LIST OF ELEMENTS WITH THEIR SYMBOLS AND ATOMIC WEIGHTS

Element	Symbol	Atomic number	Atomic weight[a]	Element	Symbol	Atomic number	Atomic weight[a]
Actinium	Ac	89	227.0278	Rhenium	Re	75	186.207
Aluminum	Al	13	26.98154	Rhodium	Rh	45	102.9055
Americium	Am	95	(243)	Rubidium	Rb	37	85.4678
Antimony	Sb	51	121.75	Ruthenium	Ru	44	101.07
Argon	Ar	18	39.948	Rutherfordium[b]	Rf	104	(261)
Arsenic	As	33	74.9216	Samarium	Sm	62	150.36
Astatine	At	85	(210)	Scandium	Sc	21	44.9554
Barium	Ba	56	137.33	Selenium	Se	34	78.96
Berkelium	Bk	97	(247)	Silicon	Si	14	28.0855
Beryllium	Be	4	9.01218	Silver	Ag	47	107.8682
Bismuth	Bi	83	208.9804	Sodium	Na	11	22.98977
Boron	B	5	10.81	Strontium	Sr	38	87.62
Bromine	Br	35	79.904	Sulfur	S	16	32.06
Cadmium	Cd	48	112.41	Tantalum	Ta	73	180.9479
Calcium	Ca	20	40.078	Technetium	Tc	43	(98)
Californium	Cf	98	(251)	Tellurium	Te	52	127.60
Carbon	C	6	12.011	Terbium	Tb	65	158.9254
Cerium	Ce	58	140.12	Thallium	Tl	81	204.383
Cesium	Cs	55	132.9054	Thorium	Th	90	232.0381
Chlorine	Cl	17	35.453	Thulium	Tm	69	168.9342
Chromium	Cr	24	51.996	Tin	Sn	50	118.71
Cobalt	Co	27	58.9332	Titanium	Ti	22	47.88
Copper	Cu	29	63.546	Tungsten	W	74	183.85
Curium	Cm	96	(247)	Unnilennium[b]	Une	109	(266)
Dysprosium	Dy	66	162.50	Unnilhexium[b]	Unh	106	(263)
Einsteinium	Es	99	(254)	Unniloctium[b]	Uno	108	(265)
Erbium	Er	68	167.26	Unnilpentium[b]	Unp	105	(262)
Europium	Eu	63	151.96	Unnilquadium[b]	Unq	104	(261)
Fermium	Fm	100	(257)	Unnilseptium[b]	Uns	107	(262)
Fluorine	F	9	18.998403	Uranium	U	92	238.0289
Francium	Fr	87	(223)	Vanadium	V	23	50.9415
Gadolinium	Gd	64	157.25	Xenon	Xe	54	131.29
Gallium	Ga	31	69.72	Ytterbium	Yb	70	173.04
Germanium	Ge	32	72.61	Yttrium	Y	39	88.9059
Gold	Au	79	196.9665	Zinc	Zn	30	65.39
Hafnium	Hf	72	178.49	Zirconium	Zr	40	91.22
Helium	He	2	4.00260				
Holmium	Ho	67	164.9304				
Hydrogen	H	1	1.0079				
Indium	In	49	114.82				
Iodine	I	53	126.9045				
Iridium	Ir	77	192.22				
Iron	Fe	26	55.847				
Krypton	Kr	36	83.80				
Lanthanum	La	57	138.9055				
Lawrencium	Lr	103	(260)				
Lead	Pb	82	207.2				
Lithium	Li	3	6.941				
Lutetium	Lu	71	174.967				
Magnesium	Mg	12	24.305				
Manganese	Mn	25	54.9380				
Mendelevium	Md	101	(258)				
Mercury	Hg	80	200.59				
Molybdenum	Mo	42	95.94				
Neodymium	Nd	60	144.24				
Neon	Ne	10	20.179				
Neptunium	Np	93	237.0482				
Nickel	Ni	28	58.69				
Niobium	Nb	41	92.9064				
Nitrogen	N	7	14.0067				
Nobelium	No	102	(259)				
Osmium	Os	76	190.2				
Oxygen	O	8	15.9994				
Palladium	Pd	46	106.42				
Phosphorus	P	15	30.97376				
Platinum	Pt	78	195.08				
Plutonium	Pu	94	(244)				
Polonium	Po	84	(209)				
Potassium	K	19	39.0983				
Praseodymium	Pr	59	140.9077				
Promethium	Pm	61	(145)				
Protactinium	Pa	91	231.0359				
Radium	Ra	88	226.0254				
Radon	Rn	86	(222)				

[a] Numbers in parentheses are mass numbers of the most stable or best-known isotope of radioactive elements.

[b] The official name and symbol have not been agreed to. The names for elements 106, 107, 108, and 109 represent their atomic numbers, as in un (1) nil (0). hex (6) = unnilhexium (Unh) for element 106.

Fundamentals
of General, Organic,
and Biological Chemistry

Fundamentals of General, Organic, and Biological Chemistry

John McMurry
Cornell University

Mary E. Castellion
Norwalk, Connecticut

Prentice Hall, Englewood Cliffs, New Jersey 07632

Library of Congress Cataloging-in-Publication Data
McMurry, John.
    Fundamentals of general, organic, and biological chemistry / John
McMurry, Mary Castellion.
        p.   cm.
    Includes index.
    ISBN 0-13-351867-1 :
    1. Chemistry.   2. Chemistry, Organic.   3. Biochemistry.
I. Castellion, Mary.   II. Title.
QD31.2.M3877   1992
540—dc20   91-12234
                                        CIP

Editor-in-Chief: Tim Bozik
Acquisitions Editor: Dan Joraanstad
Marketing Manager: Diana Farrell
Design Director: Florence Dara Silverman
Cover and Interior Designer: Peggy Kensellaer, Meryl Poweski
Prepress Buyer: Paula Massenaro
Manufacturing Buyer: Lori Bulwin
Supplements Editor: Alison Munoz
Editorial Assistant: Lynne Breitfeller
Proofreaders: Muriel Adams, Anne E. Bosch, Marie de Felice, Harriet F. Hoenie,
    Lynne C. MacAurthur, Nancy Velthaus, and Martha Williams
Illustrations by Vantage Art
Photo Research Coordinator: Lorinda Morris-Nantz
Photo Researcher: Tobi Zausner
Cover Illustration: © Howard Sochurek, All Rights Reserved/The Stock Market

*Photo credits and acknowledgments appear on pages A-35 and A-36, which
constitute a continuation of the copyright page.*

 © 1992 by Prentice-Hall, Inc.
A Simon & Schuster Company
Englewood Cliffs, New Jersey 07632

Printed in the United States of America
10 9 8 7 6 5 4 3 2

ISBN   0-13-351867-1

Prentice-Hall International (UK) Limited, *London*
Prentice-Hall of Australia Pty. Limited, *Sydney*
Prentice-Hall Canada Inc., *Toronto*
Prentice-Hall Hispanoamericana, S.A., *Mexico*
Prentice-Hall of India Private Limited, *New Delhi*
Prentice-Hall of Japan, Inc., *Tokyo*
Simon & Schuster Asia Pte. Ltd., *Singapore*
Editora Prentice-Hall do Brasil, Ltda., *Rio de Janeiro*

Contents

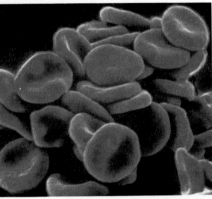

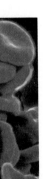

CHAPTER 3
Atoms and the Periodic Table   *47*

CHAPTER 4
Nuclear Chemistry   *71*

CHAPTER 5
Ionic Compounds *96*

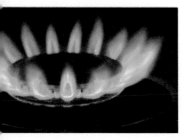

CHAPTER 6
Molecular Compounds *124*

CHAPTER 7
Chemical Reactions: Mass Relationships and Classification   *153*

CHAPTER 8
Chemical Reactions: Energy, Rates, and Equilibria   *187*

CHAPTER 9
Gases, Liquids, and Solids *216*

CHAPTER 10
Solutions *245*

CHAPTER 11
Acids, Bases, and Salts *278*

CHAPTER 12
Introduction to Organic Chemistry: Alkanes *314*

CHAPTER 13
Alkenes, Alkynes, and Aromatic Compounds 345

CHAPTER 14
Some Compounds with Oxygen, Sulfur, or Halogens 378

CHAPTER 15
Amines 407

CHAPTER 16
Aldehydes and Ketones *431*

CHAPTER 17
Carboxylic Acids and Their Derivatives *458*

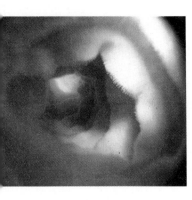

CHAPTER 23
Lipids   *642*

CHAPTER 24
Lipid Metabolism   *671*

CHAPTER **25**
Protein and Amino Acid Metabolism   *694*

CHAPTER **26**
Nucleic Acids and Protein Synthesis   *712*

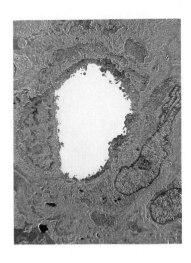

CHAPTER **27**
Body Fluids *742*

Applications and Interludes

APPLICATIONS

INTERLUDES

P r e f a c e

To provide an introduction to chemistry in general and to the chemistry of living things in particular—that is the goal of this textbook. The writing style, content, and organization are directed toward students with career goals in the allied health sciences and toward students seeking to know something about chemistry's role in our complex society.

Teaching chemistry is a challenging activity, just as learning chemistry is challenging. Teaching chemistry all the way from, "What is an atom?" to, "How do we get energy from glucose?" is especially difficult. How can that much information be covered in two semesters? Or one semester? Conversations with many of you who teach this course show that there are just about as many solutions to this problem as there are teachers. Thus, *flexibility* is of primary importance. There is ample material in this book for a thorough, two-semester introduction to general, organic, and biological chemistry. By varying the topics covered and the time devoted to them, however, each teacher can change the focus of the course to meet individual needs. Our unique biochemistry sequence, described below, allows for an unusual degree of flexibility with this material.

Another matter of primary importance is *student accessibility*. Most students in this course have their sights set well beyond academic concerns and the laboratory bench. They want to know why: Why must I study the gas laws? Why are molecular shapes important for me as a nurse, a farmer, or an informed citizen? We have therefore endeavored at every step along the way to place chemistry in the context of applications and everyday life. To meet this challenge, we have written about these matters in the mainstream of the text as well as in the Application and Interlude sections. With the intent of gaining student confidence, our writing style is relaxed and friendly, and we have included many visual and verbal study aids.

ORGANIZATION

The traditional three segments of this course have been covered in 11 chapters on general chemistry, 6 chapters on organic chemistry, and 10 chapters on biological chemistry. *General chemistry* begins with the fundamentals needed for any study of chemistry (Chapters 1–3), covers nuclear chemistry with a focus on its many applications (Chapter 4), introduces chemical compounds and chemical reactions (Chapters 5–8), and ends with specific topics essential to applied chemistry (Chapter 9, Gases, Liquids and Solids; Chapter 10, Solutions; and Chapter 11, Acids and Bases).

The six *organic chemistry* chapters provide a concise overview of the subject and focus on what students must know in order to get on with the study of biochemistry. Nomenclature rules are included with the introduction to hydrocarbons (Chapters 12, 13) and thereafter are kept to a minimum. The functional group chapters (Chapters 14–17) emphasize the structures and properties relevant to biomolecules and profile commonly encountered compounds. The chapter on amines presents their numerous biological roles and, by preceding the carboxylic acids chapter, allows amide chemistry to be covered with that of related carboxylic acid derivatives.

Our effort to allow for flexibility is especially evident in the *biological chemistry* chapters, where *structure and function are integrated*. Protein structure (Chapter 18) is followed by enzyme chemistry and biochemical energy (Chapters 19 and 20). Then come carbohydrates and their metabolism, lipids and their metabolism, and protein metabolism. If your time for biochemistry is limited, stop with Chapter 20 on biochemical energy, and your students will have an excellent preparation in the essentials of metabolism. To carry the metabolism story further, cover the next two chapters on carbohydrates and their metabolism. And if you want to cover all classes of biomolecules and their metabolism, you will find a thorough, integrated treatment in Chapters 18–25. Nutrition is not treated as yet another separate subject, but is integrated with the discussion of each type of biomolecule. In the last two chapters, we cover subjects that we've found are an essential part of the course to many of you, but optional to others—Nucleic Acids and Protein Synthesis in Chapter 26 and Body Fluids in Chapter 27. Throughout, we have made every effort to provide up-to-date coverage, recognizing that biological chemistry is a most rapidly advancing area of science.

KEY FEATURES

Applications and Interludes   A wide variety of special topics are covered in over 80 Application and Interlude sections. These sections provide thorough coverage—whether the topics are just assigned for additional reading or incorporated in the course, students can gain a reasonable understanding of each topic. Representative subjects (a complete list precedes the Preface) include *everyday chemistry* (Detergents; Minerals and Gems; The Biochemistry of Running), *environmental and societal issues* (Risk Assessment; Acid Rain; Chlorofluorocarbons and the Ozone Layer; Is It Poisonous or Isn't It?), *health and medical applications* (Homeostasis; Inhaled Anesthetics; Ethyl Alcohol as a Drug and Poison; Glucose Tolerance Test), and *modern applied chemistry* (Magnetic Resonance Imaging; Antioxidants; Prodrugs). Questions on the Applications and Interludes are provided in a separate section at the end of each chapter.

Color Photos Each chapter opens with a photo chosen to generate curiosity about the chapter subject. Then, throughout each chapter, photos are used to enhance understanding and appreciation for the subject matter.

Graphics Molecular structures, chemical and math equations, and charts and diagrams have been highlighted with color to emphasize their meaning. Many topics, especially in organic and biological chemistry, become much clearer when the reacting parts of molecules are color coded. For example, color has been used consistently to highlight phosphate groups in biomolecules and to distinguish energy-rich forms of such molecules as ATP and reduced coenzymes (red) from their lower energy counterparts, ADP and oxidized coenzymes (blue).

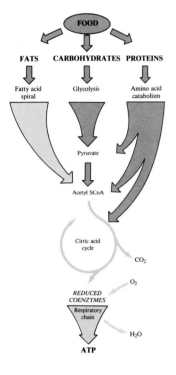

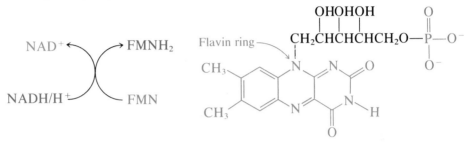

Flavin mononucleotide (FMN)

Nowhere are graphics more important than in clarifying the complexities of biochemical pathways and we believe our graphics are the best ever offered to students in this course. Our overall metabolism diagram has been adapted as a logo and is repeated as appropriate with the topic under discussion highlighted.

Computer-Generated Structures Computer-generated molecular models are used extensively in the organic and biological chapters, both for their accuracy in portraying the three-dimensional structures of molecules and for their visual appeal.

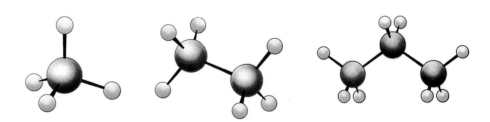

Problem Solving The problem solving skills essential to this course are illustrated with clearly explained and worked out Solved Problems. The analysis of mathematical problems for known and unknown information and the factor-label method are introduced in Chapter 2 and applied throughout. In addition, students are encouraged to find approximate Ballpark Solutions for mathematical problems as a way of testing their understanding of specific concepts and developing their analytical thinking. Solved Problems are all followed by related Practice Problems, and most text sections include additional practice problems to provide an immediate test of understanding. Practice Problems are answered at the back of the book.

PEDAGOGY

Introduction and Goals Each chapter begins with a brief introductory overview, followed by a list of goals for the student to keep in mind while studying.

Marginal Definitions Key terms are boldfaced in the text on their first use, and the definition of each term is provided in the margin nearby for easy review.

Glossary The definitions of all key terms are collected in alphabetical order in the Glossary at the back of the book.

Summaries Certain complex topics are summarized immediately after their presentation in bulleted statement lists. Each chapter ends with a clear, concise summary that reviews key points, with essential terms highlighted in boldface type.

Review Problems The end-of-chapter Review Problems provide approximately 1800 questions and problems: 800 on general chemistry, 400 on organic chemistry, and 600 on biological chemistry.

Appendices and Reference Tables The Appendices provide a review of expontential notation and useful conversion factors. The reference tables inside the cover display for easy reference the periodic table, an alphabetical list of elements, the structural features of important families of organic molecules, and a list of important tables and diagrams in the text.

Index The index is designed to be especially useful by including both general and specific citations and by the absence of cross reference entries without page numbers.

SUPPLEMENTS

Study Guide and Solutions Manual, by Susan McMurry. This companion volume answers all in-text and end-of-chapter problems and explains in detail how the answers are obtained. The solutions and data have been carefully prepared and reviewed for accuracy and coordination with the textbook. Chapter summaries, study hints, and self-test materials for each chapter are included.

Instructor's Resource Manual, by Theodore Sakano, and **Prentice Hall Test Manager** The *Manual* includes chapter overviews, lecture outlines, learning objectives, suggested readings, and 1500 multiple choice test questions. The *Test Manager* provides these questions on disk in either IBM® or MacIntosh® format and includes an editing feature that allows questions to be added or changed.

Laboratory Manual and Instructor's Manual to Laboratory Manual, by Scott Mohr and Susan Griffin The *Laboratory Manual* provides 34 laboratory experiments adaptable to either two- or three-hour laboratory periods. The *Instructor's Manual* includes detailed descriptions of all necessary chemicals, supplies, and equipment, as well as answers to pre-lab questions, typical student results, and completed report forms.

How to Study Chemistry, by Vernon Burger   This free supplement contains problem-solving strategies, helpful hints for learning and achieving success in chemistry, and a mathematics review.

Transparencies   A set of 100 two-color and four-color transparencies from this and other Prentice Hall chemistry texts is available.

Additional resources   A collection of timely news stories from *The New York Times,* described elsewhere in the opening pages of this book, and several video packages are available upon adoption. For further information please contact your local Prentice Hall sales representative.

ACKNOWLEDGMENTS

It is a pleasure to thank the many people whose help and suggestions were so valuable in preparing this book. We especially thank Leslie Kinsland, University of Southwestern Louisiana, for her assistance with questions and problems. The persons listed below provided many excellent suggestions after reviewing all or part of the manuscript. In particular, John M. Daly, Leland Harris, Larry Jackson, Gloria G. Lyle, and Les Wynston travelled across the country to make sigificant contributions for which we are very grateful, and Larry Jackson helped out with the special topic boxes for Chapter 27.

James N. Beck
McNeese State University

Richard E. Beitzel
Bemidji State University

Rodney Buyer
Hope College

John M. Daly
Bellarmine College

Lindsley Foote (Retired)
Western Michigan State University

Leland Harris
University of Arizona

Kenneth I. Hardcastle
California State University,
   Northridge

Merrill Hugo
Shasta College

Larry L. Jackson
Montana State University

Gloria G. Lyle
University of Texas, San Antonio

Frank R. Milio
Towson State University

Danny V. White
American River College

Karen Wiechelman
University of Southwestern
   Louisiana

Donald W. Williams
Hope College

Leslie Wynston
California State University,
   Long Beach

In addition, the book has benefited from the careful reading of galley proofs by Clyde Metz, College of Charleston, and by Leland Harris, University of Arizona. During production of this book, the persistence and professionalism of John Morgan, Production Editor, Prentice Hall; Tobi Zausner, Photo Researcher; and Diane Koromhas, Layout Artist were greatly appreciated. We further extend our thanks to everyone on the capable staff of Prentice Hall, both those named on the copyright page and the many others who worked and continue to work for the success of our book.

The New York Times

and

PRENTICE HALL

present

CHEMISTRY
A Contemporary View

Selected Readings

The New York Times and Prentice Hall are sponsoring *A Contemporary View,* a program designed to enhance student access to current information of relevance in the classroom.

Through this program, the core subject matter provided in the text is supplemented by a collection of time-sensitive articles from one of the world's most distinguished newspapers, *The New York Times*. These articles demonstrate the vital, ongoing connection between what is learned in the classroom and what is happening in the world around us.

To enjoy the wealth of information of *The New York Times* daily, a reduced subscription rate is available. For information, call toll-free: 1-800-631-1222.

Prentice Hall and *The New York Times* are proud to cosponsor *A Contemporary View*. We hope it will make the reading of both textbooks and newspapers a more dynamic, involving process.

A Note to the Student

Here you are, about to study chemistry, perhaps for the first time. The topics you are about to study will be useful in all health-related professions and in many business endeavors. The chemistry you are introduced to in this book will also be useful in exercising judgment in everyday life. Newspapers and magazines are filled with chemistry-related stories about protecting the environment, about new materials designed to improve the quality of life, and about drugs that promise to revolutionize medical care. The better you understand such matters, the better you will be able to function in today's society.

The following suggestions should prove helpful in your study:

Don't read the text immediately. As you begin each new chapter, look it over first. Read the introductory paragraphs and familiarize yourself with the chapter goals. Find out what topics are covered, and take a look at the illustrations—to get a feel for the topics at hand. Then turn to the end of the chapter and read the summary. You'll be in a much better position to learn new material if you first have a general idea of where you're going.

Work the problems. The problems are designed to give you practice in the skills necessary to understand and use chemistry. There are no shortcuts here. The sample problems illustrate the skills, the in-chapter practice problems provide immediate practice, and the end-of-chapter problems provide additional drill. Brief answers to in-chapter practice problems and most even-numbered review problems are given at the end of this book.

Use the study guide. Complete answers and explanations for all problems, along with chapter outlines, additional study hints, and self-tests, are given in the *Study Guide and Solutions Manual* that accompanies this text. The *Study Guide* can be extremely useful when you're working problems and when you're studying for an exam. Investigate what's there now so you'll know where to find help when you need it.

Ask questions.   Faculty members and teaching assistants are there to help you learn. Don't hesitate because you think a question might be stupid or silly. If it's something you need to know to get on with understanding chemistry, it's always a good question.

Many of the words and symbols that lie ahead in this book may at first seem strange to you. We urge you not to let their unfamiliarity cause you to lose sight of your goals: to learn about the amazing kinds of chemistry that keep us all alive and well, and to understand the impact of chemistry on everyday life.

John McMurry
Mary E. Castellion

CHAPTER

1

Matter, Energy, and Life

The eggshell is calcium carbonate, an inorganic chemical. The whipped egg white is a protein, one of nature's biomolecules. And the blue glow is from light passing through an extraordinary new product of chemical research—an almost invisible material only slightly denser than air. In this book, you're going to be learning about the chemistry of many such materials.

Look around you. Everything you see, touch, taste, and smell is made of chemicals. Many of these chemicals—those that make up rocks, trees, and your own body—occur naturally. Many others are synthetic: The plastics, the fibers, and many of the medicines that are so important a part of modern life do not occur in nature but have been created in the chemical laboratory.

Just as everything you see is made of chemicals, many of the natural changes you see taking place around you are the result of chemical reactions—the change of one chemical into another. The flowering of plants in the spring, the color change of a leaf in the fall, and the growth and aging of a human body are all the result of chemical reactions. To understand these or any other processes in life, you must have a basic understanding of chemistry.

As you might expect, the chemistry of many life processes is complex, and it's not possible to jump into their study immediately. Thus, the general plan of this book is to increase gradually in complexity, beginning in the first 11 chapters with a grounding in the scientific fundamentals that govern all of chemistry, moving in the next 6 chapters to look at the nature of carbon-containing substances like those that compose all living things, and then coming in the final 10 chapters to biological chemistry. We'll begin in this chapter by looking at the following topics:

1. *What is matter?* The goal: Be able to discuss the properties of matter and describe the three states of matter.
2. *How is matter classified?* The goal: Be able to distinguish among mixtures, pure substances, chemical elements, and chemical compounds.
3. *What kinds of change can matter undergo?* The goal: Be able to distinguish between chemical and physical changes.
4. *How are chemical elements represented?* The goal: Be able to name and give the symbols of elements.
5. *What is energy?* The goal: Be able to describe in general terms the different types of energy.

1.1 CHEMISTRY

Biochemistry The study of the chemistry of substances that occur in living matter

Chemistry The study of the nature, properties, and transformations of matter.

Matter The physical material that makes up the universe; anything that has mass and volume.

Chemistry is often referred to as the "central science," because what chemists learn is valuable to all other sciences. In fact, as more and more is learned about chemistry, biology, and physics, the old dividing lines among these disciplines become less and less clear-cut. Figure 1.1 sketches the relationship of chemistry and **biochemistry** to some other fields of study that apply to living matter. Regardless of which scientific discipline you're most interested in, a study of chemistry serves as a necessary foundation.

Chemistry, simply defined, is the science of matter. **Matter,** in turn, is a catch-all word used to describe anything physically real—anything you can see, touch, taste, or smell. In more scientific terms, matter is anything that has mass

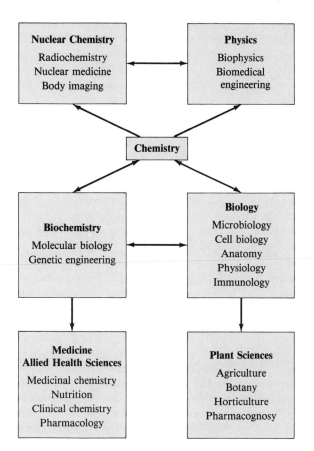

Figure 1.1
The relationship of chemistry to some other health-related scientific disciplines.

(weight) and volume (occupies space). Thus, a light bulb contains matter, but the light coming from the bulb contains no matter.

How might we describe different kinds of matter more specifically? Any characteristic that can be used to describe or identify something is called a **property.** Size, color, and temperature are all familiar properties of matter. Another, less familiar but more important property is *chemical composition:* the description of what matter is made of. Figure 1.2 shows some familiar substances—water, table sugar (sucrose), and baking soda (sodium bicarbonate)—and Table 1.1 lists their composition and some of their properties.

Property A characteristic useful for identifying a substance or object.

Figure 1.2
Samples of the pure substances water, sugar, and baking soda.

Table 1.1 Some Properties of Water, Sugar, and Baking Soda[a]

Water	Sugar (sucrose)	Baking Soda (Sodium Bicarbonate)
Colorless liquid Odorless Melting point: 0°C Boiling point: 100°C	White crystals Odorless Dissolves in water	White powder Odorless Dissolves in water
Decomposes to give hydrogen and oxygen when electric current is passed through it	Begins to decompose at 160°C, turning black and giving off water	Decomposes at 270°C, giving off water and carbon dioxide
11.2% hydrogen, 88.8% oxygen	6.4% hydrogen, 42.1% carbon, 51.5% oxygen	27.4% sodium, 1.2% hydrogen, 14.3% carbon, 57.1% oxygen
Does not burn	Burns in air	Does not burn

[a]Compositions are given by weight percent.

Physical property A property that can be determined without a change in identity.

Melting point (mp) The temperature at which a solid turns into a liquid.

Boiling point (bp) The temperature at which a liquid boils.

Physical change A change in which no change in identity occurs.

Chemical property A property that involves a change in identity of a substance.

Chemical reaction (chemical change) A process in which the identity and composition of one or more substances are changed.

Properties of matter are divided into two types: physical and chemical. **Physical properties** are those characteristics that can be determined without altering the identity of a substance. For example, when you step on a scale, you can determine your weight, but when you step off the scale, your body is unchanged. Weight is therefore a physical property, as are such other characteristics as color and odor. **Melting points** and **boiling points** like those listed in Table 1.1 for water are also physical properties because they too can be determined without changing the identity of a substance. Solid ice undergoes a **physical change** when it is converted to liquid water, but they're both *chemically* the same.

In contrast to physical properties, **chemical properties** *do* involve changes in the identity of a substance. For example, if you leave a bicycle out in the rain, the exposed metal parts are likely to rust. The iron of the bicycle changes chemically, combining with oxygen and moisture from the air to give a new substance called rust. Rusting, the chemical combination of iron with oxygen and water, is therefore a chemical property of iron. The changes described in Table 1.1 that result when an electric current is passed through water, or when sugar or baking soda is heated, are also chemical properties because new substances are produced.

Most likely, you have as one of your goals an understanding of living matter. Chemistry is applied to living matter by studying what happens at the microscopic level. When the steak and potatoes you had for dinner land in your stomach, exactly what happens to all the different kinds of substances in them? When you look out the window, how do substances in your eye send messages to your brain that register as green trees, blue sky, and red roses? Chemical reactions are responsible for these and all our other body functions. A **chemical reaction,** or **chemical change,** is a process in which one or more substances undergo a change in identity. To understand the chemistry of living things requires the study of their chemical composition and chemical reactions.

Practice Problems **1.1** Identify each of the following as a change in physical properties or in chemical properties: (a) grinding a metal surface (b) fruit ripening (c) bubbles rising in a just-opened bottle of soda (d) explosion of TNT

1.2 Which of the following is made of chemicals? (a) hairspray (b) a goldfish (c) paint (d) a watermelon

1.2 STATES OF MATTER

Solid A substance that has a definite shape and volume.

Liquid A substance that has a definite volume but that changes shape to fill its container.

Gas A substance that has neither a definite volume nor a definite shape.

State of matter The physical state of a substance as a solid, a liquid, or a gas.

Change of state The conversion of a substance from one state to another, for example, from a liquid to a gas.

Matter can be classified in a number of ways according to its properties and composition. One simple and familiar classification is based on shape and volume: Every sample of matter can be classified as a solid, a liquid, or a gas. A **solid** has a definite volume and a definite shape that doesn't change regardless of what container it is placed in. Matter in the solid form is everywhere around us. A **liquid,** in contrast, has a definite volume but an indefinite shape. The volume of a liquid doesn't change when it's poured into a different container, but its shape does. Water, gasoline, and rubbing alcohol are familiar examples of liquids. A **gas,** however, has neither a definite volume nor a definite shape. A gas expands to fill the volume and take the shape of any container it's placed in. Air, for example, is a mixture of gases, primarily oxygen and nitrogen.

Some substances, such as water, can exist in all three **states of matter**—the solid state, the liquid state, and the gaseous state—depending on the temperature and pressure. The conversion of a substance from one state to another is known as a **change of state** and is a common occurrence. The melting of a solid, the freezing or boiling of a liquid, and the condensing of a gas to a liquid are familiar to everyone.

Practice Problems **1.3** Formaldehyde is a disinfectant, a preservative, and a raw material for plastics manufacture. Its boiling point (bp) is $-2°F$, and its melting point (mp) is $-134°F$. Is formaldehyde a gas, a liquid, or a solid at room temperature (75°F)?

1.4 Acetic acid, which gives the sour taste to vinegar, has bp 244°F and mp 62°F. On a cold morning with the laboratory at 32°F, will the acetic acid bottle contain a solid or a liquid?

1.3 CLASSIFICATION OF MATTER

Pure substance Any type of matter that has unvarying properties and constant composition.

The first question a chemist asks about an unknown substance is: Is it a pure substance or is it a mixture? Every sample of matter is one or the other. Water and sugar alone are pure substances, but when you stir some sugar into a glass of water, you create a *mixture*.

What is the difference between a pure substance and a mixture? One difference is that a **pure substance** does not vary in its composition or its properties. Every sample of water, or sugar, or baking soda has the composition and properties listed in Table 1.1. A **mixture,** however, can vary in both its

Mixture A physical blend of two or more substances, each of which retains its chemical identity; can be separated by physical changes.

composition and its properties. For example, anywhere from a few grains of sugar to more than a cupful can be dissolved in a cup of water. Depending on the amounts of sugar and water present in the mixture, the sweetness, the boiling point, the mass of equal volumes, and other properties differ.

Another difference between a pure substance and a mixture is that the components of a mixture can be separated by physical methods, that is, without changing their identities. Water can be separated from a sugar–water mixture, for example, by boiling the mixture to drive off the steam and then condensing the steam to recover the pure water. The pure solid sugar is left behind in the container.

AN APPLICATION: A CHEMICAL REACTION

Nickel + hydrochloric acid ⟶ nickel chloride + hydrogen

(a)

(a) The reactants: The flat dish contains pieces of nickel, an *element* that is a typical lustrous metal. The bottle contains concentrated hydrochloric acid, a *solution* of the *chemical compound* hydrogen chloride in water. These *reactants* are about to be combined in the empty test tube.

(b) The reaction: When the *chemical reaction* occurs, the colorless solution turns green as nickel changes to soluble green nickel chloride, and bubbles slowly rise as the gaseous element hydrogen is released from hydrochloric acid.

(b)

(c)

(c) The product: Removal of water from the solution leaves behind the *product*, the solid green *chemical compound* known as nickel chloride.

Chemical compound A pure substance that can be broken down into simpler substances only by chemical reactions; composed of atoms of different elements.

Reactant A starting substance that undergoes change during a chemical reaction.

Product A substance formed as the result of a chemical reaction.

Chemical element A fundamental substance that cannot be chemically broken down into any simpler substance.

Atom The smallest and simplest particle of an element.

Pure substances are themselves classified into two groups, those that can undergo chemical breakdown to yield simpler substances and those that cannot. For example, passing electric current through pure water produces hydrogen and oxygen, chemically decomposing the water in the process:

$$\text{Water} \xrightarrow[\text{current}]{\text{electric}} \text{hydrogen} + \text{oxygen}$$

Any pure material that can be broken down into simpler substances by a chemical change is known as a **chemical compound.** Water, sugar, baking soda, and millions of other substances are examples of chemical compounds. Notice how this chemical change is written: The **reactant,** water, is written on the left; the **products,** hydrogen and oxygen, are written on the right; and an arrow connects the two to imply a chemical reaction.

Unlike water, such substances as hydrogen, oxygen, aluminum, gold, sulfur, and so forth, do not undergo chemical breakdown to yield simpler substances with different properties. Materials that can't be broken down, even by chemical reactions, are called **chemical elements.** More than 100 elements are known. **Atoms** are the smallest particles of elements, and the properties of the elements differ because each is composed of a different kind of atom. A sample of pure copper, for example, contains only copper atoms. The atoms of the different elements are the fundamental building blocks of nature from which all other substances are derived.

The classification of matter into mixtures, compounds, and elements is summarized in Figure 1.3.

Figure 1.3
A scheme for the classification of matter.

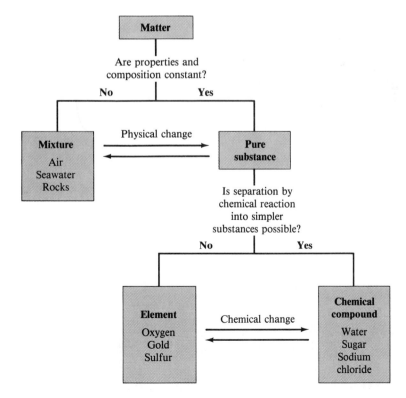

Practice Problems   **1.5**   Classify each of the following as a mixture or a pure substance:
(a) concrete   (b) helium in a balloon   (c) a lead weight   (d) wood

1.6   Classify each of the following as a physical change or a chemical change:
(a) separation by filtration
(b) production of carbon dioxide by heating limestone
(c) mixing alcohol and water

1.4   THE CHEMICAL ELEMENTS

One hundred and nine chemical elements are known today. They are listed inside the front cover of this book both alphabetically and in the form of a table called the *periodic table*. About 90 elements occur naturally; the remainder have been produced artificially by chemists and physicists. Each element has its own distinctive properties, and just about every element has been put to use in some way that takes advantage of its properties. As indicated in Figure 1.4, which shows the approximate elemental composition of the earth's crust and of the human body, the naturally occurring elements are not equally abundant. Oxygen and silicon together account for 75% of the mass in the earth's crust; oxygen, carbon, and hydrogen account for nearly all the mass of a human body.

Metal   A malleable element with a lustrous appearance that is a good conductor of heat and electricity.

Nonmetal   An element that is a poor conductor of heat and electricity.

The elements are roughly divided into metals and nonmetals. **Metals** are solids (except for mercury), usually have a lustrous appearance when freshly cut, are good conductors of heat and electricity, and are malleable rather than brittle. (A malleable substance can be beaten or rolled into different shapes or thicknesses; a brittle substance crumbles and breaks under such treatment.) Some familiar metals are aluminum, copper, and chromium. **Nonmetals,** in contrast, are poor conductors of heat and electricity and most are either gaseous or solid. For example, oxygen and nitrogen are gases present in air, and carbon is a solid present in soot, diamond, and graphite. One nonmetallic element, bromine, is a liquid.

The elements essential for human life are listed in Table 1.2. In addition to such well-known elements as carbon, hydrogen, oxygen, and nitrogen, less familiar elements such as molybdenum and selenium are also important.

Figure 1.4
Elemental composition of (a) the earth's crust, and (b) the human body.

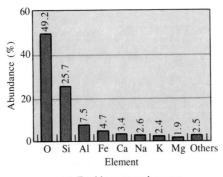

(a) Earth's crust and oceans

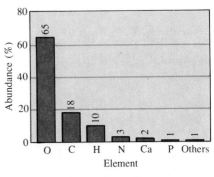

(b) Human body

AN APPLICATION: SOME CHEMICAL ELEMENTS

Nonmetals: Nitrogen, phosphorus, sulfur, and chlorine   Nitrogen, phosphorus, sulfur, and chlorine are essential to the chemical compounds in living things. Pure nitrogen (a) is a gas that doesn't condense to a liquid until it is cooled to almost $-328°F$. It makes up 80% of the air. Phosphorus (b, red solid) occurs naturally in phosphate rock, and free sulfur (b, yellow solid) is found in large underground deposits in Texas and Louisiana. Chlorine (b) is a yellow-green gas that combines so readily with other substances that it is not present in nature as the free element.

Metals: Gold and zinc   Gold (c), which we all know for its beauty, isn't very reactive and is not essential to living things, although gold compounds have been used to treat arthritis. Zinc (d), produced from zinc oxide ores, has industrial uses ranging from brass, to roofing materials, to dry cell batteries and is also an essential trace element in our diets.

(a) Nitrogen

(b) Phosphorus, sulfur, and chlorine

(c) Gold

(d) Zinc

1.5   NAMES AND SYMBOLS FOR ELEMENTS

Rather than write out the full names of elements, chemists generally use a shorthand notation in which most elements are referred to by one- or two-letter symbols. The names and symbols of some common elements are listed in Table 1.3, and other names and symbols are also included in Table 1.2. Note that all two-letter symbols have only the first letter capitalized; the second letter is always lowercase.

The symbols of the common elements are just the first one or two letters of the elements' names, such as H (hydrogen) and Al (aluminum). Pay special attention, however, to the elements grouped in the fourth column in Table 1.3.

Table 1.2   Elements Essential for Human Life[a]

Element	Symbol	Function
Carbon	C	
Hydrogen	H	These four elements are present throughout all
Oxygen	O	living organisms
Nitrogen	N	
Boron	B	Aids in the use of Ca, P, and Mg
Calcium*	Ca	Necessary for growth of teeth and bones
Chlorine*	Cl	Necessary for maintaining salt balance in body fluids
Chromium	Cr	Aids in carbohydrate metabolism
Cobalt	Co	Component of vitamin B_{12}
Copper	Cu	Necessary to maintain blood chemistry
Fluorine	F	Aids in the development of teeth and bones
Iodine	I	Necessary for thyroid function
Iron	Fe	Necessary for oxygen-carrying ability of blood
Magnesium*	Mg	Necessary for bones, teeth, and muscle and nerve action
Manganese	Mn	Necessary for carbohydrate metabolism and bone formation
Molybdenum	Mo	Component of enzymes necessary for metabolism
Nickel	Ni	Aids in the use of Fe and Cu
Phosphorus*	P	Necessary for growth of bones and teeth; present in DNA and RNA
Potassium*	K	Component of body fluids; necessary for nerve action
Selenium	Se	Aids in vitamin E action and fat metabolism
Silicon	Si	Helps form connective tissue and bone
Sodium*	Na	Component of body fluids; necessary for nerve and muscle action
Sulfur*	S	Component of proteins; necessary for blood clotting
Zinc	Zn	Necessary for growth, healing, and overall health

[a]C, H, O, and N are present in all foods. Other elements listed vary in their distribution in different foods. Those marked with an asterisk are *macronutrients,* essential in the diet at more than 100 mg/day; the rest, other than C, H, O, and N, are *micronutrients,* essential at 15 ng or less per day.

Table 1.3   Names and Symbols for Some Common Elements

Symbols Based on Modern Names						Symbols Based on Latin Names	
Al	Aluminum	Co	Cobalt	N	Nitrogen	Cu	Copper (*cuprum*)
Ar	Argon	F	Fluorine	O	Oxygen	Au	Gold (*aurum*)
Ba	Barium	He	Helium	P	Phosphorus	Fe	Iron (*ferrum*)
Bi	Bismuth	H	Hydrogen	Pt	Platinum	Pb	Lead (*plumbum*)
B	Boron	I	Iodine	Rn	Radon	Hg	Mercury (*hydrargyrum*)
Br	Bromine	Li	Lithium	Si	Silicon	K	Potassium (*kalium*)
Ca	Calcium	Mg	Magnesium	S	Sulfur	Ag	Silver (*argentum*)
C	Carbon	Mn	Manganese	Zn	Zinc	Na	Sodium (*natrium*)
Cl	Chlorine	Ni	Nickel			Sn	Tin (*stannum*)

The symbols here are derived from Latin names such as Na for sodium, once known as *natrium*. Memorizing these symbols is necessary, but fortunately there aren't many of them.

Chemical formula
Representation of a chemical compound by element symbols and subscripts that show how many atoms of each element are present.

Element symbols are combined to produce **chemical formulas,** which show by symbols and subscripts how many atoms of different elements are combined in chemical compounds. For example, the formula H_2O represents water, which contains two hydrogen atoms for each oxygen atom. Similarly, the formula CH_4 represents methane (natural gas), and the formula $C_{12}H_{22}O_{11}$ represents table sugar (sucrose).

AN APPLICATION: WHAT'S IN A CHEMICAL NAME?

A chemical name often evokes one or more of the following reactions:

"It's unpronounceable."
"It's too complicated."
"It must be something bad."

In the first place, there's no secret to pronouncing chemical names; just say every possible syllable. The word "acetaldehyde" has five syllables:

ac-et-**al**-de-hyde

and the word "phenylpropanolamine" has seven syllables:

phen-yl-pro-pa-**nol**-a-mine

But why are chemical names so complicated? Once you realize that there are more than 10 *million* known compounds, the reason is obvious: The name of a compound has to include enough information so that every chemist can recognize the composition and structure of the compound from the name. It's as if you had to devise a system for naming your friends so that each name would include information about height, hair and eye color, and other identifying characteristics.

What about chemicals being bad? These days, it seems that a different chemical gets into the news every week, usually in stories describing a threat to health or the environment. The unfortunate result is that many people conclude that everything with a chemical name is man-made and dangerous. *Neither is true*. Acetaldehyde, for example, is one of the many naturally occurring chemicals responsible for the delicious taste of fresh strawberries. In fact, small quantities of acetaldehyde are present naturally in most tart, ripe fruits and are added to artificial flavorings. *Pure* acetaldehyde, however, is also a flammable gas that can be toxic and explosive in high concentrations.

Similar comparisons of desirable and harmful properties can be made for many chemicals. To be a good scientist, and indeed a good citizen, requires looking beyond a chemical name. A large enough dose of just about anything, even water, can be dangerous. The properties of a substance and the conditions surrounding its use must be evaluated before judgments are made.

Contains the following: acetone, acetaldehyde, methyl butyrate, ethyl caproate, hexyl acetate, methanol, acrolein, crotonaldehyde, acetic acid, crotonic acid, formaldehyde, and other natural ingredients.

Practice Problems **1.7** Look at the alphabetical list on the inside front cover and find the symbols for these elements:
(a) uranium, the fuel in nuclear reactors
(b) titanium, the skin of jet fighters
(c) tungsten, the filament in light bulbs

1.8 What elements do these symbols represent?
(a) Na   (b) Ca   (c) Pd   (d) K   (e) Sr   (f) Sn

1.6 ENERGY

Energy   The capacity to do work or cause change.

Many words used in science sound familiar but have meanings different from those used in daily life. Take the word *energy* for instance. We all use the word frequently, but what exactly is energy, where does it come from, and what is it used for? In science, **energy** is formally defined as *the capacity to do work*. Put another way, energy can produce change.

Potential energy   Energy that is stored because of position, composition, or shape.

Kinetic energy   The energy of an object in motion.

Energy is classified as either *potential* or *kinetic*. **Potential energy** is stored energy. The water behind a dam, an automobile poised to coast downhill, and a coiled spring have potential energy waiting to be released. **Kinetic energy,** in contrast, is the energy of motion. When the water in the dam falls and powers a turbine, when the car rolls downhill and crashes, or when the spring uncoils and makes the hands on a clock move, the potential energy stored by each is converted to kinetic energy (Figure 1.5). Experience shows that only events leading to lower potential energy are spontaneous. We expect a car to roll spontaneously downhill and release energy; we don't expect a car to roll spontaneously uphill and gain energy.

Chemical energy   The potential energy stored in chemical compounds.

Energy has various forms, all interchangeable. The energy in your body, for instance, is produced by chemical reactions that release **chemical energy,** a form of potential energy that is stored in chemical compounds. Chemical energy is then converted by living things into many other forms. In warm-blooded animals, for instance, chemical energy is used to generate *thermal energy,* or heat, for maintaining body temperature. When walking, swimming, or crawling, you use *mechanical energy*. In speaking, you generate *sound energy*. Within nerve cells, *electrical energy* is used for sending messages. Some insects, such as fireflies, even have the ability to generate *light energy*.

All chemical reactions are accompanied by a change in energy. Most often, energy is released in the form of heat, as with the burning magnesium shown in Figure 1.6. The substances produced when heat is released in a reaction have lower potential energy than the reactants, just as the water that has fallen over a dam has less potential energy than it had at the top. In describing this result, we say that the products of a reaction are *more stable* than the reactants. The term "stable" is used frequently in chemistry to describe substances that have little natural tendency to undergo change. Understanding the role of energy in chemical reactions is an essential part of their study.

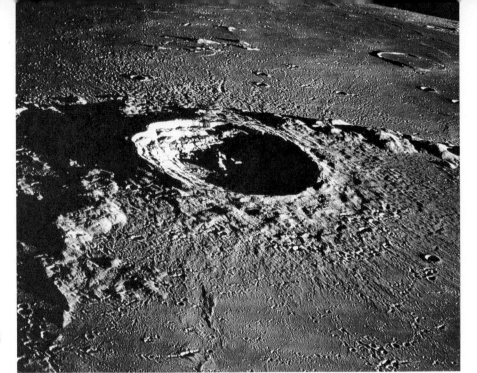

Figure 1.5
Eratosthenes, an impact crater on the Moon. Such a crater is formed by conversion of the kinetic energy of a meteor traveling 5000 miles per hour to thermal and mechanical energy. At impact, iron vaporizes, rock melts, and the molten material is pushed aside in ripples and thrown out in splashes.

Figure 1.6
A release of chemical energy. Inside the flask, magnesium metal is burning in carbon dioxide. The magnesium is converted to magnesium oxide, and carbon dioxide is converted to carbon, which can be seen as black specks clinging to the inside of the flask.

Practice Problem   ***1.9***   Decide whether each of the following has mainly kinetic energy or mainly potential energy:
(a) a book on the edge of the desk   (b) a just-pitched baseball
(c) a sandwich   (d) a flashlight battery

INTERLUDE: RISK ASSESSMENT

Living means making decisions. We decide to drive, take a train, or fly to a given destination. We choose whether to drink soda sweetened with sugar or with aspartame. Making judgments of this kind is formally known as *risk assessment,* and it's something we do every day without thinking about it.

Assessing risk to human health from exposure to chemicals can't be based on everyday experience. Evaluation has traditionally begun by exposing animals to a chemical and carefully watching for any signs of harm. To limit the large costs and long times needed for such studies, the amounts administered are vastly greater than any human is likely to encounter. Once animal tests are complete, the data must be interpreted. This is the hard part, requiring the use of complex mathematical models to arrive at a maximum concentration that will cause no harm to humans.

Any risk assessment model must include a number of assumptions. If a substance is harmful to animals, will it necessarily cause harm to humans? How should a large dose for a small animal be translated into a small dose for a large human? Should different assumptions be made for different experimental animals? What natural defenses might humans have that test animals lack, and vice versa? The uncertainties are large, and the validity of the answers is not always clear.

All decisions involve tradeoffs. Do the benefits of a pesticide that will greatly increase the availability of food outweigh a health risk to 1 person in 1 million who are exposed? We can't expect to protect ourselves from all possible threats in this complex world, and we have to ask whether it's really necessary to ban a pesticide or a food additive based on a 1-in-1-million risk of harm. One person in 1 million dies for every 10 miles traveled on a bicycle and every 100 miles traveled in a car, but we don't hesitate to do these things.

How individuals evaluate risk is strongly influenced by familiarity. Many foods contain natural ingredients far more toxic than synthetic food additives or pesticide residues, but the ingredients are ignored because the foods are familiar. For example, a peanut butter sandwich is a more potent cancer threat (because of aflatoxin, a chemical found in moldy peanut shells) than a glass of water that contains a few parts per billion of a chlorinated solvent. And most of us are fond of charcoal-broiled hamburgers. Should cooking meat over hot coals be forbidden because the process increases the concentration of cancer-causing chemicals in the meat?

Legislators proposing restrictive laws for chemicals, and citizens demanding these laws, have a responsibility to keep matters in perspective. The necessary decisions aren't easy, but your study of chemistry will be a help in such decision making.

Which is more hazardous?

SUMMARY

Chemistry is the science of **matter**. There are three **states of matter: solid, liquid,** and **gas.** Any characteristic that can be used to describe matter is called a **property. Physical properties** are those characteristics that can be measured without changing the identity of a substance, whereas **chemical properties** are those characteristics that *do* involve a change in identity. Substances are further characterized as either **pure substances** or **mixtures.** Every pure substance, in turn, is either an **element** or a **chemical compound.** Elements, which are roughly divided according to their properties into **metals** and **nonmetals,** are fundamental substances that cannot be chemically changed into anything simpler. Most elements are represented by

one- or two-letter symbols, such as H for hydrogen, Ca for calcium, and Na for sodium. **Physical changes** do not result in any changes in identity of the substances involved, whereas in chemical changes, or **chemical reactions,** the identity of one or more substances changes.

Energy is defined in science as the capacity to do work or to cause change. **Potential energy** is stored energy, and **kinetic energy** is the energy of motion. All chemical reactions are accompanied by a change in energy, most often by the release of heat. The potential energy stored in chemical compounds is known as **chemical energy.** Living things depend upon the release of chemical energy for all their energy needs.

REVIEW PROBLEMS

Chemistry and the Properties of Matter

1.10 Which of the following are composed of chemicals?
(a) a rosebush
(b) the substances that give roses their fragrance
(c) the earth in which the bush grows

1.11 Give examples of four physical properties.

1.12 Which of the following are physical properties?
(a) boiling point of water
(b) decomposition of water caused by an electric current
(c) solubility of sugar in water
(d) tendency of potassium metal to create an explosion when placed in water
(e) brittleness of glass

1.13 Which of the following are physical properties?
(a) color   (b) weight of 1 pint of water
(c) flammability of gasoline   (d) souring of milk
(e) attraction of nickel to a magnet

1.14 Identify each of the following as a physical property or a chemical property:
(a) floating of oil on water
(b) condensation of steam
(c) softness of potassium metal, allowing it to be cut with a knife
(d) ignition of matches when struck on a rough surface

1.15 Refer to Table 1.1 and classify each property of baking soda as a chemical or physical property.

States and Classification of Matter

1.16 Name and describe the three states of matter.

1.17 (a) Name two changes of state and describe what causes each to occur.
(b) Are these physical or chemical changes?

1.18 Sulfur dioxide is a compound produced when sub-

stances containing sulfur burn in air. It has mp $-76.1°C$ and bp $-10°C$. In what state does sulfur dioxide exist at room temperature (25°C)? (The symbol °C means degrees on the Celsius scale.)

1.19 Hexyl alcohol, a chemical used in the synthesis of drugs, antiseptics, and perfumes, has mp $-51.6°C$ and bp 157°C. In what state is hexyl alcohol found at 75°C?

1.20 Benzyl salicylate, which melts at 24°C and has a high boiling point (over 200°C), is a chemical compound sometimes used as a sunscreen. In what state is it found at (a) 14°C (b) 60°C?

1.21 Classify each of the following as a mixture or a pure substance:
(a) pea soup   (b) seawater
(c) the contents of a propane tank (C_3H_8)
(d) urine   (e) lead   (f) a multivitamin tablet

1.22 Classify each of the following as a mixture or a pure substance:
(a) blood   (b) silicon   (c) dishwashing liquid
(d) toothpaste   (e) gold
(f) gaseous ammonia (NH_3)

1.23 What is the difference between an element and a compound? Between a compound and a mixture?

1.24 Classify each of the following as an element, a compound, or a mixture:
(a) aluminum foil   (b) table salt   (c) water
(d) air   (e) a banana   (f) notebook paper

1.25 Which of the following terms, (i) mixture, (ii) solid, (iii) liquid, (iv) gas, (v) chemical element, (vi) chemical compound, applies to each of the following?
(a) motor oil   (b) copper   (c) carbon dioxide
(d) nitrogen   (e) sodium bicarbonate

1.26 Hydrogen peroxide, solutions of which are often used to cleanse cuts and scrapes, breaks down to yield water and oxygen:

$$\text{Hydrogen peroxide} \rightarrow \text{water} + \text{oxygen}$$

(a) Identify the reactants and products.
(b) Which of the substances are chemical compounds and which are elements?

1.27 When sodium metal is placed in water, the following change occurs:

$$\text{Sodium} + \text{water} \rightarrow \text{hydrogen} + \text{sodium hydroxide}$$

(a) Identify the reactants and products.
(b) Which of the substances are elements and which are chemical compounds?

Elements and Their Symbols

1.28 Describe the general properties of metals and nonmetals.

1.29 What symbols identify these elements?
(a) zinc   (b) mercury   (c) barium   (d) gold
(e) silicon   (f) carbon   (g) sodium   (h) lead

1.30 What elements are specified by these symbols?
(a) N   (b) K   (c) Cl   (d) Ca
(e) P   (f) Mg   (g) Mn

1.31 The symbol CO stands for carbon monoxide, a chemical compound, but the symbol Co stands for cobalt, an element. Explain how the two can be told apart.

1.32 What is wrong with these statements? Correct them.
(a) The symbol for bromine is BR.
(b) The symbol for manganese is Mg.
(c) The symbol for carbon is Ca.
(d) The symbol for potassium is po.

1.33 What is the most abundant element in the earth's crust? In the human body? List the name and symbol for each.

1.34 Small amounts of the following elements in our diets are essential for good health. What is the chemical symbol for each?
(a) iron   (b) copper   (c) cobalt
(d) molybdenum   (e) chromium   (f) fluorine
(g) sulfur

1.35 What is wrong with these statements? Correct them.
(a) Water has the formula H2O.
(b) Water is composed of hydrogen and nitrogen.

1.36 Name the elements combined in the chemical compounds represented by the following formulas:
(a) $MgSO_4$ (b) $FeBr_2$ (c) CoP
(d) AsH_3 (e) $CaCr_2O_7$

1.37 Which of the two elements described here is most likely a metal, and which a nonmetal?
(a) Osmium (mp 3000°C, bp 5500°C) is hard, shiny, very dense, and conducts electricity.
(b) Xenon (bp −108°C) is a colorless, odorless gas.

1.38 The amino acid glycine has the formula $C_2H_5NO_2$. What elements are present in glycine? What is the total number of atoms represented by the formula?

1.39 Ribose, an essential part of ribonucleic acid (RNA), has the formula $C_5H_{10}O_5$. What is the total number of atoms represented by the formula?

1.40 What is the shorthand formula for penicillin V, whose molecules contain 16 carbons, 18 hydrogens, 2 nitrogens, 5 oxygens, and 1 sulfur?

Energy

1.41 Describe the difference between kinetic and potential energy.

1.42 Decide whether each of the following has mainly kinetic energy or mainly potential energy:
(a) a tankful of gasoline
(b) a bird sitting on a telephone wire
(c) an automobile battery

1.43 Decide whether each of the following has mainly kinetic energy or mainly potential energy:
(a) a wound-up toy mouse
(b) a golf ball in flight
(c) Sterno, a fuel used for small heaters

1.44 What is the principal source of energy for human beings?

1.45 Are the following statements true or false? If false, explain why.
(a) Plants store energy from the sun as kinetic energy.
(b) Physical changes in your body produce the energy you need to run up the stairs.

Applications

1.46 Many people with high blood pressure are required to maintain a low-sodium diet. Read the ingredient labels on several foods and try to find three compounds that contain sodium. [App. Chemical Name]

1.47 For each unfamiliar substance you have seen on a food label, consult a reference book, such as the *Merck Index*, to see why the substance has been added to the food. [App: Chemical Name]

1.48 Many over-the-counter medications, beverages, and other products may have unacceptably high risks for certain classes of people. Read the labels on common products such as aspirin, artifically sweetened beverages, alcohol, and cigarettes and note the types of people to whom the product is of special risk. [Int: Risk Assessment]

Additional Questions and Problems

1.49 Choose three items in your home and classify the materials they are composed of as mixtures, elements, or chemical compounds.

1.50 Distinguish between
(a) kinetic and potential energy
(b) melting point and boiling point
(c) reactants and products
(d) metals and nonmetals

1.51 Are the following statements true or false? If false, explain why.
(a) The combination of sodium and chlorine to produce sodium chloride is a chemical reaction.
(b) The addition of heat to solid sodium chloride until it melts is a chemical reaction.

1.52 Are the following statements true or false? If false, explain why.
(a) An airplane in flight has only kinetic energy.
(b) The formula for a chemical compound that contains lead and oxygen is LiO.

(c) By stirring together salt and pepper we can create a new chemical compound to be used for seasoning food.

1.53 Which of the following are chemical compounds and which are elements?
(a) H_2O_2 (b) Mo   (c) C   (d) NO
(e) $NaHCO_3$

1.54 A white solid with a melting point of 730°C is melted. When electricity is passed through the resultant liquid, a brown gas and a molten metal are produced. Neither the metal nor the gas can be broken down into anything more simple by chemical means. Classify the white solid, the molten metal, and the brown gas as a mixture, a compound, or an element.

1.55 As a clear, red liquid sits at room temperature, evaporation occurs until only a red solid remains. Is the liquid an element, a compound, or a mixture?

1.56 Describe how you could physically separate a mixture of iron filings, table salt, and white sand.

2

Measurements in Chemistry

Paperclips and semiconductor chips must be the right size for their jobs. Everything we use or study has to be measured at some point, and this requires many kinds of units of measure.

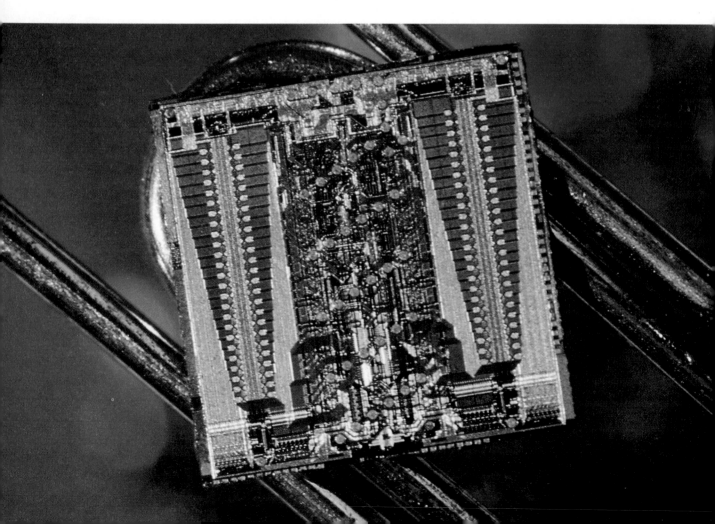

Measurements and numbers are essential in chemistry and all its applications. Whenever you use chemical compounds—whether you are adding salt to cooking water, mixing up a spray to keep bugs off your plants, or choosing a medicine dosage for a child—quantities of substances must be measured. A mistake in choosing the quantities could ruin the dinner, kill the plants, or leave the disease uncured. Though they may not be the most interesting or exciting parts of studying chemistry, the topics in this chapter, which deal with the following questions, can't be taken lightly.

1. *How are the results of measurements reported?* The goal: Be able to name the metric units of measure for mass, length, volume, and temperature.

2. *How are large and small numbers best represented?* The goal: Be able to interpret scientific notation and prefixes for units of measure and, and to convert numbers to and from scientific notation.

3. *What methods are used to indicate how good measurements are?* The goal: Be able to interpret the number of significant figures in a physical quantity and round off numbers in calculations involving physical quantities.

4. *How can a quantity be converted from one unit of measure to another?* The goal: Be able to convert physical quantities from one unit to another by using conversion factors.

5. *What methods are aids to solving problems?* The goal: Be able to apply the factor-label method, problem analysis, and the ballpark solution technique to solving problems.

6. *What are heat, specific heat, density, and specific gravity?* The goal: Be able to define these quantities and use them in calculations.

2.1 PHYSICAL QUANTITIES

To measure a physical property, you compare it with something of defined size. To measure the height of a book, for example, you place a ruler marked off in inches or centimeters next to the book and then record the height as follows:

10.0 inches   or   25.4 centimeters

Height and all other physical properties that can be measured are known as **physical quantities.**

Physical quantity A physical property that can be measured.

A person might be described by the following information:

John Doe:	Height	6.5 feet	or	2.0 meters
	Weight	135 pounds	or	61.2 kilograms

19

Comparing these quantities with the heights and weights of most people shows that John Doe is very tall and quite thin.

Notice that all physical quantities are described by both a number and a label, or **unit.**

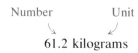

$$\overset{\text{Number}}{\searrow} \quad \overset{\text{Unit}}{\swarrow}$$
$$61.2 \text{ kilograms}$$

Unit   A defined quantity adopted as a standard of measurement.

The number alone isn't much good without a unit to define it. If you asked how much blood an accident victim had lost, the answer "three" wouldn't tell you much. Three drops? Three milliliters? Three pints? Three liters? (Let's hope it's not 3 liters; an adult human has only 5 to 6 liters of blood.)

Notice also that a given physical property can be measured in different units. For example, John Doe's height might be measured in inches, feet, or meters. In an effort to avoid confusion, scientists from many countries have agreed on a system of standard units, which has the French name *Système International d'Unités* (International System of Units). The **SI units** for the most common physical quantities are given in Table 2.1. Mass is measured in **kilograms (kg)**, length is measured in **meters (m)**, volume is measured in **cubic meters (m³),** and temperature is measured in **kelvins (K).**

SI Unit   An internationally agreed-on unit of measure (in the *Système International d'Unites*) derived from the metric system.

Kilogram (kg)   The SI unit of mass; equal to 2.205 lb.

Meter (m)   The SI unit of length; equal to 1.0936 yd.

Cubic meter (m³)   The SI unit of volume; equal to 264.2 gal.

Kelvin (K)   The SI unit of temperature; 1 K = 1C°.

Metric units   Units of a common system of measure used throughout the world.

Gram (g)   The metric unit of mass; equal to 1/1000 kg.

Liter (L)   The metric unit of volume; equal to 1.057 qt.

Celsius degree (C°)   The metric unit of temperature; equal to 1.8 Fahrenheit degrees.

SI units are closely related to the more familiar **metric units** used in all industrialized nations of the world except the United States. If you compare the SI and metric units shown in Table 2.1, you'll find that the metric unit of mass is the **gram (g)** rather than the kilogram (1 g = 1/1000 kg). Similarly, the metric unit of volume is the **liter (L)** rather than the cubic meter, and the metric unit of temperature is the **Celsius degree (C°)** rather than the kelvin (1 K = 1 C°). The meter is the unit of length in both the SI and metric systems. Although SI units are now preferred in chemical research, metric units are commonly used in biology and medicine. You'll probably find yourself working primarily with the metric system, but you may occasionally have to deal with SI units.

One problem with any system of measure is that the sizes of the units often turn out to be inconveniently large or small. A biologist describing the diameter of the red blood cell shown in Figure 2.1 (0.000006 m) would find the meter to be an inconveniently large unit, but an astronomer measuring the average distance from the earth to the sun (150,000,000,000 m) would find the meter to be inconveniently small. For this reason, metric and SI units can be modified through the use of prefixes to refer to either smaller or larger quantities.

Table 2.1   Common SI and Metric Units and Their Equivalents

Quantity	SI Unit[a] (Symbol)	Metric Unit (Symbol)	Equivalents
Mass	Kilogram (kg)	Gram (g)	1 kg = 2.205 lb
			1 kg = 1000 g
Length	Meter (m)	Meter (m)	1 m = 3.280 ft
Volume	Cubic meter (m³)	Liter (L)	1 m³ = 264.2 gal
			1 L = 0.001 m³
Temperature	Kelvin (K)	Celsius degree (C°)	1 K = 1 C°
			1 C° = 1.8 F°

[a]These are the recommended, or base, units from which other units involving these quantities are derived.

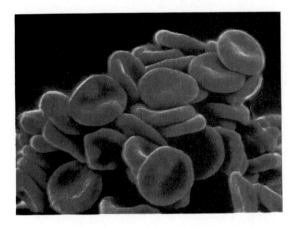

Figure 2.1
Red blood cells. The cells shown have a diameter of $6\ \mu m$, or 6×10^{-6} m. (Original magnification $\times 2000$)

The name of the SI unit for mass, the kilogram, differs by the prefix *kilo-* from the metric unit, the gram. *Kilo-* indicates that a kilogram is 1000 times larger than a gram:

$$1\ kg = 1000 \times 1\ g = 1000\ g$$

The small quantities of active ingredients in medications are often reported in milligrams. The prefix *milli-* shows that the base unit has been divided by 1000, which is the same as multiplying by 0.001:

$$1\ mg = \frac{1\ g}{1000} = \frac{1}{1000} \times 1\ g = 0.001 \times 1\ g = 0.001\ g$$

A list of prefixes is given in Table 2.2, with the most commonly used ones displayed in bold type. As an aid to remembering their meanings, note that the exponents are multiples of 3 for *mega-* (10^6), *kilo-* (10^3), *milli-* (10^{-3}), *micro-* (10^{-6}), *nano-* (10^{-9}), and *pico-* (10^{-12}). (Notation using exponents is reviewed in the next section.) The prefixes *centi-*, meaning one-hundredth of the base unit, and *deci-*, meaning one-tenth of the base unit, indicate exponents that are not multiples of 3. *Centi-* is seen most often in the length unit of *centi*meters (1 cm =

Table 2.2 Some Prefixes for Multiples of Metric and SI Units

Prefix	Symbol	Base Unit Multiplied By—[a]	Example
mega	**M**	$1,000,000 = 10^6$	1 megameter (Mm) $= 10^6$ m
kilo	**k**	$1,000 = 10^3$	1 kilogram (kg) $= 10^3$ g
hecto	h	$100 = 10^2$	1 hectogram (hg) $= 100$ g
deka	da	$10 = 10^1$	1 dekaliter (daL) $= 10$ L
deci	**d**	$0.1 = 10^{-1}$	1 deciliter (dL) $= 0.1$ L
centi	**c**	$0.01 = 10^{-2}$	1 centimeter (cm) $= 0.01$ m
milli	**m**	$0.001 = 10^{-3}$	1 milligram (mg) $= 0.001$ g
micro	**μ**	$0.000\ 001 = 10^{-6}$	1 micrometer (μm) $= 10^{-6}$ m
nano	n	$0.000\ 000\ 001 = 10^{-9}$	1 nanogram (ng) $= 10^{-9}$ g
pico	p	$0.000\ 000\ 000\ 001 = 10^{-12}$	1 picogram (pg) $= 10^{-12}$ g

[a]The scientific notation method of writing large and small numbers (for example, 10^6 for 1,000,000) is explained in Section 2.2.

0.01 m), and *deci-* is used primarily in clinical chemistry, where the concentrations of blood components are given in milligrams per *deci*liter (1 dL = 0.1 L). The term **concentration** refers to the quantity of a substance dissolved in a specific volume or mass of a solution or solvent. Like milligrams per deciliter, the units for concentration are often mass units per volume unit.

Concentration The quantity of a substance dissolved in a unit quantity of a solution or solvent.

Practice Problems

2.1 Bottles of wine sometimes carry the notation "Volume—75 cL." What does the unit cL stand for?

2.2 Give the full names of each of these units:
(a) mL   (b) kg   (c) cm   (d) km   (e) μg

2.3 Write the symbols for each of the following units:
(a) liter   (b) microliter   (c) nanometer   (d) megameter

2.4 Express each of the following in terms of the base unit (for example, 1 mL = 0.001 L):
(a) 1 nm   (b) 1 dg   (c) 1 km   (d) 1 mL   (e) 1 ng

2.2 SCIENTIFIC NOTATION

Rather than write out very large or very small numbers in their entirety, it's often convenient to express them using **scientific notation.** A number is expressed in scientific notation as the product of a number between 1 and 10, times the number 10 raised to a power. Thus, the number 215 can be written in scientific notation as 2.15×10^2:

Scientific notation Representation of a number as the product of a number (usually between 1 and 10) and 10 with an exponent.

$$215 = 2.15 \times 100 = 2.15 \times (10 \times 10) = 2.15 \times 10^2$$

Notice that in this case, where the number to be expressed in scientific notation is *larger* than 1, the decimal point has been moved to the left until it's *next to* the first digit. The exponent on the 10 tells how many places we had to move the decimal point to get it next to the first digit.

$$2\,1\,5\,. = 2.15 \times 10^2$$

Decimal point is moved two places to the left, so exponent is 2.

In order to express a number *smaller* than 1 in scientific notation, we have to move the decimal point to the right until it's just *past* the first digit. The number of places moved is the negative exponent of 10. For example, the number 0.00215 can be rewritten as 2.15×10^{-3}.

$$0.00215 = 2.15 \times \frac{1}{1000} = 2.15 \times \frac{1}{10 \times 10 \times 10} = 2.15 \times \frac{1}{10^3} = 2.15 \times 10^{-3}$$

$$0.00215 = 2.15 \times 10^{-3}$$

Decimal point is moved three places to the right, so exponent is -3.

To convert a number written in scientific notation to an ordinary number, the process is reversed. For a number with a positive exponent, the decimal point is moved to the right a number of places equal to the exponent:

$$3.7962 \times 10^4 = 37,962$$

Positive exponent of 4, so decimal point is moved to the right four places.

For a number with a negative exponent, the decimal point is moved to the left a number of places equal to the exponent:

$$1.56 \times 10^{-8} = 0.0000000156$$

Negative exponent of -8, so decimal point is moved to the left eight places.

Scientific notation is not ordinarily used for numbers that are easily written, such as 10 or 175, although sometimes it's necessary in doing arithmetic. Rules for arithmetic with numbers expressed in scientific notation are reviewed in Appendix A.

Solved Problem 2.1   There are 1,760,000,000,000,000,000,000 molecules in 1 g of sucrose (ordinary sugar). Use scientific notation to express this number.

Solution   Since we have to move the decimal point 21 places to the left to position it next to the first digit, the answer is 1.76×10^{21}.

Solved Problem 2.2   Use scientific notation to indicate the diameter of a sodium atom (0.000000000388 m).

Solution   Since we have to move the decimal point 10 places to the right to place it after the first digit, the answer is 3.88×10^{-10} m.

Solved Problem 2.3   A clinical laboratory found that a blood sample contained 0.0026 g of phosphorus and 0.000101 g of iron. (a) Give these quantities in scientific notation. (b) Give these quantities in the units in which they are usually reported—milligrams for phosphorus and micrograms for iron.

Solution   (a) 0.0026 g phosphorus $= 2.6 \times 10^{-3}$ g phosphorus

0.000101 g iron $= 1.01 \times 10^{-4}$ g iron .

(b) Since 1 mg $= 1 \times 10^{-3}$ g, 2.6×10^{-3} g phosphorus $= 2.6$ mg phosphorus. Since 1 mg $= 1 \times 10^{-6}$ g, restating the iron content with an exponent of -6 makes it easy to find the mass in micrograms:

$$1.01 \times 10^{-4} \text{ g iron} = 101 \times 10^{-6} \text{ g iron} = 101 \ \mu\text{g iron}$$

Practice Problems   **2.5**   Convert these values to scientific notation:
(a) 58 g   (b) 46,792 m   (c) 0.0006720 cm   (d) 345.3 kg

2.6   Convert these values from scientific notation back to normal notation:
(a) 4.885×10^4 mg   (b) 8.3×10^{-6} m   (c) 4.00×10^{-2} m

2.7   Ordinary table salt, or sodium chloride, is made up of small particles called ions, which we'll discuss in Chapter 5. If the distance between a sodium ion and a chloride ion is 0.000000000278 m, what is this distance in scientific notation? How many picometers (pm) is this?

2.3   MEASURING MASS

Mass   The amount of matter in an object.

Weight   The measure of the gravitational force exerted on an object by the earth, the moon, or another massive body.

The terms "mass" and "weight," although often used interchangeably, really have quite different meanings. **Mass** is a measure of the amount of matter in an object, whereas **weight** is a measure of the gravitational force that the earth, the moon, or another massive body exerts on the object. Clearly, the amount of matter (mass) in an object does not depend on its location. Whether you're standing on the earth, riding in a space shuttle, or standing on the moon, the amount of matter (mass) in your body is the same. Just as clearly, however, the weight of an object *does* depend on its location. Your weight might be 140 lb on earth, 0 lb in a space shuttle, and 23 lb on the moon, since the pull of gravity is quite different in these locations.

At the same location, two objects of identical mass have identical weights; that is, gravity pulls equally on both. Thus, the *mass* of an object can be determined on a chemical balance by comparing the *weight* of the object to the weight of a known reference standard. Much of the confusion between mass and weight is simply due to a language problem: We speak of "weighing" when we really mean that we're determining *mass* by *comparing* two weights. Figure 2.2 shows a modern electronic balance with a digital readout used for determining mass in the laboratory.

Milligram (mg)   A unit of mass equal to 1/1000 g.

Microgram (μg)   A unit of mass equal to 1/1000 mg.

One kilogram, the SI unit for mass, is just over 2 lb, which is too large for many purposes in chemistry and medicine. Thus, smaller units of mass such as the gram, the **milligram (mg),** and the **microgram (μg)** are more commonly used. Table 2.3 shows the relationships between metric and common units for mass.

Figure 2.2
A modern electronic balance that gives the weight as a digital readout to four decimal places.

Table 2.3   Units of Mass

Unit	Equivalent	Unit	Equivalent
1 kilogram (kg)	= 1000 grams = 2.205 pounds	1 ton	= 2000 pounds = 907.03 kilograms
1 gram (g)[a]	= 0.001 kilogram = 1000 milligrams = 0.03527 ounce	1 pound (lb)	= 16 ounces = 0.454 kilogram = 454 grams
1 milligram (mg)	= 0.001 gram = 1000 micrograms	1 ounce (oz)	= 0.02835 kilogram = 28.35 grams = 28,350 milligrams
1 microgram (μg)[b]	= 0.000001 gram = 0.001 milligram		

[a] Sometimes symbolized Gm in medical usage.
[b] Sometimes symbolized γ (gamma).

2.4 MEASURING LENGTH

Length The distance an object extends in a given direction.

The meter is the standard measure of **length** or distance in the metric system. One meter is 39.37 inches or about 10% longer than our usual yard, a length that is much too large for most measurements in chemistry and medicine. Other, more commonly used measures of length are the **centimeter** (**cm**; 1/100 m) and the **millimeter** (**mm**; 1/1000 m). One centimeter is a bit less than a half-inch, 0.3937 in. to be exact. A millimeter, in turn, is 0.03937 in., or about the thickness of a dime. Table 2.4 lists the relationships of these units, and Figure 2.3 shows a comparison between inches and centimeters.

Centimeter (cm) A unit of length equal to 1/100 m, or about 0.3937 in.

Millimeter (mm) A unit of length equal to 1/1000 m.

2.5 MEASURING VOLUME

Volume The amount of space occupied by an object or a sample of a substance.

Volume is the amount of space occupied by an object or substance. To measure a quantity of a chemical compound, whether in the home or garden, in the chemical industry, or for a medication, it's often more convenient to measure the volume of a solution of that compound than to weigh a specific mass. Therefore volume measurements are quite common.

Table 2.4   Units of Length

Unit	Equivalent	Unit	Equivalent
1 kilometer (km)	= 1000 meters = 0.6214 mile	1 mile (mi)	= 1.609 kilometers = 1609 meters
1 meter (m)	= 100 centimeters = 1000 millimeters = 1.0936 yards = 39.37 inches	1 yard (yd)	= 0.9144 meter = 91.44 centimeters
		1 foot (ft)	= 0.3048 meter = 30.48 centimeters
1 centimeter (cm)	= 0.01 meter = 10 millimeters = 0.3937 inch	1 inch (in.)	= 2.54 centimeters = 25.4 millimeters
1 millimeter (mm)	= 0.001 meter = 0.1 centimeter		

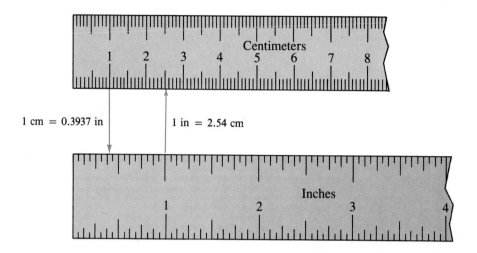

Figure 2.3
Comparison of a centimeter with an inch.

Milliliter (mL) A unit of volume equal to 1/1000 L.

Cubic centimeter (cc or cm³) An alternative name for a milliliter; volume of a 1 cm cube.

    The liter (L) is the most commonly used measure of volume in chemistry and medicine. One liter has the volume of a cube 10 cm on each edge (Figure 2.4) and is just a bit larger than one U.S. quart. Each liter is further divided into 1000 **milliliters (mL),** and one milliliter is the size of a cube one centimeter on each edge. In fact, the milliliter is often called a **cubic centimeter (cm³ or cc)** in medical work. Table 2.5 shows the relationships among measures of volume.

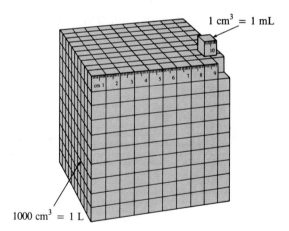

Figure 2.4
A cubic meter. A cube 10 cm on a side has a volume of 1000 cm³, which is equal to 1 liter. A cube 1.0 cm on a side has a volume of 1 cm³, or 1 mL. There are 1000 mL in a liter.

Table 2.5   Units of Volume

Unit	Equivalent	Unit	Equivalent
1 cubic meter (m³)	= 1000 liters	1 gallon (gal)	= 3.7856 liters
	= 264.2 gallons	1 quart (qt)	= 0.9464 liter
1 liter (L)	= 0.001 m³		= 946.4 milliliters
	= 1000 milliliters	1 fluid ounce (fl oz)	= 29.57 milliliters
	= 1.057 quarts		
1 deciliter (dL)	= 0.1 liter		
	= 100 milliliters		
1 milliliter (mL)	= 0.001 liter		
	= 1000 microliters		
1 microliter (μL)	= 0.001 milliliter		

AN APPLICATION: APOTHECARY UNITS

Chances are that the bottle of aspirin in your medicine cabinet gives the quantity per tablet in the mass unit of grains (gr). Occasionally a prescription may also be written for a dosage in grains.

$$1 \text{ grain} = 64.8 \text{ mg}$$

The grain is known as an *apothecary unit,* indicating that it is part of a system of measurement devised for dispensing medications and once used by all physicians and pharmacists. The system includes our customary gallons, quarts, and pints, plus some less familiar units, a few of which are still in use in certain applications. The *grain* was derived, as were many older units, from something easily available and of convenient size. Once the smallest defined unit of mass, the grain was equal to the mass of a plump grain of wheat.

The basis for liquid measurement in the apothecary system is the *minim* (a term meaning the smallest quantity of something), which was taken as the volume of one drop of water. One minim is now equal to 0.062 mL; 60 minims exactly equal one fluidram (roughly one teaspoonful); and one fluid ounce exactly equals 8 fluidrams. The fluidram is still occasionally used for fluid volume in medications and for very expensive perfume.

$$1 \text{ minim} = 0.062 \text{ mL}$$

$$1 \text{ fluidram} = 3.7 \text{ mL (60 minims)}$$

The medications in this fifteenth-century pharmacy were most likely measured out in drams and grains.

2.6 MEASUREMENT AND SIGNIFICANT FIGURES

Figure 2.5
Measuring a liquid. What is the volume of water in this graduated cylinder?

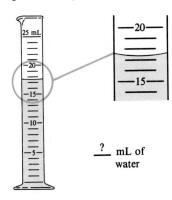

? mL of water

How much does a nickel weigh? If you put a nickel on an ordinary bathroom scale, the scale would probably register 0 lb (or 0 kg if you've gone metric). If you placed the same nickel on a common laboratory scale, however, you might get a reading of 4.95 g. Trying again by placing the nickel on an expensive analytical balance like those found in clinical and research laboratories, you might find a weight of 4.94438 g. Clearly, the exactness of your answer depends on the equipment used for the measurement.

Every experimental measurement, no matter how exact, has a degree of uncertainty to it because there's always a limit to the number of digits that can be determined. An analytical balance may be pushed to its limits in measuring a mass to the fifth decimal place, and weighing the nickel a second time might give 4.94439 g or 4.94437 g. Also, different people making the same measurement might come up with slightly different answers. (How would you record the volume of the liquid shown in Figure 2.5?)

In order to indicate the exactness of a measurement, the value recorded should use all the digits known with certainty, plus one additional estimated digit that is usually considered uncertain by plus or minus 1 (written as ± 1). The total number of digits used to express a value in this manner is called the number of **significant figures.** Thus, the quantity 4.95 g has three significant figures (4, 9, and 5), and the quantity 4.94438 g has six significant figures. Remember: All but one of the significant figures are known with certainty; the final significant figure is only an estimate accurate to plus or minus 1.

Significant figures In describing a quantity, the total of the number of digits whose values are known with certainty, plus one estimated digit.

Uncertain digit

4.95 g —a mass between 4.94 g and 4.96 g (± 0.01 g)

Uncertain digit

4.94438 g —a mass between 4.94437 g and 4.94439 g (± 0.00001 g)

Counting the number of significant figures in a measurement is usually simple, but can be troublesome when zeroes are involved. Depending on the circumstances, zeroes might be significant, or they might simply be space fillers to locate the decimal point. For example, how many significant figures does each of the following quantities have?

- 94.072 g [five significant figures (9, 4, 0, 7, 2)]
- 0.0834 g [three significant figures (8, 3, 4)]
- 23,000 g [*anywhere* from two (2, 3) to five (2, 3, 0, 0, 0) significant figures]
- 138.200 g [six significant figures (1, 3, 8, 2, 0, 0)]

At this point, you may be wondering why counting significant figures is important. Consider that when you divide 3 by 7 on your calculator, the result will look something like this,

0.428571429   Result of dividing 3 by 7 on a digital calculator

which may well be far more digits than you should use. How many of these nine digits are significant depends upon whether you divided 3.0 by 7.0 or 3.000 by 7.000, and so on. In order to use a digital calculator sensibly, understanding significant figures is very important. (Appendix A explains how to multiply and divide numbers expressed in scientific notation.) First, you must determine how many digits in an answer are significant, and then you must round off the answer shown by the calculator, as described in the next section.

The following rules are helpful for counting significant figures when zeroes are present:

1. *Zeroes in the middle of a number are always significant.* Thus, 94.072 has five significant figures.

2. *Zeroes at the beginning of a number are not significant.* They serve only to locate the decimal point. Thus, 0.0834 has three significant figures.

3. *Zeroes at the end of a number but *before* an implied decimal point may or may not be significant.* In 23,000 we can't tell whether the zeroes are part of the measurement or whether they serve only to locate the decimal point (which is implied but not written). Thus, 23,000 may have two, three, four, or five significant figures.

4. *Zeroes at the end of a number and also before a decimal point are significant.* For small numbers such as 10 or 290, the significance of the final zeros is sometimes indicated by adding a decimal point, 10. or 290. for example. To help you in learning how to use significant figures correctly, we'll follow this practice. Thus 10. has two significant figures and 200. and 290. each have three significant figures.

5. *Zeroes at the end of a number but *after* the decimal point are significant.* (We assume that these zeroes wouldn't be indicated unless they were significant.) Thus, 138.200 has six significant figures.

Scientific notation is particularly helpful for showing how many significant figures are present in a large number with zeroes at the end and avoiding the problem mentioned in rule 3. If you read that the distance from the earth to the sun is 93,000,000 mi, you don't really know how many significant figures are indicated. If, instead, the number is given as 9.3×10^7, there is no doubt that it has two significant figures. Or, it might have four significant figures and be written as 9.300×10^7 instead.

One final point about determining significant figures: Some numbers are *exact,* and for them the idea of significant figures is meaningless. Such exact numbers often occur in counting objects and in converting quantities from one unit to another. Thus, a class might have *exactly* 32 students (not 31.9, 32.0, or 32.1); similarly, there are *exactly* 100 cents to the dollar.

Solved Problem 2.4	How many significant figures does each of the following quantities have? (a) 2730.78 m (b) 0.0076 mL (c) 3400 kg (d) 3400.0 m² (e) 90. g
Solution	(a) six (rule 1) (b) two (rule 2) (c) two, three, or four (rule 3) (d) five (rule 5) (e) two (rule 4)

Practice Problems

2.8 How many significant figures does each of the following quantities have? Explain your answers.
(a) 3.45 m   (b) 0.1400 kg   (c) 10.003 L   (d) 35 cents

2.9 Rewrite the following numbers as indicated:
(a) 600,000 with two significant figures
(b) 1300 with four significant figures
(c) 794,200,000,000 with four significant figures

2.7   ROUNDING OFF NUMBERS

It often happens that a measurement appears to have more significant figures than are really justified by the precision of the work. You might read in a world almanac, for example, that the population of the United States in 1980 was 226,549,448. Because of the errors inherent in census taking, it's very unlikely that there were *exactly* this number of people. More likely, the number you read is accurate only to two or three significant figures and should be **rounded off** to 227,000,000 people.

Rounding off   The procedure for expressing values with the correct number of significant figures.

There are only two rules for rounding off numbers:

1. *If the first digit you decide to remove is 4 or less, simply drop it and all following digits.* Thus, 2.4271 becomes 2.4 when rounded off to two significant figures because the first of the dropped digits (a 2) is 4 or less.

2. *If the first digit you decide to remove is 5 or greater, round the number upward by adding a 1 to the digit to the left of the digit you drop.* Thus, 4.5832 becomes 4.6 when rounded off to two significant figures because the first of the dropped digits (an 8) is 5 or greater.

The need to round off numbers is common when doing arithmetic calculations on physical quantities, where the results often contain more digits than are significant (especially when using a hand-held calculator). There are two rules for dealing with these situations:

1. *In carrying out a multiplication or division of two numbers, the result can't have more significant figures than either of the two original numbers.* Suppose you want to calculate the mileage your car is getting. If you drove 187.3 mi and used 5.2 gal of gasoline, you could find your mileage by dividing the number of miles driven by the amount of gasoline used:

$$\text{Mileage} = \frac{\text{number of miles driven}}{\text{number of gallons used}}$$

$$= \frac{187.3 \text{ mi}}{5.2 \text{ gal}} \quad \begin{array}{l} \text{Four significant figures} \\ \\ \text{Two significant figures} \end{array}$$

$$= 36.01923077 \text{ mi/gal}$$

$$= 36 \text{ mi/gal (rounded off)}$$

Since the number of gallons is known to only two significant figures, the calculated mileage must also be rounded off to two significant figures, giving a value of 36 mi/gal.

2. *If two numbers are added or subtracted, the result can't have more digits to the right of the decimal point than either of the two original numbers.* If you add 4.672 mL of a sterile saline solution to a container already holding 34.8 mL of solution, the total volume will have to be rounded off:

Volume of solution added	4.672 mL	(three digits after decimal point)
Volume already present	34.8 mL	(one digit after decimal point)
Total volume	39.472 mL	(unrounded)
	39.5 mL	(rounded off to one digit after decimal point)

(In addition or subtraction, it's helpful to draw a line after the number with the fewest decimal places.)

Rounding off the total volume is necessary because of the difference in precision of the two measurements. Since we don't know the volume of solution in the container past the first digit after the decimal point (it could be any value between 34.7 mL and 34.9 mL), we can't know the total of the combined volumes past the same digit.

Solved Problem 2.5   Suppose that you weigh 124 lb before dinner. How much will you weigh after dinner if you eat a total of 1.884 lb of food?

Solution   Your after-dinner weight is found by adding your original weight to the weight of the food consumed:

$$124 \text{ lb} + 1.884 \text{ lb} = 125.884 \text{ lb (unrounded)}$$
$$= 126 \text{ lb (rounded)}$$

Since the value of your original weight has no significant figures after the decimal point, your after-dinner weight also must have no significant figures after the decimal point. Thus, 125.884 lb must be rounded off to 126 lb.

Solved Problem 2.6   To make currant jelly, 13.75 cups of sugar was added to 18 cups of currant juice. How much sugar was added per cup of juice?

Solution   The quantity of sugar must be divided by the quantity of juice.

$$\frac{13.75 \text{ cups sugar}}{18 \text{ cups juice}} = 0.763888889 \text{ cup sugar per cup juice (unrounded)}$$
$$= 0.76 \text{ cup sugar per cup juice (rounded)}$$

The number of significant figures in the answer is limited to two by the quantity 18 cups in the calculation. (The jelly will have just the right sweetness.)

Practice Problems   **2.10**   Round off each of the following quantities to the indicated number of significant figures:
(a) 2.304 g (three significant figures)
(b) 188.3784 mL (five significant figures)
(c) 0.00887 L (one significant figure)
(d) 1.00039 kg (four significant figures)

2.11   Carry out the following calculations, rounding each result to the correct number of significant figures:
(a) 4.87 mL + 46.0 mL   (b) 3.4 × 0.023 g   (c) 19.333 m − 7.4 m
(d) 55 mg − 4.671 mg + 0.894 mg   (e) 62,911 ÷ 611

2.8 CALCULATIONS: CONVERTING A QUANTITY FROM ONE UNIT TO ANOTHER

Many activities in the laboratory and in medicine require converting a quantity from one unit to another. For example, if there are 10 ounces of sodium bicarbonate in this box, how much is that in grams? Or, these pills contain 1.3 grains of aspirin, but I need 200 mg. Is one pill enough?

Converting between units isn't mysterious; we all do it every day. If you run 9 laps around a quarter-mile track, you have to convert between the distance unit "lap" and the distance unit "mile" to find that you have run 2.25 mi (9 laps divided by 4 laps/mi).

The simplest way to carry out calculations involving different units is to use the **factor-label method,** in which units (labels) are treated in the same way as numbers: Units can be multiplied or divided, just as numbers can. In solving a problem, the idea is to set up an equation so that the unwanted units cancel, leaving only the desired units. The units are a guide to setting the problem up correctly, for example,

Factor-label method A method of problem solving in which equations are set up so that unwanted units cancel and only the correct units remain.

$$10. \text{ oz sodium bicarbonate} \times \frac{28.35 \text{ g}}{1 \text{ oz}} = 280 \text{ g sodium bicarbonate}$$

In the factor-label method, as illustrated above, a physical quantity described in one unit is converted to a quantity in different units by using a **conversion factor:**

Conversion factor A fraction that states a relationship between two different units of measure.

(Quantity in old units) × (conversion factor) = (quantity in new units)

A conversion factor is simply any specific relationship between units, such as those given in Tables 2.3 through 2.5. For example, Table 2.4 shows that 1 km = 0.6214 mi. Restating this relationship in the form of conversion factors gives miles per kilometer or kilometers per mile:

$$\frac{0.6214 \text{ mi}}{1 \text{ km}} \qquad \frac{1 \text{ km}}{0.6214 \text{ mi}} \qquad \text{These two conversion factors express the same relationship}$$

Thus, to find out how many kilometers there are in a marathon (26.22 mi):

$$26.22 \text{ mi} \quad \times \quad \frac{1 \text{ km}}{0.6214 \text{ mi}} \quad = \quad 42.20 \text{ km}$$

Starting quantity   Conversion factor   New quantity

The mile unit (mi) cancels from the left side of the equation since it appears both above and below the division line, leaving kilometer (km) as the only remaining unit. To find out how many miles there are in a "10 K" race requires the opposite conversion factor:

$$10. \text{ km} \times \frac{0.6214 \text{ mi}}{1 \text{ km}} = 6.2 \text{ mi}$$

The factor-label method always gives the right answer if the equation is set up so that unwanted units cancel and you do the arithmetic correctly.

Solved Problem 2.7 Write the conversion factors for the following pairs of units (use Tables 2.3 through 2.5): (a) deciliters and milliliters (b) pounds and grams (c) quarts and milliliters

Solution (a) $\dfrac{1\,\text{dL}}{100\,\text{mL}} \quad \dfrac{100\,\text{mL}}{1\,\text{dL}}$ (b) $\dfrac{1\,\text{lb}}{454\,\text{g}} \quad \dfrac{454\,\text{g}}{1\,\text{lb}}$ (c) $\dfrac{1\,\text{qt}}{946.4\,\text{mL}} \quad \dfrac{946.4\,\text{mL}}{1\,\text{qt}}$

Solved Problem 2.8 (a) Convert 0.75 lb to grams. (b) Convert 0.50 qt to deciliters.

Solution (a) Set up an equation so that the pound units cancel:

$$0.75\ \cancel{\text{lb}} \times \frac{454\ \text{g}}{1\ \cancel{\text{lb}}} = 340\ \text{g}$$

(b) In this, as in many cases, it's convenient to use more than one conversion factor:

$$0.50\ \cancel{\text{qt}} \times \frac{946.4\ \cancel{\text{mL}}}{1\ \cancel{\text{qt}}} \times \frac{1\ \text{dL}}{100\ \cancel{\text{mL}}} = 4.7\ \text{dL}$$

So long as the units cancel correctly, two or more conversion factors can be strung together in the same calculation.

Practice Problems **2.12** Write the conversion factors for the following pairs of units: (a) liters and milliliters (b) grams and ounces (c) liters and quarts

2.13 (a) Convert 16.0 oz to grams.
(b) Convert 2500 mL to liters.
(c) Convert 99.0 L to quarts.

2.14 Convert 0.840 qt to milliliters in a single calculation.

2.9 PROBLEM SOLVING; DOSAGE CALCULATIONS

The main drawback to using the factor-label method is that it's easy to get the "right" answer without understanding what you're doing. It's therefore best when first approaching a problem, wherever possible, to think through a rough ballpark answer in your head before doing the exact calculation. If your rough guess isn't close to the exact answer, then there's a misunderstanding somewhere. Solved Problem 2.9 illustrates how to do this.

Solved Problem 2.9   A child is 21.5 in. long at birth. How long is this in centimeters?

Ballpark Solution   In Table 2.4 we see that 1 in. equals 2.54 cm. In other words, a centimeter is smaller than an inch, and it takes about two and one-half times more centimeters than inches to measure the same length. Thus, 21.5 in. ought to be about 2.5 times 20, or approximately 50 cm.

Exact Solution   With the factor-label method, the equation is set up so that the inch units cancel:

$$21.5 \; \cancel{\text{in.}} \times \frac{2.54 \text{ cm}}{1 \; \cancel{\text{in.}}} = 54.6 \text{ cm (rounded off from 54.61)}$$

The ballpark solution and the exact solution agree closely.

Figure 2.6
Intravenous medications. One of many situations in which careful attention to units and numbers is of crucial importance.

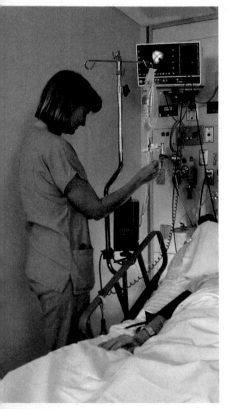

The factor-label method and the use of an estimate, or ballpark solution, are techniques that will help you solve problems of any type, not just simple unit conversions.

Many calculations include combined units, such as speed in miles per hour or concentration in milligrams per milliliter. Calculations related to dosage of medication use combined units in a variety of ways—milligrams of medication per tablet, milligrams of medication per milliliter of solution, milliliters of intravenous solution per hour, and so on. Suppose a patient is receiving an intravenous solution at the rate of 4.5 mL/min (Figure 2.6). How many minutes will it take to deliver 500. mL? The combined unit is used in exactly the same way as a conversion factor.

$$500. \; \cancel{\text{mL}} \times \frac{1 \text{ min}}{4.5 \; \cancel{\text{mL}}} = 111.111 \ldots \text{ min (unrounded)}$$
$$= 111 \text{ min (rounded)}$$

Note that calling the rate 1 min/4.5 mL expresses the same relationship as 4.5 mL/min.

Problems sometimes seem complicated, but you can sort out the complications by analyzing the problem in the following manner:

- Identify the information known.
- Identify the information needed in the answer.
- Decide what to do with the known information to convert it to the answer.

Application of the factor-label method wherever possible is an essential part of this analysis. The following dosage problems illustrate how to use analysis as an aid in problem solving. Ordinarily, though, you'd do the analysis mentally rather than writing it out. Try this with Practice Problems 2.15 and 2.16.

Solved Problem 2.10   Administration of digitalis to control atrial fibrillation in heart patients must be carefully controlled, because even a modest overdose can be fatal. To take differences between patients into account, dosages are sometimes prescribed in micrograms per kilogram of body weight (μg/kg). Thus, a child and an adult may differ greatly in weight, but both will receive the same dosage. At a dosage of 20 μg/kg body weight, how many milligrams of digitalis should a 160 lb patient receive?

Analysis   The known information is (1) the patient's weight in pounds and (2) the dosage in micrograms per kilogram of body weight. The needed information is the dose in milligrams. Two unit conversions are therefore required—from known pounds of body weight to kilograms and from known micrograms to milligrams. The problem is solved by using the conversion factors based on 1 kg = 2.205 lb and 1 mg = 1000 μg, so that all units except milligrams cancel.

Ballpark Solution   A kilogram is roughly equal to 2 lb. A 160 lb patient will therefore weigh about 80 kg. At a dosage of 20 μg/kg, an 80 kg patient should receive 80 × 20 μg or about 1600 μg of digitalis. Since 1 mg = 1000 μg, it will take only one-thousandth as many milligrams as micrograms to equal the same dose. Our ballpark answer is about 1.6 mg of digitalis.

Exact Solution
$$160 \; \cancel{lb} \times \frac{1 \; \cancel{kg}}{2.205 \; \cancel{lb}} \times \frac{20 \; \cancel{\mu g} \; \text{digitalis}}{1 \; \cancel{kg}} \times \frac{1 \; \text{mg}}{1000 \; \cancel{\mu g}} = 1.5 \; \text{mg digitalis}$$
(rounded off)

The ballpark solution and the exact solution agree.

Solved Problem 2.11   A patient requires injection of 0.16 grain of a pain killer available as a 15 mg/mL solution. How many milliliters should be administered?

Analysis   The known information is the mass of the medication in grains and the concentration of the solution in milligrams per milliliter. The unknown is the needed milliliters of solution. Before the number of milliliters can be found, grains must be converted to milligrams (1 grain = 64.8 mg).

Solution
$$0.16 \; \cancel{\text{grain}} \times \frac{64.8 \; \cancel{mg}}{1 \; \cancel{\text{grain}}} \times \frac{1 \; \text{mL}}{15 \; \cancel{mg}} = 0.69 \; \text{mL}$$

Ballpark Solution   A ballpark solution can't always be found easily before the calculation is done. In such cases, the answer should *always* be checked by finding a ballpark solution afterward. To find a ballpark solution at this point, look for an easy way to round off the numbers in the problem so that you can mentally check the math. Here, notice that 0.16 grain/15 mg is close to 0.15/15, or 0.01, times about 65, giving a ballpark solution of 0.65, which is close to the exact solution.

Practice Problems   **2.15**   Carry out the necessary conversions to answer the following questions:
(a) How many kilograms does a 7.5 lb infant weigh?
(b) How many milliliters are in a 4.0 fl oz bottle of cough medicine?

2.16   Calculate the dosage in milligrams per kilogram body weight for a 135 lb adult who takes two aspirin tablets weighing 0.324 g each. Calculate the dosage for a 40 lb child who also takes two aspirin tablets.

2.10   MEASURING TEMPERATURE

Temperature   The relative measure of how hot or cold something is; heat flows between objects or materials that are at different temperatures and are in contact.

Just when you were getting used to hearing daily **temperatures** given on the weather report in both Fahrenheit (°F) and Celsius (°C) units, along comes yet a third system. The SI unit for temperature degrees is the kelvin (K). (Note that we say only "kelvin," not "degrees kelvin.")

Both the kelvin and the Celsius degree are defined as 1/100 of the interval between the freezing point of water (0°C) and the boiling point of water (100°C). Although the size of a kelvin and the size of a Celsius degree are identical, the numbers assigned to various points on the scales differ. Whereas the Celsius scale defines the freezing point of water as 0°C, the Kelvin scale defines the coldest possible temperature, $-273.15°C$ (sometimes called *absolute zero*), as 0 K. Thus, 0 K = $-273.15°C$, and 273.15 K = 0°C as the following equations indicate (for most purposes, rounding off to 273 is sufficient):

$$\text{Temperature in K} = \text{temperature in °C} + 273$$

$$\text{Temperature in °C} = \text{temperature in K} - 273$$

For practical applications in medicine and clinical chemistry, both the Fahrenheit scale and the Celsius scale are still used almost exclusively. The Fahrenheit scale defines the freezing point of water as 32°F and the boiling point of water as 212°F. Thus, it takes 180 Fahrenheit degrees to cover the same range encompassed by only 100 Celsius degrees, and a Celsius degree is therefore exactly 180/100 = 9/5 = 1.8 times larger than a Fahrenheit degree.

Converting between the Fahrenheit and Celsius scales is similar to converting between different units of length or volume, but is a bit more complex because *two* corrections need to be made—one to adjust for the difference in degree size and one to adjust for the different zero points. The size correction is made by using the exact relationship 1C° = 1.8F°. The zero-point correction is made by remembering that the freezing point is *higher* by 32 on the Fahrenheit scale than it is on the Celsius scale. The following formulas show the conversion methods:

Celsius to Fahrenheit

$$°F = \left(\frac{1.8°F}{1°C} \times °C \right) + 32° \qquad \text{or} \qquad °F = (1.8 \times °C) + 32°$$

Fahrenheit to Celsius

$$°C = (°F - 32°) \times \frac{1°C}{1.8°F} \quad \text{or} \quad °C = \frac{(°F - 32°)}{1.8}$$

(You may already have learned to do these conversions using the fractions 9°C/5°F and 5°F/9°C. Use whichever method you find easiest to remember.)

The quickest way to become familiar with converting between the Fahrenheit and Celsius scales is to remember a few convenient points and make rough estimates of values intermediate between these remembered points. Figure 2.7 lists some common temperatures in both Celsius and Fahrenheit degrees.

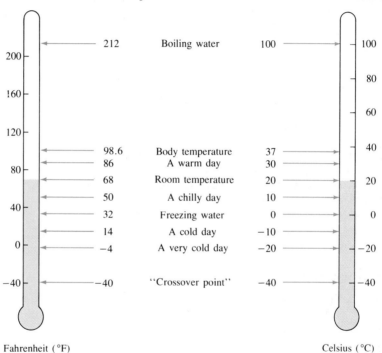

Figure 2.7
A comparison of the Fahrenheit and Celsius scales.

212	Boiling water	100
98.6	Body temperature	37
86	A warm day	30
68	Room temperature	20
50	A chilly day	10
32	Freezing water	0
14	A cold day	−10
−4	A very cold day	−20
−40	"Crossover point"	−40

Fahrenheit (°F)   Celsius (°C)

Solved Problem 2.12 The highest land temperature ever recorded was 136°F in Al Aziziyah, Libya, on September 13, 1922. What is this temperature on the Celsius scale?

Ballpark Solution A temperature of 136°F is about 100 Fahrenheit degrees above freezing (32°F). Since Celsius degrees are about twice as large as Fahrenheit degrees (1C° = 1.8F°), it takes about one-half as many of them to span the same range. Freezing on the Celsius scale is 0°C, so our rough answer is that 136°F equals about 50°C.

Exact Solution According to the equation given above,

$$(136°F - 32°) \times \frac{1°C}{1.8°F} = 57.8°C \quad \text{(rounded off from 57.77778°C)}$$

The ballpark solution and the exact solution agree.

Practice Problems   **2.17**   What is the body temperature on the Celsius scale of someone with a fever of 103°F?

2.18   The use of mercury thermometers is limited by the fact that mercury freezes at −38.9°C. What does this correspond to on the Fahrenheit scale?

2.11   HEAT AND ENERGY

Heat   A form of energy transferred from a hotter object to a colder one.

Joule (J)   The SI unit of energy.

Calorie (cal)   A common unit of energy equal to exactly 4.184 J.

Kilocalorie (kcal)   A unit of energy equal to 1000 cal.

What is heat? Everyone knows what it feels like, but few people really know what it is. In fact, **heat** is a form of energy that is transferred from a hotter object to a colder object when the two come in contact. Energy is measured in SI units by the **joule (J**; pronounced "jool"), but the metric **calorie (cal)** is still much more widely used. One calorie is just about the amount of heat necessary to raise the temperature of one gram of water by one Celsius degree. A **kilocalorie (kcal),** often called a *large calorie* by nutritionists and represented by C, equals 1000 cal.

$$1000 \text{ cal} = 1 \text{ kcal} \qquad 1000 \text{ J} = 1 \text{ kJ}$$
$$1 \text{ cal} = 4.184 \text{ J} \qquad 1 \text{ kcal} = 4.184 \text{ kJ}$$

Often when something is heated or cooled, we need to know *how much* heat is gained or lost. To find this out we measure temperature changes. The magnitude of a temperature change depends on both the mass and the identity of the material being heated or cooled. One calorie will raise the temperature of one gram of water by one Celsius degree, but the same amount of heat will raise the temperature of one gram of iron by 10 Celsius degrees. The amount of heat that will raise the temperature of one gram of a substance by one Celsius degree is called the **specific heat:**

Specific heat   The amount of heat that will raise the temperature of 1 gram of a substance by one Celsius degree.

$$\text{Specific heat} = \frac{\text{calories}}{\text{gram °C}}$$

Table 2.6
Specific Heats of Some Common Substances

Substance	Specific Heat (cal/g °C)
Water	1.00
Alcohol	0.59
Mercury	0.033
Gold	0.031
Iron	0.106
Sodium	0.293

Specific heats vary greatly from one substance to another, as shown in Table 2.6. The specific heat of water is higher than that of most other substances, which means that a large transfer of heat is required either to cool (remove heat from) or warm (add heat to) a given amount of water.

The heat added or removed and the temperature change for a substance are related as follows:

$$\text{Heat (cal)} = \text{mass (g)} \times \text{temperature change (°C)} \times \text{specific heat (cal/g °C)}$$

where the temperature change is the difference between the temperature before and the temperature after heat is added or removed. Any of the four quantities in this relationship can be calculated if the other three are known.

Solved Problem 2.13   Taking a bath might use about 95 kg water. How much energy (in calories) is needed to heat the water from a cold 15°C to a warm 40°C?

Ballpark Solution   The water is being heated 25 C° (from 15°C to 40°C), and it therefore takes 25 cal to heat each gram. Since 95 kg of water is 95,000 g, or about 100,000 g, it takes about 25 × 100,000 cal, or 2,500,000 cal, to heat the entire tub of water.

Exact Solution   Set up an equation so that unwanted units cancel:

$$25°C \times \frac{1000\ g}{kg} \times \frac{1.0\ cal}{g\ °C} \times 95\ kg = 2,400,000\ cal$$

Practice Problems **2.19** Assuming that Coca-Cola has the same specific heat as water, how much energy in calories is removed when 350 g of Coke (about the contents of one 12 oz can) is cooled from 25°C to 3°C?

2.20 If it takes 161 cal to raise the temperature of a 75 g bar of aluminum by 10 C°, what is the specific heat of aluminum?

2.12 DENSITY

We've all heard the question, Which weighs more, a pound of feathers or a pound of lead? The answer, of course, is that they both weigh the same. A pound of feathers has far more *volume* than a pound of lead, but a pound is still a pound.

Density The mass of an object per unit of volume.

The physical property that relates the mass of an object to its volume is called **density.** Density, which is simply mass divided by volume for any material or object, is usually expressed in units of grams per milliliter (g/mL) for liquids and grams per cubic centimeter (g/cm³) for solids. The densities of some common materials are listed in Table 2.7. Any substance with density less than that of water will float on water, but any substance with density greater than that of water will sink.

Table 2.7   Densities of Some Common Materials at 25°C and 1 atm Pressure

Substance	Density (g/mL)	Substance	Density (g/mL)
Ice (0°C)	0.917	Human fat	0.94
Water (3.98°C)	1.0000	Cork	0.22–0.26
Gold	19.3	Table sugar	1.59
Helium	0.000194	Balsa wood	0.12
Air	0.001185	Earth	5.54
Urine	1.003–1.030	Blood plasma	1.027

$$\text{Density} = \frac{\text{mass (g)}}{\text{volume (mL or cm}^3)}$$

Most substances, especially gases, change in volume when heated or cooled, and densities are therefore temperature-dependent. For example, at 3.98°C, a 1 mL container holds exactly 1.0000 g of water (density = 1.0000 g/mL). But, as the temperature is raised the volume occupied by the water expands so that only 0.9584 g fits in the 1 mL container at 100°C (density = 0.9584 g/mL). When reporting a density, the temperature must also be specified.

Knowing the density of a liquid can be very useful, since it's often easier to measure the volume of a liquid than its weight. Suppose, for example, that you need 1.5 g of ethyl alcohol. Rather than use a dropper to weigh out exactly the right amount, it would be much easier to look up the density of ethyl alcohol (0.7893 g/mL at 20°C) and measure out the correct volume with a syringe or graduated cylinder. As the following calculation shows, density is really just a conversion factor for changing between mass (grams) and volume (milliliters).

$$1.5 \text{ g ethyl alcohol} \times \frac{1 \text{ mL}}{0.7893 \text{ g}} = 1.9 \text{ mL ethyl alcohol}$$

Solved Problem 2.14 What volume of isopropyl alcohol (rubbing alcohol) would you use if you needed 25.0 g? The density of isopropyl alcohol is 0.7855 g/mL at 20°C.

Ballpark Solution Since 1 mL of isopropyl alcohol weighs a bit less than 1 g, it will take a bit more than 25 mL to get 25 g—perhaps about 30 mL.

Exact Solution Use density as a conversion factor and set up an equation so that the unwanted units cancel:

$$25.0 \text{ g isopropyl alcohol} \times \frac{1 \text{ mL}}{0.7855 \text{ g}} = 31.8 \text{ mL isopropyl alcohol}$$

The ballpark solution and the exact solution agree.

Practice Problems **2.21** Which of the solids whose densities are given in Table 2.7 will float on water, and which will sink?

2.22 Chloroform, once used as an anesthetic agent, has a density of 1.474 g/mL. What volume would you use if you needed 12.37 g?

2.23 A glass stopper weighing 16.8 g has a volume of 7.6 cm^3. What is the density of the glass?

AN APPLICATION: MEASURING THE PERCENTAGE OF BODY FAT

Much as we might complain about it, none of us would survive long without the layer of adipose tissue (body fat) lying just under our skin. Not only does this fat layer act as a shock absorber and as a thermal insulator to maintain body temperature, it also serves as a long-term energy storehouse. A normal adult body contains about 50% cellular protoplasm, 24% blood and other extracellular fluids, 7% bone, and 19% body fat. Overweight sedentary individuals have a higher fat percentage, while some world-class female gymnasts are reported to have as little as 3% body fat.

The percentage of body fat is measured by the underwater immersion method shown in the accompanying figure. Suppose that a person weighs 60.0 kg on land. The individual then climbs into a pool of water, exhales air from the lungs to sink below the surface, and is found to weigh 2.9 kg while totally immersed. The underwater body weight is less than the land weight because water helps support the body by giving it buoyancy. Since fat is less dense than other body components, it makes a person's body particularly buoyant. The higher the percentage of body fat, the more buoyant the person and the greater the difference between land weight and underwater body weight.

With body weight known, body volume is determined next. Physics tells us that the buoyancy of a submerged object (that is, the land weight minus the submerged weight) is equal to the weight of the water displaced by the object. Thus, our submerged person with a buoyancy of 57.1 kg (60.0 kg − 2.9 kg) must be displacing 57.1 kg of water. Since we know that water has a density of 1.0 kg/L, we can find the volume of the submerged body:

Body volume
$$= \text{weight of water displaced} \times \text{density of water}$$
$$= (\text{land weight} - \text{underwater weight}) \, \cancel{kg} \times \frac{1 \, L}{\cancel{kg}}$$
$$= (60.0 - 2.9) \, \cancel{kg} \times \frac{1 \, L}{\cancel{kg}} = 57.1 \, L$$

With both body weight and body density now known, overall body density is calculated:

$$\text{Body density} = \frac{\text{body weight}}{\text{body volume}} = \frac{60.0 \, \text{kg}}{57.1 \, \text{L}}$$
$$= 1.05 \, \text{kg/L} = 1.05 \, \text{g/mL}$$

A chart is then used to correlate body density with the percentage of body fat. Our individual with a density of 1.05 g/mL has about 20 percent body fat—not bad, but some more exercise might be a good idea.

Measurement of percentage of body fat by the underwater immersion method.

2.13 SPECIFIC GRAVITY

Specific gravity The density of a substance divided by the density of water at the same temperature.

For many purposes ranging from winemaking to medicine, it's customary to use the **specific gravity** of a substance rather than its density. The specific gravity of any material is simply a comparison of the mass of the substance to the mass of an equal volume of water at the same temperature. Since all units cancel out, specific gravity has no units:

$$\text{Specific gravity} = \frac{\text{density of substance (g/mL)}}{\text{density of water (1 g/mL)}}$$

At normal temperatures, the density of water is very close to 1 g/mL. Thus, the specific gravity of a substance is numerically equal to its density and is used in the same ways.

Hydrometer A weighted bulb used to measure the specific gravity of a liquid.

Specific gravities are usually measured only for liquids, using an instrument called a **hydrometer.** A hydrometer has a weighted bulb on the end of a calibrated glass tube, as shown in Figure 2.8. The depth to which the hydrometer sinks when placed in the fluid to be measured indicates its specific gravity.

As a general rule, water that contains dissolved particles has a specific gravity higher than 1.00, because the dissolved material is usually denser than water. In winemaking, for example, the amount of fermentation taking place is gauged by observing the change in specific gravity on going from grape juice, which contains 20% dissolved sugar (sp gr = 1.082), to dry wine, which contains 12% alcohol (sp gr = 0.984).

In medicine, a hydrometer called a urinometer is used to indicate the amount of solids in urine. The normal specific gravity of urine is about 1.005 to 1.030. Certain conditions, such as diabetes mellitus or a high fever, cause an abnormally high urine specific gravity, indicating either excessive elimination of solids or decreased elimination of water. Abnormally low specific gravity is found in individuals with an abnormally low body temperature or those using diuretics (drugs that increase water elimination).

Figure 2.8
A hydrometer for measuring specific gravity. (a) A hydrometer has a weighted bulb at the end of a calibrated glass tube. (b) The depth to which the hydrometer sinks in a fluid indicates the specific gravity of the fluid because the bulb sinks until it displaces a volume of liquid equal to its mass.

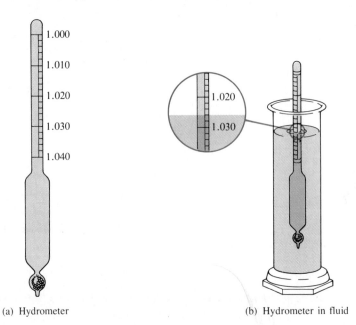

(a) Hydrometer

(b) Hydrometer in fluid

INTERLUDE: POWERS OF 10

It's not easy to grasp the enormous differences in size represented by numbers given as powers of 10 (scientific notation), but the following pictures might help. Imagine that you're looking at an ordinary household pin. Since the pin is only about 3 cm long, you can see little detail. Imagine, though, that the pin is magnified in size by about 10^2. That small change in the exponent on going from 1 to 10^2 means that the pin is now 100 times larger and would look something like part (a) in the accompanying figure.

Now imagine that the pin is magnified by about 10^4 as in part (b). You see not only the bluntness of the tip (it's not as sharp as you thought), but also that the pin is covered by tiny bacteria about 3×10^{-4} cm in length. Increasing the magnification still further to about 5×10^6 gives you the picture in part (c) of the internal details of a bacterium about to undergo cell division and shows a cell wall that is only *70 nm* thick. The magnification has changed only from 1 to 10^6, but that small change in exponent has made an extraordinary difference in your view. Powers of 10 are powerful indeed.

Bacteria on a household pin at different magnifications. The individual pictures are explained in the text.

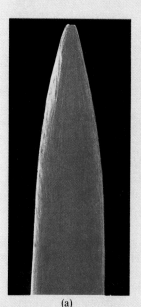
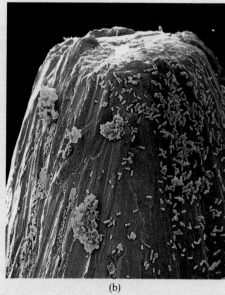
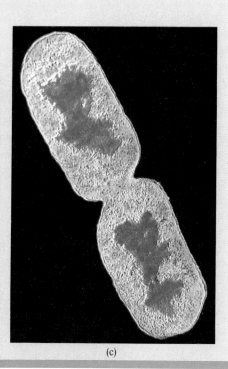

(a)   (b)   (c)

SUMMARY

Physical quantities are properties that can be measured; they are described by both a number and unit. Among the most important physical quantities are mass, length, volume, and temperature. The units used to describe these properties are either those of the International System of Units (**SI units**) or of the **metric** system. **Mass** is the amount of matter an object contains; it is measured in **kilograms (kg)** or **grams (g)**. **Length** is measured in **meters (m)**. **Volume** is measured in **cubic meters (m³)** in the SI system and in **liters (L)** or **milliliters (mL)** in the metric system. **Temperature** is measured in **kelvins (K)** in the SI system and in **Celsius degrees (°C)** in the metric system.

Measurements of small and large quantities are often expressed in **scientific notation** as the product of a number between 1 and 10 times some power of 10. For example, $3562 = 3.562 \times 10^3$. When measuring physical quantities or using them in calculations, it's important to indicate the exactness of the measurement by using the correct number of **significant figures.** All but one of the significant figures in a number should be known with certainty; the final digit is estimated to ±1.

Many applications of chemistry require numerical calculations involving the results of measurements and the use of conversion factors that allow a measurement in one unit to be given in another unit. Problems should be solved by the **factor-label method,** in which units are included and problems set up so that unwanted units cancel out and only the desired units remain. Analysis of a problem by identifying the known and needed information and then deciding how to convert the known information to the answer is always helpful. Also, it is essential to check an answer by finding a ballpark solution.

Density is the physical property that relates mass to volume; it is usually expressed in units of grams per milliliter (g/mL) or grams per cubic centimeter (g/cm³). The **specific gravity** of an object—usually a liquid—is the object's density divided by the density of water at the same temperature. Since the density of water is approximately 1 g/mL, specific gravity and density have the same numerical value.

Heat is a form of energy that is transferred from hotter to cooler objects. The **specific heat** of a substance is the amount of heat necessary to raise the temperature of 1 g of the substance by 1 C°. Water has an unusually high specific heat, which helps our bodies to maintain an even temperature.

REVIEW PROBLEMS

Definitions and Units

2.24 What is the difference between a physical quantity and a number?

2.25 What are the base units used in the SI system to measure mass, volume, length, and temperature?

2.26 What are the basic units used in the metric system to measure mass, volume, length, and temperature?

2.27 Give the full name of each of these units:
(a) cL   (b) dm   (c) mm   (d) nL   (e) mg
(f) m^3 (g) cc

2.28 Write the symbol for each of the following units:
(a) picogram   (b) centimeter   (c) deciliter
(d) microliter   (e) milliliter

2.29 How many picograms are in 1 mg? In 35 ng?

2.30 How many microliters are in 1 L? In 20 mL?

2.31 Which is smaller, a liter or a quart? By what percent?

Scientific Notation and Significant Figures

2.32 Express the following numbers in scientific notation:
(a) 9457   (b) 0.00007   (c) 20,000,000,000
(d) 0.012345   (e) 652.38

2.33 Convert the following numbers from scientific notation to normal notation:
(a) 4.87×10^3 (b) 5.501×10^6
(c) 2.540×10^{-3} (d) 3.68×10^4

2.34 How many significant figures does each of the following numbers have?
(a) 237,401   (b) 0.300   (c) 3.01   (d) 244.4
(e) 50,000   (f) 660.

2.35 Round off each of the numbers in Problem 2.34 to two significant figures and express them in scientific notation.

2.36 How many significant figures are in each of the following quantities?
(a) distance from New York City to Wellington, New Zealand, 14,397 km
(b) average body temperature of a crocodile, 25.6°C
(c) melting point of gold, 1064°C
(d) diameter of an influenza virus, 0.00001 mm
(e) radius of a phosphorus atom, 0.110 nm

2.37 The diameter of the earth at the equator is 7926.381 mi.
(a) Round off the earth's diameter to four significant figures, to two significant figures, and to six significant figures.
(b) Express the earth's diameter in scientific notation.

2.38 Carry out the following calculations, express the answers to the correct numbers of significant figures, and include units in the answers
(a) 9.02 g + 3.1 g
(b) 88.80 cm + 7.391 cm
(c) 362 mL − 99.5 mL
(d) 12.4 mg + 6.378 mg + 2.089 mg

2.39 Carry out the following calculations, express the answers to the correct numbers of significant figures, and include units in the answers.
(a) 8.08 m × 5.320 m
(b) 11.1 mi × 20. mi
(c) 1350. miles ÷ 7.311 hr
(d) 6.80 cm × 100. cm × 2.00 cm

Unit Conversions and Problem Solving

2.40 Carry out the following metric conversions.
(a) 3.614 mg to cg   (b) 12.0 kL to ML
(c) 14.4 μm to mm   (d) 6.03×10^{-6} cg to ng

2.41 Carry out the following conversions. Consult Tables 2.3 through 2.5 as needed.
(a) 56.4 mi to km; to Mm
(b) 2.0 L to quarts and to fluid ounces
(c) 7 ft 2.0 in. to cm; to m
(d) 1.35 lb to kg; to dg

2.42 Express the following quantities in more convenient units by using unit prefixes:
(a) 9.78×10^4 g   (b) 1.33×10^{-4} L
(c) 0.000 000 000 46 g

2.43 A clinical report gave the following data from a blood analysis: iron, 39 mg/dL; calcium, 8.3 mg/dL; cholesterol, 224 mg/dL. Express each of these quantities in grams per deciliter, writing the answers in scientific notation.

2.44 The speed limit in Canada is 100 km/hr.
(a) How many mi/hr is this?
(b) How many feet per second?

2.45 The speed limit in many places is 55 mi/hr.
(a) How many km/hr is this?
(b) How many m/s?

2.46 A six-pack of 12 oz soft drinks sells for $2.59, and a 2.0 L bottle of the same soft drink sells for $1.59. What is the cost in $/L for each?

2.47 According to Figure 2.1, the diameter of a red blood cell is 6×10^{-6} m. How many red blood cells are needed to make a line 1 in. long?

2.48 The World Trade Center in New York has an approximate floor area of 406,000 m^2. How many square feet of floor is this?

2.49 The longest recorded jump by a flea is 13.0 inches. How far is this in mm?

2.50 A normal value for blood cholesterol is 200 mg/dL. If a normal adult has a total blood volume of 5 L, how much total cholesterol is present?

2.51 The white blood cell concentration in normal blood is approximately 12,000 cells/mm^3. How many white blood cells does a normal adult with 5 L of blood have? Express the answer in scientific notation.

2.52 The recommended daily dose of calcium for an 18 year old male is 1200 mg. If there is 290 mg of calcium in 1.0 cup of whole milk and milk is his only calcium source, how much milk should an 18 year old male drink each day?

Temperature and Heat

2.53 What is the temperature of the normal human body (98.6°F) in °C? In K?

2.54 If you were feeling sick and a thermometer measured your body temperature at 39.3°C, what would your temperature be in °F?

2.55 The boiling point of liquid nitrogen, used in the removal of warts and in other surgical applications, is −195.8°C. What is this temperature in K and in °F?

2.56 Diethyl ether, a substance once used as a general anesthetic, has a specific heat of 0.895 cal/g °C. How many cal and how many kcal of heat are needed to raise the temperature of 30.0 g of diethyl ether from 10.0°C to 30.0°C?

2.57 Calculate the specific heat of copper if it takes 23 cal to heat a 5.0 g sample from 25°C to 75°C.

2.58 Assume that 50. cal of heat is applied to a 15 g sample of sulfur at 20°C. What is the final temperature of the sample if the specific heat of sulfur is 0.175 cal/g °C?

2.59 A 150. g sample of mercury and a 150. g sample of iron are at an initial temperature of 25.0°C. If 250. cal of heat is applied to each sample, what is the final temperature of each? Consult Table 2.6.

Density and Specific Gravity

2.60 Aspirin has a density of 1.40 g/cm^3. What is the volume in cm^3 of a tablet weighing 250 mg?

2.61 What is the density of isopropyl alcohol (rubbing alcohol) if a 5.000 mL sample weighs 3.928 g at room temperature? What is the specific gravity of isopropyl alcohol?

2.62 Ethylene glycol (automobile antifreeze) has a specific gravity of 1.1088 at room temperature. What is the volume of 1.00 kg of ethylene glycol? What is the volume of 2.00 lb of ethylene glycol?

2.63 In which liquid will a urinometer float higher, ethanol (sp gr 0.7893) or chloroform (sp gr 1.4832)?

Applications

2.64 Write an alternate conversion factor for each of the following unit equivalents from the apothecary unit system:
(a) 1 gr = 64.8 mg, 1 mg = ? gr
(b) 1 fluid ounce = 8 fluidram, 1 fluidram = ? fluid ounce
(c) 1 fluidram = 3.70 mL, 1 mL = ? fluidram
(d) 1 minim = 0.062 mL, 1 mL = ? minim
(e) 1 fluid ounce = 480 minims, 1 minim = ? fluid ounce
[App: Apothecary Units]

2.65 (a) How many milligrams of aspirin are in a 5 grain aspirin tablet?
(b) How many drams are in a 1.5 fluid ounce bottle of perfume? Use conversion factors from Problem 2.64.
[App: Apothecary Units]

2.66 What purposes does fat serve in the human body?
[App: Body Fat]

2.67 What is a person's body density if the mass is 85.0 kg on land but 5.6 kg when totally immersed in water? [App: Body Fat]

2.68 Liposuction is a technique for removing fat deposits from various areas of the body. How many liters of fat would have to be removed to result in a 5.0 lb weight loss? The density of human fat is 0.94 g/mL. [App: Body Fat]

Additional Problems

2.69 Gemstones are weighed in *carats,* where 1 carat = 200 mg. What is the mass in grams of the Hope diamond, the world's largest blue diamond at 44.4 carats?

2.70 If you were cooking in an oven calibrated in Celsius degrees, what temperature would you use if the recipe called for 350°F?

2.71 Macau, on the southern coast of China, is reported to be the most densely populated territory in the world, with 293,000 people living in an area of 6.2 square miles. What is the population density of Macau in people/mile^2? Express your answer in scientific notation with the proper number of significant figures.

2.72 What dosage in grams per kilogram of body weight does a 130 lb woman receive if she takes two 250 mg tablets of penicillin? How many 125 mg tablets should a 40 lb child take to receive the same dosage?

2.73 The density of air at room temperature is 1.3 g/L at 0°C. What is the mass of the air
(a) in grams and
(b) in pounds in a room that is 4.0 m long, 3.0 m wide, and 2.5 m high?

2.74 Approximately 75 mL of blood is pumped by a normal human heart at each beat. Assuming an average pulse of 72 beats per minute, how many milliliters of blood are pumped in one day?

2.75 A doctor has ordered that a patient be given 15 g of glucose, which is available in a concentration of 50. g glucose/100. mL of solution. What volume of solution should be given to the patient?

2.76 The density of mercury, commonly used in thermometers, is 13.6 g/mL.
(a) What is the specific gravity of mercury?
(b) Would an iron ball bearing float or sink in mercury? The density of iron is 7.86 g/mL.
(c) What is the mass in pounds of 2.00 L of mercury?

2.77 In a normal person, the level of glucose, also known as blood sugar, is about 85 mg/mL of blood. If an average body contains about 11 pt of blood, how many grams and how many pounds of glucose are present in the blood?

2.78 A patient is receiving 3000 mL/day of a solution that contains 5 g of dextrose (glucose) per 100 mL of solution. If glucose provides 4 kcal per gram of energy, how many cal/day is the patient receiving from the glucose?

2.79 A rough guide to fluid requirements based upon body weights is: 100 mL/kg for the first 10 kg of body weight, 50 mL/kg for the next 10 kg, and 20 mL/kg for weight over 20 kg. What volume of fluid per day is needed by a 55 kg woman? Give the answer with two significant figures.

2.80 Chloral hydrate, a sedative and sleep-inducing drug, is available as a solution labeled 10. gr/fluidram. What volume in milliliters should be administered to a patient who is meant to receive 7.5 gr per dose? (1 grain = 64.8 mg; 1 fluidram = 3.70 mL)

2.81 When 1.0 tablespoon of butter is burned or used by our body, it releases 100 kcal (100 food calories) of energy. If we could use all the energy provided, how many tablespoons of butter would have to be burned to raise the temperature of 3.00 L of water from 18.0°C to 90.0°C?

3 Atoms and the Periodic Table

Take a picture of an atom? Not so long ago that was impossible. But not any more. You're looking at the surface of a piece of pure germanium, the kind used in making semiconductors. The red and blue spots (falsely colored) are atoms in the top layer, and the black spots are openings to the second layer of atoms.

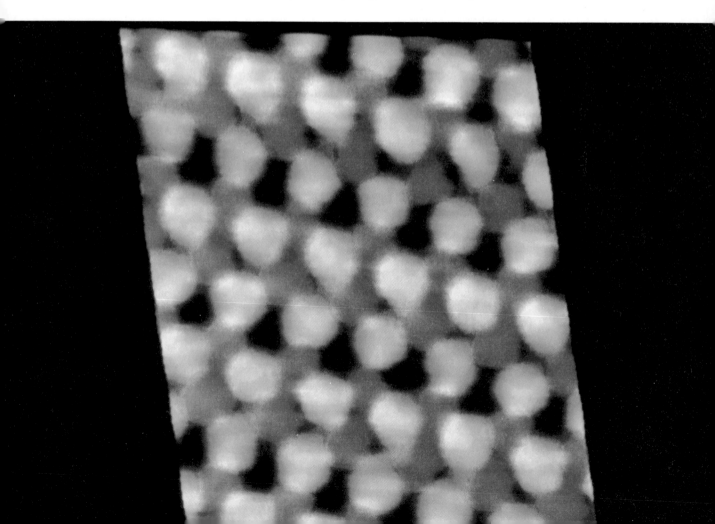

Chemistry must be studied on two levels. In Chapters 1 and 2 we've been dealing with the macroscopic view of chemistry—the properties and transformations of matter that we can see and measure numerically with ease. Now we're ready to approach the microscopic view of chemistry—individual atoms and their properties. Only within the past few years have individual atoms been seen with the help of powerful instruments, although experiments had long before convinced scientists of their existence.

In this chapter we'll first introduce modern atomic theory. Then we'll examine how the structure of atoms influences macroscopic properties. To do this requires introducing some new kinds of terms and symbols. Keep in mind that the purpose of these terms and symbols is to help you see the amazing ways that the properties of matter, including those of all living things, depend on the differences among atoms. We'll take up the following questions about atoms and atomic theory:

1. *What is the modern view of atoms?* The goal: Be able to list the major assumptions of modern atomic theory.

2. *How do atoms of different elements and their isotopes differ?* The goal: Be able to explain the composition of different atoms in terms of the number of protons, neutrons, and electrons.

3. *How are electrons arranged in atoms?* The goal: Given the number of electrons in an atom, be able to explain how they are distributed in shells and subshells and to write the electron configuration of the atom.

4. *What is the periodic table?* The goal: Be able to describe how elements are arranged in the periodic table, name the subdivisions of the periodic table, and relate the position of an element in the periodic table to its electron structure.

3.1 MODERN ATOMIC THEORY

Take a piece of aluminum foil and cut it in two. Then take one of the pieces and cut *it* in two, and so on. Assuming that you have infinitely small scissors and infinite dexterity, how long can you keep dividing the foil? Is there a limit, or is matter infinitely divisible into ever smaller and smaller pieces? In fact, there is a limit. The smallest and simplest piece that aluminum (or any other element) can be broken into and still be identifiable as aluminum is called an **atom** (derived from the Greek *atomos*, meaning "indivisible").

Chemistry is founded on atomic theory, which explains the properties of matter in terms of atoms. The Greeks, who first proposed that all matter might be made of very tiny particles, assumed that these particles must be indivisible. Today, we know that atoms themselves are composed of **subatomic particles.** Experimental evidence gathered over the years has led to the modern view of atoms and matter summarized in the following statements:

Atom The smallest and simplest particle of an element.

Subatomic particle An elementary particle smaller than an atom.

48

- **All matter is composed of atoms.**
- **Atoms of each element differ from atoms of all other elements.**
- **Atoms are composed of subatomic particles—protons, neutrons, and electrons.** Differences in the composition of their atoms are responsible for the distinctive properties of elements.
- **Chemical compounds consist of atoms combined in definite proportions.** That is, only whole atoms can combine—one A atom with one B atom, or two B atoms, and so on. The incredible diversity of kinds of matter is based on the enormous number of ways that atoms can combine with each other.
- **Chemical compounds undergo changes in their compositions and identities in chemical reactions, but atoms remain unchanged.** The result of a chemical reaction is a reorganization of the ways in which atoms are combined.
- **The identities of atoms change only in nuclear reactions.** You'll see in the next chapter that certain kinds of unstable atoms undergo nuclear reactions in which their compositions and identities change. In everyday life and in most applications of chemistry, though, nuclear reactions are of limited concern.

3.2 MASSES OF ATOMS

Atoms are extremely small, ranging from about 7.4×10^{-11} m in diameter for a hydrogen atom to 5.24×10^{-10} m for a cesium atom. Even the smallest speck of dust contains about 10^{16} atoms. In mass, atoms vary from 1.67×10^{-24} g for a hydrogen atom to 3.95×10^{-22} g for a uranium atom, one of the heaviest naturally occurring atoms. It's difficult to appreciate just how small atoms are, although it might help if you realize that a fine pencil line is about 3 million atoms across.

Comparing the masses of atoms given as very small numbers is difficult. To simplify comparisons, atomic masses are expressed on a relative mass scale. One atom is assigned a mass, and then all others are measured relative to it. The process is like deciding that a golf ball (46.0 g) will be assigned a mass of 1. On this scale a baseball (149 g), which is about three times heavier than a golf ball, would have a mass of about 3.24, a volleyball (270. g) would have a mass of 5.87, and so on.

Atomic mass unit (amu) Unit of the relative atomic mass scale, equal to $\frac{1}{12}$ the mass of a carbon atom, which is 1.6606×10^{-24} g.

The basis for the relative atomic mass scale is an atom of carbon that contains six protons and six neutrons. This type of atom is assigned a mass of exactly 12 atomic mass units. The result is that one **atomic mass unit (amu)** is equivalent to 1.6606×10^{-24} g, an equality that relates relative atomic masses to the mass unit most common in chemistry.

$$1 \text{ amu} = 1.6606 \times 10^{-24} \text{ g}$$

Dalton An alternate name for atomic mass unit.

Carbon atoms are just about 12 times heavier than hydrogen atoms, and on the relative atomic mass scale one hydrogen atom has a mass very close to 1 amu. Magnesium atoms are about twice as heavy as carbon atoms and have a mass close to 24 amu. An atomic mass unit is also sometimes known as a **dalton,** especially in biochemistry. The name honors John Dalton, who in 1808 presented the first atomic theory based on experimental evidence.

Practice Problems **3.1** What is the mass in atomic mass units of one nitrogen atom weighing 2.33×10^{-23} g?

3.2 What is the total mass in grams of 100,000 gold atoms, each weighing 197 amu?

3.3 How many atoms are in each of the following?
(a) 1.0 g of hydrogen atoms, each of mass 1.0 amu
(b) 12.0 g of carbon atoms, each of mass 12.0 amu
(c) 23.0 g of sodium atoms, each of mass 23.0 amu

3.4 What pattern do you see in your answers to Problem 3.3? (We'll return to this very important pattern in Chapter 7.)

3.3 ELECTRONS, PROTONS, AND NEUTRONS

Proton A positively charged (+1) subatomic particle found in the nuclei of atoms; mass = 1 amu.

Electron A negatively charged (−1) subatomic particle of very small mass that moves in the large volume of space surrounding nuclei.

Neutron An electrically neutral subatomic particle found in the nuclei of atoms; mass = 1 amu.

Although atoms appear indestructible in chemical reactions, they are actually composed of subatomic particles called protons, neutrons, and electrons. A **proton** is a subatomic particle with a positive (+) charge and a mass of approximately 1 amu. An **electron** has a negative charge (−) and a mass about 1/2000 that of a proton (1/1836 to be exact). In fact, electrons are so much lighter than protons and neutrons that their mass is usually ignored. A **neutron** has no electrical charge and has a mass very close to that of a proton. The properties of these particles are compared in Table 3.1.

Since the atoms in things around us rarely fly apart, we have to ask, What holds the subatomic particles together? Because they are electrically charged, protons and electrons exert forces on each other. On the one hand, electrons are negatively charged and are strongly attracted to protons, which are positively charged. That is, opposite charges attract. On the other hand, electrons repel electrons and protons repel protons. That is, identical charges repel each other. If you've handled magnets, you're familiar with forces of attraction and repulsion. When the north and south poles of magnets are brought together, they attract each other. When two identical magnetic poles are brought together, though, they repel each other.

Within atoms, electrons are attracted to protons, but at the same time are repelled by each other and must spread apart. The interplay between such attractive and repulsive forces explains many of the properties of atoms and chemical compounds.

Table 3.1 A Comparison of Subatomic Particles

| | Mass | | Charge |
	In Grams	In amu	In Charge Units
Electron	9.109×10^{-28}	5.486×10^{-4}	−1
Proton	1.673×10^{-24}	1.007	+1
Neutron	1.675×10^{-24}	1.009	0

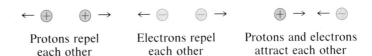

Protons repel each other   Electrons repel each other   Protons and electrons attract each other

Nucleus The dense, positively charged mass at the center of an atom where protons and neutrons are located.

All the protons and neutrons in an atom are packed closely together in a dense central region called the **nucleus,** which is very small relative to the size of the atom. (A unique, very strong force studied by physicists acts within atomic nuclei to bind the neutrons and protons together.) Surrounding the nucleus is a mostly empty, large volume of space in which the electrons move about rapidly (Figure 3.1) in a manner we'll discuss in Section 3.7. It's the behavior of electrons that is of greatest interest in chemistry.

Figure 3.1
The structure of atoms. Protons and neutrons are packed together in the nucleus, while electrons move about in the large surrounding volume. Virtually all the mass of an atom is concentrated in the nucleus.

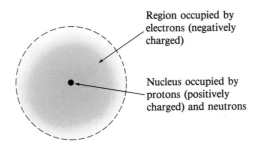

Region occupied by electrons (negatively charged)

Nucleus occupied by protons (positively charged) and neutrons

3.4 COMPOSITION OF ATOMS

Atomic number The primary characteristic that distinguishes atoms of different elements; equal to the number of protons in an atom's nucleus.

Atoms of different elements contain different numbers of protons, a value called the **atomic number** of an element. Overall, atoms are neutral and have no net charge because each atom contains an identical number of positively charged protons and negatively charged electrons. Thus, the atomic number also equals the number of electrons in every atom of a given element.

Atomic number = number of protons in an atom
= number of electrons in an atom

Atomic number
Name
Symbol
Atomic weight

20
Calcium
Ca
40.08

In the periodic table inside the front cover of this book, the elements are listed from left to right in the order of increasing atomic number. The atomic number is given just above the name and symbol for each element. (Just below the name and symbol is the atomic weight, which is discussed in Section 3.6.)

Mass number The sum of an atom's protons and neutrons.

The sum of the protons and neutrons in an atom is called the **mass number.** Most hydrogen (H) atoms have one proton and no neutrons and therefore have mass number 1; helium (He) atoms with two protons and two neutrons have mass number 4; oxygen (O) atoms with eight protons and eight neutrons have mass number 16; and so on. In general, atoms contain at least as many neutrons as protons, but there's no simple way to predict how many neutrons a given atom will have.

Mass number = number of protons + number of neutrons

Solved Problem 3.1   Phosphorus has atomic number 15. How many protons, electrons, and neutrons are there in phosphorus atoms with mass number 31?

Solution   The atomic number gives the number of protons in atoms of a given element, and the number of electrons equals the number of protons. Thus, phosphorus atoms, with atomic number 15, have 15 protons and 15 electrons. The mass number gives the total number of protons and neutrons, so to find the number of neutrons the atomic number is subtracted from the mass number:

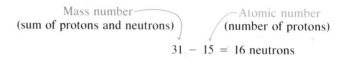

Mass number (sum of protons and neutrons)   Atomic number (number of protons)

$$31 - 15 = 16 \text{ neutrons}$$

Solved Problem 3.2   An atom contains 28 protons and has a mass number of 60. Give the number of electrons and neutrons in this atom and identify the element.

Solution   The number of protons equals both the atomic number and the number of electrons. Therefore, this atom contains 28 electrons and

$$60 - 28 = 32 \text{ neutrons}$$

Consulting the periodic table inside the front cover shows that the element with atomic number 28 is nickel (Ni).

Practice Problems   **3.5**   Identify the elements with atomic numbers 75, 3, and 52.

3.6   The uranium used in nuclear reactors has atomic number 92 and mass number 235. How many protons and neutrons are present in these uranium atoms?

3.5 ISOTOPES

Although all atoms of a given element have the same number of protons (the atomic number characteristic of that element), different atoms of an element can have different mass numbers, depending on how many neutrons they contain. Such atoms with *identical atomic numbers* but *different mass numbers* are called **isotopes.** Hydrogen, for example, has three isotopes. All hydrogen atoms have one proton, but most have no neutrons. This type of hydrogen, called *protium*, has a mass number of 1. A small percentage of hydrogen atoms have one neutron. This isotope, with mass number 2, is called *deuterium*. A third isotope, called *tritium*, has two neutrons and mass number 3.

Isotopes Atoms of the same element that have different numbers of neutrons in their nuclei and therefore different mass numbers.

Protium: 1 proton
0 neutrons
Mass number = 1

Deuterium: 1 proton
1 neutron
Mass number = 2

Tritium: 1 proton
2 neutrons
Mass number = 3

Isotopes are represented by adding the mass number as a superscript and the atomic number as a subscript to the atomic symbol. Thus protium is $^1_1$H, deuterium is $^2_1$H, and tritium is $^3_1$H.

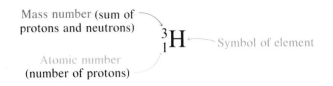

Unlike the isotopes of hydrogen, isotopes of most elements do not have distinctive names. Instead, they are referred to in writing by adding the mass number to the name of the element. In this manner, the hydrogen isotope represented by the symbol above would be named hydrogen-3.

Solved Problem 3.3	Write the symbol for carbon -12, the isotope of carbon that is the basis for the relative atomic mass scale.
Solution	Carbon has atomic number 6 (see the periodic table) and the mass number of this isotope, 12, is given in the name, so the symbol is $^{12}_{6}C$.
Solved Problem 3.4	Identify element X and give its atomic number, mass number, number of protons, and number of electrons: $^{194}_{78}X$.
Solution	Element X has atomic number 78 which shows that it is platinum. The isotope symbolized above has mass number 194. Subtracting the atomic number from the mass number gives the number of neutrons. Platinum therefore has 78 protons 78 electrons $194 - 78 = 116$ neutrons

Practice Problems   **3.7**   Chlorine, an element present in common table salt (sodium chloride), has common isotopes with mass numbers 35 and 37. How many protons and neutrons are present in each isotope?

3.8   Write the symbols for both chlorine isotopes (Problem 3.7), including the atomic numbers and mass numbers.

3.9   Complete each of the following isotope symbols: $^{11}_{5}?$, $^{56}_{?}Fe$.

3.6   ATOMIC WEIGHT

Most elements occur in nature as mixtures of isotopes. In a large sample of the natural mixture of hydrogen atoms, for example, 99.985% have mass number 1 (protium) and 0.015% have mass number 2 (deuterium). The composition of such natural mixtures of isotopes is virtually the same everywhere on earth. Therefore, it's more useful to know the *average* mass of the atoms of an element than to know the masses of individual isotopes. The average mass of the atoms in the naturally occurring mixture of isotopes of an element is called the **atomic weight** (or sometimes the *atomic mass*) of that element. Thus, because of the small percentage of deuterium atoms present in any sample of hydrogen, the atomic weight of hydrogen, 1.008 amu, is slightly higher than 1.

Atomic weight   The average mass (in amu) of atoms in the naturally occurring isotopic mixture of an element.

AN APPLICATION: CHERNOBYL AND CESIUM

Few people had heard of the element cesium until April 26, 1986, when an explosion destroyed the unit 4 reactor at the Chernobyl nuclear power plant in Russia. As a consequence of that accident, cesium came to be a household word throughout much of Europe.

Discovered in 1860 in mineral waters from the small town of Bad Dürkheim, Germany, and named for the Latin *caesius*, meaning "sky blue," naturally occurring cesium has only one isotope, $^{133}_{55}Cs$. Although this naturally occurring isotope is stable and harmless, two unstable isotopes, $^{134}_{55}Cs$ and $^{137}_{55}Cs$, are formed as by-products in nuclear reactors and were released into the atmosphere at Chernobyl. Cesium-134 and cesium-137 are radioactive, undergoing decay slowly and producing harmful radiation. As dust from the accident slowly settled to the earth, radioactive cesium became widely dispersed on agricultural land throughout northern Europe and the European part of the Soviet Union.

None of this would matter quite so much except for cesium's position in the periodic table. As the element with atomic number 55, cesium is in the same group as potassium in the periodic table. It is therefore chemically similar to potassium and can be mistaken for that element by living organisms. Potassium is a constituent of all body tissues and fluids.

When crops are grown and animals are grazed on contaminated land, radioactive cesium can move into the human food chain. The radiation emitted by cesium that has been incorporated into human tissue and fluids has the potential for causing cancer. Through careful monitoring and warnings issued by public health officials, efforts were made to limit the potential hazard to those living within reach of the materials released at Chernobyl. It is estimated, however, that an additional 28,000 cancers may occur worldwide in the next 50 years as a result of that accident.

The Chernobyl nuclear power plant after the devastating fire and explosion of April 26, 1986, that destroyed the unit 4 reactor.

Atomic weights are given for all the elements inside the front cover of this book. With just a few exceptions, atomic weight increases with atomic number. Where there are exceptions, they result from varying distributions of isotopes.

To calculate the atomic weight of an element, the individual masses of the naturally occurring isotopes and the percentage of each must be known. Chlorine, for example, occurs as a mixture of 75.53% chlorine-35 atoms (mass = 35.0 amu) and 24.47% chlorine-37 atoms (mass = 37.0 amu). The atomic weight is an average based on the percentage of atoms of each type and is found by calculating the percentage of the mass contributed by each isotope. For chlorine, the calculation is done in the following way (to three significant digits):

Contribution from ^{35}Cl: 75.53% of 35.0 amu = 26.4 amu
Contribution from ^{37}Cl: 24.47% of 37.0 amu = 9.05 amu
Atomic weight = 35.5 amu

3.7 LOCATION OF ELECTRONS IN ATOMS

How are electrons distributed in an atom? Since electrons are in constant motion around the nucleus, it's impossible to define their positions precisely. It turns out, though, that electrons are not perfectly free to move about; they're restricted to specific regions within the atom according to the amount of energy they have.

By the application of mathematics, a model that successfully explains the restrictions on the locations of electrons has been developed. For most purposes, though, it's only necessary to know that the energies of electrons in atoms are *quantized*, which means that electron energies are limited to certain values and can have no others. To understand quantization, think of riding up in an elevator. You can stop only at floor 2, floor 3, floor 4, and so on, not at floor 2.3 or floor 3.5. The level you can reach above the first floor is therefore quantized (Figure 3.2).

The main energy levels of electrons in atoms, like the floors in a building, are numbered 1, 2, 3, 4, ... They are referred to as the first main energy level, the second main energy level, and so on. These levels, or **shells**, are specific regions occupied by electrons of increasing energy at increasing distances from the nucleus. The shells are roughly like the layers in an onion: The farther a shell is from the nucleus, the larger it is, the more electrons it holds, and the greater the potential energies of its electrons because they are farther away from the positive charge of the nucleus.

Within the main energy levels, electrons are further grouped into **subshells** of four different types that are identified in order of increasing energy by the letters *s*, *p*, *d*, and *f*. Each subshell holds a different maximum number of electrons:

s subshell: 2 e^- p subshell: 6 e^-

d subshell: 10 e^- f subshell: 14 e^-

The first main energy level has only an *s* subshell and thus holds just 2 electrons. The second main energy level contains both *s* and *p* subshells, which together hold 8 electrons (2 in the *s* and 6 in the *p* subshell). Note in Table 3.2 that the

Figure 3.2
Quantized values. The energy levels of electrons in atoms, like the levels of the floors in a building, are quantized.

Shells   Specific regions surrounding a nucleus that are occupied by electrons of increasing energy; main energy levels.

Subshells   Subdivisions within shells of atoms occupied by electrons of different energies; designated by *s*, *p*, *d*, and *f*.

Table 3.2 Distribution of Electrons into Shells

Number of Shell	Electron Capacity of Shell	Subshells
First (1)	2	s
Second (2)	8	s, p
Third (3)	18	s, p, d
Fourth (4)	32	s, p, d, f
Fifth (5)	50	s, p, d, f, g^a

[a] No known atoms in their ground states contain electrons in subshells beyond *f*.

number of subshells within each shell is equal to the number of the shell. Thus, the third shell can accommodate a maximum of 18 electrons in its *s*, *p*, and *d* subshells.

Orbitals Regions of specific shapes occupied by electrons in *s*, *p*, *d*, and *f* subshells.

The regions occupied by electrons within subshells are known as **orbitals** and are identified by the subshell letters. Orbitals of each type have distinctive shapes. The *s* orbitals, for example, are spherical regions centered on the nucleus (Figure 3.3a) and the *p* orbitals are dumbbell-shaped (Figure 3.3b). Each *s* subshell contains only one *s* orbital, but each *p* subshell contains three different *p* orbitals that extend out from the nucleus at 90° angles from each other (Figure 3.3c).

Note that we haven't said anything about the *path* an electron follows as it moves about within an orbital. We can only define the general region of space in which an electron with a certain energy will most probably be found. It's helpful to think of the orbital as a time-lapse photograph of an electron's movement about the nucleus. In such a photograph, the orbital would appear as a kind of blurry cloud showing where the electron has been. In fact, orbitals are often referred to as *electron clouds*.

Each orbital may be *occupied by only two electrons*, which differ in a property known as *spin*. The maximum numbers of electrons in each shell and subshell are fixed by this requirement. Each orbital can hold only two electrons of opposite spin, as summarized in Figure 3.4.

Subshells are symbolized by writing the number of the main energy level followed by the letter for the subshell. For the *s* subshell in the first energy level this gives

1st shell⟶ ⟵*s* subshell

1*s*

Figure 3.3
The shapes of *s* and *p* orbitals: (a) *s* Orbitals are spherical and (b) *p* orbitals are dumbbell-shaped. (c) The three *p* orbitals are oriented at right angles to each other. Each orbital can hold only two electrons.

Figure 3.4
Electron capacity of shells and subshells. Each shell, or main energy level, contains subshells with different numbers of orbitals. The electron capacity of each shell is fixed by the requirement that each orbital can hold only two electrons.

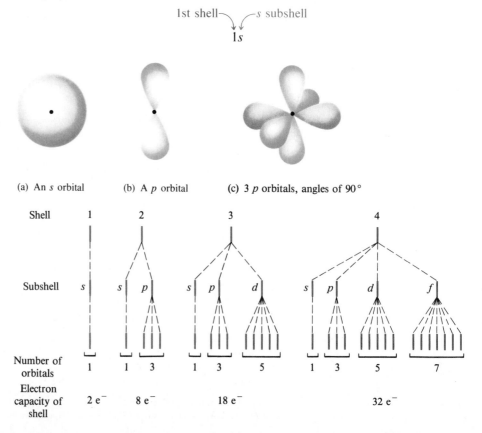

(a) An *s* orbital   (b) A *p* orbital   (c) 3 *p* orbitals, angles of 90°

The maximum numbers of electrons in each of the first three shells are accounted for as follows:

First shell—2 electrons

| 1s subshell | 2 electrons in an s orbital |

Second shell—8 electrons

| 2s subshell | 2 electrons in an s orbital |
| 2p subshell | 6 electrons in three p orbitals |

Third shell—18 electrons

3s subshell	2 electrons in an s orbital
3p subshell	6 electrons in three p orbitals
3d subshell	10 electrons in five d orbitals

Electron configuration
The specific arrangement of an atom's electrons in shells and subshells; represented by notation such as $1s^2\, 2s^2$.

Practice Problems   **3.10**   What is the maximum number of electrons that can occupy a 3p subshell? A 2s subshell? A 2p subshell?

3.11   How many electrons are present in an atom in which the 1s, 2s, and 2p subshells are filled? Name this element.

3.12   How many electrons are present in an atom in which the first and second shells are filled and also the 3s subshell? Name this element.

Figure 3.5
Order of subshell energy levels. Above the 3p subshell, the energies of subshells in different main energy levels overlap.

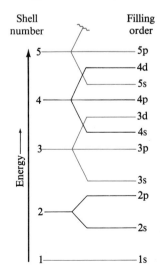

3.8   ELECTRON CONFIGURATIONS

The arrangement of electrons in an atom's shells and subshells is called an **electron configuration.** The location of the electrons in an atom can be predicted by applying three rules:

1. *Electrons occupy the lowest energy orbitals available, beginning at 1s and continuing in the order shown on the right in Figure 3.5.*
2. *A single electron occupies every orbital in a subshell before a second electron enters any of them.* In filling the three p orbitals in the same energy level, for example, all three are half-filled by one electron before any is filled by addition of the second electron.
3. *Each orbital can hold only two electrons, which must be of opposite spin.*

How electron configurations are built up according to these rules is illustrated below for the elements with from 1 to 20 electrons, many of which are among the elements most important in living things. Understanding electron configurations is helpful because, as will become apparent, the behavior of the elements depends on these configurations.
Within each shell, the subshell energies increase in the order s, p, d, f. Nevertheless, the subshells don't fill strictly in this order, apparently because the energy levels of subshells in different shells overlap as shown in Figure 3.5.

The 4*s* subshell, for example, is occupied by two electrons before the first electron enters the 3*d* subshell.

As you read the following paragraphs, check the atomic numbers and locations of the elements in the periodic table. See if you can detect the relationship between the electron configurations of different atoms and the arrangement of elements in the periodic table.

● **Hydrogen** (atomic number 1) The single electron in hydrogen is in the lowest possible energy level—the 1*s* subshell (rule 1). The configuration may be represented in either of the following ways:

$$\textbf{H} \quad 1s^1 \quad \text{or} \quad \frac{\uparrow}{1s^1}$$

In the first representation, the superscript in the notation $1s^1$ means that the 1*s* orbital is occupied by one electron. In the second representation, the 1*s* orbital is indicated by a line and the single electron is this orbital is shown by an up arrow ($\uparrow$). A single electron in an orbital is often referred to as *unpaired*.

● **Helium** (atomic number 2) The two electrons in helium are *paired* in the 1*s* subshell rather than unpaired and in different subshells (rules 1 and 3):

$$\textbf{He} \quad 1s^2 \quad \text{or} \quad \frac{\uparrow\downarrow}{1s^2}$$

● **Lithium** (atomic number 3) With the first shell full, the second begins to fill (rule 1):

$$\textbf{Li} \quad 1s^2\,2s^1 \quad \text{or} \quad \frac{\uparrow\downarrow}{1s^2}\ \frac{\uparrow}{2s^1}$$

● **Beryllium** (atomic number 4) An electron next pairs up in the 2*s* subshell (rules 1 and 3).

$$\textbf{Be} \quad 1s^2\,2s^2 \quad \text{or} \quad \frac{\uparrow\downarrow}{1s^2}\ \frac{\uparrow\downarrow}{2s^2}$$

● **Boron, carbon, nitrogen** (atomic numbers 5 through 7) The next three electrons enter the three 2*p* orbitals one at a time (rule 2).

$$\textbf{B} \quad 1s^2\,2s^2\,2p^1 \quad \text{or} \quad \frac{\uparrow\downarrow}{1s^2}\ \frac{\uparrow\downarrow}{2s^2}\quad \underbrace{\frac{\uparrow}{}\ \_\ \_}_{2p^1}$$

$$\textbf{C} \quad 1s^2\,2s^2\,2p^2 \quad \text{or} \quad \frac{\uparrow\downarrow}{1s^2}\ \frac{\uparrow\downarrow}{2s^2}\quad \underbrace{\frac{\uparrow}{}\ \frac{\uparrow}{}\ \_}_{2p^2}$$

$$\textbf{N} \quad 1s^2\,2s^2\,2p^3 \quad \text{or} \quad \frac{\uparrow\downarrow}{1s^2}\ \frac{\uparrow\downarrow}{2s^2}\quad \underbrace{\frac{\uparrow}{}\ \frac{\uparrow}{}\ \frac{\uparrow}{}}_{2p^3}$$

● **Oxygen, fluorine, neon** (atomic numbers 8 through 10) Electrons now pair up one by one to fill the $2p$ subshell and fully occupy the second energy level (rules 1 and 3).

$$\text{O} \qquad 1s^2\,2s^2\,2p^4 \qquad \text{or} \qquad \underset{1s^2\;\;2s^2}{\uparrow\downarrow\;\;\uparrow\downarrow} \qquad \underset{2p^4}{\underline{\uparrow\downarrow\;\;\uparrow\;\;\uparrow}}$$

$$\text{F} \qquad 1s^2\,2s^2\,2p^5 \qquad \text{or} \qquad \underset{1s^2\;\;2s^2}{\uparrow\downarrow\;\;\uparrow\downarrow} \qquad \underset{2p^5}{\underline{\uparrow\downarrow\;\;\uparrow\downarrow\;\;\uparrow}}$$

$$\text{Ne} \qquad 1s^2\,2s^2\,2p^6 \qquad \text{or} \qquad \underset{1s^2\;\;2s^2}{\uparrow\downarrow\;\;\uparrow\downarrow} \qquad \underset{2p^6}{\underline{\uparrow\downarrow\;\;\uparrow\downarrow\;\;\uparrow\downarrow}}$$

● **Sodium to calcium** (atomic numbers 11 through 20) The identical pattern of electron configurations for the first 10 elements is repeated in the second 10 elements. The $3s$ and $3p$ subshells are filled at argon (atomic number 18). Next comes the first variation in the order of filling. In potassium and calcium (atomic numbers 18 and 19), electrons fill the $4s$ subshell instead of the $3d$ subshell.

Complete electron configurations for the first 20 elements are shown in Table 3.3. For example, the notation $1s^2\,2s^2\,2p^6\,3s^2$ for magnesium means that magnesium atoms have two electrons in the first shell, eight electrons in the second shell, and two electrons in the third shell.

Two electrons in first shell — Eight electrons in second shell — Two electrons in third shell

Mg (atomic number 12): $1s^2 \quad 2s^2\,2p^6 \quad 3s^2$

Beyond calcium, with its full $4s^2$ orbital, 10 electrons next fill the $3d$ orbital and then electrons enter the $4p$ and $5s$ orbitals (as shown in Figure 3.5).

Table 3.3   Electron Configuration of the First 20 Elements

Element		Atomic Number	Electron Configuration	Element		Atomic Number	Electron Configuration
H	Hydrogen	1	$1s^1$	Na	Sodium	11	$1s^2\,2s^2\,2p^6\,3s^1$
He	Helium	2	$1s^2$	Mg	Magnesium	12	$1s^2\,2s^2\,2p^6\,3s^2$
Li	Lithium	3	$1s^2\,2s^1$	Al	Aluminum	13	$1s^2\,2s^2\,2p^6\,3s^2\,3p^1$
Be	Beryllium	4	$1s^2\,2s^2$	Si	Silicon	14	$1s^2\,2s^2\,2p^6\,3s^2\,3p^2$
B	Boron	5	$1s^2\,2s^2\,2p^1$	P	Phosphorus	15	$1s^2\,2s^2\,2p^6\,3s^2\,3p^3$
C	Carbon	6	$1s^2\,2s^2\,2p^2$	S	Sulfur	16	$1s^2\,2s^2\,2p^6\,3s^2\,3p^4$
N	Nitrogen	7	$1s^2\,2s^2\,2p^3$	Cl	Chlorine	17	$1s^2\,2s^2\,2p^6\,3s^2\,3p^5$
O	Oxygen	8	$1s^2\,2s^2\,2p^4$	Ar	Argon	18	$1s^2\,2s^2\,2p^6\,3s^2\,3p^6$
F	Fluorine	9	$1s^2\,2s^2\,2p^5$	K	Potassium	19	$1s^2\,2s^2\,2p^6\,3s^2\,3p^6\,4s^1$
Ne	Neon	10	$1s^2\,2s^2\,2p^6$	Ca	Calcium	20	$1s^2\,2s^2\,2p^6\,3s^2\,3p^6\,4s^2$

AN APPLICATION: ATOMS AND LIGHT

Think of light as a beam of energy moving through space in the form of a wave. The shorter the wavelength, the higher the energy of the beam.

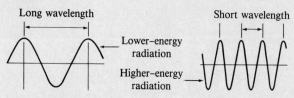

The *electromagnetic spectrum*, represented in the diagram, includes light and all other forms of radiant energy. We apply this energy for many practical purposes, from medical X rays, to microwave cooking, to radio and television.

What happens when a beam of radiant energy collides with an atom? If the amount of energy is just right, an electron can be kicked up from its usual energy level into a higher one. An electrical discharge or heat can do the same. Many practical applications, from neon lights to fireworks, take advantage of this effect.

An atom with its electrons in their usual, lowest-energy-level locations is in its *ground*

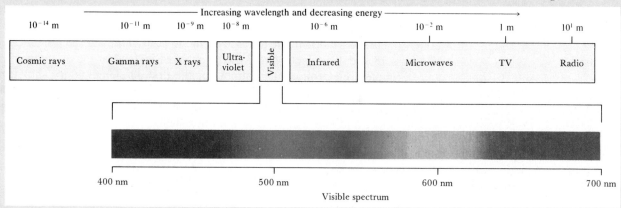

The electromagnetic spectrum with the various regions and their wavelengths, which are given in length units. From left to right wavelength increases and energy decreases.

Solved Problem 3.5   Show how the electron configuration of oxygen can be assigned.

Solution   Oxygen, atomic number 8, has eight electrons to be placed in specific orbitals. Assignments are made by putting two electrons in each orbital, according to the order shown on the right in Figure 3.5. For oxygen, the first two electrons are placed in the $1s$ orbital ($1s^2$) and the next two electrons are placed in the $2s$ orbital ($2s^2$). The next four electrons can all be placed in the three available $2p$ orbitals ($2p^4$), which can hold a maximum of six electrons. Thus, oxygen has the configuration $1s^2\,2s^2\,2p^4$.

Solved Problem 3.6   Write the complete electron configuration of phosphorus, using arrows to show how the electrons in each subshell are paired.

Solution   Phosphorus, atomic number 15, has 15 electrons to be placed in subshells. The first two electrons fill the first energy level and the next eight electrons fill

Fireworks are the most spectacular of the many applications of radiant energy.

state, which is the most stable state for the atom. With one of its electrons promoted upward in energy level, an atom is said to be *excited*. The excited state doesn't last long, though, because the electron quickly drops back to its more stable ground-state energy level, releasing the extra energy as electromagnetic radiation. If the released energy falls in the range of visible light, we can see the result.

In neon lights, noble gas atoms are excited by an electric discharge to give a variety of colors depending on the gas (red from neon, white from krypton, blue from argon). Similarly, mercury or sodium atoms excited by electrical energy are responsible for the intense bluish or yellowish light, respectively, provided by some street lamps. In the same manner, metal atoms excited by heat are responsible for the spectacular colors of fireworks (red from strontium, green from barium, and blue from copper, for example). The concentration of certain metals in body fluids is measured by sensitive instruments relying on the same principle of electron excitation that we see in fireworks. Instruments that determine the intensity of the flame color produced by lithium (red), sodium (yellow), and potassium (violet) yield the concentrations of these metals given in most clinical lab reports.

the *s* and *p* subshells of the second energy level. All electrons in these two energy levels are paired.

$$\underbrace{\underset{1s^2}{\uparrow\downarrow}\ \underset{2s^2}{\uparrow\downarrow}}\qquad \underbrace{\uparrow\downarrow\ \uparrow\downarrow\ \uparrow\downarrow}_{2p^6}$$

The remaining five electrons must enter the third energy level. Two fill the $3s$ subshell and three occupy the $3p$ subshell, one in each of the three *p* orbitals.

$$\text{P}\qquad \underset{1s^2}{\uparrow\downarrow}\ \underset{2s^2}{\uparrow\downarrow}\qquad \underbrace{\uparrow\downarrow\ \uparrow\downarrow\ \uparrow\downarrow}_{2p^6}\qquad \underset{3s^2}{\uparrow\downarrow}\qquad \underbrace{\uparrow\ \uparrow\ \uparrow}_{3p^3}$$

Practice Problems   **3.13**   Write the electron configurations (using $1s^2$-type notation) for these elements (check your answers in Table 3.3):
(a) C   (b) Na   (c) Cl   (d) Ca

3.14   Write electron configurations for the elements with atomic numbers 11 and 36.

3.15   For an atom containing 33 electrons, identify the incompletely filled subshell and show the paired and/or unpaired electrons in this subshell with arrows.

3.9   THE PERIODIC TABLE

By the early 1860s, 62 elements were known, and chemists had begun to look for ways of organizing their accumulated knowledge. Soon certain similarities among groups of elements were noticed. For example, lithium, sodium, and potassium were all known to be metals that react violently with water; chlorine, bromine, and iodine were all known to be nonmetals that have pungent odors. Furthermore, graphs of the relationships between properties of successive elements showed periodic patterns of rising and falling values.

In 1869, the Russian chemist Dmitri Mendeleev suggested a tabular way of classifying the elements according to the similarities and periodic variations of their properties. In modified form, Mendeleev's table of the elements has become the modern **periodic table**, shown in Figure 3.6.

Beginning at the upper left corner of the table, elements are arranged in horizontal rows, or **periods**, in order of increasing atomic number. The first period contains only 2 elements, hydrogen and helium; the second and third periods each contain 8 elements. The fourth and fifth periods each contain 18 elements. The sixth period contains 32 elements, 14 of which (Ce through Lu) are placed separately; and the seventh period (incomplete) contains 23 elements, with 14 also placed separately. When this arrangement is used, the elements fall into 18 vertical columns, or **groups**. The goal of the periodic table is to group together in these vertical columns elements with similar properties. By achieving this grouping, the periodic table is a valuable guide to predicting the similarities and differences between elements.

Groups are numbered in two ways, both shown in Figure 3.6. The longer groups on the left and right, called the **main groups** (or the representative element groups), are numbered 1A through 8A. Alternatively, all 18 groups are numbered from left to right with Arabic numbers. Except for hydrogen, all the nonmetals fall together on the right side of the table, separated from the metals on the left by the **metalloids**, a small number of elements with properties intermediate between those of metals and nonmetals. The shorter groups at the center of the table are known as **transition metal groups**; the elements in these groups are called **transition metals**.

What fundamental property of atoms is responsible for the periodic variations observed in so many characteristics of the elements? This question occupied the thoughts of chemists for more than 50 years after Mendeleev, and it was not until well into the 1920s that the answer was established. Today, though, we know that the properties of the elements are determined entirely by their electron configurations.

Periodic table   A table displaying the elements in order of increasing atomic number so that elements with similar properties fall into the same column.

Period   A horizontal row of elements in the periodic table; elements are listed in order of increasing atomic number across a period.

Group   A vertical column of elements in the periodic table; elements in a group have similar properties.

Main group   An element group on the far right (Groups 3A to 8A) or far left (Groups 1A to 2A) of the periodic table.

Metalloids   Elements with properties intermediate between those of metals and nonmetals.

Transition metal group   A short element group in the middle of the periodic table, where d subshells are filling.

Transition metals   Metals that fall in the 10 short groups in the center of the periodic table.

Noble gases

8A
18

Halogens

Alkali metals

1A 2A Alkaline earth metals

3A 4A 5A 6A 7A
13 14 15 16 17

1 H																	2 He
3 Li	4 Be				*GROUP B elements*							5 B	6 C	7 N	8 O	9 F	10 Ne
11 Na	12 Mg			Transition metals								13 Al	14 Si	15 P	16 S	17 Cl	18 Ar

3 4 5 6 7 8 9 10 11 12

19 K	20 Ca	21 Sc	22 Ti	23 V	24 Cr	25 Mn	26 Fe	27 Co	28 Ni	29 Cu	30 Zn	31 Ga	32 Ge	33 As	34 Se	35 Br	36 Kr
37 Rb	38 Sr	39 Y	40 Zr	41 Nb	42 Mo	43 Tc	44 Ru	45 Rh	46 Pd	47 Ag	48 Cd	49 In	50 Sn	51 Sb	52 Te	53 I	54 Xe
55 Cs	56 Ba	57 La	72 Hf	73 Ta	74 W	75 Re	76 Os	77 Ir	78 Pt	79 Au	80 Hg	81 Tl	82 Pb	83 Bi	84 Po	85 At	86 Rn
87 Fr	88 Ra	89 Ac	104 Unq	105 Unp	106 Unh	107 Uns	108 Uno	109 Une									

Lanthanides →

58 Ce	59 Pr	60 Nd	61 Pm	62 Sm	63 Eu	64 Gd	65 Tb	66 Dy	67 Ho	68 Er	69 Tm	70 Yb	71 Lu

Actinides →

90 Th	91 Pa	92 U	93 Np	94 Pu	95 Am	96 Cm	97 Bk	98 Cf	99 Es	100 Fm	101 Md	102 No	103 Lr

Figure 3.6
The periodic table of the elements. The elements cerium (Ce) through lutetium (Lu)
follow lanthanum (La) in the table, and the elements thorium (Th) through
lawrencium (Lr) follow actinium (Ac). Metallic elements are shown in yellow, metal-
like elements (metalloids) are in green, and nonmetallic elements are in blue.

Figure 3.7 shows with arrows and colors how the length of each period is
determined by the filling of shells and subshells as the atomic number increases.
By reading from left to right across the rows, you can see the order in which
orbitals are filled for all known elements.

Only two elements complete the first period—hydrogen and helium—
because only two electrons are required to fill the first shell. The second and
third periods each contain eight elements because eight electrons are needed to
fill the s and p subshells in the second and third shells.

With the fourth period, the lengths of the periods increase to 18 to make
room for 10 elements in which the d subshells are filled. In fact, there are three
series of transition elements, one each for filling the $3d$, $4d$, and $5d$ subshells.
Note that two series of elements are placed outside the periodic table and not
given group numbers. The main reason for this arrangement is simply so that the
table will fit conveniently on a page. The arrangement also makes chemical
sense, however, because the 14 elements following lanthanum (the *lanthanides*)
are all quite similar, as are the 14 elements following actinium (the *actinides*).

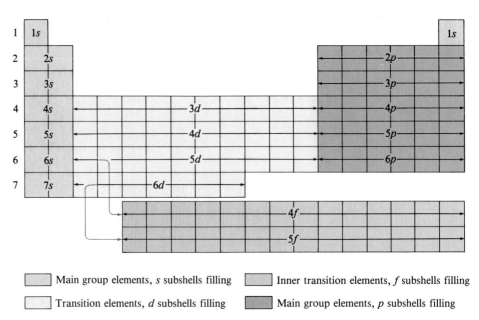

Figure 3.7
Periodic table showing subshells in the order in which they are filled. The main-group elements with their *s* subshells filling are shown in dark blue, the main-group elements with their *p* subshells filling are shown in green, the transition metals are shown in tan, and the lanthanides and actinides are shown in light blue.

Main group elements, *s* subshells filling

Transition elements, *d* subshells filling

Inner transition elements, *f* subshells filling

Main group elements, *p* subshells filling

Practice Problems **3.16** Locate aluminum in the periodic table and give its (a) group number and (b) period number.

3.17 Locate (a) krypton, (b) strontium, (c) nitrogen, and (d) cobalt in the periodic table. Indicate which of the following categories apply to each: (i) metal, (ii) nonmetal, (iii) transition element, (iv) main group element, (v) noble gas.

3.18 Using Figures 3.5 and 3.7, write the electron configuration of arsenic (atomic number 33).

AN APPLICATION: HYDROGEN

Hydrogen is the only element that doesn't fit comfortably into the periodic table. Although it's usually placed just above Group 1A because it has a single 1*s* electron, it has no resemblance to the other Group 1A elements. Hydrogen is a colorless, odorless gas that is present in a very small percentage in the air. Lithium, the first Group 1A element, is a shiny, reactive metal that must always be stored away from air because it reacts easily with oxygen and moisture.

Looking to the right in the periodic table shows that hydrogen and helium together make up the first period. Each has outer electrons only in the 1*s* subshell and both are gases. But again, these elements differ greatly in their properties. No compounds containing helium have ever been made. Hydrogen, in contrast, is part of many chemical compounds. Fifteen percent of all atoms in the earth's crust and oceans are hydrogen atoms present in chemical compounds. Even people who haven't studied chemistry know that water is H_2O—hydrogen and oxygen combined. Water and huge numbers of compounds containing hydrogen are present in all living things.

Figure 3.8
Three halogens. Chlorine (a) is a pale green gas shown condensing to a liquid on a cold tube. Bromine (b) is a volatile liquid that fills the flask with vapor, and iodine (c) is a solid that vaporizes and recondenses easily.

(a)   (b)   (c)

3.10 GROUP CHARACTERISTICS AND ELECTRON CONFIGURATIONS

Four groups of elements in the periodic table have such distinctive properties that they have been given group names:

Alkali metal An element in Group 1A of the periodic table (Li, Na, K, Rb, Cs, Fr).

● *Group 1A* **Alkali metals:** Lithium, sodium, potassium, rubidium, and cesium are shiny, soft, low-melting metals. All react violently with water to form products that are extremely alkaline, or soaplike—hence the name "alkali metals." Because of their high *reactivity*, or tendency to undergo chemical reactions with other elements, alkali metals are never found in nature in the pure state.

Alkaline earth metal An element in Group 2A of the periodic table (Be, Mg, Ca, Sr, Ba, Ra).

● *Group 2A* **Alkaline earth metals:** Beryllium, magnesium, calcium, strontium, barium, and radium are also lustrous, silvery metals, but all are less reactive than their neighbors in Group 1A.

Table 3.4 Electron Configuration for Four Groups of Elements

Group	Element	Atomic Number	Number of Electrons in Each Shell 1 2 3 4 5 6	Group	Element	Atomic Number	Number of Electrons in Each Shell 1 2 3 4 5
Group 1A **Alkali Metals**	Li (Lithium)	3	2 1	*Group 7A* **Halogens**	F (Fluorine)	9	2 7
	Na (Sodium)	11	2 8 1		Cl (Chlorine)	17	2 8 7
	K (Potassium)	19	2 8 8 1		Br (Bromine)	35	2 8 18 7
	Rb (Rubidium)	37	2 8 18 8 1		I (Iodine)	53	2 8 18 18 7
	Cs (Cesium)	55	2 8 18 18 8 1				
Group 2A **Alkaline Earths**	Be (Beryllium)	4	2 2	*Group 8A* **Noble Gases**	He (Helium)	2	2
	Mg (Magnesium)	12	2 8 2		Ne (Neon)	10	2 8
	Ca (Calcium)	20	2 8 8 2		Ar (Argon)	18	2 8 8
	Sr (Strontium)	38	2 8 18 8 2		Kr (Krypton)	36	2 8 18 8
	Ba (Barium)	56	2 8 18 18 8 2		Xe (Xenon)	54	2 8 18 18 8

Halogen   An element in Group 7A of the periodic table (F, Cl, Br, I, At).

Noble gas   An element in Group 8A of the periodic table (He, Ne, Ar, Kr, Xe, Rn).

Valence electrons
Electrons in the highest occupied main energy level of an atom.

● *Group 7A* **Halogens:** Fluorine, chlorine, bromine, and iodine are reactive, corrosive nonmetals (Figure 3.8). Halogens, like alkali metals, are found in nature only in combination with other elements, such as sodium in table salt (sodium chloride). In fact, their group name, halogen (**hal**-o-jen), derives from the Greek word *hals*, meaning ''salt.''

● *Group 8A* **Noble gases:** Helium, neon, argon, krypton, xenon, and radon are gases of extremely low reactivity. Helium, neon, and argon haven't been found to react with any other element; krypton and xenon react with very few.

Why do the elements in these and other groups of the periodic table have similar properties? The answer emerges when you look at Table 3.4 and note the numbers of **valence electrons,** those in the outer shell. *Elements within each group (column) of the periodic table have similar electron configurations in their outermost occupied electron shells.* Writing the configurations of, for example, lithium and sodium

Li $1s^2\,2s^1$ **Na** $1s^2\,2s^2\,2p^6\,3s^1$

shows that the highest occupied level in lithium is the second, which holds one electron; the highest occupied level in sodium is the third, which also holds one electron. A general notation for the outer-shell configuration of the alkali metals is ns^1, where n represents the number of the outer shell (for Li, $n = 2$; for Na, $n = 3$; for K, $n = 4$; and so on). Thus, all alkali metals (Group 1A) have a single valence electron.

Solved Problem 3.7   Using n to represent the number of the outer shell, write a general outer-shell configuration for the elements in Group 6A.

Solution   The elements in Group 6A have six valence electrons. In each, the first two of these electrons are in the outer s subshell, giving ns^2. The next four electrons are in the outer p subshell, giving np^4. For Group 6A, the general outer-shell configuration is

$$ns^2\,np^4$$

Solved Problem 3.8   How many electrons are there in a tin atom? Give the number of electrons in each shell (as in Table 3.2). How many valence electrons are there in a tin atom? Write the outer-shell configuration for tin.

Solution   Checking the periodic table shows that tin has atomic number 50 and is in Group 4A. The number of electrons in each shell is

Shell no.	1	2	3	4	5
Sn	2	8	18	18	4

As expected from the group number, there are four valence electrons. They are in the 5s and 5p subshells (see Figure 3.7) and have the configuration

$$5s^2\,5p^2$$

All alkaline earth elements (Group 2A) have two valence electrons (ns^2); all halogens (Group 7A) have seven valence electrons ($ns^2 ns^5$); and all noble gases (except helium in the first period) have eight valence electrons ($ns^2 np^6$). The same is true for every main group (Groups 1A through 8A): Atoms within each group have similar electron configurations and the same number of valence electrons. In fact, *the group numbers from 1A through 8A give the numbers of valence electrons for the elements in each group.*

Practice Problems   **3.19**   Identify the group in which all elements have the outer-shell configuration ns^2.

3.20   For chlorine, identify the group number, give the number of electrons in each occupied shell, and write the outer-shell configuration.

INTERLUDE: ARE ATOMS REAL?

All chemistry rests on the belief that matter is composed of the tiny particles we call atoms. Every chemical reaction and every physical law that governs the behavior of matter is explained by chemists in terms of atomic theory.

But how do we know that atoms are real, rather than imaginary? How do we know that our explanations have a factual basis and are not just fanciful theories? The best answer is that we can now actually *see* individual atoms through the use of an extraordinary device called a scanning tunneling microscope (STM). Invented in 1981 by a research team at the IBM Corporation, magnifications of up to 10 million have been achieved, allowing chemists for the first time to look di-

rectly at atoms themselves. To create an image like that of germanium shown at the beginning of this chapter, the STM moves a superfine tungsten needle point at a constant distance just above a surface and a computer creates an image of its ups and downs.

In 1989 the IBM scientists discovered that their new microscope can do more than take pictures of atoms—it can move them around one by one. To prove it, they created their company logo by placing 35 atoms of xenon (a noble gas) into the pattern in the accompanying photo. The actual IBM written in atoms is 660 billionths of an inch long.

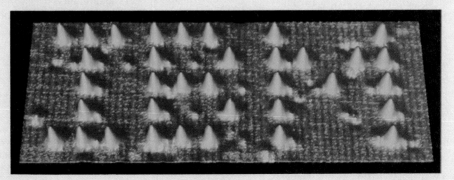

An image created by using the scanning tunneling microscope to move 35 xenon atoms into place on the surface of a supercooled nickel crystal.

SUMMARY

All matter is composed of atoms, either alone or combined in chemical compounds. In chemical reactions, the ways in which atoms are combined change, but the atoms themselves remain unchanged. An atom is the smallest and simplest unit into which an element can be broken down while maintaining the properties of the element.

Atoms are made up of three small subatomic particles called **protons**, **neutrons**, and **electrons**. Protons have a positive electrical charge; electrons have a negative electrical charge; and neutrons are electrically neutral. As a result, protons and electrons attract each other. The protons and neutrons in an atom are present in a dense, positively charged central region called the **nucleus**. Electrons are situated a great distance away from the nucleus, leaving most of the atom as empty space.

Elements differ according to the numbers of protons their atoms contain. All atoms of a given element have the same number of protons and an equal number of electrons, given by the **atomic number** characteristic of that element. Different atoms of the same element may have different numbers of neutrons, however, and thus have different **mass numbers** (the total number of protons and neutrons). Atoms with identical numbers of protons and electrons but different numbers of neutrons are called

isotopes. The **atomic weight** of an element is simply the weighted average mass of an element's naturally occurring isotopes.

The electrons surrounding an atom are grouped into layers, or **shells**, according to the amount of energy they have. Within each shell, electrons are grouped into **subshells**, and within each subshell into **orbitals**, regions of space to which they are confined. The s orbitals are spherical regions, and the p orbitals are dumbbell-shaped. Each orbital and each shell can hold only a specific number of electrons. The innermost shell can hold 2 electrons in an s orbital ($1s^2$); the second shell can hold 8 electrons in one s and three p orbitals ($2s^2\,2p^6$); the third shell can hold 18 electrons in one s, three p, and five d orbitals ($3s^2\,3p^6\,3d^{10}$); and so on. **Electron configurations** of specific elements are predicted by placing the element's electrons into orbitals, beginning with the lowest-energy orbital.

Elements are organized by atomic number into the **periodic table**. **Main-group elements** in the same **groups** in the table have the same number of **valence electrons** in their outermost shell and therefore have similar chemical properties. The lengths of the rows, or **periods**, in the table are determined by the number of electrons in each subshell.

REVIEW PROBLEMS

Atomic Theory and the Composition of Atoms

3.21 In what ways do atoms of different elements differ from each other?

3.22 Find the mass in grams of an atom of each of the following elements:
(a) Bi, atomic weight 208.9804 amu
(b) Xe, atomic weight 131.29 amu
(c) He, atomic weight 4.0026 amu

3.23 Find the mass in amu of each of the following:
(a) 5.00×10^6 C atoms of 1.99×10^{-23} g mass
(b) 1 O atom of 2.66×10^{-23} g mass
(c) 1 Br atom of 1.31×10^{-23} g mass

3.24 What is the mass in grams of 6.022×10^{23} N atoms of mass 14.01 amu?

3.25 What are the names of the three subatomic particles? What are their masses in amu, and what electrical charge does each of the three have?

3.26 Describe the location of the three types of subatomic particles in atoms.

3.27 Identify each of the following atoms:
(a) contains 19 protons
(b) contains 50 protons
(c) has atomic number 30

3.28 Give the number of neutrons in each of the naturally occurring isotopes of argon: argon-36, argon-38, argon-40.

3.29 Give the number of protons, neutrons, and electrons in each of these isotopes:
(a) aluminum-27   (b) $^{28}_{14}\text{Si}$
(c) boron-11   (d) $^{224}_{88}\text{Ra}$

3.30 Which of the following symbols represent isotopes of the same element?
(a) $^{19}_{9}\text{X}$ (b) $^{19}_{10}\text{X}$ (c) $^{21}_{9}\text{X}$ (d) $^{21}_{12}\text{X}$

3.31 Name the isotope represented by each symbol in Problem 3.30.

3.32 Write the symbols for each of the following isotopes:
(a) atoms contain 6 protons and 8 neutrons
(b) atoms have mass number 39 and contain 19 protons
(c) atoms have mass number 20 and contain 10 electrons

3.33 Complete each of the following isotope symbols:
(a) $^{206}_{84}\text{?}$ (b) $^{224}_{?}\text{Ra}$ (c) $^{197}_{?}\text{Au}$ (d) $^{84}_{36}\text{?}$

3.34 Give the number of neutrons in each isotope listed in Problem 3.33.

3.35 There are two naturally occurring isotopes of carbon with mass numbers of 12 and 13. How many neutrons does each have? Write the symbol for each isotope, indicating both atomic number and mass number.

3.36 The isotope of iodine with mass number 131 is often used in medicine as a radioactive tracer. Write the symbol for this isotope, indicating both mass number and atomic number.

3.37 Explain the role of isotopes in determining the atomic weights of the elements. ·

3.38 Naturally occurring copper is a mixture of 69.09% ^{63}Cu with a mass of 62.93 amu and 30.91% ^{65}Cu with a mass of 64.93 amu. Calculate the atomic weight of copper.

Electron Configurations

3.39 What is the maximum number of electrons that can go into an orbital?

3.40 What is the maximum number of electrons that can go into the first shell? The second shell? The third shell?

3.41 What are the shape and location within an atom of s and p orbitals?

3.42 How many electrons are present in an atom with its $1s$, $2s$, and $2p$ subshells filled? What element is this?

3.43 How many electrons are present in an atom with its $1s$, $2s$, and $3s$ subshells filled and with two electrons in the $3p$ subshell? What element is this?

3.44 Use arrows to show electron pairing in the p subshells of: (a) sulfur (b) bromine (c) silicon

3.45 Without looking at Table 3.3 or 3.4, write the electron configurations for:
(a) magnesium, atomic number 12
(b) sulfur, atomic number 16
(c) neon, atomic number 10
(d) cadmium, atomic number 48

3.46 How many electrons does the element with atomic number 20 have in its outer shell?

The Periodic Table

3.47 Why does the third period in the periodic table contain eight elements?

3.48 Why does the fourth period in the periodic table contain 18 elements?

3.49 Americium, atomic number 95, is used in household smoke detectors. What is the symbol for americium? Is americium a metal or a nonmetal?

3.50 For the elements from scandium to zinc:
(a) Are they metals or nonmetals?
(b) To what general class of elements do they belong?
(c) What subshell is being filled by electrons in these elements?

3.51 How many valence electrons do Group 4A elements have? Explain.

3.52 For (a) calcium, (b) palladium, (c) carbon, and (d) radon, choose which of the following terms apply: (i) metal, (ii) nonmetal, (iii) transition element, (iv) main-group element, (v) noble gas, (vi) alkali metal, (vii) alkaline earth metal

3.53 Identify the outermost subshell occupied by electrons in beryllium and arsenic atoms.

3.54 What group in the periodic table group has the outer-shell configuration $ns^2 \, np^4$?

3.55 What other two elements in the periodic table would you expect to be most similar to sulfur?

3.56 Give the number of valence electrons in atoms of each of the following elements:
(a) Kr   (b) C   (c) Ca   (d) K   (e) B   (f) Cl

3.57 What elements in addition to lithium make up the alkali metal family?

3.58 What elements in addition to fluorine make up the halogen family?

3.59 What elements in addition to helium make up the noble gas family?

3.60 Cesium, atomic number 55, has only one more electron than xenon, atomic number 54, yet its chemical behavior is completely different. Explain.

3.61 Using n for the number of the outer shell, write a general outer-shell configuration for the elements in Group 7A and in Group 1A.

Applications

3.62 In view of its electron configuration and chemical behavior, why is radioactive cesium especially hazardous? [App: Chernobyl and Cesium]

3.63 A strontium atom is found to have an electron in the sixth shell. (a) Why do we say that the atom is in an excited state? (b) What must happen for the atom to return to the ground state? [App: Atoms and Light]

3.64 Which type of radiation in each of the following pairs has the higher energy?
(a) infrared radiation and cosmic rays
(b) visible light and gamma rays
(c) radio waves and cosmic rays [App: Atoms and Light]

3.65 Why do you suppose ultraviolet radiation from the sun is more damaging to the skin than visible or infrared radiation? [App: Atoms and Light]

3.66 Certain kinds of chemical compounds release radiation in the 200–400 nm region when their electrons become excited. In what part of the spectrum is this radiation? [App: Atoms and Light]

3.67 Hydrogen is placed in Group 1A on many periodic charts, even though it is not an alkali metal. On other periodic charts, though, hydrogen is included with Group 7A even though it is not a halogen. Explain. [App: Hydrogen]

3.68 What is the advantage of using a scanning tunneling microscope compared to a normal light microscope? [Int: Are Atoms Real?]

Additional Problems

3.69 Give the number of electrons in each shell for lead.

3.70 Identify the highest occupied subshell in atoms of: (a) argon (b) magnesium (c) technetium (d) iron

3.71 Calculate the atomic weight of naturally occurring bromine, which contains 50.54% bromine-79 of mass 78.92 amu and 49.46% bromine-81 of mass 80.91 amu.

3.72 Naturally occurring magnesium consists of three isotopes: 78.70% ^{24}Mg with mass of 23.99 amu; 10.13% ^{25}Mg with mass of 24.99 amu; and 11.17% ^{26}Mg with mass of 25.98 amu. Calculate the atomic weight of magnesium.

3.73 If you had one atom of hydrogen and one atom of carbon, which of the two would weigh more? Explain.

3.74 If you had a pile of 10^{23} hydrogen atoms and another pile of 10^{23} carbon atoms, which of the two piles would weigh more? (See Problem 3.73.)

3.75 If your pile of hydrogen atoms in Problem 3.73 weighed about one gram, how much would your pile of carbon atoms weigh?

3.76 Based on your answer to Problem 3.75, how much would you expect a pile of 10^{23} sodium atoms to weigh?

3.77 An unidentified element is found to have an electron configuration by shell of 2 8 18 8 2. To what group and period does this element belong? Is the element a metal or a nonmetal? How many protons does an atom of the element have? What is the name of the element?

3.78 Titanium, atomic number 22, is used in building jet fighter planes because of its combination of high strength and light weight. If titanium has an electron configuration by shell of 2 8 10 2, in what orbital are the outer-shell electrons?

3.79 Zirconium, atomic number 40, is directly beneath titanium (Problem 3.78) in the periodic table. What electron configuration by shell would you expect zirconium to have? Is zirconium a metal or a nonmetal?

3.80 A blood sample is found to contain 8.6 mg/dL of Ca. How many atoms of Ca (atomic weight, 40.08 amu) are present in 8.6 mg?

3.81 What is wrong with each ground-state electron configuration listed here?
(a) Cr $1s^2\ 2s^2\ 2p^6\ 3s^2\ 3p^6\ 3d^6$
(b) N $1s^2\ 2p^5$
(c) Si $1s^2\ 2s^2\ 2p$ ⇅ __ __
(d) Mg $1s^2\ 2s^2\ 2p^6\ 3s$ ⇅

3.82 Not all elements follow precisely the electron filling order described in Figure 3.7. Atoms of which elements are represented by the following electron configurations?
(a) $1s^2\ 2s^2\ 2p^6\ 3s^2\ 3p^6\ 3d^{10}\ 4s^1$
(b) $1s^2\ 2s^2\ 2p^6\ 3s^2\ 3p^6\ 3d^{10}\ 4s^2\ 4p^6\ 4d^5\ 5s^1$

4

Nuclear Chemistry

Imagine being able to *see* where brain activity occurs in response to different kinds of stimulation. The red regions in these images of normal subjects allow exactly that. Radioactivity, the property of unstable nuclei that makes such images possible, is explored in this chapter.

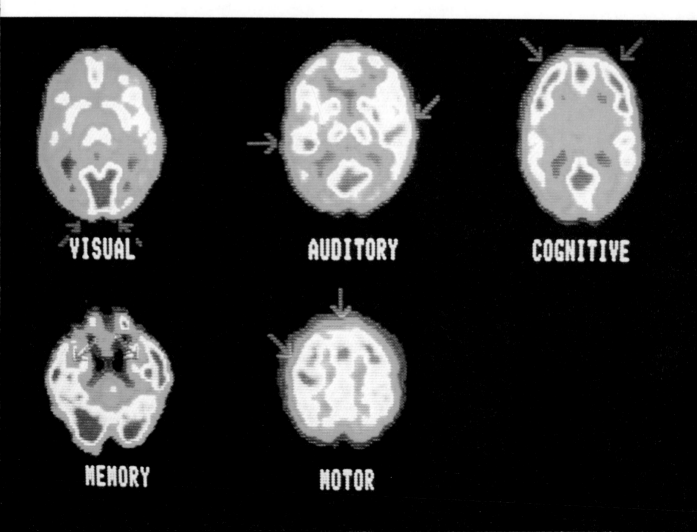

VISUAL   AUDITORY   COGNITIVE

MEMORY   MOTOR

In Chapter 3 you were introduced to atoms, the building blocks of all matter. In most areas of chemistry, you can take for granted that atoms do not change their identities in any physical or chemical change. In this chapter we're going to look at the exception, *nuclear* chemistry—chemistry in which the nuclei of atoms can change. You've seen that the identity of an isotope depends on the number of protons and neutrons in its nucleus. When nuclei change, therefore, atoms of one element can become atoms of a different element. Such changes occur when unstable nuclei break down to give more stable nuclei. You'll see that nuclear chemistry has many valuable applications in medicine, some of them quite recently developed. In this chapter we'll answer the following questions:

1. ***What is radioactivity?*** The goal: Be able to define radioactivity and the terms used to describe it.
2. ***What are the different kinds of radioactivity?*** The goal: Be able to list the characteristics of three common kinds of radiation—α, β, and γ.
3. ***What happens in nuclear decay reactions?*** The goal: Be able to describe α, β, and γ decay and to write balanced equations for nuclear decay reactions.
4. ***How is the rate of nuclear decay measured?*** The goal: Be able to explain half-life and to calculate the quantity of a radioisotope remaining after a known number of half-lives.
5. ***What is ionizing radiation?*** The goal: Be able to describe the properties of the different types of ionizing radiation, their potential for harm to living tissue, their detection, and protection from them.
6. ***How is radioactivity measured?*** The goal: Be able to list and define the common units for measuring radiation.
7. ***What is artificial transmutation?*** The goal: Be able to explain nuclear bombardment and balance equations for nuclear bombardment reactions.
8. ***What are nuclear fission and nuclear fusion?*** The goal: Be able to explain the differences between nuclear fission and nuclear fusion.

4.1 THE DISCOVERY OF RADIOACTIVITY

The discovery of radioactivity dates to the year 1896 when the French physicist Henri Becquerel made a totally unexpected observation. While investigating the nature of phosphorescence, Becquerel happened to place a sample of a uranium-containing mineral on top of a photographic plate that had been wrapped in black paper and placed in a drawer to protect it from sunlight. On developing the plate, Becquerel was surprised to find a silhouette of the mineral. He immediately concluded that the mineral was producing some kind of unknown radiation that passed through the paper and exposed the photographic plate.

Radioactivity Spontaneous emission of radiation (alpha, beta, or gamma rays) from an unstable atomic nucleus.

Marie Sklodowska Curie and her husband, Pierre, took up the challenge and began a series of investigations into this new phenomenon, which they termed **radioactivity.** They found that the source of the radioactivity was the element uranium (U) and that two previously unknown elements, which they named polonium (Po) and radium (Ra), were also radioactive. For these achievements, Becquerel and the Curies shared the 1903 Nobel Prize in physics.

4.2   THE NATURE OF RADIOACTIVITY: α, β, AND γ RADIATION

Alpha (α) **radiation** Emission of helium nuclei, $^4_2$He.

Beta (β) **radiation** Emission of electrons.

Gamma (γ) **radiation** Emission of high-energy light waves.

Further work on radioactivity by the English scientist Ernest Rutherford soon established that there were at least two types of radiation, which he named **alpha (α)** and **beta (β)** after the first two letters of the Greek alphabet. Shortly thereafter, a third type of radiation was found and named for the third Greek letter, **gamma (γ).**

Studies by Rutherford, the Curies, and other scientists showed that when the three kinds of radiation are passed between electrically charged plates, each is affected differently. α Rays bend toward the negative plate, β rays bend toward the positive plate, and γ rays continue straight ahead (Figure 4.1). These experiments demonstrated that α radiation is positively charged, that β radiation is negatively charged, and that γ radiation has no charge. Also, because the α rays were deflected less by the charged plates than the β rays, the experiments showed that α rays are heavier than β rays.

A further difference among the three kinds of radiation soon became apparent: α and β radiations are actually composed of small particles, while γ radiation consists of very-high-energy light waves (see spectrum in Application on Atoms and Light in Chapter 3). Rutherford was able to show that a β *particle* is an electron (e$^-$) and that an α *particle* is simply a helium nucleus (He^{2+}). Recall that a helium atom (atomic number 2 and atomic weight 4) consists of two protons, two neutrons, and two electrons. When the two electrons are removed, the remaining helium *nucleus,* or α particle, has only the two protons and two neutrons.

Figure 4.1
Demonstration of the electrical charges of different kinds of radiation. When passed between charged plates, α rays, since they are attracted toward the negative plate, must be positively charged; β rays, since they are attracted toward the positive plate, must be negatively charged; and γ rays, since they are not attracted to either plate, must be neutral.

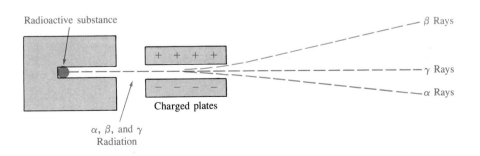

Yet a third difference among the three kinds of radiation is their penetrating power. Because of their relatively large mass, α particles move relatively slowly (up to about one-tenth the speed of light) and can be stopped by a few sheets of paper or by the top layer of skin. β Particles, because they are much lighter, move at up to nine-tenths the speed of light and have about 100 times the penetrating power of α particles. A block of wood or heavy protective clothing is necessary to stop β radiation, which can otherwise penetrate the skin and cause burns and other damage. γ Rays move at the speed of light and have about 1000 times the penetrating power of α particles. A lead block several inches thick is needed to stop γ radiation, which can otherwise penetrate and damage the body's internal organs.

The characteristics of the three kinds of radiation are summarized in Table 4.1. Even though the α particle is an ion with a $+2$ charge, it's customary to leave out the ionic charge in the written symbol, ^4_2He.

4.3 STABLE AND UNSTABLE ISOTOPES

Nuclear chemistry The study of atomic nuclei and their reactions.

Radioisotope A radioactive isotope.

The chemistry of radioactive substances is the chemistry of their nuclei—**nuclear chemistry.** It is differences in atomic nuclei that are responsible for radioactivity. Every element in the periodic table has at least one radioactive isotope (**radioisotope**), and there are nearly 2000 different radioisotopes known. Their radioactivity is the result of having unstable nuclei, although the exact causes of this instability aren't fully understood. Radiation is emitted when unstable nuclei spontaneously change into more stable nuclei.

For the elements in the first few rows of the periodic table, stability is associated with a roughly equal number of neutrons and protons. Hydrogen, for example, has stable ^1_1H and ^2_1H (deuterium) isotopes but becomes radioactive in the ^3_1H isotope (tritium). As elements get heavier, the number of neutrons relative to protons in stable nuclei increases. Lead-208, for example, the most abundant stable isotope of lead, has 126 neutrons and 82 protons in its nuclei. Nevertheless, of the 29 known isotopes of lead, only 4 are stable while 25 are radioactive. All isotopes of elements with atomic numbers higher than that of bismuth (83) are radioactive.

Table 4.1   Characteristics of α, β, and γ Radiation

Type of Radiation	Symbol	Charge	Composition	Mass (*amu*)	Velocity	Relative Penetrating Power
Alpha	α, ^4_2He	$+2$	Helium nucleus	4	Up to 10% light speed	Low (1)
Beta	β, $^0_{-1}\text{e}$	-1	Electron	1/1840	Up to 90% light speed	Medium (100)
Gamma	γ, $^0_0\gamma$	0	High-energy radiation	0	Light speed	High (1000)

Most of the known radioactive isotopes have been made in high-energy particle accelerators by reactions that we'll describe in Section 4.11. Such isotopes are called **artificial radioisotopes** and are described as having *artificial radioactivity,* since they aren't found in nature. All the elements heavier than uranium, the **transuranium elements,** are known only as artificial isotopes. The much smaller number of radioactive isotopes found in the earth's crust are called **natural radioisotopes** and are described as having *natural radioactivity.*

Aside from their radioactivity, radioisotopes have the same *chemical* properties as stable isotopes. Herein lies the great usefulness of radioisotopes as **tracers.** A chemical compound tagged with a radioactive atom undergoes all the same reactions as the untagged compound. The difference is that the tagged compound can be located with a radiation detector and its distribution determined.

Artificial radioisotope A manmade radioactive isotope.

Transuranium elements Elements of higher atomic number than uranium (atomic number 92) and known only as artificial radioisotopes.

Natural radioisotope A radioactive isotope found in nature.

Radioactive tracer A radioisotope used to track the location of a substance to which it is attached.

4.4 NUCLEAR DECAY

Think for a minute about the consequences of α and β radiation. If radioactivity involves the spontaneous emission of a small particle from an unstable atomic nucleus, then the atom itself must undergo a change. With the understanding of radioactivity came the startling discovery that atoms of one element can change into atoms of another element, something that had previously been thought impossible. The spontaneous process of particle emission from an unstable nucleus is called **nuclear decay** or radioactive decay, and the change of one element into another brought about by nuclear decay is called **transmutation.**

Nuclear decay The emission of a particle from the nucleus of an element.

Transmutation The change of one element into another brought about by a nuclear reaction.

Nuclear decay   Radioactive element $\longrightarrow$ new element + emitted particle

Alpha Emission   What happens when an atom of uranium-238 ($^{238}_{92}U$) gives off an α particle? The atom of the new element produced must have two fewer protons and two fewer neutrons after the emission. Since the number of protons in the nucleus has now changed from 92 to 90, the identity of the atom has changed from uranium to thorium, and since the total number of neutrons and protons has decreased by 4, uranium-238 has become thorium-234 ($^{234}_{90}Th$) (Figure 4.2).

$$^{238}_{92}U \;\longrightarrow\; ^{234}_{90}Th \;+\; ^{4}_{2}He$$

Figure 4.2
α Particle emission from uranium-238. Emission of an α particle from an atom of uranium-238 leaves behind an atom of thorium-234.

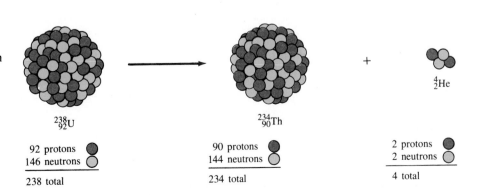

$^{238}_{92}U$	$^{234}_{90}Th$	$^{4}_{2}He$
92 protons	90 protons	2 protons
146 neutrons	144 neutrons	2 neutrons
238 total	234 total	4 total

Note that the equation for a nuclear reaction is not balanced in the chemical sense because the kinds of atoms are not the same on both sides of the arrow. Instead, *a nuclear equation is balanced when the sums of the mass numbers and the sums of the atomic numbers are the same on both sides of the equation.*

Solved Problem 4.1   Polonium-218 is one of the α emitters studied by Marie Curie. Write the equation for the α decay and identify the element formed.

Solution   First, look up the atomic number of polonium in the periodic table and write the known part of the nuclear equation using the symbol for polonium-218 in the standard format:

$$^{218}_{84}\text{Po} \longrightarrow {}^{4}_{2}\text{He} + ?$$

Next, finish the nuclear equation by calculating the mass number and atomic number of the product element. The mass number is $218 - 4 = 214$, and the atomic number is $84 - 2 = 82$. A look at the periodic table identifies the element with atomic number 82 as lead (Pb).

$$^{218}_{84}\text{Po} \longrightarrow {}^{4}_{2}\text{He} + {}^{214}_{82}\text{Pb}$$

Finally, check your answer by making sure that the mass numbers and atomic numbers on the two sides of the equation are balanced:

Mass numbers: $218 = 4 + 214$ Atomic numbers: $84 = 2 + 82$

Practice Problems   **4.1**   High levels of radioactive radon-222 ($^{222}_{86}\text{Rn}$) have been found in many homes built on radium-containing rock, leading to the possibility of health hazards. What product results from α emission by radon-222?

4.2   What isotope of radium (Ra) is converted into radon-222 ($^{222}_{86}\text{Rn}$) by α emission?

Beta Emission   Whereas α emission leads to the loss of two protons and two neutrons from the nucleus, β emission involves the *decomposition of a neutron* to yield an electron and a proton. The electron is ejected as a β particle, and the proton is retained by the nucleus. Note that the electrons of β rays come from the *nucleus* and not from the occupied orbitals surrounding the nucleus.

The net result of β emission is that the atomic number of the atom increases by 1 because there is a new proton. The mass number of the atom, however, remains the same since a neutron changed into a proton, but the total number of protons and neutrons hasn't changed. For example, iodine-131 ($^{131}_{53}\text{I}$), a radioisotope used in detecting thyroid problems, undergoes nuclear decay by β emission to yield xenon-131 ($^{131}_{54}\text{Xe}$).

$$^{131}_{53}\text{I} \longrightarrow {}^{131}_{54}\text{Xe} + {}^{0}_{-1}\text{e}$$

The symbol for a β particle ($_{-1}^{0}e$) shows that it has no mass (superscript zero); the subscript -1 for its negative charge allows the atomic numbers to balance in a nuclear equation.

Practice Problems **4.3** Carbon-14, a beta emitter, is a rare isotope of great use in dating archaeological artifacts. Write a nuclear equation for the decay of carbon-14.

4.4 Write nuclear equations for β emission of these isotopes:
(a) $_{1}^{3}H$ (b) $_{82}^{210}Pb$

Gamma Emission Emission of γ rays, unlike α and β particle emission, causes no change in mass or atomic number because γ rays are simply high-energy light rays. γ Emission almost always accompanies α and β emission because the new nucleus is produced in an excited state and must get rid of its extra energy. The process is similar to the emission of visible light when orbital electrons fall from higher to lower energy levels (Application on Atoms and Light, Chapter 3).

When cobalt-60 is used in cancer therapy, it's actually the penetrating γ rays that do the job, rather than the less penetrating β rays:

$$_{27}^{60}Co \rightarrow _{-1}^{0}e + _{28}^{60}Ni + _{0}^{0}\gamma$$

Since γ emission affects neither mass number nor atomic number, it is often omitted from nuclear equations. Nevertheless, γ rays are of great importance: Their penetrating power makes them by far the most dangerous kind of external radiation for humans. This same penetrating power also makes them useful in numerous medical applications.

Table 4.2 summarizes the three different kinds of radioactive decay.

4.5 HALF-LIFE

The rate of radioactive decay varies greatly from one isotope to another. Some isotopes such as uranium-235 decay at a barely perceptible rate, while others decay almost instantly. Rates of decay are measured in units of **half-life** ($t_{1/2}$), and one half-life is defined as the amount of time required for one-half of the radioactive sample to decay. For example, the half-life of iodine-131, a radioisotope used in thyroid testing, is 8 days. If today you have a certain amount of $_{53}^{131}I$, say 1.0 g, then 8 days from now you will have only 0.50 g of $_{53}^{131}I$ remaining because one-half of the sample will have decayed. After 8 more days (16 total) only 0.25 g of $_{53}^{131}I$ will remain; after a further 8 days (24 total) only 0.125 g will remain; and so on. Each passage of a half-life causes the decay of one-half of whatever sample remains. The half-life is the same no matter what the size of the sample, the temperature, or any other external conditions.

Half-life ($t_{1/2}$) The amount of time required for 50% of a sample to undergo radioactive decay.

$$1.0 \text{ g } _{53}^{131}I \xrightarrow[\text{days}]{8} \begin{array}{c} 0.50 \text{ g } _{53}^{131}I \\ 0.50 \text{ g } _{54}^{131}Xe \end{array} \xrightarrow[\text{days}]{16} \begin{array}{c} 0.25 \text{ g } _{53}^{131}I \\ 0.75 \text{ g } _{54}^{131}Xe \end{array} \xrightarrow[\text{days}]{24} \begin{array}{c} 0.125 \text{ g } _{53}^{131}I \\ 0.875 \text{ g } _{54}^{131}Xe \end{array} \longrightarrow$$

One Half-life Two half-lives Three half-lives

The origins of nuclear medicine reach back to 1901 when the French physician Henri Danlos first used radium in the treatment of a tuberculous skin lesion. Since that time, the use of radioactivity has become a crucial part of modern medical care, both diagnostic and therapeutic. Currently used nuclear techniques can be grouped into four classes: (1) imaging procedures (discussed in a later Application in this chapter), (2) in vivo procedures, (3) in vitro procedures, and (4) radiation therapy.

In Vivo Procedures In vivo studies—those that take place *inside* the body—are carried out to assess the functioning of a particular organ or body system. A radiopharmaceutical agent is administered, and its path in the body—whether absorbed, excreted, diluted, or concentrated—is determined by analysis of blood or urine samples.

Among the many in vivo procedures utilizing radioactive agents is a simple method for the determination of whole-blood volume by injecting a known quantity of red blood cells la-

beled with radioactive chromium-51. After a suitable interval to allow the labeled cells to be distributed evenly throughout the body, a blood sample is taken and blood volume is calculated by comparing the concentration of labeled cells in the blood with the quantity of labeled cells injected.

In Vitro Procedures: Radioimmunoassay In vitro studies—those that take place *outside* the body—are carried out on blood or urine samples and involve *radioimmunoassay (RIA) techniques*. RIA takes advantage of the ability of antibodies or other large protein molecules to selectively bind specific smaller molecules called *antigens*. The role of antibodies is to capture the antigens and protect the body against their action. In RIA, the substance to be analyzed is the antigen, and the antibody needed for each different antigen is first grown in laboratory animals.

To carry out an analysis, known amounts of the antibody and a radioactively labeled, or *tagged,* version of the antigen are combined with

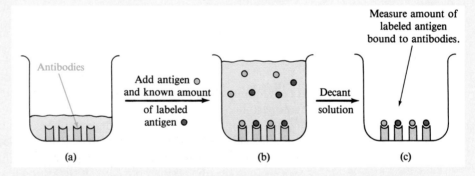

Radioimmunoassay (RIA) involves measuring the binding of antigen to an antibody against that antigen. (a) Antibodies of known binding ability are prepared and (b) added to a sample of biological fluid containing the antigen to be measured (○). A known amount of radioactively labeled antigen is also added (●). (c) The unknown amount of antigen in the fluid can be determined by counting the amount of labeled antigen bound to the antibodies. The relative numbers of labeled and unlabeled antigens that are bound are the same as the relative numbers combined (1 to 1 in this example).

a sample containing an unknown concentration of unlabeled antigen. The labeled and unlabeled antigen molecules compete with each other for the limited amount of antibody. Once equilibrium has been established so that quantities of bound antigen become constant, the relative numbers of labeled and unlabeled bound antigen molecules reflect their original concentrations. To complete the analysis, the distribution of tagged molecules is measured with a radiation detector and compared with the results on a series of known samples to determine the unknown antigen concentration. The RIA technique is used to determine very small concentrations of a wide variety of drugs, hormones, and other substances, and can often be carried out on a single drop of blood.

Therapeutic Procedures Therapeutic procedures—those in which radiation is purposely used as a weapon to kill diseased tissue—can involve either external or internal sources of radiation. External radiation therapy for the treatment of cancer can be carried out with gamma rays emanating from a cobalt-60 source. The highly radioactive source is shielded by a thick lead container and has a small opening directed toward the site of the tumor. By focusing the radiation beam on the tumor and rotating the patient's body, the tumor receives the full exposure while the exposure of surrounding parts of the body is minimized. Nevertheless, sufficient exposure occurs so that most patients treated in this manner suffer the effects of radiation sickness.

Internal radiation therapy is a much more selective technique. For example, in the treatment of thyroid disease a radioactive substance such as iodine-131 is administered. This powerful beta emitter localizes in the target tissue, the thyroid. Since β particles penetrate no further than several millimeters, the localized I-131 produces a high radiation dose that destroys only the surrounding diseased tissue. To treat certain tumors, such as those in the female reproductive system, a radioactive source is placed physically close to the tumor for a specific amount of time.

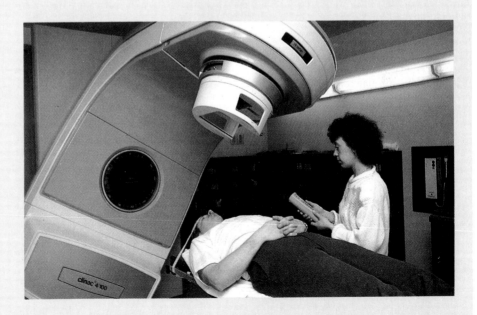

Cobalt radiation therapy

Table 4.2   Summary of Radioactive Decay

Kind of Decay	Effect on Mass Number	Effect on Atomic Number
α	Mass number decreases by 4	Atomic number decreases by 2
β	Mass number unchanged	Atomic number increases by 1
γ	Mass number unchanged	Atomic number unchanged

The percent of any radioactive isotope sample remaining after each half-life passes is represented by the curve in Figure 4.3. No matter how long the half-life period, 50% remains after one half-life, 25% remains after two half-lives, 12.5% remains after three half-lives, and so on.

		Percent remaining
One half-life	$\frac{1}{2}$	50%
Two half-lives	$\frac{1}{2} \times \frac{1}{2} = \frac{1}{4}$	25%
Three half-lives	$\frac{1}{2} \times \frac{1}{2} \times \frac{1}{2} = \frac{1}{8}$	12.5%

The half-lives of some useful radioisotopes are given in Table 4.3. As you might expect, radioisotopes used internally in medical applications have fairly short half-lives so that they decay.

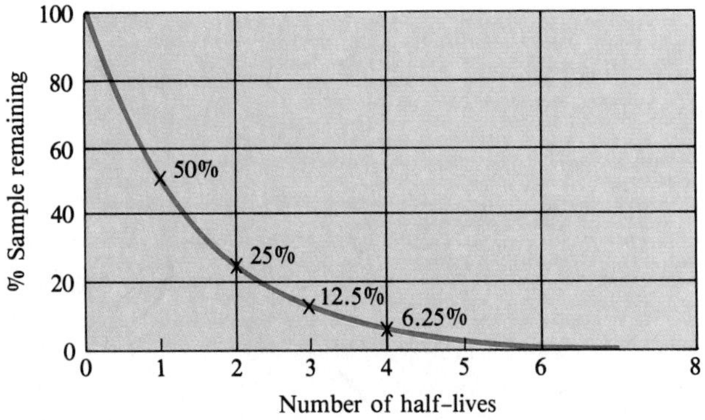

Figure 4.3
Radioactive decay curve. The decay of all radioisotopes follows this curve, whether their half-lives are measured in years, days, minutes, or seconds. The percentages of the original isotope remaining after one, two, three, and four half-lives are shown.

Table 4.3   Half-lives of Some Useful Radioisotopes

Radioisotope	Symbol	Radiation	Half-life	Use
Uranium-235	$^{235}_{92}U$	α, γ	7.1×10^8 years	Nuclear reactors
Carbon-14	$^{14}_{6}C$	β	5730 years	Archaeological dating
Cobalt-60	$^{60}_{27}Co$	β, γ	5.3 years	Cancer therapy
Tritium	$^{3}_{1}H$	β	12.26 years	Biochemical tracer
Phosphorus-32	$^{32}_{15}P$	β	14.3 days	Leukemia therapy
Iodine-131	$^{131}_{53}I$	β, γ	8.07 days	Thyroid studies
Sodium-24	$^{24}_{11}Na$	β, γ	15.0 hours	Blood studies
Iodine-123	$^{123}_{53}I$	γ	13.3 hours	Thyroid therapy
Potassium-42	$^{42}_{19}K$	β, γ	12.4 hours	Nutrition studies
[a]Technetium-99m	$^{99m}_{43}Tc$	γ	6.02 hours	Brain scans

[a]The m in technetium-99m stands for *metastable*, meaning that it decays (by γ decay) without changing its mass number or atomic number.

Solved Problem 4.2 Phosphorus-32 has a half-life of approximately 14 days. What percent of a sample remains after 8 weeks?

Solution First, determine, how many half-lives have elapsed. Since one half-life of $^{32}_{15}P$ is 14 days (2 weeks), 8 weeks represents four half-lives. Next, do the calculation. Since each half-life decreases the amount of the sample by half, multiply the starting amount (100%) by 1/2 for each of the four half-lives that has elapsed:

Four half-lives

$$\text{Final percentage} = 100\% \times (\tfrac{1}{2} \times \tfrac{1}{2} \times \tfrac{1}{2} \times \tfrac{1}{2})$$
$$= 100\% \times \tfrac{1}{16} = 6.25\%$$

Practice Problem **4.5** The half-life of carbon-14 is 5730 years. What percentage of $^{14}_{6}C$ remains in a sample estimated to be about 17,000 years old?

4.6 NATURAL RADIOACTIVE DECAY SERIES

Cosmic rays Energetic particles, primarily protons, from space.

Most natural radioisotopes were either created when the earth was formed or are products of the decay of original radioisotopes. A few radioisotopes are continually produced by the action of **cosmic rays,** energetic particles that come from interstellar space and collide with atoms and chemical compounds in the atmosphere, producing a shower of gamma rays, neutrons, and other subatomic particles that reach the earth's surface.

The decay of three very long-lived parent radioisotopes accounts for all the heavy radioactive elements found in the earth's crust. The parent of the uranium series, for example, is uranium-238, which has a half-life of 4.5×10^9 years (about 4 billion years). Decay of the unstable uranium-238 nuclei initiates a long series of α and β decay reactions. Finally, after a series of 13 decay reactions that form new radioactive elements, the stable lead-206 isotope is reached and the process is complete.

Radon-222 is a gaseous radioisotope produced in the uranium-238 decay series by the α decay of radium-226, which has a half-life of 1600 years. Rocks, soil, and building materials that contain radium are sources of radon emission, and concern has arisen over significant concentrations of radon in the air in buildings. Radon itself, a noble gas, passes in and out of the lungs without being incorporated into body tissue. The greatest threat from radon exposure is the trapping of its solid decay product, polonium-218, in the lungs and the resulting serious damage to lung tissue from α particles emitted by this radioisotope.

4.7 IONIZING RADIATION

Ionizing radiation Radiation capable of dislodging an electron from (ionizing) an atom or a chemical compound it strikes.

High-energy radiation of all kinds is usually grouped together under the name **ionizing radiation.** The interaction of any of these kinds of radiation with an atom in a chemical compound knocks out an electron, converting the compound to an extremely reactive *ion,* a charged species with an odd number of electrons. The reactive ion and the electron, known as an *ion pair,* immediately react with other chemical compounds in their surroundings, creating still other highly reactive fragments that can cause further reactions. In this manner, a large dose of ionizing radiation can destroy the delicate balance of chemical structure and chemical reactions in living cells and cause the death of an organism.

A small dose may not cause immediately visible symptoms but can be dangerous if it strikes a cell nucleus and damages its genetic machinery. The resultant changes might lead to a **genetic mutation,** to cancer, or to cell death. The nuclei of rapidly dividing cells, such as those in bone marrow, the lymph system, an embryo, or the lining of the intestinal tract, are most readily damaged in this manner. If a sperm or egg cell is affected, a genetic mutation can be passed on to offspring. It's because of the susceptibility of rapidly dividing cells to radiation effects that ionizing radiation is able to selectively destroy cancer cells, which multiply at an abnormally high rate.

Ionizing radiation includes not only α particles, β particles, and γ rays, but also **X rays** and cosmic rays. X rays are like γ rays in that they are high-energy light waves rather than particles. The principal difference between them is that γ rays originate within the nucleus and X rays originate outside the nucleus when excited inner shell electrons drop to lower energy levels.

Some properties of ionizing radiation are summarized in Table 4.4. α Rays are the most energetic, are the least penetrating, and have the greatest potential for damage, as shown by comparing the relative numbers of ion pairs produced in traveling equal distances.

The effects of ionizing radiation on the human body vary with the radiation's energy, its distance from the body, the length of the exposure, and whether the source is outside or inside the body. When coming from outside the body, γ and X radiation are potentially more harmful than α and β rays because they pass through clothing and skin and into the body's cells.

α Rays are stopped by clothing and skin, and β rays don't penetrate very far. These types of radiation are much more dangerous when emitted within the body, however, because all their radiation energy is then given up to the immediately surrounding tissue. α Emitters are especially hazardous internally and are almost never used in medical applications.

4.8 RADIATION PROTECTION

Genetic mutation An error in the code that determines inherited traits.

X rays High-energy electromagnetic radiation.

Background radiation The sum of low-level radiation to which all individuals are exposed.

No amount of radiation can be called absolutely safe. All ionizing radiation, even the small amount of **background radiation** we are all exposed to every day from naturally occurring radioisotopes and cosmic rays, has a small potential

Table 4.4   Some Properties of Ionizing Radiation

Radation	Energy Range[a]	Penetration Distance in Water	Relative No. of Ion Pairs Produced in Equal Distance Traveled
α Rays	3–9 MeV	0.02–0.04 mm	2500
β Rays	0–3 MeV	0–1 mm	100
X Rays	1 eV–100 keV	0.001 mm–1 cm	10
γ Rays	10 keV–10 MeV	1 mm–10 cm	1

[a]The small energies of subatomic particles are often measured in electron volts (eV); 1 eV = 1.602×10^{-19} J; 1 MeV is 1 million electron volts; 1 keV is 1000 electron volts.

for causing damage. It makes sense, therefore, to try to limit exposure when possible and to use protective shields when working with known radiation sources like X ray machines. The sources are surrounded with materials such as lead in layers thick enough to stop virtually all the radiation.

Protection from radiation is also afforded by controlling the distance between workers and radiation sources. The greater the intensity of radiation, the greater the potential for damage. Radiation intensity (I) decreases with the square of the distance from the radiation source. The intensities of radiation at two different distances, 1 and 2, are given by the equation

$$\frac{I_1}{I_2} = \frac{d_2^2}{d_1^2}$$

For example, suppose a source delivers 16 units of radiation at a distance of 1.0 m. Doubling the distance decreases radiation intensity by one-fourth:

$$\frac{16 \text{ units}}{I_2} = \frac{(2 \text{ m})^2}{(1 \text{ m})^2}$$

$$I_2 = 16 \text{ units} \times \frac{1 \text{ m}^2}{4 \text{ m}^2} = 4 \text{ units}$$

Practice Problem   4.6   A β-emitting radiation source gives 250 units of radiation at a distance of 4.0 m. At what distance does the radiation drop to one-tenth its original value?

4.9 RADIATION DETECTION

Small amounts of naturally occurring radiation have always been present, but people have been aware of it only within the past 100 years. The problem, of course, is that radiation is invisible. We can't see, hear, smell, touch, or taste radiation, no matter how high the dose. We can, however, detect radiation by taking advantage of its ionizing properties.

Perhaps the best known way of detecting and measuring radiation is the *Geiger counter*. A Geiger counter consists of an argon-filled tube containing two electrodes (Figure 4.4). The inner walls of the tube are coated with an electrically conducting material and given a negative charge, while a wire in the center

Figure 4.4
A Geiger counter. As radiation enters the tube through a thin window, it ionizes argon atoms, creating positively charged particles (Ar^+) which conduct a tiny electric current from the negatively charged walls to the positively charged center electrode.

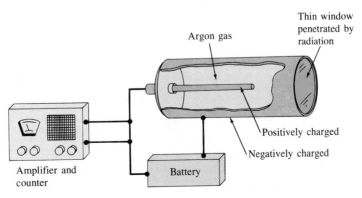

Thin window penetrated by radiation

Argon gas

Positively charged

Negatively charged

Amplifier and counter

Battery

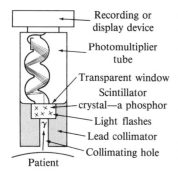

Recording or display device

Photomultiplier tube

Transparent window

Scintillator crystal—a phosphor

Light flashes

Lead collimator

Collimating hole

Patient

Figure 4.5
A γ ray detector. γ Rays from a radioisotope within the body are recorded by light from the phosphor, which is amplified by the photomultiplier tube. The collimator limits the γ rays entering the detector to those from the area under study.

of the tube is given a positive charge. As radiation enters the tube through a thin window, it strikes and ionizes argon atoms, which briefly conduct a tiny electric current from the negatively charged walls to the positively charged center electrode. The passage of the current is detected, amplified, and used to produce a clicking sound. The more radiation entering the tube, the more frequent the clicks. Geiger counters are useful for seeking out a radiation source in a large area or for gauging the intensity of emitted radiation.

The simplest device for detecting *exposure* to radiation is the photographic film badge worn by people who routinely work with radioactive materials. The film is protected from exposure to light, but any other radiation striking the badge causes the film to fog (remember Becquerel's discovery). At regular intervals, the film is developed and compared with a standard to indicate roughly the amount of radiation exposure of an individual.

The total exposure over a period of time is most often measured by taking advantage of a property of certain crystals known as *thermoluminescence,* which is the emission of light (*-luminescence*) when heated (*thermo-*). Radiation energy is trapped by electrons in the crystals within a detector (a *dosimeter*) worn by an individual. When the crystals are heated, the amount of radiation exposure is shown by the amount of light emitted.

The most versatile method for detecting radiation is based on *scintillation,* in which a *phosphor* (commonly a sodium iodide crystal activated with thallium) emits a flash of light each time it's struck by radiation. The emitted light (directly proportional in amount to the energy of the radiation) is converted to an electrical signal and amplified by a photomultiplier tube (Figure 4.5).

The expanding use of radioisotopes for imaging and diagnosis in medicine is possible because of the development of scintillation, or γ ray, cameras in which a large phosphor crystal (up to 1 m in diameter) is viewed by an array of photomultiplier tubes connected to a computer-controlled circuit. The camera produces an image of the location and intensity of radiation in a specific organ or region of the body (Figure 4.6). The image may be projected on an oscilloscope screen, recorded on film, or stored as digital data.

Figure 4.6
γ Ray camera image of spine and ribs of an individual with bone cancer. Cancerous bone concentrates an injected radionuclide and shows up as hot spots on the image, which has had false color added.

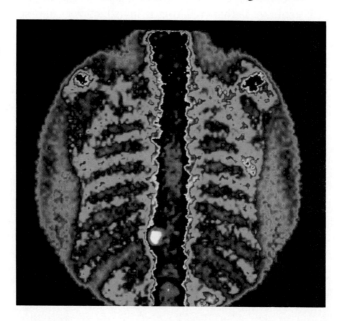

4.10 UNITS FOR RADIATION

Radiation intensity is expressed in several ways, depending on what is being measured. Some units measure the number of nuclear decay occurrences; others measure exposure to radiation or the biological consequences of radiation (Table 4.5).

Curie A unit for measuring the number of radioactive disintegrations per second.

Millicurie A unit 1/1000 the size of a curie.

Curie The **curie (Ci)** measures the number of radioactive disintegrations occurring each second in a sample, in other words, the rate of radioactive decay. One curie is an extremely large unit, and **millicuries (mCi)** are more often used.

$$1 \text{ Ci} = 3.7 \times 10^{10} \text{ disintegrations per second}$$
$$1 \text{ mCi} = 1/1000 \text{ Ci} = 3.7 \times 10^{7} \text{ disintegrations per second}$$

The curie measures how radioactive a sample is but does not measure radiation exposure or potential for damage.

The radioactivity and dosage of *radiopharmaceuticals*—radioactive substances administered orally or intravenously—is usually given in millicuries. To calculate the size of a radiopharmaceutical dose, it's necessary to determine the decay rate of the isotope solution per milliliter. Because the emitter concentration is constantly decreasing as it decays, the activity must be measured immediately before administration. Suppose, for example, that a solution containing iodine-131 for a thyroid function study is found to have a decay rate of 0.020 mCi/mL and the dose administered is to be 0.050 mCi. The amount of the solution administered must be

$$\frac{0.050 \text{ mCi}}{\text{Dose}} \times \frac{1 \text{ mL }^{131}\text{I solution}}{0.020 \text{ mCi}} = 2.5 \text{ mL }^{131}\text{I solution/dose}$$

Roentgen A unit for measuring the ionizing intensity of radiation.

Roentgen The **roentgen (R)** is a unit for the ionizing intensity of γ or X radiation. In other words, the roentgen measures the capacity of the radiation

Table 4.5 Units for Measuring Radiation

Unit	Quantity Measured	Description
Curie (Ci)	Amount of radioactivity	Amount of sample that undergoes 3.7×10^{10} disintegrations/sec
Roentgen (R)	Ionizing intensity of radiation	Amount of radiation that produces 2.1×10^{9} charges in 1 cm^3 air
Rad (D)	Amount of radiation absorbed per gram of tissue	For X and γ rays, 1 D = 1 R
Rem	Amount of tissue damage	Amount of radiation producing the same tissue damage as 1 R of X rays
Sievert (Sv)	Amount of tissue damage	Amount of radiation producing the same tissue damage as 100 R of X rays

for affecting matter. One roentgen is the amount of radiation that produces 2.1×10^9 units of charge in $1 cm^3$ of dry air at atmospheric pressure. Each collision of ionizing radiation with an atom produces one ion, or one unit of charge.

Rad A unit for measuring the amount of radiation energy absorbed per gram of tissue.

Rad The **rad** (**D**, *radiation absorbed dose*) is a unit for the energy *absorbed* per gram of material exposed to a radiation source and is defined as the absorption of 1×10^{-5} J of energy per gram. The energy absorbed varies with the type of material irradiated and the type of radiation. For most purposes, though, the roentgen and the rad are so close that they can be considered identical when used for X rays and γ rays: 1 R = 1 D.

Rem A unit for measuring the amount of tissue damage caused by radiation.

Rem The **rem** (*roentgen equivalent for man*) measures the amount of tissue damage caused by radiation; in other words, it takes into account the differences in energy of different types of radiation. One rem is the amount of radiation that produces the same effect as 1 R of X rays.

Rems are the preferred units for medical purposes because they measure equivalent doses of different kinds of radiation. For example, one rad of α rays

AN APPLICATION: BODY IMAGING

The goal of body imaging is either to look at anatomy or to visualize function. We're all familiar with the appearance of a standard X ray, produced when X radiation from an external source passes through the body and the intensity of the radiation that exits is recorded on film. Today, a host of new imaging techniques, many of them applications of radioactivity, are greatly expanding what can be learned about body structure and function. The great advantage of body imaging using radiation is that it is noninvasive: It doesn't require "invading" the body by exploratory surgery.

Development of the γ ray camera mentioned in Section 4.9 has made possible production of images that show the distribution pattern of radioactively tagged substances in the body. In essence, a radiopharmaceutical agent is injected into the body, and its distribution pattern is monitored externally, resulting in an image like that of the cancerous bone in Figure 4.6. A diseased organ might concentrate more of the radiopharmaceutical than a normal organ and thus show up as a radioactive hot spot against a cold background. Alternatively, the diseased organ might concentrate less of the radiopharmaceutical than a normal organ and thus show up as a cold spot on a hot background.

Several techniques that are revolutionizing medical diagnosis are made possible by *tomography*, in which computer processing allows production of images through "slices" of the body. In X ray tomography, now commonly known as *CT* (computer-aided tomography) scanning, an array of detectors collects up to 90,000 readings of X rays that have passed through the body as the X-ray source and detectors move rapidly in a circle around it. CT scans can detect structural abnormalities such as tumors but do not reveal function and do not require radioactive materials. (Another nonradioactive imaging technique, MRI, is discussed in Chapter 24.)

Combining tomography with radioisotope imaging gives cross-sectional views of regions that concentrate a radioactive substance and allow visualization of function in addition to structure. One such technique is known as *positron emission tomography (PET)* because it utilizes radioisotopes that emit *positrons* (a type of subatomic particle that converts to two gamma rays after it is emitted). Oxygen-15, nitrogen-13, carbon-11, and fluorine-18, the radioisotopes commonly used in PET, have the advantage that they can be used to label many physiologically active compounds. The brain images on the opening page of this chapter are PET scans made with causes 20 times more tissue damage than one rad of γ rays, but one rem of α rays and one rem of γ rays cause the same amount of damage. In effect, the rem takes *both* ionizing intensity and biological effect into account, while the rad deals only with intensity.

SI Radiation Units   In the SI system, the **becquerel (Bq)** is the unit corresponding to the curie and is defined as one disintegration per second. The SI unit for energy absorbed is the *gray (Gy)* (1 Gy = 100 rads). For radiation dose, the SI unit is the *sievert (Sv)*, which is equal to 100 rems.

The biological consequences of different radiation doses are given in Table 4.6. Although the effects listed in the table seem fearful, the average radiation dose received annually by most people is only about 0.12 rem (120 *milli*rems). About 70% of this radiation comes from natural sources (rocks and cosmic rays); the remaining 30% comes from medical procedures like X rays (Figure 4.7). The amount due to emissions from nuclear power plants and to fallout from atmospheric testing of nuclear weapons in the 1950s is barely detectable.

fluorine-18-labeled glucose. Since glucose provides the energy for brain activity, the brain regions that respond to various stimuli are made visible. The disadvantage of the radioisotopes used in PET scans is that they have such short half-lives (1 min to 2 hr) that they must be produced by bombardment at the site of their use. The cost of PET is therefore high, for a hospital must install a *cyclotron* and be able to maintain and operate it.

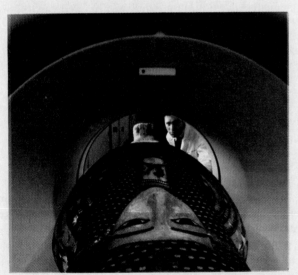
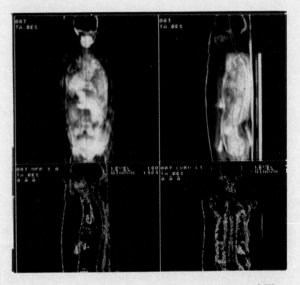

Because CT scanning is noninvasive, Egyptologists at the Boston Museum of Fine Arts were able to observe their 3000 year old mummy, Tabes, without disturbing her sarcophagus. On the left you see her being positioned in the CT scanner. On the right is a CT image of her skeleton, showing by the black space where the brain has been removed and by the colored regions that organs were preserved and replaced.

Table 4.6   Biological Effects of Short-term Radiation on Humans

Dose (rems)	Biological Effects
0–25	No detectable effects
25–100	Temporary decrease in white blood cell count
100–200	Nausea, vomiting, longer term decrease in white blood cells
200–300	Vomiting, diarrhea, loss of appetite, listlessness
300–600	Vomiting, diarrhea, hemorrhaging, eventual death in some cases
Above 600	Eventual death in nearly all cases

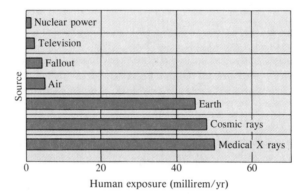

Figure 4.7
Sources of average human exposure to radiation.

Practice Problems

4.7   Initial estimates of radiation loss from the 1986 Chernobyl nuclear power plant disaster in Russia predict a worldwide increase in background radiation of about 5 mrems. By what percentage will this amount increase the annual dose of most people?

4.8   A solution of selenium-75, a radiopharmaceutical used for imaging the pancreas in the diagnosis of pancreatic disease, was found just prior to administration to have an activity of 44 μCi/mL. How many milliliters should be administered intravenously for a dose of 175 μCi?

4.11 ARTIFICIAL TRANSMUTATION

Artificial transmutation
The change of one element into another by bombardment with a high-energy particle.

Most of the nearly 2000 known radioisotopes, including all those used in medicine, don't occur naturally. Rather, they are made by **artificial transmutation** of stable isotopes brought about by nuclear bombardment reactions. An atom is bombarded with a high-energy particle such as a proton, neutron, or α particle. In the ensuing collision between particle and atom, an unstable compound nucleus is created, a nuclear change occurs, and a different element is produced. For example, transmutation occurs in the upper atmosphere when neutrons produced by bombardment of atoms with cosmic rays collide with atmospheric nitrogen, resulting in the formation of carbon-14 plus a proton. In the collision, a neutron dislodges a proton from the nitrogen nucleus as the neutron and nucleus fuse together.

$$^{14}_{7}\text{N} + ^{1}_{0}\text{n} \quad \rightarrow \quad ^{14}_{6}\text{C} + ^{1}_{1}\text{H}$$

Still other artificial transmutations can lead to the synthesis of entirely new elements never before seen on earth. In fact, all the *transuranium elements*—those elements with atomic numbers greater than 92—have been produced by bombardment reactions. For example, the element plutonium (Pu) can be made by bombardment of uranium-238 with α particles:

$$^{238}_{92}\text{U} + ^{4}_{2}\text{He} \quad \rightarrow \quad ^{241}_{94}\text{Pu} + ^{1}_{0}\text{n}$$

The plutonium-241 that results from uranium bombardment is itself radioactive with a half-life of 13.2 years, decaying by β emission to yield americium-241. Americium-241 is also radioactive, decaying by α emission with a half-life of 458 years. (If the name sounds vaguely familiar, it's because americium is used commercially in making smoke detectors, as explained in Figure 4.8.)

$$^{241}_{94}\text{Pu} \quad \rightarrow \quad ^{241}_{95}\text{Am} + ^{0}_{-1}\text{e}$$

Note that all the equations just mentioned for artificial transmutations are balanced. The sum of the mass numbers and the sum of the atomic numbers are the same on both sides of each equation.

Figure 4.8
Smoke detection by americium-241. Americium is an α emitter. Because of ionization by α particles, air within the detector becomes electrically conductive and relays a current provided by a small battery. Smoke particles cut down the flow of electrical current in the air, and the resulting decrease in the current sets off the alarm. The quantity of radiation emitted by such a detector is *so* small that it creates no hazard.

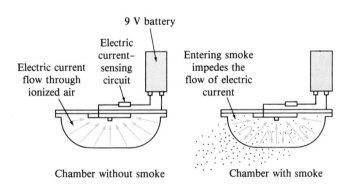

9 V battery

Electric current–sensing circuit

Electric current flow through ionized air

Entering smoke impedes the flow of electric current

Chamber without smoke

Chamber with smoke

Practice Problems

4.9 What isotope of what element results from α decay of the americium-241 in smoke detectors?

4.10 The element berkelium, first prepared at the University of California at Berkeley in 1949, is made by α bombardment of americium-241 ($^{241}_{95}\text{Am}$). Two neutrons are also produced during the reaction. What isotope of berkelium results from this transmutation? Write a balanced nuclear equation.

4.11 Write a balanced nuclear equation for the reaction of argon-40 with a proton:

$$^{40}_{18}\text{Ar} + ^{1}_{1}\text{H} \quad \rightarrow \quad ? + ^{1}_{0}\text{n}$$

4.12 NUCLEAR FISSION

Nuclear fission The splitting apart of an atom by neutron bombardment to give smaller atoms.

We saw in the previous section that particle bombardment of various elements causes artificial transmutation and results in the formation of new, usually heavier elements. With a very few isotopes, however, a somewhat different result occurs. In popular terms, it's called "splitting the atom"; in more scientific terms, it's called **nuclear fission.**

Uranium-235 is the only naturally occurring isotope that undergoes nuclear fission. When U-235 is bombarded by a stream of relatively slow neutrons, its nucleus is split into isotopes of other elements. The split can take place in any number of ways to give combinations of over 200 different isotope fragments. For example, barium-139 and krypton-94 are sometimes produced.

$$\ _{92}^{235}\text{U} + \ _0^1\text{n} \ \longrightarrow \ \ _{56}^{139}\text{Ba} + \ _{36}^{94}\text{Kr} + 3\ _0^1\text{n} + \gamma + \text{heat energy}$$

Chain reaction A self-sustaining reaction that continues unchecked once it has been started.

As indicated by the balanced nuclear equation above, *one* neutron is used to initiate fission of a U-235 nucleus, but *three* neutrons are released. Thus a nuclear **chain reaction** can be started: One neutron initiates one fission that releases 3 neutrons; the 3 neutrons initiate three new fissions that release 9 neutrons; the 9 neutrons initiate nine fissions that release 27 neutrons; and so on at an ever faster pace as long as U-235 remains.

The period of history in which we live, sometimes referred to as the Atomic Age, was ushered in by the realization that a nuclear chain reaction could release a previously undreamed of quantity of energy. The atomic bombs that in 1945 ended the World War II conflict between Japan and the United States when dropped on Hiroshima and Nagasaki were the first product of that realization.

The great quantity of heat released in nuclear fission can also be used to convert water to steam, thereby turning huge generators to produce electric power. In 1989, about 20% of the electric power generated in the United States came from nuclear power plants, and worldwide the figure was 17% in 1988. The use of nuclear power is far more advanced in some countries than in others, with France and Belgium leading the way by generating about 70% and 66% of their electricity in nuclear plants.

The major objections causing so much public debate about nuclear power plants are twofold: safety and waste disposal. Although a nuclear explosion is highly unlikely, there is a serious potential radiation hazard posed by nuclear power plants should accidents release radioactive substances to the environment.

Perhaps even more important than reactor safety in the long term is the problem posed by disposal of radioactive wastes from nuclear plants. Many of these wastes are highly radioactive and have such long half-lives that hundreds or even thousands of years must elapse before they will be safe for humans to approach. How to dispose of such hazardous materials safely is an as yet unsolved and perhaps unsolvable problem because in choosing potential storage sites, political and emotional issues can play as large a role as scientific evaluation.

Practice Problem   ***4.12***   What other isotope besides tellurium-137 is produced by nuclear fission of uranium-235 according to the following equation?

$$^{235}_{92}\text{U} + ^{1}_{0}\text{n} \rightarrow ^{137}_{52}\text{Te} + 2\,^{1}_{0}\text{n} + ?$$

4.13 NUCLEAR FUSION

Nuclear fusion The combination of two light nuclei to give a heavier nucleus.

Have you ever wondered what powers the sun and the stars? It's a different type of nuclear reaction called **nuclear fusion,** in which nuclei combine rather than split apart. One reaction in our sun, for example, is probably fusion of the hydrogen isotopes deuterium ($^{2}_{1}\text{H}$) and tritium ($^{3}_{1}\text{H}$) to form helium.

$$^{2}_{1}\text{H} + ^{3}_{1}\text{H} \rightarrow ^{4}_{2}\text{He} + ^{1}_{0}\text{n} + \text{energy}$$

Adding up the actual masses of these reactants and products would show that the products weigh less than the reactants. The difference in mass is converted into energy, in the most visible application of Einstein's now-famous equation, $E = mc^2$ (where E is energy, m is mass, and c is the speed of light). Here is the source of energy from the sun—nuclear energy released by fusion reactions.

At the temperature in the sun, about 10^8 K, all matter is reduced to a plasma in which nuclei are stripped of their electrons and collide with enough energy to fuse. Creation of such a plasma on earth was made possible once nuclear fission was achieved, but thus far its only application has been in tests of *thermonuclear* weapons, beginning with the first explosion of a "hydrogen bomb" at Bikini Atoll in the Pacific Ocean (Figure 4.9).

The dream exists that controlled nuclear fusion can provide the ultimate cheap, clean power source. The fuel is deuterium, available to all countries in virtually unlimited supply in the oceans, and there are few radioactive by-products. But providing high enough temperatures and holding the plasma together long enough for nuclei to react is a gigantic technical challenge that has yet to be met.

Figure 4.9
Thermonuclear bomb test, Bikini Atoll, 1956. Understanding of the enormous destructive potential of such weapons has so far deterred their use.

INTERLUDE: ARCHAEOLOGICAL RADIOCARBON DATING

Biblical scrolls are found in a cave near the Dead Sea; are they authentic? A mummy is discovered in an Egyptian tomb; how old is it? The burned bones of a man are dug up near Lubbock, Texas; how long ago were humans on the North American continent? Using a technique called *radiocarbon dating,* archaeologists can answer these and many other questions. (The Dead Sea Scrolls are 1900 years old and authentic, the mummy is 3100 years old, and the human remains found in Texas are 9900 years old.)

Radiocarbon dating of archaeological artifacts depends on production by cosmic ray bombardment of carbon-14, which is radioactive and has a half-life of 5730 years (Section 4.11). Carbon-14 atoms produced from nitrogen in the upper atmosphere combine with oxygen to yield $^{14}CO_2$, which slowly mixes with ordinary $^{12}CO_2$ and is then taken up by plants during photosynthesis. When these plants are eaten by animals, carbon-14 enters the food chain and is ultimately distributed evenly throughout all living organisms.

As long as a plant or animal is living, a dynamic equilibrium is established. An organism excretes or exhales the same amount of ^{14}C that it takes in, and the ratio of ^{14}C to ^{12}C that the organism contains is the same as that in the atmosphere—about 1 part in 10^{12}. When the plant or animal dies, however, it no longer takes in more ^{14}C. Thus, the $^{14}C/^{12}C$ ratio in the organism slowly diminishes as ^{14}C undergoes radioactive decay. At 5730 years (one ^{14}C half-life) after the death of the organism, the $^{14}C/^{12}C$ ratio has decreased by a factor of 2; at 11,460 years after death, the $^{14}C/^{12}C$ ratio has decreased by a factor of 4; and so on.

By simply measuring the amount of ^{14}C remaining in the traces of any once-living organism, archaeologists can determine how long ago the organism died. Human hair from well-preserved remains, charcoal or wood fragments from once-living trees, and cotton or linen from once-living plants, are all useful sources for radiocarbon dating. The accuracy of the technique lessens as samples get older and the amount of ^{14}C they contain diminishes, but artifacts with an age of 1000–20,000 years can be dated with reasonable accuracy.

Radiocarbon dating was used to determine that this painted bowl was made in ancient Persia 5000 years ago.

SUMMARY

Radioactivity is the spontaneous emission of radiation from the nucleus of an unstable atom. The three major kinds of radiation are called **alpha (α), beta (β),** and **gamma (γ).** Alpha radiation consists of helium nuclei, small particles containing two protons and two neutrons ($_2^4He$); beta radiation consists of electrons ($_{-1}^0e$); and gamma radiation consists of very-high-energy light waves. Every element in the periodic table has at least one radioactive isotope (**radioisotope**). All isotopes of elements beyond bismuth (atomic number 83) are radioactive.

Particle emission from a radioactive nucleus leads to **nuclear decay** and to the change of one element into another (**transmutation**). Thus, loss of an alpha particle from an atom leads to a new atom whose atomic number is 2 less than that of the starting atom. Loss of a beta particle arises by decomposition of a neutron and leads to an atom whose atomic number is 1 greater than that of the starting atom:

Alpha emission: $\quad {}^{238}_{92}\text{U} \quad \rightarrow \quad {}^{234}_{90}\text{Th} + {}^{4}_{2}\text{He}$

Beta emission: $\quad {}^{131}_{53}\text{I} \quad \rightarrow \quad {}^{131}_{54}\text{Xe} + {}^{0}_{-1}\text{e}$

All the heavy **natural radioisotopes** are members of natural radioactive decay series, in which an unstable parent initiates a series of α and β decay reactions that continues until a stable isotope is reached. The rate of nuclear decay is expressed in units of **half-life** ($t_{1/2}$), where one half-life is the amount of time necessary for one-half of the radioactive sample to decay.

High-energy radiation of all types—α particles, β particles, γ rays, X rays, and cosmic rays—is called **ionizing radiation.** When any of these kinds of radiation strikes an atom in a chemical compound, it dislodges an electron and gives a reactive ion. This ionizing effect can be lethal to cells. γ Rays and X rays are the most penetrating and most harmful types of external radiation, but α and β rays from internal sources are extremely dangerous because of their high energy and the resulting damage to surrounding tissue. Shielding and sufficient distance from the source provide protection from ionizing radiation.

Radiation intensity is expressed in different ways according to what property of radiation is being measured. The **curie (Ci)** measures the number of radioactive disintegrations per second in a sample; the **roentgen (R)** measures the ionizing ability of radiation; the **rad (D)** measures the amount of radiation energy absorbed per gram of tissue; and the **rem** measures the amount of tissue damage caused by radiation. Radiation effects become noticeable with a human exposure of 25 rems and become lethal at an exposure above 600 rems.

Most known radioisotopes do not occur naturally but are made by bombardment of an atom with a high-energy particle. In the ensuing collision between particle and atom, a nuclear change occurs and a new element is produced by **artificial transmutation.** With a very few isotopes, including ${}^{235}_{92}\text{U}$, the nucleus is actually split apart by neutron bombardment to give smaller fragments. A large amount of energy is released during this **nuclear fission,** leading to use of the reaction for generating electric power. **Nuclear fusion** results when small nuclei such as those of tritium and deuterium collide with sufficient energy to combine to give a heavier nucleus accompanied by conversion of lost mass into energy.

REVIEW PROBLEMS

Radioactivity

4.13 What does it mean to say that a substance is radioactive?

4.14 What word is used to describe the change of one element into another?

4.15 Describe how α, β, and γ radiation differ.

4.16 What symbol is used for an α particle in a nuclear equation?

4.17 What symbol is used for a β particle in a nuclear equation?

4.18 Which kind of radiation, α, β, or γ, has the highest penetrating power, and which has the lowest?

4.19 What happens when ionizing radiation strikes an atom in a chemical compound?

4.20 How does ionizing radiation lead to cell damage?

4.21 What are the main sources of background radiation?

4.22 How can a nucleus emit an electron during β decay when there are no electrons present in the nucleus to begin with?

4.23 What is the difference between an α particle and a helium atom?

4.24 What does it mean when we say that strontium-90, a waste product of nuclear power plants, has a half-life of 28.8 years?

4.25 What percentage of the original radioactivity remains in a sample after two half-lives have passed?

Measuring Radioactivity

4.26 Describe how a Geiger counter works.

4.27 Describe in your own words how a film badge works.

4.28 Why are rems the preferred unit for measuring the health effects of radiation?

4.29 Approximately what amount (in rems) of short-term exposure to radiation produces noticeable effects in humans?

4.30 Match each unit in the left-hand column with the property being measured in the right-hand column:

(1) curie
(2) rem
(3) rad
(4) roentgen

(a) ionizing intensity of radiation
(b) amount of tissue damage
(c) number of disintegrations per second
(d) amount of radiation per gram of tissue

Nuclear Decay and Transmutation

4.31 What does it mean to say that a nuclear equation is balanced?

4.32 What happens to the mass number and atomic number of an atom that emits an α particle? A β particle? A γ ray?

4.33 What are transuranium elements, and how are they made?

4.34 How does nuclear fission differ from normal radioactive decay?

4.35 What characteristic of uranium-235 fission causes a chain reaction?

4.36 Selenium-75, a beta emitter with a half-life of 120 days, is used medically for pancreas scans. What is the product of selenium-75 decay?

4.37 Approximately how much selenium-75 would remain from a 0.050 g sample that has been stored for 1 year? (See Problem 4.36.)

4.38 Approximately how long would it take a sample of selenium-75 to lose 99% of its radioactivity? (See Problem 4.36.)

4.39 What products result from radioactive decay of these beta emitters?
(a) $^{35}_{16}S$ (b) $^{24}_{10}Ne$ (c) $^{90}_{38}Sr$

4.40 Identify the starting radioisotopes that give these products:
(a) ? $\rightarrow$ $^{140}_{56}Ba$ + $^{0}_{-1}e$ (b) ? $\rightarrow$ $^{242}_{94}Pu$ + $^{4}_{2}He$

4.41 What products result from radioactive decay of these alpha emitters?
(a) $^{190}_{78}Pt$ (b) $^{208}_{87}Fr$ (c) $^{245}_{96}Cm$

4.42 What products are formed in the following nuclear reactions?
(a) $^{109}_{47}Ag$ + $^{4}_{2}He$ $\rightarrow$? (b) $^{10}_{5}B$ + $^{4}_{2}He$ $\rightarrow$? + $^{1}_{0}n$

4.43 Balance the following equations for the nuclear fission of $^{235}_{92}U$:
(a) $^{235}_{92}U$ + $^{1}_{0}n$ $\rightarrow$ $^{160}_{62}Sm$ + $^{72}_{30}Zn$ + ? $^{1}_{0}n$
(b) $^{235}_{92}U$ + $^{1}_{0}n$ $\rightarrow$ $^{87}_{35}Br$ + ? + 3 $^{1}_{0}n$

4.44 For centuries, alchemists dreamed of turning base metals into gold. This dream finally became reality when it was shown that mercury-198 can be converted into gold-198 on bombardment by neutrons. What small particle is produced in addition to gold-198? Write a balanced nuclear equation for the reaction.

4.45 Element 109 ($^{266}_{109}Une$), the heaviest known element, was prepared in 1982 by bombardment of bismuth-209 atoms with iron-58. What other product must also have been formed? Write a balanced nuclear equation for the transformation.

4.46 The half-life of mercury-197 is 65 hours. If a patient undergoing a kidney scan is given 5.0 ng of mercury-197, how much will remain after 6 days? After 27 days?

4.47 A Se-75 source is producing 300. rem at a distance

of 2.0 m. What distance is needed to decrease the intensity of exposure to below 25 rems (the level at which no effects would be detectable)?

4.48 If a radiation source has an intensity of 650 rems at 1.0 m, what is its intensity at 50.0 m?

4.49 Gold-198, a beta emitter that is used to treat leukemia, has a half-life of 2.7 days. The standard dosage is about 1.0 mCi/kg body weight.
(a) What is the product of the beta emission of gold-198?
(b) How long does it take a 30.0 mCi sample of gold-198 to decay so that only 3.75 mCi remains?
(c) How many mCi are required by a 70.0 kg adult?

4.50 Na-24 is used to study the circulatory system and to treat chronic leukemia. It is administered in the form of saline (NaCl) solution, with a therapeutic dosage of 180 Ci/kg body weight. How many mL of a 5.0 mCi/mL solution are needed to treat a 68 kg adult?

Applications

4.51 What is an antigen? [App: Medical Uses of Radioisotopes]

4.52 Why is radiation therapy often a last resort for cancer therapy? [App: Medical Uses of Radioisotopes]

4.53 What are the advantages of CT and PET relative to conventional X-rays? [App: Body Imaging]

4.54 Some dried beans with a $^{14}C/^{12}C$ ratio one-eighth of the current value were found in an old cave. How old are the beans? [Int: Archeological Radiocarbon Dating]

4.55 Why is ^{14}C dating useful only for samples that contain material from objects that were once alive? [Int: Archeological Radiocarbon Dating]

4.56 Why can't ^{14}C dating be used to tell if a piece of leather was made in 1850? [Int: Archeological Radiocarbon Dating]

Additional Questions and Problems

4.57 Harmful chemical spills can often be cleaned up by treatment with another chemical. For example, a spill of H_2SO_4 might be neutralized by addition of $NaHCO_3$. Why can't the harmful radioactive wastes from nuclear power plants be cleaned up just as easily?

4.58 Why are scintillation counters or Geiger counters more useful for determining the existence and source of a new radiation leak than dosimeters and film badges?

4.59 Why is it important that internally administered radioisotopes have a short half-life?

4.60 Tc-99m, used for brain scans and to monitor heart function, is formed by decay of Mo-99.
(a) By what type of decay will Mo-99 produce Tc-99m?
(b) Mo-99 is formed by neutron bombardment of a natural isotope. If one neutron is absorbed and there are no other by-products of this process, from what isotope is Mo-99 formed?

4.61 The half-life of Tc-99m (Problem 4.60) is 6.02 hours. If a sample with an initial activity of 15 Ci is injected into a patient, what is the activity in 24 hours, assuming that none of the sample is excreted?

4.62 Pu-238 is an alpha emitter used to power batteries for heart pacemakers.

(a) What is the product of this emission?

(b) Why is a pacemaker battery enclosed in a metal case before being inserted into the chest cavity?

4.63 Why do dentists place a dense apron over a patient before taking X-rays?

4.64 As a result of high exposure to radiation, such as that from the explosion at Chernobyl, there were numerous miscarriages and many instances of farm animals born with severe defects. Why are embryos and fetuses particularly susceptible to the effects of radiation?

4.65 What are the main advantages of fusion relative to fission as an energy source?

CHAPTER

5

Ionic Compounds

Ionic compounds are high-melting solids made up of crystals with characteristic shapes, such as those shown for table salt. You'll see why in this chapter.

Anywhere from a few to a huge number of atoms can be combined in a chemical compound. Atoms of different types are present in a given compound in exact numerical ratios, and all the atoms are held in exact positions relative to each other. Clearly, something holds these atoms together. The forces that do so are called *chemical bonds* and are of two major types. This chapter is concerned with the first of these—ionic bonds—and with the substances that contain such bonds.

All chemical bonds result from the attraction between positively charged nuclei and negatively charged electrons. The manner in which atoms of different elements form bonds is therefore related to their different electron configurations. In this chapter we answer the following questions:

1. *What is an ionic bond?* The goal: Be able to define an ionic bond and give the general properties of compounds that contain ionic bonds.
2. *What is the octet rule, and how does it apply to ions?* The goal: Be able to state the octet rule and use it to predict the electron configurations of ions of main-group elements.
3. *What are the relationships between the periodic table and ion formation?* The goal: Be able to predict what ions are likely to be formed by atoms of a given element.
4. *What symbols and names are used to represent ions?* The goal: Be able to name the common ions and write their symbols or formulas.
5. *What determines the chemical formulas of ionic compounds?* The goal: Be able to write formulas for ionic compounds, given the identities of the ions combined.
6. *How are ionic compounds named?* The goal: Be able to name a compound from its formula or give the formula of a named compound.
7. *What is formula weight?* The goal: Be able to calculate the formula weight of any given ionic compound and explain what the formula weight represents.

5.1 IONIC BONDING

The elements of Group 1A (the alkali metals) and Group 7A (the halogens) form many compounds with one another. Sodium chloride (table salt) is already familiar to you. The names and **chemical formulas** of some other compounds containing elements from these two groups include

Chemical formula
Representation with symbols and subscript numbers of the atoms combined in a chemical compound.

Potassium iodide	KI	Added to salt to provide iodide ion needed by the thyroid gland
Sodium fluoride	NaF	Added to the water supply to provide fluoride ion for the prevention of tooth decay
Sodium iodide	NaI	Used in scintillation crystals in gamma ray cameras

Chemical bonds   The forces that hold atoms together in chemical compounds.

Ion   An electrically charged atom or group of atoms.

Ionic bond   The force of electrical attraction between a positive ion and a negative ion in a crystal.

Crystal   A solid with planar faces meeting at fixed angles and in which atoms, ions, or molecules are arranged in a regular, repeating pattern.

Figure 5.1
Sodium + chlorine →
sodium chloride. When hot sodium metal is immersed in chlorine gas, it burns with an intense yellow flame as white sodium chloride "smoke" forms.

The composition and many properties of these compounds are similar. The two elements are combined in each case in a 1:1 ratio—one alkali metal atom for every halogen atom. Each compound has a high melting point (all are over 500°C), each is a stable, white, crystalline solid, and each is soluble in water. Furthermore, each shows a property that gives a clue to the type of **chemical bond** between the atoms: The water solution of each conducts electricity, and even the molten compounds conduct electricity. Any explanation of bonding in these compounds must account for these properties.

Electricity can flow only through a medium containing charged particles that are free to move. The electrical conductivity of metals, for example, results from the presence of electrons that can flow through the metal. What charged particles might be present in the water solutions of alkali metal–halogen compounds or in the molten compounds? To answer this question, think about the composition of atoms.

Atoms are electrically neutral because they contain equal numbers of protons and electrons. By gaining or losing one or more electrons, an atom is converted into a charged particle called an **ion.** Alkali metals and halogens are present in many compounds as ions rather than as neutral atoms.

When the elements sodium and chlorine react with each other, the product is sodium chloride, a compound quite unlike either of the elements from which it's formed (Figure 5.1). Sodium is a soft, silvery metal that reacts violently with water, and chlorine is a corrosive, poisonous green gas. Combined, they produce our familiar table salt. The salt is formed when a sodium atom transfers its valence electron to a chlorine atom to form a sodium ion (Na^+) and a chloride ion (Cl^-). Since opposite electrical charges strongly attract each other, the positive sodium ion and negative chloride ion are said to be held together by an **ionic bond.**

When a billion billion sodium atoms transfer electrons to a billion billion chlorine atoms, a barely visible **crystal** of sodium chloride results. In this crystal, equal numbers of sodium and chloride ions are packed together in a regular arrangement. Each positively charged sodium ion is surrounded by six negatively charged chloride ions, and each chloride ion is surrounded by six sodium ions (Figure 5.2). This packing arrangement allows each ion to be

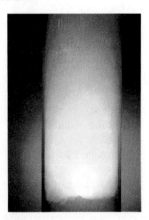

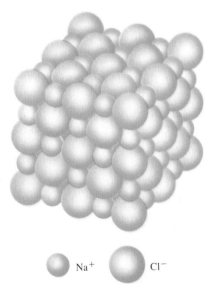

Figure 5.2
The packing arrangement of ions in a sodium chloride crystal. Each positively charged sodium ion is surrounded by six negatively charged chloride ions, and vice versa. The crystal is held together by ionic bonds—the attraction between oppositely charged ions.

Na^+ Cl^-

stabilized by the attraction of unlike charges on its six nearest-neighbor ions, while being as far as possible from like-charged ions.

Because of the three-dimensional packing arrangement of ions in a sodium chloride crystal, we can't speak of specific ionic bonds between specific pairs of ions. Rather, there are many ions attracted by ionic bonds to their nearest neighbors. We can only speak of the whole crystal as being an ionic solid and of such compounds as being **ionic compounds.** The same is true of *all* compounds composed of ions.

Ionic compound A chemical compound composed of positively and negatively charged ions held together by ionic bonds.

5.2 ELECTRON-DOT SYMBOLS AND IONS

Because valence electrons play so important a role in the behavior of atoms, including the formation of ions, a method for including them with atomic symbols is useful. In an **electron-dot symbol,** dots around the atomic symbol indicate the number of valence electrons present. For the Group 1A elements, which have a single valence electron, the electron-dot symbols are written as follows:

Electron-dot symbol An atomic symbol surrounded by dots that show the number of valence electrons.

$$Li\cdot \quad Na\cdot \quad K\cdot \quad Rb\cdot \quad Cs\cdot \quad Fr\cdot$$

For the Group 7A elements (the halogens), which have seven valence electrons, the electron-dot symbols are

$$\cdot\ddot{F}: \quad \cdot\ddot{Cl}: \quad \cdot\ddot{Br}: \quad \cdot\ddot{I}: \quad \cdot\ddot{At}:$$

The electron-dot symbols for the first few elements in each main group are given in Table 5.1. Note that helium differs from the other noble gases in having only two valence electrons. Nevertheless, it's considered a member of this group because of its similar properties. The configuration of helium also resembles those of the other noble gases because its highest occupied subshell is completely filled ($1s^2$).

Carbon, nitrogen, and oxygen are from Groups 4A, 5A, and 6A; their atoms have four, five, and six valence electrons, respectively, and they are represented by the following electron-dot symbols:

$$\cdot\dot{C}\cdot \quad \cdot\dot{N}: \quad \cdot\ddot{O}:$$

You'll be seeing these symbols frequently, for these elements are common in the chemical compounds in living things. Except for oxygen, however, these elements rarely gain or lose electrons to form ions.

Table 5.1 Electron-Dot Symbols

1A	2A	3A	4A	5A	6A	7A	Noble Gases
							He :
Li·	·Be·	·Ḃ·	·Ċ·	·N̈:	·Ö:	·F̈:	:Ne:
Na·	·Mg·	·Al·	·Si·	·P̈:	·S̈:	·Cl:	:Ar:
K·	·Ca·	·Ga·	·Ge·	·As:	·Se:	·Br:	:Kr:

When chemical compounds form, it's the valence electrons that participate in the chemical change. In the case of the combination of elements to form ionic compounds, one or more electrons shift from one atom to another and ions are formed. The loss of one electron by a sodium atom to form a sodium ion is represented as

$$\text{Na} \cdot \quad \rightarrow \quad \text{Na}^+ \quad + \quad e^-$$

| Sodium atom | Sodium ion | Electron | Electron loss produces a positively charged ion |

and the conversion of a chlorine atom to an ion is represented as

$$: \overset{\cdot\cdot}{\underset{\cdot\cdot}{\text{Cl}}} \cdot \quad + \quad e^- \quad \rightarrow \quad : \overset{\cdot\cdot}{\underset{\cdot\cdot}{\text{Cl}}} :^-$$

| Chlorine atom | Electron | Chloride ion | Electron gain produces a negatively charged ion |

Cation   A positively charged ion.

Anion   A negatively charged ion.

Ions with a positive charge, such as Na^+, are called **cations,** and ions with a negative charge, such as Cl^-, are called **anions.** Cations are smaller than the atoms from which they are derived, and anions are larger (Figure 5.3).

To symbolize a cation, either with or without electron dots, the positive charge is added as a superscript to the symbol for the element. If the charge is $+1$, the 1 is omitted, as in Na^+ and Ag^+. The same is done for anions by using a minus sign, giving Cl^- and F^-. For $+2$, -2, and larger charges, the numbers are used, for example, Ca^{2+} and S^{2-}.

A sample of sodium chloride (NaCl) is electrically neutral because the positive charge of each sodium ion is balanced by the negative charge of a chloride ion. Similarly, all other compounds between alkali metals and halogens are electrically neutral and are composed of positive metal ions such as K^+ and Li^+, and negative halogen ions such as F^- and Br^-.

Figure 5.3
Formation of Na^+ and Cl^-, showing relative sizes of atoms and ions. Loss of an electron causes a decrease in size, and addition of an electron causes an increase in size. The radii are given in picometers.

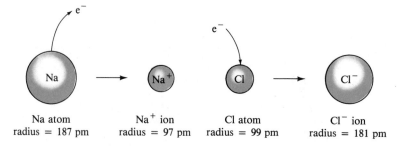

| Na atom radius = 187 pm | Na^+ ion radius = 97 pm | Cl atom radius = 99 pm | Cl^- ion radius = 181 pm |

Practice Problems

5.1   Write the electron-dot symbol for element X, which is in Group 3A.

5.2   Write the electron-dot symbols for radon, lead, bromine, and radium.

5.3   Identify the ions present in KI and write equations for their formation from atoms by electron gain and loss.

5.4   Write symbols, both with and without electron dots, for the ions formed by (a) gain of two electrons by selenium (Se), (b) loss of two electrons by barium (Ba), (c) gain of one electron by bromine (Br).

5.3   PROPERTIES OF IONIC COMPOUNDS

Like sodium chloride, ionic compounds are usually crystalline solids. Ions vary in size and charge, however, and therefore are packed together in crystals in various ways. The ions in each different compound settle into a pattern that efficiently fills space and provides low potential energy. Very pure, perfect crystals like the sodium chloride crystals shown at the beginning of this chapter reflect in their shapes the different internal arrangements of the ions (Figure 5.4).

Because the ions in ionic solids are held rigidly in place by attraction to their neighbors, they cannot move about to conduct electricity. Once an ionic solid is melted or dissolved in water, however, the ions can move freely. Thus, ionic bonding successfully explains the electrical conductivity of the compounds.

The observed high melting points and boiling points of ionic compounds are also accounted for by ionic bonding. The attractive force between oppositely charged particles is extremely strong, and for ions to loosen their grip on each other, they must gain a large amount of energy by being heated to high temperatures. For example, sodium chloride (NaCl) melts at 801°C and boils at 1413°C, and potassium iodide melts at 681°C and boils at 1330°C.

Despite the strength of ionic bonds, ionic crystals shatter if struck sharply. A blow disrupts the orderly arrangement of cations and anions, and repulsion between ions of like charge that have been pushed together aids in splitting apart the crystal.

Ionic compounds dissolve in water if the attraction between water and the ions overcomes the attraction of the ions for each other. Compounds like sodium chloride are very soluble and can be dissolved to make solutions of very high **concentration.** Don't be misled, however, by the ease with which sodium chloride and other familiar ionic compounds dissolve in water. Many ionic compounds are water-soluble, but many others are not.

Concentration   The amount of a substance dissolved in a unit quantity of a solution or solvent.

Figure 5.4
Crystals of sodium sulfate (photographed in polarized light). Note how different these crystals are in shape from those of sodium chloride on the opening page of this chapter.

AN APPLICATION: HOMEOSTASIS

The concept of **homeostasis** is central to the study of living things, much as the concept of the chemical bond is central to the study of chemistry. To understand human health and what happens when it is disrupted by disease or trauma requires understanding the meaning of homeostasis—the maintenance of a constant internal environment in the body.

''Internal environment'' is a general way of describing all the conditions within cells, organs, and body systems. Conditions such as body temperature, the availability of chemical compounds that supply energy, and the disposal of waste products must remain within very specific limits for everything to function properly. Throughout our bodies, sensors keep track of the internal environment, and when any aspect of it varies, signals are sent to restore the proper balance. For example, if oxygen is in short supply, signals are sent that make us breathe harder. When we are healthy, our bodies adjust to wide variations in our external environment, diet, or exercise level so that the conditions necessary for normal body functions are unaffected.

At the chemical level, homeostasis results in concentrations of ions and many different kinds of organic compounds that stay between normal maximum and minimum levels. The reliably predictable concentrations of such substances are the basis for **clinical chemistry**—the chemical analysis of body tissues and fluids. The blood drawn in a physician's office winds up in a clinical laboratory, and many hospitals have their own clinical laboratories. In the lab, chemical reactions and other tests are used to measure concentrations of significant ions and compounds in blood, urine, feces, spinal fluid, or other samples from a patient's body. Comparing the lab results with norms—average concentration ranges in a population of healthy individuals—shows which body systems are struggling, or failing, to maintain homeostasis. To give just one example, uric acid is a compound that helps to carry waste nitrogen from the body. A uric acid concentration higher than normal range of about 2.5–7.7 mg of uric acid per deciliter of blood can be a signal of kidney malfunction.

A copy of a clinical lab report for a routine

Homeostasis   Maintenance of unchanging internal conditions by living things.

Clinical chemistry Chemical analysis of body tissues and fluids.

Chemical reactivity   The tendency of an element or chemical compound to undergo chemical reactions.

Octet rule: Atoms of main-group elements tend to combine in chemical compounds by gaining, losing, or sharing electrons so that they attain eight outer-shell electrons.

5.4   IONS AND THE OCTET RULE

Look again at the electron configurations shown in Table 3.4 for several families of elements in the periodic table. The explanation for the formation of ions in reactions between alkali metals and halogens lies here. Alkali metal atoms have a single valence electron; halogen atoms have seven valence electrons; and atoms of elements in Group 8A (the noble gas family) have eight valence electrons. Of the three groups, both the alkali metals and the halogens are extremely reactive: They easily undergo many chemical reactions and form many compounds. The noble gases are the opposite: They are the least reactive of all the elements. Evidently there is something special about having eight outer-shell electrons (filled *s* and *p* subshells) that leads to unusual stability and lack of **chemical reactivity.**

Observations of a great many chemical compounds show that most compounds of main-group elements contain atoms combined so that each has eight outer-shell electrons, a so-called electron octet. The **octet rule** summarizes these observations: Atoms of main-group elements tend to combine in chemical blood analysis is shown here. By comparing the results and the normal ranges, you can see that this individual is free of significant variations from normal. The metal names in this report refer to *cations* that are mentioned in this chapter, and "phosphorus" (a nonmetal) refers to the polyatomic *anion*, HPO_4^{2-}.

Test	Result		Normal Range
Albumin	4.3	g/dl	3.5–5.3g/dl
Alk. Phos.*	33	U/L	25–90U/L
BUN*	8	mg/dL	8–23mg/dL
Bilirubin T.*	0.1	mg/dL	0.2–1.6mg/dL
Calcium	8.6	mg/dL	8.5–10.5mg/dL
Cholesterol	227	mg/dL	120–250mg/dL
Chol., HDL*	75	mg/dL	30–75mg/dL
Creatinine	0.6	mg/dl	0.7–1.5mg/dL
Glucose	86	mg/dl	65–110mg/dL
Iron	101	μg/dL	35–140μg/dL
LDH*	48	U/L	50–166U/L

Test	Result		Normal Range
SGOT*	23	U/L	0–28U/L
Total Protein	5.9	g/dL	6.2–8.5g/dL
Triglycerides	75	mg/dL	36–165mg/dL
Uric Acid	4.1	mg/dL	2.5–7.7mg/dL
GGT*	23	U/L	0–45U/L
Magnesium	1.7	mEq/L	1.3–2.5mEq/L
Phosphoros	2.6	mg/dL	2.5–4.8mg/dL
SGPT*	13	U/L	0–26 U/L
Sodium	137.7	mEq/L	135–155mEq/L
Potassium	3.80	mEq/L	3.5–5.5mEq/L

Clinical lab report for routine blood analysis. The abbreviations marked with asterisks are for the following tests (alternate standard abbreviations are in parentheses): Alk. Phos., alkaline phosphatase (ALP); BUN, blood urea nitrogen; Bilirubin T., total bilirubin; Chol., HDL, cholesterol, high-density lipoproteins; LDH, lactate dehydrogenase; SGOT, aspartate aminotransferase (AST); GGT, γ-glutamyl transferase; SGPT, alanine aminotransferase (ALT). Some of the units of measurement were introduced in Chapter 2, and others are discussed in later chapters. The role of many of the substances listed are also discussed in later chapters.

compounds by gaining, losing, or sharing electrons so that they attain eight outer-shell electrons. In other words, atoms tend to undergo changes that give them electron configurations like those of the noble gases just before them (for metals) or just after them (for nonmetals) in the periodic table.

Here, we're interested in the gain and loss of electrons in ionic compounds. (Electron sharing in chemical bonds is the subject of the next chapter.) Consider sodium chloride: Sodium has fully occupied first and second shells, and one valence electron in the third shell ($1s^2 2s^2 2p^6 3s^1$). Chlorine has fully occupied first and second shells and seven valence electrons ($1s^2 2s^2 2p^6 3s^2 3p^5$). In forming sodium chloride, sodium attains an outer-shell octet by *losing* the single electron in its $3s$ subshell to leave the filled second shell outermost, and chlorine attains an outer-shell octet by *gaining* an electron in its $3p$ subshell. The sodium atom acquires a positive charge, and the chlorine atom gains a negative charge.

$$Na\cdot \quad + \quad \cdot \overset{\cdot\cdot}{\underset{\cdot\cdot}{Cl}} \colon \quad \rightarrow \quad Na^+ \quad \colon \overset{\cdot\cdot}{\underset{\cdot\cdot}{Cl}} \colon^-$$

$$1s^2\,2s^2\,2p^6\,3s^1 \quad 1s^2\,2s^2\,2p^6\,3s^2\,3p^5 \quad 1s^2\,2s^2\,2p^6 \quad 1s^2\,2s^2\,2p^6\,3s^2\,3p^6$$

Solved Problem 5.1   Write the electron configuration of aluminum (atomic number 13). Show how many electrons an aluminum atom must lose to form a stable ion with an outer octet and write the configuration of the ion. Explain the reason for the ion's charge and write the ion's symbol.

Solution   Aluminum is in Group 3A and has the electron configuration

$$\text{Al} \qquad 1s^2\, 2s^2\, 2p^6\, 3s^2\, 3p^1$$

Since the second shell contains an octet of electrons ($2s^2\, 2p^6$), aluminum must lose the three electrons in the $3s$ and $3p$ subshells to achieve a stable outermost octet. The result is formation of an ion with the neon configuration:

$$\text{Al}^{3+} \qquad 1s^2\, 2s^2\, 2p^6 \qquad \text{The neon configuration}$$

A neutral aluminum atom has 13 protons and 13 electrons. With the loss of three electrons, there is an excess of three protons, accounting for the $+3$ charge of the ion, Al^{3+}.

Practice Problems   **5.5**   Write the electron configuration of potassium, atomic number 19, and show how a potassium atom can attain a noble gas configuration.

5.6   How many electrons must a magnesium atom, atomic number 12, lose to attain a noble gas configuration? Write the symbol for the ion formed.

5.7   Oxygen atoms, atomic number 8, have six valence electrons. How can an oxygen atom most easily attain a noble gas configuration? Write the electron configuration of the oxygen ion and identify the noble gas that has the same configuration.

5.5   PERIODIC PROPERTIES AND ION FORMATION

Most of the elements essential for human life (Table 1.2) fall in the first four periods of the periodic table. Since these elements play such a large role in our health and welfare, a closer look at some of their properties is worthwhile. We want to note which elements are present in body fluids as ions.

You've seen in Figure 3.6 that metals are on the left in the periodic table, and nonmetals on the right. Other related properties vary in a similar way.

Atomic radii (Figure 5.5) are a measure of the distance of the outermost electrons from the nuclei in atoms. This distance is determined by two competing effects: the greater distance of successively higher energy shells from the nucleus, and the nuclear charge, which pulls the electrons inward as it increases. From left to right across the periods, radii generally decrease because nuclear charge is increasing and pulling in electrons in the same shell. From top to bottom in the groups, the increasing nuclear charge is outweighed by the occupancy of a higher energy shell in the atoms of each element.

Atomic radius   The radius of an atom.

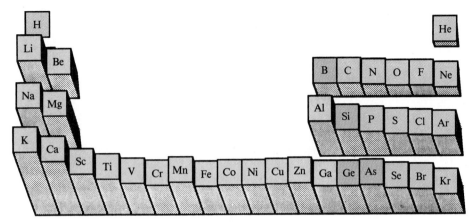

Relative atomic radii

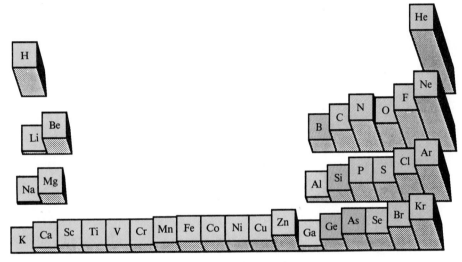

Relative ionization energies

Figure 5.5
Variation in atomic radii and ionization energy with periods and groups in the periodic table. The alkali metals Li, Na, and K have the largest atomic radii in each period and the smallest ionization energies. The reverse is true for the noble gases He, Ne, Ar, and Kr.

Ionization energy The energy required to remove one electron from a single atom in the gaseous state.

The ease with which atoms give up electrons is compared by measuring a property called **ionization energy**, defined as the energy required to remove one electron from a single atom in the gaseous state (Figure 5.5). Notice in Figure 5.5 how ionization energies and atomic radii show opposite effects. This is exactly what is to be expected—it's easier to remove an electron from a larger atom, where it's farther from the positive charge of the nucleus, than from a smaller atom. Notice in Figure 5.5 that metal atoms have larger radii and smaller ionization energies.

The elements with lighter, smaller atoms at the tops of Groups 4A, 5A, and 6A, plus all the elements in Groups 7A and 8A, are nonmetals. These elements do not ordinarily form positively charged ions for two good reasons. First, pulling even one electron away from the positive charge of the nucleus in these small atoms is very difficult. Second, to reach a stable octet, four or more outer-shell electrons would have to be removed. Since each electron after the first must be taken from a positively charged ion, it becomes more and more difficult to remove successive electrons. Ions with charges larger than $+3$ are, therefore, almost never formed.

Rather than *lose* electrons to form positively charged ions, some nonmetal atoms *add* electrons to form negatively charged ions with outer-shell octets, as we've shown for chlorine. Adding successive electrons to an already negatively charged ion, however, becomes increasingly more difficult, and anions with charges larger than -3 are also not ordinarily formed.

5.6   IONS OF THE COMMON ELEMENTS

The periodic table, as illustrated in Figure 5.6 for elements in the first four periods, is the key to understanding and remembering which elements are found in ionic compounds and which are not. Note that atoms of elements in the same group tend to form ions of the same charge.

The metals of Groups 1A and 2A form only $+1$ and $+2$ ions, respectively, and in fact are found mostly in ionic compounds. The ions of these elements all have noble gas configurations as a result of loss of electrons from their outermost s subshells:

$$\text{Group 1A} \qquad \text{M}\cdot \;\rightarrow\; \text{M}^+ + 1\,e^-$$

M = an alkali metal (Li, Na, K, Rb, or Cs)

$$\text{Group 2A} \qquad \cdot\text{M}\cdot \;\rightarrow\; \text{M}^{2+} + 2\,e^-$$

M = an alkaline earth metal (Mg, Ca, Sr, Ba, or Ra)

Four of these ions, Na^+, K^+, Mg^{2+}, and Ca^{2+}, are present in body fluids where they play extremely important roles in biochemical reactions. (The levels of

Figure 5.6
Ions most commonly formed by elements in the first four periods. Symbols for the ions that participate in body chemistry are in red.

1A	2A											3A	4A	5A	6A	7A	8A
1 H^+																	2 He
3 Li^+	4 Be^{2+}											5 B	6 C	7 N^{3-}	8 O^{2-}	9 F^-	10 Ne
11 Na^+	12 Mg^{2+}				Transition metals							13 Al^{3+}	14 Si	15 P	16 S^{2-}	17 Cl^-	18 Ar
19 K^+	20 Ca^{2+}	21 Sc^{3+}	22 Ti^{3+}	23 V^{2+}	24 Cr^{2+} Cr^{3+}	25 Mn^{2+} Mn^{3+}	26 Fe^{2+} Fe^{3+}	27 Co^{2+} Co^{3+}	28 Ni^{2+}	29 Cu^+ Cu^{2+}	30 Zn^{2+}	31 Ga	32 Ge	33 As	34 Se^{2-}	35 Br^-	36 Kr

these ions are included in the routine blood analysis illustrated in the Application on homeostasis.)

The only Group 3A element commonly encountered in ionic compounds is aluminum, which forms Al^{3+} by loss of three electrons from the outermost s and p subshells. Aluminum is not an essential element in the human diet, although it is known to be present in some organisms.

The first three elements in Groups 4A and 5A (C, Si, Ge; N, P, As) don't ordinarily form either cations *or* anions. Carbon, however, is the backbone of all **organic compounds**, and together with hydrogen, nitrogen, phosphorus, and oxygen is present in essential biological compounds that we'll be describing later.

Oxygen and sulfur are versatile elements that form large numbers of ionic compounds and nonionic compounds. Their ions have noble gas configurations achieved by gaining two electrons:

Organic compounds
Compounds containing carbon and hydrogen, and derivatives of these compounds.

$$\text{Group 6A} \qquad :\!\overset{..}{O}\!: + 2\,e^- \rightarrow :\!\overset{..}{\underset{..}{O}}\!:^{2-}$$

$$:\!\overset{.}{S}\!: + 2\,e^- \rightarrow :\!\overset{..}{\underset{..}{S}}\!:^{2-}$$

The halogens (except for astatine, which is radioactive and rare) are present in many compounds as -1 ions.

$$\text{Group 7A} \qquad :\!\overset{..}{X}\!\cdot + e^- \rightarrow :\!\overset{..}{X}\!:^-$$

X = a halogen (F, Cl, Br, I)

The chloride ion is of key importance in body fluids.

Most transition metals lose electrons to form cations in ionic compounds, and the essential transition metals are present in the body as cations. Many are part of organic compounds rather than in solution in body fluids, however. The charges of transition metal cations are not so predictable as those of main-group elements, because many transition metal atoms can lose one or more d electrons in addition to outer s electrons. For example, iron ($1s^2\,2s^2\,2p^6\,3s^2\,3p^6\,3d^6\,4s^2$) forms Fe^{2+} by losing two electrons from the $4s$ subshell and also forms Fe^{3+} by losing an additional electron from the $3d$ subshell. From the iron configuration you can see why the octet rule is limited to main-group elements: Transition metal cations, with just a few exceptions, can't have outer octets.

The major points to remember about ion formation and the periodic table are

● Metals form cations, for example, Na^+, Ca^{2+}, and Fe^{2+}.
● Nonmetals form anions, for example, O^{2-} and F^-.
● Group 1A and 2A metals form $+1$ and $+2$ ions, respectively, for example, Li^+ and Mg^{2+}.
● Group 6A nonmetals oxygen and sulfur form the anions O^{2-} and S^{2-}.
● Group 7A elements (the halogens) form -1 ions, for example, F^- and Cl^-.
● Many transition metals form ions of more than one charge, for example, Fe^{2+} and Fe^{3+}.

Solved Problem 5.2   Which of the following ions is most likely to be found in an ionic compound? S^{3-}, Co^{2+}, V^{5+}

Solution S^{3-} Sulfur is in Group 6A and therefore has six valence electrons. Gaining two electrons gives a sulfur ion with a noble gas configuration; gaining three electrons does not. This ion should not form.

Co^{2+} Cobalt is a transition element and can form ions with more than one charge. A $+2$ charge is reasonable for a metal ion. This ion is likely to be found.

V^{5+} Although vanadium is a transition metal and might form more than one ion, a charge of $+5$ is too high. This ion should not form.

Practice Problems   5.8   Is molybdenum (Mo) more likely to form a cation or an anion?

5.9   Which of the following elements can form more than one cation? strontium (Sr), chromium (Cr), or bromine (Br)

5.10   Which of the following elements are present in the human body as ions?
(a) copper   (b) magnesium   (c) phosphorus
(d) chlorine   (e) sulfur

5.7   NAMING IONS

Metal cations are named with the metal name followed by the word ''ion,'' as in

$$Mg^{2+} \qquad\qquad Al^{3+}$$

Magnesium ion   Aluminum ion

For metals that can form more than one ion, a method is needed to indicate which ion is being referred to. Two systems are used for this. The first is an old system that gives the ion with the smaller charge the word ending *-ous,* and the ion with the larger charge the ending *-ic.* The second is a newer system (meant to replace the old system) in which the charge on the ion is given as a Roman numeral in parentheses right after the metal name. For example,

	Cr^{2+}	Cr^{3+}
Old system name:	chrom*ous* ion	chrom*ic* ion
New system name:	chromium(II) ion	chromium(III) ion

We'll mostly use the new system, but it's important to understand both because the old system is often found on labels of commercially supplied chemicals and their solutions. The small differences between the names in either system

AN APPLICATION: BIOLOGICALLY IMPORTANT IONS

The human body requires many different ions for proper functioning. Several of these ions, such as Ca^{2+}, Mg^{2+}, and PO_4^{3-}, are used as structural material in the solid ionic compounds in bones and teeth, in addition to having other essential functions. Although, for example, 99% of Ca^{2+} is contained in bones and teeth, the small amounts in body fluids play a vital role in transmission of the messages that help to maintain homeostasis. Others, including most of the essential transition metal ions such as Fe^{2+}, which is found in hemo-

globin, are required for specific chemical reactions in the body. And still others, such as K^+, Na^+, and Cl^-, are present in fluids throughout the body. Just as in chemical compounds, solutions containing ions must have overall neutrality, and several polyatomic anions, especially HCO_3^- and HPO_4^{2-}, help to balance the cation charges in body fluids. Some of the most important ions and their functions are shown in the accompanying table.

Some Biologically Important Ions

Ion	Location	Function	Dietary Source
Ca^{2+}	Outside cell; 99% of Ca^{2+} is in bones and teeth as $Ca_3(PO_4)_2$ and $CaCO_3$	Bone and tooth structure; necessary for blood clotting, muscle contraction, and transmission of nerve impulses	Milk, whole grains, leafy vegetables
Fe^{2+}	Blood hemoglobin	Transports oxygen from lungs to cells	Liver, red meat, leafy green vegetables
K^+	Fluids inside cells	Maintains ion concentrations in cells; regulates insulin release and heartbeat	Milk, oranges, bananas, meat
Na^+	Fluids outside cells	Protects against fluid loss; necessary for muscle contraction and transmission of nerve impulses	Table salt, seafood
Mg^{2+}	Fluids inside cells; bone	Present in many enzymes; needed for energy generation and muscle contraction	Leafy green plants, seafood, nuts
Cl^-	Fluids outside cells; gastric juice	Maintains fluid balance in cells; helps transfer CO_2 from blood to lungs	Table salt, seafood
HCO_3^-	Fluids outside cells	Controls acid–base balance in blood	By-product of food metabolism
HPO_4^{2-}	Fluids inside cells; bones and teeth	Controls acid–base balance in cells	Fish, poultry, milk

illustrate the importance of reading the name of a chemical very carefully and more than once before using it.

The names of some common cations are listed in Table 5.2. Notice that the old names of the iron and copper ions are derived from their Latin names (*ferrum* and *cuprum*).

It can be a little confusing to use the same name for both a metal and its ion, and you sometimes have to stop and think about what's being referred to.

Table 5.2   Common Cations

Aluminum ion	Al^{3+}	Cesium ion	Cs^+
Barium ion	Ba^{2+}	Lithium ion	Li^+
Calcium ion	Ca^{2+}	Potassium ion	K^+
Lead ion	Pb^{2+}	Silver ion	Ag^+
Magnesium ion	Mg^{2+}	Sodium ion	Na^+
Strontium ion	Sr^{2+}		
Tin ion[a]	Sn^{2+}		
Zinc ion	Zn^{2+}		

Transition metals that form more than one ion

Old system name	New system name	
Chromous ion	Chromium(II) ion	Cr^{2+}
Chromic ion	Chromium(III) ion	Cr^{3+}
Cuprous ion	Copper(I) ion	Cu^+
Cupric ion	Copper(II) ion	Cu^{2+}
Ferrous ion	Iron(II) ion	Fe^{2+}
Ferric ion	Iron(III) ion	Fe^{3+}
Mercurous ion	Mercury(I) ion	Hg_2^{2+}[b]
Mercuric ion	Mercury(II) ion	Hg^{2+}

[a] Sometimes also called *stannous ion.*
[b] This unusual +2 cation is composed of two mercury atoms rather than one.

AN APPLICATION: SOME "HYPERS" AND "HYPOS"

Scientists are prone to using prefixes to pack more meaning into a single word. For example, the "extra" in *extracellular* means "outside of cells," and the "intra" in *intracellular* means "inside of cells." High concentrations of sodium ion are present in extracellular fluids, while sodium ion concentrations in intracellular fluids are low. Potassium ion concentrations are just the reverse—high inside cells and low outside cells.

Abnormal concentrations of Na^+ or K^+ ions (or other substances) in body fluids create a variety of "hypo" (meaning "too low") and "hyper" (meaning "too high") conditions, some of which are quite serious. A below-normal concentration of sodium in blood serum is described as *hyponatremia*; for potassium, it's *hypokalemia*. In each word, "hypo" means "low" or "below normal" and the ending "emia" refers to blood. The reverse conditions, in which concentrations are above normal, are *hypernatremia* and *hyper-*

kalemia. (Note that "hypo" and "low" rhyme, providing a good trick for remembering the difference between the "hypos" and the "hypers.")

Sodium ion is lost from the body mainly in urine but also in sweat and feces. Abnormal sodium loss and resulting hyponatremia may result from kidney disease; from excessive sweating due to exercise, fever, or a hot environment; from severe burns; or from vomiting and diarrhea. Ion concentrations in body fluids are dependent, of course, on the amounts of both ions and water in the body. Hyponatremia also results when the body fails to eliminate enough water, a condition that may accompany kidney or heart failure, and liver diseases. If hyponatremia continues, it can cause edema (swelling of body tissues), central nervous system problems, and muscle spasms.

The reverse condition, hypernatremia, occurs when too much water is eliminated. If more

Table 5.3
Common Anions

Bromide ion	Br^-
Chloride ion	Cl^-
Fluoride ion	F^-
Iodide ion	I^-
Oxide ion	O^{2-}
Sulfide ion	S^{2-}

For example, it's common practice in nutrition and health-related fields to talk about sodium or potassium in the bloodstream. Since both sodium and potassium *metals* react violently with water, however, they are unlikely to be present in blood. The reference is to the harmless sodium and potassium *ions*.

Anions are named by replacing the ending of the element name with the ending *-ide* and combining this name with the word "ion" (Table 5.3). For example, the ion formed by fluor*ine* is named the fluor*ide* ion, and the ion formed by sulf*ur* is the sulf*ide* ion.

$$F^- \qquad S^{2-}$$

Fluoride ion   Sulfide ion

Practice Problems

5.11   Name the following ions: Cu^{2+}, F^-, Mg^{2+}, S^{2-}

5.12   Write the symbols for the following ions: silver ion, iron(II) ion, cuprous ion, telluride ion

5.13   Ringer's solution, which is used intravenously to adjust ion concentrations in body fluids, contains the ions of sodium, potassium, calcium, and chlorine. Give the names and symbols of these ions.

water than sodium is lost from sweating, diarrhea, vomiting, or burns, hypernatremia results. The condition also results from intake of too little food and water. The symptoms are those of dehydration—severe thirst, weakness, lethargy, and ultimately shock.

Although most K^+ ions are present within cells rather than outside them, the small percentage of K^+ ions in blood serum is essential to keep the heart beating. The normal routes of K^+ loss are the same as those for Na^+, and hypokalemia may result from some of the same conditions as hyponatremia. In addition, sodium and potassium ion concentrations are interrelated. Homeostasis for these ions is maintained by their transport across cell membranes by "ion pumps." Slight shifts in the normal balance of these ions entering and leaving cells greatly change the small concentrations of potassium ion in blood serum. Hypokalemia and hyperkalemia are ex-

tremely serious conditions. Low blood potassium can cause a weak and irregular heartbeat as well as paralysis due to failure to transmit nerve impulses; high blood potassium can cause cardiac arrest.

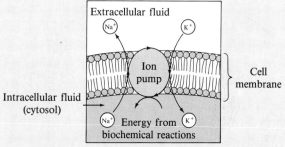

Sodium and potassium ions must be moved across cell membranes to maintain homeostasis, which in this case means high concentrations of K^+ within the cell and high concentrations of Na^+ outside the cell. When concentrations on different sides of the membrane are different, energy must be used to maintain the difference.

5.8   POLYATOMIC IONS

Polyatomic ion   An ion that contains two or more atoms.

Monatomic ion   An ion formed from a single atom.

Polyatomic ions, which are composed of more than one atom, are almost as common as the **monatomic ions** discussed in Section 5.4. Most polyatomic ions contain oxygen and another element, and their chemical formulas, like those for chemical compounds, show by subscripts how many of each type of atom are combined. Sulfate ion, for example, is composed of one sulfur atom and four oxygen atoms and has a -2 charge.

$$SO_4^{2-}$$

Sulfate ion

The atoms are held together by bonds of the type discussed in the next chapter. A polyatomic ion is charged because it contains a total number of electrons different from the total number of protons in the combined atoms.

The most common polyatomic ions are listed in Table 5.4. Note that the ammonium ion, NH_4^+, is the only one with a positive charge.

Table 5.4   Some Common Polyatomic Ions

Ammonium ion	NH_4^+	Nitrate ion	NO_3^-
Acetate ion	$CH_3CO_2^-$	Permanganate ion	MnO_4^-
Carbonate ion	CO_3^{2-}	Phosphate ion	PO_4^{3-}
Bicarbonate ion (hydrogen carbonate ion)	HCO_3^-	Hydrogen phosphate ion	HPO_4^{2-}
Hypochlorite ion	OCl^-	Dihydrogen phosphate ion	$H_2PO_4^-$
Dichromate ion	$Cr_2O_7^{2-}$	Sulfate ion	SO_4^{2-}
Cyanide ion	CN^-	Bisulfate ion (hydrogen sulfate ion)	HSO_4^-
Hydroxide ion	OH^-		
Nitrite ion	NO_2^-	Sulfite ion	SO_3^{2-}

Practice Problems   **5.14**   Write the names of the following ions: NO_3^-, CN^-, OH^-, HPO_4^{2-}

5.15   Write the formulas of the following ions: bicarbonate ion, hydroxide ion, nitrite ion, permanganate ion

5.9   CHEMICAL FORMULAS FOR IONIC COMPOUNDS

All chemical compounds are neutral, a requirement that makes finding the formulas of ionic compounds easy. Once the ions are identified, all that's needed is to figure out how many ions of each type give a total charge of zero.

If the ions have the same charge, it's clear that only one of each ion is needed:

$$K^+ \quad \text{and} \quad F^- \quad \text{form} \quad KF$$

$$Ca^{2+} \quad \text{and} \quad SO_4^{2-} \quad \text{form} \quad CaSO_4$$

What happens when the ions have different charges? The ratio of the ions must result in a total charge of zero. When potassium and oxygen combine, for example, it takes two potassium ions ($2 \times +1 = +2$) to balance the -2 charge of the oxide ion. The subscript 2 in the formula indicates that two potassium ions are present for every oxide ion.

$$2\,K^+ \quad \text{and} \quad O^{2-} \quad \text{form} \quad K_2O$$

The situation is reversed for the calcium ion and the chloride ion: There is one cation for every two anions.

$$Ca^{2+} \quad \text{and} \quad 2\,Cl^- \quad \text{form} \quad CaCl_2$$

It helps in writing the formulas for ionic compounds to remember that the number of each ion is equal to the charge on the other ion. For example, the correct formula for a compound containing Mg^{2+} and PO_4^{3-} shows three Mg^{2+} ions and two PO_4^{3-} ions:

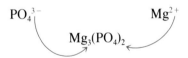

To check the formula of an ionic compound, multiply the charge of each ion by its subscript.

$$Mg_3(PO_4)_2$$

$$\text{For } Mg^{2+}: \quad 3 \times +2 = +6$$

$$\text{For } PO_4^{3-}: \quad 2 \times -3 = -6$$

The positive and negative charges are equal, so the formula is correct.

Chemical formulas for ionic compounds are written according to the following rules, as illustrated above:

- With the cation first and the anion second
- Without the charges of the ions
- Without the subscript if it would be 1
- With parentheses around a polyatomic ion formula if it must have a subscript

> **Solved Problem 5.3** Write the chemical formula for the compound formed by calcium ion and nitrite ion.
>
> **Solution** The two ions are Ca^{2+} and NO_2^-. Two nitrite ions, each with a -1 charge, will balance the $+2$ charge of the calcium ion.
>
> $$Ca^{2+} \quad\times\quad NO_2^-$$
>
> Since there must be two nitrite ions, the nitrite formula must be enclosed in parentheses. The correct formula is
>
> $$Ca(NO_2)_2$$

Practice Problems

5.16 Write the formulas for the ionic compounds formed by silver ion and (a) iodide ion, (b) oxide ion, (c) phosphate ion.

5.17 Write the formulas for the ionic compounds formed by sulfate ion with (a) sodium ion, (b) iron(II) ion, (c) chromium(III) ion.

5.18 The ionic compound containing ammonium ion and carbonate ion gives off the odor of ammonia, a property put to use in smelling salts for reviving someone who has fainted. Write the formula for this compound.

5.19 Astringents coagulate (pull out of solution) proteins in blood, sweat, and other body fluids, a property put to use in deodorants and styptic pencils that stop shaving cuts from bleeding. Two safe and effective astringents are the ionic compounds of aluminum with sulfate ion and acetate ion. Write the formulas of these compounds.

5.10 NAMING IONIC COMPOUNDS

Ionic compounds are named by giving the cation name followed by the anion name, with a space between them, as illustrated in Table 5.5. To name compounds of metals that form more than one ion, either the old (*-ous*, *-ic*) or the new system (Roman numerals) described in Section 5.7 must be used. Thus, $FeCl_2$ is called iron(II) chloride (or ferrous chloride), and $FeCl_3$ is called iron(III) chloride (or ferric chloride). Significant differences exist between compounds of the same two elements but with different charges on the cation. In treating iron-deficiency anemia, for example, iron(II) compounds are preferable, because considerably more iron is absorbed from them by the body than from iron(III) compounds.

It's customary not to use Roman numerals for ions of elements that form only one type of ion. If you're in doubt about whether to include the charge in a name, however, it's safer to put it in than leave it out, especially for transition metal ions.

Table 5.5 Some Common Ionic Compounds and Their Applications

Chemical Name (Common Name)	Formula	Applications
Ammonium carbonate	$(NH_4)_2CO_3$	Smelling salts
Calcium hydroxide (hydrated lime)	$Ca(OH)_2$	Mortar, plaster, whitewash
Calcium oxide (lime)	CaO	Lawn treatment, industrial chemical
Lithium carbonate ("lithium")	Li_2CO_3	Treatment of manic depression
Magnesium hydroxide (milk of magnesia)	$Mg(OH)_2$	Antacid
Magnesium sulfate (Epsom salts)	$MgSO_4$	Laxative, anticonvulsant
Potassium permanganate	$KMnO_4$	Antiseptic, disinfectant[a]
Potassium nitrate (saltpeter)	KNO_3	Fireworks, matches, and desensitizer for teeth
Silver nitrate	$AgNO_3$	Antiseptic, germicide
Sodium bicarbonate (baking soda)	$NaHCO_3$	Baking powder, antacid, mouthwash, deodorizer
Sodium hypochlorite	$NaOCl$	Disinfectant; active ingredient in household bleach
Zinc oxide	ZnO	Skin protection, in calamine lotion

[a] An antiseptic kills or inhibits growth of harmful microorganisms on the skin or in the body; a disinfectant kills harmful microorganisms but is generally not used on living tissue.

Solved Problem 5.4 Magnesium carbonate is used as an ingredient in Bufferin tablets. Write its formula.

Solution Look at the cation and the anion parts of the compound separately: Magnesium, a Group 2A element, forms the doubly positive Mg^{2+} cation; carbonate ion is doubly negative, CO_3^{2-}. Since magnesium carbonate must be neutral, its formula is $MgCO_3$.

Solved Problem 5.5 Sodium and calcium both form a wide variety of ionic compounds. Write the formulas for the compounds listed:
(a) sodium bromide and calcium bromide
(b) sodium sulfide and calcium sulfide
(c) sodium phosphate and calcium phosphate

Solution (a) $NaBr$ and $CaBr_2$ Note subscript 2 in $CaBr_2$ because of the $+2$ charge on Ca^{2+}.
(b) Na_2S and CaS Here subscript 2 is in Na_2S because of the -2 charge on S^{2-}.
(c) Na_3PO_4 and $Ca_3(PO_4)_2$ To balance the -3 charge of the anion requires 3 Na^+ and can be done only with 3 Ca^{2+} ions and 2 PO_4^{3-} ions.

Solved Problem 5.6 Give names for the listed compounds, using Roman numerals to indicate the charges on the cations where necessary:
(a) KF   (b) MgCl$_2$ (c) AuCl$_3$ (d) Fe$_2$O$_3$

Solution (a) Potassium fluoride No Roman numeral is necessary for a Group 1A metal, which forms only one cation.
(b) Magnesium chloride No Roman numeral is necessary since magnesium (Group 2A) forms only Mg^{2+}.
(c) Gold(III) chloride The Roman numeral III is necessary to specify the $+3$ charge on gold, a transition element that may form other ions.
(d) Iron(III) oxide Since the three oxide anions (O^{2-}) have a total negative charge of -6, the two iron cations must have a total charge of $+6$. Thus, each is Fe(III).

Practice Problems **5.20** Barium sulfate is an ionic compound swallowed by patients before their gastrointestinal tract is X-rayed. Write its formula. How many barium ions are present for each sulfate ion?

5.21 The compound Ag$_2$S is partly responsible for the tarnish found on silverware. Name Ag$_2$S and give the charge of the silver ion.

5.22 Name each of these compounds:
(a) CuO   (b) Ca(CN)$_2$ (c) NaNO$_3$ (d) Cu$_2$SO$_4$
(e) Li$_3$PO$_4$ (f) NH$_4$Cl

5.23 Write formulas for these compounds:
(a) barium hydroxide   (b) copper(II) carbonate
(c) magnesium bicarbonate   (d) cuprous fluoride
(e) ferric sulfate   (f) ferrous nitrate

5.11 FORMULA UNITS AND FORMULA WEIGHTS

The idea that atoms join together in definite ratios in chemical compounds is one of the cornerstones of chemistry. All samples of a given chemical compound contain exactly the same ratios of atoms of the combined elements. Furthermore, since atoms of different elements have different masses, masses of the combined elements have a fixed ratio in a given compound.

Simplest formula A chemical formula giving the lowest possible ratio of atoms combined in a chemical compound.

The formulas of ionic compounds are known as **simplest formulas** (or sometimes *empirical formulas*), meaning that they show the lowest possible ratio of atoms in the compound. Formulas such as K$_2$I$_2$ and K$_3$I$_3$ are not wrong for potassium iodide, but they give no different information about the compound than KI.

What exactly, then, does the chemical formula of an ionic compound represent? The smallest *particles* in ionic compounds are the individual ions.

Formula unit The group of atoms or ions represented by the formula of a chemical compound.

Formula weight The sum of the individual atomic weights for all atoms represented in the formula of a chemical compound or polyatomic ion.

There is no such thing as a single particle of sodium chloride. To identify the smallest possible neutral *unit* of an ionic compound, however, we speak of a **formula unit** (Figure 5.7). For sodium chloride, the formula unit is one sodium ion and one chloride ion. For potassium sulfate, K_2SO_4, the formula unit is two potassium ions and one sulfate ion.

Although formula units of ionic compounds ordinarily have no independent existence, the concept is useful because the weights of compounds can be compared on the same mass scale as the weights of atoms. The **formula weight** of a compound is the sum of the individual atomic weights for all the atoms in the formula of the compound.

Figure 5.7
Formula units of ionic compounds. The sum of the charges on the ions in a formula unit equals zero. For the compound on the left, $(+1) + (-1) = 0$; for the compound on the right $(+2) + (2 \times -1) = 0$.

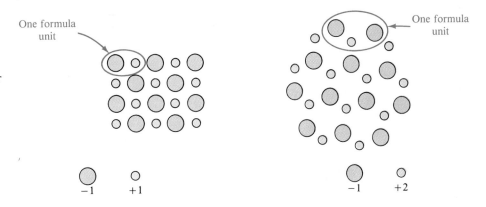

One formula unit

One formula unit

−1 +1

−1 +2

Solved Problem 5.7

Sodium nitrate ($NaNO_3$) and calcium nitrate ($Ca(NO_3)_2$) produce the brilliant yellow and red colors in fireworks. Find the formula weights of these two compounds.

Solution

A formula weight is the sum of the atomic weights of *all* atoms in a compound. In finding formula weights, be sure to account for all subscripts and remember that a subscript outside parentheses multiplies *all* the atoms inside the parentheses.

For $NaNO_3$:

$$
\begin{array}{lll}
1\ \text{Na} = & & 23.0\ \text{amu} \\
1\ \text{N} = & & 14.0\ \text{amu} \\
3\ \text{O} = 3 \times 16.0\ \text{amu} = & & \underline{48.0\ \text{amu}} \\
& & 85.0\ \text{amu}
\end{array}
$$

For $Ca(NO_3)_2$:

$$
\begin{array}{lll}
1\ \text{Ca} = & & 40.0\ \text{amu} \\
2\ \text{N} = 2 \times 14.0\ \text{amu} = & & 28.0\ \text{amu} \\
2 \times 3\ \text{O} = 6 \times 16.0\ \text{amu} = & & \underline{96.0\ \text{amu}} \\
& & 164.0\ \text{amu}
\end{array}
$$

One formula unit of calcium nitrate is almost twice as heavy as one formula unit of sodium nitrate.

Practice Problems   **5.24**   Find the formula weights of (a) Ca_3N_2, (b) $CaHPO_4$, (c) $Ca(OH)_2$, and (d) $Ca(MnO_4)_2$.

5.25   For each compound in Practice Problem 5.22, give the number of ions in one formula unit.

5.26   Find the formula weights of (a) magnesium oxide, (b) cupric oxide, (c) and manganese(II) sulfate, all of which are used as sources of their metal ions in vitamin-mineral pills.

INTERLUDE: MINERALS AND GEMS

If you have a sapphire, ruby, emerald, or zircon ring, you are walking around with a crystalline ionic compound on your finger. The gem came from the earth's crust, the source of most of our chemical raw materials. Gemstones are minerals, which to geologists means that they are naturally occurring, crystalline chemical compounds. (Nutritionists refer to the metal ions essential to human health as "minerals.")

Sapphire and ruby are forms of the mineral corundum, which is aluminum oxide (Al_2O_3); the blue of sapphire is due to traces of iron and titanium ions, and the red of ruby to traces of chromium ions. Many minerals are silicates—their anions contain silicon and oxygen, which combine in a variety of structures ranging from SiO_4^{4-} to huge chains, sheets, and three-dimensional networks. Zircon is zirconium silicate ($ZrSiO_4$), and emerald is a form of the mineral beryl ($Al_2[Be_3Si_6O_{18}]$), which is composed of Al^{3+}, Be^{2+}, and silicate chains.

Many minerals not sufficiently beautiful for use in jewelry are as valuable, if not more valuable, than gems, because they are sources for the metals so essential to our industrial civilization. Extraction of the metals requires energy and a series of chemical reactions to convert the metal ions into free elements—the reverse of ion formation from atoms.

(a)

(b)

(a) Natural malachite, a hydrated copper carbonate that is a copper ore and is also used in jewelry. (b) Topaz, a yellow-brown valuable gemstone that is a silicate containing aluminum and fluoride ions.

SUMMARY

Atoms are converted into positively charged ions (**cations**) by the loss of electrons, and into negatively charged ions (**anions**) by the gain of electrons. **Ionic compounds** are composed of cations and anions united by **ionic bonds**, which result from the attraction between opposite electrical charges of the ions. Ionic compounds conduct electricity when molten or dissolved in water. They are generally crystalline solids with high melting and boiling points; some but not all are soluble in water.

As shown by the lack of **chemical reactivity** of the noble gases (Group 8A), an outer electron configuration of eight electrons in filled s and p subshells is quite stable. In ionic compounds, main-group elements tend to form ions in which they have attained noble gas configurations by the gain or loss of electrons: the **octet rule**.

Periodic variations of **atomic radii** and **ionization energies** show that metal atoms are larger than nonmetal atoms and lose electrons more easily.

Across the periods:	Down the groups:
Metallic character decreases	Metallic character increases
Atomic radii decrease	Atomic radii increase
Ionization energy increases	Ionization energy decreases

Thus, metals are found in ionic compounds as cations, and nonmetals as anions. Metals from Groups 1A and 2A form $+1$ and $+2$ cations, respectively. Transition metals form cations without octet configurations and often of more than one charge. Oxygen and sulfur (Group 6A) form -2 anions, and halogens (Group 7A) form -1 anions. The nonmetals of Groups 4A and 5A are not common in ionic compounds.

Cations are named with the metal name, and anions are named using the name ending *-ide*. For metals that form more than one ion, a Roman numeral equal to the ion's charge (or the ending *-ous* or *-ic*) is added to the ion's name. Ionic compounds, which often contain **polyatomic ions** as well as **monatomic ions**, contain sufficient numbers of ions to give neutrality, providing a means of determining their chemical formulas. To name an ionic compound, the cation name is given first, with the charge of the metal ion indicated if necessary, and the anion name second.

The formula of an ionic compound is a **simplest formula**, representing the simplest ratio of combined atoms, for there is no single particle of an ionic compound. The weight of one **formula unit** in atomic mass units is the **formula weight** of a compound.

REVIEW PROBLEMS

Ions and Ionic Bonding

5.27 Define these terms:
(a) ion   (b) anion   (c) cation   (d) ionic bond
(e) ionic compound

5.28 Write electron-dot symbols for:
(a) hydrogen   (b) helium   (c) lithium

5.29 Write electron-dot symbols for:
(a) phosphorus   (b) sulfur   (c) chlorine
(d) barium   (e) aluminum

5.30 Write equations for loss or gain of electrons by atoms that result in:
(a) formation of Ca^{2+} (b) formation of Au^+
(c) formation of F^- (d) formation of Cr^{3+}

5.31 Write symbols for the ions formed by:
(a) gain of three electrons by nitrogen
(b) loss of one electron by lithium
(c) loss of two electrons by titanium
(d) loss of three electrons by gold

5.32 Decide whether the following statements are true or false for ionic solids:

(a) Ions are randomly arranged.
(b) All ions are the same size.
(c) Ionic solids can be shattered with a sharp blow.
(d) Ionic solids have low boiling points.

Ions and the Octet Rule

5.33 State the octet rule.

5.34 What is the charge of an ion that contains 34 protons and 36 electrons?

5.35 What is the charge of an ion that contains 11 protons and 10 electrons?

5.36 Strontium is a Group 2A element with atomic number 38. How many electrons must strontium lose to achieve a noble gas configuration?

5.37 How many electrons and how many protons does a strontium atom have (Problem 5.36)? How many electrons and protons does a strontium ion have?

5.38 Write the electron configuration for each of the following ions:
(a) Rb^+ (b) Br^- (c) S^{2-} (d) Ba^{2+} (e) Al^{3+}

5.39 Each of the following ions has a noble gas configuration. Identify the noble gas.
(a) Ca^{2+} (b) Li^+ (c) O^{2-} (d) F^- (e) Mg^{2+}

Periodic Properties and Ion Formation

5.40 Define these terms:
(a) atomic radius (b) ionization energy

5.41 Look at the periodic table and tell which member of each of the following pairs of atoms has the larger ionization energy:
(a) Li and O (b) Li and Cs
(c) K and Zn (d) Mg and N

5.42 What common features do the alkali metals have that allow them to form positively charged ions?

5.43 Write symbols for the ions most commonly formed by these elements:
(a) aluminum (b) iron (c) chlorine

5.44 Identify the main groups in the periodic table whose elements (a) all commonly form cations, (b) all commonly form anions.

5.45 Which of the following are likely to be stable ions? Explain.
(a) Li^{2+} (b) K^- (c) Mn^{3+} (d) Zn^{4+} (e) Ne^+

5.46 Which of the following elements is likely to form more than one cation?
(a) magnesium (b) silicon (c) manganese

5.47 Write the electron configurations of Cr^{2+} and Cr^{3+}.

5.48 Would you expect the ionization energy of Li^+ to be less than, greater than, or the same as the ionization energy of Li? Why?

5.49 Why does an S^{2-} ion have a larger radius than an S atom?

Symbols, Formulas, and Names for Ions

5.50 Name the following ions:
(a) S^{2-} (b) Sn^{2+} (c) Sr^{2+} (d) Mg^{2+} (e) Au^+

5.51 Name the following ions in both the old and the new systems:
(a) Cu^+ (b) Fe^{3+} (c) Hg^{2+}

5.52 Write symbols for the following ions:
(a) selenide ion (b) oxide ion (c) silver ion
(d) cobalt(II) ion

5.53 Write formulas for the following ions:
(a) hydroxide ion (b) sulfate ion
(c) acetate ion (d) permanganate ion
(e) hypochlorite ion (f) nitrate ion
(g) bicarbonate ion

5.54 Name the following ions:
(a) NH_4^+ (b) CN^- (c) CO_3^{2-} (d) SO_3^{2-}

Names and Formulas for Ionic Compounds

5.55 Write formulas for the compounds formed by nitrate ion with:
(a) aluminum (b) silver
(c) zinc (d) barium

5.56 Maalox, an over-the-counter antacid, is a mixture of magnesium hydroxide and aluminum hydroxide. Write the formula for each compound.

5.57 Write the formula for each of the following substances:
(a) sodium bicarbonate (baking soda)
(b) potassium nitrate (a backache remedy)
(c) calcium carbonate (an antacid)

5.58 Write the formula for each of the following compounds:
(a) calcium hypochlorite, used as a swimming pool disinfectant
(b) copper(II) sulfate, used to kill algae in swimming pools
(c) sodium phosphate, used in detergents to enhance cleaning action

5.59 Complete the following table by writing in the formula of the compound formed by each pair of ions.

	S^{2-}	Cl^-	PO_4^{3-}	CO_3^{2-}
copper(II)	CuS			
Ca^{2+}				
NH_4^+				
ferric ion				

5.60 Write the names of these substances:
(a) $MgCO_3$ (b) $Ca(CH_3CO_2)_2$ (c) $AgCN$
(d) $Na_2Cr_2O_7$

5.61 Which of the following structures is most likely to be correct for calcium phosphate?
(a) Ca_2PO_4 (b) $CaPO_4$ (c) $Ca_2(PO_4)_3$
(d) $Ca_3(PO_4)_2$

Formula Units and Formula Weights

5.62 What ions and how many of each type are present in one formula unit of these compounds?
(a) $AgCl$ (b) CaF_2 (c) Ag_2SO_4

5.63 Calculate the formula weights of the following compounds, all used as paint pigments:
(a) ZnO (white) (b) TiO_2 (white)
(c) $PbCrO_4$ (yellow) (d) $CdSe$ (red)

5.64 Write the formula and calculate the formula weight of each of the following compounds:
(a) ammonium dichromate
(b) chromium(III) nitrate
(c) lithium hydrogen sulfate

Applications

5.65 What is homeostasis? [App: Homeostasis]

5.66 A blood serum concentration of iron ion of 60 μg/dL falls within the normal range. Convert this concentration to g/L. [App: Homeostasis]

5.67 Where is most of the calcium ion found in the body? [App: Biologically Important Ions]

5.68 While excess sodium ion is considered hazardous, a certain amount is necessary for normal body functions. What is the purpose of sodium in the body? [App: Biologically Important Ions]

5.69 Before a person is allowed to donate blood, a drop of the blood is tested to be sure that it contains a sufficient amount of iron (men, 41 μg/dL; women, 38μg/dL). Why is this required? [App: Biologically Important Ions]

5.70 What do the prefixes "hyper" and "hypo" mean? [App: Some "Hypers" and "Hypos"]

5.71 What is the difference between a geologist's and a nutritionist's definition of a mineral? [Int: Minerals and Gems]

Additional Questions and Problems

5.72 Hydroxyapatite, a component of human bone, has the formula $Ca_5(PO_4)_3(OH)$. Name each ion in hydroxyapatite, give its charge, and show that the formula correctly represents a neutral compound.

5.73 Calculate the formula weight of hydroxyapatite, described in Problem 5.72. How many ions are present in one formula unit of hydroxyapatite?

5.74 Explain the "noble gas configuration" of the hydride ion, H^-.

5.75 Many compounds containing a metal and a non-metal are named by the Roman numeral system described in Section 5.7, even though they aren't ionic. Write the chemical formula for each of the following such compounds:
(a) chromium(VI) oxide (b) vanadium(V) chloride
(c) manganese(IV) oxide
(d) molybdenum(IV) sulfide

5.76 The arsenate ion has the formula AsO_4^{3-}. Write the formula and calculate the formula weight of lead(II) arsenate, used as an insecticide.

5.77 One commercially available calcium supplement contains calcium gluconate, a compound that is also used as an anticaking agent in instant coffee.
(a) If this compound contains one calcium ion for every two gluconate ions, what is the charge on a gluconate ion?
(b) How many gluconate ions are in one formula unit of iron(III) gluconate, a commercial iron supplement?

5.78 Each of the following names is incorrect. Write the correct name for each compound.
(a) Cu_3PO_4, copper(III) phosphate
(b) Na_2SO_3, sodium sulfide
(c) MnO_2, manganese(II) oxide
(d) $AuCl_3$, gold chloride
(e) $Pb(CO_3)_2$, lead(II) acetate
(f) Ni_2S_3, nickel(II) sulfide

5.79 Each of the following formulas is incorrect. Write the correct formula for each compound.
(a) cobalt(II) cyanide, $CoCN_2$
(b) uranium(VI) oxide, UO_6
(c) tin(II) sulfate, $Ti(SO_4)_2$
(d) potassium phosphate, K_2PO_4
(e) calcium phosphide, CaP
(f) lithium hydrogen sulfate, Li_2HSO_4

5.80 How many protons, electrons, and neutrons are in each of the following ions?
(a) $^{16}O^{2-}$ (b) $^{89}Y^{3+}$ (c) $^{133}Cs^+$ (d) $^{81}Br^-$

CHAPTER

6

Molecular Compounds

Because the shape of a molecule is essential to its properties, especially to its interaction with living systems, this chemist is scrutinizing the shape of a possible new pharmaceutical. You'll learn something in this chapter about molecular shapes.

You saw in the last chapter that ionic compounds are high-melting crystalline solids composed of positively and negatively charged ions. What are the compositions of materials that aren't ionic compounds? What kind of chemical compounds can be gases, liquids, or low-melting solids at room temperature? Searching for the smallest particles of most such materials leads not to atoms or ions but to molecules. A molecule is a group of atoms connected to each other by the second major type of chemical bond, the covalent bond.

A single molecule of water contains exactly two hydrogen atoms and one oxygen atom, no more and no less. From what is known about the size of atoms and how they're connected to each other, we can draw the following picture of a water molecule.

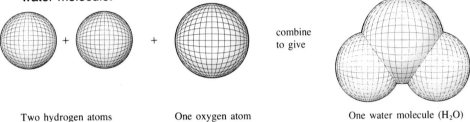

Two hydrogen atoms   One oxygen atom   One water molecule (H_2O)

Among the questions we'll answer in this chapter about the covalent bond and the properties of compounds composed of molecules are the following:

1. *What is a covalent bond?* The goal: Be able to define the covalent bond and describe its formation.

2. *How does the octet rule apply to covalent bond formation?* The goal: Be able to predict the numbers of covalent bonds usually formed by the more common main-group elements.

3. *How are molecular compounds represented by formulas?* The goal: Be able to interpret molecular formulas, structural formulas, and Lewis structures.

4. *How are electrons arranged in covalent bonds?* The goal: Be able to recognize the occurrence of single, double, and triple covalent bonds.

5. *How are electron pairs arranged in molecules, and what is their influence on molecular shape?* The goal: Be able to draw Lewis structures and use them to predict molecular geometry.

6. *When are bonds and molecules polar?* The goal: Be able to use electronegativity and molecular geometry to predict bond and molecular polarity.

7. *What are the major differences in properties between ionic and molecular compounds?* The goal: Be able to compare the structures, compositions, and melting and boiling points of ionic and molecular compounds.

8. *Under what circumstances are molecular compounds converted to ions?* The goal: Be able to describe ion formation by hydrogen loss or addition.

6.1 COVALENT BONDS

How many ionic compounds can you think of that you encounter frequently in everyday life? After you've named table salt (NaCl), baking soda ($NaHCO_3$), and maybe a few others like lime for the lawn (CaO) and epsom salts ($MgSO_4$), coming up with further examples gets difficult. Most materials around us are *not* crystalline, brittle, high-melting solids. Every day we're much more likely to encounter gases like those in the air, liquids like water and gasoline, low-melting solids like mothballs and butter, and flexible solids like plastics and wood.

Molecule A group of atoms held together by covalent bonds in a discrete unit that is capable of independent existence.

All these materials are composed mainly of **molecules** rather than ions, and most of the molecules are made of nonmetal atoms combined with each other. The gaseous element hydrogen provides the simplest example of how nonmetal atoms combine. Each hydrogen atom has one valence electron. How can hydrogen atoms form a hydrogen molecule, in which two hydrogen atoms are connected to each other? The hydrogen atoms aren't likely to gain or lose electrons, but they can *share* them. The single 1s electrons pair up in the region between the two atoms and hold them together.

Shared electron pair

$$H\cdot \; + \; \cdot H \qquad \rightarrow \qquad H \!:\! H$$

Two hydrogen atoms   A hydrogen molecule, H_2

Covalent bond A bond that results when two atoms share one or more pairs of electrons.

The shared valence electron pair forms a **covalent bond.** The electrons occupy the region between the two hydrogen nuclei, and the negative charges of the electrons are essentially the ''glue'' that holds the positively charged hydrogen nuclei together. For simplicity, the electron pair of a covalent bond is often represented by a straight line, and H—H, H_2, or H : H all represent a hydrogen molecule.

Diatomic molecule A molecule that consists of two atoms bonded together.

A molecule like H_2 composed of just two atoms is a **diatomic molecule.** Hydrogen, nitrogen, oxygen, and the halogens normally exist as such molecules: H_2, N_2, O_2, F_2, Cl_2, Br_2, I_2. There are only seven common diatomic elements (Table 6.1), and it's best to memorize them.

As a further example of covalent bonding, consider the chlorine molecule, Cl_2. The bonding in chlorine allows the two atoms to achieve stable electron octets by sharing electrons. Each individual chlorine atom has seven valence electrons in the $3s^2 3p^5$ configuration. Two of the three $3p$ orbitals are filled by two electrons each, while the third $3p$ orbital holds only one electron. When two chlorine atoms approach each other, the unpaired $3p$ electrons from each are shared by both atoms in a covalent bond. Such bond formation can be pictured as the overlap of the p orbitals containing the single electrons (Figure 6.1), with the resulting formation of a region of high electron density between the atoms.

Shared electron pair

$$: \!\ddot{C}l\cdot \; + \; \cdot \ddot{C}l\!: \qquad \rightarrow \qquad : \!\ddot{C}l\!:\!\ddot{C}l\!:$$

A chlorine molecule, Cl_2

Each chlorine atom ''owns'' six outer-shell electrons and ''shares'' two more, giving each atom a stable outer-shell octet.

Table 6.1 Elements That Exist as Diatomic Molecules

Hydrogen	H_2	Colorless, odorless, flammable and explosive gas; not toxic
Nitrogen	N_2	Colorless, odorless gas of very low reactivity; 78% (by vol) in air; not toxic
Oxygen	O_2	Colorless, odorless reactive gas; 21% (by vol) in air; supports combustion; not toxic
Fluorine	F_2	Pale yellow, highly reactive, though nonflammable, gas with pungent odor; very toxic and harmful to tissues
Chlorine	Cl_2	Dense, greenish yellow, highly reactive gas; irritating, suffocating odor; slightly water soluble; nonflammable, but supports combustion
Bromine	Br_2	Dark red-brown liquid with irritating fumes; slightly water-soluble; reacts with metals; seriously harmful to skin
Iodine	I_2	Gray-black crystalline solid with violet vapor; slightly soluble in water; moderately reactive; irritating to eyes and skin

Figure 6.1
Overlap of *p* orbitals. When two singly occupied *p* orbitals overlap, the electron density becomes concentrated in the region between the atoms, joining them together by a single covalent bond.

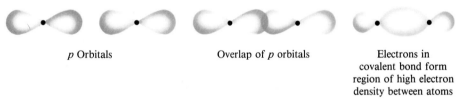

p Orbitals   Overlap of *p* orbitals   Electrons in covalent bond form region of high electron density between atoms

The best analogy for the bond formed by a shared electron pair is to think of two people who both want the same object. If both hold onto the object tightly, they are bound together. Neither person can walk away from the other as long as they both hold on. Similarly, when both chlorine atoms hold onto the shared electrons, the atoms are bonded together.

The energy changes that occur during bond formation are pictured by a diagram like that in Figure 6.2, in which potential energy is plotted against the

Figure 6.2
Change in potential energy as two approaching atoms form a covalent bond. The stable bond is formed at the point where the energy is at a minimum. The *bond length* is the distance between nuclei in a stable bond, and the *bond energy* is the amount of energy that must be supplied to break the bond and regenerate the separate atoms.

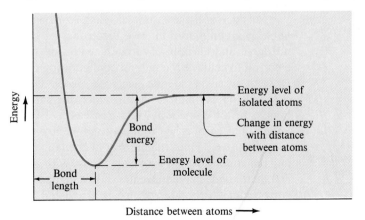

Energy level of isolated atoms

Change in energy with distance between atoms

Bond energy

Energy level of molecule

Bond length

Distance between atoms ⟶

distance between two atoms approaching each other. Beginning on the right side of the diagram where the two atoms are far apart, neither feels the effect of the other. As the atoms move closer together and begin to form a covalent bond, the electrons of each atom are attracted to the nucleus of the other and the total potential energy begins to drop. Potential energy drops still further as the atoms come closer together, until a stable molecule is formed at the point of minimum energy.

Practice Problems	*6.1*	What noble gas electron configuration do the two hydrogen atoms achieve in H_2?
	6.2	Draw the iodine molecule by combining electron-dot symbols, label the shared electron pair, and show two other ways the iodine molecule can be represented. What noble gas configuration do the iodine atoms achieve in an iodine molecule?

6.2 COVALENT BONDING, THE OCTET RULE, AND THE PERIODIC TABLE

Molecular compound A chemical compound composed of molecules.

Covalent bonds can form between unlike atoms as well as between like atoms, and **molecular compounds** are composed of atoms of two or more elements. Water molecules consist of two hydrogen atoms joined by covalent bonds to an oxygen atom, H_2O; ammonia molecules consist of three hydrogen atoms covalently bonded to a nitrogen atom, NH_3; and methane molecules consist of four hydrogen atoms covalently bonded to a carbon atom, CH_4. Notice that the different atoms in these examples—hydrogen, oxygen, nitrogen, and carbon—all form different numbers of covalent bonds when they join to other atoms in neutral molecules.

H_2	H_2O	NH_3	CH_4
Hydrogen bonds to *1* hydrogen atom	Oxygen bonds to *2* hydrogen atoms	Nitrogen bonds to *3* hydrogen atoms	Carbon bonds to *4* hydrogen atoms

In each of these examples, each atom shares enough electrons to achieve a noble gas configuration: two electrons for hydrogen, and octets for oxygen, nitrogen, and carbon. Hydrogen, with one valence electron, needs one more electron to achieve the helium configuration ($1s^2$) and forms one covalent bond. Oxygen, with six valence electrons, needs two more electrons to reach a stable octet and gains them by sharing electrons with two other atoms in two covalent bonds. In the same way, nitrogen has five valence electrons and forms three bonds, and carbon has four valence electrons and forms four bonds.

Biomolecule A molecule of a compound that occurs naturally and plays a role in the chemistry of living things.

Figure 6.3 gives the number of covalent bonds usually formed by hydrogen and the common main-group elements, with symbols for the elements present in covalent bonds in **biomolecules** shown in blue. The numbers of valence electrons surrounding the bonded atoms are also listed. You will sometimes see variations in the numbers of covalent bonds surrounding atoms of Si, P, S, Cl, and other

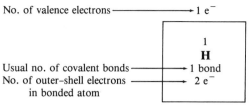

No. of valence electrons → 1 e⁻

Usual no. of covalent bonds → 1 bond
No. of outer-shell electrons → 2 e⁻
in bonded atom

Figure 6.3
Number of covalent bonds formed by common main-group elements. The usual numbers of covalent bonds are shown in green. For Si, P, S, Cl, and other elements from the third and higher periods, the number of covalent bonds may vary (partly depending on how structures are drawn). For example, chlorine and heavier halogens have more than one covalent bond in compounds with oxygen such as perchloric acid ($HClO_4$) and in compounds with other halogens such as bromine trifluoride (BrF_3). The possible numbers of outer-shell electrons, shown in red, do not vary, but those in parentheses are less common. Symbols for elements found in biomolecules are in blue.

Group 3A	Group 4A	Group 5A	Group 6A	Group 7A
3 e⁻	4 e⁻	5 e⁻	6 e⁻	7 e⁻
5 **B** 3 bonds 6 e⁻	6 C 4 bonds 8 e⁻	7 N 3 bonds 8 e⁻	8 O 2 bonds 8 e⁻	9 **F** 1 bond 8 e⁻
	14 **Si** 4 bonds 8 e⁻	15 P 3 or 5 bonds 8 e⁻ or 10 e⁻	16 S 2 or 6 bonds 8 e⁻ or 12 e⁻	17 **Cl** 1 bond 8 e⁻ (10 e⁻, 12 e⁻)
				35 **Br** 1 bond 8 e⁻ (10 e⁻, 12 e⁻)
				53 **I** 1 bond 8 e⁻ (10 e⁻, 12 e⁻)

elements from period 3 or higher, but *the numbers of electrons surrounding the atoms do not vary* from those listed in the figure.

Figure 6.3 reveals a few exceptions to the octet rule in addition to hydrogen with its one valence electron to share in one covalent bond. Boron is a similar exception: It has only three valence electrons to share in three covalent bonds and forms compounds such as BF_3.

Phosphorus and sulfur show a different type of octet rule exception because their *d* orbitals can be used by bonding electrons. In some molecules, therefore, phosphorus has five covalent bonds (with 10 outer-shell electrons) and sulfur has six covalent bonds (with 12 outer-shell electrons). For example, in combination with halogens, phosphorus and sulfur form molecules such as PCl_5 and SF_6

PCl_5

Phosphorus pentachloride
(highly reactive yellowish
solid)

SF_6

Sulfur hexafluoride
(colorless, nonflammable gas
used as insulator in electrical
equipment)

Solved Problem 6.1 Consult Figure 6.3 and decide whether the following molecules are likely to exist.

(a)
$$Br—\underset{\underset{\displaystyle Br}{|}}{\overset{\overset{\displaystyle Br}{|}}{C}}—Br$$

(b) I—Cl

(c)
$$H—\underset{\underset{\displaystyle H}{|}}{\overset{\overset{\displaystyle H}{|}}{F}}—H$$

(d) H—S—H

CBr_3 ICl FH_4 H_2S

Solution (a) No. Carbon does not have four covalent bonds in this structure.
(b) Yes. Both iodine and chlorine have one covalent bond.
(c) No. Fluorine, because it is in the second period, does not form more than one covalent bond.
(d) Yes. Sulfur, which like oxygen is in Group 6A, often forms two covalent bonds.

Practice Problems

6.3 How many covalent bonds are formed by each atom in these molecules?
(a) PH_3 (b) H_2Se (c) HCl   (d) SiF_4

6.4 Lead forms ionic compounds but also forms molecular compounds. Using Figure 6.3, explain whether a molecular compound containing lead and chlorine is most likely to be $PbCl_4$ or $PbCl_5$.

6.5 Compare Figure 6.3 with Figure 5.6 and list the elements that form both ionic and covalent bonds and those that form only covalent bonds. Why do the latter not form ionic bonds?

6.3 MULTIPLE COVALENT BONDS

Single covalent bond A covalent bond that results from sharing of two electrons between atoms.

The bonding in some molecules cannot be explained by the sharing of one electron pair in **single covalent bonds** like the Cl—Cl bond in Cl_2 and the O—H bonds in H_2O. For example, the carbon and oxygen atoms in carbon dioxide, CO_2, and the nitrogen atoms in the N_2 molecule could not have electron octets if only single bonds were present:

Combining $:\overset{..}{O}\cdot$, $\cdot\overset{.}{C}\cdot$, and $:\overset{..}{O}\cdot$ gives

$:\overset{..}{O}:C:\overset{..}{O}:$ **Unstable**

Carbon has only *six* electrons, and oxygen has only *seven* electrons.

Combining $:\overset{.}{N}\cdot$ with $:\overset{.}{N}\cdot$ gives

$:\overset{..}{N}:\overset{..}{N}:$ **Unstable**

Each nitrogen has only *six* electrons.

The only way the atoms in CO_2 and N_2 molecules can have outer-shell electron octets is if they share *more* than two electrons in each bond, resulting in the formation of *multiple* covalent bonds. If the carbon and oxygen atoms in carbon dioxide share *four* electrons, each atom then has a stable electron octet,

Double covalent bond A covalent bond that results from sharing of two electron pairs between atoms.

Triple covalent bond A covalent bond that results from sharing of three electron pairs between atoms.

and two **double bonds** result. Similarly, if two nitrogen atoms share *six* electrons, each nitrogen then has an electron octet, and a **triple bond** results.

Double bonds

:Ö::C::Ö: or   :Ö=C=Ö:

The carbon and oxygen atoms now have eight electrons each.

A triple bond

:N:::N: or   :N≡N:

Each nitrogen atom now has eight electrons.

Multiple covalent bonding is common in organic molecules, all of which contain carbon. For example, ethylene, a simple compound used commercially as a plant hormone to induce ripening in fruit, has the formula C_2H_4. The only way that all six atoms can have octets is for the two carbon atoms to share four electrons in a carbon–carbon double bond:

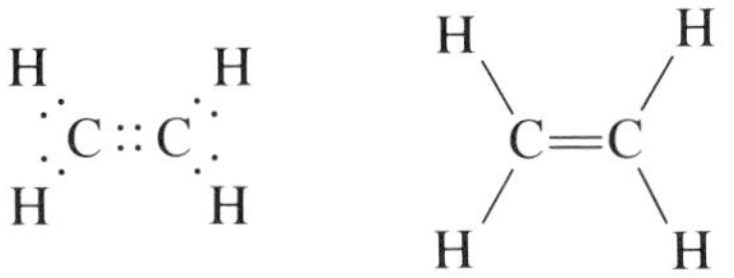

Ethylene—the carbon atoms share four electrons in a double bond.

Acetylene, the gas used in welding, has the formula C_2H_2. Thus, the two acetylene carbons must share six electrons in a carbon–carbon triple bond:

H:C:::C:H   H—C≡C—H

Acetylene—the carbon atoms share six electrons in a triple bond.

Note that in compounds with multiple bonds like ethylene and acetylene, each carbon atom still forms a total of four covalent bonds.

Carbon, nitrogen, and oxygen are the elements most often present in multiple bonds. Carbon and nitrogen form both double and triple bonds, and there is a triple bond in elemental nitrogen (N≡N). Oxygen forms double bonds, frequently with carbon and also with phosphorus and sulfur.

Most common multiple bonds in chemical compounds

Double bonds:	C=C	C=N	C=O	N=N
Triple bonds:	C≡C	C≡N		

CN cane be a polyatomic ion too

Practice Problem   **6.6**   Acetic acid is represented by combining electron-dot symbols as shown. How many outer-shell electrons are associated with each atom? Draw the structure in which dots are replaced by straight lines for covalent bonds.

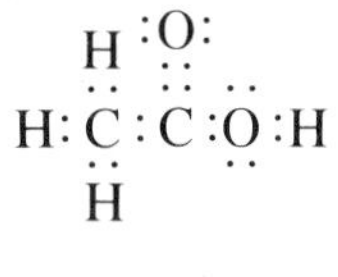

H   :Ö:
H:C:C:Ö:H
   H

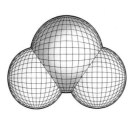
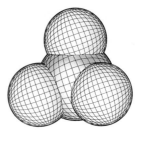
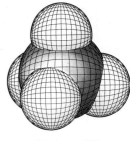

Water, H₂O   Ammonia, NH₃   Methane, CH₄

Figure 6.4
The shapes of water, methane, and ammonia molecules.

Molecular formula A formula that shows by subscripts the numbers and kinds of atoms in one molecule.

Structural formula A formula that shows how atoms are connected to each other.

6.4 STRUCTURAL FORMULAS

Formulas such as H_2O and PCl_5 that show the numbers and kinds of atoms in one molecule are called **molecular formulas.** Though useful, these formulas don't show how the atoms in a molecule are connected to each other. **Structural formulas** are used for this purpose. For example, water, ammonia, and methane (Figure 6.4) are represented as follows:

Lewis structures

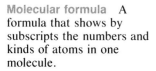

$$H\!:\!\overset{..}{\underset{..}{O}}\!:\!H \quad \text{or} \quad H\!-\!\overset{..}{\underset{..}{O}}\!-\!H \qquad H\!:\!\overset{..}{N}\!:\!H \quad \text{or} \quad H\!-\!\overset{\overset{\textstyle H}{|}}{N}\!-\!H \qquad H\!:\!\overset{H}{\underset{..}{C}}\!:\!H \quad \text{or} \quad H\!-\!\overset{\overset{\textstyle H}{|}}{\underset{\underset{\textstyle H}{|}}{C}}\!-\!H$$

Lone pairs of electrons

Lewis structure A structural formula that uses dots to represent valence electrons.

Structural formulas that include electron dots are called **Lewis structures,** after G. N. Lewis of the University of California, who devised them. The advantage of Lewis structures is that they make clear how many electrons surround each atom in a molecule. *Whenever two atoms are shown connected by a line, a covalent bond is indicated.* In a water molecule, for example, the oxygen atom is surrounded by eight electrons, four of them shared in two covalent bonds with oxygen and four others. Pairs of electrons not used in bonding, such as those on oxygen in a water molecule, are known as *unshared pairs* or **lone pairs.**

Lone pair A pair of outer-shell electrons not used by an atom for forming bonds.

The formula for a molecular compound, whether it's written as a Lewis structure or in the usual manner with subscripts, as in H_2O, NH_3, or CH_4, represents the number of atoms combined in *one molecule* of the compound. Here is a significant difference in the meaning of formulas for molecular and ionic compounds (Figure 6.5). The formulas of ionic compounds just give ratios

Figure 6.5
Distinction between ionic and molecular compounds. In ionic compounds, the smallest particles are ions; in molecular compounds, they are molecules.

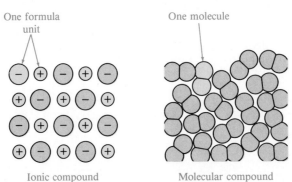

One formula unit

One molecule

Ionic compound   Molecular compound

of ions. In contrast, to be correct, molecular formulas must give the number of atoms in a molecule. The organic compounds with formulas C_2H_4 and C_4H_8 are quite different, as shown by their structural formulas.

$$\begin{array}{c} H \\ \diagdown \\ C \end{array} = \begin{array}{c} H \\ \diagup \\ C \end{array}$$

Ethylene   2-Butene

One formula unit of a molecular compound is one molecule. To find the formula weight of a molecular compound, therefore, the weights of atoms shown in the formula are combined exactly as for ionic compounds (Section 5.11). Methane (Figure 6.6), for example, has a formula weight of 16.0 amu: 12.0 amu for one C atom plus (4 × 1.0) amu for four H atoms. (The term *molecular weight,* abbreviated mol wt, is often used in place of "formula weight" when referring to molecular compounds rather than ionic compounds.)

Figure 6.6
Methane. Natural gas, which we burn for cooking, is mostly methane.

Practice Problems

6.7 Hydrogen peroxide is a reactive substance used as a bleach for hair and paper and as a cleansing agent for cuts and scrapes. Its structural formula is H—O—O—H.
(a) Write the molecular formula for hydrogen peroxide
(b) Find its molecular weight.
(c) By comparing numbers of protons and electrons, demonstrate why the peroxide ion, $[O—O]^{2-}$, formed when H—O—O—H loses two H^+ ions, has a charge of -2. (Note how different the properties of water and hydrogen peroxide are, though their structures differ by only one atom.)

6.8 Find the formula weight for chloral hydrate ($C_2Cl_3H_3O_2$), a prescription drug used as a sedative.

6.5 COORDINATE COVALENT BONDS

Coordinate covalent bond
A covalent bond in which both shared electrons are contributed by the same atom.

In the covalent bonds you've seen thus far, each electron in the shared pairs is from a different atom. Sometimes, though, both electrons are donated by the same atom to form what is known as a **coordinate covalent bond.** The ammonium ion, NH_4^+, has this kind of bond. When ammonia (Figure 6.7) reacts in water solution to add a hydrogen ion, H^+, the lone pair on ammonia and the hydrogen ion, which has no valence electrons, unite in a covalent bond.

$$H^+ + \begin{array}{c} H \\ | \\ H—N—H \\ \colon\colon \end{array} \longrightarrow \left[\begin{array}{c} H \\ | \\ H—N—H \\ | \\ H \end{array} \right]^+$$

Figure 6.7
Ammonia. Nitrogen is essential to crops. It can be supplied directly by application of ammonia or in nitrogen-containing fertilizers manufactured from ammonia.

AN APPLICATION: COORDINATE COVALENT BONDS AND YOU

An entire class of substances known as *coordination compounds* is based on the ability of metal ions, especially transition metal ions, to form coordinate covalent bonds with lone pairs on non-metal atoms. A platinum ion, for example, combines with the lone pairs on two ammonia molecules to give a positively charged ion.

$$H_3N: + Pt^{2+} + :NH_3 \rightarrow$$
$$[H_3N \rightarrow Pt \leftarrow NH_3]^{2+}$$

This ion is present in cisplatin, $[Pt(NH_3)_2]Cl_2$, a successful anticancer drug. Compounds with lone pairs that form coordinate covalent bonds to poisonous metal ions are used to remove them from the bloodstream. Many coordination compounds are brightly colored, including the copper compound that gives the color to blue jeans.

The essential metals (Table 1.2) are all present in the body as ions. The activity of magnesium, iron, cobalt, copper, and zinc ions in biochemical reactions is made possible by their formation of coordinate covalent bonds with the lone pairs on nitrogen, oxygen, or sulfur atoms in proteins or other large biomolecules. By coordinating with from two to six different atoms in the same molecule, the metal ions can influence the shapes of large molecules, pulling them into folds and bends. By coordinating with atoms in *different* molecules, they can hold the molecules in

position while chemical reactions take place. For example, without exploring the entire structures of the large protein molecules involved, it's clear that the zinc ion pictured here helps to position the bond about to be broken. The zinc is coordinated to two nitrogen atoms and one oxygen atom in one molecule, and during the reaction forms a fourth coordinate covalent bond to oxygen in the molecule undergoing a change.

Reactant molecule held in place by coordination to Zn^{2+} ion in an enzyme.

Once a coordinate covalent bond has formed, it's no different than any other covalent bond, and all four covalent bonds in $NH_4{}^+$ are identical. For this reason, the origins of bonding electrons are seldom of interest and there's little need to identify coordinate covalent bonds in structural formulas. Sometimes, though, it's done with an arrow that shows where the electron pair came from. For example, nitrous oxide, N_2O, might be written $:N\equiv N\rightarrow \overset{..}{O}:$ instead of $:N\equiv N{-}\overset{..}{\underset{..}{O}}:$. Note here and in the ammonium ion how the coordinate covalent bond results in structures that we wouldn't predict from common bonding patterns, like an N atom with four covalent bonds and an O atom with one covalent bond.

6.6 DRAWING LEWIS STRUCTURES

To draw Lewis structures, the arrangement of the atoms must first be known. Sometimes the arrangement is obvious. Water, for example, can only be H—O—H because only oxygen can be in the middle and form two covalent bonds. Sometimes knowing that hydrogen, oxygen, or halogen atoms usually surround a central atom leads to the correct arrangement. Where the arrangement of the atoms is uncertain, more information must be gathered (but you won't be asked to do this).

Once the arrangement is known, two general approaches can be used for drawing Lewis structures for simple molecules. The first is a broadly applicable stepwise procedure that distributes available electrons among bonds and lone pairs. The second approach applies to molecules in which common bonding patterns are followed by all atoms.

Stepwise Method for Drawing Lewis Structures The stepwise method for drawing Lewis structures is illustrated by finding the structure of PCl_3, in which the chlorine atoms surround the phosphorus atom.

1. *Find the total number of valence electrons in the combined atoms. For polyatomic ions, add one electron for each negative charge or subtract one electron for each positive charge.* Phosphorus (Group 5A) has 5 valence electrons, and chlorine (Group 7A) has 7 valence electrons, giving a total of 26 electrons in PCl_3

$$5 \text{ e}^- + (3 \times 7 \text{ e}^-) = 26 \text{ e}^-$$
$$P \qquad 3 \times Cl \qquad PCl_3$$

2. *Place an electron pair in each covalent bond.* Write the symbols in the predetermined arrangement and connect them with electron-pair bonds:

$$\begin{array}{c} Cl \\ | \\ Cl{-}P{-}Cl \end{array}$$

3. *Place lone pairs so that, except for hydrogen, each nonmetal atom connected to the central atom has an octet.*

$$\begin{array}{c} :\overset{..}{Cl}: \\ | \\ :\overset{..}{\underset{..}{Cl}}{-}P{-}\overset{..}{\underset{..}{Cl}}: \end{array}$$

4. *Place any remaining electrons on the central atom.* Thus far, we have used 24 of the 26 available electrons—6 electrons in three single bonds and 18 electrons in the three lone pairs on each chlorine atom. This leaves two electrons for one lone pair on phosphorus:

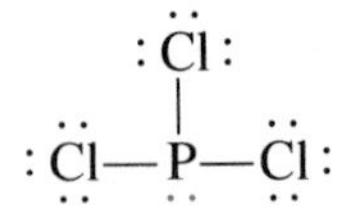

$$: \overset{\displaystyle ..}{\underset{\displaystyle ..}{Cl}} :$$

$$: \overset{..}{\underset{..}{Cl}} - \overset{|}{P} - \overset{..}{\underset{..}{Cl}} :$$

5. *Check to see that each atom has an electron octet or other correct number of valence electrons and that all available electrons have been used. If not, use one or more lone pairs to form multiple bonds to atoms that can have them.* In the structure above, each atom has an octet and all 26 available electrons have been used. The structure is correct.

Solved Problem 6.2   Draw the Lewis structure for sulfur dioxide, SO_2.

Solution   The total number of valence electrons in SO_2 is 18.

$$6\ e^- + (2 \times 6\ e^-) = 18\ e^-$$

$$\text{S} \qquad 2 \times \text{O} \qquad SO_2$$

O—S—O	Placing the sulfur atom in the middle and adding covalent bonds uses 4 e^-.
$:\overset{..}{O}—S—\overset{..}{O}:$	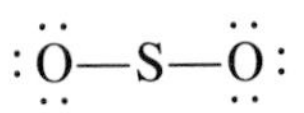 Adding lone pairs on both oxygen atoms uses 12 of the remaining 14 e^-.
$:\overset{..}{\underset{..}{O}}—\overset{..}{S}—\overset{..}{\underset{..}{O}}:$	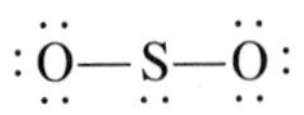 The last electron pair is placed on the central sulfur atom. Checking shows that each oxygen atom has an electron octet, but sulfur has only six electrons.
$:\overset{..}{\underset{..}{O}}—\overset{..}{S}=\overset{..}{O}:$	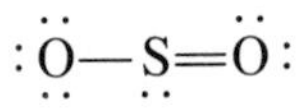 By moving one lone pair from an oxygen atom to a double bond, the oxygen maintains its share in the electron pair and sulfur achieves an octet. (It doesn't matter which side the S=O bond is written on.)

Practice Problems   **6.9**   Draw the Lewis structures of
(a) phosgene, $COCl_2$, a poisonous gas, and
(b) the hypochlorite ion, OCl^-, which is present in many swimming pool chemicals

6.10   Draw the Lewis structures of
(a) carbon monoxide, CO, and
(b) sulfur dichloride, SCl_2

6.11   Draw the Lewis structure for nitric acid, HNO_3, in which the nitrogen atom is in the middle and, as in most such acids, the hydrogen atom is bonded to an oxygen atom.

Structural Formulas Based on Common Bonding Patterns of H, C, O, N, and X (Halogen) Atoms In many molecules, especially the organic molecules of greatest interest in later chapters, hydrogen, carbon, oxygen, nitrogen, and halogen atoms can be expected to have the following common bonding patterns:

- C atoms are often bonded to each other.
- C forms four covalent bonds (and has no lone pairs in neutral molecules).
- N forms three covalent bonds and has one lone pair.
- O forms two covalent bonds and has two lone pairs.
- H forms one covalent bond.
- Halogen atoms (X = F, Cl, Br, I) form one covalent bond and have three lone pairs.

Relying on the common bonding patterns simplifies writing Lewis structures for many molecules containing only these atoms. In ethane, C_2H_6, for example, three of the four covalent bonds of each carbon atom are used in bonds to hydrogen, and the fourth is a carbon–carbon bond. There is no other arrangement in which all eight atoms can have their usual bonding patterns.

$$\begin{array}{cc} H & H \\ | & | \\ H-C-C-H \\ | & | \\ H & H \end{array}$$

Ethane

Condensed structure A structure in which central atoms and the atoms bonded to them are written as groups, for example, CH_3CH_3.

For such organic molecules, Lewis structures become awkward and **condensed structures** are often used. In its condensed form, ethane is written as CH_3CH_3, meaning that each carbon atom has three hydrogen atoms bonded to it (CH_3) and the two CH_3 units are bonded to each other. You'll get a lot more practice with condensed structures in later chapters.

Solved Problem 6.3 Draw the Lewis structure for hydrogen cyanide, HCN, in which the atoms are connected in the order shown.

Solution With the atoms connected in the order H—C—N, the only way that carbon and nitrogen can have their usual numbers of covalent bonds, four for C and three for N, is for them to be joined by a triple bond. Including the lone pair on the triply bonded nitrogen atoms gives the structure

$$H-C\equiv N:$$

To check the structure, we can confirm that all 10 valence electrons (1 e⁻ from H, 4 e⁻ from C, and 5 e⁻ from N) have been used in four covalent bonds and one lone pair.

Solved Problem 6.4   Draw the Lewis structure for chloroethylene, C_2H_3Cl.

Solution   The carbon atoms must be bonded to each other. With only four more atoms available, it's clear that the carbon atoms cannot each have four covalent bonds unless they are joined by a double bond. Putting in the double bond, adding the other three H atoms and one Cl atom, and placing three lone pairs on the chlorine atom gives

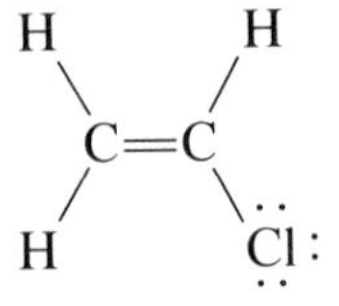

$$\begin{array}{ccc} H & & H \\ \diagdown & & \diagup \\ & C = C & \\ \diagup & & \diagdown \\ H & & \ddot{\underset{\cdot\cdot}{Cl}}\,: \end{array}$$

Checking shows that all 18 e⁻ are accounted for in six covalent bonds and three lone pairs.

Practice Problems   **6.12**   Methylamine, CH_5N, is responsible for the characteristic odor of decaying fish. Draw the Lewis structure of methylamine.

6.13   Draw the Lewis structure of propane (LP gas), used to heat homes in rural areas, which has the formula C_3H_8.

6.14   Formaldehyde, the substance used as a preservative for biological specimens, has the formula HCHO. Draw the structural formula of formaldehyde.

6.15   Add lone pairs where appropriate to the following structures:

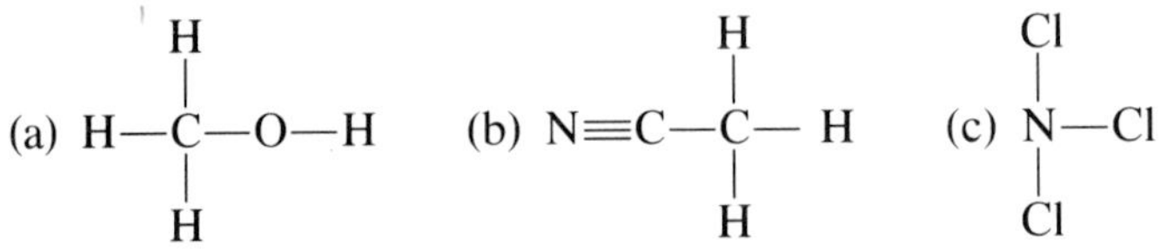

$$\text{(a)}\quad H-\underset{\displaystyle \overset{\displaystyle |}{H}}{\overset{\displaystyle \overset{\displaystyle |}{H}}{C}}-O-H \qquad \text{(b)}\quad N\equiv C-\underset{\displaystyle \overset{\displaystyle |}{H}}{\overset{\displaystyle \overset{\displaystyle |}{H}}{C}}-H \qquad \text{(c)}\quad \underset{\displaystyle \overset{\displaystyle |}{Cl}}{\overset{\displaystyle \overset{\displaystyle |}{Cl}}{N}}-Cl$$

---

† 6.7 THE SHAPES OF MOLECULES

In the introduction to this chapter, we drew a water molecule in a bent shape. What determines this shape? Why are the three atoms connected at an angle rather than in a straight line? Like so many properties, molecular shapes are related to the numbers and locations of the valence electrons in covalent bonds and lone pairs.

The prediction of molecular shape by analyzing how bonds and electron pairs surround individual atoms is based on applying what is known as the **valence-shell electron-pair repulsion** (VSEPR) model. Pairs of valence electrons in bonds and in lone pairs occupy negatively charged electron clouds that stay as far apart as possible, causing molecules to assume specific shapes.

Molecular shapes can be predicted by analyzing a Lewis structure in the following way:

Valence-shell electron-pair repulsion (VSEPR) Repulsion by valence shell electron pairs that maintains them as far apart as possible and determines molecular shape.

• *Identify the atom for which the bonding geometry is of interest.* In a simple molecule like PCl_3 or CO_2 this is usually the central atom.

• *Count the number of electron-charge clouds surrounding the atom of interest in the Lewis structure.* The number of charge clouds is simply the total number of bonds and lone pairs. Multiple bonds count the same as single bonds, because we're interested only in the *number* of charge clouds, not in how many electrons there are in each cloud.

• *Determine the molecular shape according to the number of electron-charge clouds, taking into account the effect of lone pairs on molecular shape.* The possible molecular shapes are summarized in Table 6.2. For example, an atom with two charge clouds and no lone pairs has linear geometry; an atom with three charge

Table 6.2   Molecular Geometry Around Atoms With Two, Three, and Four Charge Clouds

Number of Bonds	Number of Lone Pairs	Number of Charge Clouds on Atom	Shape of Molecule	Example
2	0	2	Linear	
$\begin{bmatrix} 3 \\ \\ 2 \end{bmatrix}$	$\begin{bmatrix} 0 \\ \\ 1 \end{bmatrix}$	3	Planar triangular Angular	
$\begin{bmatrix} 4 \\ 3 \\ 2 \end{bmatrix}$	$\begin{bmatrix} 0 \\ 1 \\ 2 \end{bmatrix}$	4	Tetrahedral Pyramidal Bent	

Linear

Planar triangular

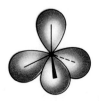

Tetrahedral

Figure 6.8
Electron-pair charge clouds surrounding a bonded atom. The drawings show the arrangements of two, three, and four charge clouds at maximum distance from each other. The charge clouds may be from single, double, or triple bonds, or lone pairs.

Bond angle   The angle formed between any two adjacent covalent bonds.

Tetrahedron   A geometrical figure with four identical triangular faces.

clouds has planar triangular geometry; an atom with three charge clouds and one lone pair has an angular geometry because the lone pair occupies space but does not include any atom that is part of the molecule.

The orientation of two, three, or four charge clouds as far apart as possible from each other is shown in Figure 6.8. If there are only two charge clouds, as occurs on the central atom in such molecules as CO_2 and $BeCl_2$, the clouds are farthest apart when they point in opposite directions. Thus, both $BeCl_2$ and CO_2 are linear molecules with **bond angles** of 180°.

These molecules are linear, with bond angles of 180°.

$$: \overset{..}{Cl} \overset{180°}{\underset{}{-}} Be \overset{..}{-} \overset{..}{Cl} :$$

$$: \overset{..}{O} \overset{180°}{=} C = \overset{..}{O} :$$

When there are three charge clouds, as occurs on the central atoms in BF_3 and SO_2, the clouds can be farthest apart if they lie in a plane and point to the corners of an equilateral triangle. Thus, a BF_3 molecule is planar triangular, with F—B—F bond angles of 120°. In the same way, an SO_2 molecule has a planar triangular arrangement of its three electron clouds, but one point of the triangle is occupied by a lone pair. The relationship of the three atoms themselves is therefore angular rather than linear, and the O—S—O bond angle is approximately 120°.

A BF_3 molecule is planar triangular, with bond angles of 120°.

Top view

Side view

An SO_2 molecule is angular, with a bond angle of approximately 120°.

Top view

Side view

When there are four charge clouds, as occurs on the central atom in CH_4, NH_3, and H_2O, the clouds can be farthest apart when they extend to the corners of a regular tetrahedron. As illustrated in Figure 6.9, a **tetrahedron** is a geometric figure with four identical faces that are equilateral triangles. The central atom lies at the center of the tetrahedron, the charge clouds point to the corners, and the angle between two lines drawn from the center to any two corners is 109.5°.

Because stable outer-shell octets are so common, a great many molecules have geometries based on the tetrahedron. In methane (CH_4), for example, the carbon atom has tetrahedral geometry with H—C—H bond angles of exactly 109.5°. In ammonia (NH_3), the nitrogen atom has a tetrahedral arrangement of its four charge clouds, but one point of the tetrahedron is occupied by a lone pair, resulting in an overall pyramidal shape for the molecule. Similarly, water,

Figure 6.9
Tetrahedral molecular geometry. The central atom is located at the center of the tetrahedron, and the four outer-shell electron pairs are oriented toward the corners. The bond angle between any two pairs is, ideally, 109.5°. The molecular geometry of water, ammonia, methane, and many other molecules is based on the tetrahedron.

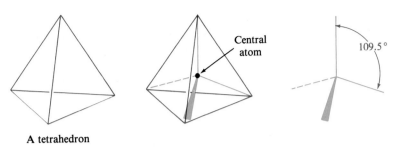

A tetrahedron

which has two points of the tetrahedron occupied by lone pairs, has an overall bent shape.

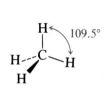

A methane molecule is tetrahedral, with bond angles of 109.5°.

An ammonia molecule is pyramidal, with bond angles of 107°.

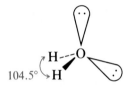

A water molecule is angular, with a bond angle of 104.5°.

Note that the H—N—H bond angle in ammonia (107°) and the H—O—H bond angle in water (104.5°) are close to, but not exactly at, the ideal 109.5° tetrahedral value. The angles are diminished somewhat from their ideal value because the lone-pair charge clouds spread out and compress the rest of the molecule.

The geometry around individual atoms in larger molecules also derives from the shapes shown in Table 6.2. For example, each of the two carbon atoms in ethylene ($H_2C\text{=}CH_2$) is attached to three other atoms, giving rise to planar triangular geometry. The molecule as a whole is planar, with H—C—C and H—C—H bond angles of approximately 120°.

The ethylene molecule is planar, with bond angles of 120°

Top view

Side view

Carbon atoms bonded to four other atoms are each at the center of a tetrahedron, as shown here for ethane, $H_3C\text{—}CH_3$.

The ethane molecule has tetrahedral carbon atoms, with bond angles of 109.5°

Solved Problem 6.5 Predict the geometry of the formaldehyde molecule, HCHO (Practice Problem 6.14).

Solution The Lewis structure of formaldehyde

$$\begin{array}{c} H \\ \diagdown \\ \qquad C{=}\ddot{\ddot{O}} \\ \diagup \\ H \end{array}$$

shows that carbon is the central atom. The carbon atom is surrounded by three charge clouds and no lone pairs. Consulting Table 6.2 shows that the molecule should have the planar triangular shape.

Solved Problem 6.6 What shape do you expect for the phosphine (PH_3) molecule?

Solution Drawing the Lewis structure for phosphine

$$\begin{array}{c} H \\ | \\ H{-}\underset{\cdot\cdot}{P}{-}H \end{array}$$

shows that the phosphorus atom has four charge clouds with one lone pair. Phosphine therefore has the same molecular shape as ammonia: It is pyramidal with bond angles somewhat less than 109.5° due to the effect of the lone pair.

Practice Problems **6.16** What shapes do you predict for the organic molecules chloroform, CH_3Cl, and dichloroethylene, $Cl_2C{=}CH_2$?

6.17 Electron-pair repulsion influences the shapes of polyatomic ions in the same manner as those of neutral molecules. What is the shape of the ammonium ion (NH_4^+)?

6.18 Hydrogen selenide (H_2Se) resembles hydrogen sulfide (H_2S)—both compounds have terrible odors and are poisonous. What are the shapes of these molecules?

6.8 ELECTRONEGATIVITY AND BOND TYPE

Polymer Very large molecule composed of identical repeating units and formed by combination of small molecules.

Electrons in a covalent bond occupy the space between the bonded atoms. If the atoms are identical, as in H_2 and Cl_2, the electrons are equally attracted to each atom and are shared equally. Usually, though, this is *not* the case. Atoms of different elements have different abilities to attract electrons.

AN APPLICATION: HOW BIG CAN A MOLECULE BE?

The answer is, "Very, very big." The really big molecules in our bodies and in many items we buy are all polymers. Like a string of beads, a **polymer** is formed of many repeating units connected in a long chain. Each "bead" in the chain comes from a simple molecule that has formed chemical bonds at both ends, linking it to other molecules. The molecules can be the same:

-a-a-a-a-a-a-a-a-a-a-a-a-a-a-a-a-a-a-a-a-a-a-a-a-a-

or they can be different and connected in an ordered pattern or a random pattern:

-a-b-a-b-a-b-a-b-a-b-a-b-a-b-a-b-a-b-a-b-a-b-

-a-a-a-b-b-b-a-b-a-b-a-b-b-b-a-a-b-b-b-a-b-a-

The polymer chains can have branches hanging down or be connected by the branches to other chains:

-a-a-a-a-a-a-a-a-a-a-
   b   b
   b   b
   b   b

-a-a-a-a-a-a-a-a-a-a-a-a-a-a-a-
   b   b   b   b
   b   b   b   b
-a-a-a-a-a-a-a-a-a-a-a-a-a-a-a-

These are just a few of the many possible variations, some of which are three-dimensional networks.

You use synthetic polymers every day—you call many of them "plastics." Solid plastics are polymers made by combining several thousand to **several hundred thousand molecules,** or "beads," connected to each other. The resulting giant molecules have formula weights ranging from a few thousand atomic mass units to **several million atomic mass units.**

Nature began to exploit the extraordinary variety of polymer properties long before humans did. In fact, despite great progress in recent years, there is still much to be learned about the polymers in living things. Carbohydrates and proteins are polymers, as are the giant molecules of deoxyribonucleic acid (DNA) that govern the reproduction of viruses, bacteria, plants, and everything alive, including us. Nature's polymer molecules are larger (some with formula weights in the billions of atomic mass units) and more complex than any that chemists have yet created.

The polymer structure decorating the edge of this box is that of polyethylene, a polymer made by combination of ethylene molecules (CH_2=CH_2). The double bond becomes a single bond in the polymer. Polyethylene has up to 50,000 repeating molecular units and is used in such items as chairs, toys, water pipes, milk bottles, and packaging films.

H—C—H
H—C—H

One repeating unit of polyethylene

Smaller atoms with higher nuclear charges, such as fluorine and chlorine, attract electrons strongly and are said to be more *electronegative* than other atoms. Fluorine has the most electronegative atoms of all the elements. In a covalent bond such as that in hydrogen fluoride (HF), electrons spend more time at the fluorine end of the bond than at the hydrogen end, creating a **polar covalent bond.** While the molecule as a whole is neutral, one end is more negative than the other. The unequal charge distribution is represented by placing a δ^- (Greek lowercase *delta*) at the more negative end of the bond and a δ^+ at the more positive end of the bond. The symbols represent *partial* charges that are equal but opposite

Polar covalent bond A covalent bond in which one atom attracts bonding electrons more strongly than the other atom.

$$\overset{\delta^+ \quad \delta^-}{H—F}$$

Electronegativity The ability of an atom in a molecule to attract electrons; a numerical value on the electronegativity scale.

The ability of an atom in a covalent bond to attract electrons is its **electronegativity.** A numerical scale that allows comparison of the electron-attracting abilities of different atoms has been established. Linus Pauling, an American chemist, devised this scale early in a distinguished career that has included the Nobel Prize in chemistry in 1954 for his work on protein structure and the Nobel Peace Prize in 1962 for his efforts to limit nuclear weapons testing.

On the Pauling electronegativity scale, fluorine is assigned a value of 4, and less electronegative atoms are assigned lower values. You can see in Figure 6.10 that electronegativity increases across the periods (as atoms get smaller) and decreases down the groups (as atoms get larger). Metals in general have lower electronegativities than nonmetals.

Comparing the electronegativity values of bonded atoms allows the polarity of bonds to be compared and the occurrence of ionic bonding to be predicted. Oxygen (electronegativity 3.5), for example, is more electronegative than nitrogen (3.0), and both are more electronegative than carbon (2.5). Both carbon–oxygen and carbon–nitrogen bonds are therefore polar, with the positive ends at the carbon atoms. The larger difference in electronegativity values shows that the carbon–oxygen bond is the more polar of the two.

Figure 6.10
Electronegativity of the main-group elements.

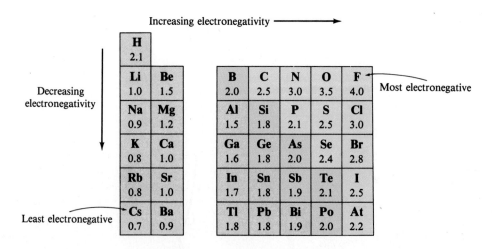

Electronegativity
difference = 0.5
$\overset{\delta^+ \ \ \delta^-}{C\!-\!N}$

$\overset{\delta^+ \ \ \delta^-}{C\!-\!O}$
Electronegativity
difference = 1.0 gives
a more polar bond

No sharp dividing line exists between covalent and ionic bonds; most bonds fall somewhere between 100% covalent and 100% ionic (Figure 6.11). As a rule of thumb, electronegativity differences from 0.4 up to 1.9 indicate increasingly polar covalent bonds, and differences of 2 or more indicate ionic bonds. The electronegativity differences show, for example, that carbon and fluorine form a highly polar covalent bond, sodium and chlorine form an ionic bond, and cesium and fluorine form an ionic bond,

	$\overset{\delta^+ \ \delta^-}{C\!-\!F}$	$Na^+ \ Cl^-$	$Cs^+ \ F^-$
Electronegativity difference =	1.5	2.1	3.3

Nonpolar covalent bond A covalent bond in which the electrons are shared equally.

Cesium and fluorine form just about the closest to 100% ionic bonds that are possible. Bonds between atoms of identical electronegativity, known as **nonpolar covalent bonds,** are 100% covalent, and bonds with electronegativity differences less than 0.4 are generally considered nonpolar.

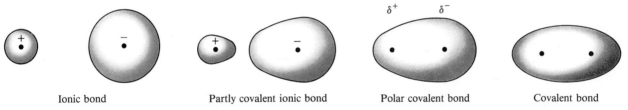

| Ionic bond | Partly covalent ionic bond | Polar covalent bond | Covalent bond |

Figure 6.11
The variation in bonding. A bond may be anywhere on the continuum from fully ionic to fully covalent.

Practice Problems **6.19** Arrange the elements most commonly bonded to carbon in organic compounds, H, N, O, P, and S, in order of increasing electronegativity.

6.20 Use electronegativity differences to classify bonds between the following pairs of atoms as ionic, covalent, or polar covalent:
(a) I and Cl   (b) Li and O   (c) Br and Br   (d) P and Br.

6.21 Using the symbols δ^+ and δ^-, identify the location of the partial charges on the polar covalent bonds formed between
(a) fluorine and sulfur,   (b) phosphorus and oxygen,
(c) arsenic and chlorine.

6.9 POLAR MOLECULES

Just as individual bonds can be polar, whole molecules can be polar if electrons are more attracted to one part of the molecule than another. A molecule that has polar covalent bonds is not necessarily a polar molecule, however. Molecular polarity depends not only on the bonds but also on the shape of the molecule. Neither carbon dioxide (CO_2) nor tetrachloromethane (CCl_4) molecules are polar, for example, despite their polar C—O and C—Cl bonds. The direction of bond polarity is often represented by arrows pointing toward where the electrons are most strongly attracted. These arrows, pointed at the negative end and crossed at the positive end, make it apparent that the individual bond polarities cancel out in the symmetrical CO_2 and CCl_4 molecules.

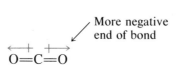

More negative end of bond

CO_2, nonpolar because individual bond polarities cancel

CCl_4, nonpolar because individual bond polarities cancel

Unlike tetrachloromethane, though, the chloromethane (CH_3Cl) molecule is polar because the individual bond polarities don't cancel each other. The C—H bonds are much less polar than the C—Cl bond.

CH_3Cl
polar molecule

Net polarity of molecule

In water and ammonia molecules, the polarities of the bonds to hydrogen plus the electron density of the lone pairs combine to make the oxygen and nitrogen ends of the molecules strongly negative.

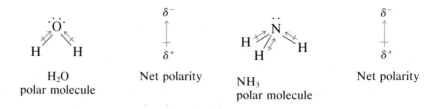

H_2O
polar molecule

Net polarity

NH_3
polar molecule

Net polarity

Practice Problem **6.22** For each of the compounds described in Solved Problems 6.5 and 6.6 (PH_3 and HCHO), decide whether the molecules are polar and show the direction of net polarity.

6.10 NAMING BINARY MOLECULAR COMPOUNDS

Binary compound A compound composed of atoms of two elements.

Atoms of two different elements combine to give what is called a **binary compound.** Whether ionic or molecular, the formulas of binary compounds are usually written with the less electronegative element first. Thus, the symbols for metals or nonmetals furthest left in the periodic table come first, as in

$$KF \qquad CrO_3 \qquad BCl_3 \qquad NO$$

The common names of binary molecular compounds are based on using the prefixes listed in Table 6.3 according to two rules:

1. *Name the first element in the formula with the element name, plus a prefix if needed.*

2. *Name the second element in the formula with the element name modified with -ide (as for ionic compounds, Section 5.7), plus a prefix if needed.*

The prefixes indicate how many atoms of each element are combined. *Mono-,* indicating one, is omitted except where needed to distinguish between two different compounds between the same elements. For example, the two oxides of carbon receive the names shown here. (Note that adjacent vowels are avoided by omitting the final vowel in the prefix before "oxide.")

$$CO \qquad\qquad CO_2$$

Carbon *mono*xide   Carbon *di*oxide

Some other examples are

$$N_2O_5 \qquad\qquad PCl_3 \qquad\qquad NI_3$$

Dinitrogen *pent*oxide   Phosphorus *tri*chloride   Nitrogen *tri*iodide

Table 6.3 Prefixes Used in Chemical Names

Number	Prefix
1	mono
2	di
3	tri
4	tetra
5	penta
6	hexa
7	hepta
8	octa
9	nona
10	deca

Practice Problems

6.23 Name
(a) S_2Cl_2,   (b) ICl,   (c) ICl_3

6.24 Write the formulas for
(a) selenium tetrafluoride,   (b) diphosphorus pentoxide, and
(c) bromine trifluoride

6.11 PROPERTIES OF MOLECULAR COMPOUNDS

You've seen that the strong force of attraction between positive and negative ions is responsible for the high melting and boiling points of ionic compounds (Section 5.3). *Molecules,* however, are neutral overall, and there is no force of comparable strength to hold groups of molecules together. Thus, many observable properties of molecular compounds arise not from covalent bonding forces, but from the weaker forces that act between molecules. The properties of ionic and molecular compounds are compared in Table 6.4.

Intermolecular forces
Forces of attraction and repulsion between partial charges in different molecules.

Intermolecular forces depend on the attraction between positive and negative ends of polar molecules and on momentary variations in electron density in neighboring molecules. Whether a substance exists as a gas, a liquid, or a solid at a given temperature depends on its intermolecular forces, as explained further in Chapter 9. Where intermolecular forces are weak, molecules are so little attracted to each other that the substances are gases at ordinary temperatures. Somewhat stronger intermolecular forces pull molecules together into liquids, and still stronger forces exist in molecular solids (Figure 6.12). Even so, the melting points and boiling points of molecular solids are usually significantly lower than those of ionic solids. Some molecular solids are soft and waxy rather than crystalline. In general, the higher the formula weight, the higher the melting point of a molecular compound.

Table 6.4   A Comparison of Ionic and Molecular Compounds

Ionic Compounds	Molecular Compounds
Smallest particles are ions	Smallest particles are molecules
Usually composed of metals combined with nonmetals	Usually composed of nonmetals with nonmetals
Crystalline solids	Gases, liquids, or low-melting solids
High melting points	Low melting points
High boiling points (above 700°C)	Low boiling points
Conduct electricity when molten or dissolved in water	Do not conduct electricity
Many are water-soluble	Few are water-soluble
Not soluble in organic liquids	Many soluble in organic liquids

Figure 6.12
Some molecular solids.

6.12   IONS FROM MOLECULAR COMPOUNDS: AN INTRODUCTION TO ACIDS AND BASES

Imagine a neutral water molecule, H—O—H. What would happen if you were to remove one hydrogen atom but leave behind the electron pair of the former O—H bond? The hydrogen you removed would bear a positive charge since it consists of only a proton (the H nucleus) without any electrons. In other words, it would be an H^+ ion. Conversely, the O—H fragment would have to bear a negative charge because an electron has been left behind, creating an OH^- ion.

$$H\!:\!\overset{..}{\underset{..}{O}}\!:\!H \;\rightarrow\; H^+ \;+\; {}^-\!:\!\overset{..}{\underset{..}{O}}\!:\!H$$

The oxygen atom has an outer-shell octet of electrons and carries a negative charge.

The hydrogen atom has *no* electrons and carries a positive charge.

Acid   A substance that is able to donate a hydrogen ion, H^+.

The hydroxide ion is, of course, one of the many polyatomic ions present in ionic compounds. Other polyatomic ions are formed in a similar manner when the molecular compounds known as **acids** give up hydrogen ions (H^+) when they're dissolved in water. The anions that remain include polyatomic ions like those listed in Table 6.5.

Hydrochloric acid, for example, is the water solution that forms when gaseous HCl is dissolved in water. The polar water molecules attract hydrogen ions away from the positive ends of the molecules to form hydronium ions, H_3O^+.

$$HCl + H_2O \;\rightarrow\; H_3O^+ + Cl^-$$

A water solution containing H_3O^+ and Cl^- is known as hydrochloric acid.

The H_3O^+ ion and acids that yield it in solution play crucial roles in all kinds of chemistry, and we'll have much more to say about them, including their

Table 6.5   Acids and Their Polyatomic Ions

Acids		Polyatomic Ions	
Acetic acid	CH_3COOH	Acetate ion	CH_3COO^{-a}
Carbonic acid	H_2CO_3	Bicarbonate ion (hydrogen carbonate ion)	HCO_3^-
Nitric acid	HNO_3	Nitrate ion	NO_3^-
Nitrous acid	HNO_2	Nitrite ion	NO_2^-
Phosphoric acid	H_3PO_4	Dihydrogen phosphate ion	$H_2PO_4^-$
		Hydrogen phosphate ion	HPO_4^{2-}
		Phosphate ion	PO_4^{3-}
Sulfuric acid	H_2SO_4	Hydrogen sulfate ion	HSO_4^-
		Sulfate ion	SO_4^{2-}
Hydrochloric acid	HCl	Chloride ion	Cl^-

^{a}Sometimes written $C_2H_3O_2^-$.

functions in living systems. You'll probably be using acids in the laboratory soon, and recognizing the formulas of the common acids in Table 6.5 will be useful.

Base A substance that can accept a hydrogen ion (H^+) from an acid.

The compounds known as **bases** are related to acids because they add the hydrogen ions released by acids. Ammonia, NH_3, is acting as a base when it adds a hydrogen ion and is converted into the ammonium ion, NH_4^+. Neutral organic compounds such as methylamine, CH_3NH_2, can also add a hydrogen ion to their nitrogen atoms in the same way to give $CH_3NH_3^+$ and are therefore bases.

Many biomolecules consist of ionic groups bonded to large organic molecules. Three especially important ions that you'll see often in biochemistry are the $—NH_3^+$ group, the $—COO^-$ group, and the $—OPO_3^{2-}$ group (derived from phosphoric acid, H_3PO_4).

$$\text{Organic molecule}-NH_2 + H^+ \longrightarrow \text{organic molecule}-NH_3^+$$

$$\text{Organic molecule}-\overset{\displaystyle O}{\overset{\|}{C}}-O-H \longrightarrow \text{organic molecule}-\overset{\displaystyle O}{\overset{\|}{C}}-O^- + H^+$$

$$\text{Organic molecule}-O-\underset{\underset{\displaystyle OH}{|}}{\overset{\overset{\displaystyle O}{\|}}{P}}-OH \longrightarrow \text{organic molecule}-O-\underset{\underset{\displaystyle O^-}{|}}{\overset{\overset{\displaystyle O}{\|}}{P}}-O^- + 2\,H^+$$

SUMMARY

Molecular compounds are composed of **molecules,** in which atoms are joined together by shared electron-pair or **covalent bonds.** Atoms that share one electron pair are joined by a **single bond** (such as C—C); atoms that share two electron pairs are joined by a **double bond** (such as C=C); and atoms that share three electron pairs are joined by a **triple bond** (such as C≡C). Depending on their numbers of valence electrons, different atoms usually form different numbers of covalent bonds (such as C, four bonds; N, three bonds; O, two bonds; H, one bond). Unshared electron pairs in molecules, called **lone pairs,** can sometimes provide both bonding electrons to form **coordinate covalent bonds.**

Structural formulas are used to show how atoms are connected in molecules, and **Lewis structures** show all valence electrons by including electron dots. Lewis structures are drawn by counting the total number of valence electrons in a molecule or polyatomic ion and then placing bonding pairs and shared pairs so that all electrons are accounted for.

Covalent bonds in molecules have specific orientations based on the repulsion between surrounding elec-tron-charge clouds (Table 6.2). Molecules with four electron pairs, for example, may be tetrahedral (no lone pairs), pyramidal (one lone pair), or bent (two lone pairs).

Electronegative elements are those that strongly attract electrons in bonds. All ionic and covalent bonds fall on a continuum from nonpolar covalent to ionic. Comparisons of **electronegativity** values allow predictions of where bonds fall on that continuum and whether bonds will be nonpolar covalent, polar covalent, or ionic. Molecules containing polar bonds can be polar overall but are nonpolar if the bonds are arranged symmetrically so that the contributions of individual bonds cancel out.

Many properties of molecular compounds are dependent on **intermolecular forces,** which are weaker than bonding forces. Molecular compounds may be gases, liquids, or low-melting solids and usually have lower melting and boiling points than ionic compounds. Molecular compounds do not conduct electricity, except that solutions of those that produce ions when dissolved in water, notably acids such as HCl and H_2SO_4, do so.

INTERLUDE: DIAMOND AND GRAPHITE

Diamond and graphite are both made of pure carbon, but what a difference there is between them. Diamond, a rare, colorless gemstone, is one of the hardest known materials. Graphite, a flaky, black substance, is so soft that it's used as a lubricant and as the "lead" in lead pencils.

The differences between diamond and graphite are due to their different structures (see the accompanying figure). A crystal of diamond is essentially one enormous molecule. Each carbon atom in the crystal is covalently bonded to four neighboring carbons with exactly the same tetrahedral geometry as that in methane. The result is an immense three-dimensional array of interlocked atoms that gives diamond its great strength and hardness. Graphite, however, consists of two-dimensional sheets of carbon atoms stacked on top of each other in layers. Each carbon atom is joined to only three other carbons and has a double bond. Because the sheets are not bonded together, they can slip and slide over one another, making graphite a good lubricant.

Scientists have known for nearly 40 years that diamonds (like those pictured here) can be made synthetically by subjecting graphite to very high pressures at temperatures around 1800°C. More recently, though, methods have been discovered for depositing an extremely thin film of diamond onto a variety of other materials to give these materials a nearly indestructible protective coat. Coated knives and scalpels should remain forever sharp, coated glass lenses should remain forever scratchproof, and coated bearings should never wear out. These and other applications may soon become reality.

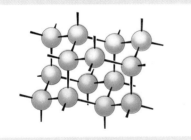

Diamond

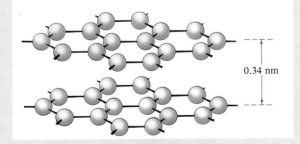

0.34 nm

Graphite

Synthetic diamonds.

REVIEW PROBLEMS

Covalent Bonds

6.25 What is a covalent bond? Describe the changes in potential energy that occur as two species approach each other to form a covalent bond.

6.26 Which of the following elements would you expect to form (i) diatomic molecules, (ii) mainly covalent bonds, (iii) mainly ionic bonds, (iv) both covalent and ionic bonds (more than one may apply)?
(a) oxygen (b) potassium (c) phosphorus
(d) iodine (e) hydrogen (f) cesium

6.27 Which of the following pairs of atoms form covalent bonds?
(a) calcium and fluorine
(b) carbon and fluorine
(c) magnesium and fluorine
(d) silicon and fluorine
(e) lithium and fluorine

6.28 Look up tellurium (atomic number 52) in the periodic table and predict how many covalent bonds it commonly forms. Explain.

6.29 Germanium (atomic number 32) is an element used in the manufacture of transistors. Judging from its position in the periodic table, how many covalent bonds does it usually form?

6.30 Identify four elements whose bonding patterns are exceptions to the octet rule and explain why for each.

6.31 How many covalent bonds are formed by each atom in the following molecules?
(a) NCl_3 (b) HI (c) $CHCl_3$ (d) PBr_5

6.32 Which of the following contains a coordinate covalent bond?
(a) $PbCl_2$ (b) $Cu(NH_3)_4^{2+}$ (c) NH_4^+
(d) H_2O (e) HOClO (f) H_3O^+

6.33 Tin forms both ionic and covalent compounds. Is the molecular compound of tin more likely to be $SnCl_3$, $SnCl_4$, or $SnCl_5$? Why?

Structural Formulas

6.34 Distinguish between:
(a) a molecular formula and a structural formula
(b) a structural formula and a condensed structure
(c) a lone pair and a shared pair of electrons

6.35 Give the total number of valence electrons in the following:
(a) N_2 (b) CO (c) CH_3CH_2CHO (d) OF_2

6.36 Add lone pairs where appropriate to each of the following structures:
(a) C≡O (b) CH_3SH

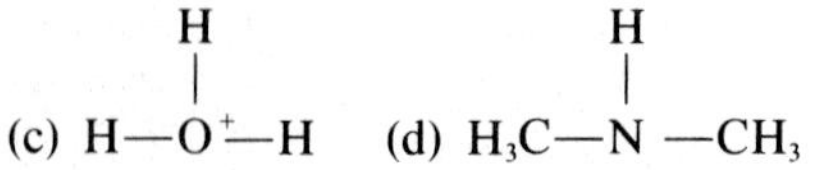

(c) H—O⁺—H (d) H_3C—N—CH_3 (with H atoms above O and N)

6.37 If a research paper appeared reporting the structure of a new molecule with formula C_2H_8, most chemists would be highly skeptical. Why?

6.38 Which of the following three possible structural formulas for $C_3H_6O_2$ is correct? Explain.

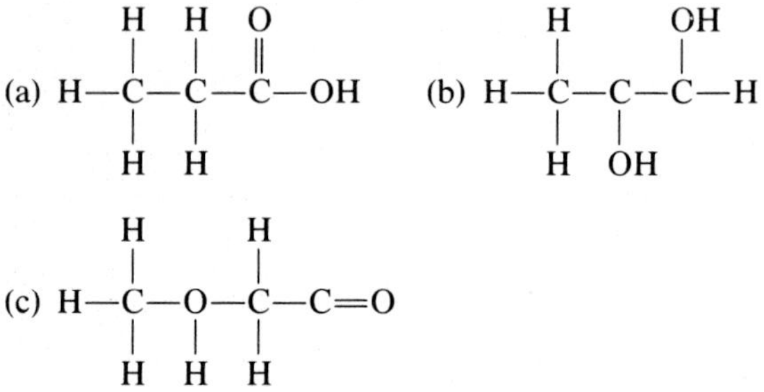

(a) structure; (b) structure; (c) structure

6.39 Convert the following Lewis structures into structural formulas in which lines replace the bonding electrons. Include the lone pairs.
(a) H:Ö:N::Ö:
(b) H:C:C:::N: (with H's on C)
(c) H:F:

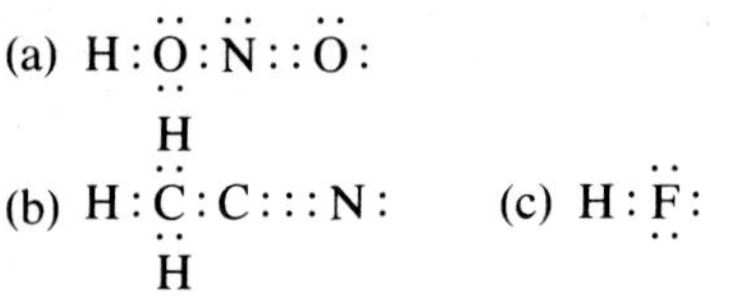

6.40 Convert the following structural formulas into condensed structures:
(a), (b), (c) structures

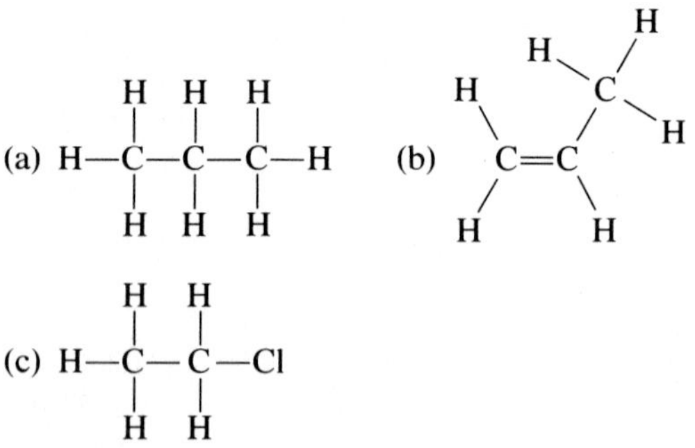

6.41 Convert the following Lewis structure for the nitrate ion into a line structure that includes the lone pairs. Why does the nitrate ion have a −1 charge?

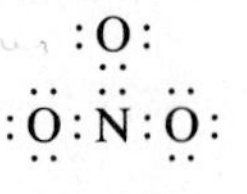

:Ö:N:Ö: with O above

Drawing Lewis Structures

6.42 Draw the Lewis structures for each of the following:
(a) SiF_4 (b) $AlCl_3$ (c) CF_2Cl_2 (d) SO_3
(e) BBr_3 (f) NF_3

6.43 Draw Lewis structures for each of the following:
(a) nitrous acid, HNO_2 (H is bonded to an O)
(b) ozone, O_3
(c) acetaldehyde, CH_3CHO (H is not bonded to O)

6.44 Ethanol, or "grain alcohol," has the formula C_2H_5OH. Propose a structure for ethanol that is consistent with common bonding patterns.

6.45 Dimethyl ether has the same molecular formula, C_2H_6O, but very different properties from ethanol (Problem 6.44). Propose a structure for dimethyl ether in which the oxygen is bonded to two carbons.

6.46 Hydrazine, a substance used to make rocket fuel, has the formula N_2H_4. Propose a structure for hydrazine.

6.47 Tetrachloroethylene, C_2Cl_4, is used commercially as a dry cleaning solvent. Propose a structure for tetrachloroethylene based on the common bonding patterns expected in organic molecules. What kind of carbon–carbon bond is present?

6.48 Draw the Lewis structure for carbon disulfide, CS_2, a foul-smelling liquid used as a solvent for fats. What kind of carbon–sulfur bonds are present?

6.49 The discovery in the 1960s that xenon and fluorine react to form a molecular compound was a surprise to most chemists, because it had been thought that noble gases could not form any bonds.
(a) Why was it thought that noble gases could not form bonds?
(b) Draw the Lewis structure of XeF_4.

6.50 Draw Lewis structures for the following polyatomic ions:
(a) formate, HCO_2^- (b) carbonate, CO_3^{2-}
(c) sulfite, SO_3^{2-} (d) thiocyanate, SCN^-

Molecular Geometry

6.51 Explain how molecular geometry can be predicted based on the numbers of shared and lone electron pairs around an atom.

6.52 Predict the geometry around atom A for molecules with the general formulas AB_4, AB_3E, and AB_2E, where E represents an unshared electron pair. What values do you expect for the bond angles in these molecules?

6.53 Sketch the three-dimensional shape of:
(a) chloroform, $CHCl_3$ (b) hydrogen sulfide, H_2S
(c) ozone, O_3 (d) carbon disulfide, CS_2
(e) nitrous acid, HNO_2

6.54 Classify each of the molecules in Problem 6.42 according to the numbers of bonds and lone pairs of electrons around the central atom and predict the shapes of the molecules.

6.55 Predict the geometry around each carbon atom in acetaldehyde (Problem 6.43c).

Polarity of Bonds and Molecules

6.56 Why are most bonds neither 100% covalent nor 100% ionic?

6.57 Which of the following bonds are polar? Identify the negative and positive ends of the bonds by using δ^- and δ^+.
(a) I—Br   (b) O—H   (c) C—F   (d) N—C
(e) C—C

6.58 Based on electronegativity differences, would you expect bonds between the following pairs of atoms to be ionic or covalent?
(a) Na and F   (b) C and Cl
(c) N and H   (d) Be and Br

6.59 Arrange the following in order of the increasing polarity of their bonds:
(a) HCl   (b) PH_3 (c) H_2O (d) CF_4

6.60 Is the planar triangular molecule boron trichloride (BCl_3) polar? Explain.

6.61 Decide whether or not each of the compounds listed in Problem 6.59 is polar and show the direction of polarity.

Names, Formulas, and Properties of Molecular Compounds

6.62 Name the following binary compounds:
(a) PI_3 (b) $AsCl_3$ (c) P_4S_3 (d)Al_2F_6
(e) NI_3 (f) IF_7

6.63 Write formulas for the following compounds:
(a) nitrogen dioxide
(b) sulfur hexafluoride
(c) bromine pentaiodide
(d) tetraphosphorus decaoxide
(e) dinitrogen tetraoxide
(f) arsenic pentachloride

6.64 For ionic and molecular compounds, compare the following:
(a) the nature of the smallest particles
(b) melting points
(c) existence as gases, liquids, or solids

Applications

6.65 What is a coordinate covalent bond? [App: Coordinate Covalent Bonds]

6.66 What metals are particularly associated with coordinate covalent bond formation in biological molecules? [App: Coordinate Covalent Bonds]

6.67 How is a polymer formed? [App: How Big?]

6.68 How are diamonds and graphite similar? How do they differ in structure and in physical properties? [Int: Diamond and Graphite]

6.69 What are two uses each of graphite and diamond? [Int: Diamond and Graphite]

Additional Questions and Problems

6.70 Acetone, a common solvent used in some nail polish removers, has the molecular formula C_3H_6O and contains a carbon–oxygen double bond.
(a) Propose two possible Lewis structures for acetone.
(b) What is the geometry around the carbon atoms in each of the structures?
(c) Which of the bonds in each structure are polar?
(d) What is the formula weight of acetone?

6.71 The following formulas are all unlikely to be correct. What is wrong with each?
(a) CCl_3 (b) N_2H_5 (c) H_3S (d) C_2OS

6.72 Count the number of protons and electrons present in the cyanide ion and show why it has a negative charge.

$$:C:::N:^-\quad\text{cyanide ion}$$

6.73 Which of the compounds (a) through (d) contain one or more of the following: (i) ionic bonds, (ii) covalent bonds, (iii) coordinate covalent bonds?
(a) $CaCl_2$ (b) $Mg(NO_3)_2$ (c) BF_4^- (d) CBr_4

6.74 The phosphonium ion (PH_4^+) is formed by reaction of phosphine (PH_3) with an acid.
(a) Draw the Lewis structure of the phosphonium ion.
(b) Predict its molecular geometry.
(c) Describe how a fourth hydrogen can be added to PH_3.
(d) Explain why the ion has a $+1$ charge.

6.75 Name the following compounds. Be sure to determine whether the compound is ionic or covalent, so that you use the proper naming rules.
(a) $CaCl_2$ (b) $TeCl_2$ (c) BF_3 (d) $MgSO_4$
(e) K_2O (f) FeF_3 (g) PF_3

6.76 Find the formula weight of each compound in Problem 6.75.

6.77 Name these common acids:
(a) H_3PO_4 (b) H_2CO_3 (c) HNO_3 (d) H_2SO_4

6.78 Draw the Lewis structure for hydroxylamine, NH_2OH. Would you expect this molecule to be polar or nonpolar? Why?

CHAPTER

7

Chemical Reactions: Classification and Mass Relationships

The chemical engineers who designed this equipment for biopharmaceutical manufacture had to know the chemical reactions that occur in the process and had to calculate the masses of materials to be handled. This is one of the many practical applications of the topics covered in this chapter.

An oyster makes a pearl, a marigold seed grows into a flowering plant, a log burns in the fireplace—these changes are all the result of chemical reactions. The study of how and why chemical reactions happen is a major part of chemistry and provides information that is both fascinating and practical. Thus far, you've seen that in a chemical reaction one or more substances change identity. In this chapter we'll begin to look more closely at chemical reactions. First we'll deal with a familiar problem in chemistry: how to represent what we're discussing on paper. Next we'll describe the mass relationships among substances involved in chemical reactions, which must be understood to put chemical reactions to practical use. Then we'll introduce you to a few easily recognized classes of chemical reactions. Among the questions we'll answer are the following:

1. **How are chemical reactions represented in writing?** The goal: Given the identities of reactants and products, be able to write a balanced chemical equation (or net ionic equation); given a chemical equation, be able to interpret it.

2. **Why is the mole such a useful unit in chemistry?** The goal: Be able to explain the meaning and uses of the mole and Avogadro's number.

3. **How are molar quantities and mass quantities related?** The goal: Be able to convert between molar and mass quantities of any element or compound.

4. **What is the quantitative significance of chemical equations?** The goal: Be able to carry out mole–mole, mass–mole, and mass–mass calculations for quantities of reactants and products, and to calculate percent yield.

5. **How are chemical reactions classified?** The goal: Be able to recognize combination, displacement, decomposition, exchange, and redox reactions.

6. **When do exchange reactions occur?** The goal: Given the reactants, be able to predict whether an exchange reaction will occur and identify the products.

7. **What are redox reactions?** The goal: Be able to explain the various ways of recognizing redox reactions and to identify the substances oxidized and reduced in a given reaction.

7.1 CHEMICAL EQUATIONS

Heating sodium bicarbonate produces water, carbon dioxide, and a white solid, sodium carbonate.

$$\text{Sodium bicarbonate} \xrightarrow{\text{heat}} \text{sodium carbonate} + \text{water} + \text{carbon dioxide}$$

Chemical equation A written expression in which symbols and formulas are used to represent reactants and products, and which is balanced.

Replacing the chemical names with chemical formulas converts the word description of this reaction into a **chemical equation:**

$$2\ NaHCO_3 \xrightarrow{\text{heat}} Na_2CO_3 + H_2O + CO_2$$

Reactant A starting substance that undergoes change in a chemical reaction.

The formulas for **reactants** are always written to the left of the arrow, and the formulas for **products** to the right of the arrow. Conditions necessary for a reaction to occur are often specified above the arrow, as is done here to show that sodium bicarbonate must be heated.

Product A substance formed as the result of a chemical reaction.

Why is the number 2 added in front of $NaHCO_3$ in the equation? Without the 2 this equation would be a wrong statement because it would represent violation of a fundamental law of nature:

Law of conservation of mass Matter can be neither created nor destroyed in any physical or chemical change.

Law of conservation of mass: Matter can be neither created nor destroyed in any physical or chemical change.

In chemical reactions, the atoms of the reactants are rearranged, but all of them must appear in the products and no new atoms can be added. (Nuclear reactions are, as you've seen, the single exception to this law.) Therefore, we must always write a **balanced chemical equation** in which the number of atoms of each kind is the same on both sides of the arrow.

Balanced chemical equation A chemical equation in which the numbers of atoms of each kind are the same in the reactants and products and, for a net ionic equation, charge is also balanced.

The numbers placed in front of formulas to balance equations are called **coefficients,** and they multiply all the atoms in a formula. Thus, 2 $NaHCO_3$ indicates two formula units of sodium bicarbonate, which contain 2 Na atoms, 2 H atoms, 2 C atoms, and 6 O atoms ($2 \times 3 = 6$, the coefficient times the subscript for O). Count the number of atoms on the right side of the equation above to convince yourself that it is indeed balanced.

Coefficient A number placed before a formula in a chemical equation to show how many formula units of that substance are required to balance the equation.

The substances in chemical reactions may be solids, liquids, or gases, or they may be in **aqueous solution,** that is, dissolved in water. Sometimes this information is added to an equation by placing the following symbols after the chemical formulas:

Aqueous solution A solution in which water is the solvent.

(g)	(l)	(s)	(aq)
Gas	Liquid	Solid	In water, or aqueous, solution

For the decomposition of solid sodium bicarbonate,

$$2\ NaHCO_3(s) \xrightarrow{\text{heat}} Na_2CO_3(s) + H_2O(l) + CO_2(g)$$

Solved Problem 7.1 Interpret in a sentence this equation for a reaction used in extracting lead from its ores and show that the equation is balanced.

$$2\ PbS(s) + 3\ O_2(g) \rightarrow 2\ PbO(s) + 2\ SO_2(g)$$

Solution The equation is read as follows:

Solid lead sulfide plus gaseous oxygen yields solid lead oxide plus gaseous sulfur dioxide.

To show that the equation is balanced, count the atoms of each type on each side of the arrow:

On the left:	2 Pb	2 S	$3 \times 2 = 6\ O$
On the right:	2 Pb	2 S	$2\ O + (2 \times 2)\ O = 6\ O$
			From 2 PbO From 2 SO$_2$

The numbers of atoms of each type are the same in the reactants and products, so the equation is balanced.

Practice Problems

7.1 Determine which of these equations are balanced:
(a) $HCl + KOH \xrightarrow{yield} H_2O + KCl$
(b) $CH_4 + Cl_2 \rightarrow CH_2Cl_2 + HCl$
(c) $H_2O + MgO \rightarrow Mg(OH)_2$
(d) $Al(OH)_3 + H_3PO_4 \rightarrow AlPO_4 + 2\ H_2O$

7.2 Interpret each of the following equations:
(a) $CoCl_2(s) + 2\ HF(g) \rightarrow CoF_2(s) + 2\ HCl(g)$
(b) $Pb(NO_3)_2(aq) + 2\ KI(aq) \rightarrow PbI_2(s) + 2\ KNO_3(aq)$

7.2 BALANCING CHEMICAL EQUATIONS

Balancing chemical equations takes a mixture of common sense and trial and error. It can be done in four steps:

1. Write an unbalanced equation with the reactants on the left and the products on the right, using the correct formulas for all substances involved. For example, hydrogen and oxygen must be written as H_2 and O_2, rather than as H and O, since we know that hydrogen and oxygen exist as diatomic molecules. **The subscripts in chemical formulas must never be changed in balancing an equation, since doing so would mean a change in the reaction itself.**

2. One by one, balance the number of atoms of each type. It helps to begin with elements that appear in only one formula on each side of the equation, which usually means leaving oxygen and hydrogen until last. If a polyatomic ion appears on both sides of an equation, it is treated like a single type of atom. For example, in the following equation,

$$H_2SO_4 + 2\,NaOH \rightarrow Na_2SO_4 + 2\,H_2O$$

the sulfate ion (SO_4^{2-}) is balanced because there is one on the left and one on the right.

3. *Check the equation to make sure the numbers and kinds of atoms on both sides of the equation are the same.*

4. *When the equation is balanced, make sure the coefficients are reduced to their lowest whole-number values.*

Solved Problem 7.2   Natural gas (methane, CH_4) burns in oxygen to yield water and carbon dioxide (CO_2). Write a balanced equation for the reaction.

Solution   *Step 1. Write the unbalanced equation using correct formulas for all substances:*

$$CH_4 + O_2 \rightarrow CO_2 + H_2O \qquad \text{(Unbalanced)}$$

Step 2. One by one, balance the atoms of each type: First, note that carbon appears in one formula on each side of the arrow. Since there is one carbon atom in each formula, the equation is already balanced with respect to carbon. Next, note that there are four hydrogen atoms on the left (in CH_4) and only two on the right (in H_2O). Placing a 2 before H_2O gives the same numbers of hydrogen atoms on both sides:

$$CH_4 + O_2 \rightarrow CO_2 + 2\,H_2O \qquad \text{(Balanced for C and H)}$$

Now look at the number of oxygen atoms. There are two on the left (in O_2) but four on the right (two in CO_2 and one in each of two H_2O's). If we place a 2 before the O_2, the numbers of oxygen atoms on both sides are the same:

$$CH_4 + 2\,O_2 \rightarrow CO_2 + 2\,H_2O \qquad \text{(Completely balanced for C, H, and O)}$$

Step 3. Check to be sure the numbers of atoms on both sides are the same.

On the left:	1 C	4 H	$(2 \times 2)\,O = 4\,O$
On the right:	1 C	$2 \times 2 = 4\,H$	$2\,O + 2\,O = 4\,O$

From CO_2 From 2 H_2O

Step 4. Make sure the coefficients are reduced to their lowest whole-number values. In fact, our answer is already correct, but we might, through trial and error, have arrived at a different balanced equation:

$$2\,CH_4 + 4\,O_2 \rightarrow 2\,CO_2 + 4\,H_2O$$

Although this equation is balanced, the coefficients are not the lowest whole numbers. Dividing all the coefficients by 2 is necessary to reach the final, correct equation.

Solved Problem 7.3   Potassium chlorate ($KClO_3$) decomposes when heated to yield potassium chloride and oxygen, a reaction used to provide oxygen for the emergency breathing masks in airliners. Write a balanced equation for this reaction.

Solution   The unbalanced equation is

$$KClO_3 \rightarrow KCl + O_2$$

The equation is already balanced for K and Cl but is unbalanced for O. In order to balance O, it's necessary to put a coefficient of 2 before $KClO_3$ and a coefficient of 3 before O_2, giving 6 O atoms on both sides of the equation:

$$2\,KClO_3 \rightarrow KCl + 3\,O_2 \quad \text{(Balanced for O)}$$

Unfortunately, when we balanced the equation for O, we *unbalanced* it for K and Cl. Thus, we have to rebalance for K and Cl by placing a coefficient of 2 before KCl.

$$2\,KClO_3 \rightarrow 2\,KCl + 3\,O_2 \quad \text{(Balanced for K, Cl, and O)}$$

Checking gives

On the left:	2 K	2 Cl	$(2 \times 3)\,O = 6\,O$
On the right:	2 K	2 Cl	$(3 \times 2)\,O = 6\,O$

Practice Problems

7.3   Write a balanced equation for the reaction of metallic sodium with chlorine to yield sodium chloride.

7.4   Ozone (O_3) is formed in the earth's upper atmosphere by the action of solar radiation on oxygen molecules. Write a balanced equation for the formation of ozone from oxygen.

7.5   Balance these equations:
(a) $Ca(OH)_2 + 2\,HCl \rightarrow CaCl_2 + 2\,H_2O$
(b) $4\,Al + 3\,O_2 \rightarrow 2\,Al_2O_3$
(c) $2\,CH_3CH_3 + 7\,O_2 \rightarrow 4\,CO_2 + 6\,H_2O$
(d) $2\,AgNO_3 + MgCl_2 \rightarrow 2\,AgCl + Mg(NO_3)_2$

7.3 AVOGADRO'S NUMBER AND THE MOLE

Imagine that you want to do a laboratory experiment by combining hydrogen and oxygen to form water. How much hydrogen and how much oxygen should you use? According to the balanced equation, it takes two molecules of hydrogen and one molecule of oxygen to make two molecules of water.

$$2\,H_2(g) + O_2(g) \rightarrow 2\,H_2O(l)$$

Figure 7.1
Counting by weighing. Two pounds of jelly beans would contain twice as many jelly beans as one pound.

In reality, of course, individual molecules are so tiny that we can see them only with very specialized instruments. Thus, a *visible* chemical reaction must involve reacting many *quadrillions* of hydrogen molecules in a 2:1 ratio with many quadrillions of oxygen molecules to yield many quadrillions of water molecules.

2:1 Ratio of H_2 to O_2

$$2 \text{ H}_2 \text{ molecules} + 1 \text{ O}_2 \text{ molecule} \rightarrow 2 \text{ H}_2\text{O molecules}$$

$$1000 \text{ H}_2 \text{ molecules} + 500 \text{ O}_2 \text{ molecules} \rightarrow 1000 \text{ H}_2\text{O molecules}$$

$$2 \times 10^{20} \text{ H}_2 \text{ molecules} + 1 \times 10^{20} \text{ O}_2 \text{ molecules} \rightarrow 2 \times 10^{20} \text{ H}_2\text{O molecules}$$

How, though, can you be sure you have the correct ratio of hydrogen and oxygen molecules in your reaction vessel? Clearly, it's impossible to count out the right number of molecules. You might decide instead to weigh out the hydrogen and oxygen. This is a common method of dealing with small objects—quantities of potatoes, jelly beans, and nails for construction are all weighed rather than counted out (Figure 7.1).

But the weighing approach leads to yet another problem. How many molecules are there in a gram of hydrogen or oxygen? You need to know this to weigh out enough hydrogen and oxygen. A convenient solution to this problem is provided by the masses of substances on the atomic mass unit scale. (You might want to review Section 3.2 at this point.)

The atomic weight of a hydrogen *atom* is 1.01 amu, and the formula weight of an H_2 *molecule* is 2×1.01 amu, or 2.02 amu. Similarly, the formula weight of an O_2 molecule is 2×16.0 amu, or 32.0 amu. Since the weight ratio of a *single* H_2 molecule to a *single* O_2 molecule is 2:32, it follows that the weight ratio of *any* given number of H_2 molecules to the same number of O_2 molecules will always be 2:32. In other words, *whenever* the weight ratio of an H_2 sample to an O_2 sample is 2:32, the samples will contain the same number of molecules (Figure 7.2).

A convenient way to use this relationship is to take amounts in grams that are numerically equal to formula weights. If, for example, you carry out your experiment with the following quantities of hydrogen and oxygen

H_2: $2 \times$ formula weight in grams $= 2 \times 2.0$ g $= 4.0$ g H_2 Two times as many H_2 molecules as O_2 molecules

O_2: $1 \times$ formula weight in grams $= 32.0$ g O_2

you could be certain that you have the correct ratio of reactants.

Figure 7.2
Weight ratios and numbers of molecules. Equal numbers of H_2 and O_2 molecules always have a weight ratio equal to the ratio of their formula weights (2:32).

O_2 molecules (heavier)

H_2 molecules (lighter)

The mass relationships between hydrogen and oxygen illustrate a general principle: *A weight in grams equal to the formula weight of any compound contains the same number of formula units as a weight in grams equal to the formula weight of any other compound.* (Perhaps you discovered this when you did Practice Problems 3.3 and 3.4.) This number, which has the value 6.02×10^{23}, is known as **Avogadro's number,** after Amadeo Avogadro, an Italian scientist who in the early 1800's recognized the relationship between mass and the number of formula units.

Avogadro's number   The number of particles (atoms, formula units, ions, or molecules) in one mole (6.02×10^{23}).

How big is Avogadro's number? Our minds can't really conceive of the magnitude of a number like 6.02×10^{23}. Consider that if you were to measure the total amount of water in all the world's oceans, the answer would be approximately Avogadro's number of milliliters.

By defining a quantity based on Avogadro's number, it's possible to count atoms, ions, or molecules by weighing. The quantity of any substance that contains Avogadro's number of units is called a **mole.** The units might be atoms, formula units, molecules, or anything else of interest. In practice, one mole of an element or compound is equal to the atomic or formula weight of the substance in grams instead of atomic mass units (Figure 7.3).

Mole (mol)   Amount of a substance containing an Avogadro's number of particles; in practice, usually a quantity in grams equal to the formula weight of the substance.

The mole, abbreviated mol, is simply a counting unit like the dozen or the gross. One mole of ping pong balls is 6.02×10^{23} ping pong balls, one mole of bowling balls is 6.02×10^{23} bowling balls, and one mole of molecules is 6.02×10^{23} molecules. A mole of bowling balls is much *heavier* than a mole of ping pong balls, but the *number* of balls in one mole of each is the same. Similarly, one mole of oxygen molecules is heavier than one mole of hydrogen molecules (32.0 g versus 2.0 g), but the number of molecules is 6.02×10^{23} in each.

7.4   GRAM–MOLE CONVERSIONS

Molar mass   The mass, usually in grams, of one mole of a substance.

The mass in grams of one mole of a substance is called the **molar mass** of that substance. To find the molar mass, the formula weight (Section 5.11) must first be known. For example, the formula weight of water is 18.0 amu, and therefore the molar mass of water is 18.0 g:

$$1 \text{ mol } H_2O = 18.0 \text{ g } H_2O \qquad \text{Molar mass of water}$$

Figure 7.3
One mole quantities of (clockwise from the top) sulfur, sugar, mercury, copper, and water.

The molar mass provides factors for converting moles of water to grams of water, or vice versa. Suppose we need to know how much 0.25 mol of water weighs:

Molar mass used as conversion factor

$$0.25 \; \text{mol H}_2\text{O} \times \frac{18.0 \; \text{g H}_2\text{O}}{1 \; \text{mol H}_2\text{O}} = 4.5 \; \text{g H}_2\text{O}$$

Alternatively, suppose we need to know how many moles of water are equal to 27 g of water:

Molar mass used as conversion factor

$$27 \; \text{g H}_2\text{O} \times \frac{1 \; \text{mol H}_2\text{O}}{18.0 \; \text{g H}_2\text{O}} = 1.5 \; \text{mol H}_2\text{O}$$

Some examples of the relationships between formula weights, molar quantities, and numbers of particles are given in Table 7.1. Because the mole is such a convenient unit in chemistry, it's often also used for concentrations. For example, a solution with a concentration of 1 mol/L of NaCl (formula weight 58.5 amu) contains 58.5 g of NaCl per 1 L of solution.

Table 7.1   Molar Quantities of Elements and Compounds

One Mole of—	Weighs—	Contains—
Sodium (Na)	23.0 g	6.02×10^{23} Na atoms
Chlorine (Cl_2)	70.9 g	6.02×10^{23} Cl_2 molecules
Sodium chloride (NaCl)	58.5 g	6.02×10^{23} NaCl formula units
		6.02×10^{23} Na^+ ions
		6.02×10^{23} Cl^- ions
Methane (CH_4)	16.0 g	6.02×10^{23} CH_4 molecules
		6.02×10^{23} C atoms
		$4 \times (6.02 \times 10^{23})$ H atoms
Water (H_2O)	18.0 g	6.02×10^{23} H_2O molecules
		$2 \times (6.02 \times 10^{23})$ H atoms
		6.02×10^{23} O atoms

Solved Problem 7.4 Many nonprescription pain relievers contain ibuprofen ($C_{13}H_{18}O_2$, formula weight 206.3 amu), which in higher doses is a prescription drug (Motrin). If the tablets in a bottle of pain reliever contain a total of 0.082 mol of ibuprofen, what is the total number of grams of ibuprofen present in the tablets?

Solution The molar mass of ibuprofen is 206.3 g. Using the molar mass as a conversion factor gives

$$0.082 \; \text{mol } C_{13}H_{18}O_2 \times \frac{206.3 \; \text{g } C_{13}H_{18}O_2}{1 \; \text{mol } C_{13}H_{18}O_2} = 17 \; \text{g } C_{13}H_{18}O_2$$

Ballpark Solution The answer should be somewhat less than $0.1 \times 200 = 20$.

Solved Problem 7.5   The maximum dose of sodium hydrogen phosphate (Na_2HPO_4, formula weight 142 amu) that should be taken in one day for use as a laxative is 3.8 g. How many moles of sodium hydrogen phosphate, how many moles of Na^+ ions, and how many moles of ions in total are in this dose?

Solution   The molar mass of Na_2HPO_4 is 142 g. Using the molar mass as a conversion factor gives

$$3.8 \text{ g } \cancel{Na_2HPO_4} \times \frac{1 \text{ mol } Na_2HPO_4}{142 \text{ g } \cancel{Na_2HPO_4}} = 0.027 \text{ mol } Na_2HPO_4$$

The chemical formula Na_2HPO_4 shows that each formula unit contains two Na^+ ions and one HPO_4^{2-}, and therefore each mole contains 2 mol of Na^+ ions, 1 mol of HPO_4^{2-} ions, and a total of 3 mol of ions.

$$\frac{2 \text{ mol } Na^+}{1 \text{ mol } \cancel{Na_2HPO_4}} \times 0.027 \text{ mol } \cancel{Na_2HPO_4} = 0.054 \text{ mol } Na^+$$

$$\frac{3 \text{ mol ions}}{1 \text{ mol } \cancel{Na_2HPO_4}} \times 0.027 \text{ mol } \cancel{Na_2HPO_4} = 0.081 \text{ mol ions}$$

Ballpark Solution   Since in finding moles of Na_2HPO_4, 3.8/142 is roughly 5/150 = 1/30, the answer should be roughly 0.03. In finding moles of ions, the answers should be close to $2 \times 0.025 = 0.050$ and $3 \times 0.025 = 0.075$.

Practice Problems   **7.6**   How many moles are in a 10.0 g sample of ethyl alcohol, C_2H_6O? How many grams are in a 0.10 mol sample of ethyl alcohol?

7.7   How many carbon atoms, hydrogen atoms, and oxygen atoms are present in 0.025 mol of aspirin ($C_9H_8O_4$)?

7.8   How many grams of NaOH are present in 1.0 L of a solution that has a concentration of 1.5 mol NaOH/L?

7.5   MOLE RELATIONSHIPS FROM CHEMICAL EQUATIONS

The mole is put to practical use in interpreting chemical equations. Because of the way the mole is defined as a counting unit for atoms, molecules, and ions, the *coefficients in chemical equations show the relative numbers of moles of reactants and products.*

According to the equation

$$2 \text{ H}_2\text{O}_2(aq) \quad \rightarrow \quad 2 \text{ H}_2\text{O}(l) \quad + \quad \text{O}_2(g)$$

$$\text{2 mol H}_2\text{O}_2 \qquad\qquad \text{2 mol H}_2\text{O} \qquad \text{1 mol O}_2$$

the decomposition of every 2 mol of hydrogen peroxide gives 2 mol of water and 1 mol of oxygen. Furthermore, for every 2 mol of water formed, 1 mol of oxygen is formed

$$\frac{2 \text{ mol } H_2O_2}{2 \text{ mol } H_2O} \qquad \frac{2 \text{ mol } H_2O_2}{1 \text{ mol } O_2} \qquad \frac{2 \text{ mol } H_2O}{1 \text{ mol } O_2}$$

Mole ratio A ratio of the molar amounts of substances shown by the coefficients in a balanced chemical equation.

By using such **mole ratios** in factor-label calculations, the molar amounts of any reactants or products are easily determined.

Mole ratios also help to solve the problem that arises when two reactants are mixed in quantities that do not match the molar amounts given by the equation coefficients. Suppose the reaction of sodium with chlorine

$$2 \text{ Na}(s) + Cl_2(g) \rightarrow 2 \text{ NaCl}(s)$$

is carried out by mixing 0.35 mol of Na and 0.75 mol of Cl_2. To find out how much NaCl can be formed, it's necessary to know which is the **limiting reactant.** Using the mole ratios gives the amounts of reactants required for complete reaction by either 0.35 mol of Na or 0.75 mol of Cl_2:

Limiting reactant The reactant present in an amount that limits the extent to which a reaction can occur.

$$0.35 \text{ mol Na} \times \frac{1 \text{ mol } Cl_2}{2 \text{ mol Na}} = 0.18 \text{ mol } Cl_2$$

$$0.75 \text{ mol } Cl_2 \times \frac{2 \text{ mol Na}}{1 \text{ mol } Cl_2} = 1.5 \text{ mol Na}$$

Sodium is clearly the limiting reactant: It requires less Cl_2 than is available, while Cl_2 requires more Na than is available. Therefore, the maximum amount of NaCl that can be formed is

$$0.35 \text{ mol Na} \times \frac{2 \text{ mol NaCl}}{2 \text{ mol Na}} = 0.35 \text{ mol NaCl}$$

Solved Problem 7.6 Iron and oxygen combine to form an iron oxide in the following manner:

$$4 \text{ Fe}(s) + 3 \text{ O}_2(g) \rightarrow 2 \text{ Fe}_2O_3(s)$$

(a) What are the mole ratios of the product to each reactant and of the reactants to each other?

(b) How many moles of iron(III) oxide are formed by the complete oxidation of 6.2 mol of iron?

Solution (a) The coefficients show the mole ratios:

$$\frac{2 \text{ mol } Fe_2O_3}{4 \text{ mol Fe}} \qquad \frac{2 \text{ mol } Fe_2O_3}{3 \text{ mol } O_2} \qquad \frac{4 \text{ mol Fe}}{3 \text{ mol } O_2}$$

(b) To find the needed moles of Fe_2O_3, begin by writing down the known information—moles of iron—and then select the mole ratio that will allow quantities to cancel, leaving the desired quantity:

$$6.20 \ \overline{\text{mol Fe}} \times \frac{2 \ \text{mol Fe}_2O_3}{4 \ \overline{\text{mol Fe}}} = 3.10 \ \text{mol Fe}_2O_3$$

Note that mole ratios are exact numbers and therefore don't limit the number of significant digits in the result of a calculation.

Ballpark Solution   The answer for part (b) should be roughly one-half of 6, or 3, in good agreement with the answer. In such problems, the cancellation of labels to give the desired quantity verifies that the answer was found in the correct manner.

Practice Problems   **7.9**   (a) Balance the following equation and determine how many moles of nickel will react with 9.81 mol of hydrochloric acid.

$$Ni(s) + HCl(aq) \rightarrow NiCl_2(aq) + H_2(g)$$

(b) What is the maximum number of moles of $NiCl_2$ that can be formed in the reaction of 6.0 mol of Ni and 6.0 mol of HCl?

7.10   In the process of photosynthesis, plants convert carbon dioxide and water to glucose ($C_6H_{12}O_6$) and oxygen. Write a balanced equation for this reaction and determine how many moles of carbon dioxide are required to produce 15.0 mol of glucose.

7.6 MASS RELATIONSHIPS FROM CHEMICAL EQUATIONS

Mole ratios permit calculation of the molar amounts of reactants and products, but actual amounts of substances must be weighed out in grams. Suppose A and B are substances involved in a chemical reaction—whether they're reactants or products doesn't matter. Often, the known information is the quantity of A and the needed information is the quantity of B.

● **Mole–mole problem** Known: moles of A. Needed: moles of B
● **Mole–mass (or mass–mole) problem** Known: moles (or mass) of A. Needed: mass (or moles) of B
● **Mass–mass problem:** Known: mass of A. Needed: mass of B

You've learned that known moles of A can be converted into moles of B by using the mole ratio (Solved Problem 7.6 is a mole–mole problem).

Mole ratio as conversion factor

Moles A   →   moles B

If the mass of B is needed, the molar mass of B must be used to complete the calculation:

Mole ratio as conversion factor

Molar mass as conversion factor

$$\text{Moles A} \;\to\; \text{moles B} \;\to\; \text{mass B}$$

But if you know the mass of A and need to know the mass of B, then *both* molar masses must be included in the calculation:

Molar mass as conversion factor

Mole ratio as conversion factor

Molar mass as conversion factor

$$\text{Mass A} \;\to\; \text{moles A} \;\to\; \text{moles B} \;\to\; \text{mass B}$$

The paths that connect known and needed information about amounts of substances in chemical reactions are diagramed in Figure 7.4. In all cases, you must begin with the balanced chemical equation and, no matter what the known information is, you must convert it to moles and use a mole ratio to find the needed information. The steps for determining mass relationships among reactants and products, even in very complex problems, are always as follows:

1. *Write the balanced chemical equation.*
2. *Choose molar masses and mole ratios to convert the known information into the needed information.*
3. *Set up the factor-label expression and calculate the answer.*
4. *Check the answer by a ballpark solution.*

Figure 7.4
Flow diagram summarizing conversions between moles and grams in finding quantities of reactants and products for chemical reactions. Mole ratios are needed as conversion factors in all these calculations, and molar masses are needed in those where the known and/ or unknown information is a mass.

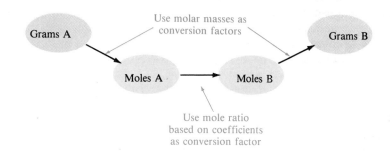

Grams A

Use molar masses as conversion factors

Grams B

Moles A

Moles B

Use mole ratio based on coefficients as conversion factor

Solved Problem 7.7   In the atmosphere, nitrogen dioxide reacts with water to produce nitric acid, which contributes to pollution by acid rain:

$$3\,NO_2(g) + H_2O(l) \;\to\; 2\,HNO_3(aq) + NO(g)$$

How many grams of HNO_3 are produced for every 1.0 mol of NO_2 that reacts in this manner?

Solution   This is a moles (1.0 mol NO_2) to mass (? g HNO_3) problem, so the molar mass of HNO_3 and the mole ratio of HNO_3 to NO_2 reacted are needed in the calculation.

$$1.0\ \text{mol NO}_2 \times \frac{2\ \text{mol HNO}_3}{3\ \text{mol NO}_2} \times \frac{63\ \text{g HNO}_3}{1\ \text{mol HNO}_3} = 42\ \text{g HNO}_3$$

Ballpark Solution   The answer is two-thirds of 63, which is exactly 42, and the labels cancel correctly.

Solved Problem 7.8 One method for determining quantities of substances in solution is by converting the substance to a solid and then weighing the solid. If the following reaction gave 0.022 g of solid calcium oxalate (CaC_2O_4), what mass of calcium chloride was present in the original sample of $CaCl_2$?

$$CaCl_2(aq) + Na_2C_2O_4(aq) \rightarrow CaC_2O_4(s) + 2\ NaCl(aq)$$

Solution Both the known information and that to be found are masses, so this is a mass–mass problem.

$$0.022\ \text{g } CaC_2O_4 \times \frac{1\ \text{mol } CaC_2O_4}{128\ \text{g } CaC_2O_4} \times \frac{1\ \text{mol } CaCl_2}{1\ \text{mol } CaC_2O_4} \times \frac{111\ \text{g } CaCl_2}{1\ \text{mol } CaCl_2} = 0.019\ \text{g } CaCl_2$$

Ballpark Solution Recognizing that 111/128 is slightly less than 1 shows that the answer should be slightly less than 0.022.

Practice Problem 7.11 Hydrogen fluoride is one of the few substances that react with quartz, which is silicon dioxide.

$$SiO_2(s) + 4\ HF(g) \rightarrow SiF_4(g) + 2\ H_2O(l)$$

(a) How many moles of HF will react completely with 9.90 mol of SiO_2?
(b) What mass in grams of water will be produced by the reaction of 23 g of SiO_2?

7.7 PERCENT YIELD

Calculations of amounts of products based on mole ratios all assume 100% of the reactants are converted to products. Only rarely is this the case in practice. The difference between the maximum possible amount of a product—the **theoretical yield**—and the amount actually obtained is given by the **percent yield** of a reaction:

Theoretical yield The calculated maximum amount of product that can be formed from a given amount of reactant.

Percent yield The percent of the theoretical yield actually obtained from a chemical reaction.

$$\text{Percent yield} = \frac{\text{actual yield}}{\text{theoretical yield}} \times 100$$

The actual yield is found by weighing the amount of product obtained. The theoretical yield is found by a mass–mass calculation like those illustrated in the preceding section.

Solved Problem 7.9 The element boron is produced by the reduction of boric oxide by magnesium.

$$B_2O_3(s) + 3\ Mg(s) \rightarrow 2\ B(s) + 3\ MgO(s)$$

If 675 g of boron was obtained from the reduction of 2350 g of boric oxide, what was the percent yield?

Solution The theoretical yield of boron from the reduction of 2350 g of boric oxide is 729 g:

$$2350 \text{ g B}_2\text{O}_3 \times \frac{1 \text{ mol B}_2\text{O}_3}{69.6 \text{ g B}_2\text{O}_3} \times \frac{2 \text{ mol B}}{1 \text{ mol B}_2\text{O}_3} \times \frac{10.8 \text{ g B}}{1 \text{ mol B}} = 729 \text{ g B}$$

The percent yield is the percent of the theoretical yield actually obtained.

$$\text{Percent yield of B} = \frac{675 \text{ g B}}{729 \text{ g B}} \times 100 = 92.6\%$$

Ballpark Solution The theoretical yield should be approximately 40 (from 2400/60) × 2 × 10, or 800. Because the theoretical and actual yield are both close to 700, it's also reasonable that the percent yield should be close to 100%.

Practice Problem 7.12 What is the percent yield in the formation of 19.4 g of ammonium chloride in the combination reaction of 15.0 g of hydrogen chloride with ammonia?

7.8 TYPES OF CHEMICAL REACTIONS

Inorganic compound A chemical compound composed of elements other than carbon; not an organic compound based on a carbon structure.

Combination reaction A reaction in which two reactants combine to give a single product: A + B → C.

Decomposition reaction A reaction in which a single compound breaks down into other substances, often after energy has been supplied as heat: C → A + B.

Displacement reaction A reaction in which one element takes the place of a less reactive element in a compound: A + BC → B + AC.

Exchange reaction A reaction in which two compounds exchange partners: AB + XY → AY + XB.

Without some guidelines on distinguishing features to look for, the variety of chemical reactions is bewildering. In this and the next few sections we're going to describe a few simple types of reactions of elements and **inorganic compounds.**

Chemistry is an experimental science, and it's important to remember that writing a balanced equation for a reaction of a common type doesn't mean that the reaction will occur. It's known from experiments, for example, that many metals combine with oxygen in reactions such as

$$2 \text{ Mg}(s) + \text{O}_2(g) \rightarrow 2 \text{ MgO}(s)$$

The equally sensible-looking equation for the reaction of gold with oxygen is *wrong*, however,

$$4 \text{ Au}(s) + 3 \text{ O}_2(g) \rightarrow 2 \text{ Au}_2\text{O}_3(s) \quad \text{Does not occur}$$

because gold is so unreactive that this reaction does not happen.

The distinctive patterns of the reactions we'll be describing are summarized in Table 7.2, where each letter in the generalized equations stands for an element, an ion, or a compound. The reactions in Table 7.2 are

- The **combination** of elements to form compounds
- The **decomposition** of single compounds to yield two or more products
- The **displacement** of an element from a compound by another element
- **Exchange reactions** (also called double decomposition or metathesis reactions) between ions in solution, in which two compounds exchange "partners."

Table 7.2   Classification of Some Reactions of Elements and Compounds

Reactants	Reaction Type	Comments
Two elements	Combination $A + B \rightarrow C$	Usually metal plus nonmetal or nonmetal plus nonmetal. O_2 and halogens combine with many other elements. Element (often O_2 or halogen) and compound can also combine.
Element plus compound	Displacement $A + BC \rightarrow B + AC$	Displacement of metal by metal and displacement of H_2 from acid by metal are common.
Single compound	Decomposition $C \rightarrow A + B$	Requires added energy unless compound is unstable.
Ions in aqueous solution	Exchange $AB + XY \rightarrow AY + XB$	Occurs if gas, solid, water (or other nonionic compound) can form. Includes acid–base (neutralization) reactions.

In addition, in Sections 7.13 and 7.14, we'll describe *redox reactions*, a large and varied class of reactions with no single reaction pattern.

Figure 7.5
Reaction of zinc with hydrochloric acid, a displacement reaction between an element and a compound.

7.9 REACTIONS OF ELEMENTS: COMBINATION AND DISPLACEMENT

All but the least reactive elements undergo combination reactions with other elements. The reactive nonmetals—oxygen and the halogens—combine with most other elements in reactions such as

$$Sn(s) + 2\,Cl_2(g) \rightarrow SnCl_4(l) \qquad \text{Combination of metal and nonmetal}$$
$$4\,P(s) + 5\,O_2(g) \rightarrow P_4O_{10}(s) \qquad \text{Combination of nonmetal and nonmetal}$$

Combination reactions are not limited to elements. For example, oxygen, fluorine, or chlorine often add to a compound, as in the conversion of sulfur dioxide to sulfur trioxide:

$$2\,SO_2(g) + O_2(g) \rightarrow 2\,SO_3(g) \qquad \text{Combination of element and compound}$$

More commonly, though, the reaction of an element with a compound is a displacement reaction, in which a reactive element takes the place of a less reactive element in the compound. Metals can be displaced by more reactive metals (Figure 7.5):

$$Cu(s) + 2\,AgNO_3(aq) \rightarrow 2\,Ag(s) + Cu(NO_3)_2(aq) \qquad \text{Copper displaces silver}$$

and the hydrogen in an acid can be displaced by reactive metals:

$$Zn(s) + 2\ HCl(aq) \rightarrow H_2(g) + ZnCl_2(aq)$$

Displacement of hydrogen from acid by metal

Practice Problems **7.13** Classify each of the following reactions as combination or displacement:
(a) $CaO(s) + H_2O(l) \rightarrow Ca(OH)_2(aq)$
(b) $Fe(s) + H_2SO_4(aq) \rightarrow FeSO_4(aq) + H_2(g)$
(c) $Fe(s) + Cl_2(g) \rightarrow FeCl_2(s)$

7.14 Write a balanced equation for the displacement reaction of magnesium with zinc chloride in aqueous solution.

7.10 REACTIONS OF COMPOUNDS: DECOMPOSITION

When there is only one reactant, the only possible simple reaction is a decomposition reaction. How easily decomposition occurs depends on the stability of the compound. Carbonic acid (H_2CO_3) is so unstable that most of it decomposes to carbon dioxide and water as soon as it's formed at room temperature.

$$H_2CO_3(aq) \rightarrow CO_2(g) + H_2O(l)$$ Decomposition

At the opposite extreme, very stable compounds such as water and sodium chloride can only be decomposed by passing an electrical current through the compound in a process called *electrolysis:*

$$2\ NaCl(l) \xrightarrow{\text{electrolysis}} 2\ Na(l) + Cl_2(g)$$ Decomposition

Heating causes decomposition of the carbonates and hydroxides of all except the most active metals:

$$FeCO_3(s) \xrightarrow{\text{heat}} FeO(s) + CO_2(g)$$ Decomposition
$$Ba(OH)_2(s) \xrightarrow{\text{heat}} BaO(s) + H_2O(l)$$ Decomposition

Often, as in the two reactions above, thermal decomposition results in the loss of a simple molecule as a gas (Figure 7.6).

Practice Problems **7.15** Write a balanced equation for the thermal decomposition of copper(II) carbonate.

7.16 Oxides of the less reactive elements decompose into their elements when heated. Write a balanced equation for the decomposition of mercury(II) oxide (HgO).

Figure 7.6
Dynamite explosion. The pressure and shock wave in a dynamite explosion are caused by the sudden release of a total of 29 mol of CO_2, H_2O, N_2, and O_2 gases for every 4 mol of nitroglycerin ($C_3H_5N_3O_9$) that decomposes.

AN APPLICATION: WHAT IS IT AND HOW MUCH OF IT IS THERE?

Analytical chemistry is concerned with answering questions like those posed in the title. Doing so today often involves some combination of chemical reactions and electronic instruments. Before chemists, physicists, and engineers learned to make such wonderful instruments, however, chemical reactions were the only tools available.

Using chemical reactions for *qualitative analysis,* which answers the What is it? question, requires that the changes caused by the reactions can be seen. Visible changes that are useful in analysis include solid formation, gas evolution, and color changes. The accompanying photo shows some changes caused by reactions so distinctive that they leave little doubt of the presence of specific ions in a sample being analyzed. Although today a complex mixture such as a metal alloy or seawater would rarely be analyzed

solely by such tests, it can be done and was relied on in the past.

Quantitative analysis is concerned with measuring the actual amounts or concentrations of substances in a sample. Food chemists must measure concentrations of contaminants, manufacturing chemists must measure concentrations of reactants and products, and much of the work in a hospital clinical laboratory is quantitative analysis. Instruments for routine blood analysis have been automated to the extent that they can carry out 400 or more tests an hour. Many of these tests rely on color changes or precipitation reactions just like the old qualitative analysis tests. The difference is that subtle increases or decreases in the intensity of color or the cloudiness of solutions are detected electronically and interpreted quantitatively.

7.11   REACTIONS OF IONS IN SOLUTION: EXCHANGE REACTIONS

Precipitation Reactions   When aqueous solutions of two ionic compounds are mixed, a reaction will occur if one product removes ions from solution. Think about what might happen when solutions of barium nitrate $(Ba(NO_3)_2)$ and sodium sulfate (Na_2SO_4) are combined. Unless some sort of reaction occurs, the Ba^{2+}, NO_3^-, Na^+, and SO_4^{2-} ions will just drift around in the new solution. Because barium sulfate $(BaSO_4)$ is insoluble in water, however, the Ba^{2+} and SO_4^{2-} ions immediately combine, and the following reaction takes place:

$$Ba(NO_3)_2(aq) + Na_2SO_4(aq) \;\rightarrow\; BaSO_4(s) + 2\,NaNO_3(aq)$$

Exchange with formation of precipitate

Precipitate   A solid that forms during a reaction in solution.

Precipitation   The formation of a solid during a reaction in solution.

The barium and sulfate ions are removed as solid barium sulfate (Figure 7.7), but the sodium and nitrate ions remain in solution.

A solid that forms during a reaction in solution is called a **precipitate.** Prediction of when **precipitation** will occur is possible if the solubilities of the compounds that might form are known. Some general solubility rules are given in Table 7.3, and learning them is very useful.

Four visible changes used to identify ions in qualitative analysis (left to right): Al^{3+} gives a red precipitate with an organic reagent; CrO_4^{2-} gives a yellow precipitate of BaCrO; CrO_4^{2-} is oxidized to deep blue CrO_5 dissolved in the organic solvent layer; Zn^{2+} gives a white precipitate of ZnS.

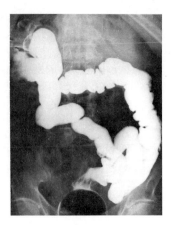

Figure 7.7
Precipitation of barium sulfate, an exchange reaction. Because barium sulfate is opaque to X rays and passes through the body unchanged, it is used to make the gastrointestinal tract visible in X ray radiography.

Table 7.3   General Rules for Water Solubility of Ionic Compounds

Mostly Soluble	Important Exceptions
Ammonium (NH_4^+) salts	None
Potassium (K^+) salts	None
Sodium (Na^+) salts	None
Acetates ($CH_3CO_2^-$)	$AgCH_3CO_2$ is only moderately soluble in hot water
Chlorides (Cl^-) ⎫ Bromides (Br^-) ⎬ Iodides (I^-) ⎭	Compounds with Ag^+, Hg_2^{2+}, Pb^{2+} are insoluble, but $PbCl_2$ is soluble in hot water
Nitrates (NO_3^-)	None
Sulfates (SO_4^{2-})	Compounds with Ba^{2+}, Sr^{2+}, Pb^{2+}, Hg_2^{2+} are insoluble

Mostly Insoluble	Important Exceptions
Carbonates	Compounds with NH_4^+, Li^+, Na^+, K^+ are soluble
Hydroxides	Compounds with Na^+, K^+, Ba^{2+} are soluble, and compounds with Ca^{2+} and Ag^+ are slightly soluble
Phosphates	Compounds with NH_4^+, Li^+, Na^+, K^+ are soluble
Sulfides	Compounds with NH_4^+, Na^+, K^+ are soluble

What might happen when solutions of sodium nitrate ($NaNO_3$) and potassium sulfate (K_2SO_4) are mixed? To answer this question, check the solubilities of both possible exchange products, Na_2SO_4 and KNO_3, in Table 7.3. Because both are water-soluble, no reaction will occur.

Acid A substance that is able to donate a hydrogen ion, H^+.

Acid–Base Reactions When an **acid** reacts with a metal hydroxide in aqueous solution, the products are a salt and water, for example,

$$HCl(aq) + NaOH(aq) \rightarrow NaCl(aq) + H_2O(l)$$

Acid   Base   Salt   Water   Exchange with formation of water

Base A substance that can accept a hydrogen ion (H^+) from an acid.

Salt An ionic substance composed of the cation of a base and the anion of an acid, such as NaCl, $BaSO_4$.

Sodium hydroxide and other compounds containing the OH^- ion are one important class of **bases.** In this type of acid–base reaction, the hydrogen of the acid combines with the OH^- to form water in an exchange reaction that removes the H^+ and OH^- ions as water molecules. Ionic compounds like NaCl composed of the cations of bases and the anions of acids are known as **salts.**

Gas-forming Reactions Production of a gas that bubbles out of solution removes ions as effectively as formation of precipitates or water. For example, reaction of hydrochloric acid with sulfides like magnesium sulfide produces hydrogen sulfide, a molecular compound that is a gas and takes H^+ and S^{2-} out of solution.

$$MgS(s) + 2\ HCl(aq) \rightarrow MgCl_2(aq) + H_2S(g)$$

Exchange with formation of a gas

[Because H_2S is toxic, reactions that produce it must be carried out in a hood.]
The reaction of carbonates with acids is also driven forward by gas formation. The exchange reaction first forms carbonic acid, H_2CO_3, but this unstable compound immediately breaks down to give carbon dioxide and water, so these are the actual reaction products, for example,

$$Na_2CO_3(aq) + 2\ HCl(aq) \rightarrow 2\ NaCl(aq) + H_2O(l) + CO_2(g)$$

In summary, exchange reactions between ionic compounds occur when one of the following occurs:

- Formation of a solid—a precipitation reaction
- Formation of water—a neutralization reaction
- Formation of a gas

Solved Problem 7.10 Will an exchange reaction occur between the following pairs of reactants in aqueous solution? If so, explain why and predict the products.
(a) KOH and H_2SO_4 (b) Na_2CO_3 and H_2SO_4
(c) $NiSO_4$ and NH_4Br (d) $ZnCl_2$ and $AgNO_3$

Solution (a) This reaction between a hydroxide base and an acid will occur; products of the acid–base reaction will be water and potassium sulfate (K_2SO_4).
(b) A gas forms when a carbonate reacts with an acid, so this reaction will occur and the products will be sodium sulfate (Na_2SO_4), carbon dioxide, and water.
(c) Since the possible products of this reaction are both water-soluble ($NiBr_2$ and $(NH_4)_2SO_4$), no reaction will occur.
(d) Of the possible exchange reaction products, AgCl is insoluble and $Zn(NO_3)_2$ is soluble, so the reaction will occur.

Solved Problem 7.11 Decide whether each of the following is a combination, decomposition, displacement, or exchange reaction.
(a) $C(s) + O_2(g) \rightarrow CO_2(g)$
(b) $NaOH(aq) + HCl(aq) \rightarrow NaCl(aq) + H_2O(l)$
(c) $2 Na(s) + 2 HCl(aq) \rightarrow 2 NaCl(aq) + H_2(g)$

Solution (a) This reaction between two elements, carbon and hydrogen, fits the pattern of a combination reaction:

$$C(s) + O_2(g) \rightarrow CO_2(g)$$

A B C

(b) Sodium hydroxide is a soluble ionic compound, and the water solution of HCl is hydrochloric acid. Therefore, this is a reaction between ions in solution that fits the exchange reaction pattern (remember that water is HOH):

$$NaOH(aq) + HCl(aq) \rightarrow NaCl(aq) + H_2O(l)$$

AB XY AY XB

(c) This is a reaction between an element and a compound; it fits the pattern of a displacement reaction

$$2 Na(s) + 2 HCl(aq) \rightarrow H_2(g) + 2 NaCl(aq)$$

A BC B AC

Practice Problems **7.17** Write balanced equations for the exchange reactions between (a) aluminum nitrate and sodium sulfide, and (b) calcium carbonate and hydrochloric acid.

7.18 Classify each of the following reactions:
(a) $Sb_2S_3(s) + 3 Fe(s) \rightarrow 2 Sb(s) + 3 FeS(s)$
(b) $Na_2S(aq) + Hg(NO_3)_2(aq) \rightarrow HgS(s) + 2 NaNO_3(aq)$
(c) $2 KClO_3(s) \rightarrow 2 KCl(s) + 3 O_2(g)$

7.19 Dinitrogen oxide (N_2O, nitrous oxide or laughing gas), which is used as an anesthetic in dentistry, is produced when ammonium nitrate is heated. Classify this reaction, identify a likely second product, and write a balanced equation.

7.12 NET IONIC EQUATIONS

In the precipitation of barium sulfate

$$Ba(NO_3)_2(aq) + Na_2SO_4(aq) \rightarrow BaSO_4(s) + 2 NaNO_3(aq)$$

the sodium and nitrate ions undergo no change. To demonstrate this, we can write the ions present during the reaction separately:

$$Ba^{2+} + 2\,NO_3^- + 2\,Na^+ + SO_4^{2-} \longrightarrow BaSO_4(s) + 2\,Na^+ + 2\,NO_3^-$$

Spectator ion An ion that undergoes no change in a reaction in aqueous solution.

Notice that the Na^+ and NO_3^- ions appear on both sides of the arrow. By canceling out these ions (known as **spectator ions**), we are left with a **net ionic equation,** which includes only species that undergo change and is balanced for both atoms and ionic charge:

Net ionic equation A chemical equation including only the species that undergo change in a reaction involving ions in solution; must be balanced for atoms and charge.

$$Ba^{2+} + \cancel{2\,NO_3^-} + \cancel{2\,Na^+} + SO_4^{2-} \longrightarrow BaSO_4(s) + \cancel{2\,Na^+} + \cancel{2\,NO_3^-}$$
$$Ba^{2+} + SO_4^{2-} \longrightarrow BaSO_4(s) \qquad \text{Net ionic equation}$$

The only chemical change here is the combination of the barium and sulfate ions to form barium sulfate.

In a correctly written net ionic equation

- No spectator ions appear.
- Gases, liquids, and insoluble compounds are represented by their usual chemical formulas.
- Coefficients are lowest possible whole numbers.
- Numbers of atoms are balanced.
- Charge is balanced.

Also, it is customary to omit (*aq*) after ions in net ionic equations, because they are always assumed to be in aqueous solution.

Solved Problem 7.12   Write net ionic equations for (a) the reaction of zinc chloride with silver nitrate, (b) the reaction of hydrochloric acid with potassium carbonate.

Solution   (a) This reaction between soluble ionic compounds will be a precipitation reaction with the formation of silver chloride:

$$2\,AgNO_3(aq) + ZnCl_2(aq) \longrightarrow 2\,AgCl(s) + Zn(NO_3)_2(aq)$$

Writing out all the ions separately and eliminating spectator ions gives the net ionic equation:

$$2\,Ag^+ + \cancel{2\,NO_3^-} + \cancel{Zn^{2+}} + 2\,Cl^- \longrightarrow 2\,AgCl(s) + \cancel{Zn^{2+}} + \cancel{2\,NO_3^-}$$
$$2\,Ag^+ + 2\,Cl^- \longrightarrow 2\,AgCl(s)$$

The coefficients can all be divided by 2 to give

$$Ag^+ + Cl^- \longrightarrow AgCl(s)$$

Checking shows that the equation is balanced for atoms (Ag and Cl on both sides) and charge (on the left, $+1$ and -1 give a total charge of zero, which is also the charge on the right).

(b) Proceeding as above and remembering to write complete formulas for carbon dioxide and water gives

$$K_2CO_3(aq) + 2\ HCl(aq) \rightarrow 2\ KCl(aq) + H_2O(l) + CO_2(g)$$

$$2\ K^+ + CO_3^{2-} + 2\ H^+ + 2\ Cl^- \rightarrow 2\ K^+ + 2\ Cl^- + H_2O(l) + CO_2(g)$$

$$CO_3^{2-} + 2\ H^+ \rightarrow H_2O(l) + CO_2(g)$$

Checking shows that atoms (2 H, 1 C, 3 O) and charge (0) are the same on both sides of the equation.

Practice Problem   7.20   Write net ionic equations for the following reactions:
(a) $2\ K(s) + Pb(NO_3)_2(aq) \rightarrow 2\ KNO_3(aq) + Pb(s)$
(b) $2\ KOH(aq) + H_2SO_4(aq) \rightarrow K_2SO_4(aq) + 2\ H_2O(l)$
(c) $CuS(s) + 2\ HCl(aq) \rightarrow CuCl_2(aq) + H_2S(g)$

7.13 REDOX REACTIONS

Oxidation–reduction (redox) reaction A reaction in which oxidation and reduction occur, shown by electron gain and loss or oxidation number changes.

Oxidation The loss of electrons or increase in oxidation number of a reactant in a chemical reaction.

Reduction The gain of electrons or decrease in oxidation number of a reactant in a chemical reaction.

Oxidized Indicates a substance containing atoms that have lost electrons or increased in oxidation number.

Reduced Indicates a substance containing atoms that have gained electrons or decreased in oxidation number.

In describing ionic compounds (Section 5.2), we discussed the loss and gain of electrons that occur during the formation of sodium chloride from sodium metal and gaseous chlorine:

$$Na \rightarrow Na^+ + e^- \quad and \quad Cl + e^- \rightarrow Cl^-$$

Electron gain and loss team up in what are known as **oxidation–reduction**, or redox reactions. Electron loss is **oxidation**, and electron gain is **reduction**.
    In the redox reaction between sodium and chlorine,

$$2\ Na(s) + Cl_2(g) \rightarrow 2\ NaCl(s)$$
0 charge   + 1 charge   − 1 charge

sodium loses electrons and is said to be **oxidized**, and chlorine gains electrons and is said to be **reduced**. One way to spot a redox reaction is by a change in the charge of two reactants. Most displacement reactions, for example, are also redox reactions:

$$Zn(s) + 2\ AgNO_3(aq) \rightarrow 2\ Ag(s) + Zn(NO_3)_2(aq)$$

or $\quad Zn(s) + 2\ Ag^+ \rightarrow 2\ Ag(s) + Zn^{2+}$

Reduced (electrons gained)   Oxidized (electrons lost)

A reactant that causes the reduction of another substance is called a **reducing agent,** and a reactant that causes the oxidation of another substance is called an **oxidizing agent.** Every redox reaction has to have one of each, because *oxidation and reduction always occur together.* In the displacement reaction above zinc is the reducing agent because it gives up electrons, and silver ion is the oxidizing agent because it accepts electrons.

Table 7.4 lists some of the most common oxidizing and reducing agents, and two are shown in Figure 7.8. Note that reactive nonmetals like chlorine are oxidizing agents, and hydrogen and reactive metals like sodium are reducing agents.

Reducing agent The reactant that causes a reduction by giving up electrons or increasing in oxidation number.

Oxidizing agent The reactant that causes an oxidation by taking electrons or decreasing in oxidation number.

Table 7.4   Some Common Oxidizing and Reducing Agents

Oxidizing Agents	Reducing Agents
Fluorine, chlorine (F_2, Cl_2)	Lithium, sodium, potassium (Li, Na, K)
Oxygen (O_2)	Magnesium, calcium (Mg, Ca)
Hydrogen peroxide (H_2O_2)	Iron, zinc (Fe, Zn)
Nitric acid (HNO_3)	Hydrogen (H_2)
Hypochlorite ion (ClO^-)	Ammonia (NH_3)
Permanganate ion (MnO_4^-)	Sulfite ion (SO_3^{2-})
Dichromate ion ($Cr_2O_7^-$)	Hydrogen phosphate ion (HPO_3^{2-})
Iron(III) ion (Fe^{3+})	Iron(II) ion (Fe^{2+})
	Iodide ion (I^-)

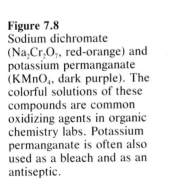

Figure 7.8
Sodium dichromate ($Na_2Cr_2O_7$, red-orange) and potassium permanganate ($KMnO_4$, dark purple). The colorful solutions of these compounds are common oxidizing agents in organic chemistry labs. Potassium permanganate is often also used as a bleach and as an antiseptic.

Practice Problems   7.21 Identify the oxidizing and reducing agents in the following reactions:
(a) $Fe(s) + Cu^{2+} \rightarrow Fe^{2+} + Cu(s)$
(b) $Mg(s) + Cl_2(g) \rightarrow MgCl_2(s)$
(c) $2\ Al(s) + Cr_2O_3 \rightarrow 2\ Cr(s) + Al_2O_3(s)$

7.22 Potassium, a silvery metal, reacts with bromine, a corrosive, reddish liquid, to yield potassium bromide, a white solid. Write the balanced equation and identify the oxidizing and reducing agents.

AN APPLICATION: ORGANIC REDOX REACTIONS

Reactions of organic compounds are likely to involve hydrogen and oxygen. Organic chemists, therefore, tend to stick with the older meanings of "oxidation" and "reduction" based on reactions of oxygen and hydrogen.

"Oxidation" in organic chemistry is usually the addition of oxygen or the removal of hydrogen. "Reduction" in organic chemistry is usually the removal of oxygen or the addition of hydrogen. To see what this means, compare the structures of acetaldehyde and ethanol.

$$
\underset{\substack{\text{Acetaldehyde}\\(C_2H_4O)}}{CH_3\overset{\displaystyle H}{\overset{|}{C}}=O}
\quad
\underset{\text{oxidation } (-2\text{ H})}{\overset{\text{reduction } (+2\text{ H})}{\rightleftharpoons}}
\quad
\underset{\substack{\text{Ethanol}\\(C_2H_6O)}}{CH_3\overset{\displaystyle H}{\underset{\displaystyle H}{\overset{|}{\underset{|}{C}}}}-OH}
$$

The compounds differ by two hydrogen atoms. Acetaldehyde, an important industrial chemical, is prepared by the *oxidation* of ethanol vapor at high temperatures in the presence of air and silver, which increases the reaction rate (a *catalyst*)

Oxidation of ethanol

$$
CH_3CH_2OH \xrightarrow{480°C,\ Ag} CH_3\overset{\displaystyle H}{\overset{|}{C}}=O
$$

As a further example, compare the structures of acetaldehyde and acetic acid:

$$
\underset{\substack{\text{Acetaldehyde}\\(C_2H_4O)}}{CH_3\overset{\displaystyle H}{\overset{|}{C}}=O}
\quad
\underset{\text{reduction } (-O)}{\overset{\text{oxidation } (+O)}{\rightleftharpoons}}
\quad
\underset{\substack{\text{Acetic acid}\\(C_2H_4O_2)}}{CH_3\overset{\displaystyle O}{\overset{\|}{C}}-OH}
$$

These two compounds differ by a single oxygen atom, and conversion of either one to the other is a redox reaction.

Often the oxidation of an alcohol does not stop with the loss of two hydrogen atoms but proceeds directly to formation of an acid by addition of an oxygen atom. The oxidation of ethanol all the way to acetic acid by bacterial fermentation is the process that converts wine stored in the presence of air to vinegar.

Oxidation of ethanol

$$
CH_3CH_2OH \xrightarrow{\text{fermentation, } O_2} CH_3\overset{\displaystyle O}{\overset{\|}{C}}-OH
$$

To make vinegar at home requires wine, some *mother*, which contains the bacterium *Acetobacter aceti*, and oxygen from the air. And patience while you wait till the bacteria finish the job of oxidizing ethanol to acetic acid.

7.14 RECOGNIZING REDOX REACTIONS; OXIDATION NUMBERS

The term "oxidation" was first applied to the addition of oxygen. In this sense the combination of sulfur and oxygen is a redox reaction, even though no ions are formed.

$$S(s) + O_2(g) \rightarrow SO_2(g)$$

Because oxygen is highly electronegative, the oxygen atoms in SO_2 attract the electrons in the S—O bonds more strongly than the sulfur does, and thus the oxygen atoms in SO_2 "own" more of the negative charge than the sulfur atom. By extending the definitions of oxidation and reduction to a loss or gain of *ownership* of electrons, the reaction of S with O_2 is classified as an oxidation–reduction reaction because sulfur has lost some electron ownership and oxygen has gained some.

A formal system has been devised for keeping track of electron ownership. Every atom in a compound is assigned an **oxidation number** (also called an *oxidation state*) according to a fixed set of rules. With the use of oxidation numbers, *every* reaction can be classified either as a redox reaction or as a nonredox reaction:

- **Redox reaction:** Any reaction in which oxidation numbers change.
- **Nonredox reaction:** Any reaction in which no oxidation numbers change.
- A substance whose oxidation number increases in a reaction has been oxidized.
- A substance whose oxidation number decreases has been reduced.

The terms used to describe redox reactions are summarized in Table 7.5. Redox reactions, a major class of chemical reactions, can be recognized in any of the ways listed below:

Recognizing redox reactions

- Hydrogen or oxygen is added to or removed from a compound (see An Application: Organic Redox Reactions)
- An element is a reactant or product.
- One of the reactants is a common oxidizing or reducing agent (Table 7.4).
- Electrons are gained and lost.
- Oxidation numbers change (in *all* redox reactions).

Oxidation number (oxidation state) The charge on an ion formed from a single atom, or a number assigned according to a fixed set of rules and indicating relative electron ownership in bonds.

Nonredox reaction A reaction in which no oxidation numbers change.

Table 7.5 Terms Used to Describe Redox Reactions

In—	Electrons Are—	Oxidation Number Becomes—
Oxidation	lost	more positive
Reduction	gained	more negative
Substance oxidized	lost	more positive
Substance reduced	gained	more negative
Oxidizing agent	gained	more negative
Reducing agent	lost	more positive

To demonstrate how oxidation numbers are used, we've given a few of the rules for assigning them. You'll have no need to use the complete set of rules, but it's helpful to see how they work. Assigning oxidation numbers to every atom in the reactants and products will always reveal whether or not oxidation and reduction occur in the specific reaction.

● *All elements have oxidation number zero.* As a result of this rule, you can immediately recognize as a redox reaction any reaction in which an element is a reactant or a product.

$$\overbrace{\text{Oxidation no. zero}}$$
$$\text{Na} \qquad \text{Cl}_2$$

● *All simple ions have oxidation numbers equal to their charges.*

$$\text{Oxidation no. } +1 \qquad \text{Oxidation no. } +2 \qquad \text{Oxidation no. } -1$$
$$\text{Na}^+ \qquad\qquad \text{Ca}^{2+} \qquad\qquad \text{Cl}$$

● *In binary molecular compounds, the more electronegative element has the oxidation number it would have if it were an anion.* As a result of this rule halogen atoms usually have oxidation number -1 and oxygen atoms usually have oxidation number -2.

$$\text{HCl} \qquad \text{Oxidation no. } -1$$
$$\text{because Cl forms a Cl}^- \text{ ion}$$

● *The sum of the oxidation numbers of all atoms in a compound equals zero.* With this rule, the oxidation number of one element in a binary compound can be found if the oxidation number of the other is known. In HCl, since Cl is -1, H must be $+1$. In SO_2, since *each* O has oxidation number -2 (the charge it would have as an anion), S must have oxidation number $+4$ to maintain a total of zero.

$$\text{Oxidation no. } +4 \qquad\qquad \text{Total oxidation no. } 2 \times -2 = -4$$
$$\text{SO}_2$$

By using oxidation numbers, the system for naming ionic compounds with Roman numerals for the charges of cations can be extended to any binary compound, ionic or not. The compound AsF_5, for example, is named arsenic(V) fluoride by this system.

Solved Problem 7.13   What is the oxidation number of the titanium atom in $TiCl_4$? Name this compound using the Roman numeral.

Solution   First look at the anion (chloride) and then work backward. Since there are four chloride ions in $TiCl_4$, each of which must have an oxidation state of -1, the oxidation state of titanium must be $+4$. The compound is named titanium(IV) chloride.

Solved Problem 7.14   Describe how you can show by using oxidation numbers that the displacement of bromine by chlorine,

$$Cl_2(g) + 2\ Br^- \rightarrow Br_2(l) + 2\ Cl^-$$

is a redox reaction. Which reactant has been oxidized, and which has been reduced? Which reactant is the oxidizing agent?

Solution   Oxidation numbers must change in redox reactions. Here the oxidation number of chlorine changes from 0 in Cl_2 to -1 in Cl^-, and that of bromine changes from -1 in Br^- to 0 in Br_2.

$$\overset{0}{Cl_2(g)} + 2\ \overset{-1}{Br^-} \rightarrow \overset{0}{Br_2(l)} + 2\ \overset{-1}{Cl^-}$$

Chlorine, whose oxidation number decreases, has been reduced, and bromide ion, whose oxidation number increases, has been oxidized. Since the oxidizing agent undergoes reduction, chlorine is the oxidizing agent.

Practice Problems   **7.23**   What are the oxidation numbers of the metal atoms in the compounds listed? Name each using the oxidation number as a Roman numeral.
(a) VCl_3
(b) MgO
(c) $SnCl_4$
(d) CrO_3
(e) $Cu(NO_3)_2$
(f) $NiSO_4$

7.24   Show by using oxidation numbers whether or not the following are redox reactions and classify the reactions according to Table 7.2.
(a) $Na_2S(aq) + NiCl_2(aq) \rightarrow 2\ NaCl(aq) + NiS(s)$
(b) $2\ Na(s) + 2\ H_2O(l) \rightarrow 2\ NaOH(aq) + H_2(g)$

7.25   Assign an oxidation number to each element of the reactants and products in the following reaction and identify which have been oxidized and which have been reduced.

$$2\ PbS(s) + 3\ O_2(g) \rightarrow 2\ PbO(s) + 2\ SO_2(g)$$

7.26   Decide which of the following are redox reactions. Give the basis for your decision.
(a) $C(s) + O_2(g) \rightarrow CO_2(g)$
(b) $CuO(s) + 2\ HCl(aq) \rightarrow CuCl_2(aq) + H_2O(l)$
(c) $CH_3CH{=}CH_2 + H_2 \rightarrow CH_3CH_2CH_3$
(d) $2\ MnO_4^- + 5\ SO_2(g) + 2\ H_2O(l) \rightarrow 2\ Mn^{2+} + 5\ SO_4^{2-} + 4\ H^+$

INTERLUDE: REDOX IN ACTION

How many redox reactions did you experience today? Quite a few, without any doubt.

As you sit reading this book, *respiration* is proceeding in your body, with oxygen entering cells where small molecules derived from food are being oxidized to produce energy in a process called *catabolism*. The crucial overall energy-yielding reaction for most living organisms is the complete oxidation of glucose to carbon dioxide and water:

Production of energy in living cells

$$C_6H_{12}O_6 + 6\ O_2\ \rightarrow\ 6\ CO_2 + 6\ H_2O + energy$$

If today happens to be a cold day, somewhere a *combustion* reaction is most likely producing heat to keep you warm. Strictly speaking, combustion is the oxidation of any fuel with the production of heat and light. Commonly, though, we refer to a fuel burning in air as "combustion." In our homes we might burn natural gas, which is mostly methane; fuel oil, a mixture of liquid carbon–hydrogen compounds; or coal, a mixture of mostly solid carbon compounds. Combustion reactions also thrust rockets into space, propel most of our vehicles, and provide energy for generating electricity.

We also depend heavily on the direct production of electricity by redox reactions in which electrons are gained and lost. What would life be like without any item that uses a cell or battery? In batteries like those in toys, flashlights, calculators, and automobiles the substances that undergo oxidation and reduction are physically separated, as they are in the simple electrochemical cell shown here. Connections between the isolated reactants permit ions and electrons to flow and complete a circuit in what is called an *electrochemical cell.*

In an automobile battery, for example, lead is oxidized to lead(II) sulfate, and lead(IV) oxide is reduced to lead(II) sulfate in the presence of sulfuric acid in a series of connected cells. The overall reaction is

Production of electricity in an automobile battery

$$Pb(s) + PbO_2(s) + 4\ H^+ + 2\ SO_4^{2-}\ \rightarrow$$
$$2\ PbSO_4(s) + 2\ H_2O(l) + energy$$

When electrical energy is supplied, the reaction above is reversed, lead and lead(IV) oxide are regenerated, and the battery is recharged.

Corrosion describes the results of a group of redox reactions that we could just as well do

SUMMARY

All chemical equations must be **balanced;** that is, the numbers of atoms of each kind must be the same in the reactants and the products. To balance an equation, coefficients are placed before formulas, and formulas must *never* be changed.

A **mole** is a counting unit that equals **Avogadro's number** (6.02×10^{23}) of atoms, ions, or molecules. One **molar mass** is the formula weight of a substance (or the atomic weight of an element) taken in grams and contains one mole of the substance. Equal molar masses of substances contain equal numbers of formula units, and molar mass provides the conversion factors that connect numbers of moles and mass in grams.

The coefficients in chemical equations represent the numbers of moles of reactants and products that participate in a reaction and provide mole ratios that relate the quantities of reactants and/or products. By appropriately using molar masses and mole ratios in factor-label calculations, unknown masses or molar amounts of reactants and products can be found from known masses or molar amounts of reactants and products. The amounts of products calculated from amounts of reactants and mole ratios represent **theoretical yields,** the amount formed if 100% of reactant is converted to product. The **percent yield** is found from the theoretical yield and the mass of product actually recovered.

without. In general, corrosion is the deterioration of materials resulting from exposure to the environment. A most common type of corrosion occurs when a metal comes in contact with moisture and various ions to create something like an electrochemical cell. A familiar and very destructive example of corrosion is the rusting of iron, which occurs in the presence of water, hydrogen ions or other ions, and oxygen. The iron is oxidized, and the freed electrons travel through the metal until they reach a spot where oxygen is available to be reduced.

Oxidation and reduction during rusting of iron

Oxidation: $2 \, Fe(s) \; \rightarrow \; 2 \, Fe^{2+} + 4 \, e^-$

Reduction: $O_2(g) + 4 \, H^+ + 4 \, e^- \rightarrow 2 \, H_2O(l)$

The end result of the complex process is production of rust, a red solid that is composed of iron(III) oxide and water, $Fe_2O_3 \cdot H_2O$. Because rust drops away from the surface as it forms, an entire object can be consumed by rusting. In contrast, the oxide formed by some metals, notably aluminum, adheres tightly to the metal surface and protects an object from deterioration.

In this electrochemical cell, a voltage is produced when electrons flow through the external circuit as the redox displacement reaction $Zn(s) + Cu^{2+} \rightarrow Zn^{2+} + Cu(s)$ occurs. At the left a copper electrode dips into a copper sulfate solution, and on the right a zinc electrode dips into a zinc sulfate solution.

Four common types of reactions can often be identified by the nature of the reactants: **combination,** in which elements or an element and a compound unite in a single product; **displacement,** in which a more reactive element displaces a less reactive element in a compound; **decomposition,** in which a single compound breaks down to two or more compounds; and **exchange,** in which ions in solutions react to form a precipitate, a gas, or water. The solubilities of ionic compounds (Table 7.3) are used to predict when **precipitation** will occur.

All reactions can be classified either as **redox reactions,** in which oxidation numbers decrease (**reduction**) or increase (**oxidation**), or as nonredox reactions, in which these changes do not occur. An **oxidizing agent** causes the oxidation of another reactant, and a **reducing agent** causes the reduction of another reactant. **Oxidation numbers** are assigned according to a set of rules (Section 7.14) so that the more electronegative element in a compound always has the more negative oxidation number. Some ways to recognize redox reactions include the presence of an element as a reactant or product, the addition or loss of hydrogen or oxygen, and the presence of a common oxidizing or reducing agent as a reactant.

REVIEW PROBLEMS

Balancing Chemical Equations

7.27 What is meant by the term "balanced equation"?

7.28 Which of the following equations are balanced? Balance those that need it.
(a) $2 C_2H_6 + 5 O_2 \rightarrow 2 CO_2 + 6 H_2O$
(b) $3 Ca(OH)_2 + 2 H_3PO_4 \rightarrow Ca_3(PO_4)_2 + 6 H_2O$
(c) $Mg + O_2 \rightarrow 2 MgO$
(d) $(NH_4)_3P + 3 NaOH \rightarrow Na_3P + NH_3 + H_2O$
(e) $K + H_2O \rightarrow KOH + H_2$

7.29 Write balanced equations for the following:
(a) Gaseous sulfur dioxide reacts with water vapor to form liquid sulfurous acid (H_2SO_3).
(b) Liquid bromine reacts with solid potassium metal to form potassium bromide.
(c) Gaseous propane (C_3H_8) burns in oxygen to form gaseous carbon dioxide and water vapor.

7.30 Balance the following equations:
(a) $Hg(NO_3)_2 + LiI \rightarrow LiNO_3 + HgI_2$
(b) $I_2 + Cl_2 \rightarrow ICl_3$
(c) $Al + O_2 \rightarrow Al_2O_3$
(d) $CuSO_4 + AgNO_3 \rightarrow Ag_2SO_4 + Cu(NO_3)_2$
(e) $Mn(NO_3)_3 + Na_2S \rightarrow Mn_2S_3 + NaNO_3$
(f) $NO_2 + O_2 \rightarrow N_2O_5$
(g) $P_4O_{10} + H_2O \rightarrow H_3PO_4$

7.31 Write a balanced equation for the reaction of sodium bicarbonate ($NaHCO_3$) with sulfuric acid (H_2SO_4) to yield CO_2, Na_2SO_4, and H_2O.

7.32 Spoilage of wine into vinegar is caused by reaction of ethyl alcohol (C_2H_6O) with oxygen to yield acetic acid ($C_2H_4O_2$) and water. Write a balanced equation for the reaction.

7.33 Photosynthesis in green leaves converts carbon dioxide and water into glucose and oxygen. Write a balanced equation for the process:

$$CO_2 + H_2O \rightarrow C_6H_{12}O_6 + O_2$$

7.34 When organic compounds are burned, they react with oxygen to form CO_2 and H_2O. Write balanced equations for the combustion of:
(a) C_4H_{10} (butane, used in lighters)
(b) C_2H_6O (ethanol, used in gasohol and as race car fuel)
(c) C_8H_{18} (octane, a component of gasoline)

Molar Masses and Moles

7.35 What is a mole of a substance? How many molecules are in one mole of a molecular compound?

7.36 How many calcium atoms are in 16.2 g of calcium?

7.37 What is the mass in grams of 2.68×10^{22} atoms of uranium?

7.38 Caffeine has the formula $C_8H_{10}N_4O_2$. If an average cup of coffee contains approximately 125 mg of caffeine, how many moles of caffeine are in one cup?

7.39 How many moles of aspirin, $C_9H_8O_4$, are in a 500 mg tablet?

7.40 What is the molar mass of diazepam (Valium), $C_{16}H_{13}ClN_2O$?

7.41 Calculate the molar masses of these substances:
(a) benzene, C_6H_6 (b) sodium bicarbonate, $NaHCO_3$
(c) chloroform, $CHCl_3$ (d) penicillin V, $C_{16}H_{18}N_2O_5S$

7.42 How many moles are present in 5.00 g samples of each of the compounds listed in Problem 7.41?

7.43 How many grams are present in 0.050 mol samples of each of the compounds listed in Problem 7.41?

7.44 Iron(II) sulfate, $FeSO_4$, is used in the clinical treatment of iron-deficiency anemia. What is the molar mass of $FeSO_4$? How many moles of $FeSO_4$ are in a standard 300 mg tablet?

Mole and Mass Relationships from Chemical Equations

7.45 What is meant by the term "limiting reactant"?

7.46 At elevated temperatures N_2 and O_2 can react to yield NO, an important cause of air pollution.
(a) Write a balanced equation for the reaction.
(b) How many moles of N_2 are needed to react with 7.50 mol of O_2?
(c) How many moles of NO can be formed when 3.81 mol of N_2 reacts?
(d) How many moles of O_2 must react to produce 0.250 mol of NO?
(e) When 12 mol of O_2 reacts with 16 mol of N_2, which is the limiting reactant, and what is the maximum number of moles of NO that can form?

7.47 Ethyl acetate reacts with H_2 in the presence of a catalyst to yield ethyl alcohol:

$$C_4H_8O_2 + H_2 \rightarrow C_2H_6O$$

(a) Write a balanced equation for the reaction.
(b) How many moles of ethyl alcohol are produced by reaction of 1.5 mol of ethyl acetate?
(c) How many grams of ethyl alcohol are produced by reaction of 1.5 mol of ethyl acetate with H_2?
(d) How many grams of ethyl alcohol are produced by reaction of 12.0 g of ethyl acetate with H_2?
(e) How many grams of H_2 are needed to react with 12.0 grams of ethyl acetate?

7.48 Large amounts of ammonia, NH_3, are prepared for use as a fertilizer by reacting N_2 with H_2.
(a) Write a balanced equation for the reaction.
(b) How many moles of N_2 are needed to react to make 16.0 grams of NH_3?
(c) How many grams of H_2 are needed to react with 75.0 grams of N_2?
(d) When 0.62 mol of N_2 reacts with 1.33 mol of H_2, which is the limiting reactant and what is the maximum mass in grams of NH_3 that can form?

Percent Yield

7.49 Methanol (CH_3OH), once made from the distillation of wood, is now made by reacting carbon monoxide and hydrogen at high pressure.

$$CO(g) + 2 H_2(g) \rightarrow CH_3OH(l)$$

(a) How many grams of CH_3OH can be made from 10.0 g of CO if it all reacts?
(b) If 9.55 g of CH_3OH was recovered when the amounts in part (a) were used, what was the percent yield?

7.50 Pure iron metal can be made by roasting the ore with carbon.

$$3 C + 2 Fe_2O_3 \rightarrow 3 CO_2 + 4 Fe$$

(a) How many kilograms of iron can be obtained from the reaction of 75.0 kg of Fe_2O_3?
(b) If 51.3 kg of iron was obtained from the reaction in part (a), what was the percent yield?

7.51 Dichloromethane, CH_2Cl_2, the solvent used to decaffeinate coffee beans, is prepared by reaction of CH_4 with Cl_2.
(a) Write the balanced reaction. (HCl is also formed.)
(b) How many grams of Cl_2 are needed to react with 50.0 g of CH_4?
(c) How many grams of dichloromethane can be prepared by reaction of 50.0 g of CH_4 and 50.0 g of Cl_2? Which is the limiting reactant?

Types of Chemical Reactions

7.52 Complete and balance the equations for the following displacement reactions:
(a) $FeCl_2(aq) + 2 Ag(s) \rightarrow ?$
(b) $Ba(s) + 2 H_2O(l) \rightarrow Ba(OH)_2(aq) + ?$
(c) $CuSO_4(aq) + Zn(s) \rightarrow ?$
(d) $Mg(NO_3)_2(aq) + Pb(s) \rightarrow ?$

7.53 Use Table 7.3 to predict whether exchange reactions will occur between the listed pairs of reactants. Write balanced equations for those reactions that should occur.
(a) NaBr and $Hg_2(NO_3)_2$ (b) $SrCl_2$ and K_2SO_4

(c) $LiNO_3$ and $Ca(C_2H_3O_2)_2$ (d) $(NH_4)_2CO_3$ and $CaCl_2$
(e) KOH and $MnBr_2$ (f) Na_2S and $Al(NO_3)_3$

7.54 Write net ionic equations for the following reactions:
(a) $Mg(s) + CuCl_2(aq) \rightarrow MgCl_2(aq) + Cu(s)$
(b) $2 KCl(aq) + Pb(NO_3)_2(aq) \rightarrow PbCl_2(s) + 2 KNO_3(aq)$
(c) $2 Cr(NO_3)_3(aq) + 3 Na_2S(aq) \rightarrow$
$$Cr_2S_3(s) + 6 NaNO_3(aq)$$
(d) $2 AuCl_3(aq) + 3 Sn(s) \rightarrow 3 SnCl_2(aq) + 2 Au(s)$
(e) $2 NaI(aq) + Br_2(l) \rightarrow 2 NaBr(aq) + I_2(s)$

7.55 For each of the following reactions, tell (i) what element is reduced; (ii) what element is oxidized; (iii) which reactant is the oxidizing agent; and (iv) which reactant is the reducing agent.
(a) $2 SO_2(g) + O_2(g) \rightarrow 2 SO_3(g)$
(b) $2 Na(s) + Cl_2(g) \rightarrow 2 NaCl(s)$
(c) $CuCl_2(aq) + Zn(s) \rightarrow ZnCl_2(aq) + Cu(s)$
(d) $2 NaCl(aq) + F_2(g) \rightarrow 2 NaF(aq) + Cl_2(g)$
(e) $BiCl_3(aq) + Cl_2(g) \rightarrow BiCl_5(aq)$

7.56 Assign an oxidation number to the metal in each of the following compounds:
(a) $CoCl_3$ (b) $FeSO_4$ (c) UO_3 (d) CuF_2
(e) TiO_2 (f) SnS

Applications

7.57 A blood sample is analyzed. Which of these are quantitative reports and which are qualitative reports on the sample? [App: What Is It?]
(a) type O+ (b) iron = 42 $\mu g/dL$
(c) elevated white cell count
(d) culture shows presence of a virus
(e) blood sugar = 92 mg/dL

7.58 It is often difficult to assign oxidation numbers in organic molecules. What are the signs that a compound has been oxidized, and what are the signs that it has been reduced in an organic reaction? [App: Organic Redox Reactions]

7.59 Look at the following organic reactants and products and decide if reduction or oxidation has occurred. [App: Organic Redox Reactions]

Reactant	Product
(a) $CH_3\overset{O}{\overset{\|}{C}}CH_3$	$CH_3\overset{OH}{\overset{\|}{C}H}CH_3$
(b) $HC\equiv CH$	CH_3CH_3
(c) CH_3CH_2OH	$CH_2{=}CH_2$
(d) $CH_3CH_2\overset{O}{\overset{\|}{C}}H$	$CH_3CH_2\overset{O}{\overset{\|}{C}}OH$

7.60 What are electrochemical cells used for? [Int: Redox in Action]

7.61 What are the symptoms of corrosion? [Int: Redox in Action]

Additional Problems

7.62 Zinc metal reacts with hydrochloric acid (HCl) according to the equation:

$$Zn + 2\ HCl \rightarrow ZnCl_2 + H_2$$

(a) How many grams of hydrogen would be produced if 15.0 grams of zinc reacted?
(b) Into which reaction category does this reaction fall?
(c) Is this a redox reaction? If so, tell what is reduced, what is oxidized, what is the reducing agent, and what is the oxidizing agent.

7.63 Lithium oxide is used aboard the space shuttle to remove water from the atmosphere according to the equation:

$$Li_2O + H_2O \rightarrow 2\ LiOH$$

(a) How many grams of Li_2O must be carried on board to remove 80.0 kg of water?
(b) Is this a redox reaction? Why or why not?

7.64 Batrachotoxin, $C_{31}H_{42}N_2O_6$, the active component of South American arrow poison, is so toxic that 0.05 μg can kill a person. How many molecules is this?

7.65 Look at Table 7.3 and predict whether a precipitate will form when $CuCl_2(aq)$ and $Na_3PO_4(aq)$ are mixed. If so, write both (a) the balanced equation and (b) the net ionic equation for the process.

7.66 When table sugar (sucrose, $C_{12}H_{22}O_{11}$) is heated, it decomposes to form C and H_2O.
(a) Write a balanced equation for the process.
(b) How many grams of carbon are formed by the breakdown of 60.0 g of sucrose?
(c) How many grams of water are formed when 6.50 g of carbon is formed?

7.67 Although Cu is not sufficiently active to displace hydrogen from acids, it can be dissolved by concentrated nitric acid, which functions as an oxidizing agent according to the following equation:

$$Cu(s) + 4\ HNO_3(aq) \rightarrow$$
$$Cu(NO_3)_2(aq) + 2\ NO_2(g) + 2\ H_2O(l)$$

(a) Write the net ionic equation for this process.
(b) Are 35.0 grams of HNO_3 sufficient to dissolve 5.00 grams of copper?

7.68 The net reaction for the Breathalyzer test used to indicate alcohol concentration in the body is:

$$16\ H^+ + 2\ Cr_2O_7{}^{2-} + 3\ C_2H_6O \rightarrow$$
$$3\ C_2H_4O_2 + 4\ Cr^{3+} + 11\ H_2O$$

(a) Is the ethanol, C_2H_6O, oxidized or reduced?
(b) How many grams of $K_2Cr_2O_7$ must be used to consume 1.50 grams of C_2H_6O?
(c) How many grams of $C_2H_4O_2$ can be produced from 80.0 g of C_2H_6O?

7.69 Ethanol is formed by enzyme action on sugars and starches during fermentation:

$$C_6H_{12}O_6 \xrightarrow{\text{enzymes}} 2\ CO_2 + 2\ C_2H_6O$$

If the density of ethanol is 0.789 g/mL, how many quarts of ethanol can be produced by the fermentation of 100.0 pounds of sugar?

7.70 Classify each of the reactions in Problem 7.30 as combination, displacement, decomposition, or exchange.

7.71 Balance the following equations:
(a) $Al(OH)_3 + HNO_3 \rightarrow Al(NO_3)_3 + H_2O$
(b) $AgNO_3 + FeCl_3 \rightarrow AgCl + Fe(NO_3)_3$
(c) $(NH_4)_2Cr_2O_7 \rightarrow Cr_2O_3 + H_2O + N_2$
(d) $Mn_2(CO_3)_3 \rightarrow Mn_2O_3 + CO_2$

7.72 Classify each of the reactions in Problem 7.71 as combination, displacement, decomposition, or exchange.

CHAPTER

8

Chemical Reactions: Energy, Rates, and Equilibrium

Combustion of the chemicals in a match head, once set off by the heat of friction, gives off heat spontaneously and happens quickly. Other reactions absorb heat and happen slowly. Such differences are the topic of this chapter.

Once reactants and products are identified and a balanced equation is written, the masses of reactants and products can be interpreted. But, as we've noted, writing a balanced equation doesn't prove a reaction will occur. We're now ready to examine the criteria for deciding whether a reaction can actually take place.

To complete the information needed about a chemical reaction, a chemist wants to know the answers to several fundamental questions. Is energy released or absorbed when the reaction occurs? Does a reaction need an energy boost to get started and, after it's started, how fast does the reaction occur? Does a reaction continue until all reactants are converted to products or is there an equilibrium point beyond which no additional product forms? Armed with answers to these questions, it's sometimes possible to control the outcome of a reaction by changing the conditions under which it is carried out. The goals of this chapter are to answer the following questions:

1. *Why is energy released in some reactions and not others?* The goal: Be able to explain the factors that influence energy changes in chemical reactions.

2. *What is the heat of reaction?* The goal: Be able to define heat of reaction and, given a heat of reaction and quantity of reactant or product, to calculate the quantity of heat absorbed or released.

3. *What is the criterion for spontaneity in chemistry?* The goal: Be able to define free energy change and entropy, and qualitatively interpret the equation that relates free energy change to heat of reaction, temperature, and entropy.

4. *What determines the rate of a chemical reaction?* The goal: Be able to explain how collisions and activation energy determine reaction rate.

5. *What is chemical equilibrium?* The goal: Be able to describe what is occurring in a reaction at equilibrium, write the equilibrium constant expression for a given reaction, and explain what its value means.

6. *What is LeChatelier's principle and how do temperature, pressure, concentrations, and catalysts influence chemical reactions?* The goal: Be able to predict the effects of changes in conditions on reaction rates and chemical equilibria.

8.1 HEAT OF REACTION

Why is it that the halogens combine easily at ordinary temperatures with many elements and compounds but nitrogen does not? Is there a difference between halogen molecules and nitrogen molecules that accounts for their different reactivities? The answer is yes: The nitrogen–nitrogen triple bond is *much stronger* than halogen–halogen single bonds and can't be broken as easily in chemical reactions.

Bond energy The energy needed to separate two covalently bonded atoms.

The characteristic strength of a bond is given by its **bond energy,** which is the amount of energy needed to break the bond. The triple bond in N_2 is four to six times stronger than any of the halogen–halogen bonds.

Some bond energies

$$N\equiv N \quad 226 \text{ kcal/mol}$$

F—F	38 kcal/mol	Br—Br	46 kcal/mol
Cl—Cl	58 kcal/mol	I—I	36 kcal/mol

Endothermic Describes a process that absorbs heat from the surroundings.

Exothermic Describes a process that gives off heat to the surroundings.

A change like bond breaking that must absorb heat to take place is described as **endothermic.** The reverse of bond breaking, bond formation, releases heat and is described as **exothermic.** Furthermore, the amount of energy released in forming a bond is the same as that absorbed in breaking it. When chlorine atoms combine to give Cl_2, 58 kcal/mol of heat is released; when Cl_2 is cleaved into Cl atoms, 58 kcal/mol of heat is absorbed.

The same energy relationships that govern bond breaking and bond formation apply to every physical or chemical change: (1) The reverse of an endothermic process is exothermic, and the reverse of an exothermic process is endothermic. (2) The amount of heat that accompanies a change in one direction is numerically equal to that for the opposite change. These relationships are required by a fundamental law of nature, the law of conservation of energy (also known as the *first law of thermodynamics*):

Law of conservation of energy Energy can be neither created nor destroyed in any physical or chemical change (first law of thermodynamics).

Law of conservation of energy: Energy can be neither created nor destroyed in any physical or chemical change.

If more energy could be released by an exothermic reaction than was consumed in its reverse, this law would be violated. Suppose the combination of two nitrogen atoms to give N_2 released more energy than was needed to break the bond in one N_2 molecule. The extra energy could be applied to breaking apart another molecule, and still more energy gained by then letting those two atoms combine. More and more new energy could be created by repeating the sequence over and over again—something we know from experience can't happen.

Heat of reaction The amount of heat released or absorbed during a chemical reaction.

Enthalpy change The amount of heat released (negative ΔH) or absorbed (positive ΔH) in a specific chemical or physical change.

In every chemical reaction some bonds are broken and some are formed. The difference between the energy absorbed and released in these two processes plus the energy of any physical changes that accompany a reaction is called the **heat of reaction** and is a quantity that can be measured. Heats of reaction at constant pressure, and also heats of physical changes such as melting and boiling, are represented by ΔH, where Δ is a general symbol used to indicate "a change in." Usually the value of ΔH represents conversion of reactants as they exist at room temperature (25°C) into products at room temperature. Although H is formally named *enthalpy* and ΔH is therefore an **enthalpy change,** it's convenient to refer to the enthalpy change for a chemical reaction as the "heat of reaction," and we will do so.

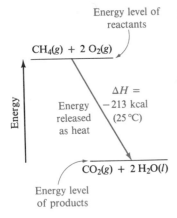

Energy level of reactants

$CH_4(g) + 2 O_2(g)$

Energy

Energy released as heat

$\Delta H = -213$ kcal (25 °C)

$CO_2(g) + 2 H_2O(l)$

Energy level of products

Figure 8.1
An exothermic reaction. In the combustion of methane, the products store less chemical energy than the reactants. The difference in energy is released as heat during the reaction.

Combustion   A chemical reaction in which heat and often light are produced; usually refers to burning in the presence of oxygen.

8.2   EXOTHERMIC AND ENDOTHERMIC REACTIONS

When the total energy of the reactants is *greater* than the total energy of the products, energy is released and a reaction is exothermic. All **combustion** reactions, for example, are exothermic—burning one mole of methane releases 213 kcal of heat (Figure 8.1). In an exothermic reaction, the heat released is essentially a reaction product and ΔH is assigned a negative value (the heat leaves):

An exothermic reaction

$$CH_4(g) + 2 O_2(g) \longrightarrow CO_2(g) + 2 H_2O(l) + 213 \text{ kcal}$$

or

$$CH_4(g) + 2 O_2(g) \longrightarrow CO_2(g) + 2 H_2O(l) \qquad \Delta H = -213 \text{ kcal}$$

Note that when ΔH is included in an equation as illustrated above, the value given is for the reaction represented by the equation as written.

The quantities of heat released in the combustion of several fuels, including natural gas, which is mostly methane, are compared in Table 8.1. You can see why there is interest in the potential of hydrogen as a fuel.

When the total energy of the reactants is *less* than the total energy of the products, energy is absorbed and a reaction is endothermic. The combination of nitrogen and oxygen to give nitrogen(II) oxide (Figure 8.2), a gas present in automobile exhaust, is such a reaction. The heat added in an endothermic reaction is like a reactant, and ΔH is given a positive sign (heat is added).

An endothermic reaction

$$N_2(g) + O_2(g) + 43 \text{ kcal} \longrightarrow 2 NO(g)$$

or

$$N_2(g) + O_2(g) \longrightarrow 2 NO(g) \qquad \Delta H = 43 \text{ kcal}$$

Some examples of heats of two kinds of reactions are given in Table 8.2.

The essential points about heat and chemical reactions covered thus far are summarized as follows:

● An exothermic reaction releases heat to the surroundings; ΔH is negative for an exothermic reaction.

● An endothermic reaction absorbs heat from the surroundings; ΔH is positive for an endothermic reaction.

● The reverse of an exothermic reaction is endothermic.

● The reverse of an endothermic reaction is exothermic.

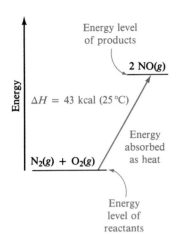

Figure 8.2
An endothermic reaction. In the combination of nitrogen and oxygen to form nitrogen(II) oxide, the products store more chemical energy than the reactants. The additional energy must be absorbed from the surroundings as the reaction takes place.

Table 8.1   Energy Values of Some Fuels

Substance	Energy Value (kcal/kg)
Wood (pine)	4,300
Anthracite coal	7,400
Charcoal	8,100
Crude oil (Texas)	10,500
Gasoline	11,500
Natural gas (mostly methane)	11,700
Hydrogen	34,000

Table 8.2   Some Heats of Reaction[a]

Compound		ΔH at 25°C (kcal/mol)
Heats of combustion (heat released per mole of compound burned completely to give carbon dioxide and water)		
Acetylene	$CH{\equiv}CH(g)$	−310
Glucose	$C_6H_{12}O_6(s)$	−671
Methanol	$CH_3OH(l)$	−174
n-Octane	$C_8H_{18}(l)$	−1308
Sucrose	$C_{12}H_{22}O_{11}(s)$	−1349
Heats of formation (heat absorbed or released per mole of compound formed by combination of the elements)		
Sulfur trioxide	$SO_3(g)$	−94
Ozone	$O_3(g)$	34
Sodium chloride	$NaCl(s)$	−98
Silicon dioxide (quartz)	SiO_2	−218
Iron(III) oxide	Fe_2O_3	−197

[a] As is common in data tables, values are given per mole of a specific reactant or product rather than for a balanced equation.

● The amount of heat absorbed or released in the reverse of a reaction is equal to that released or absorbed in the original reaction, but ΔH has the opposite sign.

8.3   CALCULATING HEATS OF REACTION

As demonstrated in Solved Problem 8.1, a balanced equation and its accompanying ΔH provide the information needed to calculate the heat for reaction of a given amount of reactant or product. Ratios and molar masses are used as was demonstrated in Sections 7.5 and 7.6 for finding masses of reactants or products.

Solved Problem 8.1 (a) How much heat is released during the combustion of 0.35 mol of methane?

(b) How much heat is released during the combustion of 8.00 g of methane?

(c) How much heat is released when 2.50 mol of O_2 reacts completely with methane?

$$CH_4(g) + 2\ O_2(g) \longrightarrow CO_2(g) + 2\ H_2O(l) \qquad \Delta H = -213 \text{ kcal}$$

Solution (a) The equation shows that 213 kcal of heat is released per mole of methane that reacts, giving the ratio 213 kcal/1 mol CH_4. To find the amount of heat released by combustion of 0.35 mol of methane requires the following factor-label calculation:

$$0.35 \text{ mol } CH_4 \times \frac{213 \text{ kcal}}{1 \text{ mol } CH_4} = 75 \text{ kcal}$$

(b) To go from the mass of a reactant to the heat of a reaction requires including the molar mass in the calculation:

$$8.00 \text{ g } CH_4 \times \frac{1 \text{ mol } CH_4}{16.0 \text{ g } CH_4} \times \frac{213 \text{ kcal}}{1 \text{ mol } CH_4} = 107 \text{ kcal}$$

(c) The ratio in this case is for 2 mol of reactant: 213 kcal/2 mol O_2.

$$2.50 \text{ mol } O_2 \times \frac{213 \text{ kcal}}{2 \text{ mol } O_2} = 266 \text{ kcal}$$

Practice Problems **8.1** In photosynthesis, green plants convert carbon dioxide and water into the carbohydrate glucose ($C_6H_{12}O_6$) according to the following equation:

$$6\ CO_2(g) + 6\ H_2O(l) + 678 \text{ kcal} \longrightarrow C_6H_{12}O_6(aq) + 6\ O_2(g)$$

(a) Is this reaction exothermic or endothermic?

(b) Give the value of ΔH for this reaction. 678 kcal

(c) Write the equation for the reverse of this reaction, including heat as a reactant or product. $C_6H_{12}O_6 + 6O_2(g) \longrightarrow 6\ CO_2(g) + 6H_2O(l) \ 678 \text{ Kcal}$

8.2 The following equation shows the conversion of aluminum oxide (from the aluminum ore known as bauxite) to aluminum.

$$2\ Al_2O_3(s) \longrightarrow 4\ Al(s) + 3\ O_2(g) \qquad \Delta H = 801 \text{ kcal}$$

(a) Is this reaction exothermic or endothermic?

(b) How many kilocalories of heat are required to produce one mole of aluminum?

(c) How many kilocalories of heat are required to produce 10. g of aluminum?

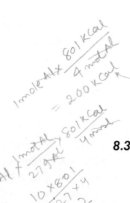
1 mole Al × $\frac{801 \text{ kcal}}{4 \text{ mol Al}}$ = 200 Kcal

10 g Al × $\frac{1 \text{ mol Al}}{27 \text{ g Al}}$ × $\frac{801 \text{ kcal}}{4 \text{ mol}}$

$\frac{10 \times 801}{27 \times 4}$

$\frac{2}{}$

8.3 How much heat is absorbed during production of 145 g of nitrogen(II) oxide by the combination of nitrogen and oxygen?

$$N_2(g) + O_2(g) \longrightarrow 2\ NO(g) \qquad \Delta H = 43 \text{ kcal}$$

$145 \times \frac{43 \text{ K cal}}{60 \text{ g NO}} = 103.9 \text{ K cal}$

8.4 ENERGY FROM FOOD

Serious efforts to lose weight usually lead to studying the caloric values of foods, and many food labels include such values for whatever is in the package. Have you ever wondered where those numbers come from?

Food is "burned" in the body to yield water, carbon dioxide, and energy, just as natural gas is burned in furnaces. In fact, the "caloric value" of a food is just the heat of reaction for complete oxidation of the food (minus a small correction factor). The value is the same whether the food is burned in the body or in the laboratory. One gram of protein releases 4 kcal, 1 g of table sugar (a carbohydrate) releases 4 kcal, and 1 g of fat releases 9 kcal (Table 8.3).

The caloric values of foods are usually given in "Calories" (note the capital C) that are actually kilocalories. To determine these values experimentally, a carefully weighed food sample together with oxygen is placed into an instrument called a *calorimeter*, the food is ignited, the temperature change is measured, and the heat given off is calculated from the temperature change. In the calorimeter, the heat from the food is released very quickly and the temperature rises immediately. Clearly, something a bit different goes on when food is burned in the body, which to remain alive must have a constant temperature of 37°C. If all the protein and fat in a large steak were burned to give heat in a minute or two, more heat would be generated (about 800 kcal) than the body could eliminate before overheating.

It's a well-established fact of chemistry (known as *Hess's law*) that the total heat of reaction in going from a given set of reactants to a given set of products is the same no matter what pathway is taken. In other words, no matter how many reactions are used to go all the way from, say, reactants to products in Figure 8.1 or 8.2, the total heat released or absorbed is the same. The body applies this principle by withdrawing energy from food bit by bit in series of interconnected reactions that proceed gradually from carbohydrate, fat, and protein to carbon dioxide, water, and energy. These and other reactions continually taking place in the body, referred to as the body's **metabolism,** will be examined in later chapters.

Metabolism The overall sum of the many reactions taking place in an organism.

Table 8.3   Caloric Values of Some Foods

Substance, Sample Size	Caloric Value (kcal)
Protein, 1 g	4
Carbohydrate, 1 g	4
Fat, 1 g	9
Alcohol, 1 g	7.1
Cola drink, 12 fl oz (369 g)	160
Apple, one medium (138 g)	80
Iceberg lettuce, 1 cup shredded (55 g)	5
White bread, 1 slice (25 g)	65
Hamburger patty, 3 oz (85 g)	245
Pizza, 1 slice (120 g)	290
Vanilla ice cream, 1 cup (133 g)	270

AN APPLICATION: NITROGEN FIXATION

The old saying about water might be modified and applied to nitrogen: Nitrogen, nitrogen everywhere, but not a bit to eat. All plants and animals need nitrogen—it's present in two major classes of biomolecules (proteins and nucleic acids, which direct the production of new organisms). But because the triple bond in the nitrogen molecule, N_2, is so strong, plants and animals can't use the free element that is 80% of the atmosphere. It's up to nature and the fertilizer industry to convert N_2 into usable nitrogen compounds, a process called *nitrogen fixation*. Plants can use ammonia (NH_3), nitrates (NO_3^-), urea (H_2NCONH_2), and other simple, water-soluble compounds. Animals get their nitrogen by eating plants or animals that feed on plants.

Most industrial fertilizer production begins with the combination of nitrogen and hydrogen to give ammonia (in the *Haber process*).

$$3 \ H_2(g) \ + \ N_2(g) \ \longrightarrow \ 2 \ NH_3(g)$$

Nature, of course, was fixing nitrogen before the fertilizer industry got started. Microorganisms such as blue-green algae and bacteria that live in the roots of plants like alfalfa, beans, and peas fix nitrogen from the atmosphere by converting it to ammonia.

Some fixed nitrogen is also created by lightning. The energy released during a lightning flash causes the endothermic combination of nitrogen and oxygen in the atmosphere to form NO, which can then combine with oxygen or ozone, ultimately producing water-soluble nitrates (NO_3^-) that plants can use. For many years, it was assumed that less than 10% of the world's supply of fixed nitrogen is created by lightning. Recent studies of the quantity of nitrogen compounds formed in each lightning strike have caused a revision of this estimate, however. There are roughly 100 lightning strikes every second, and they may be producing almost half of the world's fixed nitrogen supply.

The nitrogen cycle, showing how N_2 is fixed, used by plants and animals, and returned to the atmosphere.

8.5 WHY DO CHEMICAL REACTIONS OCCUR? FREE ENERGY

We know from observation that there is a natural tendency for events that lead to lower energy to occur spontaneously. A rock tumbles down a hill, releasing its stored (potential) energy and reaching a lower-energy, more stable position (Figure 8.3). A wound-up spring uncoils when set free. Applying the lesson of these observations to chemistry, the obvious conclusion is that exothermic processes—those that release potential energy—should be spontaneous. In many cases, this conclusion is correct because the release of energy as heat is one of the two driving forces for chemical and physical change. Endothermic processes, which absorb energy, should therefore be nonspontaneous. Often this is also correct, but not always. Some processes take place spontaneously even though they aren't exothermic.

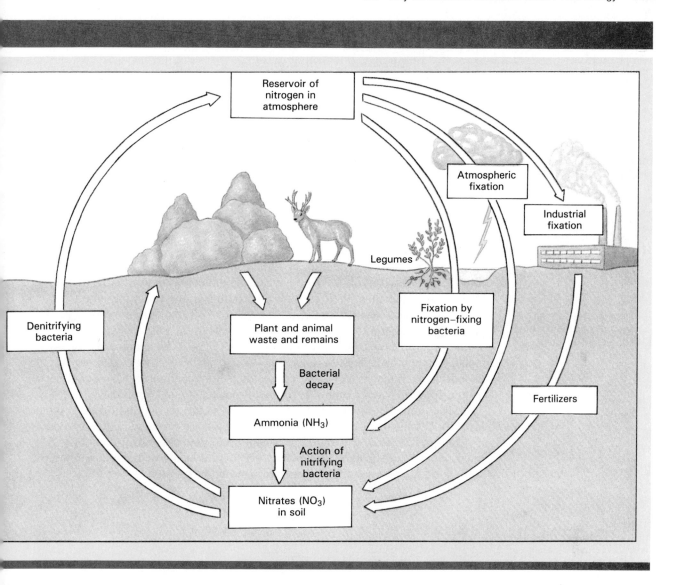

Spontaneous process A process that, once begun, proceeds without any external influence.

Before exploring the situation further, it's important to understand what "spontaneous" means in chemistry, for it's not quite the same as in everyday language. A **spontaneous process,** once started, proceeds on its own without any external influence. The change need not happen quickly, like a spring uncoiling or a rock rolling downhill. It can also happen slowly, like the gradual rusting away of an abandoned car. A *nonspontaneous process,* in contrast, takes place only in the presence of some *continuous* external influence. Energy must be expended to recoil a spring or push a car uphill. *The reverse of a spontaneous process is always nonspontaneous.*

As an example of a process that takes place spontaneously even though it doesn't give off energy, think about what happens when someone enters a room with a lighted cigarette. Before long, everyone in the room notices the smell because the smoke doesn't stay in one place. It spontaneously mixes with the air and spreads throughout the room. What this and other spontaneous proc-

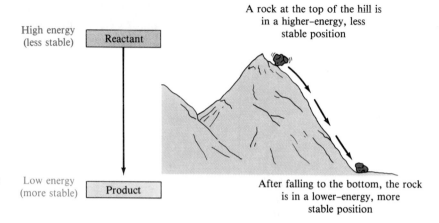

Figure 8.3
Change in potential energy. A rock rolling downhill and an exothermic reaction both result in a decrease in potential energy.

esses that don't release energy have in common is an increase in molecular disorder or randomness. As the odor-causing molecules in the smoke intermingle with air molecules, their random distribution in the room increases.

The amount of disorder in a system is called the system's **entropy,** denoted S. The greater the disorder or randomness of the particles in a substance or mixture, the larger the value of S (Figure 8.4). Gases have more disorder and higher entropy than liquids, and liquids have higher entropy than solids. In chemical reactions, entropy increases when, for example, a gas is produced from a solid or when 2 mol of reactants produces 4 mol of products.

The tendency toward increasing entropy is the second important driving force for chemical and physical change. To decide, therefore, whether a process is spontaneous, *both* the enthalpy change (ΔH) and the entropy change (ΔS) must be taken into account. When both are favorable, a process is spontaneous; when both are unfavorable, a process is not spontaneous.

Entropy (S)   A measure of the amount of molecular disorder of a substance.

Figure 8.4
Entropy. The mixture on the right has a higher entropy than that on the left and therefore has a higher value of S. The value of ΔS for converting the mixture on the left to that on the right would be positive, showing an increase in entropy.

Clearly, the two factors didn't always have to operate in the same direction. For example, it's possible to imagine a process that is *unfavored* by enthalpy (requires energy) yet favored by entropy (disorder increases). The melting of an ice cube above 0°C is just such a process. It's also possible to imagine situations in which a process can be either spontaneous or nonspontaneous depending on the conditions. The amount of disorder in a substance turns out to depend on temperature, thus making the whole question of spontaneity depend on temperature, as is the case for melting ice.

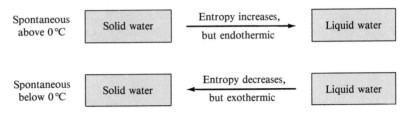

To take both heat of reaction (enthalpy) and disorder (entropy) into account when deciding about the spontaneity of a chemical reaction, a quantity called the **free energy change (ΔG)** is defined as follows:

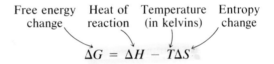

$$\Delta G = \Delta H - T\Delta S$$

Free energy change (ΔG)
A quantity that is a function of energy change and entropy change and is the criterion for spontaneous change (negative ΔG); $\Delta G = \Delta H - T\Delta S$.

The value of ΔG is negative for spontaneous processes and positive for nonspontaneous processes.

The equation for free energy change also shows how spontaneity depends on temperature. At low temperatures, $T\Delta S$ is often so small that subtracting it from ΔH makes little difference and ΔH controls spontaneity. At a high enough temperature, however, the $T\Delta S$ term can dominate by being larger than ΔH. Thus, an endothermic process that is nonspontaneous at low temperature can become spontaneous at a higher temperature. A good example is the reaction of carbon with water to yield carbon monoxide and hydrogen.

$$C(s) + H_2O(l) \longrightarrow CO(g) + H_2(g)$$

At 25°C: $\Delta H = 42$ kcal/mol $\Delta S = 60.$ cal/mol K

$$\Delta G = 42 \text{ kcal/mol} - (298 \text{ K} \times 0.061 \text{ kcal/mol K}) = 24 \text{ kcal/mol}$$

At 25°C, the reaction has an unfavorable, positive ΔH but a favorable, positive ΔS term because disorder increases when a solid and a liquid are converted into two gases. No reaction occurs if carbon and water are mixed together at 25°C (Figure 8.5a), as shown by the unfavorable, positive ΔG for the process. If the reaction is carried out at a very high temperature, however, it becomes spontaneous because above approximately 760°C the $T\Delta S$ term becomes larger than ΔH. The reaction is, in fact, the first step of an industrial process by which methanol (CH_3OH) and other organic compounds are manufactured (Figure 8.5b). As supplies of natural gas and oil are used up, this reaction may become important as a starting point for the manufacture of synthetic fuels.

(a)

(b)

Figure 8.5
Carbon and water. (a) No reaction occurs at room temperature, no matter how much we stir. (b) When superheated steam is passed over coal, carbon and water react to produce carbon monoxide and hydrogen, a mixture that can then be used in the synthesis of methanol and other organic compounds. This plant processes 900 tons of coal in this manner every day.

You won't be asked to do any calculations involving free energy and entropy in this book, but it's nevertheless helpful in understanding chemical reactions to understand the relationships among energy, entropy, temperature, and spontaneity. Knowing the free energy changes for the reactions of metabolism provides valuable insight into how metabolism works. For example, a highly favorable reaction with a large negative free energy value cannot be reversed in the body because, in reverse, the required free energy (the equal but positive ΔG) is more than the body can supply. Also, living things cannot raise temperatures to convert nonspontaneous reactions into spontaneous reactions and must resort to other strategies that we'll be examining in later chapters.

Spontaneity and free energy

● A spontaneous process, once begun, proceeds without any external assistance and has a negative ΔG.

● A nonspontaneous process requires continuous external assistance and has a positive ΔG.

● ΔG for the reverse of a reaction is equal in value to ΔG for the original reaction but has the opposite sign.

● The rate of change is unrelated to spontaneity.

● Some nonspontaneous processes become spontaneous with a change in temperature.

Practice Problems **8.4** Decide whether entropy increases or decreases in the following processes:
(a) A flower grows into a pumpkin.
(b) Gasoline fumes escape as the fuel is pumped into your car.
(c) $Mg(s) + Cl_2(g) \rightarrow MgCl_2(s)$

8.5 According to the value of ΔG, does sodium iodide dissolve spontaneously?

$$NaI(s) \xrightarrow{H_2O} Na^+ + I^- \qquad \Delta G = -1.6 \text{ kcal}$$

Does entropy increase or decrease in this change?

8.6 HOW DO CHEMICAL REACTIONS OCCUR? REACTION RATES

$$O_3 + NO \longrightarrow O_2 + NO_2$$

Effective collision

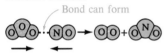

Ineffective collision

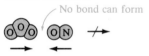

Figure 8.6
Collisions in the reaction of ozone with nitrogen(II) oxide. In an effective collision, molecules collide so that atoms that will form bonds come into contact. The collision must also occur with sufficient energy to overcome repulsion.

In considering heats of reaction and spontaneity, only the final outcome of a reaction is important. Now it's time to take a look at the changes that occur *during* a reaction.

Atoms, molecules, and ions are in constant motion. At low temperatures, these motions are relatively sluggish; at higher temperatures, motions become more vigorous and collisions between the randomly moving particles become more frequent.

For a chemical reaction to occur, reactant particles have to collide. Some chemical bonds then have to break, while new ones form. Not every collision is successful, however. In an effective collision, the colliding molecules approach so that the atoms about to form new bonds can connect (Figure 8.6), and the collision occurs with enough energy to overcome repulsion between electrons in the different reactants. It's as though a rock at the top of a hill is trapped in a small depression and isn't able to fall spontaneously. In order to fall, it first has to be lifted over the barrier and given a shove. In other words, before the rock can *release* its potential energy in a fall, something has to "collide" with it, putting enough energy into the rock to start it falling (Figure 8.7).

Similarly, just because a chemical reaction has a favorable free energy change doesn't mean the reaction will occur all by itself. Many chemical reactions need a shove to get started. We've all seen the flame from a match, but we know that the reaction doesn't begin until we strike the match. The heat of

Figure 8.7
Rock behind a barrier. Before the rock can fall, releasing its potential energy, energy must be put into it to raise it over the barrier.

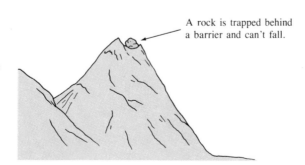

A rock is trapped behind a barrier and can't fall.

friction provides enough energy for a few molecules to react. Once started, the reaction sustains itself as the energy given off by reacting molecules pushes other molecules over the barrier.

Reaction energy diagrams like those in Figure 8.8 are helpful in visualizing the energy changes as a chemical reaction takes place. The energy levels of the reactants are shown at all stages of the reaction. At the beginning, reactants are at the energy level indicated on the left. For exothermic reactions, the energy level of the products is lower than the energy level of the reactants (Figure 8.8a). For endothermic reactions (Figure 8.8b), the situation is reversed.

The amount of energy necessary for reactants to surmount the energy barrier is called the reaction's **activation energy, E_{act}.** The activation energy determines how fast a reaction occurs—the **reaction rate,** which is given by the experimentally determined change in concentration of a reactant or product with time. The lower the activation energy, the more successful collisions there are in a given amount of time and the faster the reaction. Conversely, the higher the activation energy, the fewer successful collisions there are and the slower the reaction.

Note that the size of the activation energy and the size of the heat of reaction are *unrelated*. A reaction might take place very slowly (have a large E_{act}) even though it goes far downhill once it does react (has a large negative ΔH). Every reaction has its own characteristic activation energy and reaction rate.

Reaction energy diagram
A plot representing the energy changes that occur during a chemical reaction.

Activation energy (E_{act})
The amount of energy that reactants must have in order to surmount the energy barrier to reaction.

Reaction rate   The rate at which a chemical reaction occurs, as measured by the change of a reactant or product concentration per unit time.

Figure 8.8
Reaction energy diagrams used to show energy changes during a chemical reaction. Reaction begins on the left and proceeds to the right. In exothermic reactions, the product energy level is lower than that of reactants. In endothermic reactions, the situation is reversed. The height of the barrier between reactant energy level and the peak is the activation energy, E_{act}; the difference between reactant and product energy levels is the heat of reaction.

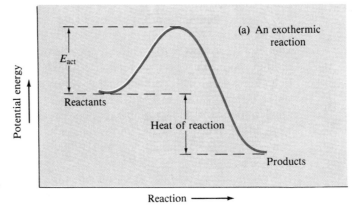

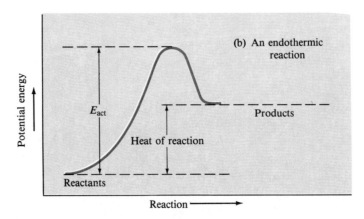

Solved Problem 8.2   Draw a reaction energy diagram for a reaction that is very fast but releases only a small amount of heat.

Solution   A very fast reaction is one that has a small E_{act}, and a reaction that releases only a small amount of heat is one that has a small heat of reaction. Thus, the diagram must show a small energy barrier and a small energy difference between starting materials and products.

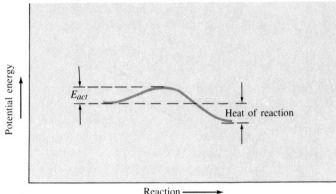

Reaction energy diagram for Solved Problem 8.2. Both E_{act} and the heat of reaction are small.

Practice Problem   **8.6**   Draw a reaction energy diagram for a reaction that is very slow and highly endothermic.

8.7 EFFECT OF TEMPERATURE, CONCENTRATION, AND CATALYSTS ON REACTION RATES

Several things can be done to help reactants over an activation energy barrier and thereby speed up a reaction. One possibility is to add energy to the reactants by raising the temperature.

Increase in temperature $\longrightarrow$ increase in reaction rate

With more energy in the system, successful collisions between reactants are more likely for two reasons. (1) The atoms, molecules, or ions move faster with increasing temperature, and the frequency of collisions therefore increases. (2) At higher temperatures the fraction of particles moving faster increases. As a result, there are more collisions forceful enough to overcome the activation energy.

In general, a 10°C rise in temperature doubles a reaction rate, although there are many exceptions to this rule of thumb. It's certain, however, that some increase in reaction rate must accompany an increase in temperature. A bicycle definitely rusts faster on a warm day in summer than on a cold day in winter.

A second possibility for speeding up a reaction is to increase the concentrations of the reactants. With reactants more crowded together, collisions become more frequent and reactions more likely. Thus, flammable materials burn much more rapidly in pure oxygen than in air because the concentration of O_2 molecules is higher (air is approximately 20% oxygen). For this reason, hospitals must take extraordinary precautions to be sure that no flames are used near patients receiving oxygen. Although different reactions respond differently to concentration changes, doubling or tripling a reactant concentration often doubles or triples the reaction rate.

Increase in reactant concentration $\longrightarrow$ increase in reaction rate

Catalyst A substance that speeds up a chemical reaction without itself undergoing any permanent chemical change.

A third possibility for speeding up a reaction is to add a catalyst. **Catalysts** are substances that accelerate chemical reactions but are unchanged when the reaction is completed. For example, natural oils contain carbon–carbon double bonds, while solid fats contain mainly carbon–carbon single bonds. Vegetable oils are converted to semisolid margerine by addition of hydrogen to some of their double bonds in the presence of a finely divided nickel catalyst:

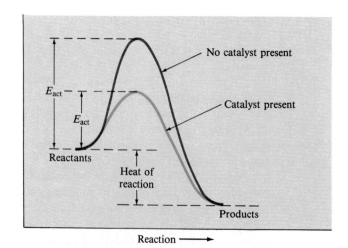

Double bonds
in natural oil

Single bonds
in margerine

Catalysts increase reaction rates by lowering the height of the activation energy barrier. To return once more to the analogy of a rock behind a barrier at the top of a hill, a catalyst functions by finding some alternative *low* point in the barrier for the rock to be lifted over. On a reaction energy diagram, the catalyzed reaction has a lower activation energy (Figure 8.9).

Addition of catalyst $\longrightarrow$ increase in reaction rate

Figure 8.9
A reaction energy diagram for a reaction in the presence (green line) and absence (red line) of a catalyst. The catalyzed reaction has a lower E_{act} because it uses an alternate pathway with a lower energy barrier. The heat of reaction is unaffected by the presence of a catalyst.

No catalyst present

E_{act}

Catalyst present

E_{act}

Reactants

Heat of
reaction

Products

Reaction $\longrightarrow$

APPLICATION: REGULATION OF BODY TEMPERATURE

Maintenance of normal body temperature is crucial if the chemical reactions of metabolism are to occur at the right rates. If the body's thermostat is unable to maintain a temperature of 37°C, the rates of the many thousands of chemical reactions that take place constantly in the body will change accordingly.

If, for example, a hiker in the mountains were to be trapped overnight by a sudden storm and unable to keep warm, **hypothermia** could result. Hypothermia is a dangerous state that occurs when the body is unable to generate enough heat to maintain normal temperature. All chemical reactions in the body slow down because of the lower temperature, the supply of energy drops, and death can result when the body is no longer able to fuel itself. The slowing of metabolism is often used to advantage during open-heart surgery, however. The heart is stopped and maintained at about 15°C, while the body, which receives oxygenated blood from an external pump, is cooled to 25–32°C.

Conversely, a marathon runner on a hot, humid day might become overheated, and **hyperthermia** could result. Hyperthermia, also called *heat stroke,* is an uncontrolled rise in temperature as the result of the body's inability to lose sufficient heat. All chemical reactions in the body are accelerated at higher temperatures, the heart struggles to pump blood faster to supply increased oxygen, and brain damage can result.

Body temperature is maintained both by the thyroid gland and by the hypothalamus region of the brain, which act together to regulate the metabolic rate (see the accompanying figure). When the body's environment changes, temperature receptors in the skin, spinal cord, and abdomen send signals to the hypothalamus, which contains both heat-sensitive neurons and cold-sensitive neurons that detect temperature changes.

Stimulation of the heat-sensitive neurons on a hot day causes a variety of effects: Impulses are sent to stimulate the sweat glands, dilate the blood vessels of the skin, decrease muscular activity, and reduce metabolic rate. Sweating cools the body through evaporation; approximately 540 cal of heat is removed by evaporation of 1 g of sweat. Dilated blood vessels cool the body by allowing more blood to flow close to the surface of the skin where heat is removed by contact with air. Decreased muscular activity and a reduced metabolic rate cool the body by lowering internal heat production.

Stimulation of the cold-sensitive neurons on a cold day also causes a variety of effects: The hormone epinephrine is released to stimulate metabolic rate; peripheral blood vessels contract to decrease blood flow to the skin and prevent heat loss; and muscular contractions increase to produce more heat, resulting in shivering and "goosebumps."

One further comment: Drinking alcohol to warm up on a cold day actually has exactly the opposite of the intended effect. Alcohol causes blood vessels to dilate, resulting in a warm feeling as blood flow to the skin increases. Although the warmth feels good temporarily, body temperature ultimately drops as heat is lost through the skin at an increased rate.

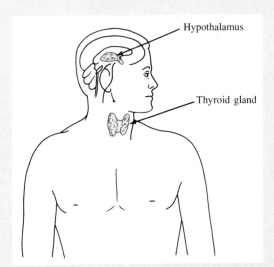

Location of the hypothalamus and the thyroid gland, the body's heat regulation centers.

We also rely on catalysts to reduce the air pollution created by exhaust from automobile engines. The catalytic converters in most automobiles are tubes packed with catalysts of two types. One accelerates the complete combustion of hydrocarbons and CO in the exhaust to give CO_2 and H_2O, and the other decomposes NO to N_2 and O_2. In general, reactants adhere to the surface of solid catalysts. In this fixed position the reactants are held in place, making successful reaction easier and lowering the activation energy.

Enzyme A protein that acts as a catalyst for biological reactions.

Enzymes, the catalysts in biochemical reactions, are very large molecules that work in essentially the same way. A biomolecule reactant can be trapped within an enzyme molecule in just the right position for easy reaction. When even one of our thousands of enzymes malfunctions, drastic consequences to health can result.

8.8 REVERSIBLE REACTIONS AND CHEMICAL EQUILIBRIUM

Many chemical reactions result in virtually complete conversion of the reactants into products. We'd be very surprised, for example, to find any unchanged sodium metal left after its reaction with water.

$$2 \text{ Na}(s) + 2 \text{ H}_2\text{O}(l) \longrightarrow 2 \text{ NaOH}(aq) + \text{H}_2(g)$$

The products are so much more stable than the reactants that, once started, the reaction keeps going until the reactants are used up. And no one has yet found a way to make the reverse reaction occur.

What happens, though, when the reactants and products are of approximately *equal* stability? This is the case in the reaction of acetic acid (vinegar) with ethyl alcohol to yield ethyl acetate, a solvent used in nail polish remover.

$$\underset{\text{Acetic acid}}{\text{CH}_3\overset{\displaystyle O}{\overset{\|}{\text{C}}}\text{OH}} + \underset{\text{Ethyl alcohol}}{\text{HOCH}_2\text{CH}_3} \underset{\text{or this direction?}}{\overset{\text{this direction?}}{\rightleftharpoons}} \underset{\text{Ethyl acetate}}{\text{CH}_3\overset{\displaystyle O}{\overset{\|}{\text{C}}}\text{OCH}_2\text{CH}_3} + \underset{\text{Water}}{\text{H}_2\text{O}}$$

Reversible reaction A reaction that can proceed in either the forward or the reverse direction.

Imagine the situation if you mixed acetic acid and ethyl alcohol. The two would begin to react to form ethyl acetate and water. But as soon as ethyl acetate and water formed, *they* would begin to react to go back to acetic acid and ethyl alcohol. Such a reaction, which can easily go in either direction, is said to be a **reversible reaction.**

Chemical equilibrium The point in a reversible reaction at which forward and reverse reactions take place at the same rate, so that the concentrations of products and reactants no longer change.

Suppose instead you mixed some ethyl acetate and water. The same thing would occur—as soon as small quantities of acetic acid and ethyl alcohol form, the reaction in the other direction would begin to take place. No matter which pair of reactants is mixed together, both reactions occur until ultimately a **chemical equilibrium** is established. Reactions that reach equilibrium are represented in equations by a double arrow:

$$\text{CH}_3\overset{\displaystyle O}{\overset{\|}{\text{C}}}\text{OH} + \text{HOCH}_2\text{CH}_3 \rightleftharpoons \text{CH}_3\overset{\displaystyle O}{\overset{\|}{\text{C}}}\text{OCH}_2\text{CH}_3 + \text{H}_2\text{O}$$

The reaction read from left to right is referred to as the *forward reaction,* and the reaction from right to left is referred to as the *reverse reaction.* At equilibrium, the reaction vessel will contain all four substances—acetic acid, ethyl acetate, ethyl alcohol, and water.

You might think that at equilibrium the forward and reverse reactions have stopped, but this is not the case. When the reaction begins, the forward and reverse reaction rates change continuously as shown at the left in Figure 8.10. Once the rates become equal, they change no further. Chemical equilibrium is a dynamic condition in which the *rates of the forward and reverse reactions are equal.* Because the substances present are continuously being formed and are reacting at the same rate, the *concentrations* of reactants and products are constant at equilibrium. So long as the reaction conditions are unchanged, the concentrations will remain the same.

You might also think that the concentrations of reactants and products at equilibrium have become equal, but this too is incorrect. *Equilibrium can be reached at any point between pure products and pure reactants.* The extent to which the forward or reverse reaction is favored over the other is a characteristic property of a given reaction under given conditions.

Figure 8.10
Reaction rates in an equilibrium reaction. Initially the forward rate decreases as the concentrations of reactants drop, and the reverse rate increases as the concentrations of products increase. At equilibrium, the forward and reverse reaction rates are equal.

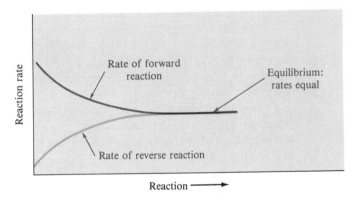

8.9 EQUILIBRIUM CONSTANT EXPRESSIONS

Experience has led to a mathematical expression that describes any chemical reaction at equilibrium. Suppose a variety of different mixtures of sulfur dioxide and oxygen are allowed to come to equilibrium

$$2\ SO_2(g)\ +\ O_2(g)\ \rightleftharpoons\ 2\ SO_3(g)$$

and the concentrations of all three gases then measured at a constant temperature. All experiments will show that no matter how the original concentrations are varied and no matter what concentrations remain at equilibrium, substituting the equilibrium concentrations into the following expression and carrying out the calculation gives the same answer when the temperature is 727°C.

$$\frac{[SO_3]^2}{[SO_2]^2[O_2]} = 429$$

Equilibrium constant expression Expression showing the concentrations of reactants and products in a reaction at equilibrium; equal to a constant

$$K = \frac{[M]^m[N]^n \cdots}{[A]^a[B]^b \cdots}$$

The square brackets are a general notation for the concentration in moles per liter of the substance shown within the brackets.

Numerous experiments like the one just described have led to what is known as the **equilibrium constant expression.** For a general reaction in which the reactants are A, B, . . ., the products are M, N, . . ., and their coefficients in the equation are a, b, . . ., and m, n, . . .

$$aA + bB + \cdots \rightleftharpoons mM + nN + \cdots$$

the expression is

$$K = \frac{[M]^m[N]^n \cdots}{[A]^a[B]^b \cdots}$$

Product concentrations raised to powers equal to coefficients

Reactant concentrations raised to powers equal to coefficients

Equilibrium constant (K) Value of the equilibrium constant expression for a given reaction.

where K is known as the **equilibrium constant.** Because the value of K varies with temperature, it's customary to either give K values for room temperature (25°C) or specify the temperature.

The value of the equilibrium constant indicates what is referred to as the *position* of a reaction at equilibrium. If the forward reaction is favored, $[M]^m[N]^n$ is larger than $[A]^a[B]^b$ and the value of K is larger than 1. If, instead, the reverse reaction is favored, $[M]^m[N]^n$ is smaller than $[A]^a[B]^b$ at equilibrium, and the value of K is smaller than 1.

For a reaction such as the combination of hydrogen and oxygen to form water vapor, the equilibrium constant is very, very large, showing how greatly the formation of water is favored over the reverse reaction:

$$2\,H_2(g) + O_2(g) \rightleftharpoons 2\,H_2O(g) \quad K = 3.1 \times 10^{81} \quad \text{Goes to completion}$$

Equilibrium is ignored for such reactions, and the reaction is described as *going to completion.* For a reaction such as the combination of nitrogen and oxygen at 25°C to give nitrogen(II) oxide, the equilibrium constant is very small, showing what we know from observation—the nitrogen and oxygen in the air don't combine noticeably at room temperature:

$$N_2(g) + O_2(g) \rightleftharpoons 2\,NO(g) \quad K = 4.7 \times 10^{-31}$$

When K is close to 1, say between 10^2 and 10^{-2}, significant amounts of both reactants and products are present at equilibrium.

Meaning of equilibrium constant values

- $K > 1$, forward reaction is favored; $K < 1$, reverse reaction is favored
- K very much larger than 1—reaction goes to completion
- K of intermediate value—significant amounts of reactants and products at equilibrium
- K very much smaller than 1—essentially no reaction.

Solved Problem 8.3   The concentrations of reactants and products in the combination of chlorine with phosphorus trichloride

$$PCl_3(g) + Cl_2(g) \rightleftharpoons PCl_5(g)$$

were determined experimentally at equilibrium and found to be 7.2 mol/L of PCl_3, 7.2 mol/L of Cl_2, and 0.050 mol/L of PCl_5. Write the equilibrium constant expression and calculate the equilibrium constant for the reaction at the temperature of these measurements. Is the forward or the reverse reaction favored?

Solution   The coefficients in this reaction are each 1, so the equilibrium constant expression is simply the concentration of the product, PCl_5, divided by the concentrations of the two reactants, PCl_3 and Cl_2.

$$K = \frac{[PCl_5]}{[PCl_3][Cl_2]}$$

Inserting the measured concentrations and calculating the value of K gives

$$K = \frac{[PCl_5]}{[PCl_3][Cl_2]} = \frac{0.050}{(7.2)(7.2)} = 9.6 \times 10^{-4}$$

The value of K is less than 1, showing that the reverse reaction is favored. (Note that it is customary to omit units in equilibrium constants and in calculations involving them.)

Practice Problems   **8.7**   Write equilibrium constant expressions for the following reactions:
(a) $N_2O_4(g) \rightleftharpoons 2\ NO_2(g)$
(b) $CH_4(g) + Cl_2(g) \rightleftharpoons CH_3Cl(g) + HCl(g)$
(c) $2\ BrF_5(g) \rightleftharpoons Br_2(g) + 5\ F_2(g)$

8.8   Do the following reactions favor reactants or products at equilibrium?
(a) $sucrose(aq) + H_2O(l) \rightleftharpoons glucose(aq) + fructose(aq)$ $K = 1.4 \times 10^5$
(b) $NH_3(aq) + H_2O(l) \rightleftharpoons NH_4^+ + OH^-$ $K = 1.6 \times 10^{-5}$
(c) $Fe_2O_3(s) + 3\ CO(g) \rightleftharpoons 2\ Fe(s) + 3\ CO_2(g)$ $K\ (at\ 727°C) = 24.2$

8.10 LECHATELIER'S PRINCIPLE: THE EFFECT OF CHANGING CONDITIONS ON EQUILIBRIA

The effect on chemical equilibria of a change in conditions can be predicted by a general rule called LeChatelier's principle:

LeChatelier's principle
When a stress is applied to a system in equilibrium, the equilibrium shifts so that the stress is relieved.

LeChatelier's principle: When a stress is applied to a system in equilibrium, the equilibrium shifts to relieve the stress.

A chemical equilibrium is stressed or disrupted by any change that causes the rates of the forward and reverse reactions to become unequal.

You've seen that reaction rates are affected by concentrations, temperature, and catalysts. What about equilibria? In the following sections, we'll look at what happens to equilibria when concentration, temperature, and pressure are changed. A catalyst has no effect except to reduce the time it takes to reach equilibrium, because the change it causes in E_{act} equally affects both forward and reverse reactions (look back at Figure 8.9). Once equilibrium is reached, concentrations will be the same as they would have been in the absence of the catalyst.

Effect of Concentration Changes on Equilibria To start with, consider the reaction of carbon monoxide with hydrogen to form methanol. Once equilibrium has been reached, the concentrations of the reactants and product are constant, and the forward and reverse reaction rates have adjusted so that they are equal.

$$CO(g) + 2 H_2(g) \rightleftharpoons CH_3OH(g)$$

How will this system respond if the concentration of CO is increased? To relieve this "stress," as expected according to LeChatelier's principle, the extra CO will be used up—in other words, the rate of the forward reaction will increase. You can think of the CO added on the left as "pushing" the equilibrium to the right:

$$CO \longrightarrow$$

$$CO(g) + 2 H_2(g) \rightleftharpoons CH_3OH(g)$$

As soon as more CH_3OH forms, the reverse reaction will speed up too and some will be converted back to CO and H_2. Once again, as when the reactants were first combined, the forward and reverse reaction rates must adjust until they are equal. When equilibrium is reestablished, the value of $[H_2]$ will be lowered because more H_2 has reacted with the added CO, and the value of $[CH_3OH]$ will be raised. The changes offset each other, however, and the new concentrations, like the original concentrations, will equal K when used in the equilibrium constant expression.

$$CO(g) + 2 H_2(g) \rightleftharpoons CH_3OH(g)$$

If this increases . . .   . . . then this decreases . . .   . . . and this increases . . .

. . . but this remains constant.
$$K = \frac{[CH_3OH]}{[CO]\,[H_2]^2}$$

What would happen if CH_3OH were added to the reaction at equilibrium? Some of the methanol would react to yield CO and H_2, and when equilibrium was reestablished, the values of [CO], $[H_2]$, and $[CH_3OH]$ would all be higher. Again, though, the value of K would be unchanged, and these new concentrations would give K when used in the equilibrium constant expression.

If this increases . . .

$$CO(g) + 2 H_2(g) \rightleftharpoons CH_3OH(g)$$

. . . then this increases . . .   . . . and this increases . . .

. . . but this remains constant.
$$K = \frac{[CH_3OH]}{[CO]\,[H_2]^2}$$

What do you suppose happens when a reactant is continuously supplied or a product is continuously removed? Since the concentrations are continuously changing, equilibrium can never be reached. It's therefore sometimes possible to force a reaction to produce large quantities of a desirable product even when the equilibrium constant is unfavorable. Take the reaction of acetic acid with ethanol to yield ethyl acetate, for example. As discussed in the previous section, the equilibrium constant K for this reaction is close to 1, meaning that substantial amounts of reactants are still present at equilibrium. If, however, the ethyl acetate is removed as rapidly as it's formed, the production of more and more product is forced to occur, in accord with LeChatelier's principle. (In practice, the removal of ethyl acetate can be done simply by boiling it away since the boiling point of ethyl acetate is 13 C° lower than that of water.)

Continuously removing this product from the reaction forces more of it to be produced.

$$CH_3CO_2H + C_2H_5OH \rightleftharpoons CH_3CO_2C_2H_5 + H_2O$$

Acetic acid   Ethyl alcohol   Ethyl acetate

Sometimes the reactions of metabolism take advantage of this effect, with one reaction prevented from reaching equilibrium by the continuous consumption of its product by the next reaction in a series.

Effect of Temperature and Pressure Changes on Equilibria You've seen that the reverse of an exothermic reaction is always endothermic (Figure 8.8). Equilibrium reactions are therefore exothermic in one direction and endothermic in the other. LeChatelier's principle predicts that an increase in temperature will cause an equilibrium to shift in favor of the reaction that absorbs energy: the endothermic reaction. Conversely, a decrease in temperature will cause an equilibrium to shift in favor of the reaction that releases energy: the exothermic reaction. Because the forward and reverse reaction rates are affected differently, the *value* of the equilibrium constant changes with temperature. In the exothermic reaction of nitrogen with hydrogen to form ammonia, for example,

Heat

$$N_2(g) + 3 H_2(g) \rightleftharpoons 2 NH_3(g) + heat$$

At 25°C, $K = 5.8 \times 10^5$ At 400°C, $K = 1.8 \times 10^{-4}$

the equilibrium constant changes from 5.8×10^5 at 25°C to 1.8×10^{-4} at 400°C—a radical change from an equilibrium position that strongly favors the formation of ammonia to one that strongly favors its decomposition. In the case of this industrially important reaction, raising the temperature will do nothing to help improve the yield of ammonia.

What about changing the pressure? Pressure influences an equilibrium if one or more of the substances involved is a gas. In such reactions, a pressure increase shifts the equilibrium in the direction that decreases the number of molecules in the gas phase (and thereby the pressure). For the ammonia synthesis, since 4 mol of gas is converted to 2 mol of gas, a pressure increase would favor increased ammonia yields.

$$\text{Pressure} \longrightarrow$$

$$N_2(g) + 3\ H_2(g) \longrightarrow 2\ NH_3(g)$$

No matter how pressure and temperature are adjusted, however, the ammonia synthesis reaction is too slow to be useful. Addition of an iron–aluminum oxide catalyst allows a practical compromise to be made between temperature and pressure, resulting in industrial production of more than 12 billion kilograms of ammonia each year in the United States.

The effects of changing reaction conditions on reaction rates and equilibria are summarized in Table 8.4.

Table 8.4 Effects of Changing Reaction Conditions on Reaction Rates and Equilibria

Change	Effect on Reaction Rate	Effect on Equilibrium
Temperature change	Increases with higher T and decreases with lower T	Value of K changes
Addition of catalyst	Increases	No change
Change in reactant concentrations	Increases with higher reactant concentrations and decreases with lower reactant concentrations	No change in value of K; concentrations of reactants and products readjust to equal K

Solved Problem 8.4 For the endothermic combination of nitrogen and oxygen to give nitrogen(II) oxide,

$$N_2(g) + O_2(g) \rightleftharpoons 2\ NO(g) \qquad \Delta H = 43\ \text{kcal}$$

explain the effects on reactant and product concentrations and the value of K of (a) increasing temperature, (b) increasing the concentration of NO, (c) adding a catalyst.

Solution (a) The reaction is endothermic (ΔH is positive), so (thinking of heat as a reactant) increasing temperature will favor the forward reaction and the concentration of NO will be higher at equilibrium. The value of K will change.

(b) Increasing the concentration of the product NO will favor the reverse reaction. At equilibrium, the concentrations of both N_2 and O_2, as well as that of NO, will be higher. K will not change.

(c) A catalyst will accelerate the rate at which equilibrium is reached, but the concentrations at equilibrium (all other conditions remaining the same) will be unchanged. K will not change.

INTERLUDE: LECHATELIER AND OXYGEN TRANSPORT

The body has many strategies for making substances available where and when they are needed. Some essential substances are kept in storage, waiting for chemical signals to trigger their release. Others are synthesized as needed. Oxygen from the atmosphere is made available in body tissues by equilibrium shifts governed by LeChatelier's principle. After entering the lungs, oxygen is transferred to arterial blood. Only about 3% of the oxygen in blood is simply dissolved; the rest is chemically bound to hemoglobin molecules (Hb), which store the oxygen until it reaches tissues where it is needed. The pickup and release of oxygen by hemoglobin is an equilibrium process. The oxygen supply in the alveoli of the lungs is so high that about 98% of the Hb molecules in blood are fully saturated with oxygen on leaving the lungs.

$$O_2 \longrightarrow$$
$$Hb(aq) + O_2(aq) \rightleftharpoons HbO_2(aq)$$

When blood reaches capillaries far from the lungs, the equilibrium is disturbed because there is less oxygen in the surrounding tissues than in the lungs. As expected according to LeChatelier's principle, HbO_2 gives up oxygen to restore equilibrium. More oxygen is released to active tissues that are consuming oxygen quickly than to inactive tissues, where the drop in the amount of oxygen is not so large.

$$O_2 \longleftarrow$$
$$Hb(aq) + O_2(aq) \rightleftharpoons HbO_2(aq)$$

The rate of oxygen release and pickup by hemoglobin is also affected (also according to LeChatelier's principle) by changes in acid (H^+) concentration, CO_2 concentration, and temperature. For example, a hydrogen ion can displace O_2 from oxygenated hemoglobin:

$$HbO_2(aq) + H^+ \rightleftharpoons HbH^+ + O_2(aq)$$

Muscles that are oxygen-starved because they are working hard produce lactic acid, providing H^+ that drives this equilibrium toward oxygen release.

A person ascending quickly to a high altitude can feel quite sick because of lack of oxygen, a condition known as *hypoxia*. After a few weeks, though, the body adjusts to high altitude by producing more hemoglobin molecules, thereby driving the Hb–O_2 reaction to the right and allowing the needed amount of oxygen to be transported. The breathing rate and heart rate also increase, so more oxygen becomes available.

The people who make their homes in the Himalayas in Nepal have a higher concentration of hemoglobin in their blood than those of us who live at lower altitudes.

Practice Problems   **8.9**   Will a high yield of sulfur trioxide at equilibrium be favored by a high or low pressure?

$$2 \, SO_2(g) + O_2(g) \; \rightleftharpoons \; 2 \, SO_3(g) \qquad \Delta H = -47 \text{ kcal}$$

8.10   What effect will each of the listed changes have on equilibrium in the reaction of carbon with hydrogen?

$$C(s) + 2 \, H_2(g) \; \rightleftharpoons \; CH_4(g) \qquad \Delta H = -18 \text{ kcal}$$

(a) increase temperature
(b) increase pressure
(c) allow CH_4 to escape continuously from reaction vessel

SUMMARY

A chemical reaction may be **exothermic** (negative ΔH) or **endothermic** (positive ΔH) depending on the relative energies of the reactants and products. The **heat of reaction** (ΔH, also known as the **enthalpy change**) is used, together with the balanced equation, to find the heat released or absorbed for a given quantity of reactant or product.

Some reactions are **spontaneous** and occur without external influence. Spontaneity is dependent on the heat of reaction, the temperature, and the change in **entropy** (ΔS), which is the increase or decrease in disorder of the products relative to the reactants. A negative **free energy change** ($\Delta G = \Delta H - T\Delta S$) indicates spontaneity, and a positive free energy change indicates nonspontaneity.

Chemical reactions occur when reactant atoms, molecules, or ions collide with sufficient energy to overcome repulsion and in an orientation that allows bond formation. The exact amount of collision energy necessary is called the **activation energy** (E_{act}). A high activation energy results in a slow reaction because few collisions occur with sufficient vigor, whereas a low activation energy results in a fast reaction. Reaction rates can be increased by raising the temperature, by raising the concentrations of reactants, or by adding a **catalyst,** which accelerates a reaction without itself undergoing any change.

A reaction that is **reversible** can go in either direction and will ultimately reach **chemical equilibrium.** At this point, the forward and reverse reactions occur at the same rate, and the concentrations of reactants and products are constant. Every reversible reaction has a characteristic value for the **equilibrium constant** (K), given by the **equilibrium constant expression** for that reaction.

$$K = \frac{[M]^m[N]^n \cdots}{[A]^a[B]^b \cdots}$$

Product concentrations raised to powers equal to coefficients

Reactant concentrations raised to powers equal to coefficients

When the forward reaction is favored, K is larger than 1; when the reverse reaction is favored, K is less than 1. **LeChatelier's principle** states that when a stress is applied to a system in equilibrium, the equilibrium shifts so that the stress is relieved. Applying this principle allows prediction of the effects of changing conditions on chemical reactions.

REVIEW PROBLEMS

Energy in Chemical Reactions

8.11   For an endothermic reaction, is the total energy of the reactants more or less than the total energy of the products?

8.12   State the law of conservation of energy.

8.13   What is meant by the term "heat of reaction"? What other name is given to the heat of a reaction?

8.14   The vaporization of Br_2 from the liquid to the gas state requires 7.4 kcal/mol.

(a) What is the sign of ΔH for this process?

(b) How many kilocalories of heat are needed to vaporize 6.5 mol of Br_2?

(c) How many kilocalories of heat are required to evaporate 75 g of Br_2?

8.15 Glucose, also known as "blood sugar," has the formula $C_6H_{12}O_6$.

(a) Write the equation for the combustion of glucose.

(b) If 4.1 kcal is provided by combustion of each gram of glucose, how many kcal are released by the combustion of 1.50 mol of glucose?

(c) What is the minimum amount of energy a plant must absorb to product 15.0 g of glucose?

8.16 During the combustion of 5.00 g of ethanol, C_2H_5OH, 35.5 kcal of energy is released.

(a) Write a balanced equation for the combustion reaction.

(b) What is the sign of ΔH for this reaction?

(c) How much energy is released by the combustion of 1.00 mol of C_2H_5OH?

(d) How many grams and how many moles of C_2H_5OH must be burned to release 450. kcal of energy?

(e) How many kcal of energy are released by the combustion of 10.0 g of C_2H_5OH?

(f) How many grams of C_2H_5OH must be burned to raise the temperature of 500.0 mL of water from 20.0°C to 100.0°C?

8.17 What does entropy measure?

8.18 Which of these processes results in an increase in disorder of the system?

(a) a drop of ink spreading out when it is placed in water

(b) steam condensing into drops on windows

(c) constructing a building from loose bricks

(d) $I_2(s) + 3\ F_2(g) \rightarrow 2\ IF_3(g)$

(e) a precipitate forming when two solutions are mixed

8.19 What is meant by a "spontaneous" process?

8.20 How is the sign of free energy related to the spontaneity of a process?

8.21 Why are most spontaneous reactions exothermic?

8.22 $NaCl(s) \xrightarrow{\text{water}} Na^+(aq) + Cl^-(aq)$

$$\Delta H = 1.00\ \text{kcal}$$

(a) Is this process endothermic or exothermic?

(b) Does entropy increase or decrease in this process?

(c) NaCl, table salt, readily dissolves in water. Explain in terms of your answers to parts (a) and (b).

(d) Using the above information, would you expect NaCl to be more soluble in hot or in cold water? Does your answer agree with what you have observed?

8.23 $2\ Hg(l) + O_2(g) \rightarrow 2\ HgO(s) \quad \Delta G = -43\ \text{kcal}$

(a) Is this process spontaneous?

(b) Does entropy increase or decrease in this process?

8.24 The combination reaction of gaseous hydrogen and liquid bromine to give gaseous hydrogen bromide has $\Delta H = -17.4$ kcal, $\Delta S = 27.2$ J/K, and $\Delta G = -25.5$ kcal.

(a) Write the balanced equation for this reaction.

(b) Does entropy increase or decrease in this process?

(c) Why is this process spontaneous at all temperatures?

Rates of Chemical Reactions

8.25 What is the activation energy of a reaction?

8.26 Which reaction is faster at the same temperature and similar concentration, one with $E_{act} = 10$ kcal/mol or one with $E_{act} = 5$ kcal/mol? Explain your answer.

8.27 Draw reaction energy diagrams for exothermic reactions that meet these descriptions.

(a) a very slow reaction that has a small heat of reaction

(b) a very fast reaction that has a high heat of reaction

8.28 Draw a reaction energy diagram for a reaction whose products have the same energies as its reactants. What is the heat of reaction in this case?

8.29 Sketch a reaction energy diagram for a reaction that has an activation energy of 12 kcal/mol and a heat of reaction of −6 kcal/mol.

8.30 Give two reasons why increasing temperature increases the rate of a reaction.

8.31 Why does increasing concentration generally increase the rate of a reaction?

8.32 What is a catalyst and what effect does it have on the activation energy of a reaction?

8.33 Superimpose on your sketch for Problem 8.29 the reaction energy diagram for the same system in the presence of a catalyst that lowers the activation energy to 3 kcal/mol. Does the heat of the reaction change?

8.34 C (s, diamond) → C (s, graphite)

$$\Delta G = -0.693\ \text{kcal}$$

(a) According to these data, do diamonds spontaneously turn into graphite?

(b) In light of your answer to part (a) what is the probable reason that diamonds can be kept for hundreds (or thousands) of years?

Chemical Equilibria

8.35 What is meant by the term "chemical equilibrium"? Must amounts of reactants and products be equal at equilibrium?

8.36 Why do the rates of most reactions decrease as the reaction progresses?

8.37 Why don't catalysts alter the amount of reactants and products present at equilibrium?

8.38 Write the equilibrium constant expressions for the following reactions.

(a) $2\ CO(g) + O_2(g) \rightleftarrows 2\ CO_2(g)$

(b) $C_2H_6(g) + 2 Cl_2(g) \rightleftharpoons C_2H_4Cl_2(g) + 2 HCl(g)$
(c) $CaCO_3(s) \rightleftharpoons CaO(s) + CO_2(g)$
(d) $HF(aq) + H_2O(l) \rightleftharpoons H_3O^+(aq) + F^-(aq)$
(e) $3 O_2(g) \rightleftharpoons 2 O_3(g)$

8.39 Do the following reactions favor reactants or products?

(a) $S_2(g) + 2 H_2(g) \rightleftharpoons 2 H_2S(g)$ $K = 2.8 \times 10^{-21}$
(b) $CO(g) + 2 H_2(g) \rightleftharpoons CH_3OH(g)$ $K = 10.5$
(c) $Br_2(g) + Cl_2(g) \rightleftharpoons 2 BrCl(g)$ $K = 58.0$
(d) $I_2(s) \rightleftharpoons 2 I(g)$ $K = 6.8 \times 10^{-3}$
(e) $SO_2(g) + Cl_2(g) \rightleftharpoons SO_2Cl_2(g)$ $K = 12.8$

8.40 For the reaction $2 NO_2(g) \rightleftharpoons N_2O_4(g)$, equilibrium concentrations at 25°C were $[NO_2] = 0.025$ M and $[N_2O_4] = 0.0869$ M.
(a) What is the numerical value of K at this temperature?
(b) What is the value of $[NO_2]$ if $[N_2O_4] = 0.40$ mol/L?

8.41 Use your answer from Problem 8.40 to calculate $[N_2O_4]$ at equilibrium when $[NO_2] = 0.12$ mol/L.

8.42 Would you expect to find relatively more reactants or more products for the reaction in Problem 8.40 if the pressure were raised?

8.43 Fluorine can react with oxygen to yield oxygen difluoride:

$$2 F_2(g) + O_2(g) \longrightarrow 2 OF_2(g)$$

If the following concentrations are found at equilibrium, what is the value of K?

$$[O_2] = 0.200 \text{ M}, [F_2] = 0.0100 \text{ M}, [OF_2] = 0.0633 \text{ M}.$$

8.44 Oxygen can be converted into ozone by the action of lightning or electric sparks: $3 O_2(g) \rightleftharpoons 2 O_3(g)$. For this process, $\Delta H = +68$ kcal and $K = 2.68 \times 10^{-29}$ at 25°C.

(a) Is the reaction exothermic or endothermic?
(b) Are the reactants or the products favored at equilibrium?
(c) Explain the effect on the equilibrium of: (1) increasing pressure (2) increasing the concentration of $O_2(g)$ (3) increasing the concentration of $O_3(g)$ (4) adding a catalyst (5) increasing the temperature

8.45 Hydrogen chloride can be made from the reaction of elemental chlorine and hydrogen: $Cl_2(g) + H_2(g) \rightleftharpoons 2 HCl(g)$. For this reaction, $K = 2.6 \times 10^{33}$ and $\Delta H = -44$ kcal at 25°C.
(a) Is the reaction endothermic or exothermic?
(b) Are the reactants or the products favored at equilibrium?
(c) Explain the effect on the equilibrium of:
(1) increasing pressure
(2) increasing the concentration of $HCl(g)$
(3) increasing the concentration of $Cl_2(g)$
(4) decreasing the concentration of $H_2(g)$
(5) adding a catalyst

Applications

8.46 What does it mean to "fix" nitrogen? [App: Nitrogen Fixation]

8.47 For the production of ammonia from its elements, $\Delta H = -22$ kcal. [App: Nitrogen Fixation]

$$N_2(g) + 3 H_2(g) \longrightarrow 2 NH_3$$

(a) Is this process endothermic or exothermic?
(b) How many kilocalories of heat are involved in the production of 0.500 mol of NH_3?

8.48 What body organs help to regulate body temperature? [Int: Body Temperature]

8.49 What is the purpose of the dilation of blood vessels? [Int: Body Temperature]

8.50 What hormone stimulates the metabolic rate? [Int: Body Temperature]

8.51 What molecule is responsible for oxygen transport in the body? [App: Oxygen Transport]

8.52 Many athletes train at high altitudes to prepare themselves for competition. Why? [App: Oxygen Transport]

Additional Questions and Problems

8.53 Hematite, an iron ore with formula Fe_3O_4, can be reduced by treatment with hydrogen gas to yield iron metal and water vapor.
(a) Write the balanced equation.
(b) This process requires 36 kcal for every 1.00 mol of Fe_3O_4 reduced. How much energy (in kilocalories) is required to produce 75 g of iron?
(c) How many grams of hydrogen are needed to produce 75 g of iron?
(d) This reaction has $K = 2.3 \times 10^{-18}$. Are the reactants or the products favored?

8.54 Hemoglobin (Hb) reacts reversibly with O_2 to form HbO_2, a substance that transfers oxygen to tissues:

$$Hb(aq) + O_2(aq) \rightleftharpoons HbO_2(aq)$$

(a) Carbon monoxide (CO) is 140 times more attracted to Hb than is O_2 and establishes another equilibrium:

$$Hb(aq) + 4 CO(aq) \rightleftharpoons Hb(CO)_4(aq)$$

Explain, using LeChatelier's principle, why inhalation of CO can cause weakening and eventual death.
(b) Still another equilibrium is established when both O_2 and CO are present:

$$Hb(CO)(aq) + O_2(aq) \rightleftharpoons HbO_2(aq) + CO(aq)$$

Explain, using LeChatelier's principle, why pure oxygen is often administered to victims of CO poisoning.

8.55 Many hospitals administer glucose intravenously to patients. If 4.0 kcal is provided by each gram of glucose, how many grams must be administered to maintain a person's normal basal metabolic needs of about 1700 kcal/day?

8.56 For the evaporation of water, $H_2O(l) \rightarrow H_2O(g)$, at 100° C, $\Delta H = 9.72$ kcal/mol.
(a) How many kilocalories of energy are needed to vaporize 10.0 g of $H_2O(l)$?
(b) How many kilocalories of energy are released when 10.0 g of $H_2O(g)$ is condensed?

8.57 Ammonia reacts slowly in air to produce nitrogen oxide and water vapor:

$$NH_3(g) + O_2(g) \rightleftharpoons NO(g) + H_2O(g) + heat$$

(a) Balance the equation.
(b) Write the equilibrium-constant expression.
(c) Explain the effect on the equilibrium of: (1) raising the

pressure, (2) adding NO(g), (3) decreasing the concentration of NH_3, (4) lowering the temperature

8.58 Methanol, CH_3OH, is used as race car fuel.
(a) Write the balanced equation for the combustion of methanol.
(b) ΔH for this process is -174 kcal/mol methanol. How many kilocalories are released by burning 50.0 g of methanol?

8.59 Sketch a reaction energy diagram for a system in which the forward reaction has an activation energy of 25 kcal/mol and the reverse reaction has an activation energy of 35 kcal/mol.
(a) Is the forward process endothermic or exothermic?
(b) What is the value of the heat of reaction?

8.60 Can a catalyst increase the amount of product formed at equilibrium from a given amount of reactants? Why or why not?

C H A P T E R

9

Gases, Liquids, and Solids

What is it that prevents oil and water from mixing with each other to form a solution? What is it that makes oil and water liquids rather than solids? Intermolecular forces make the difference. In this chapter you'll see what these forces are.

You've learned by now that all matter is composed of atoms, ions, or molecules; that these particles are in constant motion that increases with temperature; that the properties of matter vary with the different types of chemical bonds; and that physical and chemical changes are accompanied by the absorption or release of energy. Now we're going to look at how these concepts are related to the physical properties of matter.

The major questions we'll answer in this chapter are the following:

1. *What is the kinetic theory of gases?* The goal: Be able to state the assumptions of the kinetic theory of gases and use them to explain gas behavior.
2. *What is partial pressure?* The goal: Be able to define partial pressure and use Dalton's law of partial pressures.
3. *What are the gas laws?* The goal: Be able to state Boyle's law, Charles' law, and Avogadro's law and explain them in terms of the kinetic theory of gases.
4. *What is the universal gas law?* The goal: Be able to use the universal gas law equation to find the pressure, or volume, or temperature, or molar mass, or mass of a gas sample.
5. *What occurs during a change of state?* The goal: Be able to apply the concepts of heat exchange, equilibrium, and vapor pressure to changes of state.
6. *What are the major intermolecular forces?* The goal: Be able to explain dipole–dipole attraction, London forces, and hydrogen bonding, and recognize which of these forces affect a given molecule.

9.1 STATES OF MATTER REVISITED

Gases, liquids, and solids differ in their ability to maintain their shapes and volumes. The major reasons for the differences are in the distances between atoms, ions, or molecules and in their freedom to move about. In gases, molecules are very far apart, move freely, and have almost no influence on each other. (Note that in speaking of "gas molecules" we include the single-atom "molecules" of noble gases such as helium.) In liquids, molecules are close enough to be in contact but retain considerable freedom to move (like individual grains of rice as you pour them from a box). In solids, the atoms, molecules, or ions are held in specific spatial relationships to each other and can only wiggle around in place.

The attractive forces between gas molecules aren't strong enough to hold them together, and the molecules move independently of one another. When the attractive forces between atoms, molecules, or ions are stronger, however, the particles are pulled together into liquids or solids (Figure 9.1). The properties of liquids and solids are therefore much more dependent on the nature of their individual particles than are the properties of gases.

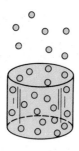

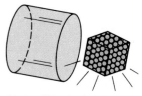

Figure 9.1
Differences between gases, liquids, and solids. The particles differ in the amount of freedom they have to move around as a result of the different strengths of attractions between them.

(a) A gas: The individual molecules feel little attraction for one another and are free to move about.

(b) A liquid: The individual molecules are attracted to one another but can slide over each other.

(c) A solid: The individual molecules are strongly attracted to one another and cannot move around.

Change of state   The conversion of a substance from one state to another, for example, from a liquid to a gas.

The transformation of a substance from one state to another is known as a **change of state.** Every change of state is reversible. In one direction the change is exothermic, in the other direction it is endothermic, and an equilibrium between the two states exists at some appropriate temperature. In the change from a solid to a liquid, for example, heat is absorbed in the forward reaction, and the solid and liquid are in equilibrium at the **melting point.**

Melting point (mp)   The temperature at which a solid turns into a liquid.

$$\text{Solid} + \text{heat} \rightleftharpoons \text{liquid} \qquad \text{Melting point}$$

The names applied to different changes of state and their exothermic or endothermic nature are summarized in Figure 9.2.

Gases, liquids, and solids vary greatly in their compressibilities. Gases are easily compressed, a property that allows them to be stored under pressure in strong tanks. Liquids are not very compressible, and solids are practically incompressible. The expansion and contraction of gases with changes in temperature are also much greater than that of liquids or solids.

Diffusion   The mixing of atoms, ions, or molecules by random motion.

Gases, liquids, and solids also differ greatly in their rates of **diffusion.** Freely mobile gas molecules diffuse rapidly. Liquids diffuse with easily observable speed, but more slowly than gases. If two metals are pressed together, some atoms at their surfaces will diffuse, but the process is vastly slower than the diffusion of liquids.

Figure 9.2
Changes of state. Each pair of changes is endothermic from left to right and exothermic from right to left. At the apropriate temperature, each pair of changes is at equilibrium. *Sublimation* is the change in which a substance changes directly from the solid state to the gas state; dry ice at 1 atm, for example, sublimes and does not melt into a liquid.

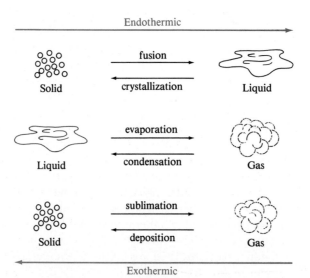

9.2 GASES AND THE KINETIC THEORY

Kinetic theory of gases
A set of assumptions for explaining the general behavior of gases.

The behavior of gases can be explained by a group of assumptions known as the **kinetic theory of gases:**

● *A gas consists of a great many molecules moving about in random straight lines with no attractive forces between them.* Because of this random motion, different gases diffuse together quickly.

● *Gas molecules move in random straight lines with an energy proportional to the Kelvin temperature.* Thus, gas molecules move faster as the temperature increases, and at a given temperature the average kinetic energy of the molecules of every gas is the same. (In fact, gas molecules move much faster than you might suspect. The average speed of a helium atom at room temperature and atmospheric pressure is approximately 1.36 km/s or 3000 mi/hr, close to that of a high-speed rifle bullet.) Gas molecules change direction only when they collide with the walls of their container or each other.

● *The amount of space actually occupied by gas molecules is much smaller than the amount of space between molecules.* Thus, most of the volume taken up by a gas is just empty space, accounting for the ease of compression and low densities of gases.

Ideal gas A gas that obeys all the assumptions of the kinetic theory of gases.

Pressure The force per unit area exerted on a surface.

● *When gas molecules collide without reacting, they spring apart elastically and their total energy is conserved.* The pressure of a gas against the walls of its container is the result of collisions of the gas molecules with the walls. The more collisions, the higher the pressure.

A gas that perfectly obeys the assumptions of the kinetic theory is called an **ideal gas.** The term ''ideal'' is used because all gases behave somewhat differently than predicted by the kinetic theory when, at very high pressures or very low temperatures, their molecules get closer and closer together. Gases usually display nearly ideal behavior, however, under the conditions at which they are most commonly encountered.

Figure 9.3
Atmospheric pressure. A column of air weighing 14.7 lb presses down on each square inch of the earth's surface at sea level, resulting in what we call atmospheric pressure.

9.3 PRESSURE

We're all familiar with the effects of air pressure. When you climb a mountain or fly in an airplane, the change in air pressure against your eardrums as you climb or descend can cause a painful ''popping.'' When you pump up a bicycle tire, you increase the pressure of air against the inside walls to keep the tire firm.

In scientific terms, **pressure** is defined as force per unit area pushing against a surface. (A ''force'' is exerted when, for example, you start something moving or stop it.) In the bicycle tire, for example, you measure the force of air molecules colliding with the inside walls of the tire. The units you probably use for tire pressure are pounds per square inch (psi), where 1 psi is equal to the pressure exerted by a 1 pound object resting on a 1 square inch surface. We on earth are under pressure from the atmosphere, the blanket of air pressing down on us, as shown in Figure 9.3. Atmospheric pressure is not constant, however; it varies slightly from day to day depending on the weather and also varies with altitude. Air pressure is about 14.7 psi at sea level but only about 4.9 psi on the summit of Mt. Everest.

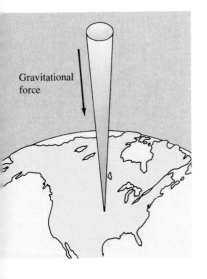

Gravitational force

Millimeter of mercury (mm Hg) A unit of pressure equal to the force exerted by a 1 mm column of mercury.

Torr Alternate name for the pressure unit, mm Hg.

Mercury barometer A mercury-filled glass tube used to measure atmospheric pressure.

Manometer A mercury-filled U-tube used to measure pressure in a closed system.

Pascal (Pa) The SI unit of pressure; equal to 0.007500 mm Hg.

The most commonly used unit of pressure is the **millimeter of mercury,** abbreviated **mm Hg** and often called the **torr** after the Italian physicist Evangelista Torricelli. This rather unusual unit dates back to the early 1600s when Torricelli made the first **mercury barometer,** a device like the one illustrated in Figure 9.4a. A thin tube, sealed at one end, is filled with mercury and then inverted in a dish of mercury. As mercury runs from the tube into the dish, a vacuum is created at the sealed end of the tube. Since the pressure in the vacuum is zero but the pressure forcing the mercury from the dish up into the tube is that of the atmosphere, a column of mercury is held in the tube. The same principle is used in the **manometer,** a device for measuring pressure in a sealed container (Figure 9.4b). The height of the mercury column varies with the pressure on the mercury surface, hence the use of mm Hg as a unit of pressure.

Pressure is also measured by an SI unit called the **pascal (Pa),** where 1 Pa = 0.007500 mm Hg (or 1 mm Hg = 133.3224 Pa). Measurements in pascals are becoming more common, and many clinical laboratories may make the switchover in the next few years. Higher pressures are often still given in atmospheres. As a unit, one atmosphere (atm) is defined as equal to exactly 760 mm Hg, which is the average atmospheric pressure at sea level.

Pressure units

$$1 \text{ atm} = 760 \text{ mm Hg} = 14.7 \text{ psi} = 101,325 \text{ Pa}$$

$$1 \text{ mm Hg} = 1 \text{ torr} = 133.3224 \text{ Pa}$$

Practice Problem **9.1** The air pressure outside a jet plane flying at 35,000 ft is about 220. mm Hg. How many atmospheres is this? How many psi? How many pascals?

Figure 9.4
Pressure measurement. (a) A mercury barometer, used to read atmospheric pressure, and (b) a mercury-filled manometer, used to read the pressure of a gas sample. Both devices are used by measuring the height of a column of mercury forced by gas pressure into a sealed tube. In the manometer the pressure of the gas forces mercury into the sealed end, and the pressure is read by measuring the difference between mercury levels at the two ends.

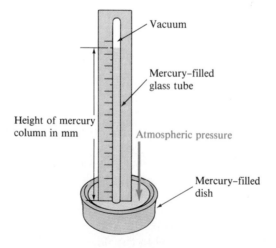

(a) Mercury barometer

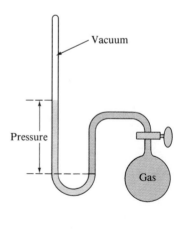

(b) Manometer

9.4 DALTON'S LAW: PARTIAL PRESSURE

According to the kinetic theory, each gas molecule acts independently of all others because they are so far apart. To any individual molecule, the chemical identity of other molecules is irrelevant. Thus, *mixtures* of gases act just the same as pure gases and obey the same laws.

Dry air, for example, is a mixture of about 21% oxygen, 78% nitrogen, and 1% argon (by volume), which means that about 21% of atmospheric air pressure is caused by oxygen molecules, 78% by nitrogen molecules, and 1% by argon molecules. We therefore say that, at a *total* air pressure of 760 mm Hg, the **partial pressure** caused by the contribution of oxygen is 0.21 × 760 mm Hg, or 160 mm Hg. Similarly, the partial pressure of nitrogen in air is 593 mm Hg, and that of argon is 6 mm Hg. The partial pressure exerted by each gas in a mixture is the same pressure that the gas would exert if it were alone. This makes sense if you think about it: The pressure exerted by each gas depends on the frequency of collisions of its molecules with the walls, and this frequency doesn't change when other gases are present because the different molecules have no influence on each other.

> **Partial pressure** The contribution to total gas pressure caused by each individual component of a mixture of gases.

According to **Dalton's law of partial pressure,** the total pressure exerted by any gas mixture is the sum of the individual pressures of the components in the mixture:

> **Dalton's law of partial pressure** The total pressure exerted by a mixture of gases is equal to the sum of the partial pressures exerted by each individual gas.

$$\text{Dalton's law of partial pressure:} \quad P_{\text{total}} = P_{\text{gas 1}} + P_{\text{gas 2}} + \cdots$$

The partial pressures of specific gases are represented by adding formulas to the symbol for pressure. You might see the partial pressure of oxygen represented as P_{O_2}, PO_2, or p_{O_2}. Air inside the lungs at 37°C and atmospheric pressure has the following average composition at sea level:

$$P_{\text{total}} = P_{N_2} + P_{O_2} + P_{CO_2} + P_{H_2O}$$

$$= 573 \text{ mm Hg} + 100 \text{ mm Hg} + 40 \text{ mm Hg} + 47 \text{ mm Hg}$$

$$= 760 \text{ mm Hg}$$

Note that the total is equal to the atmospheric pressure.

The percent composition of air does not change appreciably with altitude, although the pressure decreases. With increasing altitude, therefore, the partial pressure of oxygen in the air is lower, and it is this change that leads to difficulty in breathing.

Solved Problem 9.1   Humid air on a warm summer day is approximately 20% oxygen, 75% nitrogen, and 5% water vapor. What is the partial pressure of each component if the atmospheric pressure that day is 750 mm Hg?

Solution   The partial pressure of any gas is obtained by multiplying the percent concentration of the gas by the total gas pressure in the sample.

Oxygen partial pressure (P_{O_2}):   0.20 × 750 mm Hg = 150 mm Hg

Nitrogen partial pressure (P_{N_2}):   0.75 × 750 mm Hg = 560 mm Hg

Water vapor partial pressure (P_{H_2O}):   0.05 × 750 mm Hg = 40 mm Hg

Practice Problems

9.2 Assuming a total pressure of 9.5 atm, what is the partial pressure of each component in a mixture of 98% helium and 2.0% oxygen breathed by deep-sea divers?

9.3 How does the partial pressure of oxygen in diving gas (Practice Problem 9.2) compare with its partial pressure in normal air?

9.4 Determine the percent composition of air in the lungs from the composition in partial pressures given just before Solved Problem 9.1.

9.5 What is the partial pressure of oxygen in the lungs at an altitude where the atmospheric pressure is 685 mm Hg?

AN APPLICATION: BLOOD PRESSURE

Having your blood pressure measured is a quick and easy way to get an indication of the state of your circulatory system. Although blood pressure varies with age, a normal adult male has a reading near 120/80, and a normal adult female has a reading near 110/70. The units for these numbers are millimeters of mercury. Abnormally high values signal an increased risk of heart attack and stroke.

Pressure varies greatly in different types of blood vessels. Usually, though, measurements are carried out on arteries in the upper arm as the heart goes through a full cardiac cycle. **Systolic pressure** is the maximum pressure developed in the artery just after contraction, as the heart forces the maximum amount of blood into the artery. **Diastolic pressure** is the minimum pressure that occurs at the end of the heart cycle.

Blood pressure is measured by a *sphygmomanometer,* a device consisting of a squeeze bulb, a flexible cuff, and a mercury manometer. The cuff is placed around the upper arm over the brachial artery and inflated by the bulb to about 200 mm Hg pressure, an amount great enough to squeeze the artery shut and prevent blood flow. Air is then slowly released from the cuff, and pressure drops. As cuff pressure reaches the systolic pressure, blood spurts through the artery, creating a turbulent tapping sound that can be heard through a stethoscope. The pressure registered on the manometer at the moment the first sounds are heard is the systolic blood pressure. Sounds continue until the pressure in the cuff becomes low enough to allow diastolic blood flow. At this point, blood flow becomes smooth, no sounds are heard, and a diastolic blood pressure reading is recorded on the manometer. Readings are usually recorded as systolic/diastolic, for example, 120/80. The accompanying figure shows the sequence of events during measurement.

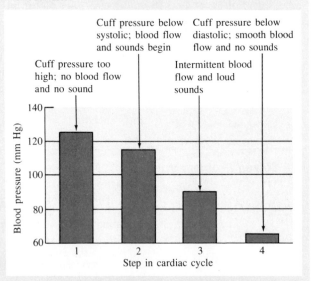

The sequence of events during blood pressure measurement, including the sounds heard.

9.5 BOYLE'S LAW: THE RELATION BETWEEN PRESSURE AND VOLUME

Gas laws A series of laws that describe the behavior of gases under conditions of differing pressure, volume, and temperature.

Observations by scientists in the 1700s led to the formulation of what are now called the **gas laws,** which allow us to predict the influence of pressure (P), volume (V), and temperature (T) on any gas or mixture of gases. In this section we'll examine Boyle's law, which shows the relation between pressure and volume.

Imagine that you have a sample of gas inside a cylinder with a plunger at one end (Figure 9.5). What would happen if you were to halve the volume of the gas by pushing the plunger halfway down? Since the gas molecules have only half as much room to move around in, the number of collisions with the walls of the cylinder increases. Thus, the pressure of the gas in the cylinder increases. According to **Boyle's law,** *the pressure of a gas at constant temperature is inversely proportional to its volume*. Inversely proportional quantities change in opposite directions. As volume goes up or down, pressure goes down or up, a relationship that can be expressed in the following way:

Boyle's law The pressure of a gas at constant temperature is inversely proportional to its volume (PV = constant, or $P_1V_1 = P_2V_2$).

For a fixed amount of gas, when temperature (T) is constant

$$\text{Boyle's law:} \quad \text{Pressure} \propto \frac{1}{\text{volume}} \quad \text{(The sign } \propto \text{ means ''proportional to'')}$$

$$\text{or} \quad \text{Pressure} \times \text{volume} = PV = \text{a constant value}$$

Since $P \times V$ is always a constant value for a given amount of gas at constant temperature, Boyle's law can be restated to show how one of these **variables** changes if the other changes. The starting pressure (P_1) times the starting volume (V_1) is equal to the final pressure (P_2) times the final volume (V_2) (because P_1V_1 and P_2V_2 are equal to the same constant):

Variable A measurable property that is not constant.

$$\text{Boyle's law restated:} \quad P_1V_1 = P_2V_2$$

The movement of air and carbon dioxide in and out of the lungs is also a consequence of the Boyle's-law behavior of gases. At the moment between each breath, the pressure inside the lungs is equal to atmospheric pressure. Inhalation takes place as the muscles of the diaphragm and rib cage contract, expanding the volume of the lungs and decreasing the pressure inside them (Figure 9.6). Air must then move into the lungs through the respiratory tract to

Figure 9.5
Boyle's law. Decreasing the volume of a gas sample by one-half increases crowding of the molecules and thereby doubles the pressure.

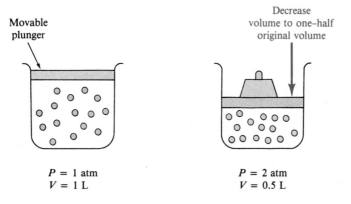

Movable plunger

Decrease volume to one–half original volume

P = 1 atm
V = 1 L

P = 2 atm
V = 0.5 L

Figure 9.6
Boyle's law in breathing.
During inhalation, the
diaphragm moves down and
the rib cage moves up and
out, thus increasing volume,
decreasing pressure in the
chest cavity, and drawing air
in to equalize pressure.
During exhalation, the
muscles relax, volume
decreases and pressure
increases, and air moves out
of the chest cavity.

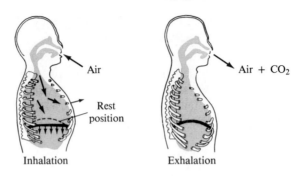

equalize the pressure with that of the atmosphere. At the end of inhalation, air pressure in the lungs is again equal to atmospheric pressure. Exhalation takes place as the muscles relax, decreasing the volume of the lungs and increasing pressure inside them. Now gases move out of the lungs until pressure is once again equalized with the atmosphere.

Scuba divers, who breathe air at higher than atmospheric pressure, must learn not to hold a breath while ascending from a dive. Otherwise, as the surrounding water pressure decreases, increasing air volume can tear the lungs, and air that escapes into the body can cause damage ranging in severity from a collapsed lung to a stroke.

Solved Problem 9.2 In a typical automobile engine, the gas mixture in a cylinder is compressed from 1 atm to 9.5 atm. If the uncompressed volume of the cylinder is 750 mL, what is the volume when fully compressed?

Solution It's helpful in using the gas laws to summarize the known and needed information:

$$P_1 = 1 \text{ atm} \qquad V_1 = 750 \text{ mL}$$

$$P_2 = 9.5 \text{ atm} \qquad V_2 = ?$$

According to Boyle's law, the pressure of a sample times its volume is constant:

$$P_1V_1 = P_2V_2$$

Solving for V_2 and substituting the known information gives

$$V_2 = \frac{P_1V_1}{P_2}$$

$$V_2 = \frac{1.0 \text{ atm} \times 750 \text{ mL}}{9.5 \text{ atm}} = 79 \text{ mL}$$

Practice Problems

9.6 A cylinder of compressed oxygen used to aid breathing has a volume of 5.0 L at 90 atm pressure. What volume would the same amount of oxygen have if the pressure were 1.0 atm, assuming constant temperature?

9.7 A sample of hydrogen gas at 273 K has a volume of 2.5 L at 5.0 atm pressure. What is its pressure if its volume is changed to 10.0 L? To 0.20 L?

9.6   CHARLES' LAW: THE RELATION BETWEEN VOLUME AND TEMPERATURE

Imagine that you again have a sample of gas inside a cylinder with a plunger at one end. What would happen if you were to double the sample's temperature while letting the plunger move freely to keep the pressure constant? The gas molecules would move with twice as much energy and collide with the walls more often and with greater force. If a constant pressure is to be maintained, the volume of the gas in the cylinder must double.

For this relationship to apply, the gas temperature must be measured in Kelvins, rather than Celsius degrees. In fact, the Kelvin scale is defined according to the temperature–volume relationship, with 0 K at absolute zero, the point at which, theoretically, gas volume diminishes to zero.

Charles' law   The volume of a gas at constant pressure is directly proportional to its Kelvin temperature ($V/T =$ constant, or $V_1/T_1 = V_2/T_2$).

According to **Charles' law,** *the volume of a gas at constant pressure is directly proportional to its Kelvin temperature*. Note the difference between directly proportional and inversely proportional as in Boyle's law. Directly proportional quantities change in the same direction. As temperature goes up or down, volume also goes up or down:

For a fixed amount of gas, when the pressure (P) is constant

$$\textbf{Charles' law:} \qquad \text{Volume} \propto \text{temperature}$$

$$\text{or} \qquad \frac{\text{Volume}}{\text{Temperature}} = \frac{V}{T} = \text{a constant value}$$

$$\text{or} \qquad \frac{V_1}{T_1} = \frac{V_2}{T_2}$$

As an example of Charles'-law behavior, think of what happens when a hot-air balloon is filled. Heating causes the air inside to expand and fill the bag (Figure 9.7).

Figure 9.7
Charles' law in hot-air ballooning. The volume of the gas increases as it is heated, causing a decrease in density and allowing the balloon to rise.

Solved Problem 9.3 An average adult inhales a volume of 0.50 L air into the lungs with each breath. If the air is warmed from room temperature (20°C = 273 K + 20° = 293 K) to body temperature (37°C = 310 K) while in the lungs, what is the volume of the air when exhaled?

Solution According to Charles' law, the volume of the gas divided by its temperature is constant. Solving for V_2, substituting, and carrying out the calculation gives

$$\frac{V_1}{T_1} = \frac{V_2}{T_2}$$

$$V_2 = \frac{V_1 \times T_2}{T_1} = \frac{(0.50 \text{ L}) \times (310 \, \cancel{K})}{293 \, \cancel{K}} = 0.53 \text{ L}$$

Practice Problem **9.8** A sample of chlorine gas has a volume of 0.30 L at 273 K and 1 atm pressure. What is its volume at 350 K and 1 atm pressure? At 500°C?

9.7 THE COMBINED GAS LAW

Combined gas law The product of the pressure and volume of a gas is inversely proportional to its temperature (PV/T = constant, or P_1V_1/T_1 = P_2V_2/T_2).

Since PV has a constant value and V/T also has a constant value for a fixed amount of gas, the two relationships can be merged into the **combined gas law:**

For a fixed amount of gas

Combined gas law: $\frac{PV}{T}$ = a constant value or $\frac{P_1V_1}{T_1} = \frac{P_2V_2}{T_2}$

If any five of the six quantities in this equation are known, the sixth can be calculated.

The combined gas law is helpful because it's the only equation you need to remember for a fixed amount of gas. If any of the three variables, T, P, or V, is constant, it drops out of the equation, leaving behind, respectively, Boyle's law, Charles' law, or the relationship showing that *at constant volume, the pressure of a gas is directly proportional to its Kelvin temperature:*

For a fixed amount of gas, when volume (V) is constant

$\frac{P_1\cancel{V}}{T_1} = \frac{P_2\cancel{V}}{T_2}$ gives $\frac{P_1}{T_1} = \frac{P_2}{T_2}$

In other words, as temperature goes up or down, pressure also goes up or down. The result of throwing an aerosol can into an incinerator is an example of the pressure–temperature relation. Pressure builds up, and the can explodes (hence the warning statement on aerosol cans).

Suppose it's necessary to compare the different volumes of gas formed in a series of laboratory experiments. Would it be sufficient to report that one experiment produced 1.0 L of gas and the next produced 1.5 L of gas? The answer is no. Since the volume of a gas changes with pressure and temperature, the experimental results must include the temperature and pressure at which the volume was measured. Only then can comparisons of the gas volumes be meaningful.

To simplify comparisons of gas samples, volumes are often reported for **standard temperature and pressure (STP)**, which is defined as follows:

Standard temperature and pressure (STP)   Standard conditions for a gas, defined as 0°C (273 K) and 1 atm (760 mm Hg) pressure.

Standard temperature and pressure (STP)

$$0°C \ (273 \ K) \quad \text{and} \quad 1 \ atm \ (760 \ mm \ Hg)$$

It's not necessary to actually make a measurement at this temperature and pressure. The combined gas law can be used to calculate the volume at STP from the measured volume, temperature, and pressure.

Solved Problem 9.4   A 6.3 L sample of helium gas stored at 25°C and 1.0 atm pressure is to be condensed into a 2.0 L tank and maintained at a pressure of 2.8 atm. What temperature is needed to maintain this pressure?

Solution   The known and needed information is as follows:

$$P_1 = 1.0 \ atm \qquad V_1 = 6.3 \ L \qquad T_1 = 25° + 273 \ K = 298 \ K$$

$$P_2 = 2.8 \ atm \qquad V_2 = 2.0 \ L \qquad T_2 = ?$$

Solving the combined gas law equation for T_2 and substituting the known information gives

$$\frac{P_1 V_1}{T_1} = \frac{P_2 V_2}{T_2}$$

$$T_2 = \frac{P_2 V_2 T_1}{P_1 V_1} = \frac{(2.8 \ \text{atm})(2.0 \ \text{L})(298 \ K)}{(1.0 \ \text{atm})(6.3 \ \text{L})} = 260 \ K \ (= -13°C)$$

Solved Problem 9.5   What would the inside pressure become if an aerosol can with an initial pressure of 4.5 atm at 20°C were heated in a fire to 600°C?

Solution   The volume is constant, while the pressure and temperature vary. This is the type of problem that can be solved by starting with the combined gas law and canceling the constant quantity.

$$T_1 = 20° + 273 \ K = 293 \ K \qquad P_1 = 4.5 \ atm$$

$$T_2 = 600° + 273 \ K = 873 \ K \qquad P_2 = ?$$

$$\frac{P_1 \cancel{V}}{T_1} = \frac{P_2 \cancel{V}}{T_2} \qquad \text{gives} \qquad \frac{P_1}{T_1} = \frac{P_2}{T_2}$$

$$P_2 = \frac{P_1 \times T_2}{T_1} = \frac{(4.5 \ atm) \times (873 \ \text{K})}{293 \ \text{K}} = 13 \ atm$$

AN APPLICATION: INHALED ANESTHETICS

William Morton's demonstration in 1846 of ether-induced anesthesia during surgery must surely rank as one of the most important medical breakthroughs of all time. Before that date, all surgery had been carried out with the patient fully conscious. Many other inhaled anesthetic agents have now been introduced, including nitrous oxide, halothane, and enflurane, to name a few.

$$H-\underset{\underset{H}{|}}{\overset{\overset{H}{|}}{C}}-\underset{\underset{H}{|}}{\overset{\overset{H}{|}}{C}}-O-\underset{\underset{H}{|}}{\overset{\overset{H}{|}}{C}}-\underset{\underset{H}{|}}{\overset{\overset{H}{|}}{C}}-H \qquad N_2O$$

Ether   Nitrous
                      oxide

$$F-\underset{\underset{F}{|}}{\overset{\overset{F}{|}}{C}}-\underset{\underset{Cl}{|}}{\overset{\overset{Br}{|}}{C}}-H \qquad H-\underset{\underset{Cl}{|}}{\overset{\overset{F}{|}}{C}}-\underset{\underset{F}{|}}{\overset{\overset{F}{|}}{C}}-O-\underset{\underset{F}{|}}{\overset{\overset{F}{|}}{C}}-H$$

Halothane   Enflurane

Despite their great importance and utility, surprisingly little is known about how inhaled anesthetics work in the body. The potency of different anesthetics correlates well with their solubility in vegetable oils, leading many scientists to believe that they act by dissolving in the fatty membranes surrounding nerve cells. The resultant change in the nerve cell membranes apparently decreases the ability of sodium ions to pass into the cells, thereby blocking the firing of nerve impulses.

The depth of anesthesia is determined by the concentration of anesthetic agent that reaches the brain. Brain concentration, in turn, depends on the solubility and transport of the anesthetic agent in blood and on its partial pressure in inhaled air. Nitrous oxide must be administered at a partial pressure of up to 500 mm Hg to induce loss of consciousness, but the more soluble halothane can be administered at a partial pressure of 20 mm Hg.

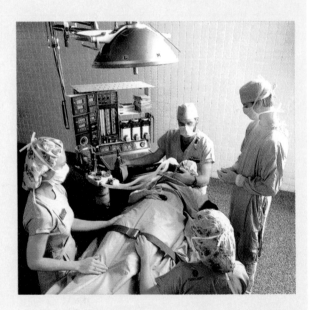

Anesthesiologists use sophisticated equipment to monitor the vital signs of the patient and administer the precise amount of anesthetic that is safe and effective.

Practice Problems

9.9 What final temperature is required for the pressure inside an automobile tire to increase from 30. psi at 0°C to 45 psi, assuming the volume remains constant?

9.10 A meteorological balloon is filled with helium to a volume of 250. L at 22°C and 745 mm Hg. The balloon ascends to an altitude where the pressure is 570. mm Hg, and its volume decreases to 232 L. What is the temperature at this altitude?

9.11   A sample of oxygen produced in a laboratory experiment had a volume of 590 mL at a pressure of 775 mm Hg and a temperature of 352 K. What is the volume of this sample at STP? (Use STP as T_2 and P_2.)

9.8   AVOGADRO'S LAW: THE RELATION BETWEEN VOLUME AND MOLES OF SAMPLE

Imagine finally that you have two equal volumes of different gases in sealed containers at the same temperature and pressure. How many molecules does each sample contain? Since molecules in a gas are so tiny compared to the empty space surrounding them, the chemical identity of the molecules doesn't matter. According to **Avogadro's law,** *equal volumes of gases at the same temperature and pressure contain equal numbers of molecules.* As long as the temperature and pressure are identical, one liter of oxygen gas contains as many molecules as one liter of chlorine gas, which contains as many molecules as one liter of carbon dioxide gas, and so on. Avogadro's law gives us another direct proportionality that applies to gases.

Avogadro's law   Equal volumes of gases at the same temperature and pressure contain equal numbers of molecules (V/n = constant, or $V_1/n_1 = V_2/n_2$).

For fixed temperature (T) and pressure (P)

$$\text{Avogadro's law:} \qquad \text{Volume} \propto n \text{ (number of moles)}$$

$$\text{or} \qquad \frac{\text{Volume}}{\text{No. of moles}} = \frac{V}{n} = \text{a constant value}$$

$$\text{or} \qquad \frac{V_1}{n_1} = \frac{V_2}{n_2}$$

Because equal volumes of gases at the same temperature and pressure contain equal numbers of molecules, they also contain equal numbers of moles. In other words, one mole (6.02×10^{23} molecules) of any gas has the same volume as one mole of any other gas at the same temperature and pressure. Thus, by comparing volumes of gases at the same temperature and pressure, it's possible to compare molar amounts. This is often done by using the **standard molar volume**—the volume of one mole of a gas at STP, which is equal to 22.4 L/mol.

Standard molar volume   The volume of one mole of a gas at standard temperature and pressure (22.4 L).

Since Avogadro's number provides the connection between the mole and numbers of molecules, molar volume can be used, for example, to find out how many molecules you inhale in an average breath, assuming a volume of 0.50 L at STP:

$$0.50 \, L \times \frac{6.02 \times 10^{23} \text{ molecules}}{22.4 \, L} = 1.3 \times 10^{22} \text{ molecules}$$

Practice Problem   **9.12**   How many moles of methane gas, CH_4, are in a 1.00×10^5 L storage tank at STP? How many grams of methane is this? How many grams of carbon dioxide gas, CO_2, could the same tank hold?

9.9 THE UNIVERSAL GAS LAW

Universal gas law A law
that relates the temperature,
pressure, volume, and molar
amount of a gas sample
($PV = nRT$).

The relations among the four variables P, V, T, and n for gases are summarized in Table 9.1. These variables can be combined into a single **universal gas law** that relates the temperature, pressure, volume, and molar amount of a gas sample. Any time we know the value of three of these four quantities, we can calculate the value of the fourth.

For a given gas sample of any molar amount (n)

$$\text{Universal gas law:} \quad \frac{PV}{nT} = R \text{ (a constant)}$$

or, as usually written

$$PV = nRT$$

P = pressure of gas (usually in mm Hg or atm)
V = volume of gas in liters (L)
n = number of moles of gas
R = the universal gas constant
T = temperature of gas in kelvins (K)

The value of the universal gas constant R varies with the units chosen for pressure. The two most commonly used values are for pressure in millimeters of mercury or in atmospheres:

$$\text{For } P \text{ in atmospheres} \quad R = 0.0821 \frac{\text{L atm}}{\text{mol K}}$$

$$\text{For } P \text{ in mm Hg} \quad R = 62.4 \frac{\text{L mm Hg}}{\text{mol K}}$$

In using the universal gas law, it's important to be sure to choose the correct value of R and, if necessary, to convert volumes into liters and temperature into kelvins.

Table 9.1 Summary of the Gas Laws

Gas Laws	Variable Quantities	Constant Quantities
Combined gas law: $P_1V_1/T_1 = P_2V_2/T_2$	Pressure, volume, temperature	Number of moles
Boyle's law: $P_1V_1 = P_2V_2$	Pressure, volume	Temperature, number of moles
Charles' law: $P_1/T_1 = P_2/T_2$	Pressure, temperature	Volume, number of moles
Avogadro's law: $V_1/n_1 = V_2/n_2$	Volume, number of moles	Temperature, pressure
Universal gas law: $PV = nRT$	Gives pressure, volume, and temperature for a fixed quantity of gas	

Solved Problem 9.6 How many moles of air are there in the lungs of an average person with a total lung capacity of 3.8 L? Assume that the person is at sea level and has a normal body temperature of 37°C.

Solution The problem asks for a value of n when P, V, and T are known. We'll take atmospheric pressure as 760 mm Hg. Volume is given in the correct units of liters, but temperature must be converted to kelvins.

$$P = 760 \text{ mm Hg (1 atm)} \qquad V = 3.8 \text{ L} \qquad T = 310 \text{ K (37°C)}$$

Solving the universal gas law for n and substituting gives

$$n = \frac{PV}{RT} = \frac{(760 \text{ mm Hg})(3.8 \text{ L})}{62.3 \frac{\text{mm Hg L}}{\text{mol K}} \times (310 \text{ K})} = 0.15 \text{ mol air}$$

There is 0.15 mol of air (that is, 0.15 mol of gas molecules) in the lungs of an average person.

Practice Problems **9.13** An aerosol spray deodorant can with a volume of 350 mL contains 3.2 g of propane gas (C_3H_8) as propellant. What is the pressure in the can at 20°C?

9.14 A helium gas cylinder of the sort used to fill balloons has a volume of 180 L and a pressure of 2200 psi (150 atm) at 25°C. How many moles of helium are in the tank? How many grams?

9.10 LIQUIDS

As they are in gases, molecules in the liquid state are in constant motion. If a molecule happens to be near the surface of a liquid and if it has enough energy, it has a chance to break free of the liquid and escape into the gas state. In an open container, the now-gaseous molecule will wander away from the liquid, and the process will continue until all the molecules escape from the container (Figure 9.8a). This, of course, is exactly what happens during **evaporation.** We're all familiar with seeing a puddle of water evaporate after a rainstorm.

Evaporation The gradual escape of molecules from a liquid into the gaseous state.

If the liquid is in a closed container, however, the situation is different. In a closed container, a gaseous molecule can never escape. Thus, the random

Figure 9.8
The transfer of molecules between liquid and gas states. (a) Molecules that escape from an open container drift away until the liquid has entirely evaporated. (b) Molecules in a closed container reach an equilibrium; that is, the rates of molecules leaving the liquid and returning to the liquid are equal and their concentration in the gas state is constant.

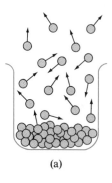
(a)

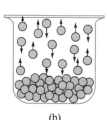
(b)

Dynamic equilibrium An equilibrium in which the rates of forward and reverse changes are equal.

motion of the molecule will occasionally bring it back into the liquid. After the concentration of molecules in the gas state has built up sufficiently high, the number of molecules reentering the liquid becomes equal to the number escaping from the liquid. At this point, a **dynamic equilibrium** exists (Figure 9.8b), exactly as in a chemical reaction at equilibrium. Evaporation and condensation are taking place at the same rate, and the concentration of vapor in the container is constant so long as the temperature does not change.

Once molecules have escaped from the liquid into the gas state, they are subject to all the gas laws previously discussed. In a closed vessel at equilibrium, for example, the gas molecules (called **vapor**) make their own contribution to the total pressure of the gas above the liquid according to Dalton's law of partial pressure. We call this contribution the **vapor pressure** of the liquid.

Vapor The gaseous state of a substance that is normally a liquid.

Vapor pressure The pressure of a vapor at equilibrium with its liquid.

Boiling point (bp) The temperature at which the vapor pressure of a liquid is equal to atmospheric pressure and the liquid boils.

Vapor pressure depends on both the temperature and the identity of a liquid. As the temperature rises, molecules become more energetic and more likely to escape into the gas state. Thus, vapor pressure rises with increasing temperature, as shown in Figure 9.9, until ultimately the **boiling point (bp)** of a liquid is reached. At the boiling point, the vapor pressure of the liquid is equal to the pressure of the atmosphere, and what we usually call the "boiling point" is the *normal boiling point*—the temperature at which a liquid boils at 760 mm Hg.

Once the vapor pressure of a liquid is equal to the pressure above the liquid, bubbles of vapor form under the surface and force their way to the top, giving rise to the violent action observed during a vigorous boil.

You might think that the more liquid in a closed container, the higher the vapor pressure will be, but this isn't the case. As long as *some* liquid is present, the vapor pressure in a closed container depends only on the temperature.

If atmospheric pressure is higher or lower than normal, the boiling point of a liquid changes accordingly. At high altitudes, for example, atmospheric pressure is lower than at sea level, and boiling points are also lower. On top of Mt. Everest (29,028 ft, or 8848 m), the boiling temperature of water is approximately 71°C. In a pressure cooker, however, pressure is higher and boiling points are also higher. The same principle is applied in strong vessels known as *autoclaves* in which steam at high pressure yields the temperatures needed for sterilizing medical and dental instruments (170°C).

Figure 9.9
Vapor pressure–temperature curves. A plot showing the change of vapor pressure with temperature for ethyl ether, ethyl alcohol, and water. At a liquid's boiling point, its vapor pressure is equal to atmospheric pressure. The commonly given *normal* boiling points are those at 760 mm Hg.

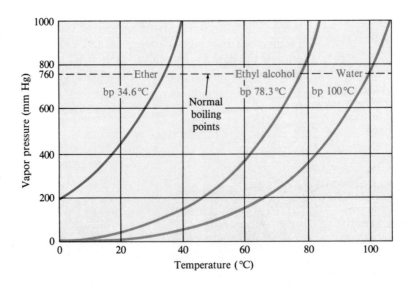

†9.11 INTERMOLECULAR FORCES: DIPOLE–DIPOLE AND LONDON FORCES

What determines whether a substance is a gas, a liquid, or a solid at a given temperature? Why is methane a gas at room temperature and water a liquid? Why does ether evaporate much more readily than water?

By examining closely what happens during the conversion of a liquid to a gas, we can answer these questions. Molecules that are close together in the liquid must gain enough energy to break away from their neighbors. The ease with which this happens depends on two opposing forces: (1) the strength of attraction between the molecules (the **intermolecular forces**), and (2) the kinetic energy of the molecules, which depends on the temperature. The stronger the intermolecular forces, the higher the boiling point and the higher the **heat of vaporization** of a liquid.

Three major types of intermolecular forces affect the properties of molecules—dipole–dipole forces and London forces, discussed below, and hydrogen bonding, discussed in the next section.

Dipole–Dipole Attraction   Recall (Sections 6.8 and 6.9) that molecules with polar covalent bonds are polar if their molecular shapes are unsymmetrical. Within a liquid, the positive and negative ends of polar molecules are attracted to each other (Figure 9.10), resulting in the type of intermolecular force known as **dipole–dipole attraction.** Its effect can be seen in the boiling point differences between nonpolar bromine and polar iodine chloride:

	Nonpolar molecule — Br_2	Polar molecule — ICl
Formula weight	160 amu	162 amu
Bp	59°C	101°C

Notice that in demonstrating dipole–dipole attraction, we've been careful to compare boiling points for compounds with similar formula weights. There's a reason for this; in fact, it's a general principle in science. To compare two properties, we must eliminate the effects of similar or competing properties. The weak intermolecular forces described in the next section act between *all* molecules and increase with increasing formula weight.

Intermolecular forces
Forces of attraction and repulsion between partial charges in different molecules.

Heat of vaporization   The amount of heat necessary to convert one gram (or one mole) of liquid into a gas.

Dipole–dipole attraction
Attraction between polar molecules.

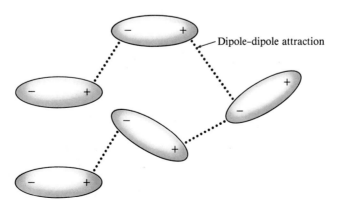

Figure 9.10
Dipole–dipole forces. The positive and negative ends of polar molecules are attracted to each other.

Dipole–dipole attraction

London forces
Intermolecular forces of attraction between short-lived, temporary dipoles.

London Forces The constant motion of electrons within molecules produces weak forces known as **London forces.** For an instant, there may be more electron density at one end of a molecule than the other, creating a short-lived polarity. At that instant, electrons in neighboring molecules are attracted by the positive end of the molecule and repelled by the negative end. The overall result of these constantly changing polarities is a slight net attraction that is independent of permanent polarity and increases with formula weight. Even the noble gases are subject to this effect and condense into liquids when kinetic energy decreases sufficiently at very low temperatures. For example, compare the boiling points of helium, at wt 4 amu, $-269°C$; and neon, at wt 20 amu, $-246°C$.

London forces also increase in effectiveness with the amount of surface contact between molecules. The more spherical and compact a molecule, the weaker its London forces compared to those of a linear molecule. Neopentane, for example, has a lower boiling point than pentane because its shape allows less contact with neighboring molecules.

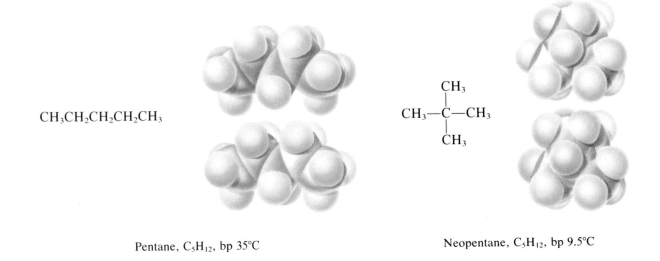

$CH_3CH_2CH_2CH_2CH_3$

$$CH_3-\overset{\overset{\displaystyle CH_3}{|}}{\underset{\underset{\displaystyle CH_3}{|}}{C}}-CH_3$$

Pentane, C_5H_{12}, bp 35°C

Neopentane, C_5H_{12}, bp 9.5°C

Practice Problem **9.15** Would you expect the boiling points to increase or decrease in the following series?
(a) Kr, Ar, Ne   (b) CH_3CH_3, $CH_3CH_2CH_3$, $CH_3CH_2CH_2CH_3$

9.12 INTERMOLECULAR FORCES: HYDROGEN BONDING

Hydrogen bonding
Intermolecular forces due to attraction between an electronegative atom (O, N, or F) and a hydrogen atom bonded to an electronegative atom (usually O or N).

Hydrogen bonding is the intermolecular force with the strongest influence on our lives. It is responsible for the high boiling point of water, for the usefulness of water as a solvent (Section 10.3), and for holding huge biomolecules in the shapes needed to play their essential roles in biochemistry.

Hydrogen bonding is the attraction between a highly electronegative atom (O, N, or F) and a hydrogen atom covalently bonded to another electronegative atom (usually O or N). In bonds between hydrogen and such electronegative atoms, electron density is pulled away from the hydrogen atom, which is left with a sizable partial positive charge. A hydrogen bond results when this positive hydrogen atom is drawn to an unshared electron pair on a neighboring electronegative atom, for example,

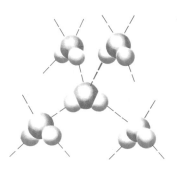

Figure 9.11
Hydrogen bonding in water. The intermolecular attraction in water is especially strong because, not only are the molecules highly polar, but each has two lone pairs and two hydrogen atoms, allowing formation of four hydrogen bonds.

The following boiling points illustrate the effects of hydrogen bonding.

	CH$_4$	NH$_3$	H$_2$O	HF	HCl
Formula weight	16 amu	17 amu	18 amu	20. amu	36.5 amu
Bp	−161°C	−33°C	100.°C	−19.5°C	−85°C

Methane is nonpolar, with no hydrogen bonding. Ammonia, water, and hydrogen fluoride are all polar molecules capable of hydrogen bonding. Hydrogen bonding is especially strong in water because each highly polar water molecule contains two hydrogen atoms and two unshared electron pairs, and can therefore form four hydrogen bonds (Figure 9.11). The table also shows for comparison, HCl, a polar molecule for which hydrogen bonding is not significant.

Surface tension Result of intermolecular forces that pull molecules at a liquid surface inward, causing the surface to act like a stretched membrane.

Intermolecular forces also determine the strength of the **surface tension** in a liquid. Because molecules at the surface are attracted by neighbors on only three sides, they are pulled inward, tightening up the surface like a stretched rubber membrane. The surface tension in water is so strong that a small object like a needle will rest on the surface if carefully placed, even though it should sink because it is denser than water. Surface tension also pulls water droplets into spheres as they fall (Figure 9.12).

Figure 9.12
Surface tension in action.

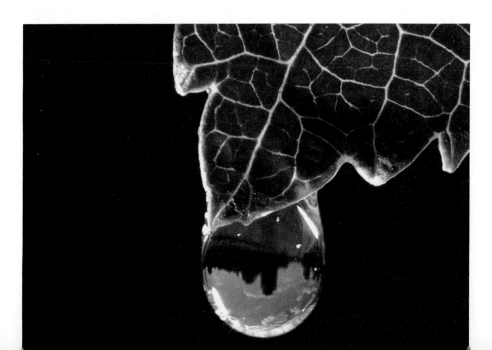

Solved Problem 9.7 The small amino acid units in protein molecules are joined together by peptide bonds, which have the structure shown here, where X and Y represent the rest of the protein molecule. Draw a structure that shows hydrogen bonding between the peptide bonds in two different protein molecules.

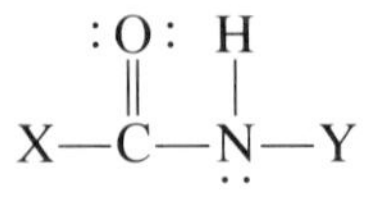

$$
\begin{array}{cc}
:\!O\!: & H \\
\| & | \\
X\!-\!C\!-\!N\!-\!Y \\
\end{array}
$$

Solution To identify a hydrogen bond site, look for a hydrogen atom bonded to O or N and a nearby O, N, or F atom with at least one unshared electron pair. The only hydrogen atom bonded to an electronegative atom in the peptide bond is the one on nitrogen. It can hydrogen-bond to the unshared pairs on the doubly bonded oxygen atom or the nitrogen atom in a neighboring peptide bond.

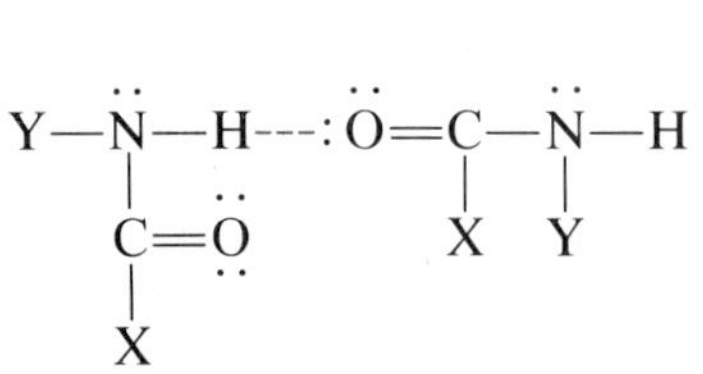

$$
\begin{array}{ccc}
Y-\overset{..}{N}-H---\!:\!\overset{..}{O}\!=\!C-\overset{..}{N}-H & & Y-\overset{..}{N}-H---\!:\!\overset{\overset{\displaystyle H}{|}}{N}-Y \\
\;\;\;\;|\;\;\;\;\;\;\;\;\;\;\;\;\;\;\;\;|\;\;\;\;\;\;\;\;\;|\; & \text{or} & \;\;\;\;|\;\;\;\;\;\;\;\;\;\;\;\;\;\;\;|\;\;\;\;\;\;\;\;\;| \\
\;\;\;\;C\!=\!\overset{..}{O}\;\;\;\;\;\;X\;\;\;\;Y & & \;\;\;\;C\!=\!\overset{..}{O}\;\;\;\;\;\;C\!=\!\overset{..}{O}\!: \\
\;\;\;\;|\; & & \;\;\;\;|\;\;\;\;\;\;\;\;\;\;\;\;\;\;\;| \\
\;\;\;\;X & & \;\;\;\;X\;\;\;\;\;\;\;\;\;\;X
\end{array}
$$

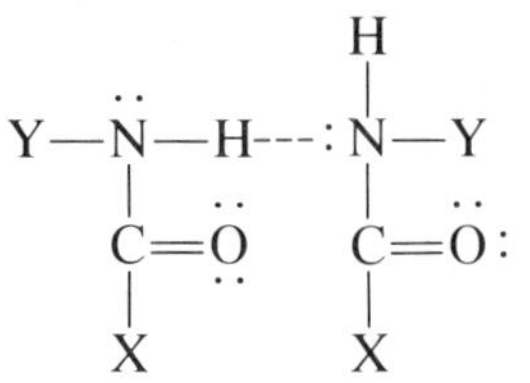

Practice Problems **9.16** Which of the following compounds will be influenced by hydrogen bonding?

(a) 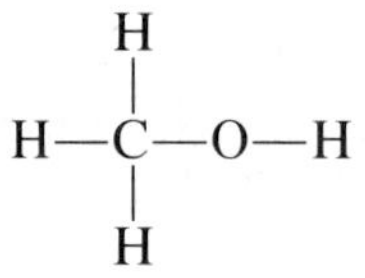
$$
\begin{array}{c}
H \\
| \\
H\!-\!C\!-\!O\!-\!H \\
| \\
H
\end{array}
$$
Methyl alcohol

(b) 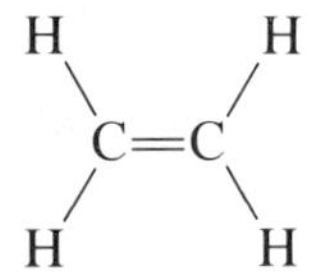
$$
\begin{array}{cc}
H & H \\
\diagdown & \diagup \\
C\!=\!C \\
\diagup & \diagdown \\
H & H
\end{array}
$$
Ethylene

(c) 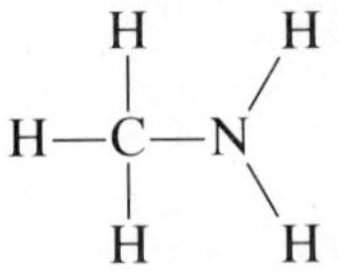
$$
\begin{array}{c}
H \;\;\;\;\; H \\
| \;\;\;\;\;\;\; \diagup \\
H\!-\!C\!-\!N \\
| \;\;\;\;\;\;\; \diagdown \\
H \;\;\;\;\; H
\end{array}
$$
Methylamine

9.17 Name the intermolecular forces that influence the properties of the following compounds and explain how these forces influence the boiling points of these compounds.
(a) ethane, CH_3CH_3, bp $-89°C$
(b) ethyl alcohol, CH_3CH_2OH, bp $78°C$
(c) ethyl chloride, CH_3CH_2Cl, bp $12.5°C$.

AN APPLICATION: BONE AND BIOMATERIALS

Until recent years, the creations of craftsmen, designers, and engineers have been limited by the properties of available materials. The times are changing because chemists and physicists are learning to control properties as they never could before. Increasingly, designers can come up with a description of a material they need and expect it to be custom-made for them. A look at new *biomaterials* in use or being considered for bone replacement illustrates the range of modern materials science.

Bone itself is a *composite*—a mixture of two or more different solid phases intimately united so that the properties of the composite depend on those of the combined materials. Fiberglass, which consists of glass fibers in a polymer matrix, is a familiar example of a composite material. Bone is roughly one-third tough, flexible fibers of collagen (a protein) and two-thirds strong but brittle crystals of an ionic compound known as calcium hydroxyapatite, $Ca_{10}(PO_4)_6(OH)_2$. The strength and flexibility of bone are intermediate between those of collagen and the ionic phosphate.

The first bone replacement materials were metal alloys—stainless steels, titanium alloys, and cobalt–chromium alloys—developed to resist attack by body fluids. Next, from polymer chemistry, came parts for hip and knee joints of high-density polyethylene with formula weights in the range of 1.5 to 2 million. A new understanding of the properties of glasses and ceramics is now also being applied to biomaterials. Ceramics for bone replacement are being made from high-density aluminum oxide (Al_2O_3). Materials known as "bioglasses," rather than being passive to their surroundings in the body, are "surface-active." Bone and bioglasses chemically bond to each other where their surfaces meet, gradually eliminating the joint between them. The closest approach to duplicating the properties of bone may be in composites with either polymer fibers in a ceramic matrix or ceramic fibers in a polymer matrix. The prosthetic arm in the accompanying photo is a similar composite made of carbon fibers in an acrylic polymer matrix.

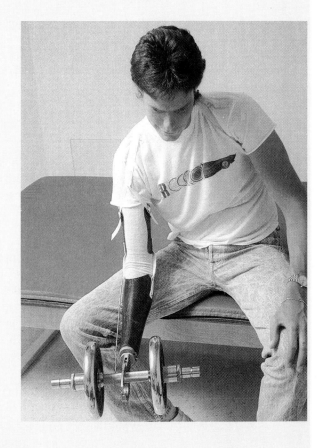

9.13 SOLIDS

Crystalline solid A solid composed of crystals, in which atoms, molecules, or ions have a regular, repeating arrangement.

A **crystalline solid** consists of atoms, molecules, or ions rigidly locked into ordered positions from which they cannot escape so long as the material remains solid. As shown in Table 9.2, ionic compounds, molecular compounds, and metals all form crystalline solids. *Network covalent substances,* also crys-

Table 9.2 Types of Solids

Substance	Smallest Unit	Interparticle Forces	Properties	Examples
Ionic compound	Ions	Attraction between positive and negative ions	Brittle and hard; high-melting; crystalline	NaCl, KI, $Ca_3(PO_4)_2$
Molecular compound	Molecules	Intermolecular forces	Soft; low- to moderate-melting; crystalline	Ice, wax, frozen CO_2, all solid organic compounds
Metal or alloy	Metal atoms	Metallic bonding (attraction between metal ions and surrounding mobile electrons)	Lustrous; soft (Na) to hard (Au); high-melting; crystalline	Elements (Fe, Cu, Sn, . . .), bronze (CuSn alloy), amalgams (Hg + other metals)
Network covalent	Atoms	Covalent bonds	Very hard; very high-melting; crystalline	Diamond, quartz (SiO_2), tungsten carbide (WC)
Amorphous solid	Atoms, ions, or molecules (including polymer molecules)	Any of the above	Noncrystalline; no sharp mp; able to flow (may be very slow); curved edges when shattered	Glasses, tar, some plastics

talline, are very hard materials in which each atom is covalently bonded to several other atoms in a huge, three dimensional network. An entire piece of a network covalent substance like diamond (Chapter 6 Interlude) is essentially one molecule.

By contrast with crystalline materials, the particles in **amorphous solids** do *not* have an orderly arrangement. Amorphous solids often result when liquids cool before they can achieve internal order, or when they can't become ordered because their molecules are large and tangled together (as happens in many polymers). All forms of glass are amorphous solids, as are tar, the gemstone opal, and some hard candies. Amorphous solids differ from crystalline solids by softening over a wide temperature range rather than having sharp melting points, and by shattering to give pieces with curved rather than planar faces.

Amorphous solid A solid in which the atoms, molecules, or ions do not have a regular, repeating arrangement as in a crystalline solid.

9.14 CHANGES OF STATE

What happens when a crystalline solid is heated? As more and more energy is put in, molecules begin to stretch, bend, and vibrate more and more vigorously; atoms or ions wiggle about with more energy. Finally, if the temperature is raised high enough and the motions become vigorous enough, the particles break free from one another, and the substance gradually becomes liquid. The temperature at which this happens is the melting point of the solid, at which the solid and liquid states are in equilibrium. Melting points, as illustrated in Table 9.3, vary with the strength of interparticle forces. The quantity of heat required to completely melt a substance is known as its **heat of fusion,** and this too is a characteristic property of chemical compounds.

The change of a liquid into a vapor proceeds in the same way as the change of a solid into a liquid. When you first put a pan of water on the stove, all the

Heat of fusion The amount of heat necessary to convert one gram (or one mole) of solid into a liquid.

Volatile Evaporates readily.

**Table 9.3
Melting Points
of Some Common
Substances[a]**

Substance	Melting Point (°C)
Sodium metal	97.8
Sodium chloride	801
Sodium hydroxide	318
Oxygen	−218
Tungsten carbide	>2850
Diamond	>3500
Glucose	146
Morphine	254
Cholesterol	148.5
Benzene	5.5
Ethyl alcohol	−117.3
Methane	−182.6

[a] Sodium is a metal; sodium chloride and sodium hydroxide are ionic compounds; and tungsten carbide and diamond are network covalent compounds. All other substances listed are molecular compounds.

added heat goes into raising the temperature of the water. Once the boiling point is reached, all the absorbed heat goes into freeing molecules from their neighbors as they escape into the gas state. Because of its strong intermolecular forces, water has a particularly high heat of vaporization (Table 9.4). Thus, water evaporates more slowly than many other liquids, takes a long time to boil away, and absorbs more heat in the process.

Liquids that evaporate very readily, like ether or rubbing alcohol (isopropyl alcohol), are referred to as **volatile.** If you spill a volatile liquid on your skin, you will feel cool as it evaporates because of the heat it is absorbing from your body. This property of liquids is apparent when an alcohol rub helps to reduce a fever, or when you perspire on a hot day. Unless it's a windy day, you may not feel so cool with water on your skin. But the large quantity of heat absorbed as water evaporates can be harmful. Individuals whose clothing becomes soaked when the temperature is low are at increased risk of hypothermia.

Figure 9.13 shows the energy changes as water is heated until it becomes a gas. You can see that the temperature rises until the melting point is reached, remains constant until melting is completed, and increases until the boiling point, when once again all the energy goes into the change of state. Reversing direction in Figure 9.13 shows that the heat of vaporization is released as steam is converted to a liquid. Thus, a burn from steam is more serious than a burn from boiling water because of the additional heat released as steam condenses.

Table 9.4   Boiling Points and Heats of Vaporization of Some Common Substances

Liquid	Boiling Point (°C)	Heat of Vaporization (cal/g)
Ammonia	−33.4	327
Butane	−0.5	91.5
Ether	34.6	84.0
Ethyl alcohol	78.3	204
Water	100.0	540

Figure 9.13
Heating curve for converting ice at −10°C to steam at 1 atm. The specific heats of ice and water govern the slanted portions of the curve where temperature is increasing. In the flat portions of the curve, all added heat is used by the change of state.

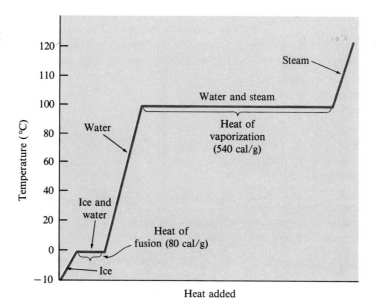

SUMMARY

The three states of matter, solid, liquid, and gas, differ principally in the distance between their particles and the freedom with which they move. According to the **kinetic theory of gases,** the physical behavior of gases can be explained by assuming that they consist of molecules moving rapidly at random, separated from other molecules by great distances, and colliding with no loss of energy. Gas pressure is the result of molecular collisions with a surface. In a mixture of gases each gas exerts the pressure, or **partial pressure,** it would if present alone.

Three **gas laws** explain the properties of gases. (1) **Boyle's law:** The pressure of a gas at constant tempera-ture is inversely proportional to its volume ($P_1V_1 = P_2V_2$). (2) **Charles' law:** The volume of a gas at constant pressure is directly proportional to its temperature in kelvins ($V_1/T_1 = V_2/T_2$). (3) **Avogadro's law:** Equal volumes of gases at the same temperature contain the same number of molecules.

Boyle's law and Charles' law give the **combined gas law** ($P_1V_1/T_1 = P_2V_2/T_2$), which applies to changing conditions for a fixed quantity of gas. The three laws are combined in the **universal gas law,** $PV = nRT$, that relates the effects of temperature, pressure, volume, and molar amount. At 0°C and 1 atmosphere pressure, called **stan-**

INTERLUDE: THE UNIQUENESS OF WATER

Largely because of its strong hydrogen bonding, many physical properties of water are very different from those of other compounds of similar structure. Planet Earth and the life that thrives upon it depend on these unique properties of water.

Water has the highest specific heat of any liquid (1 cal/g °C), giving it the capacity to absorb large quantities of heat while changing only slightly in temperature. In contrast, it takes 540 times more heat to evaporate a given quantity of water (heat of evaporation = 540 cal/g) than to raise its temperature by 1°C. Temperatures in the environment and the human body are kept within moderate ranges by these properties.

A lake stores large amounts of energy from the sun and returns some of this energy to the air and nearby land when the temperature drops. Water vapor produced over a warm body of water then travels on the winds until it reaches a cooler region, condenses, and falls as rain, releasing the 540 cal/g gained in evaporation. In this manner, water redistributes heat around the globe. When the temperature falls to 0°C and the water in a lake or river freezes, 80 cal/g of heat is released, slowing the temperature drop over nearby land.

You can feel the effect of water evaporation on your wet skin when the wind blows, and when you step out of a swimming pool at noon on a summer's day you can turn blue with cold. Even when comfortable, your body is still relying for cooling on the 540 cal/g carried away from the skin and lungs by evaporating water. The heat generated by the chemical reactions of metabolism is carried to the skin by blood, where water moves through cell walls to the surface and evaporates. When metabolism, and therefore heat generation, speeds up, blood flow increases and capillaries dilate so that heat is brought to the surface faster. We perspire, and the moisture content in exhaled air increases. If the humidity is very high, the air cannot hold much more water vapor, evaporation slows down, and we feel uncomfortable because we can't cool down.

Water is also unique in how it changes from the liquid state to the solid state. Most substances are denser as solids than as liquids, which makes sense if molecules are more closely packed in the solid than the liquid state. Indeed, like most dard temperature and pressure (STP), one mole of any gas (6.02×10^{23} molecules) occupies a volume of 22.4 L.

When molecules escape from the surface of the liquid into the gas state in a closed container, equilibrium is reached between liquid and vapor at the **vapor pressure** of the liquid, which increases with temperature. At a liquid's **boiling point,** its vapor pressure equals atmospheric pressure, and the entire liquid is converted into gas while its temperature remains constant.

There are three major types of intermolecular forces: **dipole–dipole forces,** which act between polar molecules; **London forces,** which act between all molecules and increase in strength with formula weight; and **hydrogen bonding,** the strongest of the three, which acts between a hydrogen atom bonded to O or N and a second O, N, or F atom.

Ionic compounds, molecular compounds, metals, and network covalent compounds all form **crystalline solids.** At a solid's **melting point,** particles break free from one another and the substance becomes liquid. **Amorphous solids** lack internal order and do not have sharp melting points. The amount of heat necessary to vaporize one gram (or one mole) of liquid at its boiling point is called its **heat of vaporization.**

---

solids, water becomes more dense with decreasing temperature, but only until it reaches 3.98°C. Then something unusual happens: The density *decreases* as water cools to 0°C and decreases further (to 0.917 g/mL) when it freezes. As a result, ice floats on liquid water, and lakes and rivers freeze from the top down. If the reverse were true, fish and other creatures would be wiped out as they became trapped in ice at the bottom during winter. Here also, hydrogen bonding is responsible for water's uniqueness. As water freezes, each molecule is locked into position by hydrogen bonds to four other water molecules (as shown in the accompanying figure). The resulting structure has more open space than liquid water, accounting for the lower density.

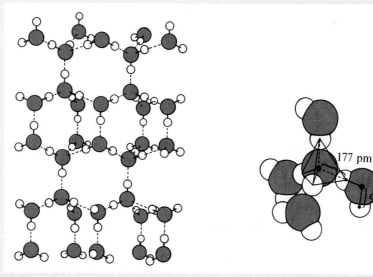

The crystal structure of ice. Because there's more open space between water molecules in ice than in liquid water, ice floats.

177 pm

99 pm

REVIEW PROBLEMS

Gases and Pressure

9.18 How is one atmosphere of pressure defined?

9.19 What is meant by partial pressure?

9.20 What is Dalton's law of partial pressure?

9.21 What are the four assumptions of the kinetic theory of gases?

9.22 How does the kinetic theory of gases explain gas pressure?

9.23 Convert these values into mm Hg:
(a) standard pressure   (b) 0.25 atm   (c) 7.5 atm
(d) 28.0 in. Hg   (e) 41.8 pascals

9.24 Atmospheric pressure at the top of Mt. Whitney is 440 mm Hg
(a) How many atm is this?
(b) How many in. Hg is this?

9.25 If the partial pressure of oxygen in air at 1.0 atm is 160 mm Hg, what is the partial pressure on the summit of Mt. Whitney (Problem 9.24)? Assume that the percent oxygen is the same.

Boyle's Law

9.26 What is Boyle's law, and what variables must be kept constant for the law to hold?

9.27 The pressure of gas in a 600.0 mL cylinder is 65.0 mm Hg. To what volume must it be compressed to increase the pressure to 385 mm Hg?

9.28 The volume of a balloon is 3.50 L at 1.00 atm. What is the pressure if the balloon is compressed to 2.00 L?

9.29 Oxygen gas is commonly sold in 50.0 L steel containers at a pressure of 150. atm. What volume would the gas occupy at a pressure of 1.00 atm if its temperature remained unchanged?

9.30 A hypodermic needle has a volume of 10.0 cc at 14.7 psi. If the tip is blocked so that air can't escape, what pressure is required to decrease the volume to 2.00 cc?

9.31 A gas tank containing N_2 has a volume of 25.0 L and a pressure of 35.0 atm. A second tank, containing O_2, has a volume of 10.0 L and a pressure of 15.0 atm. Both gases are pumped into a tank with a volume of 15.0 L.
(a) What is the partial pressure of each gas in the new tank?
(b) What is the total pressure in the new tank?

Charles' Law

9.32 What is Charles' law, and what variables must be kept constant for the law to hold?

9.33 A hot-air balloon has a volume of 960. L at 18°C. To what temperature must it be heated to raise its volume to 1200. L?

9.34 A gas sample has a volume of 125 mL at 25°C. What is its volume at 37°C?

9.35 A balloon has a volume of 43.0 L at 20°C. What is its volume at −5°C?

Combined Gas Law

9.36 For the combined gas law to be used, what variable must be kept constant?

9.37 What conditions are defined as standard temperature and pressure (STP)?

9.38 A gas has a volume of 2.84 L at STP. At what temperature does it have a volume of 7.50 L at 520. mm Hg?

9.39 A helium balloon has a volume of 3.50 L at 22.0°C and 1.14 atm. What is its volume if the temperature is increased to 30.0°C and the pressure is increased to 1.20 atm?

9.40 When H_2 gas was released by the action of HCl on Zn, the volume of H_2 collected was 55.0 mL at 26°C and 749 mm Hg. What is the volume of the H_2 at STP?

9.41 A compressed-air tank carried by scuba divers has a volume of 8.0 L and a pressure of 140 atm at 20°C. What is the volume of air in the tank at STP?

9.42 What is the partial pressure of oxygen in the scuba tank (Problem 9.41) assuming that air is 21% oxygen?

9.43 What is the effect on the pressure of a gas if you simultaneously:
(a) halve its volume and double its kelvin temperature?
(b) double its volume and halve its kelvin temperature?

Avogadro's Law and Standard Molar Volume

9.44 Explain Avogadro's law using the kinetic theory of gases.

9.45 How many liters does a mole of gas occupy at STP?

9.46 Which sample contains more molecules: 1.0 L of O_2 at STP, or 1.0 L of H_2 at STP?

9.47 Which of the two samples in Problem 9.46 weighs more?

9.48 How many mL of Cl_2 gas must you measure out to obtain 0.20 g at STP?

9.49 A typical room is 4.0 m long, 5.0 m wide, and 2.5 m high.
(a) What is the total mass of the oxygen in the room? Assume that the gas in the room is at STP and that air contains 21% oxygen and 79% nitrogen.
(b) What is the total mass of nitrogen in the room described in part (a)?
(c) What is the total mass of air in the room?

Universal Gas Law

9.50 What is the universal gas law?

9.51 Which sample contains more molecules: 2.0 L of Cl_2 at STP, or 3.0 L of CH_4 at 300 K and 1.5 atm? Which sample weighs more?

9.52 Which sample contains more molecules: 2.0 L of CO_2 at 300 K and 500 mm Hg, or 1.5 L of N_2 at 57°C and 760 mm Hg? Which sample weighs more?

9.53 If 15.0 g of CO_2 gas has a volume of 0.30 L at 300. K, what is its pressure in mm Hg?

9.54 If 20.0 g of N_2 gas has a volume of 0.40 L and a pressure of 6.0 atm, what is its temperature?

9.55 If 18.0 g of O_2 gas has a temperature of 350. K and a pressure of 550. mm Hg, what is its volume?

9.56 How many moles of a gas will occupy a volume of 0.55 L at a temperature of 347 K and a pressure of 2.5 atm?

Liquids and Intermolecular Forces

9.57 What is the vapor pressure of a liquid?

9.58 What is the value of a liquid's vapor pressure at the liquid's normal boiling point?

9.59 What is the effect of pressure on a liquid's boiling point?

9.60 What is a liquid's heat of vaporization?

9.61 Look at Figure 9.9 and answer these questions:
(a) What is the boiling point of water at an atmospheric pressure of 400 mm Hg?
(b) Is ether a liquid or a gas at 20°C and 400 mm Hg?
(c) At what pressure does ethanol boil at 40°C?

9.62 What characteristic must a compound have to experience each of the following interparticle forces?
(a) London forces
(b) dipole–dipole interactions
(c) hydrogen bonding

9.63 The observed melting points of the halogens (F_2, Cl_2, Br_2, and I_2) are: 113.5°C, −101°C, −7°C, and −220°C. Match each temperature with the appropriate molecule. Justify your answer in terms of intermolecular attractions.

9.64 In which of the following compounds are dipole–dipole attractions the most important interparticle force?
(a) N_2 (b) HCN   (c) CCl_4 (d) $MgBr_2$
(e) CH_3Cl (f) CH_3CO_2H

9.65 Both dimethyl ether (CH_3OCH_3) and ethanol (C_2H_5OH) have the same formula weight, but the boiling point of dimethyl ether is −25°C while that of ethanol is 78°C. Explain.

9.66 The heat of vaporization of water is 9.72 kcal/mol.
(a) How much heat (in kcal) is required to vaporize 3.00 mol of H_2O?

(b) How much heat (in kcal) is released when 255 g of steam condenses?

Solids

9.67 What is the difference between an amorphous and a crystalline solid?

9.68 What are the major interparticle attractions (ionic, intermolecular, or metallic bonding) in each of the following crystalline solids?
(a) KBr   (b) CO_2 (c) Ca   (d) Ne
(e) $MgCO_3$ (f) NH_3

9.69 Describe and give an example of a network covalent solid.

9.70 For most substances heats of vaporization are much larger than heats of fusion. Water, for example, has a heat of vaporization of 540 cal/g and a heat of fusion of 80 cal/g. Explain.

Applications

9.71 What is the difference between a systolic and a diastolic pressure reading? [App: Blood Pressure]

9.72 Is a blood pressure reading of 180/110 within the normal range? Which is the systolic and which is the diastolic pressure? [App: Blood Pressure]

9.73 What substance was used as the first general anesthetic? [App: Inhaled Anesthetics]

9.74 What is the advantage of a ''bioglass'' relative to metal for bone replacement? [App: Bone and Biomaterials]

9.75 Explain how the presence of a large body of water helps moderate temperature fluctuations. [Int: Uniqueness of Water]

9.76 Explain why ice has a lower density than water. [Int: Uniqueness of Water]

9.77 Sometimes when a freeze is expected, farmers turn sprinklers on their crops to coat them with water. Why? [Int: Uniqueness of Water]

Additional Questions and Problems

9.78 Use the kinetic molecular theory to explain why gas pressure increases if the temperature is raised and the volume is kept constant.

9.79 Hydrogen and oxygen react according to the equation $2 H_2 + O_2 \rightarrow 2 H_2O$. According to Avogadro's law, how many liters of hydrogen are required to react with 2.5 L of oxygen at STP?

9.80 If 3.0 L of hydrogen and 1.5 L of oxygen at STP react to yield water, how many moles of water are formed? What gas volume does the water have at a temperature of 100°C and 1 atm pressure?

9.81 Approximately 240 mL/min of CO_2 is exhaled by an average adult at rest. Assuming a temperature of 37°C and standard pressure, how many moles of CO_2 is this?

9.82   How many grams of CO_2 are exhaled by an average resting adult in 24 hours (Problem 9.81)?

9.83   Imagine that you have two identical vessels, one containing hydrogen at STP and the other containing oxygen at STP. How could you tell which was which without opening them?

9.84   One mole of any gas has a volume of 22.4 L at STP. What is the formula weight of each of the following gases, and what are the densities in g/L at STP?
(a) CH_4 (b) CO_2 (c) O_2 (d) UF_6

9.85   Gas pressure outside the space shuttle is approximately 1×10^{-14} mm Hg at a temperature of approximately 1 K. If the gas is almost entirely hydrogen atoms (H, not H_2), what volume of space is occupied by one mole of atoms? What is the density of H gas in atoms/L?

9.86   A sample containing 2.50 g of methanol (CH_3OH) and 1.50 g of H_2O is placed in a 300.0 mL container. The container is closed and heated to 150.°C, causing complete vaporization of both liquids.
(a) What is the partial pressure, in atm, of each gas in the container?
(b) What is the total pressure in the container?
(c) Why is it important to vent pressure cookers and other steam devices, allowing some of the vapors formed to escape?

9.87   Ethylene glycol, $C_2H_6O_2$, has one OH bonded to each carbon.
(a) Draw the Lewis dot structure of ethylene glycol.
(b) Draw the Lewis dot structure of chloroethane, C_2H_5Cl.
(c) Chloroethane has a slightly higher molar mass than ethylene glycol, yet has a lower melting point (-39°C for chloroethane versus -13°C for ethylene glycol) and a lower boiling point (3°C versus 198°C). Explain.

C H A P T E R

10 Solutions

The Colorado River has been carrying away an average of 400,000 tons of dissolved and suspended solids each day for millions of years. In this chapter, we'll describe how water molecules bring solids into solution.

Up to this point, we've been concerned primarily with pure substances, both elements and compounds. In day-to-day life, however, most of the materials we come in contact with are *mixtures.* Air, for example, is a gaseous mixture of oxygen and nitrogen; blood is a liquid mixture of many different components; and rocks are solid mixtures of different minerals. We'll look closely in this chapter at the characteristics and properties of solutions and other mixtures.

1. *What are mixtures, and what are solutions?* The goal: Be able to distinguish between mixtures and solutions.
2. *How does the nature of solute and solvent affect solubility?* The goal: Be able to explain "like dissolves like" and predict solubilities based on this rule of thumb.
3. *What influence do temperature and pressure have on solubility?* The goal: Be able to explain the influence of temperature and pressure, including applications of Henry's law.
4. *How is the concentration of a solution expressed?* The goal: Be able to define, use, and convert between the most common ways of expressing solution concentrations.
5. *How are dilutions carried out?* The goal: Be able to state the result of diluting a solution and explain how to make a desired dilution.
6. *What is an electrolyte?* The goal: Be able to recognize strong and weak electrolytes and nonelectrolytes, and use electrolyte concentrations in equivalents.
7. *What is vapor pressure lowering?* The goal: Be able to explain vapor pressure lowering and its associated effects.
8. *What are osmosis and dialysis?* The goal: Be able to describe osmosis, dialysis, and some of their applications.
9. *What is a colloid?* The goal: Be able to define a colloid and give some examples of colloids in various states.

10.1 MIXTURES

Heterogeneous mixture
A mixture that is visually nonuniform.

Homogeneous mixture
A mixture that is uniform to the naked eye.

Mixtures, in which combined substances retain their chemical identities, are often classified according to their visual appearance as either heterogeneous or homogeneous. **Heterogeneous mixtures** are those in which the mixing is obviously nonuniform. Most rocks, for example, show a grainy character due to the heterogeneous mixing of particles of different minerals. **Homogeneous mixtures** are those in which the mixing *is* uniform, at least to the naked eye. Seawater, a homogeneous mixture of soluble compounds in water, is an obvious example.

Mixtures are also classified according to the size of their particles as solutions, colloids, and suspensions. The divisions among the three groups aren't always clear, but the concepts are useful nonetheless. **Solutions,** the most important class of homogeneous mixtures, contain particles the size of ions and small molecules (roughly 0.1–2 nm diameter). **Colloids** are also homogeneous in appearance but contain larger particles than solutions (2–1000 nm diameter). **Suspensions** contain still larger particles, with diameters greater than 1000 nm—the size of a speck just visible to the naked eye. Generally, "suspension" refers to small particles suspended in either a gas or a liquid. Suspensions are homogeneous when well stirred, but the particles usually eventually settle out into layers.

Solutions, colloids, and suspensions can be differentiated in several ways. For example, true solutions are transparent, although they may be colored. Colloids may be as clear in appearance as true solutions but may also have a murky or opaque appearance. It's possible to distinguish between liquid colloids and solutions, though, by shining a light through them. The particles in a solution are so small that they don't affect the light beam, whereas the larger particles in a colloid scatter the light and spread out the beam (Figure 10.1), a phenomenon called the *Tyndall effect.*

Neither solutions nor colloids separate on standing and in this way differ from suspensions. In addition, the particles dispersed in solutions and colloids are too small to be removed by filtration, but those in a suspension are large enough to remain behind on a filter. Table 10.1 gives some examples of solutions, colloids, and suspensions. It's interesting to note that blood has characteristics of all three. About 45% of blood volume consists of suspended red and white cells; the rest is plasma, which contains electrolytes in solution and colloid-sized protein molecules.

Solution A homogeneous mixture in which atoms, molecules, or ions are uniformly mixed together.

Colloid (colloidal dispersion) A mixture of a dispersing medium and dispersed particles larger than most molecules but too small to be seen by the naked eye.

Suspension A mixture containing particles just large enough to be visible to the naked eye, and which settle out on standing.

Practice Problem   **10.1**   Which of the following liquid mixtures are true solutions?
(a) milk   (b) apple juice   (c) hand lotion   (d) gasoline

Figure 10.1
The Tyndall effect. Light is scattered as it passes through the colloidal dispersion in water on the left, making the path of the light visible. Because light is not scattered by solute particles in the solution on the right, the light pathway is not visible.

Table 10.1 Three Kinds of Mixtures: Solutions, Colloids, and Suspensions

Kind of Mixture	Particle Diameter	Examples	Characteristics
Solution	0.1–2.0 nm	Seawater, air, vinegar	Transparent to light; non-filterable; does not separate on standing
Colloid	2.0–1000 nm	Butter, milk, fog, pearls	Clear, murky, or opaque to light; shows Tyndall effect; nonfilterable; does not separate on standing
Suspension	>1000 nm	Paint, aerosol sprays, muddy water	Murky or opaque to light; can be separated by filtration or on standing

10.2 SOLUTIONS

State of matter The physical state of a substance as a solid, liquid, or gas.

Alloy A solution or mixture of metals.

Solute The dissolved substance in a homogeneous mixture; often a solid or gas in a liquid solution.

Solvent The major substance in a homogeneous mixture; often a liquid in which a gas or solid is dissolved.

Miscible Two substances mutually soluble in all proportions without limit.

Although we usually think of a solid dissolved in a liquid when talking about solutions, solutions occur in all three **states of matter** (Table 10.2). Even solutions of one solid with another are common, as in metal **alloys** such as 14-karat gold (58% gold with other metals). For solutions in which a gas or solid is dissolved in a liquid, the dissolved substance is called the **solute**, and the liquid is called the **solvent**. When one liquid is dissolved in another, the distinction between solute and solvent isn't always obvious.

What determines whether a substance is soluble in a given liquid? It depends on how strongly the solute and solvent particles are attracted to each other relative to the attraction holding them in the pure substances. Two liquids are **miscible**, or mutually soluble in all proportions, when the intermolecular forces within each liquid and those between molecules of the two liquids are very similar. Ethyl alcohol and water, for example, are miscible because hydrogen bonding is strong between ethyl alcohol molecules, between water molecules, and also between water and ethyl alcohol molecules.

solute -------- solute   solvent -------- solvent

Substances form solutions when these forces are similar

solute -------- solvent

Observations of solubilities have led to the rule of thumb that *like dissolves like,* meaning that substances with similar interparticle forces form solutions. Polar solvents dissolve polar substances and ionic substances; solvents that hydrogen-bond dissolve substances that hydrogen-bond; and nonpolar solvents dissolve nonpolar substances. Thus, a strongly polar, hydrogen-bonding compound like water dissolves ethyl alcohol and sodium chloride, whereas a nonpolar organic compound like hexane (C_6H_{14}) does not dissolve ionic com-

Table 10.2   Some Different Kinds of Solutions

Solution Type	Example
Gas in gas	Air (oxygen, nitrogen, other gases)
Gas in liquid	Carbonated water (CO_2 in water)
Gas in solid	Hydrogen in platinum (H_2 in Pt)
Liquid in liquid	Vinegar (acetic acid in water)
Solid in liquid	Saline solution (NaCl in water)
Solid in solid	14-karat gold

Immiscible Mutually insoluble.

pounds or polar compounds but dissolves other nonpolar organic compounds such as fats and oils. The old saying, Oil and water don't mix (which applies to more than chemical compounds), sums up what we've all observed—oil and water are **immiscible.** The intermolecular forces between water molecules are so strong that after an oil–water mixture is shaken, the water layer always reforms, squeezing out the oil molecules.

Some liquids, particularly such organic compounds as chloroform ($CHCl_3$) and diethyl ether ($CH_3CH_2OCH_2CH_3$), are slightly soluble in water because they are small, polar molecules. When mixed with water, a small amount of the organic compound dissolves, but the remainder forms a separate liquid layer. As the number of carbon atoms in organic molecules increases, though, water solubility decreases.

Practice Problem   10.2   Which of the following pairs of substances would you expect to form solutions?
(a) carbon tetrachloride (CCl_4) and water
(b) gasoline and magnesium sulfate ($MgSO_4$)
(c) hexane (C_6H_{14}) and heptane (C_7H_{16})
(d) ethyl alcohol (CH_3CH_2OH) and heptanol ($CH_3(CH_2)_5CH_2OH$)

10.3   WATER AS A SOLVENT

When NaCl crystals are put in water, ions at the crystal surface come into contact with polar water molecules. Positively charged sodium ions are attracted to the negatively polarized oxygen end of water, while negatively charged chloride ions are attracted to the positively polarized hydrogen end. The combined forces of attraction between an ion and several water molecules are enough to pull the ion away from the crystal, exposing a fresh surface, until ultimately the crystal dissolves (Figure 10.2).

Solvation The surrounding of a solute ion or molecule by solvent molecules.

Hydration The surrounding of a solute ion or molecule by water molecules.

Once in solution, Na^+ and Cl^- are completely surrounded by solvent molecules, a phenomenon called **solvation** (or, specifically for water, **hydration**). The water molecules form a loose shell around the ions, stabilizing them by electrical attraction, as shown in Figure 10.2.

Solubility is an unpredictable and complex matter that is not well understood. No one has yet devised a mathematical theory that predicts the extent to which an unknown substance will dissolve in water. As you saw in studying exchange reactions (Section 7.11), though, we do have some general rules for

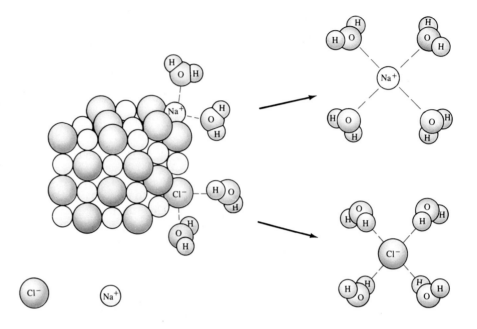

Figure 10.2
Dissolution of salt in water. A salt crystal dissolves when polar water molecules surround the individual Na^+ and Cl^- ions, pulling them from the crystal surface into solution.

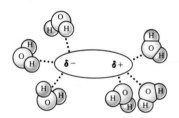

Figure 10.3
Hydration of a polar molecule in solution in water.

predicting the water solubility of ionic compounds. Water solubility is not by any means limited to ionic compounds. Many polar organic molecules, including sugars, amino acids, and even some proteins, dissolve in water. Water solvates these polar molecules in the same way it solvates ions—by surrounding the molecules and orienting to maximize electrical attractions (Figure 10.3).

10.4 HYDRATION OF SOLIDS

Some ionic compounds attract water strongly enough to hold onto water molecules even when crystalline. Such compounds are solid **hydrates.** For example, the plaster of paris used to make decorative objects, stucco walls, and casts for broken limbs is calcium sulfate monohydrate, $CaSO_4 \cdot H_2O$. The raised dot between $CaSO_4$ and H_2O in the formula indicates that one molecule of water is present for each $CaSO_4$ formula unit in the crystal.

$$CaSO_4 \cdot H_2O \qquad \text{A hydrate}$$

Hydrate A solid substance that has water molecules included in its crystals.

After being finely ground and mixed with water to make plaster, this monohydrate gradually changes into the crystalline dihydrate $CaSO_4 \cdot 2H_2O$, known as *gypsum*. During the change, the plaster hardens and expands in volume, causing it to fill a mold or shape itself closely around a broken limb. Many ionic compounds are sold and handled primarily as hydrates (Table 10.3).

Still other ionic compounds attract water so strongly that they pull water vapor from the surrounding atmosphere to become hydrated. Compounds that show this behavior, such as calcium chloride ($CaCl_2$), are called **hygroscopic;** they are often used as drying agents. You might have noticed a small bag of a hygroscopic compound (probably silica, SiO_2) included in the packing material of a new stereo or VCR to keep humidity low during shipping.

Hygroscopic Having the ability to attract water vapor from the air.

Table 10.3   Some Common Hydrates

Formula	Name	Uses
$AlCl_3 \cdot 6H_2O$	Aluminum trichloride hexa-hydrate	Antiperspirant
$CaSO_4 \cdot 2H_2O$	Calcium sulfate dihydrate (gypsum)	Cements, wallboard, molds
$CaSO_4 \cdot H_2O$	Calcium sulfate monohydrate (plaster of paris)	Casts, molds
$CuSO_4 \cdot 5H_2O$	Copper(II) sulfate penta-hydrate (blue vitriol)	Pesticide, germicide, topical fungicide
$FeSO_4 \cdot 7H_2O$	Iron(II) sulfate heptahydrate (green vitriol)	Fertilizer, feed additive
$MgSO_4 \cdot 7H_2O$	Magnesium sulfate hepta-hydrate (epsom salts)	Laxative, anticonvulsant
$Na_2B_4O_7 \cdot 10H_2O$	Sodium tetraborate deca-hydrate (borax)	Cleaning compounds, fireproofing agent
$Na_2S_2O_3 \cdot 5H_2O$	Sodium thiosulfate penta-hydrate (hypo)	Photographic fixer

Practice Problems   **10.3**   Write the formula of sodium sulfate decahydrate, known as Glauber's salt and used as a laxative.

10.4   What mass of Glauber's salt must be weighed out to provide 1.00 mol of sodium sulfate?

10.5   SOLUBILITY

Solubility   The amount of a substance that can be dissolved in a given volume of solvent.

Imagine that you are asked to prepare 100 mL of a *saline solution* (an NaCl solution). You might measure out the water, add solid sodium chloride, and stir until it dissolves. But how much NaCl should you add? Can you dissolve any amount you choose in water, or is there a limit? In fact, there is a limit. At 20°C, a maximum of 35.8 g of NaCl will dissolve in 100 mL of water. Any amount added above this limit will simply sink to the bottom of the container and sit there.

Saturated solution   A solution in which the solute has reached its solubility limit.

The maximum amount of a substance that will dissolve in a given amount of a liquid, usually expressed in grams per 100 mL (g/100 mL), is called its **solubility.** When a solution has reached its solubility limit, we say that it is **saturated.** A solution containing less than the maximum amount of solute is described as **unsaturated.** Like melting point and boiling point, solubility is a physical property characteristic of a specific substance. Different substances have greatly differing solubilities, as illustrated in Table 10.4.

Unsaturated solution   A solution that contains less than the maximum amount of solute.

In a saturated solution in contact with some undissolved solid solute, a **dynamic equilibrium** is established:

Dynamic equilibrium   An equilibrium in which the rates of forward and reverse changes are equal.

$$\text{Solid solute + solvent} \underset{\text{crystallize}}{\overset{\text{dissolve}}{\rightleftarrows}} \text{saturated solution}$$

Table 10.4   Solubilities of Some Common Substances in Water

Substance	Solubility in Water (g/100 mL)	
	At 20°C	At 100°C
Solids		
Ammonium nitrate (NH_4NO_3)	178	871
Sodium chloride (NaCl)	35.8	39.1
Sodium bicarbonate ($NaHCO_3$)	9.6	23.6
Sodium hydroxide (NaOH)	109	347
Glucose	92.3	556 (90°C)
Sucrose	204	487
Liquids		
Acetic acid ($C_2H_4O_2$)	Miscible	—
Chloroform ($CHCl_3$)	0.71	—
Diethyl ether (ether, $C_4H_{10}O$)	6.9	—
Ethyl alcohol (C_2H_6O)	Miscible	—
Phenol (carbolic acid, C_6H_6O)	9.1	—
Gases		
Ammonia (NH_3)	51.8	16.0 (60°C)
Carbon dioxide (CO_2)	0.169	0.071 (60°C)
Hydrogen chloride (HCl)	77.1	66.5 (60°C)
Oxygen (O_2)	0.0043	0.0028 (60°C)

Solute particles leave the solid surface and reenter the solid from solution at equal rates, and the concentration of solute in the saturated solution is constant (Figure 10.4).

Anything that increases contact between the solvent and the surface speeds up dissolution. In the kitchen and the lab, we instinctively stir a solution to speed up dissolving a solid—stirring brings fresh solvent into contact with the surfaces. Solids also dissolve more quickly when finely ground than in big chunks because there is more surface area. Speeding up dissolving, though, causes no change in the maximum *amount* that will dissolve at a given temperature.

Figure 10.4
A saturated solution. Solute particles enter and leave the solid surface at an equal rate. In the process, an irregularly shaped crystal is converted to a more symmetrical crystal.

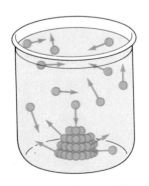

10.6   EFFECT OF TEMPERATURE ON SOLUBILITY

The effect of temperature on solubility is different for every substance and is unpredictable. The dissolution of most ionic crystalline substances is endothermic because energy is needed to pull the ions away from each other, and for such substances raising the temperature increases solubility, as illustrated in Table 10.4.

$$\text{Solid solute} + \text{solvent} + \text{heat} \xrightarrow{\text{Heat}} \rightleftharpoons \text{saturated solution}$$

AN APPLICATION: GOUT AND KIDNEY STONES—PROBLEMS IN SOLUBILITY

One of the major pathways for the breakdown of nucleic acids in the body is by their conversion to uric acid. About 0.5 g/day of uric acid is excreted in the urine of normal people. Unfortunately, the solubility of uric acid in water is fairly low—only about 7 mg/mL at 37°C. When too much uric acid is produced by the body or the mechanisms for its elimination fail, its concentration in blood and urine rises, and excess uric acid sometimes comes out of solution to be deposited in the joints and kidneys.

Gout, a disorder of nucleic acid metabolism that primarily affects middle-aged men, is characterized by excessive uric acid production leading to the deposit of uric acid crystals in soft tissue around the joints. The feet and the hands are often affected. Deposition of the sharp, needle-like crystals causes an acute and painful in-flammation that can lead ultimately to arthritis and to bone destruction.

Similarly with kidney stones. Excess uric acid in the urine can lead to the formation and deposit of crystals in the kidney. Kidney stones may also be composed of calcium or magnesium salts containing phosphate, carbonate, or oxalate anions. Although often quite small, kidney stones cause excruciating pain when passed through the ureter. In some cases, complete blockage of the ureter occurs.

Treatment of excessive uric acid production involves both dietary modification and drug therapy. Foods rich in nucleoproteins, such as liver, sardines, and asparagus, must be avoided, and drugs such as allopurinol are taken to lower production of uric acid.

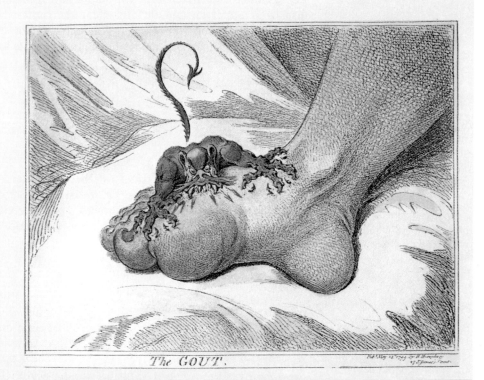

The GOUT.

This old cartoon leaves little doubt about how painful gout can be.

A few ionic substances such as calcium chloride ($CaCl_2$) do release heat when dissolved, a property taken advantage of in instant hot compresses for sports and health care. Dry $CaCl_2$ and a pouch of water are enclosed in a flexible package. Striking the package breaks the pouch, the solid and water mix, and heat is released. Ammonium nitrate (NH_4NO_3) is used in a similar fashion in cold compresses, since it absorbs heat when it dissolves in water.

Supersaturated solution A solution containing solute in greater concentration than the normal solubility limit of the solute.

Occasionally, a **supersaturated solution** can be produced, one that contains more solute than a saturated solution. Suppose a large amount of a substance is dissolved at, say, 95°C, giving a solution that would be more than saturated at room temperature. As the solution cools down, the solubility decreases and the excess solute should crystallize. But if the cooling is done very slowly, if there are no bits of solid in the solution on which crystals can start to form, and if the container stands quietly, crystallization might not occur. Instead a supersaturated solution is created. The instant crystallization is initiated—perhaps by dropping in a tiny crystal of the solute—all the excess solute rapidly crystallizes. The result may be the dramatic formation of a beautiful network of crystals (Figure 10.5).

The influence of temperature on the solubility of gases, unlike that of solids, is always predictable (Table 10.4). Heat always decreases gas solubility.

$$\xleftarrow{\hspace{3cm}}\text{Heat}$$
$$\text{Gas} + \text{solvent} \rightleftharpoons \text{saturated solution} + \text{heat}$$

For example, an increase in the temperature of natural waters can have a disastrous effect on aquatic life. Near the outflow of warm water from an industrial operation, the oxygen concentration of the water may decrease, causing fish and microorganisms that cannot tolerate lower oxygen levels to die.

Figure 10.5
The result of supersaturation. Crystals are shown at the instant they began to form when a seed crystal was dropped into a supersaturated sodium acetate ($NaCH_3COO$) solution. In a few more seconds, the beaker will be full of crystals.

10.7 EFFECT OF PRESSURE ON SOLUBILITY: HENRY'S LAW

Pressure has virtually no effect on the solubilities of solids and liquids in solvents. Gas solubilities, however, are strongly dependent on pressure. The stress of an increase in the pressure of a gas in contact with a liquid can be relieved (LeChatelier's principle again) only by more gas molecules going into solution.

$$\text{Pressure} \xrightarrow{\hspace{3cm}}$$
$$\text{Gas} + \text{solvent} \rightleftharpoons \text{solution}$$

You learned in Section 9.4 that each gas in a mixture exerts its pressure independently of the others—the *partial pressure* of that gas. According to **Henry's law,** the solubility of a gas in a liquid is directly proportional to its partial pressure over the liquid at constant temperature. If the pressure of the gas doubles, solubility doubles; if the gas pressure is halved, solubility is halved (Figure 10.6).

Henry's law The solubility of a gas in a liquid is directly proportional to its partial pressure over the liquid at constant temperature.

Figure 10.6
Henry's law. Gas solubility is directly proportional to partial pressure (for gases that do not react with the solvent). An increase in pressure causes more gas molecules to enter solution until equilibrium is restored between the gas in solution and the gas state.

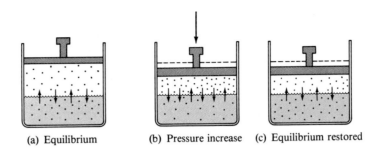

(a) Equilibrium   (b) Pressure increase   (c) Equilibrium restored

Henry's law: Gas solubility $\propto$ gas pressure

$$\frac{\text{Solubility}}{\text{Pressure}} = \begin{array}{l}\text{a constant value} \\ \text{at fixed temperature}\end{array}$$

Henry's law accounts for the fizzing in a suddenly opened bottle of soft drink or champagne. The bottle is sealed under more than one atmosphere of carbon dioxide pressure, causing some of the CO_2 to dissolve. When the bottle is opened, however, CO_2 pressure drops and dissolved gas comes bubbling out of solution.

As written below, Henry's law gives the relation between the concentration (C) of a given gas in solution and its partial pressure.

$$P_{\text{gas}} = kC \qquad \text{At constant temperature}$$

The numerical value of the Henry's law constant k is different for each gas–solvent combination, and the calculations require units that we need not explore. The important point here is that partial pressure is often used to represent gas concentration in a solution, a practice especially common in health-related sciences. Without knowing the equation written above, you might find puzzling references to partial pressures of carbon dioxide and oxygen "in" body fluids. Some of these values are given in Table 10.5 and illustrate the convenience of having the same unit for, say, oxygen concentration, in both air and blood. Compare the oxygen partial pressures in saturated alveolar air (air in

Table 10.5   Partial Pressures and Normal Gas Concentrations in Gases and Body Fluids[a]

Source of Sample	Partial Pressure of N_2 (P_{N_2}, mm Hg) (% of atm pressure)[b]	Partial Pressure of O_2 (P_{O_2}, mm Hg) (% of atm pressure)[b]	Partial Pressure of CO_2 (P_{CO_2}, mm Hg) (% of atm pressure)[b]	Partial Pressure of H_2O (P_{H_2O}, mm Hg) (% of atm pressure)[b]
Inspired air (dry)	597 (78.6%)	159 (20.8%)	0.3 (0.04%)	3.7 (0.5%)
Alveolar air (saturated)	573 (75.4%)	100 (13.2%)	40 (5.2%)	47 (6.2%)
Expired air (saturated)	569 (74.8%)	116 (15.3%)	28 (3.7%)	47 (6.2%)
Arterial blood	573	95	40	
Venous blood	573	40	45	
Peripheral tissues	573	40	45	

[a] Adapted from F. Martini, *Fundamentals of Anatomy and Physiology*, p. 659 (Prentice Hall, 1989).
[b] Percentage of average atmospheric pressure of 760 mm Hg.

the lungs) and in arterial blood. The values are almost the same because the gases dissolved in blood come to equilibrium with the same gases in the lungs. (The slight drop is caused by mixing in of some blood that doesn't pass through the lungs.)

If the partial pressure of a given gas over a solution changes while the temperature is constant, the new solubility of the gas is found by setting the Henry's law relations equal to each other. Using C for solubility in grams per 100 mL and P for partial pressure, the relation is

$$\frac{C_1}{P_1} = \frac{C_2}{P_2} \qquad \text{At constant temperature}$$

For example, at a 159 mm Hg partial pressure of oxygen in the atmosphere (P_1), the solubility of oxygen in blood is 0.44 g/100 mL (C_1). To find the solubility of oxygen in blood at 11,000 ft, where the partial pressure of O_2 is 56 mm Hg (P_2), the equation is used as follows:

$$C_2 = \frac{P_2 C_1}{P_1} = \frac{56 \text{ mm Hg} \times 0.44 \text{ g/100 mL}}{159 \text{ mm Hg}} = 0.15 \text{ g/100 mL}$$

Breathing pressurized air causes a potential problem for scuba divers because of Henry's law. Nitrogen dissolved in a diver's blood under pressure comes bubbling out of solution if a diver ascends too quickly. The result, known as the "bends" because it causes difficulty in straightening out the joints, is very painful and can be fatal. The treatment is to place the diver in a high-pressure chamber in which decompression is carried out slowly, allowing nitrogen to come out of solution gradually enough to be eliminated by natural mechanisms. Replacing nitrogen with helium in the breathing mixture decreases the problem because helium is less soluble in body fluids than nitrogen.

Practice Problem   10.5   At a partial pressure of 760 mm Hg (and 20°C), the solubility of carbon dioxide in water is 0.169 g/100 mL. What is the solubility of carbon dioxide at 2.5×10^4 mm Hg?

10.8   PERCENT CONCENTRATION

Concentration   The amount of a solute per unit quantity of solution or solvent.

Although we might speak in casual conversation of a solution as either "dilute" or "concentrated," laboratory work usually requires an exact knowledge of a solution's **concentration**. As indicated in Table 10.6, there are several common methods for expressing concentration. The units differ, but all the methods listed describe how much solute is present in a unit quantity of solution.

Because percent concentration is expressed in the three different ways listed in Table 10.6, the possibility for confusion is always present. For this reason, in many fields (as illustrated in Section 2.9 for dosage calculations) concentrations are increasingly being given in specific units, for example, 25 g/100 mL and 25 g/dL. At present, though, it's still necessary to understand all the concentration units listed in Table 10.6.

Table 10.6   Some Methods for Expressing Concentration

Solute Measure	Solution Measure	Concentration Measure
Weight (g)	Volume (100 mL)	Weight/volume percent [(w/v)%]
Weight (g)[a]	Weight (g)[a]	Weight/weight percent [(w/w)%]
Volume (mL)[b]	Volume (mL)[b]	Volume/volume percent [(v/v)%]
Number of moles	Volume (L)	Moles/liter (molarity, M)
Parts[c]	10^6 parts[c]	Parts per million (ppm)

[a] Or any other weight units, but same unit for solute and solution.
[b] Or any other volume units, but same unit for solute and solution.
[c] Any units, but same unit for solvent and solute.

Weight/Volume Percent Concentration   One of the most common methods for expressing percent concentration is to give the number of grams (weight) of the solute per 100 mL (volume) of final solution: **weight/volume percent concentration [(w/v)%].** For example, if 15 g of glucose is dissolved in enough water to give 100 mL of solution, the glucose concentration is 15 g/100 mL, or 15% (w/v).

Weight/volume percent concentration [(w/v)%] Concentration expressed as the number of grams of solute dissolved in 100 mL of solution.

$$(w/v)\% \text{ concentration} = \frac{\text{grams of solute}}{\text{mL of solution}} \times 100\%$$

$$= \frac{15 \text{ g glucose}}{100 \text{ mL solution}} \times 100\% = 15\%(w/v)$$

The weight/volume percent concentration is found from any known mass and volume of solution as illustrated in Solved Problem 10.1.

Sometimes, as a variation on weight/volume percent, concentration is given in milligram percent (mg %) for more dilute solutions. For example, 5 mg % means 5 mg per 100 mL of solution, which is equivalent to 5 mg/dL.

To prepare 100 mL of a specific weight/volume solution, the weighed solute is dissolved in just enough solvent to give a *final* volume of 100 mL—it's not dissolved in a *starting* volume of 100 mL solvent. After all, if the solute were dissolved in 100 mL of solvent, the final volume of the solution could well be a bit larger than 100 mL, since the volume of the solute would be included. In practice, solutions of specific concentrations are prepared using a **volumetric flask** as illustrated in Figure 10.7.

Volumetric flask   A flask whose volume has been precisely calibrated.

Solved Problem 10.1   A solution of heparin sodium, an anticoagulant for blood that is used intravenously, contains 1.8 g of heparin sodium dissolved to make a final volume of 15 mL of solution. What is the weight/volume percent concentration of this solution?

Solution

$$(w/v)\% \text{ concentration} = \frac{\text{grams of solute}}{\text{mL of solution}} \times 100\%$$

$$= \frac{1.8 \text{ g heparin sodium}}{15 \text{ mL}} \times 100\% = 12\% \text{ (w/v)}$$

(a)

(b)

(c)

(d)

Figure 10.7
Preparation of a solution of known concentration. (a) The calculated quantity of solute is weighed out. (b) The weighed solute is placed in a volumetric flask, and a small amount of solvent is added. (c) The solute is dissolved by swirling the flask. (d) Still more solvent is added until the flask is filled to the known calibration mark on the neck (250 mL in this case).

Solved Problem 10.2 How much sodium chloride is needed to prepare 250 mL of a 1.5% (w/v) solution?

Solution The concentration of the solution gives two conversion factors:

1.5% (w/v) means $\dfrac{1.5 \text{ g NaCl}}{100 \text{ mL of solution}}$ or $\dfrac{100 \text{ mL of solution}}{1.5 \text{ g NaCl}}$

Using the appropriate conversion factor gives

$$250 \text{ mL solution} \times \frac{1.5 \text{ g NaCl}}{100 \text{ mL solution}} = 3.8 \text{ g NaCl}$$

Practice Problems   **10.6**   In clinical lab reports (see Application on Homeostasis in Chapter 5) certain concentrations are given in milligrams per deciliter. Convert an 8.6 mg/dL concentration of Ca^{2+} to weight/volume percent.

10.7   What is the weight/volume percent concentration of a solution that contains 23 g of potassium iodide (KI) dissolved in 350 mL of aqueous solution?

10.8   Can a saturated solution be concentrated? Can it be dilute?

10.9   How many grams of solute are needed to prepare these solutions?
(a) 100. mL of 12% (w/v) glucose ($C_6H_{12}O_6$)
(b) 75 mL of 2.0% (w/v) KCl

Weight/Weight Percent Concentration   The **weight/weight percent concentration [(w/w)%]** of a solution gives the number of grams of solute per 100 g of solution.

Weight/weight percent concentration [(w/w)%]
Concentration expressed as the number of grams of solute per 100 g of solution.

$$(w/w)\% \ \text{concentration} = \frac{\text{grams of solute}}{\text{grams of solution}} \times 100\%$$

For example, if 2.5 g of NaCl is dissolved in enough water to give a final solution weighing 100 g, the NaCl concentration is (2.5 g/100 g) × 100% = 2.5% (w/w). There is no particular advantage to expressing concentrations as weight/weight percent, and the method is not often used.

Volume/Volume Percent Concentration   The concentrations of solutions of one liquid in another are often expressed as **volume/volume percent concentrations [(v/v)%]**, which give volume of liquid solute per 100 mL of solution.

Volume/volume percent concentration [(v/v)%]
Concentration expressed as the number of milliliters of solute per 100 mL of solution.

$$(v/v)\% \ \text{concentration} = \frac{\text{mL of solute}}{\text{mL of solution}} \times 100\%$$

For example, if 10.0 mL of ethyl alcohol is dissolved in enough water to give 100 mL of solution, the ethyl alcohol concentration is (10 mL/100 mL) × 100% = 10% (v/v).

The *proof* of an alcoholic beverage is two times the volume/volume percent of alcohol. A beverage that is labeled 80 proof, for example, contains 40 mL of ethyl alcohol per 100 mL of beverage. (Pure ethyl alcohol is 200 proof.)

Solved Problem 10.3   How much methyl alcohol is needed to prepare 75 mL of a 5.0% (v/v) solution?

Solution   The concentration gives the following two conversion factors:

5.0% (v/v) means $\dfrac{5.0 \text{ mL methyl alcohol}}{100 \text{ mL solution}}$ or $\dfrac{100 \text{ mL solution}}{5.0 \text{ mL methyl alcohol}}$

Setting up the equation to convert the amount of solution to the amount of methyl alcohol gives

$$75 \text{ mL solution} \times \frac{5.0 \text{ mL methyl alcohol}}{100 \text{ mL solution}} = 3.8 \text{ mL methyl alcohol}$$

Practice Problems   10.10   Describe how you would use a 500 mL volumetric flask to prepare a 7.5% (v/v) solution of acetic acid in water.

10.11   What volume of solute (in mL) is needed to prepare (a) 100 mL of 22% (v/v) ethyl alcohol (b) 150 mL of 12% (v/v) acetic acid?

Parts per million (ppm)
Number of parts per one
million (10^6) parts.

Parts per Million When concentration in percent gives numbers that are inconveniently small, scientists resort to concentrations in **parts per million (ppm)** or parts per billion (ppb). Often this occurs in dealing with trace amounts of pollutants or contaminants. For example, the maximum allowable concentration in the air of the organic solvent benzene (C_6H_6) is set by government regulation at 10 ppm by volume. Relative to percent, what does this mean? Ten parts per million (1,000,000) is

$$10 \text{ ppm} = \frac{10 \text{ mL}}{1,000,000 \text{ mL}} \times 100\% = 0.001\% \text{ (v/v)}$$

Thus, 10 ppm is equivalent to 0.001 mL per 100 mL. (It could also be 0.001 L per 100 L, or 0.001 qt per 100 qt. As long as both units are the same, it's still parts per million.)

To find ppm or ppb, where the "parts" might be mass or volume units, the relationships are

$$\text{ppm} = \frac{\text{g solute}}{\text{g solution}} \times 10^6 \qquad \text{ppb} = \frac{\text{g solute}}{\text{g solution}} \times 10^9$$

$$\text{or} \quad \text{ppm} = (\% \text{ concentration}) \times 10^4 \qquad \text{ppb} = (\% \text{ concentration}) \times 10^7$$

Practice Problems **10.12** What is the concentration of NaCl in ppm in 0.9% (w/w) NaCl solution?

10.13 What is the concentration in ppm of NaF in tap water that has been fluoridated by the addition of 32 mg of NaF for every 20 kg of solution?

√10.9 MOLE/VOLUME CONCENTRATION: MOLARITY

Although useful for many purposes, the four methods just discussed for expressing concentration are awkward for use in calculations based on chemical equations. Amounts of reactants and products in aqueous solution are most easily handled when solute concentrations are expressed as the number of moles of solute per liter of solution, known as the **molarity (M)** of a solution. For example, a solution made by dissolving 1.00 mol (36.5 g) of HCl in enough water to give 1.0 L of solution has a concentration of 1.00 mol/L, and we say it is a "one molar" (1.00 M) solution.

Molarity (M) Concentration expressed as the number of moles of solute per liter of solution.

The molarity of a solution is found by dividing the number of moles of solute by the number of liters of solution.

$$\text{Molarity (M)} = \frac{\text{moles of solute}}{\text{liters of solution}}$$

Molarity provides conversion factors that connect volume and number of moles in a solution of known molarity:

$$\text{Volume (L)} = \frac{\text{liters of solution}}{\text{moles of solute}} \times \text{moles}$$

1/molarity

$$\text{Moles} = \frac{\text{moles of solute}}{\text{liters of solution}} \times \text{volume (L)}$$

molarity

Figure 10.8 shows how molarity is used in calculating the quantities of reactants or products in chemical reactions. When the solution *volume* for a reactant or product is needed, molarity is first used to find the number of moles of solute taking part in the reaction. The number of moles is then used exactly as illustrated in Sections 7.4 and 7.5.

Note in Solved Problem 10.5 that *millimoles*, which are quite common in health care fields for solutes of low concentration such as components of body fluids, are used in the same manner as moles.

Solved Problem 10.4   A saturated solution of the strong oxidizing agent potassium permanganate ($KMnO_4$, molar mass = 158 g) contains 74.3 g per liter of solution. What is the molarity of this solution?

Solution   Since molarity is moles per liter, the 74.3 g must be converted to moles by using the molar mass as a conversion factor:

$$74.3 \ g \times \frac{1 \ \text{mol} \ KMnO_4}{158 \ g} = 0.470 \ \text{mol} \ KMnO_4$$

The molarity of the solution is

$$\frac{0.470 \ \text{mol} \ KMnO_4}{1 \ L} = 0.470 \ \text{mol} \ KMnO_4/L \ \text{or} \ 0.470 \ M \ KMnO_4$$

Figure 10.8
Flow diagram summarizing use of molarity for conversions between solution volume and moles as a step in finding quantities of reactants and products for chemical reactions in aqueous solution.

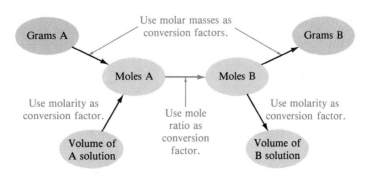

Practice Problems 10.14   What is the molarity of a solution that contains 50 g of vitamin B_1 hydrochloride (molar mass = 337 g) in 160 mL of solution?

10.15   How many grams of solute would you use to prepare these solutions?
(a) 500 mL of 1.25 M NaOH
(b) 1.5 L of 0.25 M glucose ($C_6H_{12}O_6$)

10.16   How many moles of solute are present in these solutions?
(a) 125 mL of 0.20 M $NaHCO_3$
(b) 650 mL of 2.5 M H_2SO_4

Solved Problem 10.5   A stock solution of the antibiotic tetracycline hydrochloride (tch) (molar mass 481 g) has the concentration 0.52 M. If a single dose contains 0.31 mmol of tetracycline hydrochloride, what is the volume of the dose?

Solution   A millimole is 0.001 mol (1/1000 of a mole), and a milliliter is 0.001 L (1/1000 of a liter). Therefore molarity, moles per liter, is also equal to millimoles per milliliter. The needed solution volume is found by using molarity as a conversion factor as follows:

$$0.31 \text{ mmol tch} \times \frac{1 \text{ mL}}{0.52 \text{ mmol tch}} = 0.60 \text{ mL}$$

Practice Problem 10.17   What volume, in milliliters, of a 0.200 M glucose ($C_6H_{12}O_6$) solution is needed to provide a total of 25.0 g glucose?

Solved Problem 10.6   A blood concentration of ethyl alcohol (molar mass 46.0 g) of 0.0653 M is sufficient to induce a coma. At this concentration, what is the total mass of alcohol in grams in an adult male whose total blood volume is 5.5 L?

Solution   The needed information here is a mass. Using the molar mass as a conversion factor in a single factor-label calculation gives

$$\frac{0.0653 \text{ mol alcohol}}{1 \text{ L}} \times \frac{46.0 \text{ g alcohol}}{1 \text{ mol alcohol}} \times 5.5 \text{ L} = 17 \text{ g alcohol}$$

At the serious point of coma inducement, this adult male's blood contains a total of 17 g of alcohol.

Practice Problem 10.18   The concentration of cholesterol ($C_{27}H_{46}O$) in normal blood is approximately 0.0050 M. How many grams of cholesterol are in 750 mL of blood?

Solved Problem 10.7   Within our stomachs, gastric juice that is about 0.1 M in HCl aids in digestion. What mass in grams of sodium bicarbonate ($NaHCO_3$, molar mass 84 g) is needed to react completely with the HCl in 50 mL of gastric juice according to the following equation?

$$HCl(aq) + NaHCO_3(s) \rightarrow NaCl(aq) + H_2O(l) + CO_2(g)$$

Solution   The pathway for solving this problem, which requires starting with a known volume of reactant solution and finding the mass of the other reactant, is shown in Figure 10.8 and consists of

$$\text{Volume} \xrightarrow{\text{molarity}} \text{moles} \xrightarrow{\text{mole ratio}} \text{moles} \xrightarrow{\text{molar mass}} \text{mass}$$

Since calculations based on chemical equations must be done in moles, we first have to find how many moles of HCl are in 50 mL of 0.1 M HCl.

$$50 \text{ mL} \times \frac{0.1 \text{ mol HCl}}{1000 \text{ mL}} = 0.005 \text{ mol HCl}$$

Next, check the coefficients of the balanced equation to find that each mole of HCl reacts with one mole of $NaHCO_3$. Combining the mole ratio and the molar mass of $NaHCO_3$ in a factor-label calculation gives the needed mass of $NaHCO_3$

$$0.005 \text{ mol HCl} \times \frac{1 \text{ mol NaHCO}_3}{1 \text{ mol HCl}} \times \frac{84 \text{ g NaHCO}_3}{1 \text{ mol NaHCO}_3} = 0.4 \text{ g NaHCO}_3$$

Practice Problem 10.19   What mass in grams of calcium carbonate is needed to react completely with 75 mL of 0.10 M HCl according to the following equation:

$$2 HCl(aq) + CaCO_3(s) \longrightarrow CaCl_2(aq) + H_2O(l) + CO_2(g)$$

10.10 DILUTION

Dilution A decrease in the concentration of a solution caused by addition of solvent.

Many solutions, from orange juice to chemical reagents, are stored in high concentrations and then prepared for use by **dilution.** For example, you might make up 1/2 gal of orange juice by adding water to a canned concentrate. In the same way, you might buy a medicine or chemical reagent in concentrated solution and then dilute it before use.

$$\text{Concentrated chemical solution} + \text{solvent} \longrightarrow \text{dilute solution}$$

The key fact to remember about dilution is that the amount of *solute* remains constant; only the *volume* is changed by adding more solvent. For concentrations represented as molarity

$$\text{Moles of solute} = \underset{\substack{\text{[Original}\\ \text{concentration]}}}{M_{orig}} \times \underset{\substack{\text{[Original}\\ \text{volume]}}}{V_{orig}} = \underset{\substack{\text{[Final}\\ \text{concentration]}}}{M_{final}} \times \underset{\substack{\text{[Final}\\ \text{volume]}}}{V_{final}}$$

which we can rewrite as

$$M_{final} = M_{orig} \times \frac{V_{orig}}{V_{final}} \qquad \text{where } \frac{V_{orig}}{V_{final}} \text{ is a dilution factor}$$

Dilution factor The ratio of original to final volumes of a solution being diluted.

These equations show that the concentration after dilution (M_{final}) can be found by multiplying the original concentration (M_{orig}) by a dilution factor. The **dilution factor** is simply the ratio of the original and final solution volumes (V_{orig}/V_{final}). If the solution volume *increases*, for example, by a factor of 5 from 10 mL to 50 mL, then the concentration must *decrease* to one-fifth of its original value because the dilution factor is 10 mL/50 mL, or 1:5. Solved Problem 10.8 shows how to use this relationship for calculating the effect of dilution.

Solved Problem 10.8   What is the final concentration if 75 mL of a 3.5 M glucose solution is diluted to a volume of 400. mL?

Solution   Use the ratio of original-to-final volume as a dilution factor to calculate the final concentration:

$$M_{final} = M_{orig} \times \frac{V_{orig}}{V_{final}} = \frac{3.5 \text{ mol glucose}}{1 \text{ L}} \times \frac{75 \text{ mL}}{400 \text{ mL}}$$

$$= 0.66 \text{ mol glucose/L or } 0.66 \text{ M glucose}$$

Practice Problem   **10.20**   Hydrochloric acid is normally purchased at a concentration of 12.0 M. What is the final concentration if 100 mL of 12.0 M HCl is diluted to 500 mL?

The relationship between concentration and volume can also be used to determine how to dilute a solution. We know that

$$M_{orig} \times V_{orig} = M_{final} \times V_{final}$$

Therefore

$$V_{orig} = V_{final} \times \frac{M_{final}}{M_{orig}}$$

In this case, V_{orig} is the volume that must be diluted to prepare a new solution of a given volume and lower concentration. This volume is found by multiplying the desired volume (V_{final}) by the ratio of the final and original concentrations (M_{final}/M_{orig}). For example, to decrease the concentration to one-fifth its original value, the original volume must be one-fifth the desired final volume.

Solved Problem 10.9   Sodium hydroxide (NaOH) solution can be purchased at a concentration of 1.0 M. How would you prepare 1.0 L of 0.25 M NaOH?

Solution   Using the foregoing equation, the volume of solution that must be diluted is found as follows:

$$V_{orig} = V_{final} \times \frac{M_{final}}{M_{orig}} = 1.0 \text{ L} \times \frac{0.25 \text{ M}}{1.0 \text{ M}} = 0.25 \text{ L} = 250 \text{ mL}$$

To prepare the desired solution, 250 mL of 1.0 M NaOH must be measured out and water added to bring the final volume to 1.0 L.

Practice Problem   10.21   Concentrated ammonia solution, NH_3 in H_2O, is commercially available at a concentration of 16.0 M. How much would you use to prepare 500 mL of a 1.25 M solution?

10.11 IONS IN SOLUTION: ELECTROLYTES

Try a simple experiment. Remove the lantern battery and light bulb from a large flashlight. Then, using wires, connect the battery, the part holding the light bulb, and two thin strips of metal (copper or aluminum would be good) as shown in Figure 10.9. Next take a glass of water, dissolve as much table salt in it as you can, and immerse the ends of the metal strips (which are now **electrodes**) in the salt water. The bulb lights, demonstrating that ionic compounds in aqueous solution conduct electricity.

Substances like NaCl that conduct an electric current when dissolved in water are called **electrolytes.** Conduction occurs because negatively charged Cl^- anions migrate through the solution toward the wire connected to the positive end of the battery, while at the same time positively charged Na^+ cations migrate toward the wire connected to the negative end of the battery.

As you might expect, the ability to conduct electric current depends on the concentration of ions in solution. Distilled water, which contains virtually no ions, is nonconducting; ordinary tap water contains low concentrations of dissolved ions (mostly Na^+, K^+, Mg^{2+} and Cl^-) and is very weakly conducting; and NaCl, which is 100% converted to ions in aqueous solution, is a **strong electrolyte.** Solutions of strong electrolytes become increasingly better conduc-

Electrode   Conductor through which electrical current enters or leaves a conducting medium.

Electrolyte   A substance that conducts electricity when dissolved in water.

Strong electrolyte   A substance that is completely converted to ions when dissolved in water.

Figure 10.9
A simple experiment to show that current can flow through a solution of ions. With pure water in the beaker, the light bulb does not light. With a concentrated NaCl solution in the beaker, electrical current flows from the battery and the light bulb glows.

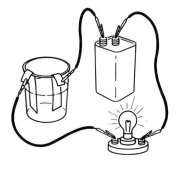

tors as their concentrations increase. Substances such as acetic acid (CH_3CO_2H) that are only partly ionized and establish equilibria between ions and nonionized molecules are known as **weak electrolytes**:

$$CH_3CO_2H(aq) \rightleftharpoons CH_3CO_2^- + H^+$$

Substances that do not conduct electricity when dissolved in water are **nonelectrolytes.**

Weak electrolyte A substance that is partially ionized when dissolved in water and establishes equilibrium between its nonionized form and ions.
Nonelectrolyte A substance that does not conduct electricity when dissolved in water.

10.12 EQUIVALENTS, MILLIEQUIVALENTS, AND BODY ELECTROLYTES

Imagine dissolving both NaCl and KBr in the same solution. Since the cations (K^+ and Na^+) and anions (Cl^- and Br^-) are all mixed together and no reactions occur among them, an identical solution could just as well be made from KCl + NaBr. Thus, we can no longer speak of having an NaCl + KBr solution; we can only speak of having a solution with four different ions in it.

A similar situation exists for blood and other body fluids, which contain many different anions and cations. Since they're all mixed together, we can't "assign" specific cations to specific anions, and we can't talk about specific ionic compounds. To discuss such mixtures, we need a new term, called *equivalents* of ions.

One **equivalent (Eq) of an ion** is the amount in grams that contains Avogadro's number, or one mole, of charges. If the ion has a charge of +1 or −1, one equivalent is simply equal to the formula weight of the ion in grams. Thus, one equivalent of Na^+ is 23 g, and one equivalent of Cl^- is 35.5 g. If the ion has a charge of +2 or −2, however, one equivalent is equal to the ion's formula weight in grams divided by 2. Thus, one equivalent of Mg^{2+} is 24.3 g/2 = 12.2 g, and one equivalent of CO_3^{2-} is (12.0 g + 16.0 g + 16.0 g + 16.0 g)/2 = 30.0 g.

Equivalent, ion (Eq) The amount of an ion in grams that contains Avogadro's number of charges.

$$\text{One equivalent of an ion} = \frac{\text{molar mass of ion (grams)}}{\text{number of charges on ion}}$$

Because ion concentrations in body fluids are often low, clinical chemists find it more convenient to talk about *milliequivalents* of ions. One **milliequivalent (mEq)** of an ion is 1/1000 of an equivalent. For example, the normal concentration of Na^+ in blood is 0.14 Eq/L, or 140 mEq/L.

Milliequivalent (mEq) The amount of an ion equal to one-thousandth of an equivalent.

$$1 \text{ mEq} = 0.001 \text{ Eq} \qquad 1 \text{ Eq} = 1000 \text{ mEq}$$

The average concentrations of the major electrolytes in blood plasma are given in Table 10.7. Of course, the *total* milliequivalents of positively and negatively charged electrolytes must be equal in order to maintain neutrality. Adding up the milliequivalents of positive and negative ions in Table 10.7 shows a higher concentration of positive ions than negative ions. The difference (the *anion gap*) is made up by negatively charged proteins and the anions of organic acids.

Table 10.7   Concentrations of Major Electrolytes in Blood Plasma

Ion	Concentration (mEq/L)	Ion	Concentration (mEq/L)
Cations		*Anions*	
Na^+	136–145	Cl^-	98–106
Ca^{2+}	4.5–6.0	HCO_3^-	25–29
K^+	3.6–5.0	SO_4^{2-} and HPO_4^{2-}	2
Mg^{2+}	3		

Solved Problem 10.10   Taking the normal concentration of Ca^{2+} in blood as 5.0 mEq/L, how many milligrams of Ca^{2+} are in 1.00 L of blood?

Solution   First calculate how many grams are in 1 Eq of Ca^{2+}. The molar mass of Ca^{2+} is 40.1 g, but because Ca^{2+} has two charges, we must divide the molar mass by 2. Therefore,

$$1 \text{ Eq } Ca^{2+} = \frac{40.1 \text{ g } Ca^{2+}}{2} = 20.1 \text{ g } Ca^{2+}$$

Next, calculate how many milligrams are in 5.0 mEq/L Ca^{2+}:

$$\frac{20.1 \text{ g } Ca^{2+}}{1 \text{ Eq } Ca^{2+}} \times \frac{1 \text{ Eq } Ca^{2+}}{1000 \text{ mEq}} \times \frac{5.0 \text{ mEq}}{1.00 \text{ L}} = 0.10 \text{ g/L } Ca^{2+}$$

Finally, convert grams into milligrams:

$$\frac{0.10 \text{ g } Ca^{2+}}{1 \text{ L}} \times \frac{1000 \text{ mg}}{1 \text{ g}} = 100 \text{ mg/L } Ca^{2+}$$

Thus, there are 100 mg of Ca^{2+} in 1 L of blood.

Practice Problems 10.22   How many grams are in one equivalent of each of the following?
(a) K^+ (b) Br^- (c) Mg^{2+} (d) SO_4^{2-}

10.23   How many grams are in 1 mEq of each ion in Problem 10.22?

10.24   Look at the data in Table 10.7 and calculate how many milligrams of Mg^{2+} are in 250 mL of blood.

10.13   VAPOR PRESSURE, FREEZING POINT, AND BOILING POINT

The vapor pressure of a liquid depends on the equilibrium between molecules entering and leaving the liquid surface (Section 9.10). If the liquid molecules are mixed with other particles that *do not* evaporate, some spots in the liquid surface are filled by these *nonvolatile* particles, leaving fewer molecules at the surface that can escape. The result is that the vapor pressure is lowered (Figure 10.10).

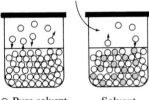

Fewer solvent molecules
leave surface, lowering
vapor pressure

○ Pure solvent   Solvent
with nonvolatile
solute ○

Figure 10.10
Vapor pressure lowering by
a nonvolatile solute.

Several other physical properties are influenced by vapor pressure lowering, and all depend only on the concentration of nonvolatile particles, not their identities. The boiling point is changed because it is the temperature at which the vapor pressure *equals* the atmospheric pressure. With a nonvolatile solute present, the vapor pressure of a liquid is lower at a given temperature, and it takes more energy (heating to a higher temperature) before the vapor pressure is high enough for boiling to begin.

The freezing point is also influenced by vapor pressure lowering, and we make practical use of this effect in the winter. A lower solution vapor pressure means that a solution must reach a lower temperature before freezing begins. "Antifreeze" for the water in automobile cooling systems is a nonvolatile solute, usually ethylene glycol ($HOCH_2CH_2OH$). The antifreeze is added in sufficient concentration to lower the freezing point below the lowest possible outdoor temperature (unless you're very unlucky). The salt sprinkled on icy roads (usually either NaCl or $CaCl_2$) works by the same principle: lowering the freezing point below the outdoor temperature.

For each mole of nonvolatile solute, the freezing point of 1 kg of water is *lowered* by 1.86 C° and the boiling point is *raised* by 0.51 C°. This means adding 1 mol of the antifreeze ethylene glycol per kilogram of water will lower the freezing point of the solution to 0°C − 1.86 C° = −1.86°C. The addition of one mole of calcium chloride ($CaCl_2$) per kilogram will lower the freezing point by 1.86 C° for *each* mole of solute ions, giving a freezing point of 0°C − (3 × 1.86 C°) = −5.58°C.

***Practice Problems* 10.25** (a) Explain why, on a molar basis, calcium chloride ($CaCl_2$) is more effective at melting ice than sodium chloride (NaCl).
(b) Compare the effect of these two compounds on the freezing point of water on a mass basis.

10.26 (a) What is the freezing point of a solution of 1.0 mol of glucose ($C_6H_{12}O_6$) in 1 kg of water?
(b) What is the boiling point of a solution of 1.0 mol of sodium chloride (NaCl) in 1 kg of water?

10.14 OSMOSIS AND OSMOTIC PRESSURE

Semipermeable membrane
A thin membrane that allows water or other small solvent molecules to pass through but blocks the flow of larger solute molecules or ions.

Osmosis The passage of solvent molecules across a semipermeable membrane from a more dilute solution to a more concentrated solution.

Certain materials, including those that make up the membranes around living cells, are *semipermeable*. A **semipermeable membrane** is one that allows small molecules or ions to pass through but blocks the passage of larger solute particles. It's as if the membrane has tiny holes in it through which only small particles can pass. Larger molecules and bulky solvated ions evidently can't fit through the holes.

Some semipermeable membranes allow the passage of only water molecules. When two solutions are separated by such a membrane, water passes from the more dilute solution to the more concentrated solution. Called **osmosis**, this process (like vapor pressure lowering) depends only on the numbers of solute particles, not their identity. The passage of water molecules continues until the concentrations of solute particles are the same on both sides of the membrane or until a high enough pressure stops the solvent flow. Note that **osmosis always results in dilution of the more concentrated solution.**

In the special case where pure water is present on one side of an **osmotic membrane** and a solution on the other, the concentrations on the two sides can never become equal no matter how much water passes from one side to the other, because all the solute remains on the one side. **Osmotic pressure** is defined as the amount of external pressure that will halt the passage of solvent molecules across an osmotic membrane separating a solution and pure water. Think of it this way: The higher the osmotic pressure of a solution, the higher the "attraction" of the solution for water molecules from across the membrane.

The experiment shown in Figure 10.11 illustrates osmotic pressure. Initially, the beaker contains pure water and the tube contains the solution. Water passes through the membrane until the column of solution is tall enough to exert a pressure that stops any more water molecules from flowing through the membrane. Water rises in plant stalks and trees in a similar manner because of the higher solute concentrations in plant cells.

Osmotic pressure varies directly with the concentration of solute particles. A 0.001 M solution of sucrose, for example, has an osmotic pressure of 0.024 atm, but the osmotic pressure of a 0.15 M NaCl solution (which contains 0.3 mol of ions) is 7.4 atm, a value high enough to support a difference in water level of approximately 230 ft!

The osmotic pressure is the same for any two solutions with equal concentrations of particles, and thus it's convenient to use the term **osmolarity (Osmol)** to describe the number of particles in solution. The osmolarity of a solution is equal to the molarity times the number of particles produced by each formula unit. Thus, a 0.2 M glucose solution has an osmolarity of 0.2 Osmol, but a 0.2 M solution of NaCl has an osmolarity of 0.4 Osmol.

Osmotic membrane A membrane that allows only the passage of solvent molecules.

Osmotic Pressure The amount of external pressure that must be applied to halt the passage of solvent molecules across an osmotic membrane.

Osmolarity (Osmol) The number of moles of dissolved solute particles (ions or molecules) per liter of solution.

$$\text{Osmolarity} = \text{molarity} \times \text{number of particles per solute formula unit}$$

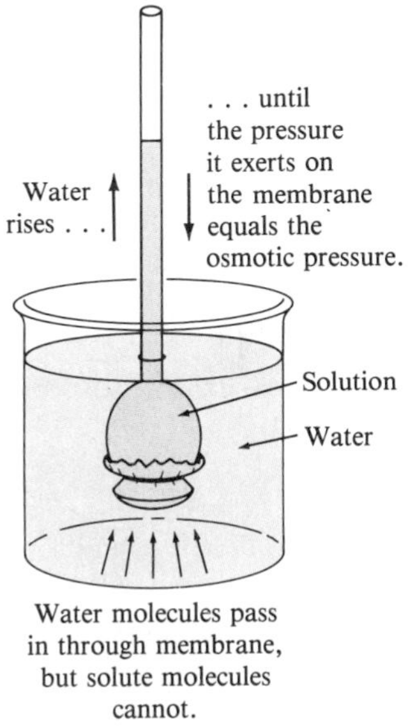

Water rises . . .
. . . until the pressure it exerts on the membrane equals the osmotic pressure.

Solution
Water

Water molecules pass in through membrane, but solute molecules cannot.

Figure 10.11
Experiment illustrating osmosis. Water molecules pass through the membrane into the solution until the pressure of the solution column prevents the passage of more water molecules.

AN APPLICATION: WHY ICE CREAM ISN'T A SOLID AND OTHER BITS OF KITCHEN CHEMISTRY

Have you ever had a corned beef sandwich, a pickle, and a dish of ice cream for lunch? If so, did you know that vapor pressure lowering and associated properties made your meal possible?

First, the corned beef: Corned beef was invented in the days before refrigeration, when one of the few ways to preserve meat was to soak it in concentrated salt solution (brine). Since the osmolarity of brine is higher than that inside the bacteria and molds that can spoil meat, water crosses their cell membranes into the brine, leaving the microorganisms dead or very damaged.

Pickles and sauerkraut also depend on osmosis for their texture and flavor. In simple pickle making, cucumbers, onions, or other vegetables are soaked in brine before being packed in vinegar. Water that would otherwise dilute the vinegar and diminish the flavor of the finished product is drawn by osmosis into the brine. Brine soaking is also used in the more complex production of dill pickles and sauerkraut. Cucumbers for dill pickles, for example, are salted to draw out water, and then more water is added to give a solution of just the right salt concentration. The characteristic flavor of the pickles is created by lactic acid and other substances produced by bacteria that thrive in a 2.25% salt solution (at about 20°C) and enter the brine from the environment.

And what about the ice cream? The starting material for making ice cream is a homogeneous mixture of sugar, cream, flavorings, and other ingredients. The distinctive texture of the finished product is partly dependent on its freezing point lowering. Dissolved sugar lowers the freezing point of water in the mixture to about −3°C.

As ice crystals form at this temperature, the concentration of sugar in the remaining liquid increases, *further* lowering the freezing point.

Ice cream never becomes a solid block of ice because the freezing point lowering keeps ahead of the dropping temperature. Tiny pockets of liquid always remain in the finished product, carrying most of the flavor and nutritional value in their high concentration of dissolved sugar and other ingredients, and their suspended milk proteins. Fine quality ice cream is a foam that consists of four phases: liquid droplets, fat globules, ice crystals, and tiny air cells trapped between the other phases. The best ice cream texture is achieved when the ice crystals are very small, which requires that they form rapidly at low temperatures.

This cucumber was soaked in concentrated sodium chloride solution for several days in the laboratory. The effects of water loss are evident and, in fact, have probably gone too far to make a good pickle.

Blood plasma The fluid surrounding blood cells.

Isotonic Referring to a solution that has the same osmolarity as another, usually blood.

Osmosis is particularly important in living organisms, because the membranes around cells are semipermeable. The fluids both inside and outside cells must therefore have the same osmolarity to prevent buildup of osmotic pressure and rupture of the cell membrane.

In blood, the fluid (**plasma**) surrounding red blood cells has an osmolarity of 0.30 Osmol and is **isotonic** with (that is, has the same osmolarity as) the cell

Hypotonic Referring to a solution that has lower osmolarity than another, usually blood.

Hemolysis Bursting of red blood cells due to a buildup of pressure in the cell.

Hypertonic Referring to a solution that has higher osmolarity than another, usually blood.

Crenation The shriveling of red blood cells due to a loss of water.

contents. If the cells are removed from plasma and placed in an isotonic saline solution (0.15 M NaCl = 0.30 Osmol), they are unharmed. If, however, red blood cells are placed in pure water or in any solution with an osmolarity *lower* than that of the cell contents (a **hypotonic** solution), water will pass through the membrane into the cell, causing the cell to swell up and burst, a process called **hemolysis.**

Finally, if red blood cells are placed in a solution having an osmolarity *greater* than the cell contents (a **hypertonic** solution), water will pass out of the cell into the surrounding solution, causing the cell to shrivel so that the cell membranes wrinkle, a process called **crenation.** Figure 10.12 shows red blood cells under all three conditions—isotonic, hypotonic, and hypertonic. Clearly, it's critical that solutions used for intravenous feeding be isotonic to prevent red blood cells from being destroyed.

Practice Problem   10.27   What is the osmolarity of these solutions?
(a) 0.35 M KBr   (b) 0.15 M glucose

Figure 10.12
Red blood cells in (a) an isotonic solution (normal in appearance), (b) a hypotonic solution (swollen), and (c) a hypertonic solution (shriveled).

(a)

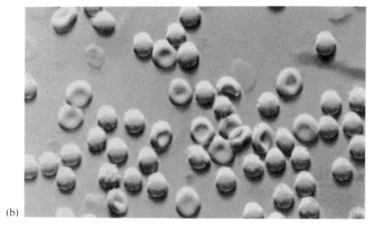

(b)

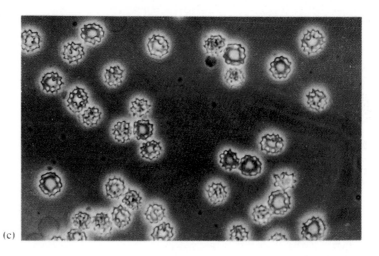

(c)

10.15 COLLOIDS

What do blood plasma, mayonnaise, and fog have in common? They are all colloids. Colloids, or colloidal dispersions, are mixtures of a *dispersing medium,* equivalent to a solvent, and a *dispersed phase* composed of particles or clusters of particles larger than those in true solutions (Section 10.1). Both the dispersing medium and the "particles" may be gases, liquids, or solids. Examples of colloids of each type are listed in Table 10.8.

In blood plasma, the dispersing medium is water and the colloidal particles are protein molecules. Other components of blood plasma, such as the electrolytes, are in true solution. Dispersions of solids in liquids may be either *sols*— individual solid particles in the liquid—or *gels.* In a gel, large molecules tangle together into a network that gives the mixture some solidity. Jello-brand dessert is a gel, for example, but if left in a warm room too long, it reverts to a sol. The sol–gel interconversion property of Jello is imparted by gelatin, a mixture of proteins extracted from animal tissue (bone, skin, connective tissue). *All proteins are of colloidal dimensions.*

Mayonnaise is an *emulsion,* a colloidal dispersion of a liquid in a liquid. In mayonnaise, the dispersing medium is water (from vinegar and lemon juice) and the particles are droplets of oil. Such an emulsion is stabilized by an *emulsifying agent,* which prevents the droplets of immiscible liquids from reuniting with each other. Egg yolk is the emulsifying agent that surrounds the oil droplets in mayonnaise. In milk, fat particles are held in emulsion by casein, a protein. Soap and detergent molecules (Chapter 23) are emulsifying agents that help remove greasy dirt from clothing and household objects.

Dialysis The passage through a semipermeable membrane of solvent and small solute molecules, but not of large solute molecules.

Solid colloidal particles pass through a normal filter but are too large to pass through semipermeable membranes. In a process called **dialysis,** which is exactly analogous to osmosis, solvent and small solute molecules pass through a semipermeable membrane from the side of higher concentration to that of lower concentration, but large colloidal particles can't pass. (The exact dividing line

Table 10.8 Types of Colloids

Dispersing Medium	Dispersed Substance	Colloid Type	Examples
Solid	Gas	Solid foam	Polyurethane foam, marshmallow
	Liquid	Solid emulsion or gel	Cheese, butter
	Solid	Solid sol	Ruby glass, certain alloys
Liquid	Gas	Foam[a]	Soapsuds, whipped cream
	Liquid	Emulsion[a]	Mayonnaise, milk, face cream
	Solid	Sols, gels	Gelatin, starch, blood serum, egg white
Gas	Liquid	Liquid aerosol	Fog, aerosol spray
	Solid	Solid aerosol	Smoke, airborne bacteria and viruses

[a]Require emulsifying agent.

between a "small" molecule and a "large" one is, of course, imprecise.) Note the difference between "osmosis," which refers to the movement of water molecules only, and "dialysis," which refers to the movement of small solute particles in addition to water molecules.

Protein molecules don't cross semipermeable membranes and thus play an essential role in determining the osmolarity of body fluids. The distribution of water and solutes across the capillary walls that separate blood plasma from the fluid surrounding cells (*interstitial fluid*) is controlled by the balance between blood pressure and osmotic pressure. The pressure of blood inside the capillary tends to push water out of the plasma (filtration), but the osmotic pressure of colloidal protein molecules tends to draw water into the plasma (reabsorption). The balance varies with location in the body. At the arterial end of a capillary, where blood pumped from the heart has a higher pressure, filtration is favored. At the venous end, where blood pressure is lower, reabsorption is favored, causing waste products from metabolism to enter the bloodstream.

In the kidneys, reabsorption of water due to osmolarity differences is one of several mechanisms that allow urine to be excreted at an osmolarity of up to four times that of normal body fluids. In the kidney tubules, sodium and chloride ions are actively pumped out of urine into surrounding tissue, raising the osmolarity of the interstitial fluid so that water then crosses the tubule membrane, leaving the urine more concentrated.

SUMMARY

Mixtures are classified according to their visual appearance as either **heterogeneous,** if the mixing is nonuniform, or **homogeneous,** if the mixing is uniform. **Solutions** contain particles the size of ions and molecules (0.1–2.0 nm diameter), whereas larger particles are present in **colloids** (particles do not settle) and **suspensions** (particles do settle).

The maximum amount of one substance (the **solute**) that can be dissolved in another (the **solvent**) is called the substance's **solubility.** Substances tend to be mutually soluble when their interparticle forces are similar (like dissolves like). In water, the most common solvent, water molecules surround solute ions and polar molecules so that opposite charges are adjacent (**hydration**). Some crystalline solids retain water of hydration. The solubility in water of many, but not all, ionic compounds increases with temperature, and the solubility of gases always decreases with temperature. Pressure significantly affects only gas solubilities, which are directly proportional to their partial pressure over the solution (**Henry's law**).

The **concentration** of a solution can be expressed in several ways, including percent composition, parts per million, and molarity. In the **weight/volume percent** (**w/v%**) method, concentration is expressed as the number of grams of solute per 100 mL of solution. In the **molarity** method, concentration is expressed as the number of moles of solute per liter of solution. Molarity is used to calculate quantities of reactants or products for reactions in aqueous solution.

Substances that form ions when dissolved in water and whose water solutions therefore conduct an electric current are called **electrolytes.** Body fluids contain small concentrations of many different electrolytes. Because a nonvolatile solute of any identity displaces a volatile solvent at a liquid surface, solvent vapor pressures are lowered, causing boiling point to be elevated and freezing point to be lowered.

Certain materials, including those that make up the membranes around living cells, are **semipermeable.** These membranes contain pores that allow water or other small solvent molecules to pass but block the passage of solute ions and molecules, a phenomenon called **osmosis** when passage of just water molecules is involved. When the number of dissolved particles on the two sides of the membrane is different, water passes from the more dilute side to the more concentrated side. **Osmotic pressure** is the pressure needed to stop osmosis between a solution and pure water. An effect similar to osmosis is noted when membranes of larger pore size are used. In **dialysis,** the membrane allows the passage of solvent and small dissolved molecules but prevents passage of colloidal and larger particles.

INTERLUDE: DIALYSIS

Proteins, because they do not dialyze, can be separated from small ions and molecules by dialysis, making dialysis a valuable procedure for purification of proteins needed for laboratory studies. Dialysis membranes include animal bladders, parchment, and cellophane. By frequently changing the solution surrounding a dialysis bag containing a protein in solution, dialysis is caused to continue until only pure protein remains in the bag.

Perhaps the most important medical use of dialysis is in artificial kidney machines, where **hemodialysis** is used to cleanse the blood of patients whose kidneys malfunction. Blood is taken from the body and pumped through a long cellophane dialysis tube suspended in an isotonic solution containing glucose, NaCl, NaHCO₃, and other essential materials as diagrammed in the accompanying figure. Small waste materials pass from the blood through the dialysis membrane where they are washed away, but cells, proteins, and other important blood components are prevented by their size from passing through the membrane. In addition, the dialysis fluid concentration can be controlled so that imbalances in electrolytes are corrected. Purified blood is then returned to the body.

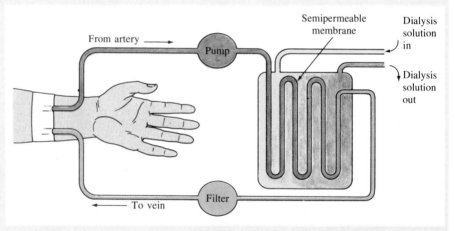

Schematic operation of a hemodialysis unit used for purifying blood. Blood is pumped from an artery through a coiled semipermeable membrane of cellophane. Small waste products pass through the membrane and are washed away by an isotonic dialysis solution.

REVIEW PROBLEMS

Solutions and Solubility

10.28 What is the difference between a homogeneous mixture and a heterogenous one?

10.29 How can you tell a solution from a colloid? A colloid from a suspension?

10.30 What characteristic of water allows it to dissolve ionic solids?

10.31 If a single 5 g block of salt is placed in water, it dissolves slowly, but if 5 g of powdered salt is placed in water, it dissolves rapidly. How can you explain this difference?

10.32 Suppose you had a mixture of salt (soluble in water) and aspirin (soluble in chloroform). How could you use the differences in solubility behavior to separate the mixture into its two components?

10.33 Which of the following are true solutions?
(a) Italian salad dressing
(b) rubbing alcohol
(c) algae in pond water
(d) black coffee

10.34 Which pairs of liquids are miscible?

(a) H_2SO_4 and H_2O (b) C_8H_{18} and C_6H_6
(c) C_2H_5OH and H_2O (d) CS_2 and CCl_4

10.35 The solubility of NH_3 in water at an NH_3 pressure of 760. mm Hg is 51.8 g/100 mL. What is the solubility of NH_3 if the partial pressure is reduced to 250. mm Hg?

10.36 At 159 mm Hg, the solubility of oxygen in blood is 0.44 g/100 mL. If pure oxygen at 760 mm Hg is administered to a patient, what is the concentration of oxygen in the blood assuming Henry's law is followed?

Concentrations of Solutions

10.37 How is weight/volume percent concentration defined?

10.38 How is molarity defined?

10.39 How is volume/volume percent concentration defined?

10.40 How much water must you add to 100. mL of orange juice concentrate if you want the final juice to be 20.0% of the strength of the original?

10.41 How would you prepare 500. mL of a 5.0% (v/v) ethyl alcohol solution?

10.42 A dilute solution of boric acid, $B(OH)_3$, is often used as an eyewash. How would you prepare 500. mL of a 0.50% (w/v) boric acid solution?

10.43 Nalorphine, a relative of morphine, is used to combat withdrawal symptoms in heroin users. How many mL of a 0.40% (w/v) solution of nalorphine must be injected to obtain a dose of 1.5 mg?

10.44 Describe how you would prepare 250 mL of a 0.10 M NaCl solution.

10.45 Describe how you would prepare 1.0 L of a 7.5% (w/v) KBr solution.

10.46 Which of the following solutions is more concentrated?
(a) 0.50 M KCl or 5.0% (w/v) KCl
(b) 2.5% (w/v) $NaHSO_4$ or 0.025 M $NaHSO_4$

10.47 What is the weight/volume percent concentration of the following solutions?
(a) 5.0 g KCl in 75 mL water
(b) 15 g sucrose in 350 mL water

10.48 How many grams of each substance are needed to prepare these solutions?
(a) 50.0 mL of 8.0% (w/v) KCl
(b) 200. mL of 7.5% (w/v) acetic acid

10.49 If you had only 23 g of KOH remaining in a bottle, how many mL of 10.0% (w/v) solution could you prepare? How many mL of 0.25 M solution?

10.50 The concentration of glucose in blood is approximately 90 mg/100 mL. What is the weight/volume percent concentration of glucose?

10.51 Assuming a blood glucose concentration of 90. mg/100 mL and a blood volume of 5.0 L, how many grams of glucose are present in the blood of an average adult?

10.52 The lethal dosage of potassium cyanide in rats is 10. mg KCN/kg body weight. What is this concentration in ppm?

10.53 The maximum concentration set by the U.S. Public Health Service for arsenic in drinking water is 0.050 mg/kg. What is this concentration in ppm and in ppb?

10.54 What is the molarity of these solutions?
(a) 12.5 g $NaHCO_3$ in 350.0 mL solution
(b) 45.0 g H_2SO_4 in 300.0 mL solution
(c) 30.0 g NaCl dissolved to make 500.0 mL solution

10.55 How many moles of solute are in these solutions?
(a) 200 mL of 0.30 M acetic acid, CH_3COOH
(b) 1.50 L of 0.25 M NaOH
(c) 750 mL of 2.5 M nitric acid, HNO_3

10.56 How many grams of solute are in each of the solutions in Problem 10.55?

10.57 How many mL of a 0.75 M HCl solution do you need to obtain 0.0040 mol of HCl?

10.58 A flask containing 450 mL of 0.50 M H_2SO_4 was accidentally knocked to the floor. How many grams of $NaHCO_3$ do you need to put on the spill to completely react with the acid according to the following equation?

$$H_2SO_4(aq) + 2\ NaHCO_3(s) \longrightarrow$$
$$Na_2SO_4(aq) + 2\ H_2O(l) + 2\ CO_2(g)$$

10.59 The fixing agent for photographic negatives, $Na_2S_2O_3$, reacts with undeveloped AgBr to convert it to a water-soluble complex that can be washed away:

$$AgBr(s) + 2\ Na_2S_2O_3(aq) \longrightarrow$$
$$Na_3Ag(S_2O_3)_2(aq) + NaBr(aq)$$

How many mL of 0.0200 M $Na_2S_2O_3$ solution are needed to dissolve 0.450 g of AgBr?

10.60 One test for vitamin C ($C_6H_8O_6$, also called ascorbic acid) is based on the oxidation of the vitamin by iodine:

$$C_6H_8O_6(aq) + I_2(aq) \longrightarrow C_6H_6O_6(aq) + 2\ HI(aq)$$

(a) If 25.0 mL of a fruit juice requires 13.0 mL of 0.0100 M I_2 solution for reaction, what is the molarity of the ascorbic acid in the fruit juice?
(b) The Food and Drug Administration recommends that 60. mg of ascorbic acid be consumed per day. How many mL of this fruit juice must a person drink to obtain the recommended dosage?

10.61 Concentrated (12.0 M) hydrochloric acid is sold for household and industrial purposes under the name "muriatic acid." How many mL of 0.500 M HCl solution can be made from 25.0 mL of 12.0 M HCl solution?

10.62 Dilute solutions of $NaHCO_3$ are sometimes used in treating acid burns. How many mL of 0.100 M $NaHCO_3$ solution are needed to prepare 750.0 mL of 0.0500 M $NaHCO_3$ solution?

Electrolytes

10.63 What is an electrolyte? Give an example.

10.64 What does it mean when we say that the concentration of Ca^{2+} in blood is 3.0 mEq/L?

10.65 Kaochlor, a 10.% KCl solution, is an oral electrolyte supplement administered for potassium deficiency. How many mEq of K^+ are in a 30. mL dose?

10.66 Look up the concentration of Cl^- ion in blood (Table 10.7) and calculate how many grams of Cl^- are in 100. mL blood.

10.67 A solution of sulfate ion (SO_4^{2-}) is 0.025 M. What is this concentration in Eq/L and in mEq/L?

Vapor Pressure, Boiling Points, and Freezing Points

10.68 Equal volumes of pure water and syrup (sugar dissolved in water) are placed in open containers at room temperature. The pure water evaporates long before all the moisture has left the syrup. Explain.

10.69 Methanol, CH_3OH, is sometimes used as an antifreeze for the water in automobile windshield washer fluids. Calculate the mass of methanol that must be added to 5.0 kg of water to lower its freezing point to $-10.0°C$. For each mole of nonvolatile solute, the freezing point of 1 kg of water is lowered 1.86 C°.

10.70 What is the boiling point of the solution produced in Problem 10.69? For each mole of nonvolatile solute, the boiling point of 1 kg of water is raised 0.51 C°.

10.71 Which lowers the freezing point of 2 kg of water more, 0.20 mol NaOH or 0.20 mol $Ba(OH)_2$? Both compounds are strong electrolytes.

Osmosis

10.72 Explain why a red blood cell will swell up and burst when placed in pure water.

10.73 What does it mean when we say that a 0.15 M NaCl solution is isotonic with blood, whereas distilled water is hypotonic?

10.74 Describe briefly what would happen if a 1.0 M NaCl solution were separated from distilled water by an osmotic membrane.

10.75 Which of the following solutions has the higher osmolarity?
(a) 0.25 M KBr or 0.40 M Na_2SO_4
(b) 0.30 M NaOH or 3.0% (w/v) NaOH

10.76 What would happen if a 0.40 M glucose solution were separated from a 0.40 M NaCl solution by an osmotic membrane?

Applications

10.77 Uric acid, the principal constituent of some kidney stones, has the formula $C_5H_4N_4O_3$. In aqueous solution, the solubility of uric acid is only 0.067 g/L. Express this concentration in (w/v)%, in ppm, and in molarity. [App: Gout and Kidney Stones]

10.78 Explain why, if you salt meat or vegetables before cooking, it draws the juices out of them. [App: Kitchen Chemistry]

10.79 What is the difference between an osmotic membrane and a dialysis membrane? [Int: Dialysis]

10.80 How does hemodialysis help to maintain a body's homeostasis? [Int: Dialysis]

Additional Questions and Problems

10.81 Emergency treatment of cardiac arrest victims sometimes involves injection of calcium chloride solutions directly into the heart muscle. How many grams of $CaCl_2$ are administered in an injection of 5.0 mL of a 5.0% solution? How many mEq of Ca^{2+} does this correspond to?

10.82 Nitric acid, HNO_3, is available commercially at a concentration of 16 M. What volume would you use to prepare 750 mL of a 0.20 M solution?

10.83 How much 16 M nitric acid (Problem 10.82) would react with 5.50 g of KOH according to the following equation?

$$HNO_3 + KOH \longrightarrow H_2O + KNO_3$$

10.84 An old bottle of 12.0 M hydrochloric acid has only 25 mL left in it. What is the HCl concentration if enough water is added so that the final volume is 525 mL?

10.85 Ringer's solution, used in the treatment of burns and wounds, is prepared by dissolving 8.6 g of NaCl, 0.30 g of KCl, and 0.33 g of $CaCl_2$ in water and diluting to a volume of 1 L. What is the molarity of each of the three components?

10.86 What is the osmolarity of Ringer's solution (Problem 10.85)? Is it hypotonic, isotonic, or hypertonic with blood plasma (0.30 Osmol)?

10.87 In most states, a person with a blood alcohol concentration of 0.10% (v/v) is considered legally drunk. What volume total alcohol does this concentration represent, assuming a blood volume of 5.0 L?

10.88 Draw the structure of an ammonia molecule, NH_3, and indicate the polarization of the N–H bonds using the δ^+/δ^- notation.

10.89 In light of your answer to Problem 10.88, show how a hydrogen bond can form between two ammonia molecules.

10.90 Ammonia (Problem 10.88) is very soluble in water (51.8 g/L at 20°C and 760 mm Hg).

(a) Show how NH_3 can hydrogen-bond to water.

(b) What is the solubility of ammonia in water in mol/L?

10.91 Cobalt(II) chloride, a blue solid, can absorb water from the air to form cobalt(II) chloride hexahydrate, a pink solid. The equilibrium is so sensitive to moisture in the air that $CoCl_2$ is used as a humidity indicator.

(a) Write a balanced equation for the equilibrium. Be sure to include water as a reactant to produce the hexahydrate.

(b) How many grams of water are released by the decomposition of 2.50 g of cobalt(II) chloride hexahydrate?

11

Acids, Bases, and Salts

The antacid tablet just dropped into the water contains both an organic acid (citric acid) and a base (sodium bicarbonate). As they dissolve they are reacting with each other and giving off carbon dioxide. Acid–base equilibria and their control are the subject of this chapter.

Acids! The word calls up images of dangerous, corrosive liquids that eat away everything they touch. Although a few well-known substances such as sulfuric acid (H_2SO_4) do indeed fit this description, most acids are relatively harmless. In fact, many acids, such as ascorbic acid (vitamin C), are necessary for life.

You've already learned the most useful definitions of acids and bases, which are based on the loss and gain of a proton, H^+. In this chapter, a further look at acids and bases will answer the following questions:

1. **What are acids and bases?** The goal: Be able to recognize acids and bases and write equations for the common acid–base reactions.
2. **How can water be either an acid or a base?** The goal: Be able to give examples of reactions in which water is either an acid or a base.
3. **What is the influence of acid and base strengths on their reactions?** The goal: Be able to predict the favored direction of acid–base equilibria.
4. **What is the ion product constant for water?** The goal: Be able to write the equation for this constant and use it to find $[H_3O^+]$ or $[OH^-]$.
5. **What is the pH scale for measuring acidity?** The goal: Be able to explain the pH scale and find pH from $[H_3O^+]$.
6. **How are acid concentrations expressed?** The goal: Be able to define and use equivalents and normality.
7. **What types of salts give acidic and alkaline solutions?** The goal: Be able to predict from its formula whether a salt solution will be neutral, acidic, or alkaline.
8. **What is a buffer?** The goal: Be able to explain how a buffer maintains pH and how the bicarbonate buffer functions in the body.

11.1 ACIDS AND BASES IN AQUEOUS SOLUTION: A REVIEW

Because of the importance of acids and bases, we couldn't have come this far into a discussion of chemistry without mentioning them. Take a moment to review what we've already said about acids and bases before we go on to a more systematic study of their properties:

- Acids are substances that donate hydrogen ions, H^+.

$$HCl(aq) \longrightarrow H^+ + Cl^-$$

Hydrochloric
acid

● Bases are substances that accept hydrogen ions.

$$NH_3(aq) + H^+ \longrightarrow NH_4^+$$

Ammonia
(a base)

● Acids are polar molecular compounds with a partial positive charge on the hydrogen atom that's donated.

$$\overset{\delta^+}{}\overset{\delta^-}{}$$
$$HCl$$

● The common polyatomic ions (Table 5.4) are the anions remaining when acids have donated their hydrogen atoms.

$$H_2SO_4(aq) \longrightarrow 2\,H^+ + SO_4^{2-}$$

Sulfuric acid   Sulfate ion

● Ionic compounds that contain the hydroxide ion (OH^-) are one important class of bases:

$$NaOH(s) \xrightarrow{\;H_2O\;} Na^+ + OH^-$$

● The exchange reaction of an acid with a base yields a salt (composed of the cation from the base and the anion from the acid) plus water:

$$HCl(aq) + NaOH(aq) \longrightarrow NaCl(aq) + H_2O(l)$$

Acid   Base   Salt   Water

11.2 PROPERTIES OF ACIDS AND BASES

Before the chemistry of acids and bases was well understood, they were recognized by certain characteristic properties (Table 11.1). Acids have a sour taste, and bases are bitter. Concentrated solutions of bases are slippery to the touch. You must always remember that no laboratory chemical should *ever* be tasted

Table 11.1   Characteristic Properties of Proton Acids and Hydroxide Bases

Acids	Bases
Sour, like vinegar	Bitter and slippery, like soap
Change indicator colors in opposite direction from base, such as blue litmus to pink	Change indicator colors in opposite direction from acid, such as pink litmus to blue
React with bases to neutralize them and form salts	React with acids to neutralize them and form salts
Liberate hydrogen in reactions with active metals	React in aqueous solution with salts of heavy metals to form insoluble hydroxides or oxides
Aqueous solutions conduct electricity	Aqueous solutions conduct electricity

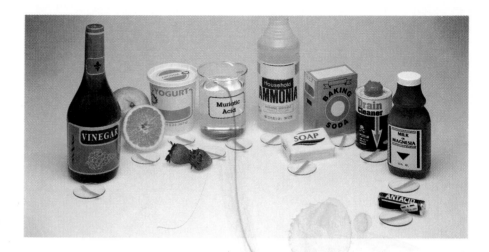

Figure 11.1
Acidic (on the left) and alkaline (on the right) everyday items. The upper ends of the litmus paper strips have been exposed to the various solutions.

Acid–base indicator A dye that changes color to indicate the acidity or alkalinity of a solution (the pH).

Strong acid An acid that is a strong electrolyte.

Strong base A hydroxide ion-containing base that is a strong electrolyte.

Strong electrolyte A substance that is completely converted to ions when dissolved in water.

or touched to experience its properties. At home, though, you're very familiar with the acid and base properties (Figure 11.1). Just about every food that's sour contains an acid—lemons, oranges, and grapefruit contain citric acid, many soft drinks contain phosphoric acid, sour milk contains lactic acid, and vinegar contains acetic acid. Bases aren't so obvious in foods, but most of us have them stored under the kitchen or bathroom sink, because they're present in many household cleaning agents, from perfumed toilet soap to the stuff you put down the drain to dissolve hair, grease, and other materials that clog it.

Acids and bases also have long been known to change the colors of dyes that are **acid–base indicators.** Red cabbage juice, for example, is a natural indicator (Figure 11.2)—it's red in acid solutions and blue to green in base solutions. Acid solutions contain hydrogen ions (H^+) and anions, and solutions of bases like sodium hydroxide contain cations and hydroxide ions (OH^-). **Strong acids** and **strong bases** are **strong electrolytes**—they are 100% converted to ions in solution. The common ones are listed in Table 11.2, and it's a good idea to learn their names and formulas.

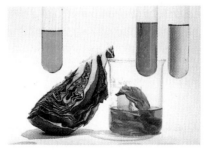

Figure 11.2
Red cabbage juice—a natural acid–base indicator. The juice changes from deep red to light red when acid is added, and to green when base is added.

Practice Problem **11.1** Which of the following are weak acids? In other words, which are *not* strong acids?
(a) hydrocyanic acid, HCN(*aq*)   (b) hydrobromic acid, HBr(*aq*)
(c) nitrous acid, HNO_2(*aq*)   (d) hydrofluoric acid, HF(*aq*)
(e) nitric acid, HNO_3(*aq*)

Table 11.2 Common Strong Acids and Bases

Name (Common Name)	Formula	Comments
Strong acids		
Perchloric acid	$HClO_4$	Strong oxidizing agent; ignites in contact with organic materials; detonates on impact
Sulfuric acid (battery acid)	H_2SO_4	Industrial chemical manufactured in largest quantity; concentrated acid is oxidizing agent; dilution generates heat, so acid must be added to water (not water to acid)
Hydriodic acid	HI	Disinfectant
Hydrobromic acid	HBr	Sensitive to light
Hydrochloric acid (muriatic acid)	HCl	Many uses, including metal cleaning, food processing, lab reagent
Nitric acid	HNO_3	Oxidizing agent; used in fertilizer and explosives manufacture; reacts with skin and other proteins to give yellow color
Strong bases, water-soluble		
Sodium hydroxide (caustic soda, lye)	NaOH	Most important commercial base; uses include peeling fruits and vegetables, manufacture of paper, detergents, cellophane, household drain cleaners; solid absorbs H_2O and CO_2 from air
Potassium hydroxide	KOH	Uses include soap manufacture, electrolyte in alkaline storage battery; absorbs H_2O and CO_2 from air
Strong bases, limited water solubility		
Barium hydroxide	$Ba(OH)_2$	Uses include dehairing hides, beet sugar refining, water softening, glass manufacture
Calcium hydroxide (hydrated lime, slaked lime)	$Ca(OH)_2$	Uses include mortar, plaster, cement, whitewash, disinfectant, water softening
Magnesium hydroxide (milk of magnesia)	$Mg(OH)_2$	Least-soluble common base; antacid, laxative; used in sugar refining, toothpaste, food additive (drying agent)

11.3 ACID DEFINITION

Acid (Brønsted-Lowry) A substance that is able to donate a hydrogen ion, H^+.

The Danish chemist Johannes Brønsted and the English chemist Thomas Lowry in 1923 simultaneously proposed the acid and base definitions we've been using. An **acid,** according to the Brønsted-Lowry definition, is a substance that is able to give away a hydrogen ion, H^+, to another molecule or ion. Since a hydrogen *atom* consists of a proton and an electron (Section 3.5), a hydrogen *ion*, H^+, is simply a proton. Thus, we refer to acids as *proton donors*.

In learning to recognize the common polyatomic ions (Table 5.4), you saw that different acids can have different numbers of hydrogen atoms available for giving away. Acids with one proton to donate, like HCl and HNO_3, are **monoprotic acids;** H_2SO_4 is a **diprotic acid** since it has two protons; and H_3PO_4 is a **triprotic acid** since it has three protons.

Monoprotic acid A substance that has one acidic hydrogen atom.

Diprotic acid A substance that has two acidic hydrogen atoms.

Triprotic acid A substance that has three acidic hydrogen atoms.

Nitric acid (monoprotic)   Sulfuric acid (diprotic)   Phosphoric acid (triprotic)

Acetic acid (CH_3COOH), an example of an organic acid, actually has a total of four hydrogens, but only the one attached to oxygen is acidic. The three hydrogens bonded to carbon are not given up easily. Most organic acids are similar: They contain many hydrogen atoms, but only the one in the —COOH group is acidic.

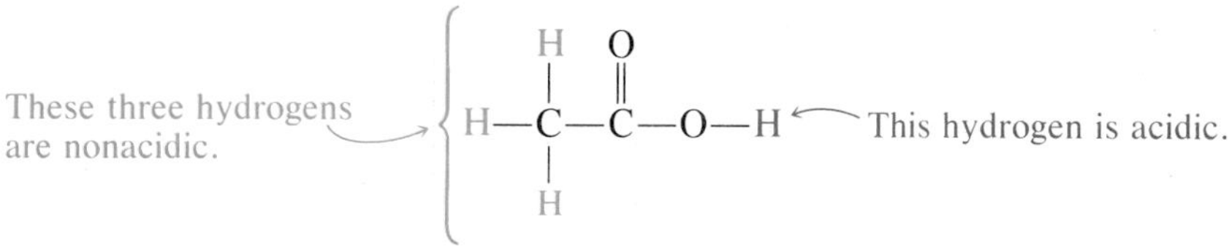

These three hydrogens are nonacidic.

This hydrogen is acidic.

Acetic acid

11.4   BASE DEFINITION

The base properties listed in Table 11.1 are those of the hydroxide ion (OH^-) in aqueous solution. Solutions with these properties are referred to as *basic* or *alkaline*. Older definitions of bases, in fact, were limited to compounds like NaOH that contain hydroxide ion. According to the broader and more useful Brønsted-Lowry definition, a **base** is any substance that accepts a proton from an acid. Thus, during an acid–base reaction, a proton is transferred.

The general reaction between proton-donor acids and proton-acceptor bases can be represented as follows:

Base (Brønsted-Lowry)   A substance that can accept a hydrogen ion (H^+) from an acid.

Electrons from the base form the new B—H bond.

$$H—A + B: \rightleftharpoons B—H^+ + A^-$$
$$H—A + B:^- \rightleftharpoons B—H + A^-$$

where H—A represents a proton-donor acid and $B:$ or $B:^-$ represents a proton-acceptor base.

Notice in these acid–base reactions that both electrons in the resulting B—H bond come from the base when a bond forms to H^+. In fact, a base must have such a **lone pair** of electrons (Section 6.4) in order to accept H^+ from an acid.

Lone pair   A pair of outer-shell electrons not used by an atom in forming bonds.

A base can be either neutral ($B:$) or negatively charged ($B:^-$). If the base is neutral, then the protonated product has a positive charge ($B—H^+$):

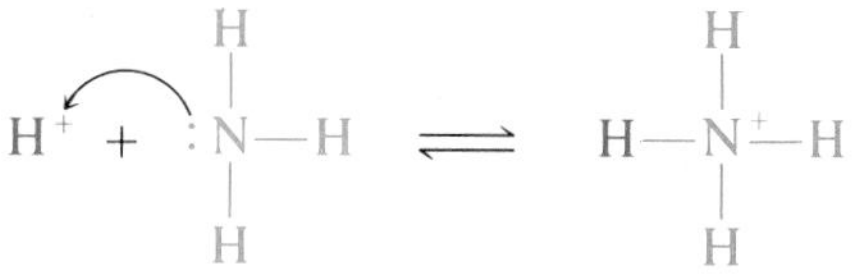

Ammonia   Ammonium ion

If the base is negatively charged to begin with, then the protonated product is neutral (B—H). According to the Brønsted-Lowry acid–base definition, therefore, it's not a compound such as NaOH that's the base but the OH^- ion, no matter what its source.

$$H^+ + :OH^- \longrightarrow H-O-H$$

The base

Practice Problems

11.2 Which of the following would you expect to be Brønsted-Lowry acids?
(a) HCOOH   (b) H_2S (c) $SnCl_2$

11.3 Which of the following would you expect to be Brønsted-Lowry bases?
(a) SO_3^{2-} (b) Ag^+ (c) F^-

11.5 WATER AS AN ACID AND A BASE

Water is not an acid in the classical sense, because it has few of the properties listed in Table 11.1. One of the most useful features of the Brønsted-Lowry definition, however, is that it includes water. When water reacts as a proton donor, it is an acid. For example, in its reaction with acetate ion, water donates a proton to the acetate ion to form acetic acid:

$$CH_3COO^- + H_2O(l) \rightleftharpoons CH_3COOH(aq) + OH^-$$

Base   Water as an acid

This equation illustrates an essential feature of the Brønsted-Lowry acid–base definitions that we'll return to later: *The anion of an acid is a base because it is a proton acceptor.*

Water also has none of the properties listed for bases in Table 11.1, but it reacts as a Brønsted-Lowry base when it accepts a proton from an acid. Although it's sometimes convenient to write a chemical equation that portrays the splitting apart (**dissociation**) of an acid into a proton and an anion, this dissociation doesn't occur spontaneously in practice. A bare proton is too unstable to exist by itself, and an acid therefore gives up a proton only when there is a base present to accept it.

Dissociation The splitting apart of a substance to yield two or more ions.

$$H-A \not\longrightarrow H^+ + A^-$$ The dissociation of a pure acid doesn't occur in isolation

$$H-A + Base \rightleftharpoons H-Base^+ + A^-$$ Instead, a base is necessary to accept the proton from the acid.

When an acid is dissolved in water, water acts as a base to accept a proton from the acid and generate the **hydronium ion** (H_3O^+). If HCl gas is dissolved in water, for example, a solution of H_3O^+ and Cl^- ions results.

Hydronium ion H_3O^+, the species formed when an acid is dissolved in water.

Water uses two of its electrons
to form a bond to H⁺

Hydrochloric acid $H—Cl(g) + :\overset{..}{O}—H(l) \longrightarrow H—\overset{+}{\underset{H}{\overset{..}{O}}}—H(aq) + Cl^-(aq)$

(Acid)   (Base)   (Aqueous solution of ions)

The acid properties listed in Table 11.1 are due to the presence of H_3O^+ in aqueous solution. Often, H_3O^+ and H^+ or $H^+(aq)$ are used interchangeably to refer to a hydrogen ion dissolved in water.

Amphoteric Able to react as both an acid and a base.

Substances like water that can react as either an acid or a base are described as **amphoteric.** When water acts as an acid, it donates a proton and becomes OH^-; when it acts as a base, it accepts a proton and becomes H_3O^+.

Water as an acid
(H⁺ donor) $:NH_3(aq) + H_2O(l) \rightleftharpoons NH_4^+ + OH^-$

Water as a base
(H⁺ acceptor) $HCl(aq) + H_2O(l) \longrightarrow H_3O^+ + Cl^-$

Practice Problem   **11.4**   Is water an acid or a base in the following reactions?
(a) $H_3PO_4(aq) + H_2O(l) \rightleftharpoons H_2PO_4^- + H_3O^+$
(b) $F^- + H_2O(l) \rightleftharpoons HF(aq) + OH^-$
(c) $NH_4^+ + H_2O(l) \rightleftharpoons H_3O^+ + NH_3(aq)$

11.6   REACTIONS OF ACIDS WITH BASES

Reactions of Acids with Metal Hydroxides Acids react with metal hydroxide bases to yield water and a salt in exchange reactions (Section 7.11):

$HCl(aq) + KOH(aq) \longrightarrow H_2O(l) + KCl(aq)$

An acid   A base   Water   A salt

All such reactions are written with a single arrow because they are driven virtually to completion by formation of water. The net ionic equation (Section 7.12) for all such reactions (unless an insoluble salt is formed) is

$$H^+ + OH^- \longrightarrow H_2O(l)$$

Neutralization The exchange reaction between an acid and a base to yield water and a salt.

and makes clear why the properties of the acid and base disappear in such **neutralization** reactions: The hydrogen ions and hydroxide ions are used up in the formation of water.

Practice Problems **11.5** Write a balanced equation for each of the following acid–base reactions and name the salt formed:
(a) $HNO_3(aq) + KOH(aq)$
(b) $H_2SO_4(aq) + LiOH(aq)$
(c) $HCl(aq) + Mg(OH)_2(aq)$

11.6 Maalox, an over-the-counter antacid, contains aluminum hydroxide, $Al(OH)_3$. Write a balanced equation for the reaction of $Al(OH)_3$ with stomach acid (HCl).

Reactions of Acids With Metal Bicarbonates and Metal Carbonates The reactions of acids with metal bicarbonates and metal carbonates are also exchange reactions. Because the exchange reaction product carbonic acid (H_2CO_3) is unstable, the actual products are carbon dioxide and water:

$$HBr(aq) + NaHCO_3(aq) \longrightarrow H_2O(l) + CO_2(g) + NaBr(aq)$$

Dropping the spectator ions, Br^- and Na^+, shows that the bicarbonate ion is indeed a base (a proton acceptor) and that this is an acid–base reaction:

$$H^+(aq) + HCO_3^-(aq) \longrightarrow [H_2CO_3(aq)] \longrightarrow H_2O(l) + CO_2(g)$$

| An acid | Bicarbonate ion (a base) | Carbonic acid (unstable) | Water | Carbon dioxide |

Acids react with metal carbonates such as Na_2CO_3 in exactly the same way that they react with metal bicarbonates, except that a carbonate ion (CO_3^{2-}) can neutralize twice as much acid as a bicarbonate ion (HCO_3^-):

$$Na_2CO_3(aq) + 2\,HCl(aq) \longrightarrow H_2O(l) + CO_2(g) + 2\,NaCl(aq)$$

Although most metal carbonates are insoluble in water (marble is almost pure calcium carbonate, $CaCO_3$), they nevertheless react easily with aqueous acid. In fact, geologists often test for carbonate-bearing rocks simply by putting a few drops of aqueous HCl on the rock and watching to see if bubbles of CO_2 form (Figure 11.3).

Figure 11.3
Calcium carbonate reacts with hydrochloric acid. Bubbles of carbon dioxide are rising in the test tube that contains marble chips. Eggshells are made of calcium carbonate, too.

Practice Problem   **11.7**   Write balanced equations for the following reactions:
(a) $KHCO_3(aq) + HNO_3(aq) \rightarrow$
(b) $MgCO_3(s) + H_2SO_4(aq) \rightarrow$
(c) $HI(aq) + LiHCO_3(aq) \rightarrow$

Reactions of Acids with Ammonia   Aqueous solutions of ammonia are sometimes labeled "ammonium hydroxide," but this is really a misnomer. Ammonia is such a weak base that only about 1% of the ammonia molecules in aqueous solution are protonated by water to yield NH_4^+ and OH^- ions; the remaining 99% are unprotonated.

$$NH_3(aq) + \text{H—O—H} \rightleftharpoons NH_4^+(aq) + OH^-(aq)$$

99% of ammonia molecules are unprotonated.

1% of ammonia molecules are protonated by water.

Acids react with ammonia to yield **ammonium salts** such as ammonium chloride, NH_4Cl, most of which are water-soluble.

$$NH_3(aq) + HCl(aq) \longrightarrow NH_4Cl(aq)$$

Ammonium salt   An ionic substance formed by reaction of ammonia with an acid and containing the ammonium ion, NH_4^+.

Living organisms contain a group of compounds called *amines,* which contain ammonia-like nitrogen atoms and undergo neutralization reactions with acids just as ammonia can. Methylamine, for example, reacts with hydrochloric acid as follows:

$$CH_3NH_2(aq) + HCl(aq) \longrightarrow CH_3NH_3Cl(aq)$$

Practice Problems   **11.8**   What products would you expect from the reaction in aqueous solution of ammonia and sulfuric acid?

$$H_2SO_4(aq) + NH_3(aq) \longrightarrow$$

11.9   Show how ethylamine ($C_2H_5NH_2$) reacts with hydrochloric acid to form an ethylammonium salt.

11.7 ACID AND BASE STRENGTH

Weak electrolyte   A substance that is partly ionized when dissolved in water and establishes equilibrium between its ionized and nonionized forms.

Different acids differ in their ability to give up a proton. At the left in Table 11.3 a number of common acids are listed in order of decreasing acid strength, that is, in order of decreasing ease of giving up a proton. The six acids at the top of the table are the strong acids. Those remaining are weak acids, that is, **weak electrolytes;** they fall roughly into the two groups shown.

Table 11.3   Relative Acid and Base Strengths

		Acid		Base		
Increasing acid strength	Strong acids: 100% ionized	Perchloric acid	$HClO_4$	ClO_4^-	Perchlorate ion	Increasing base strength
		Sulfuric acid	H_2SO_4	HSO_4^-	Hydrogen sulfate ion	
		Hydriodic acid	HI	I^-	Iodide ion	Little or no reaction as bases
		Hydrobromic acid	HBr	Br^-	Bromide ion	
		Hydrochloric acid	HCl	Cl^-	Chloride ion	
		Nitric acid	HNO_3	NO_3^-	Nitrate ion	
		Hydronium ion	H_3O^+	H_2O	**Water**	
	Weak acids	Hydrogen sulfate ion	HSO_4^-	SO_4^{2-}	Sulfate ion	Very weak bases
		Phosphoric acid	H_3PO_4	$H_2PO_4^-$	Dihydrogen phosphate ion	
		Nitrous acid	HNO_2	NO_2^-	Nitrite ion	
		Hydrofluoric acid	HF	F^-	Fluoride ion	
		Acetic acid	CH_3COOH	CH_3COO^-	Acetate ion	
	Very weak acids	Carbonic acid	H_2CO_3	HCO_3^-	Bicarbonate ion	Weak bases
		Dihydrogen phosphate ion	$H_2PO_4^-$	HPO_4^{2-}	Hydrogen phosphate ion	
		Ammonium ion	NH_4^+	NH_3	Ammonia	
		Hydrocyanic acid	HCN	CN^-	Cyanide ion	
		Bicarbonate ion	HCO_3^-	CO_3^{2-}	Carbonate ion	
		Hydrogen phosphate ion	HPO_4^{2-}	PO_4^{3-}	Phosphate ion	
		Water	H_2O	OH^-	**Hydroxide ion**	
	Little or no reaction as acids	Methyl alcohol	CH_3OH	CH_3O^-	Methoxide ion	Strong bases: 100% reaction with H_2O to give OH^-
		Ammonia	NH_3	NH_2^-	Amide ion	

Note that diprotic acids such as sulfuric acid undergo two stepwise dissociations in water. The first dissociation of H_2SO_4 to yield HSO_4^- occurs to the extent of nearly 100%. The second dissociation of HSO_4^- to yield SO_4^{2-} (sulfate ion) takes place to a much lesser extent because it involves loss of a positively charged H^+ from an ion that already carries one negative charge.

$$H_2SO_4(aq) + H_2O(l) \longrightarrow H_3O^+ + HSO_4^-$$

Hydrogen sulfate ion

$$HSO_4^- + H_2O(l) \rightleftharpoons H_3O^+ + SO_4^{2-}$$

Sulfate ion

AN APPLICATION: ULCERS, HEARTBURN, AND ANTACIDS

Although their causes are unknown, peptic ulcers are one of the hazards of fast-paced modern life—nature's way of telling us to slow down. Ulcers occur either in the stomach itself or in the upper part of the small intestine (the *duodenum*) when the protective mucosal lining is penetrated and gastric juices begin to dissolve the stomach or intestinal wall. The resultant lesion causes considerable pain and can lead to eventual perforation of the wall.

   Much of the tissue damage that produces ulcers results from the strongly acidic nature of gastric juices, which reaches a maximum acid strength approximately 60 min after eating. A less serious but nevertheless uncomfortable condition, commonly known as *heartburn*, occurs when gastric juices back up into the esophagus, causing burning in the chest and sometimes regurgitation.

   One of the simplest methods for controlling heartburn and ulcer damage is to lower stomach acidity through the use of antacids. As you might expect, all the common over-the-counter antacid remedies are bases, either hydroxides, carbon-ates, or bicarbonates. Since aluminum and calcium salts tend to cause constipation, most formulations also contain magnesium salts, which have a counteracting laxative effect. Sodium bicarbonate ($NaHCO_3$, baking soda) is less suitable than other bases because it tends to enter the bloodstream too rapidly. For complete healing of ulcers, a more effective therapeutic approach is the use of a drug such as Tagamet that interrupts the chemical signal that triggers acid secretion. By keeping the acid content of the stomach at a minimum for up to 6 weeks, the ulcer lesion is given a chance to heal.

Ingredients of Some Antacid Preparations

Trade Name	Active Antacid Ingredients
Alka-Seltzer	$NaHCO_3 + KHCO_3$
Bisodol	$CaCO_3 + Mg(OH)_2$
Gelusil	$Al(OH)_3 + Mg(OH)_2$
Maalox	$Al(OH)_3 + Mg(OH)_2$
Rolaids	$NaAl(OH)_2CO_3$
Tums	$CaCO_3$

The most striking feature of Table 11.3 is the relationship between acid and base strength. The list of corresponding anions on the right is a list of bases in the order of increasing base strength. *The weaker an acid, the stronger the base it forms by loss of a proton.*

   The pairs on the left and right in Table 11.3 are referred to as **conjugate acid–base pairs**—acid–base pairs related by the gain and loss of a proton. For example, HCl is the conjugate acid of Cl^-, and Cl^- is the conjugate base of HCl

Conjugate acid–base pair
An acid and base related to each other by the loss and gain of a proton.

The conjugate acid of Cl^- ⟶ HCl   The conjugate base of HCl ⟶ Cl^-

Weak base   A base that, because it has low affinity for a proton, establishes an equilibrium with water.

Why is there a relationship between acid and base strength? Consider what it means for an acid or base to be strong or weak. A strong acid gives up a proton readily, showing that the anion of a strong acid has little affinity for the proton. Otherwise the acid wouldn't have given up the proton in the first place. But this is exactly the definition of a **weak base**—a substance that has little affinity for a proton.

$$H\!-\!A + H_2O \longrightarrow H_3O^+ + A^-$$

If this is a strong acid because it gives up the proton readily . . .

. . . then this anion is a weak base because it has little affinity for the proton

For example,

$$HCl + H_2O \longrightarrow H_3O^+ + Cl^-$$

(HCl—a strong acid) (Cl⁻—a very weak base)

Weak acid An acid that, because it holds its proton strongly, is not fully dissociated in aqueous solution and establishes an equilibrium with water.

Similarly, a **weak acid** is one that holds a proton tightly, gives it up with difficulty, and establishes an equilibrium with water

$$H\!-\!A + H_2O \rightleftharpoons H_3O^+ + A^-$$

The anion of a weak acid is therefore a base when it reacts with water.

$$A^- + H_2O \rightleftharpoons H\!-\!A + OH^-$$

If this anion has a high affinity for a proton . . .

. . . then this is a weak acid because it does not readily give up the proton.

Practice Problems 11.10 Write formulas for
(a) the conjugate acid of HS^- (b) the conjugate acid of PO_4^{3-}
(c) the conjugate base of H_2CO_3 (d) the conjugate base of NH_4^+

11.11 Using Table 11.3, in each pair, identify the stronger acid:
(a) H_2O or NH_4^+ (b) H_2SO_4 or CH_3COOH
(c) HCN or H_2CO_3

11.12 Using Table 11.3, in each pair, identify the stronger base:
(a) F^- or Br^- (b) OH^- or HCO_3^-

11.8 PROTON-TRANSFER REACTIONS

Acid–base reaction, Brønsted-Lowry The transfer of a proton from an acid to a base.

A **Brønsted-Lowry acid–base reaction** is a proton transfer from an acid to a base:

$$HA + B \text{ (or } B^-) \rightleftharpoons BH^+ \text{ (or } BH) + A^-$$

Acid   Base   Conjugate acid   Conjugate base

Table 11.3 helps in making qualitative predictions about proton-transfer reactions. *An acid–base proton-transfer equilibrium always favors formation of the weaker acid and base.*

To try out this rule, we'll compare the reactions of acetic acid with water and hydroxide ion. In predicting the direction of such acid–base reactions, first write the equation and identify the acid and base on each side of the arrow; then decide which of each pair is stronger and weaker. Acetic acid reacts with water to give hydronium ion and acetate ion. Writing this down and consulting the table shows that this reaction will be favored in the reverse direction as written, because acetic acid and water are the weaker acid and base, respectively.

$$CH_3COOH(aq) + H_2O(l) \rightleftharpoons CH_3COO^- + H_3O^+$$

Weaker acid   Weaker base   Stronger base   Stronger acid

Reverse reaction favored

The situation is reversed for the reaction of acetic acid with hydroxide ion:

$$CH_3COOH(aq) + OH^- \rightleftharpoons CH_3COO^- + H_2O(l)$$

Stronger acid   Stronger base   Weaker base   Weaker acid

Forward reaction favored

Solved Problem 11.1   Write a balanced equation for the proton-transfer reaction between phosphate ion and water, identify each acid–base pair, and determine which direction of the equilibrium is favored.

Solution   Phosphate ion (PO_4^{3-}) is the anion of a weak acid and therefore a base, so in this reaction water will be the proton donor—the acid.

$$PO_4^{3-} + H_2O(l) \rightleftharpoons HPO_4^{2-} + OH^-$$

Base   Acid   Conjugate acid   Conjugate base

Table 11.3 shows that HPO_4^{2-} and OH^- are the stronger and therefore the reaction is favored in the reverse of the direction written.

Practice Problem 11.13   Write a balanced equation for the proton-transfer reaction between hydrogen phosphate ion and hydroxide ion, identify each acid–base pair, and determine which direction of the equilibrium is favored.

11.9 WEAK ACID DISSOCIATION CONSTANTS

The reaction of a weak acid with water, like any chemical equilibrium, can be described quantitatively by an equilibrium constant expression (Section 8.9). The equilibrium constant K_a for the reaction of a weak acid with water is called an **acid dissociation constant (K_a).**

Acid dissociation constant (K_a) The equilibrium constant for the dissociation of an acid (HA), equal to $[H^+][A^-]/[HA]$.

$$HA(aq) + H_2O(l) \rightleftharpoons H_3O^+ + A^- \qquad K_a = \frac{[H_3O^+][A^-]}{[HA][H_2O]}$$

The expression can be simplified by dropping $[H_2O]$ because, with water as the solvent present in very large excess, it's concentration is essentially constant and of no influence on the equilibrium. Writing K_a is further simplified by using $[H^+]$ instead of $[H_3O^+]$, giving the following expression that applies to any weak acid.

$$K_a = \frac{[H^+][A^-]}{[HA]}$$

The stronger an acid, the larger its value of K_a, because it is further dissociated into H^+ and A^-. In fact, the K_a values for strong acids that are 100% dissociated (such as HCl) are extremely large, difficult to measure, and not very useful.

Take a minute to look over Table 11.4, which gives acid dissociation constant values for a number of common acids. The data illustrate several important points:

- For weak acids, the equilibrium with water is always favored in the reverse direction, and the K_a values are all less than 1.
- Donation of each successive H^+ from a polyprotic acid is more difficult.
- Most simple organic acids containing the —COOH group have K_a values of about 10^{-5}.

11.10   IONIZATION OF WATER

Pure water is very slightly dissociated into ions. Since each dissociation yields one H_3O^+ ion and one OH^- ion, the concentrations of the two ions are identical; at 25°C, the concentration of each is 1.00×10^{-7} mol/L.

$$2\ H_2O(l) \ \rightleftharpoons \ H_3O^+ + OH^-$$

At 25°C: $[H_3O^+] = [OH^-] = 1.00 \times 10^{-7}$ mol/L

The equilibrium constant expression for this reaction

$$K = \frac{[H_3O^+]\,[OH^-]}{[H_2O]^2}$$

Table 11.4   Some Acid Dissociation Constants, K_a (at 25°C)

Acid	K_a	Acid	K_a
Hydrofluoric acid (HF)	6.5×10^{-4}	**Polyprotic acids**	
Hydrocyanic acid (HCN)	6.2×10^{-10}	Sulfuric acid	
Ammonium ion (NH$_4^+$)	5.6×10^{-10}	H_2SO_4	Large
		HSO_4^-	1.0×10^{-2}
Organic acids		Phosphoric acid	
Formic acid (HCOOH)	1.8×10^{-4}	H_3PO_4	7.5×10^{-3}
Acetic acid (CH$_3$COOH)	1.8×10^{-5}	$H_2PO_4^-$	6.6×10^{-8}
Propanoic acid	1.3×10^{-5}	HPO_4^{2-}	1×10^{-12}
(CH$_3$CH$_2$COOH)		Carbonic acid	
Ascorbic acid	7.9×10^{-5}	H_2CO_3	4.5×10^{-7}
(vitamin C)		HCO_3^-	4.8×10^{-11}

is simplified (like the K_a expression) by dropping the concentration of water to give an expression for the **ion product constant of water (K_w)**. Since at 25°C both $[H_3O^+]$ and $[OH^-]$ equal 1.00×10^{-7}, K_w is equal to 1.00×10^{-14}.

ion product constant of water (K_w) The product of H_3O^+ and OH^- molar concentrations in water or any aqueous solution ($K_w = [H_3O^+][OH^-] = 1 \times 10^{-14}$).

Ion product constant of water (K_w, 25°C)

$$K_w = [H_3O^+] \times [OH^-]$$
$$= (1.00 \times 10^{-7}) \times (1.00 \times 10^{-7})$$
$$= 1.00 \times 10^{-14}$$

K_w has the same value for every aqueous solution, regardless of whether an acid, base, or salt is present. If an acid is present so that the hydronium ion concentration is high, then the hydroxide ion concentration of the solution is low. By using K_w, an unknown H_3O^+ concentration can be found from a known OH^- concentration, and vice versa. For example, for a 0.10 M HCl solution, where $[H_3O^+] = 0.10$ mol/L because HCl is a strong acid and 100% dissociated, $[OH^-]$ is 1.0×10^{-13} mol/L.

$$[OH^-] = \frac{K_w}{[H_3O^+]} = \frac{1.00 \times 10^{-14}}{0.10} = 1.0 \times 10^{-13} \text{ mol/L}$$

Similarly, if a base is present so that the hydroxide ion concentration is high, then the hydronium ion concentration of the solution is low.

Solutions are identified as acidic, neutral, or alkaline (basic) by their H_3O^+ and OH^- concentrations (at 25°C):

- Acidic solution: $[H_3O^+] > 1 \times 10^{-7}$ mol/L $([H_3O^+] > [OH^-])$
- **Neutral solution:** $[H_3O^+] = 1 \times 10^{-7}$ mol/L $([H_3O^+] = [OH^-])$
- Alkaline solution: $[H_3O^+] < 1 \times 10^{-7}$ mol/L $([H_3O^+] < [OH^-])$

Solved Problem 11.2 What is $[OH^-]$ in milk if $[H_3O^+] = 4.5 \times 10^{-7}$? Is milk acidic or alkaline?

Solution Milk is slightly acidic, as shown by the value of $[H_3O^+]$, which is slightly larger than 1×10^{-7}. Since the product of $[H_3O^+]$ and $[OH^-]$ for any solution is 1.00×10^{-14}, solving for $[OH^-]$ gives

$$[H_3O^+][OH^-] = (4.5 \times 10^{-7}) \times [OH^-] = 1.00 \times 10^{-14}$$

$$[OH^-] = \frac{1.00 \times 10^{-14}}{4.5 \times 10^{-7}} = 2.2 \times 10^{-8} \text{ mol/L}$$

Practice Problem 11.14 What is $[OH^-]$ in these solutions? Identify each as either acidic or basic:
(a) beer, $[H_3O^+] = 3.2 \times 10^{-5}$ mol/L
(b) household ammonia, $[H_3O^+] = 3.1 \times 10^{-12}$ mol/L

11.11 MEASURING ACIDITY IN AQUEOUS SOLUTIONS: pH

pH A number usually between 0 and 14 that describes the acidity of an aqueous solution; mathematically, the negative common logarithm of a solution's H_3O^+ concentration.

Acidic solution A solution in which pH is less than 7 and hydronium ion concentration is larger than hydroxide ion concentration.

Neutral solution A solution in which pH is equal to 7 and hydronium ion and hydroxide ion concentrations are equal.

Alkaline solution A solution in which pH is greater than 7 and hydroxide ion concentration is greater than hydronium ion concentration.

In medicine, chemistry, and many other fields, it's often necessary to know the exact concentration of H_3O^+ or OH^- ions in solution. If, for example, the hydronium ion concentration of blood varies slightly from 4.0×10^{-8} mol/L, death can result.

Although perfectly correct, it's nevertheless awkward to refer to low concentrations of H_3O^+ by using moles per liter. If you were asked which concentration is higher, 9.0×10^{-8} mol/L or 3.5×10^{-7} mol/L, you'd probably have to stop and think for a minute before answering. Fortunately, there's an easier way to express and compare $[H_3O^+]$—the pH method.

The pH of most aqueous solutions is simply a number between 0 and 14 that indicates the solution's acid strength. A pH less than 7 corresponds to an acid solution; a pH larger than 7 corresponds to an alkaline solution; and a pH of 7 corresponds to a neutral solution. The pH scale and pH values of some common substances are shown in Figure 11.4. The smaller the pH, the more acid the solution; the larger the pH, the more alkaline the solution.

pH is defined as the negative common logarithm of the H_3O^+ concentration in moles per liter.

$$pH = -\log[H_3O^+]$$

If you've studied logarithms, you know that the common logarithm of a number is the power to which 10 must be raised to equal the number. The pH definition can therefore be restated as

$$[H_3O^+] = 10^{-pH}$$

For example, in neutral water (at 25°C) where $[H_3O^+] = 10^{-7}$, the pH is 7; in a strong acid solution, where $[H_3O^+] = 10^{-1}$, the pH is 1; and in a strong base solution, where $[H_3O^+] = 10^{-14}$, the pH is 14.

Figure 11.4
The pH scale and the pH values of some common substances. A low pH value corresponds to a strongly acidic solution; a high pH value corresponds to a strongly alkaline solution; and a pH of 7 corresponds to a neutral solution.

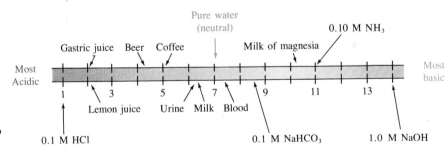

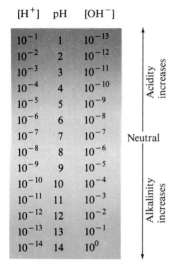

$[H^+]$	pH	$[OH^-]$	
10^{-1}	1	10^{-13}	
10^{-2}	2	10^{-12}	
10^{-3}	3	10^{-11}	Acidity increases
10^{-4}	4	10^{-10}	
10^{-5}	5	10^{-9}	
10^{-6}	6	10^{-8}	
10^{-7}	7	10^{-7}	Neutral
10^{-8}	8	10^{-6}	
10^{-9}	9	10^{-5}	
10^{-10}	10	10^{-4}	Alkalinity increases
10^{-11}	11	10^{-3}	
10^{-12}	12	10^{-2}	
10^{-13}	13	10^{-1}	
10^{-14}	14	10^{0}	

Figure 11.5
Relationship of the pH scale to hydrogen ion and hydroxide ion concentrations. Note that the sum of the $[H^+]$ and $[OH^-]$ exponents is always -14 because as one concentration increases, the other decreases so that their product equals 1×10^{-14}.

- **Acidic solution:** pH < 7   ($[H_3O^+] > 1 \times 10^{-7}$)
- **Neutral solution:** pH = 7   ($[H_3O^+] = 1 \times 10^{-7}$)
- **Alkaline solution:** pH > 7   ($[H_3O^+] < 1 \times 10^{-7}$)

To find pH, $[H_3O^+]$ can first be expressed in scientific notation. If you want to know the pH of a 0.0001 M solution of HCl, for which $[H_3O^+] = $ 0.0001 mol/L,

$$[H_3O^+] = 0.0001 = 1 \times 10^{-4} \text{ mol/L}$$
$$pH = 4.$$

It's easy to underestimate the difference in acidity or alkalinity with each pH unit. Keep in mind that the pH scale covers an enormous range of acidities because pH is a *logarithmic scale* (Figure 11.5) that involves powers of 10. A change of only 1 pH unit means a *10-fold* change in $[H_3O^+]$; a change of 2 pH units means a *100-fold* change in $[H_3O^+]$; and a change of 12 pH units, from a strong acid solution with pH 1 to a strong base solution with pH 13, means a change of 10^{12} (a million million) in $[H_3O^+]$.

It might help give you a feeling for the size of the quantities involved to think of a backyard swimming pool that contains about 100,000 L water. You would have to add only 0.10 mol of HCl (3.7 g) to lower the pH of the pool from 7.0 (neutral) to 6.0, but you would have to add 10,000 mol of HCl (370 kg!) to lower the pH of the pool to 1.0.

Solved Problem 11.3   Methyl red is an acid–base indicator that changes color at an $[H_3O^+]$ of about 1×10^{-5} mol/L. What pH is this?

Solution   Since the pH is the negative of the exponent of 10, the pH is 5.

Solved Problem 11.4   Lemon juice has a pH of about 2. What $[H_3O^+]$ is this?

Solution   Taking 2 as the negative exponent of 10 gives
$$[H_3O^+] = 1 \times 10^{-2} \text{ mol/L}$$

Solved Problem 11.5   A cleaning solution was found to have $[OH^-] = 1 \times 10^{-3}$. What is the pH of this solution?

Solution   To find pH, the value of $[H_3O^+]$ must first be found by using K_w.
$$K_w = [H_3O^+][OH^-] = 1 \times 10^{-14}$$
$$[H_3O^+][1 \times 10^{-3}] = 1 \times 10^{-14}$$
$$[H_3O^+] = \frac{1 \times 10^{-14}}{1 \times 10^{-3}} = 1 \times 10^{-11}$$
$$pH = 11$$

Practice Problems 11.15   Give the pH of solutions with the following concentrations:
(a) $[H_3O^+] = 1 \times 10^{-5}$ mol/L
(b) $[OH^-] = 1 \times 10^{-9}$ mol/L

11.16   Give the hydronium ion concentrations of solutions with (a) pH 13, (b) pH 3, (c) pH 8. Which of these solutions is most acidic? Which is most alkaline?

11.12 WORKING WITH pH

Converting between pH and $[H_3O^+]$ is simple when the pH is a whole number, but how do you find $[H_3O^+]$ of blood, which has a pH of 7.4, or the pH of a solution with $[H_3O^+] = 4.6 \times 10^{-3}$? Sometimes it's sufficient to make an estimate. The pH of blood, 7.4, is between 7 and 8, so the hydronium ion concentration of blood must be between 10^{-7} mol/L and 10^{-8} mol/L. To be exact about using pH values requires either a calculator or a log table. You're most likely to use a calculator, but we've included a brief explanation of a log table to help you understand what logs mean.

Working pH Problems with a Calculator   Converting hydronium ion concentration to pH on most scientific calculators is done with a "log" key and an "expo" or "e" key plus a "(−)" key for indicating negative exponents of 10. The number may be entered before or after pushing the "log" key. Consult your calculator instructions if you're not sure how to use these keys. The number given by the calculator must be converted to pH by changing the sign from minus to plus.

Finding the pH equivalent to $[H_3O^+] = 2 \times 10^{-3}$ and $[H_3O^+] = 4.6 \times 10^{-3}$ gives the following results:

$$pH = -[\log(2 \times 10^{-3})] = -(-2.698970) = 2.7$$

$$pH = -[\log(4.6 \times 10^{-3})] = -(-2.337242) = 2.34$$

A note about significant figures: The log is reported with the same number of digits *after* the decimal point as in the whole original number: log 2 $\times 10^{-3}$ = -2.7; log 4.6 $\times 10^{-3}$ = -2.34; log 4.65 $\times 10^{-3}$ = -2.332.

Table 11.5
Common Logarithms of Whole Numbers

log 1.0 = 0.00
log 2.0 = 0.30
log 3.0 = 0.48
log 4.0 = 0.60
log 5.0 = 0.70
log 6.0 = 0.78
log 7.0 = 0.85
log 8.0 = 0.90
log 9.0 = 0.95
log 10.0 = 1.000

Working pH Problems With a Log Table   Table 11.5 is a small table for finding common logarithms of whole numbers with one or two significant digits, results that are accurate enough for many purposes. To use a log table, remember that multiplying numbers is equivalent to adding their logs, that is,

$$\log (a \times b) = \log a + \log b$$

which for a number in scientific notation gives

$$\log (a \times 10^{-b}) = \log a + \log 10^{-b}$$

AN APPLICATION: ACID RAIN

The problem of acid rain has emerged as one of the most important environmental issues of recent times. Both the causes and the effects of acid rain are well understood. The problem is what to do about it.

As the water that evaporates from oceans and lakes condenses into raindrops, it dissolves small quantities of gases from the atmosphere. Under normal conditions, rain is slightly acidic (pH 5.6) because of dissolved CO_2. In recent decades, however, the acidity of rainwater in many industrialized areas of the world has increased by a factor of over 100 to a pH of 3–3.5.

The culprit behind acid rain is industrial and automotive pollution. Each year in the United States and Canada, large coal-burning power plants and smelters pour millions of tons of sulfur dioxide (SO_2) gas into the atmosphere, where some is oxidized by air to produce sulfur trioxide (SO_3). Sulfur oxides then dissolve in rain to form sulfurous acid and sulfuric acid:

$$SO_2 + H_2O \longrightarrow H_2SO_3 \qquad \text{Sulfurous acid}$$

$$SO_3 + H_2O \longrightarrow H_2SO_4 \qquad \text{Sulfuric acid}$$

Nitrogen oxides produced both by coal-burning plants and by automobiles make a further contribution to the problem. Nitrogen dioxide (NO_2) dissolves in water to form nitric (HNO_3) and nitrous (HNO_2) acids:

$$2\,NO_2 + H_2O \longrightarrow HNO_3 + HNO_2$$

Many processes in nature require such a fine balance that they are dramatically upset by the shift that has occurred in the pH of rain. Many thousands of lakes in the Adirondack region of upper New York State and in southeastern Canada have become so acidic that all fish life has disappeared; massive tree die-offs have occurred throughout Central Europe; and countless statues are being slowly dissolved away. What should be done about it?

These scientists are monitoring acid rain and its effects on trees.

To find the pH equivalent to $[H_3O^+] = 2 \times 10^{-3}$,

$$pH = -[\log (2 \times 10^{-3})] = -[\log 2 + \log 10^{-3}]$$

Read from the table   Exponent of 10

$$= -[0.3 + (-3)] = 2.7$$

Practice Problems ***11.17***   Identify each of the following solutions as acidic or basic and rank them in order of increasing acidity:
(a) saliva, pH = 6.5   (b) pancreatic juice, pH = 7.9
(c) orange juice, pH = 3.7   (d) wine, pH = 3.5

11.18   Estimate $[H_3O^+]$ values for each of the solutions in Problem 11.17.

11.19   Find the pH of the following solutions:
(a) seawater with $[H_3O^+] = 5.3 \times 10^{-9}$ mol/L
(b) a urine sample with $[H_3O^+] = 8.9 \times 10^{-6}$ mol/L

11.13 LABORATORY DETERMINATION OF ACIDITY

There are two common ways to measure the pH of a solution. The simpler but less accurate method is to use an acid–base indicator. For example, the well-known dye *litmus* is red below pH 4.8 but blue above pH 7.8; phenolphthalein (fee-nol-**thay**-lean) is colorless below pH 8.2 but red above pH 10; and so on, as shown in Figure 11.6. Paper strips ("pH paper") impregnated with a combination of different indicators allow one to determine pH simply by putting a drop of solution on the paper and comparing the color that appears to the color on a calibration chart (Figure 11.7).

A second and more accurate method of determining pH is to use an electronic pH meter like the one shown in Figure 11.8. An electrode is dipped into the solution, and the meter provides an accurate pH reading.

Figure 11.6
Some acid–base indicators and their color changes. The change in color with pH (written on the tubes) is shown for (left to right) phenolphthalein, bromthymol blue, and methyl red.

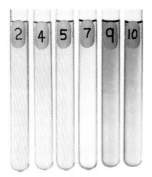

Figure 11.7
Testing pH with a paper strip. Comparison of the color with the code on the strip gives the approximate pH.

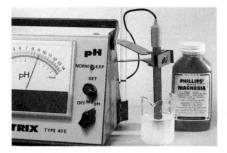

Figure 11.8
Using a pH meter to obtain an accurate reading of pH. Is milk of magnesia acidic or alkaline?

11.14 BUFFER SOLUTIONS

The pH of a solution can be kept relatively constant through the use of a **buffer:** a combination of substances that act together to prevent a drastic change in the pH of a solution. Most common buffer solutions are mixtures of a weak acid and the anion of that acid. When acid is added to the solution, it reacts with the basic anion:

$$H_3O^+ + A^- \longrightarrow HA(aq) + H_2O(l)$$

When base is added to the solution, it reacts with the acid:

$$OH^- + HA(aq) \longrightarrow A^- + H_2O(l)$$

Buffer A combination of substances, usually a weak acid and its anion, that act together to prevent a large change in the pH of a solution.

An effective buffer solution contains roughly equal concentrations of the acid and its anion. The 0.1 M acetic acid–0.1 M acetate ion buffer is a good example. At equilibrium in this buffer solution the pH is 4.74. Rearranging the equilibrium constant expression for acetic acid shows that the hydronium ion concentration, and therefore the pH, is dependent on the ratio of acid to anion concentrations:

$$K_a = \frac{[H^+][CH_3COO^-]}{[CH_3COOH]}$$

$$[H^+] = K_a \frac{[CH_3COOH]}{[CH_3COO^-]}$$

When acid is added to the acetic acid–acetate buffer

$$\overleftarrow{\hspace{3cm}} H^+$$
$$CH_3COOH \rightleftharpoons CH_3COO^- + H^+$$

the equilibrium is displaced so that the concentration of CH_3COOH increases and that of CH_3COO^- decreases. But as long as the changes in $[CH_3COOH]$ and $[CH_3COO^-]$ are small relative to their original values, the value of $[CH_3COOH]/[CH_3COO^-]$ doesn't change much and there is little change in the pH.

When base is added to this system

$$OH^- \longrightarrow$$

$$CH_3COOH \rightleftharpoons CH_3COO^- + H^+$$

it reacts with H^+ and the equilibrium is shifted to the right (H^+ is removed), causing the concentration of CH_3COO^- to increase and the concentration of CH_3COOH to decrease. Here also, as long as the concentration changes are small, there is little change in the pH.

The dramatic effect caused by addition of acetate ion to an acetic acid solution to produce a buffer solution is illustrated by the pH changes when 0.01 mol of H^+ or OH^- is added to water or the buffer solution:

	pH of 1 L of Water	pH of 1 L of 0.1 M CH_3COOH– 0.1 M CH_3COO^- Buffer
	pH 7	pH 4.74
Plus 0.01 mol H^+ gives	pH 2	pH 4.68
Plus 0.01 mol OH^- gives	pH 12	pH 4.85

The choice of acid and the concentration ratio control the pH range in which a buffer is effective. To make this apparent, we can take the negative logarithm of the general equation

$$[H^+] = K_a \frac{[HA]}{[A^-]}$$

$$pH = pK_a - \log\left(\frac{[HA]}{[A^-]}\right)$$

and rearrange the result as follows

$$pH = pK_a + \log\left(\frac{[A^-]}{[HA]}\right)$$

Henderson-Hasselbalch equation Equation that gives pH of buffer solution containing weak acid (HA) and its anion (A^-); pH = pK_a + log([A$^-$]/[HA]).

to give what is known as the **Henderson-Hasselbalch equation** in which pK_a is the negative logarithm of the K_a. Clearly, the buffer pH depends on the pK_a of the acid chosen and on the concentration ratio. (With little error, $[A^-]$ and $[HA]$ are taken as the initial concentrations in the buffer instead of the concentrations at equilibrium.)

When the $[A^-]/[HA]$ ratio is equal to 1, the pH of a buffer solution is equal to the pK_a of the acid. For example, we noted above that the pH of an acetate buffer with a concentration ratio of 1 is 4.74. Taking the negative logarithm of the K_a for acetic acid in exactly the same manner as finding a pH from $[H^+]$ shows the basis for this buffer's pH:

$$K_a = 1.8 \times 10^{-5}$$

$$pK_a = -\log K_a = -[\log(1.8 \times 10^{-5})] = -[-4.744727] = 4.74$$

In general, a buffer is expected to be effective within a pH range of its initial pH plus or minus 1. The buffering effect is overcome when such large quantities of acid or base are added that the [A⁻]/[HA] ratio of a buffer undergoes a large change.

Factors that Influence pH Maintained by Buffer

- K_a value of acid
- [A⁻]/[HA] ratio
- pH range = initial pH ± 1.

Practice Problems **11.20** A mixture of HCl and NaCl is a very poor buffer system. Propose an explanation, taking into account the base strength of the Cl⁻ ion.

11.21 Calculate the pH of a buffer solution that contains equal concentrations of dihydrogen phosphate ion ($H_2PO_4^-$) and hydrogen phosphate ion (HPO_4^{2-}) using the Henderson-Hasselbalch equation and the K_a value in Table 11.4. Write the equations for the reaction of H⁺ and OH⁻ with this buffer.

11.15 BUFFERS IN THE BODY

The pH of body fluids is maintained by interactions among buffer systems, breathing, and kidney function. Two of the three major buffer systems in the body are the bicarbonate and phosphate systems, which depend on weak acid–conjugate base interactions exactly like those of the acetate buffer system described in the preceding section,

$$H_2CO_3(aq) \rightleftharpoons H^+ + HCO_3^-$$

$$H_2PO_4^-(aq) \rightleftharpoons H^+ + HPO_4^{2-}$$

The third buffer system depends on the ability of proteins to act as either proton acceptors or proton donors at different solution pH values.

To illustrate buffers in action in the body, we'll look at the bicarbonate system—the principal buffer system in blood serum and other extracellular fluids. (The phosphate system is the major buffer within cells.)

Since carbonic acid is unstable and therefore always in equilibrium with carbon dioxide and water,

$$H_2CO_3(aq) \rightleftharpoons H_2O(l) + CO_2(g)$$

there is an extra step in the bicarbonate buffer mechanism.

$$CO_2(aq) + H_2O(l) \rightleftharpoons H_2CO_3(aq) \rightleftharpoons H^+ + HCO_3^-$$

As a result, the bicarbonate buffer system is intimately related to the elimination of carbon dioxide, which is continuously produced in cells and transported to the lungs to be exhaled. Carbonic acid, although present in very low concentration, is the intermediary in this process.

Since it's mainly CO_2 dissolved in blood rather than H_2CO_3, $[CO_2]$ is used in the Henderson-Hasselbalch equation for the bicarbonate buffer in blood:

$$pH = pK_a + \log\left(\frac{[HCO_3^-]}{[CO_2]}\right)$$

Increase in $[CO_2]$ lowers pH.
Decrease in $[CO_2]$ raises pH.

You can see that increasing $[CO_2]$ causes a decrease in pH (blood becomes more acidic), and decreasing $[CO_2]$ causes an increase in pH (blood becomes more alkaline). At the normal blood pH of 7.4, the equation is

$$7.4 = 6.1 + \log\left(\frac{[HCO_3^-]}{[CO_2]}\right)$$

and shows, on solving for $[HCO_3^-]/[CO_2]$, that the bicarbonate/carbon dioxide ratio in blood is normally 20 to 1.

Figure 11.9 represents the relationships between the bicarbonate buffer system, the lungs, and the kidneys. Under normal circumstances the reactions shown in Figure 11.9 are in equilibrium. Addition of excess H^+ (red arrows) causes formation of H_2CO_3 and results in lowering of H^+ concentration. Removal of H^+ (blue arrows) causes replacement of H^+ by dissociation of H_2CO_3. The maintenance of pH by this mechanism is supported by a reserve of bicarbonate ions in body fluids. In other words, the normal concentration of HCO_3^- ions in body fluids is larger than that of H^+. Such a buffer can accommodate larger additions of H^+ before there is a significant change in the pH, a condition that meets the body's needs because production of excess acid in the body is a more common condition than excessive loss of H^+ or production of OH^-.

A change in the breathing rate provides a quick further adjustment in the bicarbonate buffer system. Receptors monitor the concentration of CO_2 in the blood. When it rises, breathing rate increases, removing CO_2 and decreasing H^+ concentration (red arrows in Figure 11.9). When CO_2 concentration falls, breathing rate decreases and H^+ concentration increases (blue arrows in Figure 11.9).

Figure 11.9
Relationships of the bicarbonate buffer system to the lungs, the kidneys, and other buffer systems. The red and blue arrows show responses to the stresses of increased or decreased respiratory rate and removal or addition of H^+ due to metabolic changes.

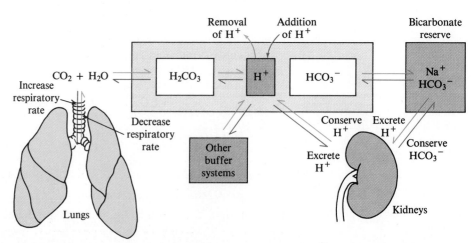

The kidneys provide additional backup to the bicarbonate buffer system. Each day a quantity of H^+ equal to the quantity produced in the body, about 100 mEq, is excreted in the urine. In the process of sending H^+ into the urine, the kidney returns HCO_3^- to the extracellular fluids, where it becomes part of the bicarbonate reserve.

The regulation of blood pH by the bicarbonate buffer system is particularly important in preventing the medical conditions called *acidosis* and *alkalosis*. **Acidosis,** which results when blood pH drops below 7.35, leads in mild cases to fainting and depression of the nervous system, and in more severe cases to coma. It can be brought on by the difficulty in breathing that accompanies asthma and emphysema, by metabolic irregularities due to fasting, and by the other conditions listed in Table 11.6. Breathing difficulties lower blood pH by drastically raising blood CO_2 concentration (see Figure 11.9), whereas fasting lowers blood pH by causing the overproduction of blood-soluble acidic substances. The body attempts to compensate for acidosis by stimulation of respiration to remove CO_2, increased excretion of H^+ in the urine, and increased return of HCO_3^- to body fluids by the kidneys.

Alkalosis, the condition that results when blood pH rises above 7.45, leads to stimulation of the nervous system and to subsequent convulsions. Alkalosis arises either from heavy breathing (**hyperventilation**) that removes large amounts of CO_2 from the blood or from conditions that raise the concentration of bicarbonate ions, which then deplete the supply of hydrogen ions. Persistent vomiting, for example, activates production of more hydrochloric acid in the stomach, which is accompanied by the "alkaline tide," the flooding of bicarbonate ions into extracellular fluids. The body attempts to compensate for alkalosis

Acidosis The medical condition that results when blood pH drops below 7.35.

Alkalosis The medical condition that results when blood pH rises above 7.45.

Hyperventilation Heavy breathing that depletes carbon dioxide from the lungs and blood.

Table 11.6   Causes of Acidosis and Alkalosis

Type of Imbalance	Causes
Respiratory acidosis	CO_2 buildup due to—
	Decreased respiratory activity (hypoventilation)
	Cardiac insufficiency (e.g., congestive failure, cardiac arrest)
	Deterioration of pulmonary function (e.g., asthma, emphysema, pulmonary obstruction, pneumonia)
Respiratory alkalosis	Loss of CO_2 due to—
	Excessive respiratory activity (hyperventilation, due, for example, to high fever, nervous condition)
Metabolic acidosis	Increased production of metabolic acids due to—
	Fasting or starvation
	Diabetes
	Excessive exercise
	Decreased acid excretion in urine due to—
	Poisoning
	Renal failure
	Decreased plasma bicarbonate concentration due to—
	Diarrhea
Metabolic alkalosis	Elevated plasma bicarbonate concentration due to—
	Vomiting
	Diuretics
	Antacid overdose

by slowing the respiration rate and increasing the excretion of bicarbonate ion. An effective treatment for hyperventilation is breathing into a paper bag, which forces the return of CO_2 to the body. When this approach fails, the body usually takes over, causing fainting and a resultant slower respiration rate.

11.16 EQUIVALENTS OF ACIDS AND BASES: NORMALITY

Equivalent, acid (base)
The amount in grams of an acid (or base) that can donate one mole of H^+ (or OH^-) ions.

You saw in the last chapter (Section 10.12) that it's sometimes useful to think in terms of ion equivalents. Since some acids have more than one H^+ per formula unit and some bases have more than one OH^- per formula unit, it's also often useful to consider **acid** or **base equivalents.**

One equivalent of an acid is equal to the molar mass of the acid divided by the number of H^+ ions produced per formula unit. Thus, one equivalent of an acid is the weight in grams that can donate one mole of H^+ ions. Similarly, one equivalent of a base is the weight in grams that can produce one mole of OH^- ions:

$$\textbf{One equivalent of acid} = \frac{\text{molar mass}}{\text{number of } H^+ \text{ ions produced}}$$

$$\textbf{One equivalent of base} = \frac{\text{molar mass}}{\text{number of } OH^- \text{ ions produced}}$$

One equivalent of the monoprotic acid HCl is 36.5 g, the molar mass of the acid, but one equivalent of the diprotic acid H_2SO_4 is 49 g, the molar mass of the acid (98 g) divided by 2.

$$\text{One equivalent } H_2SO_4 = \frac{\text{molar mass } H_2SO_4}{2} = \frac{98 \text{ g}}{2} = 49 \text{ g}$$

Since H_2SO_4 is diprotic . . .   divide by 2

Acid–base equivalents have two practical advantages: First, they are useful when only the acidity or alkalinity of a solution, rather than the identity of the acid or base, is of interest. Second, they show quantities that are chemically equivalent: 36.5 g of HCl and 49 g of H_2SO_4 are chemically equivalent quantities because they each react with one equivalent of base. *One equivalent of any acid neutralizes one equivalent of any base.*

Because acid–base equivalents are so useful, acid and base concentrations are sometimes expressed in normality (N) rather than in molarity. The **normality** of an acid or base solution is defined as the number of equivalents of acid or base per liter of solution. For example, a solution made by dissolving 1.0 equivalent (36.6 g) of HCl in 1.0 L of water has a concentration of 1.0 Eq/L, or 1.0 N.

Normality   A measure of acid (or base) concentration expressed as the number of acid (or base) equivalents per liter of solution.

$$\text{Normality (N)} = \frac{\text{equivalents of acid or base}}{1 \text{ L solution}}$$

In fields where it's necessary to express small concentrations of acids or bases in equivalents, the milliequivalent is often used.

$$1000 \text{ milliequivalents (mEq)} = 1 \text{ equivalent}$$

Especially in clinical chemistry, the concentrations of acids or bases may be reported in *milliequivalents per liter* rather than equivalents per liter. A solution that contains 0.1 Eq of acid per liter is 0.01 N and has an acid concentration of 10 mEq/L (0.01 Eq × 1000 mEq/Eq).

The values of molarity (M) and normality (N) are the same for monoprotic acids such as HCl but are not the same for diprotic or triprotic acids. A solution made by dissolving 1.0 equivalent (49 g, 0.50 mol) of the diprotic acid H_2SO_4 in 1.0 L of water has a *normality* of 1.0 Eq/L, or 1.0 N, but a *molarity* of 0.50 mol/L, or 0.50 M. For any acid (or base), normality is always equal to molarity times the number of H^+ (or OH^-) ions produced per formula unit.

Normality of acid = (molarity of acid) × (number of H^+ ions produced)

Normality of base = (molarity of base) × (number of OH^- ions produced)

Solved Problem 11.6 What is the normality of the solution made by diluting a volume of concentrated sulfuric acid containing 6.5 g of H_2SO_4 to a volume of 200 mL? What is the concentration of this solution in milliequivalents per liter?

Solution First calculate how many equivalents of H_2SO_4 are in 6.5 g by using the equivalent weight of the acid, 49 g/Eq, as a conversion factor.

$$6.5 \text{ g } H_2SO_4 \times \frac{1 \text{ Eq } H_2SO_4}{49 \text{ g } H_2SO_4} = 0.13 \text{ Eq } H_2SO_4$$

Next, determine the concentration of the acid:

$$\frac{0.13 \text{ Eq } H_2SO_4}{200 \text{ mL}} \times \frac{1000 \text{ mL}}{1 \text{ L}} = 0.65 \text{ Eq } H_2SO_4/L = 0.65 \text{ N}$$

Thus, the concentration of the sulfuric acid solution is 0.65 N, or 650 mEq/L.

Practice Problems 11.22 How many equivalents are in each of the following samples?
(a) 5.0 g HNO_3 (b) 12.5 g $Ca(OH)_2$ (c) 4.5 g H_3PO_4

11.23 What would the normalities of the solutions be if each of the samples in Problem 11.22 were dissolved in water and diluted to a volume of 300. mL?

11.24 (a) How many equivalents of HCl are in 37 mL of 0.50 N HCl?
(b) How many equivalents of NaOH will neutralize this HCl solution?

11.17 TITRATION

Determining the pH of a solution gives the solution's hydronium ion concentration but not necessarily its total acid concentration. For example, in a 0.10 M solution of acetic acid, the total acid concentration is 0.10 mol/L, yet the hydronium ion concentration is only 0.0013 mol/L (pH 2.9) because acetic acid is only about 1% dissociated.

Titration An experimental method for determining acid (or base) concentration by neutralizing a sample with a base (or acid) of known concentration.

End point The point at which neutralization is made visible in a titration.

The total acid or base concentration of a solution is found by carrying out a **titration** (Figure 11.10). A measured volume of a solution of unknown acid or base concentration is placed in a flask, and an indicator is added. If the unknown solution is an acid, then a solution of known base concentration is slowly added until the indicator changes color to signal neutralization. If the unknown solution is a base, then a solution of known acid concentration is slowly added until neutralization is complete. The point at which neutralization is made visible in a titration is known as the **end point.** From the total volume of acid (or base) solution needed to reach the end point, the concentration of the unknown base (or acid) can be calculated.

The necessary calculations for a titration can be done with moles and molarity, but equivalents and normality are often used instead and simplify the process. Since the volume of a solution times its normality gives the number of equivalents

$$\text{Volume (L)} \times \text{normality} \left(\frac{\text{equivalents}}{\text{liter}} \right) = \text{equivalents}$$

a convenient relation exists at the end point: The number of equivalents of H_3O^+ and of OH^- are equal. Letting V_A be the volume of acid and N_A its normality, and letting V_B be the volume of base and N_B its normality, gives at the end point

$$V_A N_A = V_B N_B$$

After an acid–base titration, three of these quantities will be known, and the fourth, either N_A or N_B, will be unknown. For convenience, the normalities are often stated in milliequivalents per milliliter rather than equivalents per liter as shown in Solved Problem 11.7.

Figure 11.10
Carrying out a titration. (a) A known volume of the acid (or base) solution to be analyzed is placed in the flask along with an indicator. (b) Base (or acid) of known concentration is then added drop by drop until (c) the color change of the indicator shows that neutralization is complete (the end point).

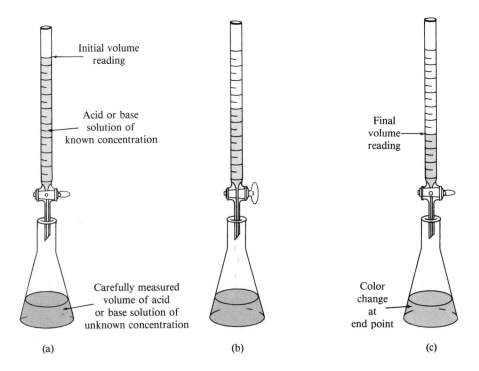

Initial volume reading

Acid or base solution of known concentration

Final volume reading

Carefully measured volume of acid or base solution of unknown concentration

Color change at end point

(a)   (b)   (c)

Solved Problem 11.7 When a 5.00 mL sample of household vinegar (dilute aqueous acetic acid) was titrated, 44.5 mL of 0.100 N NaOH solution was required to reach the end point. What is the acid concentration of vinegar, in normality, milliequivalents per liter, and molarity?

Solution The neutralization reaction is

$$CH_3COOH(aq) + NaOH(aq) \longrightarrow CH_3COO^-Na^+(aq) + H_2O(l)$$

An important consequence of using normality becomes apparent when we set out to solve this problem. The identities of the acid and base don't matter in calculating the unknown normality, because the equivalents of acid and base, whatever their identities, are equal at the end point.

$$V_A = 5.00 \text{ mL} \qquad N_A = ?$$
$$V_B = 44.5 \text{ mL} \qquad N_B = 0.100 \text{ N} = 0.100 \text{ mEq/mL}$$

$$V_A N_A = V_B N_B$$

$$(5.00 \text{ mL})(N_A) = (44.5 \text{ mL})(0.100 \text{ mEq/mL})$$

$$N_A = \frac{(44.5 \text{ mL})(0.100 \text{ mEq/mL})}{5.00 \text{ mL}} = 0.890 \text{ mEq/mL}$$

The acid is found to be 0.890 N, which means an acid concentration of 890 mEq/L. Since acetic acid is a monoprotic acid, the solution is also 0.890 M.

Practice Problems 11.25 In order to determine the concentration of the acid in an old bottle of aqueous HCl whose label had become unreadable, a titration was carried out. What is the HCl concentration if 58.4 mL of 0.250 N NaOH was required to titrate a 20.0 mL sample of the acid?

11.26 How many mL of 0.150 M NaOH are required to neutralize 50.0 mL of 0.200 M H_2SO_4?

11.27 A 21.5 mL sample of a KOH solution of unknown concentration required 16.1 mL of 0.15 M H_2SO_4 solution to reach the end point in a titration. What is the molarity of the KOH solution? (*Hint:* Solve the problem using normalities and then answer the question.)

11.18 ACIDITY AND ALKALINITY OF SALT SOLUTIONS

Dissolving sodium chloride produces a neutral solution, but this is not the case for many salts. Salt solutions may be neutral, acidic, or alkaline, depending on the ions present (Figure 11.11). To aid in predicting the acidity of salt solutions, it's convenient to classify salts according to the acid and base from which they would be formed in a neutralization reaction. The classification and some examples are given in Table 11.7.

(a)

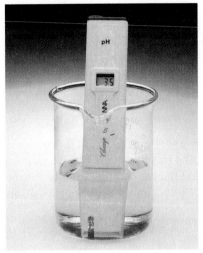
(b)

(c)

Figure 11.11
pH of salt solutions. Neutral, acidic, and basic salts are illustrated by solutions of (left to right) sodium chloride (NaCl), ammonium chloride (NH₄Cl), and sodium hypochlorite (NaClO). If you've ever spilled household bleach on your fingers and noticed that it's slippery, now you know why—it's a sodium hypochlorite solution.

Table 11.7   Acidity and Alkalinity of Salt Solutions

Base	Acid	Solution	Examples
Strong	Strong	Neutral	NaCl, K_2SO_4, $Ca(NO_3)_2$
Strong	Weak	Alkaline due to hydrolysis of anion	Na_2S, Na_2CO_3, $K(CH_3COO)$
Weak	Strong	Acidic due to hydrolysis of cation	NH_4Cl, $AlCl_3$, NH_4NO_3
Weak	Weak	Alkaline, acidic, or neutral	$NH_4(CH_3COO)$, NH_4CN

Salt of a Strong Base and a Strong Acid Sodium chloride is this type of salt because it can be formed by reaction of NaOH, a strong base, with HCl, a strong acid. Chloride ion, like the anions of the other strong acids, is so weak a base that it does not react with water. The Na^+ ion and the cations in other strong bases such as KOH and $Ca(OH)_2$ also undergo virtually no reaction with water. In general, halides salts formed by the strong hydroxide bases give neutral solutions.

Salt of a Strong Base and a Weak Acid Sodium acetate is this type of salt. The sodium ion does not react with water, but the acetate ion is a weak base that reacts with water to produce a weakly alkaline solution:

$$CH_3COO^- + H_2O(l) \rightleftharpoons CH_3COOH(aq) + OH^-$$

Hydrolysis reaction (ion)
The reaction of an ion with water.

Such reactions of ions with water are known as **hydrolysis reactions.** Any salt that contains the cation of a strong base and the basic anion from a weak acid will produce an alkaline solution. Sodium carbonate (known as *soda ash*), for example, gives such strongly alkaline solutions that it is used in cleaning powders and solutions.

Salt of a Weak Base and a Strong Acid The most common salts of this type are the ammonium salts with strong acids, for example, NH_4Cl. The ammonium ion is a weak acid and reacts with water to give an acidic solution:

$$NH_4^+ + H_2O(l) \rightleftharpoons NH_3(aq) + H_3O^+$$

The anions of strong acids (Table 11.3) undergo no hydrolysis and do not influence the acidity of a solution.

Salt of a Weak Base and a Weak Acid Both cation and anion in this type of salt react with water, so we can't predict whether the resulting solution will be acidic or alkaline without quantitative information. The ion that reacts to the greatest extent with water will govern the pH—it may be the cation *or* the anion.

Practice Problems 11.28 From the following, choose those salts that should produce alkaline solutions:
(a) K_2SO_4 (b) $AlCl_3$ (c) MgF_2 (d) NH_4Cl

11.29 From the salts listed in Practice Problem 11.28 choose those that should produce acidic solutions.

SUMMARY

Acids are substances that donate hydrogen ions (a proton, H^+); **bases** are substances that accept hydrogen ions. Thus, the generalized reaction of an acid with a base involves the reversible transfer of a proton:

$$A\!-\!H + :B \rightleftharpoons A^- + H\!-\!B^+$$

where $A\!-\!H$ = an acid and $:B$ = a base.

Reaction of an acid with a metal hydroxide such as KOH yields water and a salt; reaction with bicarbonate ion (HCO_3^-) or carbonate ion (CO_3^{2-}) yields water, a salt, and carbon dioxide gas; and reaction with ammonia yields an ammonium salt.

Acids do not spontaneously dissociate into H^+ and an anion, because a bare hydrogen ion is too unstable. In aqueous solution, however, water can react as a base and accept a proton from an acid to generate a **hydronium ion, H_3O^+.** Water can also donate a proton to a base, thereby reacting as an acid. Thus, water is **amphoteric.** The concentration of hydronium ion and hydroxide ion in any

aqueous solution is represented by the **ion product constant of water,** $K_w = [H_3O^+][OH^-] = 1 \times 10^{-14}$ (at 25°C).

Different acids and bases differ in their ability to give up or accept a proton. **Strong acids** give up a proton easily and are 100% dissociated in aqueous solution; **weak acids** give up a proton reluctantly, are only slightly **dissociated** in water, and establish equilibria between dissociated and undissociated acid. Similarly, **strong bases** accept and hold a proton readily, whereas **weak bases** have a low affinity for a proton and establish equilibria in aqueous solution. A **conjugate acid–base pair** are related by loss and gain of a proton. The stronger an acid, the weaker its conjugate base. A proton-transfer reaction always favors the formation of the weaker acid and base.

The acidity or alkalinity of an aqueous solution is given by its **pH,** which is the negative logarithm of its hydronium ion concentration, $[H_3O^+]$. A pH below 7 means an acidic solution; a pH equal to 7 means a neutral solution; and a pH above 7 means an alkaline solution.

Acid and base concentrations are often expressed in

units of **normality** (**N**) rather than molarity. The normality of a solution is the number of **equivalents** of H$^+$ or OH$^-$ ions present per liter of solution. One equivalent of an acid (or base) is the molar mass of the acid (or base), divided by the number of H$^+$ (or OH$^-$) ions that the acid (or base) produces. Acid (or base) concentrations are determined in the laboratory by **titrating** a solution of unknown concentration with a base (or acid) solution of known strength until an **indicator** signals that **neutralization** is complete.

Salt solutions can be neutral, acidic, or alkaline, depending on the extent to which their ions undergo **hydrolysis**. As shown in Table 11.7, salts can be classified according to the acids and bases from which they would form in a neutralization reaction.

The pH of a solution can be regulated through the use of a **buffer** that acts to remove either added H$^+$ ions or added OH$^-$ ions. The **carbonate buffer** present in blood and the **phosphate buffer** present in cells are particularly important examples.

INTERLUDE: pH OF BODY FLUIDS

Each body fluid is buffered to maintain a pH range suited to its function (see the accompanying table). The stability of cell membranes, the shapes of huge protein molecules that must be folded in certain ways to function, and the activities of enzymes are all dependent on appropriate hydrogen ion concentrations. The maintenance of homeostasis is, therefore, severely threatened by changes in pH.

Blood plasma and interstitial fluid, which together include one-third of body water, are both buffered to an average pH of 7.4, which is only very slightly basic. One of the functions of blood is to neutralize by its buffering action the acid by-products of cellular metabolism.

The strongly acidic environment in the stomach has three important functions. First, it aids in the digestion of proteins by causing them to *denature*, or unfold. Second, it kills most of the bacteria we consume along with our food. Third, it converts the enzyme that breaks down proteins from an inactive form to the active form.

When the acidic mixture of partially digested food (the *chyme*) leaves the stomach and enters the small intestine, it triggers secretion by the pancreas of an alkaline fluid containing bicarbonate ion. A principal function of this *pancreatic juice* and other fluids within the intestine is to dilute and neutralize the hydrochloric acid carried along from the stomach.

Urine has a wide normal pH range, depending on the diet and activities of the day. It is generally acidic, though, because one important function of urine is to eliminate a quantity of hydrogen ion equal to that produced by the body

each day. Without this elimination, the body's buffer systems would soon be overwhelmed by acid and unable to maintain homeostasis.

pH of Body Fluids

Blood plasma	7.4
Interstitial fluid (fluid surrounding cells)	7.4
Cytosol (fluid within cells)	7.0
Cerebrospinal fluid (fluid in brain and spinal cord)	7.3
Saliva	5.8–7.1
Gastric juice	1.6–1.8
Pancreatic juice	7.5–8.8
Intestinal juice	6.3–8.0
Urine	4.6–8.0
Sweat	4.0–6.8

An instrument for the electrochemical measurement of pH, P_{CO_2}, and P_{O_2} in blood. Such *blood-gas analysis* allows assessment of acid-base balance and the efficiency of oxygen transport.

REVIEW PROBLEMS

Acids and Bases

11.30 What is the molecular formula of each of the following substances?
(a) sulfuric acid   (b) nitric acid
(c) magnesium hydroxide   (d) aluminum hydroxide
(e) hydrofluoric acid   (f) potassium hydroxide

11.31 What happens when a strong acid such as HBr is dissolved in water?

11.32 What happens when a weak acid such as CH_3COOH is dissolved in water?

11.33 What happens when a strong base such as KOH is dissolved in water?

11.34 What is the difference between a monoprotic acid and a diprotic acid? Given an example of each.

11.35 Which of the following are strong acids? Consult Table 11.3 if necessary.
(a) $HClO_4$ (b) H_2CO_3 (c) H_3PO_4 (d) NH_4^+
(e) HI   (f) $H_2PO_4^-$

11.36 Which of the following are weak bases? Consult Table 11.3 if necessary.
(a) NH_3 (b) $Ca(OH)_2$ (c) HPO_4^{2-} (d) LiOH
(e) CN^- (f) NH_2^-

11.37 Write the equilibrium constant expressions for the three successive dissociations of phosphoric acid, H_3PO_4, in water.

Brønsted-Lowry Acids and Bases

11.38 For each of the following substances, tell whether it is a Brønsted-Lowry base, a Brønsted-Lowry acid, or neither.
(a) HCN   (b) $CH_3CO_2^-$ (c) $AlCl_3$ (d) H_2CO_3
(e) Mg^{2+} (f) $CH_3NH_3^+$

11.39 Label all Brønsted-Lowry acids and bases in the following equations. Show which substances make conjugate pairs.
(a) $CO_3^{2-}(aq) + HCl(aq) \rightarrow HCO_3^-(aq) + Cl^-(aq)$
(b) $H_3PO_4(aq) + NH_3(aq) \rightarrow H_2PO_4^-(aq) + NH_4^+(aq)$
(c) $NH_4^+(aq) + CN^-(aq) \rightleftharpoons NH_3(aq) + HCN(aq)$
(d) $HBr(aq) + OH^-(aq) \rightarrow H_2O(l) + Br^-(aq)$
(e) $H_2PO_4^-(aq) + N_2H_4(aq) \rightleftharpoons HPO_4^{2-}(aq) + N_2H_5^+(aq)$

11.40 Write the formulas of the conjugate acids of the following Brønsted-Lowry bases.
(a) $CH_2ClCO_2^-$ (b) C_5H_5N (c) SeO_4^{2-}
(d) $(CH_3)_3N$

11.41 Write the formulas of the conjugate bases of the following Brønsted-Lowry acids.
(a) HCN   (b) $(CH_3)_2NH_2^+$ (c) H_3PO_3
(d) $HSeO_3^-$

11.42 The hydrogen-containing anions of many polyprotic acids are amphoteric. Write equations for (a) HCO_3^- and (b) $H_2PO_4^-$ acting as bases with the strong acid HCl and as acids with the strong base NaOH.

11.43 Look at Table 11.3 and decide which in each pair is the stronger base.
(a) OH^- or PO_4^{3-} (b) Br^- or NO_2^-
(c) NH_3 or CH_3O^- (d) CN^- or H_2O
(e) I^- or HPO_4^{2-}

11.44 Write balanced equations for proton-transfer reactions between the listed pairs. Indicate the conjugate pairs and determine in which direction each equilibrium is favored.
(a) HCl and PO_4^{3-} (b) HCN and SO_4^{2-}
(c) $HClO_4$ and NO_2^- (d) CH_3O^- and HF

11.45 Tums, a drugstore remedy for acid indigestion, contains $CaCO_3$. Write an equation for the reaction of Tums with gastric juice (HCl).

11.46 Alka-Seltzer, a drugstore antacid, contains a mixture of $NaHCO_3$, aspirin, and citric acid, $C_6H_5O_7H_3$. Why does Alka-Seltzer foam and bubble when dissolved in water? Which ingredient is the antacid?

11.47 Write balanced equations for the following acid–base reactions:
(a) $LiOH + HNO_3 \rightarrow$
(b) $BaCO_3 + HI \rightarrow$
(c) $H_3PO_4 + KOH \rightarrow$
(d) $Ca(HCO_3)_2 + HCl \rightarrow$
(e) $Ba(OH)_2 + H_2SO_4 \rightarrow$

Acid and Base Strength: pH

11.48 Find K_a values in Table 11.4 and decide which acid in each of the following pairs is stronger.
(a) HCOOH or HF   (b) HSO_4^- or HCN
(c) $H_2PO_4^-$ or HPO_4^{2-}
(d) CH_3CH_2COOH or CH_3COOH

11.49 How is K_w defined? What is its numerical value (at 25°C)?

11.50 How is pH defined?

11.51 What color is phenolphthalein indicator at the following pHs? (See Figure 11.6.)
(a) pH 4.5   (b) pH 11.2   (c) pH 7.1

11.52 The electrode of a pH meter was placed in a sample of urine, and a reading of 7.9 was obtained. Is the sample acidic, basic, or neutral?

11.53 Normal gastric juice has a pH of about 2. Assuming that gastric juice is primarily aqueous HCl, what is the HCl concentration?

11.54 What is the approximate pH of a 0.10 M solution of a strong monoprotic acid?

11.55 A 0.10 N solution of the deadly poison hydrogen cyanide, HCN, has a pH of 5.2. Is HCN acidic or basic? Is it a strong or a weak acid?

11.56 What are the H_3O^+ concentrations of solutions with the following pH values?
(a) pH 4   (b) pH 11   (c) pH 0
(d) pH 1.38   (e) pH 7.96

11.57 What is the OH^- concentration of each solution in Problem 11.56?

11.58 Human spinal fluid has a pH of 7.4. Approximately what is the H_3O^+ concentration of spinal fluid?

11.59 Approximately what pHs do these H_3O^+ concentrations correspond to?
(a) fresh egg white: $[H_3O^+] = 2.5 \times 10^{-8}$ M
(b) apple cider: $[H_3O^+] = 5.0 \times 10^{-4}$ M
(c) household ammonia: $[H_3O^+] = 2.3 \times 10^{-12}$ M

11.60 What is the OH^- concentration for each solution in Problem 11.59? Rank the solutions in terms of increasing acidity.

Buffers

11.61 What are the two components of a buffer system? How does a buffer work to hold pH nearly constant?

11.62 Which system would you expect to be a better buffer: $HNO_3 + NaNO_3$, or $CH_3COOH + CH_3COO^-Na^+$? Explain.

11.63 The pH of a buffer solution containing 0.10 M acetic acid and 0.10 M sodium acetate is 5.60. Write the equations for reaction of this buffer with a small amount of HNO_3 and with a small amount of NaOH.

11.64 Blood serum generally has a pH of 7.4.
(a) Discuss a respiratory and a metabolic reason for which the pH might be greater than pH 7.45. What is this condition called?
(b) Discuss a respiratory and a metabolic reason for which the pH might be less than pH 7.45. What is this condition called?

Concentrations of Acid and Base Solutions

11.65 What does it mean when we talk about acid and base equivalents?

11.66 How is normality defined as a means of expressing acid or base concentration?

11.67 How many equivalents are in 500. mL of 0.50 M HNO_3? Of 0.50 M H_3PO_4?

11.68 How many equivalents of NaOH are needed to react with 0.035 equivalent of the triprotic acid H_3PO_4?

11.69 How many mL of 0.0050 N KOH are required to neutralize 25 mL of 0.0050 N H_2SO_4? To neutralize 25 mL of 0.0050 N HCl?

11.70 How would you prepare 250 mL of a 0.10 N HCl solution from a commercially available 12.0 M HCl solution?

11.71 What is the molarity of a 0.10 N H_3PO_4 solution?

11.72 Since hydrogen cyanide is a gas, it is easier to measure by volume than by weight. How many liters of HCN at STP are required to make 250 mL of 0.10 N solution?

11.73 How many equivalents are in each of the following?
(a) 0.25 mol $Mg(OH)_2$ (b) 2.5 g $Mg(OH)_2$
(c) 15 g CH_3COOH

11.74 What is the normality of these solutions?
(a) 0.75 M H_2SO_4 (b) 0.13 M $Ba(OH)_2$
(c) 1.4 M HF

11.75 What are the molarity and the normality of a solution made by dissolving 5.0 g $Ca(OH)_2$ in enough water to make 400. mL of solution?

11.76 What are the molarity and the normality of a solution made by dissolving 25 g of citric acid ($C_6H_5O_7H_3$, a triprotic acid) in enough water to make 750 mL of solution?

11.77 How many mL of the solution described in Problem 11.75 are needed to neutralize 25.0 mL of the solution described in Problem 11.76?

11.78 Titration of a 10.0 mL solution of KOH required 15.0 mL of 0.0250 M H_2SO_4 solution. What is the molarity of the KOH solution?

11.79 If 35 mL of a 0.10 N acid solution is needed to reach the end point in titration of 21.5 mL of a base solution, what is the normality of the base solution?

Applications

11.80 A popular antacid has $NaAl(OH)_2CO_3$ as the active ingredient. [App: Ulcers, Heartburn, and Antacids]
(a) Write a balanced equation for the reaction of this compound with HCl.
(b) How many grams of this antacid are required to neutralize 15.0 mL of 0.0955 M HCl?

11.81 Normal rain has a pH of about 5.6. What is $[H_3O^+]$ in normal rain? [App: Acid Rain]

11.82 Acid rain with a pH as low as 1.5 has been recorded in West Virginia. [App: Acid Rain]
(a) What is $[H_3O^+]$ in this acid rain?
(b) How many grams of HNO_3 must be dissolved to make 25 L of solution with a pH of 1.5?

11.83 Which body fluid is most acidic and which is most basic? [Int: pH of Body Fluids]

Additional Questions and Problems

11.84 How many mL of 0.50 M NaOH solution are required to titrate 40.0 mL of a 0.10 M H_2SO_4 solution to an end point?

11.85 How many equivalents of acid or base do each of the following samples contain? How many milliequivalents?
(a) 20.0 mL of 0.015 N HNO_3
(b) 170 mL of 0.025 N H_2SO_4

11.86 A 0.15 N solution of HCl is used to titrate 30.0 mL of a $Ca(OH)_2$ solution of unknown concentration. If 140 mL of HCl is required, what is the concentration of the $Ca(OH)_2$ solution?

11.87 Why doesn't pure water conduct electricity, even though it dissociates into H_3O^+ and OH^- ions?

11.88 Which solution contains more acid, 50 mL of a 0.20 N HCl solution, or 50 mL of a 0.20 N acetic acid solution? Which has a higher hydronium ion concentration? Which has a lower pH?

11.89 A 0.010 M solution of aspirin has pH 3.3. Is aspirin a strong or a weak acid?

11.90 Which of the following combinations produces an effective buffer solution?
(a) NaF and HF   (b) $HClO_4$ and $NaClO_4$
(c) NH_4Cl and NH_3 (d) KBr and HBr

11.91 One method of analyzing ammonium salts is to treat them with NaOH and then heat the solution to remove the NH_3 gas formed.

$$NH_4^+(aq) + OH^-(aq) \xrightarrow{\text{heat}} NH_3(g) + H_2O(l)$$

(a) Label the Brønsted-Lowry acid–base pairs.
(b) If 2.86 L of NH_3 at 60°C and 755 mm Hg is produced by the reaction of NH_4Cl, how many grams of NH_4Cl were in the original sample?

11.92 What is the pH of the solution formed by diluting 25.0 mL of 0.40 M NaOH to a final volume of 2.00 L?

C H A P T E R

12

Introduction to Organic Chemistry: Alkanes

Oil—to some it's a profitable investment, to others it's a source of energy, to still others it's a source of political power. To chemists, it's the principal raw material for creating useful organic compounds. In this chapter, you'll learn about the products of oil refineries like the one shown here.

As knowledge of chemistry slowly evolved in the 1700s, mysterious differences were noted between compounds obtained from animals and those obtained from minerals. Chemicals from animal sources were often more difficult to isolate, to purify, and to work with than those from mineral sources. To express this difference, the term *organic chemistry* was introduced to mean the study of compounds from living organisms, while *inorganic chemistry* was used to refer to the study of compounds from minerals.

Today we know there aren't any fundamental differences between organic and inorganic compounds: The same scientific principles are applicable to both. The only common characteristic of compounds from living sources is that they contain the element carbon.

Why is carbon special? Carbon atoms have the unique ability to bond together, forming long chains and rings. Of all the elements, only carbon is able to form such an immense array of compounds. In this and the next five chapters, we'll look at the chemistry of organic compounds, beginning with an exploration of these topics:

1. *How are organic molecules classified?* The goal: Be able to classify organic molecules into functional-group families.
2. *What are the structures of organic molecules?* The goal: Be able to recognize the main carbon chain in a molecule and identify constitutional isomers.
3. *How are organic molecules drawn?* The goal: Be able to convert between structural formulas and condensed or line structures.
4. *How are alkanes and cycloalkanes named?* The goal: Be able to name an alkane or cycloalkane from its structure or write the structure, given the name.
5. *What are the general properties of alkanes?* The goal: Be able to describe such properties as polarity, water solubility, flammability, toxicity, and chemical reactivity.
6. *What are the major chemical reactions of alkanes?* The goal: Be able to describe the products formed in combustion and halogenation of alkanes.

12.1 THE NATURE OF ORGANIC MOLECULES

Let's review what we've seen in earlier chapters about the structures of organic molecules:

● **Carbon is always tetravalent; it always forms four bonds** (Section 6.2). In methane, carbon is connected to four hydrogen atoms:

$$Methane, CH_4 \qquad H-\underset{\underset{H}{|}}{\overset{\overset{H}{|}}{C}}-H$$

● **Organic molecules contain covalent bonds** (Section 6.2). In ethane, the bonds result from the sharing of two electrons, either between C and C atoms or between C and H atoms:

$$Ethane, C_2H_6 \qquad H:\overset{\overset{\cdot\cdot}{H}}{\underset{\underset{\cdot\cdot}{H}}{C}}:\overset{\overset{\cdot\cdot}{H}}{\underset{\underset{\cdot\cdot}{H}}{C}}:H \quad = \quad H-\overset{\overset{H}{|}}{\underset{\underset{H}{|}}{C}}-\overset{\overset{H}{|}}{\underset{\underset{H}{|}}{C}}-H$$

● **Organic molecules contain polar covalent bonds when carbon bonds to an element on the far right or far left of the periodic table** (Section 6.8). In chloromethane, the electronegative chlorine atom attracts electrons more strongly than carbon, resulting in polarization of the carbon–chlorine bond so that carbon has a partial positive charge, δ^+:

$$Chloromethane, CH_3Cl \qquad H:\overset{\overset{H}{\cdot\cdot}}{\underset{\underset{H}{}}{C}}:\overset{\cdot\cdot}{\underset{\cdot\cdot}{Cl}} \quad = \quad H-\overset{\overset{H}{|}}{\underset{\underset{H}{|}}{C}}{}^{\delta+}-Cl^{\delta-}$$

● **Carbon can form multiple covalent bonds by sharing more than two electrons with a neighboring atom** (Section 6.3). In ethylene, the two carbon atoms share four electrons in a double bond; in acetylene, the two carbons share six electrons in a triple bond:

$$Ethylene, C_2H_4 \qquad \overset{H}{\underset{H}{\cdot}}C::C\overset{H}{\underset{H}{\cdot}} \quad = \quad \overset{H}{\underset{H}{}}C=C\overset{H}{\underset{H}{}}$$

$$Acetylene, C_2H_2 \qquad H:C:::C:H \quad = \quad H-C\equiv C-H$$

● **Covalently bonded molecules have specific three-dimensional shapes** (Section 6.7). When carbon is bonded to four atoms as in methane, CH_4, the bonds are oriented toward the four corners of an imaginary tetrahedron with carbon in the center:

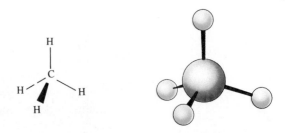

● **Hydrogen, nitrogen, and oxygen are the elements most often present in organic molecules in addition to carbon** (Section 6.6). Nitrogen and oxygen can form both single and multiple bonds to carbon.

$$C—N \qquad C—O \qquad C—H$$
$$C=N \qquad C=O$$
$$C\equiv N$$

As a result of their covalent bonding, organic compounds have properties quite different from those of many inorganic compounds. For example, inorganic salts have high melting points and boiling points because they consist of large collections of oppositely charged ions held together by strong electrical attractions (Section 5.3). Organic compounds, by contrast, consist of atoms joined by covalent bonds in individual molecules. Thus intermolecular forces, which are fairly weak, have an important influence on the properties of organic compounds. Organic compounds generally have lower melting and boiling points than inorganic salts. In fact, many simple organic compounds are liquids at room temperature, and a few are gases.

Solubility and electrical conductivity are other important differences between organic and inorganic compounds. Whereas many inorganic compounds dissolve in water to yield ions in solutions capable of conducting electricity, most organic compounds are insoluble in water and do not conduct electricity. Only small polar organic molecules, such as glucose and ethyl alcohol (both of which contain —OH groups), or large molecules with many polar groups, such as some proteins, dissolve in water.

The lack of water solubility of organic compounds has many important practical consequences, such as the difficulty in cleaning up greasy dirt and oil spilled in the ocean (Figure 12.1). In digestion, fats and other nonpolar organic molecules consumed in food must be made soluble in water before they can be used by the body.

Figure 12.1
Result of an oil spill in the Black Sea in February 1991 during the Gulf War between Iraq and a coalition of Western and Arab states. The oil washing up at Jubail, Saudi Arabia, was spilled in what was described by some as an act of "environmental terrorism."

12.2 FAMILIES OF ORGANIC MOLECULES: FUNCTIONAL GROUPS

Atoms of a relatively few different kinds of elements can bond together in a great many different ways in organic compounds. At last count, there were more than 10 million organic compounds described in the scientific literature. Each has unique physical properties such as melting point and boiling point, and each has unique chemical properties. The situation isn't as hopeless as it sounds, however, because chemists have learned through experience that organic compounds can be classified into families according to their structural features and that the chemical behavior of the members of a family is often predictable. Instead of 10 million compounds with random chemical reactivity, there are just a few general families of organic compounds whose chemistry is predictable.

Functional group A part of a larger molecule composed of an atom or group of atoms that has characteristic chemical behavior.

The structural features that allow us to class compounds together are called functional groups. A **functional group** is a part of a larger molecule and is composed of an atom or group of atoms that has characteristic chemical behavior. A given functional group undergoes the same reactions in every molecule it's a part of. For example, the carbon–carbon double bond is one of the simplest functional groups. Ethylene (C_2H_4), the simplest compound with a double bond, undergoes many chemical reactions similar to those of cholesterol ($C_{27}H_{46}O$), a far larger and more complex compound. Both, for example, react with hydrogen in the same manner, as shown in Figure 12.2. These identical reactions with hydrogen are typical: The chemistry of an organic molecule, regardless of size and complexity, is determined by the functional groups it contains.

Table 12.1 lists some of the most important families of organic molecules and their distinctive functional groups. All compounds that contain an —NH_2 group, for example, are known as amines and are members of the amine family.

Figure 12.2
The reactions of (a) ethylene and (b) cholesterol with hydrogen. The carbon–carbon double bond functional groups adds two hydrogen atoms in both cases, regardless of the complexity of the rest of the molecule.

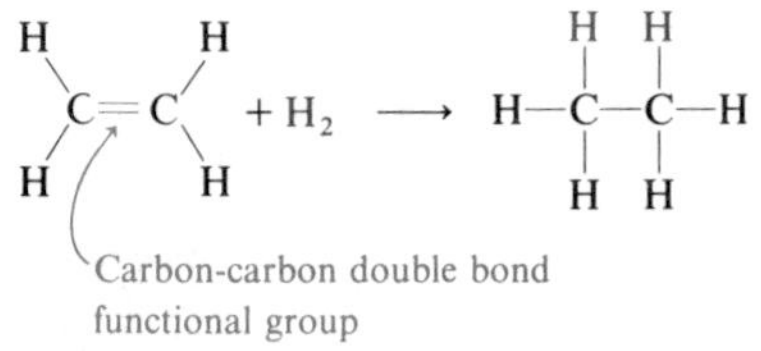

(a) Ethylene

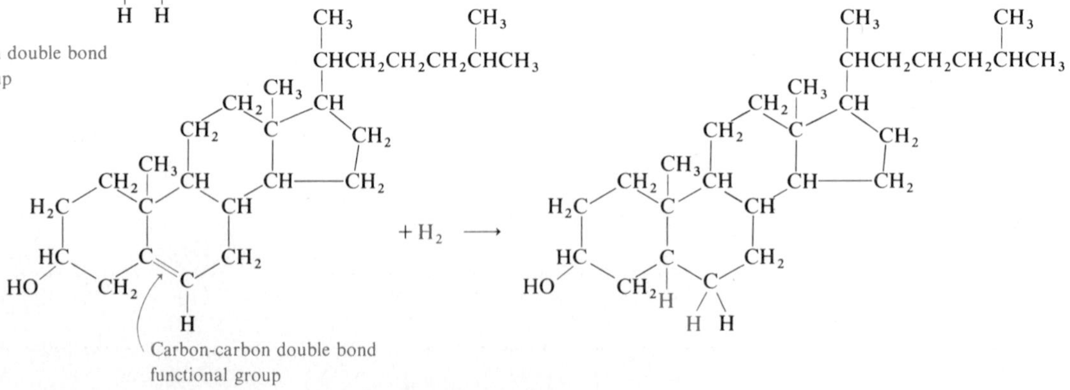

(b) Cholesterol

Table 12.1 Some Important Families of Organic Molecules

Family Name	Functional Group Structure[a]	Simple Example	Name Ending
Alkane	(contains only C—H and C—C single bonds)	CH_3CH_3 ethane	-ane
Alkene	$\underset{/}{\overset{\backslash}{C}}=\underset{\backslash}{\overset{/}{C}}$	$H_2C=CH_2$ ethylene	-ene
Alkyne	—C≡C—	H—C≡C—H acetylene (ethyne)	-yne
Arene	(benzene ring structure)	(benzene structure) benzene	none
Alkyl halide[b]	—C—X	CH_3—Cl methyl chloride	none
Alcohol	—C—O—H	CH_3—OH methyl alcohol (methanol)	-ol
Ether	—C—O—C—	CH_3—O—CH_3 dimethyl ether	none
Amine	—N—H, —N—H, —N—	CH_3—NH_2 methylamine	-amine
Aldehyde	—C(=O)—H	CH_3—C(=O)—H acetaldehyde (ethanal)	-al
Ketone	—C—C(=O)—C—	CH_3—C(=O)—CH_3 acetone	-one
Carboxylic acid	—C(=O)—OH	CH_3—C(=O)—OH acetic acid	-ic acid
Anhydride	—C(=O)—O—C(=O)—	CH_3—C(=O)—O—C(=O)—CH_3 acetic anhydride	none
Ester	—C(=O)—O—	CH_3—C(=O)—O—CH_3 methyl acetate	-ate
Amide	—C(=O)—NH_2, —C(=O)—N—H, —C(=O)—N—	CH_3—C(=O)—NH_2 acetamide	-amide

[a] The bonds whose connections aren't specified are assumed to be attached to carbon or hydrogen atoms in the rest of the molecule.
[b] X = F, Cl, Br, or I.

Hydrocarbon   A compound that contains only carbon and hydrogen.

In this and the next chapter, we'll describe the chemistry of **hydrocarbons,** organic compounds containing only carbon and hydrogen. Members of the first four families listed in Table 12.1—the alkanes, alkenes, alkynes, and arenes— are hydrocarbons when no other functional groups are present in their molecules. Alkanes are the one family of organic compounds that contain no functional groups, for alkanes are constructed entirely of carbon and hydrogen atoms joined by single bonds. As you'll see, the absence of functional groups makes alkanes relatively unreactive.

Every member of the other families is essentially an alkane in which one or more hydrogen atoms have been replaced by functional groups. Some examples of compounds formed by replacement of a hydrogen atom in methane, the simplest alkane, are

$$CH_4 \qquad CH_3OH \qquad CH_3NH_2 \qquad CH_3\overset{\overset{\displaystyle O}{\parallel}}{C}OH$$

A hydrocarbon   An alcohol   An amine   A carboxylic acid

Much of the chemistry discussed in this and the next five chapters is the chemistry of the families listed in Table 12.1, so it's best to memorize their names and structures now. Note that they fall into three groups: (1) the hydrocarbons of the first four families; (2) those whose functional groups contain only single bonds (alkyl halides, alcohols, ethers, and amines); and (3) those whose functional groups contain a carbon–oxygen double bond (aldehydes, ketones, carboxylic acids and anhydrides, esters, and amides).

Practice Problems   **12.1**   Locate and identify the functional groups in these molecules:
(a) lactic acid, from sour milk

$$CH_3-\overset{\overset{\displaystyle H}{|}}{\underset{\underset{\displaystyle OH}{|}}{C}}-\overset{\overset{\displaystyle O}{\parallel}}{C}-OH$$

(b) methyl methacrylate, used in making Lucite and Plexiglas

$$H_2C=\overset{\underset{\underset{\displaystyle CH_3}{|}}{}}{C}-\overset{\overset{\displaystyle O}{\parallel}}{C}-O-CH_3$$

12.2   Propose structures for molecules that fit these descriptions:
(a) C_2H_4O containing an aldehyde functional group
(b) $C_3H_6O_2$ containing a carboxylic acid functional group

12.3 THE STRUCTURE OF ORGANIC MOLECULES: ALKANES AND THEIR ISOMERS

Alkane A compound that contains only carbon and hydrogen and has only single bonds.

Molecules that contain only carbon and hydrogen and that have only single bonds belong to the family of organic molecules called **alkanes.** If we imagine ways that one carbon and four hydrogens can combine, there is only one possibility: methane, CH_4. If we imagine ways that two carbons and six hydrogens can combine, only ethane, CH_3CH_3, is possible; and if we imagine the combination of three carbons with eight hydrogens, only propane, $CH_3CH_2CH_3$, is possible.

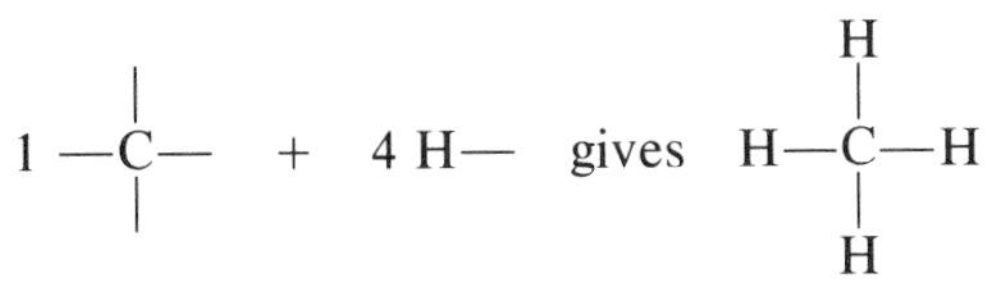

Methane

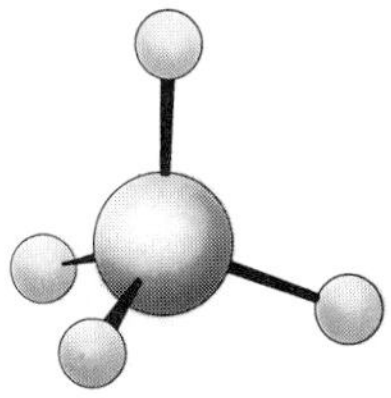

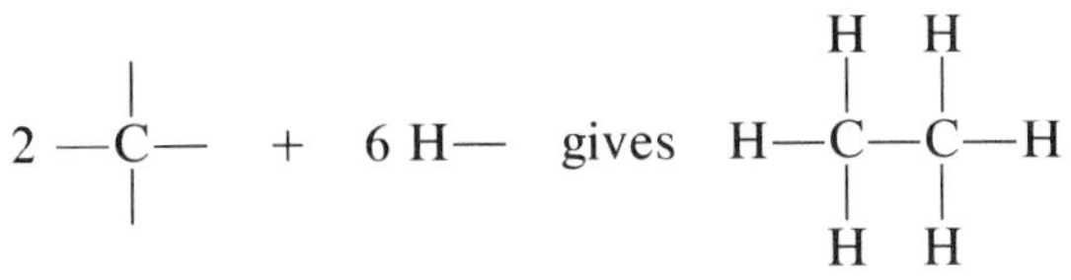

Ethane

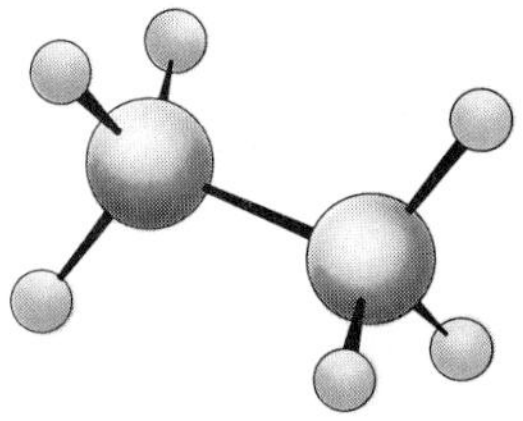

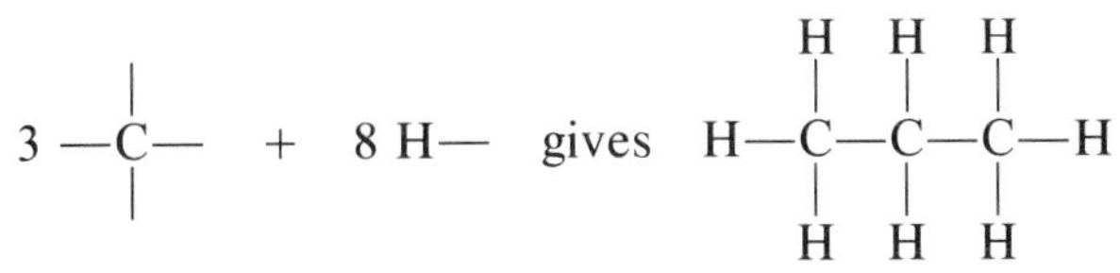

Propane

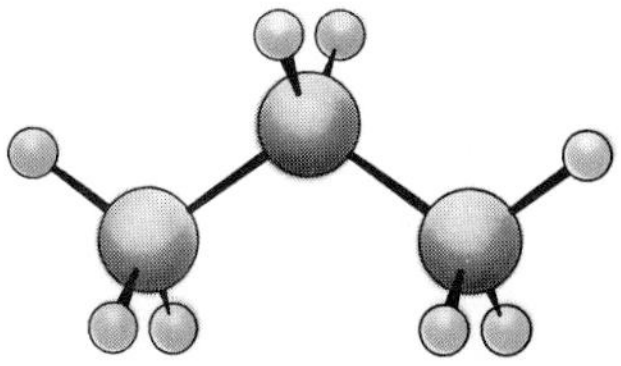

Straight-chain alkane An alkane that has all its carbon atoms connected in a row.

Branched-chain alkane An alkane that has a branching connection of carbon atoms along its chain.

If larger numbers of carbons and hydrogens combine, *more than one kind of molecule can be formed.* There are two ways in which molecules with the formula C_4H_{10} can be formed: The four carbons either can be in a row or can have a branched arrangement. Similarly, there are *three* ways in which molecules with the formula C_5H_{12} can be formed; and so on for larger alkanes. Compounds with all their carbons connected in a row are called **straight-chain alkanes,** whereas those with a branching connection of carbons are called **branched-chain alkanes.** Note that in a straight-chain alkane, you can draw a line through all the carbon atoms without lifting your pencil from the paper. In a branched-chain alkane, however, you must either lift your pencil from the paper or retrace your steps in order to draw a line through all the carbons.

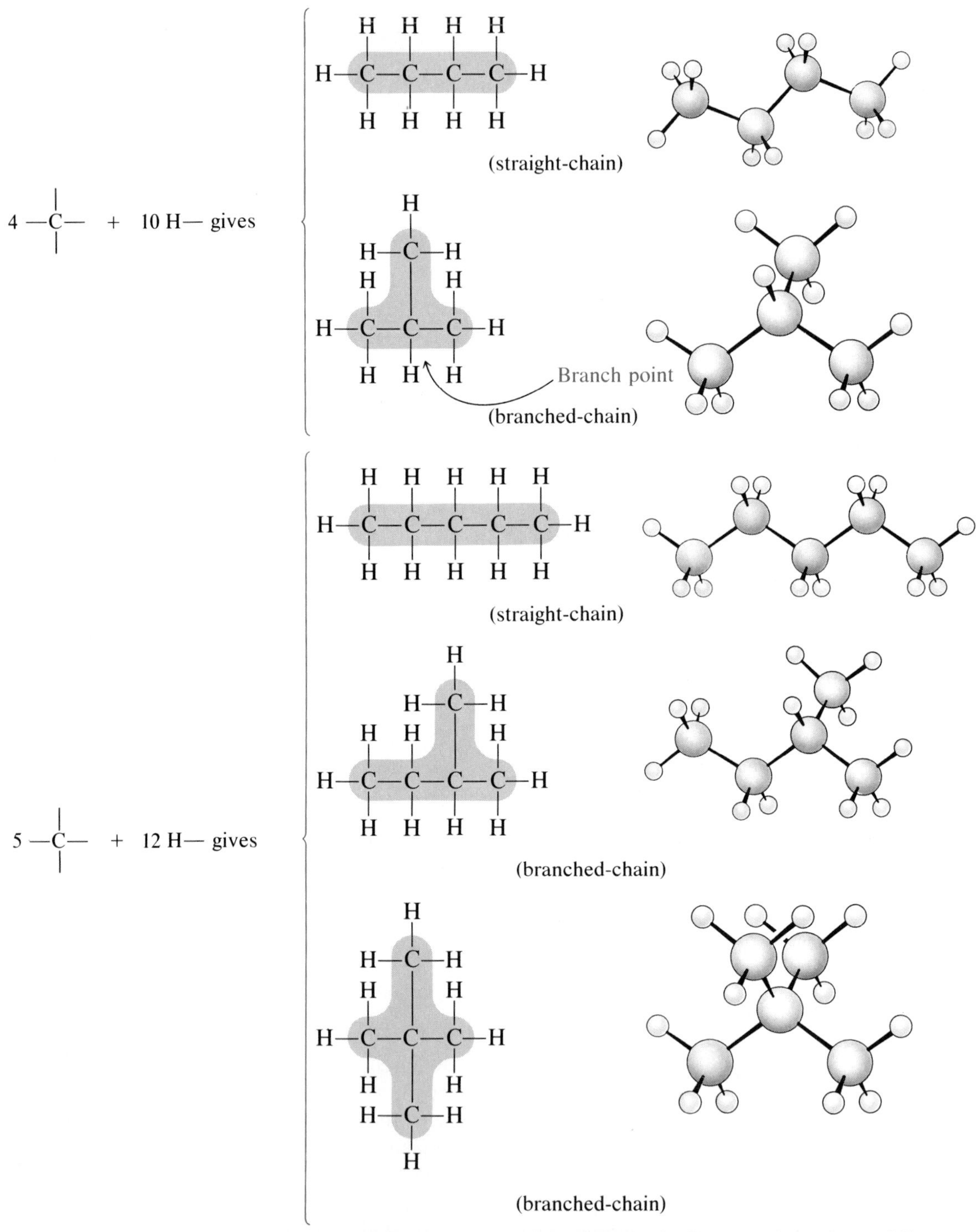

$$4 \; —\!\overset{\textstyle |}{\underset{\textstyle |}{C}}\!— \;\; + \;\; 10 \; \text{H}— \; \text{gives}$$

(straight-chain)

(branched-chain)

Branch point

$$5 \; —\!\overset{\textstyle |}{\underset{\textstyle |}{C}}\!— \;\; + \;\; 12 \; \text{H}— \; \text{gives}$$

(straight-chain)

(branched-chain)

(branched-chain)

The two different compounds with the formula C_4H_{10} and the three different compounds with the formula C_5H_{12} are called *isomers*. Specifically, these are **constitutional isomers,** compounds with the same molecular formula but with their atoms connected in different arrangements. As Table 12.2 shows, the number of possible alkane isomers grows rapidly as the number of carbon atoms increases.

Isomers, constitutional
Compounds with the same molecular formula but with different connections between atoms.

Table 12.2   Numbers of Possible Alkane Isomers

Formula	Number of Isomers	Formula	Number of Isomers
C_6H_{14}	5	$C_{10}H_{22}$	75
C_7H_{16}	9	$C_{20}H_{42}$	366,319
C_8H_{18}	18	$C_{30}H_{62}$	4,111,846,763
C_9H_{20}	35	$C_{40}H_{82}$	62,491,178,805,831

It's important to realize that different constitutional isomers are completely different chemical compounds. They have different structures, different physical properties such as melting point and boiling point, and potentially different physiological properties. For example, ethyl alcohol and dimethyl ether both have the formula C_2H_6O. Yet ethyl alcohol is a liquid central nervous system depressant with a boiling point of 78.5°C, whereas dimethyl ether is a gaseous substance with a boiling point of -23°C (Table 12.3). You can see why molecular formulas such as C_2H_6O aren't very useful in organic chemistry.

Practice Problems   **12.3**   Draw the straight-chain isomer with the formula C_7H_{16}.

12.4   Draw two branched-chain isomers with the formula C_7H_{16}.

Table 12.3   Some Properties of Ethyl Alcohol and Dimethyl Ether

Name and Molecular Formula	Structure	Boiling Point	Melting Point	Physiological Activity
Ethyl alcohol C_2H_6O		78.5°C	-117.3°C	Central-nervous-system depressant
Dimethyl ether C_2H_6O		-23°C	-138.5°C	—

AN APPLICATION: NATURAL VS. SYNTHETIC

Before organic chemistry, we had only substances from plants and animals—*natural products*—for treating diseases, perfuming ourselves, flavoring foods, gluing things together, and hundreds of other daily applications. *Essential oils,* for example, are water-insoluble mixtures distilled or extracted from plants. Their names are romantic and alluring: oil of bergamot, oil of sweet bay, oil of rose, and oil of lavender, all of which are used in making perfume. Oil of thyme, oil of sweet almond, and oil of pine needles have medical applications.

Many natural products were first used without any knowledge of their chemical compositions. Sometimes this was a mistake: Oil of bitter almond that's not purified contains hydrogen cyanide. As organic chemistry developed, chemists learned how to work out the structures of the compounds in natural products. The disease-curing miracle of penicillin, the first widely used antibiotic, was discovered in a mold in 1928, but its chemical structure was not determined until about 20 years later. Today there is a revival of interest in folk medicines and an effort to identify the chemical compounds responsible for their effectiveness.

Once a structure is known, organic chemists seek to determine how the compound might be synthesized. If the starting materials are inexpensive enough and the process is simple enough, it may be more economical to manufacture a compound than to isolate it from a natural substance. In the case of penicillin, a complete synthesis was achieved in 1958, and many structurally similar compounds that are also good antibiotics have been made in the laboratory. It is not yet practical, however, to manufacture any of the penicillins by complete synthesis. Instead, we have *semisynthetic* penicillins in which the molecular structure of a mold product is controlled by adding appropriate chemicals to the medium in which the mold grows.

12.4 DRAWING ORGANIC STRUCTURES

Structural formula A formula that shows how atoms are connected to each other.

Condensed structure A structure in which central atoms and the atoms connected to them are written as groups, e.g., $CH_3CH_2CH_3$.

Drawing **structural formulas** like those in Table 12.3 for ethyl alcohol and dimethyl ether is both time-consuming and awkward, even for relatively small molecules. **Condensed structures** (Section 6.6) are a convenient compromise because they're simpler but show the essential information about which functional groups are present and how atoms are connected. In condensed structures, carbon–hydrogen and carbon–carbon single bonds aren't shown; rather, they're understood to be there. If a carbon atom has three hydrogens bonded to it, we write CH_3; if the carbon has two hydrogens bonded to it, we write CH_2; and so on. For example, the four-carbon straight-chain compound (called *butane*) and its branched-chain isomer (called *2-methylpropane*) can be written as the following condensed structures:

Butane   2-Methylpropane

Large quantities of many of nature's chemicals, however, such as vitamin C and caffeine, are produced synthetically each year, as are large quantities of medicines nature has never produced. Some individuals demand vitamins only from natural sources, assuming that "natural" is somehow better. Do you think it would be possible to devise a chemical test to distinguish between two pure samples of vitamin C, one isolated from rose hips and one made by chemical synthesis? Does anyone ever worry about whether the caffeine in their cola drink came from tea leaves or from a chemical manufacturing plant?

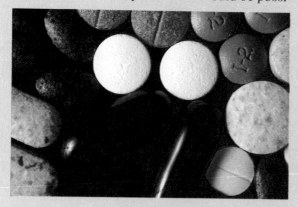

Two sources of the chemicals known as vitamins.

Note that the horizontal bonds between carbons aren't shown—the CH_3 and CH_2 units are simply placed next to each other—but that the vertical bond in 2-methylpropane is shown for clarity. Occasionally, as a further simplification, a row of CH_2 groups is shown by parentheses, with a subscript equal to the number of groups:

$$CH_3CH_2CH_2CH_2CH_2CH_3 \quad = \quad CH_3(CH_2)_4CH_3$$

Practice Problem   **12.5**   Draw the three isomers of C_5H_{12} as condensed structures.

(a)
```
    H  H  H  H  H
    |  |  |  |  |
H―C―C―C―C―C―H
    |  |  |  |  |
    H  H  H  H  H
```
Pentane

(b)
```
        H
        |
     H―C―H
     H  |  H  H
     |  |  |  |
H―C―C―C―C―H
     |  |  |  |
     H  H  H  H
```
2-Methylbutane

(c)
```
        H
        |
     H―C―H
     H  |  H
     |  |  |
H―C―C―C―H
     |  |  |
     H  |  H
     H―C―H
        |
        H
```
2,2-Dimethylpropane

12.5 THE SHAPES OF ORGANIC MOLECULES

Although every carbon atom in an alkane has its four bonds pointing toward the four corners of a tetrahedron, chemists don't usually worry about exact three-dimensional shapes when writing condensed structures. For example, the straight-chain four-carbon alkane, butane, might be represented by any of the structures shown below. These structures don't imply any particular three-dimensional shape for butane; they only show that butane has a continuous chain of four carbon atoms and indicate the *connections* between atoms without specifying geometry.

Some drawings of butane, C_4H_{10}

$$CH_3 \quad CH_2 \atop CH_2 \quad CH_3 \quad \text{or} \quad {CH_3 \atop CH_2CH_2CH_3} \quad \text{or} \quad {CH_2CH_3 \atop CH_2CH_3} \quad \text{or} \quad {CH_3CH_2CH_2 \atop CH_3}$$

Conformation The exact three-dimensional shape of a molecule at any given instant.

In fact, butane has no one single shape because *rotation* is possible around carbon–carbon single bonds. The two parts of a molecule joined by a carbon–carbon single bond are free to spin around the bond, giving rise to an infinite number of possible three-dimensional structures, or **conformations.** A given butane molecule might be fully extended at one instant (Figure 12.3a) but be twisted an instant later (Figure 12.3b). An actual sample of butane contains a great many molecules that are constantly changing shape. At any given instant, however, most of the molecules have the less crowded extended conformation shown in Figure 12.3a. The same is true for all other alkanes: At any given instant most molecules are in the least-crowded conformation.

So long as any two structures show identical connections between atoms, they represent identical compounds, no matter how the structures are drawn. Sometimes you have to mentally rotate structures to see whether they're the same or different. To "see" that the following two structures represent the same compound rather than two isomers, picture one of them turned end for end (picture the red CH_3 groups on the same end).

$$CH_3CHCH_2CH_2CH_3 \qquad CH_3CH_2CH_2CHCH_3 \atop {CH_2 \atop OH} \qquad\qquad {CH_2 \atop OH}$$

Figure 12.3
Some possible conformations of butane. There are many other conformations as well.

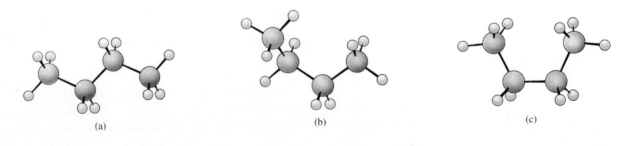

(a) (b) (c)

Solved Problem 12.1   The following molecules have the same formula, C_7H_{16}. Which of the structures represent the same molecule?

$$\overset{\displaystyle CH_3}{\underset{|}{}}$$
(a) $CH_3CHCH_2CH_2CH_2CH_3$ (b) $CH_3CH_2CH_2CH_2\overset{CH_3}{\underset{|}{C}HCH_3}$

$$\overset{\displaystyle CH_3}{\underset{|}{}}$$
(c) $CH_3CH_2CH_2CHCH_2CH_3$

Solution   The important point in determining whether two structures are identical is to pay attention to the order of connection between atoms. Don't get confused by the apparent differences caused by writing a structure right to left versus left to right. In this example, molecule (a) has a straight chain of six carbons with a —CH_3 branch on the second carbon from the end. Molecule (b) also has a straight chain of six carbons with a —CH_3 branch on the second carbon from the end and is therefore identical to (a). The only difference between (a) and (b) is that one is written "forward" and one is written "backward." Molecule (c), by contrast, has a straight chain of six carbons with a —CH_3 branch on the *third* carbon from the end and is therefore an isomer of (a) and (b).

Solved Problem 12.2   Are the following pairs of compounds isomers, the same compound, or different compounds?

(a) $CH_3\overset{CH_3}{\underset{|}{C}H}CH_2\overset{|}{\underset{CH_3}{C}H_2}$ $CH_3\overset{CH_3}{\underset{|}{C}H}CH_2CH_2CH_3$

(b) $CH_3CH_2\overset{}{\underset{CH_2CH_3}{C}HCH_3}$ $CH_3\overset{CH_2CH_3}{\underset{CH_3}{C}HCH_2}$

(c) $CH_3CH_2OCH_3$ $CH_3CH_2\overset{O}{\overset{||}{C}}H$

Solution   (a) First, compare molecular formulas to determine whether the compounds are the same or different. In this case, both compounds have the same molecular formula (C_6H_{14}). Next, examine their structures to see if they are the same compound or isomers. To do this, locate the longest straight carbon chain in each and then compare the locations of the **substituents.**

Substituent A group attached to a root compound.

$CH_3\overset{CH_3}{\underset{|}{C}H}CH_2CH_2$ $CH_3\overset{CH_3}{\underset{|}{C}H}CH_2CH_2CH_3$

Since the CH_3— group is on the second carbon from the end of a five-carbon chain in both cases, these compounds are identical.
(b) Both compounds have the same molecular formula (C_6H_{14}), and the

longest chain in each is five carbon atoms. A comparison shows, however, that the CH_3 group is on the middle carbon atom in one structure and on the second carbon atom in the other. These compounds are isomers.

$$CH_3CH_2CHCH_3$$
$$|$$
$$CH_2CH_3$$

$$CH_2CH_3$$
$$|$$
$$CH_3CHCH_2$$
$$|$$
$$CH_3$$

(c) Each of these compounds has three carbon atoms and one oxygen atom, but the numbers of hydrogen atoms differ (eight H atoms and six H atoms, respectively). These are different compounds, but they are not isomers.

AN APPLICATION: DISPLAYING MOLECULAR SHAPES

Molecular shape is critical to the proper functioning of all biological molecules. The tiniest difference in shape between two compounds can cause them to behave differently or to have different physiological effects in the body. It's therefore critical that chemists have techniques available both for determining molecular shapes with great precision and for visualizing these shapes in useful and manageable ways.

Three-dimensional shapes of molecules are determined by *X ray crystallography,* a technique that allows us to "see" molecules in a crystal using X ray waves rather than light waves. The molecular "picture" obtained by X ray crystallography looks at first like a series of regularly spaced dark spots on a photographic film. After computerized manipulation of the data, however, recognizable molecules can be drawn. Relatively small molecules like morphine are usually displayed on paper, but enormous biological molecules like immunoglobulins are best displayed on computer terminals where their structures can be enlarged, rotated, and otherwise manipulated for the best view.

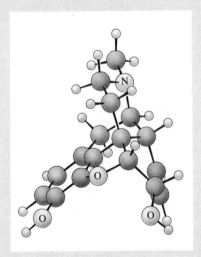
(a)

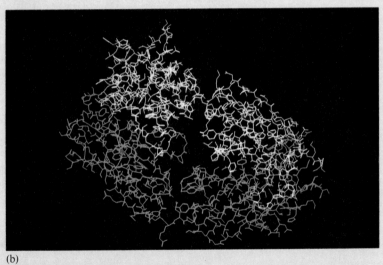
(b)

Computer-generated shapes of (a) morphine and (b) an immunoglobulin, one of the antibodies in blood that protect us from harmful invaders such as bacteria and viruses.

Practice Problems **12.6** Which of the following three structures are identical?

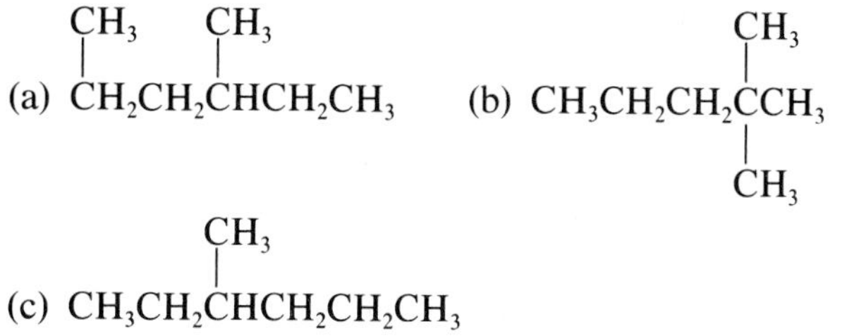

$$\underset{\substack{\big| \\ CH_3}}{\text{(a) } CH_2CH_2} \underset{\substack{\big| \\ CH_3}}{CHCH_2CH_3} \qquad \text{(b) } CH_3CH_2CH_2\underset{\substack{\big| \\ CH_3}}{\overset{\overset{\textstyle CH_3}{\big|}}{C}}CH_3$$

$$\text{(c) } CH_3CH_2\underset{\substack{\big| \\ CH_3}}{\overset{\overset{\textstyle CH_3}{\big|}}{C}}HCH_2CH_2CH_3$$

12.7 According to Table 12.2, there are five isomers with the formula C_6H_{14}. Draw as many as you can.

12.8 Using toothpicks for bonds and two marshmallows to represent the carbon atoms, build a model of an ethane molecule, CH_3CH_3. (Hydrogen atoms can be "understood" to be at the ends of the toothpicks coming out of the marshmallows.) Rotate the model around a C—C bond and note the changing relationships between hydrogens on neighboring carbons.

12.6 NAMING ALKANES

In earlier times when relatively few pure organic chemicals were known, new compounds were named at the whim of their discoverer. Thus, urea is a crystalline substance first isolated from urine, and morphine is a plant extract that's a sedative and is named after Morpheus, the Greek god of dreams. As more and more compounds became known, however, the need for **nomenclature**—a systematic method of naming compounds—became apparent.

Nomenclature A system for naming chemical compounds.

The nomenclature now used is that devised by the International Union of Pure and Applied Chemistry: the IUPAC system (pronounced **eye**-you-pac). In the IUPAC system for organic compounds, a chemical name has three parts: prefix, root, and suffix. The root specifies the overall size of the molecule by telling how many carbon atoms are present in the longest continuous chain, the suffix identifies what family the molecule belongs to, and the prefix specifies the location of functional groups and other substituents on the chain:

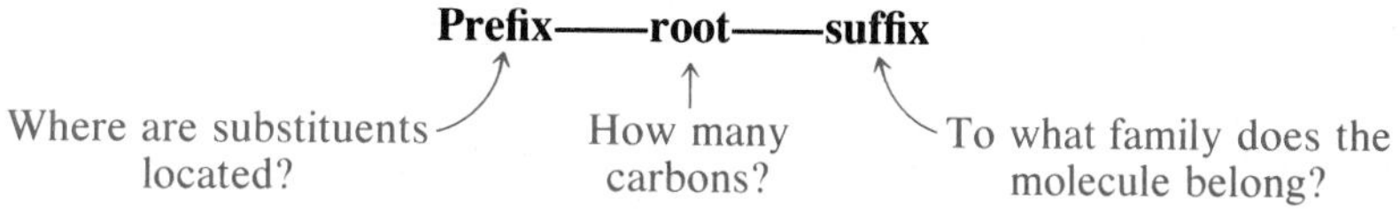

Prefix——root——suffix

Where are substituents located?   How many carbons?   To what family does the molecule belong?

Straight-chain alkanes are named simply by counting the number of carbon atoms in the chain and adding the family suffix *-ane* to the root name to give the name of the parent compound. With the exception of the first four compounds—*meth*ane, *eth*ane, *prop*ane, and *but*ane—whose root names have historical origins, the alkanes are named from Greek numbers according to the number of carbons present. Thus, *pent*ane is the five-carbon alkane, *hex*ane is

Alkyl group The part of an alkane that remains when one hydrogen atom is removed.

Methyl group —CH₃, the alkyl group derived from methane.

Ethyl group —CH₂CH₃, the alkyl group derived from ethane.

the six-carbon alkane, and so on. The roots shown by italics in Table 12.4 are used in naming many types of organic compounds. These first 10 alkane names should be memorized.

Substituents that branch off a main chain are called **alkyl groups.** Each different alkyl group can be thought of as the part of an alkane that remains when one hydrogen atom is removed. For example, removal of one hydrogen from methane gives the **methyl group,** —CH₃, and removal of one hydrogen from ethane gives the **ethyl group,** —CH₂CH₃. Notice that these alkyl groups are named simply by replacing the *-ane* ending of the parent alkane with a *-yl* ending.

Methane $H-\overset{\overset{\displaystyle H}{|}}{\underset{\underset{\displaystyle H}{|}}{C}}-H \xrightarrow{\text{remove one H}} -\overset{\overset{\displaystyle H}{|}}{\underset{\underset{\displaystyle H}{|}}{C}}-H = -CH_3$ (A methyl group)

Ethane $H-\overset{\overset{\displaystyle H}{|}}{\underset{\underset{\displaystyle H}{|}}{C}}-\overset{\overset{\displaystyle H}{|}}{\underset{\underset{\displaystyle H}{|}}{C}}-H \xrightarrow{\text{remove one H}} -\overset{\overset{\displaystyle H}{|}}{\underset{\underset{\displaystyle H}{|}}{C}}-\overset{\overset{\displaystyle H}{|}}{\underset{\underset{\displaystyle H}{|}}{C}}-H = -CH_2CH_3$ (An ethyl group)

Methane and ethane are special because each has only one kind of hydrogen. It doesn't matter which of the four methane hydrogens is removed because all four are equivalent. Thus, there is only one possible kind of methyl group. Similarly, it doesn't matter which of the six ethane hydrogens is removed, and only one kind of ethyl group is possible.

n-Propyl group —CH₂CH₂CH₃, the alkyl group derived by removing a hydrogen atom from an end carbon of propane.

Isopropyl group —CH(CH₃)₂, the alkyl group derived by removing a hydrogen atom from the central carbon of propane.

The situation is more complex for larger alkanes that contain more than one kind of hydrogen. For example, propane has two different kinds of hydrogens. Removal of any one of the six hydrogens attached to an end carbon yields a straight-chain propyl group called *n-propyl,* whereas removal of one of the two hydrogens attached to the central carbon yields a branched-chain propyl group called **isopropyl.** (The "*n*" prefix in *n*-propyl stands for *normal,* meaning straight-chain, and is used with other alkyl groups, too.)

Table 12.4   Names of Straight-Chain Alkanes

No. of Carbons	Structure	Name
1	CH₄	*Meth*ane
2	CH₃CH₃	*Eth*ane
3	CH₃CH₂CH₃	*Prop*ane
4	CH₃CH₂CH₂CH₃	*But*ane
5	CH₃CH₂CH₂CH₂CH₃	*Pent*ane
6	CH₃CH₂CH₂CH₂CH₂CH₃	*Hex*ane
7	CH₃CH₂CH₂CH₂CH₂CH₂CH₃	*Hept*ane
8	CH₃CH₂CH₂CH₂CH₂CH₂CH₂CH₃	*Oct*ane
9	CH₃CH₂CH₂CH₂CH₂CH₂CH₂CH₂CH₃	*Non*ane
10	CH₃CH₂CH₂CH₂CH₂CH₂CH₂CH₂CH₂CH₃	*Dec*ane

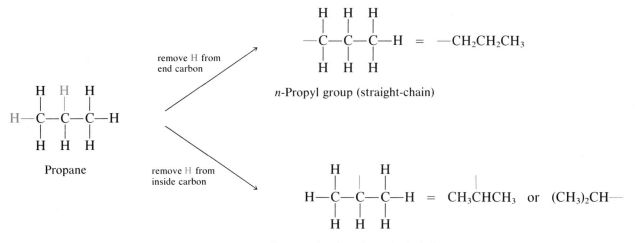

n-Propyl group (straight-chain)

Propane

Isopropyl group (branched-chain)

It's important to realize that alkyl groups themselves are not compounds and that the "removal" of a hydrogen from an alkane is just a way of looking at things, not a chemical reaction. The names of some common alkyl groups are listed in Table 12.5.

Branched-chain alkanes are named by following four steps:

Step 1. Name the main chain. Find the *longest continuous chain of carbons* present in the molecule and name the chain using the root for the number of carbons it contains. The longest chain may not always be obvious from the manner of writing; you may have to "turn corners" to find it.

$$CH_3—CH_2$$
$$CH_3—CH—CH_2—CH_3$$

Name as a substituted pentane, not as a substituted butane, because the *longest* chain has five carbons.

Step 2. Number the carbon atoms in the main chain. Beginning at the end nearer the first branch point, number each carbon atom in the main chain:

$$\underset{1}{CH_3}—\underset{2}{CH}—\underset{3}{CH_2}—\underset{4}{CH_2}—\underset{5}{CH_3}$$
with CH₃ branch on C2

The first (and only) branch occurs at C2 if we start numbering from the left, but would occur at C4 if we started from the right by mistake.

Table 12.5   Some Common Alkyl Groups*[a]*

			CH₃
CH₃—	CH₃CH₂—	CH₃CH₂CH₂—	CH₃CH—
Methyl	Ethyl	*n*-Propyl	Isopropyl
		CH₃	
CH₃CH₂CH₂CH₂—	CH₃CHCH₂CH₃	CH₃CHCH₂—	CH₃CCH₃
			CH₃
n-Butyl	*sec*-Butyl	Isobutyl	*tert*-Butyl

[a] The red bond shows the connection to the rest of the molecule.

Step 3. Identify and number the branching substituents. Assign a number to each branching substituent on the main chain according to its point of attachment.

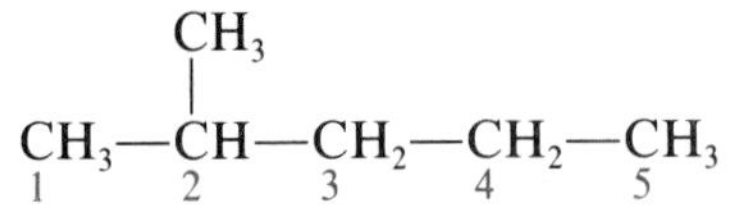

$$CH_3$$
$$|$$
$$\underset{1}{CH_3}-\underset{2}{CH}-\underset{3}{CH_2}-\underset{4}{CH_2}-\underset{5}{CH_3}$$

The main chain is a pentane. There is one —CH_3 substituent group connected to C2 of the chain.

If there are two substituents on the same carbon, assign the same number to both of them. There must always be as many numbers in the name as there are substituents.

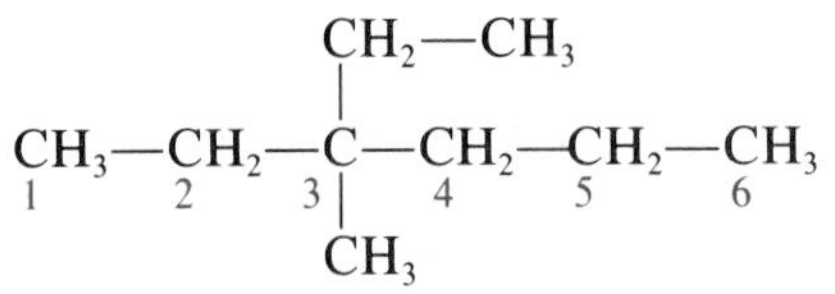

$$CH_2-CH_3$$
$$|$$
$$\underset{1}{CH_3}-\underset{2}{CH_2}-\underset{3}{C}-\underset{4}{CH_2}-\underset{5}{CH_2}-\underset{6}{CH_3}$$
$$|$$
$$CH_3$$

The main chain is a hexane. There are two substituents, a —CH_3 and a —CH_2CH_3, both connected to C3 of the chain.

Step 4. Write the name as a single word. Use hyphens to separate the numbers from the different prefixes and use commas to separate numbers if necessary. If two or more different substituent groups are present, cite them in alphabetical order. If two or more identical substituents are present, use one of the prefixes *di-*, *tri-*, *tetra-*, and so forth, but don't use these prefixes for alphabetizing purposes.

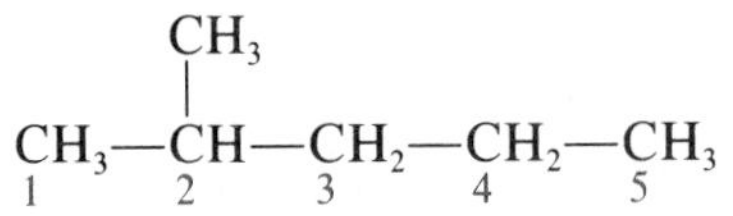

$$CH_3$$
$$|$$
$$\underset{1}{CH_3}-\underset{2}{CH}-\underset{3}{CH_2}-\underset{4}{CH_2}-\underset{5}{CH_3}$$

2-Methylpentane (a five-carbon main chain with a 2-methyl substituent)

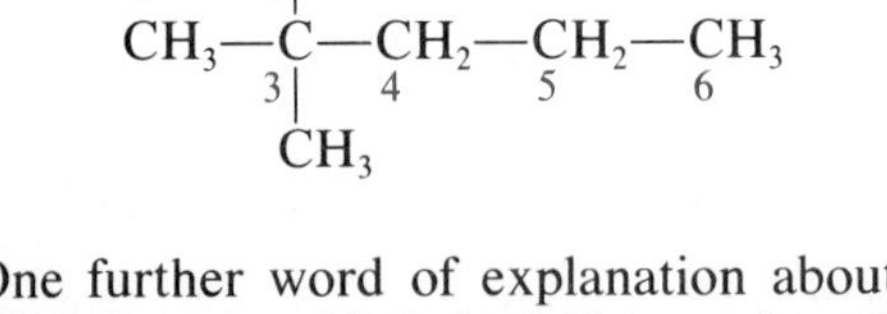

$$CH_2-CH_3$$
$$|$$
$$\underset{1}{CH_3}-\underset{2}{CH_2}-\underset{3}{C}-\underset{4}{CH_2}-\underset{5}{CH_2}-\underset{6}{CH_3}$$
$$|$$
$$CH_3$$

3-Ethyl-3-methylhexane (a six-carbon main chain with 3-ethyl and 3-methyl substituents cited alphabetically)

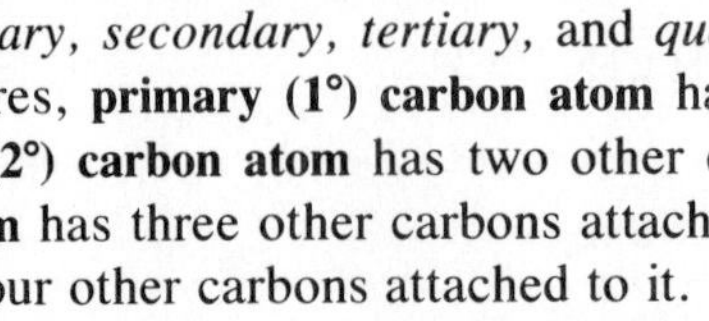

$$\underset{1}{CH_3}-\underset{2}{CH_2}$$
$$|$$
$$\underset{3}{CH_3}-\underset{}{C}-\underset{4}{CH_2}-\underset{5}{CH_2}-\underset{6}{CH_3}$$
$$|$$
$$CH_3$$

3,3-Dimethylhexane (a six-carbon main chain with two 3-methyl substituents)

Primary (1°) carbon A carbon atom that is bonded to one other carbon atom.

Secondary (2°) carbon A carbon atom that is bonded to two other carbons.

Tertiary (3°) carbon A carbon atom that is bonded to three other carbons.

Quaternary (4°) carbon A carbon atom that is bonded to four other carbons.

One further word of explanation about alkanes and alkyl groups: It's sometimes useful to think about the *number of other carbon atoms attached* to a given carbon atom. There are four possible substitution patterns for carbon, called *primary, secondary, tertiary,* and *quaternary.* As indicated by the following structures, **primary (1°) carbon atom** has one other carbon attached to it, a **secondary (2°) carbon atom** has two other carbons attached to it, a **tertiary (3°) carbon atom** has three other carbons attached to it, and a **quaternary (4°) carbon atom** has four other carbons attached to it.

$$R—\overset{\displaystyle H}{\underset{\displaystyle H}{C}}—H \qquad R—\overset{\displaystyle R}{\underset{\displaystyle H}{C}}—H \qquad R—\overset{\displaystyle R}{\underset{\displaystyle R}{C}}—H \qquad R—\overset{\displaystyle R}{\underset{\displaystyle R}{C}}—R$$

Primary carbon (1°) has one other carbon attached.　　*Secondary* carbon (2°) has two other carbons attached.　　*Tertiary* carbon (3°) has three other carbons attached.　　*Quaternary* carbon (4°) has four other carbons attached.

R— The general symbol for an alkyl group.

The symbol **R** is used here and in later chapters to represent a *generalized* alkyl group, where R may stand for any alkyl group. For example, the generalized formula R—OH for an alcohol might refer to CH_3OH, CH_3CH_2OH, or any of a great many other possibilities.

Removal of the hydrogen atom from a secondary or a tertiary carbon atom produces a branched alkyl group. Note in Table 12.5 that two of the four-carbon alkyl groups are distinguished from each other by prefixes showing that one bonds through a secondary (*sec-*) carbon atom and the other through a tertiary (*tert-*) carbon atom.

Solved Problem 12.3　　What is the IUPAC name of this alkane?

$$CH_3—\overset{\displaystyle CH_3}{\underset{}{CH}}—CH_2—CH_2—\overset{\displaystyle CH_3}{\underset{}{CH}}—CH_2—CH_3$$

Solution　　First, find and name the longest continuous chain of carbon atoms (in this case seven carbons, or *hept*ane). Second, number the main chain beginning at the end nearer the first branch (on the left in this case).

$$\underset{1}{CH_3}—\underset{2}{\overset{\displaystyle CH_3}{CH}}—\underset{3}{CH_2}—\underset{4}{CH_2}—\underset{5}{\overset{\displaystyle CH_3}{CH}}—\underset{6}{CH_2}—\underset{7}{CH_3}$$

Name as a heptane.

Third, identify and number the substituents (a 2-methyl and a 5-methyl in this case). Fourth, write the name as one word using the prefix *di-* since there are two methyl groups. Separate the two numbers by a comma and use a hyphen between the numbers and the word:

$$\underset{1}{CH_3}—\underset{2}{\overset{\displaystyle CH_3}{CH}}—\underset{3}{CH_2}—\underset{4}{CH_2}—\underset{5}{\overset{\displaystyle CH_3}{CH}}—\underset{6}{CH_2}—\underset{7}{CH_3}$$

Substituents: 2-methyl, 5-methyl　　Name: 2,5-dimethylheptane

Solved Problem 12.4　　Identify each of the carbon atoms in this molecule as primary, secondary, tertiary, or quaternary.

$$CH_3\overset{\displaystyle CH_3}{\underset{}{CH}}CH_2CH_2\overset{\displaystyle CH_3}{\underset{\displaystyle CH_3}{C}}CH_3$$

Solution   Look at each carbon atom in the molecule, count the number of other carbon atoms attached, and make the assignment.

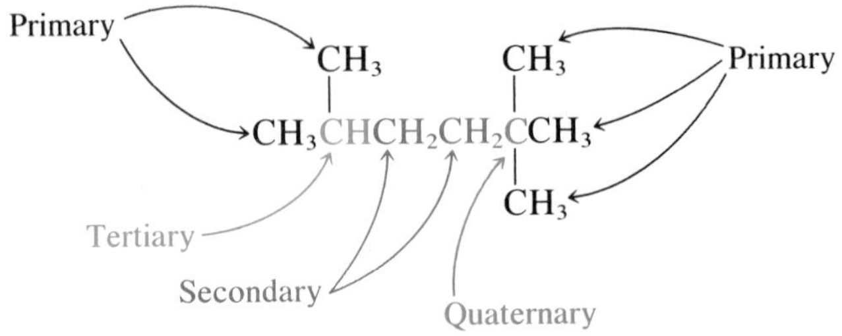

Primary

CH_3 CH_3 Primary

$CH_3CHCH_2CH_2CCH_3$

Tertiary CH_3 Primary

Secondary   Quaternary

Practice Problems   **12.9**   What are the IUPAC names of the following alkanes?

CH_2—CH_3 CH_3

(a) 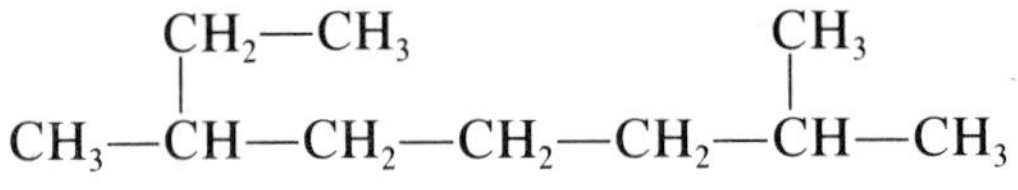 CH_3—CH—CH_2—CH_2—CH_2—CH—CH_3

$C\,H_2$—CH_3

(b) 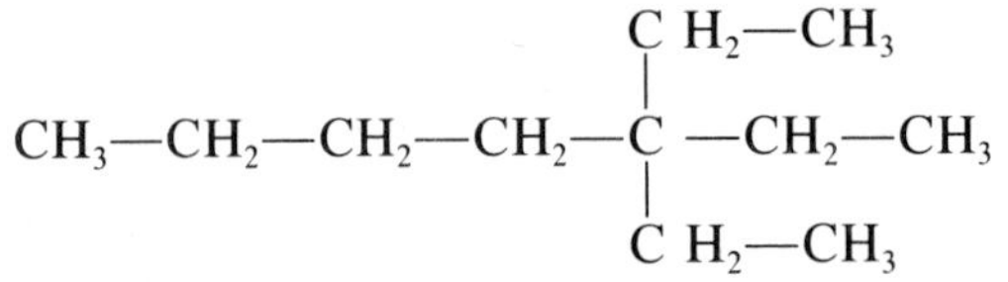 CH_3—CH_2—CH_2—CH_2—C—CH_2—CH_3

$C\,H_2$—CH_3

12.10   Draw structures corresponding to these IUPAC names:
(a) 3-methylhexane   (b) 3,4-dimethyloctane   (c) 2,2,4-trimethylpentane

12.11   Identify the carbon atoms in the molecules shown in Problem 12.9 as primary, secondary, tertiary, or quaternary.

12.12   Draw and name alkanes that meet these descriptions:
(a) an alkane with a tertiary carbon atom
(b) an alkane that has both a tertiary and a quaternary carbon atom

12.7   PROPERTIES OF ALKANES

Alkanes contain only nonpolar carbon–carbon and carbon–hydrogen bonds, so the only intermolecular forces influencing their properties are the weak London forces (Section 9.11). The effect of these forces is shown in the regularity with which the melting points and boiling points of straight-chain alkanes increase with molecular size (Figure 12.4). The first four alkanes—methane, ethane, propane, and butane—are gases at room temperature and pressure. Alkanes with from 5 to 15 carbon atoms are liquids, and those with 16 or more carbon atoms are generally low-melting, waxy solids.

In keeping with their low polarity, alkanes are insoluble in water but soluble in nonpolar organic solvents, including other alkanes. Since alkanes are generally less dense than water, they float on its surface. The liquid alkanes are volatile and must be handled with care because their vapors are flammable, as are the gaseous alkanes. Mixtures of alkane vapors and air can explode when detonated by a single spark.

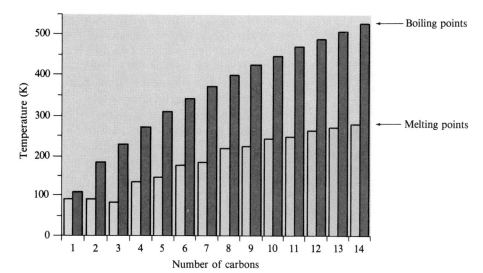

Figure 12.4
Boiling and melting points for the C_1–C_{15} straight-chain alkanes. There is a regular increase with molecular size.

The physiological effects of alkanes are limited. Methane, ethane, and propane gases are nontoxic, and the danger of inhaling them lies in suffocation due to lack of oxygen. Breathing the vapor of higher alkanes in large concentrations can induce sleep followed by loss of consciousness. There's also a danger in breathing droplets of liquid alkanes because they dissolve nonpolar substances in lung tissue and cause pneumonia-like symptoms.

Petrolatum (a jelly) and mineral oil (a thick liquid) are both mixtures of higher alkanes produced directly from petroleum. They're so harmless to tissue that both are approved for use in eye ointments and are used as vehicles for medications of many kinds. Petrolatum (sold as Vaseline) softens, lubricates, and protects the skin. Mineral oil is sometimes used as a laxative; it passes through the body unchanged.

Properties of Alkanes

- Odorless or mild odor, colorless, tasteless
- Nonpolar; insoluble in water and soluble in nonpolar organic solvents; less dense than water
- Flammable
- Regular increase in melting and boiling points with molecular weight; those with up to four C atoms are gases
- Not very reactive.

12.8 CHEMICAL REACTIONS OF ALKANES

If you've ever done any home canning, you might have used paraffin wax to seal the jars. Chemically, what you've used is a mixture of straight-chain alkanes in the C_{20}–C_{36} range. The word **paraffin,** derived from the Latin *parum affinis* meaning "slight affinity," is often applied to alkanes because of their lack of chemical reactivity. Alkanes show little chemical affinity for other molecules and are inert to acids, bases, and most other common laboratory reagents. Their major reactions are those with oxygen and the halogens.

Paraffin   A mixture of waxy alkanes having 20 to 36 carbon atoms.

Combustion A chemical reaction in which heat and often light are produced; usually refers to burning in the presence of oxygen.

Combustion   The chemical reaction of alkanes with oxygen occurs during the **combustion** of fuels in engines and furnaces (see Section 8.2). Carbon dioxide and water are the products of complete combustion of any hydrocarbon (or any organic compound containing carbon, hydrogen, and oxygen), and a large amount of heat is released. (Some examples were given in Table 8.1.)

When hydrocarbon combustion is less than complete due to faulty engine or furnace performance, carbon monoxide (CO) and carbon-containing soot are among the combustion products. Carbon monoxide is a highly toxic and dangerous substance. It's especially hazardous because it has no odor and can go undetected. Breathing air containing as little as 2% carbon monoxide for 1 hr can cause either respiratory and nervous system damage or death. The supply of oxygen to the brain is cut off by carbon monoxide because it binds strongly to hemoglobin at the site where oxygen is normally bound. By contrast with carbon monoxide, carbon dioxide is nontoxic and causes harm only by suffocation when it replaces oxygen in the atmosphere.

***Practice Problem*   12.13**   Write a balanced equation for the complete combustion of ethane.

Halogenation, alkane The substitution of one or more hydrogen atoms in an alkane by halogen atoms.

Substitution reaction An organic reaction in which two reactants exchange atoms or groups, AB + XY → AY + XB.

Halogenation   The second notable reaction of alkanes is the replacement of hydrogen atoms by chlorine or bromine atoms, known as **halogenation.** Such a reaction in which a hydrogen atom or other substituent in an organic compound is replaced by a different substituent is called a **substitution reaction.** Halogenation of alkanes takes place at a useful rate only when initiated by heat or light. Complete chlorination of methane, for example, yields carbon tetrachloride

$$CH_4 + 4\ Cl_2 \xrightarrow{\text{heat or light}} CCl_4 + 4\ HCl$$

Although we've written a neatly balanced equation for the reaction of methane with chlorine, it doesn't represent well what actually happens because this reaction, like many organic reactions, usually yields a mixture of products:

$$CH_4 + Cl_2 \longrightarrow CH_3Cl + HCl$$
$$\xrightarrow{Cl_2} CH_2Cl_2 + HCl$$
$$\xrightarrow{Cl_2} CHCl_3 + HCl$$
$$\xrightarrow{Cl_2} CCl_4 + HCl$$

CH_3Cl, methyl chloride
CH_2Cl_2, dichloromethane
$CHCl_3$, chloroform
CCl_4, carbon tetrachloride

In considering an organic reaction, attention is usually focused on converting a particular starting material into a desired product. Because they aren't of interest, possible minor side products are often ignored in writing organic equations. Also, nonorganic products such as the HCl formed in the chlorination of methane, are often of little interest. Therefore, it's frequently not

Reagent A reactant used to bring about a specific chemical reaction.

necessary to balance the equation for an organic reaction as long as the reactant, the major product, and any necessary **reagents** and conditions are shown. A chemist who plans to convert methane into methyl bromide might therefore write the equation as follows:

$$CH_4 \xrightarrow[\text{light, heat}]{Br_2} CH_3Br$$

Like many equations for organic reactions, this equation isn't balanced.

Practice Problem **12.14** Write the structures of all possible products with one or two chlorine atoms that could be formed in the substitution reaction of propane with chlorine.

12.9 CYCLOALKANES

Acyclic alkane An alkane that contains no rings.

Cycloalkane An alkane that contains a ring of carbon atoms.

The organic compounds described thus far have all been open-chain or **acyclic alkanes. Cycloalkanes,** which contain rings of carbon atoms, are also well known and are widespread throughout nature. Cholesterol, for example (Figure 12.2), has three rings of 6 carbon atoms and one of 5 carbon atoms. Compounds of all ring sizes from 3 through 30 carbon atoms and beyond have been prepared in the laboratory. The two simplest cycloalkanes contain 3 (cyclopropane) and 4 carbon atoms (cyclobutane) joined in rings:

Cyclopropane
(mp −128°C, bp −33°C)

Cyclobutane
(mp −50°C, bp −12°C)

Cyclic and acyclic alkanes are similar in many of their properties, for the cycloalkanes are also nonpolar molecules. Cyclopropane and cyclobutane are gases, and those with more carbon atoms are liquids and solids. Like the alkanes, the cycloalkanes are insoluble in water and flammable.

Because of their cyclic structures, cycloalkane molecules are more rigid and less flexible than their open-chain counterparts. Rotation is not possible around the carbon–carbon bonds in cycloalkanes without breaking open the ring. To maintain closed rings, the carbon bond angles in the small cyclopropane and cyclobutane rings are compressed, causing these compounds to be less stable and more reactive than other cycloalkanes. The six-membered cyclohexane ring exists in a puckered, nonplanar shape known as the *chair conformation* (Figure 12.5), in which the carbon atoms have nearly tetrahedral bond angles. The cyclohexane ring is therefore very stable, and many naturally occurring and biochemically active molecules contain cyclohexane rings.

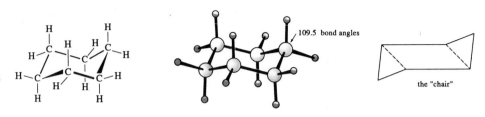

Figure 12.5
The chair conformation of cyclohexane. All bond angles are close to the 109.5° tetrahedral value.

12.10 DRAWING AND NAMING CYCLOALKANES

Even condensed structures become cluttered and awkward when working with large molecules that contain rings. Thus, a more streamlined way of drawing structures is often used in which cycloalkanes are represented simply by polygons. A triangle represents cyclopropane, a square represents cyclobutane, a pentagon represents cyclopentane, and so on.

Cyclopropane   Cyclobutane   Cyclopentane   Cyclohexane

Line structure A shorthand way of representing ring structures as polygons without showing individual carbon atoms.

Notice that carbon and hydrogen atoms aren't even shown in these **line structures.** A carbon atom is simply "understood" to be at every intersection of two lines, and the proper number of hydrogen atoms necessary to give each carbon atom four covalent bonds is supplied mentally. Methylcyclohexane looks like this in a line structure:

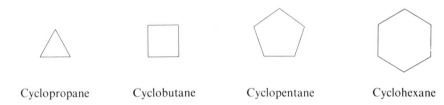

is the same as

These intersections represent CH_2 groups.

This three-way intersection is a CH group.

Cycloalkanes are named by a straightforward extension of the rules for open-chain alkanes. The carbon ring is equivalent to the main chain. In most cases, only two rules are needed:

1. *Use the cycloalkane name as the parent.* That is, compounds should be named as alkyl-substituted cycloalkanes rather than as cycloalkyl-substituted alkanes. If there is only one substituent on the ring, it's not even necessary to assign a number since all ring positions are equivalent.

Parent compound: Cyclohexane
Name: Methylcyclohexane
(not cyclohexylmethane)

2. ***Number the substituents.*** Start numbering at the group that has alphabetical priority and proceed around the ring in the direction that gives the second substituent the lowest possible number.

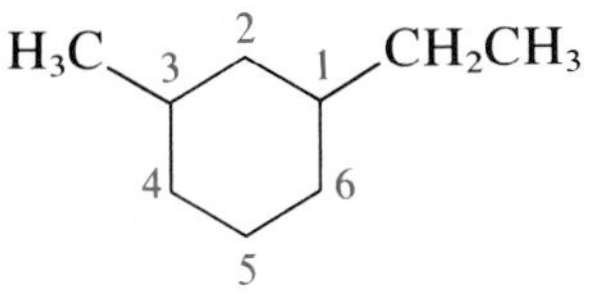
H3C
CH2CH3
1
2
3
4
5
6

1-Ethyl-3-methylcyclohexane
(not 1-ethyl-5-methylcyclohexane or
1-methyl-3-ethylcyclohexane or
1-methyl-5-ethylcyclohexane)

Solved Problem 12.5 What is the IUPAC name of this cycloalkane?

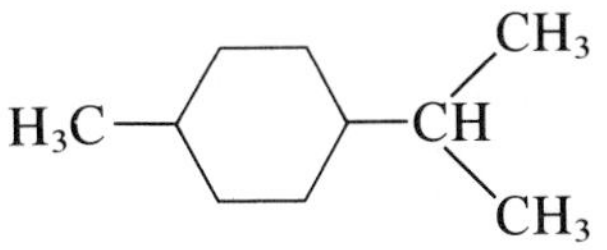
H3C
CH
CH3
CH3

Solution First, identify the parent cycloalkane (cyclohexane in this case). Second, identify the two substituents (a methyl group and an isopropyl group). Third, number the compound beginning at the group having alphabetical priority (isopropyl rather than methyl) and proceed around the ring in a direction that gives the second group the lowest possible number.

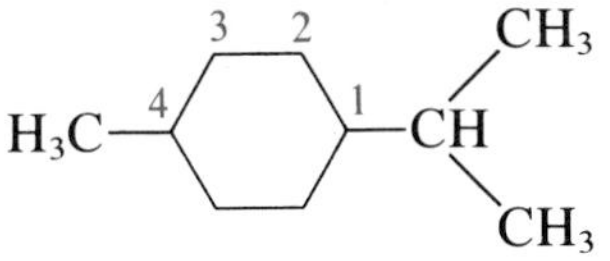
H3C
CH
CH3
CH3
1
2
3
4

1-Isopropyl-4-methylcyclohexane

Solved Problem 12.6 Draw a line structure for 1,4-dimethylcyclohexane.

Solution First, draw a hexagon to represent a cyclohexane ring and then attach a —CH_3 (methyl) group at an arbitrary position that becomes C1. Then count around the ring to C4 and attach another —CH_3 group (written here as H_3C— since it is attached on the left side of the ring).

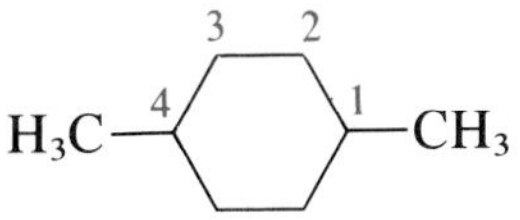
H3C
CH3
1
2
3
4

1,4-Dimethylcyclohexane

Practice Problems 12.15 What are the IUPAC names of these cycloalkanes?

(a) 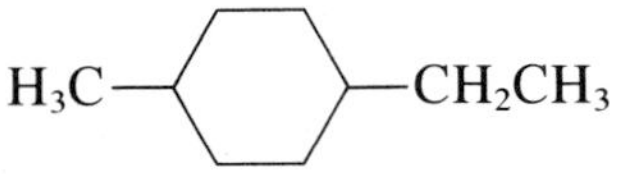
H3C
CH2CH3

(b) 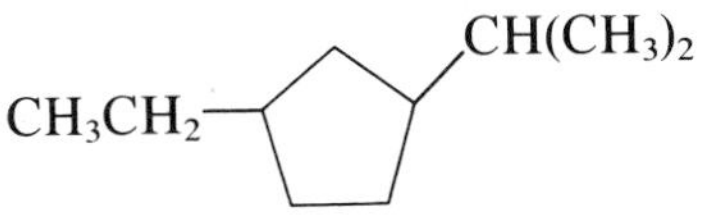
CH3CH2
CH(CH3)2

12.16 Draw structures representing these IUPAC names. Use simplified line structures rather than condensed structures.

(a) 1,1-diethylcyclohexane (b) 1,3,5-trimethylcycloheptane

INTERLUDE: PETROLEUM

Although many alkanes occur naturally throughout the plant and animal world, natural gas and petroleum deposits provide the most abundant supply. Laid down eons ago, these deposits are largely derived from the decomposition of marine organic matter.

Natural gas consists chiefly of methane, with smaller amounts of ethane, propane, and butane also present. **Petroleum** is a complex mixture of hydrocarbons that must be separated, or *refined,* into different fractions before it can be used. Petroleum refining begins with **distillation** to separate the crude oil into three main fractions according to boiling points: straight-run gasoline (bp 30–200°C), kerosene (bp 175–300°C), and gas oil (bp 275–400°C). The residue is then further distilled under reduced pressure to recover lubricating oils, waxes, and asphalt, as shown in the diagram.

Distillation of petroleum is just the beginning of the process for making automobile fuel. It's long been known that straight-chain alkanes burn far less smoothly than branched-chain compounds, a quality measured by determining a compound's **octane number.** Heptane, a particularly bad fuel, is assigned an octane rating of 0, and 2,2,4-trimethylpentane (known as isooctane) is given a rating of 100. Straight-run gasoline,

with its high percentage of unbranched alkanes, is thus a poor fuel. Petroleum chemists, however, have devised sophisticated methods to remedy the problem. One of these methods, **catalytic cracking,** involves taking the kerosene cut (C_{11}–C_{14}) and "cracking" it into smaller C_3–C_5 molecules at high temperature. These small hydrocarbons are then catalytically recombined to yield C_7–C_{10} branched-chain molecules that are perfectly suited for use as automobile fuel.

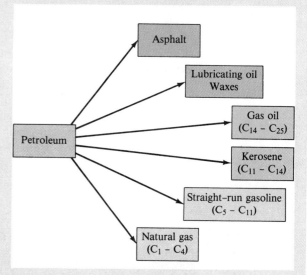

The principal products of petroleum refining.

SUMMARY

Carbon has the unique ability to form strong bonds to other carbon atoms, resulting in the formation of rings or long chains and giving rise to many millions of possible structures. Organic compounds can be classified into various families according to the functional groups they contain. A **functional group** is a part of a larger molecule and is composed of an atom or group of atoms that has characteristic chemical reactivity. A given functional group undergoes the same chemical reactions in every molecule where it occurs.

Compounds of carbon and hydrogen that contain only single bonds are called **alkanes. Straight-chain alkanes** have all their carbons connected in a row; **branched-chain alkanes** have a branching connection of atoms some-

where along their chains; and **cycloalkanes** have a ring of carbon atoms. **Isomerism** is possible in alkanes having four or more carbons. **Constitutional isomers** are compounds that have the same molecular formula but have different structures and properties because of different connections between atoms. Organic compounds may be represented by **structural formulas** that show all atoms and bonds, by condensed structures in which not all bonds are drawn, or by **line structures** in which the carbon skeleton is represented by lines and the locations of C and H atoms are understood.

Alkanes are named in the **IUPAC system** by applying a set of **nomenclature** rules. Straight-chain alkanes are named by adding the family ending *-ane* to a root that tells how many carbon atoms are present. Branched-chain alkanes are named by using the longest continuous chain of carbon atoms for the root and then identifying the alkyl groups present as branches off the main chain. An **alkyl group** is the part of an alkane that remains when one hydrogen is removed. Isomerism is possible in alkyl groups just as in alkanes themselves. Cycloalkanes are named in the same system by adding *cyclo-* as a prefix to

the name of the alkane with the same number of carbon atoms.

Alkanes are generally soluble only in nonpolar organic solvents, have weak intermolecular forces, increase in melting and boiling points with formula weight, and are nontoxic. Their principal chemical reactions are **combustion** and **halogenation,** a **substitution reaction** in which halogen atoms are substituted for hydrogen atoms.

REVIEW PROBLEMS

Organic Molecules and Functional Groups

12.17 What special characteristic of carbon makes possible the existence of so many different organic compounds?

12.18 What are functional groups, and why are they important?

12.19 Why are most organic compounds insoluble in water and nonconducting?

12.20 If you were given two unlabeled bottles, one containing hexane and one containing water, how could you tell them apart?

12.21 What is meant by the term *polar covalent bond?* Give an example of such a bond.

12.22 Give examples of compounds that are members of the following families:
(a) alcohol   (b) amine   (c) carboxylic acid
(d) ether

12.23 Locate and identify the functional groups in these molecules:
(a)

Menthol

(b)

Aspirin (acetylsalicylic acid)

12.24 Propose structures for molecules that fit these descriptions:
(a) a ketone with the formula $C_5H_{10}O$
(b) an ester with the formula $C_6H_{12}O_2$

(c) a compound with the formula $C_2H_5NO_2$ that is both an amine and a carboxylic acid
(d) an amide with the formula C_4H_9NO

12.25 Identify the functional groups in these molecules:

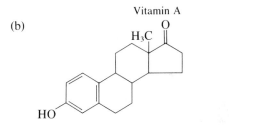

(a)

Vitamin A

(b)

Estrone, a female sex hormone

Alkanes and Isomers

12.26 What structural feature distinguishes a straight-chain alkane from a branched-chain alkane?

12.27 What requirement must be met in order for two compounds to be isomers?

12.28 If one compound has the formula C_5H_{10} and another has the formula C_4H_{10}, are the two compounds isomers? Explain.

12.29 What is the difference between a secondary carbon and a tertiary carbon? Between a primary carbon and a quaternary carbon?

12.30 Why can't a compound have a *quintary* carbon (five R groups attached to C)?

12.31 Give examples of compounds that meet these descriptions:
(a) an alkane that has two tertiary carbons
(b) a cycloalkane that has only secondary carbons

12.32 There are three isomers with the formula C_3H_8O. Draw their structures.

12.33 Write condensed structures for each of the following molecular formulas. You may have to use rings and/or multiple bonds in some instances.
(a) C_2H_7N (b) C_4H_8 (c) C_2H_4O (d) CH_2O_2

12.34 If someone reported the preparation of a compound with the formula C_3H_9, most chemists would be skeptical. Why?

12.35 How many isomers can you write that fit these descriptions?
(a) alcohols with formula $C_4H_{10}O$
(b) amines with formula C_3H_9N
(c) ketones with formula $C_5H_{10}O$
(d) aldehydes with formula $C_5H_{10}O$
(e) esters with formula $C_4H_8O_2$
(f) carboxylic acids with formula $C_4H_8O_2$

12.36 Which of the following pairs of structures are identical, which are isomers, and which are unrelated?
(a) $CH_3CH_2CH_3$ and $\underset{\overset{|}{C}H_2CH_3}{CH_3}$

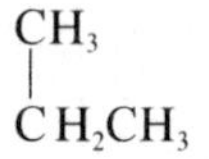

(b) 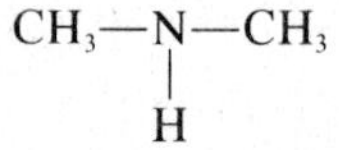 $CH_3{-}\underset{\overset{|}{H}}{N}{-}CH_3$ and $CH_3CH_2{-}\underset{\overset{|}{H}}{N}{-}H$

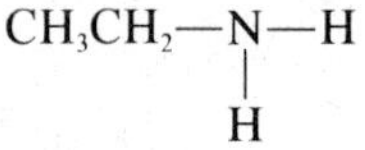

(c) $CH_3CH_2CH_2{-}O{-}CH_3$ and
$CH_3CH_2CH_2{-}\overset{\overset{O}{\|}}{C}{-}CH_3$

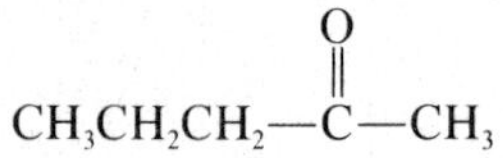

(d) 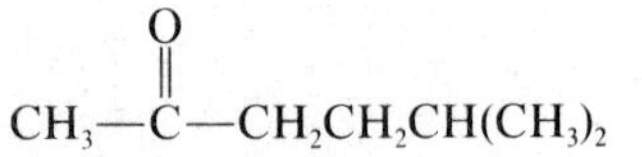 $CH_3{-}\overset{\overset{O}{\|}}{C}{-}CH_2CH_2CH(CH_3)_2$ and
$CH_3CH_2{-}\overset{\overset{O}{\|}}{C}{-}CH_2CH_2CH_2CH_3$

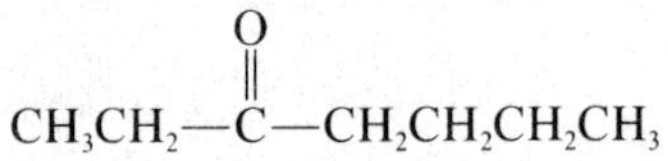

(e) $CH_3CH{=}CHCH_2CH_2{-}O{-}H$ and
$CH_3CH_2\underset{\overset{|}{CH_3}}{CH}{-}\overset{\overset{O}{\|}}{C}{-}H$

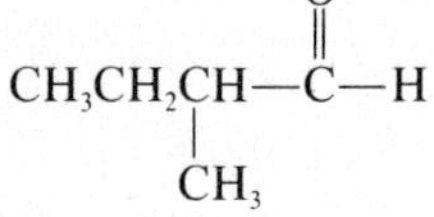

12.37 What is wrong with each of the following structures?
(a) $CH_3{=}CHCH_2CH_2OH$

(b) 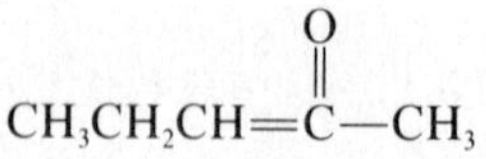 $CH_3CH_2CH{=}\overset{\overset{O}{\|}}{C}{-}CH_3$

(c) $CH_2CH_2CH_2C{\equiv}\underset{\overset{|}{CH_3}}{C}CH_3$

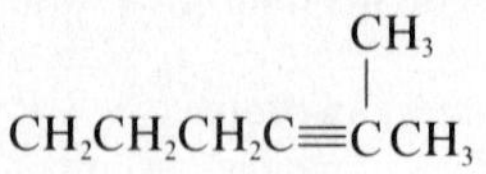

12.38 Which of the structures in each group represent the same compound, and which represent isomers?

(a)

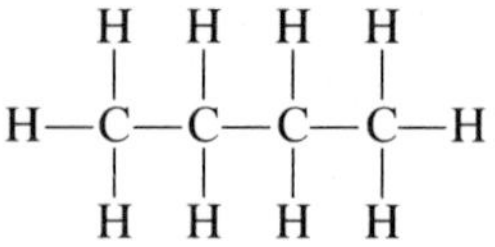

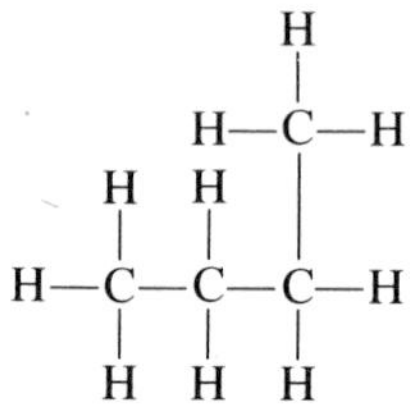

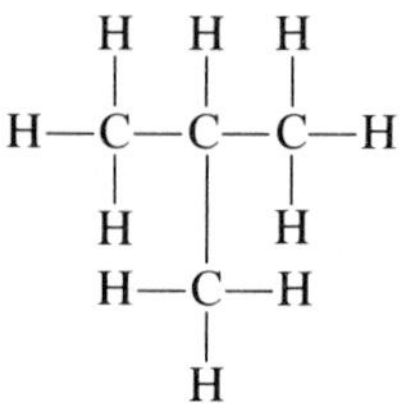

(b) 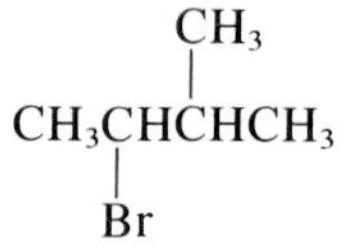 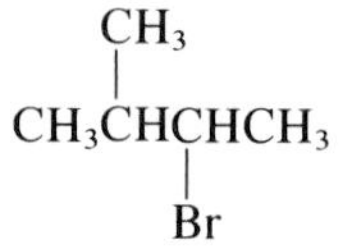 $\underset{\overset{|}{Br}}{CH_3\underset{\overset{|}{CH_3}}{CH}CHCH_3}$ $\underset{\overset{|}{Br}}{CH_3\underset{\overset{|}{CH_3}}{CH}CHCH_3}$

$\underset{\overset{|}{Br}}{\underset{CH_2CHCH_2CH_3}{CH_3}}$

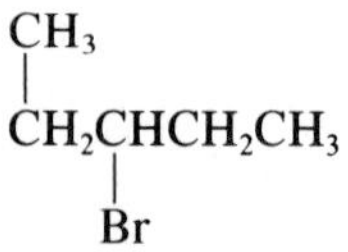

Alkane Nomenclature

12.39 What are the IUPAC names of these alkanes?
(a) 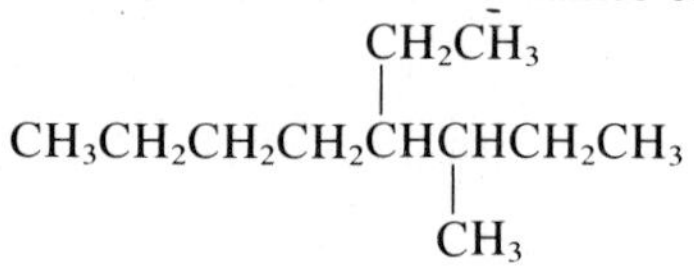 $CH_3CH_2CH_2CH_2\underset{\overset{|}{CH_3}}{\overset{\overset{CH_2CH_3}{|}}{C}H}CHCH_2CH_3$

(b) $CH_3CH_2CH_2\underset{\overset{|}{CH_2CH_3}}{CH}CH_2\overset{\overset{CH_3CHCH_3}{|}}{C}HCH_3$

(c) $CH_3\underset{\overset{|}{CH_3}}{\overset{\overset{CH_3}{|}}{C}}CH_2CH_2CH_2\overset{\overset{CH_3}{|}}{C}HCH_3$

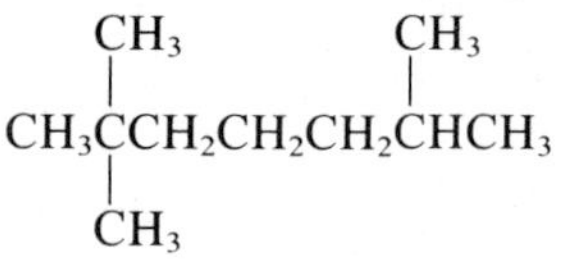

(d) 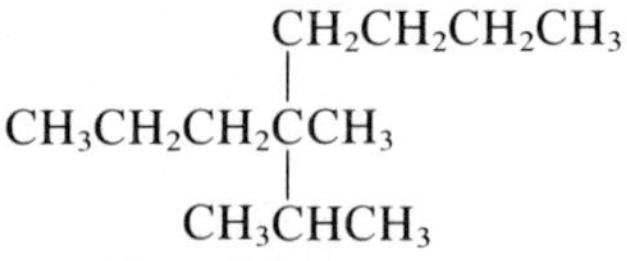 $CH_3CH_2CH_2\underset{\overset{|}{CH_3CHCH_3}}{\overset{\overset{CH_2CH_2CH_2CH_3}{|}}{C}}CH_3$

(e) 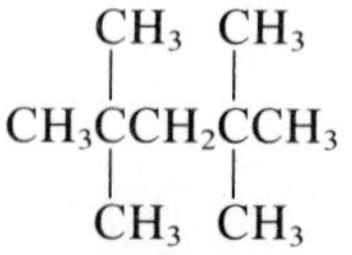 $CH_3\underset{\overset{|}{CH_3}}{\overset{\overset{CH_3}{|}}{C}}CH_2\underset{\overset{|}{CH_3}}{\overset{\overset{CH_3}{|}}{C}}CH_3$

(f) 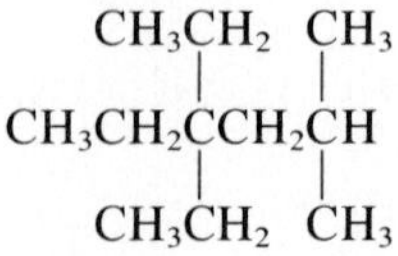 $CH_3CH_2\underset{\overset{|}{CH_3CH_2}}{\overset{\overset{CH_3CH_2\ \ CH_3}{}}{C}}CH_2\underset{\overset{|}{CH_3}}{CH}$ (g) 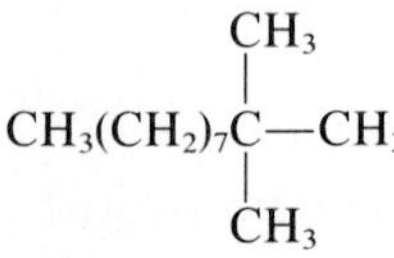 $CH_3(CH_2)_7\underset{\overset{|}{CH_3}}{\overset{\overset{CH_3}{|}}{C}}{-}CH_3$

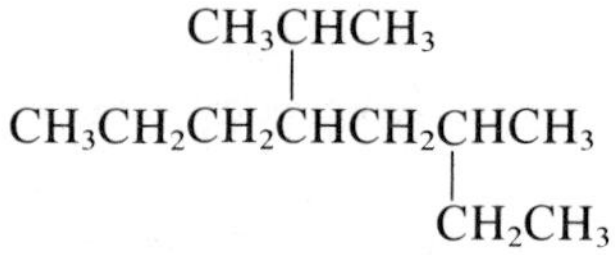

12.40 Write condensed structures for each of these compounds:
(a) 3-ethylhexane   (b) 2,2,3-trimethylpentane
(c) 3-ethyl-3,4-dimethylheptane
(d) 5-isopropyl-2-methyloctane
(e) 2,2,6,6-tetramethyl-4-propylnonane

12.41 The following compound, known trivially as isooctane, is important as a reference substance for determining the "octane" rating of gasoline. What is the proper IUPAC name of isooctane?

$$\text{Isooctane} \quad \underset{\underset{\text{CH}_3}{|}}{\overset{\overset{\text{CH}_3 \quad \text{CH}_3}{|\quad\quad|}}{\text{CH}_3\text{CCH}_2\text{CHCH}_3}}$$

12.42 Provide IUPAC names for each of the five isomers with the formula C_6H_{14}.

12.43 Draw structures corresponding to these IUPAC names:
(a) cyclooctane   (b) 1,1-dimethylcyclopentane
(c) 1,2,3,4-tetramethylcyclobutane
(d) 4-ethyl-1,1-dimethylcyclohexane
(e) ethylcycloheptane   (f) 1,3,5-triethylcyclohexane
(g) 2-isopropyl-1,1-dimethyl-3-propylcyclobutane
(h) 1,2,4,5-tetramethylcyclooctane

12.44 Provide IUPAC names for these cycloalkanes:

(a)
(b)
(c) $CH_3CH_2CH_2-$
(d) $CH_3CH_2CH_2CH_2-$
(e)
(f)
(g)

12.45 The following names are incorrect. Tell what is wrong with each and provide the correct names.

(a) 2,2-Methylpentane

(b) 5-Ethyl-3-methylhexane

(c) 1-Cyclobutyl-2-methylpropane

12.46 Draw structures and give IUPAC names for the nine isomers of C_7H_{16}.

Reactions of Alkanes

12.47 Propane, more commonly known as LP gas, burns in air to yield CO_2 and H_2O. Write a balanced equation for the reaction.

12.48 Write the balanced equation for the combustion of isooctane (Problem 12.41).

12.49 Write the balanced equation for the combustion of a component of paraffin, $C_{25}H_{52}$.

12.50 In a substitution reaction in alkanes, what atom does the halogen replace?

12.51 Write the formulas of the three singly substituted isomers formed when 2,2-dimethylbutane reacts with chlorine in the presence of light.

12.52 Write the formulas of the seven doubly substituted isomers formed when 2,2-dimethylbutane reacts with bromine in the presence of light.

Applications

12.53 Why does a synthetically produced version of a compound have exactly the same properties as a "natural" compound? [App: Natural vs. Synthetic]

12.54 What does it mean when a compound is semisynthetic? [App: Natural vs. Synthetic]

12.55 Must "natural" products be beneficial? Try to find examples of five "natural" chemicals that are hazardous or toxic to human beings. [App: Natural vs. Synthetic]

12.56 Why is it important to know the shape of a molecule? [App: Displaying Molecular Shapes]

12.57 How does petroleum differ from natural gas? [Int: Petroleum]

12.58 What types of hydrocarbons burn most efficiently in an internal combustion engine? [Int: Petroleum]

12.59 Discuss how different alkanes can be separated by the process of distillation. [Int: Petroleum]

12.60 Methane and ethane have boiling points of $-164°C$ and $-89°C$, respectively. Hexadecane ($C_{16}H_{34}$) and heptadecane ($C_{17}H_{36}$) have boiling points of $287°C$ and $303°C$, respectively. Suggest a reason why the boiling points of the latter compounds are so much closer to each other and so much higher than those of the first members of the alkane family? [Int: Petroleum]

Additional Questions and Problems

12.61 Label the functional groups present in the following molecules:

(a) Testosterone, a male sex hormone

(b) Aspartame (an artificial sweetener)

12.62 Label each carbon in Problem 12.61 as primary, secondary, tertiary, or quaternary.

12.63 An analysis proves that two different pure samples each have the molecular formula C_5H_{12}. At atmospheric pressure one boils at $36°C$, while the other boils at $30°C$. How can this be?

12.64 Most lipsticks are about 70% castor oil and wax. Why is lipstick more easily removed with petroleum jelly than with water?

12.65 Write the balanced combustion reaction for dodecane ($C_{12}H_{26}$), a component of kerosene.

12.66 When cyclohexane is exposed to bromine in the presence of light, substitution occurs. Write the formulas of (a) all possible monosubstituted products, and (b) all possible disubstituted products.

12.67 The following names are incorrect. Write the structural formula that agrees with the apparent name and then write the correct name of the compound.
(a) 2-ethylbutane
(b) 2-isopropyl-2-methylpentane
(c) 5-ethyl-1,1-methylcyclopentane
(d) 3-ethyl-3,5,5-trimethylhexane
(e) 1,2-dimethyl-4-ethylcyclohexane
(f) 2,4-diethylpentane
(g) 5,5,6,6-methyl-7,7-ethyldecane

12.68 Draw the structural formulas and name all the cyclic isomers with the formula C_5H_{10}.

12.69 Which should have a higher boiling point, pentane or neopentane? Why?

C H A P T E R

13

Alkenes, Alkynes, and Aromatic Compounds

A fennel plant is an aromatic herb; a phenyl group, pronounced exactly the same way, is the characteristic structural feature of "aromatic" organic compounds. And the connection is more than a pun because anethole ($CH_3OC_6H_4CH=CHCH_3$), the compound that gives fennel its characteristic licorice aroma and flavor, contains a phenyl group.

In this and the remaining four chapters on organic chemistry, we'll examine families of organic compounds whose functional groups give them characteristic properties. Compounds in the three families described in this chapter all contain carbon–carbon multiple bonds. Alkenes, such as ethylene, contain a carbon–carbon double bond; alkynes, such as acetylene, contain a carbon–carbon triple bond; and aromatic compounds, such as benzene, contain a six-membered ring of carbon atoms with three double bonds or other rings with similar arrangements of alternating double bonds. Note that the computer-generated representations of these compounds shown below indicate only the connection between atoms. The multiple bonds themselves are understood rather than specifically represented.

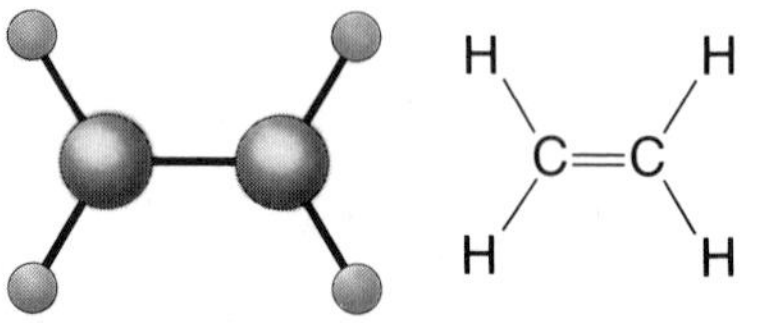

Ethylene
(an *alkene*—has
a C—C double bond)

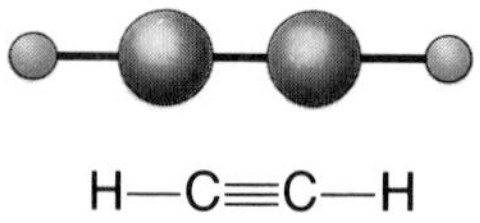

Acetylene
(an *alkyne*—has
a C—C triple bond)

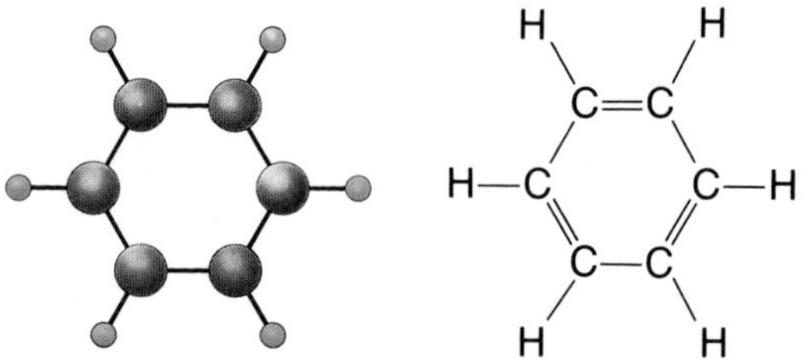

Benzene
(an *aromatic compound*—
has a six-membered ring
and three double bonds)

In this chapter, we'll look at answers to these questions:

1. *What are alkenes, alkynes, and aromatic compounds?* The goal: Be able to recognize the structures of members of these three families of unsaturated organic compounds and give examples of each.

2. *How are alkenes, alkynes, and aromatic compounds named?* The goal: Be able to name an alkene, alkyne, or simple aromatic compound from its structure or write the structure, given the name.

3. *What are cis–trans isomers?* The goal: Be able to identify cis–trans isomers and predict their occurrence.

4. *What are the general properties of alkenes, alkynes, and aromatic compounds?* The goal: Be able to describe and compare such properties as polarity, water solubility, flammability, bonding, and chemical reactivity.

5. *What are the common chemical reactions of alkenes, alkynes, and aromatic compounds?* The goal: Be able to predict the products of reactions of alkenes, alkynes, and aromatic compounds.

6. *How do organic reactions take place?* The goal: Be able to show how addition reactions occur and describe the difference between addition and substitution reactions.

13.1   SATURATED AND UNSATURATED HYDROCARBONS

Saturated   Containing only single bonds between carbon atoms and thus unable to accommodate additional hydrogen atoms.

The alkanes, introduced in Chapter 12, are often referred to as **saturated** hydrocarbons because each carbon atom is fully "saturated" by bonds to the maximum number of hydrogen atoms: No more hydrogen atoms can be added. Alkenes and alkynes, however, contain carbon–carbon multiple bonds to which more hydrogen atoms can be added and are therefore known as **unsaturated** hydrocarbons. **Alkenes** are hydrocarbons that contain carbon–carbon double bonds and **alkynes** are hydrocarbons that contain carbon–carbon triple bonds.

Unsaturated   Containing one or more double or triple bonds between carbon atoms and thus able to add more hydrogen atoms.

$$CH_3CH_2CH_3 \qquad CH_3CH{=}CH_2 \qquad CH_3C{\equiv}CH$$

An alkane   An alkene   An alkyne

Alkene   A compound that contains only carbon and hydrogen and has a carbon–carbon double bond.

13.2   ALKENES

Alkyne   A compound that contains only carbon and hydrogen and has a carbon–carbon triple bond.

Simple alkenes are made in huge quantities by *thermal cracking* of natural gas and petroleum and are the basic building blocks of the vast *petrochemical industry,* which produces organic chemicals from petroleum. Alkanes are rapidly heated to such high temperatures (750–900°C) that they "crack" apart, forming reactive fragments that then reunite or rearrange into lower-molecular-weight molecules such as ethylene ($CH_2{=}CH_2$) and propylene ($CH_2{=}CHCH_3$).

$$CH_3(CH_2)_nCH_3 \xrightarrow{\text{750–900°C}} H_2 + CH_4 + CH_2{=}CH_2 + CH_3CH{=}CH_2 + CH_3CH_2CH{=}CH_2$$

By far the majority of organic chemicals used in making drugs, explosives, paints, plastics, pesticides, and many other products are synthesized by routes that depend on alkene starting materials. More ethylene is produced in the United States each year than any other organic chemical: 18.7 million tons in 1990. Much of it goes into making polyethylene.

Curiously enough, ethylene is also a natural product, being formed in the leaves, flowers, and roots of plants. At each stage in the development of a plant, ethylene has a role to play as a hormone: It controls seedling growth, stimulates root formation, and regulates fruit ripening. During ripening, a fruit produces ethylene that then speeds up ripening in other nearby fruits.

The ability of ethylene to hasten ripening was discovered indirectly by citrus growers who at one time ripened green oranges in rooms heated with kerosene stoves. When the stoves were replaced, the fruit, to their surprise, no longer ripened. Ethylene from incomplete combustion of the kerosene had caused the ripening, not heat from the stoves. Today, fruit is intentionally exposed to ethylene to continue the ripening process during shipment to market.

13.3   NAMING ALKENES AND ALKYNES

Alkenes and alkynes are named systematically by a series of three rules similar to those used for alkanes. The roots indicating the number of carbon atoms in the main chain are the same as those for alkanes (Table 12.4). For an alkene, the

-*ene* suffix is used in place of -*ane*, and for an alkyne the -*yne* suffix is used to give the name of the parent compound.

Step 1. Name the parent compound. Find the longest chain *containing the double or triple bond* and name the parent compound by adding the suffix -*ene* or -*yne* to the root for this main chain.

$CH_3CH_2CH_2CH{=}CH_2$ Name as a *pentene*—a five-carbon chain containing a double bond.

$CH_3CH_2CH_2C{\equiv}CCH_3$ Name as a *hexyne*—a six-carbon chain containing a triple bond.

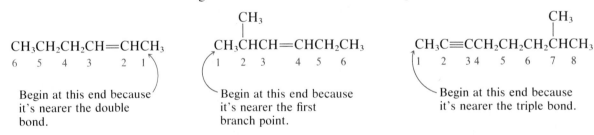

Name as a *hexene*—a six-carbon chain containing a double bond—

not

as a heptene, since the double bond is not contained in the seven-carbon chain.

Step 2. Number the carbon atoms in the main chain. Beginning at the end nearer the multiple bond, number each carbon atom in the chain. If the multiple bond is an equal distance from both ends, as in the middle structure shown here, begin numbering at the end nearer the first branch point.

$$CH_3CH_2CH_2CH{=}CHCH_3$$
$$\,6\quad\,5\quad\,4\quad\,3\quad\quad2\quad\,1$$

Begin at this end because it's nearer the double bond.

$$\overset{\displaystyle CH_3}{\overset{|}{CH_3CHCH{=}CHCH_2CH_3}}$$
$$\,1\quad\,2\quad\,3\quad\quad4\quad\,5\quad\,6$$

Begin at this end because it's nearer the first branch point.

$$\overset{\displaystyle CH_3}{\overset{|}{CH_3C{\equiv}CCH_2CH_2CHCH_3}}$$
$$\,1\quad\,2\quad\,3\,4\quad\,5\quad\,6\quad\,7\quad\,8$$

Begin at this end because it's nearer the triple bond.

For cyclic alkenes, the -*ane* ending of the cycloalkane is replaced by -*ene*. No numbers are needed for an unsubstituted cycloalkene. In substituted cycloalkenes, the double-bond carbon atoms are numbered 1 and 2 so that the first substituent has the lowest number:

Don't begin here.

Name as a cyclohexene.

Begin here so that the substituent has the lowest number.

Step 3. Write out the full name. Identify and number branching substituents in the same way as for alkanes: Assign numbers to the branching substituents on the chain and list them alphabetically. Use commas to separate numbers and use hyphens to separate words from numbers. Indicate the position of the multiple bond in the chain by giving the number of the *first* multiple-bonded carbon. If more than one double bond is present, identify the position of each and use the name endings *-diene, -triene, -tetraene,* and so forth, to give the exact number of double bonds. In naming cycloalkenes, it isn't necessary to include the number 1 to locate the double bond. For example,

$$\underset{5\quad4\quad3\quad2\quad1}{CH_3CH_2CH_2CH{=}CH_2}$$

1-Pentene

$$\underset{6\quad5\quad4\quad3\quad2\ 1}{CH_3CH_2CH_2C{\equiv}CCH_3}$$

2-Hexyne

$$\underset{6\quad5\quad4}{CH_3CH_2CH_2}\!\!\underset{3}{\diagdown}\!\underset{2\ 1}{C{=}CHCH_3}$$
$$\underset{}{CH_3CH_2CH_2}\!\diagup$$

3-Propyl-2-hexene

$$\underset{1\ \ 2\ \ \ 3\,4\ \ \ 5\ \ \ 6\ \ \ 7\ \ 8}{CH_3C{\equiv}CCH_2CH_2CH_2\overset{\displaystyle CH_3}{\overset{|}{C}}HCH_3}$$

7-Methyl-2-octyne

$$\underset{1\ \ \ 2\ \ \ \ 3\ \ \ \ 4}{H_2C{=}\overset{\displaystyle CH_2CH_3}{\overset{|}{C}}{-}CH{=}CH_2}$$

2-Ethyl-1,3-butadiene

4-Methylcyclohexene

For historical reasons, there are a small number of alkenes and alkynes whose names don't conform strictly to the rules. The two-carbon alkene $H_2C{=}CH_2$ should properly be called *ethene,* but the name *ethylene* has been used for so long that it's now accepted by IUPAC. Similarly, the three-carbon alkene *propene* ($CH_3CH{=}CH_2$) is often called *propylene.* The simplest alkyne, $HC{\equiv}CH$, should be known as *ethyne* but is more often called *acetylene.*

Solved Problem 13.1 What is the IUPAC name of the following alkene?

$$CH_3CH_2CH_2{-}\overset{\displaystyle H_3C}{\overset{|}{C}}{=}\overset{\displaystyle CH_2CH_3}{\overset{|}{C}}{-}CH_3$$

Solution First, find and circle the longest chain containing the double bond. In this case, we have to turn a corner to find that it's a heptene:

$$CH_3CH_2CH_2{-}\overset{\displaystyle H_3C}{\overset{|}{C}}{=}\overset{\displaystyle CH_2CH_3}{\overset{|}{C}}{-}CH_3 \qquad \text{Name as a }heptene.$$

Next, number the chain from the end nearer the double bond. The first double-bond carbon is C4 starting from the left-hand end, but C3 starting from the right:

$$\underset{7\ \ \ 6\ \ \ 5}{CH_3CH_2CH_2}{-}\underset{4}{\overset{\displaystyle \overset{2}{H_3C}}{\overset{|}{C}}}{=}\underset{3}{\overset{\displaystyle \overset{1}{CH_2CH_3}}{\overset{|}{C}}}{-}CH_3 \qquad \text{Name as a substituted }3\text{-}heptene.$$

Finally, identify the substituents and write the name.

$$\underset{7\ \ \ 6\ \ \ 5}{CH_3CH_2CH_2}{-}\underset{4}{\overset{\displaystyle \overset{2}{H_3C}}{\overset{|}{C}}}{=}\underset{3}{\overset{\displaystyle \overset{1}{CH_2CH_3}}{\overset{|}{C}}}{-}CH_3 \qquad \begin{array}{l}\text{Substituents: 3-methyl, 4-methyl}\\[4pt]\text{Name: }3,4\text{-}dimethyl\text{-}3\text{-}heptene.\end{array}$$

Solved Problem 13.2   Draw the structure of 3-ethyl-4-methyl-2-pentene.

Solution   First, identify the root name (*pent*) and write the number of carbon atoms it indicates in a straight line. Then, counting from one end, put in the double bond between the number 2 and 3 carbon atoms:

$$\overset{}{C} \ \ \overset{}{C} \ \ \overset{3}{C}=\overset{2}{C} \ \ \overset{1}{C} \qquad \text{2 Pentene}$$

Next, add the ethyl and methyl substituents on carbons 3 and 4 and write in the additional hydrogen atoms so that each carbon atom has four bonds.

$$\overset{5}{CH_3}-\overset{4}{CH}-\overset{3}{\underset{|}{C}}=\overset{2}{CH}-\overset{1}{CH_3} \qquad \text{3-Ethyl-4-methyl-2-pentene}$$

with CH_2CH_3 on C-3 and CH_3 on C-4.

Practice Problems   **13.1**   What are the IUPAC names of the following compounds?
(a) $CH_3CH_2CH_2CH=CHCH(CH_3)_2$
(b) $H_2C=CHCH_2CH_2C=CH_2$ with CH_3 substituent

13.2   Draw structures corresponding to these IUPAC names:
(a) 3-methyl-1-heptene   (b) 4,4-dimethyl-2-pentyne
(b) 2-methyl-2-hexene   (d) 3-ethyl-2,2-dimethyl-3-hexene

13.4 THE STRUCTURE OF ALKENES: CIS–TRANS ISOMERISM

Alkenes and alkynes differ from alkanes in shape because of their different kinds of bonds. Thus, whereas methane is tetrahedral, ethylene is flat (planar) and acetylene is linear (straight).

Methane—a tetrahedral molecule with bond angles of 109.5°.

Ethylene—a flat molecule with bond angles of 120°.

Acetylene—a straight molecule with bond angles of 180°.

The two carbons and four attached atoms that make up the double-bond functional group always lie in a plane. Unlike the situation in alkanes where free rotation around the C—C single bond occurs (Section 12.5), there is no free rotation around double bonds. As a consequence, a new kind of isomerism is possible for alkenes.

To see this new kind of isomerism, look at the following list of C_4H_8 compounds. When written as condensed structures, there appear to be three alkene isomers of formula C_4H_8: 1-butene, 2-butene, and 2-methylpropene. In fact, though, there are *four*. 1-Butene and 2-butene are isomers because their double bonds occur at different positions along the chain, while 2-methylpropene is isomeric with both because it has a different connection of carbon atoms. But because rotation can't occur around carbon–carbon double bonds, *there are two different 2-butenes*. In one isomer, the two methyl groups are close together on the same side of the double bond, but in the other isomer, the two methyl groups are far apart on opposite sides of the double bond.

1-Butene $\quad H_2C=CHCH_2CH_3 \quad =$

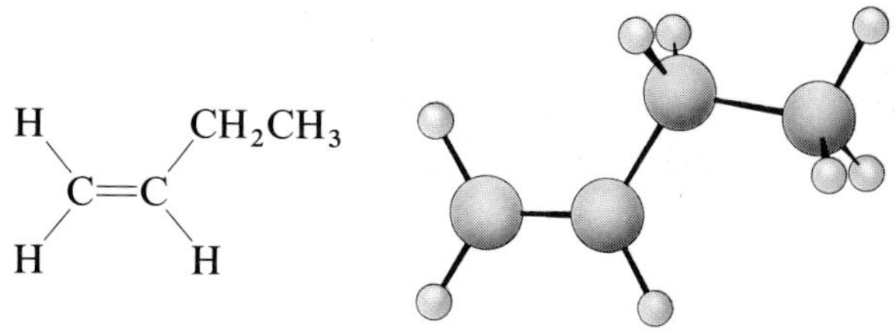

2-Butene $\quad CH_3CH=CHCH_3 \quad =$

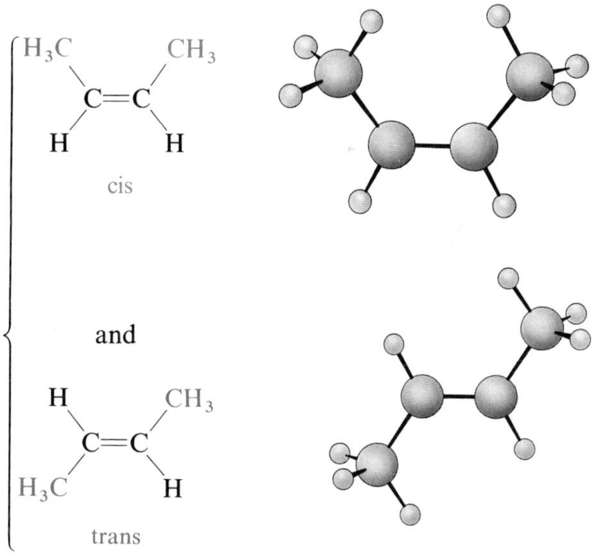

cis

and

trans

2-Methylpropene $\quad H_2C=C(CH_3)_2 \quad =$

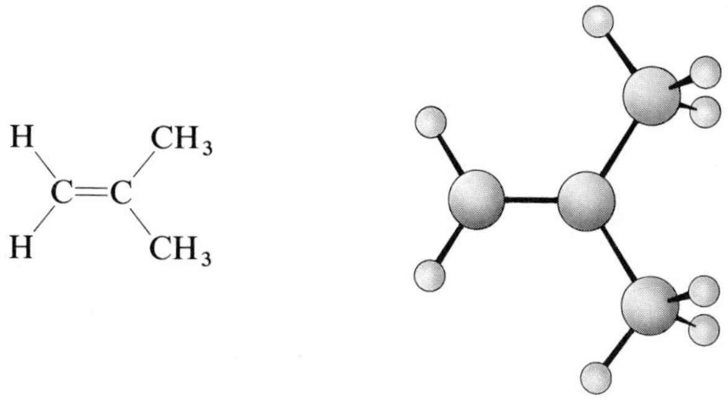

Isomers, cis–trans
Alkenes that have the same formula and connections between atoms but differ in having pairs of groups on opposite sides of the double bond.

Cis isomer Isomer with a specific pair of atoms or groups on the same side of the double bond.

Trans isomer Isomer with a specific pair of atoms or groups on opposite sides of the double bond.

The two 2-butenes are called **cis–trans isomers;** they have the same formula and connections between atoms but have different structures because of the way that groups are attached to different sides of the double bond. The **cis isomer,** the one with its methyl groups on the same side of the double bond, is named *cis*-2-butene. The **trans isomer,** the one with its methyl groups on opposite sides of the double bond, is named *trans*-2-butene. Notice above the considerable difference in shape between cis and trans isomers.

Side-view condensed formulas also illustrate the different geometries of cis and trans isomers. The following formulas are shown in a manner frequently used in organic chemistry: Bonds in the plane of the paper are solid lines; bonds that extend behind this plane are dashed lines; and bonds that come toward you from the paper are wedges.

(Top view)	(Side view)	(Top view)	(Side view)

cis-2-Butene   *trans*-2-Butene

Cis–trans isomerism occurs in an alkene whenever each double-bond carbon is bonded to two different substituent groups. If either double-bond carbon is attached to two identical groups, however, cis–trans isomerism is not possible. The two following compounds are identical, and cis–trans isomerism is impossible for them. You can convince yourself of this by mentally flipping one of the structures and seeing that it becomes identical to the other structure.

These compounds are identical. Because the left-hand carbon of the double bond has two —H's attached, cis–trans isomerism is impossible.

In the next two compounds, however, the structures do not become identical when flipped.

These compounds are not identical. Neither carbon of the double bond has two identical groups attached to it.

Cis isomer   Trans isomer

Solved Problem 13.3   Draw structures for the cis–trans isomers of 2-hexene.

Solution   First, draw a condensed structure of 2-hexene to see what groups are attached to the double-bond carbons.

$$CH_3CH\!=\!CHCH_2CH_2CH_3$$
$$1 \quad 2 \quad\quad 3 \quad 4 \quad 5 \quad 6$$

Attached to C2: —H and —CH₃
Attached to C3: —H and —CH₂CH₂CH₃

Next, draw two double bonds. Choose one end of each double bond and attach its groups in the same way to generate two identical part-structures:

Finally, attach the proper groups to the other end in the two possible ways:

cis-2-Hexene   trans-2-Hexene

The compound with the two hydrogens on the same side of the double bond is named the cis isomer, and that with the two hydrogens on opposite sides is named the trans isomer.

Practice Problems   **13.3**   Which of these substances exist as cis–trans isomers?
(a) 3-heptene   (b) 2-methyl-2-hexene   (c) 5-methyl-2-hexene

13.4   Draw the cis–trans isomers of 3,4-dimethyl-3-hexene.

13.5   PROPERTIES OF ALKENES

Alkenes resemble alkanes in most general properties except for the chemical reactivity of their double bonds. The bonds in alkenes are essentially nonpolar, and the physical properties of alkenes, like those of alkanes, are influenced mainly by weak London forces. Alkenes with one to four carbon atoms are gases, and boiling points increase with the size of the molecules, as illustrated in Table 13.1.

Like alkanes, alkenes are insoluble in water, soluble in nonpolar solvents, and less dense than water. Notice in the table that the different geometrical arrangement in cis and trans isomers is reflected in different physical properties. The alkenes are flammable, and those that are gases present explosion hazards when mixed with air. Gaseous alkenes are not toxic but cause suffocation because of lack of oxygen. As you will see in the next section, alkenes react by addition across their double bonds.

Properties of Alkenes

● Nonpolar; insoluble in water; soluble in nonpolar organic solvents; less dense than water

● Alkane-like in melting points, boiling points, and other physical properties

● Flammable; gaseous alkenes not toxic

● Display cis–trans isomerism when each double-bond C atom has different substituents

● Chemically reactive at double bond, which undergoes addition reactions.

AN APPLICATION: THE CHEMISTRY OF VISION

Does eating carrots really improve your vision? Although carrots probably don't do much to help someone who's already on a proper diet, it's nevertheless true that the chemistry of carrots and the chemistry of vision are related.

Carrots, peaches, sweet potatoes, and other yellow vegetables are rich in β-carotene, an orange-colored alkene that provides our main dietary source of vitamin A. The conversion of β-carotene to vitamin A takes place in the liver, where enzymes first cut the molecule in half and then change the geometry of the C11-C12 double bond to produce a compound named 11-*cis*-retinal. After transport from the liver to the eye, 11-*cis*-retinal reacts with the protein *opsin* to produce the light-sensitive substance *rhodopsin*.

The human eye has two kinds of light-sensitive cells: *rod cells,* which are responsible for seeing in dim light, and *cone cells,* which are responsible for color vision. When light strikes the rod cells of the eye, cis–trans isomerization of the C11-C12 double bond occurs and 11-*trans*-rhodopsin, also called metarhodopsin II, is produced. This cis–trans isomerization is accompanied by a change in molecular geometry, which in turn causes a nerve impulse to be sent to the brain where it is perceived as vision. Metarhodopsin II is then changed back to 11-*cis*-retinal for use in another vision cycle.

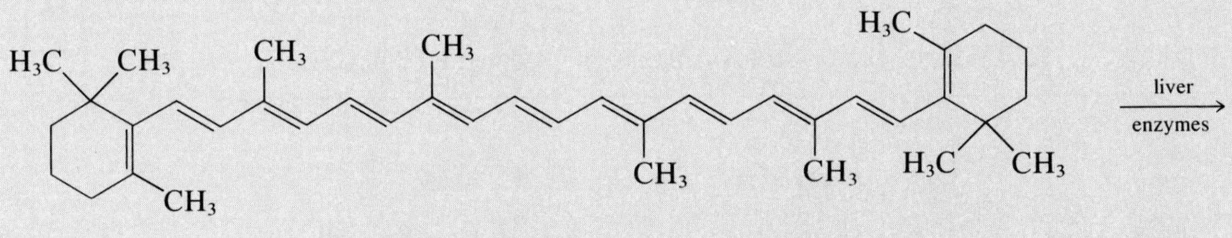

β-Carotene (a polyalkene)

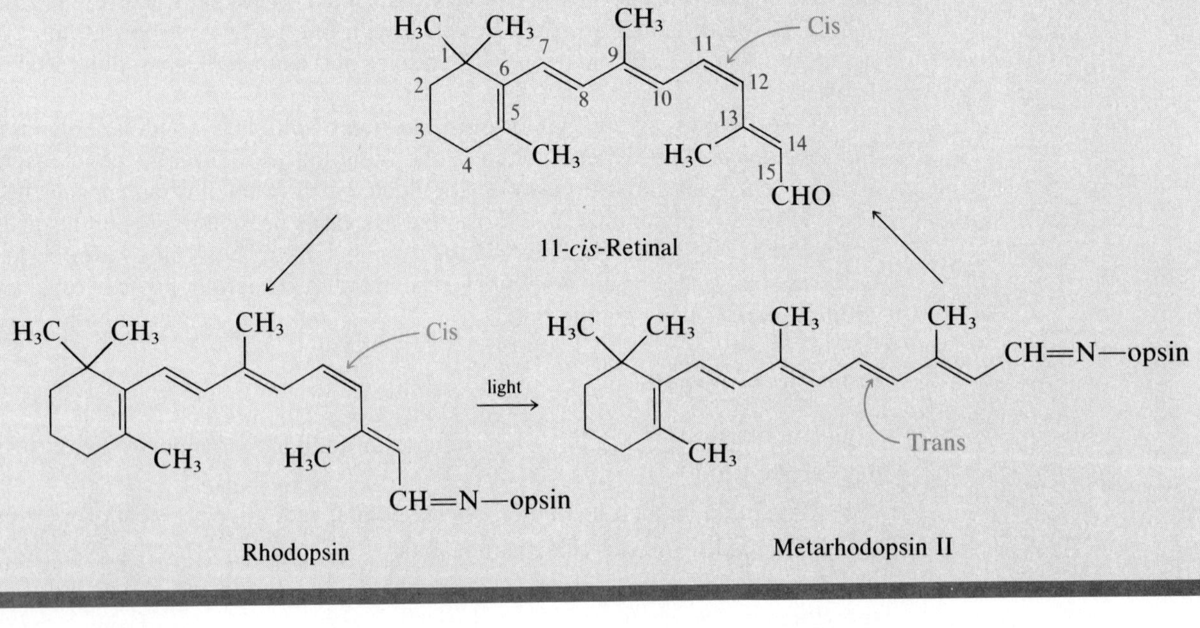

11-*cis*-Retinal

Rhodopsin

Metarhodopsin II

Table 13.1   Physical Properties of Simple Alkenes

Structure	Name	Boiling Point (°C)
$CH_2{=}CH_2$	Ethylene	-104
$CH_3CH{=}CH_2$	Propylene	-47
$CH_3CH_2CH{=}CH_2$	1-Butene	-6
(structure: H₃C, H on left carbon; H, CH₃ on right carbon across C=C)	*trans*-2-Butene	1
(structure: H₃C, CH₃ on top; H, H on bottom across C=C)	*cis*-2-Butene	4
$CH_3CH_2CH_2CH{=}CH_2$	1-Pentene	30
(structure: CH₃CH₂, H on left carbon; H, CH₃ on right carbon across C=C)	*trans*-2-Pentene	36
(structure: CH₃CH₂, CH₃ on top; H, H on bottom across C=C)	*cis*-2-Pentene	37

13.6 CHEMICAL REACTIONS OF ALKENES AND ALKYNES

Most of the reactions of carbon–carbon multiple bonds can be grouped under the category of **addition reactions,** which have the same general pattern as the inorganic *combination* reactions you saw in Section 7.9 (A + B → C). A reagent we might write as X—Y adds to the multiple bond in the unsaturated starting material to yield a product that has only single bonds. Alkenes and alkynes react similarly in many ways, but we'll look mainly at alkenes in this chapter because they're more common and more important.

Half of this double bond breaks.   This single bond breaks.   These two single bonds form.   An addition reaction

Addition of Hydrogen to Alkenes and Alkynes: Hydrogenation   Alkenes and alkynes react with hydrogen in the presence of a metal catalyst to yield the corresponding alkane product:

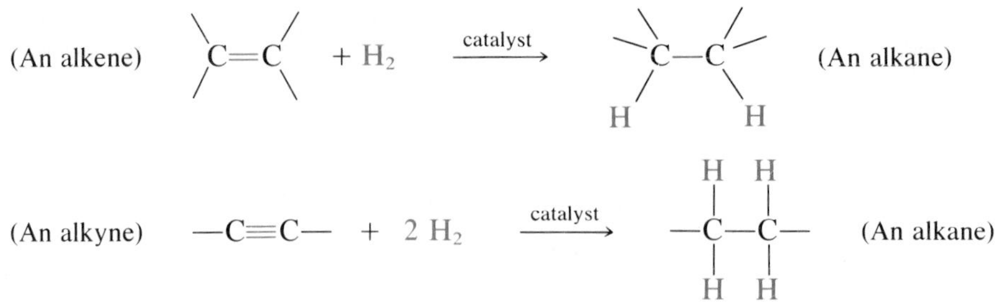

(An alkene) $\overset{\diagdown}{\underset{\diagup}{C}}=\overset{\diagup}{\underset{\diagdown}{C}}$ + H_2 $\xrightarrow{\text{catalyst}}$ $\overset{\diagdown}{\underset{\diagup}{C}}\overset{}{\underset{H}{|}}-\overset{}{\underset{H}{|}}\overset{\diagup}{\underset{\diagdown}{C}}$ (An alkane)

(An alkyne) $-C\equiv C-$ + 2 H_2 $\xrightarrow{\text{catalyst}}$ $-\overset{H}{\underset{H}{\overset{|}{\underset{|}{C}}}}-\overset{H}{\underset{H}{\overset{|}{\underset{|}{C}}}}-$ (An alkane)

For example,

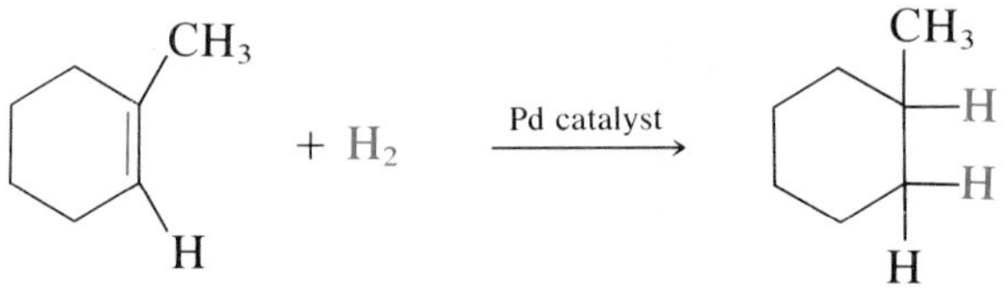

1-Methylcyclohexene   Methylcyclohexane (85% yield)

Hydrogenation The reaction of an alkene (or alkyne) with H_2 to yield an alkane product.

The addition of hydrogen to an alkene, often called **hydrogenation,** is used commercially to convert unsaturated vegetable oils, which contain numerous double bonds, to the saturated fats used in margarine and cooking fats. We'll see exact structures for these fats and oils in Chapter 23.

Solved Problem 13.4   What product would you obtain from the following reaction?

$$CH_3CH_2CH_2CH=CHCH_3 + H_2 \xrightarrow{Pd} ?$$

Solution   First, rewrite the starting material, showing a single bond and two partial bonds in place of the double bond:

$$CH_3CH_2CH_2\overset{|}{C}H-\overset{|}{C}HCH_3$$

Next, add one hydrogen to each carbon atom of the double bond and rewrite the product in condensed form:

$$CH_3CH_2CH_2\overset{|}{\underset{H}{C}}H-\overset{|}{\underset{H}{C}}HCH_3 \ = \ CH_3CH_2CH_2CH_2CH_2CH_3 \qquad \text{Hexane}$$

The reaction is

$$CH_3CH_2CH_2CH=CHCH_3 + H_2 \xrightarrow{Pd} CH_3CH_2CH_2CH_2CH_2CH_3$$

Practice Problem   13.5   Write the structures of products of these hydrogenation reactions:

(a) $CH_3CH_2CH=CH_2 + H_2 \xrightarrow{Pd}$?   (b) *cis*-2-butene + $H_2 \xrightarrow{Pd}$?

(c) *trans*-2-butene + $H_2 \xrightarrow{Pd}$?   (d) 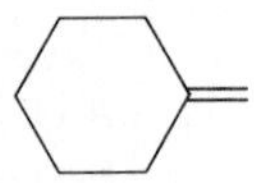 $=CH_2 + H_2 \xrightarrow{Pd}$?

Halogenation, alkene The reaction of an alkene with a halogen (Cl_2 or Br_2) to yield a 1,2-dihaloalkane product.

Addition of Cl_2 and Br_2 to Alkenes: Halogenation Alkenes react with the halogens Br_2 and Cl_2 to give 1,2-dihaloalkane addition products in a **halogenation** reaction:

$$\diagdown C=C \diagdown + X_2 \longrightarrow -C-C- \qquad \text{(A 1,2-dihaloalkane where X = Br or Cl)}$$

For example,

Ethylene + Cl_2 ⟶ 1,2-Dichloroethane

Approximately 7 million tons per year of 1,2-dichloroethane are manufactured by this reaction as the first step in making PVC [poly(vinyl chloride)] plastics.

Bromination provides a convenient test for the presence in a molecule of a carbon–carbon double or triple bond (Figure 13.1). A drop of the reddish-brown solution of bromine in carbon tetrachloride (CCl_4) is added to a sample of an unknown compound. Immediate disappearance of the color as the bromine reacts to form a colorless dibromide reveals the presence of the multiple bond.

Practice Problem 13.6 What products would you expect from the following halogen addition reactions?
(a) 2-methylpropene + Br_2 → ?   (b) 1-pentene + Cl_2 → ?

Figure 13.1
Testing for unsaturation with bromine. (a) No color change results when the bromine solution is added to hexane (C_6H_{14}). (b) Disappearance of the bromine color when it is added to 1-hexene (C_6H_{12}) indicates the presence of a double bond.

(a)   (b)

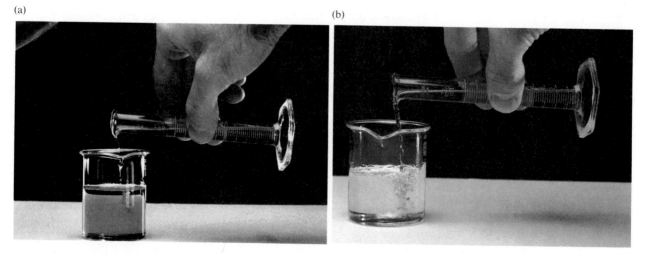

Hydrohalogenation The reaction of an alkane with HCl or HBr to yield an alkyl halide product.

Addition of HBr and HCl to Alkenes Alkenes react with hydrogen bromide (HBr) to yield *alkyl bromides* (R—Br) and with hydrogen chloride (HCl) to yield *alkyl chlorides* (R—Cl), in what are called **hydrohalogenation** reactions:

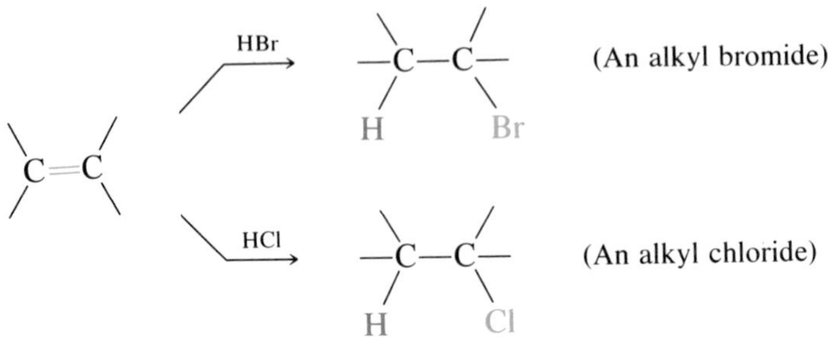

The addition of HBr to 2-methylpropene is typical of such reactions:

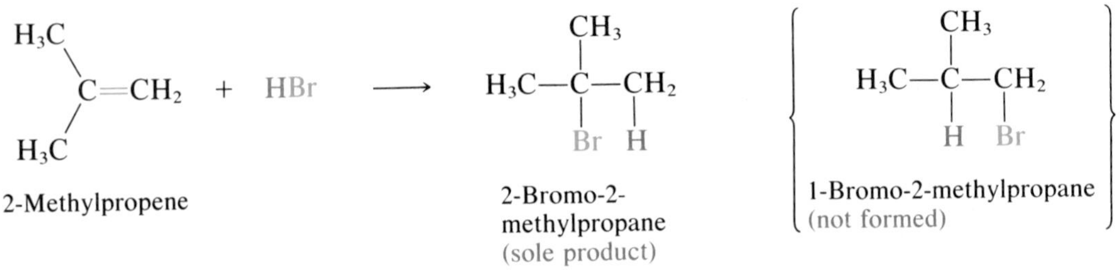

Look carefully at the previous example. Only one of the two possible addition products is obtained. 2-Methylpropene *could* add HBr to give 1-bromo-2-methylpropane, but it doesn't; it gives only 2-bromo-2-methylpropane. This is what usually happens when HBr and HCl (or other reagents) add to alkenes with unsymmetrically substituted double bonds, that is, double bonds in which one carbon is bonded to more hydrogens than the other. Such addition takes place according to a rule formulated in 1869 by the Russian chemist Vladimir Markovnikov:

Markovnikov's rule In the addition of HX to an alkene, the H becomes attached to the carbon that already has the most H's, and the X becomes attached to the carbon that has fewer H's.

Markovnikov's rule: In the addition of HX to an alkene, the H becomes attached to the carbon that already has the most H's, and the X becomes attached to the carbon that has fewer H's.

This rule allows prediction of the results of hydrohalogenation:

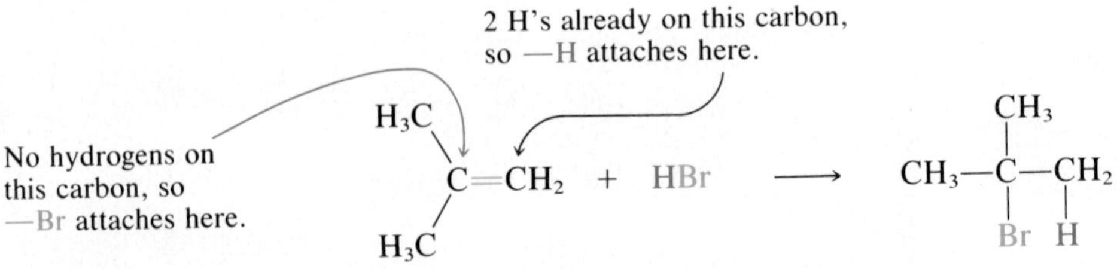

Solved Problem 13.5 What product would you expect from the following reaction?

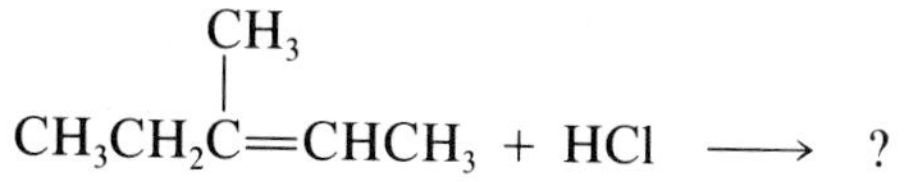

$$\underset{\overset{\displaystyle CH_3}{|}}{CH_3CH_2C}=CHCH_3 + HCl \longrightarrow \quad ?$$

Solution We know that reaction of an alkene with HCl leads to formation of an alkyl chloride addition product according to Markovnikov's rule. To make a prediction, look at the starting alkene and count the number of hydrogens already attached to each double-bond carbon. Then write the product by attaching H to the carbon that already has more hydrogens and attaching Cl to the carbon that has fewer hydrogens.

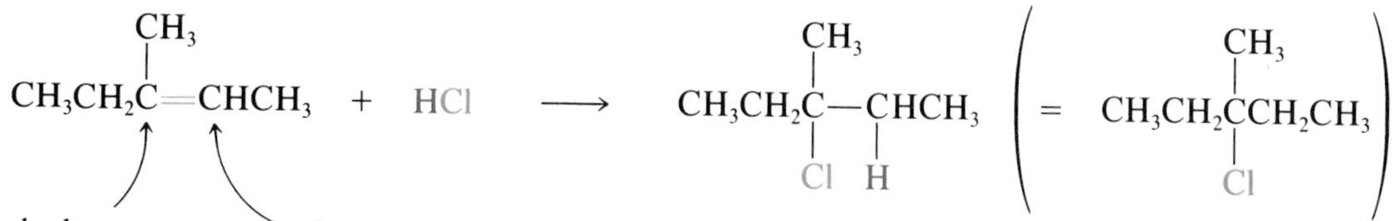

$$\underset{\overset{\displaystyle CH_3}{|}}{CH_3CH_2C}=CHCH_3 \ + \ HCl \longrightarrow \ \underset{\overset{\displaystyle CH_3}{|}}{CH_3CH_2C}-\underset{|}{\underset{H}{C}}HCH_3 \underset{Cl}{} \quad \left(= \ \underset{\overset{\displaystyle CH_3}{|}}{CH_3CH_2C}\underset{\underset{Cl}{|}}{C}H_2CH_3 \right)$$

No hydrogens on this carbon, so —Cl attaches here.　One hydrogen already on this carbon, so —H attaches here.　3-Chloro-3-methylpentane

Practice Problems 13.7 What products would you expect from these reactions?
(a) cyclohexene + HBr → ?　(b) 1-hexene + HCl → ?
(c) $(CH_3)_2CHCH=CH_2$ + HI → ?

13.8 What alkenes are the following alkyl halides likely to have been made from? (Careful, there may be more than one answer.)
(a) 3-chloro-3-ethylpentane　(b) $(CH_3)_2CHCBr(CH_3)_2$

Addition of Water to Alkenes: Hydration An alkene doesn't react with pure water if the two are just mixed together, but if the right experimental conditions are used, an addition reaction takes place to yield an *alcohol* (R—OH). This **hydration** reaction occurs on treatment of the alkene with water in the presence of a strong acid catalyst such as H_2SO_4. In fact, nearly 300 million gallons of ethyl alcohol (ethanol) are produced each year in the United States by this method.

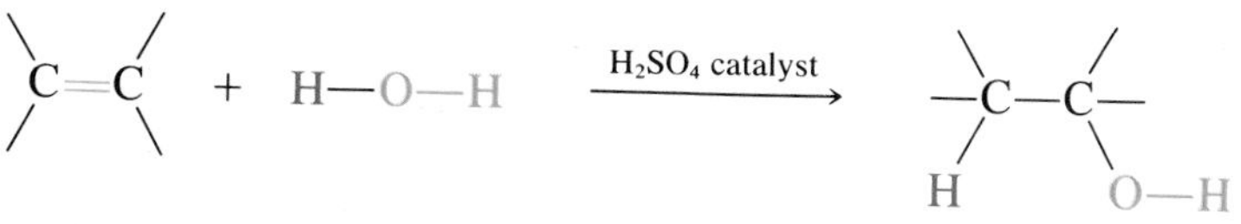

Hydration, alkene The reaction of an alkene with water to yield an alcohol.

$$\underset{/}{\overset{\backslash}{C}}=\underset{\backslash}{\overset{/}{C}} \ + \ H-O-H \ \xrightarrow{H_2SO_4 \text{ catalyst}} \ -\underset{\underset{H}{|}}{C}-\underset{\underset{O-H}{|}}{C}- \quad \text{(An alcohol)}$$

For example,

$$H_2C=CH_2 \ + \ H_2O \ \xrightarrow{H_2SO_4, \ 250°C} \ \underset{\underset{H}{|} \ \underset{O-H}{|}}{H_2C-CH_2} \quad (= \ CH_3CH_2OH)$$

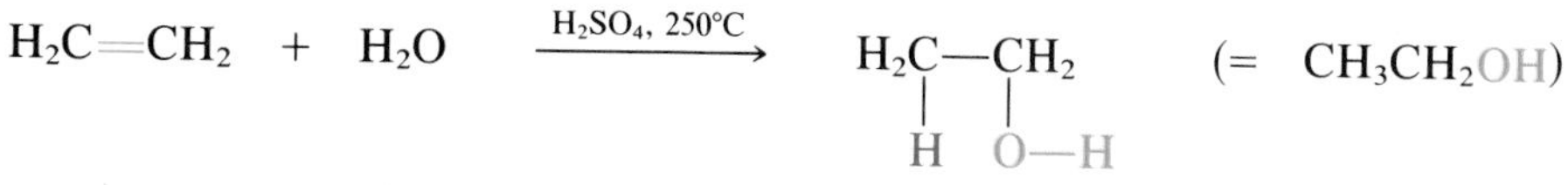

Ethylene　Ethyl alcohol

As with the addition of HBr and HCl, Markovnikov's rule can be used to predict the product when water adds to an unsymmetrically substituted alkene:

No hydrogens on this carbon, so —OH attaches here.

Two hydrogens already on this carbon, so —H attaches here.

$$H_3C \underset{H_3C}{\overset{}{C}} = CH_2 \ + \ H-O-H \ \xrightarrow[250°C]{H_2SO_4} \ H_3C-\underset{\underset{OH}{|}}{\overset{\overset{CH_3}{|}}{C}}-\underset{\underset{H}{|}}{CH_2} \ \left(= \ H_3C\underset{\underset{OH}{|}}{\overset{\overset{CH_3}{|}}{C}}CH_3 \right)$$

2-Methyl-2-propanol

Practice Problems   **13.9**   What products would you expect from these hydration reactions?

(a) [cyclohexane ring]=CH₂ + H₂O ⟶ ?   (b) [cyclohexene ring with CH₃] + H₂O ⟶ ?

13.10   What alkene starting material might 3-methyl-3-pentanol be made from?

$$CH_3CH_2\underset{\underset{OH}{|}}{\overset{\overset{CH_3}{|}}{C}}CH_2CH_3 \qquad \text{3-Methyl-3-pentanol}$$

13.7 HOW AN ALKENE ADDITION REACTION OCCURS

How do addition reactions take place? Do two molecules, say ethylene and HBr, simply collide and immediately form a product molecule of bromoethane, or is the process more complex? Studies have shown that alkene addition reactions take place in two distinct steps, as illustrated in Figure 13.2 for the addition of HBr to ethylene.

In the first step, the alkene reacts with H⁺ from the acid HBr. The carbon–carbon double bond partially breaks, and two electrons go to form a new covalent bond between one of the carbons and the incoming hydrogen. The other double-bond carbon, having had electrons removed from it, now has only six electrons in its outer shell and bears a positive charge. Unlike sodium ion (Na^+) and other metal cations, which are stable and easily isolated in salts like NaCl, such **carbocations** are quite unstable. As soon as it's formed by reaction of an alkene with H^+, the carbocation immediately reacts with bromide ion to form a neutral product.

An accounting like that in Figure 13.2 of the individual steps by which old bonds are broken and new bonds formed in a reaction is known as a **reaction mechanism.** Although we won't examine many reaction mechanisms in this book, understanding mechanisms is an important part of organic chemistry.

Carbocation A polyatomic ion with a positively charged carbon atom.

Reaction mechanism A complete description of how a reaction occurs, including the details of each individual step in the overall process.

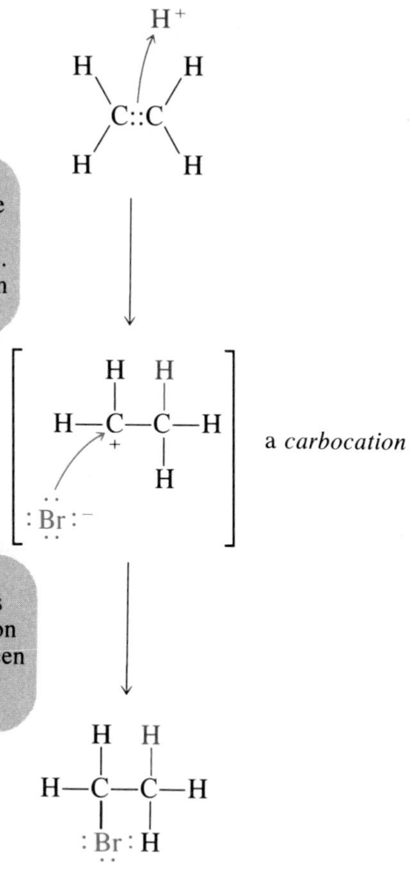

Step 1
Two electrons from the C—C double bond are used to form a new single bond between the incoming hydrogen ion and one of the carbons. The other carbon now has only six electrons in its outer shell and has a positive charge.

a *carbocation*

Step 2
The positively charged *carbocation* then reacts with the negative bromide ion, using an electron pair from bromide to form a single bond between carbon and bromine. The second carbon thus regains an outer-shell octet.

Figure 13.2
How the addition reaction of HBr to an alkene occurs: a reaction mechanism. The reaction takes place in two steps and involves a carbocation intermediate.

Their study is, for example, essential to our ever-expanding ability to understand biochemistry and the physiological effects of drugs. If you continue your study of chemistry, you'll see reaction mechanisms often.

Practice Problem 13.11 Draw the structure of the carbocation intermediate formed during the reaction of 2-methylpropene with HCl.

13.8 ALKENE POLYMERS

Polymer A very large molecule composed of identical repeating units and formed by the combination of small molecules.

Monomer A small molecule combined to form a polymer.

Polymers are large molecules formed by the bonding together of many smaller molecules called **monomers.** As we'll see in later chapters, *biological polymers* occur throughout nature. Cellulose and starch, for example, are polymers built from sugars. Although the basic idea is the same, synthetic polymers are much simpler than biopolymers, since the starting monomer units are usually small and simple organic molecules.

AN APPLICATION: ISOPRENE, TERPENES, AND NATURAL RUBBER

Nature is a clever architect, constructing many complex biomolecules by combination of smaller, simpler units. The isoprene structure and the molecules that contain it provide a beautiful example of this principle.

A branched alkene with two double bonds

$$CH_2{=}\overset{\displaystyle \overset{CH_3}{|}}{C}{-}CH{=}CH_2$$

Isoprene

Isoprene itself is not a natural product. Nevertheless, an amazing array of molecules that contain the five-carbon isoprene skeleton are found in plants and animals, where they perform a wide variety of functions. Some variations on the isoprene unit are drawn here as line structures. In some, a double bond is preserved, and in others it is lost.

Isoprene   Some isoprene units in terpenes

Many essential oils owe their distinctive aromas to simple *terpenes,* compounds composed entirely of isoprene units and found in all plants, for example,

Myrcene—two isoprene units
(oil of bay)

Limonene—two isoprene units
(oil of lemon, orange, caraway)

More complex terpenes and compounds with various functional groups on their isoprene units are also widely found in nature. The hydrocarbon squalene, for example, is an intermediate used by organisms in synthesizing steroids, an important class of biomolecules (Section 23.10). Retinal, a key substance in vision, is also a terpene.

Retinal—four isoprene units

The raw material from rubber trees is a polyisoprene with an average of 5000 isoprene units per molecule. Interestingly, the double bonds that remain all have cis geometry.

Natural rubber (*cis*-polyisoprene)

Rubber collected in the Amazon region in Brazil is being smoked to dry it. Natural rubber is sticky and not too useful until the polymer molecules are connected (*vulcanized*) by cross links that pull them back together when rubber is stretched.

Squalene—six isoprene units

Vinyl monomer   A compound with the structure $CH_2{=}CHZ$ that undergoes polymerization to give a polymer with the repeating unit $-CH_2-CHZ-$.

Polymerization   A reaction in which monomers combine to form a polymer.

Chain-growth polymer   A polymer formed by the addition of monomer molecules one by one to the end of a growing chain.

Many simple alkenes (called **vinyl monomers** because $H_2C{=}CH_2$ is known as a *vinyl group*) undergo **polymerization** reactions when treated with the proper catalysts. Ethylene yields polyethylene on polymerization, propylene yields polypropylene (Figure 13.3), and styrene yields polystyrene. The polymer product might have anywhere from a few hundred to a few thousand monomer units incorporated into a long chain.

The fundamental reaction in the polymerization of a vinyl monomer resembles the additions to double bonds described in the previous section. The addition to an alkene of a species called an *initiator* yields a reactive intermediate, which in turn adds to a second alkene molecule to produce another reactive intermediate, which adds to a third alkene molecule, and so on. Because the result is continuous addition of one monomer after another to the end of the growing polymer chain, polymers made in this way are classified as **chain-growth polymers** (also known as addition polymers).

$$H_2C{=}\overset{\overset{\displaystyle Z}{|}}{C}H \quad \xrightarrow{\text{polymerization}} \quad -(-CH_2-\overset{\overset{\displaystyle Z}{|}}{C}H-CH_2-\overset{\overset{\displaystyle Z}{|}}{C}H-CH_2-\overset{\overset{\displaystyle Z}{|}}{C}H-)_n-$$

Vinyl monomer (where the colored Z represents a substituent group)

Polymer

Variations in the Z group impart very different properties, as illustrated by the familiar alkene polymers listed in Table 13.2. Some vinyl polymers, like styrene–butadiene rubber shown in Table 13.2, are *copolymers:* polymers made from more than one kind of monomer.

Polymer properties also vary depending on the average size of the huge molecules (*macromolecules*) in a particular sample and on how much they are branched. The long molecules in straight-chain polyethylene pack close together, giving a rigid material (*high-density polyethylene*) mainly used in bottles for products such as milk and motor oil (Figure 13.4a). When polyethylene molecules contain many branches, they can't pack together so tightly and instead form a flexible material (*low-density polyethylene*) used mainly in packaging films (Figure 13.4b).

Figure 13.3
Plastics and creative endeavor. In 1983, the artist Christo surrounded 11 islands in Biscayne Bay with 6.5 million square feet of floating pink fabric to create a striking work of environmental art. The fabric used was woven polypropylene, an alkene polymer. (© Christo, 1983.)

Table 13.2   Some Alkene Polymers and Their Uses

Monomer Name	Monomer Structure	Polymer Name	Uses
Ethylene	$H_2C{=}CH_2$	Polyethylene	Packaging, bottles
Propylene	$H_2C{=}CH{-}CH_3$	Polypropylene	Bottles, rope, pails, medical tubing
Vinyl chloride	$H_2C{=}CH{-}Cl$	Poly(vinyl chloride)	Insulation, plastic pipe
Styrene	$H_2C{=}CH{-}\bigcirc$	Polystyrene	Foams and molded plastics
Styrene and butadiene	$H_2C{=}CH{-}\bigcirc$ and $H_2C{=}CHCH{=}CH_2$	Styrene–butadiene rubber (SBR)	Synthetic rubber for tires
Acrylonitrile	$H_2C{=}CH{-}C{\equiv}N$	Orlon, Acrilan	Fibers, outdoor carpeting
Methyl methacrylate	$H_2C{=}\overset{O}{\overset{\|}{\underset{\underset{CH_3}{\|}}{C}}}COCH_3$	Plexiglas, Lucite	Windows, contact lenses, fiber optics
Tetrafluoroethylene	$F_2C{=}CF_2$	Teflon	Nonstick coatings, bearings, replacement heart valves and blood vessels

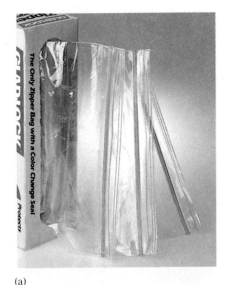

Figure 13.4
Polyethylene items. (a) Low-density polyethylene (LDPE) is moisture-proof but flexible and is used in plastic bags and packaging of all kinds. (b) High-density polyethylene (HDPE) is strong and rigid enough to be used in many kinds of bottles.

(a)

(b)

Polymer technology has come a long way since the development of synthetic rubber, nylon, Plexiglas, and Teflon, which was hastened by the need for materials during World War II. The use of polymers has changed the nature of activities from plumbing and carpentry to clothing and auto manufacture. In the health care fields one result has been the proliferation of inexpensive, disposable equipment (Figure 13.5).

Figure 13.5
Disposable syringes. These syringes have been removed from their packaging, used once, and discarded.

Practice Problem **13.12** Write the structure of the repeating units in (a) polystyrene and (b) polyacrylonitrile (Acrilan).

13.9 ALKYNES

Alkynes are named systematically in the same manner as alkenes and are like alkenes in physical and chemical properties.

$$HC\equiv CH \qquad CH_3C\equiv CH \qquad \underset{\displaystyle CH_3}{\overset{\displaystyle CH_3}{CH_3CHC\equiv CH}}$$

Acetylene   Propyne   3-Methyl-1-butyne
(ethyne)

Acetylene (ethyne), the commercially most important alkyne, is a colorless gas with a light odor like that of ether. Industrial-grade acetylene used in cutting and welding, however, often has a bad odor because of impurities. When mixed with oxygen in an oxyacetylene welding torch, burning acetylene produces a clear, light-blue flame with temperatures up to 3300°C.

Alkynes are not plentiful in nature, but they have been found in a number of plants and some animals. The compound known as *ichthyothereol* was identified in a natural poison used by Amazonian indians on their arrowheads. Not many alkynes play essential roles in human biochemistry, but one group of alkynes is indeed significant: oral contraceptives.

Some physiologically active alkynes

Ichthyothereol
(a poison of plant origin)

Norethynodrel
(a contraceptive, used in Enovid and others)

13.10 AROMATIC COMPOUNDS AND THE STRUCTURE OF BENZENE

In the early days of organic chemistry, the word *aromatic* was used to describe certain fragrant substances from fruits, trees, and other natural sources. It was soon realized, however, that substances grouped as aromatic behaved in a chemically different manner from most other organic compounds. Today, the term *aromatic* is used to refer to the class of compounds containing benzene-like rings. The association with fragrance has long been lost.

Aromatic compound A compound that contains a six-membered ring of carbon atoms with three double bonds or a similarly stable ring.

Benzene, the simplest **aromatic compound,** is a flat, symmetrical molecule with the molecular formula C_6H_6. It's often represented as cyclohexatriene: a six-membered carbon ring with three double bonds. The problem with this representation is that it gives the wrong impression about benzene's chemical reactivity and bonding. Since benzene appears to have three double bonds, you might expect it to react with H_2, Br_2, HCl, and H_2O to give the same kinds of addition products that alkenes do. But this expectation is wrong. Benzene and other aromatic compounds are much less reactive than alkenes and don't normally undergo addition reactions.

Benzene's remarkable stability is a consequence of its structure. Forming a ring of six carbons joined by single bonds to each other and to six hydrogen atoms (Figure 13.6a) leaves six electrons left over. Placing these electrons in three alternating double bonds gives each carbon atom an electron octet. But where are these double bonds located in the ring? There are two equivalent possibilities, as shown in Figure 13.6b.

The problem with the cyclohexatriene representations for benzene is that neither of the two equivalent structures in Figure 13.6b is fully correct by itself. Experimental evidence shows that all six carbon–carbon bonds in benzene are identical, so a picture with three double bonds and three single bonds can't be right.

Resonance The existence of a molecule in a single structure intermediate among two or more correct possible double-bond-containing structures that can be drawn.

The stability of benzene can best be explained by assuming it to be a completely symmetrical molecule with the six ''double-bond'' electrons circulating around the *entire* ring (Figure 13.6c). Such a structure is hard to represent by the standard conventions using lines for covalent bonds. Sometimes it's done by representing the six electrons as a circle inside the six-membered ring (Figure 13.6d). Usually, though, we draw the ring with three double bonds, with the understanding that it's an aromatic ring with equivalent bonding all around. Similar limitations arise in placing double bonds in the structures of many molecules. The name **resonance** is given to this phenomenon: the existence of a molecule in a single structure intermediate among two or more possible correct double-bond-containing structures that can be drawn.

Figure 13.6
Some representations of benzene. The planar molecular structure is shown in (a), and an electron density cloud of bonding electrons distributed around the ring is shown in (c). Benzene is usually represented by the two equivalent structures in (b) or by the symmetrical structure in (d).

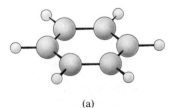

(a)

and

(b)

(c)

(d)

The simple aromatic hydrocarbons are nonpolar, insoluble in water, volatile, and flammable. Unlike the alkanes and alkenes, however, several are toxic. Benzene exposure must be limited because it is carcinogenic, and the dimethyl-substituted benzenes are central-nervous-system depressants.

Everything that we've said about the structure and stability of the benzene ring (C_6H_6) also applies to the ring when it has substituents, for example, in compounds such as hexachlorophene and vanillin:

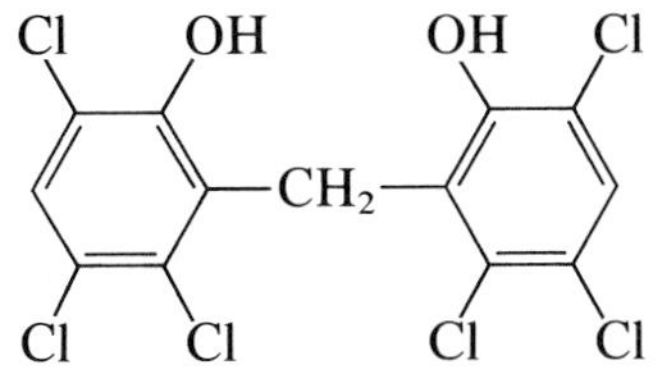

Hexachlorophene
(a germicide)

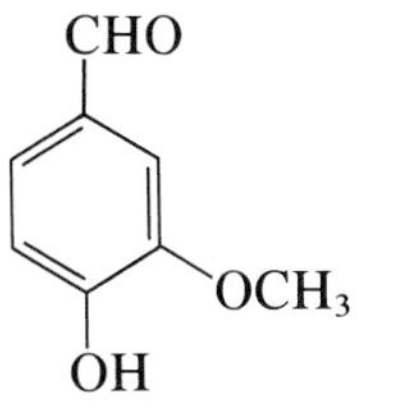

Vanillin
(vanilla flavoring)

The benzene ring is also present in many biomolecules and retains its characteristic properties in these compounds as well. "Aromaticity" is not limited to rings containing only carbon. Many aromatic compounds contain nitrogen atoms, as shown here for indole and adenine. These and all compounds that contain a substituted benzene ring, or a similar stable ring in which electrons are equally shared around the ring, are classified as aromatic compounds.

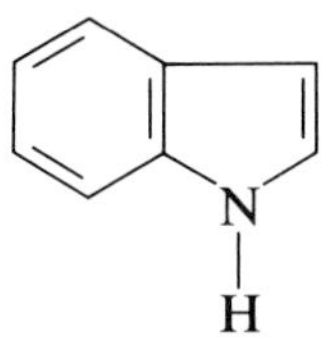

Indole
(from indigo plant and coal tar,
used in perfume)

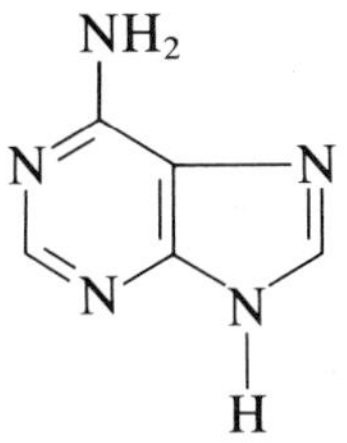

Adenine
(a nucleic acid component)

13.11 NAMING AROMATIC COMPOUNDS

Compounds in which one or more hydrogen atoms in the benzene ring are replaced by other groups use *-benzene* (or *-benz-*) as the root for their systematic, IUPAC names. Thus, C_6H_5Br is bromobenzene, $C_6H_5CH_2CH_3$ is ethylbenzene, and so on. No number is needed for monosubstituted benzenes because all the ring positions are equivalent.

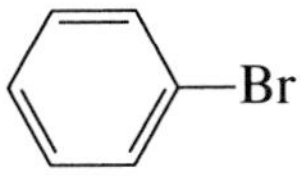

Bromobenzene

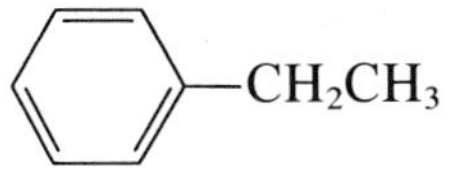

Ethylbenzene

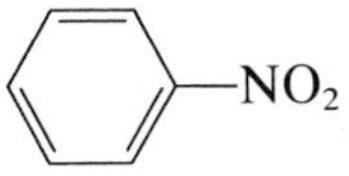

Nitrobenzene

Ortho Indicates 1,2 substituents on a benzene ring.

Meta Indicates 1,3 substituents on a benzene ring.

Para Indicates 1,4 substituents on a benzene ring.

Disubstituted aromatic compounds are named using one of the prefixes **ortho-, meta-,** or **para-.** An *ortho-* or *o*-disubstituted benzene has its two substituents in a 1,2 relationship on the ring; a *meta-* or *m*-disubstituted benzene has its two substituents in a 1,3 relationship; and a *para-* or *p*-disubstituted benzene has its substituents in a 1,4 relationship.

ortho-Dibromobenzene   *meta*-Chloronitrobenzene   *para*-Dimethylbenzene

Although all substituted aromatic compounds can be named systematically, many have common, nonsystematic names as well. For example, methylbenzene is familiarly known as *toluene*, hydroxybenzene as *phenol*, aminobenzene as *aniline*, and so on, as shown in Table 13.3. Frequently, these common names are used together with *ortho, para-,* or *meta-,* for example,

p-Chlorotoluene   *m*-Nitrophenol   *o*-Bromoaniline

Numbers must be included in the name when a benzene ring has three or more substituents. In the same manner as for alkanes, the numbers are chosen so that substituents have the lowest possible numbers and the substituents are named in alphabetical order. Here too, the common names for monosubstituted benzenes are often used as the parent names:

1-Chloro-2,4-dinitrobenzene   1-Chloro-2-ethyl-3-nitrobenzene   2,4,6-Trinitrotoluene (TNT)

Table 13.3   Common Names of Some Aromatic Compounds

Structure	Name	Structure	Name
⬡—CH₃	Toluene	H₃C—⬡—CH₃	*para*-Xylene
⬡—OH	Phenol	⬡—C(=O)—OH	Benzoic acid
⬡—NH₂	Aniline	⬡—C(=O)—H	Benzaldehyde

Occasionally, the benzene ring itself might be considered a substituent group attached to another parent compound. When this happens, the name **phenyl** is used for the C_6H_5— unit:

Phenyl group The name of the C_6H_5— unit when a benzene ring is considered a substituent group.

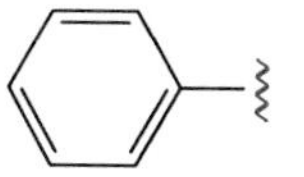

A phenyl group
(C_6H_5—⌇)

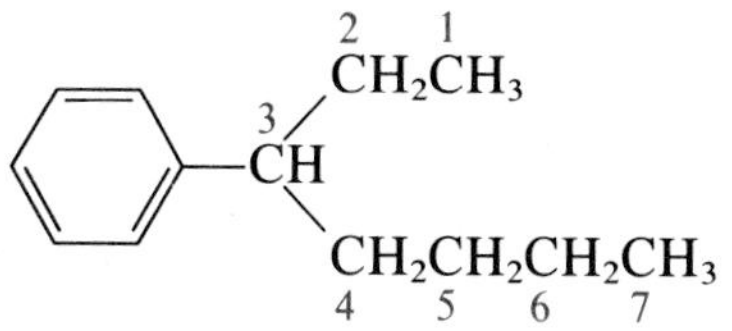
2 1
CH₂CH₃
3
CH
CH₂CH₂CH₂CH₃
4 5 6 7

3-Phenylheptane

Aryl is a general term for any substituent group that contains an aromatic ring.

Aryl group The general name for any substituent derived from an aromatic compound.

Solved Problem 13.6 Draw the structure of *m*-chloroethylbenzene.

Solution First, look at the name and determine that the root is benzene. *m*-Chloroethylbenzene must have a benzene ring with two substituents, chloro and ethyl, in a meta relationship. Next, draw a benzene ring and attach one of the substituents, say chloro, to any position:

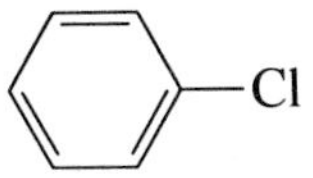
Cl

Now, go the meta position two carbons away from the chloro-substituted carbon and attach the second (ethyl) substituent:

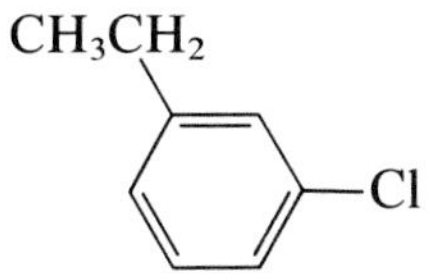
CH₃CH₂
Cl

m-Chloroethylbenzene

Practice Problems 13.13 What are the correct IUPAC names for these compounds?

(a)

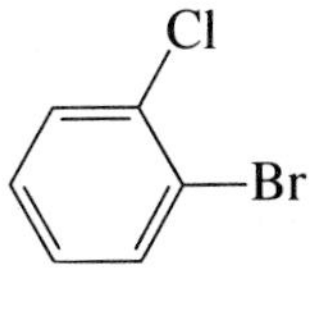
Cl
Br

(b)

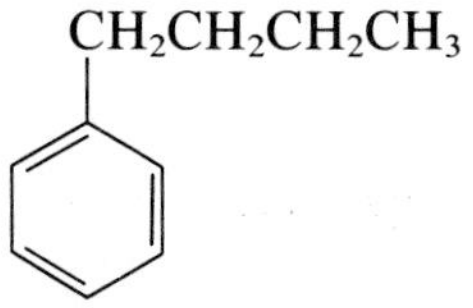
CH₂CH₂CH₂CH₃

(c)

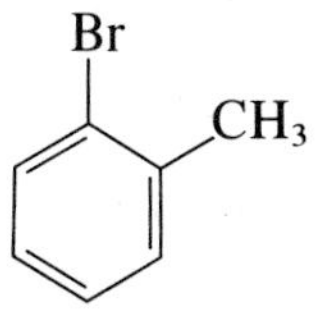
Br
CH₃

13.14 Draw structures corresponding to these names:
(a) *o*-dibromobenzene (b) *p*-nitrotoluene (c) *m*-diethylbenzene
(d) 2-chloro-4-isopropylphenol

13.12 CHEMICAL REACTIONS OF AROMATIC COMPOUNDS

Substitution reaction, aromatic Substitution of an atom or group for one of the hydrogens on an aromatic ring.

Unlike alkenes, which undergo addition reactions, aromatic compounds usually undergo **aromatic substitution reactions.** That is, a group Y *substitutes* for one of the hydrogen atoms on the aromatic ring without changing the highly stable ring itself. Of course, it doesn't matter which of the six ring hydrogens is replaced since all six are equivalent.

$$\text{(benzene)} - \text{H} + \text{Y}-\text{X} \longrightarrow \text{(benzene)} - \text{Y} + \text{H}-\text{X}$$

Nitration Substitution of a nitro group (—NO₂) for one of the ring hydrogens occurs when benzene reacts with nitric acid in the presence of sulfuric acid as catalyst:

$$\text{(benzene)}-\text{H} + \text{HO}-\text{NO}_2 \xrightarrow{\text{H}_2\text{SO}_4} \text{(benzene)}-\text{NO}_2 + \text{H}_2\text{O}$$

Benzene   Nitric acid   Nitrobenzene

Nitration of aromatic rings is a key step in the synthesis both of explosives like TNT (trinitrotoluene) and of many important pharmaceutical agents. Nitrobenzene itself is the industrial starting material for preparation of many of the brightly colored dyes used in clothing (Figure 13.7).

Figure 13.7
A variety of aniline dyes. Many dyes are derivatives of aminobenzene, or aniline. Aniline itself is made by the reduction of nitrobenzene.

Halogenation Substitution of a bromine or chlorine for one of the ring hydrogens occurs when benzene reacts with Br_2 or Cl_2 in the presence of iron as catalyst:

Benzene   Chlorine   Chlorobenzene

Sulfonation Substitution of a sulfonic acid group ($-SO_3H$) for one of the ring hydrogens occurs when benzene reacts with concentrated sulfuric acid and sulfur trioxide:

Benzene   Sulfuric acid   Benzenesulfonic acid

Aromatic-ring sulfonation is a key step in the synthesis of such compounds as the sulfa-drug family of antibiotics.

Sulfanilamide—a sulfa antibiotic

Practice Problems 13.15 Write the products from reaction of these reagents with *para*-xylene (*p*-dimethylbenzene).
(a) Br_2 and Fe   (b) HNO_3 and H_2SO_4 (c) SO_3 and H_2SO_4

13.16 Reaction of Br_2 and Fe with toluene (methylbenzene) can lead to a mixture of *three* substitution products. Show the structure of each.

13.13 POLYCYCLIC AROMATIC COMPOUNDS AND CANCER

Polycyclic aromatic compound A substance that has two or more benzene-like rings fused together along their edges.

The definition of the term *aromatic* can be extended beyond simple monocyclic (one-ring) compounds to include **polycyclic aromatic compounds:** substances that have two or more benzene-like rings joined together by sharing a common bond. Naphthalene, familiar for its use in mothballs, is the simplest and best known polycyclic aromatic compound.

INTERLUDE: COLOR IN UNSATURATED COMPOUNDS

The compounds below are brightly colored, as is β-carotene (look at its structure in the earlier Application in this chapter on vision). What do all these compounds have in common?

Have you noticed that each has numerous alternating double and single bonds?

In describing benzene, we said that the double-bond electrons are spread out over the whole

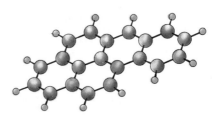

Cyanidin
(bluish-red color in flowers
and cranberries)

Mauve
(the first synthetic dye)

In addition to naphthalene, there are many more-complex polycyclic aromatic compounds. 1,2-Benzpyrene, for example, contains five benzene-like rings joined together; ordinary graphite (the "lead" in pencils) consists of enormous two-dimensional sheets of benzene rings.

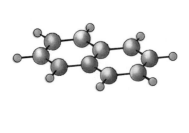

Naphthalene

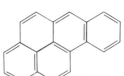

Benz[a]pyrene

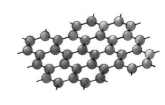

A graphite segment

Carcinogenic Cancer-causing.

Benz[a]pyrene is particularly well known because it is one of the **carcinogenic** (cancer-causing) substances found in chimney soot and cigarette smoke. When taken into the body by eating or inhaling, benz[a]pyrene is converted by enzymes into an oxygen-containing metabolite. This metabolite is able to bind to cellular DNA, causing mutations and interfering with the normal flow of genetic information.

molecule. The same phenomenon occurs whenever there are many alternating double and single bonds. The double-bond electrons form a uniform region of electron density that is capable of absorbing light. Organic compounds such as benzene, with small numbers of delocalized (spread-out) electrons, absorb in the ultraviolet region of the electromagnetic spectrum which our eyes can't detect. Compounds with long stretches of alternating double and single bonds absorb in the visible region. The color that we see is complementary to the color that's absorbed; that is, we see what's left of the white light after certain colors have been absorbed. For example, β-carotene absorbs blue light and appears yellow-orange. The same principle applies to almost all colored organic compounds: They contain large regions of delocalized electrons and absorb some

portion of the visible spectrum. When the delocalized electron system is broken up, for example by a bleach that breaks double bonds, the color is lost.

The sodium hypochlorite bleach has oxidized the dyes in the colored papers.

Benz[*a*]pyrene → enzymes → An oxygenated metabolite

Even benzene itself can cause certain types of cancer on prolonged exposure. Breathing the fumes of benzene and other volatile aromatic compounds in the laboratory should therefore be strictly avoided.

SUMMARY

Alkenes contain a carbon–carbon double bond, and **alkynes** contain a carbon–carbon triple bond. Both families are said to be **unsaturated** since they have fewer hydrogens than corresponding alkanes. Alkenes are named using the family ending *-ene*, while alkynes use the family ending *-yne*.

The planar arrangement of doubly bonded carbons and their substituents and the lack of rotation around carbon–carbon double bonds lead to **cis–trans isomerism** for disubstituted alkenes. The cis isomer has a specific

pair of atoms or groups on the same side of the double bond; the trans isomer has them on opposite sides.

Alkenes and alkynes resemble alkanes in their melting and boiling points and in being nonpolar, flammable, and generally nontoxic. They differ, however, in readily undergoing chemical reactions at their multiple bonds. Alkenes undergo **addition reactions** readily. Addition of hydrogen to an alkene (**hydrogenation**) yields an alkane product; addition of Cl_2 or Br_2 (**halogenation**) yields a 1,2-dihaloalkane product; addition of HBr and HCl (**hydrohalogenation**) yields an alkyl halide product; and addition of water (**hydration**) yields an alcohol product. **Markovnikov's rule** predicts that in the addition of HX or HOH to a double bond, the H becomes attached to the carbon that has the most H's, and the X or OH to the other carbon atom. **Polymers** such as polyethylene are formed when vinyl monomers undergo **chain-growth polymerization** (also known as addition polymerization).

Aromatic compounds contain a six-membered ring of carbon atoms (or one like it). Benzene rings are usually written with three double bonds but actually have equal bonding between neighboring carbon atoms because the double-bond electrons are delocalized around the entire ring. Simple aromatic compounds are named using the root **-benzene**. Disubstituted benzenes are named with one of the prefixes **ortho-** (1,2 substitution), **meta-** (1,3 substitution), or **para-** (1,4 substitution). Aromatic compounds are unusually stable but can be made to undergo **substitution reactions,** in which one of the ring hydrogens is replaced by some other group ($C_6H_6 \rightarrow C_6H_5Y$). Among these substitutions are **nitration** (substitution of $-NO_2$ for $-H$), **halogenation** (substitution of $-Br$ or $-Cl$ for $-H$), and **sulfonation** (substitution of $-SO_3H$ for $-H$).

REVIEW PROBLEMS

Naming Alkenes, Alkynes, and Aromatic Compounds

13.17 Why are alkenes, alkynes, and aromatic compounds said to be unsaturated?

13.18 Not all compounds that smell nice are called "aromatic," and not all compounds called "aromatic" smell nice. Explain.

13.19 Circle the aromatic portions of these molecules:

(a)

Epinephrine

(b)

Penicillin V

13.20 What family-name endings are used for alkenes, alkynes, and substituted benzenes?

13.21 Write structural formulas for compounds that meet these descriptions:
(a) an alkene with five carbons
(b) an alkyne with four carbons
(c) a substituted aromatic hydrocarbon with eight carbons

13.22 What are the IUPAC names of these compounds?
(a) $CH_3CH_2CH_2CH{=}CH_2$
(b) $CH_3CHCH_2C{\equiv}CCH_3$ with CH_3 branch
(c) $(CH_3)_2C{=}C(CH_3)_2$

(d) $CH_3CH{=}C-C{=}CH_2$ with CH_3 and CH_2CH_3 substituents

(e)

(f)

13.23 Draw structures corresponding to these IUPAC names:
(a) *cis*-2-hexene   (b) 2-methyl-3-hexene
(c) 2-methyl-1,3-butadiene   (d) *trans*-3-heptene
(e) 3,3-diethyl-6-methyl-4-nonene
(f) 3-isopropyl-3,4-dimethyl-1,4-hexadiene

13.24 There are only three alkynes with the formula C_5H_8. Draw and name them.

13.25 Provide correct IUPAC names for these aromatic compounds:

(a)   (b) $Br-\!\!-\!\!NO_2$

(c)

13.26 Draw structures corresponding to these names:
(a) aniline   (b) phenol   (c) *o*-xylene
(d) toluene   (e) benzoic acid

13.27 Draw structures corresponding to these names:
(a) *p*-nitrophenol   (b) *o*-chloroaniline
(c) *m*-bromotoluene   (d) 1,3,5-trimethylbenzene
(e) *o*-ethylphenol   (f) *m*-dipropylbenzene

13.28 Draw and name all aromatic compounds with the formula C_7H_7Br.

13.29 The following names are incorrect by IUPAC rules. Draw the structures represented by these names and write correct names.
(a) 2-methyl-4-hexene   (b) 5,5-dimethyl-3-hexyne
(c) 2-butyl-1-propene   (d) 1,5-dibromobenzene
(e) 1,2-dimethyl-3-cyclohexene
(f) 3-methyl-2,4-pentadiene
(g) 1,3,4-trichlorobenzene

13.30 How many dienes (compounds with two double bonds) are there with the formula C_5H_8? Draw and name structures of as many as you can.

Alkene Cis–Trans Isomers

13.31 What requirement must be met for an alkene to show cis–trans isomerism?

13.32 Why don't alkynes show cis–trans isomerism?

13.33 Excluding cis–trans isomers, there are five alkenes with the formula C_5H_{10}. Draw structures for as many as you can and give their IUPAC names.

13.34 Which of the alkenes in Problem 13.33 can exist as cis–trans isomers?

13.35 Draw structures of the double-bond isomers of these compounds:
(a)   (b)

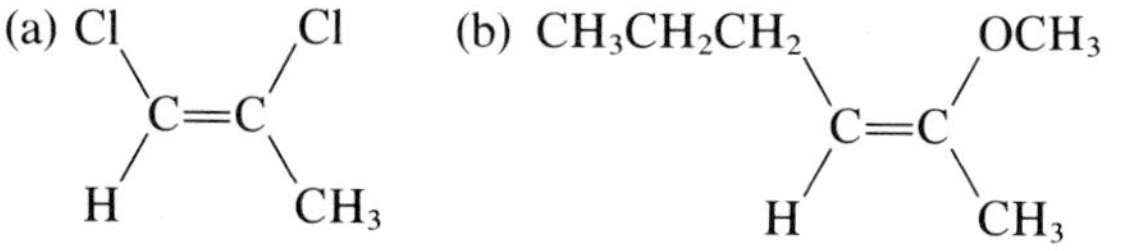

13.36 Which of the following pairs are isomers, and which are identical?
(a)

(b)

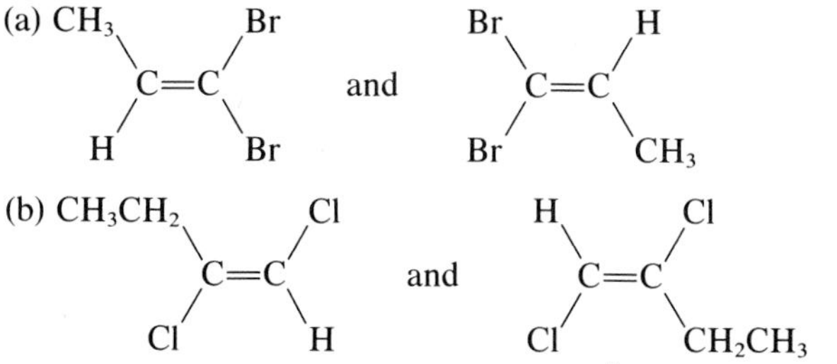

13.37 Why do you suppose small-ring cycloalkenes like cyclohexene don't exist as cis–trans isomers, whereas large-ring cycloalkenes like cyclodecene *do* show isomerism?

Reactions of Alkenes and Alkynes

13.38 What is meant by the term *addition reaction?*

13.39 Give a specific example of an alkene addition reaction.

13.40 Write equations for the reaction of 2,3-dimethyl-2-butene with these reagents:

(a) H_2 and Pd catalyst   (b) Br_2 (c) HBr
(d) H_2O and H_2SO_4 catalyst

13.41 Write equations for the reaction of 2-methyl-2-butene with the reagents shown in Problem 13.40.

13.42 Write the equations for the reaction of 3-methyl-1-butyne with
(a) excess H_2 with a Pd catalyst   (b) excess Cl_2
(c) excess HBr

13.43 What alkene could you use to make the following products. Draw the structural formula of the alkene and tell what inorganic reagent is also required for the reaction to occur.

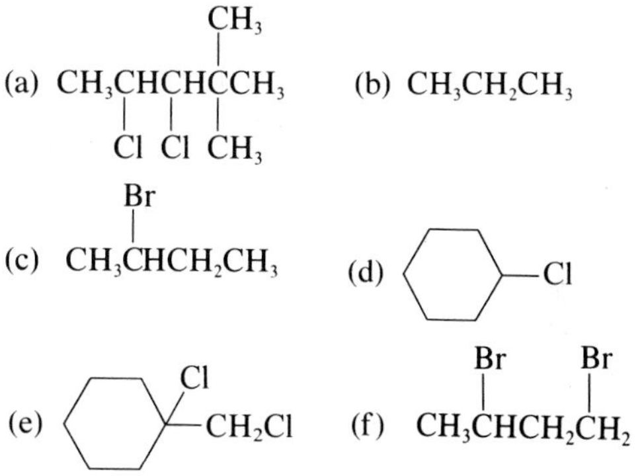

(a)   (b) $CH_3CH_2CH_3$

(c)   (d)

(e)   (f)

13.44 Write a balanced equation for the combustion of acetylene, used in high-temperature oxyacetylene torches.

13.45 Draw the carbocation formed as an intermediate when HCl adds to styrene (Table 13.2).

13.46 Polyvinylpyrrolidine (PVP) is often used in hair sprays as the substance that holds hair in place. Draw a few units of the PVP polymer. The vinylpyrrolidine monomer unit has the structure

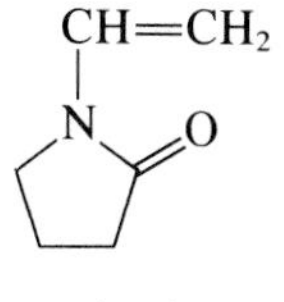

13.47 Saran, used as a plastic wrap for foods, is a polymer with the following structure. What is the monomer unit of Saran?

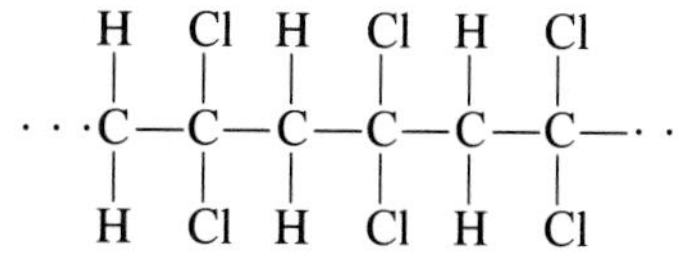

Reactions of Aromatic Compounds

13.48 What is meant by the term *substitution reaction?*

13.49 Give a specific example of a substitution reaction of an aromatic compound.

13.50 Benzene reacts with only one of the following four reagents. Which of the four is it, and what is the structure of the product?
(a) H_2 and Pd catalyst   (b) Br_2 and catalyst
(c) HBr   (d) H_2O and H_2SO_4 catalyst

13.51 Write equations for the reaction of p-dichlorobenzene with these reagents:
(a) Br_2 and Fe catalyst   (b) HNO_3 and H_2SO_4 catalyst
(c) H_2SO_4 and SO_3 (d) Cl_2 and Fe catalyst

13.52 Benzene and other aromatic compounds don't normally react with hydrogen in the presence of a palladium catalyst. If very high pressures (200 atm) and high temperatures are used, however, benzene will add three molecules of H_2 to give an addition product. What is a likely structure for the product?

13.53 How can you account for the fact that reaction of benzene with Br_2 and an iron catalyst yields a single substitution product, but the similar reaction of o-xylene (o-dimethylbenzene) with Br_2 and iron yields a mixture of two different substitution products? What are their structures?

13.54 The explosive trinitrotoluene, or TNT, is made by carrying out three successive nitration reactions on toluene. If these nitrations take place in the ortho and para positions relative to the methyl group, what is the structure of TNT?

Applications

13.55 What is the difference in the purpose of the rod and the cone cells in the eye? [App: Chemistry of Vision]

13.56 Describe the isomerization that occurs when light strikes the rhodopsin in the eye. [App: Chemistry of Vision]

13.57 Circle the isoprene units in the structure of β-carotene shown in Chemistry of Vision. [App: Terpenes]

13.58 Carvone is a component of spearmint oil. Circle the isoprene units in carvone. [App: Terpenes]

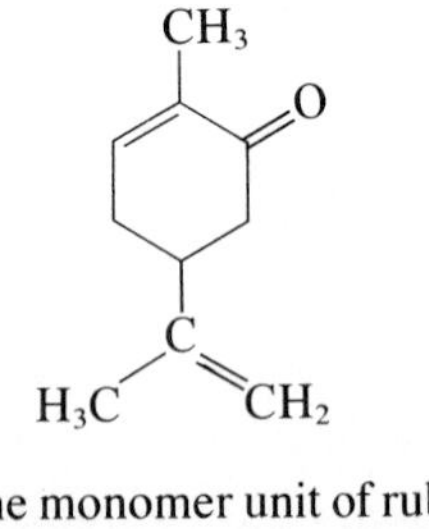

13.59 Isoprene, the monomer unit of rubber, is not capable of cis–trans isomerism, yet the natural rubber polymer exhibits cis–trans isomerism (primarily cis). What is responsible for this difference? [App: Terpenes]

13.60 Naphthalene is a white solid. Does it absorb light in the visible or in the ultraviolet range? [App: Color in Unsaturated Compounds]

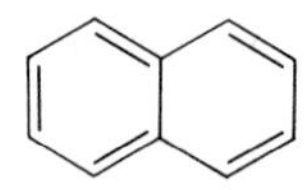

13.61 Tetrabromofluorescein, a purple dye often used in lipsticks, has the following structure. What structural feature makes it colored? If the dye is purple, what color must it absorb? [App: Color in Unsaturated Compounds]

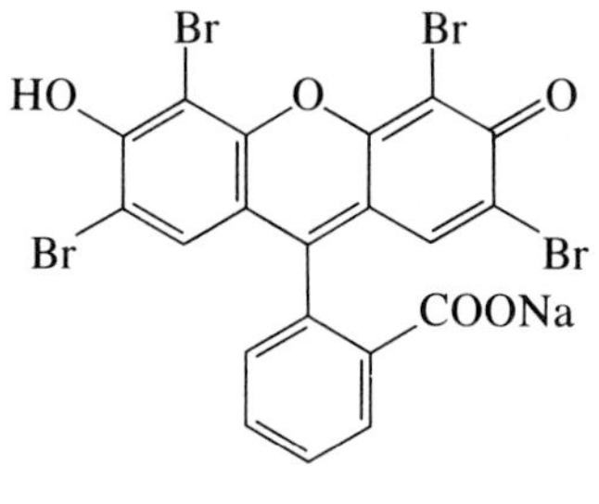

Additional Problems

13.62 Why can't alkanes form addition polymers?

13.63 Salicylic acid, or o-hydroxybenzoic acid, is used as starting material for the industrial preparation of aspirin. Draw the structure of salicylic acid.

13.64 Assume that you have two unlabeled bottles, one with cyclohexane and one with cyclohexene. How could you tell them apart by carrying out chemical reactions?

13.65 Assume you have two unlabeled bottles, one with cyclohexene and one with benzene. How could you tell them apart by carrying out chemical reactions?

13.66 The compound p-dichlorobenzene has been used as an insecticide. Draw its structure.

13.67 Menthene, a compound found in mint plants, has the formula $C_{10}H_{18}$ and the IUPAC name 1-isopropyl-4-methylcyclohexene. What is the structure of menthene?

13.68 Cinnamaldehyde, the pleasant-smelling substance found in cinnamon oil, has the following structure. What product would you expect to obtain from reaction of cinnamladehyde with hydrogen and a palladium catalyst?

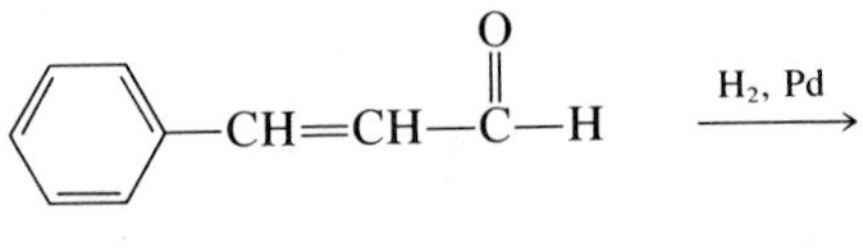

13.69 Write the products of these reactions:

(a)

(b)

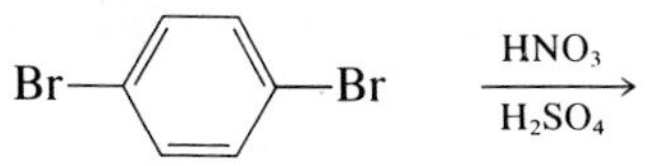

$$Br\!\!-\!\!\bigcirc\!\!-\!\!Br \quad \xrightarrow[\text{H}_2\text{SO}_4]{\text{HNO}_3}$$

(c)

$$\bigcirc \quad \xrightarrow[\text{H}_2\text{SO}_4]{\text{H}_2\text{O}}$$

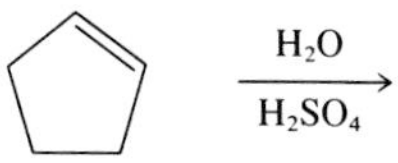

(d) CH_3CHCH_3
     $|$
     $CH_3CHCH_2CH\!=\!CH_2 \quad \xrightarrow{\text{HBr}}$

(e) $CH_3C\!\equiv\!CCH_2CH_3 \quad \xrightarrow{\text{H}_2,\ \text{Pd}}$

(f)

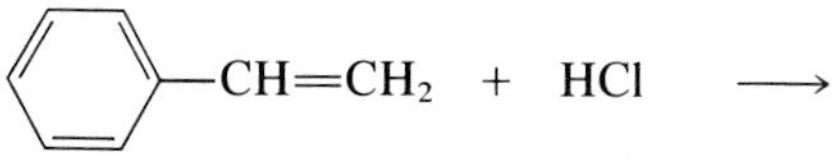

$$\bigcirc\!\!-\!\!CH\!=\!CH_2 \ + \ HCl \quad \longrightarrow$$

13.70 Two different products are possible when 2-pentene is treated with HBr. Write the structural formulas of the possible products and discuss why they are made in about equal amounts.

13.71 Benzene is a liquid at room temperature, while naphthalene, another aromatic compound (Problem 13.60), is a solid. Account for this difference in physical properties.

13.72 The compound 4-chloro-3,5-dimethylphenol is used as an antiseptic, as a germicide, and for mildew prevention. Draw its structure.

13.73 Ocimene, a compound isolated from the herb basil, has the IUPAC name 3,7-dimethyl-1,3-6-octatriene.
(a) Draw its structure.
(b) Circle the isoprene units in the formula.
(c) Draw the structure of the compound formed if enough HBr is added to react with all the double bonds in ocimene.

13.74 Describe how you could prepare the following compound from an alkene. Draw the formula of the alkene, name it, and list the inorganic reactants needed for the conversion.

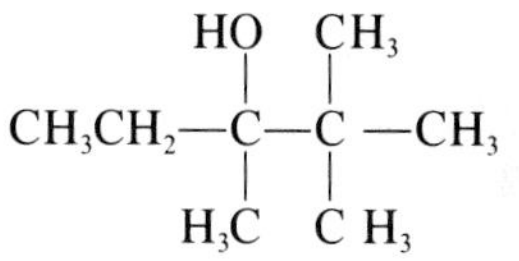

$$\begin{array}{ccc} & HO & CH_3 \\ & | & | \\ CH_3CH_2\!-\!\!&C\!-\!C&\!-\!CH_3 \\ & | & | \\ & H_3C & C\,H_3 \end{array}$$

CHAPTER

14

Some Compounds with Oxygen, Sulfur, or Halogens

Nighttime balloon launching in January in Sweden. The helium-filled balloon carries instruments that will measure ozone and pollutants in polar stratospheric clouds at 16–25 km altitude. The pollutants responsible for ozone depletion include chlorofluorocarbons, one of a wide variety of compounds introduced in this chapter.

Having introduced the hydrocarbons that form the skeletons of all organic compounds, we're ready to move on to compounds with functional groups containing other elements. In this chapter we'll concentrate on functional groups with single bonds to oxygen, sulfur, and the halogens. In future chapters you'll be introduced to groups with bonds to nitrogen and those containing the $C\!=\!O$ group.

At this point in studying organic chemistry, students often begin to wonder what they're meant to "learn" about organic compounds. There are so many different kinds. First, it's important to recognize the structures of the functional groups, whether they're present in simple or complex compounds. Then, it's useful to know the general properties imparted by these groups and a few of their most important chemical reactions. It's also valuable to learn something about some representative members of each family of compounds, such as their properties, hazards, occurrences in nature, and uses.

Questions we'll answer in this chapter include the following:

1. *How do alcohols, phenols, and ethers compare with water and alkanes?* The goal: Be able to describe the major differences in properties among members of these families.

2. *What are the distinguishing features of ethers, sulfur-containing compounds, and halogen-containing compounds?* The goal: Be able to describe the structures and outstanding properties of compounds with these functional groups.

3. *How are alcohols, phenols, ethers, sulfur-containing compounds, and halogen-containing compounds named?* The goal: Be able to give systematic names for the simple members of these families and write their structures, given the names.

4. *What are the general properties of alcohols and phenols?* The goal: Be able to describe such properties as polarity and hydrogen bonding, water solubility, and boiling points.

5. *Why are alcohols and phenols weak acids?* The goal: Be able to show why alcohols and phenols are acids.

6. *What are the major chemical reactions of alcohols?* The goal: Be able to describe and predict the products of the dehydration and oxidation of alcohols.

7. *What are some of the significant applications of alcohols, phenols, ethers, sulfur-containing compounds, and halogen-containing compounds?* The goal: Be able to identify major applications of each family of compounds and describe some important members of each family.

14.1 ALCOHOLS, PHENOLS, AND ETHERS

Alcohol A compound that contains an —OH functional group covalently bonded to an alkane-like carbon atom, R—OH.

Hydroxyl group A name for the —OH group in an organic compound.

Phenol A compound that has an —OH functional group bonded directly to an aromatic, benzene-like ring.

Ether A compound that has an oxygen atom bonded to two carbon atoms, R—O—R.

Alcohols are compounds that have an —OH group (a **hydroxyl group**) covalently bonded to a saturated, alkane-like carbon atom; **phenols** have an —OH group bonded directly to an aromatic, benzene-like ring (an aryl group); and **ethers** have an oxygen atom bonded to two organic groups. Compounds in all three families can be thought of as organic derivatives of water in which one or both of the water hydrogens have been replaced by an organic substituent. Using R for any organic group and Ar for one that is specifically aromatic, the general formulas for these families are

$$H—O—H \qquad R—O—H \qquad Ar—O—H \qquad R—O—R$$

Water   Alcohol   Phenol   Ether

An alcohol, a phenol, and an ether are shown here. Note that each functional group contains an oxygen atom with single bonds to its neighbors.

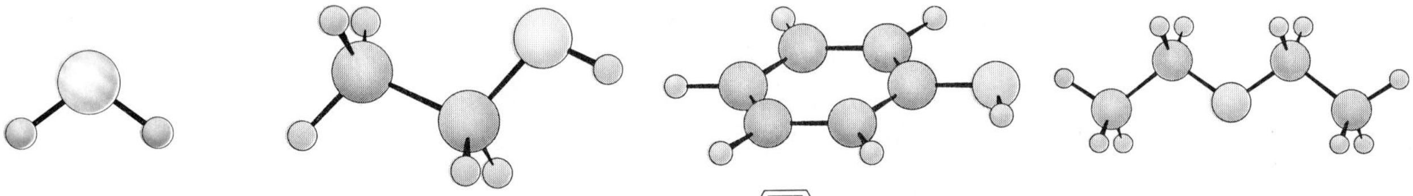

$$H—O—H \qquad CH_3CH_2—O—H \qquad \bigcirc—O—H \qquad CH_3CH_2—O—CH_2CH_3$$

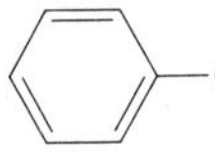

Water   Ethyl *alcohol*   *Phenol*   Diethyl *ether*

The structural similarity between alcohols and water extends to many of their properties. For example, compare the boiling points of ethyl alcohol, dimethyl ether, propane, and water:

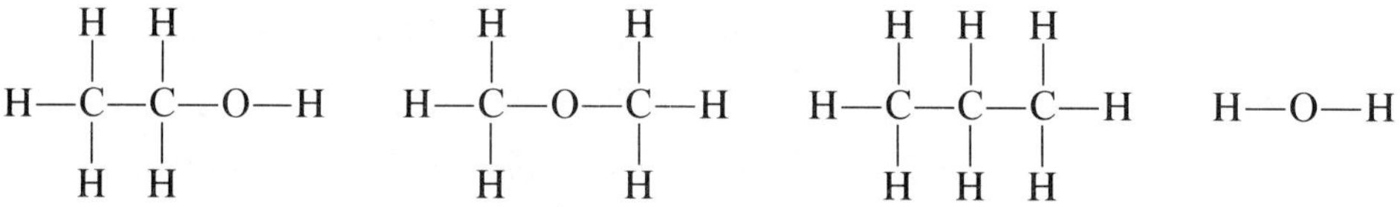

Ethyl alcohol	Dimethyl ether	Propane	Water
(mol wt 46, bp 78.5°C)	(mol wt 46, bp −23°C)	(mol wt 44, bp −42°C)	(mol wt 18, bp 100°C)

Ethyl alcohol, dimethyl ether, and propane have similar molecular weights, yet ethyl alcohol boils more than 100° higher than the other two. In fact, the boiling point of ethyl alcohol is close to that of water. Why should this be?

You saw in Section 9.12 that water has a high boiling point because of the formation of hydrogen bonds. These hydrogen bonds cause an attraction between individual molecules that prevents their easy escape into the vapor phase. In a similar manner, hydrogen bonds form between alcohol (or phenol) molecules (Figure 14.1). Since alkanes and ethers don't have hydroxyl groups, they can't form hydrogen bonds and therefore have much lower boiling points. In many properties ethers resemble alkanes, while alcohols resemble water.

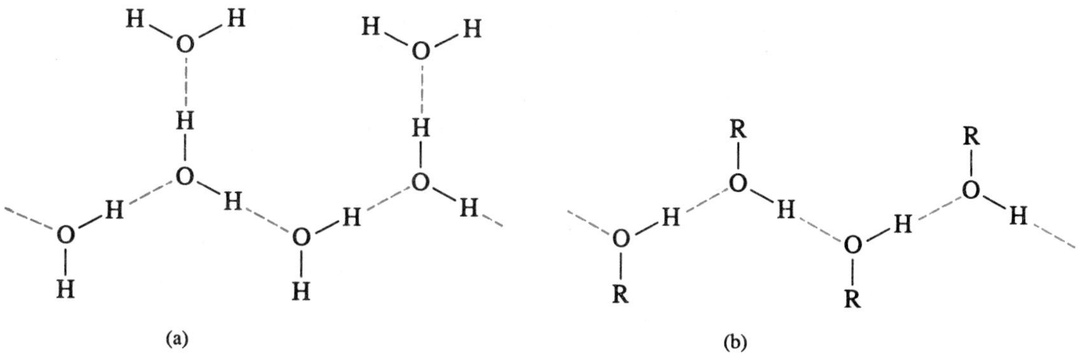

Figure 14.1
The formation of hydrogen bonds in water (a) and in alcohols (b). Due to the attractive forces of the hydrogen bonds, the easy escape of molecules into the vapor phase is prevented, resulting in high boiling points.

Practice Problems   **14.1**   Identify each of these compounds as an alcohol, a phenol, or an ether.

(a) $CH_3CH_2CHCH_3$
       |
      OH

(b) 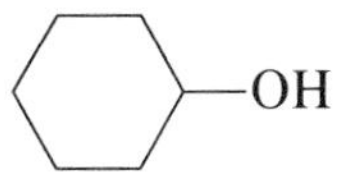 [cyclohexane ring]—OH

(c) 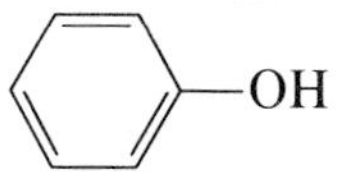 [benzene ring]—OH

(d) 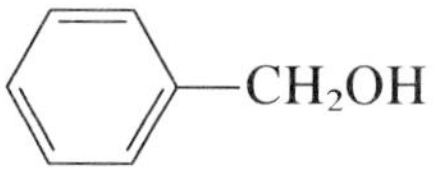 [benzene ring]—CH_2OH

(e) 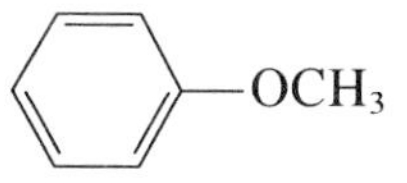 [benzene ring]—OCH_3

(f) $CH_3CHOCH_2CH_3$
       |
      CH_3

   14.2   Explain the difference between a hydroxyl group and a hydroxide ion.

14.2   SOME COMMON ALCOHOLS

Simple alcohols are among the most commonly encountered of all organic chemicals (Figure 14.2). They are useful as solvents, antifreeze agents, and disinfectants, and they are valuable in the laboratory and in industry because compounds containing many other functional groups can be synthesized from alcohols. In nature, lower-molecular-weight alcohols are formed during the fermentation of organic matter. "Higher alcohols," those with six or more carbon atoms, are obtained by chemical breakdown of natural fats and oils and are used in the production of plastics and detergents.

Wood alcohol   A common name for methyl alcohol, CH_3OH.

Methyl Alcohol (CH_3OH, Methanol)   Methyl alcohol, the simplest alcohol, is commonly known as **wood alcohol** because it was once prepared by heating wood in the absence of air. Today it is made in large quantities from a mixture of carbon monoxide and hydrogen,

$$CO(g) + 2\ H_2(g) \xrightarrow[\text{high pressure, 250°C}]{\text{Cu catalyst}} CH_3OH(l)$$

Coal and natural gas are sources for the mixture of CO and H_2, and there is great interest in methyl alcohol as a fuel and a raw material for organic chemicals now produced from petroleum. Methyl alcohol is colorless, miscible

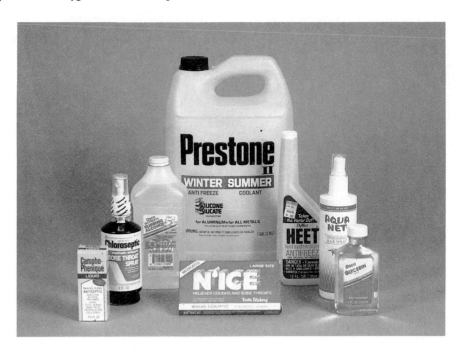

Figure 14.2
Some common household products made entirely or mostly of alcohols.

with water, and toxic to humans whether ingested or inhaled, causing blindness in low doses (about 15 mL) and death in larger amounts (about 30 mL).

Ethyl Alcohol (CH₃CH₂OH, Ethanol)   Ethyl alcohol is one of the oldest known pure organic chemicals, and its production by fermentation of grain and sugar goes back for many thousands of years. Sometimes called **grain alcohol,** ethyl alcohol is the "alcohol" present in all table wines (10–13%), beers (3–5%), and distilled liquors (35–90%). During fermentation, starches or complex sugars are converted to simple sugars ($C_6H_{12}O_6$) which are then converted to ethyl alcohol:

Grain alcohol   A common name for ethyl alcohol, CH_3CH_2OH.

$$C_6H_{12}O_6 \xrightarrow{\text{yeast enzymes}} 2\ CH_3CH_2OH\ +\ 2\ CO_2$$

About 14% (by volume) is the maximum alcohol concentration produced by fermentation. Higher alcohol concentrations in beverages are produced by distillation of the fermentation product or addition of distilled products such as brandy. The production of ethyl alcohol and alcoholic beverages in the United States is supervised by the government. To avoid the government tax on the sale of consumable alcohol, a toxic substance such as methyl alcohol or benzene is added to ethyl alcohol to produce *denatured alcohol* for nonbeverage uses.

Industrially, most ethyl alcohol is made by the hydration of ethylene (Section 13.6). Distillation yields a 95% ethyl alcohol–5% water mixture useful for any application with which water does not interfere. Reaction of the mixture with a dehydrating agent gives 100% ethyl alcohol, known as *absolute alcohol*. In some states, a blend of ethyl alcohol and gasoline known as *gasohol* is commercially available, and numerous experiments are under way with cars that burn methyl or ethyl alcohol. These fuels are desirable because they

Figure 14.3
Ethylene glycol, a poisonous dialcohol used in antifreeze, and glycerol, a sweet trialcohol used as a food additive. Can you tell by their appearance which is which?

Glycol A dialcohol; a compound that contains two —OH groups.

Primary alcohol An alcohol in which the OH-bearing carbon atom is bonded to one other carbon (and two hydrogens), RCH_2OH.

Secondary alcohol An alcohol in which the OH-bearing carbon atom is bonded to two other carbons (and one hydrogen), R_2CHOH.

Tertiary alcohol An alcohol in which the OH-bearing carbon atom is bonded to three other carbons (and no hydrogens), R_3COH.

produce fewer air pollutants than gasoline; also, ethyl alcohol can be produced from corn rather than from dwindling fossil fuel reserves.

Isopropyl alcohol ($(CH_3)_2CHOH$, Isopropanol) Isopropyl alcohol, or *rubbing alcohol,* is used as a 70% mixture with water for rubdowns because it cools the skin through evaporation and causes pores to close. It's also used as a solvent for medicines, as a sterilant for instruments, and to cleanse the skin before drawing blood or giving injections. The "medicinal" odor we associate with doctors' offices is often that of isopropyl alcohol. Although not as toxic as methyl alcohol, isopropyl alcohol is much more toxic than ethyl alcohol.

Ethylene Glycol ($HOCH_2CH_2OH$) A dialcohol, ethylene glycol is a colorless liquid miscible with water and insoluble in nonpolar solvents. Its two major uses are in antifreeze and in making polymer films and fibers such as Dacron. In humans it's a central nervous system depressant with a lethal dose of about 100 mL (Figure 14.3).

Glycerol ($HOCH_2CH(OH)CH_2OH$) Like ethylene glycol, the trialcohol glycerol is a colorless liquid that is miscible with water. Unlike ethylene glycol, it's not toxic but has a sweet taste that's put to use in candy and prepared foods. Glycerol, often called glycerin, is also used in cosmetics and tobacco as a moisturizer, in plastics manufacture, in antifreeze and shock absorber fluids, and as a solvent. The glycerol molecule provides the structural backbone of natural fats and oils (Section 23.1).

14.3 NAMING ALCOHOLS

Alcohols with one —OH group are often referred to by common names like those used in the preceding section, which consist of the name of the alkyl group followed by the word "alcohol." Thus, the two-carbon alcohol is ethyl alcohol, a four-carbon alcohol is a butyl alcohol, and so on. Both the common and the systematic names of some representative alcohols are included in Table 14.1. Notice that *dialcohols* are often called **glycols.** Ethylene glycol is the simplest glycol; propylene glycol is often used as a solvent for medicines that need to be inhaled or rubbed onto the skin.

Alcohols are classified according to the number of carbon substituents bonded to the hydroxyl-bearing carbon. Alcohols with one substituent are classified as **primary** (1°), those with two substituents as **secondary** (2°), and those with three substituents as **tertiary** (3°). Of course, the substituent groups don't have to be the same, so we'll use the representations R, R′, and R″ to indicate different groups.

$$\begin{array}{ccc}
\quad\;\;\text{H} & \quad\;\;\text{R}' & \quad\;\;\text{R}' \\
\quad\;\;| & \quad\;\;| & \quad\;\;| \\
\text{R---C---OH} & \text{R---C---OH} & \text{R---C---OH} \\
\quad\;\;| & \quad\;\;| & \quad\;\;| \\
\quad\;\;\text{H} & \quad\;\;\text{H} & \quad\;\;\text{R}''
\end{array}$$

A *primary* alcohol (one R group on OH-bearing carbon)   A *secondary* alcohol (two R groups on OH-bearing carbon)   A *tertiary* alcohol (three R groups on OH-bearing carbon)

Table 14.1 Physical Properties of Some Simple Alcohols[a]

Structure	Common Name	IUPAC Name	Boiling Point (°C)
CH_3OH	Methyl alcohol	Methanol	65
CH_3CH_2OH	Ethyl alcohol	Ethanol	78
$CH_3CH_2CH_2OH$	n-Propyl alcohol	1-Propanol	97
$\begin{array}{c} OH \\ | \\ CH_3CHCH_3 \end{array}$	Isopropyl alcohol	2-Propanol	82
$CH_3CH_2CH_2CH_2OH$	n-Butyl alcohol	1-Butanol	118
$\begin{array}{c} CH_3 \\ | \\ CH_3-C-OH \\ | \\ CH_3 \end{array}$	tert-Butyl alcohol	2-Methyl-2-propanol	83
$\begin{array}{c} CH_2-CH_2 \\ | \quad | \\ OH \quad OH \end{array}$	Ethylene glycol	1,2-Ethanediol	198
$\begin{array}{c} CH_2-CH-CH_3 \\ | \quad | \\ OH \quad OH \end{array}$	Propylene glycol	1,2-Propanediol	187
$\begin{array}{c} CH_2-CH-CH_2 \\ | \quad | \quad | \\ OH \quad OH \quad OH \end{array}$	Glycerol (glycerin)	1,2,3-Propanetriol	290

[a] All the alcohols listed are liquids at room temperature and completely miscible with water, except 1-butanol (7 g/100 mL water).

Alcohols are named in the IUPAC system as derivatives of alkanes, using the *-ol* ending for the parent compound, according to the following steps.

Step 1. *Name the parent compound.* Find the longest chain *containing the hydroxyl group* and name the chain by replacing the *-e* ending of the corresponding alkane (Table 12.4) with *-ol*:

$$CH_3CH_2CH_2CH_2OH$$

Name as a *butanol*—a four-carbon chain containing a hydroxyl group.

If the compound is a cycloalkane, its name provides the root name, for example, *cyclohexan*ol.

Step 2. *Number the carbon atoms in the main chain.* Begin at the end nearer the hydroxyl group, ignoring the location of other substituents:

$$\begin{array}{c} CH_3 \quad\quad OH \\ | \quad\quad\quad | \\ CH_3CHCH_2CHCH_2CH_3 \\ {\scriptstyle 6 \quad 5 \quad 4 \quad 3 \quad 2 \quad 1} \end{array}$$

— Begin at this end because it's nearer the OH group.

In an OH-substituted cycloalkane, begin with the —OH group.

Step 3. Identify and number the hydroxyl group and the substituents.

A 1 hydroxyl

A 5 methyl   A 3 hydroxyl

                          CH_3 OH

$CH_3CH_2CH_2CH_2OH$ $CH_3CHCH_2CHCH_2CH_3$
    4   3   2   1   6   5   4   3   2   1

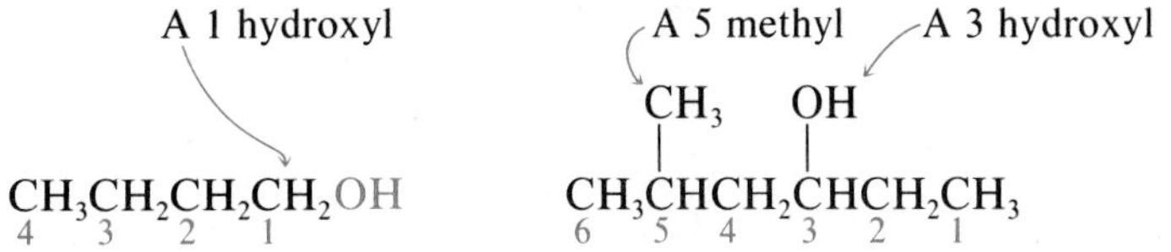

Step 4. Write the full name. Place the number that locates the hydroxyl group immediately before the parent compound name (with the *-ol* ending). Number all other substituents according to their positions and list them alphabetically:

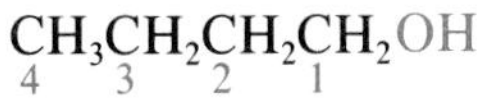

                                   CH_3 OH

$CH_3CH_2CH_2CH_2OH$ $CH_3CHCH_2CHCH_2CH_3$
    4   3   2   1   6   5   4   3   2   1

      1-Butanol 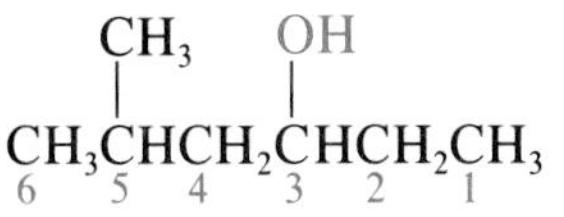   5-Methyl-3-hexanol

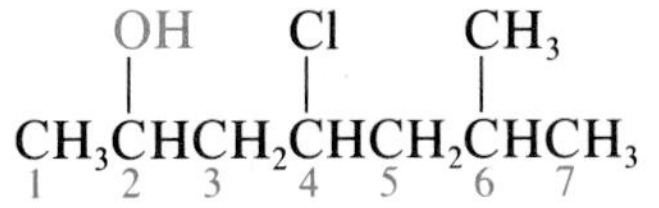

  OH Cl CH_3 2 OH

$CH_3CHCH_2CHCH_2CHCH_3$ Br
 1   2   3   4   5   6   7

4-Chloro-6-methyl-2-heptanol 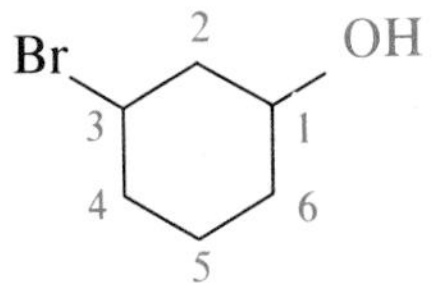   3-Bromocyclohexanol

---

Solved Problem 14.1   Give the systematic name of the following alcohol and classify it as primary, secondary, or tertiary.

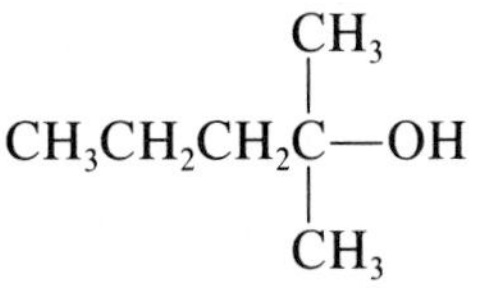

$$CH_3$$
$$CH_3CH_2CH_2C—OH$$
$$CH_3$$

Solution   First, identify the longest carbon chain and number the carbon atoms beginning at the end nearest —OH. The longest of the two chains attached to the —OH here has five carbon atoms.

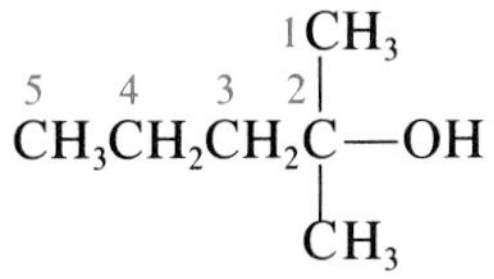

               1CH_3 Name as a pentanol.
   5 4 3 2|
$CH_3CH_2CH_2C—OH$
            CH_3

Next, identify and number the hydroxyl group and the substituents. Finally, write the full name of the compound.

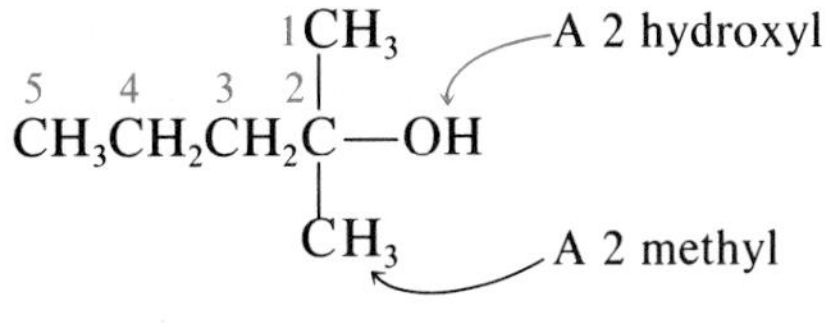

            1CH_3 A 2 hydroxyl
   5 4 3 2|
$CH_3CH_2CH_2C—OH$
            CH_3 A 2 methyl

      2-Methyl-2-pentanol

    The —OH group is bonded to a carbon atom that has three alkyl substituents, so this is a tertiary alcohol.

Practice Problems   **14.3**   Draw structures corresponding to these names:
(a) 3-methyl-3-pentanol   (b) cyclohexanol
(c) 2-methyl-4-heptanol   (d) 2-butanol   (e) 3-chloro-1-propanol

14.4   Give systematic names for these compounds:

$$\underset{\substack{| \\ \text{(a) CH}_3\text{CH}_2\text{CHCH}_2\text{CH}_3}}{\overset{\text{OH}}{}}$$

(a) CH₃CH₂CHCH₂CH₃ with OH group

(b) CH₃CH₂CHCH₂CH₂CH₃ with CH₂OH group

(c) CH₃CH₂CHCH₂CH₂CH₂Br with CH₂OH group

(d) H₃C, H₃C substituted cyclohexanol —OH

14.5   Identify each of the alcohols in Problems 14.3 and 14.4 as primary, secondary, or tertiary.

14.4   PROPERTIES OF ALCOHOLS

Hydrogen bonding (Section 9.12) is the major influence on the physical properties of alcohols. Straight-chain alcohols with up to 12 carbon atoms are liquids, and each boils at a higher temperature than similar alkanes. Most familiar alcohols, in fact, are liquids.

Solubility behavior, like hydrogen bonding, is a property shared by water and alcohols. Water, because of its polar bonds and its ability to solvate ions, is an excellent solvent for ionic compounds and polar molecules but a poor solvent for nonpolar organic molecules. Low-molecular-weight alcohols such as methanol and ethanol are water-like in their solubility behavior. Methanol and ethanol are both infinitely soluble in water, with which they can form hydrogen bonds. Both can also dissolve small amounts of many salts, and both are miscible with many organic solvents. Higher-molecular-weight alcohols, such as 1-heptanol, however, are much more alkane-like and less water-like. 1-Heptanol is nearly insoluble in water and can't dissolve salts but does dissolve alkanes.

$$\text{CH}_3\text{—OH} \qquad\qquad \text{CH}_3\text{CH}_2\text{CH}_2\text{CH}_2\text{CH}_2\text{CH}_2\text{CH}_2\text{—OH}$$

Methanol—has a small organic   1-Heptanol—has a large organic part
part and is therefore water-like.   and is therefore alkane-like.

Alcohols with two or more —OH groups can form more than one hydrogen bond. They are therefore higher boiling and more water-soluble than similar alcohols with one —OH group. For example, compare 1-butanol and 1,4-butanediol,

CH₃CH₂CH₂CH₂OH   Bp 117°C,   HOCH₂CH₂CH₂CH₂OH   Added —OH
                                        water solubility   raises bp to
    1-Butanol   of 7 g/100 mL   1,4-Butanediol   230°C and gives
                                                                                                    miscibility with
                                                                                                    water

Properties of Alcohols

- Alcohols are hydrogen-bonded and higher boiling than similar alkanes.
- Common alcohols are liquids.
- Lower alcohols are miscible with water; higher dialcohols are more water-soluble than similar monoalcohols.
- Alcohols are very weak acids (proton donors); water solutions are neutral (Section 14.7).
- Common alcohols are flammable; can be toxic.
- Dehydration and oxidation are major alcohol reactions.

Practice Problems **14.6** Which of the following compounds has the highest boiling point?
(a) $CH_3CH_2CH_2OH$ (b) $CH_3CH_2OCH_3$ (c) $CH_3CH_2CH_3$

14.7 Which of the following compounds is most water-soluble?
(a) $CH_3(CH_2)_{10}CH_2OH$ (b) $CH_3CH_2CHCH_3$ (c) $CH_3CH_2OCH_3$
                                                        OH

14.5 CHEMICAL REACTIONS OF ALCOHOLS

Dehydration Loss of water, as from an alcohol to yield an alkene.

Dehydration of Alcohols   Alcohols undergo loss of water (**dehydration**) on treatment with a strong acid catalyst. The —OH group is lost from one carbon, and an —H is lost from an adjacent carbon to yield an alkene product:

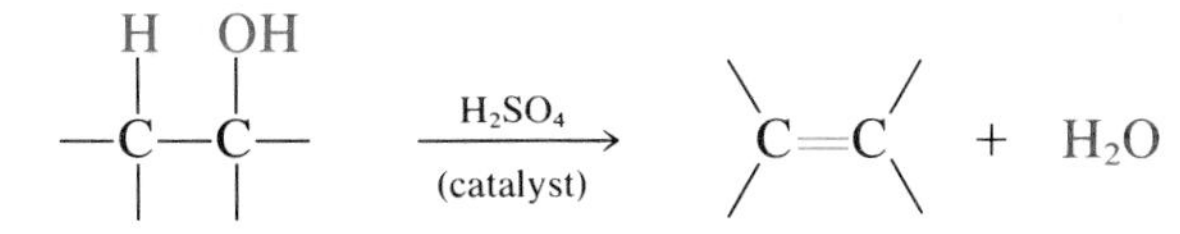

An alcohol   An alkene

For example,

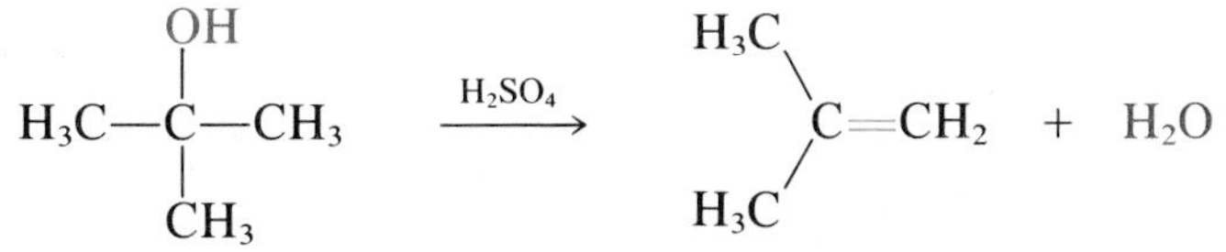

tert-Butyl alcohol   2-Methylpropene

In cases where more than one alkene can result from dehydration, a mixture of products is usually formed. A good rule of thumb is that the *major* product is the one that has the greater number of alkyl groups attached to the double-bond carbons. For example,

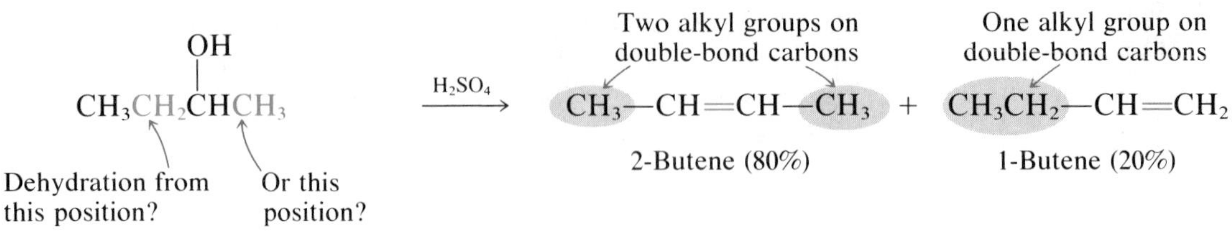

$$CH_3CH_2CHCH_3 \xrightarrow[]{H_2SO_4}$$

OH (above CHCH₃ carbon)

Dehydration from this position?   Or this position?

Two alkyl groups on double-bond carbons → $CH_3-CH=CH-CH_3$ ← + $CH_3CH_2-CH=CH_2$ ← One alkyl group on double-bond carbons

2-Butene (80%)   1-Butene (20%)

Solved Problem 14.2   What products would you expect from the following dehydration reaction? Which product will be major, and which minor?

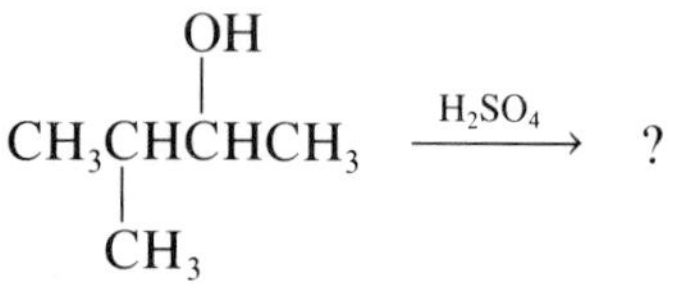

$$CH_3CHCHCH_3 \xrightarrow[]{H_2SO_4} \ ?$$

OH (on second carbon), CH₃ (below third carbon)

Solution   First, find the hydrogens on carbons next to the OH-bearing carbon and rewrite the structure to emphasize these hydrogens:

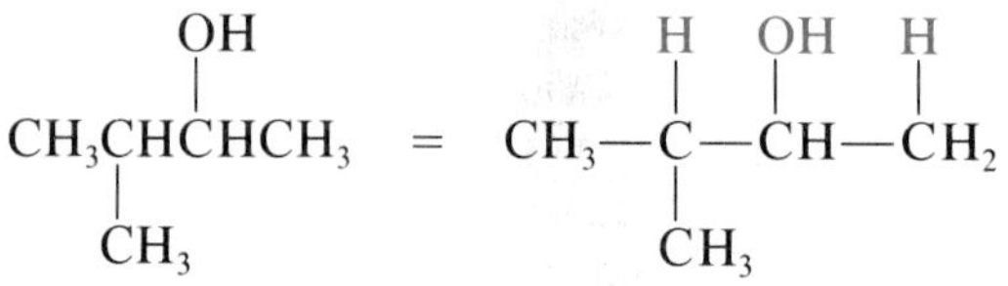

$$CH_3CHCHCH_3 \quad = \quad CH_3-\overset{\displaystyle OH}{\underset{\displaystyle CH_3}{C}}-\overset{\displaystyle H}{CH}-\overset{\displaystyle H}{CH_2}$$

with OH on the CHCHCH₃ and CH₃ below; on the right H OH H above with CH₃ below

Next, circle and remove the possible combinations of H and OH, drawing a double bond where H and OH have come from:

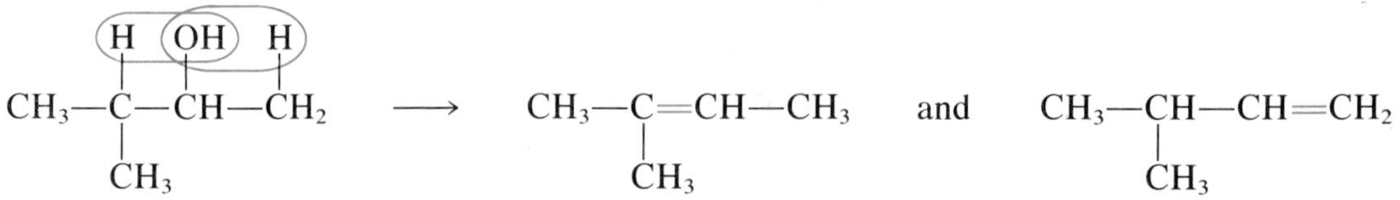

$$CH_3-\overset{\displaystyle \boxed{OH}}{\underset{\displaystyle CH_3}{C}}-CH-CH_2 \longrightarrow CH_3-\overset{}{\underset{\displaystyle CH_3}{C}}=CH-CH_3 \quad \text{and} \quad CH_3-\overset{}{\underset{\displaystyle CH_3}{CH}}-CH=CH_2$$

(H (OH) H circled above)

Finally, determine which of the alkenes has the larger number of alkyl substituents on its double-bond carbons and is therefore the major product:

$$\underset{H_3C}{\overset{H_3C}{>}}C=C\underset{CH_3}{\overset{H}{<}} \qquad \text{and} \qquad \underset{H}{\overset{(CH_3)_2CH}{>}}C=C\underset{H}{\overset{H}{<}}$$

2-Methyl-2-butene
major product (three alkyl groups)

3-Methyl-1-butene
minor product (one alkyl group)

Practice Problems   14.8   What alkenes might be formed by dehydration of the following alcohols? If more than one product is possible, indicate which is major.

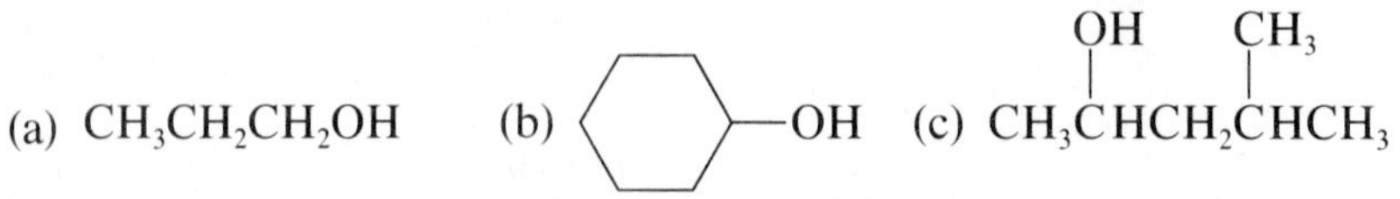

(a) $CH_3CH_2CH_2OH$ (b) ⬡—OH   (c) $CH_3CHCH_2CHCH_3$ (with OH and CH₃ substituents)

14.9   Dehydration of what alcohols would yield these alkenes?
(a) $(CH_3)_2C=C(CH_3)_2$ (b) $CH_3CH_2CH=CH_2$

Carbonyl group A functional group that has a carbon atom joined to an oxygen atom by a double bond, C=O.

Oxidation of Alcohols Primary and secondary alcohols are converted into carbonyl-containing compounds on treatment with an oxidizing agent. A **carbonyl group** (pronounced car-bo-**neel**) is a functional group that has a carbon atom joined to an oxygen atom by a double bond, C=O. Many different oxidizing agents can be used—potassium permanganate ($KMnO_4$), sodium dichromate ($Na_2Cr_2O_7$), or nitric acid (HNO_3), for example—and it usually doesn't matter which specific reagent is chosen. Thus, we'll simply use the symbol [O] to indicate a generalized reagent.

Oxidation In organic chemistry, the removal of hydrogen from a molecule or the addition of oxygen to a molecule.

The net effect of an alcohol **oxidation** is the removal of two hydrogen atoms. One hydrogen comes from the —OH group, and the other comes from the carbon atom bonded to the —OH group. These hydrogens are converted into water during the reaction by the oxidizing agent [O].

An alcohol   A carbonyl compound

Aldehyde A compound with the —CH=O functional group.

Carboxylic acid A compound with the —COOH functional group.

Different kinds of carbonyl-containing products are formed depending on the structure of the starting alcohol and on the exact reaction conditions. Thus, primary alcohols (RCH_2OH) are converted either into **aldehydes** (RCH=O) if carefully controlled conditions are used, or into **carboxylic acids** (RCOOH) if an excess of oxidant is used.

A primary alcohol   An aldehyde   A carboxylic acid

For example,

1-Butanol   Butanal   Butanoic acid

Ketone A compound that has a carbonyl group bonded to two carbon atoms, $R_2C=O$.

Secondary alcohols (R_2CHOH) are converted into **ketones** ($R_2C=O$) on treatment with oxidizing agents:

A secondary alcohol   A ketone

For example,

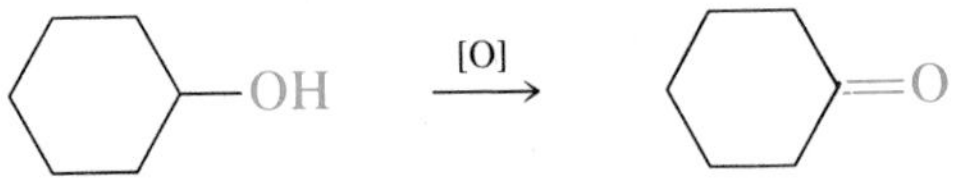

Cyclohexanol   Cyclohexanone

Tertiary alcohols don't normally react with oxidizing agents because they don't have a hydrogen on the carbon atom next to the —OH group:

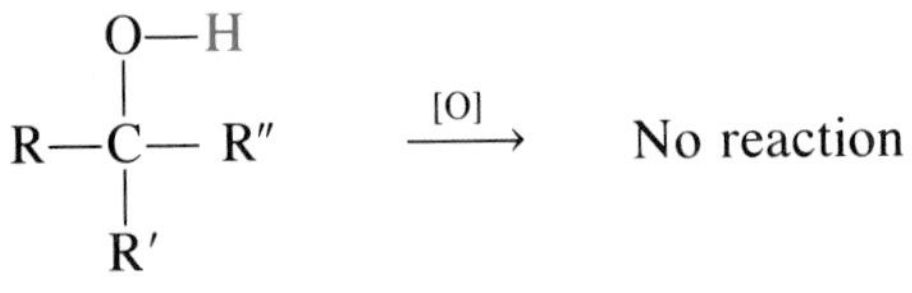

$$R-\underset{\underset{R'}{|}}{\overset{\overset{O-H}{|}}{C}}-R'' \quad \xrightarrow{[O]} \quad \text{No reaction}$$

A tertiary alcohol

Alcohol oxidations are critically important steps in many biological processes. For example, when lactic acid builds up in tired, overworked muscles, the liver removes it by oxidizing it to pyruvic acid. Our bodies, of course, don't use $Na_2Cr_2O_7$ or $KMnO_4$ for the oxidation; they use specialized, highly selective enzymes to carry out their chemistry. Regardless of the details, though, the net chemical transformation is the same whether carried out in a laboratory flask or in a living cell. (In writing biochemical reactions, we'll write acids in the ionized forms that exist in body fluids.)

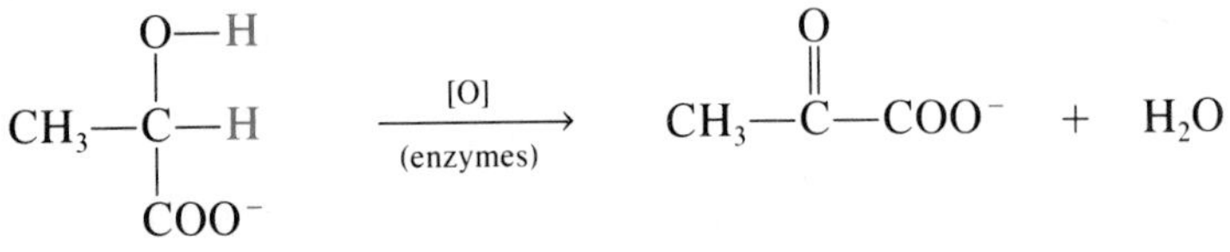

$$CH_3-\underset{\underset{COO^-}{|}}{\overset{\overset{O-H}{|}}{C}}-H \quad \xrightarrow[\text{(enzymes)}]{[O]} \quad CH_3-\overset{\overset{O}{\|}}{C}-COO^- \ + \ H_2O$$

Lactic acid anion   Pyruvic acid anion
(an alcohol-acid)

Solved Problem 14.3   What are the products of the following reaction?

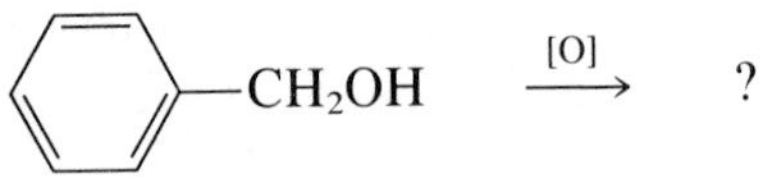

Benzyl alcohol

Solution   Since the starting material is a primary alcohol, it will be converted first to an aldehyde and then to a carboxylic acid. To find the structures of these products, first rewrite the starting alcohol to emphasize the hydrogen atoms on the OH-bearing carbon:

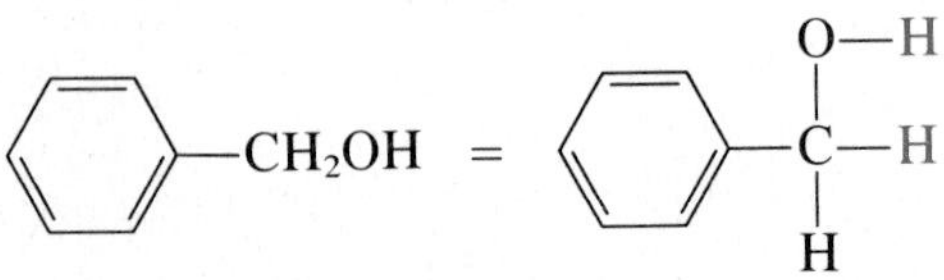

Next, circle and remove two hydrogens, one from the —OH group and one from the neighboring carbon. In their place, make a C—O double bond. This is the aldehyde product that will form initially:

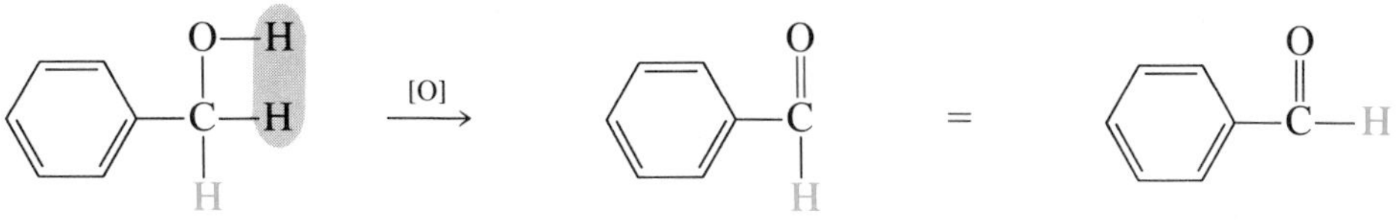

Finally, convert the aldehyde to a carboxylic acid by removing the —CH=O hydrogen and replacing it with an —OH group:

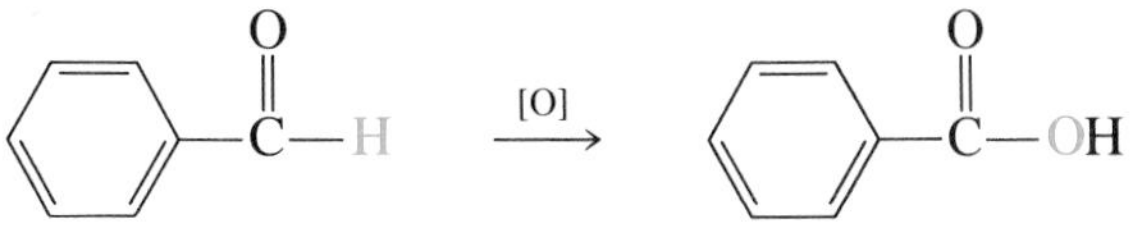

Practice Problems **14.10**   What products would you expect from oxidation of these alcohols?

(a) $CH_3CH_2CH_2OH$ (b) $CH_3\overset{\displaystyle OH}{\overset{|}{C}}HCH_2CH_2CH_3$

(c)

—CH(OH)CH₃

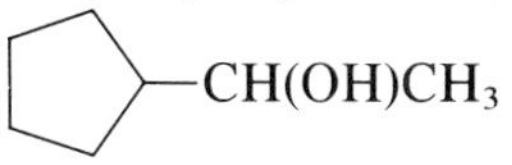

14.11   What alcohols might these carbonyl products have come from?

(a) $CH_3\overset{\displaystyle O}{\overset{||}{C}}CH_3$ (b)

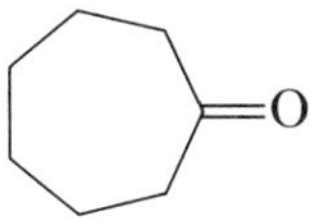

=O   (c) $CH_3\overset{\displaystyle CH_3}{\overset{|}{C}}HCH_2COOH$

14.6 PHENOLS

Antiseptic An agent that can be used to destroy or prevent the growth of harmful microorganisms on or in the body.

The word *phenol* is the name both of a specific compound (hydroxybenzene, C_6H_5OH) and of a family of compounds. Phenol itself, also called *carbolic acid*, is a medical **antiseptic** first used by Joseph Lister in 1867. Lister showed that the occurrence of postoperative infection dramatically decreased when phenol was used to cleanse the operating room and the patient's skin, and in dressings for surgical wounds. Because phenol numbs the skin, it also became popular in topical drugs for pain and itching, and in treating sore throats.

The medical use of phenol is now restricted because it can cause severe skin burns and has been found to be toxic, both by ingestion and by absorption through the skin and inhalation. The once-common use of phenol for treating diaper rash is especially hazardous because phenol is more readily absorbed through a rash. Only water or alcohol solutions containing a maximum of 1.5% phenol and lozenges containing a maximum of 50 mg of phenol are now allowed in nonprescription drugs. Many mouthwashes and throat lozenges contain alkyl-substituted phenols as active ingredients for pain relief, for example,

OH

CH₂CH₂CH₂CH₂CH₂CH₃

4-Hexylresorcinol
(a topical anesthetic)

CH₃

CH₃CHCH₃

Thymol
(a topical anesthetic; occurs
naturally in the herb thyme)

Disinfectant An agent that can be used to destroy or prevent the growth of harmful microorganisms on inanimate objects only.

Phenols and substituted phenols such as the cresols (methylphenols) are common as **disinfectants** in hospitals and elsewhere. By contrast with an anti-

AN APPLICATION: ETHYL ALCOHOL AS A DRUG AND POISON

Ethyl alcohol as a drug is classified as a central-nervous-system (CNS) depressant. Its direct effects (being "drunk") resemble the response to anesthetics and are quite predictable. At first, there is an appearance of stimulation—excitability and outgoing sociable behavior—but this results from depression of inhibition rather than stimulation. At a blood alcohol concentration of 100–300 mg/dL, motor coordination and pain perception are affected, accompanied by loss of balance, slurred speech, and amnesia. At the next stage (300–400 mg/dL), voluntary responses to stimuli are affected, and there may be nausea and loss of consciousness. Further increases in blood alcohol levels cause progressive loss of protective reflexes in stages like those of surgical anesthesia. Above 600 mg/dL of blood alcohol, spontaneous respiration and cardiovascular regulation are affected, and the ultimate result can be death.

The pathway of ethyl alcohol through the body begins with its ready absorption in the stomach and small intestine, followed by rapid distribution to all body fluids and organs. In the pituitary gland, alcohol inhibits the production of

a hormone that regulates urine flow, causing increased urine production and dehydration. In the stomach, it stimulates production of acid. Throughout the body, it causes blood vessels to dilate, resulting in flushing of the skin and a sensation of warmth as blood moves into capillaries beneath the surface. The result, though, is not a warming of the body but an increased loss of heat at the surface, making alcoholic beverages a poor choice for help in enduring cold weather.

The metabolism of alcohol occurs mainly in the liver and proceeds by oxidation in two steps, first to acetaldehyde and then to acetic acid. One of the hydrogen atoms lost in the oxidation at each stage binds to the biochemical oxidizing agent NAD^+ (nicotinamide adenine dinucleotide), and the other leaves as a hydrogen ion. When continuously present in the bodies of chronic alcoholics, alcohol and acetaldehyde are toxic, leading to devastating physical and metabolic deterioration. The liver usually suffers the worst damage, as it is the major site of alcohol metabolism.

Other alcohols are oxidized in the same manner as ethyl alcohol. The toxicity of methyl

Alcohol metabolism

$$CH_3CH_2OH + NAD^+ \xrightarrow{\substack{\text{alcohol} \\ \text{dehydrogenase} \\ \text{enzyme}}} CH_3\overset{\overset{\displaystyle O}{\|}}{C}-H + NADH/H^+ \quad CH_3\overset{\overset{\displaystyle O}{\|}}{C}-H + NAD^+ + H_2O \xrightarrow{\substack{\text{aldehyde} \\ \text{dehydrogena} \\ \text{enzyme}}}$$

septic, which safely kills microorganisms on living tissue, a disinfectant should be used only on inanimate objects. The germicidal properties of phenols can be partially explained by their ability to disrupt the permeability of cell walls of microorganisms.

Phenols are usually named with the ending -*phenol* rather than -*benzene*. For example,

o-Chlorophenol   p-Methylphenol

alcohol is due to the formation of formaldehyde (HCHO), a more toxic chemical than acetaldehyde. The danger in ingesting ethylene glycol is from its oxidation product, oxalic acid, which is not soluble in body fluids.

Because ethyl alcohol competes with other alcohols in binding to alcohol dehydrogenase, it slows their conversion to harmful oxidation products, giving the body a chance to eliminate the alcohols before damage is done. Therefore, ethyl alcohol is often administered as an antidote in methyl alcohol or ethylene glycol poisoning.

The quick and uniform distribution of ethyl alcohol in body fluids, the ease with which it crosses lung membranes, and its ready oxidizability provide the basis for tests for blood alcohol concentration. The Breathalyzer test measures alcohol concentration in expired air by the color change that occurs when the chromium in the bright yellow-orange oxidizing agent potassium dichromate ($K_2Cr_2O_7$) is reduced to blue-green chromium(III). The color change can be interpreted by instruments to give an accurate measure of alcohol concentration in the blood. In most states, a blood alcohol level above 0.10% (100 mg/dL) is evidence for legal charges of driving while intoxicated (DWI).

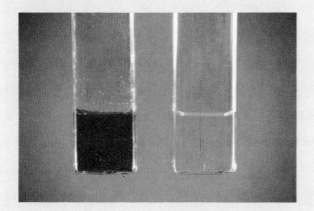 

The Breathalyzer test is based on the chemical reaction between orange potassium dichromate ($K_2Cr_2O_7$) and ethyl alcohol (in the vessels at the left). When they are mixed, the alcohol is oxidized to acetaldehyde and chromium is reduced to the green Cr^{3+} ion (on the right).

$CH_3\overset{O}{\overset{\|}{C}}OH + NADH + H^+$

The properties of phenols, like those of alcohols, are influenced by hydrogen bonding. Most phenols are water-soluble to some degree and have higher melting and boiling points than similarly substituted alkylbenzenes.

Among biomolecules, the amino acid tyrosine contains a hydroxyl-substituted benzene ring, as do many compounds that give plants their characteristic properties, such as eugenol and urushiol (Figure 14.4).

Some naturally occurring phenols

Tyrosine
(an *amino acid*)

Eugenol
(in cloves, bananas, and
other fruits; used for
toothache pain)

A urushiol
(skin irritant in
poison ivy)

Practice Problems 14.12 Draw structures for (a) *m*-bromophenol, (b) 4-ethylphenol.

14.13 Name the following compounds:
(a) Cl—⟨benzene⟩—OH   (b) HO—⟨benzene⟩—Br with CH₃

Figure 14.4
Do you recognize this plant? If you don't know what poison ivy looks like, you may experience firsthand some chemical interactions between urushiol and your body.

AN APPLICATION: ANTIOXIDANTS

If you're prone to reading food ingredient labels, the names "butylated hydroxytoluene" and "butylated hydroxyanisole" or their abbreviations BHT and BHA are familiar to you. You can see them on most cereals, cookie, and cracker boxes. Both compounds are substituted phenols.

Butylated hydroxytoluene (BHT)

Butylated hydroxyanisole (BHA)
(a mixture of two isomers)

Foods that contain unsaturated fats (those with carbon–carbon double bonds) become rancid when oxygen from the air reacts with their double bonds, breaking them apart and producing bad-smelling and bad-tasting compounds. The oxidation reactions proceed by formation of extremely reactive intermediates that contain unpaired electrons and are known as **free radicals.** Each free radical reacts with a stable molecule to form another radical, which then reacts with another molecule, and so on, in long series of *chain reactions.*

BHT and BHA are called *antioxidants* because they interrupt a free radical chain reaction by combining with the free radicals to form stable compounds. The chain reaction is thus terminated rather than carried forward. Vitamin E is a natural antioxidant that traps free radicals within the body.

Free radical formation is suspected of playing a role in both cancer and normal aging of living tissue. Although there is no conclusive evidence, questions have been raised about the possible effectiveness of antioxidants in slowing the progress of these two conditions.

Free radical An atom or group that has an unpaired electron.

14.7 ACIDITY OF ALCOHOLS AND PHENOLS

Alcohols and phenols are weakly acidic. They dissociate only slightly in aqueous solution and therefore establish equilibria.

Water: $H-O-H + H_2O \rightleftharpoons H-O^- + H_3O^+$

An alcohol: $CH_3-O-H + H_2O \rightleftharpoons CH_3-O^- + H_3O^+$

A phenol: $C_6H_5-O-H + H_2O \rightleftharpoons C_6H_5-O^- + H_3O^+$

Methanol and ethanol are no more acidic than water itself and are so slightly dissociated in water that their aqueous solutions are neutral (pH 7). The anion of an alcohol, RO^-, known as an *alkoxide ion*, is a strong base, as expected for the anion of a weak acid. Alkoxides are prepared by reaction of an alcohol with an alkali metal. For example,

$$2\,CH_3OH \; + 2\,Na \; \longrightarrow \; 2\,CH_3O^-Na^+ \; + \; H_2(g)$$

Methyl alcohol   Sodium methoxide

Phenols are considerably more acidic than water. Phenol itself ($K_a = 1 \times 10^{-10}$, for example, is close to HCN ($K_a = 6 \times 10^{-10}$) and HCO_3^- ($K_a = 5 \times 10^{-11}$) in acidity. Phenols react as weak Brønsted-Lowry acids with hydroxide ion and are soluble in dilute aqueous sodium hydroxide.

A phenol:

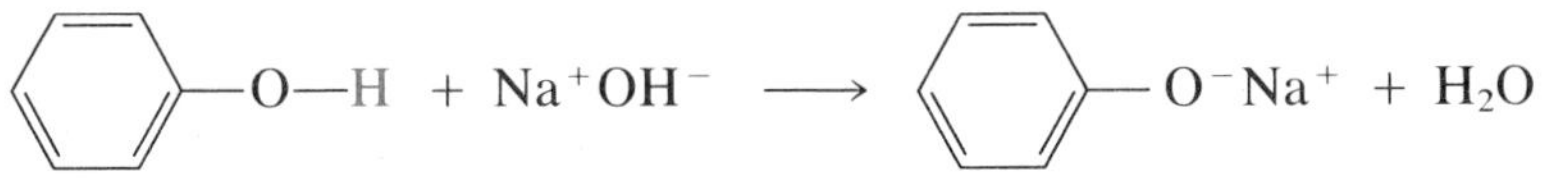

$$\text{C}_6\text{H}_5{-}O{-}H \; + \; Na^+OH^- \; \longrightarrow \; \text{C}_6\text{H}_5{-}O^-Na^+ \; + \; H_2O$$

Sodium phenoxide

14.8 NAMES AND PROPERTIES OF ETHERS

Ethers with simple alkyl or aryl groups are named just by identifying the two groups bonded to oxygen and adding the word *ether:*

$$CH_3{-}O{-}CH_3 \qquad CH_3{-}O{-}CH_2CH_3 \qquad CH_3CH_2{-}O{-}CH_2CH_3$$

Dimethyl ether   Ethyl methyl ether   Diethyl ether
(bp $-24.5°C$)   (bp $10.8°C$)   (bp $34.5°C$)

If both groups are the same, the "di-" is often left out and we refer to "methyl ether" or "ethyl ether." A reference to just "ether" usually means diethyl ether. Compounds containing the C—O—C group in a ring are classified as cyclic ethers but are not named as such. Many have common names.

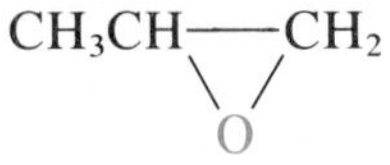

$CH_3CH{-}CH_2$ with O bridge	Reactive intermediate in polymer synthesis	A solvent	A solvent
Propylene oxide		Dioxane	Tetrahydrofuran

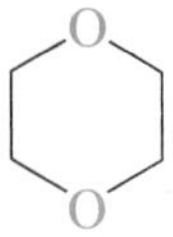

Alkoxy group An RO— group.

The RO— group is referred to as an **alkoxy group;** CH_3O- is a *methoxy group*, CH_3CH_2O- is an *ethoxy group*, and so on. These names are used when the ether functional group is present with other functional groups. For example,

$$CH_3CH_2OCH_2CH_2OH$$

2-Ethoxyethanol

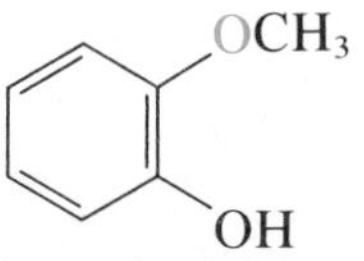

o-Methoxyphenol

Though polar, ethers lack the hydroxyl group of water and alcohols, and ether molecules do not hydrogen-bond to each other. Thus, the simple ethers are higher boiling than comparable alkanes but lower boiling than alcohols. The ether oxygen can hydrogen-bond with water, causing dimethyl ether to be water-soluble and diethyl ether to be partially miscible with water. Higher ethers are only slightly soluble or insoluble in water. Ethers are good solvents for most organic molecules.

Ethers are alkane-like in many chemicals properties and don't react with most acids, bases, or other chemical reagents. Ethers do, however, react readily with oxygen, and the simple ethers are highly flammable. On standing in air, many ethers form explosive *peroxides*, compounds which contain the unstable —O—O— group. Thus, ethers must be handled with care and stored in the absence of oxygen.

Properties of Ethers

● No hydrogen bonding occurs between ether molecules, but they are polar; lower boiling than alcohols, but higher boiling than alkanes.

● Lower ethers are volatile, flammable liquids.

● Ethers are slightly soluble or insoluble in water (except dimethyl ether, which is water-soluble).

● Are good solvents for organic compounds.

● Are not very reactive.

● Form explosive peroxides on standing in air.

14.9 SOME COMMON ETHERS

Diethyl ether, the most common ether, is best known as a solvent and anesthetic. Its value as an inhalation anesthetic was discovered in the 1840s, and until around 1930, diethyl ether, nitrous oxide (N_2O), and chloroform ($CHCl_3$) were the mainstays of the operating room. The ideal general anesthetic should act quickly, produce a deep sleep, allow a quick and smooth return to consciousness, produce few side effects, exit the body unchanged, and be safe to handle.

Ether acts quickly and is a very effective anesthetic, but it is far from ideal because recovery is not quick and it often induces nausea. Moreover, its effectiveness is strongly counterbalanced by the hazards of handling it. Diethyl ether is a highly volatile, flammable liquid whose vapor forms explosive mixtures with air that are ignited by the slightest spark. Ether has now been replaced by safer, less flammable anesthetics, two of which are halogenated ethers. It's interesting to note that these two compounds were products of an intensive effort during the 1960s in which more than 400 halogenated ethers were synthesized in a search for improved anesthetics (Figure 14.5).

Anesthetics

Enflurane   Isoflurane

Figure 14.5
Inhalation anesthetics.
Enflurane (trade name
Ethrane) and isoflurane
(trade name Forane)
are widely used nonflammable,
nonexplosive anesthetics that
produce a rapid anesthesia
from which recovery is also
rapid. They are both low-
boiling liquids administered
as vapors (enflurane, bp
56.5°C; isoflurane, bp
48.5°C).

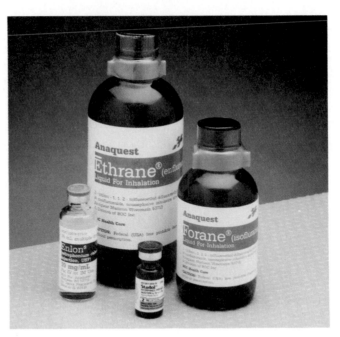

Ethers exist throughout the plant and animal kingdoms. Some are present in essential oils and are used in perfumes; others have a variety of biological roles. Juvenile hormone, for example, is a cyclic ether that helps govern the growth of insects and has attracted some interest as an insecticide. The three-membered *epoxide ring* in juvenile hormone is quite reactive because the oxygen and carbon bond angles are highly strained in the small ring.

Some biomolecules containing ether groups

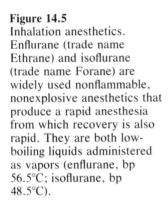

From anise;
a flavoring
and perfume in
soap and
toothpaste

Anethole

Controls insect
growth; as an insecticide
prevents formation
of the adult

Silkmoth juvenile hormone

14.10 SULFUR-CONTAINING COMPOUNDS: THIOLS AND DISULFIDES

Thiol A compound that contains the —SH functional group, R—SH.

Mercaptan An alternate name for a thiol, R—SH.

Sulfur is the element just below oxygen in the periodic table, and many oxygen-containing compounds have sulfur analogs. For example, **thiols (R—SH)**, also called **mercaptans,** are sulfur analogs of alcohols. The systematic parent name of a thiol is formed by adding *-thiol* to the parent hydrocarbon name. Otherwise, thiols are named in the same way as alcohols.

$$CH_3CH_2SH$$

$$\underset{|}{CH_3}$$
$$CH_3CHCH_2CH_2SH$$

$$CH_3CH{=}CHCH_2SH$$

Ethanethiol   3-Methyl-1-butanethiol   2-Butene-1-thiol

The most outstanding characteristic of thiols is their appalling odor. Skunk scent is caused by two of the simple thiols shown above, 3-methyl-1-butanethiol and 2-butene-1-thiol. Thiols are also in the air whenever garlic and onions are being sliced, or when there's a natural gas leak. Natural gas itself is odorless, so a low concentration of methanethiol (CH_3SH) is added as a safety measure, making it easy to detect the leak with the slightest sniff.

Disulfide A compound that contains a sulfur–sulfur single bond, R—S—S—R.

Thiols react with mild oxidizing agents to yield **disulfides,** R—S—S—R. Two thiols join together in this reaction, the hydrogen from each is lost, and the two sulfurs bond together.

$$RSH + HSR \xrightarrow{[O]} RSSR$$

Two thiol molecules   A disulfide

For example,

$$H_3C{-}S{-}H + H{-}S{-}CH_3 \xrightarrow[\text{(oxidizing agent)}]{[O]} CH_3{-}S{-}S{-}CH_3 + H_2O$$

Methanethiol   Dimethyl disulfide

The reverse of this reaction occurs easily in the presence of many reducing agents, represented here by [H],

$$RSSR \xrightarrow{[H]} RSH + RSH$$

Thiols are important biologically because they occur as a functional group in the amino acid cysteine, which is part of many proteins.

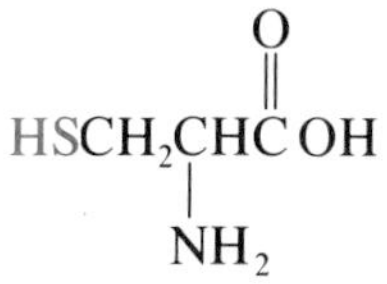

$$\underset{\underset{NH_2}{|}}{HSCH_2CHCCC}\overset{\overset{O}{\|}}{}OH \qquad \text{An amino acid}$$

Cysteine

The easy formation of —S—S— bonds between cysteine groups helps to pull large protein molecules into the shapes they need to function. Hair protein, for example, is unusually rich in —SH groups. When hair is "permed," a mild oxidizing agent causes disulfide bonds to form between —SH groups, resulting in the introduction of bends and kinks into the hair (Figure 14.6).

Practice Problem   **14.14**   What disulfides would you obtain from oxidation of these thiols?
(a) $CH_3CH_2CH_2SH$ (b) 3-methyl-1-butanethiol (skunk scent)

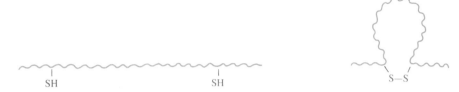

Figure 14.6
Sulfide bridges in hair. Chemistry can give you curly hair. A permanent wave results when disulfide bridges are formed between —SH groups in hair protein molecules.

14.11 HALOGEN-CONTAINING COMPOUNDS

Halogen-containing organic compounds result from the replacement of hydrogen atoms by halogen atoms. The simplest such compounds are the **alkyl halides,** RX, where R is an alkyl group and X is a halogen. Their common names are formed by giving the name of the alkyl group followed by the halogen name with an -*ide* ending. Systematic names for alkyl halides are derived like those of other families of compounds, by identifying the hydrocarbon root, using the alkane ending -*ane*, numbering the halogen substituents, and naming them as fluoro-, chloro-, bromo-, or iodo- groups. Some halogenated compounds, such as chloroform and halothane, are also known by common names.

Alkyl halide A compound that contains an alkyl group bonded to a halogen atom, RX.

CH_3CH_2Cl

Ethyl chloride, bp 12.5°C
(chloroethane)

Halothane, bp 50°C
(1-bromo-1-chloro-2,2,2-
trifluoroethane)

Chloroform, bp 61°C
(trichloromethane)

Chloral hydrate, bp 97.5°C
(1,1,1-trichloro-2,2-ethanediol)

A Halon, bp ‑‑58°C
(bromotrifluoromethane)

Halogenated organic compounds have a variety of medical and industrial uses. Ethyl chloride is used as a topical anesthetic because it cools the skin through rapid evaporation; halothane is the most important non-ether anesthetic. Chloroform was once employed as an anesthetic (often by criminals in detective stories) and as a solvent for cough syrups and other medicines, but is now considered too toxic for such uses. Chloral hydrate is a sleep-inducing prescription drug (the "knockout" drops of detective stories). Bromotrifluoromethane is useful for extinguishing fires in aircraft and electronic equipment because it is not flammable, is nontoxic, and evaporates without a trace.

Although numerous halogen-containing organic compounds are found in nature, especially in marine organisms, few play significant roles in human biochemistry. One exception is thyroxine, an iodine-containing hormone secreted by the thyroid gland. A deficiency of iodine in the human diet leads to a low thyroxine level, which causes a condition called *goiter* and a resulting swelling of the thyroid gland. To ensure adequate iodine in the diet and prevent goiter, potassium iodide is added to table salt.

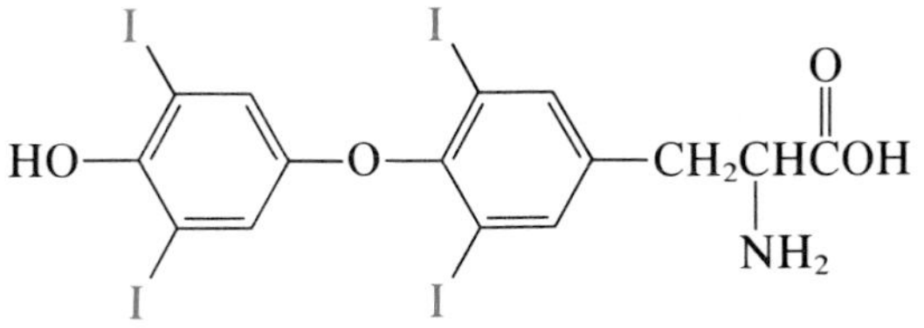

Thyroid gland hormone; deficiency causes goiter

Thyroxine

Halogenated compounds are also of great importance in industry and agriculture. Dichloromethane (CH_2Cl_2, methylene chloride) and trichloromethane ($CHCl_3$, chloroform) are common solvents, and trichloroethylene ($Cl_2C=CHCl$) is used to degrease machined metal parts. Because these substances are excellent solvents for the greases in skin, continued exposure often causes dermatitis.

Agricultural use of herbicides such as 2,4-D and fungicides such as Captan has resulted in vastly increased crop yields in recent decades, and the widespread application of chlorinated insecticides such as DDT is largely responsible for the progress made toward worldwide control of malaria and typhus. Despite their enormous benefits, chlorinated pesticides present problems because they are not broken down by natural processes and persist in the environment. They remain in the fatty tissues of organisms and accumulate up the food chain as larger organisms consume smaller ones. Eventually the concentration in some animals becomes high enough to cause harm. In an effort to maintain a balance between the value of halogenated pesticides and the harm they can do, the use of many has been restricted and some have been banned.

Some chlorinated pesticides

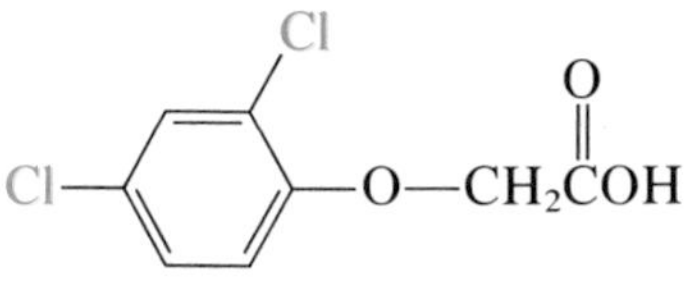

2,4-D

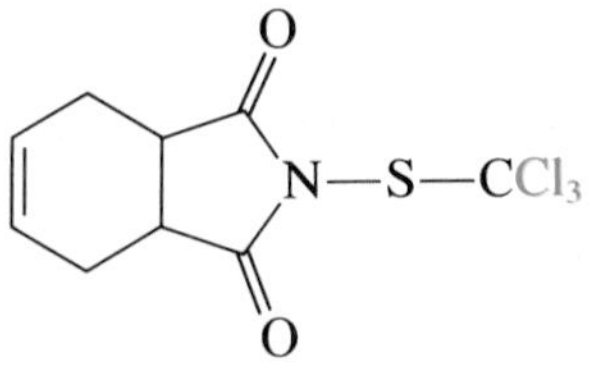

Captan

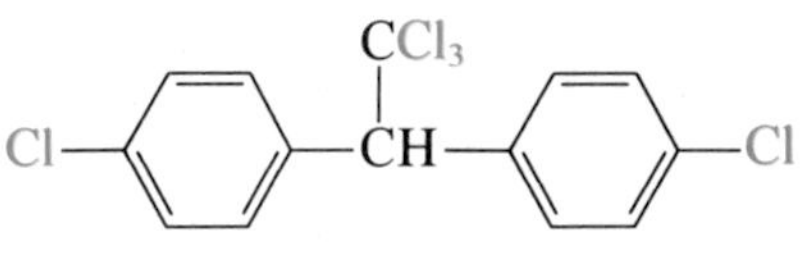

DDT

INTERLUDE: CHLOROFLUOROCARBONS AND THE OZONE HOLE

In recent years, newspaper stories about a "hole" in the ozone layer have appeared with regularity. What began as speculation about potential problems is now accepted as fact: Up to 75% of the ozone over the South Pole disappears in the fall when the so-called *ozone hole* develops. Furthermore, the overall ozone concentration in the region of the atmosphere extending from about 20 to 40 km above the earth's surface has decreased 2.5% in the last 10 years.

Although toxic to all life forms at high concentrations, ozone (O_3) is nevertheless critically important in the upper atmosphere because it acts as a shield to protect the earth from intense solar radiation. If the ozone layer were depleted, a great deal more solar radiation would reach the earth, causing an increase in the incidence of skin cancer and eye cataracts. The effects on microscopic plants in the oceans and crop plants are less predictable but could also be troublesome.

The causes of ozone depletion, although not fully understood, almost certainly involve a group of halogen-substituted alkanes called chlorofluorocarbons, familiar to most as *Freons* (DuPont). The chlorofluorocarbons (CFCs) are simple alkanes in which all the hydrogens have been replaced by either chlorine or fluorine. Fluorotrichloromethane (CCl_3F, Freon 11) and dichlorodifluoromethane (CCl_2F_2, Freon 12) are two of the most common CFCs in industrial use.

Because they are inexpensive and highly stable, yet not toxic, flammable, or corrosive, CFCs are ideal for use as propellants in aerosol cans, refrigerants, solvents, and fire extinguishers, and for blowing bubbles into foamed plastics such as those in insulation and mattresses. Unfortunately, the stability that makes CFCs so useful results in their persistence in the environment. The molecules slowly find their way into the upper atmosphere where they undergo a complex series of reactions that ultimately result in ozone destruction. When ultraviolet (UV) light strikes a CFC molecule, a carbon–chlorine bond breaks, producing a chlorine atom, which is highly reactive. The chlorine atom then reacts with ozone to yield oxygen and $ClO\cdot$, another reactive intermediate (a free radical):

$$CCl_2F_2 \xrightarrow{\text{UV light}} \cdot CClF_2 + Cl\cdot$$

$$Cl\cdot + O_3 \longrightarrow O_2 + ClO\cdot$$

Worldwide efforts to protect the ozone layer began in 1987 with an agreement, ultimately ratified by 36 countries and the European Economic Community, to gradually phase out CFC production. These efforts have since accelerated beyond the terms of the 1987 agreement as evidence for the serious nature of the problem has accumulated. Major producers of the chemicals are pooling their research efforts to find alternative, safe chemicals, and many nations have pledged to cease use of CFCs and other chemicals that release chlorine to the atmosphere by the year 2000. According to some predictions, even with these stringent efforts chlorine concentration will continue to rise due to molecules already in the atmosphere and will not return to current levels until the year 2045.

The atmosphere scientist is reviewing data on the ozone hole gathered by flights over the Antarctic.

SUMMARY

Alcohols have an —OH group (**hydroxyl**) bonded to a saturated alkane-like carbon atom and are given the family-name ending -*ol*. **Phenols** have an —OH group bonded directly to an aromatic ring and are named as phenols. **Ethers** have an oxygen atom bonded to two groups that may be either alkyl or aryl groups or may be in a ring, and their common names combine both group names with the word "ether." Both alcohols and phenols are "water-like" in their ability to form hydrogen bonds. The lower alcohols are miscible with water. Also like water, alcohols and phenols are weak acids that can donate their —OH protons to strong bases. Phenols are acidic enough to dissolve in aqueous NaOH. Ethers do not hydrogen-bond and are more alkane-like in their properties.

Alcohols undergo loss of water (**dehydration**) to yield alkenes when treated with a strong acid, and undergo **oxidation** to yield carbonyl-group-containing compounds. **Primary alcohols** (RCH$_2$OH) are oxidized to yield either aldehydes (RCH=O) or carboxylic acids (RCOOH); **secondary alcohols** (R$_2$CHOH) are oxidized to yield ketones (R$_2$C=O); and **tertiary** alcohols are not oxidized.

The —OH group is present in all carbohydrates and many other biochemically active molecules. Some common alcohols are methyl, ethyl, and isopropyl alcohol and ethylene glycol and glycerol. Except for glycerol, all are toxic in varying degrees. Phenols are notable for their use as disinfectants and antiseptics, and ethers for their solvent properties, flammability, and use as anesthetics.

Thiols, or **mercaptans** (RSH), are sulfur analogs of alcohols, many of which have unpleasant odors. They react with mild oxidizing agents to yield **disulfides** (RSSR), a reaction of importance in protein chemistry.

Alkyl halides, RX, contain a halogen atom bonded to an alkyl group. With increasing numbers of halogen atoms, compounds become higher boiling and less flammable. Halogenated compounds are rare in human biochemistry except for iodine-containing thyroxine but are widely used in industry as solvents and in agriculture as herbicides, fungicides, and insecticides. They are noted for their persistence in the environment.

REVIEW PROBLEMS

Alcohols, Ethers, and Phenols

14.15 How do alcohols, ethers, and phenols differ structurally?

14.16 What is the structural difference between primary, secondary, and tertiary alcohols?

14.17 Why do alcohols have higher boiling points than ethers of the same formula weight?

14.18 Which is the stronger acid, ethanol or phenol?

14.19 The steroidal compound prednisone is often used to treat poison ivy and poison oak inflammations. Identify the functional groups present in prednisone.

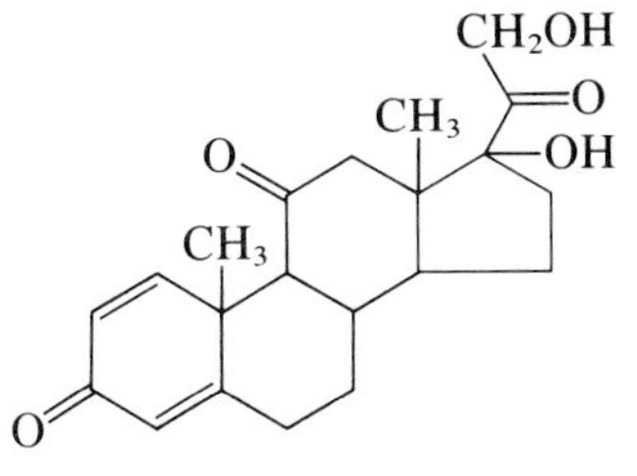

Prednisone

14.20 Vitamin E has the following structure. Identify the functional-group class to which each oxygen belongs.

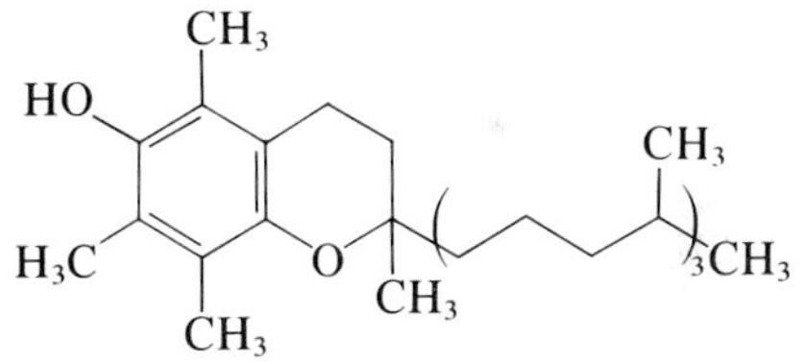

14.21 Give systematic names for these alcohols:

(a) CH$_3$CH$_2$CHCH$_2$CH$_2$CH$_3$ with CH$_2$OH substituent

(b) (CH$_3$)$_2$CHCH$_2$CH$_2$OH

(c) HOCH$_2$CH$_2$CHCH$_2$OH with OH substituent

(d)

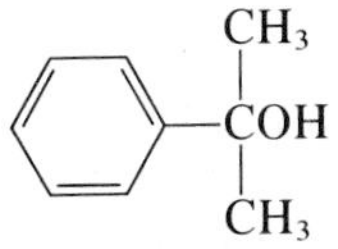

(e)

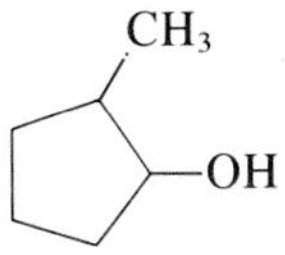

(f) CH$_3$CH$_2$CCH$_2$OH with CH$_2$CH$_3$ and CH$_3$ substituents

14.22 Draw structures corresponding to these names:
(a) 2,4-dimethyl-2-pentanol

(b) 2,2-dimethylcyclohexanol
(c) 5,5-diethyl-1-heptanol
(d) 3-ethyl-3-hexanol
(e) 2,3,7-trimethylcyclooctanol
(f) 3,3-diethyl-1,6-hexanediol

14.23 Identify each of the alcohols named in Problem 14.22 as primary, secondary, or tertiary.

14.24 Give systematic names for these compounds:
(a)

(b) CH₃—CH—O—CH₃ with CH₃

(c) O₂N—⟨⟩—O—CH₃   (d)

(e)

(f) CH₃CH₂CH₂OCH₂CH₂CH₃

14.25 Draw structures corresponding to these names:
(a) ethyl phenyl ether
(b) *o*-dihydroxybenzene (catechol)
(c) *tert*-butyl *p*-bromophenyl ether
(d) *p*-nitrophenol
(e) 2,4-diethoxy-3-methylpentane
(f) 4-methoxy-3-methyl-1-pentene

14.26 Arrange the following six-carbon compounds in order of their expected boiling points and explain your ranking:
(a) hexane   (b) 1-hexanol   (c) dipropyl ether

Reactions of Alcohols

14.27 Give a specific example of an alcohol dehydration reaction.

14.28 What product is formed on oxidation of a secondary alcohol?

14.29 What structural feature prevents tertiary alcohols from undergoing oxidation reactions?

14.30 What product(s) can form on oxidation of a primary alcohol?

14.31 Which of these three compounds would you expect to be the most soluble in water, and which the least soluble? Explain.
(a) ethane   (b) 1-pentanol   (c) 1,2,3-propanetriol

14.32 Assume that you have samples of the following two compounds, both with formula C₇H₈O. Both compounds dissolve in ether, but only one of the two dissolves

in aqueous NaOH. How could you use this information to distinguish between them?

14.33 Assume that you have samples of the following two compounds, both with formula C₆H₁₂O. What simple chemical reaction will allow you to distinguish between them? Explain.

14.34 The following alkenes can be prepared by dehydration of an appropriate alcohol. Show the structure of the alcohol in each case. If the alkene can arise from dehydration of more than one alcohol, show all possibilities.

(a)   (b) CH₃CH₂, CH₃CH₂ C=CH₂

(c) 3-hexene   (d) CH₃C=CHCH₂CH₃ with CH₃

(e) 1,3-butadiene   (f)

14.35 Phenols undergo the same kind of substitution reactions that other aromatic compounds do (Section 13.12). Formulate the reaction of *p*-methylphenol with bromine to give a mixture of two substitution products.

14.36 What carbonyl-containing products would you obtain from oxidation of these alcohols? If no reaction occurs, write "NR."

(a)   (b) CH₃CHCH₂OH with CH₃

(c) 3-methyl-3-pentanol   (d)

(e) CH₃CH₂CHCCH₃ with H₃C OH and CH₃   (f)

14.37 What alkenes might be formed by dehydration of

these alcohols? If more than one product is possible, indicate which you expect to be major.

(a)

(b)

$$\underset{\substack{| \\ OH}}{CH_3CH_2CHCHCH_3}\overset{\substack{CH_3}}{}$$

HO CH₃ on the CHCH positions

(c)

(d)

—CHCH₂CH₃
  |
  OH

(e)

$$\underset{\substack{| \\ CH_2CH_3}}{CH_3CH_2CCH_2CH_3}\overset{\substack{OH}}{}$$

Thiols and Disulfides

14.38 What is the most noticeable characteristic of thiols?

14.39 What is the structural relationship between a thiol and an alcohol?

14.40 The amino acid cysteine forms a disulfide when oxidized. What is the structure of this disulfide?

$$\underset{\substack{| \\ NH_2}}{HSCH_2CHCOH}\overset{\substack{O \\ \|}}{}$$

14.41 Name these compounds.

(a) CH_3CHCH_3
        |
        SH

(b)

SH

14.42 The boiling point of propanol is 97°C, while that of ethanethiol, with about the same molar mass, is 37°C. Chloroethane, with a similar molar mass, has a boiling point of 13°C. Explain.

14.43 Propanol is very soluble in water, while ethanethiol and chloroethane are only very slightly soluble. Explain.

Applications

14.44 Is ethanol a stimulant or a depressant? Justify your answer. [App: Ethyl Alcohol]

14.45 At what blood alcohol concentration does speech begin to be slurred? What is the approximate lethal concentration of ethyl alcohol in the blood? [App: Ethyl Alcohol]

14.46 Why does alcohol consumption cause dehydration? [App: Ethyl Alcohol]

14.47 Cirrhosis of the liver is a common disease of alcoholics. Why is the liver particularly affected by alcohol consumption? [App: Ethyl Alcohol]

14.48 Ethyl alcohol is toxic in high concentrations, yet it is administered to counteract the effects of both methanol and ethylene glycol poisoning. Name the enzyme responsible for alcohol metabolism and explain the rationale for this treatment. [App: Ethyl Alcohol]

14.49 Describe the basis of the Breathalyzer test for alcohol concentration. [App: Ethyl Alcohol]

14.50 Old westerns often show whiskey being consumed before bullets or arrows are removed from the body. Often the whiskey is poured into the wounds as well. What purpose does the whiskey serve? [App: Ethyl Alcohol]

14.51 What is a free radical? [App: Antioxidants]

14.52 What vitamin appears to be an antioxidant? [App: Antioxidants]

14.53 Ozone is considered to be an air pollutant at the earth's surface. Why is it of great benefit in the upper atmosphere? [Int: Chlorofluorocarbons]

14.54 Chlorofluorocarbons (CFCs) are still widely used as coolants in refrigerators and air conditioners, but states are beginning to legislate methods of transfer for CFCs and disposal of CFC-containing appliances. Why? [Int: Chlorofluorocarbons]

Additional Problems

14.55 Name the ether and alcohol isomers with formula $C_4H_{10}O$ and write their structural formulas.

14.56 What are the advantages and the disadvantages associated with the use of chlorinated pesticides?

14.57 Thyroxine (Section 14.11) is synthesized in the body by reaction of thyronine with iodine. Formulate the reaction and tell what kind of process is occurring.

HO—⬡—O—⬡—CH₂CH(NH₂)C—OH
                                          ‖
                                          O

Thyronine

14.58 Neither 1-nonanol nor *n*-decane is water-soluble. Explain.

14.59 Thymol, mentioned in the chapter, has the following structure. Provide an IUPAC name for the compound.

14.60 What is the difference between an antiseptic and a disinfectant?

14.61 Write the formulas and IUPAC names for the following common alcohols:
(a) rubbing alcohol (b) wood alcohol
(c) grain alcohol (d) diol used as antifreeze

14.62 Why is diethyl ether no longer ordinarily used as a general anesthetic?

14.63 Name the following compounds.

(a)

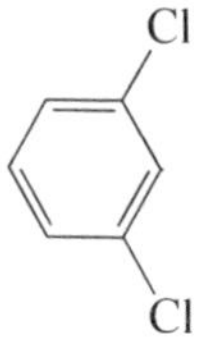

(b) $BrCH=CCH_2CH_3$
  |
  Br

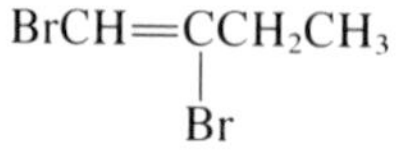

(c)

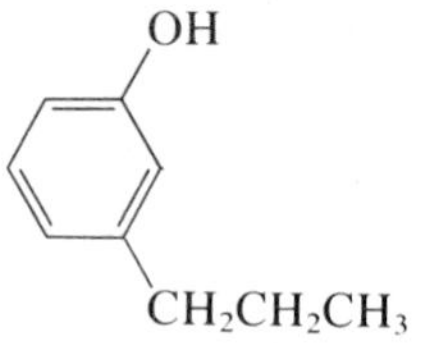

(d)

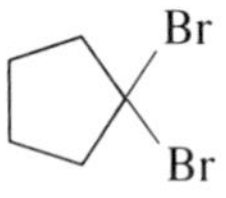

(e)
$$CH_3CCH_2CCH_3$$
with Cl, OH above and CH_3, CH_2CH_3 below

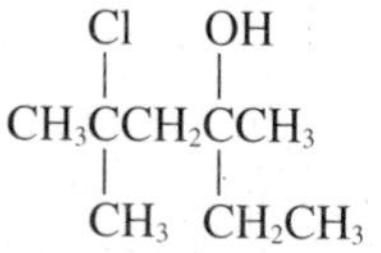

(f)
$$CH_3CH_2CHCHCH_2CHCH_3$$
with OH, OH above and CH_3 below

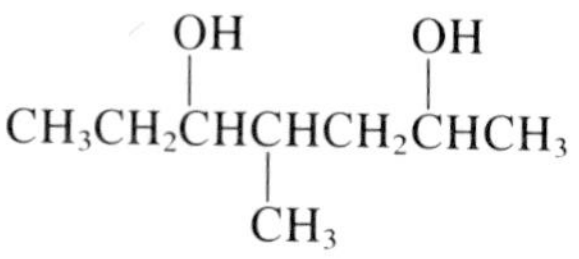

(g)
$$CH_3C\equiv CCHCH_2CCH_3$$
with Br, CH_3 above and CH_3 below

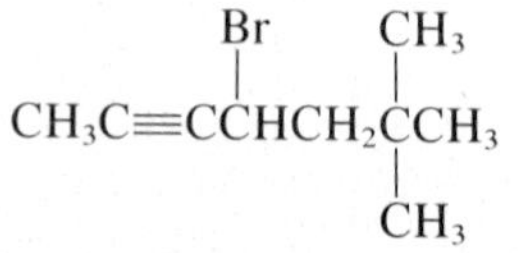

(h)

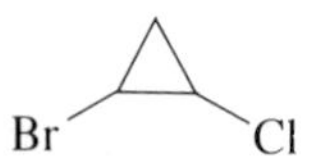

14.64 Complete these reactions.

(a) $CH_3C=CHCH_3 + HBr \longrightarrow$
      |
      CH_3

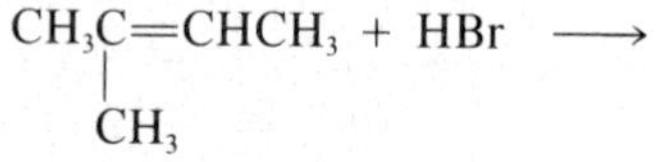

(b) $CH_3CH_2CH_2C-CHCH_3 \xrightarrow{[O]}$
  with H_3C, OH above and H_3C below

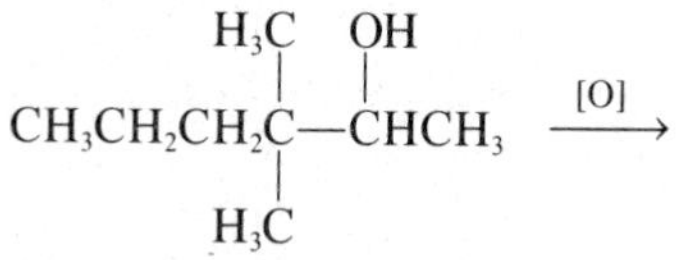

(c) $CH_3CH_2CH_2C-CHCH_3 \xrightarrow{H_2SO_4}$
  with H_3C, OH above and H_3C below

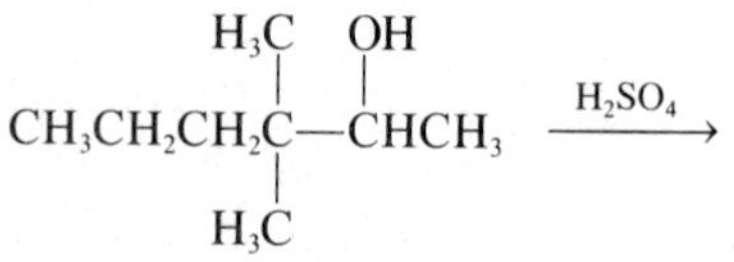

(d) $CH_3-C-C=C-CH_3 + Br_3 \longrightarrow$
  with OH above C, and H_3C, CH_3, CH_3 below

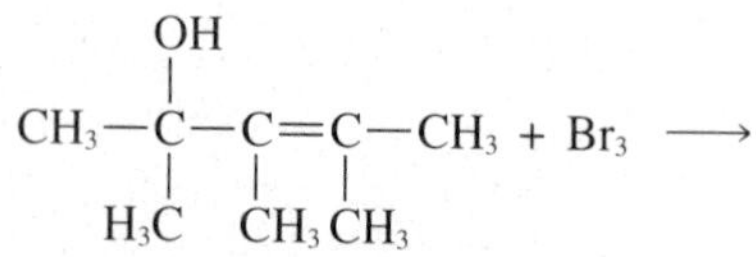

(e) $2\ (CH_3)_3CSH \xrightarrow{[O]}$

(f) $CH_3CH_2CH=CCH_3 \xrightarrow[H_2SO_4]{H_2O}$
      |
      CH_3

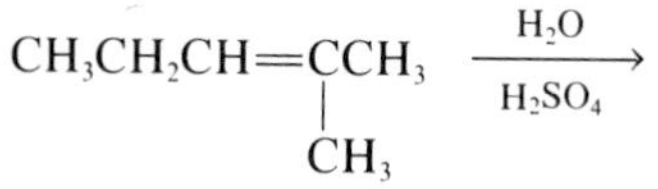

(g)
$-CH_2CHCH_3 \xrightarrow{[O]}$
  |
  OH

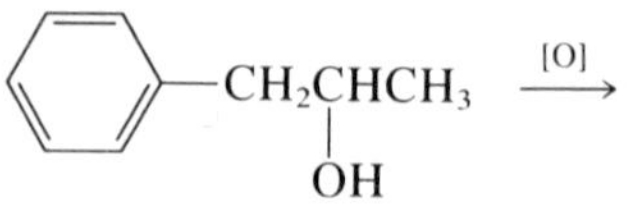

14.65 The odor of roses is due to geraniol.

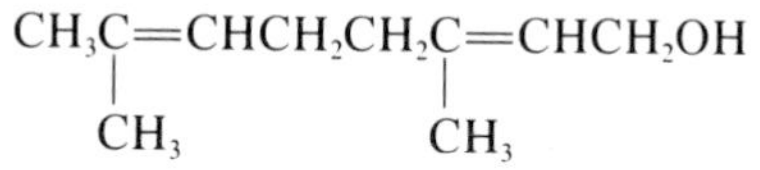

$$CH_3C=CHCH_2CH_2C=CHCH_2OH \quad \text{Geraniol}$$
with CH_3 and CH_3 below

(a) Name this compound by the IUPAC system.
(b) Circle the isoprene groups in this compound.
(c) When the alcohol group is oxidized to the aldehyde, citral, one of the compounds responsible for lemon scent is formed. Write the structure of citral.

14.66 Concentrated ethanol solutions can be used to kill microorganisms. However, at low concentrations, such as in some wines, the microorganisms can survive and cause oxidation of the alcohol. What is the structure of the acid formed?

14.67 "Flaming" desserts, such as cherries jubilee, use the ethanol in brandy or other distilled spirits as the flame carrier. Write the equation for the combustion of ethanol.

14.68 Simple sugars such as fructose, shown below, are very soluble in water in spite of having a long carbon chain. Why?

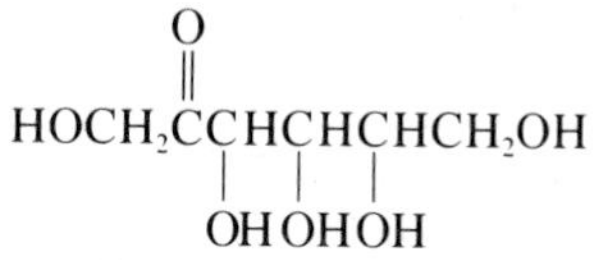

$$\overset{O}{\overset{\|}{HOCH_2CCHCHCHCH_2OH}}$$
with OH OH OH below

14.69 In Chapter 13, you saw that H_2SO_4 is a catalyst in the addition of water to alkenes to form alcohols. In this chapter, you found that H_2SO_4 is also used to dehydrate alcohols to make alkenes. Why do you think that sulfuric acid can serve two purposes—aiding in both hydration and dehydration?

CHAPTER

15 Amines

The active ingredients of many herbs in this early American apothecary shop in Shelburne, Vermont, were undoubtedly amines, the subject of this chapter.

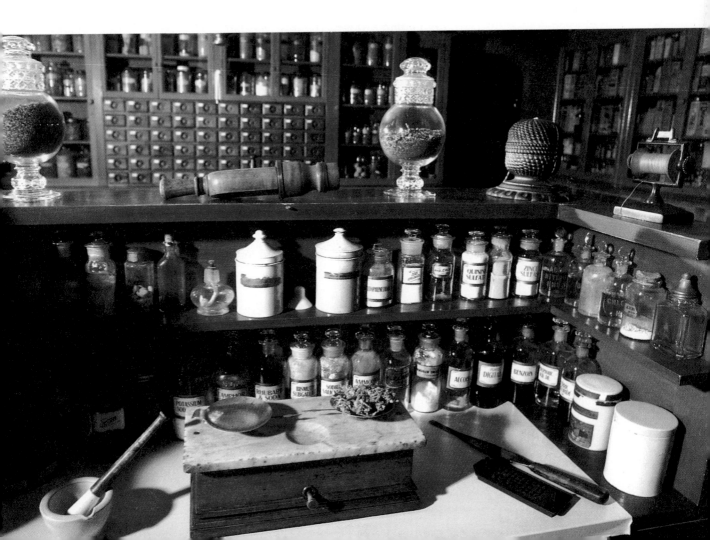

In the preceding chapter we discussed organic compounds with single bonds between oxygen, sulfur, or halogen atoms and carbon atoms. Amines are organic compounds with single bonds between nitrogen and carbon atoms other than carbon atoms in C=O groups (amides, Section 17.9). There are many more kinds of amines in nature and in use in medicine than we can explore, but this chapter will introduce you to their diversity and numerous natural functions.

1. **What are the different types of amines?** The goal: Be able to recognize primary, secondary, tertiary, and heterocyclic amines.
2. **How are amines named?** The goal: Be able to name simple amines and write their structures, given the names.
3. **What are the general properties of amines?** The goal: Be able to describe amine properties such as hydrogen bonding, solubility, boiling point, and basicity.
4. **How do amines react with water and acids?** The goal: Be able to predict the structures of the ammonium ions and of the salts formed by reactions of amines with water and acids.
5. **What are some naturally occurring types of amines and some amine-containing drugs?** The goal: Be able to describe some types of amines found in biomolecules, plants, and drugs.

15.1 AMINES

Amine A compound that has one or more organic groups bonded to nitrogen, RNH_2, R_2NH, or R_3N.

Primary amine An amine that has one organic group bonded to nitrogen, RNH_2.

Secondary amine An amine that has two organic groups bonded to nitrogen, R_2NH.

Tertiary amine An amine that has three organic groups bonded to nitrogen, R_3N.

Amines are compounds that contain one or more organic groups bonded to nitrogen: RNH_2, R_2NH, or R_3N. Thus, they are organic derivatives of ammonia in the same way that alcohols and ethers are organic derivatives of water. Amines are classified as **primary, secondary,** or **tertiary,** depending on how many organic substituents are bonded to the nitrogen atom.

$$H—N—H$$
$$|$$
$$H$$

Ammonia

$$R—N—H \qquad R—N—H \qquad R—N—R''$$
$$| \qquad\qquad | \qquad\qquad |$$
$$H \qquad\qquad R' \qquad\qquad R'$$

A *primary* amine (one R group on nitrogen) A *secondary* amine (two R groups on nitrogen) A *tertiary* amine (three R groups on nitrogen)

The groups bonded to the amine nitrogen atom may be alkyl or aryl groups. For example,

CH_3NH_2

Methylamine
(a primary alkyl amine)

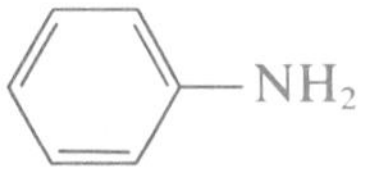

—NH_2

Aniline
(a primary
aromatic amine)

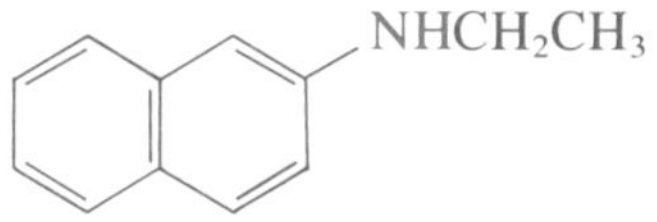

—$NHCH_2CH_3$

N-Ethylnaphthylamine
(a secondary aromatic amine)

15.2 NAMING AMINES

Primary alkyl amines, RNH_2, are named simply by identifying the alkyl group attached to nitrogen and adding the suffix -*amine* to the alkyl group name.

$CH_3CH_2NH_2$

$\underset{\underset{CH_3}{|}}{CH_3}CH NH_2$

—NH_2

Ethylamine   Isopropylamine   Cyclohexylamine

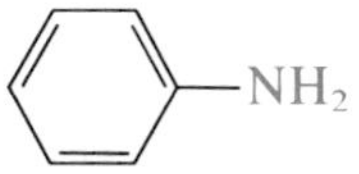

Secondary and tertiary amines with two or three identical groups are named by adding the appropriate prefix, *di-* or *tri-*, to the alkyl group name.

$\underset{\underset{H}{|}}{CH_3CH_2CH_2}NCH_2CH_2CH_3$

$\underset{\underset{CH_2CH_3}{|}}{CH_3CH_2}NCH_2CH_3$

Dipropylamine   Triethylamine

Secondary and tertiary amines with nonidentical R groups are named as N-substituted derivatives of a primary amine. The largest of the organic groups bonded to nitrogen is chosen as the parent compound, and the other groups are considered N-substituents (N because they're attached directly to nitrogen). The following compounds, for example, are named as propylamines because the propyl group in each is the largest alkyl group.

$\underset{\underset{H}{|}}{CH_3CH_2}NCH_2CH_2CH_3$

$\underset{\underset{CH_3}{|}}{CH_3}NCH_2CH_2CH_3$

N-Ethylpropylamine   N,N-Dimethylpropylamine

Amino group The —NH_2 functional group.

The simplest aromatic amine is known by the common name *aniline*. When the —NH_2 group must be named as a substituent, **amino-** is used as a prefix.

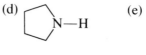 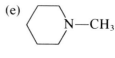

Aniline   *N*-Methylaniline   3-Aminopropanoic acid

Practice Problems   **15.1**   Identify these compounds as primary, secondary, or tertiary amines.
(a) $CH_3CH_2CH_2NH_2$ (b) $CH_3CH_2NHCH_2CH_3$

(c)
$$CH_3—\underset{\underset{CH_3}{|}}{\overset{\overset{CH_3}{|}}{C}}—NH_2$$

(d) N—H (pyrrolidine ring)

(e) N—CH₃ (piperidine ring)

      15.2   What are the names of these amines?
(a) $CH_3CH_2CH_2NH_2$ (b) $H—\underset{\underset{CH_3}{|}}{N}—CH_3$ (c) —NHCH₂CH₃ (benzene ring)

      15.3   Draw structures corresponding to these names:
(a) butylamine   (b) *N*-methylethylamine
(c) *N,N*-dimethylaniline   (d) 2-aminobutanol

15.3   HETEROCYCLIC NITROGEN COMPOUNDS

Heterocycle   A ring that contains nitrogen or some other atom in addition to carbon.

In many nitrogen-containing compounds, the nitrogen atom is in a ring with carbon atoms. Such **heterocyclic** nitrogen compounds may be nonaromatic or aromatic. Piperidine, for example, is a saturated heterocyclic amine with a six-membered ring, and pyridine is an aromatic heterocyclic amine which, like other aromatic compounds, is usually represented on paper by showing alternating double and single bonds in the ring.

Piperidine   Pyridine

      The names and structures of several heterocyclic nitrogen compounds are given in Table 15.1. You need not memorize all these names and structures but should realize that nitrogen heterocycles are very common in both plants and animals. Nicotine, found in tobacco leaves, contains two heterocyclic rings; quinine, an antimalarial drug isolated from the bark of the South American *Cinchona* tree, contains a quinoline ring system plus a nitrogen ring with a two-carbon bridge across it. The amino acid tryptophan contains an indole ring system.

Table 15.1   Some Heterocyclic Nitrogen Compounds

Pyrrolidine

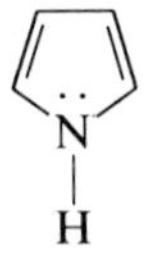

Pyrrole

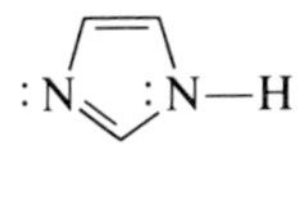

Imidazole

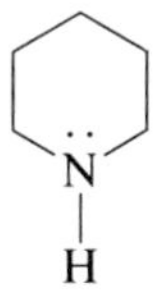

Piperidine

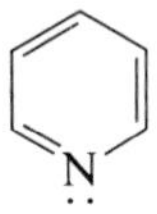

Pyridine

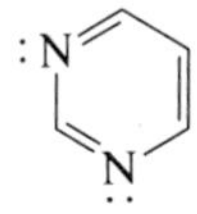

Pyrimidine

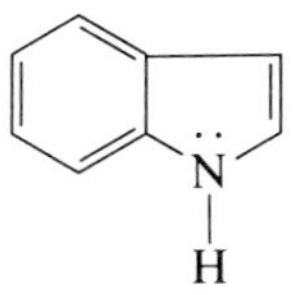

Indole

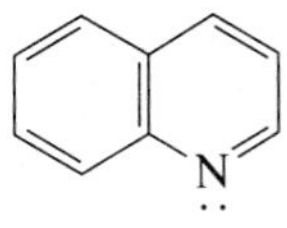

Quinoline

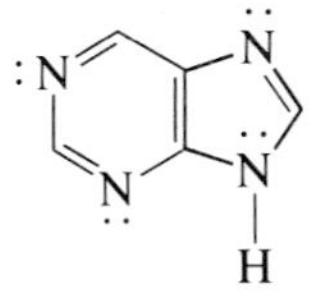

Purine

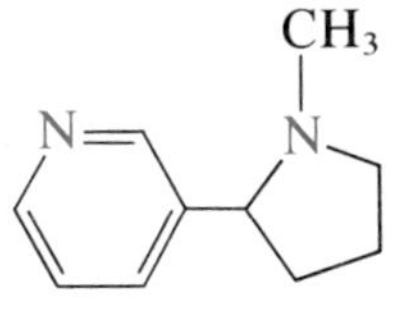

Nicotine
from tobacco
(an insecticide)

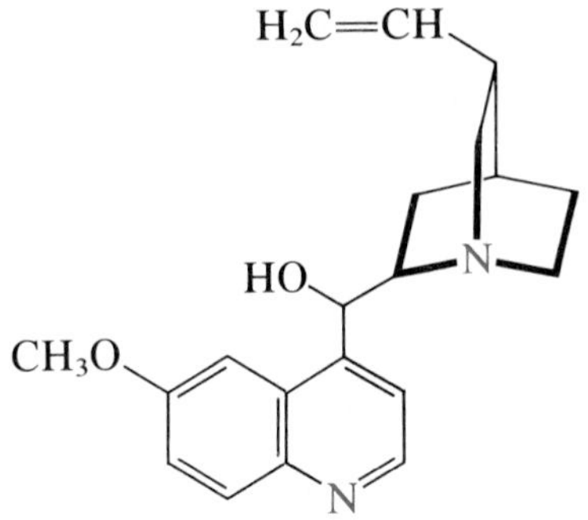

Quinine
from the *Cinchona* tree
(an antimalarial drug)

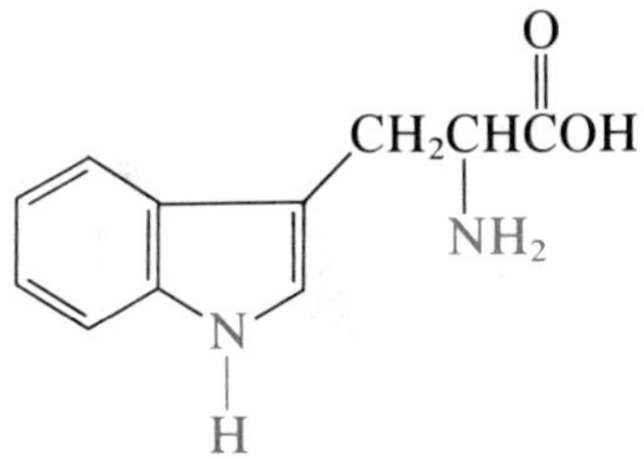

Tryptophan
(an amino acid)

15.4 PROPERTIES OF AMINES

Amines are analogous to ammonia in many of their physical properties, just as alcohols are analogous to water. Like ammonia, amines have an unshared electron pair on the nitrogen atom and have polar bonds because the nitrogen atom is electronegative. Thus, amines form hydrogen bonds in the same way that ammonia does (Figure 15.1) and are higher boiling than alkanes of similar size. Comparison of the following boiling points suggests, though, that intermolecular forces in amines are weaker than in alcohols:

$$CH_3CH_2CH_2CH_3 \qquad CH_3CH_2CH_2NH_2 \qquad CH_3CH_2CH_2OH$$

Butane, bp 0°C   Propylamine, bp 48°C   Propanol, bp 97°C

AN APPLICATION: CHEMICAL INFORMATION

Suppose you're reading an article in a magazine about caffeine in soft drinks and you decide you'd like to know something about the chemical called "caffeine." Where would you look for information?

The first place might be a chemical handbook, where you would find the structure and an entry like the one shown at the bottom of page 413 from *Lange's Handbook of Chemistry* which includes some physical properties and a reference to Beilstein's *Handbook of Organic Chemistry,* a publication that would give more chemical information (but it's written in German). A second standard chemical handbook is the *CRC Handbook of Chemistry and Physics* (CRC Press, Inc., Boca Raton, Fla.).

With your curiosity not yet satisfied, you could look next in another generally available and reliable source of chemical information, *The Merck Index: An Encyclopedia of Chemicals and Drugs,* first published in 1889 as a list of products of the Merck pharmaceutical company. The index has since grown way beyond that to a standard reference work giving information on the preparation, properties, and uses of over 10,000 chemical compounds. It has a strong medical emphasis and, where appropriate, lists toxicity data, therapeutic uses in both human and veterinary medicine, and the physiological effects and cautions associated with hazardous chemicals.

The entry for caffeine in *The Merck Index,* reproduced below, begins with a list of alternate chemical names and a capitalized entry that's the name of a medication containing caffeine. One of the joys of using *The Merck Index* is that every one of the alternate names of every substance in the book, including the drug names, appears in the index, so no matter what name you come across, you can discover exactly what substance it refers to.

Caffeine information from *The Merck Index.* Reproduced from *The Merck Index,* 11th Ed. (1989), S. Budavari, M. J. O'Neil, A. Smith, P. E. Heckelman, Eds., by permission of the copyright owner, Merck & Co., Inc., Rahway, N.J., U.S.A., © Merck & Co., Inc., 1989.

1635. Caffeine. *3,7-Dihydro-1,3,7-trimethyl-1H-purine-2,6-dione;* 1,3,7-trimethylxanthine; 1,3,7-trimethyl-2,6-dioxopurine; coffeine; thein; guaranine; methyltheobromine; No-Doz. $C_8H_{10}N_4O_2$; mol wt 194.19. C 49.48%, H 5.19%, N 28.85%, O 16.48%. Occurs in tea, coffee, maté leaves; also in guarana paste and cola nuts: Shuman, U.S. pat. **2,508,-545** (1950 to General Foods). Obtained as a by-product from the manuf of caffeine-free coffee: Barch, U.S. pat. **2,817,588** (1957 to Standard Brands); Nutting, U.S. pat. **2,802,739** (1957 to Hill Bros. Coffee); Adler, Earle, U.S. pat. **2,933,395** (1960 to General Foods). Crystal structure: Sutor, *Acta Cryst.* **11,** 453 (1958). Synthesis: Fischer, Ach, *Ber.* **28,** 2473, 3135 (1895); ....

Hexagonal prisms by sublimation, mp 238°. Sublimes 178°. Fast sublimation is obtained at 160-165° under 1 mm press. at 5 mm distance. d_4^{18} 1.23. pH of 1% soln 6.9. Aq solns of caffeine salts dissociate quickly. Absorption spectrum: Hartley, *J. Chem. Soc.* **87,** 1802 (1905). One gram dissolves in 46 ml water, 5.5 ml water at 80°, 1.5 ml boiling water, 66 ml alcohol, 22 ml alcohol at 60°, 50 ml acetone, 5.5 ml chloroform, 530 ml ether, 100 ml benzene, 22 ml boiling benzene. Freely sol in pyrrole; in tetrahydrofuran contg about 4% water; also sol in ethyl acetate; slightly in petr ether. Soly in water is increased by alkali benzoates, cinnamates, citrates or salicylates. LD_{50} orally in mice, hamsters, rats, rabbits (mg/kg): 127, 230, 355, 246 (males); 137, 249, 247, 224 (females) (Palm).

Monohydrate, felted needles, contg 8.5% H_2O. Efflorescent in air; complete dehydration takes place at 80°. ....

THERAP CAT: CNS stimulant.
THERAP CAT (VET): Has been used as a cardiac and respiratory stimulant and as a diuretic.

Next comes information about the molecular formula and sources of the compound, followed by a list of patent and journal references to articles about caffeine, including an account of its first synthesis in 1895. After the structure is a paragraph of information about the physical properties of caffeine, followed by a listing of the properties of some important derivatives of caffeine. The final lines give the therapeutic uses of caffeine, which is a central-nervous-system (CNS) stimulant.

Having learned that caffeine is present in No-Doz, you can next turn to the *Physicians Desk Reference*, the *PDR*, to learn about the use of caffeine in drugs. While intended primarily for physicians, the *PDR* is readily available in bookstores and libraries. For each drug, it contains the full product labeling information provided by the manufacturers. You can go directly to the entry for No-Doz via the product name index and learn more about the effects and cautions accompanying the intended use of this medication. By using the Generic and Chemical Name Index in the *PDR* you can also discover that caffeine is present in 27 other medications ranging from nonprescription analgesics like Anacin to a potent prescription drug for migraine headaches that combines caffeine with belladonna and ergotamine (both alkaloids, Section 15.8), and sodium pentobarbital.

NO DOZ® Tablets
[nō'dōz]

COMPOSITION
Each tablet contains 100 mg. Caffeine. Other Ingredients: Cornstarch, Flavors, Mannitol, Microcrystalline Cellulose, Stearic Acid, Sucrose.

INDICATIONS
Helps restore mental alertness or wakefulness when experiencing fatigue or drowsiness.

DOSAGE AND ADMINISTRATION
Adults and children 12 years of age and over: One or two tablets not more often than every 3 to 4 hours.

CAUTION
Do not take without consulting physician if under medical care. No stimulant should be substituted for normal sleep in activities requiring physical alertness.

WARNING
For occasional use only. Not intended for use as a substitute for sleep. If fatigue or drowsiness persists or continues to recur, consult a doctor. The recommended dose of this product contains about as much caffeine as a cup of coffee. Limit the use of caffeine-containing medications, foods, or beverages while taking this product because too much caffeine may cause nervousness, irritability, sleeplessness and, occasionally, rapid heart beat. Do not give to children under 12 years of age. **KEEP THIS AND ALL MEDICINES OUT OF THE REACH OF CHILDREN. IN CASE OF ACCIDENTAL OVERDOSE, SEEK PROFESSIONAL ASSISTANCE OR CONTACT A POISON CONTROL CENTER IMMEDIATELY. As with any drug, if you are pregnant or nursing a baby, seek the advice of a health professional before using this product.**

OVERDOSE
Typical of caffeine.

HOW SUPPLIED
NO DOZ® is supplied as:
A circular white tablet with "NoDoz" debossed on one side.
. . . .

Entry on No-Doz® from *The Physicians Desk Reference*. Reproduced from *The Physicians Desk Reference* with permission of Medical Economics Company, Inc., Oradell, N.J.

Caffeine information from *Lange's Handbook of Chemistry*. Reproduced from *Dean/Lange's Handbook of Chemistry*, 13th Ed. (1987), Entry C1, p. 7-194, with permission of McGraw-Hill, Inc.

Name	Formula	Formula weight	Beilstein reference	Density	Refractive index	Melting point	Boiling point	Flash point	Solubility in 100 parts solvent
Caffeine		194.19	26, 461	1.23_4^{18}		238	subl 178		2.1 aq; 1.5 alc; 18 chl; 0.19 eth; 1 bz

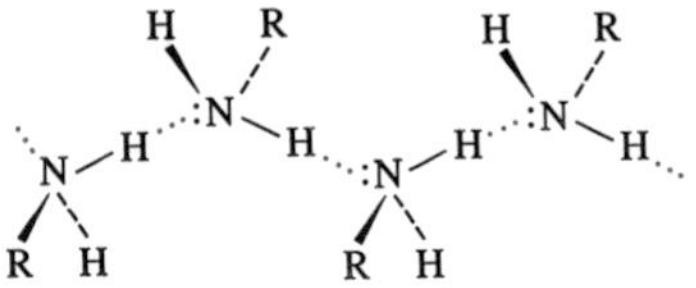

Figure 15.1
The formation of hydrogen bonds (red) in amines.

As in so many other cases, the differences in properties between amines and alcohols are accounted for by hydrogen bonding. Because nitrogen atoms are less electronegative than oxygen atoms, they form weaker hydrogen bonds, and amines are therefore lower boiling than comparable alcohols. In fact, mono-, di-, and trimethylamine and ethylamine are gases at room temperature. Other common amines with molecules of moderate size are liquids (Table 15.2).

Tertiary amine molecules have no hydrogen atoms attached to nitrogen and, because they cannot hydrogen-bond with each other, are lower boiling than either amines or alcohols of similar molecular weight. Compare the boiling point of trimethylamine, which is 3°C, with those of the compounds shown above.

All amines, however, can hydrogen-bond to water molecules through the lone electron pair on their nitrogen atoms. As a result, amines with up to about six carbon atoms are quite soluble in water. Also like ammonia, amines are weak Brønsted-Lowry bases and raise the pH of aqueous solutions (Section 15.5).

One noticeable difference from alcohols is that many volatile amines have strong odors, some like ammonia and others like fish or decaying meat. The protein in flesh contains amine groups, and the smaller, volatile amines pro-

Table 15.2   Physical Properties of Some Simple Amines

Structure	Name	Melting Point (°C)	Boiling Point (°C)	Water Solubility (g/100 mL)	Base Ionization Constant K_b
NH_3	Ammonia	−77.7	−33.3	90	1.7×10^{-5}
Primary amines					
CH_3NH_2	Methylamine	−94	−6.3	Very soluble	1.8×10^{-5}
$CH_3CH_2NH_2$	Ethylamine	−81	16.6	Miscible	4.4×10^{-4}
$(CH_3)_3CNH_2$	*tert*-Butylamine	−67.5	44.4	Miscible	2.8×10^{-4}
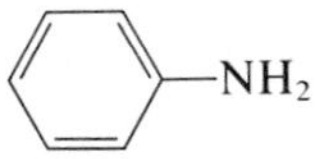 (aniline structure)—NH_2	Aniline	−6.3	184.1	4	3.8×10^{-10}
Secondary amines					
$(CH_3)_2NH$	Dimethylamine	−93	7.4	Miscible	3.8×10^{-10}
$(CH_3CH_2)_2NH$	Diethylamine	−48	56.3	Miscible	9.6×10^{-4}
$[(CH_3)_2CH]_2NH$	Diisopropylamine	−61	84	11	—
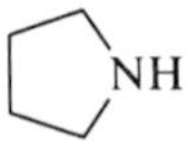 (pyrrolidine structure) NH	Pyrrolidine	2	89	Miscible	—
Tertiary amines					
$(CH_3)_3N$	Trimethylamine	−117	3	41	5×10^{-5}
$(CH_3CH_2)_3N$	Triethylamine	−114	89.3	14	5.7×10^{-4}
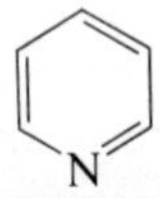 (pyridine structure) N	Pyridine	−42	115	Miscible	1.8×10^{-9}

duced during decay and protein breakdown are responsible for the odor of rotten meat. Two of the worst offenders have been given common names that are self-explanatory.

$$H_2NCH_2CH_2CH_2CH_2NH_2 \qquad H_2NCH_2CH_2CH_2CH_2CH_2NH_2$$

<table>
<tr><td>Putrescine</td><td>Cadaverine</td></tr>
</table>

Another significant property of amines is that many are physiologically active. The simpler amines are irritating to the skin, eyes, and mucous membranes and are toxic by ingestion. Some of the more complex amines from plants (Section 15.8) are among the most poisonous substances known, and some heterocyclic amines are believed to be carcinogens. On the other hand, all living things contain a wide variety of amines, and many useful drugs are amines.

Properties of Amines

● Primary and secondary amines are hydrogen-bonded and higher boiling than alkanes, but lower boiling than alcohols.

● Tertiary amines are lower boiling than secondary or primary amines because hydrogen bonding is not possible.

● The simplest amines are gases; other common amines are liquids.

● Volatile amines have unpleasant odors.

● Simple amines are water-soluble because of hydrogen bonding.

● Amines are weak Brønsted-Lowry bases (Section 15.5).

● Many amines are physiologically active, and many are toxic.

15.5 BASICITY OF AMINES

Recall from Section 11.4 that ammonia is a weak Brønsted-Lowry base. That is, ammonia can use its lone pair of electrons to accept a hydrogen ion (a proton) from an acid and form the ammonium ion, NH_4^+.

$$:NH_3 + H—O—H \rightleftharpoons NH_4^+ \quad + \quad OH^-$$

<table>
<tr><td>Ammonium ion</td><td>Hydroxide ion</td></tr>
</table>

$$:NH_3 + HCl(aq) \longrightarrow NH_4^+ \, Cl^-(aq)$$

<table>
<tr><td>Ammonium chloride</td></tr>
</table>

Amines behave similarly. The aqueous solutions of many amines are basic because of the following equilibrium:

$$RNH_2 + H_2O \rightleftharpoons RNH_3^+ \quad + OH^-$$

<table>
<tr><td>Amine</td><td>Ammonium ion</td></tr>
</table>

Ammonium salt An ionic compound composed of an ammonium cation and an anion; an amine salt.

Also, like ammonia, amines react with strong acids such as hydrochloric acid to yield **ammonium salts** according to the general equation

$$RNH_2 + HX \longrightarrow RNH_3^+ X^-$$

Amine $\qquad\qquad$ Ammonium salt

For example,

$$CH_3-\overset{..}{\underset{\underset{H}{|}}{N}}-H \ + \ HCl(aq) \ \rightleftharpoons \ CH_3-\overset{\overset{H}{|}}{\underset{\underset{H}{|}}{N^+}}-HCl^-(aq)$$

Methylamine $\qquad$ Hydrochloric acid $\qquad$ Methylammonium chloride

$$CH_3CH_2-\overset{..}{\underset{\underset{CH_2CH_3}{|}}{N}}-H \ + \ HCl(aq) \ \rightleftharpoons \ CH_3CH_2-\overset{\overset{H}{|}}{\underset{\underset{CH_2CH_3}{|}}{N^+}}-HCl^-(aq)$$

Diethylamine $\qquad$ Hydrochloric acid $\qquad$ Diethylammonium chloride

The reaction between amines and acids is reversible (like many acid–base reactions) and can be made to take place in either direction depending on the reaction conditions. An amine is protonated to yield an ammonium ion when treated with acid, and an ammonium ion is deprotonated to yield a neutral amine when treated with base. Thus, the amount of pronation of the amine depends on the pH of the medium.

Acidic conditions: $\qquad RNH_2 \ + \ H_3O^+ \longrightarrow RNH_3^+ \ + \ H_2O$
(pH < 7) $\qquad\quad$ An amine $\quad$ An acid $\qquad$ An ammonium ion

Basic conditions: $\qquad R-NH_3^+ \ + \ OH^- \longrightarrow R-NH_2(aq) \ + \ H_2O$
(pH > 7) $\qquad$ An ammonium ion $\quad$ A base $\qquad\quad$ An amine

Alkylamines are weak bases, and aromatic and heterocyclic amines are weaker bases still. Base strengths are usually expressed in terms of base ionization constants (K_b) included in Table 15.2. The value of K_b is the equilibrium constant for the ionization of a base in water in the same way that K_a is the equilibrium constant for the ionization of an acid (Section 11.9).

$$RNH_2 + H_2O \rightleftharpoons RNH_3^+ + OH^- \qquad\qquad K_b = \frac{[RNH_3^+][OH^-]}{[RNH_2]}$$

The smaller the value of K_b, the weaker the base (Table 15.2).

The positive ions formed by protonation of alkylamines are named by replacing the ending -*amine* by the ending -*ammonium*. To name the ions of heterocyclic amines, the amine name is modified by replacing the -*e* with -*ium*. For example,

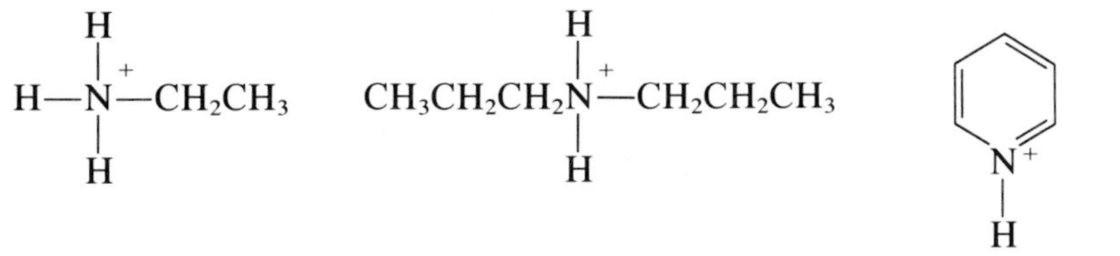

$$H-\overset{\overset{H}{|}}{\underset{\underset{H}{|}}{N^+}}-CH_2CH_3$$

$$CH_3CH_2CH_2\overset{\overset{H}{|}}{\underset{\underset{H}{|}}{N^+}}-CH_2CH_2CH_3$$

Pyridinium ion

Ethylammonium ion
(from ethylamine)

Dipropylammonium ion
(from dipropylamine)

Pyridinium ion
(from pyridine)

Any nitrogen atom with four bonds has a positive charge and is an ammonium ion, even though the charge may not be written in the formula.

Practice Problems

15.4 Write an equation for the acid–base equilibrium of dimethylamine and water.

15.5 Complete the following equations:
(a) $CH_3CH_2CHNH_2 + HBr(aq) \longrightarrow$
      $\overset{\qquad|}{\qquad CH_3}$

(b)  ⬡—NH_2 + $HCl(aq)$ $\longrightarrow$

(c) $CH_3CH_2NH_2 + CH_3COOH(aq) \longrightarrow$
(d) $CH_3NH_3{}^+Cl^- + NaOH(aq) \longrightarrow$

15.6 Name the organic ions produced in reactions (a) through (c) in Problem 15.5.

15.7 Which is the stronger base?
(a) ammonia or ethylamine   (b) triethylamine or pyridine

15.6 AMMONIUM SALTS

Like any salt, an amine salt is composed of a cation and an anion and is usually named by combining the cation and anion names. In methylammonium chloride ($CH_3NH_3{}^+Cl^-$), for example, the methylammonium ion, $CH_3NH_3{}^+$, is the cation, and the chloride ion is the anion. In an older system, ammonium salts were written and named by combining the structures and names of the amine and the acid. Methylammonium chloride, in this system, is written $CH_3NH_2 \cdot HCl$ and named methylamine hydrochloride. You'll often see this system used with drugs that are ammonium salts. For example,

$(C_6H_5)_2CHOCH_2CH_2N(CH_3)_2 \cdot HCl$ or $(C_6H_5)_2CHOCH_2CH_2\overset{+}{N}H(CH_3)_2\ Cl^-$

Diphenhydramine hydrochloride (Benadryl), an antihistamine

Ammonium salts are generally odorless, white, crystalline solids that are much more water-soluble than neutral amines because they're ionic. Thus, ammonium salt formation, like carboxylate salt formation, provides a means for converting an insoluble compound into a water-soluble derivative.

$$CH_3CH_2-\overset{\displaystyle |}{\underset{\displaystyle CH_2CH_3}{N}}-CH_2CH_3 \;+\; HCl(aq) \;\longrightarrow\; CH_3CH_2-\overset{\displaystyle +}{\underset{\displaystyle CH_2CH_3}{\overset{\displaystyle |}{N}H}}-CH_2CH_3 \; Cl^-(aq)$$

| Triethylamine (water-insoluble) | Hydrochloric acid | Triethylammonium chloride (water-soluble) |

Also, because many amines are unstable in air, salt formation allows the amine to be stored in a more stable form. When the free amine is needed, it is easily regenerated by treatment with a base:

$$CH_3NH_3^+ \; Cl^-(aq) + NaOH(aq) \;\rightarrow\; CH_3NH_2(aq) + NaCl(aq) + H_2O(l)$$

Ammonium salts can be formed by reaction of an amine with an acid, as shown above, and also by reaction of an alkyl halide with an amine:

$$RNH_2 \;+\; R'X \;\longrightarrow\; RR'NH_2^+ \; X^-$$

| Amine | Alkyl halide | Alkyl ammonium halide |

For example,

$$CH_3CH_2NH_2 \;+\; CH_3Cl \;\longrightarrow\; CH_3CH_2\overset{\displaystyle CH_3}{\overset{\displaystyle |}{N}}H_2^+ \; Cl^-$$

| Ethylamine | Methyl chloride | Ethylmethylammonium chloride |

The reaction of a tertiary amine with an alkyl halide produces a **quaternary ammonium salt**—an ammonium salt with *four* R groups on the nitrogen atom:

Quaternary ammonium salt An ammonium salt with four organic groups bonded to the nitrogen atom.

$$R-\overset{\displaystyle R}{\underset{\displaystyle R}{\overset{\displaystyle |}{\underset{\displaystyle |}{N}}}} \;+\; R'X \;\longrightarrow\; R-\overset{\displaystyle R}{\underset{\displaystyle R}{\overset{\displaystyle |}{\underset{\displaystyle |}{N}}}}{}^+{-}R' \; X^-$$

| A tertiary amine | An alkyl halide | A quaternary ammonium salt |

For example,

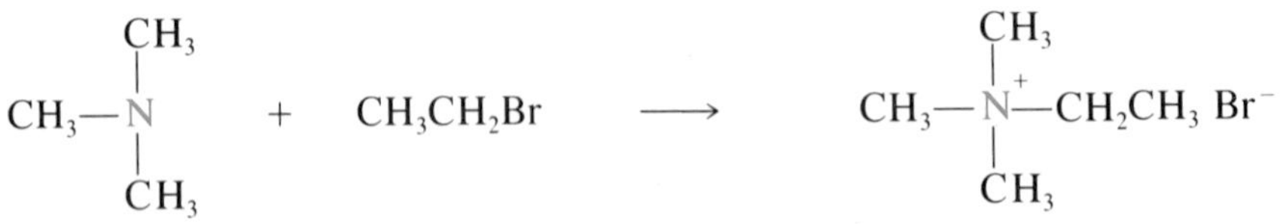

$$CH_3-\overset{\displaystyle CH_3}{\underset{\displaystyle CH_3}{\overset{\displaystyle |}{\underset{\displaystyle |}{N}}}} \;+\; CH_3CH_2Br \;\longrightarrow\; CH_3-\overset{\displaystyle CH_3}{\underset{\displaystyle CH_3}{\overset{\displaystyle |}{\underset{\displaystyle |}{N}}}}{}^+{-}CH_2CH_3 \; Br^-$$

| Trimethylamine | Ethyl bromide | Ethyltrimethylammonium bromide |

Quaternary ammonium salts have no H atom that can be removed by a base and no lone pair that can be protonated, so their structure in solution is unaffected by changes in pH. One commonly encountered quaternary ammonium salt has the following structure, where R represents a range of C_8 to C_{18} alkyl groups.

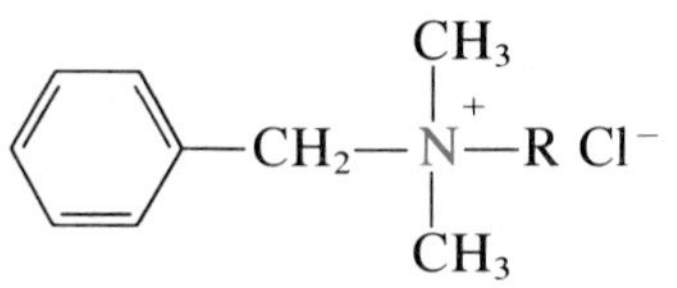

$$R = -C_8H_{17} \text{ to } -C_{18}H_{37}$$

Benzalkonium chloride (an antiseptic and disinfectant)

Practice Problems   **15.8**   Write the structures of the following compounds:
(a) hexyldimethylammonium chloride   (b) ethylammonium bromide

15.9   Identify each compound in Problem 15.8 as the salt of a primary, secondary, or tertiary amine.

15.10   Complete the following equation and name the product:

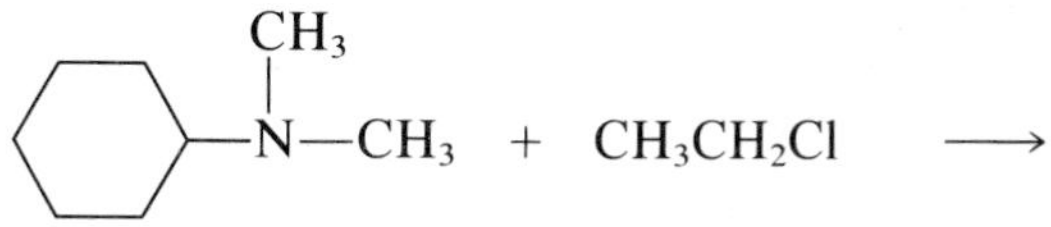

15.7 AMINES IN BIOMOLECULES

Proteins, we've noted, are polymers of amino acids, which all contain an amino group and a carboxylic acid group.

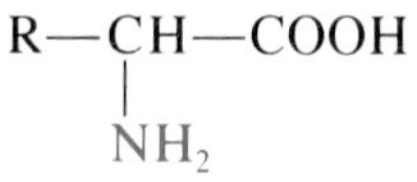

General formula of
α-amino acids found in proteins

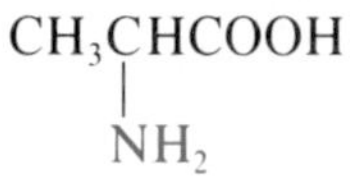

Alanine
(an α-amino acid)

A second major class of biomolecules, the *nucleotides* (Chapter 26), all contain nitrogen heterocyclic rings derived from either purine or pyrimidine (see Table 15.1). Nitrogen heterocycles are also part of several vitamins, including the B vitamins:

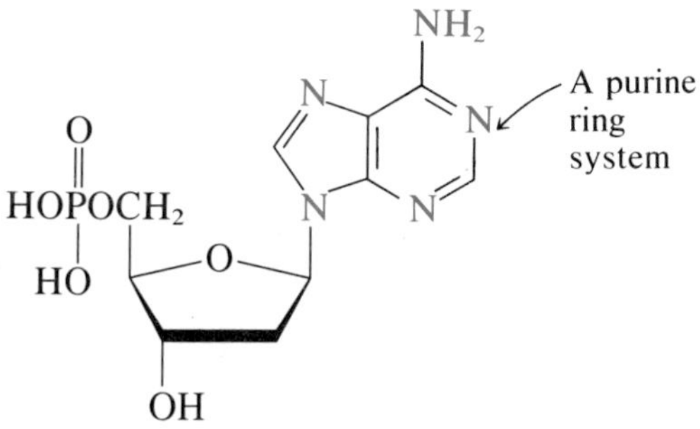

A DNA nucleotide containing
a purine heterocyclic system

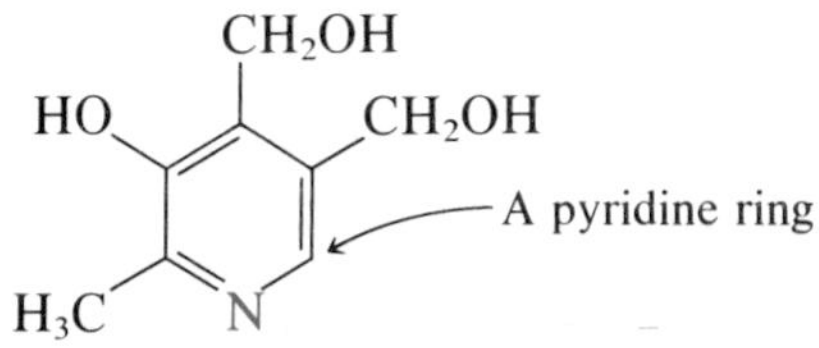

Pyridoxine
(one of the water-soluble B_6 vitamins)

AN APPLICATION: ORGANIC COMPOUNDS IN BODY FLUIDS

The chemical reactions that keep us alive and functioning occur in the aqueous solutions known as "body fluids"—blood, digestive juices, urine, and the fluid inside cells. But you've seen that for organic compounds of all classes, water solubility decreases as the hydrocarbon-like portions of the molecules become larger and molecular weight increases. How does the body manage its reactions in water solution, especially when large and complex biomolecules are involved?

Acidic and basic functional groups are part of many biomolecules, which then exist as soluble ions at the pH values of the various fluids. The most common ionized groups in biomolecules are carboxylate ions, phosphate ions, and ammonium ions:

$$\text{Carboxylate ion} \quad \text{Phosphate ion} \quad \text{Ammonium ion}$$

For example, NAD is an oxidizing agent that participates in a great many biochemical reactions:

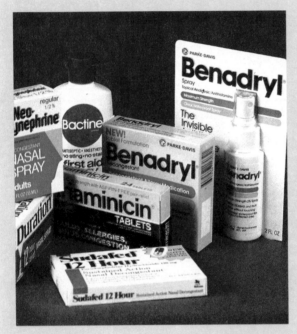

An ammonium chloride salt is the active ingredient in each of these over-the-counter medications.

NAD (nicotinamide adenine dinucleotide)
(a coenzyme and biochemical oxidizing agent)

In addition, several members of the diverse class of compounds known as *neurotransmitters* (Section 19.15), which transmit nerve impulses throughout the body, are amines, including serotonin and dopamine. Neurotransmitters have a variety of physiological effects including the ability to dilate small airways in the lung (bronchodilation), to contract capillaries, and to increase blood pressure.

Serotonin   Dopamine

Because carboxylic acids (RCOOH) are ionized in body fluids, biochemists often refer to them by their carboxylate ion names rather than their acid names. You are just as likely, for example, to see pantothenic acid referred to as *pantothenate,* or pyruvic acid as *pyruvate,* or citric acid as *citrate.* Either way, the reference is to the same substance in solution. Also, you may see the same biomolecule structures written with —COOH or —COO⁻, —PO₃H₂ or —PO₃²⁻, and —NH₂ or —NH₃⁺.

Drugs too must be soluble in body fluids and must be transported in fluids from their entry point in the body to their site of action. Most drugs are weak acids or bases and therefore are in equilibrium with their ions in body fluids. For example,

Aspirin
(an acid)

Amphetamine
(a base)

The extent of ionization of a drug helps determine how it's distributed in the body, because ions are less likely to cross cell membranes than uncharged molecules. Weak acids like aspirin are less dissociated in the acidic environment in the stomach and are absorbed readily there. Weak bases are not significantly absorbed in the stomach but are better absorbed in the more basic environment in the small intestine. It's even possible to control where a drug tablet dissolves by applying an *enteric coating,* which remains intact in the strongly acidic gastric juice but dissolves when it reaches the small intestine. This approach is taken for drugs that would be attacked and broken down by acid.

Many drugs must be delivered to the body in their more water-soluble forms as salts. Penicillin G, for example, is usually administered as its sodium or potassium salt to increase its water solubility.

Penicillin G potassium

Also, many amine-containing drugs are not very soluble in water. By converting such amines as phenylephrine, the decongestant in Neo-Synephrine, and phenylpropanolamine, the decongestant in Triaminic cough medicine, to ammonium hydrochlorides, however, their solubility increases to the point where delivery in solution is possible.

15.8 AMINES IN PLANTS

Alkaloid A naturally occurring nitrogen-containing compound isolated from a plant; usually basic, bitter, and poisonous.

The roots, leaves, and fruits of flowering plants (angiosperms) are a rich source of amines, including many that are physiologically active. These compounds, once called "vegetable alkali" because their water solutions are basic, are now referred to as **alkaloids.** The molecular structures of approximately 5500 alkaloids have been determined. Most are bitter-tasting and toxic to human beings and other animals in sufficiently high doses. Quinine, in fact, is used as a standard for bitterness: Even a 1×10^{-6} M solution tastes bitter.

The bitterness and poisonous nature of alkaloids probably evolved to protect plants from being devoured by animals. It's interesting to speculate on why, of the four basic taste sensations—sweet, bitter, salty, and sour—bitterness is the one we like least in foods. The three poisonous compounds described here—coniine, atropine, and solanine—illustrate some of the many types of alkaloid structures:

Coniine

Atropine

Solanine
(X = a group of three sugar molecules)

Figure 15.2
A potato plant. The potato grows in the absence of sunlight and should also be stored away from sunlight, which causes formation of poisonous solanine (along with green chlorophyll) under the skin.

● **Coniine** is extracted from poison hemlock (*Conium maculatum*). It was the poison with which Socrates ended his life after being convicted of corrupting Greek youth with philosophical discussions.

● **Atropine** is the toxic substance in the herb known as deadly nightshade or *belladonna* (*Atropa belladonna*). In Meyerbeer's opera, *L'Africaine*, the heroine sings of the peaceful death this plant can bring before committing suicide over her lost love. Like many other alkaloids, atropine acts on the central nervous system, a property sometimes applied (in appropriately low dosage!) in medications to reduce cramping of the digestive tract.

● **Solanine,** an even more potent poison than atropine (Figure 15.2), is found in potatoes and tomatoes, both of which belong to the same botanical family as the deadly nightshade (Solanaceae). If you've ever been warned that you must peel green potatoes and wondered why, herein lies the reason and it's a good one. The tiny amount of solanine in properly stored potatoes only contributes to their characteristic flavor. But when potatoes are exposed to sunlight, the production of solanine is increased to levels that can be dangerous. The alkaloids are formed under the skin and aren't destroyed by heating, but they can be removed by peeling. Sunlight fortunately also stimulates the formation of chlorophyll, and its green color provides a warning that potatoes must be peeled so that all the green flesh is removed. Potato sprouts also contain solanine and should be cut out before potatoes are cooked.

Some alkaloids are notable not as poisons but for their pain-relieving ability. The opium poppy, *Papaver somniferum* (Figure 15.3), has been used for this purpose at least since the seventeenth century. Morphine was the first pure compound to be isolated from the poppy, but several close relatives including codeine are also present in poppies. Heroin, another close relative of morphine, does not occur naturally but is easily synthesized in chemical laboratories. Within the body, hydrolysis of the $CH_3C{=}O$ groups converts heroin back to morphine.

Figure 15.3
The opium poppy, source of morphine and other addictive alkaloids.

Morphine

Codeine

Heroin

15.9 AMINES IN DRUGS

Given the variety of nitrogen-containing compounds that are physiologically active, it's not surprising that many drugs are amines or nitrogen heterocycles. The relationship between molecular structure and physiological activity is the key to understanding drug action and to discovering new drugs. We'll give just a few more examples of the many classes of drugs that are amines.

Histamine is the biomolecule responsible for the symptoms of the allergic reaction familiar to hay fever sufferers. It's also the chemical that causes an itchy bump when an insect bites you (Figure 15.4). The *antihistamines* are a

Figure 15.4
A mosquito in action. The mosquito injects antigens, substances that trigger a response by the body's immune system. The immune system in turn triggers release of histamine, which causes swelling, itchiness, and redness around the bite.

family of drugs that counteract histamine, and members of this family have in common a disubstituted ethylamine side chain, usually with two *N*-methyl groups.

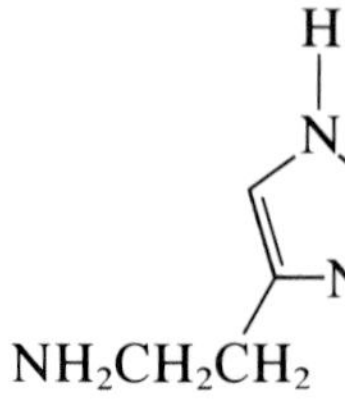

Histamine

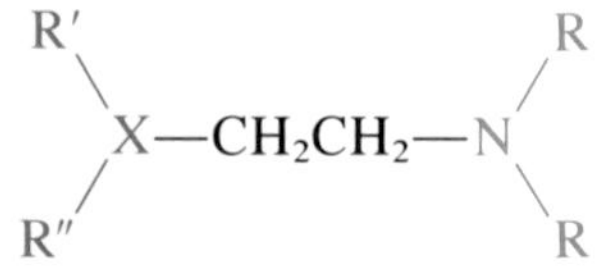

General antihistamine structure

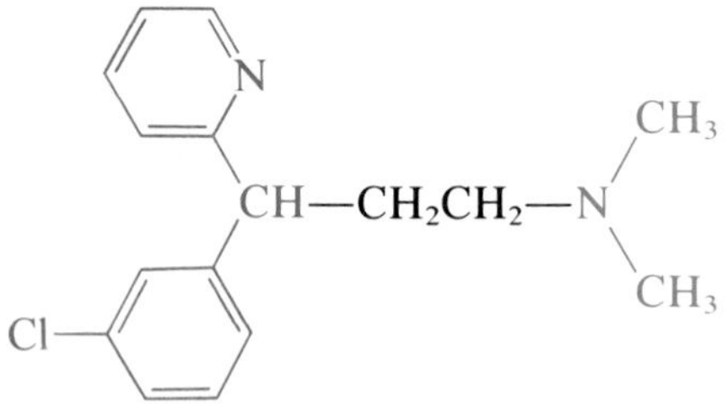

Chlorpheniramine
(an antihistamine)

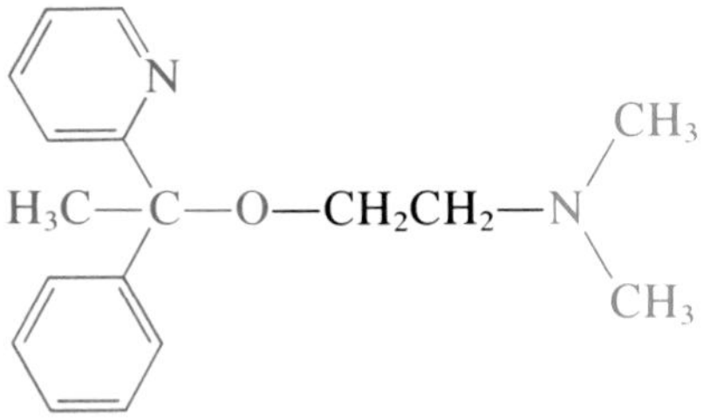

Doxylamine
(an antihistamine)

The R′ and R″ groups in the generalized structure tend to be bulky, and they may be bonded to carbon, nitrogen, or oxygen. Antihistamines act not by causing a physiological effect but by preventing one. They physically block the attachment of histamine to sites with which it must connect to create its unpleasant effects.

As another example of structure–activity relationships, compare the naturally occurring neurotransmitters on the left in Table 15.3 with the synthetic drugs on the right. Each is a variation on the same general structure, and each acts on the nervous system. The drugs are known as *sympathomimetics* because they mimic the action of the sympathetic nervous system. In the body, dopamine is a precursor of norepinephrine, a principal messenger in the nervous system. Epinephrine (adrenaline) is the compound specifically responsible for causing the surge of energy we feel in a frightening situation.

Ephedrine and its salts are used in over-the-counter medications for asthma and nasal congestion because they dilate bronchial and nasal passageways. Isoproterenol is a prescription drug for bronchial asthma. Amphetamine is the parent compound of a family of central-nervous-system stimulants that produce an array of effects like those of epinephrine, including excitement, increased blood pressure, quickened reflexes, and appetite suppression. The amphetamines are all hazardous because they are habit-forming.

Although enormously important as medicines, morphine and its alkaloid relatives pose a great problem because of their addictive properties. Much effort has therefore been devoted to understanding how morphine works and to

Table 15.3   Neurotransmitters and Drugs That Mimic Their Action (Sympathomimetics)

A general ephedrine-like structure
(groups in parentheses may or may not be present)

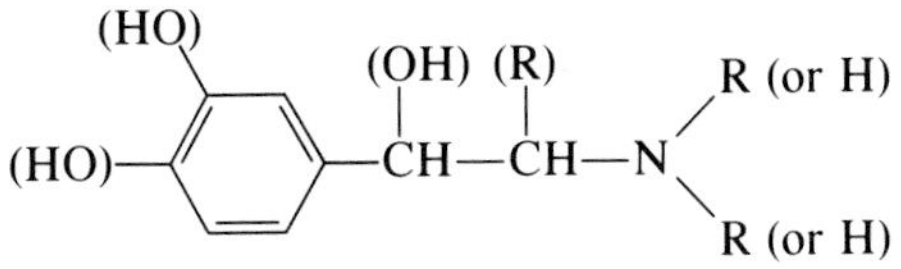

Natural Neurotransmitters

Dopamine

Norepinephrine

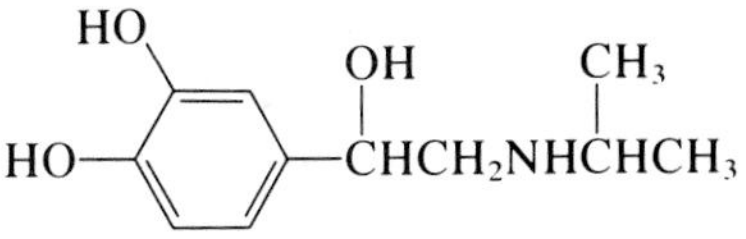

Epinephrine
(adrenaline)

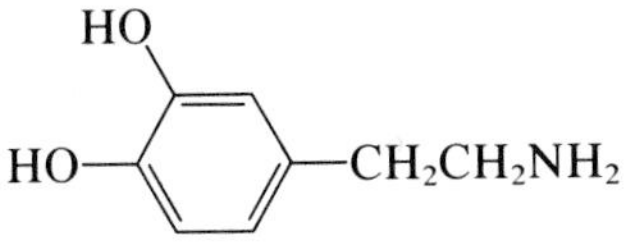

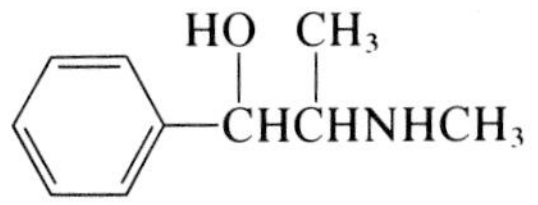

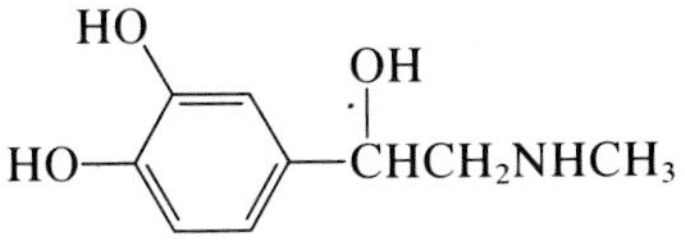

Sympathomimetic Drugs

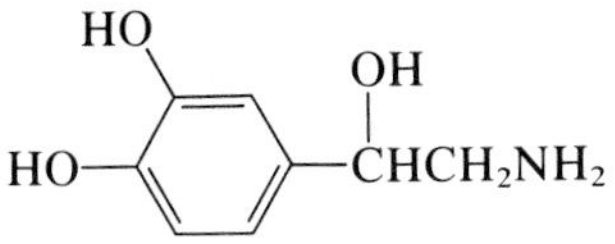

Ephedrine

Isoproterenol

Amphetamine

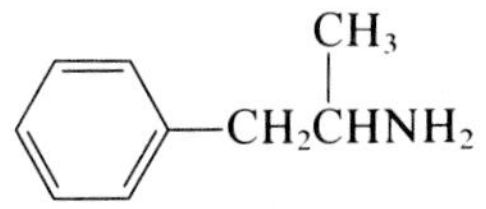

developing modified morphine derivatives that retain the desired painkilling activity but don't cause addiction. Many compounds with similar painkilling properties, including meperidine (Demerol) and methadone, have been synthesized, but no fully satisfactory morphine substitutes have yet been found. Demerol is widely used as a painkiller, and methadone is used in the treatment of heroin addiction.

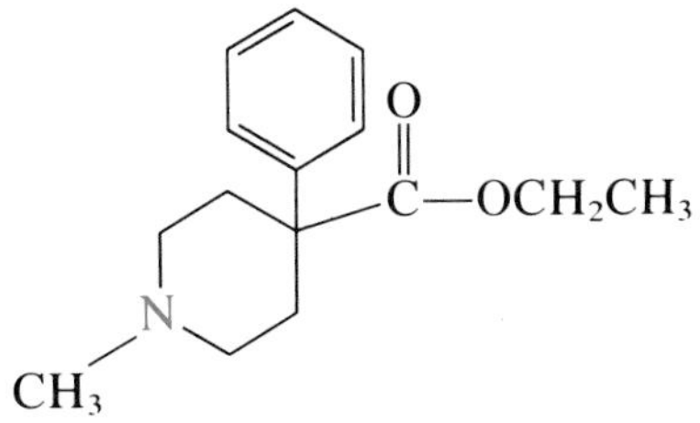

Demerol

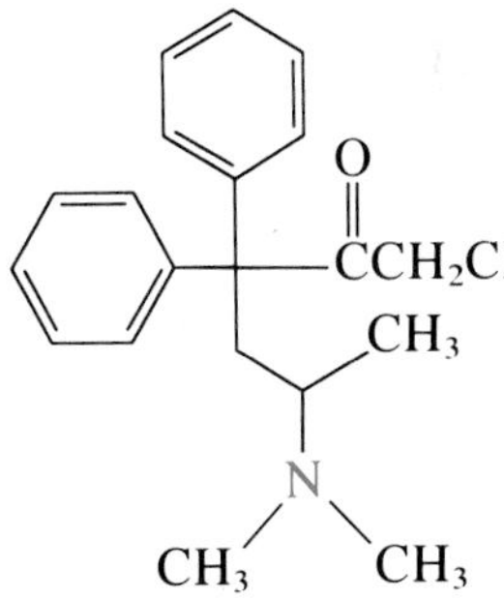

Methadone

Practice Problem   15.11   Compare the structures of Demerol and methadone with the structures in Section 15.8 of the morphine alkaloids and identify the structural unit that all have in common. (*Hint:* It includes the nitrogen atom.)

The modern approach to drug development has come to be called *drug design*, meaning the design of a drug molecule with a structure destined to meet a specific need. Problems arise in drug design when a molecule with the desired therapeutic effect can't be administered as a drug. This can happen for a variety of reasons: The compound might be unstable or insoluble in body fluids or taste terrible, it might have harmful side effects, or it might fail to reach the site where it must act.

A successful solution to such problems is the design of a *prodrug:* an inactive compound that's converted to the active drug *after* it's administered. Sulfasalazine, a sulfa antibacterial for treating ulcerative colitis, is an example of a site-specific prodrug. The active parent drug, sulfapyridine, does not reach the colon in effectively high concentrations because it is absorbed from the intestinal tract above the colon. Sulfasalazine remains unchanged until it is broken down by enzymes in the colon to generate the active drug (as shown below).

A prodrug based on epinephrine allows for safer and more effective use of this compound in relieving fluid pressure in the eye due to glaucoma. Relatively large concentrations of epinephrine must be used in eyedrops because the polar molecule does not easily cross the eye membrane. Reaction of epinephrine with a carboxylic acid yields dipivefrin, a prodrug that is less polar; it enters the eye more easily, is less

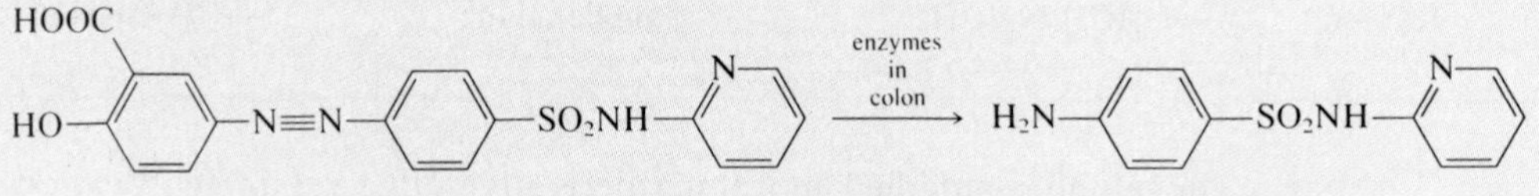

Sulfasalazine
(a prodrug)

Sulfapyridine
(an antibacterial drug)

SUMMARY

Amines are organic derivatives of ammonia in the same sense that alcohols and ethers are organic derivatives of water. Amines are classified according to the number of organic groups bonded to nitrogen as **primary amines,** RNH_2; **secondary amines,** R_2NH; and **tertiary amines,** R_3N. Nitrogen is also found in many types of **heterocyclic compounds,** both nonaromatic and aromatic.

Amines are polar compounds and can form hydrogen bonds with water through the lone pairs on nitrogen, making the simpler amines water-soluble. Primary and secondary amines are higher boiling than alkanes, but lower boiling than alcohols because they form weaker hydrogen bonds. Tertiary amines don't hydrogen-bond with each other and are lower boiling than other amines or alcohols of similar molecular weight. Volatile amines have bad odors, and many amines are physiologically active.

The most significant property of amines is their basicity. Like ammonia, amines can act as bases to accept protons from acids and are weak bases. Protonation of an amine by an acid yields an **ammonium salt** ($RNH_2 + HCl \rightarrow RNH_3^+Cl^-$). The protonation reaction is reversible, and ammonium salts can be reconverted to amines by treatment with OH^-. Formation of an ammonium salt is often used to obtain a more stable, more water-soluble derivative of an amine. Reaction of tertiary amines with alkyl halides gives **quaternary ammonium salts** ($R_4N^+X^-$), which are unaffected by changes in the pH of a solution.

Amino groups and nitrogen heterocycles are present in many biomolecules, including amino acids, nucleotides, and neurotransmitters. The **alkaloids,** which include many poisons and many drugs, are a large family of amines and nitrogen heterocycles found in plants. Many drugs are amines or nitrogen heterocycles.

irritating, and also reduces the danger of epineph-rine entering the cardiovascular system where it can have undesirable side effects. Once dipi-vefrin has entered the eye tissue, enzymes release the epinephrine.

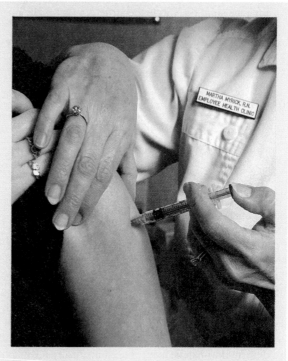

Some drugs can reach their sites of action through the bloodstream.

$(CH_3)_3C\overset{O}{\underset{\|}{C}}-O-$ ... $\overset{OH}{\underset{|}{C}}HCH_2NHCH_3$ Dipivefrin (a prodrug)

$(CH_3)_3C\overset{O}{\underset{\|}{C}}-O-$

enzymes in eye ↓

$2\ (CH_3)_3C\overset{O}{\underset{\|}{C}}-OH$ + HO... $\overset{OH}{\underset{|}{C}}HCH_2NHCH_3$

HO

Epinephrine (an antiglaucoma drug)

REVIEW PROBLEMS

Amines and Ammonium Salts

15.12 What family-name ending is used for amines?

15.13 What is the structural difference between primary, secondary, and tertiary amines?

15.14 Draw structures corresponding to these names:
(a) propylamine
(b) diethylamine
(c) N-methylpropylamine
(d) N-butyl-N-methylhexylamine
(e) N-ethylcyclopentylamine
(f) 2-methyl-3-propoxyaniline

15.15 Name these compounds:

(a) $CH_3CH_2CH_2CH_2NH_2$

(b) $CH_3\overset{NH_2}{\underset{|}{C}}HCH_3$

(c) ☐—NHCH₃

(d) ⬠ ☐—NCH₃ (phenyl-cyclopentyl N-methyl)

15.16 Identify each of the amines in Problems 15.14 and 15.15 as primary, secondary, or tertiary.

15.17 Propose structures for amines that fit these descriptions:
(a) a secondary amine with formula $C_5H_{13}N$
(b) a tertiary amine with formula $C_6H_{15}N$

15.18 Is pyridine a weaker or a stronger base than ammonia? (See Table 15.2).

15.19 Which solution has a higher pH, 0.20 M aniline or 0.20 M methylamine? (See Table 15.2).

15.20 Assume you have an aqueous solution of ethylamine.

(a) Write the equilibrium reaction for this weak base in aqueous solution. Circle the structure that is the predominant species.

(b) HNO_3 is added until the pH is 5.00. What is the predominant species now?

15.21 Name these ammonium salts. (See Table 15.2.)

(a) $CH_3NH_3^+$ Cl^-

(b)

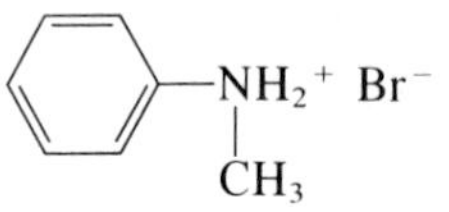

(c)

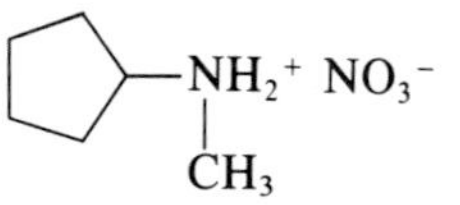

15.22 Draw the structures of these ammonium salts. (See Table 15.2.)

(a) anilinium chloride

(b) N-propylbutylammonium bromide

(c) N-ethyl-N-isopropylhexylammonium chloride

15.23 Identify each salt in Problems 15.21 and 15.22 as the salt of a primary, secondary, or tertiary amine.

15.24 Which is the stronger base, diethyl ether or diethylamine?

15.25 The compound L-dopa (2-amino-3-(3,4-dihydroxyphenyl)propanoic acid) is used medically for its potent activity against Parkinson's disease, a chronic disease of the central nervous system. Identify the functional groups present in L-dopa.

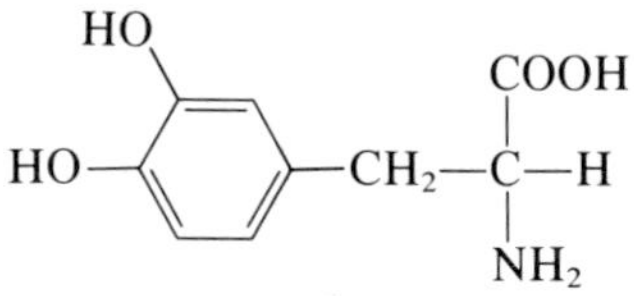

15.26 Cocaine has the structure indicated. Is cocaine a primary, secondary, or tertiary amine?

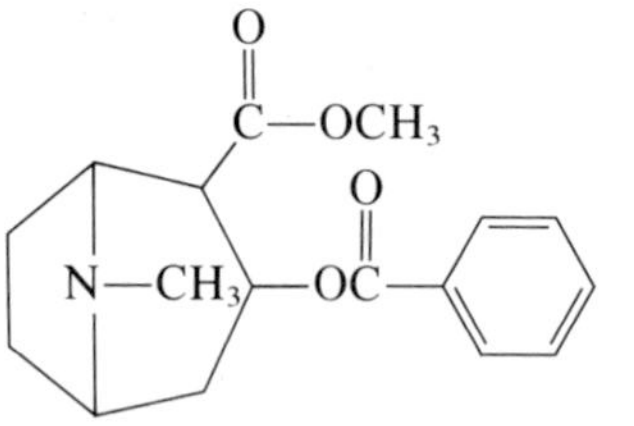

Reactions of Amines

15.27 Most illicit cocaine is actually cocaine hydrochloride—the product from the reaction of cocaine (Problem 15.26) with HCl. Show the structure of cocaine hydrochloride.

15.28 Assume that you have samples of quinine, an amine, and menthol, an alcohol. What simple chemical test could you do to distinguish between them?

15.29 Complete the following equations:

(a)

(b)

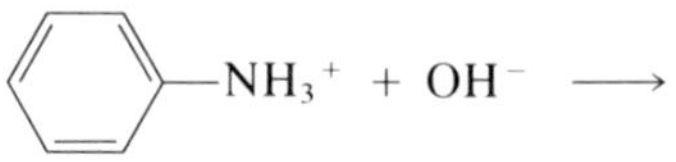

(c)

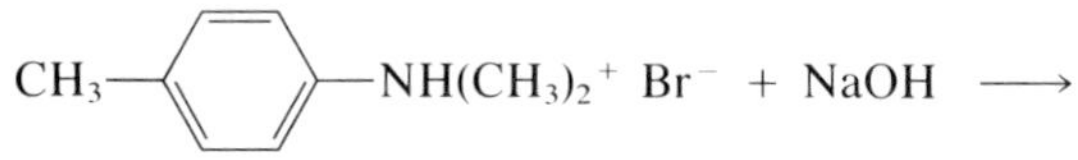

(d) $CH_3CHNH_2 + H_2O \rightleftharpoons$
    CH_3

(e) $CH_3CH_2N(CH_3)_2 + CH_3CH_2CH_2CH_2Cl \longrightarrow$

(f) $CH_3CH_2NH + H_3O^+ \longrightarrow$
    CH_3

(g)

15.30 Many hair conditioners contain an ammonium salt such as the following to help prevent "flyaway" hair. Will this salt react with acids or bases? Why or why not?

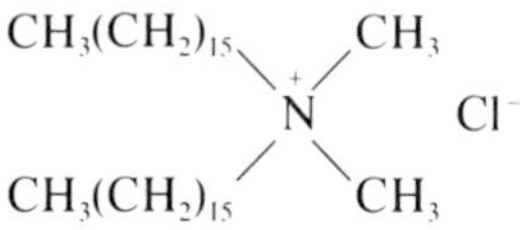

15.31 Zectran, a pesticide, has the following structure. Do you think that this substance reacts with aqueous hydrochloric acid? If so, what are the product(s)?

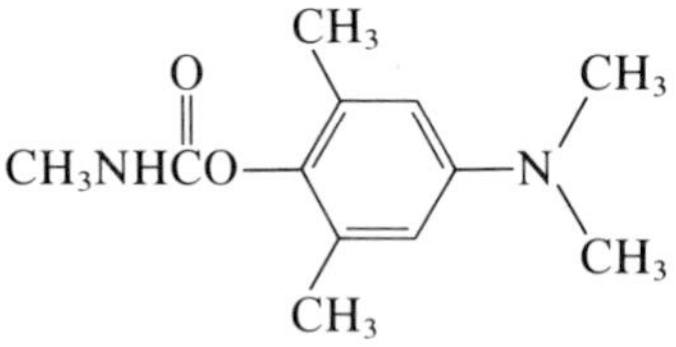

Applications

15.32 Acetaminophen is advertised as a mild pain reliever, superior to aspirin. Check the references listed in the Application on chemical information to determine (a) the structure of this compound; (b) its melting point, boiling point, and relative solubilities; (c) a few of the trade names under which it is sold; (d) uses for which it is prescribed; and (e) hazards associated with overdose. [App: Chemical Information]

15.33 The pH of most bodily fluids is about 7. Why do phosphoric, carbonic, and most organic acids exist in the ionized form at pH 7? [App: Organic Compounds in Body Fluids]

15.34 Why is it often necessary to use the salts of compounds rather than the neutral compound as drugs? [App: Organic Compounds in Body Fluids]

15.35 Salol, an ester of salicylic acid, is used as an enteric coating for other drugs. What is meant by the term "enteric coating"? [App: Organic Compounds in Body Fluids]

15.36 Promazine, a potent antipsychotic tranquilizer, is administered as the hydrochloride salt. Write the formula of this salt. [App: Organic Compounds in Body Fluids]

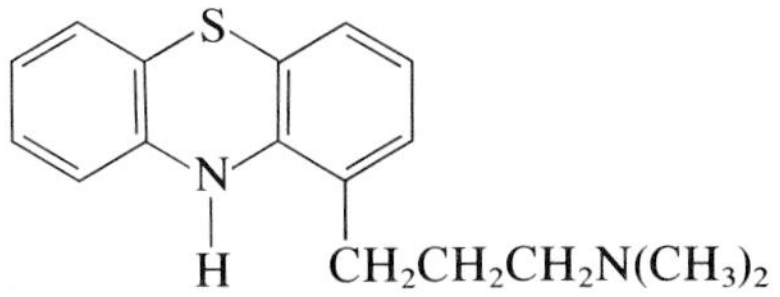

Promazine

15.37 What is a prodrug? [App: Prodrugs]

15.38 Explain why prodrugs are needed for certain applications. [App: Prodrugs]

Additional Questions and Problems

15.39 Propylamine, propanol, acetic acid, and *n*-butane have about the same molar masses. Which would you expect to have the (a) highest boiling point, (b) lowest boiling point, (c) least solubility in water, and (d) least chemical reactivity? Explain?

15.40 Explain why decylamine is much less soluble in water than ethylamine.

15.41 Each amino acid has a characteristic pH at which it exists in a form analogous to a salt where the acid group donates a proton to the amine group in the same molecule. Two naturally occurring amino acids are glycine and threonine. Show the salt form of each.

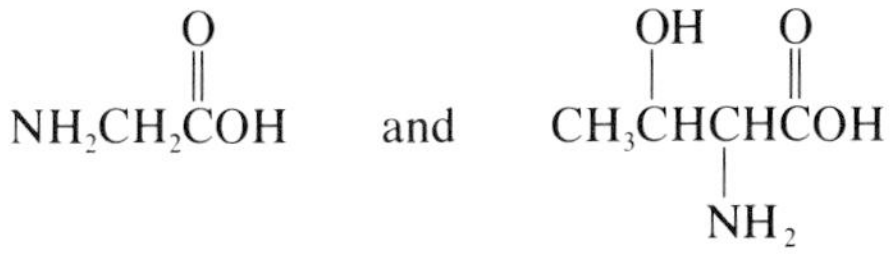

Glycine   Threonine

15.42 *para*-Aminobenzoic acid (PABA) is a common ingredient in sunscreens. Draw the structure of PABA.

15.43 PABA (Problem 15.42) is used by certain bacteria as a starting material from which folic acid (a necessary vitamin) is made. Sulfa drugs such as sodium sulfanilimide work because they resemble PABA. The bacteria try to metabolize the sulfa drug, fail to do so, and die due to lack of folic acid.

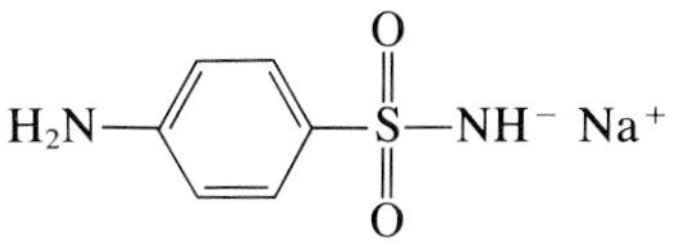

(a) Describe how this structure is similar to that of PABA.

(b) Why do you think the sodium salt, rather than the neutral compound, is used as the drug?

15.44 Acyclovir is an antiviral drug with the following structure:

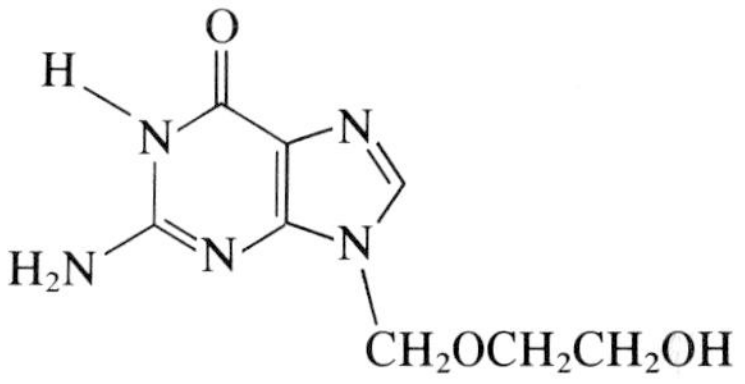

(a) What heterocyclic base (Table 15.1) is the parent of this compound?

(b) Label the other functional groups present.

(c) Label each nitrogen as primary, secondary, or tertiary.

15.45 Which is the stronger base, trimethylamine or pyridine? In which direction will the following reaction proceed?

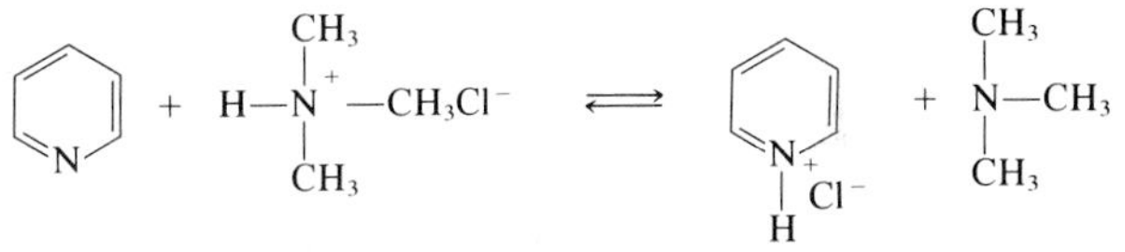

15.46 Mescaline, a powerful hallucinogen derived from the peyote cactus, has the systematic name 3,4,5-trimethoxyphenylethylamine and is similar in structure to epinephrine. Draw it.

15.47 Many illegal drugs are sympathomimetic to epinephrine. What does this term mean?

15.48 How do amines differ from analogous alcohols in (a) odor, (b) basicity, and (c) boiling point?

15.49 What two characteristics are often associated with alkaloids?

15.50 Describe how antihistamines function.

15.51 Name these compounds.

(a) $CH_3CHCH_2CH_2CH=CHCH_3$ with CH_3 on the second carbon

(b) HO—⟨benzene ring with CH₃CH₂ substituent⟩—CH(CH₃)₂

(c) (CH₃CH₂CH₂CH₂)₂NH

15.52   Complete these equations.

(a) $CH_3CH_2CCH_2CH=CCH_3$ + HCl →
      (with CH₃ above second carbon, CH₃ below that carbon, CH₂CH₃ below the =C)

(b) $CH_3CH_2\overset{OH}{C}HCH(CH_3)_2$ + H_2SO_4 →

(c) 2 CH_3CH_2SH $\xrightarrow{[O]}$

(d) ⟨benzene ring⟩—$CH_2\overset{OH}{C}HCH_2CH_3$ $\xrightarrow{[O]}$

(e) $CH_3CH_2NH_2$ + CH_3CH_2Cl →
(f) $(CH_3)_3N$ + H_2O →
(g) $(CH_3)_3N$ + HCl →
(h) $(CH_3)_3NH^+$ + OH^- →

CHAPTER

16

Aldehydes and Ketones

This bombardier beetle is spraying a potential predator with boiling-hot benzoquinones, produced in a fraction of a second by a chemical reaction that the insect carries out within its body. Benzoquinones are ketones, members of a class of compounds you'll learn about in this chapter. (For another example of insect chemical warfare, see the Interlude.)

In this and the next chapter, we'll discuss several families of compounds that are widespread in organic and biological chemistry: those that have a carbonyl group. Every carbonyl group contains a carbon–*oxygen* double bond in the same way that an alkene grouping contains a carbon–*carbon* double bond.

$$\underset{\displaystyle \diagdown\diagup}{\overset{\displaystyle \overset{O}{\|}}{C}} \qquad \text{A carbonyl group}$$

Here we'll look at the two most simple types of carbonyl compounds— aldehydes, in which the carbonyl group has one R group and one H atom on the carbonyl carbon atom, and ketones, in which there are two R groups on the carbonyl carbon.

$$\underset{R \diagup \diagdown H}{\overset{\displaystyle \overset{O}{\|}}{C}} \qquad \underset{R \diagup \diagdown R'}{\overset{\displaystyle \overset{O}{\|}}{C}}$$

$$\text{Aldehyde} \qquad\qquad \text{Ketone}$$

We'll answer the following questions in this chapter:

1. ***What are the kinds of carbonyl groups, and how do they differ?*** The goal: Be able to recognize and draw the structures of the important families of carbonyl compounds.

2. ***What are the general properties of aldehydes and ketones?*** The goal: Be able to describe such properties as polarity, hydrogen bonding, and water solubility.

3. ***How are ketones and aldehydes named?*** The goal: Be able to name the simple members of these families and write their structures, given the names.

4. ***What are some of the significant applications and occurrences of aldehydes and ketones?*** The goal: Be able to identify major applications and occurrences of each family of compounds and describe some important members of each family.

5. ***What are the major chemical reactions of aldehydes and ketones?*** The goal: Be able to describe and predict the products of the oxidation and reduction of aldehydes and ketones, and their addition of alcohols.

6. ***What are hemiacetals and acetals?*** The goal: Be able to recognize hemiacetals and acetals, describe the conditions under which they are formed, and predict the products of acetal hydrolysis.

7. ***What is the aldol reaction and why is it important in biochemistry?*** The goal: Be able to describe an aldol reaction, predict the products, and explain why it is important in biochemistry.

16.1 KINDS OF CARBONYL COMPOUNDS

There are many different kinds of carbonyl compounds, depending on what other substituents are bonded to the carbonyl-group carbon atom. As indicated in Table 16.1, *aldehydes* (often written as RCHO) have a carbon substituent and a hydrogen bonded to the carbonyl group; *ketones* (RCOR') have two carbon substituents; *carboxylic acids* (RCOOH) have a carbon and a hydroxyl (—OH); *anhydrides* (RCO_2COR') have a carbon and an oxygen bonded to another R—C=O group; *esters* (RCOOR') have a carbon and an alkoxyl (—OR); and *amides* ($RCONH_2$) have a carbon and a nitrogen group (—NH_2, NHR, or NR_2).

Since oxygen attracts electrons more strongly than carbon, carbonyl groups are strongly polarized, with a partial positive charge on carbon and a partial negative charge on oxygen. As we'll see in subsequent sections, this polarity of the carbonyl group helps to explain how many of its reactions take place.

$$O^{\delta -} \longleftarrow \text{Partial negative charge here}$$
$$C^{\delta +} \longleftarrow \text{Partial positive charge here}$$

Another property common to all carbonyl groups is planarity. The carbonyl carbon atom is surrounded by three regions of electron density, and the bond angles between the three substituents on carbon are 120° or close to it.

120° angles in a planar triangle

Table 16.1   Some Kinds of Carbonyl Compounds

Name	Structure	Example	
Aldehyde	R—C(=O)—H	H_3C—C(=O)—H	Acetaldehyde
Ketone	R—C(=O)—R'	H_3C—C(=O)—CH_3	Acetone
Carboxylic acid	R—C(=O)—O—H	H_3C—C(=O)—O—H	Acetic acid
Anhydride	R—C(=O)—O—C(=O)—R	H_3C—C(=O)—O—C(=O)—CH_3	Acetic anhydride
Ester	R—C(=O)—O—R'	H_3C—C(=O)—O—CH_3	Methyl acetate
Amide	R—C(=O)—N	H_3C—C(=O)—NH_2	Acetamide

It's useful to classify carbonyl compounds into two groups based on their chemical properties. In one group are aldehydes and ketones. Since the carbonyl group in these compounds is bonded to atoms (H and C) that don't attract electrons strongly, the bonds to the carbonyl-group carbon of aldehydes and ketones are not strongly polar. Thus, the compounds behave similarly.

In the second group are carboxylic acids, esters, and amides. Since the carbonyl-group carbon in these compounds is bonded to an atom (O or N) that *does* attract electrons strongly, the bonds are strongly polar. Thus, carboxylic acids, esters, anhydrides, and amides behave similarly in many ways. This second group of carbonyl-containing compounds is discussed in Chapter 17.

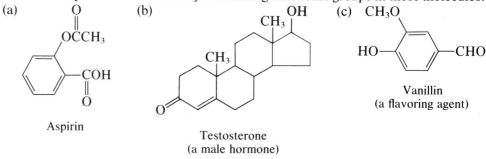

| Ketone | Aldehyde | Carboxylic acid | Anhydride | Ester | Amide |

| Less polar bonds | More polar bonds |

Practice Problem **16.1** Identify the kinds of carbonyl-containing functional groups in these molecules:

(a)

Aspirin

(b)

Testosterone
(a male hormone)

(c)

Vanillin
(a flavoring agent)

(d) $C_4H_9COCH_3$ (e) C_4H_9CHO (f) $C_4H_9COOCH_3$

16.2 NAMING ALDEHYDES AND KETONES

Aldehyde A compound that has a carbonyl group bonded to one carbon atom and one hydrogen atom, RCHO.

Aldehydes are named systematically by replacing the final *-e* of the corresponding alkane name with *-al*. Thus, the three-carbon aldehyde derived from propane is propanal, the four-carbon aldehyde is butanal, and so on. When substituents are present, the chain is numbered beginning at the —CHO end. In addition to these systematic names, some simple aldehydes have common names that end in *-aldehyde*. The one-carbon aldehyde derived from methane (systematic name, *methanal*) is often called *formaldehyde,* the two-carbon aldehyde derived from ethane (systematic name, *ethanal*) is often called *acetaldehyde,* and the simplest aromatic aldehyde is *benzaldehyde.*

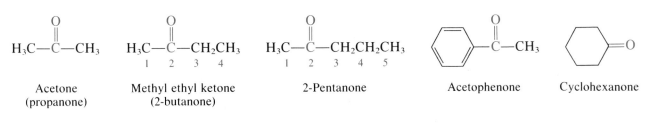

Formaldehyde   Acetaldehyde   3-Methylbutanal   Benzaldehyde
(methanal)   (ethanal)

Ketone A compound that has a carbonyl group bonded to two carbon atoms, $R_2C=O$.

Ketones are named systematically by replacing the final *-e* of the corresponding alkane name with *-one* (pronounced "own"). The numbering of the chain begins at the end nearer the carbonyl group, as shown for 2-pentanone, a five-carbon ketone derived from pentane. As with aldehydes, some simple ketones such as acetone and acetophenone also have common names. Often, the common name of a ketone gives the names of the two alkyl groups followed by the word *ketone*.

Acetone   Methyl ethyl ketone   2-Pentanone   Acetophenone   Cyclohexanone
(propanone)   (2-butanone)

Practice Problems **16.2** Draw structures corresponding to these names:
(a) hexanal (b) *p*-bromoacetophenone (c) 4-methyl-2-hexanone
(d) methyl isopropyl ketone

16.3 Give systematic names for these compounds:

(a) $CH_3CH_2CH_2CH_2CH$ (b) $CH_3CH_2CCH_2CH_3$ (c) $CH_3CH_2CHCH_2CH_2CH$

16.3 PROPERTIES OF ALDEHYDES AND KETONES

The polarity of the carbonyl group makes aldehydes and ketones moderately polar compounds. As a result, they are higher boiling than alkanes with similar molecular weights. Because they have no hydrogen atoms bonded to oxygen or nitrogen, however, their molecules don't hydrogen-bond with each other, and this makes aldehydes and ketones lower boiling than alcohols. In a series of compounds with similar molecular weights, the alkane is lowest boiling, the alcohol is highest boiling, and the aldehyde and ketone fall in between.

$CH_3CH_2CH_2CH_3$ CH_3CH_2CH CH_3CCH_3 $CH_3CH_2CH_2OH$

Butane, bp $-43°C$ Propanal, bp 50°C   Acetone, bp 56°C   Propanol, bp 97°C

Formaldehyde (HCHO), the simplest aldehyde, is a gas. The other simple aldehydes and ketones are liquids (Table 16.2), and those with more than 12 carbon atoms are solids.

Aldehydes and ketones are soluble in common organic solvents, and those with fewer than five carbon atoms are also soluble in water because they're able to hydrogen-bond with water molecules (Figure 16.1). These simple aldehydes and ketones are excellent solvents because they dissolve both polar and nonpolar compounds. With increasing numbers of carbon atoms, aldehydes and ketones become more alkane-like and less water-soluble. Many aldehydes and ketones, including several that are found naturally, have distinctive odors (Figure 16.2).

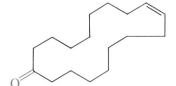

| Cinnamaldehyde (cinnamon flavor in foods, drugs; from the Chinese cinnamon plant) | Camphor (cools skin in liniment and itch remedies, but is not safe at over 2.5% concentration; from the camphor tree) | Civetone (musky odor in perfumes; from the scent gland of the civet cat) |

Camphor has been used in remedies for thousands of years but has been judged ineffective in most situations by the U.S. Food and Drug Administration. Although camphor does cool the skin, its major benefit may be that it has a "medicinal" smell.

The lower-boiling aldehydes and ketones are flammable and form explosive mixtures with air. The simple ketones have generally low toxicity, while the simple aldehydes, especially formaldehyde, are toxic.

Table 16.2   Physical Properties of Some Simple Aldehydes and Ketones[a]

Structure	Name	Boiling Point (°C)	Water Solubility (g/100 mL H_2O)
HCHO	Formaldehyde	−21	55
CH_3CHO	Acetaldehyde	21	Miscible
CH_3CH_2CHO	Propanal	49	16
$CH_3CH_2CH_2CHO$	Butanal	76	7
$CH_3CH_2CH_2CH_2CHO$	Pentanal	103	1
⬡—CHO	Benzaldehyde	178	0.3
CH_3COCH_3	Acetone	56	Miscible
$CH_3CH_2COCH_3$	2-Butanone	80	26
$CH_3CH_2CH_2COCH_3$	2-Pentanone	102	6
⬡=O	Cyclohexanone	156	2

[a] All the compounds listed are liquids at room temperature.

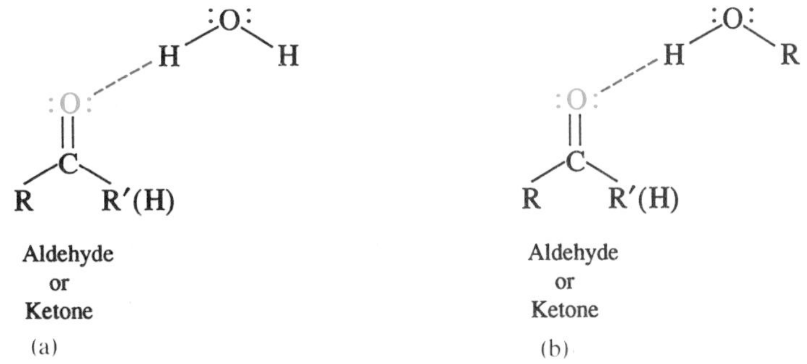

Figure 16.1
Hydrogen bonding between aldehyde or ketone molecules and (a) water and (b) alcohol molecules.

Figure 16.2
A civet, one of a large family of carnivores that live in Asia, Africa, and southern Europe. They mark their territory with musk, a potent perfume ingredient.

Properties of Aldehydes and Ketones

● No hydrogen bonding between aldehyde or ketone molecules, but they are polar and lower boiling than alcohols, but higher boiling than alkanes.

● Common aldehydes and ketones are liquids (except gaseous formaldehyde).

● Simple aldehydes and ketones are water-soluble due to hydrogen bonding with water molecules.

● Volatile aldehydes and ketones are flammable.

● Many have distinctive odors.

● Simple ketones are less toxic than simple aldehydes

16.4 SOME COMMON ALDEHYDES AND KETONES

Ketones are used in industry primarily as solvents, and aldehydes are valuable as reactants in organic synthesis and polymer formation. In living things, aldehyde and ketone functional groups are present in a great many compounds. Natural aldehydes and ketones range from simple compounds, such as those responsible for the odors and flavors of foods, spices, and flowers, to large molecules that contain numerous functional groups. Glucose and most other sugars contain —CHO groups in addition to their —OH groups. Take a good look at the glucose structure, for it plays a crucial role in the production of biochemical energy and we'll have much more to say about it. Testosterone, progesterone, and many other steroid hormones contain keto groups.

Some biomolecules containing —CHO or —C=O groups

Glucose
(a pentahydroxyhexanal;
a sugar and a biochemical
energy source)

Progesterone
(a female sex hormone)

α-Ketoglutaric acid
(a key intermediate
in metabolism)

Formaldehyde and acetaldehyde, the two simplest aldehydes, and acetone, the simplest ketone, are described in the following paragraphs.

Formaldehyde

Acetaldehyde

Acetone

Formaldehyde (HCHO) At room temperature, pure formaldehyde is a colorless gas with a pungent, suffocating odor. Low concentrations in the air cause eye, throat, and bronchial irritation. Because formaldehyde is formed during incomplete combustion of coal and other organic substances, it's partly responsible for the irritation caused by smog-laden air. Higher concentrations of formaldehyde in the air cause bronchial pneumonia, and skin contact can produce dermatitis. Formaldehyde is highly toxic by ingestion, causing serious kidney damage, coma, and death. As noted in the Application on ethyl alcohol in Chapter 14, formaldehyde is responsible for the poisonous effects of methanol.

Formaldehyde is commonly sold as a 37% (w/w) aqueous solution under the name *formalin* for use as a disinfectant, antiseptic, and preservative for biological specimens. The activity of formaldehyde in these applications results from its reaction with the —NH_2 groups in proteins. Some other aldehydes are also good disinfectants, including the four-carbon dialdehyde succinaldehyde ($OHCCH_2CH_2CHO$), a chemical sterilant used for delicate surgical instruments that can't be sterilized by heat.

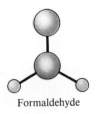
Formaldehyde

In the chemical industry, the major use of formaldehyde is in polymers with applications such as adhesives for binding plywood, foam insulation used in houses, textile finishes, and hard and durable manufactured objects such as telephone parts. *Bakelite,* one of the first commercial plastics, is a copolymer of phenol and formaldehyde. *Melamine,* familiar for its use in dinnerware and decorative laminated sheets for countertops, is a copolymer of formaldehyde and urea (H_2NCONH_2). Once the final polymerization of such a copolymer is carried out, no further melting and reshaping is possible because of the cross-linked, three-dimensional structure. In these general structures, wavy lines indicate bonds to the rest of the polymer; the CH_2 groups are from the formaldehyde molecules.

Phenol–formaldehyde polymer

Urea–formaldehyde polymer

Because concern has arisen over the toxicity and possible carcinogenicity of formaldehyde released from formaldehyde-based products, especially during fires, their use in some applications has been limited.

Acetaldehyde (CH_3CHO) Acetaldehyde is a sweet-smelling, flammable liquid present in ripe fruits. It is less toxic than formaldehyde, and small amounts are produced in the normal breakdown of carbohydrates. Acetaldehyde is, however, a general narcotic, and large doses can cause respiratory failure. Chronic exposure produces symptoms like those of alcoholism. In industry, the major use of acetaldehyde is as an intermediate in the synthesis of other compounds.

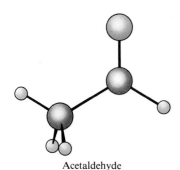

Acetaldehyde

Acetone (CH_3COCH_3) Acetone, a liquid at room temperature, is perhaps the most widely used of all organic solvents. You may have seen cans of acetone sold in paint stores for general-purpose cleanup work, and it is the solvent in many varnishes, lacquers, and nail polish removers. Not only can acetone dissolve most organic compounds, it is also miscible with water.

Acetone is highly volatile and is a serious fire and explosion risk when allowed to evaporate in a closed space. No chronic health risk has been associated with acetone exposure, however, and it is not very toxic. When the biochemical breakdown of fats and carbohydrates to yield energy is out of balance, acetone is produced in the liver, a condition that can occur during starvation or diabetes and in severe cases leaves the odor of acetone on a patient's breath.

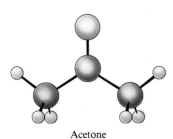

Acetone

AN APPLICATION: IS IT POISONOUS OR ISN'T IT?

In its broadest meaning, the term "drug" refers to any chemical agent other than food that affects living organisms. Mostly we refer either to substances used in preventing or treating disease, or to addictive substances obtained illegally, as "drugs." By contrast, a "poison" or "toxic substance" is any chemical agent that harms living organisms. How do we tell one from the other? Paracelsus, a Swiss physician who lived in the 1500s, understood the problem: "It is only the dose which makes a thing a poison."

The standard method for evaluating toxicity, in use since the 1920s, is the LD_{50}—lethal dose—test, and LD_{50} data are cited frequently. The LD_{50} is a measure of the toxicity of a single dose, known as *acute toxicity*. A substance is fed in varying doses to laboratory animals. The result of the test is reported as the LD_{50}, the dose that kills 50% of the animals in a uniform population. By comparing LD_{50} values, relative toxicities can be evaluated. The LD_{50} values in the accompanying table, reported as the dose in milligrams or grams per kilogram of body weight of the test animal, show that in this group of compounds formaldehyde is the most toxic and acetone the least toxic by ingestion.

	LD_{50}
Formaldehyde, 37% aqueous solution	800 mg/kg
Acetaldehyde	1.9 g/kg
Butyraldehyde	5.8 g/kg
Acetone	8.4 g/kg

To put these values in further perspective, compare them with the LD_{50} of 17.2 mg/kg for *ouabain*, an arrow poison extracted from plants by natives in Africa, or the LD_{50} of 1.0 μg/kg for the even more toxic *batrachotoxin*, an arrow poison extracted from frogs in South America.

For therapeutic use in humans, a compound must show a comfortably wide margin between the dose that produces the desired effect and the dose that produces an acute toxic effect. Aspirin,

16.5 OXIDATION OF ALDEHYDES

Oxidation In organic chemistry, the removal of hydrogen from a molecule or addition of oxygen to a molecule.

We've already seen one important reaction of aldehydes: their **oxidation** to yield carboxylic acids (Section 14.5). In this reaction, the $-CH=O$ hydrogen attached to the carbonyl carbon is removed and replaced by $-OH$. Ketones don't have this hydrogen, however, and thus don't react with most oxidizing agents.

$$\underset{\text{Aldehyde}}{R-\overset{\displaystyle\overset{O}{\|}}{C}-H} \quad \xrightarrow{[O]} \quad \underset{\text{Carboxylic acid}}{R-\overset{\displaystyle\overset{O}{\|}}{C}-OH} \qquad \underset{\text{Ketone}}{R-\overset{\displaystyle\overset{O}{\|}}{C}-R'} \quad \xrightarrow{[O]} \quad \text{No reaction}$$

Tollens' reagent A reagent ($AgNO_3$ in aqueous NH_3) that converts an aldehyde into a carboxylic acid and deposits a silver mirror on the inside surface of the reaction flask.

Reaction with a mild oxidizing agent that doesn't affect other functional groups in a molecule can be used as a test for the presence of an aldehyde group. Among the many mild oxidizing agents that cause the aldehyde → carboxylic acid conversion, Tollens' reagent is the most visually appealing. Treatment of an aldehyde with **Tollens' reagent,** a dilute solution of silver nitrate in aqueous ammonia, rapidly yields the ammonium salt of the carboxylic acid and silver for example, has an LD_{50} of 1.75 g/kg in rats. An average therapeutic dose of two aspirin tablets is only 650 mg, and the amount taken in 24 hr should not exceed 4000 mg.

The testing that precedes bringing a new drug to market includes many procedures in addition to the LD_{50} test. There must be *subacute toxicity* studies, in which the effects of less than lethal doses administered over several months are evaluated, and there must be *chronic toxicity* studies on the effects of low, medium, and high doses over the course of 6 months to 2 years. Additional studies include tests for effects on fertility and in pregnant animals, tests for behavioral effects, and tests for allergic reactions.

In the desire to decrease the enormous cost in time (over 8 years average) and money (millions of dollars) required to evaluate a new drug and to decrease the sacrifice of laboratory animals, the search for alternative methods of drug evaluation is under way. Significant progress is being made in procedures carried out *in vitro*, meaning literally "in glass," that is, not in ani-

mals. In the test pictured, for example, a value comparable to the LD_{50} is found by determining the concentration that kills 50% of the living cells in a cultured medium.

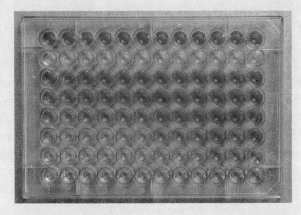

Each indentation in the test plate contains a red dye along with cells that were cultured in the presence of various concentrations of a chemical being tested for toxicity. Since only living cells take up the dye, the effect of the chemical is determined by the presence or absence of color.

metal as by-product. Addition of an acid such as HCl then yields the carboxylic acid. If the reaction is done in a clean flask or test tube, metallic silver deposits on the inner walls of the vessel, producing a beautiful, shiny mirror (Figure 16.3). In fact, a slight modification of this reaction is used in the industrial manufacture of glass mirrors.

$$\underset{\text{Aldehyde}}{R-\overset{\overset{\displaystyle O}{\|}}{C}-H} + 2\ AgNO_3(aq) + 3\ NH_3 + H_2O \longrightarrow \underset{\substack{\text{Ammonium salt}\\\text{of acid}}}{R-\overset{\overset{\displaystyle O}{\|}}{C}-O^-\ NH_4^+} + \underset{\substack{\text{Silver}\\\text{metal}}}{2\ Ag} + 2\ NH_4NO_3(aq)$$

For example,

Benzaldehyde $\xrightarrow[\text{2. } H_3O^+]{\text{1. Tollens' reagent}}$ Benzoic acid $+\ 2\ Ag$

(a)

(b)

(c)

Figure 16.3
The Tollens' test for aldehydes: RCHO + 2 Ag(NH$_3$)$_2^+$ + 3 OH$^-$ → RCOO$^-$ + 2 Ag + 4 NH$_3$ + 2 H$_2$O. In a clean container, the silver deposits on the walls to form a mirror.

Benedict's reagent A reagent (Cu^{2+} in basic solution) that converts an aldehyde into a carboxylic acid and yields a brick-red precipitate of Cu$_2$O.

Another mild oxidizing agent, known as **Benedict's reagent,** also relies on reduction of a metal to produce a visible test for aldehydes. The reagent solution contains blue copper(II) ion that is reduced to give a precipitate of red copper(I) oxide in the reaction with an aldehyde (Figure 16.4):

$$\underset{\substack{\text{Deep-blue}\\\text{solution}}}{R-\overset{\displaystyle O}{\overset{\|}{C}}-H} \;+\; 2\,Cu^{2+} \;+\; 5\,OH^- \;\longrightarrow\; R-\overset{\displaystyle O}{\overset{\|}{C}}-O^- \;+\; \underset{\substack{\text{Brick-red}\\\text{precipitate}}}{Cu_2O(s)} \;+\; 3\,H_2O$$

Because many sugars contain aldehyde groups, Benedict's reagent has been used as a test for sugars in the urine (although more specific tests are now preferred; Chapter 21).

Figure 16.4
The Benedict's test for aldehydes: The Cu^{2+}-containing reagent solution is on the left. A few drops of glucose produce the greenish color in the center. A few milliliters produce the brick-red precipitate of Cu$_2$O on the right.

Practice Problem   ***16.4***   Draw structures of the products you would obtain by treating the following compounds with Tollens' reagent. If no reaction occurs, write "NR."

$$\underset{\displaystyle \overset{|}{CH_3}}{}$$

(a) $CH_3\overset{\displaystyle CH_3}{\underset{\displaystyle |}{C}}HCH_2CH_2CH_2CHO$ (b) 2,2-dimethylpentanal

(c) 2-methyl-3-pentanone

16.6   REDUCTION OF ALDEHYDES AND KETONES

Reduction In organic chemistry, the addition of hydrogen, for example, to a carbon–oxygen double bond or a carbon–carbon double bond, or the removal of oxygen.

Just as an alcohol can be converted into a carbonyl compound by *removing* hydrogen (an *oxidation*), a carbonyl compound can be converted back into an alcohol by *adding* hydrogen to the C=O double bond (a *reduction*). **Reduction** in organic chemistry usually refers to the addition of hydrogen to a molecule.

1° or 2° Alcohol   Aldehyde or ketone

The reduction of aldehydes and ketones is usually accomplished by treatment with sodium borohydride, $NaBH_4$. When mixed with a ketone or aldehyde, sodium borohydride transfers one of its hydrogens as a hydride ion ($:H^-$). The hydride ion contributes both electrons to a covalent bond to the carbonyl carbon, forming an anion. Aqueous acid is then added, H^+ adds to the oxygen, and a neutral alcohol product results. Aldehydes are converted into primary alcohols ($RCHO \rightarrow RCH_2OH$) by reaction with $NaBH_4$.

Note that the two new hydrogen atoms in the alcohol product come from different sources. The hydrogen atom attached to carbon comes from reaction with $:H^-$ (hydride ion) in the first step, whereas that attached to oxygen comes from reaction with H^+ (acid) in the second step.

Ketones are converted into secondary alcohols ($R_2CO \rightarrow R_2CHOH$) by reaction with $NaBH_4$.

$$R-\overset{\overset{\displaystyle O}{\|}}{C}-R' \quad \xrightarrow[\text{2. } H_3O^+]{\text{1. NaBH}_4} \quad R-\overset{\overset{\displaystyle O-H}{|}}{\underset{\underset{\displaystyle R'}{|}}{C}}-H$$

From H_3O^+

From : H

Ketone   2° Alcohol

For example, the aldehyde propanal gives a primary alcohol, and the ketone cyclohexanone gives a secondary alcohol.

$$CH_3CH_2\overset{\overset{\displaystyle O}{\|}}{CH} \quad \xrightarrow[\text{2. } H_3O^+]{\text{1. NaBH}_4} \quad CH_3CH_2CH_2OH$$

Propanal   1-Propanol

Cyclohexanone   Cyclohexanol

The fact that the reduction of aldehydes and ketones takes two steps, addition of : H$^-$ and addition of H$^+$, makes good sense when you think about the polarity of the carbonyl group. Since the carbonyl-group carbon is polarized δ^+, the negatively charged hydride ion is naturally drawn to it. Similarly, a positively charged hydrogen ion is drawn to the negatively polarized carbonyl-group oxygen.

$$\overset{\overset{\displaystyle O^{\delta-}}{\|}}{C^{\delta+}}$$

$O^{\delta-} \longleftarrow$ H$^+$ attracted here

$C^{\delta+} \longleftarrow$:H$^-$ attracted here

The reduction of aldehydes and ketones to yield alcohols occurs in living cells as well as in the laboratory. Although the body doesn't use NaBH$_4$ as its ''reagent,'' there is nevertheless a remarkable similarity in the way the two kinds of reductions are done. Living organisms use *nicotinamide adenine dinucleotide*, abbreviated *NADH*, as a biological source of hydride ion. (For the structure of NAD$^+$, the oxidized form of NADH, see Application on organic compounds in body fluids, Chapter 15.) NADH serves exactly the same role in the cell that NaBH$_4$ serves in the laboratory: It donates : H$^-$ to an aldehyde or ketone to yield an anion that picks up H$^+$ from surrounding aqueous fluids. The net result is reduction.

An example of biochemical reduction occurs when pyruvic acid, an intermediate in glucose metabolism, is converted into lactic acid by active skeletal muscles. Vigorous exercise causes a buildup of lactic acid in muscles, leading to a tired, flat feeling.

$$H_3C-\overset{\overset{\displaystyle O}{\|}}{C}-COOH \quad \xrightarrow{\text{NADH}} \quad H_3C-\overset{\overset{\displaystyle O^-}{|}}{\underset{\underset{\displaystyle H}{|}}{C}}-COOH \quad \xrightarrow{H_3O^+} \quad H_3C-\overset{\overset{\displaystyle O-H}{|}}{\underset{\underset{\displaystyle H}{|}}{C}}-COOH$$

Pyruvic acid   Lactic acid

Solved Problem 16.1   What product would you obtain by reduction of benzaldehyde?

Solution   First, draw the structure of the starting material, showing the double bond in the carbonyl group. Then rewrite the structure showing only a *single* bond between C and O, along with *partial* bonds to both C and O:

Benzaldehyde   rewrite as   Partial bonds

Finally, attach hydrogen atoms to the two part-bonds and rewrite the product:

= ⟨benzene ring⟩—CH₂OH   (Benzyl alcohol)

Practice Problems   **16.5**   What product would you obtain from reduction of the following ketones and aldehydes?

(a) $(CH_3)_2CHCHO$ (b) *m*-chlorobenzaldehyde   (c) cyclopentanone

16.6   What ketones or aldehydes might be reduced to yield the following alcohols?

(a)   (b) $HOCH_2CH_2CH_2CHCH_3$
                                                              $|$
                                                              CH_3

(c) 2-methyl-1-pentanol

16.7 REACTION WITH ALCOHOLS: HEMIACETALS AND ACETALS

Addition reaction, aldehydes and ketones
Addition of an alcohol or other reagent to the carbon–oxygen double bond to give a carbon–oxygen single bond.

Hemiacetal A compound that has both an alcohol-like —OH group and an ether-like —OR group bonded to the same carbon atom.

Hemiacetal Formation   Aldehydes and ketones undergo **addition reactions** with alcohols. Just as the C=C bond is converted to a single C—C bond when hydrogen adds to an alkene, the C=O bond is converted to a single C—O bond when an alcohol adds to an aldehyde or ketone. During the reaction, the H from the alcohol bonds to the carbonyl-group oxygen, and the OR from the alcohol bonds to the carbonyl-group carbon. The addition products, called **hemiacetals,** are compounds that have both an alcohol-like —OH group and an ether-like —OR group bonded to what was a carbonyl carbon atom.

Aldehyde or ketone   Alcohol   Hemiacetal (unstable)

Two important features of this reaction need to be pointed out. It is the *negatively* polarized alcohol oxygen atom that adds to the *positively* polarized carbonyl carbon, just as negatively charged hydride ions adds to the positively polarized carbonyl oxygen during reduction (Section 16.6). Almost all carbonyl-group reactions follow this same polarity pattern. Also, note that the reaction is reversible. Hemiacetals rapidly revert back to aldehydes or ketones by loss of alcohol. Equilibrium is therefore always established between an aldehyde or ketone and its hemiacetal. For example, ethanol (CH_3CH_2OH) forms hemiacetals with acetaldehyde and acetone as follows:

Acetaldehyde + HOCH₂CH₃ ⇌ Hemiacetal (unstable)

Acetone + HOCH₂CH₃ ⇌ Hemiacetal (unstable)

(In older nomenclature, a distinction was made by calling the compounds from aldehydes "hemiacetals," and those from ketones "hemiketals"; thus you may see "hemiketal" or "ketal" in some publications.)

In practice, hemiacetals are often too unstable to be isolated. When the equilibrium position for the reaction is reached, very little hemiacetal is present. The one major exception to this rule occurs in the case of sugars such as glucose in which the —OH and —CHO functional groups that react are part of the *same* molecule. The resulting cyclic hemiacetal is more stable than a noncyclic hemiacetal. As you'll see in Chapter 21, most simple sugars contain such a stable hemiacetal link.

Glucose → Cyclic hemiacetal form of glucose

Solved Problem 16.2   Which of the following compounds is a hemiacetal?

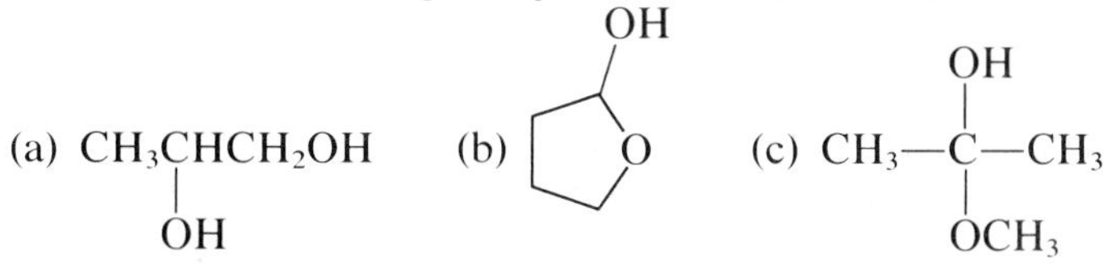

(a) CH_3CHCH_2OH
      |
      OH

(b) [cyclic structure with OH and O on ring]

(c) $CH_3-\overset{\overset{OH}{|}}{\underset{\underset{OCH_3}{|}}{C}}-CH_3$

Solution   To identify a hemiacetal, look for a carbon atom attached to two oxygen atoms, one in an —OH group and one in an —OR group. Compound (a) contains two O atoms, but they are bonded to different C atoms; it is not a hemiacetal. Compound (b) has one ring C atom bonded to two oxygen atoms, one in the substituent —OH group and one bonded to the rest of the ring, the same as an R group; it is a cyclic hemiacetal. Compound (c) also contains a C atom bonded to one —OH group and one —OR group, so it too is a hemiacetal.

Acetal A compound that has two ether-like —OR groups bonded to the same carbon atom.

Acetal Formation   If a small amount of acid catalyst is added to the reaction of an alcohol and a carbonyl compound, the initially formed hemiacetal is converted into an acetal. An **acetal** is a compound that has *two* ether-like —OR groups bonded to what was a carbonyl carbon atom.

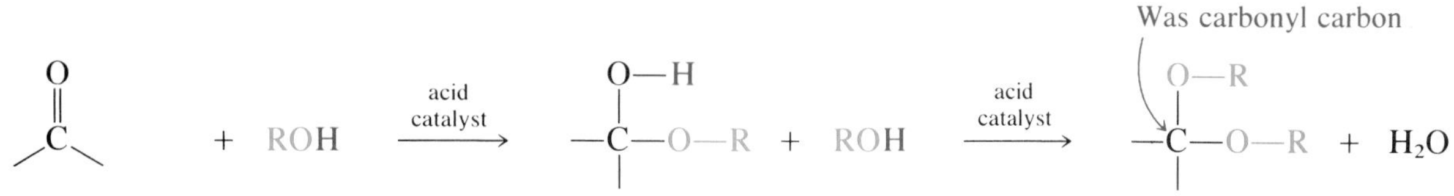

| Aldehyde or ketone | | Hemiacetal | | Acetal |

For example,

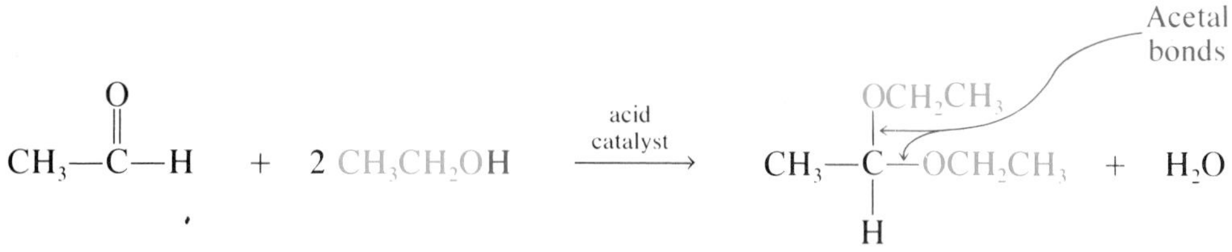

Acetaldehyde   Acetal

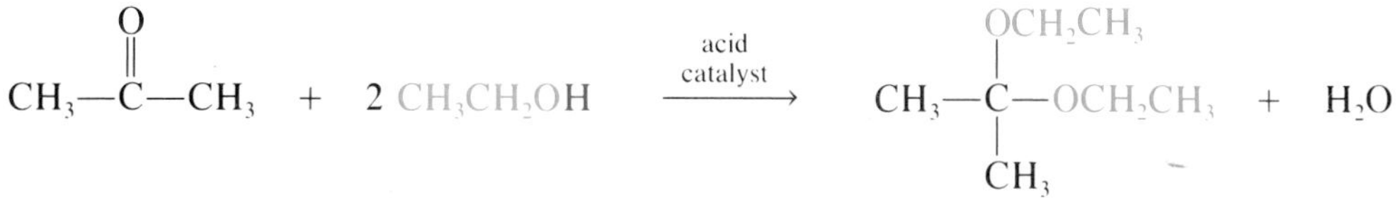

Acetone   Acetal

Unlike hemiacetals, acetals are quite stable and are easily isolated. They are so stable, in fact, that some sugars and complex carbohydrates like cellulose and starch are held together simply by acetal groupings between individual sugar units. The sugar maltose, for example, is the product of acetal formation between the hemiacetal —OH group of one glucose molecule and an alcoholic —OH group of another glucose molecule. You'll learn more about carbohydrate acetals in Chapter 21.

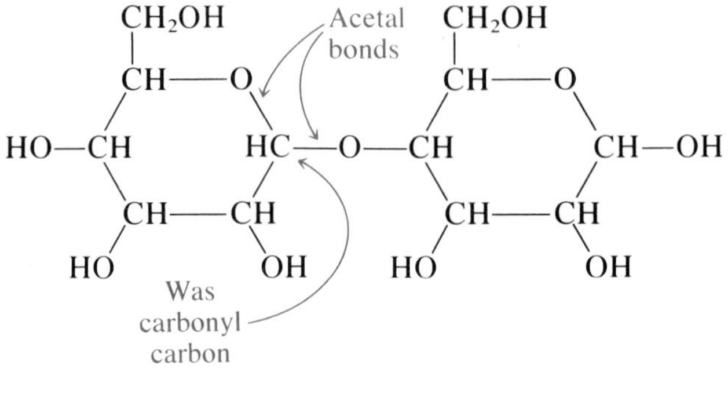

Maltose (a sugar)

Solved Problem 16.3 Write the structure of the intermediate hemiacetal and the acetal final product formed in the following reaction:

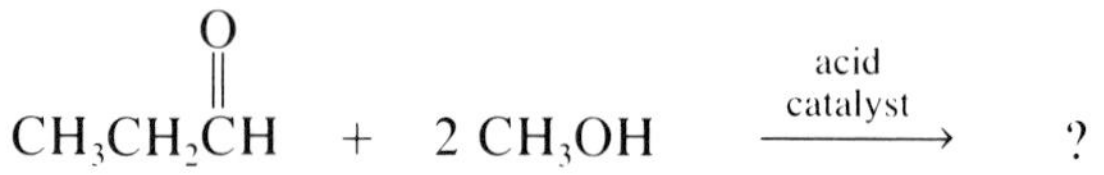

$$CH_3CH_2CH{=}O \ + \ 2\ CH_3OH \xrightarrow[\text{catalyst}]{\text{acid}} \ ?$$

Solution First, rewrite the structure showing only a single bond between C and O, along with partial bonds to both C and O:

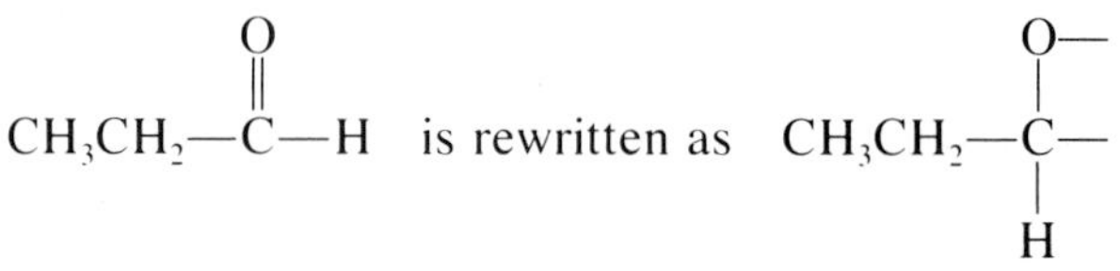

$$CH_3CH_2{-}\overset{O}{\underset{}{C}}{-}H \quad \text{is rewritten as} \quad CH_3CH_2{-}\overset{O-}{\underset{H}{C}}{-}$$

Next, add one molecule of the appropriate alcohol (CH_3OH in this case) by attaching —H to the oxygen part-bond and —OCH_3 to the carbon part-bond. This yields the hemiacetal intermediate:

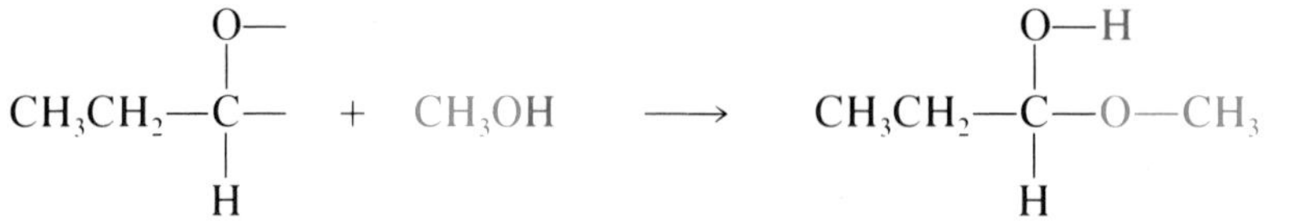

$$CH_3CH_2{-}\overset{O-}{\underset{H}{C}}{-} \ + \ CH_3OH \longrightarrow CH_3CH_2{-}\overset{O-H}{\underset{H}{C}}{-}O{-}CH_3 \quad \text{Hemiacetal}$$

Finally, replace the —OH group of the hemiacetal with an —OCH_3 from a second molecule of alcohol. This yields the acetal product and water.

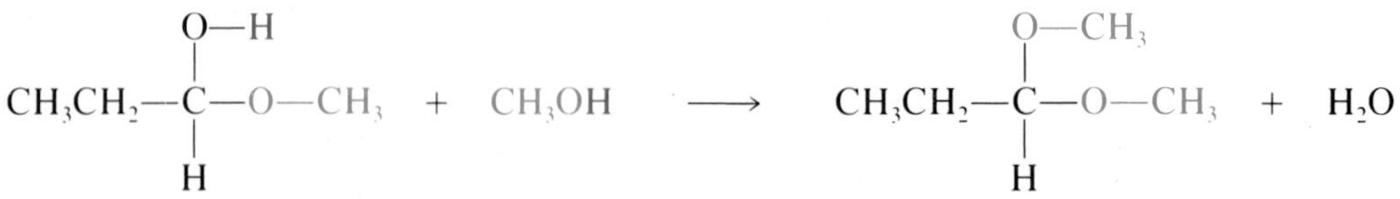

$$CH_3CH_2{-}\overset{O-H}{\underset{H}{C}}{-}O{-}CH_3 \ + \ CH_3OH \longrightarrow CH_3CH_2{-}\overset{O-CH_3}{\underset{H}{C}}{-}O{-}CH_3 \ + \ H_2O$$

Acetal product

Practice Problems   16.7   Draw the structures of the hemiacetals formed in these reactions:
(a) $CH_3CH_2CH_2CHO + CH_3CH_2OH \longrightarrow$?

(b) $CH_3CH_2\overset{\overset{\displaystyle O}{\|}}{C}CH_2CH(CH_3)_2 + CH_3OH \longrightarrow$?

16.8   Draw the structure of each acetal final product formed in the reactions shown in Problem 16.7.

16.9   Which of the following compounds are hemiacetals or acetals?
(a)
$$CH_3-\underset{\underset{\displaystyle OCH_3}{|}}{\overset{\overset{\displaystyle OCH_3}{|}}{C}}-CH_3$$
(b) $HOCHCH_2CH_2CH_3$
$\quad\;\; OCH_2CH_3$

(c)
$$CH_3\overset{\overset{\displaystyle O}{\|}}{C}CH_2OH$$
(d) a cyclic structure with $-O-$ and $-OCH_2CH_3$

Acetal Hydrolysis   Although stable to most reagents, acetals react with aqueous acid to regenerate the original ketone or aldehyde plus two molecules of alcohol. This reaction, called a **hydrolysis** because water is used to break down the starting material, is exactly what happens in the mouth and small intestine when carbohydrates are digested.

Hydrolysis   The breakdown of a compound, such as an acetal, by reaction with water; the H's and O of water add to the atoms of the broken bond.

$$\underset{\text{Acetal}}{-\overset{\overset{\displaystyle O-R}{|}}{\underset{|}{C}}-O-R} + H_2O \xrightarrow{\text{acid catalyst}} \underset{\text{Aldehyde or ketone}}{\overset{\overset{\displaystyle O}{\|}}{C}} + \underset{\text{Alcohol}}{2\,R-O-H}$$

For example,

$$CH_3\overset{\overset{\displaystyle OCH_3}{|}}{\underset{\underset{\displaystyle CH_3}{|}}{C}}-OCH_3 + H_2O \xrightarrow{\text{acid catalyst}} CH_3\overset{\overset{\displaystyle O}{\|}}{C}CH_3 + 2\,CH_3OH$$

Solved Problem 16.4   Write the structure of the aldehyde or ketone that would be formed by hydrolysis of the following acetal:

$$(CH_3)_2CHCH_2CH(OCH_2CH_3)_2 \xrightarrow{H_3O^+} ?$$

Solution   First, rewrite the starting acetal so that the two ether-like acetal bonds are more evident:

$$(CH_3)_2CHCH_2CH(OCH_2CH_3)_2 \ = \ \begin{array}{c} CH_3 \qquad\quad O{-}CH_2CH_3 \\ | \qquad\qquad\quad | \\ CH_3CHCH_2{-}C{-}O{-}CH_2CH_3 \\ | \\ H \end{array} \quad \text{Acetal bonds}$$

Next, remove the two —OR groups from the acetal carbon, converting each into a molecule of alcohol and leaving two part-bonds on carbon:

$$\begin{array}{c} CH_3 \qquad O{-}CH_2CH_3 \\ | \qquad\quad | \\ CH_3CHCH_2{-}C{-}O{-}CH_2CH_3 \\ | \\ H \end{array} = \begin{array}{c} CH_3 \\ | \qquad\quad | \\ CH_3CHCH_2{-}C{-} \\ | \\ H \end{array} + \ 2\ CH_3CH_2{-}O{-}H$$

Ethanol

Finally, add an oxygen to the two part-bonds on the acetal carbon to form a carbonyl group and rewrite the structure of the product. In this example, the product is an aldehyde. Note that the carbonyl carbon is the one that was bonded to two oxygen atoms.

$$\begin{array}{c} CH_3 \\ | \qquad\quad | \\ CH_3CHCH_2{-}C{-} \\ | \\ H \end{array} \longrightarrow \begin{array}{c} CH_3 \qquad O \\ | \qquad\qquad \| \\ CH_3CHCH_2{-}C{-}H \end{array} = (CH_3)_2CHCH_2CHO$$

3-Methylbutanal

Practice Problem 16.10 What aldehydes or ketones result from the following acetal hydrolysis reactions? What alcohol is formed in each case?

(a)

$$\text{C}_6\text{H}_5{-}CH_2C(OCH_3)_2CH_2CH_3 \xrightarrow{\ H_3O^+\ } \ ?$$

(b) $CH_3CH_2CH_2OCH_2OCH_2CH_2CH_3 \xrightarrow{\ H_3O^+\ } \ ?$

16.8 ALDOL REACTION OF ALDEHYDES AND KETONES

All the organic reactions we've covered up to this point have been interconversions of one functional group with another. For example, alcohols can be oxidized to yield aldehydes and ketones, which in turn can be reduced to give back alcohols. Although important, these functional-group interconversion reactions are of limited use because they don't change the size or carbon framework of molecules.

In contrast to functional-group interconversions, certain other reactions form new carbon–carbon bonds. By using such reactions, it's possible to take two small pieces, join them together, and thereby make a larger molecule. Many of the biochemical processes in living cells do exactly this.

The **aldol reaction** occurs when an aldehyde or ketone is treated with a base. In the reaction, a bond forms between the carbon atom next to the carbonyl group of one molecule and the carbonyl-group carbon of the second molecule. The product, formed by joining together two molecules of starting material, is a hydroxy ketone or a hydroxy aldehyde.

Aldol reaction The reaction of a ketone or aldehyde to form a hydroxy ketone product on treatment with a base catalyst.

AN APPLICATION: A BIOLOGICAL ALDOL REACTION

Aldol reactions are routinely used by living organisms as a key step in the biological synthesis of many different molecules. One particularly important example takes place in green leaves during the photosynthesis of carbohydrates when two three-carbon molecules are joined to form the six-carbon sugar, fructose, which is in turn converted into glucose, sucrose, and all other sugars.

As in most biochemical reactions, this transformation relies on enzymes as catalysts. As in many biochemical reactions involving sugars, some of the sugar —OH groups are replaced by phosphate ester groups. Although we haven't shown all the details, the reaction of glyceraldehyde 3-phosphate with dihydroxyacetone phosphate to yield the fructose diphosphate is clearly an aldol reaction. A hydrogen next to the dihydroxyacetone carbonyl group bonds to the carbonyl-group oxygen of glyceraldehyde, and a carbon–carbon bond forms between the two partners:

$$^{2-}O_3POCH_2CH-\overset{\displaystyle O}{\overset{\displaystyle \|}{C}}-H \;+\; \overset{\displaystyle H}{\underset{\displaystyle OH}{\overset{\displaystyle |}{C}}}H-\overset{\displaystyle O}{\overset{\displaystyle \|}{C}}-CH_2OPO_3{}^{2-} \;\longrightarrow\; {}^{2-}O_3POCH_2CHCH-\overset{\displaystyle OH}{}CHCCH_2OPO_3{}^{2-}$$

Glyceraldehyde
3-phosphate

Dihydroxyacetone
phosphate

Fructose 1,6-diphosphate
(a sugar phosphate)

This oxygen and this hydrogen form a bond.

This new C—C bond is formed.

$$-\overset{\displaystyle O}{\underset{\displaystyle |}{\overset{\displaystyle |}{C}}}-\overset{\displaystyle |}{\underset{\displaystyle |}{C}}- \;+\; -\overset{\displaystyle H\;\;O}{\underset{\displaystyle |}{\overset{\displaystyle |\;\;\|}{C}}}-C- \;\underset{\text{catalyst}}{\overset{\text{NaOH}}{\rightleftharpoons}}\; -\overset{\displaystyle OH}{\underset{\displaystyle |}{\overset{\displaystyle |}{C}}}-\overset{\displaystyle |}{\underset{\displaystyle |}{C}}-\overset{\displaystyle O}{\underset{\displaystyle |}{\overset{\displaystyle \|}{C}}}-C-$$

Two ketones or aldehydes

This carbon and this carbon form a bond.

A hydroxy ketone or aldehyde

For example,

$$\overset{\displaystyle H}{\underset{\displaystyle H}{\overset{\displaystyle |}{H-C}}}-\overset{\displaystyle O}{\overset{\displaystyle \|}{C}}-H \;+\; \overset{\displaystyle H}{\underset{\displaystyle H}{\overset{\displaystyle |}{H-C}}}-\overset{\displaystyle O}{\overset{\displaystyle \|}{C}}-H \;\underset{\text{catalyst}}{\overset{\text{NaOH}}{\rightleftharpoons}}\; \overset{\displaystyle H}{\underset{\displaystyle H}{\overset{\displaystyle |}{H-C}}}-\overset{\displaystyle OH}{\underset{\displaystyle H}{\overset{\displaystyle |}{C}}}-\overset{\displaystyle H}{\underset{\displaystyle H}{\overset{\displaystyle |}{C}}}-\overset{\displaystyle O}{\overset{\displaystyle \|}{C}}-H$$

Two acetaldehydes

3-Hydroxybutanal

One limitation of the aldol reaction is that the aldehyde or ketone starting material must have a hydrogen atom on the carbon next to the carbonyl group. If there is no such hydrogen present, an aldol reaction can't take place. For example, acetone can easily undergo an aldol reaction since it has six available hydrogens next to its carbonyl group, but benzaldehyde can't react in this way because it has no appropriately positioned hydrogen.

$$H_3C-\overset{\overset{\displaystyle O}{\|}}{C}-CH_3 \ + \ H_2\overset{\overset{\displaystyle H}{|}}{C}-\overset{\overset{\displaystyle O}{\|}}{C}-CH_3 \ \underset{\longleftarrow}{\overset{NaOH}{\longrightarrow}} \ H_3C-\overset{\overset{\displaystyle OH}{|}}{\underset{\underset{\displaystyle CH_3}{|}}{C}}-CH_2-\overset{\overset{\displaystyle O}{\|}}{C}-CH_3$$

Two acetone molecules—
each with six hydrogens
next to the carbonyl group

4-Hydroxy-4-methyl-2-pentanone

$$\text{(benzaldehyde structure)} \overset{\overset{\displaystyle O}{\|}}{C}-H \quad \overset{NaOH}{\longrightarrow} \quad \text{No aldol reaction}$$

Benzaldehyde—
has no hydrogens on carbon
next to the carbonyl group

Solved Problem 16.5 What aldol product would you obtain from this reaction:

$$CH_3CH_2CHO \overset{NaOH}{\longrightarrow} \ ?$$

Solution First, rewrite the reaction to emphasize the carbonyl group of one molecule and the C—H bond next to the carbonyl group of a second molecule:

$$CH_3CH_2-\overset{\overset{\displaystyle O}{\|}}{\underset{\underset{\displaystyle H}{|}}{C}} \ + \ \overset{\overset{\displaystyle H}{|}}{\underset{\underset{\displaystyle CH_3}{|}}{C}}H-\overset{\overset{\displaystyle O}{\|}}{\underset{\underset{\displaystyle H}{|}}{C}} \ \overset{NaOH}{\longrightarrow}$$

Next, draw the carbonyl group of the first molecule showing a C—O single bond and two part-bonds. Remove the appropriate hydrogen from the second molecule and connect it to the oxygen part-bond of the first molecule. Then connect the appropriate carbon from the second molecule to the carbon part-bond of the first. The structure that results is the final product.

Move this hydrogen to oxygen.

$$CH_3CH_2-\overset{\overset{\displaystyle O}{}}{\underset{\underset{\displaystyle H}{|}}{C}} \ + \ \overset{\overset{\displaystyle H}{}}{\underset{\underset{\displaystyle CH_3}{|}}{C}}H-\overset{\overset{\displaystyle O}{\|}}{\underset{\underset{\displaystyle H}{|}}{C}} \ \rightleftharpoons \ CH_3CH_2-\overset{\overset{\displaystyle O-H}{|}}{\underset{\underset{\displaystyle H}{|}}{C}}-\overset{\overset{\displaystyle }{}}{\underset{\underset{\displaystyle CH_3}{|}}{C}}H-\overset{\overset{\displaystyle O}{\|}}{\underset{\underset{\displaystyle H}{|}}{C}}$$

Connect these
carbons.

Practice Problems 16.11 Draw the aldol products from treatment of these aldehydes or ketones with base:

(a)

$$CH_3CH_2\overset{\overset{\displaystyle O}{\|}}{C}CH_2CH_3$$

(b)

$$\text{(phenyl ring)}-CH_2CHO$$

16.12 Which of the following compounds *can't* undergo aldol reactions?
(a) $CH_2=O$ (b) $(CH_3)_3CCHO$ (c) cyclopentanone

INTERLUDE: CHEMICAL WARFARE AMONG THE INSECTS

Life in the insect world is a jungle. Predators abound, just waiting to make a meal of any insect that happens along. Without missiles to protect themselves, many insects have evolved extraordinarily effective means of *chemical* protection. Take the humble millipede *Apheloria corrugata*, for example. When attacked by ants, the millipede protects itself by discharging a compound called *benzaldehyde cyanohydrin*.

Cyanohydrins [RCH(OH)C≡N] are compounds formed by addition of the toxic gas HCN (hydrogen cyanide) to ketones or aldehydes. Like the reaction of a ketone or aldehyde with an alcohol to yield a hemiacetal (Section 16.7), the reaction with HCN to yield a cyanohydrin is reversible. Thus, the benzaldehyde cyanohydrin secreted by the millipede can decompose to yield benzaldehyde and HCN. The millipede actually protects itself by discharging poisonous hydrogen cyanide at would-be attackers: a remarkably clever and very effective kind of chemical warfare.

Benzaldehyde cyanohydrin

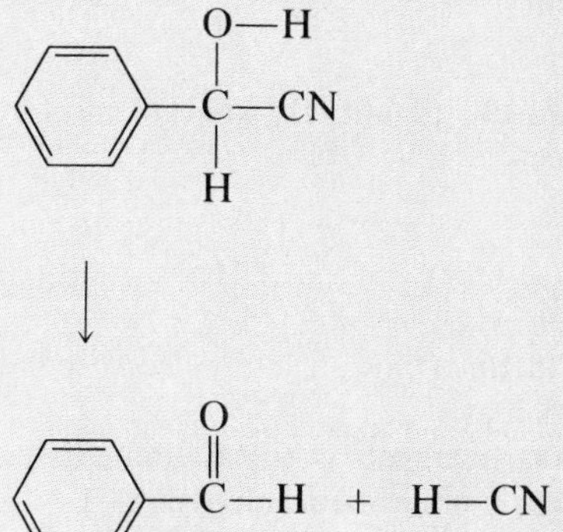

Benzaldehyde

Using a strategy similar to that of millipedes, apricots and peaches protect their seeds with a group of substances called *cyanogenic glycosides*. These compounds consist of benzaldehyde cyanohydrin bonded to glucose or another sugar. When eaten, the sugar unit is cleaved off by enzymes, and HCN is released. One of the cyanogenic glycosides, amygdalin, called Laetrile, was once well known because of its widely touted but never scientifically proven anticancer activity.

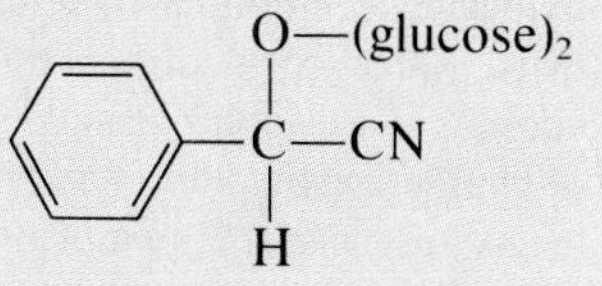

Amygdalin (Laetrile)

The potent chemical weapon shown on the opening page of this chapter is benzoquinone, the simplest member of a class of compounds that are cyclohexadienediones (cyclohexene rings with two double bonds and two carbonyl groups). When threatened, the bombardier beetle initiates the enzyme-catalyzed oxidation of a dihydroxybenzene by hydrogen peroxide. A cloud of irritating benzoquinone vapor shoots out of the beetle's defensive organ at up to 100°C and with such force that it sounds like a pistol shot.

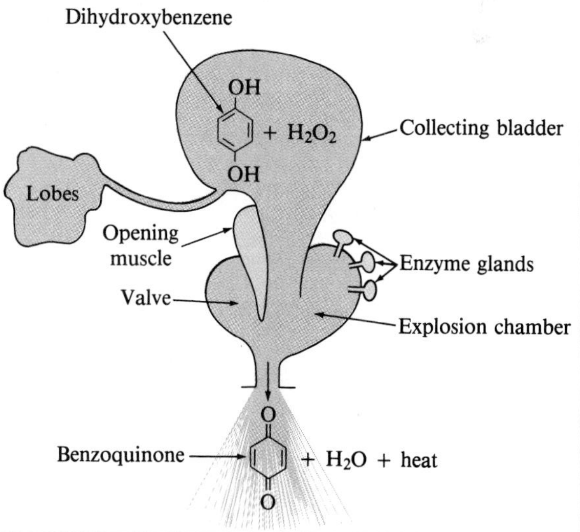

The defensive organ of the bombardier beetle.

SUMMARY

Carbonyl compounds contain the carbon–oxygen double bond, C=O. Different kinds of carbonyl compounds exist, depending on what other groups are bonded to the carbonyl carbon. **Aldehydes** (RCHO) and **ketones** (RCOR′) behave similarly because the atoms (—H and —C) bonded to the carbonyl-group carbon don't attract electrons strongly, and their bonds have similar polarity. They have no hydrogen bonding with each other, but hydrogen-bond with water or alcohols, making simple aldehydes and ketones water-soluble and excellent solvents. Many have characteristic odors and flavors; sugars contain aldehyde groups, and many hormones contain keto groups.

The major reaction of aldehydes and ketones is the **addition** of various reagents to the carbon–oxygen double bond. For example, aldehydes and ketones both add hydrogen to yield alcohol products. This **reduction** is carried out by treating the carbonyl compound first with NaBH₄

to add :H⁻, and then with aqueous acid to add H⁺. Aldehydes and ketones also add alcohols to yield **hemiacetals** or **acetals,** depending on the reaction conditions. A hemiacetal has an alcohol-like —OH group and an ether-like —OR group bonded to what was the carbonyl carbon; an acetal has two ether-like —OR groups bonded to what was the carbonyl carbon. Acetals react with aqueous acid (**hydrolysis**) to regenerate carbonyl compounds.

Aldehydes can be distinguished from ketones by their oxidation to carboxylic acids; the Tollens' and Benedict's reagents rely on metal reduction for this test.

Aldehydes and ketones that have a hydrogen atom at the position next to the carbonyl group can undergo the **aldol reaction.** Because this reaction results in formation of a carbon–carbon bond between two molecules of starting material, it is useful for building larger molecules from smaller pieces.

REVIEW PROBLEMS

Aldehydes and Ketones

16.13 What is the structural difference between an aldehyde and a ketone?

16.14 What family-name endings are used for aldehydes and ketones?

16.15 Use δ^+ and δ^- to show how the carbonyl group is polarized.

16.16 Draw structures for compounds that meet these descriptions:
(a) a ketone, C_5H_8O
(b) an aldehyde with eight carbons
(c) a keto aldehyde, $C_6H_{10}O_2$
(d) a hydroxy ketone, $C_5H_8O_2$

16.17 Which of these molecules contain carbonyl groups?

(a) CH₃CH₂C̈=O (with OH above)   (b) O=CCH₂CH₂CHCH₃ (with NH₂ and CH₃ below)

(c) CH₃CH₂—O—CH=CH₂   (d) CH₃CH₂C(OCH₃)₃
(e) CH₃CHCOOH (with CH₃ below)   (f) CH₃COCH₂CH₂OH

16.18 Which of these molecules contain carbonyl groups?
(a) CH₃CH₂CHO   (b) CH₃CH₂COCH₂CH₃
(c) (CH₃)₂C(OH)CH₂CH₂CH₃   (d) CH₃COOCH₃

(e) CH₃—⟨benzene ring⟩—CONH₂   (f) CH₃CHCH₂CHCH₃ (with OH and OCH₃ below)

16.19 Identify the kinds of carbonyl groups in these molecules:

(a) O=⟨cyclohexane ring⟩—CHO   (b) ⟨cyclohexane ring⟩—COCH₃ (with O above C)

16.20 Draw structures corresponding to these aldehyde names:
(a) heptanal   (b) 4,4-dimethylpentanal
(c) o-chlorobenzaldehyde
(d) 3-hydroxy-2,2,4-trimethylpentanal
(e) 4-ethyl-2-isopropylhexanal
(f) p-bromobenzaldehyde

16.21 Draw structures corresponding to these ketone names:
(a) 3-heptanone   (b) 2,4-dimethyl-3-pentanone
(c) m-nitroacetophenone   (d) cyclohexanone
(e) 1,1,1-trichloro-2-butanone
(f) 2-methyl-3-hexanone

16.22 Give systematic names for these aldehydes:

(a) CH₃CH₂CHCHO (with CH₃ below)   (b) CH₃CH₂CH₂CHCH₃ (with CHO above)

(c) $(CH_3)_3CCHO$

(d) [structure: benzene ring with CH₃, NO₂, and CHO substituents]

(e) $CH_3CH_2\overset{\underset{\displaystyle CH_3}{|}}{C}\overset{\displaystyle O}{\overset{||}{C}}H$ with CH_2CH_3 below

(f) $CH_3\overset{\underset{\displaystyle Br}{|}}{C}CH_2\overset{\displaystyle O}{\overset{||}{C}}H$ with Br below

16.23 Give systematic names for these ketones:

(a) $CH_3\overset{\displaystyle O}{\overset{||}{C}}CH_2CH_3$

(b) $CH_3\overset{\displaystyle O}{\overset{||}{C}}CH_2CH_2\overset{\underset{\displaystyle CH_3}{|}}{C}HCH_3$

(c) $(CH_3)_3C\overset{\displaystyle O}{\overset{||}{C}}C(CH_3)_3$

(d) [cyclopentanone ring with CH(CH₃)₂ substituent and =O]

16.24 The following names are incorrect. What is wrong with each?
(a) 1-butanone   (b) 4-butanone   (c) 3-butanone
(d) cyclohexanal   (e) 2-butanal

Reactions of Aldehydes and Ketones

16.25 What kind of compound is produced when an aldehyde reacts with an alcohol in a 1:1 ratio?

16.26 What kind of compound is produced when an aldehyde reacts with an alcohol in a 1:2 ratio in the presence of an acid catalyst?

16.27 Why does reduction of an aldehyde give a primary alcohol, whereas reduction of a ketone gives a secondary alcohol?

16.28 Give specific examples of these reactions:
(a) reduction of a ketone with $NaBH_4$
(b) oxidation of an aldehyde with Tollens' reagent
(c) formation of an acetal from an aldehyde and an alcohol

16.29 Which of the following compounds react with Tollens' reagent? Draw structures of the reaction products.
(a) cyclopentanone   (b) hexanal

(c) $CH_3CH_2CH_2\overset{\underset{\displaystyle}{|}}{C}HCH_2CH_3$ with CHO above

(d) [benzene ring]—CHO

(e) $CH_3CH_2\overset{\displaystyle O}{\overset{||}{C}}CH_3$

(f) $Cl_2CH\overset{\displaystyle O}{\overset{||}{C}}H$

16.30 Draw structures of the products obtained by reaction of the compounds in Problem 16.29 with $NaBH_4$ followed by acid treatment.

16.31 What is the difference between an acetal and a hemiacetal?

16.32 Assume that you are given two unlabeled bottles, one containing pentanal and the other containing 2-pentanone. What simple chemical test would allow you to distinguish between the contents of the two bottles?

16.33 Draw structures of the aldehydes that might be oxidized to yield these carboxylic acids:

(a) H_3C—[benzene ring]—COOH

(b) $CH_3CH_2\overset{\underset{\displaystyle COOH}{|}}{C}HCH_2\overset{\underset{\displaystyle CH_3}{|}}{C}HCH_3$

(c) $CH_3CH{=}CHCOOH$

(d) [benzene ring with OH]—COOH

16.34 Draw structures of the primary alcohols that might be oxidized to yield the carboxylic acids shown in Problem 16.33.

16.35 What ketones or aldehydes might be reduced to yield these alcohols?

(a) $CH_3CH_2CH_2\overset{\underset{\displaystyle}{|}}{C}HCH_3$ with CH₂OH above

(b) 2,2-dimethyl-1-hexanol

(c) HO—[cyclohexane ring]—OH

(d) $CH_3CH_2\overset{\underset{\displaystyle CH_3}{|}}{C}H\overset{\underset{\displaystyle}{|}}{C}HCH_3$ with OH above

16.36 Write the structures of the hemiacetals and acetals that result from the following reactions:
(a) 2-butanone + 1-propanol
(b) butanal + isopropyl alcohol
(c) acetone + ethanol
(d) hexanal + 2-butanol

16.37 Acetals are usually made by reaction of an aldehyde or ketone with two molecules of a monoalcohol. If an aldehyde or ketone reacts with *one* molecule of a dialcohol, however, a *cyclic* acetal results. Draw the structure of the cyclic acetal formed in the following reaction:

[cyclohexanone]=O + HO—CH_2CH_2—OH $\xrightarrow{\text{H}}$?

16.38 What products result from hydrolysis of these acetals?

(a) $CH_3CH_2CH_2\overset{\underset{\displaystyle O-CH_3}{|}}{C}H$ with O—CH₂CH₃ above

(b) [structure: central C with H₃C and O—CH₂ groups, cyclic]

(c)

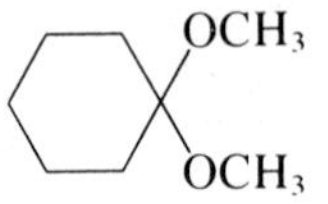

(d)

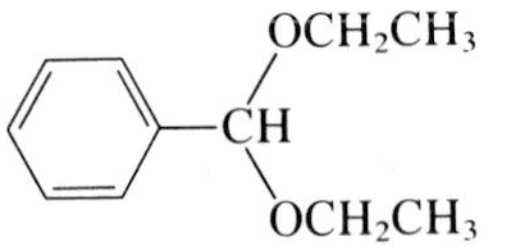

16.39 Cyclic hemiacetals sometimes form if an alcohol group in one part of a molecule adds to a carbonyl group elsewhere in the same molecule. What is the structure of the open-chain hydroxy aldehyde from which the following hemiacetal might form?

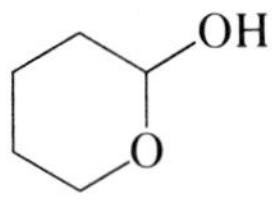

A cyclic hemiacetal

16.40 Like the cyclic hemiacetal in Problem 16.39, cyclic acetals are also known. What products would you expect on hydrolysis of the following cyclic acetal?

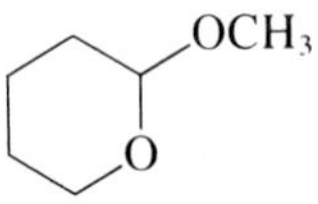

A cyclic acetal

16.41 Glucose exists largely in the cyclic hemiacetal form shown. Draw the structure of glucose in its open-chain hydroxy aldehyde form.

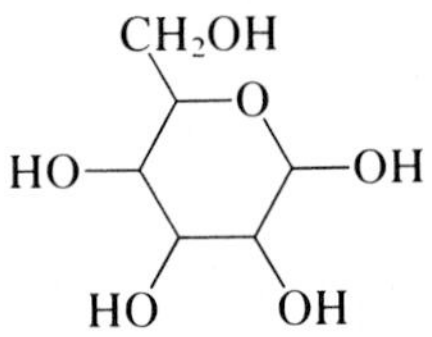

Glucose

16.42 In many respects, glucose acts as if it were an aldehyde, even though the structure shown in Problem 16.41 contains no —CHO group. For example, glucose reacts with Tollens' reagent to produce a silver mirror. Explain.

16.43 What products result from hydrolysis of the following cyclic acetal?

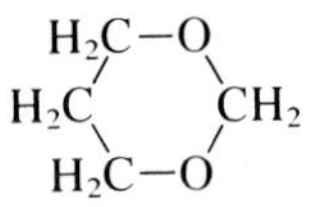

16.44 Aldosterone is a key steroid involved in controlling the sodium–potassium salt balance in the body. Identify the carbonyl groups in aldosterone.

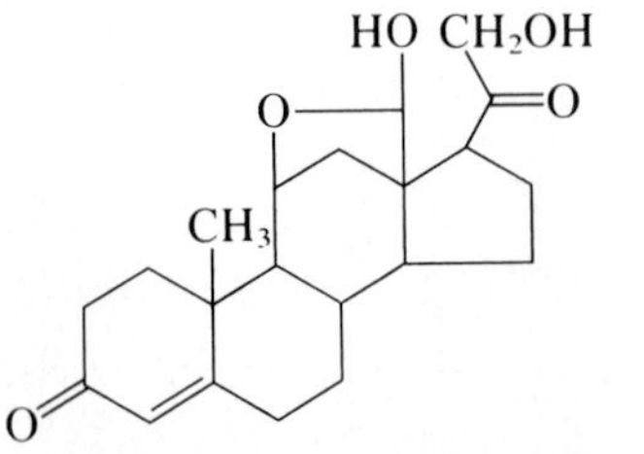

16.45 Aldosterone (Problem 16.44) also contains a cyclic hemiacetal linkage. Identify the linkage and tell whether it's derived from an aldehyde or a ketone.

16.46 3-Cyclohexenone is an example of a *difunctional* molecule, a compound that has two different functional groups. What products would you expect from treatment of 3-cyclohexenone with the reagents listed?

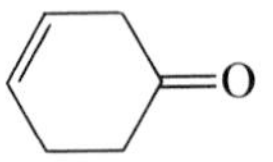

3-Cyclohexenone

(a) H_2 and a Pd catalyst   (b) $NaBH_4$, then H_3O^+
(c) Br_2

16.47 What structural requirement must be met in order for an aldehyde or ketone to undergo an aldol reaction?

16.48 Write the structures of the aldol products that result from base treatment of these ketones or aldehydes.
(a) CH_3CHCH_2CHO (b) cyclohexanone
        |
       CH_3

(c) 3-pentanone   (d)

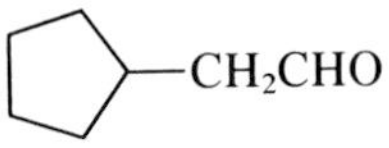

16.49 When 2-butanone is treated with base, a mixture of two different aldol products results. What are their structures?

16.50 The aldol reaction is reversible; that is, aldol products can sometimes break apart to yield aldehydes or ketones. What products would result if the following hydroxy aldehyde broke apart in a reverse aldol reaction?

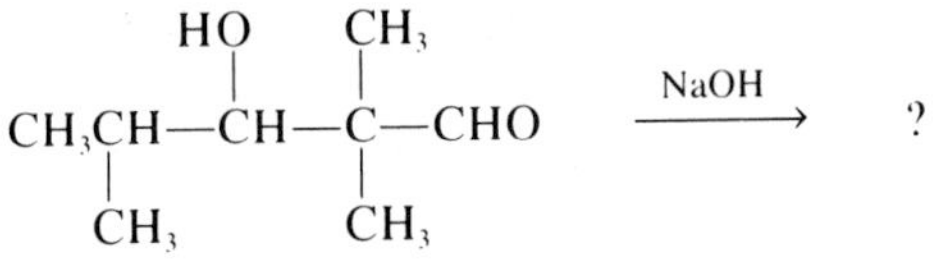

16.51 The following compound can be made by a mixed aldol reaction between two different carbonyl compounds. What two aldehyde or ketone starting materials would you use?

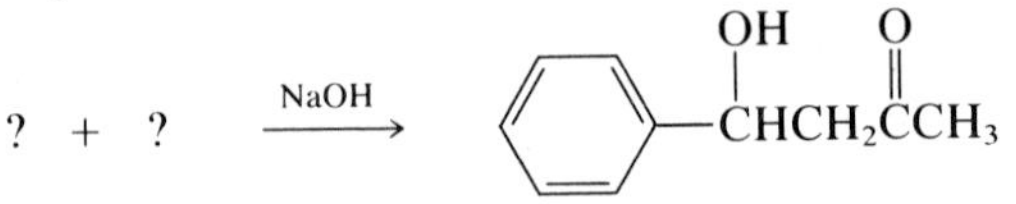

Applications

16.52 Discuss the differences among *acute, subacute,* and *chronic* toxicities of a drug. [App: Is It Poisonous?]

16.53 The LD_{50} value for the acute toxicity of botulism toxin, which is associated with improperly canned foods, is less than 0.01 mg/kg. This is less than 1 drop for an average adult. If 1 drop of botulism toxin is ingested, does this guarantee death? Discuss the significance of an LD_{50} rating. [App: Is It Poisonous?]

16.54 While many substances, such as ethanol, are only moderately toxic to adults, they can pose special dangers to embryos and fetuses. Why do you think this special susceptibility occurs? [App: Is It Poisonous?]

16.55   In a metabolism pathway in the body, oxaloacetic acid reacts with acetyl coenzyme A (acetyl SCoA) in the presence of an enzyme and water to form citric acid and free coenzyme A.

$$\underset{\text{Oxaloacetic acid}}{HOCCH_2-C-C-OH} \; + \; \underset{\text{Acetyl SCoA}}{CH_3CSCoA} \longrightarrow$$

$$\underset{\text{Citric acid}}{HOCCH_2-C-COH} \; + \; \underset{\text{Coenzyme A}}{CoASH}$$

How is this reaction similar to an aldol reaction? [App: Biological Aldol Reaction]

16.56   Both HCN and benzaldehyde have the aroma of almonds. If a liquid had a faint almond odor, how would you test it to determine whether the aldehyde or the acid was present? [Int: Chemical Warfare]

16.57   Old recipes for peach jam suggested grinding up the peach pits and adding them to the jam for more flavor and a crunchy texture. Why is this not a good idea? [Int: Chemical Warfare]

16.58   HCN is quite toxic: How do you suppose that a millepede can use this weapon without killing himself? [Int: Chemical Warfare]

Additional Questions and Problems

16.59   Name vanillin (Practice Problem 16.1) systematically.

16.60   Can the following alcohol be formed by the reduction of an aldehyde or ketone? Why or why not?

$$(CH_3)_3COH$$

16.61   Many flavorings and perfumes are partially based on fragrant aldehydes and ketones. Why do you think the portion of the odor due to the ketone is more stable than that due to the aldehyde?

16.62   One problem with burning some plastics is the release of formaldehyde. What are some of the physiological effects of exposure to formaldehyde?

16.63   Chloral hydrate, a potent sedative and component in "knockout" drops, is formed by reacting 2,2,2-trichloroethanal with water in a reaction analogous to hemiacetal formation. Draw the formula of chloral hydrate.

16.64   Name these compounds.

(a) $CH_3CH_2CCH(CH_3)_2$

(b) $CH_2{=}CHCHCH_2CH{=}CH_2$

(c)

(d) $(CH_3)_3CCHCCH_2CH_3$
   $\quad\quad\quad\;\; CH_3$

(e)

(f) $CH_3CH_2C{\equiv}CC(CH_2CH_3)_3$

16.65   Draw the structural formulas of these compounds.
(a) o-nitrobenzaldehyde
(b) 2,3-dichlorocyclopentanone
(c) methyl n-butyl ether   (d) 2,4-dimethyl-3-pentanol
(e) 2,3,3-triiodobutanal   (f) 1,1,3-tribromoacetone

16.66   Complete these equations.

(a) $CH_3CH{=}C(CH_3)_2 + H_2 \xrightarrow{Pd}$

(b) $CH_3CH_2\overset{\displaystyle CH_2OH}{\underset{\displaystyle CH_2CH_3}{C}}CH_2CH_3 \xrightarrow{[O]}$

(c) $CH_3CCH_2CH_2CH_3 \xrightarrow[H_3O^+]{NaBH_4}$

(d) $\text{C}_6\text{H}_5CH_2CH +$

   $HOCH_2CH_2CH_3 \longrightarrow$ (hemiacetal)

(e) $\text{C}_6\text{H}_5CH_2CH +$

   $2\; HOCH_2CH_2CH_3 \longrightarrow$ (acetal)

(f) $CH_3CH{=}\overset{\displaystyle CH_2CH_3}{C}CH_2CH_2CH_3 + HCl \longrightarrow$

(g) $CH_3{-}\text{C}_6\text{H}_4{-}CH_2CH_2OH \xrightarrow{H_2SO_4}$

16.67   How could you differentiate between 3-hexanol and hexanal using a simple chemical test?

CHAPTER

17

Carboxylic Acids and Their Derivatives

Every now and then, we should all take a moment to enjoy the aroma of some carboxylic acid esters, one of the kinds of compounds described in this chapter.

We said in the last chapter that carbonyl compounds can be classified into two groups, based on their structural and chemical similarities. In one group are aldehydes and ketones, and in the other group are carboxylic acids and their derivatives: esters, anhydrides, and amides.

$$R—\overset{\overset{\displaystyle O}{\|}}{C}—OH \qquad R—\overset{\overset{\displaystyle O}{\|}}{C}—O—R' \qquad R—\overset{\overset{\displaystyle O}{\|}}{C}—O—\overset{\overset{\displaystyle O}{\|}}{C}—R \qquad R—\overset{\overset{\displaystyle O}{\|}}{C}—NH_2$$

Carboxylic acid   Ester   Acid anhydride   Amide

We'll look at the chemistry of this second group of carbonyl compounds in this chapter and answer the following questions:

1. **What are the general properties of carboxylic acids and their derivatives?** The goal: Be able to describe and compare such properties as hydrogen bonding, water solubility, boiling point, and acidity or basicity.

2. **How are carboxylic acids, anhydrides, esters, and amides named?** The goal: Be able to name the simple members of these families and write their structures, given the names.

3. **What are some of the significant applications and occurrences of carboxylic acids, anhydrides, esters, and amides?** The goal: Be able to identify major applications and occurrences of each family of compounds and describe some important members of each family.

4. **How are esters and amides synthesized from carboxylic acids and converted back to carboxylic acids?** The goal: Be able to describe and predict the products of the ester- and amide- forming reactions of carboxylic acids and the hydrolysis of esters and amides.

5. **What is the Claisen condensation reaction, and why is it important in biochemistry?** The goal: Be able to describe how the Claisen condensation reaction occurs, predict products of the reaction, and explain why it is important in biochemistry.

6. **What are phosphate esters and anhydrides, and why are they important?** The goal: Be able to recognize and write the structures of phosphate esters and anhydrides and explain why they are important in biochemistry.

17.1 PROPERTIES OF CARBOXYLIC ACIDS AND THEIR DERIVATIES

Carboxylic acid A compound that has a carbonyl group bonded to one carbon substituent and one —OH group, RCOOH.

Acid anhydride A compound that has a carbonyl group bonded to a carbon substituent and an —OCOR group, RCO₂COR.

Ester A compound that has a carbonyl group bonded to one carbon atom and one —OR group, RCOOR'.

Amide A compound that has a carbonyl group bonded to one carbon atom and one nitrogen atom group, RCONR'₂, where R' may be an H atom.

Unlike the aldehydes and ketones described in the last chapter, **carboxylic acids, acid anhydrides, esters,** and **amides** all have their carbonyl groups bonded to an atom (O or N) that strongly attracts electrons. Thus, the carbonyl groups are more strongly polarized in these families of compounds than in ketones and aldehydes.

Carboxylic acid
(RCOOH or RCO₂H)

Acid anhydride
(RCO₂COR)

Ester
(RCOOR' or RCO₂R')

Amide
(RCONH₂)

The structural similarity of these families also leads to similarities in their chemistry. All undergo **carbonyl-group substitution reactions,** in which a group we can represent as —X replaces (substitutes for) the —OH, —OCOR, —OR, or —NH₂ group of the starting material.

A carbonyl-group substitution reaction

$$R-\overset{\overset{O}{\|}}{C}-OH \ + \ H-X \ \longrightarrow \ R-\overset{\overset{O}{\|}}{C}-X \ + \ H-OH$$

$$(-OR', -NH_2, -O\overset{\overset{O}{\|}}{C}R) \qquad\qquad (H-OR', H-NH_2, H-O\overset{\overset{O}{\|}}{C}R)$$

Carbonyl-group substitution reaction A reaction in which a new group X replaces (substitutes for) a group Y attached to a carbonyl-group carbon.

We'll give numerous examples of these carbonyl-group substitution reactions in the sections that follow. Also, we'll show that amides and esters can easily be either made from acids or converted back to acids. Acid anhydrides are the most reactive of these carboxylic acid derivatives, and you'll see that they cannot be used in aqueous solution because of their substitution reaction with water (Section 17.11).

Since carboxylic acids and their derivatives all contain polar functional groups, all are higher boiling than comparable alkanes. Like alcohols, carboxylic acids can form hydrogen bonds with each other. Molecules pair off to form **dimers** (Figure 17.1), and even formic acid (HCOOH), the lightest and simplest carboxylic acid, is a liquid at room temperature with a boiling point of 101°C.

Dimer A unit formed by the joining of two identical molecules.

Figure 17.1
Acetic acid hydrogen-bonded dimer. The ability to form hydrogen bonds makes simple carboxylic acids high boiling and soluble in water.

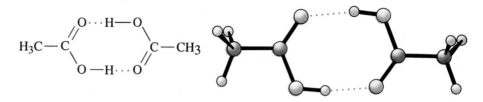

Acids with saturated straight-chain R groups containing up to nine carbon atoms are volatile liquids with strong, sharp odors, and those with up to four carbons are water-soluble. Higher saturated acids are waxy, odorless solids, and water solubility falls off as the size of the alkane-like portion increases relative to the size of the water-soluble —COOH portion.

Carboxyl group   The —COOH functional group.

When the —OH of the **carboxyl group** (—COOH) is converted to the —OR of an ester, the ability of the molecules to hydrogen-bond with each other is lost. Simple esters are therefore lower boiling than the acids from which they are derived and are also lower boiling than comparable alcohols.

$$\underset{\text{Acetic acid, bp 118°C}}{CH_3\overset{\displaystyle O}{\overset{\|}{C}}OH} \qquad \underset{\text{Methyl ester, bp 57°C}}{CH_3\overset{\displaystyle O}{\overset{\|}{C}}OCH_3} \qquad \underset{\text{Ethyl ester, bp 77°C}}{CH_3\overset{\displaystyle O}{\overset{\|}{C}}OCH_2CH_3}$$

The simple esters are colorless, volatile liquids with pleasant odors, and many of them are responsible for the natural fragrance of flowers and ripe fruits. The lower esters have some solubility in water. Esters are neither acids nor bases in aqueous solution.

Replacement of the —OH group of an acid by an —NH$_2$ group produces an *unsubstituted amide (RCONH$_2$)*, that is, an amide with two hydrogen atoms on the nitrogen atom. Such amides can form three strong hydrogen bonds to other amide molecules and are thus higher melting and higher boiling than the acids from which they are derived. For example,

$$\underset{\text{Propanoic acid, mp }-21°C,\text{ bp }141°C}{CH_3CH_2\overset{\displaystyle O}{\overset{\|}{C}}OH} \qquad\qquad \underset{\text{Propanamide, mp 81°C, bp 213°C}}{CH_3CH_2\overset{\displaystyle O}{\overset{\|}{C}}NH_2}$$

In fact, except for the simplest one (formamide, HCONH$_2$), all low-molecular-weight unsubstituted amides are relatively low-melting solids that are soluble in both water and organic solvents.

Note the distinction between amines, described in Chapter 15, and amides. The nitrogen atom is bonded to a carbonyl-group carbon atom in an amide, but not in an amine.

An amide
(RCONH$_2$)   An amine
(RNH$_2$)

As a result, amides are not basic like amines, mainly because the electron-withdrawing ability of the carbonyl group causes the unshared pair of electrons on nitrogen to be held tightly and prevents it from bonding to hydrogen.

The common carboxylic acids share the concentration-dependent corrosive properties of all acids but are not generally considered hazardous to human health. The vapors of volatile esters can be irritating and have a narcotic effect, while individual esters vary widely in their acute toxic effects. Amides can also be irritating to the skin, but the simple amides are not significant health hazards. Because anhydrides form acids on contact with water or water vapor, their corrosive and toxic properties are those of their parent acids.

Properties of Carboxylic Acid Derivatives

- All undergo carbonyl-group substitution reactions.
- Acid anhydrides react with water to form acids.
- Esters and amides are made from acids and easily converted back to acids.
- Hydrogen bonding is strong in pure acids and unsubstituted or monosubstituted amides; none in esters.
- Simpler acids and esters are liquids; all unsubstituted amides except formamide are solids.
- Acids give acidic aqueous solutions; esters and amides do not change pH.
- Volatile acids have strong odors; volatile esters have pleasant, fruity odors.

Practice Problems

17.1 Which of the following compounds has the highest boiling point and which has the lowest boiling point?
(a) CH_3OCH_3 (b) CH_3COOH (c) CH_3CH_2OH

17.2 Which of the following compounds has the highest melting and boiling points? Which has the lowest boiling point? Why?

(a) $CH_3\overset{\overset{\text{O}}{\|}}{C}NH_2$ (b) $CH_3\overset{\overset{\text{O}}{\|}}{C}NHCH_2CH_3$ (c) $CH_3\overset{\overset{\text{O}}{\|}}{C}N(CH_2CH_3)_2$

17.3 In each of the following pairs of compounds, which would you expect to be more soluble in water?
(a) $CH_3CH_2CONH_2$ or $CH_3COOCH_2CH_3$
(b) $C_8H_{17}COOH$ or $CH_3CH_2CH_2COOH$
(c) $CH_3\underset{\underset{\text{CH}_3}{|}}{C}HCOOH$ or $CH_3CH_2COO\underset{\underset{\text{CH}_3}{|}}{C}HCH_3$

17.2 NAMING CARBOXYLIC ACIDS AND THEIR DERIVATIVES

Carboxylic Acids   Carboxylic acids (RCOOH) are named systematically by replacing the final *-e* of the corresponding alkane name with *-oic acid*. The three-carbon acid is propanoic acid; the straight-chain four- carbon acid is butanoic

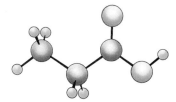

Propanoic acid

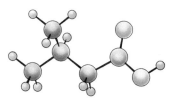

3-Methylbutanoic acid

acid; and so on. If alkyl substituents or other functional groups are present, the chain is numbered beginning at the —COOH end, as in 3-methylbutanoic acid.

$$CH_3CH_2 \overset{\overset{O}{\|}}{-C} -OH \qquad \underset{4}{CH_3}\underset{3}{CHCH_2} \overset{\overset{O}{\|}}{-\underset{1}{C}} -OH$$

Propanoic acid   3-Methylbutanoic acid

Because many simple carboxylic acids were among the first organic compounds to be isolated and purified, a number of them also have common names. For example, the one-carbon compound (HCOOH) is always called *formic acid* rather than methanoic acid; the two-carbon compound (CH₃COOH) is always called *acetic acid* rather than ethanoic acid; and the three-carbon compound is sometimes called *propionic acid* rather than propanoic acid. The names and physical properties of some common carboxylic acids are given in Table 17.1.

Dicarboxylic acids, which contain two —COOH groups, are named systematically by adding the ending *-dioic* to the alkane name (the *-e* is retained), but many are usually referred to by their common names. Oxalic acid, for example, is found in plants of the genus *Oxalis*, which includes rhubarb and

Table 17.1   Physical Properties of Some Simple Carboxylic Acids

Structure	Common Name	IUPAC Name	Melting Point (°C)	Boiling Point (°C)	Water Solubility (g/100 mL)
HCOOH	Formic	Methanoic	8	101	Miscible
CH₃COOH	Acetic	Ethanoic	17	118	Miscible
CH₃CH₂COOH	Propionic	Propanoic	−22	141	Miscible
CH₃CH₂CH₂COOH	Butyric	Butanoic	−4	163	Miscible
CH₃CH₂CH₂CH₂COOH	Valeric	Pentanoic	−34	185	4
CH₃(CH₂)₁₆COOH	Stearic	Octadecanoic	70	383	Insoluble
HOOCCOOH	Oxalic	Ethanedioic	190	Decomposes	10
HOOCCH₂COOH	Malonic	Propanedioic	135	Decomposes	154
HOOCCH₂CH₂COOH	Succinic	Butanedioic	188	Decomposes	8
HOOCCH₂CH₂CH₂COOH	Glutaric	Pentanedioic	98	Decomposes	64
H₂C=CHCOOH	Acrylic	Propenoic	13	141	Miscible
CH₃CH=CHCOOH	Crotonic	2-Butenoic	72	185	55
⬡—COOH	Benzoic	Benzoic	122	249	0.3
⬡—COOH (OH)	Salicylic	*o*-Hydroxy-benzoic	159	Decomposes	0.2

Figure 17.2
Oxalic acid. The rhubarb leaves contain oxalic acid which, when pure, is a white crystalline solid. Oxalate ion is added to blood samples to prevent clotting because it forms a precipitate with Ca^{2+} needed for the clotting process.

spinach, and is rarely called by any other name (Figure 17.2). Malonic acid is an intermediate in the synthesis of pharmaceuticals.

$$
\underset{\substack{\text{Oxalic acid}\\\text{(ethanedioic acid)}}}{\text{HO}\overset{\overset{\displaystyle O}{\|}}{C}-\overset{\overset{\displaystyle O}{\|}}{C}\text{OH}} \qquad\qquad \underset{\substack{\text{Malonic acid}\\\text{(propanedioic acid)}}}{\text{HO}\overset{\overset{\displaystyle O}{\|}}{C}-CH_2-\overset{\overset{\displaystyle O}{\|}}{C}\text{OH}}
$$

Many other varieties of compounds containing —COOH groups are also common. There are unsaturated acids, which are named systematically with the ending -*enoic*, and there are acids containing other functional groups such as hydroxy acids and keto acids, several of which are important biomolecules.

$$
\underset{\substack{\text{Acrylic acid}\\\text{(propenoic acid)}}}{H_2C{=}CH\overset{\overset{\displaystyle O}{\|}}{C}OH} \qquad \underset{\substack{\text{Lactic acid}\\\text{(2-hydroxypropanoic acid)}}}{\underset{\underset{\displaystyle OH}{|}}{CH_3CH}\overset{\overset{\displaystyle O}{\|}}{C}OH} \qquad \underset{\substack{\text{Pyruvic acid}\\\text{(2-oxopropanoic acid)}}}{CH_3-\overset{\overset{\displaystyle O}{\|}}{C}-\overset{\overset{\displaystyle O}{\|}}{C}OH}
$$

The carbon atoms *next to* the —COOH group are sometimes identified by Greek letters, beginning with α. In this system, pyruvic acid would be referred to as an α-keto acid.

$$
\underset{\substack{\text{Pyruvic acid}\\\text{(an }\alpha\text{-keto acid)}}}{\underset{\substack{\beta\quad\ \alpha}}{CH_3-\overset{\overset{\displaystyle O}{\|}}{C}-\overset{\overset{\displaystyle O}{\|}}{C}OH}} \qquad \underset{\substack{\text{Alanine}\\\text{(an }\alpha\text{-amino acid)}}}{\underset{\substack{\beta\qquad\ \alpha}}{\underset{\underset{\displaystyle NH_2}{|}}{CH_3-CH}-\overset{\overset{\displaystyle O}{\|}}{C}OH}}
$$

In alanine, as in all common amino acids, the —NH₂ group is on the alpha carbon atom

---

Solved Problem 17.1   What is the IUPAC name for this acid?

$$
\underset{\substack{\ \ |\ \ \ \ \ |}}{\underset{\text{HO}\ \ CH_3}{CH_3CHCHCOOH}}
$$

Solution   First, identify the longest chain containing the —COOH group and number it starting with the carboxyl-group carbon atom:

$$
\overset{4\quad 3\quad 2\quad 1}{\underset{\substack{\ \ |\ \ \ \ \ |}}{\underset{\text{HO}\ \ CH_3}{CH_3CHCHCOOH}}}
$$

The parent compound is the four-carbon acid, butanoic acid. It has a methyl group on position 2 and a hydroxy group on position 3, giving the name 3-hydroxy-2-methylbutanoic acid.

Acid Anhydrides The word "anhydride" means "without water," and acid anhydrides have the structure that results from loss of a molecule of water by two acid molecules (whether or not the anhydride is made that way). Anhydride names are derived by replacing *acid* in the name of the original acid with *anhydride*.

$$CH_3\overset{O}{\underset{\|}{C}}{-}OH \ + \ HO{-}\overset{O}{\underset{\|}{C}}CH_3 \longrightarrow CH_3{-}\overset{O}{\underset{\|}{C}}{-}O{-}\overset{O}{\underset{\|}{C}}{-}CH_3 \ + \ H_2O$$

2 Acetic acid   Acetic anhydride

Esters Esters (RCOOR′) are named by first specifying the alkyl group attached to the ether-like oxygen (—O—R′) and then identifying the related carboxylic acid. The family-name ending *-ic acid* is replaced by *-ate*.

This part is from *acetic* acid.   This part is an *ethyl* group.

$$H_3C{-}\overset{O}{\underset{\|}{C}}{-}O{-}CH_2CH_3$$

This part is from *benzoic* acid.   This part is a *methyl* group.

$$\langle\text{C}_6\text{H}_5\rangle{-}\overset{O}{\underset{\|}{C}}{-}O{-}CH_3$$

Ethyl acetate   Methyl benzoate

Ester names always consist of two words—the name of the alkyl group in —COOR followed by the name of the acid with an *-ate* ending. Both common and systematic names are derived in this manner. An ester of the straight-chain, four-carbon carboxylic acid, for example, can be named as a butanoate or a butyrate:

$$CH_3CH_2CH_2\overset{O}{\underset{\|}{C}}OCH_3$$

An ester used as a food flavoring to give the taste of apples

Methyl butyrate
(Methyl butanoate)

The names and properties of some common esters are given in Table 17.2.

Solved Problem 17.2 What is the structure of butyl acetate?

Solution The two-word name consisting of an alkyl group name followed by an acid name with an *-ate* ending shows that the compound is an ester. The name "acetate" shows that the RCOO part of the molecule is from acetic acid (CH$_3$COOH). A butyl group is attached to the ether-like oxygen atom in the carboxyl group. Therefore the structure is

From acetic acid   A butyl group

$$CH_3\overset{O}{\underset{\|}{C}}OCH_2CH_2CH_2CH_3$$

Table 17.2   Physical Properties of Some Simple Esters[a]

Structure	Name	Boiling Point (°C)	Water Solubility (20°C) (g/100 mL)
HCOOCH₃	Methyl formate	32	30
CH₃COOCH₂CH₃	Ethyl acetate	77	9
CH₃COOCH₂(CH₂)₃CH₃	Pentyl acetate	149	Slightly soluble
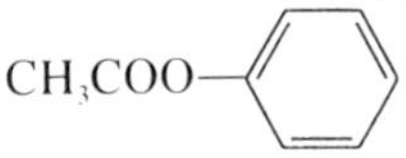 CH₃COO—⬡	Phenyl acetate	196	Insoluble
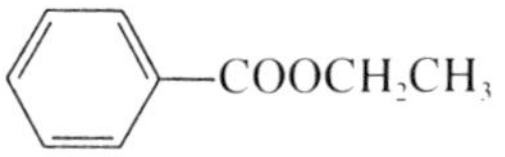 ⬡—COOCH₂CH₃	Ethyl benzoate	212	Slightly soluble
CH₃(CH₂)₁₆COOCH₃	Methyl stearate	mp 39	Insoluble
CH₃OOCCH₂CH₂COOCH₃	Dimethyl succinate	mp 19.5	Slightly soluble

[a] All but methyl stearate and dimethyl succinate are liquids.

Solved Problem 17.3   What is the name of the compound shown?

$$\overset{\displaystyle O}{\overset{\|}{HCOCH_2CH_2CH_3}}$$

Solution   The compound has the general formula RCOOR′, so it is an ester in which R = H, showing that it is derived from formic acid, HCOOH. The R′ group has three carbon atoms and is therefore a propyl group. The compound is propyl formate.

Amides   Amides (RCONH₂) with an unsubstituted —NH₂ group are named by replacing the *-oic acid* of the corresponding carboxylic acid name with *-amide*. For example, the two-carbon amide derived from acetic acid is called *acetamide*. If the nitrogen atom of the amide has alkyl substituents on it, the compound is named by first specifying the alkyl group and then identifying the amide name. The alkyl substituents are preceded by the letter *N* to identify them as being attached directly to nitrogen. Some common amides are illustrated in Table 17.3.

Table 17.3   Physical Properties of Some Simple Amides[a]

Structure	Name	Melting Point (°C)	Water Solubility (g/100 mL)
HCONH₂	Formamide	3	Miscible
CH₃CONH₂	Acetamide	81	70
CH₃CONHCH₃	N-Methylacetamide	31	Soluble
CH₃CON(CH₃)₂	N,N-Dimethylacetamide	−20	Miscible
CH₃CH₂CONH₂	Propanamide	81	Soluble
CH₃CH₂CH₂CONH₂	Butanamide	116	16
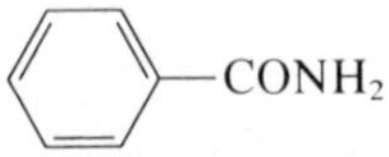 ⬡—CONH₂	Benzamide	127	1.3

[a] All but formamide and N,N-dimethylacetamide are solids.

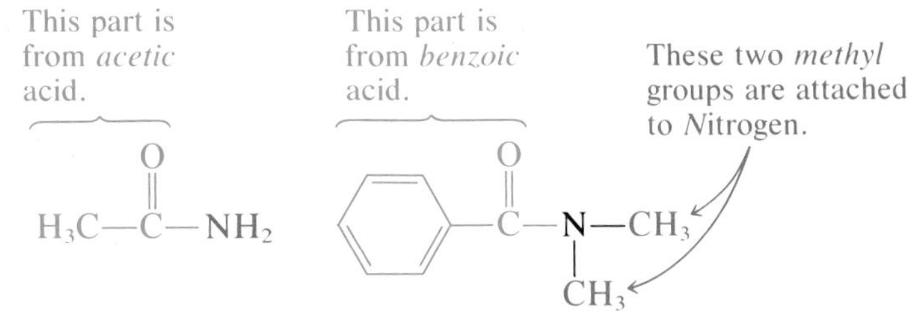

This part is from *acetic* acid.

This part is from *benzoic* acid.

These two *methyl* groups are attached to *Nitrogen*.

$$\underset{H_3C-\overset{\displaystyle O}{\overset{\displaystyle \|}{C}}-NH_2}{}$$

Acetamide

N, *N*-Dimethylbenzamide

Practice Problems   **17.4**   What are the names of these compounds?

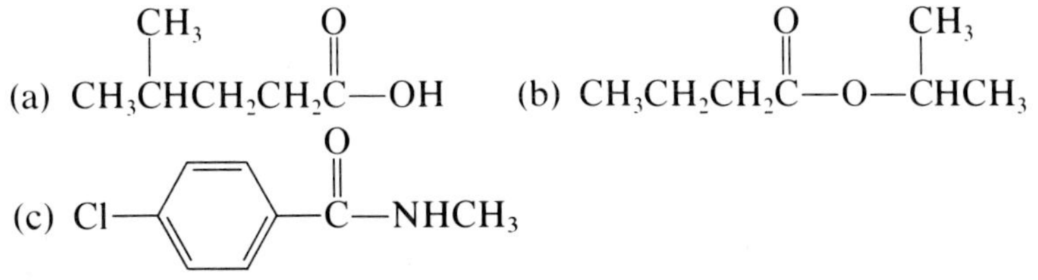

(a) $CH_3\overset{\overset{\displaystyle CH_3}{|}}{CH}CH_2CH_2\overset{\overset{\displaystyle O}{\|}}{C}-OH$

(b) $CH_3CH_2CH_2\overset{\overset{\displaystyle O}{\|}}{C}-O-\overset{\overset{\displaystyle CH_3}{|}}{C}HCH_3$

(c) $Cl-$⟨benzene ring⟩$-\overset{\overset{\displaystyle O}{\|}}{C}-NHCH_3$

17.5   Draw structures corresponding to these names:
(a) 3-methylhexanoic acid   (b) 4-methylpentanamide
(c) propyl benzoate   (d) *o*-nitrobenzoic acid
(e) *N*-methylbutanamide   (f) ethyl propanoate
(g) propanoic anhydride

17.3 SOME COMMON CARBOXYLIC ACIDS

Carboxylic acids occur widely throughout the plant and animal kingdoms. For example, butanoic acid (from the Latin, *butyrum*, "butter") is responsible for the odor of rancid butter, and cinnamic acid is partially responsible for the odor of cinnamon. Long-chain carboxylic acids such as stearic acid are components of all animal fats and vegetable oils, and many organic acids are formed in the body during the breakdown of foods.

Some naturally occurring carboxylic acids

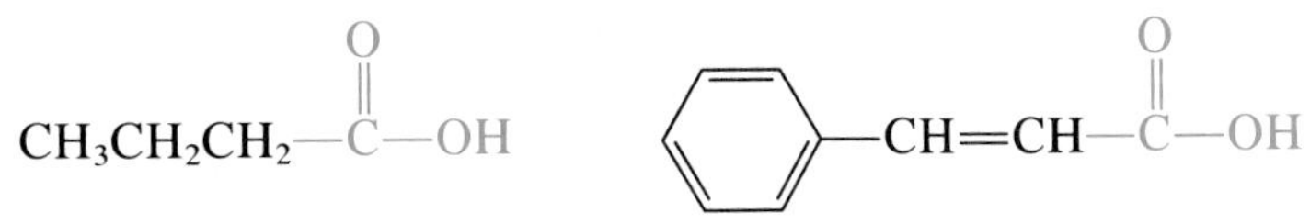

$$CH_3CH_2CH_2-\overset{\overset{\displaystyle O}{\|}}{C}-OH \qquad \langle\text{benzene ring}\rangle-CH=CH-\overset{\overset{\displaystyle O}{\|}}{C}-OH$$

Butanoic acid (from butter)

Cinnamic acid (from oil of cinnamon)

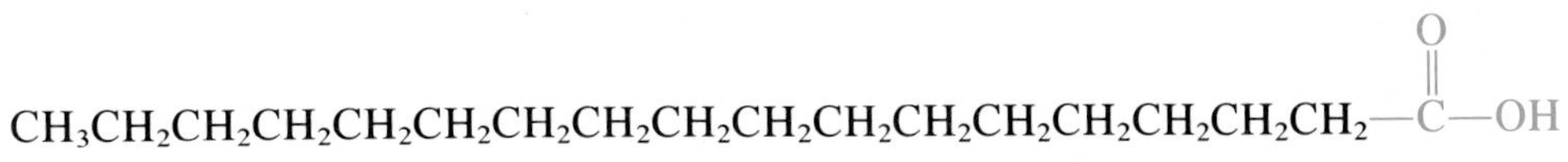

$$CH_3CH_2CH_2CH_2CH_2CH_2CH_2CH_2CH_2CH_2CH_2CH_2CH_2CH_2CH_2CH_2CH_2-\overset{\overset{\displaystyle O}{\|}}{C}-OH$$

Stearic acid ($C_{18}H_{36}O_2$, from animal fat and vegetable oil)

$$H_3C-\overset{\overset{\displaystyle O}{\|}}{C}-OH$$

$$HO\overset{\overset{\displaystyle O}{\|}}{C}CH_2\overset{\overset{\displaystyle OH}{|}}{\underset{\underset{\displaystyle COOH}{|}}{C}}CH_2\overset{\overset{\displaystyle O}{\|}}{C}OH$$

Acetic acid
(from vinegar)

Citric acid
(from citrus fruits)

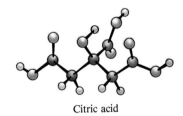

Acetic acid

Acetic Acid (CH₃COOH)   Everyone recognizes the sour taste of acetic acid, the most common carboxylic acid, because vinegar is a solution of 4–8% acetic acid in water (with various other flavoring agents). When fermentation of grapes, apples, and other fruits proceeds in the presence of ample oxygen, oxidation proceeds beyond the formation of ethanol to the formation of acetic acid (named from the Latin *acetum*, meaning "vinegar"). The acid in various concentrations in water solution is also a common laboratory reagent.

In concentrations over 50%, acetic acid is corrosive and can damage the skin, eyes, nose, and mouth. There is no pain when the concentrated acid is spilled on unbroken skin, but painful blisters form 30 min or more later. Pure acetic acid is known as *glacial* acetic acid because with just a slight amount of cooling the liquid forms icy-looking crystals at its melting point of 17°C.

Acetic acid is a reactant in many industrial processes and is sometimes used as a solvent, especially for other carboxylic acids. As a food additive, it is used to adjust pH. Earlier we noted that carboxylic acids often undergo carbonyl-group substitution reactions in which the —OH group is replaced by some other group, —X. The portion of acetic acid that bonds to X in any compound CH₃COX is known as an **acetyl group.**

Acetyl group   The CH₃C—
$$\overset{\overset{\displaystyle O}{\|}}{}$$
group.

$$CH_3\overset{\overset{\displaystyle O}{\|}}{C}-\xi$$ An acetyl group

The acetyl group and its transfer from one molecule to another play a crucial role in biochemical reactions.

Citric Acid (A Tricarboxylic Hydroxy Acid)   Citric acid is produced by almost all plants and animals during metabolism, and its normal concentration in human blood is about 2 mg/100 mL. Citrus fruits owe their tartness to citric acid; for example, lemons contain 4–8% and oranges about 1% citric acid. Pure citric acid is a white, crystalline solid (mp 153°C) that is very soluble in water. Citric acid and its salts are extensively used in pharmaceuticals, foods, and cosmetics. It serves to buffer pH in shampoos and hair-setting lotions; it adds tartness to candies and soft drinks; and it is the acid that reacts with bicarbonate ion to produce the fizz in AlkaSeltzer.

Citric acid

17.4   ACIDITY OF CARBOXYLIC ACIDS

Carboxylate anion   The anion that results from dissociation of a carboxylic acid, RCOO⁻.

Carboxylic acids are generally weak acids and establish equilibria in aqueous solution with **carboxylate anions,** RCOO⁻. Many carboxylic acids have about the same acid strength as acetic acid, as shown by the acid ionization constants in Table 17.4. There are some exceptions, though. Trichloracetic acid, used in preparing microscope slides and in precipitating protein in the analysis of body   Table

Table 17.4   Carboxylic Acid Ionization Constants[a]

Structure	Name	K_a
CH_3COOH	Acetic acid	1.8×10^{-5}
$ClCH_2COOH$	Chloroacetic acid	1.4×10^{-3}
$HCOOH$	Formic acid	1.8×10^{-4}
CH_3CH_2COOH	Propanoic acid	1.3×10^{-5}
$HOOCCOOH$	Oxalic acid	5.4×10^{-2}
		5.2×10^{-5}
$HOOCCH_2CH_2COOH$	Succinic acid	6.4×10^{-5}
		2.3×10^{-6}
$H_2C{=}CHCOOH$	Acrylic acid	5.6×10^{-5}
—COOH	Benzoic acid	6.5×10^{-5}

[a]The K_a is the equilibrium constant for the ionization of an acid; the smaller its value, the weaker the acid (Section 11.9).

$$RCOOH + H_2O \rightleftharpoons RCOO^- + H_3O^+ \qquad K_a = \frac{[RCOO^-][H_3O^+]}{[RCOOH]}$$

fluids, is a very strong acid that must be handled with the same respect as sulfuric acid.

Carboxylate ions are named by replacing the -*ic* ending in the carboxylic acid name with -*ate*. (The same names and endings are used in naming esters).

Acetic acid + H_2O ⇌ Acetate ion + H_3O^+

Propanoic acid + H_2O ⇌ Propanoate ion + H_3O^+

Pyruvic acid + H_2O ⇌ Pyruvate ion + H_3O^+

Carboxylic acid salt An ionic compound containing a carboxylate ion.

Carboxylic acids react completely with strong bases such as sodium hydroxide to give water and a **carboxylic acid salt.** Like any salt, a carboxylic acid salt is named with cation and anion names. For example,

Acetic acid (a weak acid) + Sodium hydroxide ⟶ Sodium acetate + H_2O

Acetate ion, as the anion of a weak acid, is itself a base that can accept a proton from HCl or some other strong acid (Section 11.18). Thus, carboxylate anions can be converted back into free carboxylic acids by reaction with acids.

$$H_3C-\overset{\overset{\displaystyle O}{\|}}{C}-O^-\ Na^+(aq)\ +\ HCl(aq)\ \longrightarrow\ H_3C-\overset{\overset{\displaystyle O}{\|}}{C}-O-H\ +\ Na^+\ Cl^-(aq)$$

Sodium acetate   Acetic acid

The interconversion of a carboxylic acid and its carboxylate ion is easily reversible depending on the conditions. Which form is present depends on the pH of the medium and can be controlled by adjusting the pH. At low pH (acidic solution), carboxylic acid is present; at high pH (basic solution), carboxylate ion is present:

Acidic conditions: $RCOO^-\ +\ H_3O^+\ \longrightarrow\ RCOOH\ +\ H_2O$
(pH < 7)   Carboxylate ion   Acid   Carboxylic acid

Basic conditions: $RCOOH\ +\ OH^-\ \longrightarrow\ RCOO^-\ +\ H_2O$
(pH > 7)   Carboxylic acid   Base   Carboxylate ion

AN APPLICATION: ACID SALTS AS FOOD ADDITIVES

Run your eye over the ingredients listed on the labels of soft drinks, cookies or cakes, dried sauce mixes, or preserved meats. Chances are you'll find the name of one or more acid salts.

Scanning the label on a package of strawberry jam-filled cookies, for example, turns up *sodium benzoate, sodium propionate, potassium sorbate,* and *sodium citrate.* What's the purpose of all these *food additives?* The first three are preservatives. Sodium benzoate prevents the growth of microorganisms, especially in acidic foods. It's present in many soft drinks, is one of the most common food additives, and has been used for over 70 years. Sodium propionate, also very common, prevents the growth of mold in baked goods. Potassium sorbate, the salt of an unsaturated acid that occurs naturally in many plants (sorbic acid, $CH_3CH{=}CHCH{=}CHCOOH$), is a good mold and fungus inhibitor. The fourth ingredient is the trisodium salt of citric acid. In the

cookies, it's combined with citric acid to buffer the acidity of the strawberry jam.

A package of dehydrated cream sauce with bacon bits and noodles includes a group of less familiar acid salts. *Disodium guanylate, disodium inosinate,* and *monosodium glutamate* are salts of acids that occur naturally in meats. As food additives, these salts serve as flavor enhancers by imparting a "meaty" flavor and are sometimes found in packaged foods that should taste "meaty" but don't contain much meat.

Many foods also contain two other additives that are salts, not of carboxylic acids, but of compounds with acidic hydroxyl groups: *sodium ascorbate* and *sodium erythorbate.* Ascorbic acid is vitamin C, and erythorbic acid is an isomer of vitamin C. Both salts act as antioxidants and help to maintain the color of cured meat such as bacon, though only the ascorbate has value as a vitamin.

Sodium or potassium salts of carboxylic acids are ionic solids that are usually far more soluble in water than the carboxylic acids themselves and are used when acid solubility must be increased. For example, benzoic acid has a water solubility of only 3.4 g/L at 25°C, whereas sodium benzoate has a water solubility of 550 g/L.

Practice Problems **17.6** Write the products of these reactions:
(a) $CH_3CH_2CH_2COOH + KOH \longrightarrow$?
(b) 2-methylpentanoic acid + $Ba(OH)_2 \longrightarrow$?

17.7 Write the formulas of calcium formate and sodium acrylate. (See Table 17.1.)

17.5 REACTIONS OF CARBOXYLIC ACIDS: ESTER FORMATION

Esterification reaction The reaction between an alcohol and a carboxylic acid to yield an ester plus water.

Both in chemical laboratories and in living organisms, the conversion of carboxylic acids into esters is a commonly encountered reaction. In the laboratory, an **esterification reaction** is carried out by warming a carboxylic acid with an alcohol in the presence of a strong-acid catalyst such as H_2SO_4. In so doing, an —H is lost from the alcohol, an —OH is lost from the acid, and water is formed as a by-product. The net effect is substitution of —OR' for —OH:

Two common food additives

Sodium ascorbate
(an antioxidant)

Monosodium glutamate
(flavor enhancer)

All the additives mentioned are salts of acids that occur naturally in plants and animals, and all have been OK'd by the U.S. Food and Drug Administration (FDA). Some additives are essential—without preservatives certain foods would harbor disease-causing microorganisms. Other additives make convenience foods possible or make food more appealing by enhancing fla-vor, color, or consistency. It could be argued that such additives are not essential, although that doesn't necessarily mean they are harmful. Do we really need dry powders that turn into cream sauce when water, butter, and milk are added? Each of us must decide this for ourselves.

Without certain acid salts, this will happen to your bread more quickly than with them.

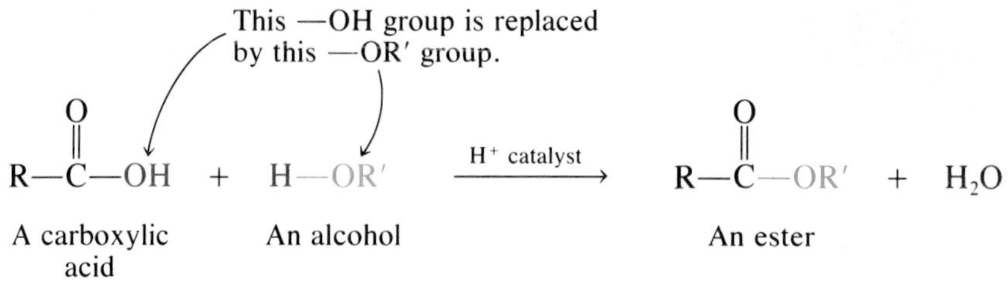

This —OH group is replaced
by this —OR' group.

$$R-\overset{\overset{\displaystyle O}{\|}}{C}-OH \quad + \quad H-OR' \quad \xrightarrow{\text{H}^+ \text{ catalyst}} \quad R-\overset{\overset{\displaystyle O}{\|}}{C}-OR' \quad + \quad H_2O$$

A carboxylic   An alcohol   An ester
acid

For example,

$$CH_3CH_2CH_2-\overset{\overset{\displaystyle O}{\|}}{C}-OH \quad + \quad H-OCH_2CH_3 \quad \underset{}{\overset{\text{H}^+}{\rightleftharpoons}} \quad CH_3CH_2CH_2-\overset{\overset{\displaystyle O}{\|}}{C}-OCH_2CH_3 \quad + \quad H_2O$$

Butanoic acid   Ethanol   Ethyl butanoate
                                                                (in pineapple oil)

Esterification reactions are reversible and often reach equilibrium with substantial amounts of both reactants and products present. In ester synthesis, good yields are obtained either by using a large excess of the alcohol or by continuously removing one of the products (for example, by distilling off a low-boiling ester). Both techniques are applications of LeChatelier's principle.

Solved Problem 17.4   The flavor ingredient in oil of wintergreen can be made by reaction of *o*-hydroxybenzoic acid with methanol. What is its structure?

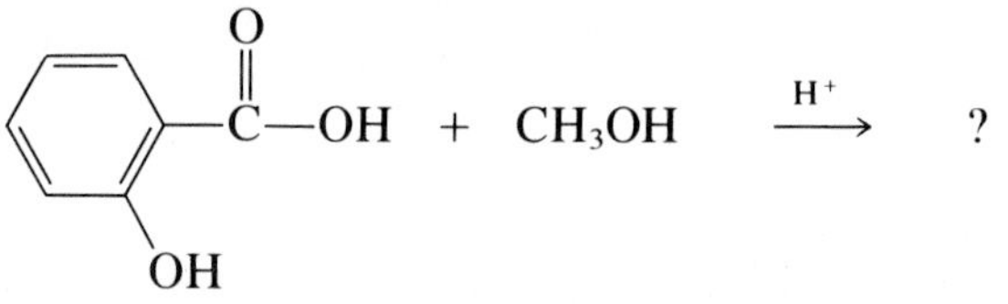

Solution   First, write the two reaction partners so that the —COOH group of the acid and the —OH group of the alcohol are facing each other:

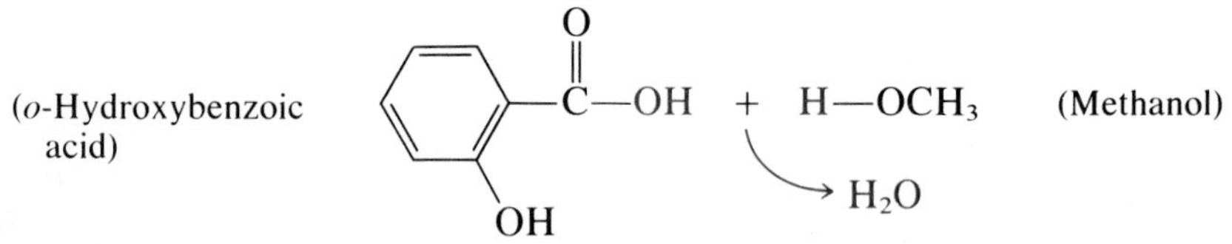

(*o*-Hydroxybenzoic acid)   (Methanol)

Next, remove —OH from the acid and —H from the alcohol to form water and then join the two resulting organic fragments. The product is the ester.

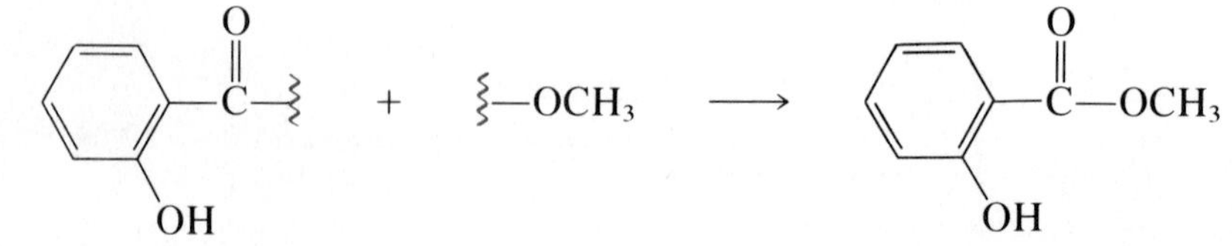

Methyl *o*-hydroxybenzoate
(in oil of wintergreen)

Practice Problems   **17.8**   Raspberry oil contains an ester that can be made by reaction of formic acid with 2-methyl-1-propanol. What is its structure?

$$HCOOH + (CH_3)_2CHCH_2OH \longrightarrow \quad ?$$

17.9   What carboxylic acid and what alcohol are needed to make each of the following esters?

(a)

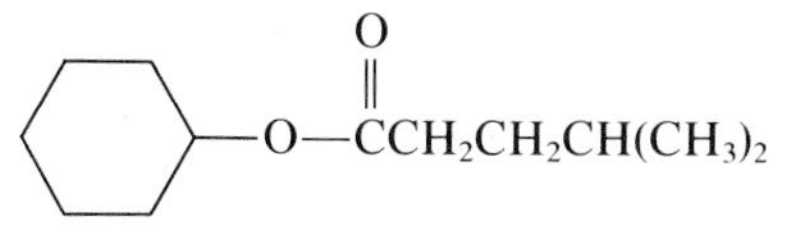

(b) $CH_3CH_2CH_2CH_2\overset{\overset{\displaystyle O}{\|}}{C}-O-CH(CH_3)_2$

17.6 SOME COMMON ESTERS

Esters have many uses in medicine, in industry, and in living systems. In medicine, a number of important pharmaceutical agents, including aspirin and the local anesthetic benzocaine, are esters. In industry, the formation of polymers by esterification is an important reaction (see Interlude in this chapter). In nature, many simple esters are responsible for the fragrant odors of fruits and flowers, such as isopentyl acetate in bananas.

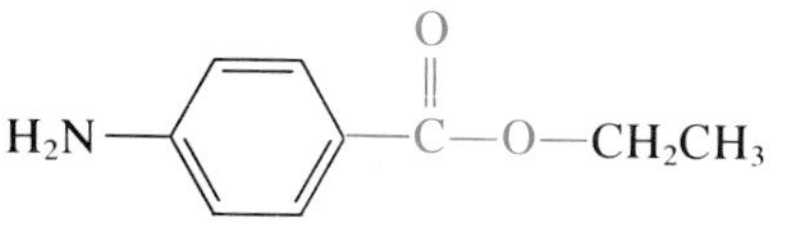

Benzocaine
(a local anesthetic)

Isopentyl acetate
(from bananas)

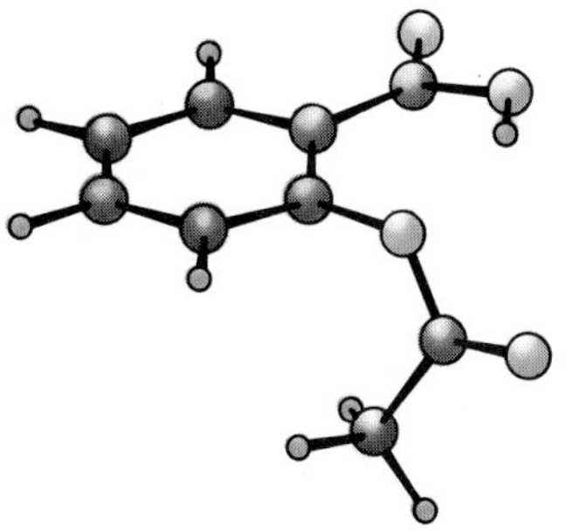

Acetylsalicylic acid

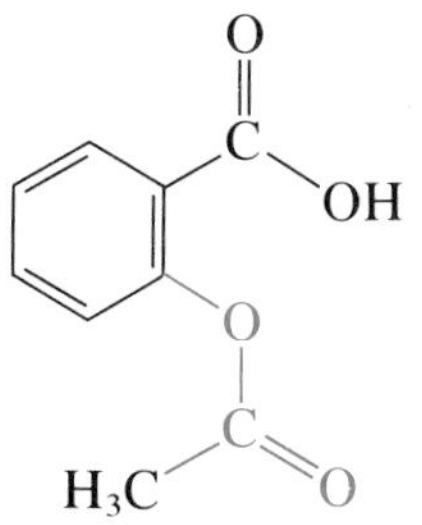

Acetylsalicylic acid
(aspirin)

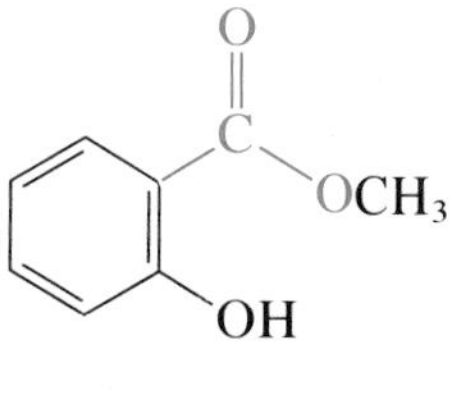

Methyl salicylate
(methyl *o*-hydroxybenzoate)

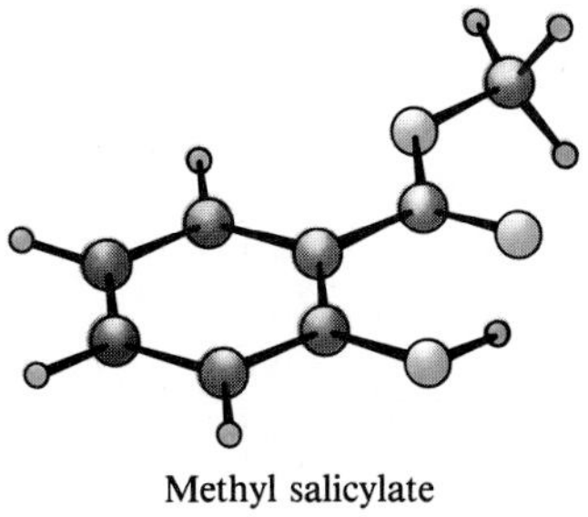

Methyl salicylate

Aspirin and Methyl Salicylate   Aspirin is a white, crystalline solid (mp 135°C) that's a member of the group of drugs known as *salicylates:* esters of salicylic acid. Methyl salicylate, the methyl ester formed at the carboxylic acid group, is a *counterirritant*, a substance that relieves internal pain by stimulating nerve endings in the skin for warmth, coolness, or pain. Poisonous if swallowed in significant amounts, methyl salicylate is one of the active ingredients in liniments such as Ben Gay and Heet. In low concentration it is the flavoring *oil of wintergreen.*

AN APPLICATION: THIOL ESTERS—BIOLOGICAL CARBOXYLIC ACID DERIVATIVES

Although the principles remain the same, many of the carbonyl-group substitution reactions that take place in living organisms use *thiol esters*, or *thioesters*, in place of carboxylic acid esters. A thiol ester, which is simply a sulfur-containing analog of a carboxylic acid ester, has its carbonyl group bonded to one —SR group (like that in a *thiol*, *RSH*) and has the general formula, RCOSR′. Many of the reactions used in the body to extract energy from food involve thiol esters.

Acetyl coenzyme A, usually abbreviated as acetyl SCoA, is the most common thiol ester in nature. Although it has a much more complex structure than esters such as ethyl acetate, it reacts in almost exactly the same way that such esters do. To give just one example of the use of acetyl coenzyme A by living organisms, *N*-acetylglucosamine, an important constituent of cell-surface membranes in mammals, is synthesized in nature by reaction of acetyl SCoA with glucosamine. Note how the result of the reaction is transfer of the acetyl group from one molecule to another.

Acetyl coenzyme A—a thiol ester

Glucosamine   Acetyl coenzyme A   *N*-Acetylglucosamine

Aspirin is only very slightly ionized in the acidic environment in the stomach but causes trouble once inside the stomach lining. There it ionizes to give acetylsalicylate ion and hydrogen ion, which do not easily exit across cell membranes. When sufficient ions have collected, which happens with even one aspirin tablet, the cell membranes are damaged and bleeding occurs. The usual blood loss of a few milliliters per tablet is not harmful, but in susceptible individuals more extensive bleeding can occur.

Aspirin was introduced as a medication in Europe in about 1900 by the Bayer Chemical Company after it was found to be as effective a pain reliever as other salicylates but with fewer unpleasant side effects (Figure 17.3). In addition to providing pain relief, aspirin lowers fever and acts to reduce inflammation. For a drug that has been in use so long and in such large quantity (about 100 aspirin tablets per U.S. resident per year), it's amazing that discoveries about the physiological effects of aspirin are still being made. Since 1982, aspirin has been known to cause a liver disorder called *Reye's syndrome* in children and is no longer recommended for use in childhood diseases such as chicken pox. Evidence is accumulating, however, that regular doses of aspirin may reduce the incidence of heart attacks in adults because of its ability to slow down blood coagulation.

Glycerol Esters   Glycerol is a trihydroxy alcohol that forms the backbone of natural fats and oils when it is esterified with carboxylic acids that have long hydrocarbon chains.

$$3 \ RCOOH \quad + \quad \begin{array}{c} HO-CH_2 \\ | \\ HO-CH \\ | \\ HO-CH_2 \end{array} \quad \longrightarrow \quad \begin{array}{c} RCOOCH_2 \\ | \\ RCOOCH \\ | \\ RCOOCH_2 \end{array}$$

A fatty acid   Glycerol   A triglyceride
(a fat or oil)

Such esters, generally known as *triglycerides,* are members of the lipid class of biomolecules. The R groups in a given fat or oil may be the same or different. The digestion of fats and oils, as you'll see in Chapter 24, is essentially a reverse of the esterification reaction.

Figure 17.3
Pharmaceutical ad from the 1900s. The ad points out that aspirin was a substitute for salicylates. Substitutes have since been found for some of the other pharmaceuticals listed, too.

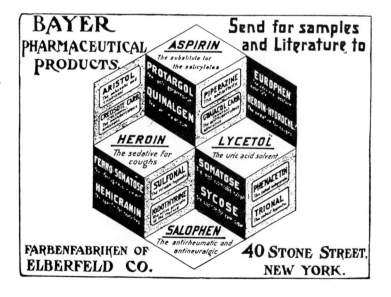

17.7 REACTIONS OF ESTERS: HYDROLYSIS

Hydrolysis The breakdown of a compound by reaction with water; the H's and O from water usually add to the atoms in the broken bond in one way or another.

Esters undergo a **hydrolysis** reaction with water that splits the ester molecule into the carboxylic acid and alcohol combined in the ester. The net effect of the hydrolysis is a substitution of —OH for —OR'.

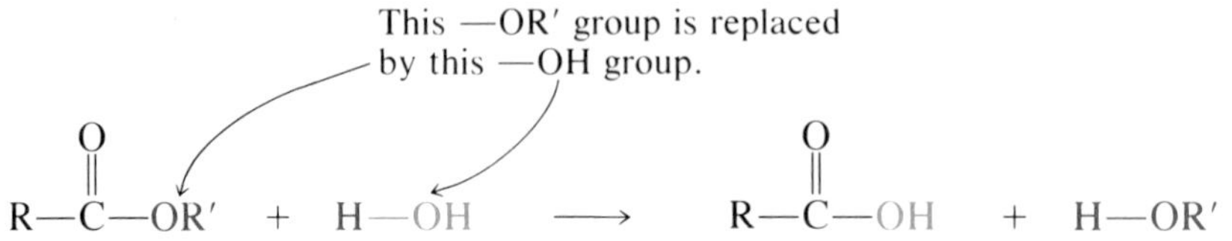

This —OR' group is replaced by this —OH group.

$$R-\overset{O}{\underset{\|}{C}}-OR' + H-OH \longrightarrow R-\overset{O}{\underset{\|}{C}}-OH + H-OR'$$

An ester   Carboxylic acid   Alcohol

Ester hydrolysis is catalyzed both by acids and by bases. Acid-catalyzed hydrolysis is simply the reverse of the esterification reaction discussed in Section 17.5. An ester is treated with water in the presence of a strong acid such as H_2SO_4, and hydrolysis takes place. For example,

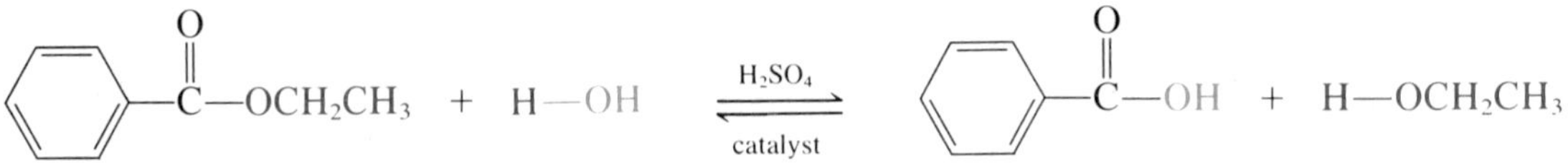

Ethyl benzoate $+ H-OH \underset{\text{catalyst}}{\overset{H_2SO_4}{\rightleftharpoons}}$ Benzoic acid $+ H-OCH_2CH_3$ Ethanol

Saponification reaction The reaction of an ester with aqueous hydroxide ion to yield an alcohol and the metal salt of a carboxylic acid.

Base-catalyzed hydrolysis, often called a **saponification reaction** (after the Latin word *sapo*, "soap"), takes place when an ester is treated with water in the presence of a base such as NaOH or KOH. The main difference between the acid- and base-catalyzed hydrolysis methods is that the base-catalyzed reaction yields the carboxylate anion as a product rather than a free carboxylic acid. In order to isolate the acid, the anion has to be protonated by treatment with HCl or H_2SO_4. (The use of saponification in making soap is discussed in Section 23.4.)

Saponification

$$R-\overset{O}{\underset{\|}{C}}-OR' + NaOH(aq) \longrightarrow R-\overset{O}{\underset{\|}{C}}-O^- Na^+ + R'O-H$$

Ester   Carboxylate salt   Alcohol

For example,

$$CH_3CH_2CH_2-\overset{O}{\underset{\|}{C}}-OCH_3 + NaOH(aq) \longrightarrow CH_3CH_2CH_2-\overset{O}{\underset{\|}{C}}-O^- Na^+ + CH_3OH$$

Methyl butanoate   Sodium butanoate   Methanol

Solved Problem 17.5   What product would you obtain from acid-catalyzed hydrolysis of ethyl formate, a flavor constituent of rum?

$$
\underset{\substack{\|\\ O}}{H-C}-O-CH_2CH_3 \ + \ H_2O \ \longrightarrow \ ?
$$

Solution   First, look at the name of the starting ester. Usually, the name of the ester gives a good indication of the names of the two products. Thus, *ethyl formate* yields *ethyl* alcohol and *form*ic acid. To find the product structures in a more systematic way, write the structure of the ester and locate the bond between the carbonyl-group carbon and the —OR′ group.

This bond is the one that breaks

$$
\underset{\substack{\|\\ O}}{H-C}-OCH_2CH_3 \ \longrightarrow \ \underset{\substack{\|\\ O}}{H-C}\!\!\!\xi \ + \ \xi\!-OCH_2CH_3
$$

Next, carry out a substitution reaction on paper. First form the carboxylic-acid product by connecting an —OH to the carbonyl-group carbon. Then add an —H to the —OCH₂CH₃ group to form the alcohol product.

Connect —OH here.

Connect —H here.

$$
\underset{\substack{\|\\ O}}{H-C}\!\!\!\xi \ + \ \xi\!-OCH_2CH_3 \ \xrightarrow{H_2O} \ \underset{\substack{\|\\ O}}{H-C}-OH \ + \ H-OCH_2CH_3
$$

Practice Problem   17.10   What products would you obtain from acid-catalyzed hydrolysis of these esters?

(a)

$$
\underset{\substack{|\\ CH_3}}{CH_3CH}-\underset{\substack{\|\\ O}}{C}-O-\underset{\substack{|\\ CH_3}}{CHCH_3}
$$

(b) $CH_3CH{=}CHCOOCH_2CH_3$

(c) propyl *p*-bromobenzoate

17.8 REACTIONS OF ESTERS: CLAISEN CONDENSATION

Claisen condensation reaction   A reaction that joins two ester molecules to yield a keto ester product.

Just as the aldol reaction (Section 16.8) joins two aldehyde or ketone molecules together, the **Claisen condensation reaction** joins two ester molecules together. A Claisen condensation takes place when an ester is treated with a strong base such as sodium methoxide, $Na^+OCH_3^-$, the sodium salt of methanol. In the reaction, the —OR′ group is lost from the carbonyl-group carbon of one ester molecule, and a bond forms between that carbon and the carbon atom next to the carbonyl group of the second molecule. The product is a ketone-ester, or *keto* ester.

This —OR′ and this —H split out.

This new C—C bond is formed.

$$R-\overset{O}{\overset{\|}{C}}-OR' \;+\; H-\overset{}{\underset{|}{C}}-\overset{O}{\overset{\|}{C}}-OR' \;\xrightarrow[\text{catalyst}]{Na^+ \;^-OCH_3}\; R-\overset{O}{\overset{\|}{C}}-\overset{}{\underset{|}{C}}-\overset{O}{\overset{\|}{C}}-OR' \;+\; H-OR'$$

This carbon and this carbon form a bond.

A keto ester   An alcohol

For example,

$$H-\overset{H}{\underset{H}{\overset{|}{\underset{|}{C}}}}-\overset{O}{\overset{\|}{C}}-O-CH_3 \;+\; H-\overset{H}{\underset{H}{\overset{|}{\underset{|}{C}}}}-\overset{O}{\overset{\|}{C}}-O-CH_3 \;\xrightarrow{NaOCH_3}\; H-\overset{H}{\underset{H}{\overset{|}{\underset{|}{C}}}}-\overset{O}{\overset{\|}{C}}-\overset{H}{\underset{H}{\overset{|}{\underset{|}{C}}}}-\overset{O}{\overset{\|}{C}}-O-CH_3 \;+\; H-OCH_3$$

2 Methyl acetate

Methyl 3-oxobutanoate
(methyl acetoacetate)

Methanol

Claisen condensation reactions are used by living organisms for the biological synthesis of many different kinds of molecules. Fats, sugars, steroid hormones, and many other classes of compounds are synthesized in the body by joining small ester molecules by Claisen condensations.

Solved Problem 17.6   What product would you obtain from the following Claisen condensation reaction?

$$2 \; CH_3CH_2\overset{O}{\overset{\|}{C}}OCH_3 \;\xrightarrow{NaOCH_3}\; ?$$

Solution   First, rewrite the reaction to emphasize the ester group of one molecule and a C—H bond next to the ester group of the second molecule:

$$CH_3CH_2\overset{O}{\overset{\|}{C}}-OCH_3 \;+\; H-\underset{\underset{CH_3}{|}}{CH}\overset{O}{\overset{\|}{C}}OCH_3 \;\longrightarrow\; ?$$

Next, remove the —OCH₃ group from the first ester and the appropriate —H from the second ester to yield methanol. Then connect the remaining fragments to yield the keto ester product.

Remove this —OCH₃ and this —H.

$$CH_3CH_2\overset{O}{\overset{\|}{C}}-OCH_3 \;+\; H-\underset{\underset{CH_3}{|}}{CH}\overset{O}{\overset{\|}{C}}OCH_3 \;\longrightarrow\; H-OCH_3 \;+\; CH_3CH_2\overset{O}{\overset{\|}{C}}-\underset{\underset{CH_3}{|}}{CH}\overset{O}{\overset{\|}{C}}OCH_3$$

Connect these carbons.

Practice Problem   **17.11**   Draw the products from Claisen condensation of these esters:

(a) [cyclopentyl]—CH$_2$COCH$_3$ (b) methyl butanoate

17.9   REACTIONS OF CARBOXYLIC ACIDS: AMIDE FORMATION

The reaction of a carboxylic acid with ammonia or an amine yields an amide, just as the reaction of a carboxylic acid with an alcohol yields an ester. In both cases, water is formed as a by-product, and the —OH part of the carboxylic acid is replaced.

$$R-\overset{O}{\overset{\|}{C}}-OH \; + \; H-NR'_2 \; \longrightarrow \; R-\overset{O}{\overset{\|}{C}}-NR'_2 \; + \; H_2O$$

Acid   Amine   Amide

The reaction of a carboxylic acid with ammonia or an amine to form an amide doesn't take place, however, unless the carboxyl —OH group is first replaced with a more reactive group. In the laboratory, a compound called *DCC* (dicyclohexylcarbodiimide) is often used.

DCC   =   [cyclohexyl]—N=C=N—[cyclohexyl]

Although the exact way in which DCC works is a bit too complex to discuss in detail, its function is to *activate* the carboxylic acid by making it much more reactive. Once activated, the carboxylic acid then reacts rapidly with an amine to generate an amide.

$$R-\overset{O}{\overset{\|}{C}}-OH \; \xrightarrow{\text{DCC}} \; R-\overset{O}{\overset{\|}{C}}-\boxed{\text{activator}} \; \xrightarrow{\text{NH}_3} \; R-\overset{O}{\overset{\|}{C}}-NH_2$$

For example,

[benzene ring]—$\overset{O}{\overset{\|}{C}}$—OH   +   H—$\overset{H}{\underset{|}{N}}$—CH$_3$ $\xrightarrow{\text{DCC}}$ [benzene ring]—$\overset{O}{\overset{\|}{C}}$—$\overset{H}{\underset{|}{N}}$—CH$_3$ +   H$_2$O

Benzoic acid   Methylamine   *N*-Methylbenzamide

CH$_3$CH$_2$—$\overset{O}{\overset{\|}{C}}$—OH   +   H—$\overset{CH_3}{\underset{|}{N}}$—CH$_3$ $\xrightarrow{\text{DCC}}$ CH$_3$CH$_2$—$\overset{O}{\overset{\|}{C}}$—$\overset{CH_3}{\underset{|}{N}}$—CH$_3$ +   H$_2$O

Propanoic acid   Dimethylamine   *N,N*-Dimethylpropanamide

In living organisms, complex biomolecules function as activating reagents to allow amide formation from a carboxylic acid and an amine. Nevertheless, the principle behind a chemist's use of DCC in a laboratory and an organism's use of complex molecules in a cell to accomplish amide formation is the same.

The **amide bond** between nitrogen and a carbonyl-group carbon is the fundamental link in protein molecules (Figure 17.4). In addition, some synthetic polymers such as nylon contain amide groups (see the Interlude in this chapter), and important pharmaceutical agents such as acetaminophen, the aspirin substitute in Tylenol and Excedrin, are amides. Glutamic acid (glutamine), one of the natural amino acids, contains an unsubstituted amide group, and urea, which carries waste nitrogen from the body, is a diamide.

Amide bond The bond between a carbonyl group and the nitrogen atom in an amide.

Some common amides

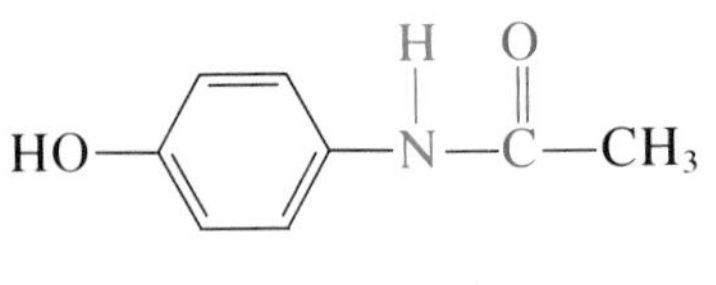

Acetaminophen 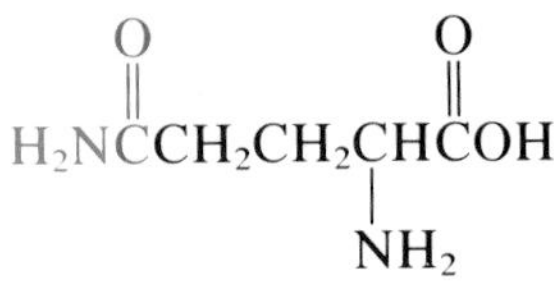   Glutamine 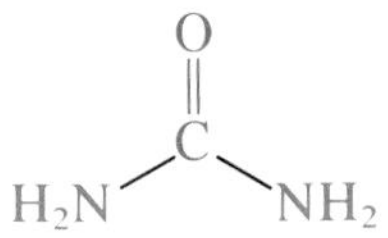   Urea

---

Solved Problem 17.7   The mosquito repellent DEET (diethyltoluamide) can be prepared by reaction of diethylamine with *p*-methylbenzoic acid (toluic acid) in the presence of DCC. What is the structure of DEET?

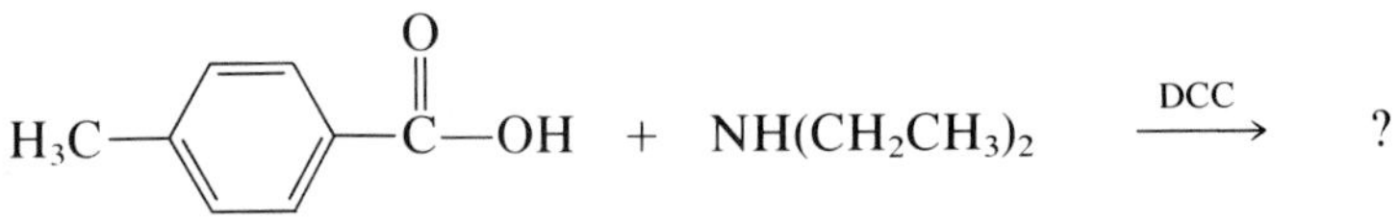

Solution   First, rewrite the equation so that the —OH of the acid and the —H of the amine are facing each other:

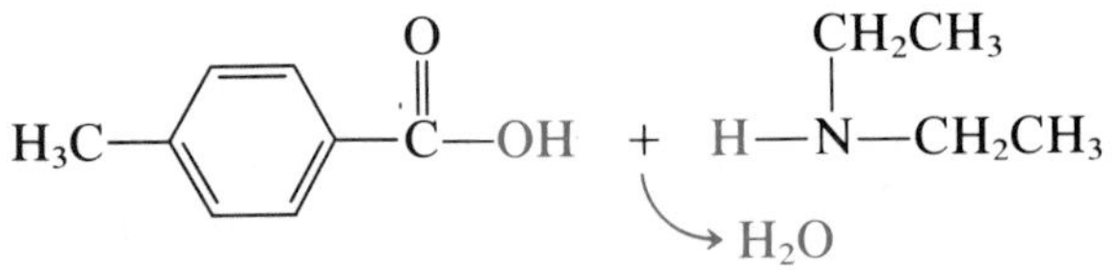

Next, remove the —OH from the acid and the —H from the nitrogen atom of the amine to form water. Then join the two resulting fragments together to form the amide product.

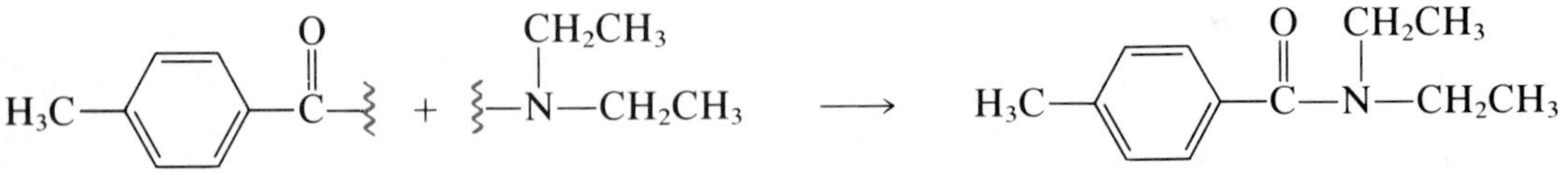

N,N-Diethyltoluamide (DEET)

Figure 17.4
Normal human striated muscle fibers. Protein molecules, which run lengthwise in each fiber, consist of amino acids held together by amide bonds.

***Practice Problems* 17.12** Draw structures of the amides formed in these reactions:

(a) CH_3NH_2 + $(CH_3)_2CHCOOH$ $\xrightarrow{\text{DCC}}$?

(b) ⬡—NH_2 + ⬠—$COOH$ $\xrightarrow{\text{DCC}}$?

17.13 What carboxylic acid and what amine would you use if you wanted to prepare the headache remedy phenacetin?

$$CH_3CH_2O-\text{⬡}-NHCCH_3 \quad \overset{\text{O}}{\overset{\|}{}} \qquad \text{Phenacetin}$$

17.10 REACTIONS OF AMIDES: HYDROLYSIS

Amides undergo a hydrolysis reaction with water in the same way that esters do. Just as an ester yields a carboxylic acid and an alcohol when hydrolyzed (Section 17.7), an amide yields a carboxylic acid and an amine. The net effect of amide hydrolysis is a substitution of —OH for NH_2 or the substituted amide nitrogen. As we'll see in Chapter 25 the cleavage of amide bonds by hydrolysis is the key process that occurs in the stomach during digestion of proteins.

This —NR′R″ group is replaced
by this —OH group from water

$$\underset{\text{An amide}}{R-\overset{\overset{\text{O}}{\|}}{C}-\underset{\underset{R''}{|}}{N}R'} \ + \ H-OH \ \longrightarrow \ \underset{\text{Carboxylic acid}}{R-\overset{\overset{\text{O}}{\|}}{C}-OH} \ + \ \underset{\underset{R''}{|}}{\underset{\text{Amine}}{H-N-R'}}$$

Amide hydrolysis is catalyzed both by acids and by bases. If an acid catalyst such as HCl is used, the amine product is converted into its ammonium salt as soon as it's formed. If a basic catalyst such as NaOH is used, the carboxylic acid product is converted into its carboxylate ion. For simplicity, however, it's easiest to write the hydrolysis products as a free carboxylic acid and a free amine. For example,

N-Methylbenzamide · Benzoic acid · Methylamine

Solved Problem 17.8 What products result from hydrolysis of N-ethylbutanamide?

$$CH_3CH_2CH_2-\overset{\overset{\displaystyle O}{\|}}{C}-NHCH_2CH_3 \; + \; H_2O \; \longrightarrow \; ?$$

Solution First, look at the name of the starting amide. Often, the name of the amide indicates the names of the two products. Thus, N-ethylbutanamide will yield ethylamine and butanoic acid. To be more systematic about finding the product structures, write the amide and locate the bond between the carbonyl-group carbon and the nitrogen. Then break this amide bond and write the two fragments:

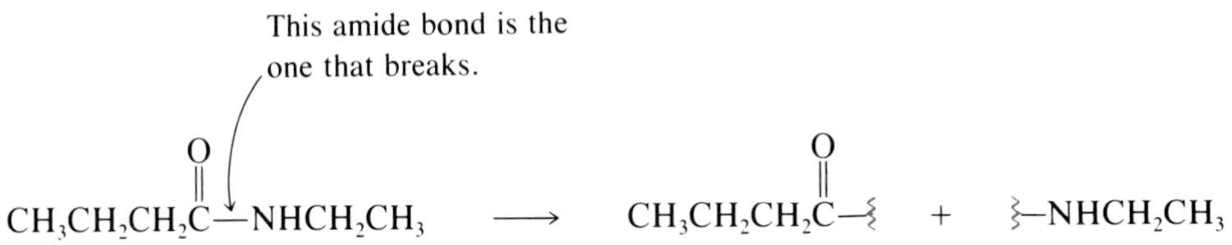

This amide bond is the one that breaks.

Next, carry out a hydrolysis reaction on paper and form the products by connecting an —OH to the carbonyl-group carbon and an —H to the nitrogen:

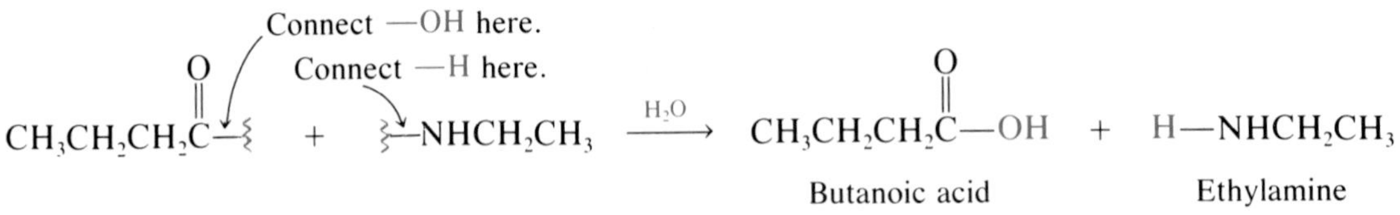

Connect —OH here.
Connect —H here.

Butanoic acid · Ethylamine

Practice Problem 17.14 What products result from hydrolysis of these amides?

$$\text{(a)} \; CH_3CH=CH\overset{\overset{\displaystyle O}{\|}}{C}NHCH_3 \qquad \text{(b)} \; N,N\text{-diethyl-}p\text{-chlorobenzamide}$$

17.11 ACID ANHYDRIDES

Acid anhydrides contain the R—C=O, or acyl group, portion of two carboxylic acids, both bonded to a central oxygen atom. The acyl groups may be from the same or from different acids, and the anhydride group may form between two carboxyl groups in the same molecule. For example,

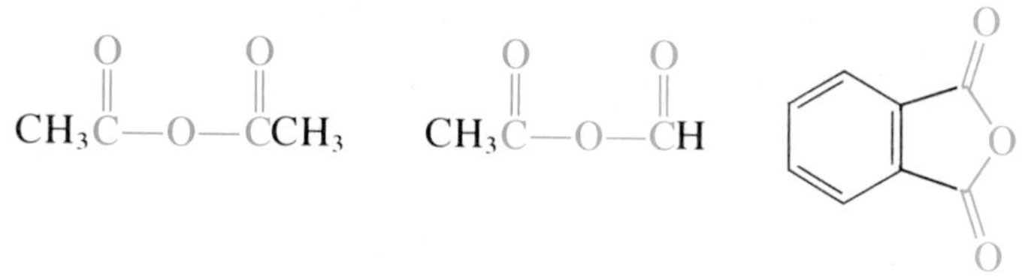

Acetic anhydride   Acetic formic anhydride   Phthalic anhydride

Because acid anhydrides (RCO_2OCR) react rapidly with water to regenerate acids, they are not found in plants and animals.

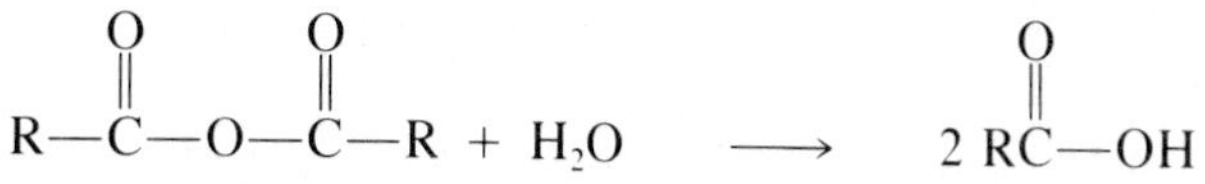

Their reactivity, however, makes anhydrides useful in the synthesis of other carboxylic acid derivatives. The R—C=O portion of an anhydride combines with an alcohol to give an ester or with an amine to give an amide, while the remainder of the anhydride adds hydrogen to give an acid:

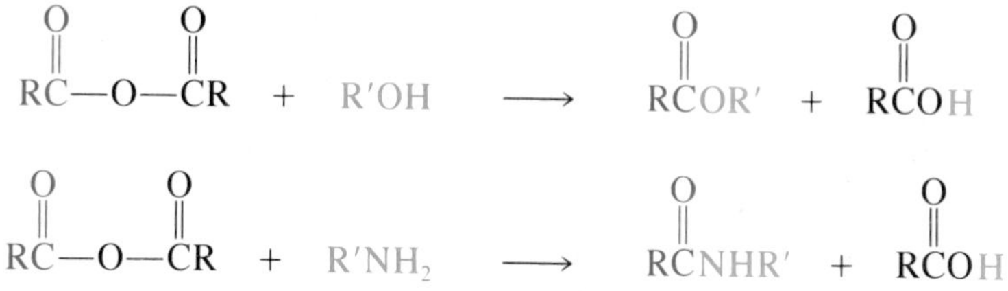

Aspirin, an ester in which salicylic acid provides the —OH group, is synthesized by the reaction of salicylic acid and acetic anhydride:

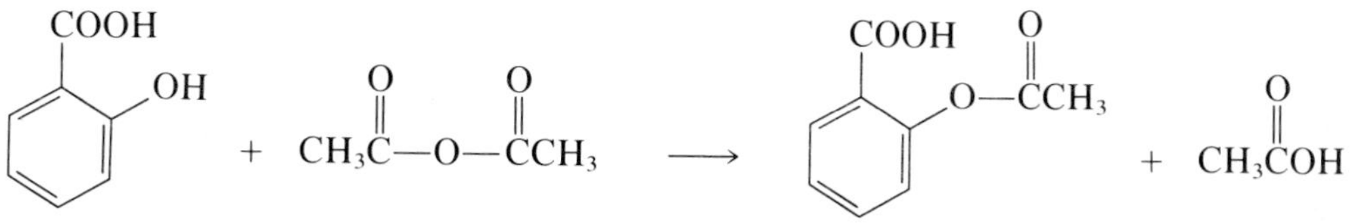

Salicylic acid   Acetic anhydride   Aspirin   Acetic acid

17.12 PHOSPHATE ESTERS AND ANHYDRIDES

Certain inorganic acids such as phosphoric acid, H_3PO_4, and nitric acid, HNO_3, have structures that are similar in many respects to the structures of carboxylic acids. All three contain an atom (C, P, or N) that is singly bonded to an —OH group and doubly bonded to another oxygen. Thus, it's not surprising to find that both phosphoric acid and nitric acid react with alcohols to form **phosphate esters** and **nitrate esters.**

Phosphate ester A compound formed by reaction of an alcohol with phosphoric acid.

Nitrate ester A compound formed by reaction of an alcohol with nitric acid.

A carboxylic acid	Phosphoric acid	Nitric acid

| R'OH | R'OH | R'OH |

A carboxylic ester	A phosphate ester	A nitrate ester

Nitrate esters don't occur naturally, although nitroglycerin, a triester between glycerol (glycerin) and three molecules of nitric acid, is well known for its use in the treatment of heart disease.

| Glycerin | Nitric acid | Nitroglycerin (a nitrate triester) |

A nitroglycerin tablet placed beneath the tongue provides relief within a few minutes for the pain (*angina pectoris*) caused by a brief interference in the flow of blood to the heart. By contrast, a mixture of nitroglycerin with oxygen-supplying salts and oxidizable materials is what we know as *dynamite*.

Phosphoric acid may be esterified at one, two, or all three of its —OH groups.

| Methyl phosphate (a phosphate monoester) | Dimethyl phosphate (a phosphate diester) | Trimethyl phosphate (a phosphate triester) |

Because they still contain acidic hydrogen atoms, the monoester and diester are also acidic. Thus, in neutral or alkaline solutions, including body fluids that are slightly alkaline, they are present as ions:

$$\overset{\displaystyle O}{\underset{\displaystyle O_-}{\overset{\displaystyle \|}{^-O-\!\!\overset{\displaystyle |}{\underset{\displaystyle |}{P}}\!\!-O-CH_3}}} \qquad \overset{\displaystyle O}{\underset{\displaystyle O^-}{\overset{\displaystyle \|}{CH_3-O-\!\!\overset{\displaystyle |}{\underset{\displaystyle |}{P}}\!\!-O-CH_3}}}$$

$$CH_3OPO_3^{2-} \qquad\qquad (CH_3O)_2PO_2^-$$

Phosphate esters are widespread throughout all living organisms and are key substances in nearly all metabolic pathways. Glyceraldehyde 3-phosphate, for example, is one of several phosphate esters that are intermediates in the breakdown of carbohydrates to yield energy.

$$\begin{array}{l} CHO \\ | \\ CHOH \\ | \\ CH_2-O-\overset{\displaystyle O}{\underset{\displaystyle O^-}{\overset{\displaystyle \|}{P}}}-O^- \end{array} \quad \text{or} \quad \begin{array}{l} CHO \\ | \\ CHOH \\ | \\ CH_2-O-PO_3^{2-} \end{array}$$

Glyceraldehyde 3-phosphate

Reaction of two molecules of phosphoric acid results in loss of water and produces a phosphorus anhydride that resembles carboxylic acid anhydrides but unlike them still contains four acidic hydrogens that will be ionized in solutions of appropriate pH. Reaction of this compound, known as *pyrophosphoric acid* (sometimes called diphosphoric acid), with yet another phosphoric acid molecule produces *triphosphoric acid,* also an acid anhydride with acidic hydrogens.

$$\underset{\substack{\displaystyle \,\\ \displaystyle OH \quad OH}}{HO-\overset{\displaystyle O}{\overset{\displaystyle \|}{P}}-O-\overset{\displaystyle O}{\overset{\displaystyle \|}{P}}-OH} \qquad \underset{\substack{\displaystyle \,\\ \displaystyle OH \quad OH \quad OH}}{HO-\overset{\displaystyle O}{\overset{\displaystyle \|}{P}}-O-\overset{\displaystyle O}{\overset{\displaystyle \|}{P}}-O-\overset{\displaystyle O}{\overset{\displaystyle \|}{P}}-OH}$$

Pyrophosphoric acid   Triphosphoric acid

Esters of these acids are referred to as diphosphates and triphosphates. In the following two methyl phosphate esters, note the difference between the P—O—C ester linkage and the P—O—P phosphorus anhydride linkages.

Anhydride

$$\underset{\substack{\displaystyle \,\\ \displaystyle OH \quad OH}}{HO-\overset{\displaystyle O}{\overset{\displaystyle \|}{P}}-O-\overset{\displaystyle O}{\overset{\displaystyle \|}{P}}-O-CH_3} \qquad \underset{\substack{\displaystyle \,\\ \displaystyle OH \quad OH \quad OH}}{HO-\overset{\displaystyle O}{\overset{\displaystyle \|}{P}}-O-\overset{\displaystyle O}{\overset{\displaystyle \|}{P}}-O-\overset{\displaystyle O}{\overset{\displaystyle \|}{P}}-O-CH_3}$$

Ester

Methyl diphosphate   Methyl triphosphate

Phosphoric acid anhydride formation, like carboxylic acid anhydride formation, can be reversed by hydrolysis to give two acids. The diphosphate ADP and the triphosphate ATP, shown here ionized as they are in body fluids, are critical intermediates in the storage and use of energy in the body. Addition of the third phosphate group to ADP absorbs, or "stores," energy, and hydrolysis of ATP to give back ADP releases energy, a reaction we'll describe further in Chapter 20.

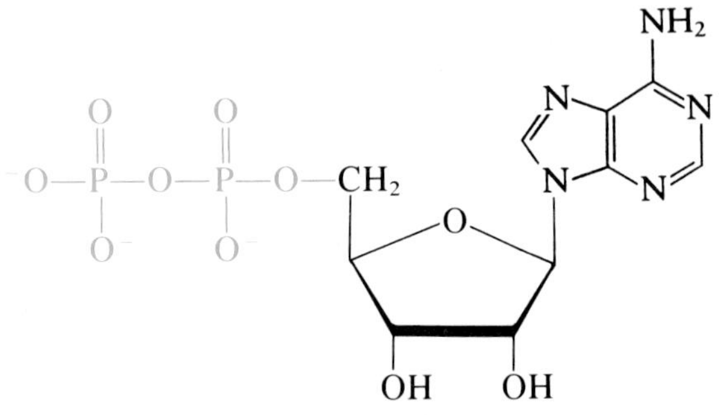

Adenosine diphosphate (ADP)

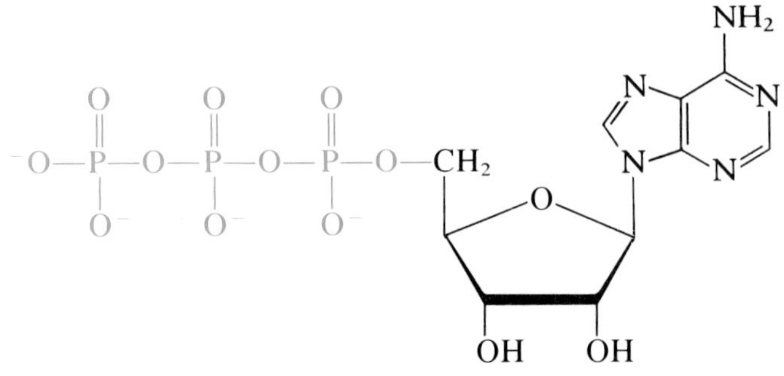

Adenosine triphosphate (ATP)

Practice Problems   **17.15**   Identify the functional group in each of the following compounds and give the structures of the products of hydrolysis of these compounds.

$$\text{(a) } CH_3\overset{\displaystyle O}{\overset{\displaystyle \|}{C}}NH_2 \qquad \text{(b) } CH_3CH_2OPO_3{}^{2-} \qquad \text{(c) } CH_3CH_2\overset{\displaystyle O}{\overset{\displaystyle \|}{C}}OCH_3$$

17.16   Draw the structure of ATP and identify its purine ring, ester group, and anhydride group.

17.13   ORGANIC REACTIONS

Equations for organic reactions tend to appear complicated. The bigger the molecules reacting or the longer the series of reactions, the more intimidating the equations. Remembering a few simple facts can help you to "see" what's going on, however. You've now been introduced to a variety of organic reactions. Let's examine what they have in common.

First, look for reaction patterns, just as you did with the inorganic reactions described in Section 7.8. Many organic reactions have the same patterns, although organic chemists tend to call them by different names. For example, you've seen the *addition reactions* of alkenes, in which water (or halogens or hydrogen halides) add across the double bond:

$$A + B \longrightarrow C$$

$$H_2C{=}CH_2 + H_2O \longrightarrow CH_3CH_2OH$$

Addition reactions have the same pattern as "combination" reactions.

Elimination reactions, such as the elimination of water to give an alkene, are essentially the reverse of addition reactions.

$$C \longrightarrow A + B$$

$$CH_3CHCH_3 \longrightarrow CH_3CH{=}CH_2 + H_2O$$
$$\quad\quad |$$
$$\quad\ OH$$

The "decomposition" reactions of inorganic compounds have the same pattern.

In this chapter you've seen several *substitution reactions*, which have the pattern called "exchange" when the reactants are ions.

$$A{-}B + X{-}Y \longrightarrow A{-}Y + X{-}B$$

The formation of esters and amides are substitution reactions, and so is the Claisen condensation.

$$\underset{\displaystyle RCOH}{\overset{\displaystyle O \atop \displaystyle \|}{}} + R'OH \longrightarrow \underset{\displaystyle RCOR'}{\overset{\displaystyle O \atop \displaystyle \|}{}} + HOH$$

$$\underset{\displaystyle RCOH}{\overset{\displaystyle O \atop \displaystyle \|}{}} + R'NH_2 \longrightarrow \underset{\displaystyle RCNHR'}{\overset{\displaystyle O \atop \displaystyle \|}{}} + HOH$$

$$\underset{\displaystyle RCH_2COR'}{\overset{\displaystyle O \atop \displaystyle \|}{}} + \underset{\displaystyle RCH_2COR'}{\overset{\displaystyle O \atop \displaystyle \|}{}} \longrightarrow \underset{\displaystyle RCH_2CCHCOR'}{\overset{\displaystyle O \ \ O \atop \displaystyle \| \ \ \|}{}} + R'OH$$
$$\quad\quad\quad\quad\quad\quad\quad\quad\quad\quad\quad\quad |$$
$$\quad\quad\quad\quad\quad\quad\quad\quad\quad\quad\quad\ R$$

In organic substitution reactions, one product is often a small molecule, such as water or an alcohol, that splits out during the formation of a larger molecule. Sometimes, the small molecule isn't even written in the equation because it's unimportant for the result of greatest interest. In subsequent chapters you'll see many examples of substitution reactions in the buildup of biomolecules.

When looking at organic equations, in addition to the reaction patterns, remember that reactions take place at functional groups. We've shown here, as it might appear in a biochemistry textbook, the sequence of reactions for synthesis of the amino acid serine. You can ignore for now the other reactants and products above the arrows. Exactly what changes take place in these reactions?

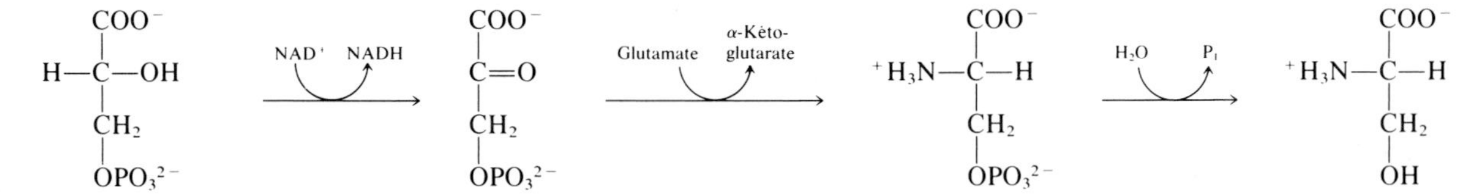

3-Phosphoglycerate   Serine

By comparing the functional groups in each product, you can immediately see that the first two reactions occur at the same carbon atom. One is the conversion of an —OH group to a C=O group, an oxidation reaction that is the elimination of two H atoms. The next reaction is the substitution of an —NH$_3$$^+$ group for the C=O. In the final step, the —NH$_3$$^+$ is unchanged, but an —OH group is substituted for the phosphate group. Whenever you come across a sequence of reactions, mentally highlight the changes as we have done here in color.

$$
\begin{array}{ccccc}
\text{COO}^- & & \text{COO}^- & & \text{COO}^- & & \text{COO}^- \\
| & & | & & | & & | \\
\text{H}-\text{C}-\text{OH} & \xrightarrow[\text{NAD}^+\ \text{NADH}]{} & \text{C}=\text{O} & \xrightarrow[\text{Glutamate}\ \alpha\text{-Keto-glutarate}]{} & {}^+\text{H}_3\text{N}-\text{C}-\text{H} & \xrightarrow[\text{H}_2\text{O}\ \text{P}_i]{} & {}^+\text{H}_3\text{N}-\text{C}-\text{H} \\
| & & | & & | & & | \\
\text{CH}_2 & & \text{CH}_2 & & \text{CH}_2 & & \text{CH}_2 \\
| & & | & & | & & | \\
\text{OPO}_3{}^{2-} & & \text{OPO}_3{}^{2-} & & \text{OPO}_3{}^{2-} & & \text{OH}
\end{array}
$$

INTERLUDE: POLYAMIDES AND POLYESTERS

When a reaction takes place between a carboxylic acid and an amine, the two molecules link together to form an amide. Imagine what would happen, though, if a molecule with *two* carboxylic acid groups were to react with a molecule having *two* amino groups. An initial reaction would join two molecules together, but further reactions would then link more and more molecules together until a giant chain resulted. This is exactly what happens when certain kinds of synthetic polymers, known as *condensation polymers*, are made.

Nylons are *polyamides* that are prepared industrially by reaction of a diamine with a diacid. For example, nylon 66 is prepared by heating adipic acid (hexanedioic acid) with hexamethylenediamine (1,6-hexanediamine) at 280°C (see equation below).

Nylons have a great many uses, both in engineering applications and in fibers. High impact strength and abrasion resistance make nylon an excellent material for bearings and gears, and high tensile strength makes it suitable as fibers for a range of applications from clothing to mountaineering ropes to carpets.

Just as diacids and diamines react to yield polyamides, diacids and dialcohols react to yield *polyesters*.

The most industrially important polyester, made from reaction of terephthalic acid (1,4-benzenedicarboxylic acid) with ethylene glycol, is used under the trade name Dacron to make clothing fiber and under the name Mylar to make plastic film and recording tape. This same polymer, poly(ethylene terephthalate) or PET, is the material in clear, flexible soft-drink bottles.

A polyester
(Dacron, PET)

Nylon 6,10 (like nylon 6,6, but with $(CH_2)_8$ replacing $(CH_2)_4$) is shown forming at the interface between layers of hexamethylenediamine (in NaOH) and a diacid chloride, $ClCO(CH_2)_8COCl$.

$$
\left.\begin{array}{c}
\text{HOOC}-(\text{CH}_2)_4-\text{COOH} \\
\text{Adipic acid} \\
+ \\
\text{H}_2\text{N}-(\text{CH}_2)_6-\text{NH}_2 \\
\text{Hexamethylenediamine}
\end{array}\right\} \xrightarrow{280°\text{C}}
$$

Nylon 6,6, a polyamide

SUMMARY

Carboxylic acids (RCOOH), **esters** (RCOOR′), and **amides** (RCONH$_2$) occur widely throughout all living organisms. Structurally, these three families are related in that all have a carbonyl group bonded to a strongly electron-attracting atom (O or N). Compounds in all three families undergo **substitution reactions** in which a group we can represent by —**X** (for example, —OR, —NR$_2$′, or —OH) substitutes for (replaces) the —OH, —OR′, or —NH$_2$ group of the starting material.

A carbonyl-group substitution reaction

$$
\underset{\underset{\displaystyle}{}}{\overset{\overset{\displaystyle O}{\parallel}}{R-C}}-OH\ (-OR',\ -NH_2)\ +\ H-X\ \longrightarrow
$$

$$
\overset{\overset{\displaystyle O}{\parallel}}{R-C}-X\ +\ H-OH\ (H-OR',\ H-NH_2)
$$

Acid anhydrides (RCO$_2$COR) resemble other carboxylic acid derivatives in undergoing carbonyl-group substitution reactions but are not found in living things because they are hydrolyzed by water.

Carboxylic acids exist as hydrogen-bonded dimers and are higher boiling than comparable alcohols. The simpler acids are water-soluble liquids with strong odors. Diacids, unsaturated acids, and aromatic acids are all common. The simpler esters are volatile liquids with fruit-like odors and some water solubility, but esters are not hydrogen-bonded. Unsubstituted amides are hydrogen-bonded, and all but formamide are solids and are water-soluble. Esters and amides are neither acidic nor basic.

Carboxylic acids are weak acids and therefore react with bases like NaOH to form water-soluble salts that contain **carboxylate anions** (RCOO$^-$). Carboxylic acids undergo conversion into esters by reaction with an alcohol in the presence of a strong-acid catalyst. The net effect of the **esterification reaction** is a substitution of —OR′ for —OH (RCOOH → RCOOR′).

Esters also undergo substitution reactions. For example, they undergo a **hydrolysis reaction** with water to yield a carboxylic acid and an alcohol. The reaction is catalyzed by both acids and, in what is known as **saponification,** by bases. Esters also undergo the **Claisen condensation reaction,** which joins two ester molecules together and forms a keto ester product.

Amides are usually prepared by reaction of a carboxylic acid with an amine in the presence of an activator. Like esters, amides undergo acid- and base-catalyzed hydrolysis, yielding carboxylic acid and amine products.

Certain inorganic acids such as nitric acid and phosphoric acid are analogous to organic carboxylic acids in that they react with alcohols to form esters. **Phosphate esters, diphosphate esters,** and **triphosphate esters** are particularly important in many biological processes.

Despite their complex appearance, many organic reactions fall into the categories known as **addition, elimination,** and **substitution.**

REVIEW PROBLEMS

Carboxylic Acids and Anhydrides

17.17 What are the structural differences among carboxylic acids, acid anhydrides, esters, and amides?

17.18 In what general way do carboxylic acids, acid anhydrides, esters, and amides differ from ketones and aldehydes?

17.19 Write the equation for the ionization of benzoic acid in water.

17.20 Show how two molecules of a carboxylic acid can hydrogen-bond to each other.

17.21 There are four carboxylic acids with the formula C$_5$H$_{10}$O$_2$. Draw and name each one.

17.22 Assume that you have a sample of propanoic acid dissolved in water.

(a) Draw the structure of the major species present in the water solution.

(b) Now assume that aqueous HCl is added to the propanoic acid solution until pH 2 is reached. Draw the structure of the major species present.

(c) Finally, assume that aqueous NaOH is added to the propanoic acid solution until pH 12 is reached. Draw the structure of the major species present.

17.23 Give IUPAC names for these carboxylic acids:

(a) $\overset{\overset{\displaystyle O}{\parallel}}{CH_3CH_2CH_2CH_2CH_2C}OH$

(b) $CH_3CH_2CH_2\underset{\underset{\displaystyle COOH}{|}}{CH}CH_3$

(c) $CH_3CH_2\underset{\underset{\displaystyle CH_3CH_2\!\!}{}}{\overset{\overset{\displaystyle COOH}{|}}{CH}}CH_2CH_3$

(d) $\triangleright\!\!-CH_2CH_2\overset{\overset{\displaystyle O}{\parallel}}{C}OH$

(e) $BrCH_2CH_2\underset{\underset{\displaystyle CH_3}{|}}{\overset{\overset{\displaystyle O}{\parallel}}{CH}C}OH$

(f) aromatic ring with CH$_3$ and —COOH

(g) (CH$_3$CH$_2$)$_3$CCOOH

(h) CH$_3$(CH$_2$)$_5$COOH

17.24 Give IUPAC names for these carboxylic acid salts:

(a)
$$CH_3CH_2CHCH_2CO^-\ K^+$$
with $\overset{O}{\overset{\|}{C}}$ and CH_2CH_3 substituent

(b)
phenyl—$CO^-\ NH_4^+$ (with C=O)

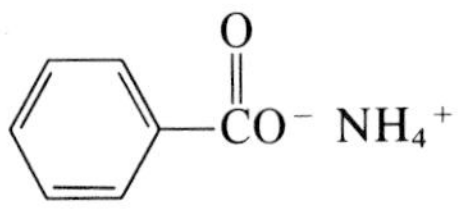

(c)
$$[CH_3CH_2CO^-]_2\ Ca^{2+}$$
(with C=O)

17.25 Draw structures corresponding to these names:
(a) 3,4-dimethylpentanoic acid
(b) triphenylacetic acid
(c) *m*-ethylbenzoic acid
(d) methylammonium butanoate
(e) 2,2-dichlorobutanoic acid
(f) 3-hydroxyhexanoic acid
(g) 3,3-dimethyl-4-phenylpentanoic acid
(h) benzoic anhydride

17.26 Draw and name three different carboxylic acids with the formula $C_7H_{14}O_2$.

17.27 Malic acid, a dicarboxylic acid found in apples, has the IUPAC name *hydroxybutanedioic acid*. Draw its structure.

17.28 What is the formula of the disodium salt of malic acid (Problem 17.27)?

17.29 Aluminum acetate is used as an antiseptic ingredient in some skin-rash ointments. Draw its structure.

17.30 How many grams of KOH does it take to neutralize these acids?
(a) 10.0 g of acetic acid
(b) 250 mL of 2.0 M propanoic acid

17.31 Write the reaction for the neutralization of citric acid (Section 17.3) by NaOH. What volume in milliliters of 0.040 M NaOH is required to neutralize 60.0 mL of 0.020 M citric acid?

17.32 What volume of 0.20 M acetic acid is needed to react with 1.4 g of $Ca(OH)_2$?

17.33 What is the molarity of an oxalic acid solution if 15.0 mL is needed to react with (a) 0.100 g of NaOH (b) 20.0 mL of 0.300 M NaOH?

17.34 Look at Table 17.4 and tell which is the stronger acid, chloroacetic acid or acetic acid?

Esters and Amides

17.35 Draw and name compounds that meet these descriptions:
(a) three different amides with the formula $C_5H_{11}NO$
(b) three different esters with the formula $C_6H_{12}O_2$

17.36 Give IUPAC names for these esters:

(a) $CH_3\overset{O}{\overset{\|}{C}}OCH_2CH_2\overset{CH_3}{\overset{|}{C}}HCH_3$ (b) $CH_3\overset{CH_3}{\overset{|}{C}}HCH_2CH_2\overset{O}{\overset{\|}{C}}OCH_3$

(c) $(CH_3)_3C\overset{O}{\overset{\|}{C}}OCH_2CH_3$

(d) phenyl—$\overset{O}{\overset{\|}{C}}OCH_2CH_3$

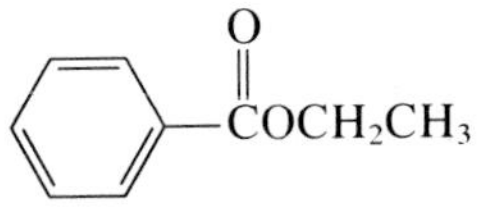

(e) $CH_3CH_2\overset{O}{\overset{\|}{C}}O$—cyclopentyl

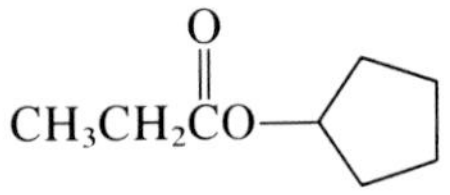

17.37 Draw structures corresponding to these IUPAC names:
(a) methyl pentanoate
(b) isopropyl 2-methylbutanoate
(c) cyclohexyl acetate   (d) phenyl *o*-hydroxybenzoate

17.38 Show the structures of the carboxylic acids and alcohols you would use to prepare each of the esters in Problem 17.37.

17.39 Provide IUPAC names for these compounds:
(a) $CH_3CH_2CHCH_2CH_3$ (b)
$\quad\quad\quad\underset{CONH_2}{|}$

(b) phenyl—$\overset{O}{\overset{\|}{C}}NH$—phenyl

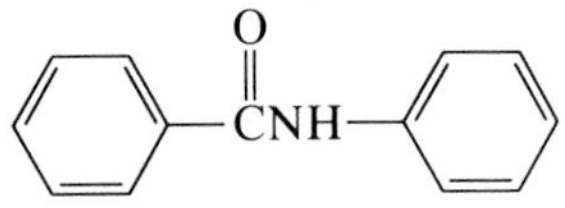

(c) $H\overset{O}{\overset{\|}{C}}N(CH_3)_2$ (d) $CH_3CH_2\overset{O}{\overset{\|}{C}}NH\overset{CH_3}{\overset{|}{C}}HCH_3$

17.40 Show how you could prepare each of the amides in Problem 17.39 from an appropriate carboxylic acid and amine.

17.41 Draw structures corresponding to these IUPAC names:
(a) 3-methylpentanamide   (b) *N*-phenylacetamide
(c) *N*-ethyl-*N*-methylbenzamide
(d) 2,3-dibromohexanamide

17.42 What compounds would result from hydrolysis of each of the amides listed in Problem 17.41?

Reactions of Carboxylic Acids and Their Derivatives

17.43 What general kind of reaction do carboxylic acids and their derivatives undergo? Write the general equation for this type of reaction.

17.44 Methyl *o*-aminobenzoate, commonly called *methyl anthranilate*, is used as a flavoring agent in grape drinks. Write an equation for the preparation of methyl anthranilate from the appropriate alcohol and carboxylic acid.

17.45 Procaine, a local anesthetic whose hydrochloride is Novocain, has the following structure. Identify the functional groups present and show the structures of the alcohol and carboxylic acids you would use to prepare it.

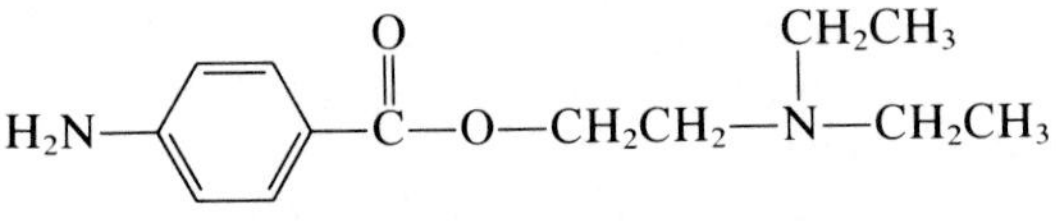

$$H_2N—\text{phenyl}—\overset{O}{\overset{\|}{C}}—O—CH_2CH_2—\overset{CH_2CH_3}{\overset{|}{N}}—CH_2CH_3$$

Procaine

17.46 *Lactones* are cyclic esters in which the carboxylic acid part and the alcohol part are connected to form a ring. What product(s) would you expect to obtain from hydrolysis of butyrolactone?

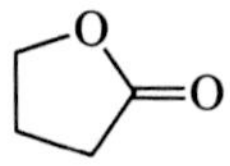

Butyrolactone (a cyclic ester)

17.47 Lidocaine (Xylocaine) is a local anesthetic closely related to procaine. Identify the functional groups present in lidocaine and show how you might prepare it from a carboxylic acid and an amine.

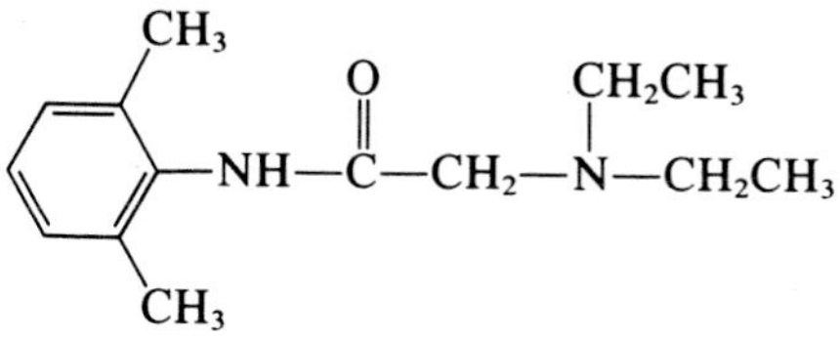

Lidocaine

17.48 Cocaine, an alkaloid isolated from the leaves of the South American coca plant, *Erythroxylon coca,* has the structure indicated. Identify the functional groups present and give the structures of the products you would obtain from hydrolysis of cocaine.

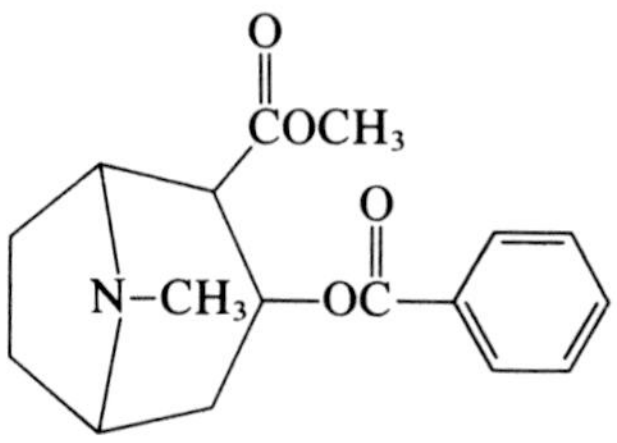

Cocaine

17.49 Household soap is a mixture of the sodium or potassium salts of long-chain carboxylic acids that arise from saponification of animal fat. Draw the structures of soap molecules produced in the following reaction:

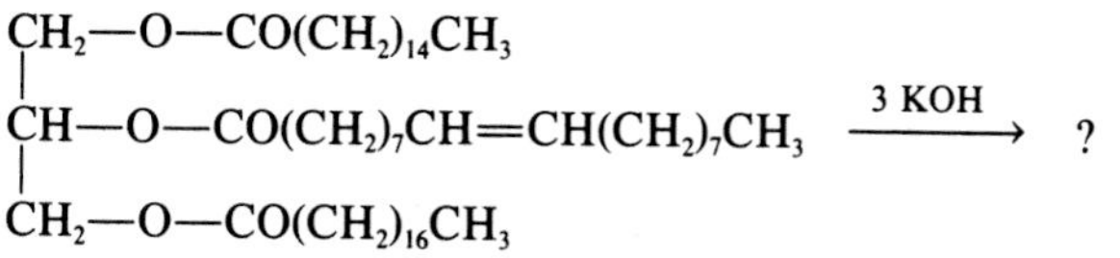

$$CH_2-O-CO(CH_2)_{14}CH_3$$
$$CH-O-CO(CH_2)_7CH=CH(CH_2)_7CH_3 \xrightarrow{\text{3 KOH}} ?$$
$$CH_2-O-CO(CH_2)_{16}CH_3$$

(A fat)

17.50 A *lactam* is a cyclic amide in which the carboxylic acid part and the amine part are connected. Draw the structure of the product(s) from hydrolysis of caprolactam, an industrial precursor of nylon.

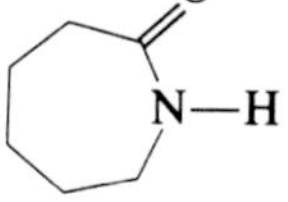

Caprolactam

17.51 Assume that you're given samples of pentanoic acid and methyl butanoate, both of which have the formula $C_5H_{10}O_2$. Describe how you can tell them apart.

17.52 Show how phenacetin, a drug once used in headache remedies, can be prepared from the reaction of an anhydride and an amine.

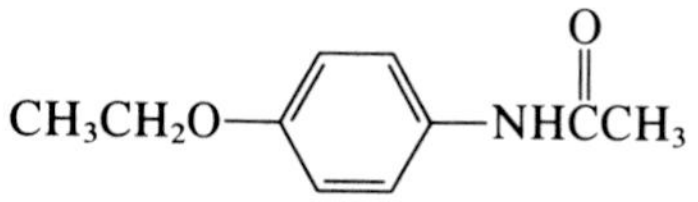

Phenacetin

17.53 The following phosphate ester is an important intermediate in carbohydrate metabolism. What products would result from hydrolysis of this ester?

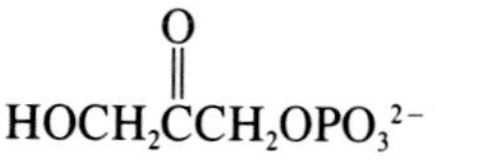

$$HOCH_2CCH_2OPO_3{}^{2-}$$

Claisen Condensation Reaction

17.54 What structural feature must an ester have in order to undergo a Claisen condensation reaction?

17.55 Which of the following esters can't undergo Claisen condensation reactions? Explain.

(a) 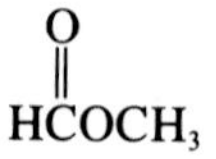 $HCOCH_3$ (b) 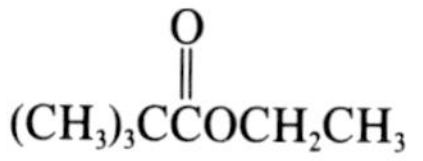 $(CH_3)_3CCOCH_2CH_3$

(c) 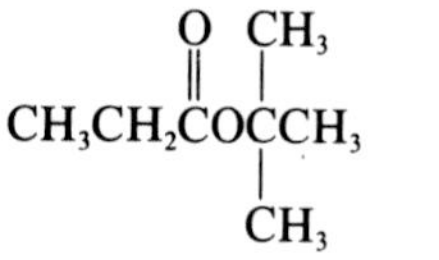 $CH_3CH_2COCCH_3$ with CH_3 groups

17.56 Draw the Claisen condensation products of the following esters:

(a) 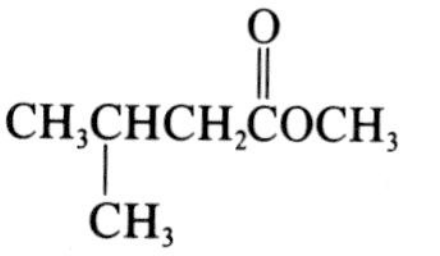 $CH_3CHCH_2COCH_3$ with CH_3 (b) isopropyl acetate

17.57 When a mixture of methyl acetate and methyl propanoate is treated with sodium methoxide in a Claisen condensation reaction, four keto ester products are formed. What are their structures?

Phosphate Esters

17.58 Why are pyrophosphoric acid and triphosphoric acid also acid anhydrides?

17.59 What is the result of complete acid hydrolysis of this diphosphate ester?

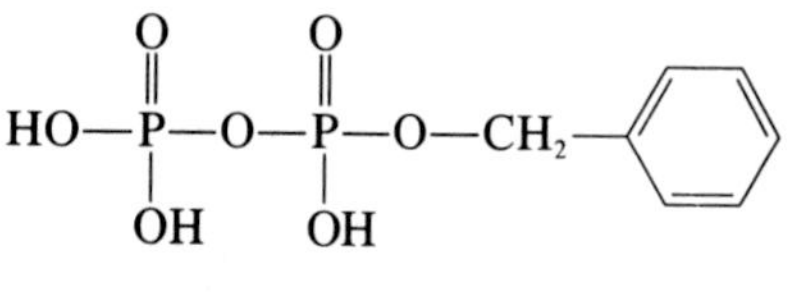

Organic Reactions

17.60 Categorize each of the following reactions as elimination, substitution, or addition.

(a)
$$CH_3\overset{\displaystyle O}{\overset{\|}{C}}H + HOCH_3 \longrightarrow CH_3\underset{\underset{\displaystyle OH}{|}}{C}HOCH_3$$

(b) $CH_3CH{=}CH_2 + HCl \longrightarrow CH_3CHClCH_3$

(c)
$$CH_3\underset{\underset{\displaystyle OH}{|}}{C}H\overset{\displaystyle O}{\overset{\|}{C}}H_2CH \overset{\Delta}{\longrightarrow} CH_3CH{=}CH\overset{\displaystyle O}{\overset{\|}{C}}H$$

17.61 Describe in words the chemical changes that take place in the following series of biochemical reactions.

Succinate   Fumarate

Malate   Oxaloacetate

Applications

17.62 Against what type of microorganisms are sodium benzoate and potassium sorbate particularly effective? [App: Acid Salts as Food Additives]

17.63 What is the general formula of a thiol ester? [App: Thiol Esters]

17.64 Predict the structure of the thiol ester that could be formed from the reaction of acetic acid and methanethiol. [App: Thiol Esters]

17.65 Baked-on paints used for automobiles and many appliances are often based on alkyds, such as can be made from terephthalic acid and glycerol. Sketch a section of the resultant polyester polymer. Note that the glycerol can be esterified at any of the three alcohol groups, providing *cross linking* to form a very strong surface. [Int: Polyamides and Polyesters]

17.66 A simple polyamide can be made from ethylenediamine and oxalic acid. Draw a few units of the polymer formed. [Int: Polyamides and Polyesters]

$$H_2NCH_2CH_2NH_2 \qquad \text{Ethylenediamine}$$

Additional Questions and Problems

17.67 Three amide isomers, *N,N*-dimethylformamide, *N*-methylacetamide, and propanamide, have respective boiling points of 153°C, 202°C, and 213°C. Explain these boiling points in light of their structural formulas.

17.68 Salol, the phenyl ester of salicylic acid, is used as an intestinal antiseptic. Draw the structure of phenyl salicylate.

17.69 Sketch the hydrogen-bonded dimer form of benzoic acid.

17.70 Propanamide and methyl acetate have about the same molar mass, both are quite soluble in water, and yet the boiling point of propanamide is 213°C while that of methyl acetate is 57°C. Explain.

17.71 Mention at least two simple chemical tests by which you could distinguish between benzaldehyde and benzoic acid.

17.72 What ester(s) would you need to produce this condensation product.

$$CH_3CH_2CH_2{-}\overset{\displaystyle O}{\overset{\|}{C}}{-}\underset{\underset{\displaystyle CH_2CH_3}{|}}{C}H{-}\overset{\displaystyle O}{\overset{\|}{C}}OCH_2CH_3$$

17.73 Write the formula of the triester formed from glycerol and stearic acid (Table 17.1).

17.74 Name these compounds.

(a)
$$CH_3CH_2\overset{\overset{\displaystyle H_3C}{|}}{C}{=}\underset{\underset{\displaystyle CH_3}{|}}{\overset{\overset{\displaystyle Cl}{|}}{C}}HCH_3$$

(b)
$$CH_3CH_2\overset{\displaystyle O}{\overset{\|}{C}}NCH_3$$

(c)
$$(CH_3CH_2)_3C\overset{\displaystyle O}{\overset{\|}{C}}O{-}$$

(d)
$${-}\overset{\displaystyle O}{\overset{\|}{C}}NHCH_2CH_3$$
$$NO_2$$

17.75 Complete these reactions.

(a)
$$CH_3CH_2{-}$$
$$+ \text{ HBr} \longrightarrow$$

(b) $CH_3\overset{O}{\overset{\|}{C}}H$ + $HOCH_2CH_2CH_3$ $\longrightarrow$ (hemiacetal)

(e) $2\ CH_3CH_2\overset{O}{\overset{\|}{C}}CH_2CH_3$ $\xrightarrow{\text{NaOH}}$ (aldol)

(c) $CH_3CH_2\overset{O}{\overset{\|}{C}}O\overset{CH_3}{\overset{|}{C}}HCH_3$ + H_2O $\xrightarrow{\text{H}_2\text{SO}_4}$

(f) $CH_3CH_2CCl_2\overset{O}{\overset{\|}{C}}H$ $\xrightarrow{[O]}$

(d)

$\overset{Cl}{\underset{}{}}$ benzene ring with $\overset{O}{\overset{\|}{C}}OH$ + pyrrolidine N—H $\xrightarrow{\text{DCC}}$

(g) $CH_3CH_2\overset{O}{\overset{\|}{C}}C(CH_3)_3$ $\xrightarrow[\text{H}_3\text{O}^+]{\text{NaBH}_4}$

CHAPTER

18

Amino Acids
and Proteins

Muscle proteins are easy to see. Many other kinds of proteins that we'll be describing in this chapter are less visible, but equally important to the efforts of these athletes.

The word *protein* is familiar to everyone. Taken from the Greek *proteios,* meaning "primary," the name "protein" is an apt description for the group of biological molecules that are of primary importance to all living organisms. Approximately 50% of your body's dry weight is protein, and every reaction that occurs in your body is catalyzed by proteins. In fact, a human body contains well over *100,000* different kinds of proteins.

Proteins have many different biological functions. Some proteins, such as the *keratin* in skin, hair, and fingernails, and the *collagen* in connective tissue, serve a structural purpose. Other proteins, such as the *insulin* that controls glucose metabolism, serve as hormones to regulate specific body processes. And still other proteins, such as *DNA polymerase,* serve as enzymes, the biological catalysts that carry out all body chemistry.

In this chapter, we'll look at the following questions about proteins:

1. **What are the structures of amino acids?** The goal: Be able to recognize amino acid structures and give some representative structures of amino acids.

2. **What are the properties of amino acids?** The goal: Be able to describe how the properties of amino acids vary with their side chains and how their ionic charges vary with pH.

3. **Why do amino acids have "handedness"?** The goal: Be able to explain what is responsible for handedness in certain molecules and recognize simple molecules that display this property.

4. **What types of interactions occur between amino acids?** The goal: Be able to give examples of the disulfide bridge and the noncovalent interactions between backbone peptide links and between side chains.

5. **What are the primary structures of proteins?** The goal: Be able to use structural formulas or symbols to show the primary structures of proteins.

6. **What are the secondary, tertiary, and quaternary structures of proteins?** The goal: Be able to distinguish among these types of structures and explain the most common kinds of secondary protein structure.

7. **How are proteins classified?** The goal: Be able to describe three different ways of classifying proteins and give examples of each.

8. **What chemical properties do proteins have?** The goal: Be able to describe protein hydrolysis and denaturation and give some examples of agents that cause denaturation.

18.1 AN OVERVIEW OF PROTEIN STRUCTURE

Protein A large biological molecule made of many amino acids linked together through amide bonds.

Amino acid A molecule that contains both an amino group and a carboxylic acid functional group; proteins are polymers of amino acids.

Regardless of their differing biological functions, all proteins are chemically similar. **Proteins** are made up of many amino acids linked together to form a long chain. As their name implies, **amino acids** are molecules that contain two functional groups, an acidic carboxyl group (—COOH) and a basic amino group (—NH₂). Glycine is the simplest example of an amino acid.

Acidic carboxyl group

Basic amino group

$$H_2N-CH_2-\overset{\overset{\textstyle O}{\|}}{C}-OH$$

Glycine—an amino acid

Peptide bond An amide bond that links two amino acids together.

Two or more amino acids can link together by forming amide bonds (Section 17.9), usually called **peptide bonds.** A *dipeptide* results from the linking together of two amino acids by formation of a peptide bond between the —NH$_2$ group of one and the —COOH group of the second; a *tripeptide* results from linkage of three amino acids via two peptide bonds; and so on. Any number of amino acids can link together to form a long chain. Chains with between about 10 and 100 amino acids are usually called **polypeptides,** while the term *protein* is usually reserved for larger chains.

Polypeptide A molecule composed of roughly 10–100 amino acids linked by peptide bonds.

Peptide bond

$$H_2N-\underset{\underset{\textstyle R}{|}}{CH}-\overset{\overset{\textstyle O}{\|}}{C}-OH \;+\; H-NH-\underset{\underset{\textstyle R'}{|}}{CH}-\overset{\overset{\textstyle O}{\|}}{C}-OH \;\longrightarrow\; H_2N-\underset{\underset{\textstyle R}{|}}{CH}-\overset{\overset{\textstyle O}{\|}}{C}-NH-\underset{\underset{\textstyle R'}{|}}{CH}-\overset{\overset{\textstyle O}{\|}}{C}-OH \;+\; H_2O$$

Two amino acids   A dipeptide

(R and R′ may be the same or different)

18.2 AMINO ACIDS

Alpha (α) amino acid An amino acid in which the amino group is bonded to the carbon atom next to the —COOH group.

There are 20 different amino acids commonly found in proteins. As shown in Table 18.1, all 20 are **alpha (α) amino acids;** that is, the amino group in each is connected to the carbon atom *alpha to* (next to) the carboxylic acid group. The 20 amino acids differ only in the nature of the R group (called the *side chain*) attached to the α carbon).

The —NH$_2$ group is on the carbon alpha to (next to) the —COOH.

$$H_2N-\underset{\underset{\textstyle R}{|}}{\overset{\alpha}{C}H}-\overset{\overset{\textstyle O}{\|}}{C}-OH$$

This side-chain R group is different for each amino acid.

An α-amino acid

Nineteen of the 20 common amino acids are primary amines, RNH$_2$, and the remaining one (proline) is a secondary amine whose nitrogen and α-carbon atoms are joined together in a five-membered ring. Each amino acid is usually referred to by a three-letter shorthand code, such as Ala (alanine), Gly (glycine), Pro (proline), and so on. In addition, a space-saving one-letter code is often used. These codes are shown in Table 18.1.

The 20 common amino acids can be classified as *neutral, basic,* or *acidic,* depending on the structure of their side chains. Fifteen of the 20 have neutral side chains; 2 (aspartic acid and glutamic acid) have an additional carboxylic

Hydrophobic Water-fearing; a hydrophobic substance does not dissolve in water.

Hydrophilic Water-loving; a hydrophilic substance dissolves in water.

Nonessential amino acid An amino acid that is synthesized by the body.

Essential amino acid An amino acid that can't be synthesized by the body and so must be obtained in the diet.

acid group in their side chains and are classified as acidic amino acids; and 3 (lysine, arginine, and histidine) have an additional amine nitrogen atom in their side chains and are classified as basic amino acids. The 15 neutral amino acids can be further divided into those with nonpolar hydrocarbon side chains and those with polar functional groups such as amide or hydroxyl groups. Nonpolar side chains are often described as **hydrophobic** (water-fearing) because they are repelled by water, while polar side chains are described as **hydrophilic** (water-loving) because they are attracted to water.

All 20 amino acids are needed to make proteins, but our bodies can synthesize only 10, known as **nonessential amino acids** because we don't have to get them from food. The remaining 10 are called **essential amino acids** (in red in Table 18.1) because they must be obtained from our diet. Failure to receive an adequate dietary supply of the essential amino acids leads to retarded growth and development in children and to disease and body deterioration in adults.

Practice Problems   **18.1**   Name the common amino acids that contain an aromatic ring. Name those that contain sulfur. Name those that are alcohols. Name those that have alkyl-group side chains.

18.2   Draw alanine showing the tetrahedral geometry of its α carbon.

18.3 DIPOLAR STRUCTURE OF AMINO ACIDS

Zwitterion A neutral dipolar compound that contains both + and − charges in its structure.

Since amino acids contain both acidic and basic groups in the same molecule, they undergo an *internal* acid–base reaction and exist primarily as dipolar ions called **zwitterions** (German *zwitter*, "hybrid"). Because a zwitterion has one plus charge and one minus charge, it is electrically neutral overall. Although you'll often see amino acids written in the nonionized form for convenience, they're actually never in this form in either the solid state or aqueous solution.

Amino acid—nonionized form   Amino acid—zwitterion form

Because they're zwitterions, amino acids have many of the physical properties we associate with salts (Section 5.3). Thus, amino acids are crystalline, have high melting points, and are soluble in water but not in hydrocarbon solvents. In addition, amino acids can react either as acids or as bases depending on the circumstances. In acid solution (low pH), amino acid zwitterions *accept* a proton on their basic —COO^- group to yield a cation. In base solution (high pH), amino acid zwitterions *lose* a proton from their acidic —NH_3^+ group to yield an anion. The predominant structure and charge of an amino acid at any given time depend on the particular amino acid and on the pH of the medium.

Table 18.1 Structures of the 20 Common Amino Acids[a]

Name	Abbreviations	Molecular Weight	Structure	Isoelectric Point
Basic amino acids				
Arginine[a]	Arg (R)	174	$H_2NC-NHCH_2CH_2CH_2-\overset{\overset{\displaystyle NH_2}{\mid}}{\underset{\underset{\displaystyle H}{\mid}}{C}}-COOH$ (with $\overset{\displaystyle NH}{\parallel}$ on the left carbon)	10.8
Histidine	His (H)	155	imidazole ring$-CH_2-\overset{\overset{\displaystyle NH_2}{\mid}}{\underset{\underset{\displaystyle H}{\mid}}{C}}-COOH$	7.6
Lysine	Lys (K)	146	$H_2NCH_2CH_2CH_2CH_2-\overset{\overset{\displaystyle NH_2}{\mid}}{\underset{\underset{\displaystyle H}{\mid}}{C}}-COOH$	9.7
Neutral Amino Acids—Nonpolar Side Chains				
Alanine	Ala (A)	89	$CH_3-\overset{\overset{\displaystyle NH_2}{\mid}}{\underset{\underset{\displaystyle H}{\mid}}{C}}-COOH$	6.0
Glycine	Gly (G)	75	$H-\overset{\overset{\displaystyle NH_2}{\mid}}{\underset{\underset{\displaystyle H}{\mid}}{C}}-COOH$	6.0
Isoleucine	Ile (I)	131	$CH_3CH_2CH-\overset{\overset{\displaystyle NH_2}{\mid}}{\underset{\underset{\displaystyle H}{\mid}}{C}}-COOH$ (with CH_3 on the CH)	6.0
Leucine	Leu (L)	131	$CH_3CHCH_2-\overset{\overset{\displaystyle NH_2}{\mid}}{\underset{\underset{\displaystyle H}{\mid}}{C}}-COOH$ (with CH_3 on the CH)	6.0
Methionine	Met (M)	149	$CH_3SCH_2CH_2-\overset{\overset{\displaystyle NH_2}{\mid}}{\underset{\underset{\displaystyle H}{\mid}}{C}}-COOH$	5.7
Phenylalanine	Phe (F)	165	benzene ring$-CH_2-\overset{\overset{\displaystyle NH_2}{\mid}}{\underset{\underset{\displaystyle H}{\mid}}{C}}-COOH$	5.5
Proline	Pro (P)	115	pyrrolidine ring$-C-COOH$ (ring with NH)	6.3
Valine	Val (V)	117	$CH_3CH-\overset{\overset{\displaystyle NH_2}{\mid}}{\underset{\underset{\displaystyle H}{\mid}}{C}}-COOH$ (with CH_3 on the CH)	6.0

Table 18.1   Structures of the 20 Common Amino Acids[a] (continued)

Name	Abbreviations	Molecular Weight	Structure	Isoelectric Point
Neutral Amino Acids—Polar Side Chains				
Asparagine	Asn (N)	132	$\underset{H}{\overset{NH_2}{H_2NCCH_2-C-COOH}}$ ($\overset{O}{\parallel}$)	5.4
Cysteine	Cys (C)	121	$\underset{H}{\overset{NH_2}{HSCH_2-C-COOH}}$	5.0
Glutamine	Gln (Q)	146	$\underset{H}{\overset{NH_2}{H_2NCCH_2CH_2-C-COOH}}$ ($\overset{O}{\parallel}$)	5.7
Serine	Ser (S)	105	$\underset{H}{\overset{NH_2}{HOCH_2-C-COOH}}$	5.7
Threonine	Thr (T)	119	$\underset{H}{\overset{OH\quad NH_2}{CH_3CH-C-COOH}}$	5.6
Tryptophan	Trp (W)	204	$\underset{H}{\overset{NH_2}{\text{(indole)}-CH_2-C-COOH}}$	5.9
Tyrosine	Tyr (Y)	181	$\underset{H}{\overset{NH_2}{HO-\text{(phenyl)}-CH_2-C-COOH}}$	5.7
Acidic Amino Acids				
Aspartic acid	Asp (D)	133	$\underset{H}{\overset{NH_2}{HOCCH_2-C-COOH}}$ ($\overset{O}{\parallel}$)	3.0
Glutamic acid	Glu (E)	147	$\underset{H}{\overset{NH_2}{HOCCH_2CH_2-C-COOH}}$ ($\overset{O}{\parallel}$)	3.2

[a] Amino acids essential to the human diet are shown in red. Some lists, including that of Recommended Dietary Allowances, omit arginine as an essential amino acid because it is synthesized in the body; others include it as essential because, although synthesized, most is broken down and therefore not available for protein synthesis.

$$H_3N^+-CH-\overset{\overset{\displaystyle O}{\|}}{C}-O-H \qquad H_3N^+-CH-\overset{\overset{\displaystyle O}{\|}}{C}-O^- \qquad H_2N-CH-\overset{\overset{\displaystyle O}{\|}}{C}-O^-$$

R	R	R

Predominant form at low pH (acidic) +1 charge   Predominant form at intermediate pH 0 charge   Predominant form at high pH (basic) −1 charge

Isoelectric point The pH at which a large sample of amino acid molecules has equal numbers of + and − charges.

The pH at which the numbers of positive and negative charges in a large sample are equal and an amino acid exists mainly in its neutral dipolar form is called the amino acid's **isoelectric point (pI).** As indicated in Table 18.1, the 15 amino acids with neutral side chains have isoelectric points near neutrality in the pH range 5.0–6.5. Alanine, for example, has pI = 6.0. At the physiological pH of 7.4 a sample of an amino acid like alanine has a slight excess of negative charges because some acidic —NH_3^+ groups have lost their protons.

Alanine at pH 2.0; positively charged   Alanine at isoelectric point, pH 6; dipolar ion   Alanine at pH 12; negatively charged

The two amino acids with acidic side chains, aspartic acid and glutamic acid, have isoelectric points at more acidic (lower) pH values than those with neutral side chains. Since the side chain —COOH groups of these compounds are substantially ionized at physiological pH of 7.4, they are often referred to as *aspartate* and *glutamate*.

The three amino acids with basic side chains have isoelectric points at more basic (higher) pH values than those with neutral side chains. In near-neutral solutions, lysine and arginine carry positive charges on their side chains, and the side chain in histidine is about 50% ionized.

Lysine   Arginine   Histidine

Because most proteins contain large numbers of both acidic and basic side chains, they can react with both H_3O^+ and OH^- ions, allowing them to act as buffers and help maintain the constant pH of body fluids.

Solved Problem 18.1 Draw the zwitterion forms of phenylalanine and serine. Which of these two acids has a hydrophobic side chain and which has a hydrophilic side chain?

Solution The zwitterions are shown here. The side chain in phenylalanine ($C_6H_5CH_2$—) is a hydrocarbon and therefore is nonpolar and hydrophobic. The side chain in serine ($HOCH_2$—) contains a polar hydroxyl group and is therefore hydrophilic.

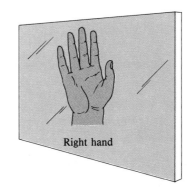

Phenylalanine

Serine

Practice Problems **18.3** Look up the structure of valine in Table 18.1 and draw it at low pH, at its isoelectric point, and at high pH.

18.4 Draw glutamic acid in its fully ionized form. Is this amino acid hydrophobic or hydrophilic?

18.4 HANDEDNESS

Are you right-handed or left-handed? Although you may not often think about it, handedness affects almost everything you do. Anyone who has played much softball knows that the last available glove always fits the wrong hand; any left-handed person taking notes in a lecture knows that it's awkward to do so in a right-handed chair. The reason for these difficulties is that your hands aren't identical. Rather, they're **mirror images.** When you hold your left hand up to a mirror, the image you see looks like your right hand (Figure 18.1). Try it.

Not all objects are handed, of course. There's no such thing as a right-

Mirror image The reverse image produced when an object is reflected in a mirror.

Figure 18.1
The meaning of *mirror image:* If you hold your left hand up to a mirror, the image you see looks like your right hand.

Left hand

Right hand

AN APPLICATION: THE HENDERSON-HASSELBALCH EQUATION AND AMINO ACIDS

In Section 11.14 we derived the *Henderson-Hasselbalch* equation, which relates the pH of a weak-acid solution to the acid's pK_a (negative logarithm of the K_a) and to the concentrations of the acid and its anion. [You might want to review Chapter 11 to brush up on acid dissociation constants and pH.]

The Henderson-Hasselbalch equation:

$$pH = pK_a + \log \frac{[A^-]}{[HA]}$$

The Henderson-Hasselbalch equation is particularly valuable for estimating $[A^-]/[HA]$ for amino acids at varying pHs. For example, when pH = pK_a, then $\log [A^-]/[HA] = 0$ and $[A^-]/[HA] = 1$. Thus, half of a weak acid is present as HA and half as A^-:

If $pH = pK_a$

then $\log [A^-]/[HA] = 0$

so $[A^-]/[HA] = 1/1$ and $[A^-] = [HA]$

At the point where pH = $pK_a - 2$, then $[A^-]/[HA] = 1/100$ and the acid is only about 1% ionized:

If $pH = pK_a - 2$

then $\log [A^-]/[HA] = -2$

so $[A^-]/[HA] = 1/100$

The Henderson-Hasselbalch equation thus gives the following relationships:

$pH = pK_a + 2$	$[A^-]/[HA] = 100/1$
$pH = pK_a + 1$	$[A^-]/[HA] = 10/1$
$pH = pK_a$	$[A^-]/[HA] = 1/1$
$pH = pK_a - 1$	$[A^-]/[HA] = 1/10$
$pH = pK_a - 2$	$[A^-]/[HA] = 1/100$

The pK_a of any acid can be experimentally determined by a titration in which a base is slowly added to the acid solution and the change in pH is recorded. The pH at which one-half the acid present has reacted is equal to the pK_a. If the acid has two acidic hydrogens, however, then two pK_a's are measured. Take alanine, for example. The accompanying *titration curve* shows the results of starting at low pH with fully protonated alanine (structure at the bottom) and titrating with 2 equivalents of NaOH until the alanine is fully deprotonated (structure at the top).

The two "legs" of the alanine titration curve—the first leg from pH 1 to 6 and the second leg from pH 6 to 11—show first the titration of the more acidic —COOH group and second the titration of the —NH$_3^+$ group. When 0.5 equivalent of NaOH is added, the first deprotonation is half complete and the pH equals the first pK_a; when 1.0 equivalent of NaOH is added, the first deprotonation is fully complete and the isoelectric point is reached; when 1.5 equivalents of NaOH is added, the second deprotonation is half complete and the pH equals the second pK_a; and when 2.0 equivalents of NaOH is added, the second deprotonation is complete. Of course, the procedure can also be run in reverse. That is,

Chiral Having (right or left) handedness; able to have two different mirror image forms.

Achiral The opposite of chiral; not having (right or left) handedness.

handed tennis ball or a left-handed coffee mug. When a tennis ball or a coffee mug is held up to a mirror, the image reflected is identical to the ball or mug itself. Objects that have handedness are said to be **chiral** (pronounced **ky-ral**, from the Greek *cheir*, meaning "hand"), and objects like the coffee mug that lack handedness are said to be nonchiral, or **achiral.**

Why is it that some objects are chiral (handed) but others aren't? In general, an object is not chiral if it is symmetrical. Conversely, an object *is*

once the pK_a's of an amino acid are known, a titration curve can be calculated with the Henderson-Hasselbalch equation.

The titration curve shows that the isoelectric point of an amino acid with two acidic groups is halfway between its two pK_a's. The value of pI, the pH at the isoelectric point, is therefore one-half the sum of the pK_a's, as illustrated here for alanine:

$$pI = \frac{pK_{a1} + pK_{a2}}{2} = \frac{2.3 + 9.8}{2} = 6.1$$

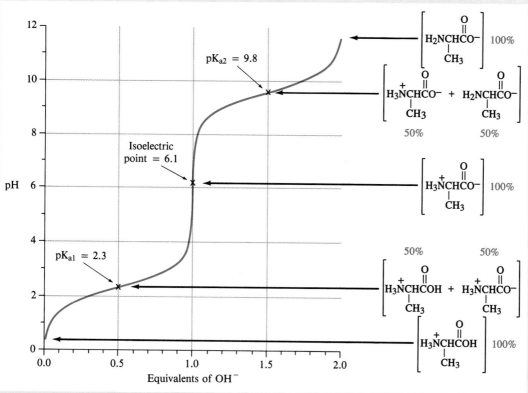

A titration curve for alanine. At pH < 1, alanine is entirely protonated; at pH 2.3, alanine is a 50:50 mix of protonated and neutral forms; at pH 6.1, alanine is entirely neutral; at pH 9.8, alanine is a 50:50 mix of neutral and deprotonated forms; at pH > 11, alanine is entirely deprotonated.

Symmetry plane An imaginary plane cutting through the middle of an object so that one half of the object is a mirror image of the other half.

chiral if it is *not* symmetrical. Symmetrical objects are those like the coffee mug that have an imaginary plane (a **symmetry plane**) cutting through their middle so that one half of the object is an exact mirror image of the other half. If you were to cut the mug in half, one half of the mug would be the mirror image of the other half. A hand, however, has no symmetry plane and is therefore chiral. If you were to cut a hand in two, one half of the hand would not be a mirror image of the other half (Figure 18.2).

Figure 18.2
The meaning of *symmetry plane:* An achiral object like the coffee mug has a symmetry plane passing through it, making the two halves mirror images. A chiral object like the hand has no symmetry plane because the two halves of the hand are not mirror images.

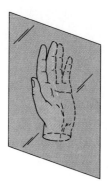

Practice Problems   **18.5**   Which of the following objects are handed?
(a) a glove   (b) a baseball   (c) a screw   (d) a nail

18.6   List three common objects that are handed and another three that aren't.

18.5 MOLECULAR HANDEDNESS AND AMINO ACIDS

Just as certain objects like a hand are chiral, certain *molecules* are also chiral. For example, compare propane and alanine (Figure 18.3). An alanine molecule has no symmetry plane because its two halves aren't mirror images. Like a hand, alanine is chiral and can exist in two forms: a "right-handed" form known as D-alanine and a "left-handed" form known as L-alanine. The two forms are not identical; they are related in the same way that your left and right hands are related. Propane, however, is achiral. It has a symmetry plane cutting through the three carbons such that one half of the molecule is a mirror image of the other half. Thus, propane exists in a single form.

Figure 18.3
Symmetry in molecules. Alanine (2-aminopropanoic acid) has no symmetry plane because the two halves of the molecule are not mirror images. Thus, alanine can exist in two forms—a "right-handed" form, referred to as D-alanine, and a "left-handed" form, referred to as L-alanine. Propane, however, has a symmetry plane and is achiral.

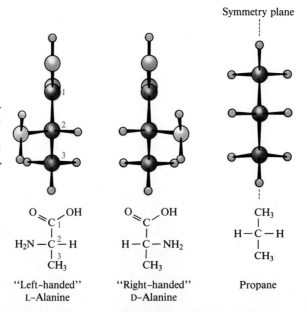

Symmetry plane

"Left-handed"
L-Alanine

"Right-handed"
D-Alanine

Propane

Why are some molecules chiral but others aren't? The answer has to do with the three-dimensional nature of organic compounds. As you saw in Section 6.7, carbon forms four bonds that are oriented to the four corners of an imaginary tetrahedron. **Whenever a carbon atom is bonded to four *different* groups, chirality results.** If a carbon is bonded to two or three of the same groups, however, no chirality is present. In alanine, for example, carbon 2 is bonded to four different groups: a —COOH group, an —H atom, an —NH₂ group, and a —CH₃ group. Thus, alanine is chiral. In propane, however, each of the three carbons is bonded to at least two groups—the H atoms—that are identical. Thus, propane is achiral.

	Groups attached to carbon 2			Groups attached to carbon 2

COOH

H₂N—C—H

CH₃

Alanine
(chiral)

1. —COOH
2. —H
3. —NH₂
4. —CH₃ } different

CH₃

H—C—H

CH₃

Propane
(achiral)

1. —CH₃ } identical
2. —CH₃
3. —H } identical
4. —H

The easiest way to see how tetrahedral geometry leads to chirality is to make paper models of the sort shown in Figure 18.4. Cut two large equilateral triangles out of stiff paper, fold each one so that its three corners come together to form a tetrahedron, and then color each corner as indicated. When four different colors (groups) are used for the four corners of the tetrahedrons, the two models are not identical but have a right-hand–left-hand relationship.

The two mirror-image forms of a chiral molecule like alanine are called **optical isomers,** or **enantiomers:** "optical" because of their effect on polarized

Optical isomers, enantiomers The two mirror-image forms of a chiral molecule.

Figure 18.4
Paper molecular models. The two tetrahedrons are mirror images (that is, are chiral) when the four corners have four different colors.

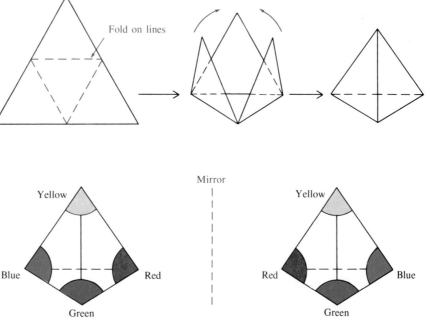

Right-hand model is a mirror image of left-hand model — *chiral.*

Stereoisomers Isomers that have the same molecular and structural formulas but different arrangements of their atoms in space.

Chiral carbon atom A carbon atom bonded to four different groups.

light, as will be explained in Chapter 21. Like other isomers (Section 12.3), optical isomers have the same formula but have different arrangements of their atoms. More specifically, optical isomers are one kind of **stereoisomer,** compounds that have the same formula and the same connections of their atoms but have different arrangements of their atoms in space.

Although they are different compounds, optical isomers are closely related and have most of the same physical properties. Both optical isomers of alanine, for example, have the same melting point, the same solubility in water, the same density, and the same chemical reactivity. They differ only in their effect on polarized light. The mirror-image relationship of the optical isomers of a compound with four different groups on a **chiral carbon atom** is illustrated in Figure 18.5. Compare these structures with the tetrahedrons in Figure 18.4.

What about the amino acid structures in Table 18.1? Are any of them chiral? Of the 20 common amino acids, 19 are chiral because they have four different groups bonded to their α carbons, —H, —NH₂, —COOH, and —R (the side chain). Only glycine, H₂NCH₂COOH, is achiral. Even though the naturally occurring chiral α-amino acids can exist as pairs of optical isomers, nature uses only a single isomer of each for making proteins. For historical reasons that we'll discuss in Chapter 21, the naturally occurring isomers are all classified as left-handed or ʟ-amino acids. The isomeric right-handed ᴅ-amino acids occur very rarely in nature.

Mirror

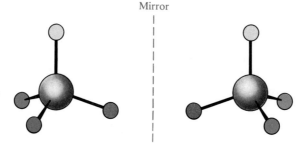

Figure 18.5
Mirror-image molecules. The central atom is chiral because it is bonded to four different groups.

Solved Problem 18.2 Lactic acid can be isolated from sour milk. Is lactic acid chiral?

$$\underset{3}{CH_3}-\underset{2}{\underset{|}{\overset{\overset{OH}{|}}{CH}}}-\underset{1}{\overset{\overset{O}{\|}}{C}OH} \qquad Lactic\ acid$$

Solution To find out if lactic acid is chiral, list the groups attached to each carbon:

Groups on carbon 1	Groups on carbon 2	Groups on carbon 3
1. —OH	1. —COOH	1. —CH(OH)COOH
2. =O	2. —OH	2. —H
3. —CH(OH)CH₃	3. —H	3. —H
	4. —CH₃	4. —H

Next, look at the lists to see if any carbon is attached to four *different* groups. Of the three carbons, carbon 2 has four different groups, and lactic acid is therefore chiral.

Practice Problems   **18.7**   2-Aminopropane is an achiral molecule, but 2-aminobutane is chiral. Explain.

18.8   Which of these molecules are chiral?

$$\begin{array}{cc} CH_3 & CH_3 \\ | & | \end{array}$$

(a) 3-chloropentane   (b) 2-chloropentane   (c) 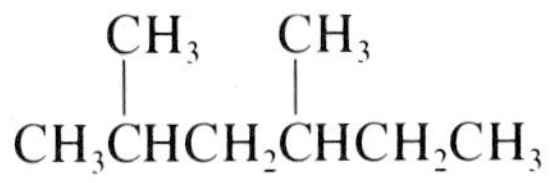 $CH_3CHCH_2CHCH_2CH_3$

18.9   Two of the 20 common amino acids have two chiral carbon atoms in their structures. Identify them.

18.6   THE PEPTIDE BOND AND PRIMARY STRUCTURE

Residue (amino acid) An alternative name for an amino acid unit in a polypeptide or protein.

Peptides and proteins are polymers in which individual amino acids, called **residues,** are linked together by peptide bonds. An amino group from one residue forms a peptide bond with the carboxylic acid group of a second residue; the amino group of the second forms a peptide bond with the carboxylic acid of a third; and so on. The repeating chain of amide linkages is called the *backbone* of the protein, and the amino acid side chains are substituents on the backbone.

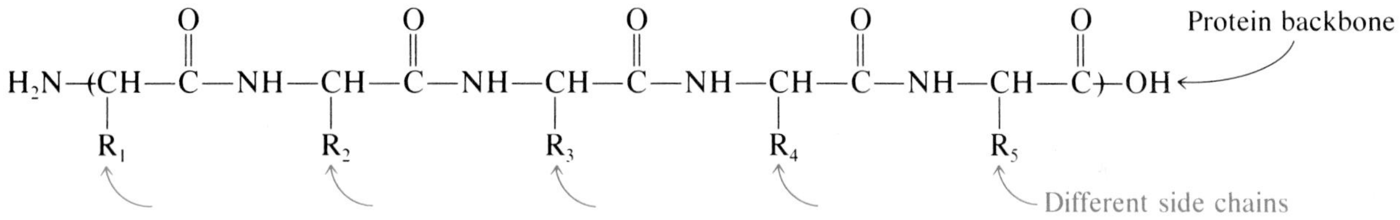

As a simple example, the dipeptide alanylserine results when a peptide bond is made between the alanine —COOH and the serine —NH₂:

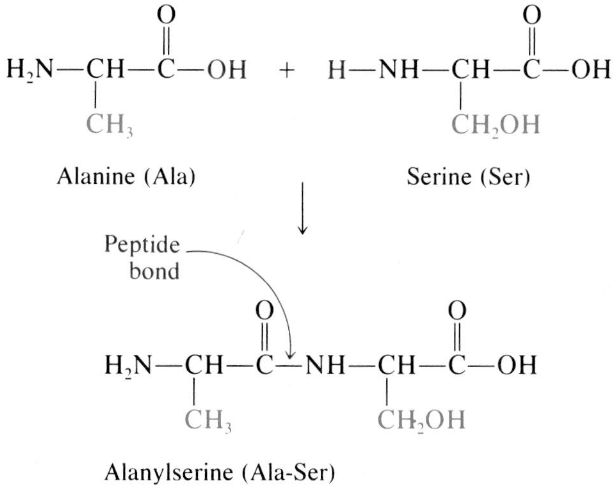

Alanine (Ala)   Serine (Ser)

Peptide bond

Alanylserine (Ala-Ser)

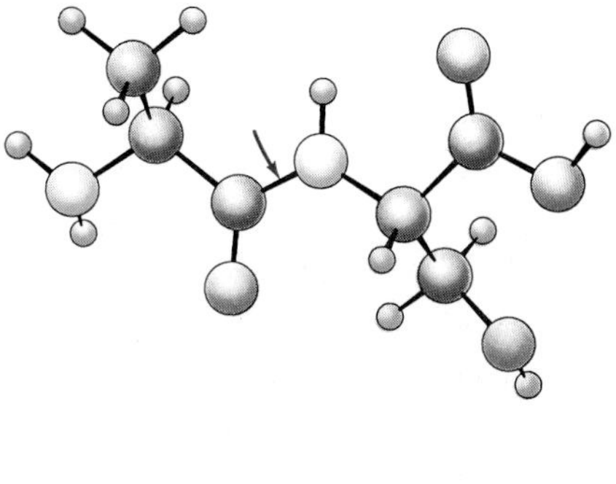

Note that two *different* dipeptides can result from reaction of alanine with serine depending on which —COOH group reacts with which —NH$_2$ group. If the alanine —NH$_2$ reacts with the serine —COOH, serylalanine results.

Serine (Ser)   Alanine (Ala)

Peptide
bond

Serylalanine (Ser-Ala)

The peptide bond is marked with an arrow in the molecular model above, and the entire amide linkage between the two amino acids is circled. Notice that the C and N of the peptide bond and the four attached atoms all lie in the same plane with approximately 120° angles between them. If another amino acid is connected by a peptide bond to the nitrogen at the left end in the model above, that new linkage is also planar, as are all peptide bonds along the entire protein backbone.

By convention, peptides and proteins are always written with the **N-terminal amino acid** (the one with the free —NH$_2$ group) on the left, and the **C-terminal amino acid** (the one with the free —COOH group) on the right. A peptide is named by citing the amino acids in order starting at the N-terminal acid and ending with the C-terminal acid. All residues except the C-terminal one have the *-yl* ending instead of *-ine*, as in alanylserine (abbreviated Ala-Ser) or serylalanine (Ser-Ala). As a further example, the abbreviations listed at the bottom of Figure 18.6 are combined to show the structure of the blood-pressure-regulating hormone angiotensin II.

The number of possible isomeric peptides increases rapidly as the number of amino acid residues increases. There are 6 ways in which 3 different amino acids can be joined, more than 40,000 ways in which the 8 amino acids present in angiotensin II could be joined, and an inconceivably vast number of ways in which the *1800* amino acids in myosin, the major component of muscle filaments, could be arranged (approximately 10^{1800}—a far larger number than there are atoms in the universe).

The size and complexity of protein chains built from up to 20 different amino acids is the basis for their ability to carry out so many different functions in living things. With molecular weights of up to 1/2 *million*, many proteins are so large that the word *structure* takes on a broader meaning when applied to these immense molecules than it does with simple organic molecules. In fact, chemists usually speak of four levels of structure when describing proteins. The **primary structure** of a protein is the sequence in which the various amino acids

N-terminal amino acid The amino acid with the free —NH$_2$ group at the end of a protein.

C-terminal amino acid The amino acid with the free —COOH group at the end of a protein.

Primary protein structure The sequence in which amino acids are linked together in a protein.

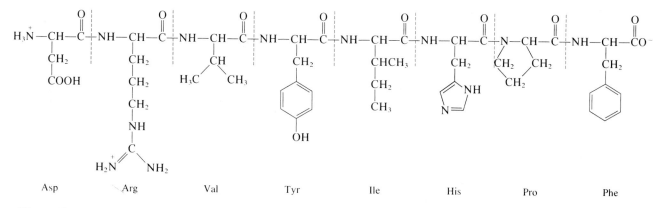

Figure 18.6
The structure of angiotensin II, a blood-pressure-regulating hormone present in blood plasma. Its primary structure is abbreviated Asp-Arg-Val-Tyr-Ile-His-Pro-Phe.

are linked together. The secondary, tertiary, and quaternary structures, as you'll see in later sections of this chapter, are various ways that proteins fold into the shapes essential for their functions.

Primary structure is the most important of the four structural levels because it is a protein's amino acid sequence that determines its overall shape, function, and properties. So crucial is primary structure to function that the change of only one amino acid out of several hundred can drastically alter a protein's biological properties. Sickle-cell anemia, for example, is caused by a genetic defect in blood hemoglobin whereby valine (with a hydrophobic side chain) is substituted for glutamic acid (with a hydrophilic side chain) at only one position, six amino acids from the N-terminal end in the chain of 146 amino acids.

Sickle-cell anemia is named for the sickle shape of affected red blood cells, which do not carry oxygen efficiently (Figure 18.7). Sickled cells are fragile and tend to collect and block capillaries, causing inflammation and pain. The percentage of individuals carrying the genetic trait for sickle-cell anemia is highest among ethnic groups with origins in tropical regions. The ancestors of these individuals survived because carriers of the sickle-cell trait are more resistant to malaria than those without this defect.

Figure 18.7
Sickled red blood cells of a patient with sickle-cell anemia. Compare with the normal red blood cells in Figure 2.1.

Solved Problem 18.3   Draw the structure of the dipeptide Ala-Gly.

Solution   First look up the names and structures of the two amino acids, Ala (alanine) and Gly (glycine).

$$H_2N-\underset{\underset{\displaystyle CH_3}{|}}{CH}-\overset{\overset{\displaystyle O}{\|}}{C}-OH \qquad H_2N-CH_2-\overset{\overset{\displaystyle O}{\|}}{C}-OH$$

Alanine (Ala)   Glycine (Gly)

Since alanine is N-terminal and glycine is C-terminal, Ala-Gly must have a peptide bond between the alanine —COOH and the glycine —NH$_2$.

Peptide bond

$$Free \ —NH_2 \ group \qquad H_2N-\underset{\underset{\displaystyle CH_3}{|}}{CH}-\overset{\overset{\displaystyle O}{\|}}{C}-NH-CH_2-\overset{\overset{\displaystyle O}{\|}}{C}-OH \qquad Free \ —COOH \ group$$

Alanine   Ala-Gly   Glycine

AN APPLICATION: PROTEIN ANALYSIS BY ELECTROPHORESIS

By taking advantage of their overall positive or negative charge, protein molecules can be separated from each other. When an electric field is created between two electrodes, a positively charged particle moves toward the negative electrode and a negatively charged particle moves toward the positive electrode. This movement is called *electrophoresis*, and its amount varies with the strength of the electric field, the charge of the particle, and the size and shape of the particle. Since the overall charge on a protein depends on pH, the electrophoresis of a protein also depends on the pH of the solution or other medium through which the particle moves.

Electrophoresis is routinely used in the clinical laboratory for determination of protein concentrations in blood serum. In the example of an electrophoresis apparatus shown here, a strip of buffer-soaked cellulose acetate is held between

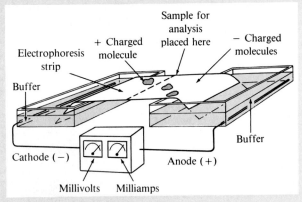

Electrophoresis apparatus for separation of blood serum proteins by overall charge.

buffer reservoirs in contact with the electrodes. The serum to be analyzed is placed in the center of the strip, and different proteins in the serum migrate different distances when the electric field

Practice Problems 18.10 Use the three-letter shorthand notations to name the two isomeric dipeptides that could be made from valine and cysteine. Draw the structure of each.

18.11 Name the six tripeptides that contain valine, tyrosine, and glycine.

18.12 Identify the amino acids in the following dipeptide and tripeptide and write the abbreviated forms of the peptide names.

(a)

$$H_2N-\underset{\underset{\underset{CH_3CHCH_3}{|}}{CH_2}}{CH}-\overset{\overset{O}{||}}{C}-NH-\underset{\underset{CH_2COOH}{|}}{CH}-\overset{\overset{O}{||}}{C}OH$$

(b)

$$H_2N-\underset{\underset{CH_2}{|}}{CH}-\overset{\overset{O}{||}}{C}-NH-\underset{\underset{CH_2OH}{|}}{CH}-\overset{\overset{O}{||}}{C}-NH-\underset{\underset{CH_2(CH_2)_3NH_2}{|}}{CH}-\overset{\overset{O}{||}}{C}OH$$

OH

18.13 Draw the structures of the dipeptide and tripeptide shown in Practice Problem 18.12 with ions wherever ionization is possible.

is applied. After a standard time has passed, the electrophoresis strip is removed and a dye is added to make the pattern of the protein bands visible. Interpretation of the differences between bands is aided by an instrument that converts the intensity of the dye color into density plots like those shown here.

Blood serum contains over 125 proteins, but clinically useful electrophoresis need not separate them all. The distribution pattern of five or six different protein fractions (albumin, two α-globulins, one or two β-globulins, and the γ-globulins) is sufficient for diagnostic purposes, with abnormally high or low readings in each region indicative of different clinical conditions. More elaborate variations on electrophoresis are extensively used in biochemical research and allow separation of individual proteins according to· their molecular weights and isoelectric points.

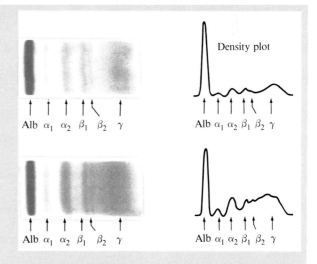

Normal electrophoresis pattern at top and abnormal pattern at bottom showing elevated γ-globulin, which indicates possibility of chronic liver disease, collagen disorder, or chronic infection.

18.7 DISULFIDE BRIDGES IN PROTEINS

Disulfide bridge An S—S bond formed between two cysteine residues that can join two peptide chains together or cause a loop in a peptide chain.

Although the peptide bond is the fundamental link between amino acid residues, a second kind of covalent bonding sometimes occurs when a **disulfide bridge** forms between two cysteine residues. Recall from Section 14.10 that thiols, RSH, react with mild oxidizing agents to yield disulfides, RS—SR. If the thiol groups are on the side chains of two cysteine amino acids, then disulfide bond formation links the two cysteine residues together. The linkage is sometimes indicated by writing CyS with a capital "S" (for sulfur) and then drawing a line from one CyS to the other: CyS—CyS.

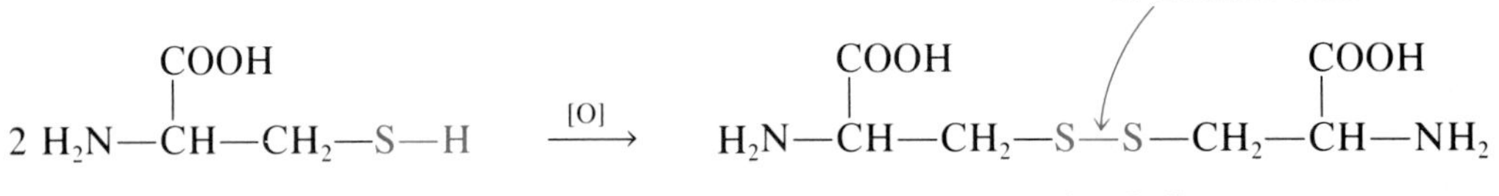

A disulfide bond

$$2 \text{ H}_2\text{N}-\underset{\underset{\text{COOH}}{|}}{\text{CH}}-\text{CH}_2-\text{S}-\text{H} \xrightarrow{[\text{O}]} \text{H}_2\text{N}-\underset{\underset{\text{COOH}}{|}}{\text{CH}}-\text{CH}_2-\text{S}-\text{S}-\text{CH}_2-\underset{\underset{\text{COOH}}{|}}{\text{CH}}-\text{NH}_2$$

Cysteine (Cys) CyS—CyS

If a disulfide bond forms between two cysteine residues in different peptide chains, the otherwise separate chains are linked together. Alternatively, if a disulfide bond forms between two cysteines in the same chain, a loop is formed in the chain. Insulin, for example, consists of two polypeptide chains linked together by disulfide bridges in two places. One of the chains also has a loop in it caused by a third disulfide bridge (Figure 18.8).

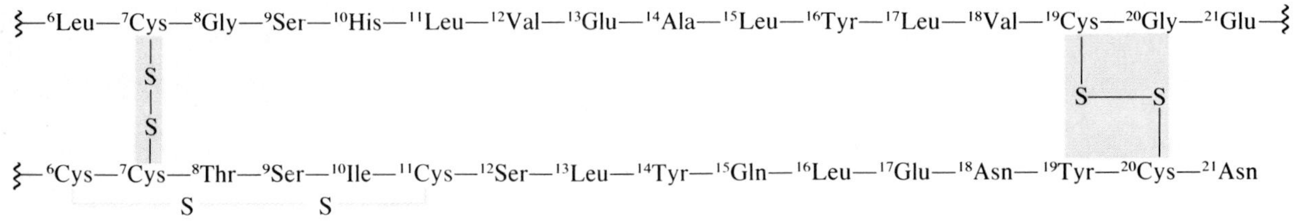

Figure 18.8
The structure of human insulin. There are two disulfide bridges linking the 21 amino acid A chain and the 30 amino acid B chain together and there is a disulfide loop in the A chain.

18.8 SECONDARY STRUCTURE OF PROTEINS

When looking at the primary structure of insulin in Figure 18.8, you might get the idea that the polypeptide chains are simply long threads, stretching from the N-terminal amino acids at one end to the C-terminal amino acids at the other end. The fact is, though, that proteins are not thread-like. Most proteins fold in such a way that nearby segments of the chain orient into a regular pattern called a **secondary structure.**

Secondary protein structure The way in which nearby segments of a protein chain are oriented into a regular pattern, for example, an α helix or a β-pleated sheet.

There are two common kinds of secondary structure patterns: the α helix and the β-pleated sheet. Many proteins have regions of α helix, regions of β-pleated sheet, and still other less organized regions where there is just a random coil. There are, in addition, other less common kinds of secondary structure such as the triple helix found in tropocollagen.

α **Helix**   A common secondary protein structure in which a protein chain wraps into a coil stabilized by hydrogen bonds between peptide links in the backbone.

The α Helix: Secondary Structure of α-Keratin   α-Keratin is a fibrous structural protein found in wool, hair, fingernails, and feathers. Studies have shown that α-keratin exists mostly as a helical coil, much like the coiled cord on a telephone (Figure 18.9a). Called an **α helix,** the coil is stabilized by the formation of hydrogen bonds between the N—H group of one peptide linkage and the carbonyl group of another peptide linkage four residues farther along the chain:

A hydrogen bond connecting two peptide linkages

The location of a hydrogen bond in the α helix is shown in red in Figure 18.9b. Although the strength of each individual hydrogen bond is low (about 5 kcal/mol), the large number present in the helix results in an extremely stable secondary structure. Each coil of the helix contains 3.6 amino acid residues, with a distance between coils of 0.54 nm (5.4 Å).

Figure 18.9
α-Helix secondary structure in α-keratin. (a) The amino acid backbone in the helical secondary structure of α-keratin winds in a spiral much like that in a telephone cord. (b) One example of the hydrogen bonds that hold the spiral in place is shown.

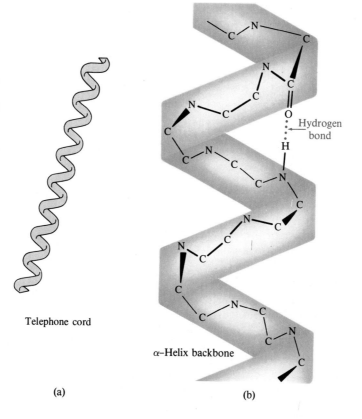

Telephone cord

α–Helix backbone

(a)   (b)

AN APPLICATION: PROTEINS IN THE DIET

Protein is often thought of as "muscle food" because of its popularity with weightlifters and other athletes. Although it's true that protein is needed for growing muscles, it's also true that we *all* need substantial amounts of protein daily because our bodies don't store protein as they do carbohydrates and fats. Furthermore, our bodies can synthesize only 10 of the 20 common amino acids from simple precursor molecules; the remaining 10 amino acids must be obtained by digestion of edible proteins.

Children need large amounts of protein for proper growth, and adults need protein to replace what is lost each day by normal biochemical reactions in the body. The accompanying table shows the estimated daily requirements of essential amino acids for an infant and an adult. The *total* recommended daily amount of protein for an adult with ideal body weight is 0.8 g per kilogram of body weight, which converts to about 9% of daily calories from protein. For a 70 kg male, the recommended total protein is 56 g, and for a 55 kg female, it's 44 g. For reference, a McDonald's Big Mac contains 25 g of protein (along with 34 g of fat). Pregnant or lactating women and individuals recovering from illness or surgery need additional protein. Infants require 2.2 g of protein per kilogram of body weight.

Not all foods are equally good sources of protein. A high-quality protein source is one that provides the 10 essential amino acids in sufficient amount to meet our minimum daily needs. Most meat and dairy products meet this requirement, but many vegetable sources such as wheat and corn don't. The term *incomplete protein* refers to protein in which one or more of the 10 essential amino acids is present in too low a quantity to sustain the growth of laboratory animals. For example, wheat is low in lysine, and corn is low in both lysine and tryptophan. Eating only food with incomplete protein can cause nutritional deficiencies, particularly in growing children.

Some of the limiting amino acids found in various foods are listed below.

Limiting Amino Acids in Some Foods

Food Category	Limiting Amino Acid
Wheat, other grains	Lysine, threonine
Peas, beans, other legumes	Methionine, tryptophan
Nuts and seeds	Lysine
Leafy green vegetables	Methionine

Vegetarians must be certain to adopt a varied diet that provides proteins from several different sources. Legumes and nuts, for example, are particularly valuable in overcoming the deficiencies of wheat and grains. In some regions, the traditional cuisine includes combinations that provide complementary proteins, for example, rice and lentils in India, corn tortillas and beans in Mexico, and rice and black-eyed peas in the southern United States.

Estimated Amino Acid Requirements

Amino Acid	Daily Requirement (mg/kg body weight)	
	Infant	Adult
Arginine	?	?
Histidine	33	10
Isoleucine	83	12
Leucine	135	16
Lysine	99	12
Methionine[a]	49	10
Phenylalanine[b]	141	16
Threonine	68	8
Tryptophan	21	3
Valine	92	14

[a] Also supplied by synthesis from cysteine in diet.
[b] Also supplied by synthesis from tyrosine in diet.

The β-Pleated Sheet: Secondary Structure of Fibroin

Fibroin, the fibrous
protein found in silk, contains mainly the **β-pleated sheet** type of secondary
structure. In this structure, polypeptide chains line up next to each other and are
held together by hydrogen bonds between peptide linkages (Figure 18.10a). The
bond angles along the backbone are such that the sheet has a pleated contour,
with side chains extending above and below the sheet (Figure 18.10b). Small
β-pleated sheet regions are found in many proteins where sections of peptide
chains double back on themselves.

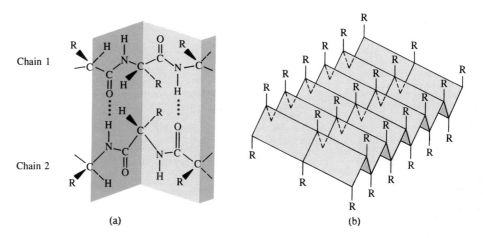

(a)   (b)

Figure 18.10
β-Pleated sheet secondary
structure. (a) Hydrogen
bonding between chains in
the β-pleated sheet. (b) The
β-pleated sheet secondary
structure of silk fibroin,
showing how the chains
line up.

The Triple Helix: Secondary Structure of Tropocollagen

Collagen is the most
abundant of all proteins in mammals, making up 30% or more of the total
protein. A fibrous protein, collagen is the major constituent of skin, tendons,
bones, blood vessels, and connective tissues. The basic structural unit of
collagen is *tropocollagen,* a large protein that consists of three loosely coiled
chains of about 1000 amino acids each, wrapped around each other to form a
stiff, rod-like **triple helix** (Figure 18.11) held together by hydrogen bonds along
the backbone. *Each* tropocollagen rod has a length of about 300 nm and a width
of 1.5 nm. (For comparison, a simple organic molecule like propane is only
about 0.5 nm long.)

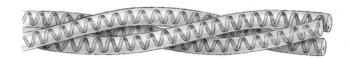

Figure 18.11
The triple helix in
tropocollagen. Three
individual protein strands
wrap around each other to
form a stiff helical cable.

18.9 NONCOVALENT INTERACTIONS IN PROTEINS

Noncovalent interactions such as the hydrogen bonds between backbone peptide bonds (which stabilize α helices and β-pleated sheets) are essential to the
structure and function of proteins. In addition, various other noncovalent interactions are possible between the amino acid side chains along a protein backbone. For instance, attractions between ionized acidic and basic side chains

form what are known as *salt bridges*. In the imaginary protein shown in Figure 18.12, hydrogen bonds cause the chain to fold, and a salt bridge between an acidic glutamate and a basic lysine side chain pulls the chains together in the middle.

In addition to the salt bridges and hydrogen bonding interactions shown in Figure 18.12, the protein is also folded so that hydrophobic hydrocarbon side chains cluster together to exclude water. Here, the same intermolecular forces that hold an oil droplet together on the surface of a puddle are at work. Such a hydrophobic region within an enzyme can, for example, hold in place the hydrocarbon portion of a reactant while a chemical reaction occurs at another spot in the molecule.

To summarize the kinds of noncovalent interactions that occur between amino acids:

● **Hydrogen bonds** form between —OH or —NH$_2$ groups and oxygen or nitrogen atoms with unshared electron pairs.

● **Salt bridges,** or ionic attractions, occur between positively and negatively charged side chains.

● **Hydrophobic interactions** pull nonpolar side chains together and exclude water.

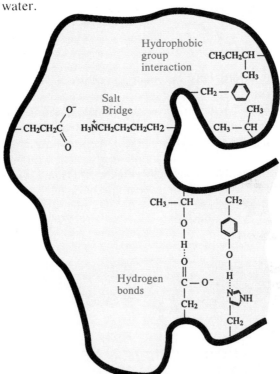

Figure 18.12
Noncovalent interactions of amino acid side chains in proteins.

Practice Problem   18.14   Look at Table 18.1 and identify the type of noncovalent interaction expected between each pair of amino acids:
(a) glutamine and tyrosine   (b) leucine and proline
(c) aspartate and arginine   (d) isoleucine and phenylalanine
(e) threonine and glutamine

18.10 TERTIARY AND QUATERNARY STRUCTURES OF PROTEINS

Tertiary protein structure The way in which an entire protein chain is coiled and folded into its specific three-dimensional shape.

The overall three-dimensional shape that results from the coiling and folding of a protein chain is called the protein's **tertiary structure.** In contrast to secondary structure, which depends on relationships between amino acids close to each other in the chain, tertiary structure can be determined by interactions of amino acid side chains that are far apart.

Tertiary Protein Structure: Myoglobin Myoglobin, a globular protein with a single chain of 153 amino acid residues, provides a good example of tertiary structure. A relative of hemoglobin, myoglobin is found in skeletal muscles, where it stores O_2. Sea mammals, for example, rely on myoglobin for oxygen to sustain them during long dives. Structurally, myoglobin consists of a number of straight segments, each of which adopts an α-helical secondary structure. These helical sections then fold up further to form a compact, nearly spherical, tertiary structure (Figure 18.13) in which a molecule of heme for binding the O_2 is embedded.

Although the bends appear to be irregular and the three-dimensional structure appears to be random, this is not the case. By bending and twisting in precisely this way, myoglobin can achieve maximum stability. All myoglobin molecules adopt this same shape because it's more stable than any other.

The most important forces stabilizing a protein's tertiary structure are the hydrophobic interactions of the nonpolar side chains, which congregate together in the hydrocarbon-like interior of a protein molecule, away from the aqueous medium. The acidic or basic amino acids with charged side chains and those with polar side chains that can form hydrogen bonds, by contrast, are usually found on the exterior of the protein where they can be solvated by water. Myoglobin has many such polar groups on its surface and is therefore water-soluble. Also important for stabilizing a protein's tertiary structure is the

Figure 18.13
Secondary and tertiary structure of myoglobin. The sausage-like shape is often used alone to represent the helical portions of a globular protein. The red structure embedded in the protein is a molecule of heme, to which O_2 binds.

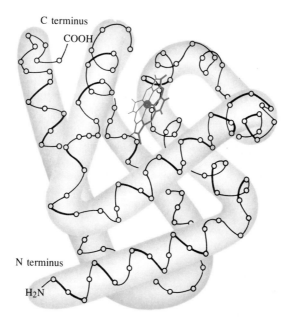

C terminus

COOH

N terminus

H_2N

Figure 18.14
Example of a globular protein, drawn with α helices as coils and β-pleated sheets as arrows pointing toward the C-terminal end. Ribonuclease is an enzyme that catalyzes hydrolysis of ribonucleic acid.

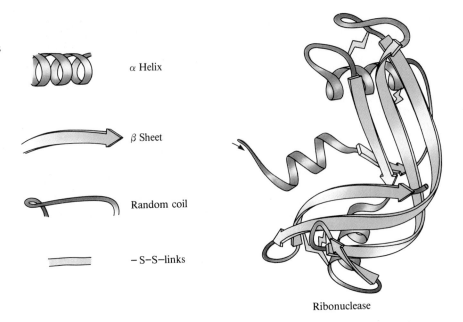

α Helix

β Sheet

Random coil

— S—S—links

Ribonuclease

formation of disulfide bridges between cysteine residues. Another example of tertiary protein structure is given in Figure 18.14. Note the combination of α-helix and β-pleated sheet regions as well as randomly coiled regions.

Quaternary Protein Structure: Collagen and Hemoglobin   When two or more protein chains combine in a single functional unit, the shape of the resulting protein is referred to as **quaternary structure.** We saw in the previous section, for example, that tropocollagen is a stiff, rod-like protein with a triple-helix secondary structure. Collagen itself, the actual protein present in skin, teeth, bones, and connective tissues, has a complex quaternary structure formed when a great many tropocollagen strands aggregate together by overlapping lengthwise in a *quarter-stagger arrangement,* as shown in Figure 18.15.

Depending on the exact purpose the collagen serves in the body, further structural modifications also occur. In connective tissue like tendons, chemical bonds form between strands to give collagen fibers with a rigid, cross-linked structure. In teeth and bones, a mineral called *calcium hydroxyapatite,* $Ca_5(PO_4)_3OH$, deposits in the gaps between chains to further harden the overall assembly.

Figure 18.16 shows another example of quaternary structure: that of hemoglobin, the oxygen carrier in blood, which consists of four polypeptide chains held together primarily by hydrophobic interactions.

Quaternary protein structure The way in which two or more protein chains aggregate to form large, ordered structures.

Figure 18.15
The quarter-stagger arrangement of tropocollagen triple helix units aggregating into a collagen fiber. Each tropocollagen chain is offset from its next neighbor by about one-fourth of its length, and the individual chains are connected by covalent bonds.

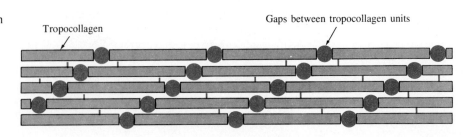

Tropocollagen

Gaps between tropocollagen units

Figure 18.16
Computer-generated model of the oxygen-carrying protein hemoglobin. The hemoglobin molecule consists of four polypeptide chains (yellow, blue, red, and green, with helical regions shown as cylinders). Within each polypeptide chain is embedded a planar heme molecule (gray carbon atoms and red oxygen atoms). Oxygen is carried by bonding to the single Fe^{2+} in the center of each of the four heme molecules.

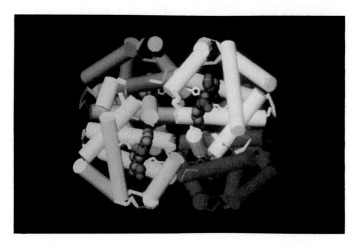

Protein structure can be summarized as follows:

● **Primary structure**—the amino acid sequence, for example,

Asp-Arg-Val-Tyr

● **Secondary structure**—the regular patterns of chain segments such as an α helix, a β-pleated sheet, or a triple helix, held together by hydrogen bonds between backbone peptide links near each other in the chain.

α Helix β–Pleated sheet

● **Tertiary structure**—the folding of a protein molecule into a specific three-dimensional shape, held together mainly by noncovalent interactions between side-chain groups that can be quite far apart.

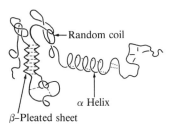

Random coil

α Helix

β–Pleated sheet

● **Quaternary structure**—the combination of several protein chains into a larger three-dimensional structure held together by noncovalent interactions or covalent cross-links.

Simple protein A protein that yields only amino acids when hydrolyzed.

Conjugated protein A protein that yields one or more other substances in addition to amino acids when hydrolyzed.

Fibrous protein A tough, insoluble protein whose peptide chains are arranged in long filaments.

18.11 CLASSIFICATION OF PROTEINS

Proteins can be classified in several ways, the simplest of which is to group them according to composition. **Simple proteins,** such as blood-serum albumin, are those that yield only amino acids and no other compounds on hydrolysis. **Conjugated proteins,** which are far more common than simple proteins, yield nonprotein substances in addition to amino acids on hydrolysis. Table 18.2 lists some examples of conjugated proteins.

Another way to classify proteins is according to their three-dimensional shape as either *fibrous* or *globular*. **Fibrous proteins,** such as collagen and the keratins, consist of polypeptide chains arranged side by side in long filaments. Because these proteins are tough and insoluble in water, nature uses them for

Table 18.2   Some Different Kinds of Conjugated Proteins

Name	Nonprotein Part	Examples
Glycoproteins	Carbohydrates	Glycoproteins in cell membranes
Lipoproteins	Lipids	High- and low-density lipoproteins that transport cholesterol and other lipids through the body
Metalloproteins	Metal ions	The enzyme cytochrome oxidase, necessary for biological energy production, and many other enzymes
Phosphoproteins	Phosphate groups	Milk casein, which serves to store nutrients for a growing embryo
Hemoproteins	Heme	Hemoglobin and myoglobin, which transport and store oxygen, respectively
Nucleoproteins	RNA (ribonucleic acid)	Found in cell ribosomes, where they take part in protein synthesis

Table 18.3   Some Common Fibrous and Globular Proteins

Name	Occurrence and Function
Fibrous proteins (insoluble)	
Keratins	Found in skin, wool, feathers, hooves, silk, fingernails
Collagens	Found in animal hide, tendons, bone, eye cornea, and other connective tissue
Elastins	Found in blood vessels and ligaments, where ability of the tissue to stretch is important
Myosins	Found in muscle tissue
Fibrin	Found in blood clots
Globular proteins (soluble)	
Insulin	Regulatory hormone for controlling glucose metabolism
Ribonuclease	Enzyme controlling RNA hydrolysis
Immunoglobulins	Proteins involved in immune response
Hemoglobin	Protein involved in oxygen transport
Albumins	Proteins that perform many transport functions in blood; protein in egg white

Globular protein A water-soluble protein that adopts a compact, coiled-up shape.

structural materials like tendons, hair, ligaments, and muscle. **Globular proteins,** by contrast, are usually coiled into compact, nearly spherical shapes like myoglobin (Figure 18.13) and ribonuclease (Figure 18.14). Globular proteins, which might have anywhere from 100 to well over 1000 amino acids in their chains, include most of the 2000 or so known enzymes. Like myoglobin, they are mobile within cells and are generally soluble in water, with their hydrophobic groups folded in toward the center and their hydrophilic groups on the outside. Table 18.3 lists some common examples of both fibrous and globular proteins.

Yet a third way to classify proteins is to group them according to biological function. As indicated in Table 18.4, proteins have an extraordinary diversity of roles in the body.

Table 18.4   Some Biological Functions of Proteins

Type	Function and Example
Enzymes	Proteins such as alcohol dehydrogenase that act as biological catalysts
Hormones	Proteins such as insulin and growth hormone that regulate body processes
Storage proteins	Proteins such as ferritin that store nutrients
Transport proteins	Proteins such as hemoglobin that transport oxygen and other substances through the body
Structural proteins	Proteins such as keratin, elastin, and collagen that form an organism's structure
Protective proteins	Proteins such as the antibodies that help fight infection
Contractile proteins	Proteins such as actin and myosin found in muscles
Toxic proteins	Proteins such as the snake venoms that serve a defensive role for the plant or animal

18.12   CHEMICAL PROPERTIES OF PROTEINS

Protein Hydrolysis   Just as a simple amide can be hydrolyzed to yield an amine and a carboxylic acid (Section 17.10), a protein can be hydrolyzed to yield many amino acids. In fact, digestion of proteins involves nothing more than breaking the numerous peptide bonds that link amino acids together. For example,

Alanine   Glycine   Cysteine   Aspartic Acid

Although a chemist in the laboratory might choose to hydrolyze a protein by heating it with a solution of hydrochloric acid, most digestion of proteins in the body takes place in the stomach and small intestine, where the process is catalyzed by enzymes. Once formed, individual amino acids are absorbed through the wall of the intestine and transported in the bloodstream to tissues.

Practice Problem   18.15   Look up the structure of angiotensin II in Figure 18.6 and draw the products that would be formed during digestion.

Denaturation The loss of secondary, tertiary, or quaternary protein structure due to disruption of noncovalent interactions that leaves peptide bonds intact.

Protein Denaturation   Since the overall shape of a protein is determined by a delicate balance of noncovalent forces, it's not surprising that a change in protein shape often results when the balance is disturbed. Such a disruption in shape without affecting the protein's primary structure is known as **denaturation.** When denaturation of a globular protein occurs, the structure unfolds from a well-defined globular shape to a randomly looped chain (Figure 18.17).

Denaturation is accompanied by changes in both physical and biological properties. Solubility is often decreased by denaturation, as occurs when egg white is cooked and the albumins coagulate into an insoluble white mass. In addition, enzymes lose their catalytic activity and other proteins are no longer able to carry out their biological functions when their shapes are altered by denaturation.

The agents that cause denaturation include heat, mechanical agitation, detergents, organic solvents, pH change, inorganic salts, and oxidizing agents:

● **Heat**   The weak side-chain attractions in globular proteins are easily disrupted by heating, in many cases only to temperatures above 50°C. Cooking meat, for example, converts some of the insoluble collagen into soluble gelatin, which can be used in glue and Jell-O, and for thickening sauces.

● **Mechanical agitation**   The most familiar example of denaturation by agitation is the foam produced by beating egg whites. Denaturation of proteins at the surface of the air bubbles stiffens the protein and causes the bubbles to be held in place.

● **Detergents**   Even very low concentrations of detergents can cause denaturation by disrupting the association of hydrophobic side chains.

● **Organic compounds**   Polar solvents such as acetone and ethanol interfere with hydrogen bonding by competing for hydrogen-bonding sites. The disinfectant action of ethanol, for example, results from its ability to denature bacterial protein.

● **pH change**   Excess H^+ or OH^- ions react with the basic or acidic side chains in amino acids and disrupt salt bridges. One familiar example of denaturation by pH change is the protein coagulation that occurs when milk turns sour.

● **Inorganic salts**   Ions can disturb salt bridges, and heavy metal ions such as those of lead, mercury, and silver react with —SH groups to form precipitates. Many heavy metals are poisons because they denature and inactivate crucial enzymes by such reactions.

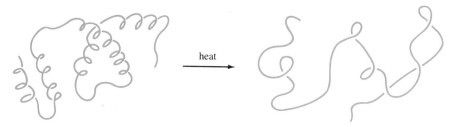

Figure 18.17
Denaturation. Disruption of its secondary, tertiary, or quaternary structure without breaking any peptide bonds results in denaturation of a protein.

Most denaturation is irreversible: Hard-boiled eggs don't soften when their temperature is lowered. Many cases are known, however, in which unfolded proteins spontaneously undergo *renaturation*—a return to their natural state. Renaturation is accompanied by full recovery of biological activity, indicating that the protein has completely returned to its stable secondary and tertiary structure. By refolding into their active shapes, proteins demonstrate that all the information needed to determine these shapes is present in the primary structure.

SUMMARY

Proteins are large biomolecules consisting of α-amino acid residues linked together by amide or **peptide bonds.** Twenty amino acids are commonly found in proteins. All exist as **zwitterions** containing both $-NH_3^+$ and $-COO^-$ ions and in solution establish equilibria between neutral and charged forms. At the **isoelectric point** the numbers of positive and negative charges in an amino acid sample in solution are equal. Amino acids and proteins vary in structure and properties according to the amino acid side chains: Some side chains are nonpolar, some are polar, some are acidic, and some are basic.

Certain organic molecules have handedness and are said to be **chiral;** others have no handedness and are **achiral.** The difference is one of symmetry: Achiral molecules have an imaginary **symmetry plane** cutting through the middle so that one half of the molecule is an exact **mirror image** of the other half. Chiral molecules lack this symmetry plane. The two right- and left-handed forms of a chiral substance are called **optical isomers,** or **enantiomers.** Nineteen of the 20 common amino acids are chiral, and the naturally occurring optical isomers belong to the L **family.** Amino acids of the L **family** occur much more rarely in nature.

Large protein molecules are held in their characteristic shapes by **disulfide bridges, salt bridges,** hydrogen bonds, and **hydrophobic interactions.** Proteins are so large that the word *structure* has several meanings. A protein's **primary structure** is its amino acid sequence. Its **secondary structure** is the way in which segments of the protein chain are oriented into a regular pattern, such as an α helix, a β-pleated sheet, or a triple helix. Its **tertiary structure** is the way in which the entire protein molecule is coiled or folded into a three-dimensional shape that may include α helix, β-pleated sheet, and randomly coiled regions. Its **quaternary structure** is the way in which several protein chains aggregate to form a larger structure. When the structure of a protein is disrupted by heating or other means, the protein is said to be **denatured.**

Proteins can be classified by composition, shape, or biological function. By composition, proteins are either simple or complex. **Simple proteins** yield only amino acids on hydrolysis; **conjugated proteins** yield other substances in addition to amino acids. By shape, proteins are either fibrous or globular. **Fibrous proteins** such as α-keratin are tough and water-insoluble; **globular proteins** such as myoglobin are water-soluble and mobile within cells. By biological function, proteins have an enormous diversity of roles: Some are enzymes, some are hormones, and some serve to store nutrients. Others act as structural, protective, or transport agents.

The chemistry of proteins is similar to that of simple amides. Hydrolysis, either by reaction with aqueous acid or by digestive enzymes, breaks a protein into its individual amino acid constituents.

INTERLUDE: DETERMINING PROTEIN STRUCTURE

Determining the primary structure of a peptide or protein requires answers to three questions: What amino acids are present? How many of each are present? In what order does each occur in the peptide chain? The answers to these questions are provided by two remarkable instruments, the amino acid analyzer and the protein sequenator.

An *amino acid analyzer* is an automated instrument for determining the identity and amount of each amino acid in a protein. The protein is first broken down into its constituent amino acids by reducing all disulfide bonds and hydrolyzing all amide bonds. The amino acid mixture that results is then separated by placing it at the top of a glass column filled with a special adsorbent material and pumping a series of aqueous buffer solutions through the column. Different amino acids pass down the column at different rates depending on their structures and are thus separated.

As each different amino acid passes from the end of the glass column, it is mixed with a solution of *ninhydrin*, a reagent that forms a deep purple color on reaction with α-amino acids, or with a solution of a reagent that forms a fluorescent compound on reaction with the amino acids. The purple color or the fluorescence is detected by an electronic sensor, and its intensity is measured. Since the amount of time required for a given amino acid to pass through a standard column is reproducible, the identity of all amino acids present in the sample can be determined simply by noting the time at which each comes off. The amount of each amino acid is measured by determining the intensity of the purple color or the fluorescence. The figure at the bottom of the page shows the results of amino acid analysis of a standard equimolar mixture of 17 amino acids.

With the identity and amount of each amino acid known, the next task is to *sequence* the peptide: to find out in what order the amino acids are linked together. The general idea of peptide sequencing is to cleave selectively one amino acid residue at a time from the end of the peptide chain, separate and identify that amino acid, and then repeat the process on the chain-shortened peptide until the entire structure is known.

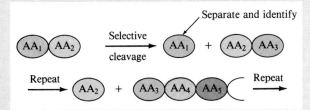

Although the chemistry of the cleavage process is complex, automated protein sequenators are available that allow a series of 30 or more repetitive sequencing steps to be carried out.

Amino acid analysis of an equimolar amino acid mixture. Each peak represents a different amino acid passing through the analysis column. The identities of the amino acids are determined by noting the times at which the peaks appear (horizontal axis), and the amounts are determined by measuring the area of each peak.

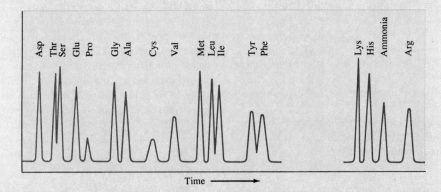

REVIEW PROBLEMS

Amino Acids

18.16 What is an amino acid?

18.17 What does the prefix "α-" mean when referring to α-amino acids?

18.18 What amino acids do these abbreviations stand for?
(a) Ser   (b) Thr   (c) Pro   (d) Phe   (e) Cys

18.19 Draw the structures of the amino acids listed in Problem 18.18.

18.20 Name and draw the structures of amino acids that fit these descriptions:
(a) contains an isopropyl group
(b) contains a secondary alcohol group
(c) contains a thiol group   (d) contains a phenol group

18.21 What do the following terms mean as they apply to amino acids?
(a) zwitterion   (b) essential amino acid
(c) isoelectric point

18.22 Classify these amino acids as neutral, basic, or acidic:
(a) lysine   (b) phenylalanine   (c) glutamic acid
(d) proline

18.23 At what pH would you expect aspartic acid to have the following structures, pH 3, pH 7, or pH 13?

(a) 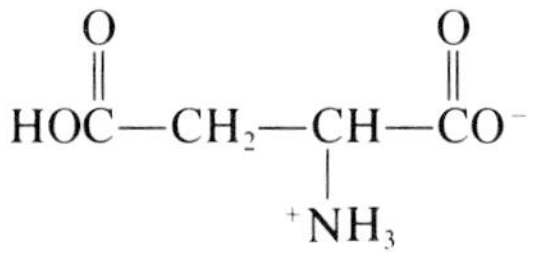 HOC—CH$_2$—CH—CO$^-$ (with =O on first C, =O on last C, $^+$NH$_3$ below middle C)

(b) 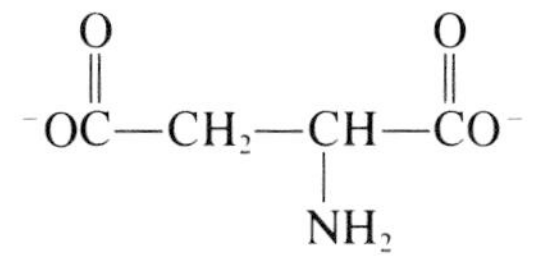 $^-$OC—CH$_2$—CH—CO$^-$ (with =O on first C, =O on last C, NH$_2$ below middle C)

18.24 Draw the structures of proline at pH 1.0, pH 6.3, and pH 9.7.

18.25 Draw the structures of lysine at pH 1.0, pH 6.3, and pH 9.7.

Handedness in Molecules

18.26 What do the terms *chiral* and *achiral* mean?

18.27 Give two examples of chiral objects and two examples of achiral objects.

18.28 Which of the following objects are chiral?
(a) a shoe   (b) a bed   (c) a light bulb
(d) a flower pot   (e) a house key
(f) a pair of scissors

18.29 2-Bromo-2-chloropropane is an achiral molecule, but 2-bromo-2-chlorobutane is chiral. Explain.

18.30 Draw the structures of the following compounds. Which of them is chiral? Put an asterisk by each chiral carbon.
(a) 2-bromo-2-chloro-3-methylbutane
(b) 3-chloropentane   (c) cyclopentanol
(d) 2-methylpropanol

18.31 Which of the carbon atoms marked with arrows in the following compounds are chiral?

(a) 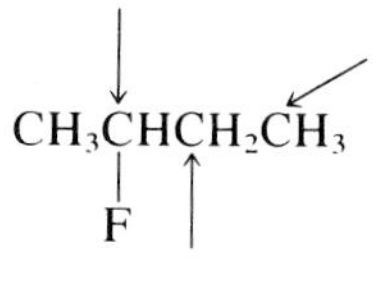 CH$_3$CHCH$_2$CH$_3$ (with F below)   (b)

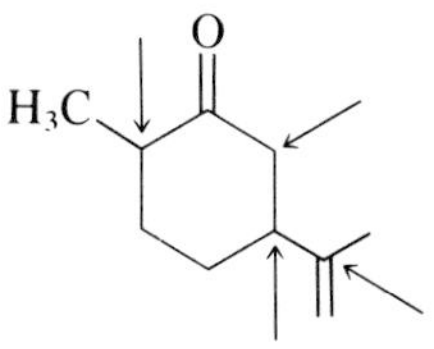

Peptides and Proteins

18.32 What is a protein?

18.33 What is the difference between a peptide and a protein?

18.34 What is the difference between a simple protein and a conjugated protein?

18.35 What kinds of molecules are found in these kinds of conjugated proteins in addition to the protein part?
(a) nucleoproteins   (b) lipoproteins
(c) glycoproteins

18.36 What is the difference between fibrous and globular proteins?

18.37 Name three different biological functions that proteins have in the body.

18.38 What is meant by the following terms as they apply to proteins?
(a) primary structure   (b) secondary structure
(c) tertiary structure   (d) quaternary structure

18.39 What kind of bonding stabilizes helical and β-pleated sheet secondary protein structures?

18.40 Why is cysteine such an important amino acid for defining the tertiary structure of proteins?

18.41 How can cross-links form between cysteine residues in a protein?

18.42 How do the following noncovalent interactions help to stabilize the tertiary and quaternary structure of a protein?
(a) hydrophobic interactions   (b) salt bridges
(c) hydrogen bonding

18.43 For each noncovalent interaction in Problem 18.42, list one pair of amino acids that could give rise to that interaction.

18.44 What kind of change takes place in a protein when it is denatured?

18.45 Explain how a protein is denatured by the following:
(a) heat (b) addition of a strong base (c) mercury ions

18.46 Look at the structure of angiotensin II in Figure 18.6 and identify both the N-terminal and C-terminal amino acids.

18.47 Use the three-letter abbreviations to name all tripeptides containing methionine, isoleucine, and lysine.

18.48 Write structural formulas for the two dipeptides containing phenylalanine and glutamic acid.

18.49 Which of the following amino acids are most likely to be found on the outside of a globular protein and which on the inside? Explain.
(a) valine (b) leucine (c) aspartic acid
(d) asparagine

18.50 Why do you suppose diabetics must receive insulin subcutaneously by injection rather than orally?

18.51 The *endorphins* are a group of naturally occurring neurotransmitters that act in a manner similar to morphine to control pain. Research has shown that the biologically active part of the endorphin molecule is a simple pentapeptide called an *enkephalin*, with the structure Tyr-Gly-Gly-Phe-Met. Draw the complete structure of this enkephalin.

18.52 Identify the N-terminal and C-terminal amino acids in enkephalin (Problem 18.51).

Properties and Reactions of Amino Acids and Proteins

18.53 Much of the chemistry of amino acids is the familiar chemistry of carboxylic acid and amine functional groups. What products would you expect to obtain from these reactions of glycine?

(a) $H_2N-CH_2-\overset{\overset{\displaystyle O}{\|}}{C}OH$ + CH_3OH $\xrightarrow{H^+ \text{ catalyst}}$

(b) $H_2N-CH_2-\overset{\overset{\displaystyle O}{\|}}{C}OH$ + HCl $\longrightarrow$

18.54 How can you account for the fact that glycine is a solid that decomposes without melting when heated to 262°C, whereas the ethyl ester of glycine is a liquid at room temperature?

18.55 We saw in Section 17.9 that amides can be prepared by reaction between a carboxylic acid and an amine in the presence of DCC. What problem would you expect to encounter if you tried to prepare the simple dipeptide glycylglycine by

$2\ H_2N-CH_2-\overset{\overset{\displaystyle O}{\|}}{C}OH$ + $\xrightarrow{DCC}$

$H_2N-CH_2-\overset{\overset{\displaystyle O}{\|}}{C}-NH-CH_2-\overset{\overset{\displaystyle O}{\|}}{C}OH$ + many other products

18.56 (a) Identify the amino acids present in the hexapeptide shown at the bottom of the page.
(b) Identify the N-terminal and C-terminal amino acids of the hexapeptide.
(c) Show the structures of the products that would be obtained on digestion of the hexapeptide.

18.57 Which would you expect to be more soluble in water, a peptide rich in aspartic acid and lysine, or a peptide rich in valine and alanine? Explain.

18.58 Proteins are generally least soluble in water at their isoelectric points. Explain.

18.59 Proteins provide buffering action in cellular fluids. Explain or write equations to show they can do this.

18.60 *Aspartame*, marketed under the trade name Nutra-Sweet for use as a nonnutritive sweetener, is the methyl ester of a simple dipeptide.

$H_2N-CH-\overset{\overset{\displaystyle O}{\|}}{C}-NH-CH-\overset{\overset{\displaystyle O}{\|}}{C}-OCH_3$
with CH_2/COOH and CH_2/phenyl groups — Aspartame

Identify the two amino acids present in aspartame and show all the products of digestion, assuming both amide and ester bonds are hydrolyzed in the stomach.

Applications

18.61 What is pK_a? [App: Henderson-Hasselbalch Equation]

18.62 Calculate what percent of the molecules in alanine are in the zwitterion form at pH 3.5. [App: Henderson-Hasselbalch Equation]

18.63 Why is it more important to have a daily source of protein than a daily source of fats or carbohydrates? [App: Proteins in the Diet]

18.64 What is an incomplete protein? [App: Proteins in the Diet]

$H_2N-CH-\overset{\overset{\displaystyle O}{\|}}{C}-NH-CH-\overset{\overset{\displaystyle O}{\|}}{C}-NH-CH-\overset{\overset{\displaystyle O}{\|}}{C}-NH-CH-\overset{\overset{\displaystyle O}{\|}}{C}-NH-CH-\overset{\overset{\displaystyle O}{\|}}{C}-NH-CH-\overset{\overset{\displaystyle O}{\|}}{C}-OH$
$CH_3CHCH_3 \quad\quad H \quad\quad CH_2OH \quad\quad CH_2CH_2SCH_3 \quad CH_3 \quad\quad CH_2COOH$

18.65 In general, which is more likely to contain complete protein—food from plant sources or from animal sources? [App: Proteins in the Diet]

18.66 What information can be obtained from an amino acid analyzer? [Int: Determining Protein Structure]

18.67 Ninhydrin is often used by forensic scientists to find fingerprints on paper and other surfaces. How do you think this technique works? [Int: Determining Protein Structure]

Additional Questions and Problems

18.68 Fresh pineapple can't be used in gelatin desserts because it contains an enzyme that hydrolyzes the proteins in gelatin, destroying the gelling action. Canned pineapple can be added to gelatin with no problem. Why?

18.69 In persons with sickle-cell anemia, a valine is sub-stituted for a glutamic acid. What amino acid, if substituted for glutamic acid, might not cause a health problem?

18.70 Both α-keratin and tropocollagen have helical secondary structure. How do they differ?

18.71 Bradykinin, a peptide that helps to regulate blood pressure, has the primary structure Arg-Pro-Pro-Gly-Phe-Ser-Pro-Phe-Arg. Draw the complete structural formula of bradykinin.

18.72 Estimate the isoelectric pH of bradykinin (Problem 18.71).

18.73 For each amino acid listed, tell whether its influence on tertiary structure is largely through hydrophobic interactions, hydrogen bonding, formation of salt bridges, covalent bonding, or some combination of these effects.
(a) glutamic acid   (b) methionine   (c) glutamine
(d) threonine   (e) histidine   (f) phenylalanine
(g) cysteine   (h) valine

CHAPTER

19

Enzymes, Vitamins, and Chemical Messengers

Messages cannot flow along nerve cells like these without the aid of neurotransmitters—one of several types of biomolecules that help to keep body chemistry under control.

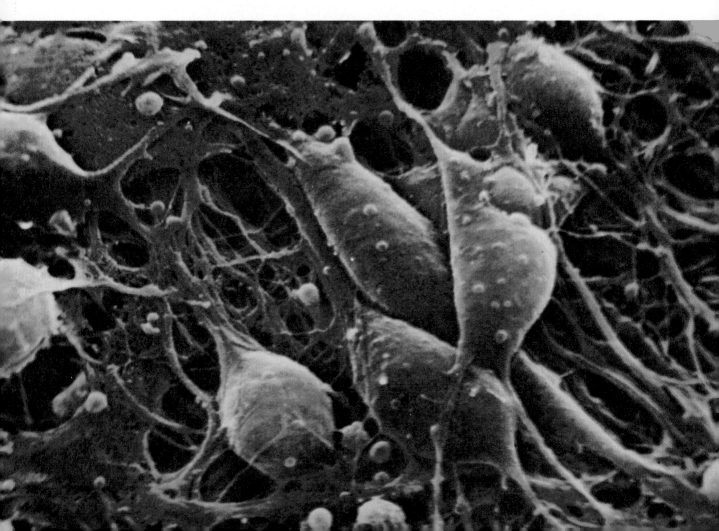

Think of your body as a walking chemical laboratory. Although the analogy isn't perfect, there's a good deal of truth to it. In a laboratory, chemical reactions are carried out one at a time by mixing pure chemicals in test tubes or flasks. In your body, chemical reactions take place in cells rather than in test tubes, and many thousands of reactions take place simultaneously.

The main difference between chemistry in a laboratory and chemistry in a living organism is *control.* In a laboratory, the speed of a reaction is controlled by adjusting experimental conditions such as temperature, solvent, reagent concentrations, and pH. In an organism, though, these conditions can't be adjusted. The human body must maintain a temperature of 37.0°C, the solvent must be water, and the pH must be 7.4.

How then does an organism control its thousands of different reactions so that all occur to the proper extent? The answer is that all reactions in living organisms are governed by biological catalysts called *enzymes.* It has been estimated that the body has at least 50,000 enzymes for regulating the multitude of reactions that take place inside us.

The activities of enzymes are controlled by several families of chemical messengers, including hormones and neurotransmitters. These messengers coordinate responses to changes in our environment and in conditions within our bodies. In this chapter, we'll explore how enzymes control biological reactions and how enzymes are in turn regulated by chemical messengers. Along the way we'll answer the following questions:

1. ***What are enzymes?*** The goal: Be able to describe the general structure of enzymes and their function in biological reactions.

2. ***What kinds of enzymes are there and how are they classified?*** The goal: Be able to name the classes of enzymes and describe the type of reaction catalyzed by each.

3. ***How do enzymes work and why are they so specific?*** The goal: Be able to use the lock-and-key and induced-fit models to explain why enzymes are specific and how they speed up reactions.

4. ***What effect do temperature, pH, enzyme concentration, and substrate concentration have on enzyme activity?*** The goal: Be able to describe what happens to enzyme activity when temperature, pH, enzyme concentration, and substrate concentration gradually change.

5. ***How is enzyme activity regulated?*** The goal: Be able to describe enzyme inhibition, feedback control, and allosteric control.

6. ***What are vitamins, and how do they function?*** The goal: Be able to describe the types of vitamins and their roles in biochemical reactions.

7. ***What are hormones, and how do they function?*** The goal: Be able to describe the origin, role in biochemical reactions, and mechanism of action of hormones.

8. ***What are neurotransmitters, and how do they function?*** The goal: Be able to describe the origin, role in biochemical reactions, and mechanism of action of neurotransmitters.

19.1 ENZYMES

Enzyme A protein or other molecule that acts as a catalyst for a biological reaction.

Enzymes are catalysts for biological reactions. Recall from Section 8.7 that a **catalyst** is a substance that speeds up the rate of a chemical reaction but that itself undergoes no permanent chemical change during the reaction. Sulfuric acid, for example, catalyzes the reaction of a carboxylic acid with an alcohol to yield an ester (Section 17.5). The reaction would occur very slowly if the catalyst were not present.

Reaction would occur very slowly without catalyst

$$CH_3-\overset{O}{\underset{||}{C}}-OH \ + \ HOCH_3 \ \xrightarrow[\text{catalyst}]{H_2SO_4} \ CH_3-\overset{O}{\underset{||}{C}}-OCH_3 \ + \ H_2O$$

Catalyst A substance that speeds up a chemical reaction without itself undergoing any permanent chemical change.

Enzymes are similar to sulfuric acid in that they catalyze reactions that might otherwise occur very slowly, but they differ in two important respects. First, enzymes are far larger, more complicated molecules than simple inorganic catalysts since, with just a few exceptions, enzymes are proteins. Second, enzymes are far more specific in their action. Whereas sulfuric acid catalyzes the reaction of nearly *every* carboxylic acid with nearly *every* alcohol, many enzymes catalyze only a single reaction of a single reactant.

The reactant in an enzyme-catalyzed reaction is known as the enzyme's **substrate,** and the **specificity** of an enzyme is the extent to which it reacts with different substrates.

Substrate The reactant in an enzyme-catalyzed reaction.

Specificity The extent to which an enzyme reacts with different substrates.

$$\text{Substrate} \ \xrightarrow{\text{enzyme catalyst}} \ \text{product}$$

Enzymes differ greatly in their substrate specificity. For example, the enzyme *amylase* found in human digestive systems is able to catalyze the hydrolysis of starch to yield glucose but has no effect on cellulose, which is also composed of glucose. By contrast, the enzyme *papain*, a globular protein of 212 amino acids isolated from papaya fruit, catalyzes the hydrolysis of the peptide bonds in a variety of different polypeptides. It's this ability to hydrolyze peptides, in fact, that accounts for the use of papain in meat tenderizers and in contact-lens cleaners.

Bond that breaks

$$\xi-NH-\underset{R}{CH}-\overset{O}{\underset{||}{C}}-NH-\underset{R'}{CH}-\overset{O}{\underset{||}{C}}\xi \ \xrightarrow[\text{papain}]{H_2O} \ \xi-NH-\underset{R}{CH}-\overset{O}{\underset{||}{C}}-OH \ + \ H_2N-\underset{R'}{CH}-\overset{O}{\underset{||}{C}}\xi$$

Like all catalysts, an enzyme doesn't affect the equilibrium point of a reaction and can't bring about a reaction that is energetically unfavorable. What an enzyme *does* do is decrease the time it takes to reach equilibrium by lowering the activation energy. Catalase, for example, accelerates the rate of decomposition of hydrogen peroxide about 10^4 times more than a metal catalyst does (Figure 19.1).

Figure 19.1
The action of catalase on hydrogen peroxide (H_2O_2). When ground beef liver is added, the hydrogen peroxide foams up as it rapidly decomposes to water and oxygen. In the body, catalase catalyzes the same reaction, needed to destroy hydrogen peroxide produced by oxidase enzymes that remove hydrogen from substrates by combination with oxygen.

Turnover number The number of substrate molecules acted on by one molecule of enzyme per unit time.

The catalytic activity of an enzyme is measured by its **turnover number,** the number of substrate molecules acted on by one molecule of enzyme per unit time. As indicated in Table 19.1, enzymes vary greatly in their turnover number. Most enzymes have values in the 1–10,000 range, but some are much higher. Carbonic anhydrase in red blood cells, for example, is able to catalyze the reaction of *600,000* substrate molecules per second.

$$CO_2 + H_2O \xrightarrow{\text{carbonic anhydrase}} H_2CO_3 \longrightarrow HCO_3^- + H^+$$

19.2 ENZYME STRUCTURE

Cofactor A small, nonprotein part of an enzyme that is essential to the enzyme's catalytic activity.

Apoenzyme The protein portion of an enzyme.

Holoenzyme The combination of apoenzyme and cofactor that is active as a biological catalyst.

Many enzymes are globular proteins and thus have specific three-dimensional shapes determined by their secondary and tertiary structure. Many other enzymes consist of more than one protein chain and therefore have quaternary structure also.

In addition to their protein part, many enzymes are associated with small, nonprotein portions called **cofactors,** which may be held within the enzyme by either covalent bonds or noncovalent interactions. In such enzymes, the protein part is called an **apoenzyme,** while the entire assembly of apoenzyme plus cofactor is called a **holoenzyme.** Only holoenzymes are active as catalysts; neither apoenzyme nor cofactor alone can catalyze a reaction. An enzyme cofactor can be either an inorganic ion (usually a metal ion) or a small organic

Table 19.1   Turnover Numbers of Some Enzymes

Enzyme	Turnover Number (per second)
Carbonic anhydrase	600,000
Acetylcholinesterase	25,000
β-Amylase	18,000
Penicillinase	2,000
DNA polymerase I	15

Coenzyme A small, organic molecule that acts as an enzyme cofactor.

molecule called a **coenzyme.** Some enzymes need both a metal ion and a coenzyme to become active.

$$\text{Holoenzyme} = \text{apoenzyme} + \text{cofactor} \quad \begin{array}{l}\text{Metal ion and/or small}\\ \text{organic molecule}\\ \text{(coenzyme)}\end{array}$$

Why are cofactors necessary? The functional groups in proteins are limited to those of the amino acid side chains. By joining up with cofactors, however, enzymes can acquire chemically reactive groups not available from their amino acids, for example, groups needed to catalyze redox reactions.

The requirement that many enzymes have for metal ion cofactors is the main reason behind our dietary need for trace minerals. Iron, zinc, copper, manganese, molybdenum, cobalt, nickel, vanadium, and selenium are all known to function as enzyme cofactors. Every molecule of carboxypeptidase A, for example, contains one Zn^{2+} ion that serves as a catalyst in the hydrolysis of a peptide bond. Like the trace minerals, many vitamins are a necessary part of our diets because they are needed for enzyme activity (Section 19.11).

Practice Problem 19.1 Check the label on a bottle of vitamin-mineral supplements and look at the correlation between the metals listed and the metals known to be present in the body as enzyme cofactors.

19.3 ENZYME CLASSIFICATION

Enzymes are arranged into six main classes according to the general kind of reaction they catalyze, and each main class is further subdivided (Table 19.2). Most of the main-class names are self-explanatory; they're listed here with some examples:

● *Oxidoreductases* catalyze oxidation–reduction reactions of substrate molecules, which may involve addition or removal of oxygen or hydrogen:

$$\text{A(reduced)} + \text{B(oxidized)} \longrightarrow \text{A'(oxidized)} + \text{B'(reduced)}$$

$$\text{CH}_3\text{—CH}_2\text{—OH} + \text{NAD}^+ \xrightarrow[\text{dehydrogenase}]{\text{alcohol}} \text{CH}_3\overset{\overset{\text{O}}{\|}}{\text{—C}}\text{—H} + \text{NADH} + \text{H}^+$$
(Reduced) (Oxidized) (Oxidized) (Reduced)

● *Transferases* catalyze the transfer of a group from one substrate to another:

$$\text{A} + \text{B—C} \longrightarrow \text{A—B} + \text{C}$$

Glucose + adenosine triphosphate (ATP) $\xrightarrow{\text{hexose kinase}}$

glucose 6-phosphate + adenosine diphosphate (ADP)

Table 19.2   Classification of Enzymes

Main Class	Some Subclasses	Type of Reaction Catalyzed
Oxidoreductases	Oxidases	Oxidation of a substrate
	Reductases	Reduction of a substrate
	Dehydrogenases	Introduction of double bond (oxidation) by formal removal of H_2 from substrate
Transferases	Transaminases	Transfer of an amino group between substrates
	Kinases	Transfer of a phosphate group between substrates
Hydrolases	Lipases	Hydrolysis of ester groups in lipids
	Proteases	Hydrolysis of amide groups in proteins
	Nucleases	Hydrolysis of phosphate groups in nucleic acids
Lyases	Dehydrases	Loss of H_2O from substrate
	Decarboxylases	Loss of CO_2 from substrate
Isomerases	Epimerases	Isomerization of chiral center in substrate
Ligases	Synthetases	Formation of new bond between two substrates, with participation of ATP
	Carboxylases	Formation of new bond between substrate and CO_2, with participation of ATP

• *Hydrolases* catalyze the hydrolysis of substrates, that is, the breaking of bonds with addition of water:

$$A\!-\!B + H_2O \longrightarrow A\!-\!OH + B\!-\!H$$

Polypeptide   Shortened polypeptide

• *Isomerases* catalyze the isomerization (rearrangement of atoms) of substrates:

$$A \longrightarrow B$$

Maleate   Fumarate

• *Lyases* (from the Greek *lein*, meaning "to break") catalyze the addition of a small molecule such as H_2O or NH_3 (usually to a double bond) and the reverse reaction, the elimination of a small molecule (usually to leave behind a double bond).

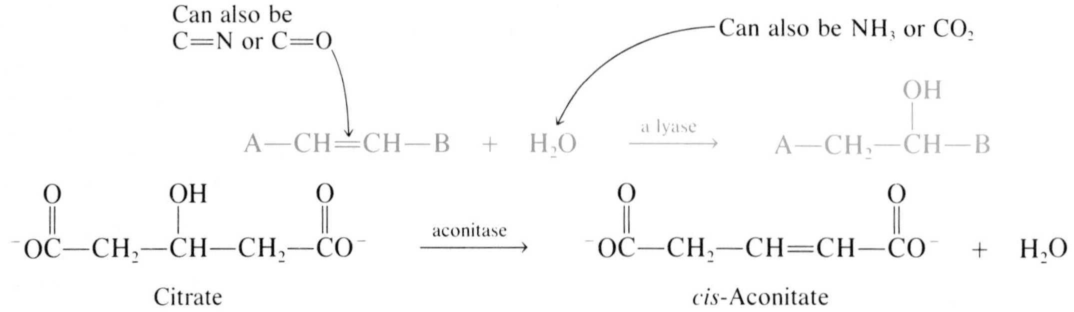

Can also be C=N or C=O

Can also be NH_3 or CO_2

$$A—CH{=\!=}CH—B \;+\; H_2O \xrightarrow{\text{a lyase}} A—CH_2—\overset{\overset{\displaystyle OH}{|}}{CH}—B$$

$$\overset{\overset{\displaystyle O}{\|}}{^-OC}—CH_2—\overset{\overset{\displaystyle OH}{|}}{CH}—CH_2—\overset{\overset{\displaystyle O}{\|}}{CO^-} \xrightarrow{\text{aconitase}} \overset{\overset{\displaystyle O}{\|}}{^-OC}—CH_2—CH{=}CH—\overset{\overset{\displaystyle O}{\|}}{CO^-} \;+\; H_2O$$

Citrate   *cis*-Aconitate

● *Ligases* (from the Latin *ligare*, meaning "to tie together") catalyze the bonding together of two substrate molecules, with the participation of ATP.

$$A + B + \text{adenosine triphosphate (ATP)} \longrightarrow A—B + \text{adenosine diphosphate (ADP)} + PO_4^{3-} \;(P_i)$$

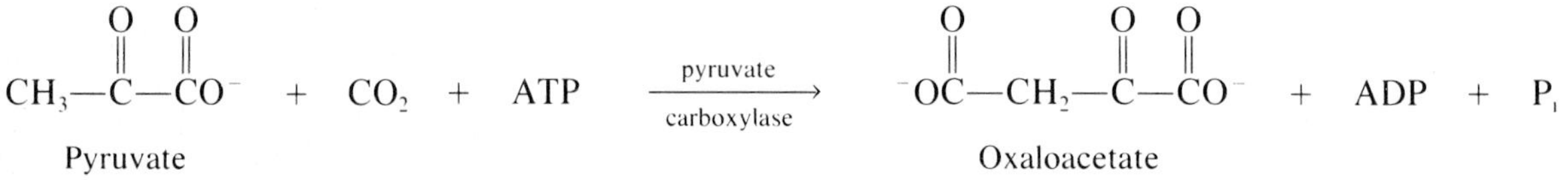

$$CH_3—\overset{\overset{\displaystyle O}{\|}}{C}—\overset{\overset{\displaystyle O}{\|}}{CO^-} + CO_2 + ATP \xrightarrow[\text{carboxylase}]{\text{pyruvate}} \overset{\overset{\displaystyle O}{\|}}{^-OC}—CH_2—\overset{\overset{\displaystyle O}{\|}}{C}—\overset{\overset{\displaystyle O}{\|}}{CO^-} + ADP + P_i$$

Pyruvate   Oxaloacetate

Note in the examples given above that all enzymes have the family-name ending *-ase*. Although many enzymes like papain and trypsin were given uninformative common names in the past, the modern systematic names have two parts: The first part identifies the substrate molecule on which the enzyme operates, and the second part is an enzyme subclass name like those shown in Table 19.2. For example, *alcohol dehydrogenase* is an oxidoreductase enzyme that oxidizes an alcohol to yield an aldehyde by removal of two hydrogens that are transferred to a coenzyme.

Solved Problem 19.1 To what class does the enzyme that catalyzes the following reaction belong?

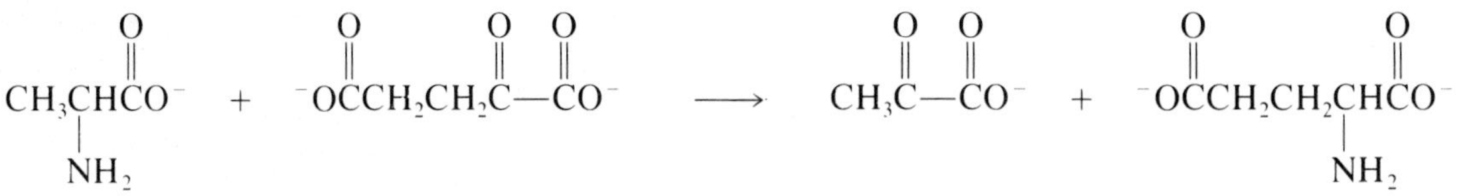

$$CH_3\overset{\overset{\displaystyle O}{\|}}{\underset{\underset{\displaystyle NH_2}{|}}{CH}}CO^- + {}^-OCCH_2CH_2\overset{\overset{\displaystyle O}{\|}}{C}—\overset{\overset{\displaystyle O}{\|}}{CO^-} \longrightarrow CH_3\overset{\overset{\displaystyle O}{\|}}{C}—\overset{\overset{\displaystyle O}{\|}}{CO^-} + {}^-OCCH_2CH_2\underset{\underset{\displaystyle NH_2}{|}}{\overset{\overset{\displaystyle O}{\|}}{CH}}CO^-$$

Solution Comparing the functional groups in the reactants and products shows that an amino group and a carbonyl group have changed places. The reaction is therefore a functional-group transfer, and the enzyme is a transferase.

Practice Problems **19.2** What reactions would you expect these enzymes to catalyze?
(a) fumarate hydrase (b) squalene oxidase (c) glucose kinase
(d) cellulose hydrolase

19.3 To what class of enzymes does pyruvate decarboxylase, which acts with the participation of ATP, belong?

19.4   ENZYME SPECIFICITY

Any theory of how enzymes work must explain two facts: It must explain why enzymes are so specific, and it must explain how enzymes speed up reactions. Our current picture of enzymes relies on two complementary models to explain these facts. In the first, called the **lock-and-key model,** an enzyme is pictured as a large, irregularly shaped molecule with a cleft or crevice in its middle. Inside the crevice is an **active site,** a small, three-dimensional region of the enzyme with the specific shape, side chains, and cofactors necessary to bind the substrate, usually by noncovalent interactions, and catalyze the appropriate reaction. In other words, the shape of the active site is like a lock into which only a specific key, the substrate, can fit (Figure 19.2).

The lock-and-key model requires that enzyme and substrate have un- changing shapes, which means that only one substrate can fit a given enzyme. Although it's true that some enzymes are so specific they operate on only a single substrate molecule, other enzymes are less selective. For example, the lipase enzymes secreted by the pancreas hydrolyze all dietary fats and oils, regardless of their exact structure. To account for the broader specificity of many enzymes, the lock-and-key model has been expanded.

According to the **induced-fit model,** some enzymes are flexible enough to change the shapes and sizes of their active sites to fit the spatial requirements of different substrates. As the enzyme and substrate come together, their interac- tion *induces* exactly the right fit (Figure 19.3). Changes in the shape of the

Lock-and-key model   A model for enzyme specificity that pictures an enzyme as a large molecule with a cleft into which only specific substrate molecules can fit.

Active site   A small, three- dimensional portion of an enzyme with the specific shape and structure necessary to bind a substrate.

Induced-fit model   A model that pictures an enzyme with a conformationally flexible active site that can change shape to accommodate a range of different substrate molecules.

Figure 19.2
The lock-and-key model of enzymes. Enzymes are large, three-dimensional molecules containing a crevice with a well-defined active site. Only a substrate whose shape and chemical nature is complementary to that of the active site can fit into the enzyme.

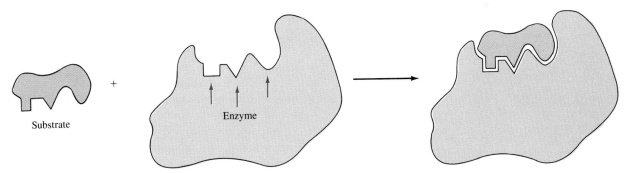

Substrate   +   Enzyme

Figure 19.3
The induced-fit model of enzyme action. Some enzymes have a conformation that is flexible enough to allow them to adapt the size and shape of their active site to the shapes of several different substrates.

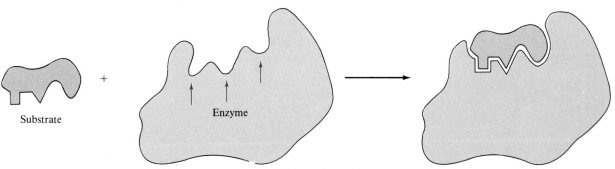

Substrate   +   Enzyme

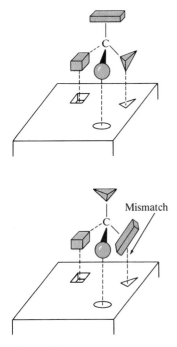

Figure 19.4
A chiral reactant and a chiral reaction site. The optical isomer on the left will fit the reaction site like a hand in a glove, but the optical isomer on the right can't fit and therefore can't react.

Enzyme–substrate complex A complex of enzyme and substrate in which the two are bound together by noncovalent interactions with the substrate in position to react.

enzyme allow it to conform to the substrate, and changes in the shape of the substrate prepare it to react.

Just as an enzyme is often specific for a given substrate, it's also specific for one of a pair of optical isomers if the substrate is a chiral molecule (Section 18.5). For example, the enzyme lactate dehydrogenase catalyzes the removal of hydrogen from "left-handed" L-lactate but not from "right-handed" D-lactate.

$$
\begin{array}{c}
\underset{\text{L-Lactate}}{\overset{\displaystyle \overset{\text{O}}{\underset{\text{C}}{\parallel}}\text{—O}^-}{\underset{\text{CH}_3}{\underset{|}{\text{HO—C—H}}}}} + \text{NAD}^+
\xrightarrow[\text{dehydrogenase}]{\text{lactate}}
\underset{\text{Pyruvate}}{\overset{\displaystyle \overset{\text{O}}{\underset{\text{C}}{\parallel}}\text{—O}^-}{\underset{\text{CH}_3}{\underset{|}{\text{C=O}}}}} + \text{NADH} + \text{H}^+
\end{array}
$$

$$
\begin{array}{c}
\underset{\text{D-lactate}}{\overset{\displaystyle \overset{\text{O}}{\underset{\text{C}}{\parallel}}\text{—O}^-}{\underset{\text{CH}_3}{\underset{|}{\text{H—C—OH}}}}} + \text{NAD}^+
\xrightarrow[\text{dehydrogenase}]{\text{lactate}}
\text{no reaction}
\end{array}
$$

As we saw in the previous chapter, amino acids and proteins have handedness. Thus, the specificity of an enzyme for one of two optical isomers is a matter of fit. As illustrated schematically in Figure 19.4, a left-handed enzyme can't fit with a right-handed substrate any more than a left-handed glove can fit on a right hand.

19.5 HOW ENZYMES WORK

Now that you know why enzymes are so specific, we need to examine how they speed up reactions. Enzyme-catalyzed reactions begin with migration of the substrate into the active site to form an **enzyme–substrate complex.** Before complex formation, the substrate molecule is in its most stable, lowest-energy form. Within the complex, the molecule is forced into a higher-energy form in which certain bonds have been weakened. The result is to make the energy barrier between substrate and product smaller.

Within the enzyme–substrate complex, atoms that will form new bonds must connect with each other, and groups needed for catalysis must be close to the necessary locations in the substrate. Many organic reactions, for example, require acidic, basic, or metal ion catalysts. The active site can provide acidic or basic groups without disrupting the constant-pH environment in body fluids. Once the chemical reaction is completed, enzyme and product molecules separate from each other and the enzyme becomes available for another substrate.

$$
\underset{\text{Enzyme}}{\text{E}} + \underset{\text{Substrate}}{\text{S}} \;\rightleftharpoons\; \underset{\text{Complex}}{[\text{E—S}]} \longrightarrow \underset{\text{Enzyme}}{\text{E}} + \underset{\text{Product}}{\text{P}}
$$

The hydrolysis of a peptide bond by chymotrypsin shown in Figure 19.5 illustrates how an enzyme functions. In the enzyme–substrate complex (Figure

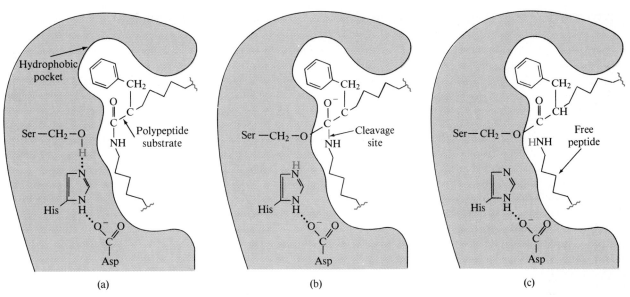

Figure 19.5
Hydrolysis of a peptide bond by chymotrypsin. (a) The polypeptide enters the active
site with its hydrophobic side chain in the hydrophobic pocket and the peptide bond
opposite serine and histidine residues. (b) Hydrogen transfer from serine to histidine
allows bonding to the peptide. (c) The peptide bond is broken. Subsequent steps, not
shown, release the other piece of the hydrolyzed polypeptide and restore the H to
serine.

19.5a), attraction of a hydrophobic side chain into a hydrophobic pocket in the
active site positions the substrate and places the bond to be broken next to the
catalytic site. The enzyme has not only brought the two reactants together but
has also oriented them correctly. Next (Figure 19.5b), the peptide bond carbon
is temporarily bonded to serine in the active site, making it easier for the peptide
bond to break (Figure 19.5c). In further steps not shown, the active site returns
to its original condition and the rest of the polypeptide is set free.

In summary, enzymes act as catalysts because of their ability to

● Bring reactants together (*proximity effect*)
● Hold reactants at the exact distance and in the exact orientation necessary
for reaction (*orientation effect*)
● Lower the energy barrier by inducing strain in the substrate (*energy effect*)
● Provide acidic, basic, or other types of sites required for catalysis
(*catalytic effect*).

19.6 INFLUENCE OF TEMPERATURE AND pH ON ENZYMES

Enzymes have been finely tuned through evolution so that their maximum
catalytic ability is highly dependent on pH and temperature. As you might
expect, optimum conditions vary slightly for each enzyme but are generally
near neutral pH and body temperature.

AN APPLICATION: MEDICAL USES OF ENZYMES AND ISOENZYMES

In a healthy person, most enzymes are found mainly within cells. When some diseases occur, however, enzymes are released from dying cells into the blood, where their increased levels can be measured by sensitive clinical instruments. The accompanying table lists some of the presently available enzyme tests, along with the medical conditions indicated by abnormal blood levels. Enzymes marked with an asterisk in the table are included in the representative routine blood analysis given in the Application on homeostasis in Chapter 5.

Enzyme analysis relies on measuring the *activity* of an enzyme rather than its concentration. Because activity is influenced by pH, temperature, and substrate concentration, it is measured in international units at standard conditions. One unit (U) is defined as the amount of the enzyme that converts one micromole of substrate to product under defined standard conditions of pH, temperature, and substrate concentration. The analytical results are thus reported in units per liter (U/L).

Among the most useful enzyme assays are those done for diagnosis of heart disease. Three enzymes, creatine phosphokinase (CPK), aspartate transaminase (AST), and lactate dehydrogenase (LDH), are found in the muscle cells of a healthy heart, as well as in other tissues. When a heart attack, or *myocardial infarction* (MI), occurs, some cells are damaged and their enzymes leak into the bloodstream. As shown in the first figure, the blood levels of CPK, AST, and LDH all increase markedly in the hours immediately following a heart attack. The CPK level rises almost immediately following an MI, reaching a sixfold increase over normal values after about 30 hr; the AST level triples after about 40 hr; and the LDH level doubles after about 4 days.

Confirmation that a heart attack has occurred can be gained by a careful analysis of the individual enzyme levels. Recent work has shown that a number of enzymes, including CPK

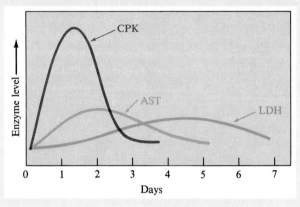

Blood levels of creatine phosphokinase (CPK), aspartate transaminase (AST), and lactate dehydrogenase (LDH) in the days following a heart attack (myocardial infarction).

and LDH, are actually mixtures of several closely related compounds called *isoenzymes* that have slightly different structures but that catalyze the same reaction. For example, creatine phosphokinase is a mixture of three isoenzymes, denoted CK(MM), CK(MB), and CK(BB). Brain tissue is rich in CK(BB), skeletal muscles are rich in CK(MM), and heart tissue is rich in CK(MB). Similarly, lactate dehydrogenase is a mixture of five isoenzymes, denoted LDH_1, LDH_2, LDH_3, LDH_4, and LDH_5. Of the five, heart tissue contains primarily LDH_1.

Since heart tissue contains primarily the CK(MB) and LDH_1 isoenzymes, their levels increase the most following a heart attack. Thus, separation and analysis of the five individual LDH isoenzymes shows that the LDH_2 level is higher than that of LDH_1 in a normal profile but that the two levels "flip" following a heart attack (facing page). Similarly, an analysis of the three CK isoenzymes shows that the CK(MB) level increases to a much greater extent than that of the other two following a heart attack.

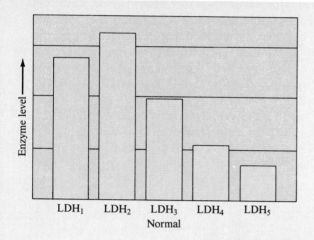 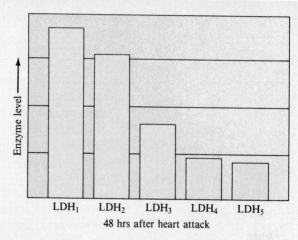

Blood levels of the five lactate dehydrogenase (LDH) isoenzymes in a normal person and in a heart attack victim after 48 hr. Note the flip of LDH_1 and LDH_2 levels following a heart attack.

Some Enzyme Assays in Body Fluids

Enzyme	Condition Indicated by Abnormal Level
Lactate dehydrogenase (LDH)[a]	Heart disease, liver diseases
Creatine phosphokinase (CPK)	Heart disease
Aspartate transaminase (AST)[a]	Heart disease, liver diseases, muscle damage
Alanine transaminase (ALT)[a]	Heart disease, liver diseases, muscle damage
Glutamyl transferase (GMT)	Liver diseases
Alkaline phosphatase (ALP)[a]	Bone disease, liver diseases
Amylase	Pancreatic diseases
Lipase (LPS)	Pancreatic diseases
Acid phosphatase (ACP)	Prostate cancer
Renin	Hypertension
Glucose-6-phosphate dehydrogenase (GPD)	Hemolytic anemia
γ-Glutamyl transferase (GGT)[a]	Liver disease, alcoholism

[a] Included in routine blood analysis illustrated in the Application on homeostasis in Chapter 5 (see footnote there on abbreviations).

Effect of Temperature   An increase in temperature leads to an increase in rate for most chemical reactions, and enzyme-catalyzed reactions are no exception. Unlike many simple reactions, however, enzyme-catalyzed processes always reach an optimum temperature, after which their rate again falls as shown in Figure 19.6. The falloff in reaction rate with heating beyond the optimum temperature occurs because enzymes begin to denature when heated too strongly (Section 18.12). The noncovalent attractions between protein side chains are disrupted, the delicately maintained three-dimensional shape of the enzyme begins to come apart, and as a result the active site needed for catalytic activity disappears.

Most enzymes denature and lose their catalytic activity above 50–60°C, a fact that explains why medical instruments and laboratory glassware can be sterilized by heating with steam in an autoclave. The high temperature of the autoclave denatures the enzymes of any bacteria present, thereby killing the organisms.

Effect of pH   The catalytic activity of many enzymes depends on the pH of the surroundings and often has a well-defined optimum point. For example, trypsin, a protease enzyme that aids digestion of proteins in the small intestine, has optimum activity at pH 8.0, an activity that falls dramatically at a pH either higher or lower than 8.0 (Figure 19.7). Changes in pH can change enzyme structure because of denaturation and can also change the ability of the active site to bind a substrate because of changes in the charges of acidic and basic side chains.

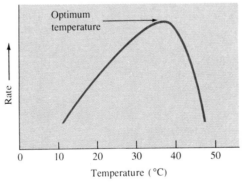

Figure 19.6
The effect of temperature on the rate of an enzyme-catalyzed reaction. There is always an optimum temperature at which the reaction rate reaches its maximum.

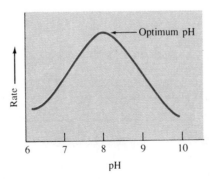

Figure 19.7
The effect of pH on the catalytic activity of trypsin, a protease digestive enzyme of the small intestine.

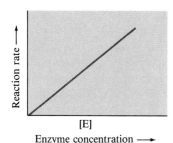

Figure 19.8
Change in reaction rate with enzyme concentration when [E] < [S] at constant temperature, pH, and substrate concentration. The reaction rate in the presence of excess substrate is directly proportional to the enzyme concentration. Doubling the enzyme concentration, for example, doubles the reaction rate.

19.7 EFFECT OF ENZYME AND SUBSTRATE CONCENTRATION ON ENZYME ACTIVITY

For a reaction to occur, the enzyme and substrate molecules must come together and form the enzyme–substrate complex. Therefore, some variation in the reaction rate can be expected if enzyme or substrate concentrations change.

First, consider what happens when the substrate concentration is high relative to the enzyme concentration. Under these conditions, all the enzymes are working to capacity. With *increasing enzyme concentration*, the additional enzyme molecules immediately start taking up substrate and the reaction rate increases. If the enzyme concentration is doubled, for example, the rate also doubles, a directly proportional relationship as shown in Figure 19.8.

Next, consider what happens when the substrate concentration is low relative to the enzyme concentration. Not all the enzymes are being utilized, and the rate is dependent on the concentration of substrate. With *increasing substrate concentration*, the rate increases because more of the enzyme molecules are put to work. Eventually, however, the substrate concentration reaches a point where all the available active sites are occupied. Since the reaction rate is now determined by how fast the enzyme–substrate complex is converted to product, the reaction rate levels off and becomes constant, and the enzyme is said to be *saturated*. Beyond this point, increasing substrate concentration has no effect on the rate (Figure 19.9).

Under most conditions, an enzyme is not likely to be saturated and the reaction rate is controlled by the overall efficiency of the enzyme. If the enzyme–substrate complex is rapidly converted to product, the rate at which enzyme and substrate combine to form the complex becomes the limiting factor. Calculations show there's an upper limit to this rate: Enzyme and substrate molecules moving at random in solution can collide with each other no more often than about 10^8 collisions per mole per second. Thus, no reaction can take place faster than this limit. Remarkably, a few enzymes actually operate with this maximum efficiency—every one of those 10^8 collisions results in the formation of product! One example is *triose phosphate isomerase*, which catalyzes a step in the breakdown of glucose.

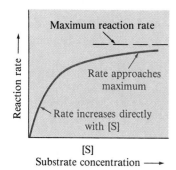

Figure 19.9
Change in reaction rate with substrate concentration at constant temperature, pH, and enzyme concentration. At low substrate concentration, the reaction rate is directly proportional to the concentration. At high substrate concentration, the rate approaches a maximum.

$$
\begin{array}{ccc}
\mathrm{CH_2OH} & & \mathrm{H}\diagdown\mathrm{C}{=}\mathrm{O} \\
| & \xrightarrow[\text{isomerase}]{\text{triose phosphate}} & | \\
\mathrm{C}{=}\mathrm{O} & & \mathrm{H{-}C{-}OH} \\
| & & | \\
\mathrm{CH_2OPO_3{}^{2-}} & & \mathrm{CH_2OPO_3{}^{2-}}
\end{array}
$$

Dihydroxyacetone phosphate   Glyceraldehyde 3-phosphate

19.8 ENZYME INHIBITION

The control of biological reactions by enzymes is only half the overall picture. It's equally important that the enzymes themselves be regulated. After all, if an enzyme were continually functioning at full speed, it would soon run out of substrate and the body would soon have a huge oversupply of the enzyme's product.

Living things control their enzymes by a variety of strategies. Although we'll describe them separately, you should keep in mind that a number of control strategies operate simultaneously in most series of related biochemical reactions. Considering that a cell contains thousands of protein molecules and hundreds of other kinds of biomolecules, all in the concentrations required to maintain **homeostasis,** the achievement of enzyme control is awe-inspiring.

Any process that starts up or increases the action of an enzyme is called **activation.** Any process that slows down or stops the action of an enzyme is called **inhibition.** Inhibition is important for natural enzyme control and can also be put to work in medications that modify enzyme activity. Some inhibitors, however, are poisons because they prevent an enzyme from carrying out a necessary function. In this and the next two sections, we'll describe three types of inhibition and some of the major strategies for regulating enzyme activity.

Homeostasis Maintenance of unchanging internal conditions by living things.

Activation (of an enzyme) Any process that initiates or increases the action of an enzyme.

Inhibition (of an enzyme) Any process that slows down or stops the action of an enzyme.

Competitive enzyme inhibition Enzyme regulation in which an inhibitor competes with a similarly shaped substrate for binding to the enzyme active site.

Competitive Inhibition What would happen if an enzyme were to encounter a molecule with a shape and size similar to that of its normal substrate? The imposter molecule could bind to the enzyme's active site and thereby prevent a normal substrate molecule from binding to the same site. As a result, the enzyme would be inactivated (Figure 19.10) and the overall effect would be equivalent to decreasing the enzyme concentration. Inhibition of this sort is called **competitive inhibition** because the inhibitor competes with substrate for binding to the active site. Of course, if the inhibitor later happens to migrate *out* of the active site, the enzyme will once again be able to bind with substrate and be fully active. Thus, competitive inhibition is reversible. As shown in Figure 19.11 (middle curve), the maximum reaction rate is unchanged by competitive inhibition, but a higher substrate concentration is required to reach that rate.

Competitive inhibition is used to good advantage in the treatment of methanol poisoning. Although not harmful itself, methanol (wood alcohol) is oxidized in the body to formaldehyde, which is highly toxic ($CH_3OH \rightarrow H_2C=O$). Because of its similarity to methanol, ethanol is able to act as a competitive inhibitor of the methanol oxidase enzyme, thereby blocking oxidation and allowing methanol to be excreted harmlessly. Thus, the medical treatment of methanol poisoning includes administering high levels of ethanol.

Figure 19.10
Competitive inhibition. A competitive inhibitor blocks an enzyme's active site by mimicking the shape and size of the normal substrate, thereby preventing the enzyme from functioning.

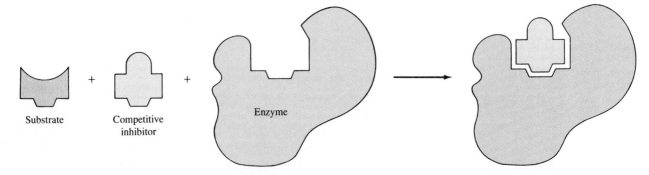

Substrate + Competitive inhibitor + Enzyme →

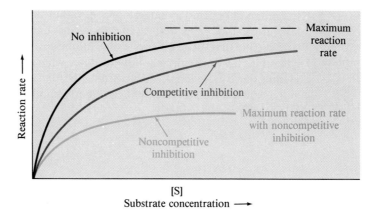

Figure 19.11
Change in reaction rate with inhibitors. The rate increases more slowly with a competitive inhibitor (red) but at high enough [S] reaches the maximum rate for the enzyme. A noncompetitive inhibitor (blue) lowers the maximum rate.

Noncompetitive Inhibition   A second kind of reversible inhibition occurs when an inhibitor binds to an enzyme at some place other than the active site. Such binding can lead to a change in the shape of the enzyme, with a resultant change in the shape and binding ability of the active site. Binding of substrate thus becomes more difficult, enzyme activity diminishes, and the maximum reaction rate decreases (Figure 19.11, bottom curve). Called **noncompetitive inhibition** because the inhibitor does not compete with substrate for the active site, this kind of effect is illustrated in Figure 19.12.

> **Noncompetitive enzyme inhibition** Enzyme regulation in which an inhibitor binds to an enzyme elsewhere than at the active site, thereby changing the shape of the enzyme's active site.

   Lead, mercury, and other heavy metals are toxic because of their ability to act as noncompetitive inhibitors. These metals bind strongly to thiol groups (—SH) on cysteine units in enzymes, thus altering the enzymes' shapes.

Irreversible Inhibition   Yet a third type of inhibition occurs when a molecule enters an enzyme's active site and forms a covalent bond to the enzyme. Since

Figure 19.12
Noncompetitive inhibition. By binding elsewhere on the enzyme, a noncompetitive inhibitor changes the shape of the active site and decreases the ability of the enzyme to bind with substrate.

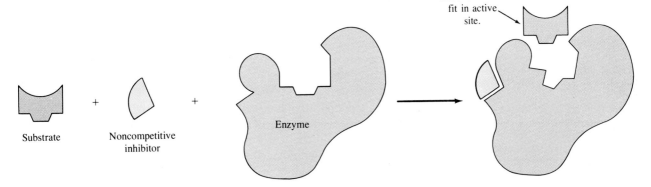

Irreversible enzyme inhibition A mechanism of enzyme deactivation in which an inhibitor forms covalent bonds to the active site.

the inhibitor is firmly bonded in place, rather than just loosely held by noncovalent attractions, the active site is permanently blocked and the enzyme is irreversibly inactivated. **Irreversible inhibition** accounts for the toxicity of nerve gases like diisopropylphosphofluoridate, which reacts with the enzyme acetylcholinesterase to block nerve transmission. Acetylcholinesterase has at its active site a serine residue that covalently bonds to the inhibitor. Most irreversible inhibitors are poisons.

Serine amino acid at active site of acetylcholinesterase

Diisopropylphosphofluoridate

Covalent bond that irreversibly binds the inhibitor to the enzyme

19.9 ENZYME REGULATION: FEEDBACK AND ALLOSTERIC CONTROL

"Feedback" is a general term applied whenever the result of a process feeds information back to affect the beginning of the process. For example, any device such as an oven that maintains a constant temperature has a sensor that detects temperature and feeds back that information to turn the heating elements on or off.

Consider a series of biochemical reactions in which A is converted to B, then B is converted to C, and so on.

$$A \xrightarrow{\text{enzyme 1}} B \xrightarrow{\text{enzyme 2}} C \xrightarrow{\text{enzyme 3}} D \xrightarrow{\text{enzyme 4}} E$$

Feedback control Activation or inhibition of the first reaction in a sequence by a product of the sequence.

Allosteric control Cooperative interaction by which binding a substrate or a regulator at one site in an enzyme influences shape and therefore binding at other sites.

Allosteric enzyme Enzyme with two or more protein chains (quaternary structure) and two or more interactive binding sites for substrate and regulators.

What will happen if product E is an inhibitor for enzyme 1? Since the amount of A converted to B will decrease, the amounts of B, C, and D will also decrease. The effect of this mechanism, known as **feedback control,** is to keep the concentration of E in a cell constant. When more E than needed is present, its synthesis is slowed down or stopped and no energy is wasted making the unneeded intermediates B, C, and D. When eventually there's not enough E, it leaves the binding sites in enzyme 1, the enzyme is no longer inhibited, and the production of E starts up again.

Generally, control strategies are more elaborate than simple feedback control because feedback by molecules with structures totally different than that of the substrate is necessary. Most biochemical pathways are regulated by **allosteric control**—a cooperative interaction in which binding a molecule at one site in an enzyme influences binding at other sites. All known **allosteric enzymes** have more than one protein chain and have two kinds of binding sites: those for substrate and those for regulators. Binding a *positive regulator* changes the shapes of active sites so that they accept substrate more readily and the rate accelerates. Binding a *negative regulator* (a noncompetitive inhibitor) changes the shapes of active sites so they accept substrate less readily, and the rate accelerates more slowly with increasing substrate concentration. The changes in enzyme shape that accompany binding either substrate or regulator are likely to be changes in quaternary structure. Because allosteric enzymes usually have several substrate and several regulator binding sites, and because there is cooperative interaction among them all, very fine control is achieved.

19.10 ENZYME REGULATION: ZYMOGENS AND GENETIC CONTROL

So far, we've discussed enzyme control by changes in enzyme shape. There are also two other important types of regulation.

Zymogen (proenzyme) A compound that becomes an active enzyme after undergoing a chemical change.

Some enzymes are synthesized in inactive forms that differ from the active forms in composition. Activation of such enzymes, known as **zymogens** or **proenzymes,** requires a chemical reaction that either adds or splits off some part of the molecule (Figure 19.13). Some of the enzymes that digest proteins, for example, are produced in the pancreas as the zymogens *trypsinogen, chymotrypsinogen,* and *proelastase.* These enzymes must not be active when they're synthesized, because they would attack the pancreas. One of the dangers of traumatic injury to the pancreas, in fact, is premature activation of these zymogens, resulting in potentially fatal *pancreatitis.* Each protease zymogen has a short polypeptide segment not present in the active enzymes, *trypsin, chymotrypsin,* and *elastase.* The unnecessary polypeptide segments are snipped off when the zymogens reach the small intestine, where protein digestion occurs.

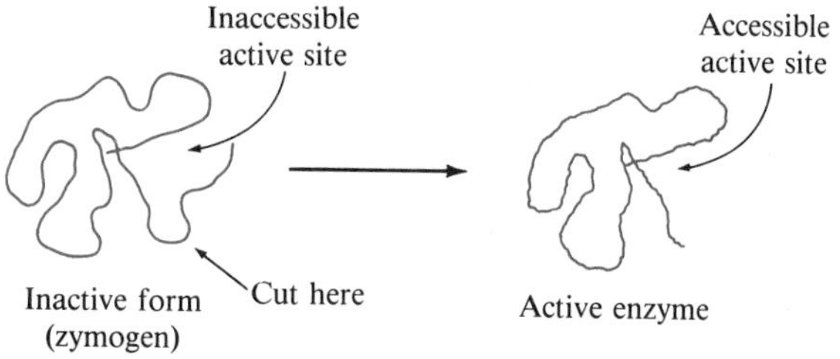

Figure 19.13
Zymogens. Some enzymes are produced initially in inactive forms called zymogens, which are then activated at the proper time by snipping off a small segment.

Inaccessible active site

Accessible active site

Inactive form (zymogen)   Cut here   Active enzyme

Yet another enzyme control strategy goes back to the supply of the enzyme itself. The synthesis of enzymes, like that of all proteins, is regulated by genes (Chapter 26), and mechanisms exist that turn enzyme synthesis on and off. The **genetic control** strategy is especially useful for enzymes needed only at certain stages in an organism's growth.

Genetic enzyme control Regulation of enzyme activity by control of the synthesis of enzymes.

19.11 VITAMINS

Vitamin A small, organic molecule that must be obtained in the diet and that is essential in trace amounts for proper biological functioning.

It has been known since antiquity that there is a critical relationship between diet and health. Lime and other citrus juices cure scurvy, meat and milk cure pellagra, and cod-liver oil cures rickets. The active substances in all three cases are **vitamins,** small, organic molecules that must be obtained through the diet and that are required in trace amounts.

Precursor A compound necessary for the synthesis of another compound.

Vitamins are grouped by solubility into two classes: water-soluble and fat-soluble. The water-soluble vitamins (Table 19.3) are found in the aqueous environment inside cells where they function either as coenzymes or as **precursors** of coenzymes. Vitamin C, for example, is biologically active without any change in structure, but thiamine must be converted to the coenzyme thiamine pyrophosphate.

Vitamin C
(ascorbic acid)

Thiamine, X = —OH

Thiamine pyrophosphate, X = $-OP_2O_6^{3-}$
(a coenzyme)

The fat-soluble vitamins A, D, E, and K are stored in the body's fat deposits, but their diverse functions are less well understood than those of the water-soluble vitamins.

Vitamin A, which is essential for normal development of epithelial tissue, night vision, and healthy eyes, has three active forms: retinol, retinal, and retinoic acid. The all-cis form of retinoic acid, the drug known as Retin-A, is

Table 19.3   Vitamins and Their Functions

Vitamin	Function (Coenzyme Example)	Deficiency Syndrome
Water-soluble vitamins		
Ascorbic acid (vitamin C)	Hydroxylases	Scurvy: bleeding gums, bruising
Thiamin (vitamin B_1)	Oxidoreductases (thiamine pyrophosphate)	Beriberi: fatigue, depression, heart disease
Riboflavin (vitamin B_2)	Oxidoreductases (flavin adenine dinucleotide, FAD)	Cracked lips, scaly skin, sore tongue
Pyridoxine (vitamin B_6)	Aminotransferases (pyridoxal phosphate)	Anemia, irritability, skin lesions, retarded growth
Niacin	Oxidoreductases (nicotinamide adenine dinucleotide, NAD)	Pellagra: dermatitis, dementia, breakdown of central nervous system, gastrointestinal function
Folic acid (vitamin M)	Methyltransferases	Megaloblastic anemia, retarded growth
Vitamin B_{12}	Isomerases (cobamide coenzymes)	Megaloblastic anemia, neurodegeneration
Pantothenic acid	Acyltransferases (coenzyme A)	Weight loss, irritability, retarded growth
Biotin (vitamin H)	Carboxylases	Dermatitis, anorexia, depression
Fat-soluble vitamins		
Vitamin A	Visual system	Night blindness, dry skin
Vitamin D	Calcium metabolism	Rickets, osteomalacia
Vitamin E	Antioxidant	Hemolysis of red blood cells
Vitamin K	Blood clotting	Hemorrhage, delayed blood clotting

used for treating skin diseases and has been reported to counteract the wrinkles of aging.

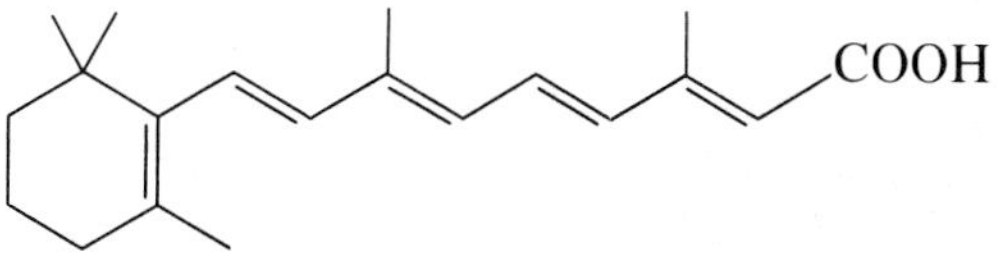

Retinoic acid

Vitamin D is related in structure to cholesterol and is synthesized when ultraviolet light from the sun strikes a cholesterol derivative in the skin. In the kidney, vitamin D is converted to a hormone that regulates calcium absorption and bone formation. Vitamin D deficiencies are most likely to occur in malnourished individuals living where there is little sunlight.

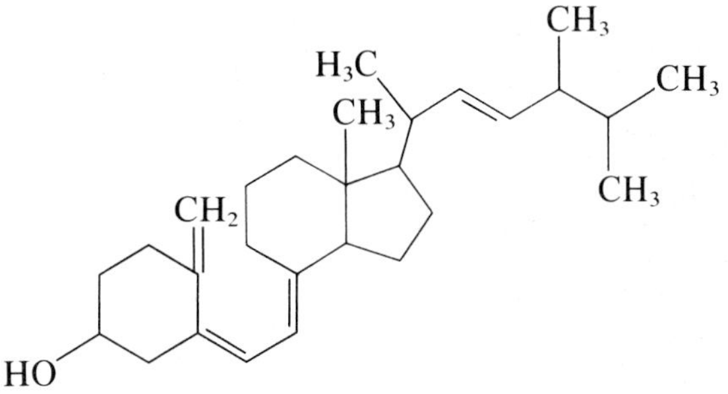

Vitamin D

Vitamin E refers to a group of structurally similar compounds called *tocopherols*, the most active of which is α-tocopherol. The best-understood function of vitamin E is as an antioxidant: It prevents the breakdown by oxidation of vitamin A and fats. The need for vitamin E apparently increases with the proportion of polyunsaturated fats in the diet.

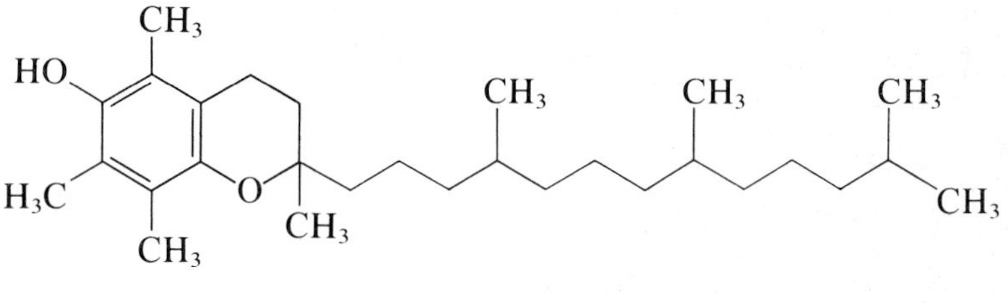

Vitamin E

Vitamin K includes the following structure and a number of related compounds with side chains of varying length. The letter K comes from the name *koagulations vitamin* bestowed by its Danish discoverer for the role of vitamin K in blood clotting. The vitamin is essential to the synthesis of several blood-clotting factors.

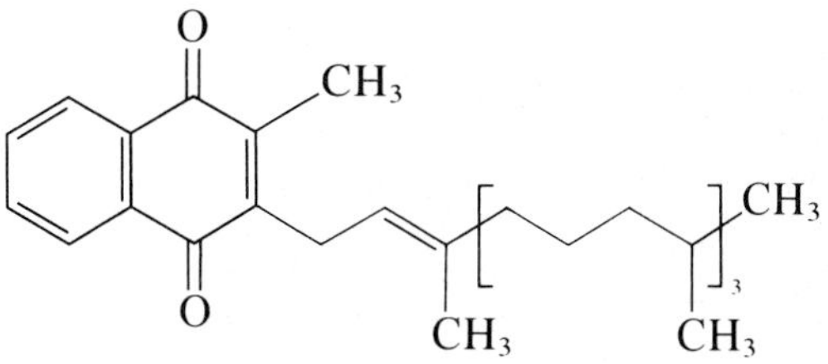

Vitamin K

AN APPLICATION: VITAMINS AND MINERALS IN THE DIET

It's not uncommon to encounter incomplete or incorrect information about vitamins and minerals in the diet. For instance, we've been frightened by the possibility that aluminum will cause Alzheimer's disease and tantalized by the possibilities that vitamin E will cure cancer and that vitamin C will defeat the common cold. Sorting out fact from fiction or distinguishing preliminary research results from scientifically proven relationships is especially difficult in this area of nutrition. Much is yet to be learned about the various functions of vitamins and minerals in the body, and new research results are continuously being reported. It's tempting for health-conscious individuals to look for guaranteed ways to better health by taking vitamins, and it's tempting to profit-making organizations to take advantage of this motivation.

One consistent source of information on nutrition is the Food and Nutrition Board of the National Academy of Sciences–National Research Council. Every 5 years they survey the latest nutritional information and publish Recommended Daily Dietary Allowances that are "designed for the maintenance of good nutrition of the majority of healthy persons in the United States." They also publish a list of estimated "safe and adequate" daily dietary ranges for selected vitamins and minerals for which there's not enough information to establish recommended allowances. Currently, biotin, pantothenic acid, copper, manganese, fluoride, chromium, molybdenum, sodium, potassium, and chloride are on this second list. The "percent of RDA" values given on food labels refers to U.S. RDAs derived from the National Academy of Sciences data by the Food and Drug Administration.

Daily vitamin intake way above the RDAs has become popular in recent years. The benefits of megadoses of any vitamins are doubted by many nutritionists, but there's no doubt about the potential for harm in megadoses of fat-soluble vitamins. Because they dissolve in fatty tissue, cell membranes, and the liver, excesses can build up in the body. Excess vitamin A, for example, can cause liver damage, skin peeling, nausea, and even death. Continuous overdoses of vitamin D lead to buildup of excess calcium in bones and also in soft tissues such as the kidneys (kidney stones), lungs, and inner ear. While overdoses of water-soluble vitamins can also disrupt body functions, this happens infrequently because water-soluble vitamins circulate in body fluids and excesses are eventually excreted in the urine.

Recommended Daily Dietary Allowances (Revised 1989)[a]

Fat-Soluble Vitamins	Male	Female	Water-Soluble Vitamins	Male	Female	Minerals	Male	Female
Vitamin A (μg RE)[b]	1000	800	Vitamin C (mg)	60	60	Ca (mg)	1200	1200
Vitamin D (μg)	10	10	Thiamin (mg)	1.5	1.1	P (mg)	1200	1200
Vitamin E (mg TE)[c]	10	8	Riboflavin (mg)	1.7	1.3	Mg (mg)	350	280
Vitamin K (μg)	70	60	Niacin (mg NE)[d]	19	15	Fe (mg)	10	15
			Vitamin B_6 (mg)	2.0	1.6	Zn (mg)	15	12
			Folate (μg)	200	180	I (μg)	150	150
			Vitamin B_{12} (μg)	2.0	2.0	Se (μg)	70	55

[a] From *Recommended Dietary Allowances*, 10th ed., Food and Nutrition Board, National Research Council–National Academy of Sciences, 1989.

[b] 1 RE = 1 retinol equivalent = 1 μg retinol.

[c] 1 TE = 1 α-tocopherol equivalent = 1 mg α-tocopherol.

[d] 1 NE = 1 niacin equivalent = 1 mg of niacin or 60 mg of dietary tryptophan.

Practice Problem   ***19.4***   Compare the structures of vitamin A and vitamin C. What structural features does each have that make one water-soluble but the other fat-soluble?

19.12   CHEMICAL MESSENGERS

At this point, you've seen some of the many kinds of enzyme-catalyzed reactions that take place in cells: oxidations, reductions, bond formations, and so forth. What hasn't been discussed yet is how the individual reactions are tied together. Clearly, the many thousands of separate reactions that take place in cells throughout the body don't occur randomly; there must be an overall control mechanism that coordinates them and keeps the entire organism in chemical balance.

Strategies much like those that inhibit or activate individual enzymes are also put to work in coordinating biochemical reactions. Messenger molecules unite with receptor molecules in induced-fit-type interactions, often in response to feedback control mechanisms, and are held together by noncovalent interactions long enough for messages to be delivered.

Hormone   A chemical messenger, usually secreted by an endocrine gland and transported through the bloodstream to elicit response from a specific target tissue.

One group of chemical messengers is composed of the **hormones** produced in the endocrine glands. Another group is made up of **neurotransmitters,** compounds that help to transmit chemical messages in the nervous system. Together the endocrine and nervous systems bear the major responsibility for keeping track of internal and external conditions and for making changes when necessary. Chemical messengers also include such diverse compounds as the *pheromones,* which carry messages between different organisms, and *endorphins* and *enkephalins,* which act in the brain to relieve pain.

Neurotransmitter   A chemical messenger that transmits a nerve impulse between neighboring nerve cells.

With increased understanding of biochemical interactions, the traditional definition of hormones as substances produced by the endocrine glands and acting on cells distant from the glands is being expanded. The prostaglandins and leukotrienes (Section 23.11), for example, don't meet the strict definition of hormones because they neither are synthesized by endocrine glands nor act on distant cells. They do, however, exert hormone-like influence on nearby cells. Similarly, the distinction between hormones and neurotransmitters is not absolute. It's been found that epinephrine (the fight-or-flight hormone discussed in Section 19.14) and norepinephrine meet both definitions—as hormones because they're secreted by endocrine glands and as neurotransmitters because they're also produced at nerve endings in the brain.

Endocrine system   A system of specialized cells, tissues, and ductless glands that excretes hormones and shares with the nervous system the responsibility for maintaining constant internal body conditions and responding to changes in the environment.

19.13   HORMONES

The glands of the **endocrine system** are located in various parts of the body and secrete hormones that are transported through the bloodstream to their target tissues. Once at the target tissue, hormones interact with specific receptors to elicit a biological response. In no case, however, do hormones themselves carry out any chemical reactions; they act solely by turning existing biological mechanisms on or off.

The endocrine system is organized into successive stages under the overall control of the *hypothalamus,* a section of the brain just above the pituitary gland. The hypothalamus secretes tiny amounts of peptide hormones called *releasing factors* that in turn stimulate the release of other hormones by the pituitary gland. Several of these hormones then activate other endocrine glands or organs to produce yet a third round of hormones, which ultimately act on their target tissues, as diagrammed in Figure 19.14.

Endocrine hormones vary greatly in structure. Some, such as insulin and the releasing factors secreted by the hypothalamus, are polypeptides. Others, such as thyroxine and epinephrine, are small, organic molecules derived from amino acids. And still others, such as estrone and the other sex hormones, are steroids (Section 23.10).

Hormone receptors may be on the outer surface of the cell, in the cytoplasm inside the cell, or in the nucleus of the cell. Receptors on the surface act in several ways: Some release a substance within the cell that regulates an enzyme; some open up channels that quickly let needed ions or molecules into the cell; some extend through the cell membrane to catalytic sites within the cell that are activated when the hormone is bound to the receptor.

Figure 19.14
The primary organization of the endocrine system, together with the names and functions of some major hormones. The *hypothalamus* is a region in the brain that monitors information from the nervous system and coordinates the activities of the endocrine and nervous systems. Releasing factors from the hypothalamus act on the pituitary gland, which then produces hormones that act on either secondary targets (blue) or final target tissues (green).

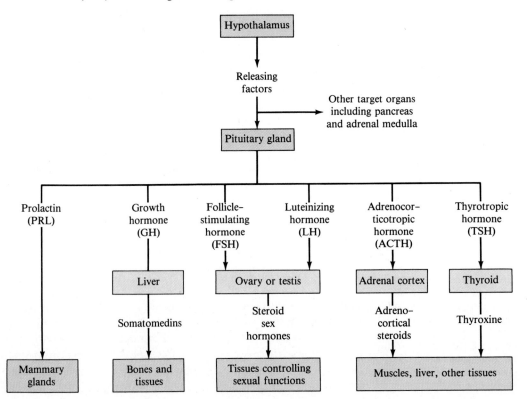

19.14   HOW HORMONES WORK: THE ADRENALINE RESPONSE

Let's look at a specific example of hormone action to see how the overall process works. *Epinephrine,* better known as *adrenaline,* is often called the fight-or-flight hormone because its release prepares the body for instant response to danger (Figure 19.15).

Figure 19.15
A grizzly bear exhibits the fight-or-flight reaction to a threatening situation. The hormone epinephrine produces a variety of physiological responses, such as increased heart rate and blood sugar level, and behavioral changes, such as these bared fangs.

Epinephrine

We've all felt the rush of adrenaline that accompanies a near-miss accident or a sudden loud noise. Chemically, epinephrine is a simple aromatic amine that is secreted into the bloodstream by the adrenal medulla within the adrenal glands located just above each kidney. Although it produces a variety of responses from numerous target tissues, epinephrine's main function is to raise the body's blood sugar level by increasing the rate of its release from storage in muscles and the liver.

Epinephrine, like many of the body's chemical messengers, doesn't cross the target cell's membrane. Instead, epinephrine binds to a receptor site on the outside of a liver cell and stimulates an enzyme (adenylate cyclase) in the cell membrane to begin production of what is called a *second messenger*. The second messenger, cyclic adenosine monophosphate (cyclic AMP or cAMP) in the present case, is released into the cell's interior. There it interacts in a series of steps to activate *glycogen phosphorylase enzymes,* the enzymes needed to release glucose from storage. These enzymes convert glucose molecules from the glucose storage polymer, glycogen, to glucose phosphate. The phosphate is then converted to glucose and released to the bloodstream (Figure 19.16).

Figure 19.16
The series of events by which release of the hormone epinephrine raises blood glucose level. Interaction of the hormone with a cell membrane receptor causes the second messenger cyclic AMP to be produced inside the target cell from ATP. The cyclic AMP in turn activates a series of steps that initiates the phosphorylase enzymes needed to break down glycogen to glucose phosphate, which is then converted to glucose for transport in the bloodstream. Epinephrine also acts to inhibit the enzymes that synthesize glycogen from glucose.

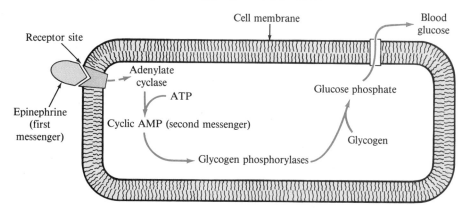

Muscle cells make energy directly from the phosphate by glycolysis (Section 22.3). When enough glucose has been produced, an enzyme called *phosphodiesterase* catalyzes the hydrolysis of cAMP, and glucose production ends. The entire series of events from initial stimulus to blood glucose production takes only seconds.

A number of hormones and neurotransmitters act via cyclic AMP as the second messenger, with the responses of different cells dependent on the different enzymes present within them. Ca^{2+} ion is also often a second messenger.

19.15 NEUROTRANSMITTERS

Synapse Narrow gap between nerve cells across which a signal is carried by a neurotransmitter.

Neurotransmitters are substances that mediate the flow of nerve impulses by transmitting a signal between neighboring nerve cells, or **neurons** (Figure 19.17). Structurally, nerve cells have a bulb-like body connected to a long, thin stem called an *axon*. Short, tentacle-like appendages called *dendrites* protrude from the bulbous end of the neuron, while numerous feathery filaments protrude from the axon at the opposite end. When bundled together in a nerve fiber, the axon of one neuron butts up against the dendrites of the next neuron, separated only by a narrow gap called a **synapse.**

Transmission of a nerve impulse between cells occurs when neurotransmitter molecules are released from the axon of a *presynaptic* neuron, cross the synaptic cleft, and fit into appropriately shaped receptor sites on the dendrites of the next, *postsynaptic* neuron (Figure 19.18). Once the message has been

Figure 19.17
A typical nerve cell.

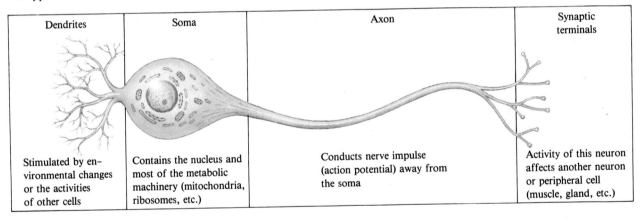

Dendrites	Soma	Axon	Synaptic terminals
Stimulated by environmental changes or the activities of other cells	Contains the nucleus and most of the metabolic machinery (mitochondria, ribosomes, etc.)	Conducts nerve impulse (action potential) away from the soma	Activity of this neuron affects another neuron or peripheral cell (muscle, gland, etc.)

Figure 19.18
Transmission of a nerve signal. Transmission occurs between neurons when a neurotransmitter molecule is released by the presynaptic neuron, crosses the synaptic cleft, and fits into a receptor site on the postsynaptic neuron.

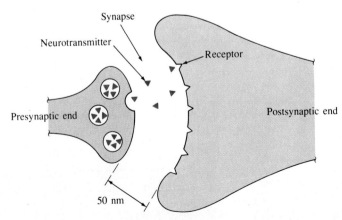

received, the cell transmits an electrical signal down its own axon and further passes along the nerve impulse.

The human nervous system has two main parts: the *central nervous system* (CNS), which consists of nerves in the brain and spinal cord, and the *peripheral nervous system* (PNS), which consists of sensory and motor nerves. The PNS, in turn, is divided into *autonomic* and *somatic* parts. The autonomic part deals with unconsciously controlled functions like blood circulation and digestion, while the somatic part deals with consciously controlled functions like movement.

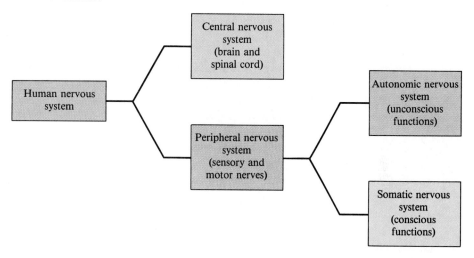

Neurons in the PNS can be divided into *cholinergic* and *adrenergic* types, depending on what kind of neurotransmitter molecules they release. Cholinergic neurons, which include all those in the somatic part and many in the autonomic part, act by releasing *acetylcholine*. Adrenergic neurons, which are found only in the autonomic part, act by releasing *norepinephrine*.

Acetylcholine

Norepinephrine

Unlike the PNS, the CNS uses a surprising variety of chemicals as neurotransmitters. Among the most interesting are *dopamine* and *serotonin*, which appear to be vital to such higher brain functions as perception and emotion. In addition, several amino acids and small peptides have been found to act as neurotransmitters in certain regions of the brain.

Dopamine

Serotonin

INTERLUDE: PENICILLIN, THE FIRST ANTIBIOTIC

Scarlet fever, diphtheria, bubonic plague, and rheumatic fever are all easily controlled diseases that are now almost unknown in modern, industrialized countries. But if you had the bad fortune to contract one of these diseases before 1940, you might well not have survived. The difference between then and now is due to the discovery in 1928 and the subsequent isolation in 1940 of penicillin, the first successful antibiotic.

Antibiotics are substances that are produced by one microorganism and are toxic to other microorganisms. Penicillin was discovered by the British bacteriologist Alexander Fleming as the result of a chance observation made when several culture plates of staphylococcus bacteria he was growing became contaminated. Fleming noticed that the contaminant, a mold subsequently identified as *Penicillium notatum,* produced a substance that appeared to kill the staphylococcus.

Chemically, the penicillins are known as *beta-lactam* antibiotics because they contain a four-membered amide ring. More than 30 naturally occurring penicillins with closely related structures are known, and several thousand synthetic derivatives have been prepared in the laboratories of pharmaceutical companies.

Different penicillins have different acyl side chains (in blue) attached to nitrogen. Penicillin G, the most common one, is shown.

The penicillins kill bacteria by acting as irreversible inhibitors of a critical transpeptidase enzyme that is involved in the synthesis of bacterial cell walls. Bacterial cells, unlike those of higher organisms, are enveloped by a protective coating called a *cell wall*. In the final stages of cell wall synthesis, strands of glycoprotein known as

Once a neurotransmitter has done its job, it must be rapidly removed from the receptor site so that the postsynaptic neuron is ready to receive another impulse. In cholinergic neurons, for example, this removal is accomplished by the enzyme *acetylcholinesterase*. Nerve gases (as illustrated in Section 19.8) and some insecticides act as cholinesterase inhibitors to block the activity of the enzyme, thereby allowing acetylcholine to accumulate at the receptor sites. Convulsions, paralysis of respiratory muscles, and death can result.

Malfunctions in the nervous system are involved as either symptoms or causes in a number of serious diseases. Parkinson's disease and schizophrenia, for example, have both been linked to defects in the brain's use of dopamine. Alzheimer's disease, a progressive disorder involving memory loss and deterioration of mental function, appears to involve a deficiency of the enzyme that synthesizes acetylcholine.

SUMMARY

Enzymes are large proteins that function as biological catalysts in converting **substrate** to product. Unlike typical inorganic catalysts, enzymes are large, complex molecules that are quite specific in their action. They are classified into six groups according to the kind of reaction they catalyze: **oxidoreductases, transferases, hydrolases, isomerases, lyases,** and **ligases.**

In addition to their protein part, many enzymes contain **cofactors** which can be either metal ions or small, organic molecules known as **coenzymes.** Often, the peptidoglycans are cross-linked together to form the final three-dimensional web. The crucial cross-linking step involves enzyme-catalyzed reaction of an Ala-Ala terminus on one strand with a Gly terminus on another strand.

Peptidoglycan—Gly—NH$_2$ D-Ala—Peptidoglycan
                                            |
                              D-Ala

$$\downarrow \text{transpeptidase}$$

Peptidoglycan—Gly—D-Ala—Peptidoglycan + D-Ala

Penicillin's three-dimensional shape is evidently similar enough to that of the Ala-Ala end of the peptidoglycan side chain that it is able to fit into the active site of the transpeptidase enzyme. Once in the active site, penicillin binds irreversibly by covalent bond formation between the enzyme and the carbonyl group of the β-lactam ring. With bacterial cell wall synthesis thus halted, the cell contents leak out through the weakened wall and the cell dies. Since the cells of

higher organisms have no cell walls, penicillin is completely specific for bacteria and is nontoxic to all other organisms.

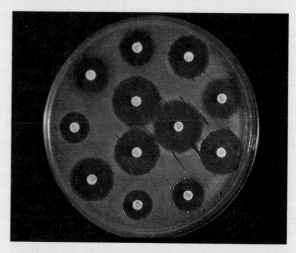

Testing the sensitivity of bacteria to different antibiotics. The clear circular areas on the culture dish are regions free of bacterial growth. The bacteria have been killed or prevented from reproducing by antibiotics diffusing outward from the smaller spots at the center of each circle.

coenzyme comes from a water-soluble **vitamin.** There are 13 known human vitamins, 9 of which are water-soluble and 4 of which are fat-soluble.

Enzymes contain an **active site,** a small, three-dimensional region with the specific shape and structure needed to bind a substrate in the **enzyme–substrate complex,** often by noncovalent interactions. Enzyme specificity is partly explained by the **lock-and-key model,** in which the substrate is thought to fit the enzyme like a key in a lock. Variations in specificity and the catalytic action of enzymes are further explained by the **induced-fit model,** in which both enzyme and substrate change shape in forming the complex. Overall, enzymes work by exercising effects on proximity, orientation, energy, and catalysis.

Enzymes function best at an optimum pH and temperature. Their action can be suppressed by **competitive inhibitors,** which compete with substrate for binding to the active site; by **noncompetitive inhibitors,** which act by changing the shape of the active site by binding elsewhere in the enzyme; or by **irreversible inhibitors,** which act by covalently bonding to and blocking the active site.

At low substrate concentration, reaction rate in-

creases with the substrate concentration; at high substrate concentration, the rate increases more slowly and eventually reaches a maximum. Competitive inhibitors slow the approach to the maximum rate, but noncompetitive inhibitors decrease the maximum rate. Enzyme activity is controlled by **feedback,** in which a reaction product is an inhibitor or activator for the first enzyme in a series of reactions, and by **allosteric control,** in which an enzyme's quaternary structure changes shape in response to binding of activators or inhibitors. Control is also achieved by **zymogen** synthesis and **genetic control.**

Enzyme-catalyzed reactions throughout the body are coordinated by chemical messengers. **Hormones** from the **endocrine system** are produced under the overall control of the hypothalamus and pituitary gland and are carried in the bloodstream to target tissues. **Neurotransmitters** are secreted by nerve cells and transmit nerve impulses from cell to cell. Many hormones and neurotransmitters connect to receptors on cell surfaces, which then produce second messengers that carry the message to the cell's interior.

REVIEW PROBLEMS

Structure and Classification of Enzymes

19.5 What is the family-name ending for an enzyme?

19.6 What is an enzyme, and what general kind of structure does it have?

19.7 There are approximately 50,000 different enzymes in the body. Why are so many needed?

19.8 How do catalysts speed up reactions?

19.9 What are the two primary differences between an inorganic catalyst and an enzyme?

19.10 What general kinds of reactions do these classes of enzymes catalyze?
(a) hydrolases   (b) isomerases   (c) lyases

19.11 What is a holoenzyme, and what are its two parts called?

19.12 What is the difference between a cofactor and a coenzyme?

19.13 What feature of enzymes makes them so specific in their action?

19.14 Describe in general terms how enzymes act as catalysts.

19.15 What classes of enzymes would you expect to catalyze these reactions:

(a) $H_2N-\underset{\underset{R}{|}}{CH}-\overset{\overset{O}{||}}{C}-NH-\underset{\underset{R'}{|}}{CH}-\overset{\overset{O}{||}}{C}OH + H_2O \longrightarrow$

$H_2N-\underset{\underset{R}{|}}{CH}-\overset{\overset{O}{||}}{C}OH + H_2N-\underset{\underset{R'}{|}}{CH}-\overset{\overset{O}{||}}{C}OH$

(b) $HO\overset{\overset{O}{||}}{C}OH \longrightarrow H_2O + CO_2$

(c) $HO\overset{\overset{O}{||}}{C}-CH_2-CH_2-\overset{\overset{O}{||}}{C}OH \longrightarrow$

$HO\overset{\overset{O}{||}}{C}-CH=CH-\overset{\overset{O}{||}}{C}OH$

19.16 What kind of reaction does each of the following enzymes catalyze?
(a) a protease   (b) a DNA ligase
(c) a transmethylase

19.17 The following reaction is catalyzed by an enzyme called *urease*. To what class of enzyme does urease belong?

$H_2N-\overset{\overset{O}{||}}{C}-NH_2 + 2 H_2O \xrightarrow{\text{urease}} 2 NH_3 + H_2CO_3$

19.18 The catalytic activity of urease (Problem 19.17) can be inhibited by adding dimethylurea to the reaction. What kind of inhibition is probably occurring?

$CH_3NH-\overset{\overset{O}{||}}{C}-NHCH_3$ Dimethylurea

Enzyme Function and Regulation

19.19 What is the difference between the lock-and-key model of enzyme action and the induced-fit model?

19.20 Fats are esters that are broken down during digestion to glycerol (1,2,3-propanetriol) and long-chain carboxylic acids. What class of enzyme would accomplish this?

19.21 Which enzyme is more specific in its action, a lock-and-key enzyme or an induced-fit enzyme?

19.22 Draw an energy diagram for the exothermic enzyme-catalyzed hydrolysis of urea (Problem 19.17). Label the energy levels of reactants and products, the activation energy, and the overall energy difference in the reaction.

19.23 Which part of a holoenzyme would you expect to be affected by denaturation, the coenzyme or the apoenzyme? Explain.

19.24 Prior to the development of modern antibiotics, a drop of dilute $AgNO_3$ was often placed in the eyes of newborn infants to prevent gonorrheal eye infections. How might $AgNO_3$ kill bacteria?

19.25 Discuss how the rate of reaction changes as enzyme concentration is increased.

19.26 Discuss how the rate of reaction changes as substrate concentration is increased.

19.27 Why don't increases in concentration of substrate or enzyme change reaction rates in exactly the same way?

19.28 What general effects would you expect the following changes to have on the speed of an enzyme-catalyzed reaction?
(a) lowering the reaction temperature from 37°C to 27°C
(b) raising the pH from 7.5 to 10.5
(c) adding a heavy-metal salt like $Hg(NO_3)_2$
(d) raising the reaction temperature from 37°C to 40°C

19.29 How can you explain the observation that pepsin, a digestive enzyme found in the stomach, has a high catalytic activity at pH 1.5 while trypsin, a digestive enzyme of the small intestine, has no activity at that pH?

19.30 What are the three kinds of enzyme inhibitors?

19.31 How does each of the three kinds of enzyme inhibitors work?

19.32 What kinds of bonds are formed between an enzyme and each of the three kinds of enzyme inhibitors?

19.33 Poisoning by which of the three types of enzyme inhibitors is probably the most difficult to treat medically? Why?

19.34 The meat tenderizer used in cooking is primarily *papain*, a protease enzyme isolated from the papaya tree. Why do you suppose papain is so effective at tenderizing meat?

19.35 Papain (Problem 19.34) is also used to help relieve the pain from bee stings. Why do you suppose that works?

19.36 What is feedback inhibition?

19.37 Why do allosteric enzymes have two types of binding sites?

19.38 Discuss the purpose of positive and negative regulators.

19.39 What is a zymogen? Why must some enzymes be secreted as zymogens?

19.40 A substantial number of calories in regular beer are due to the presence of complex carbohydrates called *amylopectins*. By adding an amyloglucosidase enzyme during brewing, the amylopectins can be hydrolyzed into glucose and then fermented. What two characteristics does the resultant light beer have? (*Hint:* The first one is that it tastes great.)

Vitamins, Hormones, and Neurotransmitters

19.41 What is the difference between a hormone and a vitamin?

19.42 What is the difference between a hormone and a neurotransmiter?

19.43 What is the relationship between vitamins and enzymes?

19.44 Which of the following three are required in the diet: hormones, enzymes, or vitamins?

19.45 What are the four fat-soluble vitamins?

19.46 Why is daily ingestion of vitamin C more critical than daily ingestion of vitamin A?

19.47 What is the general purpose of the body's endocrine system?

19.48 What gland is primarily responsible for controlling the endocrine system?

19.49 How is a hormone transported from its secretory gland to its target tissue?

19.50 Name as many endocrine glands as you can.

19.51 Describe in general terms how a hormone works.

19.52 What is the structural difference between an enzyme and a hormone?

19.53 What is the relationship between enzyme specificity and tissue specificity of hormones?

19.54 What is the cellular role of cyclic AMP?

19.55 What biological effect does the hormone epinephrine have?

19.56 What are the two parts of the human nervous system?

19.57 What is a synapse, and what role does it play in nerve transmission?

19.58 Name two classes of neurotransmitters.

19.59 Describe in general terms how a nerve impulse is passed from one neuron to another.

19.60 How do the autonomic and somatic parts of the nervous system differ?

Applications

19.61 Why must enzyme activity be monitored under standard conditions? [App: Medical Uses of Enzymes and Isoenzymes]

19.62 What three enzymes show marked concentration increases in the blood after a heart attack? Why? [App: Medical Uses of Enzymes and Isoenzymes]

19.63 What are isoenzymes? [App: Medical Uses of Enzymes and Isoenzymes]

19.64 Why are overdoses less common with water-soluble than with fat-soluble vitamins? [App: Vitamins and Minerals in the Diet]

19.65 Read the labels of food that you eat for a day or look the foods up in a nutrition table and determine what percent of your daily dosage of vitamins and minerals you get from each. Are you getting the recommended amounts from the food you eat, or should you be taking a vitamin or mineral supplement? [App: Vitamins and Minerals in the Diet]

19.66 How does penicillin kill bacteria? [Int: Penicillin, the First Antibiotic]

19.67 Why is penicillin toxic to bacteria but not to higher organisms? [Int: Penicillin, the First Antibiotic]

Additional Questions and Problems

19.68 Look up the structures of vitamin C and vitamin E in Section 19.11 and identify the functional groups in these vitamins.

19.69 Thyrotropin-releasing factor, one of the primary hormones of the hypothalamus, is the C-terminal amide of the simple tripeptide PyroGlu-His-Pro, where PyroGlu is pyroglutamic acid. Draw the full structure of thyrotropin-releasing factor.

PyroGlu

19.70 What is the chemical relationship of pyroglutamic acid (Problem 19.69) to glutamic acid, one of the 20 common amino acids?

19.71 The adult recommended daily allowance (RDA) of riboflavin is 1.6 mg. If one glass (100 mL) of red wine contains 0.014 mg of riboflavin, how much wine would an adult have to consume to obtain the RDA?

19.72 Some bacteria require *p*-aminobenzoic acid (PABA) to make folic acid, which they then use to produce a necessary enzyme. Sulfa drugs, which have the general structure shown here, interfere with folic acid synthesis. Draw the structure of PABA and propose a reason for the success of these drugs.

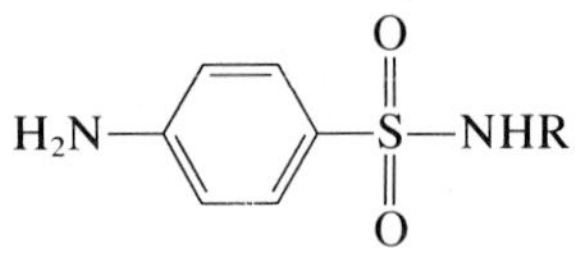

A sulfa drug

19.73 The active site of an enzyme is a very small portion of the molecule. What is the function of the rest of the huge molecule?

20

The Generation of Biochemical Energy

These athletes need a well-practiced strategy to pass the baton successfully. The energy they're using is produced with the aid of the strategies of metabolism discussed in this chapter.

All organisms need to draw energy from their surroundings to stay alive. In animals, the energy comes from food and is released in the exquisitely interconnected reactions of **metabolism.** We are powered by the oxidation in cells of biomolecules containing mainly carbon, hydrogen, and oxygen. The end products are carbon dioxide, water, and energy.

$$C, H, O \text{ (food molecules)} + O_2 \longrightarrow CO_2 + H_2O + \text{energy}$$

The principal food molecules—fats, proteins, and carbohydrates—differ in structure and are broken down by distinctive pathways that are examined in later chapters. For the present, though, we're going to concentrate on the final, common pathways by which energy is released from all types of food molecules. The questions to be answered include the following:

1. *What is the general structure of a eukaryotic cell?* The goal: Be able to identify the major parts of a eukaryotic cell, describe the structure of a mitochondrion, and tell how eukaryotic and prokaryotic cells differ.
2. *How are the reactions of catabolism organized?* The goal: Be able to list the stages of catabolism and describe the role of each.
3. *What are the major strategies of metabolism?* The goal: Be able to explain and give examples of the roles of ATP, coupled reactions, and oxidized and reduced coenzymes in metabolic pathways.
4. *What is the citric acid cycle?* The goal: Be able to describe what happens in the citric acid cycle and what its role is in energy production.
5. *How is ATP generated in the final stage of catabolism?* The goal: Be able to describe the respiratory chain, oxidative phosphorylation, and their coupling according to the chemiosmotic theory.

20.1 ENERGY AND LIFE

Metabolism The overall sum of the many reactions taking place in an organism.

Living things must do mechanical work—microorganisms engulf food, plants bend toward the sun, humans walk about. All organisms must also do the chemical work of synthesizing biomolecules needed for energy storage, growth, repair, and replacement (Figure 20.1). In addition, cells need energy for the work of moving certain molecules and ions across cell membranes. It is the energy released from food that allows these various kinds of work to be done.

Energy, we've noted previously, can be converted from one form to another but can be neither created nor destroyed. Ultimately, the energy used by all but a few living things on earth comes from the sun (Figure 20.2). Plants convert sunlight to potential energy stored mainly in the chemical bonds of carbohydrates. Plant-eating animals then convert some of this energy according to their own needs for staying alive and store the rest of it, mainly in the chemical bonds of fats. Other animals, including humans, are able to eat plants or animals and make use of the chemical energy these organisms have stored.

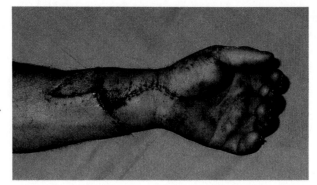

Figure 20.1
Energy at work in the body. The biomolecules needed to heal this wound will be synthesized using energy from the catabolism of food molecules.

Our bodies can't simply produce energy by burning up a large steak all at once because the large quantity of heat released would be harmful to us. Furthermore, it's difficult to capture energy for storage once it has been converted to heat. What we need is energy that can be stored and then released in the right amounts, whether we're running away from an angry dog or sleeping peacefully.

We therefore have some specific requirements for energy:

- Energy must be released from food gradually.
- Energy must be stored in a readily accessible form.
- The rate of energy release must be finely controlled.
- Just enough energy must be released as heat to maintain constant body temperature.
- Energy in a form other than heat must be available to drive reactions that aren't spontaneous at body temperatures.

In this chapter we'll look at some of the ways these requirements for energy management are met. We'll begin by reviewing some basic concepts about energy. Then we'll take an overview of metabolism and the strategies it

Figure 20.2
Flow of energy through the biosphere. Energy comes from the sun and is ultimately stored in chemical bonds, used for work, or lost as heat.

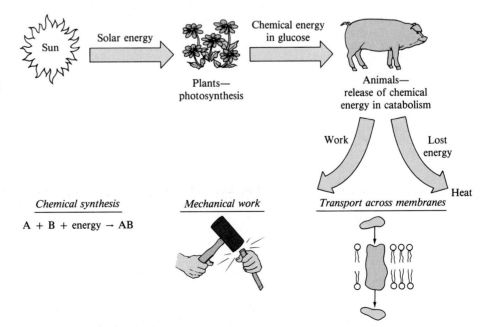

relies on, and in the rest of the chapter we'll look more closely at the generation of *adenosine triphosphate* (ATP), the molecule that carries stored energy to where it's needed.

20.2 FREE ENERGY AND BIOCHEMICAL REACTIONS

We said in Chapter 8 that a chemical reaction is favorable, or *spontaneous,* if free energy is released during the process. Conversely, a reaction is not favorable if free energy is absorbed. The change in free energy (ΔG) depends on two factors: the release or absorption of heat (ΔH) and the increase or decrease in disorder that result from the reaction (ΔS). The release of heat and the increase in disorder both favor spontaneity, and the combined effect on spontaneity of ΔH and ΔG is given by

$$\Delta G = \Delta H - T\Delta S$$

Reactions in living organisms are no different from laboratory reactions in test tubes. Both follow the same laws, and both have the same kinds of energy requirements. The free energy change therefore applies equally well to biochemical reactions and laboratory reactions.

For a chemical reaction to occur spontaneously, energy must be given off, which is indicated by a negative value of ΔG. In other words, the products of a reaction must have less energy and be more stable than the reactants. Such a reaction is described as **exergonic.** As shown by the energy diagram in Figure 20.3, the favorable, exergonic reaction whose curve is on the left has its products farther "downhill" on the energy scale than the reactants.

Oxidation reactions are often downhill reactions that give off large amounts of energy. For example, oxidation of glucose, the principal source of energy for animals, releases 686 kcal per mole of glucose.

$$C_6H_{12}O_6 + 6\ O_2 \longrightarrow 6\ CO_2 + 6\ H_2O \qquad \Delta G = -686\ \text{kcal}$$

Glucose

The more negative the free energy change, the greater the amount of energy released and the further a reaction proceeds toward products before reaching equilibrium. In the combustion of glucose, virtually no reactant remains at equilibrium.

Exergonic Describes a process that releases free energy to the surroundings (negative δg).

Figure 20.3
Energy diagrams for two reactions. In the favorable reaction, diagrammed on the left, the products have less energy than the reactants. In the unfavorable reaction, diagrammed on the right, the products have more energy than the reactants.

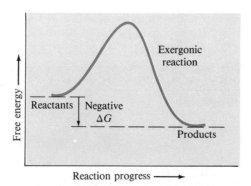

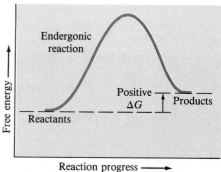

Endergonic Describes a process that absorbs free energy from the surroundings (positive ΔG).

Reactions in which the products are *higher* in energy than the reactants can also take place under the right circumstances, but such reactions can't be spontaneous. In other words, energy has to be added to the reactants for an energetically "uphill" change to occur (on the right in Figure 20.3). Such reactions have positive values of ΔG and are described as **endergonic.** An example is the conversion of glucose to glucose 6-phosphate, an important step in the breakdown of glucose from carbohydrates in the diet:

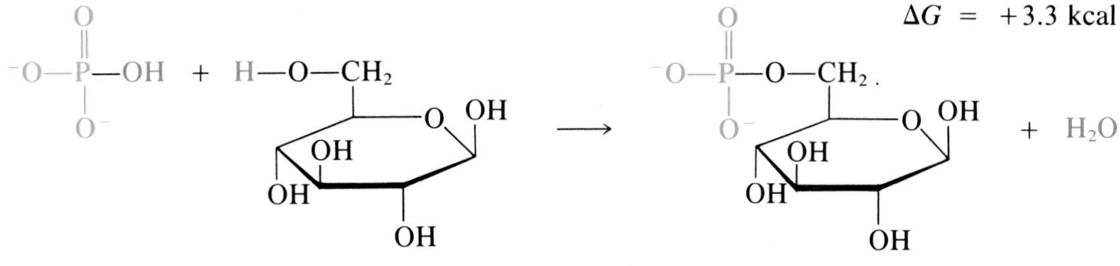

$$\Delta G \;=\; +3.3 \text{ kcal}$$

| Hydrogen phosphate ion | Glucose | Glucose 6-phosphate |

The more positive the free energy change, the greater the amount of energy required and the less product present when a reaction reaches equilibrium.

Like the value of ΔH, the value of ΔG changes sign for the reverse of a given reaction. For photosynthesis, the process whereby plants store solar energy by converting CO_2 and H_2O to glucose plus O_2, ΔG is therefore positive and of the same numerical value as for the oxidation of glucose.

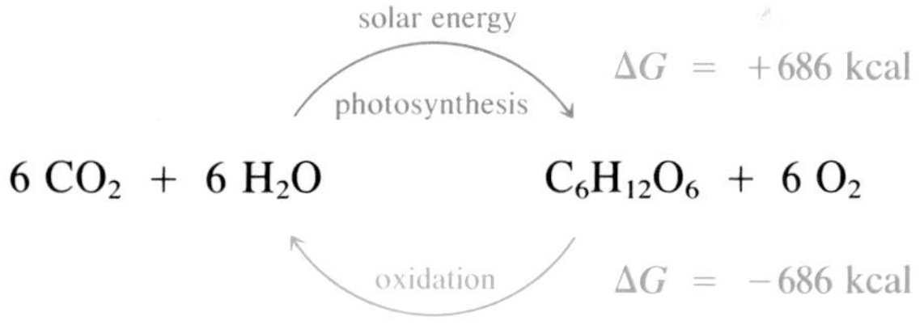

$$6\,CO_2 \;+\; 6\,H_2O \qquad\qquad C_6H_{12}O_6 \;+\; 6\,O_2$$

solar energy
photosynthesis $\Delta G \;=\; +686 \text{ kcal}$
oxidation $\Delta G \;=\; -686 \text{ kcal}$

Photosynthesis and glucose oxidation show how energy stored in the products of an *endergonic* reaction is released in the *exergonic* reverse of that reaction. Living systems make constant use of this principle.

Practice Problem **20.1** Each of the following reactions occurs in the citric acid cycle, an energy-producing sequence of reactions that we'll discuss later in this chapter. (Recall that organic acids are usually referred to in biochemistry with the -*ate* ending because they exist as ions in body fluids.) Which of the reactions listed is exergonic? Which is endergonic? Which will proceed farthest toward products at equilibrium?

(a) acetyl coenzyme A + oxaloacetate + H_2O $\longrightarrow$
                             citrate + coenzyme A $\Delta G = -9\text{ kcal}$

(b) citrate $\longrightarrow$ isocitrate $\Delta G = +3\text{ kcal}$

(c) fumarate + H_2O $\longrightarrow$ L-malate $\Delta G = -0.9\text{ kcal}$

Prokaryotic cell Cell that has no nucleus; found in bacteria and algae.

Eukaryotic cell Cell with a membrane-enclosed nucleus; found in all higher organisms.

Cell membrane (plasma membrane) The membrane surrounding a cell.

Cytoplasm The region between the cell membrane and the nuclear membrane in a eukaryotic cell.

Organelle A small, organized unit in the cell that performs a specific function.

Cytosol The contents of the cytoplasm surrounding the organelles.

Mitochondrion An egg-shaped organelle where small molecules are broken down to provide the energy to power an organism.

Cristae The inner folds of a mitochondrion.

20.3 CELLS AND THEIR STRUCTURE

Before we take an overview of metabolism, it's important to see where the energy-generating reactions take place. There are two main categories of cells: **prokaryotic** cells, found in bacteria and algae, and **eukaryotic** cells, found in higher organisms, both plant and animal.

Prokaryotic cells are simple in structure: They have no nucleus, they are quite small, and the DNA that governs reproduction is dispersed throughout their contents. Eukaryotic cells, by contrast, have a membrane-enclosed nucleus containing their DNA and are about 1000 times larger than bacterial cells. These cells are separated from their environment by a **cell membrane,** sometimes called a *plasma membrane*. Everything between the cell membrane and the nuclear membrane in a eukaryotic cell is referred to as the **cytoplasm.**

A further important difference between cell types is that eukaryotic cells contain in their cytoplasm a large number of subcellular structures called **organelles**—small, functional units that perform specialized tasks. In addition to the nucleus, other organelles include ribosomes, endoplasmic reticulum, lysosomes, Golgi complex, and mitochondria. All these organelles are surrounded by the **cytosol,** the material that fills the interior of the cell and contains electrolytes, nutrients, and enzymes. A diagram of a typical eukaryotic cell is shown in Figure 20.4.

Among the organelles shown in Figure 20.4, the **mitochondria,** often called the cell's "power plants," are the most important for energy production. It is in the mitochondria that 90% of the body's energy-carrying molecule, ATP, is produced. A typical liver cell contains about 800 mitochondria, totaling about 20% of the cellular volume.

A mitochondrion is a roughly egg-shaped structure composed of a smooth outer membrane and a folded inner membrane (Figure 20.5). The folds in the inner membrane are called **cristae,** and the space surrounded by the inner

Figure 20.4
A typical eukaryotic cell, showing some of its organelles.

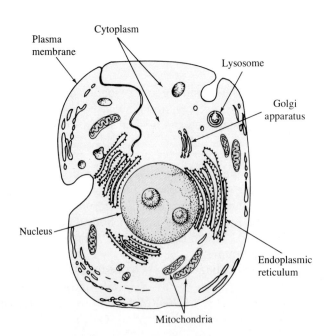

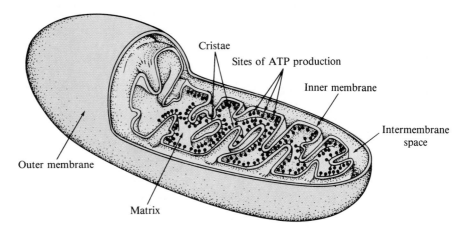

Figure 20.5
A mitochondrion. The citric acid cycle takes place in the matrix. The respiratory chain and oxidative phosphorylation take place at the inner membrane.

Mitochondrial matrix The space surrounded by the inner membrane.

membrane is called the mitochondrial **matrix.** The intermembrane space lies between the inner and outer membranes. You'll see in the following sections that some energy-producing reactions take place in the matrix, and others on the small knob-like protuberances on the cristae.

20.4 AN OVERVIEW OF METABOLISM AND ENERGY PRODUCTION

Most of the chemical reactions of metabolism occur in sequences called *metabolic pathways.* Such a pathway may be linear, in which the product of one reaction serves as the starting material for the next, or it may be cyclic, in which a series of reactions regenerates the first reactant.

A linear sequence $A \longrightarrow B \longrightarrow C \longrightarrow D \longrightarrow \cdots$ A cyclic sequence $D \overset{A}{\underset{C}{\circlearrowright}} B$

Catabolism Metabolic reactions that break down molecules into smaller pieces.

Anabolism Metabolic reactions that build larger biological molecules from smaller pieces.

Those pathways that break molecules apart are known collectively as **catabolism,** while those that put building blocks back together to assemble larger molecules are known collectively as **anabolism.** The purpose of catabolism is to release energy from food, and the purpose of anabolism is to synthesize new biomolecules, including those that store energy.

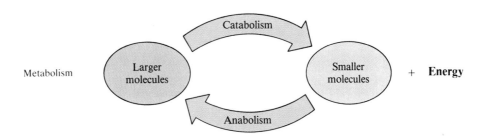

The overall picture of catabolism and energy production is simple: Eating provides fuel, breathing provides oxygen, and our bodies oxidize the fuel to extract energy. The process can be roughly divided into the four stages shown in Figure 20.6.

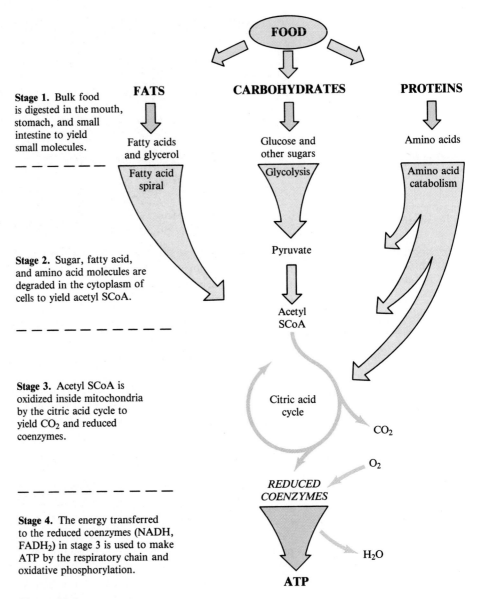

Stage 1. Bulk food is digested in the mouth, stomach, and small intestine to yield small molecules.

Stage 2. Sugar, fatty acid, and amino acid molecules are degraded in the cytoplasm of cells to yield acetyl SCoA.

Stage 3. Acetyl SCoA is oxidized inside mitochondria by the citric acid cycle to yield CO_2 and reduced coenzymes.

Stage 4. The energy transferred to the reduced coenzymes (NADH, $FADH_2$) in stage 3 is used to make ATP by the respiratory chain and oxidative phosphorylation.

Figure 20.6
An overview of catabolic pathways for the degradation of food and the production of biochemical energy stored in ATP.

Acetyl group The CH_3C=O group.

Coenzyme A Coenzyme that functions as a carrier of acetyl groups and other acyl (R—C=O) groups.

The first catabolic stage, *digestion,* takes place in the mouth, stomach, and small intestine when bulk food is broken down into individual small molecules. Carbohydrates are broken down to glucose and other sugars; proteins are broken down into amino acids; lipids (commonly called "fats") are broken down into long-chain carboxylic acids called *fatty acids.* In stage 2, these small molecules are further degraded to yield two-carbon **acetyl groups** (CH_3C=O) that are attached to·a large carrier molecule called **coenzyme A.** The resultant compound, *acetyl coenzyme A,* is an intermediate in the breakdown of all main classes of food molecules.

AN APPLICATION: BASAL METABOLISM

The minimum amount of energy expenditure required per unit of time to stay alive—to breathe, maintain body temperature, circulate blood, and keep all body systems functioning—is referred to as the **basal metabolic rate.** Ideally, it is measured in a person who is awake, is lying down at a comfortable temperature, has fasted and avoided strenuous exercise for 12 hr, and is not under the influence of any medications. The rate can be measured by monitoring respiration and finding the rate of oxygen consumption, which is proportional to the energy used.

An average basal metabolic rate is 70 kcal/hr, or about 1700 kcal/day. The rate varies with many factors, including sex, age, weight, and physical condition. A rule of thumb used by nutritionists to estimate basal energy needs per day is the requirement for 1 kcal/hr per kilogram of body weight by a male and 0.95 kcal/hr per kilogram of body weight by a female. For example, a 50. kg (110 lb) female has an estimated basal metabolic rate of (50. kg)/(0.95 kcal/kg hr) = 48 kcal/hr, giving a daily basal requirement of 1200 kcal.

The total calories a person needs each day is determined by his or her basal requirements plus the energy used in additional physical activities. The caloric consumption rates associated with some activities are listed in the following table. A relatively inactive person requires about 30% above basal requirements per day, a lightly active person requires about 50% above basal, and a very active person such as an athlete or construction worker can use 100% above basal requirements in a day. Each day that you consume food with more calories than you use, the excess calories are stored as potential energy in the chemical bonds of substances in your body and your weight rises. Each day that you consume food with fewer calories than you burn, some chemical energy in your body is taken out of storage and your weight drops.

Calories Used in Various Activities

Activity	Kilocalories used per minute
Sleeping	1.2
Sitting reading	1.3
Listening to lecture	1.7
Weeding garden	5.6
Walking, 3.5 mph	5.6
Pick-and-shovel work	6.7
Recreational tennis	7.0
Soccer, basketball	9.0
Walking up stairs	10.0–18.0
Running, 12 min/mi (5 mph)	10.0
Running, 5 min/mi (12 mph)	25.0

The cola drink contains 160 Calories (kcal) and the hamburger contains 500 Calories. How long would you have to play tennis to burn off these calories?

Acetyl groups are oxidized in the third stage, the *citric acid cycle,* to yield carbon dioxide, water, and a great deal of energy. Some of this energy is lost as heat, and some is carried by reduced coenzymes to the fourth stage, the *respiratory chain* and *oxidative phosphorylation,* in which molecules of ATP, the primary energy carrier, are produced.

In this chapter, we'll discuss stages 3 and 4, because that's where energy production mainly occurs. We'll discuss stages 1 and 2 along with other aspects of the metabolism of carbohydrates, lipids, and proteins in later chapters.

20.5 STRATEGIES OF METABOLISM: ATP AND ENERGY TRANSFER

Adenosine triphosphate (ATP) The triphosphate of adenosine; the "energy currency" molecule of metabolism.

The storage of energy in a readily accessible form is achieved in most organisms by the synthesis of **adenosine triphosphate (ATP)** (Figure 20.7). Recall from Section 17.12 that two molecules of phosphoric acid (H_3PO_4), which we'll write as the hydrogen phosphate ions present in body fluids (HPO_4^{2-}) and symbolize as P_i, can condense to form an anhydride. Pyrophosphoric acid, the resulting anhydride, exists in body fluids as pyrophosphate ion ($P_2O_7^{4-}$ or PP_i).

Hydrogen phosphate ions   Pyrophosphate ion

Phosphoric anhydride A —P—O—P— group, like those in ADP and ATP.

Note that the **phosphoric anhydride** link lies between two negatively charged groups. For this and other reasons, it's a relatively strained link that's often formed in an endergonic reaction and broken in an exergonic reaction.

Phosphate group A —PO_3^{2-} group in an organic molecule.

ATP, which has two phosphoric anhydride links, can be produced by addition of a **phosphate group** to ADP, which has one phosphoric anhydride link (Figure 20.7). The **phosphorylation** of ADP requires 7.3 kcal of energy per mole of ATP, and the hydrolysis of ATP releases this much energy.

Phosphorylation A reaction that results in addition of a phosphate group (—PO_3^{2-}).

$$ADP + HPO_4^{2-} + H^+ \longrightarrow ATP + H_2O \qquad \Delta G = +7.3 \text{ kcal}$$

$$ATP + H_2O \longrightarrow ADP + HPO_4^{2-} + H^+ \qquad \Delta G = -7.3 \text{ kcal}$$

As the molecule that stores the energy needed to power all life processes, ATP has been called the "energy currency of the living cell." Catabolic reactions pay off in the endergonic synthesis of ATP, and anabolic reactions spend ATP by transferring a phosphate group to other molecules in exergonic reactions. The entire process of energy production and use thus revolves around the ATP $\rightleftarrows$ ADP interconversion.

Since the primary metabolic function of ATP is to store energy, it is often also referred to as being a "high-energy" molecule or as containing "high-energy" P—O bonds. These terms, while useful, are misleading because they promote the idea that ATP is somehow different from other compounds. The

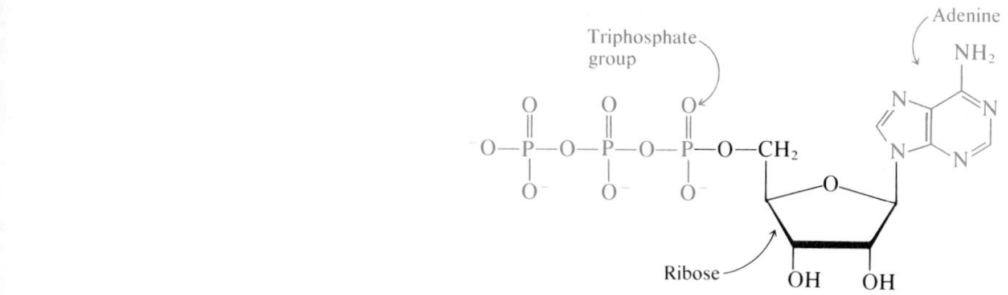

Adenosine triphosphate (ATP)

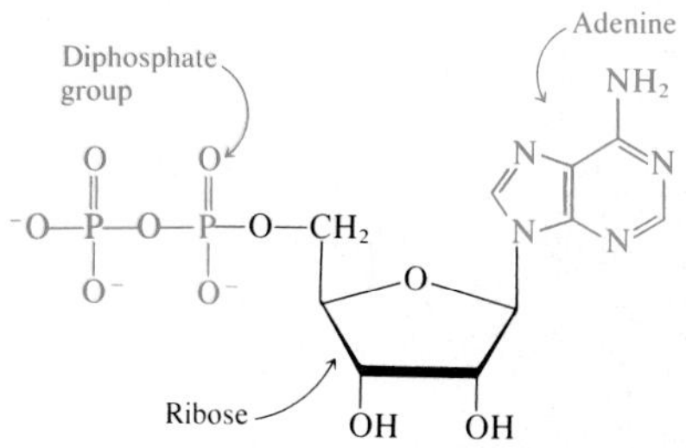

Adenosine diphosphate (ADP)

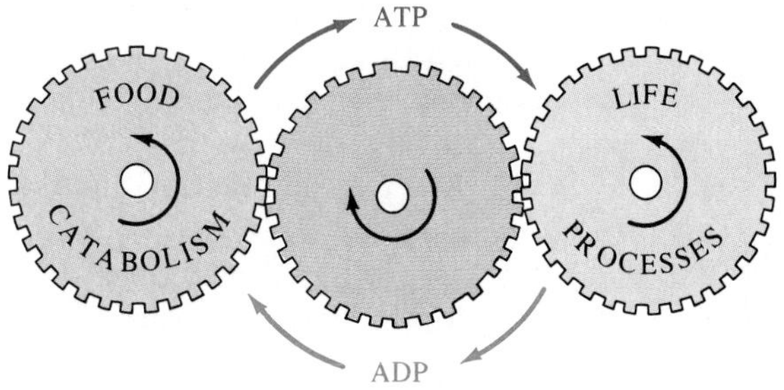

Figure 20.7
ATP and ADP. The interconversion of ATP and ADP serves as the central cog in the overall process of producing energy from food.

terms mean only that ATP is more reactive than many other biomolecules and releases a useful amount of energy when it transfers a phosphate group. In fact, if the hydrolysis of ATP released *unusually* large amounts of energy, the body wouldn't be able to provide enough energy to convert ADP back to ATP. ATP is a convenient energy carrier in metabolism *because* its hydrolysis is of intermediate energy, as illustrated in Table 20.1.

Table 20.1 Free Energies of Hydrolysis of Some Phosphates

$$R{-}O{-}P(=O)(O^-){-}O^- + H_2O \rightleftharpoons ROH + HO{-}P(=O)(O^-){-}O^-$$

Compound Name	ΔG (kcal/mol)
Phosphoenol pyruvate	−14.8
1,3-Bisphosphoglycerate	−11.8
Creatine phosphate	−10.3
ATP (→ ADP)	−7.3
Glucose 1-phosphate	−5.0
Fructose 6-phosphate	−3.3
Glucose 6-phosphate	−3.3

20.6 STRATEGIES OF METABOLISM: METABOLIC PATHWAYS AND COUPLED REACTIONS

Now that you're acquainted with ATP, we'll explore how stored chemical energy is gradually released and how it can be used to drive endergonic, or uphill, reactions. We've noted before that our bodies can't burn up a steak all at once. As shown by the diagram on the left in Figure 20.3, however, the energy difference between a reactant (the steak) and the ultimate products of its catabolism (mainly carbon dioxide and water) is a fixed quantity. *The same amount of energy is released no matter what pathway is taken between reactants and products.* The metabolic pathways of catabolism take advantage of this fact by releasing energy bit by bit in series of reactions, somewhat like the stepwise release of potential energy as water flows down an elaborate waterfall (Figure 20.8).

The overall reaction and the overall free energy change for any series of reactions can be found by summing up the equations and the free energy changes for each individual step. For example, the total free energy change (ΔG) for the 10 reactions in the *glycolysis* pathway by which one mole of glucose is converted to pyruvate (part of stage 2, Figure 20.6) is about -8 kcal, showing that the pathway is downhill and favorable. The reactions of all metabolic pathways *add up* to downhill processes with negative free energy changes.

Figure 20.8
Stepwise release of potential energy. No matter what the pathway from the top to the bottom of this waterfall, the amount of potential energy released as the water falls is the same.

Coupled reactions
Combination of an energy-requiring and energy-releasing reaction that occur together to give an overall energy-releasing reaction.

Unlike the waterfall, however, not every step in a metabolic pathway is downhill. The metabolic strategy for dealing with an energetically unfavorable, uphill reaction is to *couple* it with an energetically favorable, downhill reaction so that the overall energy change for the two reactions is favorable. For example, take the reaction of glucose with hydrogen phosphate ion (HPO_4^{2-}) to yield glucose 6-phosphate plus water, for which $\Delta G = +3.3$ kcal. The reaction doesn't take place spontaneously because the two products are about 3.3 kcal higher in energy than the starting materials.

But now let's see what the energy picture looks like when this reaction is coupled with the favorable hydrolysis of ATP to give ADP:

(*Unfavorable*) Glucose + HPO_4^{2-} ⟶ glucose 6-phosphate + H_2O $\Delta G = 3.3$ kcal

(*Favorable*) ATP + H_2O ⟶ ADP + HPO_4^{2-} $\Delta G = -7.3$ kcal

(*Favorable*) Glucose + ATP ⟶ glucose 6-phosphate + ADP $\Delta G = -4.0$ kcal

The net energy change for these two **coupled reactions** for the synthesis of glucose 6-phosphate is favorable by about 4 kcal. Enough energy is transferred directly from the hydrolysis of the phosphate group in ATP to drive the addition of a phosphate group to glucose. The excess energy is released as heat and contributes to maintaining body temperature (Figure 20.9).

Although we've written the reactions separately to show how their energies combine, coupled reactions don't necessarily take place separately. The net change occurs all at once and, in fact, *it's only by such coupling that the energy stored in one chemical compound can be transferred to the reaction of other compounds.* ATP is used in this manner to drive many different kinds of unfavorable biochemical reactions.

What about the synthesis of ATP from ADP, which has $\Delta G = +7.3$ kcal and is an unfavorable reaction? The same principle of coupling is put to use. For this endergonic reaction to occur, it must be coupled with a reaction that releases *more* than 7.3 kcal. In a different reaction of glycolysis, the formation of ATP is coupled with the hydrolysis of phosphoenolpyruvate, a high-energy phosphate of higher energy than ATP (Table 20.1):

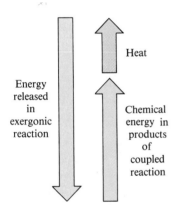

Figure 20.9
Energy exchange in coupled reactions.

$H_2C=\overset{\overset{\displaystyle O-PO_3^{2-}}{|}}{C}-COO^-$ + H_2O ⟶ $CH_3-\overset{\overset{\displaystyle O}{\|}}{C}-COO^-$ + HPO_4^{2-} $\Delta G = -14.8$ kcal

Phosphoenolpyruvate   Pyruvate

ADP + HPO_4^{2-} ⟶ ATP + H_2O $\Delta G = +7.3$ kcal

$H_2C=\overset{\overset{\displaystyle OPO_3^{2-}}{|}}{C}-COO^-$ + ADP ⟶ $CH_3\overset{\overset{\displaystyle O}{\|}}{C}-COO^-$ + ATP $\Delta G = -7.5$ kcal

To simplify matters, such coupled biochemical reactions are often written in a way that highlights the important transformation. The curved arrow intersecting the normal straight arrow shows that ADP is a reactant and ATP is a product:

$H_2C=\overset{\overset{\displaystyle O-PO_3^{2-}}{|}}{C}-COO^-$ $\overset{ADP \quad ATP}{\curvearrowright} \longrightarrow$ $CH_3-\overset{\overset{\displaystyle O}{\|}}{C}-COO^-$

Practice Problems

20.2 One of the steps in lipid metabolism is the reaction of glycerol [1,2,3-propanetriol; $HOCH_2CH(OH)CH_2OH$] with ATP to yield glycerol 1-phosphate. Write the equation for this reaction using the curved arrow symbolism.

20.3 Why must a metabolic pathway that synthesizes a given molecule occur by a different series of reactions than a pathway that breaks down the same molecule?

20.4 The hydrolysis of acetyl phosphate to give acetate has $\Delta G = -10.3$ kcal. Combine the equations and ΔG values to show whether or not coupling of this reaction with phosphorylation of ADP is favorable. (You need only give compound names or abbreviations in the equations.)

20.7 STRATEGIES OF METABOLISM: OXIDIZED AND REDUCED COENZYMES

Many metabolic reactions are oxidation–reduction (redox) reactions, which means that a steady supply of oxidizing and reducing agents must be available. To deal with this requirement, a few coenzymes continuously cycle between their oxidized and reduced forms, just as adenosine continuously cycles between ATP and ADP.

To review briefly the basics of oxidation and reduction, you might want to reread Sections 7.13 and 7.14. The important points are that in inorganic chemistry oxidations are the gain of electrons, reductions are the loss of electrons, and oxidation and reduction always occur together.

$$
\begin{aligned}
\text{Oxidation:} && Zn &\longrightarrow Zn^{2+} + 2\,e^- \\
\text{Reduction:} && 2\,Cu^+ + 2\,e^- &\longrightarrow 2\,Cu \\
\hline
\text{Redox reaction:} && 2\,Cu^+ + Zn &\longrightarrow 2\,Cu + Zn^{2+}
\end{aligned}
$$

Remember also that in organic chemistry, oxidation is often the addition of oxygen or the loss of hydrogen, and reduction is often the removal of oxygen or the addition of hydrogen.

Reduced form	Oxidized form
$R-\overset{\displaystyle O}{\overset{\|}{C}}-H$	$R-\overset{\displaystyle O}{\overset{\|}{C}}-OH$
RCH_2CH_2R	$RCH{=}CHR$

With these points in mind, let's use a reaction from the citric acid cycle to illustrate the action of a coenzyme as an oxidizing agent (Table 20.2). The entire cycle is described in Section 20.8.

Table 20.2   Coenzymes as Oxidizing and Reducing Agents

Coenzyme	As Oxidizing Agent	As Reducing Agent
Nicotinamide adenine dinucleotide	NAD^+	NADH (or $NADH/H^+$)
Nicotinamide adenine dinucleotide phosphate	$NADP^+$	NADPH (or $NADPH/H^+$)
Flavin adenine dinucleotide	FAD	$FADH_2$

Oxidation of 2° alcohol by NAD^+

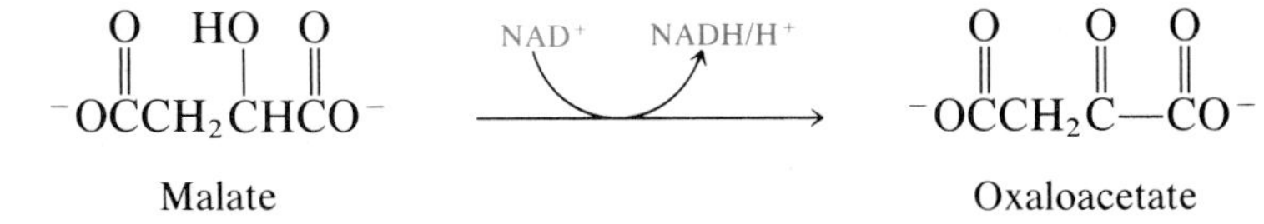

Malate → Oxaloacetate

The oxidation of malate to oxaloacetate amounts to the removal of two hydrogen atoms. The oxidizing agent is NAD^+, the oxidized form of **nicotinamide adenine dinucleotide,** whose structure is shown in Figure 20.10. NAD^+ is a coenzyme for the dehydrogenase enzymes that often catalyze removal of hydrogen atoms from secondary alcohols to produce ketones. The structure of $NADP^+$, a related coenzyme that plays a role in biosynthesis, is indicated in red in Figure 20.10.

It's important in considering enzyme-catalyzed redox reactions to recognize that a hydrogen *atom* is equivalent to a hydrogen *ion,* H^+, plus an electron, e^-. Thus, for the two hydrogen atoms removed in the oxidation of malate,

$$2 \text{ H atoms} = 2 \text{ H}^+ + 2 \text{ e}^-$$

When NAD^+ is reduced, both electrons accompany one H^+, in effect adding a hydride ion, $H:^-$, to the molecule to give NADH. The second hydrogen re-

Nicotinamide adenine dinucleotide (NAD⁺)
Coenzyme that functions as an oxidizing agent and forms $NADH/H^+$.

Figure 20.10
Oxidizing agents common in catabolism. NAD^+ is reduced to NADH plus H^+; $NADP^+$ is reduced to NADPH plus H^+; FAD is reduced to $FADH_2$.

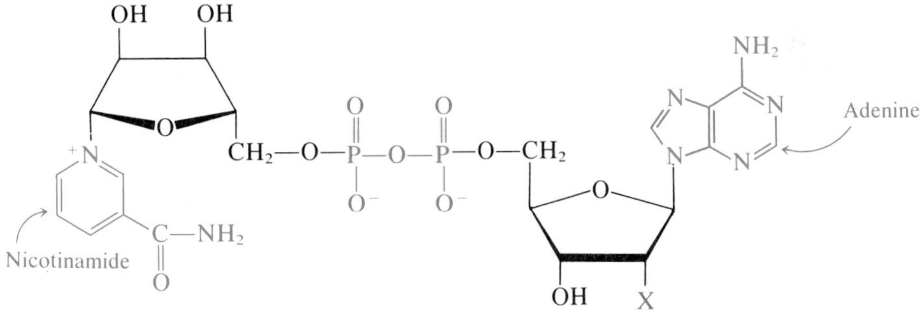

Nicotinamide adenine dinucleotide (NAD^+), X = —OH
Nicotinamide adenine dinucleotide phosphate ($NADP^+$), X = $—OPO_3^{2-}$

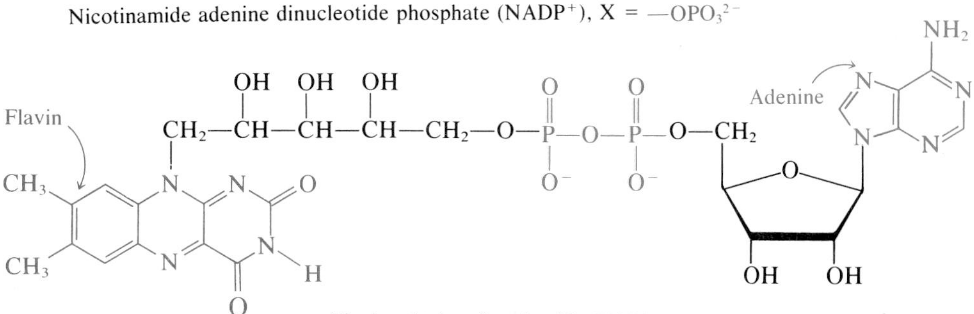

Flavin adenine dinucleotide (FAD)

moved from the oxidized substrate enters the solution as a hydrogen ion, H^+. The product of NAD^+ reduction is therefore often represented as $NADH/H^+$. $NADP^+$ is reduced in a similar manner to $NADPH/H^+$.

Reduction of NAD^+

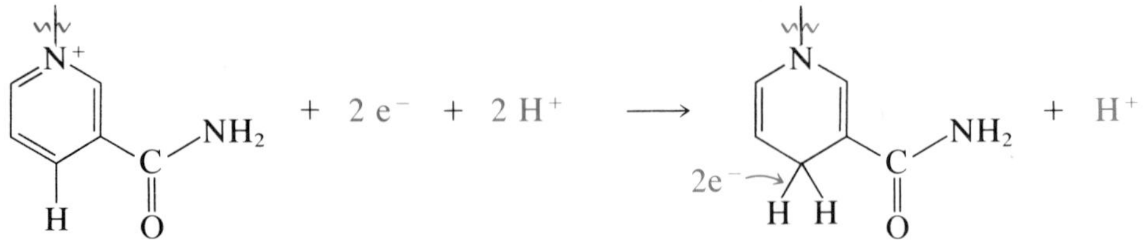

NAD^+ and **flavin adenine dinucleotide (FAD),** also shown in Figure 20.10, are the most common oxidizing agents in the reactions of catabolism. FAD, which usually removes two hydrogen atoms from a carbon chain to form an alkene double bond, adds two hydrogen ions and two electrons to give $FADH_2$ when it is reduced. Because NADH and $FADH_2$ have picked up electrons, they are often referred to as "electron carriers."

Flavin adenine dinucleotide (FAD) Coenzyme that functions as an oxidizing agent and forms $FADH_2$.

Reduction of FAD

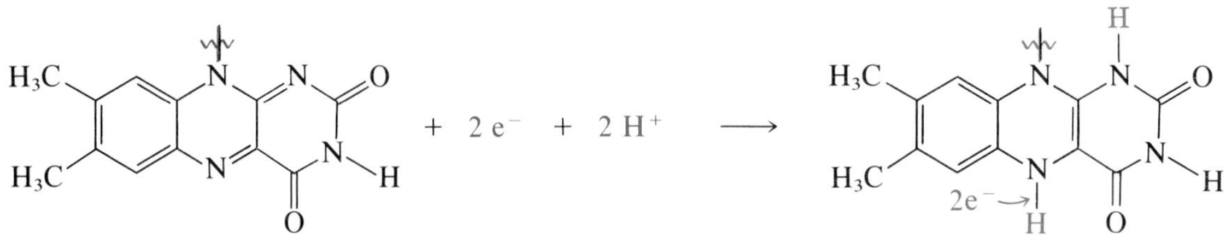

Oxidation by FAD to give alkene

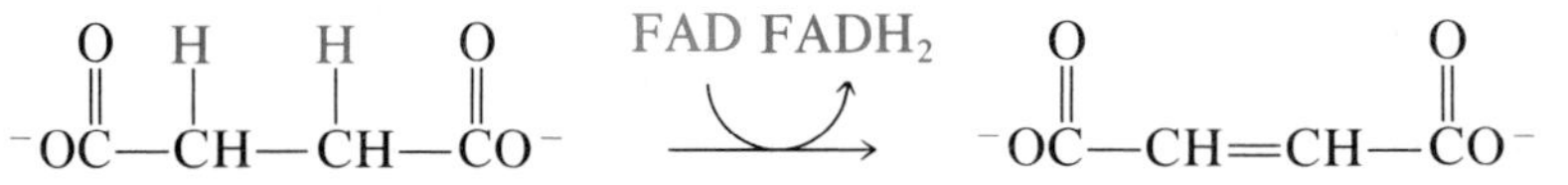

During the reduction of NAD^+ or FAD, some of the chemical energy released from the oxidized substrate is captured in NADH and $FADH_2$. As these coenzymes cycle through their oxidized and reduced forms, they are therefore able to carry energy along from reaction to reaction. Ultimately, this energy is passed on to ATP for redistribution throughout an organism.

Practice Problems   **20.5** Which of these reactions probably requires NAD^+ as a coenzyme and which requires FAD?

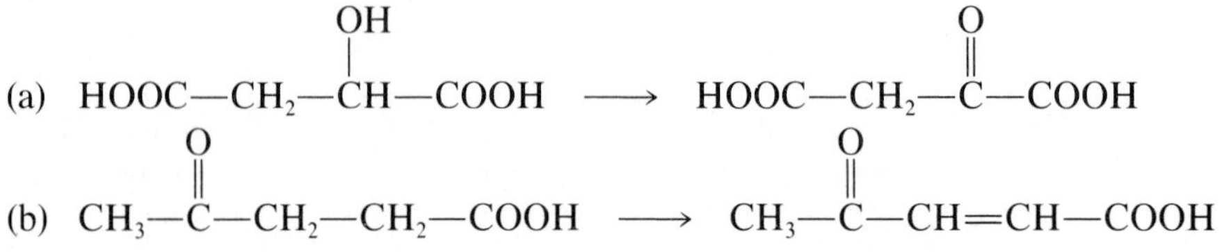

20.6 Look ahead to Figure 20.11 for the citric acid cycle. In steps 3, 6, and 8, draw the structure of the reactant and indicate which hydrogen atoms are removed.

20.8 THE CITRIC ACID CYCLE

Acetyl coenzyme A (acetyl SCoA) Acetyl-substituted coenzyme A that is the common intermediate for carrying acetyl groups into the citric acid cycle.

The first two stages of catabolism, shown earlier in Figure 20.6, result in the conversion of lipids, carbohydrates, and proteins into acetyl groups. These acetyl groups then enter the third stage, the citric acid cycle. The cycle takes place within cell mitochondria, and its reactions are coupled to the fourth stage of catabolism. Stages 1 and 2 of catabolism converge on **acetyl coenzyme A (acetyl SCoA)** when acetyl groups from the breakdown of food molecules become bonded to the thiol group of coenzyme A.

Coenzyme A, X = H—   Acetyl coenzyme A, X = CH_3C—

Like the phosphate groups in ATP, the acetyl group in acetyl SCoA is a high-energy group because it is readily removed in an exergonic hydrolysis reaction.

$$CH_3-\overset{O}{\overset{||}{C}}-S-CoA + H_2O \longrightarrow CH_3-\overset{O}{\overset{||}{C}}-O^- + H-S-CoA + H^+ \qquad \Delta G = -7.5 \text{ kcal}$$

Acetyl coenzyme A   Coenzyme A

Citric acid cycle The series of biochemical reactions that breaks down acetyl groups to carbon dioxide and energy stored in reduced coenzymes.

Breakdown of the two-carbon acetyl groups to give CO_2 occurs in the **citric acid cycle,** also called the *tricarboxylic acid cycle (TCA)* or the *Krebs cycle* after Hans Krebs who unraveled its complexities in 1937. The eight steps of the citric acid cycle and a brief description of each reaction are given in Figure 20.11. The names of the enzymes for each of the steps are listed in Table 20.3.

As its name implies, the citric acid *cycle* is a closed loop of reactions in which the *product* of the final step (step 8) is a *reactant* in the first step. The intermediates are constantly regenerated and flow continuously through the cycle, which takes place in the mitochondrial matrix where the enzymes are either free or embedded in the inner mitochondrial membrane. Some of the intermediates also serve as precursors for the synthesis of other biomolecules.

The cycle operates as long as (1) acetyl groups are available from acetyl SCoA and (2) the oxidizing agents NAD^+ and FAD are available. To meet

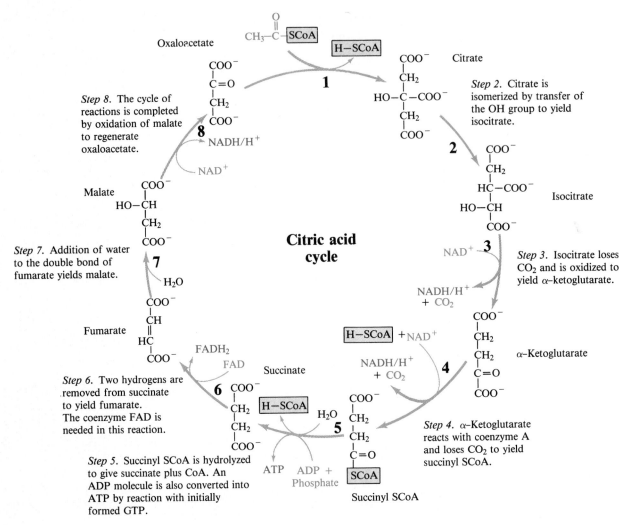

Figure 20.11
The citric acid cycle. The net effect of this eight-step series of reactions is the metabolic breakdown of acetyl groups (from acetyl SCoA) into two molecules of carbon dioxide plus reduced coenzymes. Here and throughout this and the following chapters, energy-rich forms (ATP, reduced coenzymes) are shown in red and their lower energy counterparts (ADP, oxidized coenzymes) are shown in blue.

Table 20.3   Enzymes of the Citric Acid Cycle

Step no. (Fig. 20.11)	Enzyme Name	Reaction Product
1	Citrate synthetase	Citrate
2	Aconitase	Isocitrate
3	Isocitrate dehydrogenase	α-Ketoglutarate
4	α-Ketoglutarate dehydrogenase complex	Succinyl SCoA
5	Succinyl CoA synthetase	Succinate
6	Succinate dehydrogenase	Fumarate
7	Fumarase	Malate
8	Malate dehydrogenase	Oxaloacetate

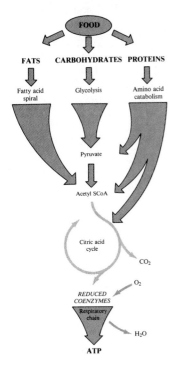

condition 2, the reduced coenzymes NADH and FADH$_2$ must be reoxidized via the respiratory chain, which in turn relies on oxygen as the final electron acceptor. Thus, the cycle is also dependent on the availability of oxygen and on the operation of the respiratory chain.

Steps 1 and 2 of the Citric Acid Cycle: Preparation   The first two steps set the stage for oxidation. Acetyl groups from acetyl SCoA enter the cycle at step 1 by addition to four-carbon oxaloacetate to give citrate, a six-carbon intermediate. Citrate is a tertiary alcohol and is converted in step 2 to its isomer, isocitrate, an easier-to-oxidize secondary alcohol. The isomerization occurs in two steps catalyzed by the same enzyme, aconitase; water is removed and then added back so that the —OH is on a different carbon atom.

$$
\begin{array}{ccc}
\begin{array}{c}
\text{COO}^- \\
| \\
\text{CH}_2 \\
| \\
\text{HO—C—COO}^- \\
| \\
\text{CH}_2 \\
| \\
\text{COO}^-
\end{array}
&
\xrightarrow[\text{aconitase}]{-\,\text{H}_2\text{O}}
\begin{array}{c}
\text{COO}^- \\
| \\
\text{CH}_2 \\
\| \\
\text{C—COO}^- \\
| \\
\text{CH} \\
| \\
\text{COO}^-
\end{array}
&
\xrightarrow[\text{aconitase}]{+\,\text{H}_2\text{O}}
\begin{array}{c}
\text{COO}^- \\
| \\
\text{CH}_2 \\
| \\
\text{H—C—COO}^- \\
| \\
\text{HO—CH} \\
| \\
\text{COO}^-
\end{array}
\\
\text{Citrate} & \text{Aconitate} & \text{Isocitrate}
\end{array}
$$

Steps 3 and 4 of the Citric Acid Cycle: Oxidative Decarboxylation   One molecule of CO$_2$ leaves at step 3 and one at step 4 as energy is released to the reduced coenzyme NADH. The reactions are oxidations in which first one and then another of the three carboxyl groups of isocitrate are removed and are thus known as *oxidative decarboxylations*. Step 4 requires coenzyme A and produces succinyl SCoA, which carries four carbon atoms along to the remaining reactions of the cycle.

Steps 5–8 of the Citric Acid Cycle: Oxidation of Succinate and Regeneration of Oxaloacetate   In step 5, the exergonic hydrolysis of succinyl SCoA is coupled with phosphorylation of guanosine diphosphate (GDP) to give guanosine triphosphate (GTP), a high-energy compound that transfers its phosphate directly to ADP. Such a reaction is known as a **substrate-level phosphorylation** to distinguish it from ATP formation in stage 4 of catabolism. Step 5 is the only step in the cycle that directly generates ATP instead of passing energy along to reduced coenzymes. In steps 6–8, two more reduced coenzyme molecules are produced as the four-carbon succinate is converted back to oxaloacetate, ready for the cycle to turn again.

Substrate-level phosphorylation
Formation of ATP by transfer of a phosphate to ADP from another phosphate-containing compound.

Net result of citric acid cycle

Acetyl SCoA + 3 NAD$^+$ + FAD + ADP + P$_i$ + 2 H$_2$O $\longrightarrow$

HSCoA + 3 NADH + 3 H$^+$ + FADH$_2$ + ATP + 2 CO$_2$

- Production of four reduced coenzyme molecules
- Conversion of an acetyl group to two CO$_2$ molecules
- Production of one energy-rich ATP molecule.

The citric acid cycle as a whole is exergonic but includes several endergonic reactions, notably step 8 which regenerates oxaloacetate. Each endergonic reaction, however, is pulled along by subsequent exergonic reactions that consume their products.

The rate at which the citric acid cycle turns is controlled by the body's need for reduced coenzymes and for the energy produced from them. When the body's supply of NADH is high, for example, NADH inhibits the enzyme for step 3. When energy is being used at a high rate, ADP accumulates and activates the enzyme for step 3. When energy is in good supply, ATP accumulates and inhibits the enzyme for step 1. In other words, the cycle is activated when energy is needed and inhibited when energy is in good supply.

Practice Problems

20.7 Which of the substances in the citric acid cycle are tricarboxylic acids (thus giving the cycle its alternate name)?

20.8 Balance the equation for the conversion of acetyl SCoA into CO_2 and coenzyme A.

$$CH_3-\overset{\overset{\displaystyle O}{\|}}{C}-SCoA \;+\; O_2 \;\longrightarrow\; CO_2 \;+\; H-SCoA \;+\; H_2O$$

20.9 Why would you expect step 5 of the citric acid cycle to be sufficiently energy-releasing to be coupled with the phosphorylation of ADP?

20.9 THE RESPIRATORY CHAIN

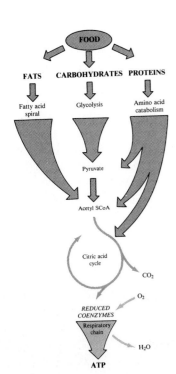

With carbon dioxide already released, reduced coenzymes from the citric acid cycle and other metabolic pathways are now ready to participate in forming the remaining two products of cellular respiration: water and energy stored in ATP. The end product of the **respiratory chain,** or *electron transport chain,* is water assembled from oxygen that we breathe, hydrogen ions, and electrons passed along from reduced coenzymes:

$$\tfrac{1}{2} O_2 + 2\,e^- + 2\,H^+ \longrightarrow H_2O$$

The reactions of the respiratory chain are coupled to *oxidative phosphorylation,* the conversion of ADP to ATP discussed in the next section.

The enzymes of the respiratory chain are embedded in the inner membrane of the mitochondrion and are arranged so that electrons pass directly from one to the next. Several types of electron carriers participate in the successive reactions. Some are located in fixed enzyme complexes and two, coenzyme Q and cytochrome c, are free to move as indicated in the schematic diagram in Figure 20.12.

The order of electron transfer, described in the following paragraphs, is from weaker to increasingly stronger electron acceptors, with energy released at each transfer, much like the waterfall shown in Figure 20.8. The coupled reactions of the pathway add up to the strongly exergonic oxidations of NADH (Figure 20.13) and FADH₂:

Figure 20.12
The respiratory chain. Symbols for the major electron carriers show their locations in the mitochondrial inner membrane. Coenzyme Q and cytochrome c are mobile, and the others are fixed within enzyme complexes. The red line shows the path of electrons along the chain. Electrons from $FADH_2$ join the chain at CoQ via an enzyme complex not shown in the figure.

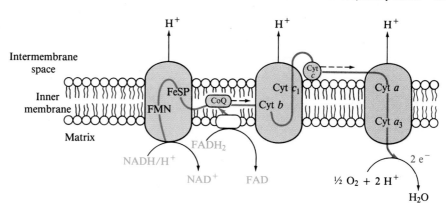

Figure 20.13
Energy released in the respiratory chain. The energy is used to pump protons across the mitochondrial inner membrane and into the intermembrane space.

$$NADH + H^+ + \tfrac{1}{2} O_2 \longrightarrow NAD^+ + H_2O + 52.7 \text{ kcal}$$
$$FADH_2 + \tfrac{1}{2} O_2 \longrightarrow FAD + H_2O + 43.4 \text{ kcal}$$

Transfer of 2 H Atoms to Flavin Mononucleotide (FMN)   The chain begins when $NADH/H^+$ from the matrix is oxidized to NAD^+ as it passes two hydrogen atoms to *flavin mononucleotide (FMN)* (Figure 20.14). This coenzyme contains the same heterocycle as FAD and is reduced in the same way to $FMNH_2$ (Section 20.7).

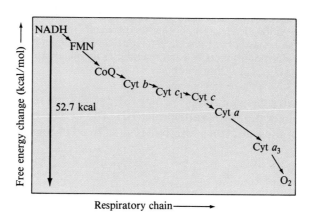

Flavin mononucleotide (FMN)

Respiratory chain (electron transport chain)   The series of biochemical reactions that passes electrons from reduced coenzymes to oxygen and is coupled to ATP formation.

Transfer of 2 e^- to Iron/Sulfur Protein (FeSP)   Following the formation of $FMNH_2$, electrons only are passed along to iron/sulfur protein (FeSP) within the

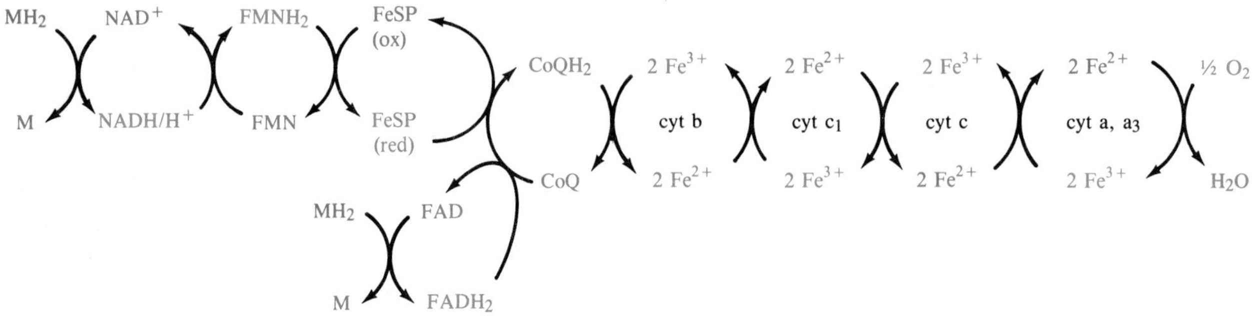

Figure 20.14
An overview of reactions in the respiratory chain. A complex series of coupled reactions leads to the formation of water, the generation of 52.7 kcal/mol energy, and the release of H^+ ions that migrate through the mitochondrial membrane. The oxidized products of each reaction in the chain are shown in red, and the reduced products are shown in blue.

same enzyme complex. As FeSP cycles from reduced to oxidized forms, its iron alternates from Fe(II) to Fe(III). The two hydrogens of $FMNH_2$, now stripped of their electrons, are released as two H^+ ions into the intermembrane space.

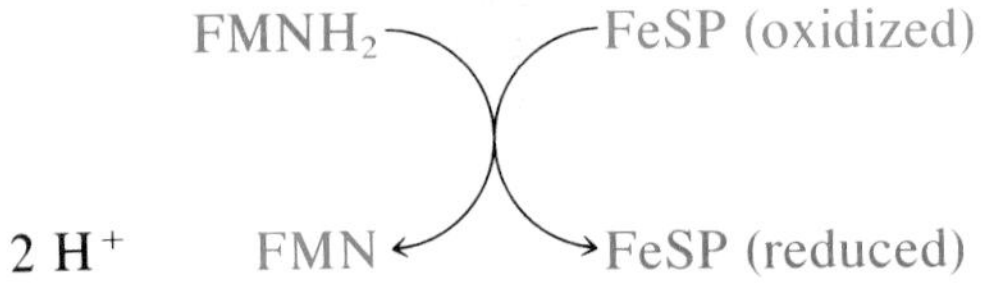

Transfer of Electrons from FeSP and FADH$_2$ to Coenzyme Q   The iron/sulfur protein next passes its electron pair to *coenzyme Q (CoQ)*, sometimes known as *ubiquinone* because of its ubiquitous occurrence in most organisms. Electrons from $FADH_2$ also enter the chain here by transfer to CoQ. The CoQ picks up two H^+ ions to give reduced coenzyme Q ($CoQH_2$), and $FADH_2$ and FeSP are returned to their oxidized forms, ready for further reactions.

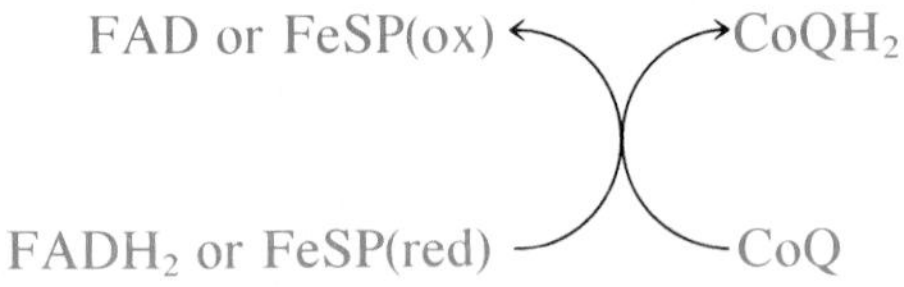

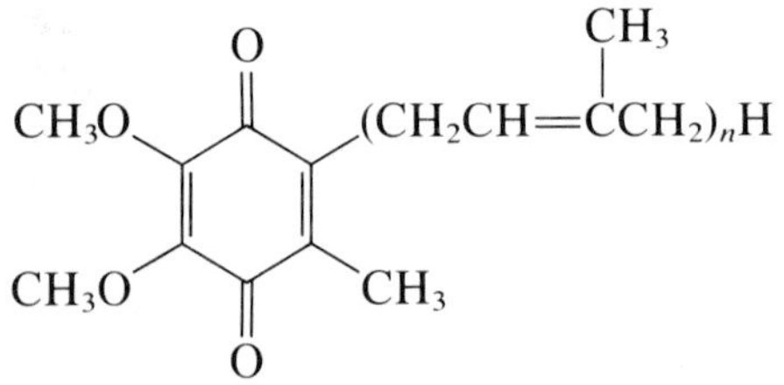

Coenzyme Q (CoQ)
(CoQH$_2$ has two —OH's
on a benzene ring)

Cytochromes Heme-containing coenzymes that function as electron carriers.

Cytochromes   Following the formation of $CoQH_2$, hydrogen ions no longer participate directly in the reductions of the respiratory chain. Instead, electrons are transferred to a series of enzymes called the **cytochromes**. There are a number of different cytochromes, abbreviated cyt a, cyt b, and so forth. All are closely related in structure, and each contains a heme coenzyme (Figure 20.15) with an iron ion that cycles between the +2 and +3 oxidation states.

Electrons are shuttled along one by one as an iron atom in one cytochrome gives up an electron to an iron atom in the next cytochrome and is oxidized from Fe(II) to Fe(III) in the process. At the same time, the iron atom in the cytochrome that accepts the electron is reduced from Fe(III) to Fe(II). Ultimately, cytochrome a₃, also called *cytochrome oxidase,* passes electrons along to an oxygen atom, which combines with two H⁺ to yield water.

$$\tfrac{1}{2}\,O_2 + 2\,e^- + 2\,H^+ \longrightarrow H_2O$$

Now that we've described the most important parts of the respiratory chain, you should look back at the entire sequence shown in Figure 20.12. As the figure indicates, H⁺ ions are released for transport through the inner mitochondrial membrane at each fixed enzyme complex. Evidence indicates that the total number of H⁺ ions transported for each NADH oxidized may be 10, but the numbers transported at each location, exactly how they are transported, and the total are still questions for active investigation. Whatever the details, the transfer of hydrogen ions across the mitochondrial inner membrane as electrons move along is essential to the synthesis of ATP via oxidative phosphorylation, as described in the next section.

Practice Problems 20.10 Without looking at Figure 20.12 or 20.13, identify the missing substances in the following series of coupled reactions:

NAD⁺ ⤵ ? ⤴ FeSP (ox) ⤵ CoQH₂
? ⤴ FMN ? ⤴ ? ⤴ ?

20.11 Look at the structure of coenzyme Q and explain whether you would expect it to be soluble in water or in organic solvents.

Figure 20.15
(a) The structure of heme, the iron-containing coenzyme present in the cytochromes.
(b) Computer-generated structure of two molecules of cytochrome c. The heme groups are shown in light green.

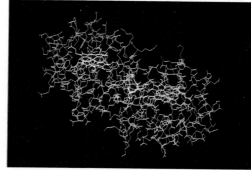

(a)   (b)

AN APPLICATION: BARBITURATES

The barbiturates, discovered in 1864 by the German chemist Adolph von Baeyer, were among the first pharmaceutical agents to come from the chemical laboratory rather than from nature. Different barbiturates are extremely easy to synthesize, and more than 2500 analogs have been prepared. All are cyclic amides that differ only in the nature of the two groups attached to the ring. Among the more common ones are barbital (sold as the drug Veronal, named after the Italian city of Verona), phenobarbital (Luminal), secobarbital (Seconal), and amobarbital (Amytal).

Barbiturates are usually classified into two groups: short-acting ones like secobarbital and long-acting ones like phenobarbital. All act as tranquilizers at low doses and as sleep inducers at somewhat higher doses. At one point several years ago it was estimated that enough barbiturates were being produced in the United States to put 10 million people to sleep every night.

Although medically useful at doses in the 50–100 mg range, barbiturates are toxic in higher amounts. At doses above 1.5 g, barbiturates depress the respiratory system, leading to severe oxygen deficiency in the brain and ultimately to death. These toxic effects appear to arise from an enzyme inhibition that blocks the participation of NADH/H$^+$ in the respiratory chain. Barbiturates somehow prevent electron transfer in the FMN/FeS/CoQ coenzymes, thereby preventing the major part of the chain from functioning.

Veronal

Phenobarbital

Secobarbital

Amobarbital

20.10 OXIDATIVE PHOSPHORYLATION: THE SYNTHESIS OF ATP

Chemiosmotic theory An explanation of how the establishment of a pH difference across a mitochondrial membrane by the respiratory chain is coupled to ATP synthesis.

How does the complex sequence of reactions that make up the respiratory chain lead to the synthesis of ATP? Despite many years of work, the full answer to this question is still not known. It seems clear, however, that the key point is the establishment of a difference in hydrogen ion concentration between the inner and outer sides of the mitochondrial membrane.

According to the **chemiosmotic theory** introduced in 1961 by Peter Mitchell, a British biochemist, hydrogen ions are "pumped" across the inner mito-

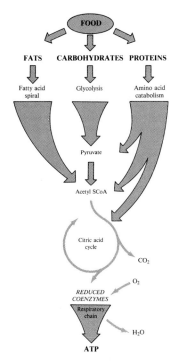

chondrial membrane during electron transport and are unable to diffuse back inside spontaneously. In other words, membrane passage of H^+ is easier in one direction than in the other, and a concentration difference is therefore maintained. The H^+ ion concentration difference between the inner and outer membranes of the mitochondrion is about 1.4 pH units, with the outside being more acidic. It's this pH difference that drives ATP formation.

By a mechanism whose details are unknown, the protons are able to cross back into the mitochondrial matrix only at certain sites. These sites, which are visible at very high magnification as knobby protrusions on the inner membrane, evidently house the ATP synthase enzyme necessary for combining ADP with phosphate to yield ATP. As the protons return to the matrix, the energy expended to maintain the concentration difference is released, and this proton flow activates the enzyme. Chemiosmotic coupling has now been demonstrated in numerous experiments, and Mitchell's achievement in proposing it was recognized in 1973 by award of a Nobel Prize.

Figure 20.16 summarizes schematically the relationships between the respiratory chain and **oxidative phosphorylation:** the synthesis of ATP from ADP. The passage of electrons from one mole of NADH down the respiratory chain releases 52.7 kcal and is coupled to the production of three molecules of ATP:

$$NADH + H^+ + \tfrac{1}{2} O_2 + 3\ ADP + 3\ P_i \longrightarrow NAD^+ + 3\ ATP + H_2O$$

Oxidative phosphorylation
The synthesis of ATP from ADP using energy released in the respiratory chain.

Since 7.3 kcal is required for each $ADP \rightarrow ATP$ conversion, about 22 kcal, or about 40% of the energy stored in NADH, is captured by oxidative phosphorylation. The oxidation of $FADH_2$, which enters the chain farther along than NADH, is coupled to the production of only two molecules of ATP.

Figure 20.16
Synthesis of ATP. The respiratory chain pumps H^+ ions from the matrix to the intermembrane space, creating a concentration imbalance. The ions can return only through channels in knobs that house the ATP synthetase enzyme complex. As the H^+ ions return, the energy expended to maintain the concentration imbalance is released and harnessed to synthesize ATP.

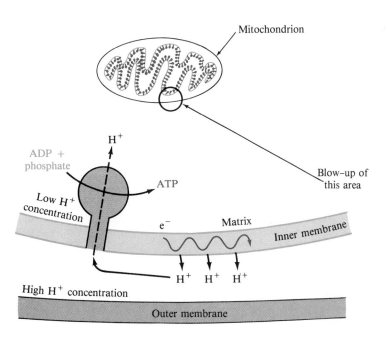

INTERLUDE: DIETS, BABIES, AND HIBERNATING BEARS

Most warm-blooded animals generate enough heat to maintain body temperature through normal metabolic reactions. In certain cases, however, normal metabolism is not able to satisfy the body's need for warmth, and a supplemental method of heat generation is needed. This occurs, for example, in newborn infants and in hibernating animals that must survive for long periods without food. Nature has therefore provided babies, bears, and many other creatures with *brown fat tissue*. Brown fat tissue, so called because its color contrasts with that of normal white fatty tissue, contains high concentrations of blood vessels and mitochondria. In addition, it contains a special protein that acts as an *uncoupler* of ATP synthesis.

Normally, the electron transport reactions of the respiratory chain are coupled to the synthesis of ATP. In other words, the purpose of the respiratory chain is to harness the energy released during catabolism to synthesize ATP. In the presence of an uncoupling agent like the protein in brown fat, however, the electron-transport reactions continue to take place, but no ATP is synthesized. As a result, the energy that would otherwise be used for ATP synthesis is released simply as heat, and the temperature of the organism rises. In essence, brown fat acts as a furnace to produce heat energy rather than stored energy.

The uncoupling of ATP synthesis from the respiratory chain can also be caused by chemical agents. It has been found, for example, that 2,4-dinitrophenol can cause uncoupling even in normal white fatty tissue of adults.

$$HO-\underset{NO_2}{\underset{|}{C_6H_3}}-NO_2$$

2,4-Dinitrophenol

Imagine the possibilities for a simple chemical agent that causes food to be converted directly into heat rather than ATP. The body, in critical need of ATP to drive its many reactions, would continue to burn fat at a frantic rate. Pound after pound of fat would literally melt away, giving every previously unsuccessful dieter the figure of a movie star. In fact, the idea works, and weight control pills containing 2,4-dinitrophenol were marketed for a brief period in the 1930s. Unfortunately, the dose needed to *slow* ATP synthesis is very close to the lethal dose that *stops* ATP synthesis, a fact that was discovered only after several fatalities had occurred. Thus, there's still no magic pill for dieters.

Brown fat and the uncoupling of the respiratory chain and ATP synthesis keep this harp seal warm.

SUMMARY

Metabolism is the sum total of all chemical reactions going on in the body. Those reactions that break down large molecules into smaller fragments are called **catabolism,** while those that build up large molecules from small pieces are called **anabolism.** Lipids, carbohydrates, and proteins are all catabolized to yield acetyl SCoA, which is then further degraded in the citric acid cycle. In the final stage of catabolism, energy is stored in ATP.

There are two broad categories of cells: **prokaryotic** cells, found in algae and bacteria, and **eukaryotic** cells, found in higher organisms. Eukaryotic cells contain a large number of small subcellular structures called **organelles** that perform specialized tasks in the cell. Among the most important organelles are the **mitochondria,** which house the enzymes necessary for producing energy from the breakdown of food-derived molecules.

Energy transfer is managed in metabolism by (1) storage in **adenosine triphosphate (ATP),** a molecule suitable for energy storage because it contains reactive P—O—P (**phosphoric anhydride**) bonds that allow it to transfer a phosphate group; (2) coupled reactions, in which the energy from an exergonic reaction is used to make possible an endergonic reaction; and (3) the cycling between their oxidized and reduced forms of enzymes, notably **nicotinamide adenine dinucleotide (NAD)** and **flavin adenine dinucleotide (FAD),** that act as electron and hydrogen carriers.

The **citric acid cycle** is a series of eight enzyme-catalyzed steps whose primary role is releasing energy from **acetyl SCoA,** which carries acetyl groups from earlier stages of catabolism. The acetyl groups are converted into two molecules of carbon dioxide plus a large amount of energy. The energy output of the citric acid cycle is carried by reduced coenzymes to the **respiratory chain,** in which electrons from the enzymes are carried in stepwise fashion through a series of enzymes ending with the **cytochromes** and with the reaction with oxygen to form water. According to the **chemiosmotic theory,** the protons released in the respiratory chain migrate back into the matrix at certain knob-like sites on the membrane, thereby activating the enzymes that carry out ATP synthesis (**oxidative phosphorylation**).

REVIEW PROBLEMS

Free Energy and Biochemical Reactions

20.12 What is the difference between an endergonic and an exergonic process?

20.13 What energy requirement must be met in order for a reaction to be spontaneous?

20.14 Why is ΔG a useful quantity for predicting the spontaneity of biochemical reactions?

20.15 Many biochemical reactions are catalyzed by enzymes. Do enzymes have an influence on the magnitude or sign of ΔG? Why or why not?

20.16 Each of the following reactions occurs during the catabolism of glucose (Chapter 22). Which is exergonic? Which is endergonic? Which proceeds farthest toward products at equilibrium?
(a) phosphoenol pyruvate + H_2O → pyruvate + phosphate $\Delta G = -14.8$ kcal/mol
(b) glucose + phosphate → glucose 6-phosphate + H_2O $\Delta G = 3.3$ kcal/mol
(c) 1,3-bisphosphoglycerate + H_2O → 3-phosphoglycerate + phosphate $\Delta G = -11.8$ kcal/mole

Cells and Their Structure

20.17 List several differences between prokaryotic and eukaryotic cells.

20.18 What kinds of organisms have prokaryotic cells, and what kinds have eukaryotic cells?

20.19 What is an organelle, and what is its general function?

20.20 What is the difference between the cytoplasm and the cytosol?

20.21 Describe in general terms the structural makeup of a mitochondrion.

20.22 Why are mitochondria called the body's "power plant"?

20.23 What are cristae, and why are they important?

Metabolism

20.24 What is the difference between digestion and metabolism?

20.25 What is the difference between catabolism and anabolism?

20.26 What key metabolic substance is formed from the catabolism of all three major classes of foods: carbohydrates, fats, and proteins?

20.27 Put the following events in the correct order of their occurrence: respiratory chain, digestion, oxidative phosphorylation, citric acid cycle.

Strategies of Metabolism

20.28 What is the full name of the substance formed during catabolism to store chemical energy?

20.29 What is the chemical difference between ATP and ADP?

20.30 Why is ATP called a high-energy molecule?

20.31 What general kind of chemical reaction does ATP carry out?

20.32 Adenosine *mono*phosphate is an important intermediate in certain biochemical pathways. Draw its structure.

20.33 What does it mean when we say that two reactions are coupled?

20.34 Show why coupling the reaction for the hydrolysis of 1,3-bisphosphoglycerate (Problem 20.16) to the phosphorylation of ADP is energetically favorable. Combine the equations and calculate ΔG for the coupled process. You need only give names or abbreviations, not chemical structures.

20.35 Write the reaction in Problem 20.34 with the curved arrow symbolism.

20.36 Would the hydrolysis of fructose 6-phosphate (Table 20.1) be favorable for phosphorylating ADP? Why or why not?

20.37 Both NAD^+ and FAD are coenzymes for dehydrogenation.
(a) When a molecule is dehydrogenated is it oxidized or reduced?
(b) Are NAD^+ and FAD oxidizing agents or reducing agents?
(c) What type of substrate is each coenzyme associated with, and what is the type of product molecule after dehydrogenation?
(d) What is the form of each coenzyme after the dehydrogenation process?
(e) Write a generalized equation for the operation of each coenzyme with the curved arrow symbolism.

The Citric Acid Cycle

20.38 Where in the cell does the citric acid cycle take place?

20.39 By what other names is the citric acid cycle known?

20.40 What substance acts as the starting point of the citric acid cycle, reacting with acetyl SCoA in the first step and being regenerated in the last step? Draw its structure.

20.41 Look at the eight steps of the citric acid cycle (Figure 20.11) and answer these questions:
(a) Which steps involve oxidation reactions?
(b) Which steps involve decarboxylations (loss of CO_2)?
(c) Which step involves a hydration reaction?

20.42 Consider step 5 of the citric acid cycle. What is substrate-level phosphorylation and what is the function of GTP in this process?

20.43 How many ATPs are directly formed as a result of the citric acid cycle?

20.44 How many molecules of NADH and $FADH_2$ are formed in the citric acid cycle?

20.45 What is the final fate of the carbons in the acetyl CoA after several turns of the citric acid cycle?

The Respiratory Chain

20.46 By what other name is the respiratory chain known?

20.47 What does the term "oxidative phosphorylation" mean? How does it differ from substrate-level phosphorylation?

20.48 In oxidative phosphorylation, what is oxidized and what is phosphorylated?

20.49 What two coenzymes initiate the events of the respiratory chain?

20.50 What are the ultimate products of the respiratory chain?

20.51 What do the following abbreviations stand for?
(a) FAD   (b) CoQ   (c) $NADH/H^+$ (d) cyt

20.52 What atom in the cytochromes undergoes oxidation and reduction in the respiratory chain?

20.53 Put the following substances in the correct order of their action in the respiratory chain: iron/sulfur protein, cytochrome a, coenzyme Q, NAD^+.

20.54 Fill in the missing substances in these coupled reactions:

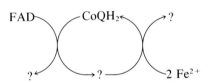

20.55 In the respiratory chain, how many ATPs are released by reoxidation of each NADH? Each $FADH_2$?

20.56 What would happen to the citric acid cycle if NADH and $FADH_2$ were not reoxidized?

20.57 According to the chemiosmotic theory, across what membrane is there a pH differential caused by the release of H^+ ions? On which side of the membrane are there more H^+ ions?

Applications

20.58 How is basal metabolic rate defined? [App: Basal Metabolism]

20.59 Estimate your basal metabolic rate using the guidelines in the application. [App: Basal Metabolism]

20.60 Why do activities such as walking raise a body's needs above the basal metabolic rate? [App: Basal Metabolism]

20.61 What is the medical function of a barbiturate? [App: Barbiturates]

20.62 By what mechanism do barbiturates become lethal at high doses? [App: Barbiturates]

20.63 Other than color, what is the difference between brown and white fat? [Int: Diets, Babies, and Hibernating Bears]

20.64 What compound has been used as an uncoupler of ATP synthesis? How can the uncoupling be used as a diet aid and why can it be dangerous? [Int: Diets, Babies, and Hibernating Bears]

Additional Questions and Problems

20.65 Why must the breakdown of molecules for energy in the body occur in several steps, rather than in one step?

20.66 The first step of the citric acid cycle involves an aldol condensation of acetyl CoA and oxaloacetic acid. Show the product of the aldol condensation before the hydrolysis to yield citrate.

20.67 The fumaric acid produced in step 6 of the citric acid cycle must have a trans double bond in order to continue on in the cycle. Suggest a reason why the correspondg cis double-bond isomer can't continue in the cycle.

20.68 With what class of enzymes (Section 19.3) are the coenzymes NAD^+ and FAD associated?

20.69 Considering both substrate-level phosphorylation and oxidative phosphorylation, how many molecules of ATP are generated by one turn of the citric acid cycle?

20.70 We talk of burning food in a combustion process, producing CO_2 and H_2O from food and oxygen. Explain how oxygen is involved in the process since there is no oxygen directly involved in the citric acid cycle. What enzyme is associated with the use of O_2?

C H A P T E R

21

Carbohydrates

Carbohydrates are stored in the liver and muscles in the polymer known as glycogen. This falsely colored micrograph of liver tissue shows a liver cell (at top), its large nucleus (green), packages of stored glycogen (pink-red), and vacuoles containing triacylglycerols (yellow), the stored lipids described in Chapter 23.

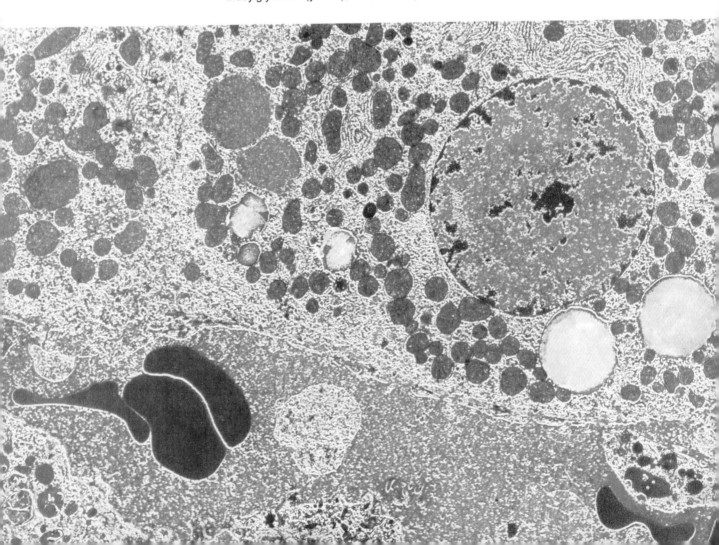

Carbohydrates occur in every living organism. The starch in food and the cellulose in grass are pure carbohydrate; modified carbohydrates form part of the coating around all living cells; other carbohydrates are found in the DNA that carries genetic information from one generation to the next; and still other carbohydrates such as streptomycin are valuable as medicines.

The word *carbohydrate* was used originally to describe glucose, the simplest and most readily available sugar. Because glucose has the formula $C_6H_{12}O_6$, it was once thought to be a "hydrate of carbon"—$C_6(H_2O)_6$. Although this view was soon abandoned, the name "carbohydrate" persisted until it is now used to refer to a large class of polyhydroxylated aldehydes and ketones.

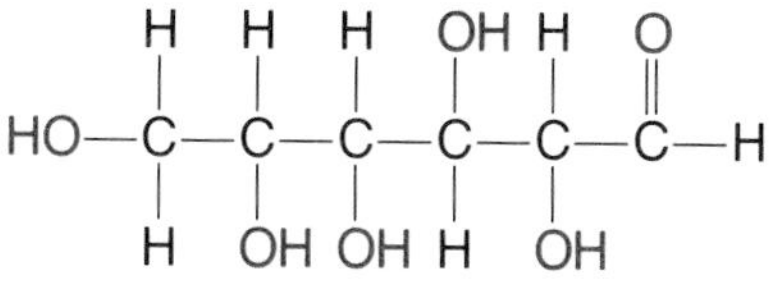

Glucose (a pentahydroxyhexanal)

Carbohydrates are synthesized in green leaves by the conversion of carbon dioxide into glucose during photosynthesis. Many molecules of glucose are then linked together to form either cellulose or starch. When starch is eaten and digested, the freed glucose provides the major source of energy required by living organisms. Thus, carbohydrates act as the intermediaries by which energy from the sun is converted into chemical energy.

In this chapter, we'll look for answers to these questions about carbohydrates:

1. ***What are the different kinds of carbohydrates?*** The goal: Be able to define the main classifications of carbohydrates and classify specific examples.

2. ***Why do carbohydrates have handedness?*** The goal: Be able to explain why carbohydrates have handedness and be able to show that handedness using Fischer projections.

3. ***What are the structures of glucose and other important simple sugars?*** The goal: Be able to describe the structural features of monosaccharides and distinguish between D and L sugars.

4. ***How are monosaccharide molecules drawn?*** The goal: Be able to draw the Fischer projection and cyclic formulas of D and L sugars.

5. ***What is a reducing sugar?*** The goal: Be able to recognize a reducing sugar and describe its oxidation.

6. ***What is a glycoside?*** The goal: Be able to predict the product of a given glycoside-forming reaction and classify the glycosidic bond formed.

7. ***What are the structures of some important disaccharides?*** The goal: Be able to draw the structures of the disaccharides maltose, sucrose, and lactose.

8. ***What are the structures of some important polysaccharides?*** The goal: Be able to describe how cellulose and starch are constructed and how they differ.

21.1 CLASSIFICATION OF CARBOHYDRATES

Carbohydrate A member of a large class of naturally occurring polyhydroxy ketones and aldehydes.

Monosaccharide (simple sugar) A carbohydrate that can't be chemically broken down into a smaller sugar by hydrolysis with aqueous acid.

Disaccharide A carbohydrate that yields two monosaccharides on hydrolysis.

Acetal A compound that has two —OR groups bonded to the same carbon atom.

Polysaccharide (complex carbohydrate) A carbohydrate composed of many monosaccharides bonded together.

Carbohydrates, a large class of naturally occurring polyhydroxylated aldehydes and ketones, are classified according to whether or not they can be broken down into smaller units. **Monosaccharides,** sometimes called **simple sugars,** are carbohydrates that can't be broken down into smaller molecules by hydrolysis with aqueous acid. Glucose, the pentahydroxyhexanal shown in the introduction to this chapter, is the most important monosaccharide. Naturally occurring monosaccharides have from three to seven carbon atoms, with an —OH group on each carbon atom except that with the aldehyde or ketone carbonyl group. The family-name ending *-ose* indicates a sugar, and specific sugars are known by common rather than systematic names.

Disaccharides are carbohydrates composed of two monosaccharides linked by C—O—C **acetal** bonds, while **polysaccharides** are compounds made of many simple sugar molecules bonded together. On hydrolysis, polysaccharides such as cellulose and starch are cleaved to yield many molecules of simple sugars. (Recall that in hydrolysis, larger molecules split apart with addition of the H— and —OH from water to the atoms in the broken bond.)

A hydrolysis reaction

Hydrolysis of a polysaccharide

$$\text{Cellulose or starch} + H_2O \xrightarrow[\text{catalyst}]{H^+} \text{thousands of glucose molecules}$$

Aldose A monosaccharide that contains an aldehyde carbonyl group.

Ketose A monosaccharide that contains a ketone carbonyl group.

Monosaccharides are classified as either aldoses or ketoses according to the kind of carbonyl group they have. An **aldose** contains an aldehyde carbonyl group; a **ketose** contains a ketone carbonyl group. The number of carbon atoms in an aldose or ketose is specified by using one of the prefixes *tri-, tetr-, pent-,* or *hex-*. Thus, glucose is an aldo-*hex*-ose (aldo- = aldehyde, -*hex* = six carbon, -ose = sugar); fructose is a keto*hex*ose (a six-carbon ketone sugar); and ribose is an aldo*pent*ose (a five-carbon aldehyde sugar). Most naturally occurring simple sugars are either aldopentoses or aldohexoses.

Glucose
(an aldohexose)

Fructose
(a ketohexose)

Ribose
(an aldopentose)

Practice Problems   **21.1**   Classify each of these monosaccharides:

(a) $HOCH_2\!-\!\underset{\underset{OH}{|}}{CH}\!-\!\underset{\underset{OH}{|}}{CH}\!-\!\underset{\underset{OH}{|}}{CH}\!-\!\underset{\overset{O}{\parallel}}{C}\!-\!H$ (b) $HOCH_2\!-\!\underset{\overset{O}{\parallel}}{C}\!-\!CH_2OH$

(c) $HOCH_2\!-\!\underset{\underset{OH}{|}}{CH}\!-\!\underset{\underset{OH}{|}}{CH}\!-\!\underset{\overset{O}{\parallel}}{C}\!-\!H$

21.2   Draw the structures of an aldohexose and a ketotetrose.

21.2   HANDEDNESS OF CARBOHYDRATES

Glyceraldehyde, the simplest naturally occurring carbohydrate, has the structure shown in Figure 21.1. Is glyceraldehyde chiral? As explained in Section 18.5, chiral compounds have a carbon atom bonded to four different atoms or groups of atoms. Such compounds lack a plane of symmetry and can exist as a pair of optical isomers (enantiomers) in either a "right-handed" D form or a "left-handed" L form. Glyceraldehyde is therefore chiral because it has four different groups bonded to C2: —CHO, —H, —OH, and —CH₂OH. Like all optical isomers, the two forms of glyceraldehyde have the same physical properties except for the way in which they interact with polarized light. This interaction is discussed in the Application on polarized light.

Compounds like glyceraldehyde have only one chiral carbon atom and can exist as two optical isomers. But what about compounds with more than one chiral carbon atom? How many isomers are there of compounds that have two, three, four, or more chiral carbons? Aldotetroses, for example, have two chiral carbon atoms and can exist in the four isomeric forms shown in Figure 21.2.

Figure 21.1
Glyceraldehyde (2,3-dihydroxypropanal), a chiral molecule. Glyceraldehyde is chiral because it has four different groups attached to carbon 2. Thus, glyceraldehyde can exist in two forms: a "right-handed" form referred to as D-glyceraldehyde and a "left-handed" form referred to as L-glyceraldehyde.

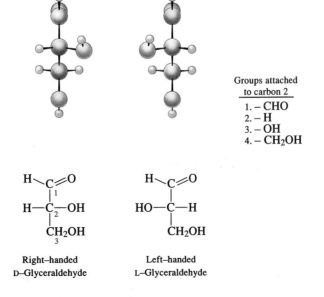

Groups attached
to carbon 2
1. – CHO
2. – H
3. – OH
4. – CH₂OH

Right–handed
D–Glyceraldehyde

Left–handed
L–Glyceraldehyde

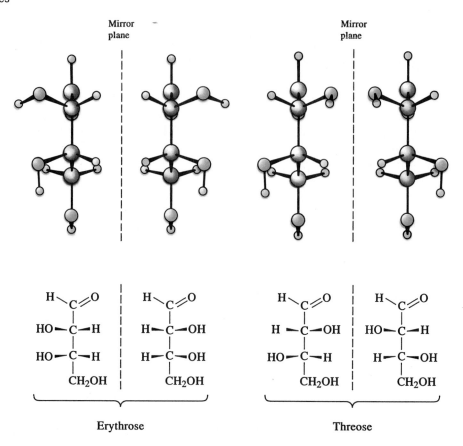

Figure 21.2
The four isomeric
aldotetroses (2,3,4-
trihydroxybutanais).
Erythrose and threose differ
in arrangement at C2.

Erythrose

Threose

The four stereoisomeric aldotetroses can be classifed into two mirror-image pairs of optical isomers, one pair named *erythrose* and one pair named *threose*. Erythrose and threose, however, are not mirror images of each other but are instead a different kind of stereoisomer. Non-mirror-image stereoisomers like erythrose and threose are called **diastereomers.**

Diastereomers
Stereoisomers that are not
mirror images of each other.

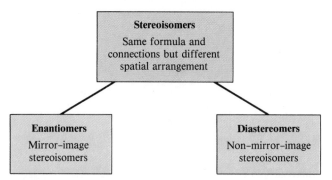

In general, a compound with n chiral carbon atoms has a maximum of 2^n possible stereoisomers and half that number of optical isomer pairs. The aldotetroses, for example, have $n = 2$ so that $2^n = 2^2 = 4$, showing that 4 stereoisomers are possible. Glucose, an aldohexose, has four chiral carbon atoms and a total of $2^4 = 16$ possible stereoisomers (8 pairs of optical isomers) that differ in the spatial arrangements of the substituents around chiral carbon atoms. All 16 are known.

AN APPLICATION: POLARIZED LIGHT AND OPTICAL ACTIVITY

Louis Pasteur, the French chemist best known for discovering the value of "pasteurizing" milk, was the first to propose that some compounds could exist as mirror-image pairs of molecules. In 1849, many years before molecular structure was understood, he was working with the crystalline salts of tartaric acid when he discovered two visibly different kinds of asymmetric crystals in a sample of sodium ammonium tartrate. By using tweezers, he patiently separated them and found that crystals of one type were *mirror images* of crystals of the second type. To explore the properties of these two tartrates, Pasteur placed their solutions in the path of *plane-polarized light* beams.

Light as we usually see it consists of electromagnetic waves oscillating in all planes at right angles to the direction of travel of the light beam. When ordinary light is passed through a *polarizer,* only the waves in one plane get through, producing what is known as *plane-polarized light*. From earlier work by another French scientist, Jean Baptiste Biot, Pasteur knew that solutions of certain organic compounds change the plane in which the light is polarized. Because of their interaction with light, such compounds are described as *optically active*. The angle of rotation of plane-polarized light by an optically active compound is measured in an instrument known as a *polarimeter,* illustrated schematically below.

Pasteur discovered that solutions of his two types of crystals rotated plane-polarized light by equal amounts, but in *opposite directions*. One kind of crystal rotated the light in a right-handed direction, denoted (+), and the other kind of crystal rotated the light in a left-handed direction, denoted (−). To explain this phenomenon he proposed the existence of two types of tartaric acid molecules with opposite asymmetric arrangements that are, like the visible crystals, mirror images of each other. Although it could not be proven until many years later, Pasteur had indeed discovered what we now know as *optical isomers*.

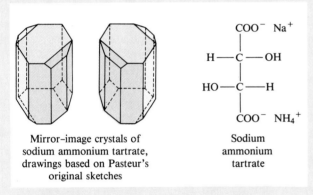

Mirror-image crystals of sodium ammonium tartrate, drawings based on Pasteur's original sketches

Sodium ammonium tartrate

Mirror-image crystals, drawings based on Pasteur's original sketches

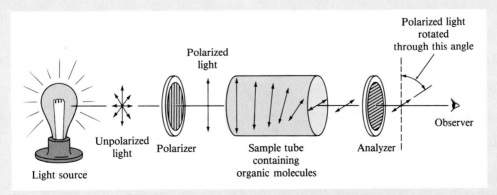

Diagram of a polarimeter, which measures the optical activity of a molecule

Note that we've waffled a bit in the above discussion by saying that 2^n represents the maximum number of "possible" stereoisomers. In some cases, variations around chiral atoms produce molecules that have symmetry planes and are thus identical to their mirror images. In such cases, there are fewer than the maximum possible number of stereoisomers.

Practice Problems   **21.3**   Draw tetrahedral representations of the two glyceraldehyde optical isomers using the standard method of wedged, dashed, and normal lines to show three-dimensionality.

   21.4   Aldopentoses have three chiral carbon atoms. What is the maximum possible number of aldopentose stereoisomers?

Fischer projection
Structure that represents a chiral carbon atom as the intersection of two lines. The horizontal line represents bonds pointing out of the page, and the vertical line represents bonds pointing behind the page. For sugars, the aldehyde or ketone is at the top.

21.3 DRAWING SUGAR MOLECULES: THE D AND L FAMILIES OF SUGARS

A standard method of representation called a **Fischer projection** has been adopted for drawing stereoisomers on a flat page so that we can tell one from another. A chiral carbon atom is represented in Fischer projections as the intersection of two crossed lines. Bonds that point out of the page are shown as horizontal lines, and bonds that point behind the page are shown as vertical lines.

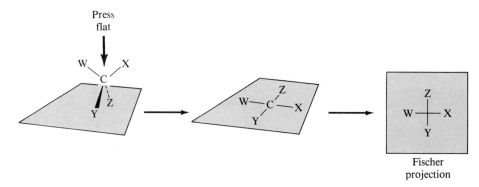

For consistency, the aldehyde or ketone carbonyl group is shown at or near the top when drawing a carbohydrate. Thus, one optical isomer of glyceraldehyde can be drawn as follows:

Mirror

H—C≷O
H—C—OH
CH₂OH

D–Glyceraldehyde

H—C≷O
HO—C—H
CH₂OH

L–Glyceraldehyde

D **Sugar** Monosaccharide with —OH group on chiral carbon atom farthest from the carbonyl group pointing to the right in the Fischer projection.

L **Sugar** Monosaccharide with —OH group on chiral carbon atom farthest from the carbonyl group pointing to the left in the Fischer projection.

Monosaccharides are divided into two families—the **D sugars** and the **L sugars**—based on their structural relationships to glyceraldehyde. Note how the D and L forms of glyceraldehyde were shown back in Figure 21.1: In the D form, the —OH group on carbon 2 comes out of the plane of the paper and points to the right when the —CHO group is at the top; in the L form, the —OH group at carbon 2 comes out of the plane of the paper and points to the left when the —CHO group is at the top. If you mentally place a mirror plane between the Fischer projections of these molecules, you can see that they're mirror images.

Nature has a strong preference for one type of handedness in carbohydrates, just as it does in amino acids (Section 18.5) and also in snail shells (Figure 21.3). It happens, however, that carbohydrates and amino acids have opposite handedness. Most naturally occurring α-amino acids belong to the L family, but most carbohydrates belong to the D family. L-Carbohydrates occur only rarely in nature.

The designations D and L derive from the Latin *dextro* for "right" and *levo* for "left." In Fischer projections, the D form of a monosaccharide has *the hydroxyl group on the chiral carbon atom farthest from the carbonyl group* pointing toward the right, while the mirror-image L form has the hydroxyl group on this same carbon pointing toward the left. The D and L designations don't, however, have any relationship to the direction in which an isomer rotates plane-polarized light.

Fischer projections of molecules with more than one chiral carbon atom are written by stacking the chiral atoms one above the other. The following structures show two of the eight pairs of optical isomers of the aldohexoses written in this manner. Given the Fischer projection of one isomer, you can draw the other isomer by reversing the substituents on each chiral atom. Note that each pair of optical isomers has a different name.

Two pairs of aldohexose optical isomers

O≷C—H
H—C—OH
H—C—OH
H—C—OH
H—C—OH
CH₂OH

D-Allose

O≷C—H
HO—C—H
HO—C—H
HO—C—H
HO—C—H
CH₂OH

L-Allose

O≷C—H
H—C—OH
HO—C—H
H—C—OH
H—C—OH
CH₂OH

D-Glucose

O≷C—H
HO—C—H
H—C—OH
HO—C—H
HO—C—H
CH₂OH

L-Glucose

Chiral C atom farthest from C=O

Figure 21.3
Nature's preference. Not only molecules, but also snail shells have a preferrred handedness. Most snail shells, like the one shown, are right-handed.

Practice Problems **21.5** Identify the following monosaccharides as (a) D-ribose or L-ribose, (b) D-mannose or L-mannose.

(a)

$$
\begin{array}{c}
\text{H} \diagdown\!\!\!\diagup \text{O} \\
\text{C} \\
| \\
\text{H—C—OH} \\
| \\
\text{H—C—OH} \\
| \\
\text{H—C—OH} \\
| \\
\text{CH}_2\text{OH}
\end{array}
$$

(b)

$$
\begin{array}{c}
\text{H} \diagdown\!\!\!\diagup \text{O} \\
\text{C} \\
| \\
\text{H—C—OH} \\
| \\
\text{H—C—OH} \\
| \\
\text{HO—C—H} \\
| \\
\text{HO—C—H} \\
| \\
\text{CH}_2\text{OH}
\end{array}
$$

21.6 Draw the optical isomers of the following monosaccharides and identify which is in the D family of sugars and which is in the L family in each pair.

(a)

$$
\begin{array}{c}
\text{H} \diagdown\!\!\!\diagup \text{O} \\
\text{C} \\
| \\
\text{HO—C—H} \\
| \\
\text{H—C—OH} \\
| \\
\text{H—C—OH} \\
| \\
\text{CH}_2\text{OH}
\end{array}
$$

(b)

$$
\begin{array}{c}
\text{CH}_2\text{OH} \\
| \\
\text{C}=\text{O} \\
| \\
\text{H—C—OH} \\
| \\
\text{HO—C—H} \\
| \\
\text{HO—C—H} \\
| \\
\text{CH}_2\text{OH}
\end{array}
$$

21.4 STRUCTURE OF GLUCOSE AND OTHER MONOSACCHARIDES

Hemiacetal A compound that has an —OH and —OR group bonded to the same carbon atom.

D-Glucose, sometimes called *dextrose* or *blood sugar,* is the most widely occurring of all monosaccharides. It is found in nearly all foods and in all living organisms, where it serves as a source of energy to fuel biochemical reactions. Before looking further at the chemical structure of glucose, glance back briefly at Section 16.7. We said there that aldehydes and ketones react reversibly with alcohols to yield **hemiacetal** addition products in which a carbon atom is bonded to both an —OH and an —OR group.

$$
\underset{\text{An aldehyde}}{\text{R—}\overset{\displaystyle \text{O}}{\overset{\|}{\text{C}}}\text{—H}} \quad + \quad \underset{\text{An alcohol}}{\overset{\text{H}}{\underset{}{|}}\ \text{O—R}'} \quad \rightleftharpoons \quad \underset{\text{A hemiacetal}}{\text{R—}\overset{\text{O—H}}{\underset{\text{H}}{\overset{|}{\text{C}}}}\text{—O—R}'}
$$

Now look at the Fischer projection structure of D-glucose shown in Figure 21.4. Since glucose is an aldohexose—a pentahydroxy aldehyde—it has alcohol

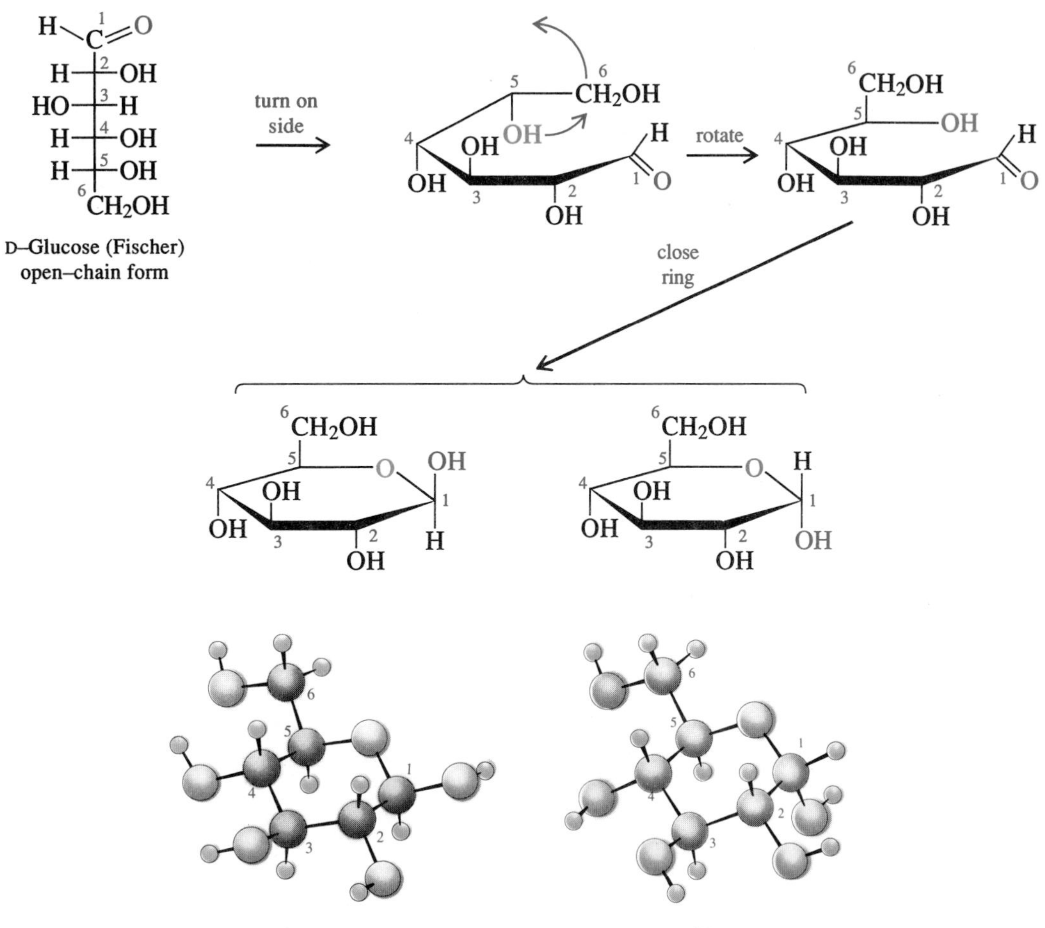

Figure 21.4
The structure of D-glucose. A glucose molecule can exist either in an open-chain
hydroxy aldehyde form or in a cyclic hemiacetal form. There are two cyclic
hemiacetal forms, called α-glucose and β-glucose, that differ in whether the
hemiacetal hydroxyl group at C1 is on the opposite side of the six-membered ring
from the CH₂OH (α) or on the same side (β). To convert the Fischer formula into
the six-membered ring formula, the Fischer formula is laid down with C1 to the
right in front and the other end curled around at the back. Then the single bond
between C4 and C5 is rotated so that the —CH₂OH group is vertical. Finally, the
hemiacetal O—R bond is formed by connecting oxygen from the —OH group on
C5 to C1 and the hemiacetal O—H group is placed on C1.

hydroxyl groups and an aldehyde carbonyl group *in the same molecule*. Thus,
an *internal* addition reaction can take place between the aldehyde carbonyl
group at carbon 1 and one of the hydroxyl groups in the chain to yield a *cyclic*
hemiacetal. In fact, it's the hydroxyl group at carbon 5 that reacts in glucose,
leading to a six-membered-ring hemiacetal. The three structures at the top in
Figure 21.4 show schematically how the 5-hydroxyl and the aldehyde group
approach for acetal formation.

Actually, there are *three* forms of D-glucose—an open-chain form, a cyclic α form, and a cyclic β form—all shown in Figure 21.4. To see the difference between the α and β forms, compare the locations of the hemiacetal —OH groups on carbon 1, which is now a chiral carbon atom. In the β form, the hydroxyl at carbon 1 points *up* on the same side of the ring as the —CH₂OH group connected to carbon 5; in the α form, the hydroxyl at carbon 1 points *down* on the opposite side of the ring from the —CH₂OH group at carbon 5.

Cyclic sugars that differ only in the positions of substituents at carbon 1 are known as **anomers,** and carbon 1 is said to be an **anomeric carbon atom.** (Remember—such a carbon atom is bonded to two oxygen atoms. In the hemiacetal, they're in an —OH group and an —OR group in which R is the ring.) Note that anomers are not optical isomers because they aren't mirror images. Although the structural difference between anomers might seem small, it has enormous biological consequences. You'll see in Section 21.8, for instance, that this one small change in structure accounts for the vast difference between starch and cellulose.

The ordinary crystalline glucose you might take from a bottle is usually entirely in the cyclic α form. In solution, however, equilibria are established by both anomers with the open-chain form. As a result, the optical effect of the solution on polarized light changes, a phenomenon known as **mutarotation.** The equilibria favor the hemiacetal forms, and either anomer when placed in solution produces the identical equilibrium mixture containing mainly the two hemiacetals. For glucose at room temperature in water, at equilibrium the mixture contains 0.02% open chain, 36% α form, and 64% β form.

$$\alpha\text{-D-glucose} \quad \rightleftharpoons \quad \text{open-chain D-glucose} \quad \rightleftharpoons \quad \beta\text{-D-glucose}$$
$$(36\%) \qquad\qquad\qquad (0.02\%) \qquad\qquad\qquad (64\%)$$

To convert Fischer projections into cyclic structures so that the same relative arrangements are maintained at the chiral carbon atoms requires following the procedure illustrated in Figure 21.4. When cyclic structures, called *Haworth projections,* are drawn in this manner, the —CH₂OH group in D sugars is always above the plane of the ring. Thus, in β-D-glucose, the hydroxyl at carbon 1 points up, on the same side of the ring as the —CH₂OH on carbon 5. In α-D-glucose, the hydroxyl at carbon 1 points down, on the opposite side of the ring from the —CH₂OH group at carbon 5.

To summarize—monosaccharide structures have the following characteristics:

● Monosaccharides are polyhydroxy aldehydes or ketones.

● Monosaccharides have three to seven carbon atoms, and 2^n possible stereoisomers, where n is the number of chiral carbon atoms.

● D and L Optical isomers differ in the location of the —OH group on the chiral carbon atom farthest from C1; in Fischer projections, D sugars have —OH on the right, and L sugars have —OH on the left.

● α and β Anomers differ in the location of the —OH on the hemiacetal carbon atom in cyclic forms; α has the —OH on the opposite side from the —CH₂OH, and β has the —OH on the same side as —CH₂OH.

Anomers Cyclic sugars that differ only in positions of substituents at the hemiacetal carbon (the anomeric carbon); the α form has the —OH on the opposite side from the —CH₂OH; the β form has the —OH on the same side as the —CH₂OH.

Anomeric carbon atom The hemiacetal C atom in a cyclic sugar; the C atom bonded to an —OH group and an —OR group (R is the ring).

Mutarotation Change in rotation of plane-polarized light resulting from the equilibrium between cyclic anomers and the open-chain form of a sugar.

Solved Problem 21.1 The open-chain form of D-altrose, an aldohexose isomer of glucose, has the following structure. Draw D-altrose in its cyclic hemiacetal form.

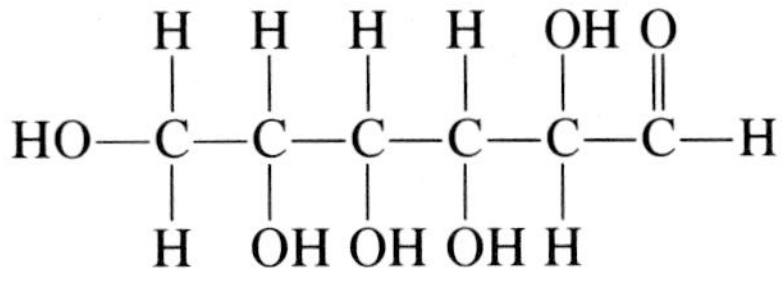

$$
\begin{array}{c}
\quad\ \ \text{H}\ \ \text{H}\ \ \text{H}\ \ \text{H}\ \ \text{OH O}\\
\quad\ \ |\ \ \ |\ \ \ |\ \ \ |\ \ \ |\ \ \parallel\\
\text{HO}-\text{C}-\text{C}-\text{C}-\text{C}-\text{C}-\text{C}-\text{H}\\
\quad\ \ |\ \ \ |\ \ \ |\ \ \ |\ \ \ |\\
\quad\ \ \text{H}\ \ \text{OH OH OH H}
\end{array}
\qquad \text{D-Altrose}
$$

Solution First, coil D-altrose into a circular shape by mentally grasping the end farthest from the carbonyl group and bending it backward into the plane of the paper:

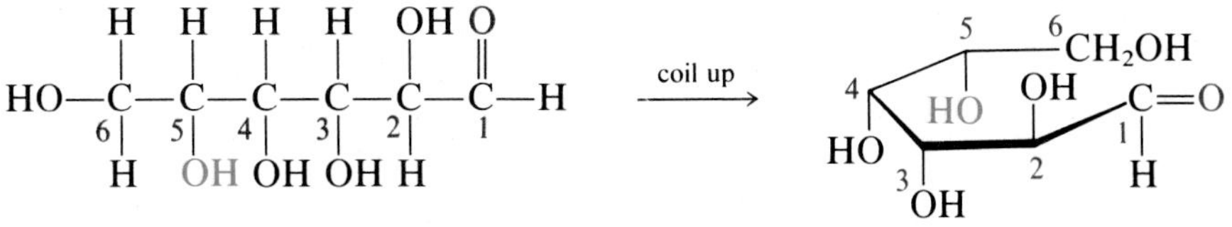

Next, rotate around the single bond between C4 and C5 so that the —CH$_2$OH group at the end of the chain is pointing up and the —OH group on C5 is pointing toward the aldehyde carbonyl group on the right:

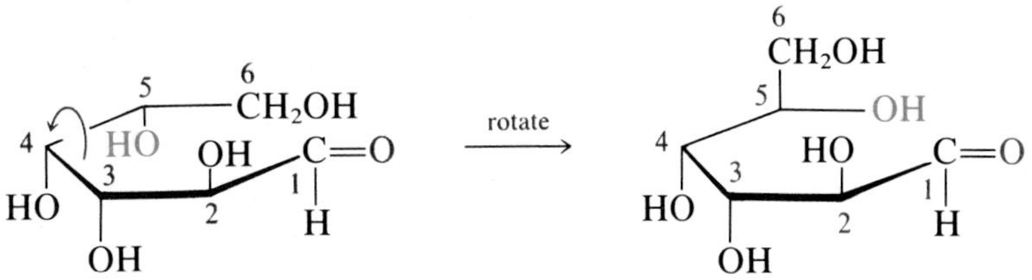

Finally, add the —OH group at C5 to the carbonyl C=O to form a hemiacetal ring. The new —OH group formed on C1 can be either up (β) or down (α).

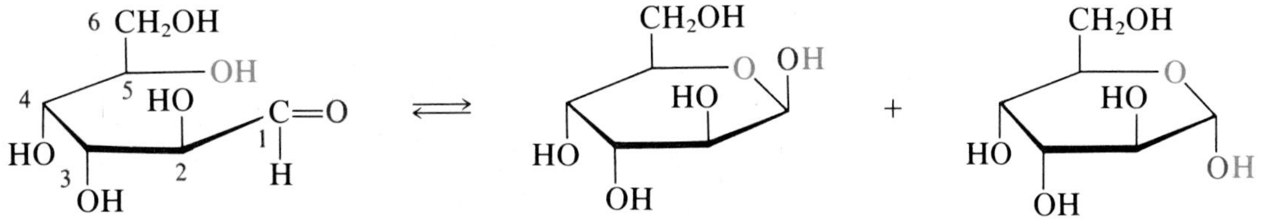

Practice Problem 21.7 D-Talose, a constituent of certain antibiotics, has the following open-chain structure. Draw D-talose in its cyclic hemiacetal form.

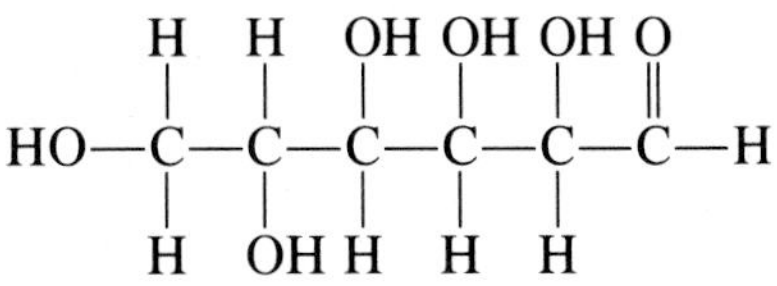

$$
\begin{array}{c}
\quad\ \ \text{H}\ \ \text{H}\ \ \text{OH OH OH O}\\
\quad\ \ |\ \ \ |\ \ \ |\ \ \ |\ \ \ |\ \ \parallel\\
\text{HO}-\text{C}-\text{C}-\text{C}-\text{C}-\text{C}-\text{C}-\text{H}\\
\quad\ \ |\ \ \ |\ \ \ |\ \ \ |\ \ \ |\\
\quad\ \ \text{H}\ \ \text{OH H}\ \ \text{H}\ \ \text{H}
\end{array}
\qquad \text{D-Talose}
$$

AN APPLICATION: CARBOHYDRATES IN THE DIET

The major monosaccharides in our diets are fructose and glucose from fruits and honey. Sucrose (common table sugar) and lactose from milk are the major disaccharides. In addition, our diets contain large amounts of the digestible polysaccharide starch, present in grains such as wheat and rice, root vegetables such as potatoes, and legumes such as beans and peas. Nutritionists, it should be noted, often refer to polysaccharides as *complex carbohydrates*.

The body's major use of digestible carbohydrates is to provide energy, about 4 kcal per gram of carbohydrate. It's necessary to have some carbohydrate in our daily diet, even though it's not our only source of energy. A small amount of any excess carbohydrate is converted to glycogen for storage, but most dietary carbohydrate in excess of our immediate needs for energy is converted into fat.

As Americans have become increasingly health- and diet-conscious in recent years, many have modified their carbohydrate intake. On the one hand, the desire for weight loss has fostered the development of low-calorie sweeteners to replace sucrose. On the other hand, replacement of meats that have a high fat and high cholesterol content by polysaccharides is viewed as desirable, and a flurry of studies suggesting health benefits has created a desire for "high-fiber" foods.

Cereal boxes now list percentages of "dietary fiber" and even distinguish between "soluble" and "insoluble" fiber. Dietary fiber is the type of polysaccharide that can't be hydrolyzed to monosaccharides and absorbed into the bloodstream. Thus, "fiber" includes not only cellulose, the indigestible polysaccharide in vegetable stalks and leaves, but a variety of noncellulose polysaccharides. For example, pectins, which are present in fruits and provide the "gel" in jelly, contain polymers derived from a polyhydroxy compound with an aldehyde group at one end and a carboxylic acid group at the other [$HOOC(CHOH)_4CHO$]. Pectin and vegetable gums are either soluble or dispersible in water and make up the soluble portion of dietary fiber.

They are often added to prepared foods to retain moisture, to thicken sauces, or to give a creamier texture. Foods high in soluble fiber include fruits, carrots, barley, and oats. Foods high in insoluble fiber include wheat, bran cereals, and brown rice. Beans and peas have both types of fiber.

Fiber functions in the body to soften and add bulk to solid waste. Studies have shown that increased fiber in the diet may reduce the risk of colon and rectal cancer, hemorrhoids, diverticulosis, and cardiovascular disease. Cancer reduction may occur because potential carcinogenic substances are absorbed on fiber surfaces and eliminated before doing any harm. Pectin may also absorb and carry away bile acids, causing an increase in their synthesis from cholesterol in the liver and a resulting decrease in blood cholesterol levels. The mechanisms of action and the interrelated effects of different fibers are far from being understood.

Beans are an excellent source of dietary fiber, both soluble and insoluble.

21.5 SOME IMPORTANT MONOSACCHARIDES

The monosaccharides, with their many opportunities for hydrogen bonding through hydroxyl groups, are generally high-melting, white, crystalline solids that are soluble in water and insoluble in nonpolar solvents. Most monosaccharides (and also disaccharides) are sweet-tasting. The natural monosaccharides of interest in human biochemistry, except for glyceraldehyde (a triose) and fructose (a ketohexose), are aldopentoses and aldohexoses. Most are of the D family and in solution are in equilibrium with both cyclic hemiacetal forms. Glucose, which we've already discussed extensively, is the most important monosaccharide in human metabolism.

Galactose   D-Galactose is widely distributed as a constituent of many plant gums and pectins. It's also a component of lactose (milk sugar) and is produced by lactose hydrolysis. Like glucose, galactose is an aldohexose. It's described as an *epimer* of glucose at carbon 4, meaning that it differs from glucose only in having the —OH on this one carbon atom on the opposite side of the ring from that in glucose. Also like glucose, galactose exists in solution as a mixture of cyclic α and β forms along with a small amount of open-chain form. In the body, galactose is converted to glucose to provide energy and is synthesized from glucose for use in lactose for milk and compounds needed in brain tissue.

α-D-Galactose   Open-chain D-galactose   β-D-Galactose

Fructose   D-Fructose, often called *levulose* or *fruit sugar,* occurs in honey and in a large number of fruits and is one of the monosaccharides in the disaccharide sucrose. Unlike glucose and galactose, fructose is a *keto*hexose: a six-carbon ketone sugar. Like glucose and galactose, however, fructose can exist in solution both in open-chain form and in cyclic forms. The cyclic hemiacetal form of fructose is a five-membered ring rather than a six-membered ring.

α-D-Fructose   Open-chain D-Fructose   β-D-Fructose

Ribose and Deoxyribose   Ribose and its relative deoxyribose are both aldopentoses: five-carbon aldehyde sugars. Both occur in the cells of all living organisms as constituents of the nucleic acids RNA (ribonucleic acid) and DNA (deoxyribonucleic acid), substances that direct the synthesis of proteins and maintain genetic information.

As its name implies, *deoxy*ribose differs from ribose in that it is missing one oxygen atom. Since it's the hydroxyl at C2 that is missing, it's more accurate to name the compound 2-deoxyribose. Both ribose and 2-deoxyribose exist in the usual mixture of open-chain and cyclic hemiacetal forms.

α-D-Ribose   Open-chain D-ribose   β-D-Ribose

α-D-2-Deoxyribose   Open-chain D-2-deoxyribose   β-D-2-Deoxyribose

Practice Problems   **21.8**   In the following monosaccharide hemiacetal, identify the anomeric carbon atom, number all the carbon atoms, and identify it as the α or β anomer.

21.9   Identify the chiral carbons in α-D-fructose, α-D-ribose, and β-D-2-deoxyribose.

21.6 REACTIONS OF MONOSACCHARIDES

Reaction With Oxidizing Agents: Reducing Sugars   You saw in Section 16.5 that aldehydes can be oxidized to yield carboxylic acids (RCHO ⟶ RCOOH). Because they too are aldehydes, aldoses like glucose undergo exactly the same oxidation reaction. Even though only a small amount of open-chain aldehyde form is present at any one time, the hemiacetal ring opening takes place so rapidly that the entire sample is eventually oxidized. Carbohydrates that react with oxidizing agents are called **reducing sugars** because they reduce the oxidizing agent.

Reducing sugar   A carbohydrate that reacts with an oxidizing agent such as Benedict's reagent.

In basic solution, ketoses are also reducing sugars because equilibrium is established between the ketose and an *enol* (alk*ene* + alcoh*ol*), which then further equilibrates with an aldose. The keto form, the enol form, and the aldose form are isomers that differ only in the position of a hydrogen atom. Oxidation of the aldose to an acid drives the equilibrium toward the right, and complete oxidation of the ketose occurs. Thus, *in basic solution all monosaccharides, whether aldoses or ketoses, are reducing sugars.*

A ketose   An enol   An aldose   A carboxylic acid

Reaction With Alcohols: Glycoside and Disaccharide Formation   Hemiacetals, as shown in Section 16.7, react with alcohols with the loss of H_2O to yield acetals, compounds that have two —OR groups bonded to the same carbon:

A hemiacetal   An alcohol   An acetal

Glycoside A cyclic acetal formed by reaction of a monosaccharide with an alcohol, accompanied by loss of H_2O.

Because glucose and other monosaccharides are cyclic hemiacetals, they react with alcohols to yield cyclic acetals called **glycosides.** In a glycoside, the —OH group on the anomeric carbon atom is replaced by an —OR group. For example, glucose reacts with methanol to produce methyl glucoside, which has an —OCH$_3$ group on carbon 1. (Note that a *gluc*oside is a cyclic acetal formed by glucose. A *glyc*oside is a cyclic acetal derived from *any* sugar.)

α-D-Glucose   Methyl α-D-glucoside, an acetal

Glycosidic bond Bond between the anomeric carbon atom of a monosaccharide and an —OR group.

The bond between the anomeric carbon atom of the monosaccharide and the oxygen atom of the —OR group is called a **glycosidic bond.** Such bonds may be either α or β according to the same definitions for monosaccharide anomers: α points below the ring and β points above the ring when the ring is drawn as described in Section 21.4. Since acetals are stable and not in equilibrium with an open-chain form, acetals like methyl α-D-glucoside are not reducing sugars.

When a monosaccharide forms a glycosidic bond to one of the —OH groups of a second monosaccharide, a disaccharide is produced. To describe the link between the two monosaccharides, the α or β orientation of the glycosidic bond (shown in red) and the numbers of the connected carbon atoms are specified. For example, the following two general structures show a **1,4 link,** a link between C1 of one monosaccharide and C4 of the second monosaccharide.

1,4 Link An acetal link between the hydroxyl group at C1 of one sugar and the hydroxyl group at C4 of another sugar.

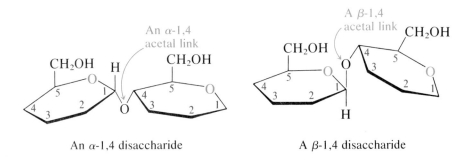

An α-1,4 disaccharide   A β-1,4 disaccharide

In some biomolecules, such as the glycoproteins (Figure 21.5) found in cell membranes, a carbohydrate and a protein are joined by a glycosidic bond to nitrogen:

Glycosidic bond in a glycoprotein

Figure 21.5
Glycoproteins at cell surfaces. At the top in the photo is the innermost layer of the small intestine (known as the *glycocalyx*). The layer consists of polysaccharide chains extending out from cell membrane proteins to which they are bonded. Such polysaccharide chains are associated with enzymes that aid in the digestion of food macromolecules.

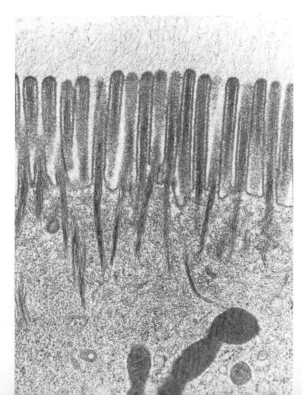

AN APPLICATION: GLUCOSE IN BLOOD AND URINE

Glucose measurements are essential in the diagnosis of *diabetes mellitus* and in the management of diabetic patients, either in a clinical setting or on a day-to-day basis by the patients themselves. One form of diabetes results when an individual has an insufficient supply of the hormone insulin. Without insulin, glucose does not move from the bloodstream into cells, where it is needed as a source of energy. As a result, glucose levels rise in blood and urine.

Most tests for glucose in urine or blood rely on detecting a color change that accompanies the oxidation of glucose. Because glucose and its oxidation product, gluconic acid, are colorless, the oxidation must be tied chemically to the color change of a suitable indicator.

D-Glucose $\xrightarrow{\text{oxidizing agent}}$ D-Gluconic acid

Benedict's reagent is one of the oxidizing agents mentioned in Section 16.5 as a test for aldehydes. The detectable color change with Benedict's reagent is the reduction of blue copper(II) ion to a brick-red precipitate of copper(I) oxide. Tablets containing the necessary Benedict's test chemicals in solid form (Clinitest tablets) are dropped into a diluted urine sample, and the color of the resulting mixture is compared to a standard chart to estimate glucose concentration. Unfortunately, Benedict's test is flawed because it is positive for any reducing sugar, for example, lactose in the urine of a pregnant woman. Although Clinitest tablets are still available in drugstores for home use by diabetics, there are now better methods.

Modern methods for glucose detection rely on the action of an enzyme specific to glucose, *glucose oxidase*. The reagent mixture includes a second enzyme called a *peroxidase* that catalyzes the reaction of hydrogen peroxide (H_2O_2) with a dye that gives a detectable color change.

$$\text{Glucose} + O_2 \xrightarrow{\text{glucose oxidase}} \text{gluconic acid} + H_2O_2$$

$$H_2O_2 + \begin{array}{c}\text{reduced dye}\\\text{(colorless)}\end{array} \xrightarrow{\text{peroxidase}} H_2O + \begin{array}{c}\text{oxidized dye}\\\text{(colored)}\end{array}$$

The glucose oxidase test is available for urine and blood. Increasingly, diabetic individuals are monitoring their glucose levels in blood, often with a modestly priced instrument that reads the color change electronically. The blood test is desirable because it is more specific and it detects rising glucose levels earlier than the urine test.

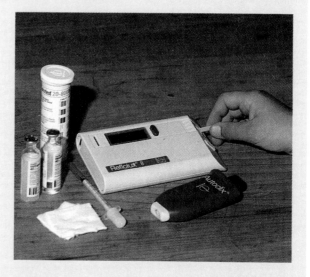

Equipment used by a diabetic individual for monitoring blood glucose and administering insulin.

Practice Problem **21.10** Look at the structure of α-D-galactose (Section 21.5) and write the two glycoside products you would expect to obtain by reaction of galactose with methanol. Are these compounds reducing sugars?

21.7 SOME IMPORTANT DISACCHARIDES

Three of the most important naturally occurring disaccharides, maltose, lactose, and sucrose, illustrate three different ways monosaccharides can be linked together.

Maltose Maltose, often called *malt sugar,* is present in fermenting grains and can be prepared in about 80% yield by enzyme-catalyzed degradation of starch. In the body, it is produced in starch digestion. Two α-D-glucose molecules are joined in maltose by an α-1,4 link. A careful look at maltose shows that it is both an acetal and a hemiacetal. The α-D-glucose on the left contains an acetal grouping (C1 is bonded to two —OR groups), while that on the right contains a hemiacetal (C1 is bonded to an —OH and an —OR). Since the acetal ring on the left doesn't open and close spontaneously, the bond linking the two glucose units in maltose is not easily cleaved. The hemiacetal group of the glucose unit on the right, however, establishes equilibrium with the aldehyde, making maltose a reducing sugar.

Maltose

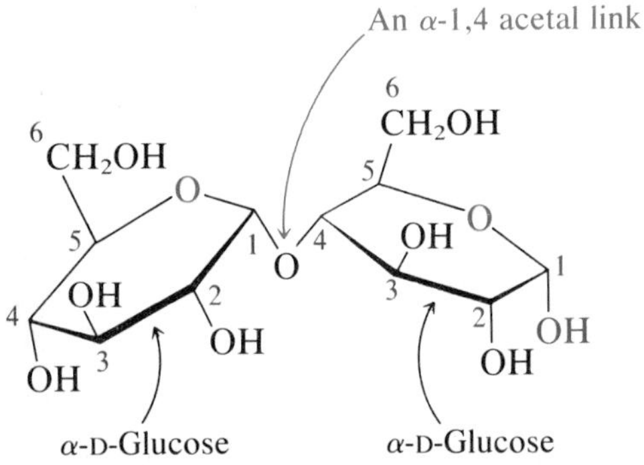

Practice Problem **21.11** Draw maltose in the form in which the glucose unit on the right is an open-chain aldehyde.

Lactose Lactose, or *milk sugar,* is the major carbohydrate present in mammalian milk. Human milk, for example, is about 7% lactose. Structurally, lactose is a disaccharide composed of galactose and glucose. The two sugars are joined in lactose by a β-1,4 acetal link between C1 of β-D-galactose and C4 of β-D-glucose. Like maltose, lactose is a reducing sugar because one glucose ring (on the right in the following structure) is a hemiacetal that is in equilibrium with an oxidizable aldehyde.

A β-1,4 acetal link

6
CH₂OH

Lactose

β-D-Galactose   β-D-Glucose

Sucrose   Sucrose—plain table sugar—is probably the most common pure organic chemical in the world. Although it's found in many plants, sugar beets (20% by weight) and sugarcane (15% by weight) are the most common sources of sucrose. Hydrolysis of sucrose yields one molecule of glucose and one molecule of fructose. The 50:50 mixture of glucose and fructose that results, often referred to as *invert sugar,* is commonly used as a food additive because it's sweeter than sucrose.

Sucrose differs from maltose and lactose in that it has no hemiacetal group because a 1,2 link joins *both* anomeric carbon atoms. Since it has no hemiacetal group that opens to expose a free aldehyde, sucrose is not a reducing sugar.

6
CH₂OH

α-D-Glucose

Sucrose

β-D-Fructose

HOCH₂

CH₂OH

Practice Problems   21.12   The disaccharide cellobiose can be obtained by enzyme-catalyzed hydrolysis of cellulose. Would you expect cellobiose to be a reducing or a nonreducing sugar? Explain.

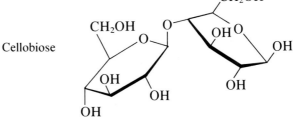

Cellobiose

21.13   How would you classify the link between the monosaccharides in cellobiose?

21.14   Show the structures of the two monosaccharides that are formed on hydrolysis of cellobiose (Practice Problem 21.12). What are their names?

AN APPLICATION: CELL SURFACE CARBOHYDRATES AND BLOOD TYPE

It was discovered more than 80 years ago that human blood can be classified into four blood group types, called A, B, AB, and O. If a transfusion becomes necessary, blood from a donor of one type can't be given to a recipient with blood of another type unless the two types are compatible. If an incompatible mix is made, the red blood cells clump together, or *agglutinate,* and death can result. Agglutination indicates that the

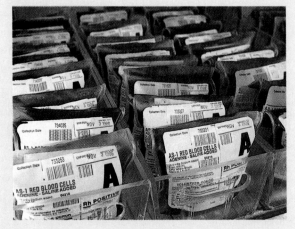

Blood to be used for transfusion must be labeled by type so that compatibility with the recipient's blood type can be assured.

recipient's immune system has recognized foreign cells in the body and has formed antibodies to them.

Cells of types A, B, and O each have on their surfaces characteristic structural features called *antigenic determinants,* which can provoke an immune response that results in the production of antibodies. Cells of type AB have both A and B markers. As shown at the bottom of this box, the structures of the three blood group determinants are known. The marker for blood group O is a trisaccharide whose constituent sugars are common to all three types, whereas the markers for blood groups A and B have one additional sugar unit.

Human Blood Group Compatibilities

Donor Blood Type	Acceptor Blood Type			
	A	**B**	**AB**	**O**
A	o	x	o	x
B	x	o	o	x
AB	x	x	o	o
O	o	o	o	o

o = compatible; x = incompatible

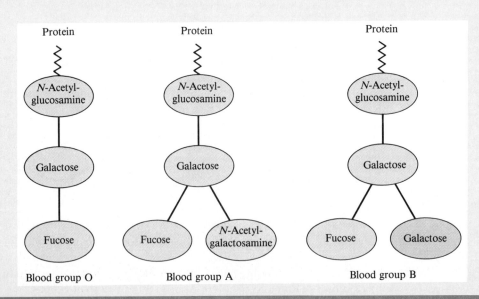

21.8 SOME IMPORTANT POLYSACCHARIDES

Polysaccharides have tens, hundreds, or even thousands of monosaccharides linked together through glycosidic bonds of the same sort as in maltose and lactose. Three of the most important polysaccharides are cellulose, starch, and glycogen.

Cellulose   Cellulose (Figure 21.6), the fibrous substance used by plants as a structural material in the cell walls of leaves, stems, and tree trunks, consists entirely of several thousand β-D-glucose units joined by β-1,4 links to form one large molecule. Cows and termites are able to digest cellulose because microorganisms living in their digestive tracts produce enzymes that hydrolyze the β acetal bonds. Humans, however, can't hydrolyze cellulose.

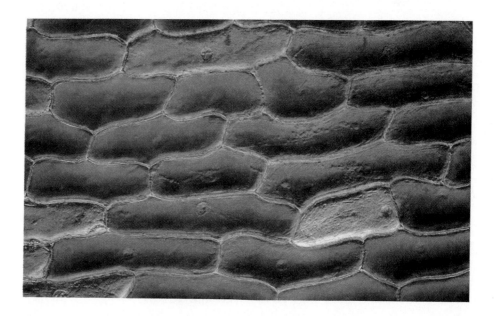

Starch   Starch, like cellulose, is a polymer of glucose. There is, however, a big difference between the two because starch, unlike cellulose, can be digested by human beings. Indeed, the starch in such vegetables as beans, wheat, rice, and potatoes is an essential part of the human diet. Structurally, starch differs from cellulose in that its individual glucose units are joined by α-1,4 links, as in maltose, rather than by the β-1,4 links of cellulose (Figure 21.7).

Figure 21.6
Cellulose structure in the skin of a red onion.

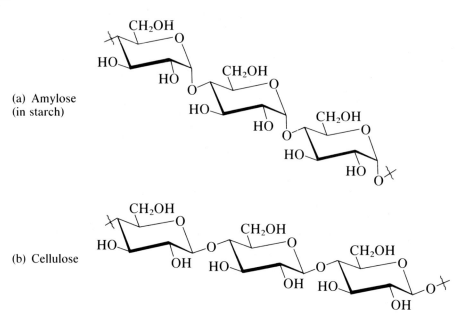

Figure 21.7
Cellulose and the amylose form of starch. These structures represent the molecular shapes more correctly than the flat rings by showing the six-membered glucose rings in their chair forms. Note the difference between the α-1,4 links in amylose and the β-1,4 links in cellulose. In cellulose several thousand β-D-glucose units are joined by 1,4 acetal links between C1 of one sugar and C4 of its neighbor.

(a) Amylose (in starch)

(b) Cellulose

Unlike cellulose, which has only one form, there are two kinds of starch, called *amylose* and *amylopectin*. Amylose, which accounts for about 20% of starch, consists of several hundred to a thousand α-D-glucose units linked together in a long chain by α-1,4 acetal bonds (Figure 21.8a). Amylopectin, which accounts for about 80% of starch, is similar to amylose but is much larger (up to 100,000 glucose units per molecule) and has α-1,6 branches approximately every 25 units along its chain. A glucose molecule at one of these branch points uses *two* of its hydroxyl groups (those at C4 and C6) to form acetal links to two other sugars (Figure 21.8b).

When eaten, starch molecules are digested mainly in the small intestine by α-amylase, which catalyzes hydrolysis of the α-1,4 links so that the starch chain is broken down. As is usually the case in enzyme-catalyzed reactions, α-amylase is highly specific in its action. It hydrolyzes only α acetal links between monosaccharides (as in starch) while leaving β acetal links (as in cellulose) untouched. Thus, starch is easily digested, but cellulose is unaffected by these digestive enzymes.

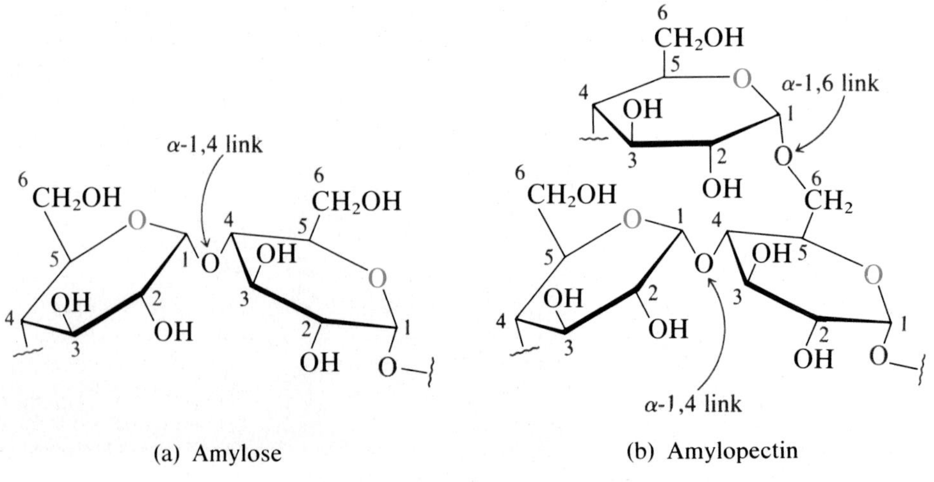

Figure 21.8
The glucose polymers in starch. Amylose consists only of linear chains of α-D-glucose units linked by 1,4 acetal bonds, whereas amylopectin has branch points about every 25 sugars in the chain. A glucose unit at a branch point uses two of its hydroxyls (at C4 and C6) to form 1,4 and 1,6 acetal links to two other sugars.

(a) Amylose

(b) Amylopectin

Practice Problem 21.15 An individual starch molecule contains thousands of glucose units but has only a single hemiacetal group at the end of the long polymer chain. Would you expect starch to be a reducing carbohydrate? Explain.

Glycogen   Glycogen, sometimes called *animal starch,* serves the same food storage role in animals that starch serves in plants. After we eat starch and the body breaks it down into simple glucose units, some of the glucose is used immediately as fuel and some is stored in the body as glycogen for later use. Any additional excess is stored as fat.

Structurally, glycogen is similar to amylopectin in being a long polymer of α-D-glucose with branch points in its chain. Glycogen has many more branches than amylopectin, however, and is much larger: up to 1 million glucose units per molecule (Figure 21.9).

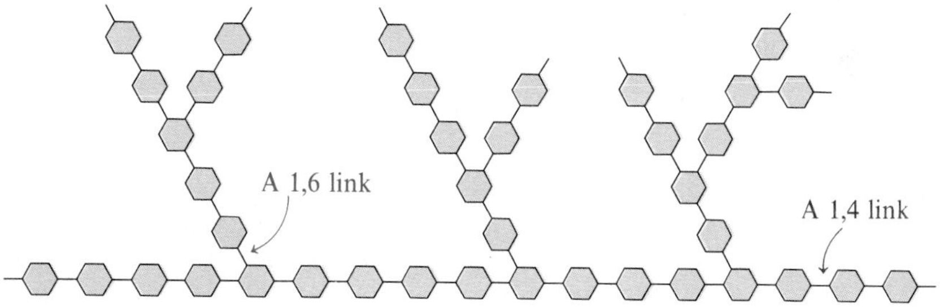

A 1,6 link

A 1,4 link

Figure 21.9
The structure of glycogen. The hexagons represent α-D-glucose units linked by 1,4 acetal bonds and (at branch points) 1,6 acetal bonds.

SUMMARY

Carbohydrates are polyhydroxy aldehydes and ketones. They are classified according to the number of carbon atoms and the kind of carbonyl group they contain. Glucose, for example, is an *aldohexose*—a six-carbon aldehyde sugar. **Monosaccharides** such as glucose can't be hydrolyzed to smaller molecules. All monosaccharides are chiral, and the naturally occurring optical isomers belong to the D **family.** Monosaccharides of the L **family** occur much more rarely in nature. **Polysaccharides** such as starch and cellulose contain many monosaccharides linked together.

Monosaccharides such as glucose, galactose, fructose, ribose, and 2-deoxyribose exist as a mixture of open-chain and cyclic hemiacetal forms in equilibrium with each other. There are two cyclic hemiacetal forms, called the α **form** and the β **form,** which differ in the orientation of the hemiacetal hydroxyl group, that is, the hydroxyl group on the **anomeric carbon.** Monosaccharides undergo many of the same reactions that other aldehydes do. Thus, they react with alcohols to yield cyclic acetals called **glycosides,** and they react with oxidizing agents to yield carboxylic acids. Sugars that can be oxidized in this way are called **reducing sugars.** All monosaccharides are reducing sugars because they establish equilibria with aldehyde forms that can be oxidized.

Disaccharides such as maltose, lactose, and sucrose contain two simple sugars joined by an acetal link classified according to whether the **glycosidic bond** is α or β and according to the carbon atoms it connects. Important polysaccharides include cellulose, starch, and glycogen. Cellulose is a linear polymer of up to 1000 glucose units bonded together by β**-1,4 acetal links.** Starch and glycogen are glucose polymers in which the individual sugar units are connected by α acetal links. The fraction of starch called amylose consists of linear chains, and the fraction called amylopectin has branched chains.

INTERLUDE: SWEETNESS

Mention the word *sugar* to most people and they'll automatically think of sweet-tasting foods or candies. In fact, most simple mono- and disaccharides *do* taste sweet, although the degree of sweetness varies from one compound to another. Using sucrose (table sugar) as a reference point, fructose is nearly twice as sweet, whereas glucose, galactose, lactose, and others are much less so. Exact comparisons between sugars are impossible, because sweetness is not a physical property and can't be accurately measured. Sweetness is simply a taste perception, and the ranking of different sugars is a matter of personal opinion. Nevertheless, the ordering shown in the accompanying table is generally agreed on.

Dietary concerns and the desire of many people to cut their caloric intake have led to widespread use of the artificial sweeteners aspartame, saccharin, and cyclamate. All are far sweeter than natural sugars, but doubts have been raised as to the long-term safety of all three. As their structures indicate, none of the three bear the slightest chemical resemblance to carbohydrates.

Relative Sweetness of Some Sugars and Sugar Substitutes

Name	Type	Sweetness
Lactose	Disaccharide	16
Galactose	Monosaccharide	30
Maltose	Disaccharide	33
Glucose	Monosaccharide	75
Sucrose	Disaccharide	100
Fructose	Monosaccharide	175
Cyclamate	Artificial	3,000
Aspartame	Artificial	15,000
Saccharin	Artificial	35,000

Aspartame

Saccharin

Sodium cyclamate

Sugarcane contains a high concentration of the disaccharide sucrose. The stems of the plant derive much of their strength and stiffness from another carbohydrate that is not digestible by humans: the polysaccharide cellulose.

REVIEW PROBLEMS

Classification and Structure of Carbohydrates

21.16   What is a carbohydrate?

21.17   What is the family-name ending for a sugar?

21.18   What is the structural difference between an aldose and a ketose?

21.19   Classify each of the following carbohydrates by indicating the nature of its carbonyl group and the number of carbon atoms present. For example, glucose is an aldohexose.

(a)

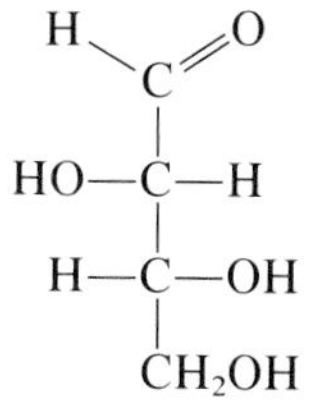

H—C=O
HO—C—H
H—C—OH
CH₂OH

(b)

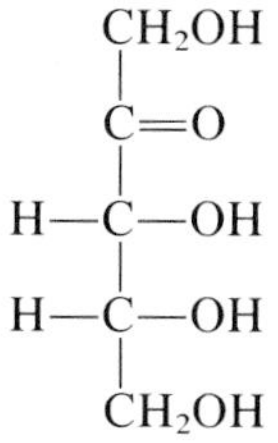

CH₂OH
C=O
H—C—OH
H—C—OH
CH₂OH

Threose   Ribulose

(c)

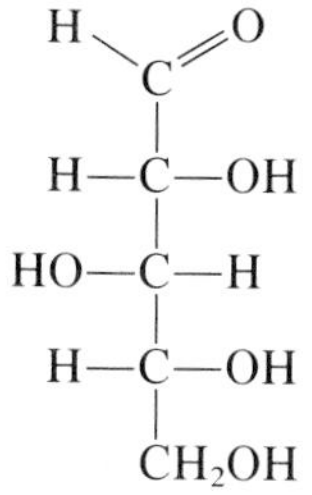

H—C=O
H—C—OH
HO—C—H
H—C—OH
CH₂OH

(d)

CH₂OH
C=O
HO—C—H
HO—C—H
H—C—OH
CH₂OH

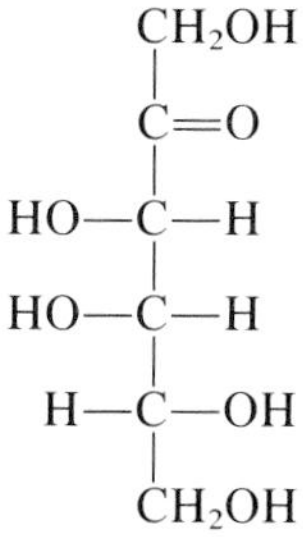

Xylose   Tagatose

21.20 Write the open-chain structure of a ketotetrose.

21.21 Write the open-chain structure of a four-carbon deoxy sugar.

21.22 Give the names of three important monosaccharides and tell where each occurs in nature.

21.23 "Dextrose" is an alternative name for what sugar?

Handedness in Carbohydrates

21.24 What is the difference between enantiomers and diastereomers?

21.25 What is the structural relationship of L-glucose to D-glucose?

21.26 Would you expect L-glucose to be a good food source in the human diet in the same way that D-glucose is?

21.27 Only three stereoisomers are possible for 2,3-dibromo-2,3-dichlorobutane. Draw them, indicating which pair are enantiomers (optical isomers). Why does the other isomer not have an enantiomer?

21.28 There are two D-aldotetroses, whose structures are shown. One of the two reacts with NaBH₄ to yield a chiral product, but the other yields an achiral product. Explain.

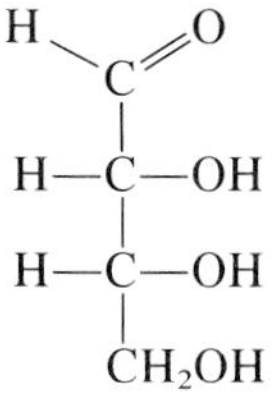

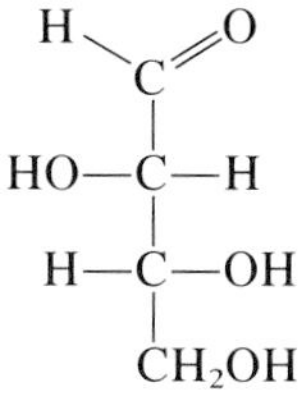

H—C=O
H—C—OH
H—C—OH
CH₂OH

D-Erythrose

H—C=O
HO—C—H
H—C—OH
CH₂OH

D-Threose

21.29. Draw the enantiomer of each molecule in Problem 21.19. Label each as a D or an L sugar.

Reactions of Carbohydrates

21.30 What does the term *reducing sugar* mean?

21.31 What is the structural difference between the α hemiacetal form of a carbohydrate and the β form?

21.32 D-Gulose, an aldohexose isomer of glucose, has the following cyclic structure. Which is shown, the α form or the β form?

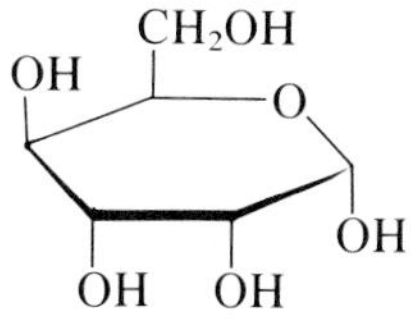

CH₂OH
OH   O

OH
OH   OH

D-Gulose

21.33 Draw D-gulose (Problem 21.32) in its open-chain aldehyde form, both coiled and uncoiled.

21.34 D-Mannose, an aldohexose found in orange peels, has the following structure in open-chain form. Coil mannose around and draw it in cyclic hemiacetal α and β forms.

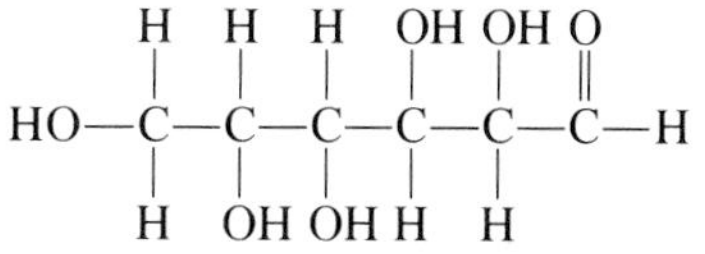

H H H OH OH O
| | | | | ||
HO—C—C—C—C—C—C—H
| | | | |
H OH OH H H

D-Mannose

21.35 D-Ribulose, a ketopentose related to ribose, has the following structure in open-chain form. Coil ribulose around and draw it in its five-membered cyclic β hemiacetal form.

H H H O H
| | | || |
HO—C—C—C—C—C—OH
| | | |
H OH OH   H

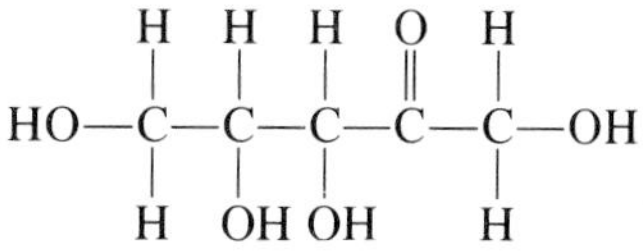

D-Ribulose

21.36 D-Allose, an aldohexose, is identical with D-glucose except that the hydroxyl group at C3 points down rather than up in the cyclic hemiacetal form. Draw the β form of this cyclic form of D-allose.

21.37 Draw D-allose (Problem 21.36) in its open-chain form.

21.38 We saw in Section 16.5 that aldehydes react with reducing agents like NaBH₄ to yield primary alcohols (RCH=O → RCH₂OH). Treatment of D-glucose with NaBH₄ yields *sorbitol*, a substance used as a sugar substitute by diabetics. Draw the structure of sorbitol.

21.39 Reduction of D-fructose with NaBH₄ (Problem 21.38) yields a mixture of D-sorbitol along with a second, isomeric product. What is the structure of the second product?

21.40 Refer to Problems 21.38 and 21.39 and explain why two products result from the reduction of D-fructose, while only one results from reduction of D-glucose.

21.41 Treatment of an aldose with an oxidizing agent like Tollens' reagent (Section 16.5) yields a carboxylic acid. Gluconic acid, the product of glucose oxidation, is used as its magnesium salt for the treatment of magnesium deficiency. Draw the structure of gluconic acid.

21.42 What is the structural difference between a hemiacetal and an acetal?

21.43 What are glycosides, and how can they be formed?

21.44 Look at the structure of D-mannose (Problem 21.34) and draw the two glycosidic products that you would expect to obtain by reacting D-mannose with methanol.

21.45 Show two D-mannose molecules (Problem 21.34) attached by an α-1,4 glycosidic linkage.

Disaccharides and Polysaccharides

21.46 Give the names of three important disaccharides. Tell where each occurs in nature. From which two monosaccharides is each made?

21.47 Starch and cellulose are both polymers of glucose. What is the main structural difference between them, and what different roles do they serve in nature?

21.48 How are amylose and amylopectin similar and how are they different?

21.49 Starch and glycogen are both α-linked polymers of glucose. What is the structural difference between them, and what different roles do they serve in nature?

21.50 Lactose and maltose are reducing disaccharides, but sucrose is a nonreducing disaccharide. Explain.

21.51 Gentiobiose, a rare disaccharide found in saffron, has the following structure. What simple sugars would you obtain on hydrolysis of gentiobiose?

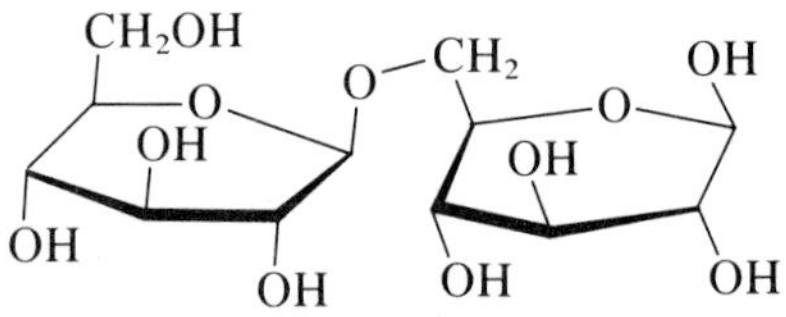

Gentiobiose

21.52 Look carefully at the structure of gentiobiose (Problem 21.51). Does gentiobiose have an acetal grouping? A hemiacetal grouping? Would you expect gentiobiose to be a reducing or a nonreducing sugar? How would you classify the linkage (α or β and carbon numbers) between the two monosaccharides?

21.53 Trehalose, a disaccharide found in the blood of insects, has the following structure. What simple sugars would you obtain on hydrolysis of trehalose?

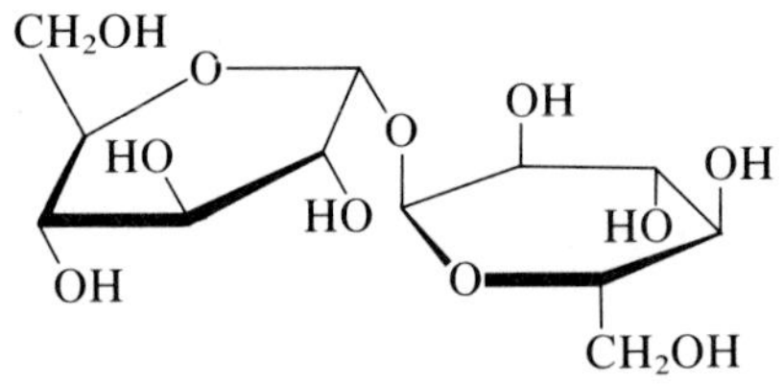

Trehalose

21.54 Does trehalose (Problem 21.53) have an acetal grouping? A hemiacetal grouping? Would you expect trehalose to be a reducing or a nonreducing sugar?

21.55 Amygdalin, or Laetrile, is a glycoside isolated in 1830 from almond and apricot seeds. It is called a *cyanogenic glycoside* because hydrolysis with aqueous acid liberates hydrogen cyanide (HCN) along with benzaldehyde and two molecules of glucose. Structurally, amygdalin is a glycoside between the hemiacetal, gentiobiose (Problem 21.51), and the alcohol, mandelonitrile. Draw the structure of amygdalin.

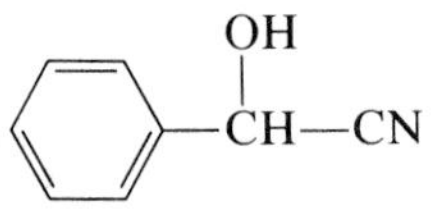

Mandelonitrile

Applications

21.56 What is an optically active compound? [App: Polarized Light and Optical Activity]

21.57 What does a polarimeter measure? [App: Polarized Light and Optical Activity]

21.58 Sucrose and D-glucose rotate plane-polarized light to the right; D-fructose rotates light to the left. When sucrose is hydrolyzed, the glucose–fructose mixture rotates light to the left. (a) What does this indicate about the relative degrees of rotation of light of glucose and fructose? (b) Why do you think that the mixture is called invert sugar? [App: Polarized Light and Optical Activity]

21.59 What generalization can you make about the direction and degree of rotation of light by enantiomers? [App: Polarized Light and Optical Activity]

21.60 Our bodies don't have the enzyme required to digest cellulose, yet it is a necessary addition to a healthy diet. Why? [App: Carbohydrates in the Diet]

21.61 What are two types of soluble fiber and two sources of this type of fiber? [App: Carbohydrates in the Diet]

21.62 What hormone deficiency gives rise to diabetes? What purpose does this hormone serve? [App: Glucose in the Blood and Urine]

21.63 Why is Benedict's test not infallible in testing for glucose? [App: Glucose in the Blood and Urine]

21.64 Briefly describe the enzymatic process for determination of glucose. [App: Glucose in the Blood and Urine]

21.65 Describe what happens when incompatible blood types are mixed. Why does this occur? [App: Cell Surface Carbohydrates]

21.66 What is the role of an antigenic determinant on a blood cell? [App: Cell Surface Carbohydrates]

21.67 Look at the structures of the blood group antigenic determinants. What groups do all blood types have in common? Why do you think that blood type O is the "universal donor"? [App: Cell Surface Carbohydrates]

21.68 Sugar substitutes have caloric value, yet are still prescribed for diabetics and for lowering dietary calorie consumption. Look at the sweetness table and propose a justification for this substitution. [Int: Sweetness]

Additional Problems

21.69 What is the relationship between D-ribose and L-ribose? What generalizations can you make about D-ribose and L-ribose with respect to (a) melting point (b) rotation of plane-polarized light (c) density (d) solubility in water (e) chemical reactivity?

21.70 What is the relationship between D-ribose and D-xylose (Problem 21.19). What generalizations can you make about D-ribose and D-xylose with respect to (a) melting point (b) rotation of light (c) density (d) solubility in water (e) chemical reactivity?

21.71 L-Sorbose, which is used in the commercial production of vitamin C, differs from D-fructose only at carbon 5. Draw the open-chain structure of D-sorbose.

21.72 D-Fructose can form a six-membered cyclic hemiacetal as well as the more prevalent five-membered cyclic form. Draw the α isomer of D-fructose in the six-membered ring.

21.73 Are the α and β forms of monosaccharides enantiomers? Why or why not?

21.74 Raffinose, found in sugar beets, is the most prevalent trisaccharide. It is formed by an α-1,6 linkage of D-galactose to the glucose portion of sucrose. Draw the structure of raffinose.

21.75 Does raffinose (Problem 21.74) have a hemiacetal grouping? An acetal grouping? Is raffinose a reducing sugar?

21.76 When you chew a cracker for several minutes, it begins to taste sweet. What do you think that the saliva in your mouth does to the starch in the cracker?

C H A P T E R

22
Carbohydrate Metabolism

The tiny surface projections (villi), where digested food is absorbed, are clearly visible in this cross section of the small intestine. In this chapter we'll begin the story of what happens to food molecules during digestion.

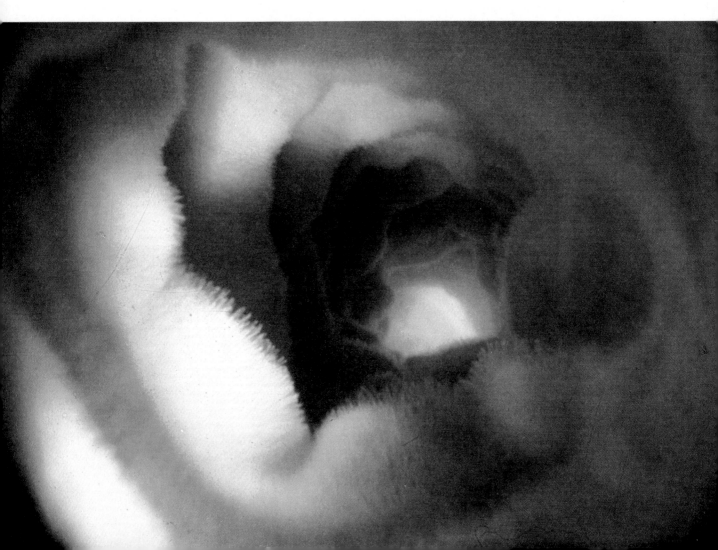

The story of carbohydrate metabolism is essentially the story of glucose: how it is broken down to pyruvate for entrance into the citric acid cycle, how it is stored and then released, and how it is synthesized when carbohydrates are in short supply. Because of the importance of glucose in metabolism, the body has several alternative strategies for maintaining glucose's concentration in blood and providing it to cells that depend on it. In this chapter, we'll answer the following questions about carbohydrate metabolism:

1. **What happens during digestion of carbohydrates?** The goal: Be able to describe where carbohydrates are digested and what the major products are.

2. **What are the major pathways in the metabolism of glucose?** The goal: Be able to identify the alternative pathways available for the synthesis and breakdown of glucose and describe their interrelationships.

3. **What is glycolysis?** The goal: Be able to list the major products of glycolysis and give an overview of the pathway.

4. **What happens to pyruvate once it is formed?** The goal: Be able to describe when each of the alternative pathways occurs and what its product is.

5. **What is the result of complete catabolism of glucose?** The goal: Be able to list the major products of the catabolism of glucose and explain what determines the total amount of ATP produced.

6. **Which hormones influence glucose metabolism?** The goal: Be able to name three hormones that influence glucose metabolism and describe their roles.

7. **What are the roles of glycogen and the pentose phosphate pathway in metabolism?** The goal: Be able to describe the anabolism and catabolism of glycogen, and the major products of the pentose phosphate pathway.

22.1 DIGESTION OF CARBOHYDRATES

Digestion A general term for the breakdown of food into small molecules.

The first stage in catabolism, previously summarized in Figure 20.6, is known as **digestion,** a catch-all term used to describe the breakdown of bulk food into individual small molecules. Digestion entails the physical grinding, softening, and mixing of food, as well as the enzyme-catalyzed hydrolysis of carbohydrates, proteins, and fats. Digestion begins in the mouth when foods are chewed; it continues in the stomach; and it concludes in the small intestine (Figure 22.1a).

The products of food digestion—glucose, fatty acids, glycerol, and amino acids—are mostly relatively small molecules that are absorbed from the intestinal tract. The absorption happens through millions of tiny hair-like projections called *microvilli* (Figure 22.1b) that provide a total surface area as big as a

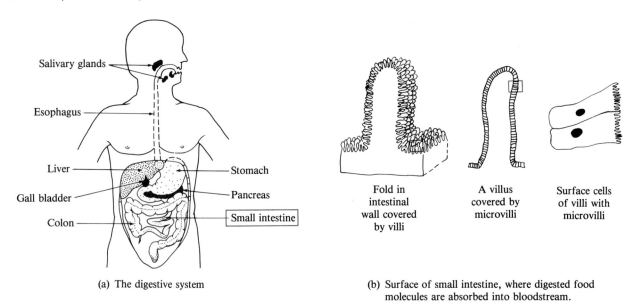

(a) The digestive system

(b) Surface of small intestine, where digested food molecules are absorbed into bloodstream.

Figure 22.1
The digestive system and details of the small intestine surface. The end products of digestion are absorbed into the bloodstream at the microvilli. Enzymes from the mucous membrane of the small intestine complete hydrolysis of disaccharides to monosaccharides.

Figure 22.2
Digestion of carbohydrates.

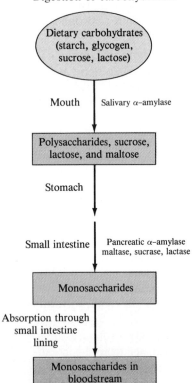

football field. Once in the bloodstream, the small molecules are transported into target cells where many are further broken down to produce energy. Some of the fragments produced by cellular breakdown of food molecules are exhaled as carbon dioxide, some are excreted, and some are used as building blocks to synthesize new biomolecules.

The digestion of carbohydrates, summarized in Figure 22.2, begins in the mouth as the α-amylase in saliva catalyzes hydrolysis of the glycosidic bonds in sugar molecules. Baceteria in the mouth do the same thing, usually not to our benefit (Figure 22.3). Starch from plants and glycogen from meat are hydrolyzed by α-amylase into smaller polysaccharides and the disaccharide maltose. Salivary α-amylase continues to act on dietary polysaccharides in the stomach until, after an hour or so, it is inactivated by stomach acid. No further carbohydrate digestion takes place in the stomach.

$$(\text{Glucose})_n \ + \ H_2O \ \xrightarrow{\alpha\text{-amylase}} \ (\text{glucose})_{<n} \ + \ (\text{glucose})_2$$

Starch and glycogen   Maltose

α-Amylase is also secreted by the pancreas and enters the small intestine, where conversion of polysaccharides to maltose continues. Other enzymes from the mucous lining of the small intestine hydrolyze maltose and the dietary disaccharides sucrose and lactose to the monosaccharides glucose, fructose, and galactose, which are then transported across the intestinal wall.

A non-life-threatening but troublesome condition known as *lactose intolerance* develops in many individuals when the enzyme lactase ceases to be

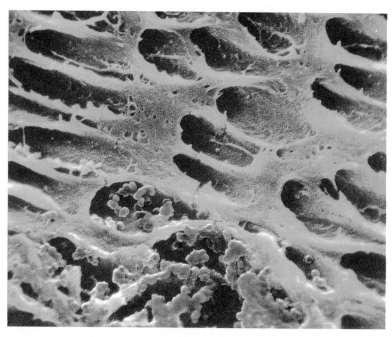

Figure 22.3
Dentine and bacteria. A cavity gets started when bacteria (orange particles) in dentine (blue) below the outer tooth enamel hydrolyze sucrose and then make the polysaccharide dextran from the resulting glucose. The tooth becomes coated with plaque, which is composed of dextran, microorganisms, and proteins from saliva. Further breakdown of sugars within the plaque produces acids that attack the tooth enamel.

made in the small intestine after the age of about 4. Lactose then proceeds unchanged to the large intestine, where it increases the osmolarity of the intestine contents and causes diarrhea. Bacterial attack on lactose in the intestine also produces methane and other gases that can cause nausea and vomiting.

Practice Problem **22.1** Complete the following word equations:

(a) $\text{lactose} + \text{H}_2\text{O} \xrightarrow{\text{lactase}}$

(b) $\text{sucrose} + \text{H}_2\text{O} \xrightarrow{\text{sucrase}}$

22.2 GLUCOSE METABOLISM: AN OVERVIEW

The principal role of glucose is as a fuel to yield the energy carried by ATP. All cells require a certain amount of glucose, and there are several metabolic strategies for maintaining a normal glucose concentration in the blood. Some cells, notably red blood cells and cells in the brain, central nervous system, and muscles, rely on glucose for energy but can't synthesize it.

In Chapter 20, we described the final stages of ATP production beginning with acetyl SCoA, a common intermediate in the catabolism of all foods. Now we'll go back and further examine the central position of glucose in metabolism, as summarized in Figure 22.4. Look back at this diagram and at Table 22.1 often as you read this chapter for help in sorting out the pathways that have similar names.

When glucose enters a cell, it is immediately converted in an irreversible reaction to glucose 6-phosphate. At this point, several pathways are available. Its most likely fate when energy is needed is *glycolysis*, the pathway leading to pyruvate, acetyl SCoA, and subsequent energy production. When cells are already well supplied with glucose, some is converted to glycogen for storage, a pathway known as *glycogenesis*. A supply of glucose beyond what can be stored as glycogen is diverted to the synthesis of fatty acids, which we'll describe in Chapter 24. In addition, and depending on the types of cells, varying amounts of glucose enter the *pentose phosphate pathway,* which supplies NADPH and sugars needed for biosynthesis. These pathways are summarized in Table 22.1.

The pyruvate produced in glycolysis also has several alternative fates awaiting it. Most likely, the pyruvate will be converted to acetyl SCoA. This pathway, however, is short-circuited in some tissues, especially where there's not enough oxygen. Under these **anaerobic** conditions, pyruvate is instead converted to lactate. The lactate may, in turn, be cycled back to pyruvate and reconverted to glucose by *gluconeogenesis,* a pathway that also allows glucose

Anaerobic In the absence of oxygen.

Figure 22.4
Glucose metabolism. Synthetic pathways (anabolism) are shown in blue, pathways that break down biomolecules (catabolism) are shown in yellow, and connections to lipid and protein metabolism are shown in green.

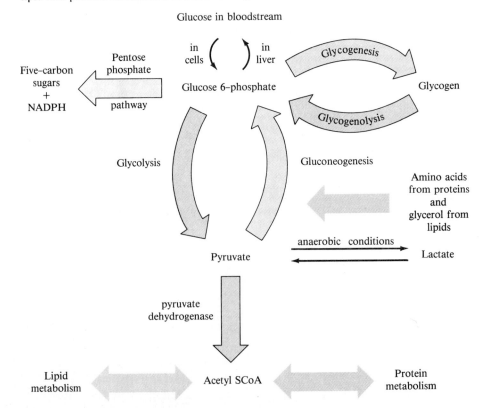

Table 22.1   Metabolic Pathways of Glucose

Name	Derivation of Name	Function
Glycolysis	*glyco-*, glucose (from the Greek, meaning "sweet") *-lysis*, decomposition	Conversion of glucose to pyruvate
Gluconeogenesis	*gluco-*, glucose *-neo-*, new *-genesis*, creation	Synthesis of glucose from amino acids, pyruvate, and other noncarbohydrates
Glycogenesis	*glyco (gen)-*, glycogen *-genesis*, creation	Synthesis of glycogen from glucose
Glycogenolysis	*glycogen-*, glycogen *-lysis*, decomposition	Breakdown of glycogen to glucose
Pentose phosphate pathway	*pentose*, a five-carbon sugar *phosphate*	Conversion of glucose to five-carbon sugar phosphates

to be synthesized from pyruvate, amino acids, or glycerol when the body is starved for glucose. In yeasts, pyruvate is metabolized differently than in human beings, and ethyl alcohol is the end product.

Practice Problems   22.2   Identify each of the following pathways:
(a) pathway for release of glucose from glycogen
(b) pathway for synthesis of glucose from lactate
(c) pathway for synthesis of glycogen

22.3   Name the synthetic pathways that have glucose as their first reactant.

22.3 GLYCOLYSIS

Glycolysis The biochemical pathway that breaks down a molecule of glucose into two molecules of pyruvate plus energy.

Glycolysis is a series of 10 enzyme-catalyzed reactions that break down each glucose molecule into two pyruvate molecules. The steps of glycolysis, also called the *Embden-Meyerhoff pathway* after its discoverers, are summarized in Figure 22.5, where the reactions and the structures of intermediates should be looked at as you proceed through the description of glycolysis in the following paragraphs. Almost all organisms carry out glycolysis; in humans it occurs in the cytosol of all cells, in contrast with the further oxidation of pyruvate, which occurs in mitochondria.

Steps 1–3 of Glycolysis: Phosphorylation   Glucose from the breakdown of food is carried in blood to cells that it enters by facilitated diffusion across cell membranes. As soon as it enters the cell, glucose is phosphorylated in *step 1* of glycolysis, which requires an energy investment from ATP. From here on, all intermediates in glycolysis are phosphates and are trapped within the cells because phosphates do not cross cell membranes. Glucose 6-phosphate, the product of step 1, inhibits hexokinase, an enzyme that plays an important role in the elaborate and delicate control of glucose metabolism.

Figure 22.5
The glycolysis pathway for converting glucose to pyruvate.

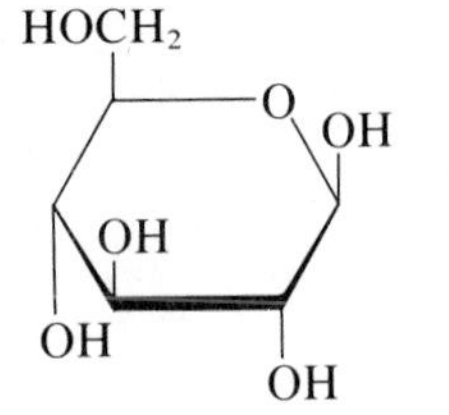
HOCH$_2$
O
OH
OH
OH
OH

Glucose

Exergonic and not reversible **1** ATP ADP

Step 1. Glucose undergoes reaction with ATP to yield glucose 6-phosphate plus ADP in a reaction catalyzed by *hexokinase.*

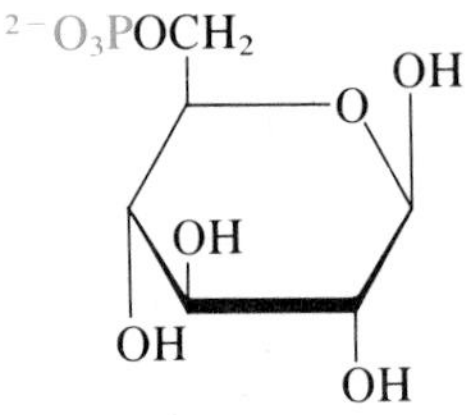
$^{2-}O_3POCH_2$
O
OH
OH
OH
OH

Glucose
6-phosphate

2

Step 2. Isomerization of glucose 6-phosphate yields fructose 6-phosphate. The reaction is catalyzed by the mutase enzyme, *phosphohexoseisomerase.*

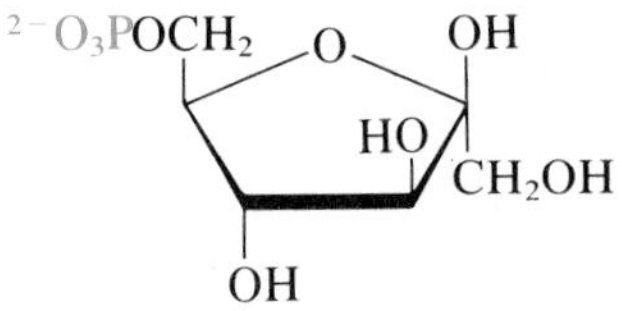
$^{2-}O_3POCH_2$
O
OH
HO
CH_2OH
OH

Fructose
6-phosphate

Exergonic and not reversible **3** ATP ADP

Step 3. Fructose 6-phosphate reacts with a second molecule of ATP to yield fructose 1,6-bisphosphate plus ADP. *Phosphofructokinase,* the enzyme for step 3, is a major control point in glycolysis.

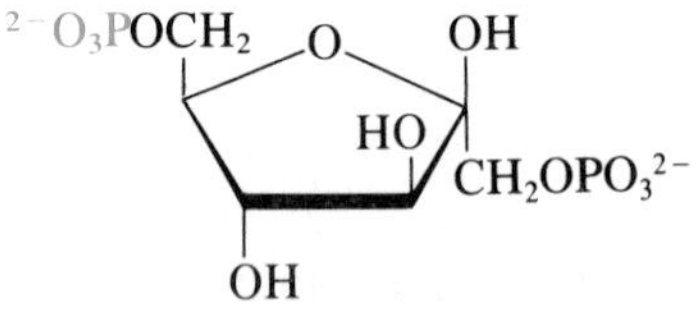
$^{2-}O_3POCH_2$
O
OH
HO
$CH_2OPO_3^{2-}$
OH

Fructose
1,6-bisphosphate

4

Step 4. The six-carbon chain of fructose 1,6-bisphosphate is cleaved into two three-carbon pieces by the enzyme *aldolase.* (Continued on next page.)

$$\underset{\substack{\text{Dihydroxyacetone} \\ \text{phosphate}}}{{}^{2-}O_3POCH_2-\overset{\displaystyle O}{\overset{\|}{C}}-CH_2OH} \quad \underset{5}{\rightleftharpoons} \quad \underset{\substack{\text{Glyceraldehyde} \\ \text{3-phosphate}}}{{}^{2-}O_3POCH_2-\overset{\displaystyle OH}{\underset{|}{C}H}-\overset{\displaystyle O}{\overset{\|}{C}}-H}$$

Step 5. The two products of step 4 are both three-carbon sugars, but only glyceraldehyde 3-phosphate can continue in the glycolysis pathway. Dihydroxyacetone phosphate must first be isomerized by the enzyme *triose phosphate isomerase*.

$$\mathbf{6} \overset{\displaystyle \text{NAD}^+, \text{P}_i}{\underset{\displaystyle \text{NADH/H}^+}{\Bigg\lgroup}}$$

Step 6. Two reactions occur as glyceraldehyde 3-phosphate is first oxidized to a carboxylic acid and then phosphorylated by the enzyme *glyceraldehyde 3-phosphate dehydrogenase*. The coenzyme nicotinamide adenine dinucleotide (NAD) and inorganic phosphate ion (P_i) are required.

$$\underset{\text{1,3-Bisphosphoglycerate}}{{}^{2-}O_3POCH_2-\overset{\displaystyle OH}{\underset{|}{C}H}-\overset{\displaystyle O}{\overset{\|}{C}}-OPO_3{}^{2-}}$$

$$\mathbf{7} \overset{\displaystyle \text{ADP}}{\underset{\displaystyle \text{ATP}}{\Bigg\lgroup}}$$

Step 7. A phosphate group from 1,3-bisphosphoglycerate is transferred to ADP, resulting in synthesis of ATP, and catalyzed by *phosphoglycerate kinase*.

$$\underset{\text{3-Phosphoglycerate}}{{}^{2-}O_3POCH_2-\overset{\displaystyle OH}{\underset{|}{C}H}-\overset{\displaystyle O}{\overset{\|}{C}}-O^-}$$

$$\mathbf{8}$$

Step 8. A phosphate group is next transferred from carbon 3 to carbon 2 of phosphoglycerate in a step catalyzed by the enzyme *phosphoglyceromutase*.

$$\underset{\text{2-Phosphoglycerate}}{HO-CH_2-\overset{\displaystyle {}^{2-}O_3PO}{\underset{|}{C}H}-\overset{\displaystyle O}{\overset{\|}{C}}-O^-}$$

$$\underset{\displaystyle H_2O}{\mathbf{9}}\Bigg\lgroup$$

Step 9. Loss of water from 2-phosphoglycerate produces phosphoenolpyruvate (PEP). The dehydration is catalyzed by the enzyme *enolase*.

$$\underset{\text{Phosphoenolpyruvate}}{H_2C=\overset{\displaystyle {}^{2-}O_3PO}{\underset{|}{C}}-\overset{\displaystyle O}{\overset{\|}{C}}-O^-}$$

Exergonic and not reversible $\mathbf{10} \overset{\displaystyle \text{ADP}}{\underset{\displaystyle \text{ATP}}{\Bigg\lgroup}}$

Step 10. Transfer of the phosphate group from phosphoenolpyruvate to ADP yields pyruvate and generates ATP, catalyzed by *pyruvate kinase*.

$$\underset{\text{Pyruvate}}{CH_3-\overset{\displaystyle O}{\overset{\|}{C}}-\overset{\displaystyle O}{\overset{\|}{C}}-O^-}$$

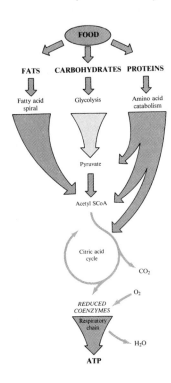

Step 2 is an isomerization that converts glucose 6-phosphate to fructose 6-phosphate. As the following open-chain formulas show, the reaction is conversion of an aldose to a ketose, which prepares the molecule for addition of a second phosphate group on the end carbon.

$$
\begin{array}{ccc}
\underset{\displaystyle \text{H}}{\overset{\displaystyle \text{O}}{\text{C}}} & & \text{CH}_2\text{OH} \\
\text{H}{-}\text{C}{-}\text{OH} & & \text{C}{=}\text{O} \\
\text{HO}{-}\text{C}{-}\text{H} & \rightleftharpoons & \text{HO}{-}\text{C}{-}\text{H} \\
\text{H}{-}\text{C}{-}\text{OH} & & \text{H}{-}\text{C}{-}\text{OH} \\
\text{H}{-}\text{C}{-}\text{OH} & & \text{H}{-}\text{C}{-}\text{OH} \\
\text{CH}_2\text{OPO}_3{}^{2-} & & \text{CH}_2\text{OPO}_3{}^{2-}
\end{array}
$$

Glucose 6-phosphate   Fructose 6-phosphate
(an aldohexose)   (a ketohexose)

Step 3 makes a second energy investment as fructose 6-phosphate is converted to fructose 1,6-bisphosphate by reaction with ATP. (''Bis-'' means ''two.'') Step 3 is another major control point for glycolysis. When the cell is short of energy, ADP (adenosine diphosphate) and AMP (adenosine monophosphate) concentrations build up and activate the step 3 enzyme, phosphofructokinase. When energy is in good supply, ATP and citrate build up and inhibit the enzyme. The outcome of steps 1–3 is formation of a molecule ready to be split into the two three-carbon intermediates that will ultimately become two molecules of pyruvate.

Steps 4 and 5 of Glycolysis: Cleavage and Isomerization　·*Step 4* converts the six-carbon bisphosphate from step 3 into two three-carbon monophosphates, one an aldose and one a ketose. The bond between carbons 3 and 4 in fructose 1,6-bisphosphate breaks, and a C=O group is formed. If you look back at Section 16.8, you'll see that this is the reverse of an aldol reaction.

$$
\begin{array}{ccc}
\text{CH}_2\text{OPO}_3{}^{2-} & & \text{CH}_2\text{OPO}_3{}^{2-} \\
\text{C}{=}\text{O} & & \text{C}{=}\text{O} \\
\text{HO}{-}\text{C}{-}\text{H} & & \text{CH}_2\text{OH} \\
\text{H}{-}\text{C}{-}\text{OH} & \rightleftharpoons & + \\
\text{H}{-}\text{C}{-}\text{OH} & & \underset{\displaystyle \text{H}}{\overset{\displaystyle \text{O}}{\text{C}}} \\
\text{CH}_2\text{OPO}_3{}^{2-} & & \text{H}{-}\text{C}{-}\text{OH} \\
 & & \text{CH}_2\text{OPO}_3{}^{2-}
\end{array}
$$

Dihydroxyacetone phosphate

Glyceraldehyde 3-phosphate

Fructose 1,6-bisphosphate

The two three-carbon sugars produced in step 4 are isomers that are interconvertible in an aldose–ketose equilibrium (*step 5* in Figure 22.5), but only glyceraldehyde 3-phosphate can continue on the glycolysis pathway. The overall result of steps 4 and 5 is therefore the production of *two* glyceraldehyde 3-phosphate molecules.

Steps 1–5 are sometimes referred to as the *energy investment* part of glycolysis. So far, two ATPs have been invested and no income has been earned, but the stage has been set for a small profit. Note that since one glucose molecule gives two glyceraldehyde 3-phosphates that each pass separately down the rest of the pathway, *steps 6–10 of glycolysis each take place twice for every glucose molecule that enters at step 1.*

Steps 6–10 of Glycolysis: Energy Generation   The second half of glycolysis is devoted to alternately generating molecules with high-energy phosphate groups that can be removed in exergonic reactions and then capturing this energy in ATP.

Step 6 is the combined oxidation of glyceraldehyde 3-phosphate to a carboxylic acid and phosphorylation of the acid. NAD^+ is the oxidizing agent, making this the first energy-generating step of glycolysis. Some of the energy from the exergonic oxidation is captured in NADH, and some is devoted to forming the high-energy phosphate.

Step 7 generates the first ATP of glycolysis by transferring a phosphate group to ADP from 1,3-bisphosphoglycerate. Because this step occurs twice for each glucose molecule, the energy balance sheet in glycolysis is even after step 7. Two ATPs were spent in steps 1–5, and now they've been replaced.

Steps 8 and 9 generate the second high-energy phosphate by an isomerization followed by dehydration to give phosphoenolpyruvate. Transfer of the phosphate group of this molecule to ADP in *step 10* then generates ATP in a highly exergonic reaction. The two ATPs formed by the two occurrences of step 10 are pure profit, and the overall results of glycolysis are as follows:

Net Result of Glycolysis

$$C_6H_{12}O_6 + 2\ NAD^+ + 2\ P_i + 2\ ADP \longrightarrow 2\ CH_3\overset{\overset{O}{\|}}{C}\overset{\overset{O}{\|}}{C}-O^- + 2\ NADH + 2\ ATP + 2\ H_2O + 2\ H^+$$

Glucose   Pyruvate

- Conversion of glucose to two pyruvates
- Production of two ATPs
- Production of two reduced coenzyme molecules, NADH.

Practice Problems   **22.4**   Identify the two pairs of steps in glycolysis in which high-energy phosphate intermediates are synthesized and their energy then harvested in ATP.

22.5   Identify each step in glycolysis that is an isomerization.

22.4 ENTRY OF OTHER SUGARS INTO GLYCOLYSIS

Fructose from fruits and sucrose, galactose from the lactose in milk, and mannose from plant polysaccharides are the major simple sugars other than glucose produced by digestion. Each eventually joins the glycolysis pathway.

Fructose is converted to glycolysis intermediates in two ways: In muscle, it is phosphorylated to fructose 6-phosphate, and in the liver it is converted to glyceraldehyde 3-phosphate. Mannose, like glucose, is converted by hexose kinase to a 6-phosphate, which then undergoes a multistep, enzyme-catalyzed rearrangement and enters glycolysis as fructose 6-phosphate. A five-step pathway is needed to convert galactose to glucose 6-phosphate for entry into the glycolysis pathway. In a genetic disease known as *galactosemia,* one of the enzymes of this pathway is missing. The resulting buildup of galactose can cause cataracts, mental retardation, liver damage, and death. The condition can be treated with a lactose- and galactose-free diet, which means strict, lifelong avoidance of all milk and milk products.

Practice Problems **22.6** Use the curved arrow symbolism to write an equation for the conversion of fructose to fructose 6-phosphate by ATP. At what step does fructose 6-phosphate enter glycolysis?

22.7 Compare glucose and galactose (see Section 21.5) and tell how their structures differ.

22.5 THREE REACTIONS OF PYRUVIC ACID

The breakdown of glucose to pyruvate is a central metabolic pathway in most living systems. The further reactions of pyruvate, however, depend on conditions and on the nature of the organism.

Aerobic In the presence of oxygen.

Under normal oxygen-rich (**aerobic**) conditions, pyruvate is converted into acetyl SCoA as part of the central pathway of glucose metabolism. Under oxygen-poor (anaerobic) conditions, it is converted into lactate, a reaction that takes place in bacteria, as when milk turns sour, and in muscle tissue of higher organisms.

Fermentation The breakdown of glucose to ethanol plus carbon dioxide by the action of yeast enzymes.

When pyruvate undergoes **fermentation** by yeast, it is converted into ethanol plus carbon dioxide, the process used to produce beer, wine, and other alcoholic drinks (Figure 22.6). Fermentation is also used to make bread: The carbon dioxide causes the bread to rise, and the alcohol evaporates. The first leavened, or raised, bread was probably made by accident when airborne yeasts got into the dough.

aerobic conditions

$$CH_3-\overset{\overset{O}{\|}}{C}-\overset{\overset{O}{\|}}{C}-O^-$$

Pyruvate

Pyruvate dehydrogenase

NADH

NAD$^+$

Anaerobic conditions

Fermentation with yeast

$$CH_3-\overset{\overset{O}{\|}}{C}-S-CoA \quad + \quad CO_2$$

Acetyl SCoA

$$CH_3-\overset{\overset{OH}{|}}{CH}-\overset{\overset{O}{\|}}{C}-O^-$$

Lactate

$$CH_3-CH_2-OH \ + \ CO_2$$

Ethanol

Oxidation of Pyruvate to Acetyl SCoA   Under aerobic conditions, pyruvate formed in the cytosol crosses the mitochondrial membrane and enters the matrix, where acetyl SCoA is produced with the aid of an enzyme embedded in the mitochondrial membrane. The overall reaction is simple in appearance:

Net Result of Pyruvate Oxidation

$$\text{Pyruvate} + \text{NAD}^+ + \text{H-S-CoA} \xrightarrow[\text{dehydrogenase}]{\text{pyruvate}} \text{acetyl SCoA} + CO_2 + \text{NADH}$$

- Transformation of two carbon atoms from pyruvate to acetyl SCoA
- Release of one molecule of CO_2
- Production of one reduced coenzyme molecule (NADH) in the mitochondrion.

   Pyruvate dehydrogenase complex, the enzyme responsible for pyruvate oxidation, contains about 60 subunits of three different enzymes with a total molecular weight of over 4 million, and the conversion of pyruvate to acetyl SCoA is actually far from simple. It requires NAD^+, CoA, FAD, and two other coenzymes (lipoic acid and thiamine pyrophosphate) derived from thiamine (vitamin B_1). The enzyme subunits, which are adjacent to each other, swing into position one after the other as pyruvate loses CO_2 and is converted to an acetyl group which is then transferred to coenzyme A.

Figure 22.6
Fermentation of grapes. Inside these barrels, yeast is at work, converting the sugar in grape juice into wine.

Practice Problem   22.8   Complete oxidation of glucose produces six molecules of carbon dioxide. Describe the stages of catabolism at which all six are formed.

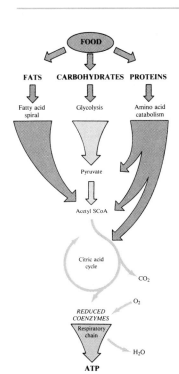

Conversion of Pyruvate to Lactate   Why does pyruvate take an alternate pathway in the absence of oxygen? Since no oxygen has been used in glucose catabolism thus far, what's the connection? The problem lies with the NADH formed in step 6 of glycolysis (Figure 22.5). Under aerobic conditions, NADH is continually reoxidized during electron transport, and fresh NAD^+ is in good supply. If the respiratory chain shuts down because of lack of oxygen, however, NADH builds up in concentration, no NAD^+ is available, and glycolysis can't continue. An alternative way to reoxidize NADH is therefore essential, because without the ATP from oxidative phosphorylation, glycolysis is needed as a source of fresh ATP.

   The reduction of pyruvate to lactate solves the problem. NADH serves as the reducing agent and is reoxidized to NAD^+. Red blood cells have no mitochondria and therefore always form lactate as the end product of glycolysis. We'll pursue the fate of the lactate a bit later in this chapter.

22.6   ENERGY OUTPUT IN COMPLETE CATABOLISM OF GLUCOSE

The total energy output from oxidation of glucose is the combined result of (1) glycolysis, (2) conversion of pyruvate to acetyl SCoA, (3) the passage of the two acetyl groups through the citric acid cycle, and (4) the passage of reduced coenzymes from each of these pathways through the respiratory chain.

To determine the total number of ATPs generated from one glucose molecule, we can first add up the net equations for each pathway. Since each glucose yields two pyruvates and two acetyl SCoAs, the net equations for pyruvate oxidation and the citric acid cycle must be multiplied by 2:

Net Result of Catabolism of One Glucose Molecule

Glycolysis

$$\text{Glucose} + 2\,\text{NAD}^+ + 2\,\text{P}_i + 2\,\text{ADP} \longrightarrow 2\,\text{pyruvate} + 2\,\text{NADH} + 2\,\text{ATP} + 2\,\text{H}_2\text{O} + 2\,\text{H}^+$$

Pyruvate oxidation

$$2\,\text{Pyruvate} + 2\,\text{NAD}^+ + 2\,\text{HSCoA} \longrightarrow 2\,\text{acetyl SCoA} + 2\,\text{CO}_2 + 2\,\text{NADH}$$

Citric acid cycle

$$2\,\text{Acetyl SCoA} + 6\,\text{NAD}^+ + 2\,\text{FAD} + 2\,\text{ADP} + 2\,\text{P}_i + 4\,\text{H}_2\text{O} \longrightarrow$$
$$2\,\text{HSCoA} + 6\,\text{NADH} + 6\,\text{H}^+ + 2\,\text{FADH}_2 + 2\,\text{ATP} + 4\,\text{CO}_2$$

$$\text{Glucose} + 10\,\text{NAD}^+ + 2\,\text{FAD} + 2\,\text{H}_2\text{O} + 4\,\text{ADP} + 4\,\text{P}_i \longrightarrow$$
$$10\,\text{NADH} + 8\,\text{H}^+ + 2\,\text{FADH}_2 + 4\,\text{ATP} + 6\,\text{CO}_2$$

For each NADH that passes through electron transport in the mitochondrion, 3 ATPs are produced, and for each $FADH_2$, 2 ATPs are produced (Section 20.10). Translating the net equation for glucose catabolism into ATPs according to these relationships gives a total of 38 ATPs, the maximum number that can be produced from each glucose molecule.

$$10\,\text{NADH} \left(\frac{3\,\text{ATP}}{\text{NADH}} \right) + 2\,\text{FADH}_2 \left(\frac{2\,\text{ATP}}{\text{FADH}_2} \right) + 4\,\text{ATP} = 38\,\text{ATP, maximum per glucose}$$

There's one step in the pathway from glucose to ATP that we've not mentioned yet, but that in some cells reduces the maximum number of ATPs produced by 2—from 38 to 36. The additional step arises because the two NADHs from glycolysis are produced in the *cytosol,* but oxidative phosphorylation takes place in the *mitochondrion* and NADH can't cross the inner mitochondrial membrane. This problem is solved by shuttle mechanisms that carry hydrogen atoms from NADH across the membrane, producing $FADH_2$ and decreasing the ATP yield.

Practice Problem **22.9** Why does the FAD-requiring shuttle decrease the ATP yield from 38 to 36?

22.7 REGULATION OF GLUCOSE METABOLISM AND ENERGY PRODUCTION

Hypoglycemia Lower-than-normal blood glucose concentration.

Normal blood glucose concentration ranges roughly from 65 to 110 mg/dL. When departures from normal occur, we're in trouble (Figure 22.7). Low blood glucose (**hypoglycemia**) causes weakness, sweating, and rapid heartbeat, and in

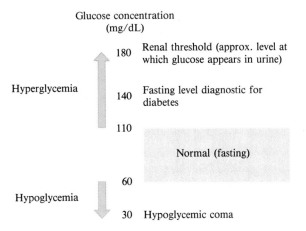

Figure 22.7
Blood glucose
concentrations.

severe occurrences, mental confusion, convulsions, coma, and death. At a
blood glucose level of 30 mg/dL consciousness is impaired or lost, and prolonged hypoglycemia can cause permanent dementia. Excess blood glucose
(**hyperglycemia**) causes increased urine flow as water is drawn from cells because of the higher osmolarity of urine containing glucose. Prolonged hyperglycemia can cause low blood pressure, coma, and death.

Hyperglycemia Higher-
than-normal blood glucose
concentration.

Two hormones from the pancreas have the major responsibility for blood
glucose regulation (Figure 22.8). The first and most important is *insulin*, which
is released when blood glucose concentration rises (Figure 22.9). Although

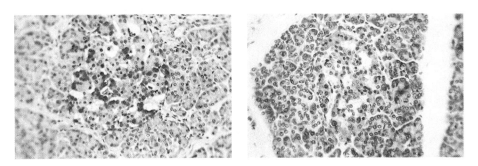

Figure 22.8
Pancreatic islets. These cell
clusters contain β cells that
produce insulin (shown by
the stain at the left) and α
cells that produce glucagon
(shown by the stain at the
right).

Figure 22.9
Regulation of glucose concentrations by insulin and glucagon from the pancreas.

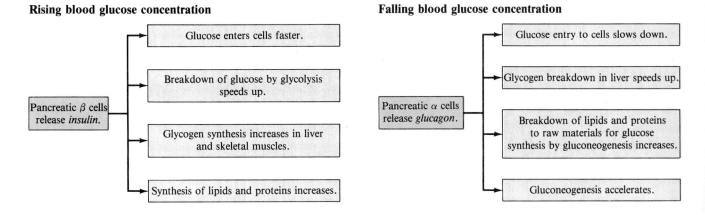

AN APPLICATION: GLYCOGEN STORAGE DISEASES

A group of genetic diseases known as *glycogen storage diseases* result from defects in enzymes associated with glycogen metabolism. Clinical symptoms are of two types: those due to hypoglycemia because of reduced availability of glucose from glycogen, and those due to accumulation of excessive amounts of glycogen in liver and muscle tissue.

von Gierke's disease results from a deficiency of glucose-6-phosphatase, the enzyme that removes phosphate from glucose 6-phosphate so that glucose can be released from the liver into the bloodstream. Symptoms include (1) a greatly enlarged liver containing excessive normal glycogen and (2) dangerously low blood sugar after nothing has been eaten for a few hours. Management of von Gierke's disease requires maintenance of adequate blood glucose levels around the clock by techniques such as nasogastric feeding in infants, drugs that inhibit uptake of glucose by the liver, or overnight delivery of carbohydrates directly to the stomach through a gastrostomy tube.

In *McArdle's disease,* an individual develops normally but as a young adult experiences rapid fatigue and muscle cramps on exertion. The defect is in the enzyme for breakdown of muscle glycogen to glucose. Moderate buildup of normal muscle glycogen occurs, but cramps arise because insufficient ATP is available from glycolysis due to a shortage of glucose 6-phosphate.

insulin has multiple physiological effects, its primary role is to aid the passage of blood glucose across cell membranes. When insulin concentration rises, cells produce energy from glucose and excess glucose is converted to glycogen, proteins, or lipids.

The second, *glucagon,* is released when blood glucose concentration drops. In a reversal of insulin's effects, glycogen is broken down in the liver rather than synthesized, and proteins and lipids are broken down into simple molecules that can be converted to glucose in the liver by gluconeogenesis. Epinephrine, the fight-or-flight hormone, also accelerates the breakdown of glycogen, but its primary target is muscle tissue where energy is needed for quick action, rather than the liver. A closer look at glycogen metabolism and gluconeogenesis is provided in Sections 22.10 and 22.11.

22.8 METABOLISM IN FASTING AND STARVATION

Imagine that you're lost in the woods. You've had no carbohydrates and very little else to eat for hours, and you're exhausted. The glycogen stored in your liver and muscles will soon be used up, but your brain must still rely on glucose to keep functioning. What happens next? Fortunately, your body's not yet ready to give up. The gluconeogenesis pathway can make glucose from proteins. And if you're lost for a long time, there's a further backup system that extracts energy from compounds other than glucose.

The metabolic changes in the absence of food begin with a gradual decline in blood glucose concentration accompanied by an increased release of glucose from glycogen. All cells contain glycogen, but most is stored in liver cells (about 70 g) and muscle cells (about 200 g). Free glucose and glycogen represent less than 1% of our energy reserves and are used up in 15–20 hr of normal activity (3 hr in a marathon). Fats are our largest energy reserve, but adjusting to dependence on them for energy takes time. In the meantime, as glucose and glycogen are exhausted, metabolism turns to breakdown of proteins in muscles and glucose production from amino acids via gluconeogenesis.

During the first few days of starvation, protein is used up at a rate as high as 75 g/day. Meanwhile, lipid catabolism is being mobilized and acetyl SCoA molecules are being manufactured from fats at an ever-increasing rate. Eventually, the citric acid cycle is overloaded and can't degrade acetyl SCoA as rapidly as it's produced. Acetyl SCoA therefore builds up inside cells and begins to be removed by a new series of metabolic reactions that transform it into substances called *ketone bodies,* which we'll describe further in our discussion of lipid metabolism in Chapter 24.

The ketone bodies enter the bloodstream, and, as starvation continues, the brain and other tissues are able to switch over to producing up to 50% of their ATP from ketone bodies instead of glucose. By the fortieth day of starvation, metabolism has stabilized at the use of about 25 g of protein and 180 g of fat each day, a condition in which glucose and protein are conserved as much as possible. An average person can survive in this state for several months—those with more fat can survive longer.

22.9 METABOLISM IN DIABETES MELLITUS

Diabetes mellitus A chronic condition due to either insufficient insulin or failure of insulin to cross cell membranes.

Metabolic disease A disease due to disruption of metabolism caused by a genetically determined enzyme defect.

Diabetes mellitus, one of the most common **metabolic diseases,** is estimated to affect up to 4% of the population. Diabetes is not a single disease but is divided into two major types, *insulin-dependent* and *non-insulin-dependent*. The insulin-dependent type, also called *juvenile-onset* or Type I diabetes because it often appears in childhood, is caused by an insufficient supply of insulin due to damage to the pancreatic cells that produce it. By contrast, in the non-insulin-dependent type, also called *adult-onset* or Type II diabetes because it usually occurs in obese individuals over about 40 years of age, insulin is in good supply but fails to promote the passage of glucose across cell membranes. Although often thought of only as a disease of glucose metabolism, diabetes affects protein and fat metabolism as well, and in some ways the metabolic response resembles starvation.

The symptoms by which diabetes is usually detected—excessive thirst accompanied by frequent urination, abnormally high glucose concentrations in urine and blood, and wasting of the body despite a good diet—result when available glucose does not enter cells where it is needed. Glucose builds up in the blood, causing the symptoms of hyperglycemia and spilling over into the urine. In untreated diabetes, metabolism responds to the glucose shortage within cells by proceeding through the same stages from depletion of glycogen stores to breakdown of proteins and fats.

Adult-onset diabetes is thought to result when cell membrane receptors fail to recognize insulin. Drugs that increase insulin levels are an effective treatment because more of the undamaged receptors are put to work. To treat juvenile-onset diabetes, the missing insulin must be supplied by injection. Commercially available insulin (Figure 22.10) is now produced by bacteria modified by recombinant DNA techniques (Chapter 26) and is the first product of genetic engineering approved for use in humans.

Figure 22.10
Synthetic insulin. The drop of insulin pictured is one of the first to be made by genetic engineering techniques (Chapter 26). Before the availability of human insulin from genetically engineered bacteria, diabetic individuals had to rely on insulin extracted from the pancreas of cows, which is expensive and differs in three amino acids from human insulin.

AN APPLICATION: GLUCOSE TOLERANCE TEST

The glucose tolerance test is among the clinical laboratory tests usually done to pin down a diagnosis of diabetes mellitus. The patient must fast for 10–16 hr, must avoid a diet high in carbohydrates prior to the fast, and must not be taking any of a long list of drugs that can interfere with the test.

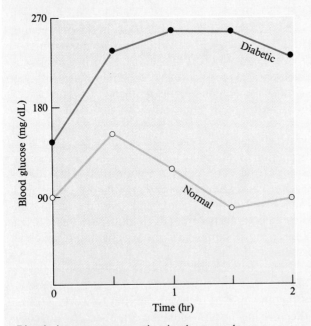

Blood glucose concentration in glucose tolerance test for normal and diabetic individuals.

First, a blood sample is drawn for determination of the fasting glucose concentration. Then, the patient drinks a solution containing 75 g of glucose, and further blood samples are taken at regular intervals after ingesting the glucose. The accompanying figure compares the changes in diabetic and nondiabetic individuals. In both, there is a rise in blood glucose concentration in the first hour. The difference becomes apparent after 2 hr, when the concentration in a nondiabetic individual has dropped to close to the fasting level but that in a diabetic individual remains high.

As listed below, a fasting blood glucose concentration of 140 mg/dL or higher, and/or a glucose tolerance test concentration that remains above 200 mg/dL beyond 1 hr, are considered diagnostic criteria for diabetes. For a firm diagnosis, the glucose tolerance test is usually given more than once.

Key Diagnostic Features of Diabetes Mellitus

Classic symptoms such as frequent urination, excessive thirst, rapid weight loss
Random blood glucose concentration > 200 mg/dL
Fasting blood glucose ≥ 140 mg/dL
Sustained blood glucose concentration ≥ 200 mg/dL after glucose administration in glucose tolerance test

An insulin-dependent diabetic is at risk for two types of medical emergencies: *Acidosis* as the result of buildup of acidic ketones in the late stages of uncontrolled diabetes may lead to coma and diminished brain function but can be reversed by timely insulin administration. Hypoglycemia or "insulin shock," by contrast, may be due to an overdose of insulin or failure to eat. If untreated, diabetic hypoglycemia can cause nerve damage or death. The arrival at the emergency room of a diabetic patient in a coma requires quick determination of whether the condition is due to ketoacidosis or excess insulin. Observations are backed up by bedside tests for glucose and ketones in blood and urine.

22.10   GLYCOGEN METABOLISM: GLYCOGENESIS AND GLYCOGENOLYSIS

Glycogen, the storage form of glucose in animals, is a branched polymer of glucose (Section 21.8). In the muscles, it is a source of energy for the muscle cells themselves; in the liver, it is a source of glucose needed to maintain normal blood glucose concentrations.

Glycogenesis The biochemical pathway for synthesis of glycogen.

To synthesize glycogen via **glycogenesis,** a strategy like that used for the transfer of phosphate or acetyl groups is needed. Glucose 6-phosphate from the first step of glycolysis is converted to glucose 1-phosphate, which then bonds through its phosphate group to form the high-energy diphosphate of uridine (UDP). One by one, UDP carries glucose molecules to the growing glycogen chain (Figure 22.11).

Figure 22.11
Glycogenesis and glycogenolysis. The pathways to and from glycogen, (glucose)$_n$, are shown schematically.

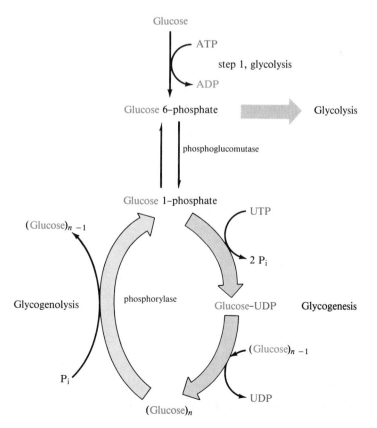

Glucose-UDP, the activated carrier of glucose in glycogen synthesis

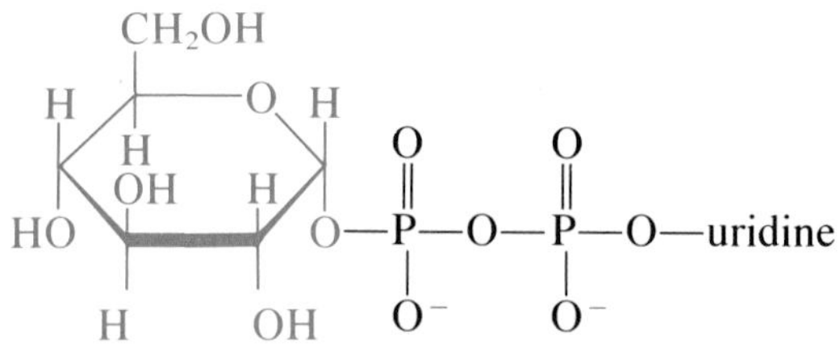

$$CH_2OH$$

(structure of glucose-UDP) ...—O—P—O—P—O—uridine

Glycogenolysis The biochemical pathway for breakdown of glycogen to free glucose.

Glycogenolysis, the release of glucose from glycogen, is not the reverse of glycogen synthesis, because it does not require uridine triphosphate (UTP). First, the enzyme phosphorylase catalyzes the conversion of a glucose at the end of the glycogen chain to glucose 1-phosphate by reaction with a phosphate ion. The glucose 1-phosphate is then converted to glucose 6-phosphate in the reverse of the equilibrium by which it is formed in glycogenesis.

Glycogenolysis in muscle cells occurs when there is an immediate need for energy from glucose and may be triggered by release of epinephrine, the fight-or-flight hormone. The glucose 6-phosphate produced in muscle cells goes directly into glycolysis. Glycogenolysis in the liver occurs when there is a need for glucose in the blood. Because phosphates can't cross cell membranes, liver cells contain the enzyme for hydrolysis to yield free glucose for release into the bloodstream.

$$\text{Glucose 6-phosphate} + H_2O \longrightarrow \text{glucose} + P_i$$

Practice Problem 22.10 The following reaction is a good example of coupling of endergonic and exergonic reactions:

$$\text{UTP} + \text{glucose 1-phosphate} + H_2O \longrightarrow \text{glucose-UDP} + 2\ P_i$$

The first of the coupled reactions is formation of glucose-UDP and pyrophosphate ion (PP_i) by combination of UTP and glucose 1-phosphate, for which $\Delta G = +1.1$ kcal. The second is hydrolysis of the pyrophosphate ion to give two phosphate ions ($2\ P_i$), for which $\Delta G = -8.0$ kcal. What is the common intermediate in these coupled reactions? What is ΔG for the coupled reactions? Is the change favorable or unfavorable?

22.11 GLUCONEOGENESIS: GLUCOSE FROM NONCARBOHYDRATES

Gluconeogenesis The biochemical pathway for the synthesis of glucose from noncarbohydrates, such as lactate, amino acids, or glycerol.

Gluconeogenesis in the liver is a pathway for making glucose from a variety of small molecules. For one, there's lactate, which is in especially abundant supply when muscles are working hard and which is also the normal product of glycolysis in red blood cells. The lactate is converted to pyruvate, which is the substrate for the first step of gluconeogenesis. The cycling of lactate from muscles and red blood cells to the liver and the return of glucose via the bloodstream to muscle cells is shown in Figure 22.12. Amino acids and glycerol from lipid catabolism are also able to enter the gluconeogenesis pathway for conversion to glucose 6-phosphate.

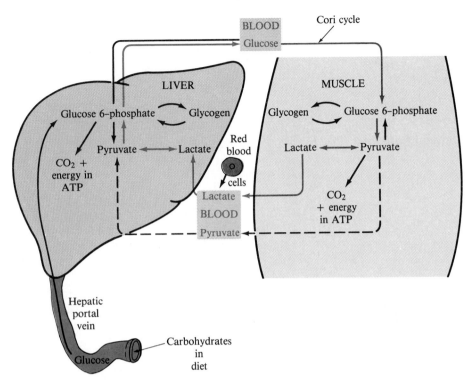

Figure 22.12
Major pathways of carbohydrate metabolites. Lactate produced in muscle during exercise or in red blood cells is sent in the bloodstream to the liver, where it is converted back into glucose 6-phosphate that can be broken down to give energy, stored as liver glycogen, or returned to the bloodstream. The cycle marked in red is known as the *Cori cycle* after the husband-and-wife team Carl and Gerti Cori who first recognized it and who received the Nobel Prize for this work.

We noted earlier that, for metabolic pathways to be favorable, they must be exergonic. As a result, most are not reversible because the amount of energy required by the reverse, endergonic pathway is too large to be supplied by cellular metabolism. Glycolysis and gluconeogenesis provide a good example of this relationship and of the way around it.

Three reactions in glycolysis—steps 1, 3, and 10 in Figure 22.5—are too exergonic to be reversed. Gluconeogenesis therefore uses different enzymes and coenzymes, with different energy requirements, to bypass these three reactions. The other 7 reactions of the 10-step gluconeogenesis pathway are the reverse of those in glycolysis and are catalyzed by the same enzymes running in reverse.

We're not going to examine all the reactions of gluconeogenesis in detail, but the first provides an interesting example. In what adds up to a reversal of the last step of glycolysis, pyruvate is first converted to oxaloacetate and then to phosphoenolpyruvate. Recall that phosphoenolpyruvate (PEP) is a high-energy compound whose conversion to the more stable pyruvate in glycolysis releases about 6 kcal. To bypass the endergonic reversal of this reaction requires the investment of two energy-carrying molecules, ATP and GTP (guanosine triphosphate, a high-energy phosphate similar in structure to ATP), in a process that releases about 5 kcal.

Oxaloacetate is another significant intermediate in gluconeogenesis because it provides a connection to the first step of the citric acid cycle. If, for example, energy is needed, oxaloacetate from pyruvate can go directly into the citric acid cycle instead of continuing on to phosphoenolpyruvate and the production of glucose.

Practice Problems   **22.11**   The net reaction for gluconeogenesis is

$$2 \text{ Pyruvate} + 4 \text{ ATP} + 2 \text{ GTP} + 2 \text{ NADH} + 2 \text{ H}^+ + 2 \text{ H}_2\text{O} \longrightarrow$$
$$\text{glucose} + 4 \text{ ADP} + 2 \text{ GDP} + 6 \text{ P}_i + 2 \text{ NAD}^+$$

Compare this result with the net reaction for glycolysis in Section 22.3. If glucose is converted to two pyruvates and both pyruvates are recycled back to glucose, is there an overall energy cost or profit?

22.12   Which would be the appropriate coenzyme in conversion of lactate to pyruvate, NADH, NAD$^+$, or FADH$_2$? Is this oxidation or reduction?

$$\underset{\text{Lactate}}{\text{CH}_3-\overset{\overset{\text{OH}}{|}}{\text{CH}}-\overset{\overset{\text{O}}{\|}}{\text{C}}-\text{O}^-} \xrightarrow{\quad ? \quad} \underset{\text{Pyruvate}}{\text{CH}_3-\overset{\overset{\text{O}}{\|}}{\text{C}}-\overset{\overset{\text{O}}{\|}}{\text{C}}-\text{O}^-}$$

22.12   THE PENTOSE PHOSPHATE PATHWAY, AN ALTERNATIVE TO GLYCOLYSIS

Pentose phosphate pathway   The biochemical pathway that produces ribose (a pentose), NADPH, and other sugar phosphates from glucose; an alternative to glycolysis.

Depending on the type of cell, varying amounts of glucose enter the **pentose phosphate pathway** as an alternative to glycolysis. One major function of this alternative is to provide NADPH, a coenzyme needed in the synthesis of lipids. Thus, the pathway occurs mainly where lipids are produced: in fatty tissue, the liver, the adrenal cortex, and mammary glands. Another major function is production of ribose, a pentose needed for the synthesis of nucleic acids. When the need for NADPH or ribose is low, intermediates from the pentose phosphate pathway can continue on through glycolysis and the citric acid cycle.

Glucose 6-phosphate formed in step 1 of glycolysis is the starting material for the pentose phosphate pathway, which, like glycolysis, takes place in the cytosol. The pathway is divided into two stages. The first, the *oxidative stage*, results in the conversion of glucose 6-phosphate to ribulose 5-phosphate and the reduction of two NADP$^+$s to two NADPHs:

Oxidative stage of the pentose phosphate pathway

$$\text{Glucose 6-phosphate} + 2 \text{ NADP}^+ + \text{H}_2\text{O} \longrightarrow$$
$$\text{ribulose 5-phosphate} + \text{CO}_2 + 2 \text{ NADPH} + 2 \text{ H}^+$$

NAD$^+$/NADH and NADP$^+$/NADPH differ in structure by only one phosphate group (Figure 20.10), but they differ significantly in function and are not interchangeable. The first of these coenzymes is needed mainly in its oxidized form, NAD$^+$, as an oxidizing agent in the exergonic reactions of catabolism. The second is needed mainly in its reduced form, NADPH, as a reducing agent in the endergonic reactions of biosynthesis.

Next after the oxidative phase, ribulose 5-phosphate is isomerized to ribose 5-phosphate, the sugar phosphate essential for nucleic acid and coenzyme synthesis.

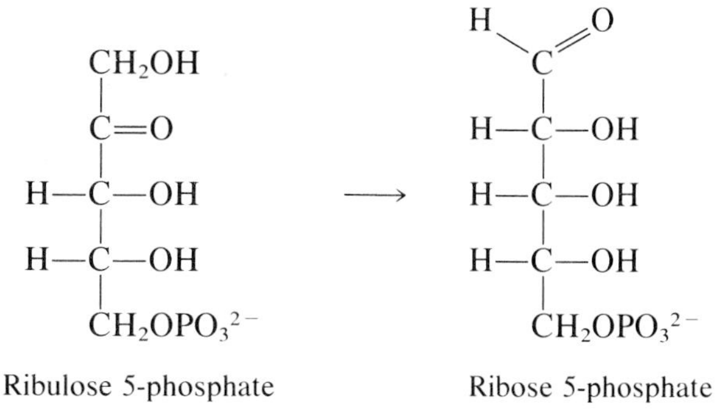

Ribulose 5-phosphate   Ribose 5-phosphate

At this point, ribose might be converted to one or more of several three-, four-, five-, or seven-carbon sugars that are also intermediates in various biosyntheses. Ultimately, glyceraldehyde 3-phosphate and fructose 6-phosphate are formed:

Net result of the pentose phosphate pathway

$$3 \text{ Glucose 6-phosphate } + 6 \text{ NADP}^+ \longrightarrow 2 \text{ fructose 6-phosphate } + 3 \text{ CO}_2 + \text{glyceraldehyde 3-phosphate } + 6 \text{ NADPH } + 6 \text{ H}^+$$

The pentose phosphate pathway is remarkably versatile. The details aren't important to us here, but consider some of the possible variations: When NADPH demand is high, intermediates are recycled to glucose 6-phosphate for further NADPH production by the oxidative phase. When ATP demand is high, fructose 6-phosphate and glyceraldehyde 3-phosphate can enter glycolysis, and the overall result is complete oxidation of glucose 6-phosphate to CO_2 and H_2O. When nucleic acid synthesis is the priority, most of the nonoxidative phase is skipped and ribose 5-phosphate is the major product.

Practice Problem   **22.13**   Eighteen carbon atoms enter the pentose phosphate pathway in three glucose 6-phosphate molecules. Describe the fate of these carbon atoms according to the net result of the pathway.

INTERLUDE: BIOCHEMISTRY OF RUNNING

A runner is tense and ready, waiting for the signal to start the race. With the training period over, biochemistry is in control of providing the energy needed to run. The flood of epinephrine and the fight-or-flight reaction take control first, but only during the opening seconds of the race. The next energy resource to be tapped is creatine phosphate, a high-energy amino acid phosphate that stands ready in muscles for instant reaction. Although it's quick because it goes in one step to

ATP, there's enough creatine phosphate to last no more than 30 seconds.

Next, muscles start to use their stores of glycogen, freeing glucose for glycolysis, pyruvate production, and under these anaerobic conditions, lactate formation instead of acetyl SCoA formation (see Fig. 22.4). In a 100-m sprint, probably all the energy comes from creatine phosphate and anaerobic glycolysis, but anaerobic breakdown of glucose from glycogen suffices for

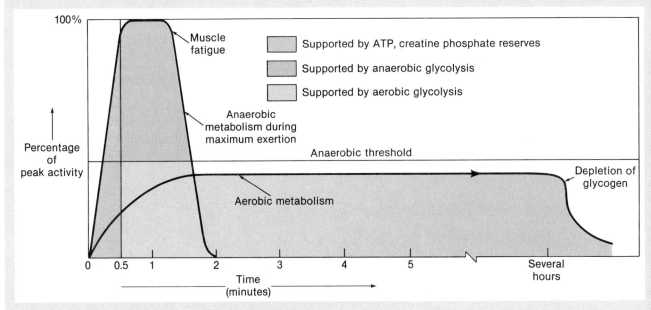

Changes in metabolism with muscle activity. The anaerobic threshold varies with factors such as age, weight, and level of athletic training.

SUMMARY

Carbohydrate **digestion** begins with α-amylase in the mouth, continues with pancreatic α-amylase and other enzymes in the small intestine, and yields monosaccharides that enter the bloodstream by absorption through the intestinal wall. The major route for glucose catabolism is **glycolysis** to yield pyruvate, followed by pyruvate oxidation to acetyl SCoA, which then yields ATP via the citric acid cycle and oxidative phosphorylation. For the com-

plete catabolism of one glucose molecule, 38 (or 36) ATPs are produced.

Glycolysis is a 10-step pathway in which two ATPs are first invested to form active phosphate intermediates that ultimately yield two pyruvates, two ATPs, and two NADHs for each glucose. Under **anaerobic** conditions, NADH serves as the reducing agent for conversion of pyruvate to lactate, thereby providing the NAD^+ needed only a minute or two of maximum exertion since a buildup of lactate causes muscle fatigue (see the curve, p. 638). Beyond this, other pathways come into action. As breathing and heart rate speed up and oxygen-carrying blood flows more quickly to muscles, the aerobic pathway is activated, and ATP is once again generated by oxidative phosphorylation. The trick is to run at a speed just under that relying on anaerobic ATP generation, the "anaerobic threshold."

Now the question is, what fuel will metabolism rely on, carbohydrate or fat? Burning fatty acids from fats is more efficient: More than twice as many calories are generated by burning a gram of fat than by burning a gram of carbohydrate. When we're sitting quietly, in fact, our muscle cells are burning mostly fat, and the fat in storage could support the exertion of marathon running for several days. By contrast, glycogen alone can

provide enough glucose to fuel only 2–3 hr of such running.

The difficulty is that fatty acids can't be delivered to muscle cells fast enough to maintain the ATP level needed for running, so metabolism compromises and glycogen remains the limiting factor for the marathon runner. Once glycogen is gone, extreme exhaustion and mental confusion set in—the condition known as "hitting the wall." Running speed is then limited to that sustainable by fats only. To delay this point as long as possible, glycogen synthesis is encouraged by diets high in carbohydrates prior to and during a race. In the hours just before the race, however, carbohydrates are avoided. Their effect of triggering insulin release is undesirable at this point because the resulting faster use of glucose will hasten depletion of glycogen.

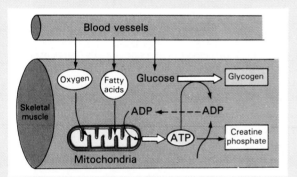

ATP generation in resting muscle.

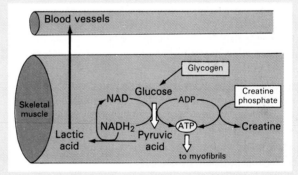

ATP generation during maximum exertion.

for glycolysis to continue producing ATP. The pancreatic hormones insulin and glucagon regulate blood glucose concentrations and help to guard against the ill effects of **hypoglycemia** and **hyperglycemia.**

Glucose is stored as glycogen made by **glycogenesis,** especially in liver and muscle tissue, and is released via **glycogenolysis.** In the liver, glycogenolysis occurs when glucose is needed in the blood. In muscle cells that need quick energy, glycogenolysis is triggered by epinephrine. The synthesis of glucose from lactate and other

noncarbohydrates, **glyconeogenesis,** is a pathway of especial importance in starvation or glucose shortage within cells because of **diabetes mellitus.** Seven of the reactions of glycolysis run in reverse in gluconeogenesis, while alternate enzymes carry out the reverse of the three glycolysis reactions that are too exergonic to be reversed.

Varying amounts of glucose are also metabolized in the **pentose phosphate pathway,** which provides ribose (a pentose), NADPH, and other sugars needed for biosynthesis.

REVIEW PROBLEMS

Digestion and Metabolism

22.14   Where in the body does digestion occur, and what kinds of chemical reactions does it involve?

22.15   What is meant by the words *aerobic* and *anaerobic?*

22.16   What three products are formed from pyruvate under aerobic, anaerobic, and fermentation conditions?

22.17   Complete the following word equation:

$$\text{Maltose} + \text{H}_2\text{O} \longrightarrow \quad ? + ?$$

Where in the digestive system does this process occur?

22.18   Four of the processes in glucose metabolism have similar-sounding names. Differentiate among glycolysis, gluconeogenesis, glycogenesis, and glycogenolysis.

Glycolysis

22.19   By what other name is the glycolysis pathway known?

22.20   What is the name and structure of the final product of carbohydrate glycolysis?

22.21   Where in the cell does glycolysis occur? How does that compare to where the citric acid cycle occurs?

22.22   Look at the 10 steps in glycolysis and then answer these questions:
(a) Which steps involve phosphorylation?
(b) Which step is an oxidation?
(c) Which step is a dehydration?

22.23   Lactate can be converted into pyruvate by the enzyme lactate dehydrogenase and the coenzyme NAD^+. Write the reaction in the standard biochemical format, using a curved arrow to show the involvement of NAD^+.

22.24   How many moles of ATP are produced by
(a) glycolysis of one mole of glucose?
(b) aerobic conversion of one mole of pyruvate to one mole of acetyl SCoA?
(c) catabolism of one mole of acetyl SCoA in the citric acid cycle?

22.25   How many moles of ATP are released by complete catabolism of one mole of glucose with the FAD shuttle in operation in glycolysis?

22.26   If fructose is isomerized to yield glucose prior to glycolysis, how many moles of ATP would you expect to obtain from complete catabolism of 1.0 mol of sucrose?

22.27   How many moles of acetyl SCoA are produced by catabolism of 1.0 mol of sucrose (Problem 22.26)?

22.28   How many moles of CO_2 are produced by the complete catabolism of sucrose (Problem 22.27)?

22.29   How many moles of ATP are produced by complete catabolism of one mole of maltose to yield CO_2?

22.30   What is the purpose of the formation of lactate under anaerobic conditions?

Glucose Levels in the Blood

22.31   Differentiate between blood sugar levels and resulting symptoms in hyperglycemia and hypoglycemia.

22.32   What is the effect of epinephrine on the glucose metabolic pathways?

22.33   From what molecules is glucose initially produced during starvation or fasting?

22.34   As starvation continues, to what is acetyl SCoA converted to prevent buildup in the cells?

22.35   What is the difference between juvenile- and adult-onset diabetes?

22.36   Where is most of the glycogen in the body stored?

22.37   Of what use is UDP in the formation of glycogen from glucose?

22.38   Why is glycogenolysis not the exact reverse of the process of glycogenesis?

Glucose Anabolism

22.39   What is the name of the anabolic pathway for making glucose?

22.40   What two molecules serve as starting materials for glucose synthesis?

22.41   To what molecule is pyruvate initially converted in the anabolism of glucose? To what substance is that molecule in turn converted?

22.42   Why can't pyruvate be converted to glucose in an exact reverse of the glycolysis pathway?

22.43   What is the major purpose of the pentose phosphate pathway?

22.44   Depending on the body's needs, into what type of compounds is glucose converted in the pentose phosphate pathway?

Applications

22.45   What genetic deficiency gives rise to von Gierke's disease and what are the symptoms? [App: Glycogen Storage Diseases]

22.46   What deficiency gives rise to McArdle's disease and what are the symptoms? [App: Glycogen Storage Diseases]

22.47   How do fasting levels of glucose in a diabetic person compare to those in a nondiabetic person? [App: Glucose Tolerance Test]

22.48   Discuss the differences in the response of a diabetic person compared to those of a nondiabetic person after drinking a glucose solution. [App: Glucose Tolerance Test]

22.49 Why is creatine phosphate a better source of quick energy than glucose? [Int: Biochemistry of Running]

22.50 Why is it not possible for a person to sprint for miles? [Int: Biochemistry of Running]

Additional Questions and Problems

22.51 When NADH is reoxidized in the citric acid cycle, 3 mol of ATP is produced. In most cells, only 2 mol of ATP is produced from the reoxidation of the NADH produced during glycolysis. What causes this difference in energy yield?

22.52 What are the symptoms and cause of the genetic disease galactosemia?

22.53 Why can pyruvate cross the mitochondrial membrane while no other molecule after step 1 in glycolysis can?

22.54 Look at the glycolysis pathway. With what type of process are kinase enzymes usually associated?

22.55 Is the same net production of ATP observed in the complete combustion of fructose as is observed in glucose? Why or why not?

22.56 Explain why one more ATP is produced when glucose is obtained from glycogen rather than used directly from the blood?

CHAPTER

23

Lipids

Soap bubbles and cell membranes are both composed of lipids, and some of their properties are quite similar.

Lipids are less well known to most people than carbohydrates and proteins, yet they're just as essential to our diet and well-being. Lipids have many important biological roles, serving as sources of fuel, as a protective coat around many plants and insects, and as a major component of the membranes that surround every living cell.

Chemically, *lipids* are naturally occurring organic molecules that dissolve in nonpolar organic solvents when a sample of plant or animal tissue is crushed or ground. For example, if a plant is placed in a kitchen blender, finely ground, and then extracted with ether, anything that dissolves in the ether is called a lipid, while anything that remains insoluble (including carbohydrates, proteins, and inorganic salts) is not a lipid.

In this chapter, we'll answer the following questions about lipids:

1. **How are lipids classified?** The goal: Be able to recognize the names and structures of the different kinds of lipids.
2. **What are fats, waxes, and oils?** The goal: Be able to describe the general structures and general properties of waxes, fats, and oils.
3. **What reactions do triacylglycerols undergo?** The goal: Be able to list the most important reactions of these lipids and, given the reactants, predict the products.
4. **What are cell membranes?** The goal: Be able to give the general structure of cell membranes and describe the lipids they contain.
5. **What are steroids, prostaglandins, and leukotrienes?** The goal: Be able to describe the general structures of these biomolecules, describe some of their functions, and recognize specific examples.

23.1 STRUCTURE AND CLASSIFICATION OF LIPIDS

Lipid A naturally occurring molecule found in plant or animal sources that is soluble in nonpolar organic solvents.

Look at the structures of some representative **lipids** shown in Figure 23.1. Since lipids are defined by solubility in nonpolar solvents (a physical property) rather than by chemical structure, it's not surprising that there are a great many different kinds. Note in all the structures shown that the molecules contain large hydrocarbon portions and not many polar groups, which accounts for their solubility behavior.

Figure 23.2 summarizes the important classes of lipids that we'll be looking at in this chapter. *Waxes* are esters of fatty acids; *triacylglycerols* and *glycerophospholipids* are esters of glycerol; and *sphingolipids* are derivatives of an 18-carbon amino alcohol. *Steroids* (cyclic hydrocarbon derivatives) and *prostaglandins* and *leukotrienes* (both derivatives of a 20-carbon acid) differ from the other lipids in having no ester groups that can be hydrolyzed. Most of the lipid class names in Figure 23.2 are probably unfamiliar to you at this point, so you might want to check back occasionally as you read this chapter.

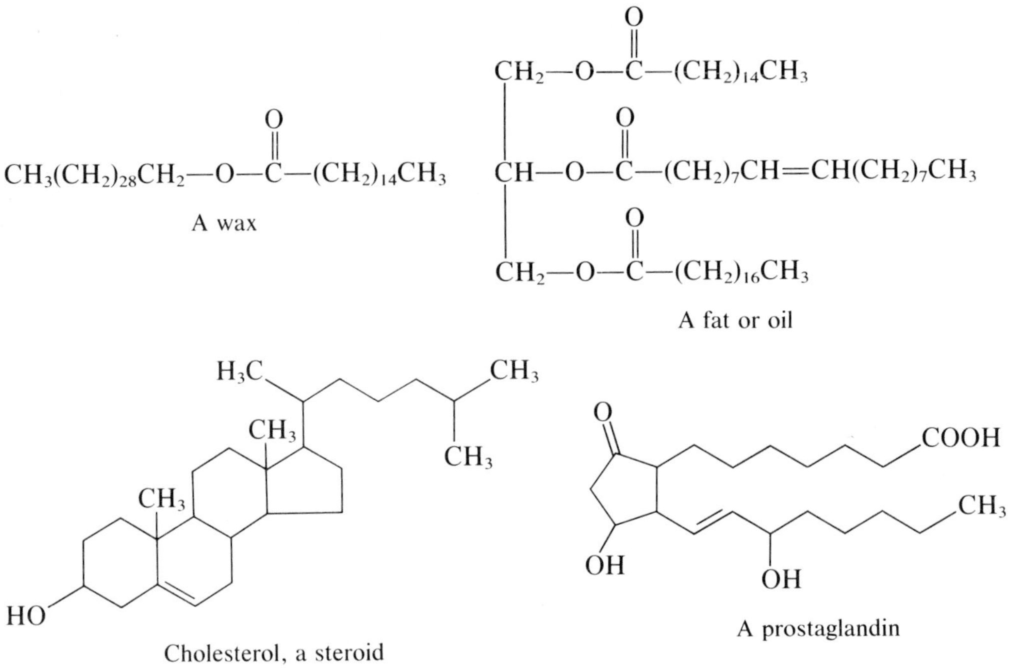

$$CH_3(CH_2)_{28}CH_2-O-\overset{\overset{\displaystyle O}{\|}}{C}-(CH_2)_{14}CH_3$$

A wax

$$CH_2-O-\overset{\overset{\displaystyle O}{\|}}{C}-(CH_2)_{14}CH_3$$
$$CH-O-\overset{\overset{\displaystyle O}{\|}}{C}-(CH_2)_7CH=CH(CH_2)_7CH_3$$
$$CH_2-O-\overset{\overset{\displaystyle O}{\|}}{C}-(CH_2)_{16}CH_3$$

A fat or oil

Cholesterol, a steroid

A prostaglandin

Figure 23.1
Structures of some representative lipids. Compounds that can be isolated from plant and animal tissue by extraction with nonpolar organic solvents are classified as lipids.

Figure 23.2
Some important families of lipids. Red shaded boxes show those derived from glycerol, and blue shaded boxes show those derived from sphingosine, an amino alcohol. Phospholipids are marked with an asterisk, and boxes outlined in red show lipids that are not esters and cannot by hydrolyzed into components.

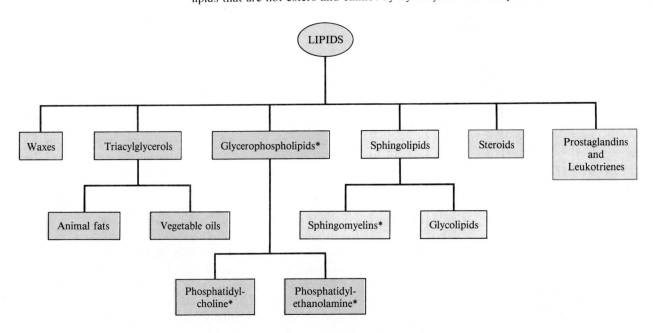

23.2 FATTY ACIDS, WAXES, FATS, AND OILS

Fatty acid A long-chain carboxylic acid; those in animal fats and vegetable oils often have 12–22 carbon atoms.

Saturated Containing only single bonds between carbon atoms and thus unable to accommodate additional hydrogen atoms.

Unsaturated Containing one or more double or triple bonds between carbon atoms and thus able to add more hydrogen atoms.

Polyunsaturated fatty acid (PUFA) A long-chain fatty acid that has two or more carbon–carbon double bonds.

The carboxylic acids present in naturally occurring fats and oils are referred to as **fatty acids.** All consist of long, straight hydrocarbon chains with a carboxylic acid group at one end, and most have even numbers of carbon atoms. Fatty acids may or may not contain double bonds. Those without double bonds are known as **saturated** fatty acids; those with double bonds are known as **unsaturated** fatty acids. If double bonds are present, they are usually cis rather than trans.

$$\text{A fatty acid} \qquad \qquad \overset{\displaystyle O}{\overset{\|}{CH_3(CH_2)_n C}} \text{OH}$$
(n is usually an even number)

About 40 different fatty acids are known to occur naturally, with some of the common ones listed in Table 23.1. Palmitic acid (16 carbons) and stearic acid (18 carbons) are the most abundant saturated acids; oleic and linoleic acids (both with 18 carbons) are the most abundant unsaturated ones. Oleic acid is *monounsaturated,* because it has only one carbon–carbon double bond, but linoleic, linolenic, and arachidonic acids are **polyunsaturated fatty acids** (called **PUFAs**), because they have more than one carbon–carbon double bond.

Two of the polyunsaturated fatty acids, linoleic and linolenic, are essential in the human diet because the body can't synthesize them. Infants grow poorly and develop severe skin lesions if fed a diet of nonfat milk, which lacks these acids; adults usually have sufficient reserves of body fat to avoid such problems. Arachidonic acid also plays an essential role in metabolism (Section 23.11), but the body is able to synthesize it from linolenic acid.

$$CH_3CH_2CH_2CH_2CH_2CH_2CH_2CH_2CH_2CH_2CH_2CH_2CH_2CH_2CH_2\overset{\displaystyle O}{\overset{\|}{C}}\text{—OH}$$

Palmitic acid—a saturated fatty acid

Linolenic acid—a polyunsaturated fatty acid (PUFA)

Table 23.1   Structures of Some Common Fatty Acids

Name	Number of Carbons	Number of Double Bonds	Structure	Melting Point (°C)
Saturated				
Lauric	12	0	$CH_3(CH_2)_{10}COOH$	44
Myristic	14	0	$CH_3(CH_2)_{12}COOH$	58
Palmitic	16	0	$CH_3(CH_2)_{14}COOH$	63
Stearic	18	0	$CH_3(CH_2)_{16}COOH$	70
Unsaturated				
Oleic	18	1	$CH_3(CH_2)_7CH{=}CH(CH_2)_7COOH$ (cis)	4
Linoleic	18	2	$CH_3(CH_2)_4CH{=}CHCH_2CH{=}CH(CH_2)_7COOH$ (all cis)	−5
Linolenic	18	3	$CH_3CH_2CH{=}CHCH_2CH{=}CHCH_2CH{=}CH(CH_2)_7COOH$ (all cis)	−11
Arachidonic	20	4	$CH_3(CH_2)_4(CH{=}CHCH_2)_4CH_2CH_2COOH$ (all cis)	−50

Wax A mixture of esters of long-chain carboxylic acids with long-chain alcohols.

Most natural **waxes** are mixtures of esters, formed in nature by reaction of fatty acids with long-chain alcohols. The carboxylic acid portion usually has an even number of carbons from 16 through 36, while the alcohol portion contains an even number of carbons in the range from 24 through 36. For example, one of the major components in beeswax is triacontyl hexadecanoate, the ester formed from the 30-carbon alcohol (triacontanol) and the 16-carbon acid (hexadecanoic acid, commonly known as palmitic acid). The waxy protective coatings on most fruits, berries, leaves, and animal furs have similar structures. Aquatic birds have a water-repellent waxy coating on their feathers and are in danger when caught in an oil spill because the waxy coating dissolves (Figure 23.3).

A wax:

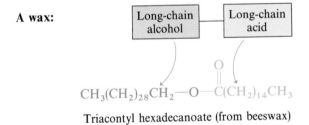

$$CH_3(CH_2)_{28}CH_2{-}O{-}\overset{\displaystyle O}{\overset{\|}{C}}(CH_2)_{14}CH_3$$

Triacontyl hexadecanoate (from beeswax)

Figure 23.3
Bird caught in a petroleum spill. The oil-soaked feathers have lost their waxy coating and also their insulating ability.

Fat A mixture of triacylglycerols that is solid because it contains a high proportion of saturated fatty acids.

Oil A mixture of triacylglycerols that is liquid because it contains a high proportion of unsaturated fatty acids.

Triacylglycerol (triglyceride) A triester of glycerol with three fatty acids.

Animal **fats** and vegetable **oils** are the most plentiful lipids in nature. Although they appear different—animal fats like butter and lard are usually solid, while vegetable oils like corn, olive, soybean, and peanut oil are liquid—their structures are closely related. All fats and oils are **triacylglycerols,** or *triglycerides;* that is, they are esters of glycerol (1,2,3-propanetriol, *glycerin*) with three fatty acids.

A triacylglycerol

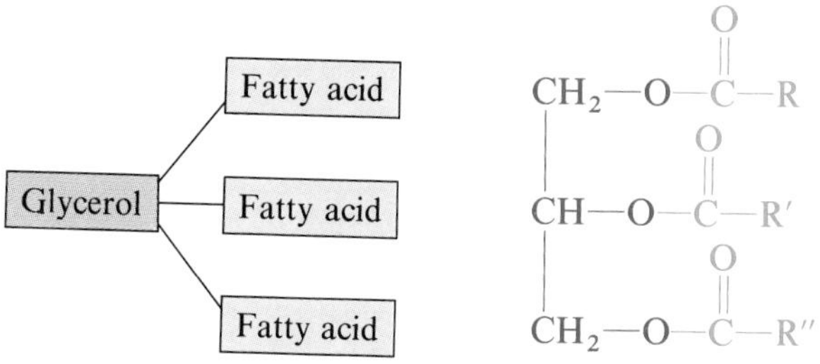

A fat or oil

For example, a typical triglyceride might have the following structure:

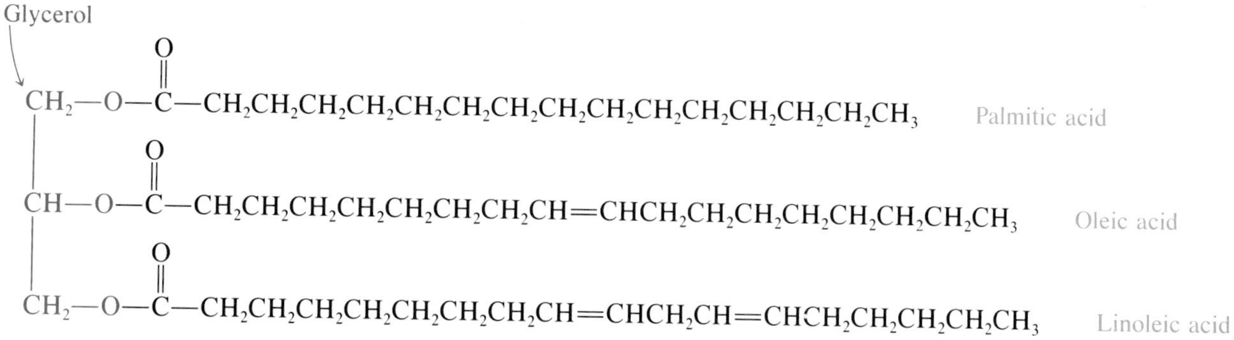

As shown in the triacylglycerol structure above, the three fatty acids of any specific fat or oil molecule are not necessarily the same. Furthermore, the fat or oil from a given natural source is a complex mixture of many different triacylglycerols. Table 23.2 lists the average composition of fats and oils from several different sources. Note particularly in Table 23.2 that vegetable oils consist almost entirely of unsaturated fatty acids, while animal fats contain a much larger percentage of saturated fatty acids. We'll see in the next section that this difference in composition is the primary reason for the different melting points of fats and oils.

Practice Problems   **23.1**   One of the constituents of the carnauba wax used in floor and furniture polish is an ester of a C_{32} straight-chain alcohol with C_{20} straight-chain carboxylic acid. Draw the structure of this ester.

23.2   Show the structure of glyceryl trioleate, a fat molecule whose components are glycerol and three oleic acid units.

Table 23.2 Approximate Composition of Some Common Fats and Oils[a]

	Saturated Fatty Acids (%)				Unsaturated Fatty Acids (%)	
Source	C_{12} Lauric	C_{14} Myristic	C_{16} Palmitic	C_{18} Stearic	C_{18} Oleic	C_{18} Linoleic
Animal Fat						
Lard	—	1	25	15	50	6
Butter	2	10	25	10	25	5
Human fat	1	3	25	8	46	10
Whale blubber	—	8	12	3	35	10
Vegetable Oil						
Corn	—	1	8	4	46	42
Olive	—	1	5	5	83	7
Peanut	—	—	7	5	60	20
Soybean	—	—	7	4	34	53

[a]Where totals are less than 100%, small quantities of several other acids are present.

23.3 PROPERTIES OF FATS AND OILS

The melting points of the various fatty acids listed in Table 23.1 show that as a general rule, the more double bonds a fatty acid has, the lower its melting point. Thus, the saturated C_{18} acid (stearic) melts at 70°C, the monounsaturated C_{18} acid (oleic) melts at 4°C, and the diunsaturated C_{18} acid (linoleic) melts at −5°C. This same trend also holds true for triacylglycerols: The more highly unsaturated a triacylglycerol, the lower its melting point. The difference in melting points between fats and oils is a consequence of differences in their chemical structures. As shown in Table 23.2, vegetable oils generally have a higher proportion of unsaturated to saturated fatty acids than animal fats.

How do the double bonds make such a significant difference in properties? Compare the structures of palmitic and linoleic acids shown in Section 23.2, for example. The hydrocarbon chains in saturated acids like palmitic are uniform in shape with identical angles of each carbon atom, allowing the chains to nestle easily together in a crystal. By contrast, the carbon chain in an unsaturated acid like linoleic has a kink wherever it passes through a cis double bond. The kinks make it difficult for such chains to fit next to each other in a uniform fashion to form crystals. The more double bonds there are in a triacylglycerol, the harder it is for it to crystallize, an effect illustrated in Figure 23.4 with molecular models.

We are accustomed to the characteristic yellow color and flavors of cooking oils, but these are contributed by natural materials carried along during production of the oils from plants; pure oils are colorless and odorless. Overheating, or exposure to air, oxidizing agents, or water, causes decomposition to products with unpleasant odors or flavors, creating what we call a *rancid* oil.

Practice Problem **23.3** Draw the full structure of arachidonic acid (Table 23.1) in a way that shows the *cis* stereochemistry of its four double bonds.

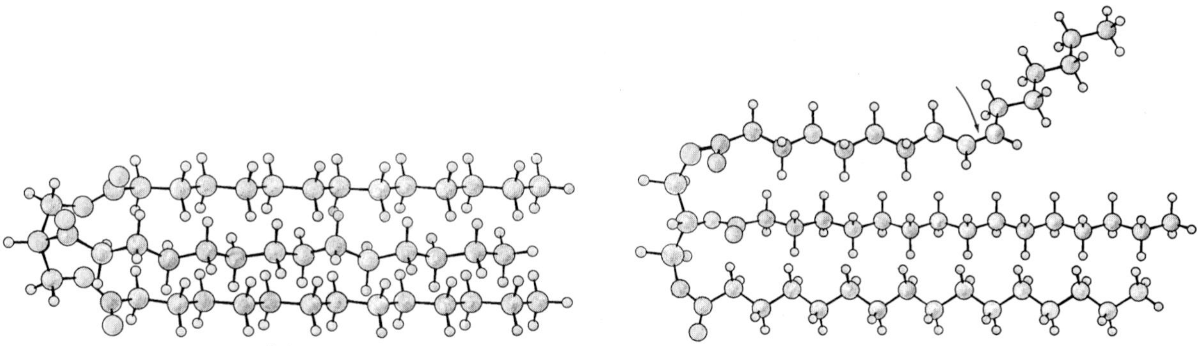

Figure 23.4
Molecular models of (a) a saturated and (b) an unsaturated triacylglycerol. The double bond (red arrow) in (b) prevents the molecule from adopting a regular shape and crystallizing easily. Fats are solids because of their high proportion of saturated hydrocarbon chains, and oils are liquids because of their high proportion of unsaturated hydrocarbon chains.

23.4 CHEMICAL REACTIONS OF TRIACYLGLYCEROLS

Hydrogenation   The carbon–carbon double bonds in vegetable oils can be hydrogenated to yield saturated fats in the same way that any alkene can react with hydrogen to yield an alkane (Section 13.6). Margarine and solid cooking fats like Crisco are produced commercially by hydrogenation of vegetable oils to give a product chemically similar to that found in animal fats.

Part-structure of an unsaturated vegetable oil

$$-O-\overset{\overset{\displaystyle O}{\|}}{C}-CH_2CH_2CH_2CH_2CH_2CH_2CH_2CH=CHCH_2CH=CHCH_2CH_2CH_2CH_2CH_3$$

2 H$_2$, Pd catalyst

Part-structure of a saturated fat

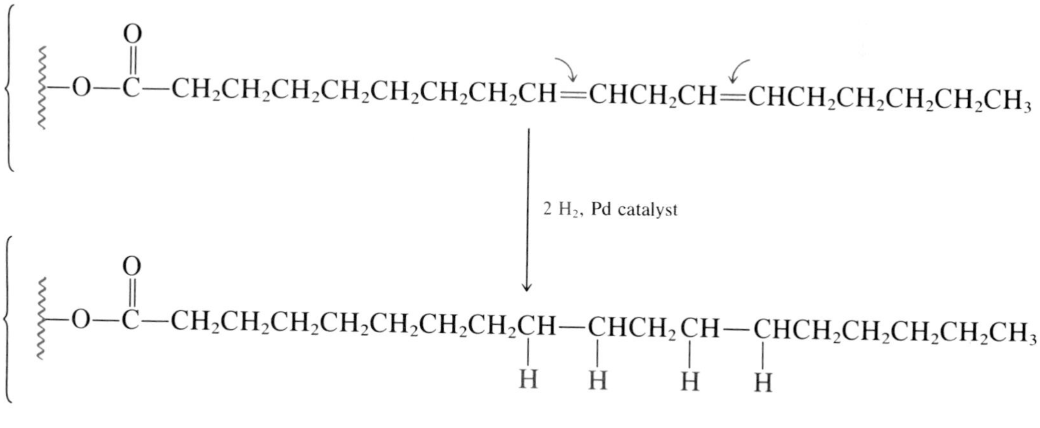

$$-O-\overset{\overset{\displaystyle O}{\|}}{C}-CH_2CH_2CH_2CH_2CH_2CH_2CH_2CH-CHCH_2CH-CHCH_2CH_2CH_2CH_2CH_3$$
$$\qquad\qquad\qquad\qquad\qquad\qquad\qquad\quad \underset{H}{|}\quad \underset{H}{|}\qquad \underset{H}{|}\quad \underset{H}{|}$$

By carefully controlling the exact extent of hydrogenation, the final product can be made to have any desired consistency. Margarine, for example, is prepared so that only about two-thirds of the double bonds present in the starting vegetable oil are hydrogenated. The remaining double bonds are left unhydrogenated, so the margarine has exactly the right consistency to keep soft in the refrigerator and to melt on warm toast.

AN APPLICATION: LIPIDS IN THE DIET

The major recognizable sources of fats and oils in our diet are butter and margarine, lard and vegetable oils, the visible fat in meat, and chicken skin. In addition, there are "invisible" lipids in milk and milk products, meat, fish, poultry, nuts and seeds, whole-grain cereals, and prepared foods. As illustrated in Table 23.2, different foods have characteristic distributions of unsaturated and saturated triacylglycerols. In meat, poultry, fish, dairy products, and eggs, the triacylglycerols are mostly saturated and are accompanied by small quantities of cholesterol. Vegetable oils have a higher unsaturated fatty acid content and contain no cholesterol. They never have, so the salad oil labels that proclaim, "No cholesterol," are stating a fact, not announcing something special the manufacturer has done for us.

Fats and oils are a popular component of our diet: they taste good, give a pleasant texture to food, and, because they are digested slowly, give a feeling of satisfaction after a meal. The percentage of calories from fats and oils in the average U.S. diet in the early 1980s was 40–45%, considerably more than needed. Remember that only two fatty acids are designated as essential because our bodies can't make them. Fats and oils in our diet beyond what we burn are mostly stored as fat in *adipose tissue* that lies in layers under the skin and around internal organs.

Concern for the relationships among saturated fats, cholesterol levels, and various diseases caused a modest decrease to 37% average calories from fats and oils in the U.S. diet by 1990. We have significantly decreased our consumption of butter, eggs, beef, and whole milk (all containing relatively high proportions of saturated fat and cholesterol) but at the same time have increased our consumption of cheese and the oils in salad dressings and baked goods. An interesting development, largely the result of public pressure, has been the agreement by several large commercial bakeries to decrease their

Practice Problem **23.4** Write an equation showing the complete hydrogenation of glyceryl trioleate (Practice Problem 23.2) to yield glyceryl tristearate.

Hydrolysis The breakdown of a compound by reaction with water.

Saponification Reaction of a fat or oil with aqueous hydroxide ion to yield glycerol and carboxylate salts of fatty acids.

Hydrolysis: Soap Triacylglycerols, like all esters, can be hydrolyzed to yield carboxylic acid and alcohol components. In the body, this **hydrolysis** is carried out by enzymes as the first step in lipid metabolism. In the laboratory, the same reaction is usually carried out with aqueous NaOH or KOH, where it is called **saponification** (Section 17.7). The initial products of fat (or oil) saponification are one molecule of glycerol and three molecules of fatty acid carboxylate salts.

A fat or oil   Glycerol   Fatty acid salts (soap)

use of the "tropical oils" palm oil and coconut oil. Although called "oils" because of their plant origin, these substances are solids at room temperature and contain significant proportions of saturated fats.

Several organizations concerned with health maintenance have recommended a diet with no more than 30% of its calories from fats and oils. Furthermore, it's recommended that no more than 10% of daily calories come from saturated fat and no more than 300 mg of cholesterol per day be included. In a daily diet of 3000 Calories, 30% from fats and oils is about 100 g, the amount in a bit more than 1/4 lb of butter.

Triacylglycerols from the seeds of these sunflowers will soon appear on super-market shelves as vegetable oil and margarine.

Soap The mixture of carboxylate salts of fatty acids formed on saponification of animal fat.

The complex mixture of fatty acid salts produced by saponification of animal fat with NaOH or KOH is known as **soap**. Crude soap curds, which contain glycerol and excess alkali as well as soap, are first purified by boiling with water and adding NaCl to precipitate the pure sodium carboxylate salts. The smooth soap that precipitates is then dried, perfumed, and pressed into bars for household use. Dyes are added for colored soaps; antiseptics are added for medicated soaps; pumice is added for scouring soaps; and air is blown in for soaps that float.

Hydrophilic Water-loving; a hydrophilic substance dissolves in water.

Hydrophobic Water-fearing; a hydrophobic substance does not dissolve in water.

Soaps act as cleaning agents because the two ends of a soap molecule are so different. The sodium-salt end of the long-chain molecule is ionic and therefore **hydrophilic** (water-loving); it tends to dissolve in water. The long organic chain portion of the molecule, however, is nonpolar and therefore **hydrophobic** (water-fearing); it tends to avoid water and to dissolve in grease. Because of these two opposing tendencies, soaps are attracted to *both* grease and water.

Micelle A spherical cluster formed by the aggregation of soap molecules (or other molecules with hydrophilic and hydrophobic ends) in water.

When soap molecules are dispersed in water, very few of them actually dissolve. Instead, large numbers of soap molecules aggregate together, with their long, hydrophobic hydrocarbon tails clustering in a ball to exclude water. At the same time, their hydrophilic ionic heads on the surface of the cluster stick out into the water. These clusters, called **micelles,** are diagrammed in Figure 23.5. Grease and dirt are made soluble in water when they are coated by the nonpolar tails of the soap molecules in the center of micelles. Once solubilized, the grease and dirt can be washed away.

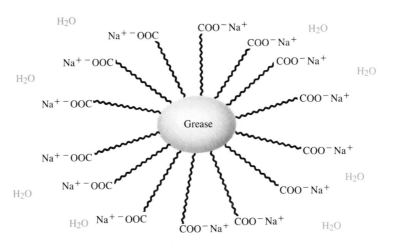

Figure 23.5
A soap micelle. The hydrophobic ends of the soap molecules extend into the grease particle, while the hydrophilic ends keep the micelle soluble in water. Lipids are transported in the bloodstream in similar micelles, as described in Chapter 24.

Practice Problems   **23.5**   In hard water, which contains iron, magnesium, and calcium ions, soap often leaves a scummy residue composed of precipitated acid salts. Draw the structure of calcium oleate, one of the components of bathtub scum.

23.6   Write the complete equation for the hydrolysis of a triacylglycerol in which the fatty acids are two molecules of stearic acid and one of oleic acid (see Table 23.1).

AN APPLICATION: DETERGENTS

Strictly speaking, anything that washes away dirt is a *detergent*. The term is usually applied, however, to synthetic materials made from petroleum chemicals that began to replace soaps in the 1950s. The goal was to overcome the problems caused by precipitation of soap scum in hard water, which include waste of soap, residues left in washed clothing, and difficulty in cleaning bathtubs and washing machines.

Like soaps, detergent molecules have hydrophobic hydrocarbon tails and hydrophilic heads, and they cleanse by the same mechanism as soap—forming micelles around greasy dirt. Materials that function in this manner are described as *surface-active agents* or *surfactants*. The hydrophilic heads may be anionic, cationic, or nonionic. Anionic detergents are commonly used in home laundry products.

An anionic detergent

$$Na^+ \ ^-O-\overset{\displaystyle O}{\underset{\displaystyle O}{S}}-\!\!\!\!-\!\!\!\!\text{—}CH_2CH_2CH_2CH_2CH_2CH_2CH_2CH_2CH_2CH_2CH_2CH_3$$

Sodium dodecylbenzenesulfonate

23.5 MEMBRANE LIPIDS

The major function of lipids in the body, other than energy storage, is as components of cell membranes. Lipids are ideal for this function because membranes must separate the aqueous solutions inside cells from the aqueous extracellular fluid. The molecules of membrane lipids are similar to soap and detergent molecules in having hydrophobic tails and hydrophilic heads, but they differ in having two tails instead of one.

Membrane lipids are built up from alcohols, amino alcohols, fatty acids, and phosphate groups. Some are esters of glycerol, like the triacylglycerols, but with a phosphate ester in place of one fatty acid ester group. Others are esters of the amino alcohol *sphingosine*.

$$_1CH_2OH$$
$$_2CHOH$$
$$_3CH_2OH$$

Glycerol

$$_1CH_2OH$$
$$_2CHNH_2$$
$$_3CHOH$$
$$_4CH{=}CH(CH_2)_{12}CH_3$$

Sphingosine

The general structures and names of the major types of membrane lipids are summarized in Figure 23.6. After describing these membrane lipids, we'll take a look at how cell membranes are organized and function.

Cationic detergents are used in fabric softeners and in disinfectant soaps.

A cationic detergent

$$\text{Ph}-CH_2-\overset{CH_3}{\underset{CH_3}{\overset{|}{\underset{|}{N^+}}}}-R\ Cl^-$$

A benzalkonium chloride; R = C_8H_{17} to $C_{18}H_{37}$

Nonionic detergents are low-sudsing and are effective at low temperatures.

A nonionic detergent

$$CH_3(CH_2)_{10}CH_2OCH_2CH_2(OCH_2CH_2)_7OH$$

A polyether

Note that the hydrocarbon chains in these detergents are unbranched. Some of the first detergents contained branched-chain hydrocarbons, but this was soon discovered to be a mistake. The bacteria in natural waters and sewage treatment plants are slow to eat branched-chain hydrocarbons, so detergents containing them went undecomposed and produced suds in streams and lakes.

The average washing machine consumes about 20 lb of detergents per year, but the use of surfactants goes way beyond the home laundry. Of the 7.6 billion pounds of surfactants used in 1988, 29% went into household products, 54% went into industrial uses, and 17% went into personal care products such as shampoos and cosmetics.

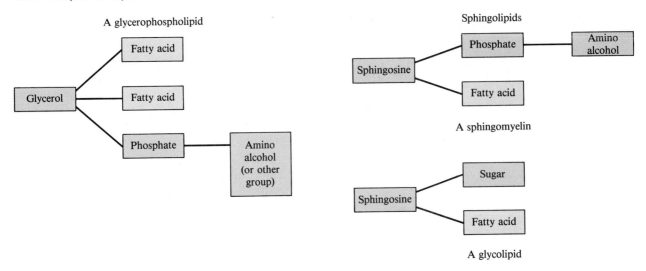

Figure 23.6
The major membrane lipids. The polar head groups are shown in the green boxes.

23.6 GLYCEROPHOSPHOLIPIDS

Glycerophospholipid
(phosphoglyceride)   A lipid
in which glycerol is linked by
ester bonds to two fatty
acids and one phosphate,
which is in turn linked by
another ester bond to an
amino alcohol (or other
alcohol); a derivative of
phosphatidic acid.

Glycerophospholipids, also known as **phosphoglycerides,** resemble triacylglycerols in having a glycerol unit linked by ester bonds to two fatty acids (the two hydrophobic tails). They differ, however, at the third position of glycerol, which is linked to a phosphate instead of a fatty acid. (Recall from Section 17.12 how phosphoric acid can form phosphate esters with alcohols.) The fatty acids may be any of the C_{12} to C_{20} acids normally present in fats, with the acid bonded to C1 of glycerol usually saturated, while that at C2 is usually unsaturated. The phosphate group, which is always bonded to C3, is also connected by a separate ester link to another compound, often a small amino alcohol. The glycerophospholipids, some of which are listed in Table 23.3, are named as derivatives of *phosphatidic acid*. For example, phosphatidylcholine:

Phosphatidic
acid

Phosphatidylcholine (a lecithin)

Table 23.3   Some Glycerophospholipids

General Structure	X =	X—OH Name	Type of Glycerophospholipid
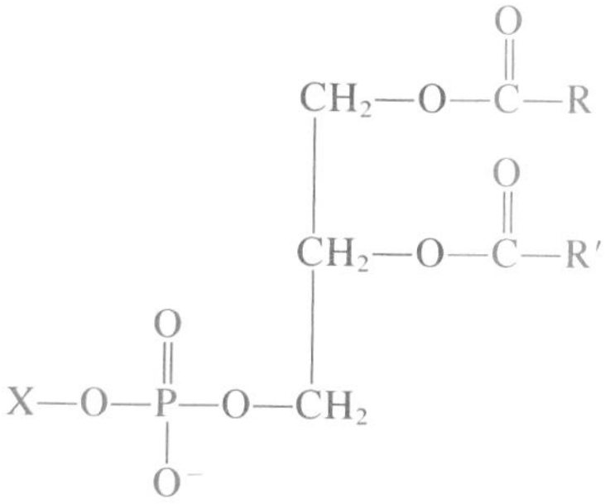	H—	Water	Phosphatidic acid (parent compound for glycerophospholipids)
	$H_3\overset{+}{N}CH_2CH_2$—	Ethanolamine	Phosphatidylethanolamine
	$(CH_3)_3\overset{+}{N}CH_2CH_2$—	Choline	Phosphatidylcholine (lecithin; most abundant cell membrane phospholipid)
	$^-OOCCHCH_2$— with $NH_3{}^+$	Serine	Phosphatidylserine (found in most tissues)
	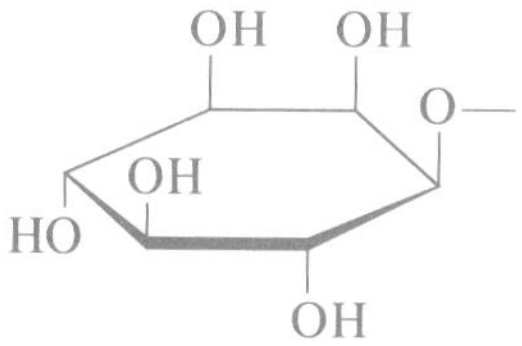 Inositol ring structure	Inositol	Phosphatidylinositol (relays chemical signals from outside cell to inside cell)

General Structure:

$$CH_2-O-\overset{\overset{O}{\|}}{C}-R$$
$$CH_2-O-\overset{\overset{O}{\|}}{C}-R'$$
$$X-O-\overset{\overset{O}{\|}}{\underset{\overset{\|}{O^-}}{P}}-O-CH_2$$

Phosphatidylcholine (lecithin) A glycerophospholipid containing choline.

The **phosphatidylcholines,** also known as **lecithins,** are major components of cell membranes in all higher organisms and are also important as a source of the choline needed in transmission of nerve signals. In solution at physiological pH, the hydrophilic head group in a phosphatidylcholine has a positive charge on the quaternary nitrogen in choline and a negative charge on the ionized phosphate group, as shown in green in the structure above. A compound referred to in the singular as phosphatidycholine or lecithin is often a mixture of molecules with different R and R' tails.

Phosphatidylcholine and other glycerophospholipids are *emulsifying agents,* substances that can coat droplets of nonpolar liquids and hold them in suspension in water (Section 10.15). You can find lecithin, usually obtained from soybean oil, listed as an ingredient in chocolate bars and other foods in which it serves as an emulsifier. It's also the lecithin in egg yolk that emulsifies the oil droplets in mayonnaise. Lecithin can be purchased at many vitamin counters and health food stores, and it has been touted as able to lower blood pressure and improve memory, although firm evidence for these effects is yet to be presented.

Phosphatidylethanolamine and phosphatidylserine are next after phosphatidylcholine in occurrence in cell membranes and are particularly abundant in brain tissue. Note in Table 23.3 that the hydrophilic head groups in these glycerophospholipids (shown in green) also contain ionic charges in solution at physiological pH.

Practice Problem   **23.7**   Show the structure of the glycerophospholipid that contains a stearic acid unit, an oleic acid unit, and a phosphate bonded to ethanolamine.

23.7 SPHINGOLIPIDS

Sphingolipid A lipid derived from the amino alcohol sphingosine (or related amino alcohols) rather than glycerol.

Sphingomyelin A sphingolipid with a fatty acid bonded to the C2 —NH₂ and a phosphate bonded to the C1 —OH group of sphingosines.

Phospholipid A lipid that has an ester link between phosphoric acid and an alcohol.

Sphingolipids are derivatives of the amino alcohol *sphingosine* rather than glycerol. The most common sphingolipids are the **sphingomyelins,** which contain phosphorus and are therefore also **phospholipids.** In sphingomyelins, a long-chain fatty acid is bonded to the C2 —NH₂ group by an amide link, and a phosphoric acid unit is bonded to the C1 —OH group. The phosphate group is also linked by an ester bond to choline. The sphingomyelins are a major constituent of the coating around nerve fibers (the *myelin sheath;* Figure 23.7) and are found in large quantities in brain tissue. A diminished amount of sphingomyelins in the brain myelin is associated with multiple sclerosis, a disease whose cause is not understood.

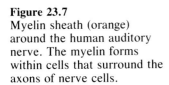

Sphingomyelin
(a sphingolipid)

Glycolipid A sphingolipid with a fatty acid bonded to the C2 —NH₂ and a sugar bonded to the C1 —OH group of sphingosines.

Yet another class of sphingosine-based lipids are the **glycolipids,** or *sphingoglycolipids:* compounds that contain both carbohydrate and lipid parts but no phosphorus. Glycolipids differ from sphingomyelins in having a sugar at C1 instead of a phosphate ester group. In the *cerebrosides,* for example, an amide link connects a fatty acid and the —NH₂ group at C2, and a glycoside link connects a monosaccharide and the —OH group at C1. Cerebrosides are particularly abundant in nerve cell membranes in the brain, where the sugar unit is D-galactose. They are also found in nonneural cell membranes, where the sugar unit is D-glucose.

Figure 23.7
Myelin sheath (orange) around the human auditory nerve. The myelin forms within cells that surround the axons of nerve cells.

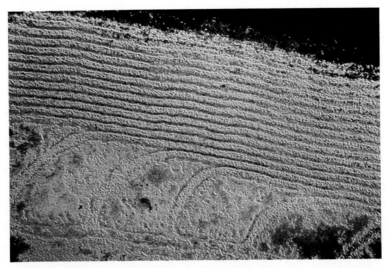

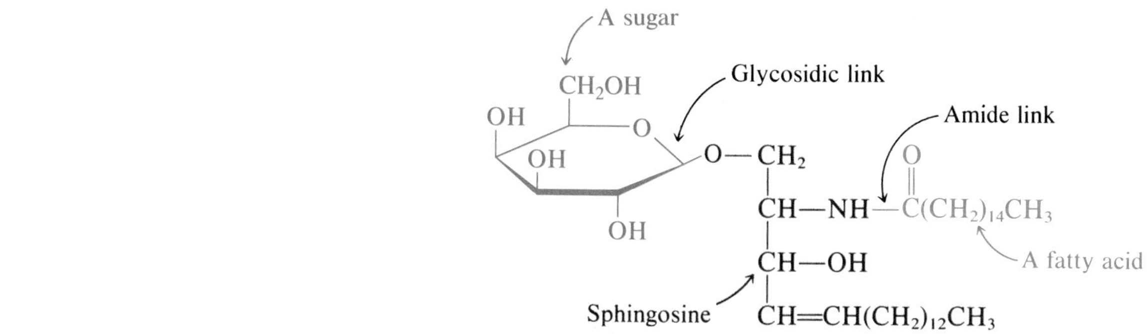

A sugar

CH_2OH

OH

OH

O

OH

Glycosidic link

O—CH_2

Amide link

O
‖
CH—NH—$C(CH_2)_{14}CH_3$

A fatty acid

CH—OH

Sphingosine

CH=$CH(CH_2)_{12}CH_3$

A cerebroside
(a glycolipid)

Closely related to the cerebrosides are the *gangliosides:* glycolipids in which the carbohydrate component linked to sphingosine is a small polysaccharide rather than a monosaccharide. Over 60 gangliosides are known. Their saccharide portions extend out of the cell membrane and apparently play important roles as receptors for hormones and neurotransmitters and in recognizing toxins and other cells.

Tay-Sachs disease, a fatal genetic disorder, is due to a defect in lipid metabolism that causes a greatly elevated concentration of a particular ganglioside in the brain. An infant born with this defect suffers mental retardation and liver enlargement and often succumbs by age 3. Tay-Sachs is one of a group of *sphingolipid storage diseases*, metabolic diseases that result from deficiencies in the supply of enzymes that break down sphingolipids.

Solved Problem 23.1

A class of membrane lipids known as *plasmalogens* has the general structure shown here. Identify the component parts of this lipid and choose the terms that apply to it: phospholipid, glycerophospholipid, sphingolipid, glycolipid. Is it most similar to a phosphatidylethanolamine, a phosphatidylcholine, a cerebroside, or a ganglioside?

CH_2—O—CH=CHR

A plasmalogen

O
‖
CH—O—C—R'

O
‖
CH_2—O—P—O—$CH_2CH_2\overset{+}{N}H_3$
|
O^-

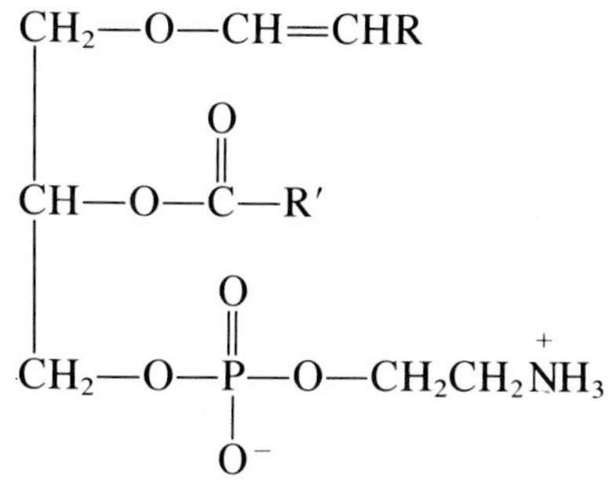

Solution

The molecule contains a phosphate group and thus is a *phospholipid*, The glycerol backbone of three carbon atoms bonded to three oxygen atoms is also present, so the compound is a *glycerophospholipid*, but one in which there is an ether linkage (CH_2—O—CH=CHR) instead of one ester linkage. The phosphate group is bonded to ethanolamine ($HOCH_2CH_2NH_2$). The compound is not a sphingolipid or a glycolipid because it is not derived from sphingosine; for the same reason it is not a cerebroside or a ganglioside. Except for the ether group in place of an ester group, the compound has the same structure as a *phosphatidylethanolamine*.

Practice Problem   **23.8**   Show the structure of the sphingomyelin that contains a myristic acid unit. Identify the hydrophilic head group and the hydrophobic tails in this molecule.

23.8 CELL MEMBRANES

Phospholipids form a major part of the membranes that surround all living cells. The lipid membranes serve not only to separate the interior of cells from their outside environment but also act as a kind of semipermeable barrier to allow the selective passage of nutrients and other components into and out of cells.

Phospholipids aggregate in a closed, sheet-like membrane structure called a **lipid bilayer** (Figure 23.8). Two parallel layers of lipids are oriented in the bilayer so that ionic head groups protrude into the aqueous environment on the inner and outer faces of the bilayer. The nonpolar tails cluster together in the middle of the bilayer where they can interact with each other and avoid water.

Cell membranes are a good deal more complex than the simple bilayer picture indicates, leading to an overall membrane structure called the *fluid-mosaic model* shown in Figure 23.9. One complexity, for example, is that the inner and outer faces of cell membranes are not identical. In human red blood cells (erythrocytes), for example, the outer layer consists primarily of sphingomyelin and phosphatidylcholine, while the inner layer is mainly phosphatidylethanolamine and phosphatidylserine.

Another complexity is that cell membranes contain many other kinds of molecules embedded in the bilayer. Glycolipids and cholesterol are present, and at least 20% of the weight of a membrane consists of protein globules (visible in Figure 23.10). Some proteins are only partially embedded (*peripheral* proteins), but others extend completely through the membrane (*integral* proteins). Lipid components of the membrane serve a primarily structural purpose, while proteins act to mediate the interactions of the cell with outside agents. Some proteins, for example, act as channels to allow specific molecules or ions to enter or leave the cell. The carbohydrate portions of glycolipids and glycoproteins (proteins bonded to carbohydrates) interact with molecules in the extracellular fluid.

Lipid bilayer The basic structural unit of cell membranes; composed of two parallel sheets of membrane lipid molecules arranged tail to tail.

Figure 23.8
Aggregation of phospholipids into the bilayer structure of a cell membrane.

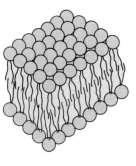

Where   is

$$CH_2-CH-CH_2-O-\overset{\overset{\displaystyle O}{\|}}{P}-O-CH_2CH_2\overset{+}{N}(CH_3)_3$$

Hydrophilic head

Hydrophobic tails

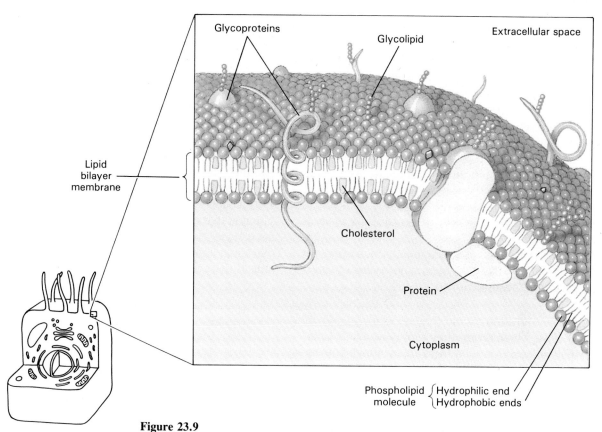

Glycoproteins

Glycolipid

Extracellular space

Lipid bilayer membrane

Cholesterol

Protein

Cytoplasm

Phospholipid molecule { Hydrophilic end / Hydrophobic ends

Figure 23.9
Fluid mosaic model of a cell membrane. Cholesterol forms part of the membrane, proteins are embedded in the bilayer, and the carbohydrate chains of glycoproteins and glycolipids extend into the extracellular space, where they interact with other substances.

Figure 23.10
A fractured cell membrane. The lipid bilayer has been split between the two lipid layers. The smoother, hydrophilic side of the top layer—the outside of the cell—is at the top in the photo. The rougher, hydrophobic interior side of the bottom layer—the inside of the bilayer—is on the bottom. The globular particles (6–9 nm in diameter) are membrane proteins.

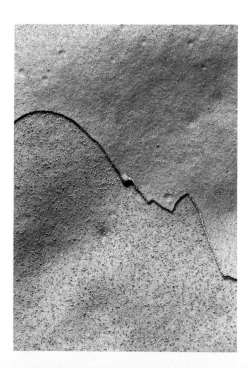

AN APPLICATION: TRANSPORT ACROSS CELL MEMBRANES

On the one hand, the membrane surrounding a living cell can't be impermeable, because nutrients must enter and waste products must leave the cell. On the other hand, the membrane can't be completely permeable, or substances would move back and forth until their concentrations were equal on both sides. In actuality, some substances move across the membranes freely in what is known as *passive transport,* and others do so only by using energy, a process known as *active transport.*

Transport across cell membranes is not yet fully understood, but three major types of transport have been identified.

Simple Diffusion   Some solutes *diffuse* in and out of cells, which means they simply wander by normal molecular motion from areas of high concentration to areas of low concentration. Simple diffusion is one of the two types of passive transport. To cross the cell membrane by simple diffusion, a solute must leave the aqueous solution on one side of the membrane, dissolve in the nonpolar region in the bilayer, and return to the aqueous solution on the other side. The lipid bilayer is just about impermeable to ions and polar molecules, which aren't soluble in the nonpolar hydrocarbon region, but oxygen molecules, for example, easily diffuse through.

Facilitated Diffusion   Like simple diffusion, facilitated diffusion is passive transport and requires no energy input. The difference is that in facilitated diffusion solutes are helped across the membrane by proteins, allowing their transport at a faster rate than could be accomplished by simple diffusion. Glucose is believed to be transported into red blood cells by facilitated diffusion. One possible way facilitated diffusion might occur is diagrammed at the bottom of this box.

Active Transport   It's essential to life that the concentrations of some solutes be different inside and outside cells. Such differences are opposite the natural tendency of solutes to move about until their concentration equalizes and therefore can be maintained only by the expenditure of energy. A notable example of active transport is the constant movement of Na^+ and K^+ ions across cell membranes. Only in this manner is it possible to maintain low Na^+ concentrations within cells and high Na^+ concentrations in extracellular fluids, and the opposite for K^+. It's thought that energy is used to change the shape of an integral protein, simultaneously bringing three Na^+ ions into the cell and moving two K^+ ions out of the cell.

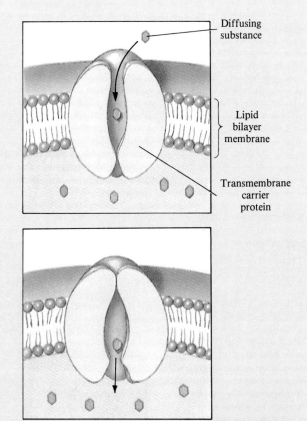

Diffusing substance

Lipid bilayer membrane

Transmembrane carrier protein

Facilitated diffusion. Molecules from the region of higher concentration diffuse into a carrier protein, which then changes conformation and releases molecules into the cell. Facilitated diffusion requires no energy and can only move species from higher- to lower-concentration regions.

One of the central features of the fluid-mosaic model is that the membrane is *fluid* rather than rigid. As a consequence, the membrane is not easily ruptured, because the lipids in the bilayer simply flow back together to repair any small hole or puncture. The effect is similar to what's observed in cooking when a thin film of oil or melted butter lies on top of water. The film can be punctured and broken, but it immediately flows back together when left alone. Still other consequences of bilayer fluidity are that small nonpolar molecules can move easily through the membrane and that individual lipid or protein molecules can diffuse rapidly from place to place within the membrane.

23.9 STEROIDS: CHOLESTEROL

Steroid A lipid whose structure is based on a tetracyclic (four-ring) carbon skeleton.

In addition to triacylglycerols and phospholipids, plant and animal cells also contain steroids. A **steroid** is a lipid whose structure is based on the following tetracyclic (four-ring) system. Three of the rings are six-membered, while the fourth is five-membered. The rings are designated A, B, C, and D, beginning at the lower left-hand corner, and the carbon atoms are numbered from 1 to 19, beginning in ring A.

Steroid skeleton

Steroids have many diverse roles throughout both the plant and animal kingdoms. Some, such as the plant steroid digitoxigenin isolated from the purple foxglove (*Digitalis purpurea*), are used in medicine as heart stimulants; others are hormones; and still others, such as cholic acid, which is synthesized in the liver from cholesterol, act as emulsifying agents to aid in the digestion of fats.

Digitoxigenin
(from purple foxglove)

Cholic acid, a bile acid
(from liver bile)

662 Chapter 23   Lipids

Cholesterol, an unsaturated alcohol, is the most abundant animal steroid. It's been estimated that a 130 lb person has a total of about 175 g of cholesterol distributed throughout the body. Much of this cholesterol is esterified with fatty acids, but some is also found as the free alcohol. Gallstones, for example, are nearly pure cholesterol.

Cholesterol, an unsaturated steroidal alcohol

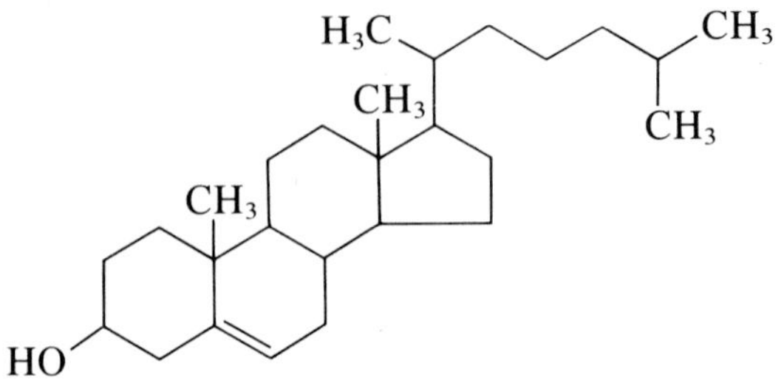

Cholesterol serves two important functions in the body. First, it is a component of cell membranes, where it acts as a regulator of membrane fluidity. Cholesterol is able to fit in between the fatty acid chains of the lipid bilayer as shown in Figure 23.9, restricting their motion and making the bilayer more rigid. Second, cholesterol serves as the body's starting material for the synthesis of all other steroids, including all the sex hormones and bile acids.

The human body obtains its cholesterol both by synthesis in the liver and by ingestion of food. Even on a strict no-cholesterol diet, the body of an adult is able to synthesize approximately 800 mg of cholesterol per day.

23.10 STEROID HORMONES

Hormones are chemical messengers that mediate biochemical events in target tissues (Chapter 19). There are two major classes of steroid hormones: the sex hormones, which control maturation and reproduction, and the adrenocortical hormones, which regulate a variety of metabolic processes.

Sex Hormones   *Testosterone* and *androsterone* are the two most important male sex hormones, or *androgens*. Androgens are responsible both for the development of male secondary sex characteristics during puberty and for promoting tissue and muscle growth. Both are synthesized in the testes from cholesterol and to a lesser extent in the adrenal cortex.

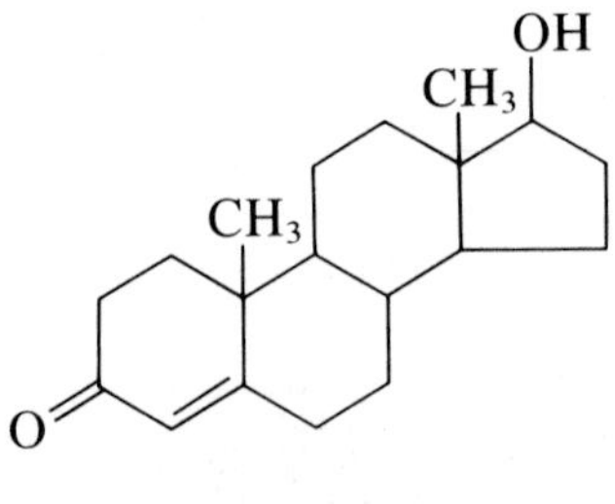

Testosterone
(an androgen)

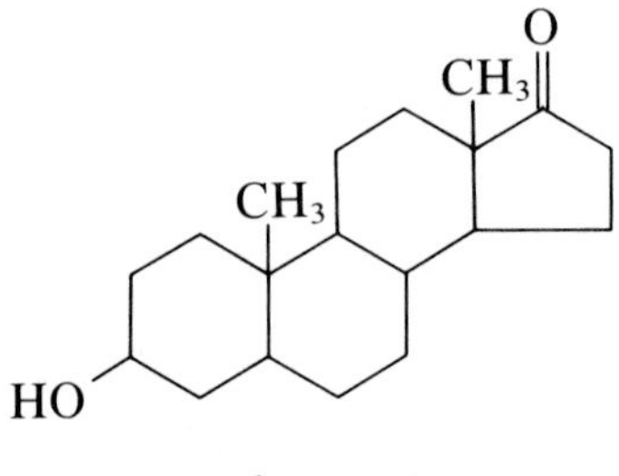

Androsterone
(an androgen)

Estrone and *estradiol* are the two most important female sex hormones, or *estrogens*. These substances, which are synthesized in the ovaries from testosterone and also to a small extent in the adrenal cortex, are responsible for the development of female secondary sex characteristics and for regulation of the menstrual cycle. Note that both have a benzene-like aromatic A ring. In addition, another kind of sex hormone called a *progestin* is essential for preparing the uterus for implantation of a fertilized ovum during pregnancy. *Progesterone* is the most important progestin.

Estrone   Estradiol   Progesterone (a progestin)

(estrogens)

Adrenocortical Hormones   Steroid hormones are secreted from the cortex (outer part) of the adrenal glands, small organs about half the size of a thumb, located near the upper end of each kidney. These steroids are of three types: *mineralocorticoids, glucocorticoids,* and the sex hormones mentioned above. Mineralocorticoids, such as *aldosterone,* function to control tissue swelling by regulating the delicate cellular salt balance between Na^+ and K^+ (hence the "mineral" in their name). Glucocorticoids, such as *hydrocortisone* and its close relative *cortisone,* are involved in the regulation of glucose metabolism and in the control of inflammation. You may well have used an antiinflammatory ointment containing cortisone to reduce the swelling and itching if you've ever been exposed to poison oak or poison ivy.

Aldosterone   Hydrocortisone   Cortisone
(a mineralocorticoid)   (a glucocorticoid)

Synthetic Steroids   In addition to the several hundred known steroids isolated from plants and animals, a great many more have been synthesized in the laboratory in the search for new drugs. The general idea is to start with a natural hormone, carry out chemical modifications of the structure, and then see what biological properties the modified steroid might have.

Among the best known of the synthetic steroids are oral contraceptive and anabolic agents. Most birth control pills are a mixture of two compounds, a synthetic estrogen such as *ethynylestradiol* and a synthetic progestin such as *norethindrone*. These synthetic steroids appear to function by tricking the body into thinking it's pregnant and therefore temporarily infertile. Anabolic steroids such as stanozolol, detected in several athletes during the 1988 Olympics, are synthetic androgens that mimic the tissue-building effect of testosterone.

Ethynylestradiol
(a synthetic estrogen)

Norethindrone
(a synthetic progestin)

Stanozolol
(a synthetic androgen)

Practice Problems   23.9   Look at the structure of progesterone and identify the functional groups in the molecule.

23.10   Look carefully at the structures of estradiol and ethynylestradiol and point out the differences. What common structural feature do they share that makes both estrogens?

Nandrolone,
an anabolic steroid

23.11   *Nandrolone* is an anabolic, or tissue-building, steroid sometimes taken by athletes seeking to build muscle mass. Among its side effects is a high level of androgenic activity. Compare the structures of nandrolone and testosterone and point out their structural similarities.

23.11   PROSTAGLANDINS AND LEUKOTRIENES

Few groups of compounds have caused as much excitement among medical researchers in the past 15 years as the prostaglandins and related compounds. Although the name *prostaglandin* reflects the original isolation of these compounds from the male prostate gland, they have subsequently been found in small amounts in all body tissues and fluids.

Prostaglandin A lipid derived from a C_{20} carboxylic acid and containing a cyclopentane ring with two long side chains.

Chemically, *prostaglandins* are C_{20} carboxylic acids that contain a five-membered ring with two long side chains. They differ only in the number of oxygen atoms they contain and in the number of double bonds present. Prostaglandin E_1 (abbreviated PGE_1) and prostaglandin $F_{2\alpha}$ are examples:

Prostaglandin E_1 (PGE_1)

Prostaglandin $F_{2\alpha}$ ($PGF_2\alpha$)

AN APPLICATION: RESEARCH REPORT ON A LEUKOTRIENE INHIBITOR

An article appeared recently about a new drug that apparently inhibits leukotriene synthesis. The story provides a good illustration of how drug development proceeds.

In a clinical trial, a new compound prepared by chemists at a pharmaceutical company had given "significant" relief to 48 victims of ulcerative colitis, an inflammatory disease that causes pain and bloody diarrhea and can eventually require removal of the colon. There is at present no cure for ulcerative colitis, nor any understanding of its cause. The new drug is also undergoing tests for treating asthma, hay fever, psoriasis (a skin disease), and rheumatoid arthritis.

The researchers began their studies by focusing on known information about arachidonic acid and the compounds synthesized from it. It was known that aspirin reacts as illustrated here with an enzyme that initiates the series of reactions leading from arachidonic acid to the prostaglandins. Transfer of the acetyl group from aspirin inactivates the enzyme.

same manner that aspirin prevents prostaglandin synthesis. They appear to have been successful—their experimental drug is reported to bind solely to the enzyme for the first step in leukotriene synthesis.

The clinical results, though preliminary, are exciting because they provide the first link between leukotrienes and inflammatory bowel disease. If all goes well, in a few years we may have a drug that's effective against a most unpleasant disease that formerly had no cure. The drug might even be more broadly useful and alleviate the symptoms of many diseases, as does aspirin. On the other hand, this particular drug may not pass the rigorous testing needed to bring it into general clinical use. Should that happen, you can be sure that structure modifications designed to lead to a safe drug with the same leukotriene-locking action will be under way.

Aspirin + HOCH₂—(enzyme) ⟶

Aspirin   Active enzyme

Salicylic acid + CH₃—C(=O)—O—CH₂—(enzyme)

Salicylic acid   Inactive enzyme

A similar series of reactions, catalyzed by a different set of enzymes, leads from arachidonic acid to leukotrienes, which are potent initiators of allergic and inflammatory responses. The researchers set their sights on finding a compound that would shut down leukotriene synthesis in the

Everyone working with science and technology must keep in touch with new developments in their specialty.

The several dozen known prostaglandins have an extraordinary range of biological effects. They can lower blood pressure, influence platelet aggregation during blood clotting, stimulate uterine contractions, and lower the extent of gastric secretions. In addition, they are responsible for some of the pain and swelling that accompanies inflammation.

In addition to the prostaglandins themselves, closely related compounds have still other important physiological actions. Interest has centered particularly on the **leukotrienes,** which differ from prostaglandins by the absence of the ring. Leukotriene release in the body has been found to trigger the asthmatic response, severe allergic reactions, and inflammation. Like all hormones, prostaglandins and leukotrienes control the body's responses to various stimuli. Unlike most other hormones, however, these compounds are produced in the cells where they act.

Prostaglandins and leukotrienes are synthesized in the body from the 20-carbon unsaturated fatty acid arachidonic acid. Arachidonic acid, in turn, is synthesized in the body from linolenic acid, helping to explain why linolenic is one of the two essential fatty acids. To illustrate the relationships among arachidonic acid, the prostaglandins, and the leukotrienes, we've shown arachidonic acid and a leukotriene so that they're ''bent'' into similar shapes.

Leukotriene A simple lipid derived from a C_{20} carboxylic acid like a prostaglandin but with no cyclopentane ring.

Arachidonic acid

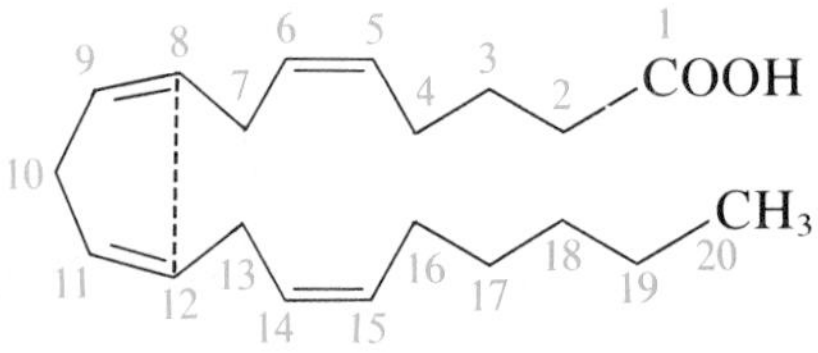

Arachidonic acid (bent)

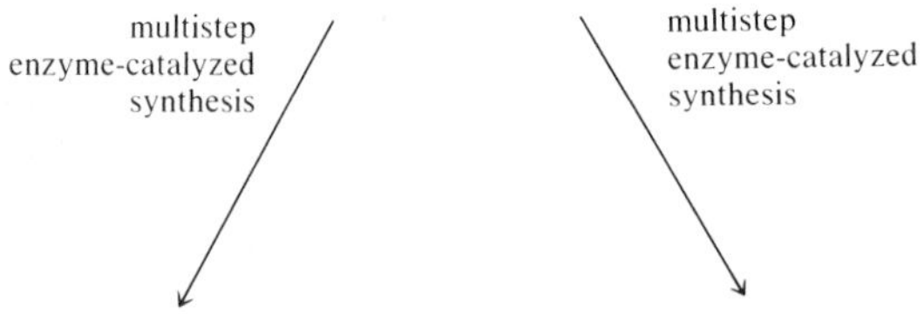

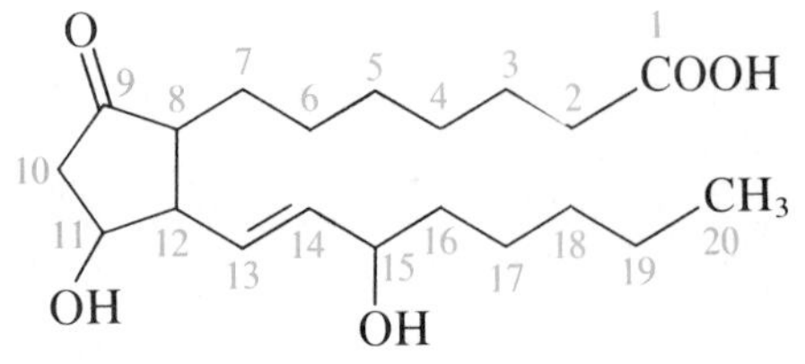

multistep enzyme-catalyzed synthesis

multistep enzyme-catalyzed synthesis

PGE_1, a prostaglandin

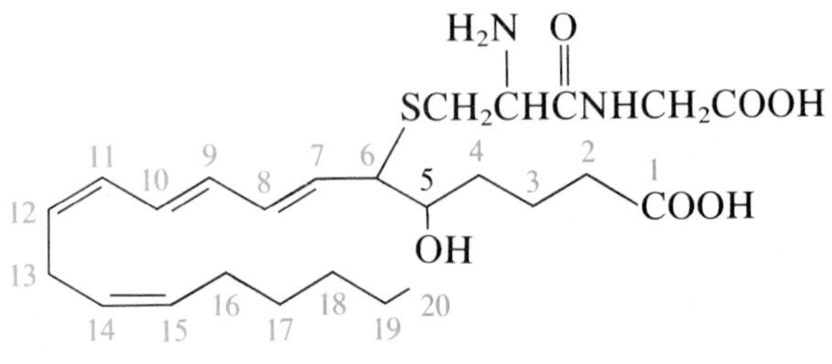

Leukotriene D_4

INTERLUDE: CHEMICAL COMMUNICATION

To be honest about it, a lot of scientific work is done for the fun of it, without great concern for immediate practical application. Practical applications often *do* follow, but the applications are not themselves the initial goal. Take the study of insects, for example. Have you ever wondered how ants are able to follow a precise path from their nests to a food source, or how the male and female of an insect species find each other for mating? Such questions are interesting to the curious, but it's not clear what value the answers might have.

Insects and many other organisms deal with each other and with their surroundings primarily by sending and receiving chemical messages using substances called *semiochemicals.* Among the most important semiochemicals are *pheromones,* chemicals that are released by one member of a species to evoke a specific response in another member of the same species. Ants and termites, for example, lay down trail pheromones that mark a path from nest to food. Similarly, butterflies, moths, and other flying insects release sex pheromones to indicate their location to other interested parties.

A good semiochemical must be volatile, so that it disappears after its job has been done, and must be structurally unique, so that it is species-specific. After all, release of a sex pheromone wouldn't be of much use to a butterfly if it attracted moths, bees, houseflies, and every other flying insect within range. Low-molecular-weight lipids are ideally suited for use as insect semiochemicals, as shown by the examples given at the bottom of this box.

These mating gypsy moths probably didn't meet just by chance. Volatile lipids used as pheromones enable insects to attract members of the opposite sex, often from great distances.

What about practical applications? An understanding of the chemical messages used by insects offers hope for developing environmentally safe, species-specific means of insect control. For example, gypsy moth infestations in hardwood forests of the northeast United States are now being fought with *pheromone traps* rather than with broad-range insecticides. A tiny amount of the gypsy moth sex pheromone is used to lure the moths into a trap without harming insects of any other species.

Finally, what about *human* sex pheromones? That's another story, but one that is actively being worked on.

$$CH_3C-OCH_2CH_2CHCH_3$$

Honeybee alarm pheromone

A termite trail pheromone

Gypsy moth sex pheromone

SUMMARY

Lipids are the naturally occurring organic molecules that dissolve when a plant or animal sample is extracted with a nonpolar solvent. Because they are defined by a physical property rather than by structure, there are a great many different kinds of lipids.

Waxes are esters between long-chain alcohols and long-chain carboxylic acids, whereas fats and oils are **triacylglycerols**—triesters of glycerol with three long-chain **fatty acids.** The fatty acids are unbranched; most have an even number of carbon atoms; and they may be either **saturated** or **unsaturated.** Vegetable oils contain a higher proportion of unsaturated fatty acids than animal fats and have consequently lower melting points.

Fats and oils can be **saponified** by treatment with aqueous NaOH or KOH to yield **soap,** a mixture of long-chain fatty acid carboxylate salts. Soaps act as cleansers because their two ends are so different. The negatively charged carboxylate end is ionic and **hydrophilic,** while the hydrocarbon chain end is nonpolar and **hydrophobic.** Thus, a soap molecule is attracted to both grease and water.

Phospholipids are lipids that contain a phosphate group. **Glycerophospholipids** are derived from glycerol by esterification with two fatty acids and one phosphoric acid. The phosphate group, in turn, is also bonded through an ester link to another molecule such as choline, ethanolamine, or serine. **Sphingolipids,** the second main class of phospholipids, are based on the alcohol sphingosine rather than on glycerol. The sphingolipids known as **sphingomyelins** and **glycolipids** are important components of cell membranes.

Phospholipids oriented into a **lipid bilayer** form a major part of the membranes that surround living cells. The ionic phosphate head groups orient on the outside of the bilayer toward the aqueous environment, while the nonpolar hydrocarbon chains cluster in the middle of the bilayer to avoid water.

Steroids are lipids based on a structure with four rings joined together, whereas **prostaglandins** are C_{20} carboxylic acids that have a single five-membered ring connected to two long side chains. **Leukotrienes** are like prostaglandins but have no ring.

REVIEW PROBLEMS

Waxes, Fats and Oils

23.12 What is a lipid?

23.13 Why are there so many different kinds of lipids?

23.14 What is a fatty acid?

23.15 What does it mean to say that fats and oils are triacylglycerols?

23.16 Draw the structure of glycerol trilaurate made from glycerol and three lauric acid molecules.

23.17 How does animal fat differ chemically from vegetable oil?

23.18 What function does a fat serve in an animal?

23.19 What function does a wax serve in a plant or animal?

23.20 Spermaceti, a fragrant substance isolated from sperm whales, was commonly used in cosmetics until it was banned in 1976 to protect the whales from extinction. Chemically, spermaceti is cetyl palmitate, the ester of palmitic acid with cetyl alcohol (the straight-chain C_{16} alcohol). Show the structure of spermaceti.

23.21 What kind of lipid is spermaceti (Problem 23.20)—a fat, a wax, or a steroid?

23.22 There are two isomeric fat molecules whose components are glycerol, one palmitic acid, and two stearic acid units. Draw the structures of both and explain how they differ.

23.23 One of the two molecules in Problem 23.22 is chiral (Section 21.3). Which molecule is chiral and why?

23.24 Write the structures of these molecules:
(a) sodium palmitate (b) decyl oleate
(c) glyceryl palmitodioleate

Chemical Reactions of Lipids

23.25 How would you convert a vegetable oil like soybean oil into a solid cooking fat?

23.26 Draw the structures of all products you would obtain by saponification of the following lipid with aqueous KOH. What are the names of the products?

$$CH_2-O-\overset{\displaystyle O}{\overset{\displaystyle \|}{C}}(CH_2)_{16}CH_3$$

$$CH-O-\overset{\displaystyle O}{\overset{\displaystyle \|}{C}}(CH_2)_7CH{=}CH(CH_2)_7CH_3$$

$$CH_2-O-\overset{\displaystyle O}{\overset{\displaystyle \|}{C}}(CH_2)_7CH{=}CHCH_2CH{=}CHCH_2CH{=}CHCH_2CH_3$$

23.27 Draw the structure of the product you would obtain on hydrogenation of the lipid in Problem 23.26. What is its name?

23.28 Would the product in Problem 23.27 have a higher or a lower melting point than the original lipid? Why?

23.29 What products would you obtain by treatment of oleic acid with these reagents?
(a) Br$_2$ (b) H$_2$, Pd catalyst
(c) CH$_3$OH, HCl catalyst

Phospholipids, Glycolipids, and Cell Membranes

23.30 What is the difference between a fat and a phospholipid?

23.31 How do sphingomyelins and cerebrosides differ structurally?

23.32 Why is it that phosphoglycerides rather than triacylglycerols are found in cell membranes?

23.33 Why are phosphoglycerides more soluble in water than triacylglycerols?

23.34 How does a soap micelle differ from a membrane bilayer?

23.35 What constituents besides phospholipids are present in a cell membrane?

23.36 What are the names of the two different kinds of sphingosine-based lipids?

23.37 Show the structure of a cerebroside made up of D-galactose, sphingosine, and myristic acid.

23.38 Draw the structure of a phosphoglyceride that contains palmitic acid, oleic acid, and the phosphate bonded to propanolamine.

23.39 Draw the structure of a sphingomyelin that contains a stearic acid unit.

23.40 *Cardiolipin,* a compound found in heart muscle, has the following structure. What products would be formed if all ester bonds in the molecule were saponified by treatment with aqueous NaOH?

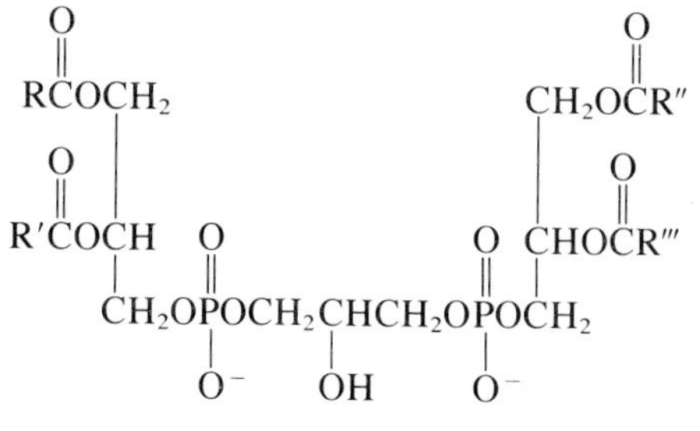

Cardiolipin

Steroids and Prostaglandins

23.41 What functional groups are present in the two sex hormones estradiol and testosterone?

23.42 How do the sex hormones estradiol and testosterone differ structurally?

23.43 The female sex hormone estrone has four chiral centers. Identify them.

23.44 What function does cholesterol serve in cell membranes?

23.45 Draw the products you would expect from treatment of cholesterol with these reagents:
(a) Br$_2$ (b) H$_2$, Pd catalyst
(c) [O], an oxidizing agent

23.46 Diethylstilbestrol (DES) exhibits estradiol-like hormonal activity even though it is not itself a steroid. Once used widely as an animal food additive, DES has been implicated as a causative agent in several types of cancer. Show how DES can be drawn so that it is structurally similar to estradiol.

Diethylstilbestrol

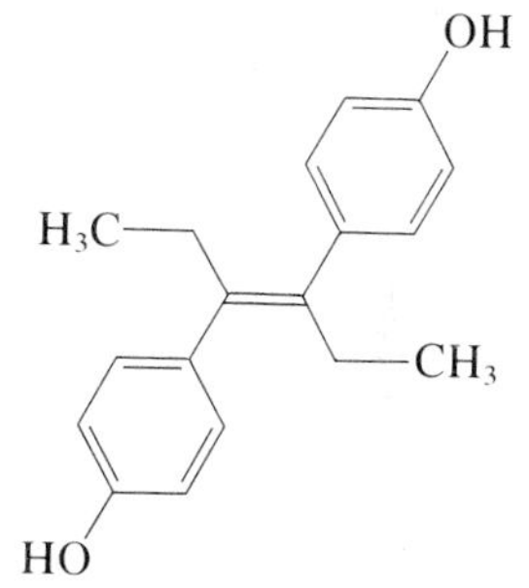

23.47 Thromboxane A$_2$ is a lipid involved in the blood-clotting process. To what category of lipids does thromboxane A$_2$ belong?

Thromboxane A$_2$

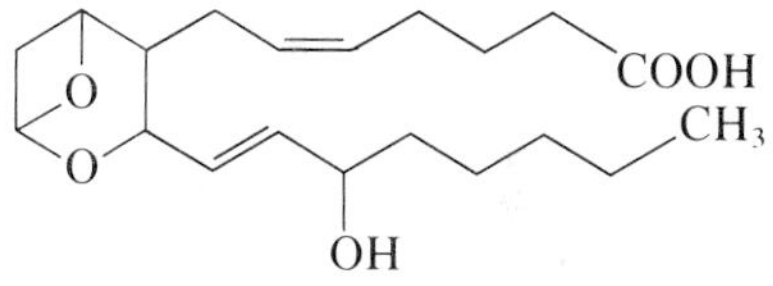

23.48 What fatty acid do you think serves as a biological precursor of thromboxane A$_2$? (Problem 23.47.)

Applications

23.49 What is the advantage of a detergent relative to a soap as a cleaning agent? [App: Detergents]

23.50 Describe the mechanism by which soaps and detergents provide cleaning action. [App: Detergents]

23.51 Cationic detergents are rarely used for cleaning. For what purposes are they used? [App: Detergents]

23.52 Why are branched-chain hydrocarbons no longer used for detergents? [App: Detergents]

23.53 Why must cells be semipermeable? [App: Transport Across Cell Membranes]

23.54 Which process requires energy—passive or active transport? Why is energy sometimes required to move solute across the cell membrane? [App: Transport Across Cell Membranes]

23.55 Why is it desirable to inhibit the production of leukotrienes? [App: Leukotriene Inhibitor]

23.56 How does aspirin function to inhibit the formation of prostaglandins from arachidonic acid? [App: Leukotriene Inhibitor]

23.57 What is a semiochemical? [Int: Chemical Communication]

23.58 What are the advantages of using pheromone traps for insect control compared to using insecticides? [Int: Chemical Communication]

Additional Questions and Problems

23.59 Which of the following are saponifiable lipids?
(a) prostaglandin E_1 (b) a lecithin
(c) progesterone   (d) a sphingomyelin
(e) a cerebroside   (f) glyceryl trioleate

23.60 Identify the component parts of each saponifiable lipid listed in Problem 23.59.

23.61 Draw the structure of the fat made from two molecules of linolenic acid and one molecule of myristic acid.

23.62 Would the fat described in Problem 23.61 have a higher or lower melting point than the fat made from one molecule each of linolenic, myristic, and stearic acids?

23.63 Why is cholesterol not saponifiable?

23.64 Jojoba wax, used in candles and cosmetics, is partially composed of the ester of stearic acid and a straight-chain 22-carbon alcohol. Draw the structure of this wax component.

23.65 What three types of lipids are particularly abundant in brain tissue?

23.66 What is the general purpose of a hormone?

23.67 What are some of the functions that prostaglandins serve in the body?

23.68 Lecithins are often used as food additives to provide emulsification. How do they accomplish this purpose?

23.69 If the average molecular weight of a sample of soybean oil is 1500, how many grams of NaOH are needed to saponify 5.0 g of the oil?

23.70 The concentration of cholesterol in the blood serum of a normal adult is approximately 200 mg/dL. How much total blood cholesterol does a person with a blood volume of 5.75 L have?

C H A P T E R

24

Lipid Metabolism

Cholesterol! It's a relatively simple molecule (carbons in green, hydrogens in blue, and oxygen in red), but understanding how it affects body chemistry is not so simple.

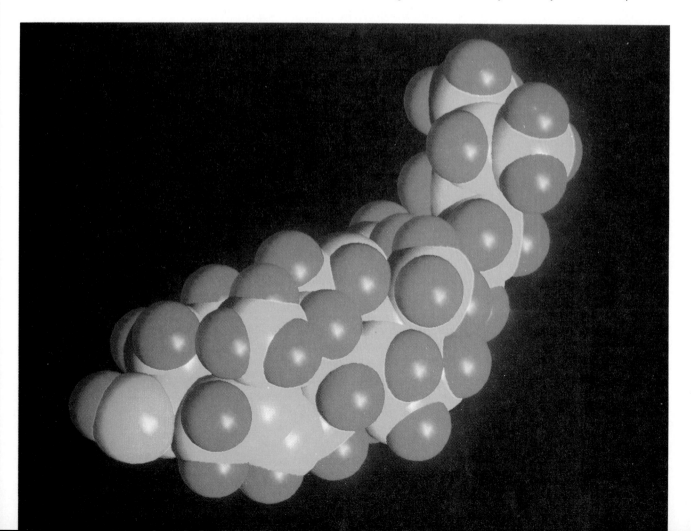

In the kitchen, we make a distinction between fats and oils based on melting point. In biochemistry, however, this distinction is unimportant because both are *triacylglycerols*—triesters of glycerol with a variety of fatty acids. Triacylglycerols make up 98–99% of the lipids in our diet, and about 90% of the lipids in our bodies.

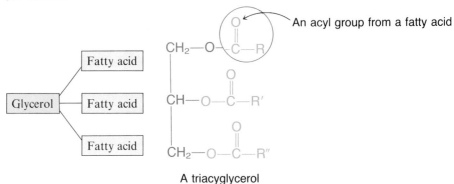

A triacyglycerol

Having introduced the chemistry of triacylglycerols in Chapter 23, we'll concentrate in this chapter on their digestion and metabolism. The numerous other types of lipids mentioned in Chapter 23 follow similar metabolic pathways. Some questions we'll answer include

1. ***What happens during the digestion of triacylglycerols?*** The goal: Be able to list the sequence of events in the digestion of triacylglycerols and their transport into the bloodstream.
2. ***What are the major pathways in the metabolism of triacylglycerols?*** The goal: Be able to name the major pathways for the synthesis and breakdown of triacylglycerols and fatty acids and be able to show their interrelationships.
3. ***How are triacylglycerols moved in and out of storage in adipose tissue?*** The goal: Be able to explain the regulation and reactions of the storage and mobilization of triacylglycerols.
4. ***How are fatty acids oxidized?*** The goal: Be able to show what happens to a fatty acid from its entry into a cell to its conversion to acetyl SCoA.
5. ***What is the energy yield from fatty acid catabolism?*** The goal: Be able to calculate the number of molecules of ATP produced from a given fatty acid.
6. ***What is the function of ketogenesis?*** The goal: Be able to identify ketone bodies, describe their synthesis, and explain their role in metabolism.
7. ***How are fatty acids synthesized?*** The goal: Be able to compare the pathways for fatty acid synthesis and oxidation, and be able to describe the reactions of the synthesis pathway.

24.1 DIGESTION OF TRIACYLGLYCEROLS

Chyme The semisolid mixture produced by digestion in the stomach.

Bile Fluid secreted by the liver and released into the small intestine during digestion; contains bile salts, bicarbonate ion, and other electrolytes.

Bile acids Steroid acids derived from cholesterol that are secreted in bile.

When eaten, triacylglycerols (TAGs) pass through the mouth unchanged and enter the stomach, where their physical breakdown takes place (Figure 24.1). The heat and churning action of the stomach break lipids into smaller droplets, a process that takes longer than the physical breakdown and digestion of other foods. To be sure that there's time for this breakdown, the presence of lipids in consumed food slows down the rate at which the mixture of partially digested foods known as **chyme** leaves the stomach. One of the reasons lipids are a pleasing part of the diet is that the stomach feels "full" for a longer time after a fatty meal.

When chyme leaves the stomach, it enters the upper end of the small intestine, where its arrival triggers the release of *pancreatic lipases*, enzymes for the hydrolysis of lipids (Figure 24.2a). (The pancreas also produces enzymes for carbohydrate and protein hydrolysis, and bicarbonate ion for neutralization of stomach acid in the chyme.) The gallbladder simultaneously releases **bile,** which is manufactured in the liver and stored in the gallbladder until needed (Figure 24.2b). Enzymes can't attack the lipids inside water-insoluble droplets, and it's the job of bile to emulsify them. Steroids called **bile acids,** or *bile salts,* solubilize lipid droplets by forming micelles much like the soap micelle shown previously in Figure 23.5. You can see from the structure of cholic acid, the major bile acid, that it resembles soaps and detergents in containing both hydrophilic and hydrophobic regions. When freed of the TAGs they transport during digestion, bile acids are mostly recycled. The small amount that escapes recycling is excreted.

Figure 24.1
Digestion of triacylglycerols.

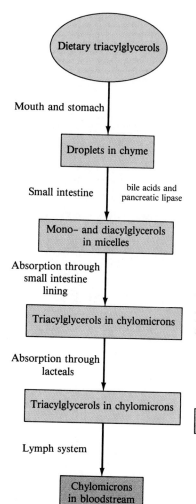

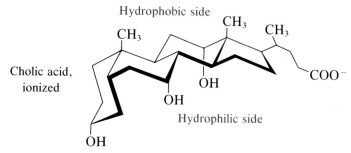

Once micelles have formed, the digestion of triacylglycerols proceeds. Within the small intestine, pancreatic lipase acts at the micelle surface to partially hydrolyze the triacylglycerols, producing mainly mono- and di-acylglycerols, plus fatty acids and a small amount of glycerol.

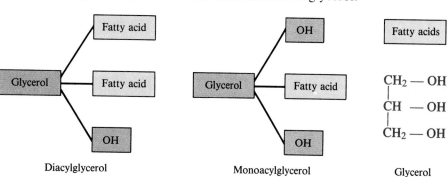

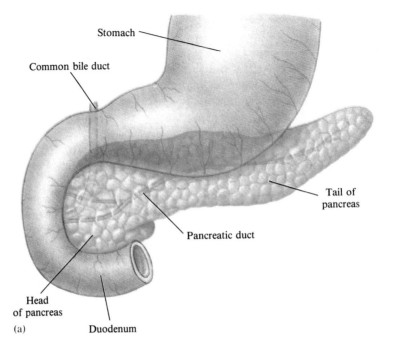

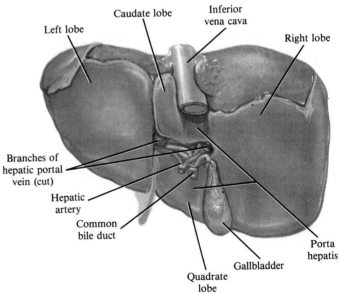

Figure 24.2
Anatomy of the pancreas and liver. (a) The pancreas secretes enzymes for the digestion of lipids, carbohydrates, and proteins, plus bicarbonate ion. Note the bile duct through which bile from the gallbladder enters the small intestine at its upper end (the duodenum). (The pancreas also secretes the hormones insulin and glucagon, from a different group of cells.) (b) Blood carrying metabolites from the digestive system enters the liver through the hepatic portal vein. The gallbladder is the site for storage of bile.

The smaller fatty acids and glycerol are water-soluble and are absorbed directly through the surface of the villi that line the small intestine. Within the villi, they enter the bloodstream through capillaries (Figure 24.3a) and are carried to the liver via the hepatic portal vein. The larger acylglycerols and fatty acids are resolubilized in the intestine by the bile salts. The resulting micelles carry these water-insoluble molecules to the intestinal lining. There, the acylglycerols and fatty acids are released from the micelles and absorbed.

Next, within the intestinal lining, the still-insoluble fatty acids and acylglycerols are repackaged by a roundabout route to enter the bloodstream (Figure 24.3b). First, the fatty acids and acylglycerols are reconverted to triacylglycerols. Then, the triacylglycerols are bound into **chylomicrons,** one of several kinds of **lipoproteins** that aid in lipid transport. The general structure of a lipoprotein is a globule of lipid surrounded by a phospholipid-based membrane with the hydrophilic regions toward the outside (Figure 24.4).

Chylomicron   Spherical, low-density lipoprotein particles that transport triacylglycerols

Lipoprotein   A lipid–protein complex that transports lipids.

Figure 24.3
Site of lipid absorption in the small intestine. Each villus (a) is lined with huge numbers of microvilli, which provide the large surface area needed for absorption. After absorption (b), smaller molecules go directly into the bloodstream via capillaries. Larger lipid molecules are repackaged into chylomicrons for entry into lacteals and delivery to the bloodstream via the lymphatic system.

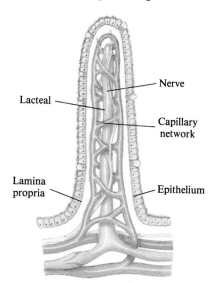

Nerve

Lacteal

Capillary network

Lamina propria

Epithelium

(a) A single villus

Acylglycerols

Free fatty acids

Triacylglycerols

Partly hydrolyzed phospholipids

Phospholipids

Chylomicrons

Lacteal (to bloodstream via lymphatic system)

Cholesterol

Smaller free fatty acids and acylglycerols, glycerol

Capillary (to bloodstream via hepatic portal vein)

(b) Cell at surface of villus

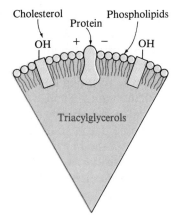

Cholesterol   Phospholipids
     Protein

Triacylglycerols

Figure 24.4
The structure of a TAG-carrying lipoprotein. The lipoproteins that transport cholesterol and other lipids have similar structures.

Because chylomicrons are too large to pass through capillary walls, they instead enter the lymphatic system by absorption into lacteals (Figure 24.3a). The chylomicrons are then carried to the thoracic duct (just below the collarbone), where the lymphatic system empties into the bloodstream. At this point, the lipids from foods are ready to be put to use.

24.2 TRIACYLGLYCEROL METABOLISM: AN OVERVIEW

Triacylglycerols and fatty acids enter metabolism from three different sources: the diet, storage in adipose tissue, and synthesis in the liver. Whatever their origin, these lipids must be made soluble by association with water-soluble proteins. The various ways this is accomplished are summarized in Figure 24.5.

The chylomicrons that carry dietary TAGs through the lymphatic system into the blood are the lowest density lipoproteins because they carry the highest ratio of lipids to proteins. The slightly more dense ones, the *VLDLs* (very-low-density lipoproteins), carry TAGs from the liver where they have been synthesized to tissues where they will be utilized or stored. When TAGs stored in adipose tissue are needed for energy, they are first hydrolyzed and then the resulting fatty acids are carried in the bloodstream by proteins known as *albumins*.

As summarized in Figure 24.6, the first step in utilizing TAGs is complete hydrolysis to glycerol and fatty acids:

$$
\begin{array}{ccc}
\underset{\text{Triacylglycerol (TAG)}}{
\begin{array}{c}
\mathrm{CH_2-O-\overset{\overset{\displaystyle O}{\|}}{C}-R} \\[6pt]
\mathrm{CH-O-\overset{\overset{\displaystyle O}{\|}}{C}-R'} \\[6pt]
\mathrm{CH_2-O-\overset{\overset{\displaystyle O}{\|}}{C}-R''}
\end{array}}
&
\xrightarrow[\mathrm{H_2O}]{\text{lipase}}
&
\underset{\text{Glycerol}}{
\begin{array}{c}
\mathrm{CH_2-OH} \\[6pt]
\mathrm{CH-OH} \\[6pt]
\mathrm{CH_2-OH}
\end{array}}
\;+\;
\underset{\text{Fatty acids}}{
\begin{array}{c}
\mathrm{R-\overset{\overset{\displaystyle O}{\|}}{C}-O^-} \\[6pt]
\mathrm{R'-\overset{\overset{\displaystyle O}{\|}}{C}-O^-} \\[6pt]
\mathrm{R''-\overset{\overset{\displaystyle O}{\|}}{C}-O^-}
\end{array}}
\end{array}
$$

Figure 24.5
Transport of TAGs and fatty acids. The carriers are shown in blue.

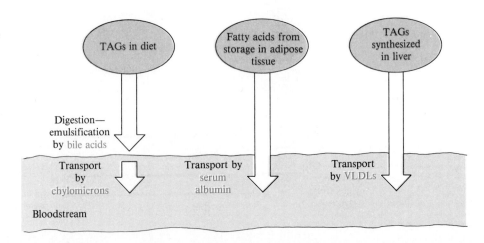

Mobilization (of triacylglycerols)
Hydrolysis of triacylglycerols in adipose tissue and release of fatty acids into bloodstream.

TAGs that are released from storage, or **mobilized,** are hydrolyzed in fat cells known as *adipocytes* by a lipase enzyme that is activated by epinephrine, glucagon, and several other hormones. TAGs from digested foods are hydrolyzed when the lipoproteins carrying them in the bloodstream encounter the *lipoprotein lipase,* which is anchored in capillary walls in adipose tissue and elsewhere. The resulting free fatty acids then travel through the bloodstream in association with albumins and enter cells for further metabolism.

The glycerol from TAG hydrolysis is carried in the bloodstream to the liver or kidneys where it is converted in the following series of reactions to dihydroxyacetone phosphate (DHAP). Isomerization of DHAP gives glyceraldehyde 3-phosphate, which can enter either the glycolysis pathway or the glyconeogenesis pathway. Thus, dihydroxyacetone phosphate represents one of several points at which carbohydrate and lipid metabolism are connected.

$$\underset{\text{Glycerol}}{\begin{array}{c}CH_2OH\\|\\HO-C-H\\|\\CH_2OH\end{array}} \quad \xrightarrow[\text{ATP ADP}]{} \quad \underset{\substack{\text{Glycerol}\\\text{3-phosphate}}}{\begin{array}{c}CH_2OH\\|\\HO-C-H\\|\\CH_2-O-PO_3{}^{2-}\end{array}} \quad \xrightarrow[\text{NAD}^+\text{ NADH/H}^+]{} \quad \underset{\substack{\text{Dihydroxyacetone}\\\text{phosphate (DHAP)}}}{\begin{array}{c}CH_2OH\\|\\C=O\\|\\CH_2-O-PO_3{}^{2-}\end{array}}$$

Figure 24.6
Metabolism of triacylglycerols. Pathways that break down molecules (catabolism) are shown in yellow, and synthetic pathways (anabolism) are shown in blue. Connections to other pathways or intermediates of metabolism are shown in green.

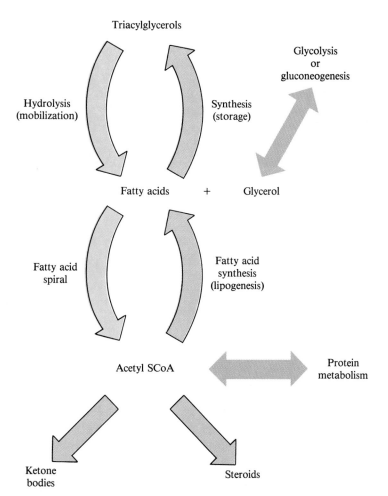

AN APPLICATION: CHOLESTEROL, LIPOPROTEINS, AND HEART DISEASE

A debate about the relationship between cholesterol and heart disease has been going on for over three decades. Newspaper and television ads trumpet the idea that cholesterol is a killer and that we'd all live much healthier lives if only we would switch our brand of cooking oil.

What are the facts? Several points seem clear. The first is that a diet rich in saturated animal fats generally leads to an increase in blood-serum cholesterol. The second is that a diet lower in saturated fat and higher in unsaturated fat leads to a lowering of the serum cholesterol level. Since the body manufactures much of its own cholesterol, however, the serum level can't be controlled entirely by diet. Remember also that since cholesterol is an essential component of cell membranes and the precursor of all other steroids in the body, a certain amount of serum cholesterol is necessary for life.

The third point is that a level of serum cholesterol greater than 200 mg/dL (a normal value is 150–200 mg/dL) has been correlated with *atherosclerosis*, a condition in which yellowish deposits composed of cholesterol and other lipid-containing materials form within the larger arteries, notably those of the brain and heart. Numerous studies have also correlated high cholesterol levels with the risk of heart attack brought on by blockage of blood flow to heart muscles.

It's recently been found that a better indication of a person's risk of heart disease comes from a measurement of blood lipoprotein levels. As shown in Figure 24.4, lipoproteins are complex assemblages of lipids and proteins that serve to transport water-soluble lipids throughout the body. They can be somewhat arbitrarily divided into four major types distinguishable by their density, as shown in the accompanying table. Since lipids are generally less dense than proteins, the chylomicrons and the very-low-density lipoproteins (VLDLs) are richer in lipid and lower in protein than the low-density lipoproteins (LDLs) and high-density lipoproteins (HDLs).

The four lipoprotein fractions have different roles. Chylomicrons and VLDL act primarily as carriers of triacylglycerols as described in Sections 24.1 and 24.2. LDL and HDL act as carriers of cholesterol to and from the liver. Present evidence suggests that LDL transports cholesterol as its linolenate ester *to* peripheral tissues where it is deesterified and can accumulate in the deposits of atherosclerosis. HDL transports choles-

The fatty acids from TAG hydrolysis have two possible fates. When energy is in good supply, fatty acids are converted back to TAGs for storage. When cells are in need of energy, carbon atoms are removed two at a time from fatty acids via the *fatty acid spiral*. The result is complete conversion of fatty acids with even numbers of carbon atoms to acetyl SCoA, and subsequent ATP production via the citric acid cycle and oxidative phosphorylation. In resting muscle, fatty acids are the major energy source.

Acetyl SCoA, which is at a crossroads in metabolism (Figure 24.6), serves as the starting material for the biosynthesis of fatty acids (*lipogenesis*) in the liver. Another pathway available to acetyl SCoA is *ketogenesis*, the production from acetyl SCoA of the three compounds known as *ketone bodies*. Normally acetyl SCoA takes this route to only a small extent. Under the stress of glucose shortage, however, ketogenesis becomes increasingly important because ketone bodies can be used as fuel instead of glucose. Acetyl SCoA is also the starting material for the synthesis of cholesterol, from which all other steroids are made.

terol as its stearate ester *from* dead or dying cells back to the liver where it is converted to bile acids and excreted.

If LDL delivers more cholesterol than is needed, and if not enough HDL is present to remove it, the excess is deposited in cells and arteries. Thus, the higher the HDL level, the less the likelihood of deposits and the lower the risk of heart disease. A rule of thumb is that a person's risk drops about 24% for each increase of 5 mg/dL in HDL concentration. Normal HDL values are about 45 mg/dL for men and 55 mg/dL for women, perhaps helping to explain why women are generally less susceptible than men to heart disease.

If high serum levels of HDL are good, then how do we get them? The answer is simple enough: As common sense tells you, the most important factor is a generally healthy lifestyle. Obesity, smoking, lack of exercise, and heavy drinking appear to lead to low HDL levels, while regular exercise, a prudent diet, and moderate alcohol consumption lead to high HDL levels. Runners and other endurance athletes, in particular, have HDL levels nearly 50% higher than those of the general population.

Serum Lipoproteins

Name	Density (g/mL)	% Lipid	% Protein
Chylomicrons	<0.94	98	2
VLDL (very-low-density)	0.940–1.006	90	10
LDL (low-density)	1.006–1.063	75	25
HDL (high-density)	1.063–1.210	60	40

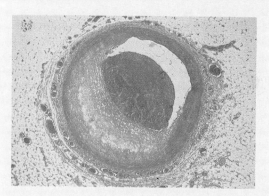

This coronary artery is severely narrowed by atherosclerosis, the buildup of cholesterol deposits in vessel walls (pink). The resulting roughening of artery walls also promotes the formation of clots (dark red) that can cut off blood flow completely, causing a heart attack.

Practice Problem **24.1** Look back at Figure 22.5 and tell where dihydroxyacetone phosphate can enter the glycolysis or gluconeogenesis pathway.

24.3 STORAGE AND MOBILIZATION OF TRIACYLGLYCEROLS

We've noted that adipose tissue is the storage depot for TAGs and that TAGs are our primary energy storage form. TAGs don't just sit unused until needed, however. The passage of fatty acids in and out of storage in adipose tissue is a continuous process essential to maintaining homeostasis.

To see how storage and mobilization are regulated, look back at Figure 22.9 which shows the effects of the hormones insulin and glucagon on metabolism. After a meal, blood glucose levels are high, insulin levels rise, and glucagon levels drop. Glucose is entering cells, and glycolysis is proceeding actively. In addition, insulin stimulates the synthesis of TAGs for storage.

The reactants in TAG synthesis are glycerol 3-phosphate and fatty acyl derivatives of coenzyme A. The glycerol 3-phosphate is made from dihydroxyacetone phosphate (DHAP), which you've seen is an intermediate in glycolysis and also a product of glycerol metabolism in the liver. Since adipocytes don't have the enzyme needed to convert glycerol to glycerol 3-phosphate, they can't make triacylglycerols unless some glucose is undergoing glycolysis and producing glycerol 3-phosphate.

$$\text{CH}_2\text{OH} \quad \text{C}=\text{O} \quad \text{CH}_2\text{OPO}_3^{2-} \quad \xrightarrow{\text{NADH/H}^+ \;\; \text{NAD}^+} \quad \text{CH}_2\text{OH} \quad \text{CHOH} \quad \text{CH}_2\text{OPO}_3^{2-}$$

DHAP   Glycerol 3-phosphate

Fatty acyl SCoAs consist of a fatty acid acyl group (like an acetyl group, but with a long carbon chain) bonded to coenzyme A. The fatty acids may have come from digestion or from biosynthesis. TAG synthesis begins with the conversion of glycerol 3-phosphate to phosphatidic acid by addition of two fatty acid groups:

Glycerol 3-phosphate → Phosphatidate

Next, the phosphate group is removed and the third fatty acid group is added to give a triacylglycerol.

Phosphatidate → A triacylglycerol

After a meal has been digested, blood glucose levels are low so that insulin levels drop and glucagon levels rise, activating the enzymes in adipose tissue responsible for mobilizing TAGs. Also, with glycerol 3-phosphate in short supply, fewer fatty acids can be recycled into storage as TAGs. Now, fatty acids and glycerol are produced in adipose tissue and released into the bloodstream.

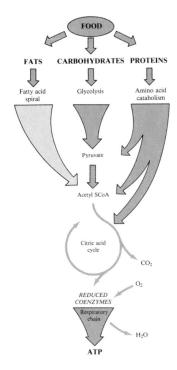

24.4 ACTIVATION, MEMBRANE TRANSPORT, AND OXIDATION OF FATTY ACIDS

Once a fatty acid crosses the cell membrane into the cytosol of a cell that needs energy (Figure 24.7), three successive processes must occur.

1. The fatty acid must be *activated* by conversion to fatty acyl SCoA.
2. The fatty acyl SCoA must be *transported* into the mitochondrial matrix.
3. The fatty acyl SCoA must be *oxidized* by enzymes in the matrix to produce acetyl SCoA plus reduced coenzymes (FADH$_2$ and NADH).

Fatty Acid Activation   Activation of a fatty acid serves the same purpose as the first few steps in oxidation of glucose by glycolysis. That is, some energy must be initially invested in conversion of the fatty acid to a form that can be broken down more easily.

The reaction of a fatty acid with coenzyme A to form a fatty acyl SCoA is endergonic. In order to take place, the reaction is coupled with the exergonic hydrolysis of ATP to yield pyrophosphate ion ($P_2O_7^{4-}$) and adenosine monophosphate (AMP), and with the even more exergonic subsequent hydrolysis of the pyrophosphate, to yield hydrogen phosphate ion (HPO_4^{2-}). The overall reaction breaks one high-energy phosphate bond in ATP and one in pyrophosphate, which is equivalent to "spending" two ATPs.

$$\underset{\displaystyle R-\overset{\displaystyle O}{\overset{\displaystyle \|}{C}}-O^-}{} \ + \ HSCoA \ + \ ATP \ + \ H_2O \ \longrightarrow$$

$$R-\overset{\displaystyle O}{\overset{\displaystyle \|}{C}}-SCoA \ + \ AMP \ + \ 2\ HPO_4^{2-} \ + \ H^+$$

Figure 24.7
A glandular cell. The nucleus and its contents appear in yellow, and the mitochondria in dark red. The mitochondria are the site of energy production: Fatty acid oxidation, amino acid breakdown, the citric acid cycle, and oxidative phosphorylation occur there. The cytosol is the site of fatty acid biosynthesis, glycolysis, the pentose phosphate pathway, and gluconeogenesis.

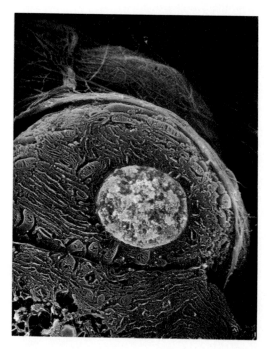

Transport Into Mitochondrion The fatty acyl SCoA molecule is too large to cross the inner mitochondrial membrane and must be transferred instead by a shuttle mechanism that uses carnitine as a temporary carrier molecule. The fatty acyl SCoA reacts with carnitine to yield a fatty acyl carnitine; transport across the membrane occurs; the fatty acyl group is picked up by HSCoA; and carnitine crosses back to begin another cycle.

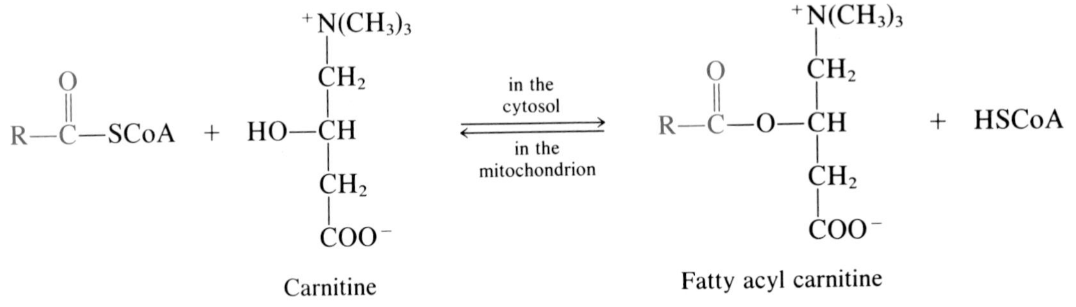

$$R-\overset{\overset{\displaystyle O}{\|}}{C}-SCoA \;+\; HO-\underset{\underset{\displaystyle COO^-}{|}}{\underset{\underset{\displaystyle CH_2}{|}}{\underset{\underset{\displaystyle CH}{|}}{\overset{\overset{\displaystyle ^+N(CH_3)_3}{|}}{\overset{\overset{\displaystyle CH_2}{|}}{}}}}} \quad \underset{\substack{\text{in the} \\ \text{mitochondrion}}}{\overset{\substack{\text{in the} \\ \text{cytosol}}}{\rightleftharpoons}} \quad R-\overset{\overset{\displaystyle O}{\|}}{C}-O-\underset{\underset{\displaystyle COO^-}{|}}{\underset{\underset{\displaystyle CH_2}{|}}{\underset{\underset{\displaystyle CH}{|}}{\overset{\overset{\displaystyle ^+N(CH_3)_3}{|}}{\overset{\overset{\displaystyle CH_2}{|}}{}}}}} \quad +\; HSCoA$$

Carnitine   Fatty acyl carnitine

Fatty Acid Oxidation Fatty acids are oxidized in the mitochondrial matrix by the series of four reactions shown in Figure 24.8 that make up the **fatty-acid spiral.** Each reaction cycle results in cleavage of a two-carbon acetyl group from the end of a fatty acyl group. The pathway is referred to as a *spiral* because a long-chain fatty acyl group must continue to return to the pathway until each pair of carbon atoms is removed as acetyl SCoA. The alternative name for the pathway, **β oxidation,** reflects the oxidation in two of the steps of the carbon atom β to the SCoA group.

Fatty acid spiral (β oxidation) A repetitive series of biochemical reactions that degrade fatty acids to acetyl SCoA.

Figure 24.8
The fatty acid spiral. Passage of acyl SCoA through these four steps leads to the cleavage of one acetyl group from the end of the fatty acid chain.

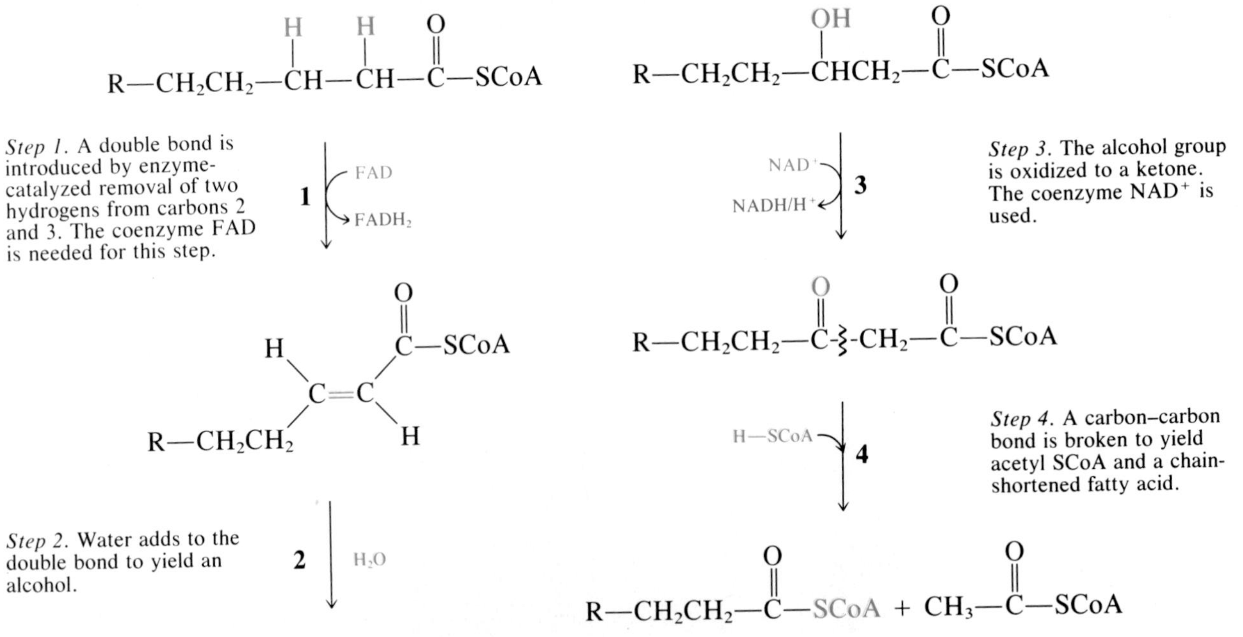

Step 1. A double bond is introduced by enzyme-catalyzed removal of two hydrogens from carbons 2 and 3. The coenzyme FAD is needed for this step.

Step 2. Water adds to the double bond to yield an alcohol.

Step 3. The alcohol group is oxidized to a ketone. The coenzyme NAD⁺ is used.

Step 4. A carbon–carbon bond is broken to yield acetyl SCoA and a chain-shortened fatty acid.

Steps 1 and 2 of fatty acid oxidation: The first β oxidation and hydrolysis. In step 1, the oxidizing agent FAD removes two hydrogen atoms from the carbon atoms α and β to the C=O group in the fatty acyl SCoA, forming a double bond. In step 2, H_2O adds across this newly created double bond to give an alcohol with the —OH group on the β carbon.

Step 3 of fatty acid oxidation: The second β oxidation. NAD^+ serves as an oxidizing agent for conversion of the β —OH group to a carbonyl group.

Step 4 of fatty acid oxidation: Cleavage to remove an acetyl group. In the final step of the fatty acid spiral, an acetyl group is split off and attached to a new coenzyme A molecule, leaving behind an acyl SCoA that is two carbon atoms shorter. If you look carefully, you might recognize this step as the reverse of a Claisen condensation reaction (Section 17.8).

To see how the fatty acid spiral works, look at the catabolism of palmitic acid shown in Figure 24.9. One turn of the spiral converts the 16-carbon palmitoyl SCoA into the 14-carbon myristyl SCoA plus acetyl SCoA; a second turn of the spiral converts myristyl SCoA into the 12-carbon lauryl SCoA plus acetyl SCoA; a third turn converts lauryl SCoA into the 12-carbon capryl SCoA plus acetyl SCoA; and so on.

You can predict how many molecules of acetyl SCoA will be obtained from a given fatty acid simply by counting the number of carbon atoms and dividing by 2. For example, the 16-carbon palmitic acid yields eight molecules of acetyl SCoA after seven turns of the spiral. The number of turns of the spiral is always one less than the number of acetyl SCoA molecules produced, because the last turn cleaves a 4-carbon chain into two acetyl SCoAs.

Figure 24.9
Passage of palmitoyl SCoA (16 C atoms) through the fatty acid spiral. Each repetition of the spiral cleaves two carbons from the end of the chain, yielding a molecule of acetyl CoA and a fatty acid chain shortened by two carbon atoms.

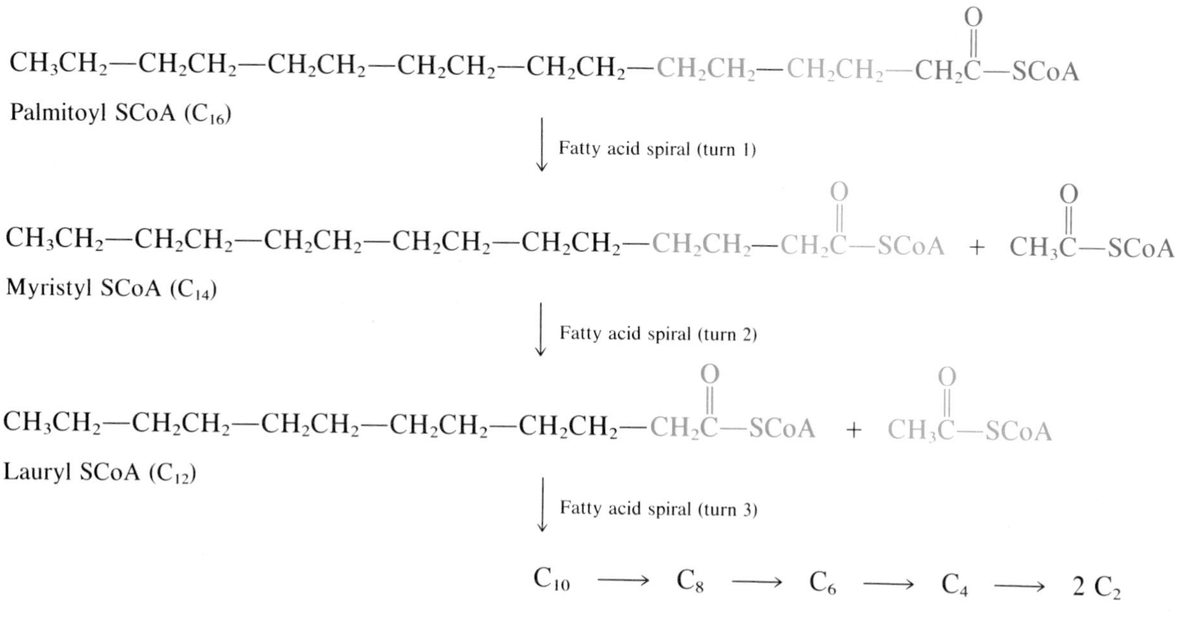

Since fatty acids usually have an even number of carbon atoms, none are left over. Additional oxidation steps deal with fatty acids with odd numbers of carbon atoms and those with double bonds, ultimately also releasing all carbon atoms for further oxidation in the citric acid cycle.

Practice Problems   **24.2**   How many molecules of acetyl SCoA are produced by catabolism of these fatty acids, and how many turns of the fatty acid spiral are needed?
(a) lauric acid, $CH_3(CH_2)_{10}COOH$
(b) arachidic acid, $CH_3(CH_2)_{18}COOH$

24.3   Write the equations for the next three turns of the fatty acid spiral following those in Figure 24.9.

24.4   Look back at the reactions of the citric acid cycle (Figure 20.11) and identify the three reactions that are similar to the first three reactions of the fatty acid spiral.

AN APPLICATION: THE LIVER, CLEARINGHOUSE FOR METABOLISM

The liver is the largest reservoir of blood in the body and is the largest internal organ, making up about 2.5% of the body's mass. Blood carrying the end products of digestion enters the liver through the hepatic portal vein before going on into general circulation, and the liver is therefore ideally situated to regulate the concentrations of nutrients and other substances in the blood.

Various functions of the liver have been previously described in scattered sections of this book, but it's only by taking an overview that the central role of the liver in metabolism can be appreciated. Among its many functions, the liver synthesizes glycogen from glucose, glucose from noncarbohydrates, triacylglycerols from mono- and diacylglycerols, fatty acids from acetyl SCoA, phospholipids, cholesterol, bile acids, and the majority of the serum proteins. In addition, liver cells can catabolize glucose, fatty acids, and amino acids to yield carbon dioxide and energy stored in ATP. The *urea cycle,* by which nitrogen from amino acids is converted to urea for excretion, also takes place in the liver.

Because reserves of glycogen, certain lipids and amino acids, iron, and fat-soluble vitamins are held in storage in the liver, it is able to release them as needed to maintain homeostasis. In addition, only liver cells have the enzyme needed to convert glucose 6-phosphate from glycogenolysis and gluconeogenesis to glucose that can enter the bloodstream.

Given its central role in metabolism, the liver is subject to a number of pathologic conditions based on excessive accumulation of various metabolites. One example is *cirrhosis,* the development of fibrous tissue that is preceded by excessive triacylglycerol buildup. Cirrhosis occurs in alcoholism, uncontrolled diabetes, and metabolic conditions in which the synthesis of lipoproteins from triacylglycerols is blocked. Another example is *Wilson's disease,* a genetic defect in copper metabolism. In Wilson's disease, copper accumulates in the liver rather than being excreted or recycled for use in a number of coenzymes. Chronic liver disease, as well as brain damage and anemia, are symptoms of Wilson's disease. The disease is treated by a low-copper diet and drugs that enhance the excretion of copper.

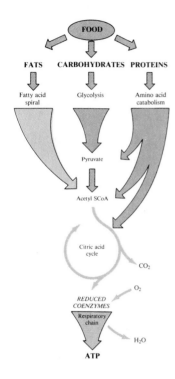

24.5 ENERGY OUTPUT DURING FATTY ACID CATABOLISM

The total energy output during fatty acid catabolism, like that from glucose catabolism, is measured by the total number of ATPs produced. In the case of fatty acids, this total is the sum of the ATPs from acetyl SCoA and those from reduced coenzymes from fatty acid oxidation. Recall from Section 20.8 that for each acetyl SCoA that passes through the citric acid cycle, the overall reaction is

$$\text{Acetyl SCoA} + 3\,NAD^+ + FAD + ADP + P_i + 2\,H_2O \longrightarrow$$
$$HSCoA + 3\,NADH + 3\,H^+ + FADH_2 + ATP + 2\,CO_2$$

Since in electron transport each NADH generates 3 ATPs and each $FADH_2$ generates 2 ATPs, the total number of ATPs produced by each acetyl SCoA is 12. Thus, a fatty acid that yields n molecules of acetyl SCoA releases $12n$ ATPs, where n is one-half the number of carbon atoms in a given fatty acid.

In addition, each turn of the fatty acid spiral (Figure 24.8) releases two reduced coenzymes: one $FADH_2$ and one NADH. The $FADH_2$ coenzyme generated in step 1 yields two molecules of ATP when it is recycled, and the NADH coenzyme generated in step 3 yields three ATPs. The total is thus five ATPs for each turn of the spiral, and the spiral turns $n - 1$ times for a given acid.

To arrive at the final total of ATPs, however, we have to subtract the two molecules of ATP that are required for the initial bonding of the free fatty acid with coenzyme A. Thus the overall energy scorecard looks like this:

	Change in Number of ATPs per Fatty Acid
Initial bonding to coenzyme A	-2
(Molecules of acetyl SCoA) $\times \dfrac{12\ ATP}{\text{acetyl SCoA}}$	$+12n$
(Number of turns of spiral) $\times \dfrac{5\ ATP}{\text{turn}}$	$+5(n-1)$
Net change for complete fatty acid catabolism	$17n - 7$

where

$$n = 1/2 \text{ number of carbons in the fatty acid chain}$$
$$= \text{number of molecules of acetyl SCoA produced}$$
$$n - 1 = \text{number of turns of the fatty acid spiral}$$

Comparing the amount of ATP produced by catabolism of a fatty acid to the amount produced by catabolism of glucose illustrates why our bodies use triacylglycerols rather than carbohydrates for long-term energy storage. One mole of glucose (180 g) generates 36 mol of ATP, whereas one mole of lauric acid (200 g) generates 95 mol of ATP. Thus, fatty acids yield more than twice as much energy per gram as carbohydrates (and do twice as much damage to diets). In addition, stored fats have a greater ''energy density'' because they are hydrophobic. Glycogen is hydrophilic, however, and each gram of stored glycogen is associated with about 2 g of water.

Solved Problem 24.1 Calculate the number of molecules of ATP produced during the catabolism of lauric acid, $CH_3(CH_2)_{10}COOH$.

Solution First, count the number of carbons in the lauric acid chain and then divide by 2 to find how many molecules of acetyl SCoA are produced:

$$CH_3(CH_2)_{10}COOH \longrightarrow C_{12} \longrightarrow 12/2 = 6 \text{ molecules of acetyl SCoA}$$

Next, find how many turns of the fatty acid spiral are needed to degrade lauric acid by subtracting 1 from the number of acetyl SCoA molecules:

$$6 \text{ Acetyl SCoA} - 1 \longrightarrow 5 \text{ turns of the fatty acid spiral}$$

Finally, do the arithmetic:

Initiation	$= -2$ ATP
6 Acetyl SCoA × (12 ATP/acetyl SCoA) =	72 ATP
5 Turns × (5 ATP/turn)	= 25 ATP
Net change	= 95 ATP

Checking shows that with $n = 6$, $(17 \times 6) - 7 = 95$ ATPs.

Practice Problems **24.5** Calculate the amount of ATP produced by catabolism of one mole of palmitic acid, $CH_3(CH_2)_{14}COOH$.

24.6 How many grams of ATP (mol wt 507 amu) are produced by catabolism of 1.0 g of palmitic acid (mol wt 256 amu)?

Ketone bodies Compounds produced in the liver that can be used as fuel by muscle and brain tissue; β-hydroxybutyrate, acetoacetate, and acetone.

Ketogenesis The synthesis of ketones bodies from acetyl SCoA.

24.6 PRODUCTION OF KETONE BODIES; KETOACIDOSIS

Glucose and fatty acids are our major fuel molecules, but small amounts of alternative fuels called **ketone bodies** are also produced from acetyl SCoA in liver mitochondria, a process known as **ketogenesis.** The first step in the production of ketone bodies is the combination of two acetyl SCoAs to give acetoacetyl SCoA. Hydrolysis then yields acetoacetate, and further reactions yield acetone and 3-hydroxybutyrate.

Ketone bodies

Under well-fed, healthy conditions, skeletal and heart muscles derive a small portion of their daily energy needs from one of the ketone bodies, acetoacetate, which is converted back to two acetyl SCoAs for oxidation via the citric acid cycle. But consider the situation when glucose levels are abnormally low. The body must then respond by providing other sources for energy. For example, the synthesis of glucose by gluconeogenesis and the breakdown of stored fat to give acetyl SCoA are accelerated. Meanwhile, however, glucose metabolism and the citric acid cycle are slowing down. The result is production from fatty acids of more acetyl SCoA than can be processed by the citric acid cycle. This condition leads to diversion of more and more acetyl SCoA into ketone bodies that can be used to produce energy. During the early stages of starvation, heart and muscle tissues burn larger quantities of acetoacetate, thereby preserving glucose for use in the brain. In prolonged starvation, even the brain can switch to ketone bodies to meet up to 75% of its energy needs.

The condition in which ketone bodies are produced faster than they are utilized (*ketosis*) is recognized by the smell of acetone on the breath and the presence of ketone bodies in the urine (*ketonuria*). Because two of the ketone bodies are carboxylic acids, continued ketosis such as might occur in untreated diabetes leads to the potentially serious condition known as **ketoacidosis:** acidosis resulting from increased concentrations of ketone bodies in the blood. The blood's buffers are overwhelmed, and blood pH drops. An individual experiences dehydration due to increased urine flow, labored breathing because acidic blood is a poor oxygen carrier, and depression. Ultimately, the condition can lead to coma and death.

Ketoacidosis Lowered blood pH due to accumulation of ketone bodies.

Practice Problem 24.7 Which of the following classifications apply to the formation of 3-hydroxybutyrate from acetoacetyl SCoA: (i) hydrolysis, (ii) oxidation, (iii) reduction, (iv) condensation?

24.7 BIOSYNTHESIS OF FATTY ACIDS

The biosynthesis of fatty acids from acetyl SCoA, a process known as **lipogenesis,** provides a link between carbohydrate, lipid, and protein metabolism. Because the catabolism of both carbohydrates and amino acids produces acetyl SCoA, fatty acid synthesis followed by triacylglycerol synthesis allows excess carbohydrates and amino acids to be stored in adipose tissue. As you've seen for glycolysis and gluconeogenesis, biochemical pathways that run in reverse directions don't usually occur by reverse routes because the reverse of an energetically favorable pathway is likely to be energetically unfavorable. This principle applies to oxidation of fatty acids by the fatty acid spiral and to its reverse, the biosynthesis of fatty acids (Table 24.1).

Fatty acid synthesis begins when an acetyl SCoA is converted to malonyl SCoA, and the malonyl group is then transferred to the **acyl carrier protein (ACP).** One malonyl SACP is required for each turn of the biosynthesis spiral, because it carries the carbon atoms to be added to the growing chain. At the same time, the acetyl group from another acetyl SCoA is bonded to a second acyl carrier protein to generate acetyl SACP.

Lipogenesis The biochemical pathway for synthesis of fatty acids from acetyl SCoA.

Acyl carrier protein Protein that carries acyl groups during fatty acid synthesis.

Table 24.1 Comparison of Fatty Acid Oxidation and Synthesis

Oxidation	Synthesis
Occurs in mitochondria.	Occurs in cytosol.
Enzymes different from synthesis.	Enzymes different from oxidation.
Intermediates carried by coenzyme A.	Intermediates carried by acyl carrier protein.
Coenzymes: FAD, NAD$^+$.	Coenzyme: NADPH.
Spiral pathway, carbon atoms removed two at a time.	Spiral pathway, carbon atoms added to at a time.

$$CH_3-\overset{\overset{\text{O}}{\|}}{C}-SCoA \ + \ HCO_3^- \ \longrightarrow \ H_2O \ + \ {}^-O-\overset{\overset{\text{O}}{\|}}{C}-CH_2-\overset{\overset{\text{O}}{\|}}{C}-SCoA \ \longrightarrow \ {}^-O-\overset{\overset{\text{O}}{\|}}{C}-CH_2-\overset{\overset{\text{O}}{\|}}{C}-SACP$$

Acetyl SCoA   Malonyl SCoA   Malonyl SACP

$$CH_3-\overset{\overset{\text{O}}{\|}}{C}-SCoA \ \longrightarrow \ CH_3-\overset{\overset{\text{O}}{\|}}{C}-SACP$$

Acetyl SCoA   Acetyl SACP

Once malonyl SACP and acetyl SACP have been generated, a repeating cycle of four reactions takes place to lengthen the growing fatty acid chain by two carbon atoms with each repetition (Figure 24.10).

Step 1 of fatty acid synthesis: Condensation. The malonyl group from malonyl ACP transfers to acetyl SACP with the loss of CO_2, resulting in addition of two carbon atoms. Note that malonyl SACP is the source of the added carbon atoms.

Steps 2–4 of fatty acid synthesis: Reduction, dehydration, and reduction. These three reactions accomplish the reverse of steps 3, 2, and 1 in the fatty acid spiral (Figure 24.8). The carbonyl group is reduced to an —OH group, dehydration yields a carbon–carbon double bond, and the double bond is reduced by addition of hydrogen.

The result of the first cycle in fatty acid synthesis is the addition of 2 carbon atoms to an acetyl group to give a 4-carbon acyl group. The next cycle then adds 2 more carbon atoms to give a 6-carbon acyl group.

$$CH_3CH_2CH_2\overset{\overset{\text{O}}{\|}}{C}-SACP \ + \ {}^-O\overset{\overset{\text{O}}{\|}}{C}CH_2\overset{\overset{\text{O}}{\|}}{C}-SACP \ \xrightarrow[\text{step 1}]{\text{repeat}} \ CH_3CH_2CH_2\overset{\overset{\text{O}}{\|}}{C}CH_2\overset{\overset{\text{O}}{\|}}{C}-SACP \ + \ CO_2$$

$$\xrightarrow[\text{steps 2-4}]{\text{repeat}} \ CH_3CH_2CH_2CH_2CH_2\overset{\overset{\text{O}}{\|}}{C}-SACP$$

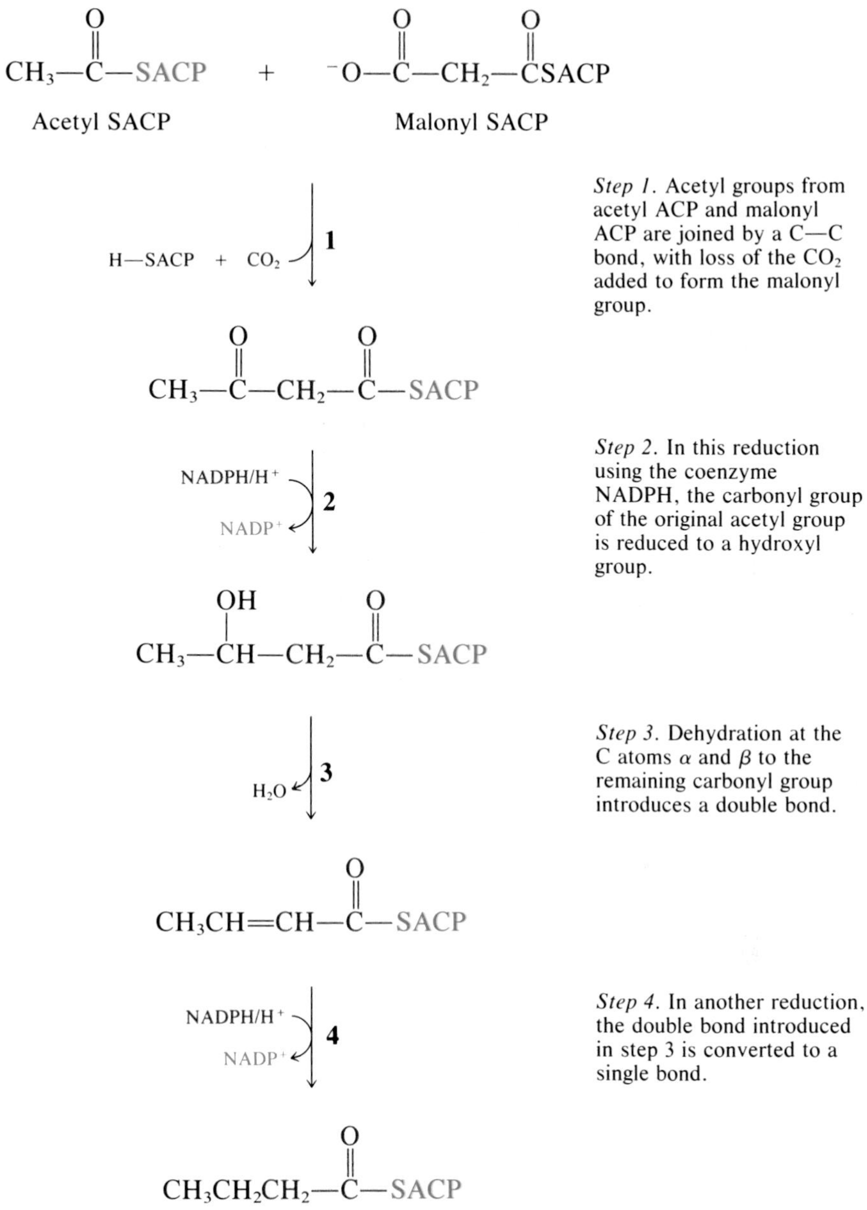

$$CH_3-\overset{\overset{\text{O}}{\|}}{C}-SACP \quad + \quad {}^-O-\overset{\overset{\text{O}}{\|}}{C}-CH_2-\overset{\overset{\text{O}}{\|}}{C}SACP$$

Acetyl SACP   Malonyl SACP

H—SACP + CO₂ ⟍⟍ **1**

Step 1. Acetyl groups from acetyl ACP and malonyl ACP are joined by a C—C bond, with loss of the CO_2 added to form the malonyl group.

$$CH_3-\overset{\overset{\text{O}}{\|}}{C}-CH_2-\overset{\overset{\text{O}}{\|}}{C}-SACP$$

NADPH/H⁺ ⟍
                2
NADP⁺ ⟍

Step 2. In this reduction using the coenzyme NADPH, the carbonyl group of the original acetyl group is reduced to a hydroxyl group.

$$CH_3-\overset{\overset{\text{OH}}{|}}{CH}-CH_2-\overset{\overset{\text{O}}{\|}}{C}-SACP$$

H₂O ⟍ **3**

Step 3. Dehydration at the C atoms α and β to the remaining carbonyl group introduces a double bond.

$$CH_3CH=CH-\overset{\overset{\text{O}}{\|}}{C}-SACP$$

NADPH/H⁺ ⟍
                4
NADP⁺ ⟍

Step 4. In another reduction, the double bond introduced in step 3 is converted to a single bond.

$$CH_3CH_2CH_2-\overset{\overset{\text{O}}{\|}}{C}-SACP$$

Figure 24.10
Chain elongation in the biosynthesis of fatty acids. The steps shown begin with acetyl acyl carrier protein (acetyl SACP), the reactant in the first cycle of palmitic acid synthesis. Each new pair of carbon atoms is carried into the next cycle by a fresh malonyl SACP. The growing chain remains attached to the carrier protein from the original acetyl SACP.

After seven trips through the elongation cycle, a 16-carbon palmitoyl group has been produced. Hydrolysis then breaks the thioester bond to give palmitic acid, $CH_3(CH_2)_{14}COOH$. All other fatty acids are synthesized from palmitic acid.

The major transport and biochemical pathways of acylglycerols and fatty acids are summarized in Figure 24.11.

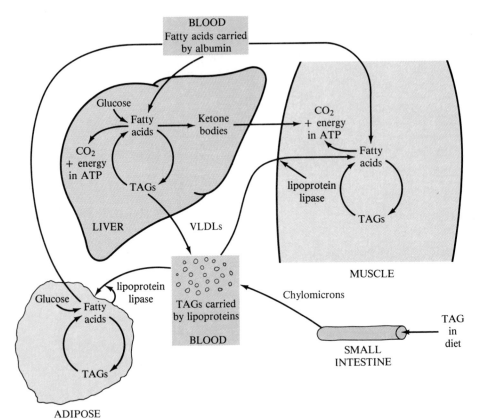

Figure 24.11
Major pathways of lipid
metabolites. Note that
synthesis of fatty acids from
glucose takes place via
acetyl SCoA mainly in
adipose tissue and the liver.

SUMMARY

The digestion and metabolism of triacylglycerols (TAGs) and other lipids is complicated by their insolubility in water. **Bile acids** therefore emulsify dietary TAGs in the small intestine, where they are partially hydrolyzed by pancreatic lipase and absorbed through the intestinal lining. After absorption of the hydrolysis products, TAGs are resynthesized and repackaged in **lipoproteins** known as **chylomicrons** that are carried in the lymphatic system to the bloodstream. Lipoproteins known as VLDLs similarly transport TAGs synthesized in the liver, and albumins in blood serum transport fatty acids released from storage in adipose tissue.

The major catabolic pathway of TAGs is hydrolysis to fatty acids plus glycerol, followed by oxidation of fatty acids in the **fatty acid spiral** to yield acetyl SCoA. The spiral proceeds by removal of two carbon atoms at a time as acetyl SCoA. A fatty acid undergoes $n - 1$ turns of the spiral and yields n acetyl SCoAs, where n is one-half the number of carbon atoms in the fatty acid. The acetyl SCoAs may then either enter the citric acid cycle for conversion into energy or be used as starting material for the synthesis of steroids and **ketone bodies.**

The **lipogenesis** pathway synthesizes fatty acids from acetyl SCoA. Fatty acid synthesis occurs in the cytosol, and the reactants are acyl groups bonded to **acyl carrier protein.** Carbon atoms are added two at a time to produce palmitic acid, from which other fatty acids are synthesized. TAG synthesis also requires dihydroxyacetone phosphate, which is made in the liver from glycerol or comes from glycolysis. Because the acetyl SCoA needed for lipogenesis is a product of carbohydrate and protein metabolism, lipogenesis followed by synthesis of TAGs is a pathway for converting excess protein and carbohydrate to fat.

Ketogenesis in the liver produces ketone bodies from acetyl SCoA. When glucose is in short supply, ketogenesis accelerates and ketones are burned for energy. An excess of ketone bodies, two of which are acids, produces the dangerous condition known as **ketoacidosis.**

INTERLUDE: MAGNETIC RESONANCE IMAGING (MRI)

The recent development in body imaging known as *magnetic resonance imaging (MRI)* is creating considerable excitement in the medical community because of its advantages over techniques that use X rays or radioactive materials. In MRI, the patient is not exposed to any ionizing radiation, there is no need to introduce contrast materials into the patient's body, and soft-tissue structures that are obscured by bone in X-ray images are visible.

Essentially, MRI takes advantage of the magnetic properties of certain atomic nuclei and looks at the location of those nuclei in the body. By placing the nuclei in a magnetic field and irradiating them with electromagnetic radiation in the radiofrequency range, the nuclei are made to absorb and release specific amounts of energy, which are measured. Body images are produced by computer processing of the resulting data.

Most MRI images currently look at hydrogen nuclei, present in abundance wherever there is water or fat in the body. The signals produced vary with the density of hydrogen atoms and with the nature of their surroundings, allowing identification of different types of tissue and the visualization of motion. For example, the volume of blood leaving the heart in a single stroke can be measured, allowing observation of the heart in motion. Several types of atoms in addition to hydrogen can also be detected by MRI, and the applications of images based on the phosphorus atoms in ATP and other phosphates are actively being explored. The technique holds great promise for studies of metabolism.

MRI has quickly become the method of choice for detecting brain tumors not seen with other techniques; stories abound of dramatic cures of patients with motor difficulties caused by tiny brain tumors. The technique is also valuable in diagnosing knee damage because it is a painless alternative to arthroscopy, in which an endoscope is physically introduced into the knee joint.

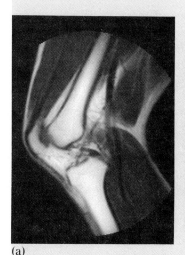 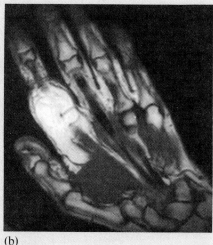 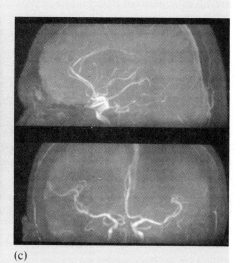

(a)   (b)   (c)

MRI images of (a) a normal knee, (b) a hand with a bone abnormality on the index finger, and (c) a normal brain from the side (top image) and front (bottom image) showing blood flow in arteries. In (a) and (b) the dark lines are blood vessels and the white areas are bone. In (a) the cruciate ligaments, often injured in sporting accidents, show faintly as dark lines between the two bones. The signal intensity in (c) has been modified to emphasize blood flow; such an image might be used to detect an aneurism or observe blood flow to a tumor.

REVIEW PROBLEMS

Digestion and Catabolism of Lipids

24.8 Where does digestion of lipids occur?

24.9 Why do lipids make one feel full for a long time after a meal?

24.10 What is the purpose of bile in lipid catabolism?

24.11 Lipases break down triacylglycerols by catalyzing hydrolysis. What are the products of this hydrolysis?

24.12 What are chylomicrons, and how are they involved in lipid metabolism?

24.13 What is the origin of the triacylglycerols transported by very-low-density lipoproteins?

24.14 How are the fatty acids from adipose tissue transported?

24.15 The glycerol derived from digestion of dietary triacylglycerols is converted into glyceraldehyde 3-phosphate, which then enters into step 6 of the glycolysis pathway. What further transformations are necessary to convert glyceraldehyde 3-phosphate into pyruvate?

24.16 If the conversion of glycerol to glyceraldehyde 3-phosphate releases one molecule of ATP, how many molecules of ATP are released during the conversion of glycerol to pyruvate? (Problem 24.15.)

24.17 How many molecules of ATP are released in the overall catabolism of glycerol to acetyl SCoA?

24.18 How many molecules of ATP are released in the complete catabolism of glycerol to CO_2 and H_2O? (Problem 24.17.)

24.19 What is an adipocyte?

24.20 Where in the cell does the fatty acid spiral occur?

24.21 What is the alternative name of the fatty acid spiral? Why is this name appropriate?

24.22 What initial chemical transformation takes place on a fatty acid to activate it for catabolism?

24.23 What is the function of carnitine in fatty acid catabolism?

24.24 Why do you suppose the sequence of reactions that catabolize fatty acids is called the fatty acid *spiral* rather than the fatty acid *cycle?*

24.25 How many moles of ATP are produced by one turn of the fatty acid spiral?

24.26 Arrange these four molecules in order of their biological energy content per mole:
(a) glucose
(b) capric acid, $CH_3(CH_2)_8COOH$
(c) sucrose
(d) myristic acid, $CH_3(CH_2)_{12}COOH$

24.27 Show the products of each step in the following fatty acid spiral on hexanoic acid:

(a)
$$CH_3CH_2CH_2CH_2CH_2\overset{\overset{\displaystyle O}{\|}}{C}SCoA \xrightarrow[\substack{\text{acetyl SCoA} \\ \text{dehydrogenase}}]{\text{FAD} \quad \text{FADH}_2} \ ?$$

(b) product of (a) $+ H_2O \xrightarrow[\text{hydratase}]{\text{enoyl SCoA}} \ ?$

(c) product of (b) $\xrightarrow[\substack{\beta\text{-hydroxyacyl SCoA} \\ \text{dehydrogenase}}]{\text{NAD}^+ \quad \text{NADH/H}^+} \ ?$

(d) product of (c) $+$ HSCoA $\xrightarrow[\text{transferase}]{\text{acetyl SCoA}} \ ?$

24.28 Write the equation for the final step in the catabolism of any fatty acid with an even number of carbons.

24.29 How many molecules of acetyl SCoA result from complete catabolism of these compounds?
(a) caprylic acid, $CH_3(CH_2)_6COOH$
(b) myristic acid, $CH_3(CH_2)_{12}COOH$

24.30 How many turns of the fatty acid spiral are necessary to completely catabolize caprylic and myristic acids? (Problem 24.29.)

Fat Anabolism

24.31 What is the name of the anabolic pathway for synthesizing fatty acids?

24.32 What is the starting material for fatty acid synthesis?

24.33 Why can't the fatty acid spiral be run backward to produce triacylglycerols?

24.34 Why are fatty acids generally composed of an even number of carbons?

24.35 What is the fatty acid from which all other fatty acids are synthesized?

24.36 How many rounds of the lipigenesis cycle are needed to synthesize palmitic acid, $C_{15}H_{31}COOH$?

Applications

24.37 What is a normal blood cholesterol range? [App: Cholesterol, Lipoproteins, and Heart Disease]

24.38 What is atherosclerosis? [App: Cholesterol, Lipoproteins, and Heart Disease]

24.39 What is the difference in the roles of LDL and HDL? [App: Cholesterol, Lipoproteins, and Heart Disease]

24.40 Which is better, a high or low HDL/LDL level? Why? [App: Cholesterol, Lipoproteins, and Heart Disease]

24.41 What is cirrhosis of the liver and what can trigger it? [App: Liver, Clearinghouse for Metabolism]

24.42 What type of atomic nucleus is primarily monitored in MRI? [Int: Magnetic Resonance Imaging (MRI)]

24.43 What is the advantage of MRI compared to radiation techniques of imaging the body? [Int: Magnetic Resonance Imaging (MRI)]

Additional Questions and Problems

24.44 Consuming too many carbohydrates causes deposition of fats in adipose tissue. How can this happen? Why aren't the molecules stored in glycogen instead?

24.45 Are any of the intermediates in the fatty acid spiral chiral? Explain.

24.46 What three compounds are classified as ketone bodies? Why are they so designated? What process in the body produces them? Why do they form?

24.47 What is ketosis? What condition results from prolonged ketosis? Why is it dangerous?

24.48 How many molecules of acetyl SCoA result from catabolism of glyceryl trimyristate?

24.49 How many ATPs are released in the complete catabolism of glyceryl trimyristate? (Problem 24.48.)

24.50 Compare fats and carbohydrates as energy sources in terms of the amount of energy released per mole and account for the observed energy difference.

C H A P T E R

25

Protein and Amino Acid Metabolism

Urea, an end product of protein metabolism, forms beautiful crystals, as seen here under polarized light. Nitrogen not needed for making new amino acids must be excreted as urea, which is formed by a metabolic pathway described in this chapter.

To understand how we function, we need to understand proteins. The complexities of their structure and their role as enzymes have already been discussed. Their biosynthesis is determined by the genetic code and is such a broad subject that it requires an entire chapter itself. What remains for this chapter is a look at the metabolism of amino acids. We'll answer the following questions:

1. **What happens during the digestion of proteins?** The goal: Be able to list the sequence of events in the digestion of proteins and the transport of amino acids into the bloodstream.

2. **What are the major strategies in the catabolism of proteins?** The goal: Be able to explain the general catabolic pathways of amino acids.

3. **What is transamination?** The goal: Be able to describe where transamination occurs in amino acid metabolism and be able to draw the structures of the products of transamination.

4. **What is the urea cycle?** The goal: Be able to list the major reactants and products of the reactions in the urea cycle and explain how the cycle is linked to the citric acid cycle.

5. **What are glucogenic and ketogenic amino acids?** The goal: Be able to define these terms and tell how these kinds of amino acids are metabolized.

6. **What is the strategy for amino acid synthesis?** The goal: Be able to describe in general the compounds used in amino acid synthesis.

25.1   PROTEIN DIGESTION

The end result of protein digestion is simple—the hydrolysis of all peptide bonds to produce a collection of amino acids:

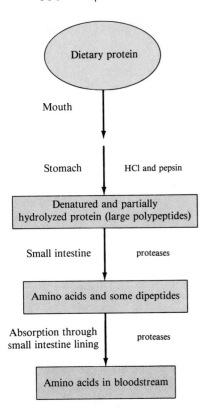

Dietary protein

Mouth

Stomach · HCl and pepsin

Denatured and partially hydrolyzed protein (large polypeptides)

Small intestine · proteases

Amino acids and some dipeptides

Absorption through small intestine lining · proteases

Amino acids in bloodstream

Figure 25.1
Digestion of proteins.

The digestion of dietary proteins, as summarized in Figure 25.1, begins with their denaturation in the strongly acidic environment of the stomach (pH 1–2). In addition to hydrochloric acid, gastric secretions include *pepsinogen,* a zymogen that is activated by acid to give the enzyme *pepsin* (Figure 25.2). Some protein peptide bonds made accessible by the unfolding of denaturation are hydrolyzed by pepsin, giving large polypeptides that then enter the small intestine.

As the polypeptides enter the small intestine, the pH is raised to about 7–8, pepsin is inactivated, and a group of pancreatic zymogens is secreted. A small amount of *trypsinogen* is activated by an enzyme from the intestinal lining to give *trypsin,* one of the **protease** enzymes that carry out peptide hydrolysis. This trypsin then activates the remaining trypsinogen and other pancreatic zymogens to give *chymotrypsin, carboxypeptidase,* and *elastase,* proteases that further hydrolyze polypeptides.

The combined action of the pancreatic proteases in the small intestine and other proteases in the intestinal lining converts all polypeptides to amino acids. After transport across cell membranes lining the intestine, the amino acids are absorbed directly into the bloodstream.

25.2 AMINO ACID METABOLISM: AN OVERVIEW

The entire collection of free amino acids throughout the body—the **amino acid pool**—occupies a central position in protein and amino acid metabolism (Figure 25.3). Since living organisms are dynamic rather than static, all tissues and biomolecules in the body are constantly being degraded, repaired, and replaced.

Figure 25.2
Digestion in cheese making. (a) The first step in cheese making is the addition of bacteria and enzymes that sour the milk and denature the milk protein (casein). (b) After excess water is removed, the remaining solid curd is shaped and stored while enzymes from microorganisms hydrolyze proteins, lactose, and triacylglycerols.

(a)

(b)

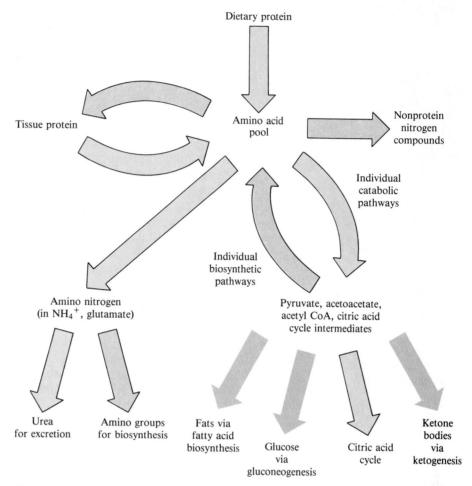

Protease An enzyme for hydrolysis of the peptide bonds in proteins.

Amino acid pool The entire collection of free amino acids in the body.

Figure 25.3
Protein and amino acid metabolism. Synthetic pathways are shown in blue, pathways that break down molecules are shown in yellow, and connections to fat and glucose metabolism are shown in green. The amino groups are used in the biosynthesis of amino acids and nonprotein nitrogen compounds.

Thus, amino acids enter the pool not only by digestion of proteins and synthesis from other compounds, but also from breakdown of tissues. In fact, a normal adult breaks down about 2% of their protein, or approximately 400 g of protein, every day.

The enzymes for hydrolysis of waste protein to form amino acids are located in cells throughout the body. Also, cells throughout the body call on amino acids from the pool for protein synthesis. In addition, as the principal nitrogen-containing compounds of the body, amino acids are needed for the synthesis of nonprotein nitrogen-containing compounds.

Amino acid catabolism is quite complex since each of the 20 different amino acids is degraded through its own unique pathway. The general idea of all the pathways, however, is that the amino nitrogen atom is removed and the amino acids are converted into compounds that can enter the citric acid cycle. For some amino acids, a single step is sufficient; for others, a multistep pathway is needed.

In keeping track of amino acid metabolism, it helps to think of each amino acid as having two parts—the amino group and the carbon atoms—that take separate courses. The amino group may be used in the synthesis of other nitrogen-containing compounds, including nucleic acids and other amino acids. Since there is no storage form for nitrogen compounds equivalent to glycogen or triacylglycerols, however, any nitrogen not immediately needed must be excreted by conversion to urea, as discussed in the next section.

Once the carbon atoms of amino acids are converted to compounds associated with the citric acid cycle, they are available for several alternative pathways. For instance, they can continue through the cycle in the body's main energy-generating pathway to give CO_2 and energy stored in ATP. About 10–20% of our energy is normally produced from amino acids. If not needed for energy, however, the carbon-carrying intermediates produced from amino acids enter the gluconeogenesis pathway for glucose synthesis (Section 22.11), the ketogenesis pathway for the synthesis of ketone bodies from acetyl SCoA (Section 24.6), or the lipogenesis pathway for the synthesis of fatty acids from acetyl SCoA (Section 24.7).

25.3 AMINO ACID CATABOLISM: REMOVAL OF THE AMINO GROUP

Transamination The interchange of the amino group of an amino acid and the keto group of an α-keto acid.

Transamination As the first step in their catabolism, most amino acids undergo **transamination,** a reaction in which the amino group of the amino acid and the keto group of an α-keto acid change places.

$$\underset{\substack{| \\ NH_3^+ \\ \text{Amino acid}}}{R'-CH-COO^-} + \underset{\text{α-Keto acid}}{R''-\overset{\overset{\textstyle O}{\|}}{C}-COO^-} \underset{\text{transaminase}}{\rightleftharpoons} \underset{\text{α-Keto acid}}{R'-\overset{\overset{\textstyle O}{\|}}{C}-COO^-} + \underset{\substack{| \\ NH_3^+ \\ \text{Amino acid}}}{R''-CH-COO^-}$$

There are a number of *transaminase* enzymes. Most are specific for α-ketoglutarate as the amino group acceptor and work with several different amino acid donors, resulting in formation of an α-keto acid and glutamate as diagrammed in Figure 25.4. For example, alanine is converted to pyruvate by transamination. The enzyme for this conversion, *alanine aminotransferase (ALT),* is especially abundant in the liver, and above-normal ALT concentrations are taken as an indication of liver damage, which allows ALT to leak into the bloodstream.

Figure 25.4
Major pathway of nitrogen from an amino acid to urea. The red pathway shows the complete catabolism of amino acid nitrogen to give urea. The green pathway shows the reverse route used in the biosynthesis of some amino acids.

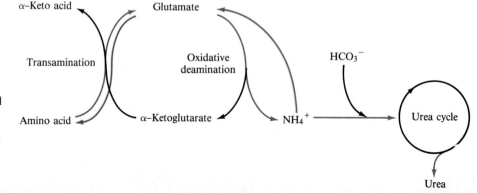

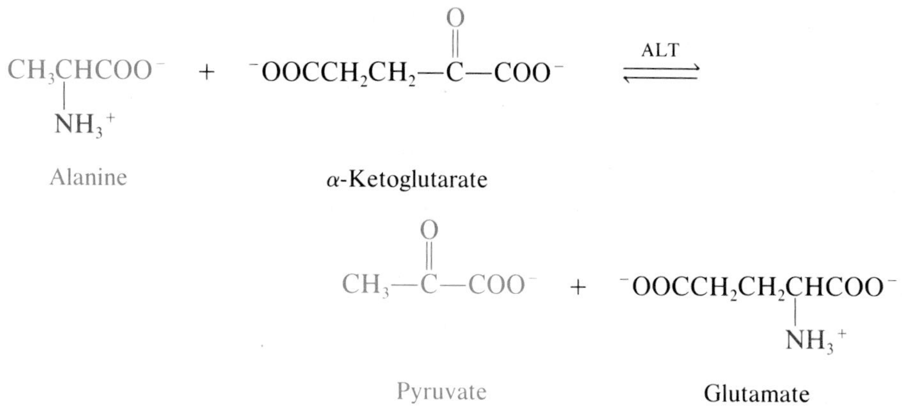

$$CH_3CHCOO^- \ + \ ^-OOCCH_2CH_2-\overset{\overset{\displaystyle O}{\|}}{C}-COO^- \ \rightleftharpoons^{ALT}$$

$$\underset{NH_3^+}{|}$$

Alanine   α-Ketoglutarate

$$CH_3-\overset{\overset{\displaystyle O}{\|}}{C}-COO^- \ + \ ^-OOCCH_2CH_2\underset{\underset{NH_3^+}{|}}{C}HCOO^-$$

Pyruvate   Glutamate

Each transaminase uses the coenzyme pyridoxal phosphate, a derivative of vitamin B$_6$ (pyridoxine) that bonds to amino acids as shown in Figure 25.5.

Transamination is a key part of many biochemical pathways involving amino acids and α-keto acids, interconverting amino groups and carbonyl groups as necessary. The transamination reactions are equilibria that go easily in either direction depending on the concentrations of the reactants. In this way, they regulate amino acid concentrations by keeping synthesis and breakdown in balance.

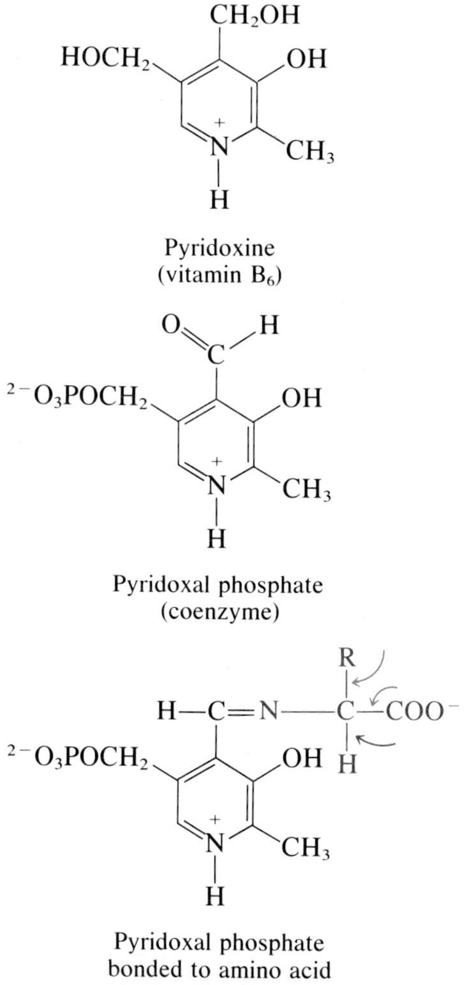

Figure 25.5
Pyridoxal phosphate (PLP), a vitamin B$_6$ derivative and a coenzyme for amino acid reactions. In transamination, the bond marked by the red arrow breaks, the resulting intermediate rearranges, and the coenzyme structure is restored after the products have formed. PLP is a coenzyme for other types of amino acid reactions and also facilitates breaking of the bonds marked with blue arrows.

Pyridoxine
(vitamin B$_6$)

Pyridoxal phosphate
(coenzyme)

Pyridoxal phosphate
bonded to amino acid

Solved Problem 25.1 The serum concentration of the transaminase from heart muscle, *aspartate aminotransferase (AST)*, is used in the diagnosis of heart disease because the enzyme escapes into the serum from damaged heart cells. AST catalyzes transamination of aspartate with α-ketoglutarate. What are the products of this reaction?

Solution The reaction is the interchange of an amino group from aspartate with the keto group from α-ketoglutarate. We know that α-ketoglutarate always gives glutamate in transamination, so one product is glutamate. The product from the amino acid will have a keto group instead of the amino group. Consulting Table 18.1 shows that the structure of aspartate is

$$\text{Aspartate} \qquad {}^-\text{OOCCH}_2\overset{\alpha}{\text{C}}\text{HCOO}^-$$
$$\underset{\text{NH}_3^+}{|}$$

Removing the —NH$_3^+$ and —H groups bonded to the α carbon and replacing them by a =O gives the desired α-keto acid, oxaloacetate, a compound you've seen in the citric acid cycle.

$$\text{Oxaloacetate} \qquad {}^-\text{OOC}-\text{CH}_2-\overset{\overset{\text{O}}{\|}}{\text{C}}-\text{COO}^-$$

The reaction is

$$\text{Aspartate} + \alpha\text{-ketoglutarate} \xrightarrow{\text{AST}} \text{oxaloacetate} + \text{glutamate}$$

Practice Problems **25.1** What are the structure and name of the α-keto acid formed by transamination of the amino acid leucine?

25.2 What is the product of the following reaction?

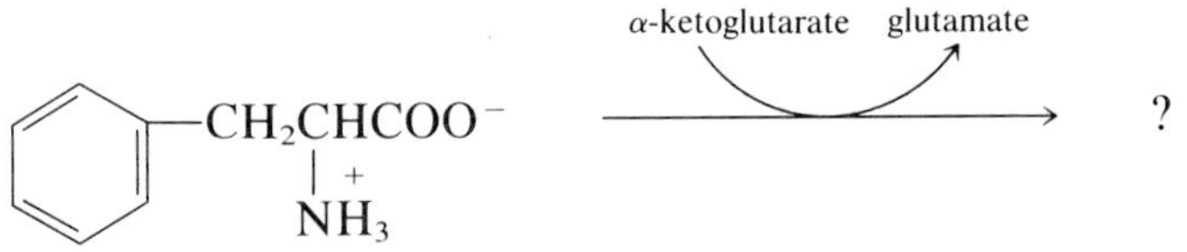

$$\text{—CH}_2\text{CHCOO}^- \quad \xrightarrow[\text{α-ketoglutarate}\quad\text{glutamate}]{} \quad ?$$
$$\underset{\text{NH}_3}{\overset{+}{|}}$$

Oxidative deamination
Conversion of an amino acid —NH$_2$ group to an α-keto group, with removal of NH$_4^+$.

Oxidative Deamination The glutamate produced from transamination of α-ketoglutarate acts as an amino group carrier. Although it's a reactant in the synthesis of several nonessential amino acids (Section 25.6), most of it undergoes recycling to fresh α-ketoglutarate plus ammonia that is eliminated (see Figure 25.4). In a process known as **oxidative deamination,** the glutamate amino group is removed as ammonia (*deamination*) and replaced by a carbonyl group (*oxidation*) to give back α-ketoglutarate. The enzyme for this reaction, *glutamate dehydrogenase,* is unique in being able to use either NAD$^+$ or NADP$^+$ as its oxidizing coenzyme. Oxidative deamination, like transamination, is a reaction that is used in the reverse direction (called *reductive amination*) in biosynthesis.

$$^-OOCCH_2CH_2CH-COO^- \;+\; H_2O \xrightarrow[\substack{\text{glutamate}\\\text{dehydrogenase}}]{\substack{NAD^+(NADP^+) \quad NADH(NADPH)}}$$

$$\underset{NH_3{}^+}{|}$$

Glutamate

$$NH_4{}^+ \;+\; {}^-OOCCH_2CH_2\overset{\displaystyle O}{\overset{\|}{C}}-COO^-$$

α-Ketoglutarate

25.4   THE UREA CYCLE

Urea cycle   The cyclic biochemical pathway that produces urea for excretion.

Ammonia is highly toxic to living things and must be eliminated efficiently. Fish are able to excrete ammonia into their watery surroundings without undergoing any harm (Figure 25.6), but mammals must first convert the ammonia to nontoxic urea by the **urea cycle.** By so doing, they avoid elimination of ammonia in the urine, which could only be done safely if it were dissolved in a large volume of water.

The conversion of ammonia to urea takes place in the liver, from which it is transported to the kidney and transferred to urine for excretion. Like many other biochemical pathways, urea formation begins with an energy investment. Ammonia from oxidative deamination (actually, ammonium ion at physiological pH), carbon dioxide from the citric acid cycle (in solution as bicarbonate ion), and ATP combine to form carbamoyl phosphate. Two ATPs are invested in this reaction, and one high-energy phosphate is transferred to the carbamoyl phosphate.

$$NH_4{}^+ \;+\; HCO_3{}^- \;+\; 2\,ATP \;\longrightarrow\; H_3\overset{+}{N}-\overset{\displaystyle O}{\overset{\|}{C}}-O-PO_3{}^{2-} \;+\; 2\,ADP \;+\; HPO_4{}^{2-} \;+\; H_2O$$

Carbamoyl phosphate

Carbamoyl phosphate next reacts in the first step of the four-step urea cycle, shown in Figure 25.7.

Figure 25.6
An orange-fin anemone fish in its natural environment in the Fiji Islands. Because fish are surrounded by water, their metabolism need not bother to convert ammonia to urea.

Figure 25.7
The urea cycle.

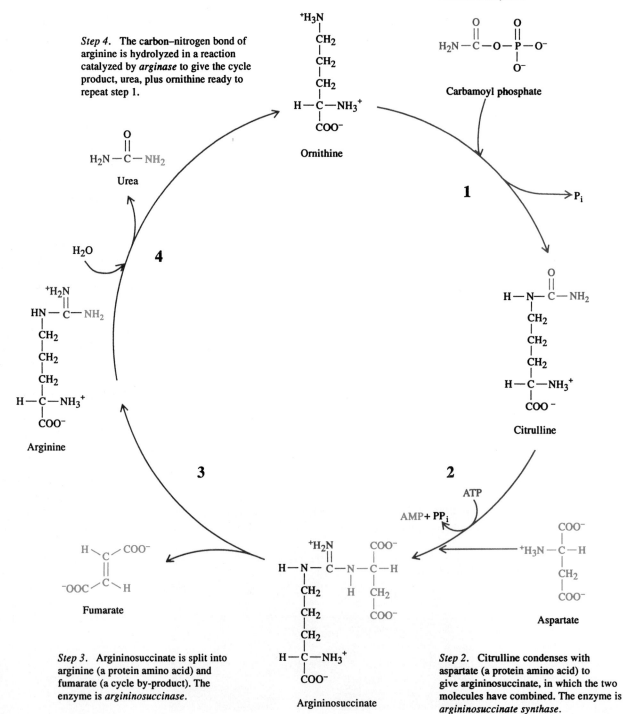

Step 1. **Carbamoyl phosphate transfers its H₂NC═O group to ornithine (a nonprotein amino acid) to give citrulline in a reaction catalyzed by** *ornithine transcarbamoylase.*

Step 4. **The carbon–nitrogen bond of arginine is hydrolyzed in a reaction catalyzed by** *arginase* **to give the cycle product, urea, plus ornithine ready to repeat step 1.**

Step 3. Argininosuccinate is split into arginine (a protein amino acid) and fumarate (a cycle by-product). The enzyme is *argininosuccinase.*

Step 2. Citrulline condenses with aspartate (a protein amino acid) to give argininosuccinate, in which the two molecules have combined. The enzyme is *argininosuccinate synthase.*

Steps 1 and 2 of the urea cycle: Building up a reactive intermediate. The first step of the urea cycle transfers the carbamoyl group, $H_2NC\!=\!O$, from carbamoyl phosphate to *ornithine*, an amino acid not found in proteins, to give *citrulline*, another nonprotein amino acid. Note that the bond broken in carbamoyl phosphate is an energetic anhydride link. The reaction introduces the first nitrogen of the end-product urea into the cycle.

Next, a molecule of aspartate, a standard protein amino acid, condenses with citrulline in a reaction driven by conversion of ATP to AMP and pyrophosphate (PP_i), followed by the additional exergonic hydrolysis of pyrophosphate. Both of what will become the urea nitrogen atoms are now bonded to the same carbon atom in argininosuccinate.

Steps 3 and 4 of the urea cycle: Cleavage and hydrolysis of the step 2 product. Step 3 cleaves argininosuccinate into two pieces: arginine, an amino acid, and fumarate, an intermediate in the citric acid cycle. Now all that remains, in step 4, is the hydrolysis of arginine to give urea and to restore ornithine, the reactant in step 1 of the cycle.

Net result of the urea cycle

$$CO_2 + NH_4^+ + 3\,ATP + {}^-OOC-CH_2-\underset{\underset{NH_3^+}{|}}{CH}-COO^- + 2\,H_2O \longrightarrow$$

Aspartate

$$\underset{Urea}{H_2N-\overset{\overset{O}{\|}}{C}-NH_2} + 2\,ADP + AMP + 4\,P_i + \underset{Fumarate}{{}^-OOC-CH\!=\!CH-COO^-}$$

- Elimination as urea of carbon from CO_2, nitrogen from NH_4^+, and nitrogen from the amino acid aspartate
- Breaking of four high-energy phosphate bonds to provide energy
- Production of the citric acid cycle intermediate, fumarate.

Fumarate links the citric acid cycle with the urea cycle. Fumarate travels around the citric acid cycle to give oxaloacetate, which can be diverted to form aspartate by a reverse transamination, thereby replenishing this reactant for step 2 of the urea cycle.

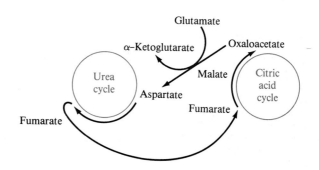

Practice Problem   25.3   Look at the citric acid cycle (Figure 20.11) and then write the structure of each reaction product in the pathway connecting fumarate and aspartate and describe the structural change in each reaction.

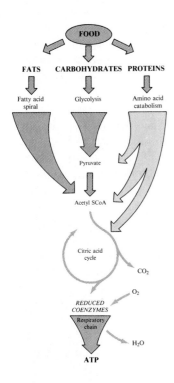

Glucogenic amino acid
An amino acid catabolized to a compound that can enter the citric acid cycle for conversion to glucose.

Ketogenic amino acid   An amino acid catabolized to a compound that can be converted to ketone bodies (also fatty acids).

25.5   AMINO ACID CATABOLISM: FATE OF THE CARBON ATOMS

The 20 amino acids found in proteins can be grouped according to the carbon-containing products of their catabolism. The amino acids listed in the blue boxes in Figure 25.8 can, instead of being completely catabolized, be converted to glucose. Recall that the first two reactions in glucose synthesis from noncarbohydrates (gluconeogenesis, Section 22.11) convert pyruvate to phosphoenolpyruvate, the high-energy intermediate needed to continue on the pathway.

$$\text{Pyruvate} \rightarrow \text{oxaloacetate} \rightarrow \text{phosphoenolpyruvate}$$

All amino acids catabolized to pyruvate or to citric-acid-cycle intermediates that precede oxaloacetate can enter gluconeogenesis via oxaloacetate. They are referred to as **glucogenic amino acids.**

Amino acids that are catabolized to acetyl SCoA or acetoacetyl SCoA can be used in the synthesis of ketone bodies and are therefore referred to as **ketogenic amino acids.** Ketone bodies, you may recall, are made from acetoacetyl SCoA synthesized from acetyl SCoA (Section 24.6). The ketogenic amino acids are also able to enter fatty acid biosynthesis via acetyl SCoA (Section 24.7). Of the ketogenic amino acids, there are only two—leucine and lysine—that aren't also glucogenic.

A few of the carbon-containing end products of amino acid catabolism result from simple transformations. Alanine, for example, gives pyruvate on transamination, and serine is converted to pyruvate in a single step that results in loss of the amino group.

$$\underset{\text{Serine}}{\underset{\underset{NH_3^+}{|}}{HOCH_2CHCOO^-}} \xrightarrow[\text{dehydratase}]{\text{serine}} \underset{\text{Pyruvate}}{\overset{O}{\overset{\|}{CH_3CCOO^-}}} + NH_4^+$$

Asparagine is hydrolyzed to aspartate, followed by transamination of aspartate to give oxaloacetate.

$$\underset{\text{Asparagine}}{\overset{O}{\overset{\|}{H_2NCCH_2}}\underset{\underset{NH_3^+}{|}}{CHCOO^-}} + H_2O \longrightarrow \underset{\text{Aspartate}}{\overset{-}{OOCCH_2}\underset{\underset{NH_3^+}{|}}{CHCOO^-}} + NH_4^+$$

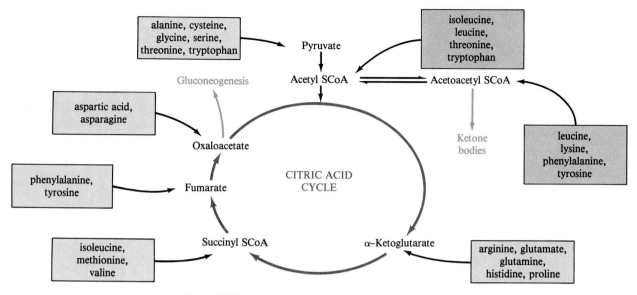

Figure 25.8
Amino acid catabolism. The carbon atoms of different amino acids enter the citric acid cycle as the indicated compounds. Glucogenic amino acids are listed in blue boxes, and ketogenic amino acids are listed in pink boxes.

The amino acids that enter the citric acid cycle as α-ketoglutarate are first transformed by one- to four-step pathways to glutamate, which is then converted by transamination to α-ketoglutarate containing the amino acid carbon atoms. The larger amino acids arrive at their end products through more complex pathways.

Practice Problem   25.4   Arginine is converted to ornithine in the last step of the urea cycle (Figure 25.7). To carry arginine carbon atoms to the citric acid cycle, ornithine undergoes transamination at its terminal amino group to give an aldehyde, followed by oxidation to glutamate and conversion to α-ketoglutarate. Write the structures of the five molecules in the pathway beginning with arginine and ending with α-ketoglutarate. Circle the region of structural change in each.

25.6 BIOSYNTHESIS OF NONESSENTIAL AMINO ACIDS

Nonessential amino acid
One of 10 amino acids that are synthesized in the body and are therefore not necessary in the diet.

Essential amino acid   One of 10 amino acids that cannot be synthesized by the body and so must be obtained in the diet.

We said in Section 18.2 that humans are able to synthesize only 10 of the 20 amino acids found in proteins, known as the **nonessential amino acids** because they don't have to be in our diets. The remaining 10 are **essential amino acids** that must be obtained from dietary sources (Table 25.1). In evolutionary terms, it seems that when the diet could provide an amino acid that would otherwise have to be synthesized by a complex route, our metabolism gave up on making it. The nonessential amino acids can all be made in 1–3 steps, but the essential amino acids would require 7–10 steps based on observations of their synthesis in microorganisms.

Table 25.1   Amino Acids Essential and Nonessential in the Diet

Essential Amino Acids	Nonessential Amino Acids
Histidine	Alanine
Isoleucine	Arginine
Leucine	Asparagine
Lysine	Aspartate
Methionine	Cysteine
Phenylalanine	Glutamate
Threonine	Glutamine
Tryptophan	Glycine
Tyrosine[a]	Proline
Valine	Serine

[a] Sometimes classified as nonessential because it is synthesized from phenylalanine.

AN APPLICATION: NITROGEN BALANCE AND KWASHIORKOR

A healthy adult is normally at *nitrogen equilibrium,* meaning that the amount of nitrogen taken in each day is equal to the amount excreted. The recommended daily dietary allowance for the average 70 kg adult is 56 g of protein per day, equivalent to 0.8 g of protein per kilogram of ideal body weight. Total protein needs are increased when the diet is low in calories from fats and carbohydrates, when the foods consumed have a poor distribution of essential amino acids, or when physical activity increases. The average protein intake in the United States is about 110 g/day, well above what most of us need.

Infants and children, pregnant women, those recovering from starvation, and those with healing wounds are usually in *positive nitrogen balance,* that is, they're excreting less nitrogen than they consume, a condition to be expected when new tissue is growing. The reverse condition, *negative nitrogen balance,* occurs when more nitrogen is excreted than consumed. This happens when protein intake is inadequate, during starvation, and in a number of pathologic conditions including malignancies, malabsorption syndromes, and kidney disease.

Health and nutrition professionals group all disorders caused by inadequate protein intake as *protein energy malnutrition (PEM).* Children, because of their higher protein needs, suffer most from such a diet. The problem is rampant where meat and milk are in short supply and where the dietary staples are vegetables or grains.

Protein deficiency alone is rare, and its symptoms are often accompanied by those of vitamin deficiencies, infectious diseases, and starvation. At one end of the spectrum is *kwashiorkor,* in which protein is deficient although caloric intake may be adequate. Children with kwashiorkor have edema (swelling due to water retention), have an enlarged, fatty liver, and are underdeveloped. The word "kwashiorkor" is from the language of Ghana and is translated as "the sickness the older gets when the next child is born," which is the time when weaning from mother's milk means conversion to a high-carbohydrate, low-protein diet. At the other end of the spectrum is *marasmus,* a condition caused mainly by starvation. As distinguished from kwashiorkor, marasmus is identified with severe muscle wasting, below-normal stature, and poor response to treatment. As many as 1 billion people worldwide suffer from PEM, most in the poor populations of developing countries.

Reductive amination
Conversion of an α-keto acid to an amino acid by reaction with NH_4^+.

All of the 10 nonessential amino acids derive their amino groups from glutamate. The glutamate itself can be made from ammonia and α-ketoglutarate by **reductive amination,** the reverse of the oxidative deamination so common in amino acid catabolism (Section 25.3). The same glutamate dehydrogenase enzyme carries out the reaction.

$$NH_4^+ \;+\; {}^-OOCCH_2CH_2\overset{\displaystyle O}{\overset{\|}{C}}-COO^- \;\xrightarrow[\substack{\text{glutamate}\\\text{dehydrogenase}}]{\text{NADH(NADPH)}\quad\text{NAD}^+\text{(NADP}^+\text{)}}\; {}^-OOCCH_2CH_2\underset{\underset{NH_3^+}{|}}{C}HCOO^- \;+\; H_2O$$

α-Ketoglutarate   Glutamate

The following four common metabolic intermediates, which you've seen by now in many roles, are the precursors for synthesis of the nonessential amino acids. As examples of how the intermediates are converted to amino acids, a few syntheses are shown in Table 25.2.

Table 25.2 Some Examples of Amino Acid Biosynthesis

(1) $CH_3\overset{\displaystyle O}{\overset{\|}{C}}COO^- \xrightarrow{\text{transamination}} CH_3\underset{\underset{NH_3^+}{|}}{C}HCOO^-$

Pyruvate   Alanine

(2) ${}^-OOCCH_2CH_2\overset{\displaystyle O}{\overset{\|}{C}}COO^- \xrightarrow{\text{transamination}} {}^-OOCCH_2CH_2\underset{\underset{NH_3^+}{|}}{C}HCOO^- \xrightarrow{\text{phosphorylation}}$

α-Ketoglutarate   Glutamate

$$\left[\, {}^{2-}{}_3OPO\overset{\displaystyle O}{\overset{\|}{C}}CH_2CH_2\underset{\underset{NH_3^+}{|}}{C}HCOO^- \,\right] \longrightarrow H_2N\overset{\displaystyle O}{\overset{\|}{C}}CH_2CH_2\underset{\underset{NH_3^+}{|}}{C}HCOO^-$$

Glutamine

(3) ${}^-OOC\underset{\underset{OH}{|}}{C}HCH_2OPO_3^{2-} \xrightarrow{\text{oxidation}} {}^-OOC\overset{\displaystyle O}{\overset{\|}{C}}CH_2OPO_3^{2-} \xrightarrow{\text{transamination}}$

3-Phosphoglycerate   3-Phosphohydroxypyruvate

${}^-OOC\underset{\underset{NH_3^+}{|}}{C}HCH_2OPO_3^{2-} \xrightarrow{\text{hydrolysis}} {}^-OOC\underset{\underset{NH_3^+}{|}}{C}HCH_2OH$

3-Phosphoserine   Serine

Precursors in syntheses of nonessential amino acids

$CH_3\overset{O}{\overset{\|}{C}}{-}COO^-$ $\quad$ $^-OOC{-}\overset{O}{\overset{\|}{C}}CH_2COO^-$ $\quad$ $^-OOCCH_2CH_2\overset{O}{\overset{\|}{C}}{-}COO^-$ $\quad$ $^{2-}O_3POCH_2\underset{\underset{OH}{|}}{C}HCOO^-$

Pyruvate $\qquad$ Oxaloacetate $\qquad$ α-Ketoglutarate $\qquad$ 3-Phosphoglycerate

The amino acid tyrosine is sometimes classified as nonessential rather than essential because it is synthesized from phenylalanine. Whatever the classification, we have a high nutritional requirement for phenylalanine and

INTERLUDE: XENOBIOTICS

How do our bodies tackle a chemical compound they've never met before? The compound might be a drug, a pesticide, or an environmental pollutant. It could be addictive, it could be a carcinogen, or it could be a harmless compound that need only be excreted. In medical terminology, such compounds are known as *xenobiotics*—chemical compounds foreign to the body.

A xenobiotic is likely to encounter a two-pronged attack when it enters the body. First, it undergoes chemical reactions that make it hydrophilic and water-soluble, so that it doesn't accumulate in fatty tissues. Second, enzymes further modify its structure, converting it to a compound that is even more water-soluble and is likely to be excreted.

In the first reaction for many xenobiotics, a hydrogen atom is replaced by a hydroxyl group in a *hydroxylation* reaction carried out by a *monooxygenase* enzyme complex known as *cytochrome P-450*. (The enzyme name derives from a peak in its absorption spectrum at 450 nm.) The xenobiotic (RH), oxygen, and NADPH react as follows:

$$RH + O_2 \xrightarrow{\text{NADPH} \quad \text{NADP}^+} R{-}OH + H_2O$$

About half of all drugs are metabolized by cytochrome P-450. For example, hydroxylation

of phenobarbital solubilizes it for excretion. As you might expect, this protective action takes place mainly in the liver as blood passes through on its way to general circulation.

The second phase of xenobiotic metabolism is usually bonding through an oxygen, nitrogen, or sulfur atom to an even more polar group, in many cases the glucose derivative glucuronic acid or the amino acid derivative glutathione:

Glucuronide group

Glutathione group

several metabolic diseases are associated with defects in the enzymes needed to convert it to tyrosine. The best known of these diseases is *phenylketonuria (PKU),* the first inborn error of metabolism for which the biochemical cause was recognized. In 1947 it was found that failure of the tyrosine synthesis leads to PKU.

$$\text{Phenylalanine} \longrightarrow \text{Tyrosine}$$

Phenylalanine

Tyrosine

Cytochrome P-450 has both good and bad effects: good in detoxifying some poisons; bad in converting some compounds, especially polycyclic aromatic hydrocarbons like those in cigarette smoke, to carcinogenic compounds that react with and damage DNA. Like many biomolecules, the cytochrome is *inducible,* meaning that when more of it is needed, more of it is made.

It's well known that smokers have more cytochrome P-450 than nonsmokers. In fact, the wide variability of xenobiotic enzyme activity with age, sex, genetic makeup, and previous exposure to xenobiotics is thought to play a role in the wide variability in individual responses to carcinogens.

Smoke swirling down the windpipe and into the lungs. This person probably has more cytochrome P-450 than a nonsmoker.

PKU results in elevated blood serum and urine concentrations of phenylalanine, phenylpyruvate, and several other metabolites produced from excess phenylalanine. Undetected PKU causes mental retardation by the second month of life. Estimates are that, prior to the 1960s, 1% of those institutionalized for mental retardation were PKU victims. The only defense against PKU and similar treatable metabolic disorders that take their toll early in life is widespread screening of newborn infants. In the 1960s a test for PKU was introduced, and virtually all hospitals in the United States now routinely screen for PKU. Treatment consists of a restricted phenylalanine diet, which is maintained in infants with special formulas and in older individuals by eliminating meat and using low-protein grain products.

Practice Problem   25.5   Look at the reactions in Table 25.2 and identify those that are
(a) transaminations   (b) hydrolyses   (c) oxidations

SUMMARY

The digestion of dietary proteins results in complete enzyme-catalyzed hydrolysis of the proteins to yield amino acids. Digestion takes place in the stomach, where the enzyme is *pepsin,* and in the small intestine under the influences of several other **proteases.** The catabolism of amino acids is complex because different pathways are needed for each of the 20 amino acids in proteins. The pathways have in common the initial removal of the amino group, followed by conversion of the carbon atom part of the molecule to a compound that can enter the citric acid cycle.

The amino group is usually removed by **transamination,** in which the amino group of an amino acid and the keto group of α-ketoglutarate are interchanged to give glutamate and a new α-keto acid. In **oxidative deamination** the amino group of glutamate is removed as ammonium ion for reaction with bicarbonate ion to form carbamoyl phosphate. The carbamoyl phosphate is then converted to urea for excretion via the **urea cycle.** The urea cycle requires aspartate and is connected to the citric acid cycle by the production of fumarate. Fumarate is converted in the citric acid cycle to oxaloacetate from which the necessary aspartate is produced by transamination.

Glucogenic amino acids are those that are catabolized to pyruvate or to intermediates in the citric acid cycle. **Ketogenic amino acids** are catabolized to acetyl SCoA or acetoacetyl SCoA, from which ketone bodies can be synthesized. Ketogenic amino acids can also enter fatty acid synthesis via acetyl SCoA.

Nine of the 10 nonessential amino acids are biosynthesized from four precursor molecules (pyruvate, oxaloacetate, α-ketoglutarate, and 3-phosphoglycerate); their amino groups come from glutamate. Tyrosine is synthesized from the essential amino acid phenylalanine.

REVIEW PROBLEMS

Amino Acid Pool

25.6   In what part of the digestive tract does the digestion of proteins begin?

25.7   What is the body's amino acid pool?

25.8   What are the sources of the body's amino acid pool?

25.9   What are the fates of amino acids in the amino acid pool?

Amino Acid Catabolism

25.10   What is meant by transamination?

25.11   How is vitamin B_6 associated with the process of transamination?

25.12   What is the structure of the α-keto acid formed by transamination of the following amino acids with α-ketoglutarate? (See Table 18.1 for structures.)
(a) isoleucine   (b) cysteine   (c) phenylalanine

25.13 Pyruvate and oxaloacetate, as well as α-keto-glutarate, can be used as receptors for the amino group in transamination. Write the structures of the products formed when these two substances react in the transamination process.

25.14 What is meant by an oxidative deamination reaction?

25.15 What coenzymes are associated with oxidative deamination?

25.16 What type of enzyme is associated with oxidative deamination?

25.17 Write the structures of the α-keto acids produced by oxidative deamination of
(a) phenylalanine   (b) tryptophan

25.18 What is the other product formed in oxidative deamination besides an α-keto acid?

25.19 What is a glucogenic amino acid?

25.20 What is a ketogenic amino acid?

The Urea Cycle

25.21 Why does the body convert NH_4^+ to urea for excretion?

25.22 Why might the urea cycle be called the ornithine cycle?

25.23 What are the sources of the carbon and each of the two nitrogens in urea?

Amino Acid Anabolism

25.24 How do essential and nonessential amino acids differ in the number of steps required for their synthesis?

25.25 What substance serves as the starting material for amino acid anabolism?

25.26 What is the name of the process by which amino acids are made from common nonnitrogen metabolites? What process is this the reverse of?

25.27 How is tyrosine made in the body?

25.28 What causes PKU, and what are its symptoms?

Applications

25.29 What is positive nitrogen balance? [App: Nitrogen Balance and Kwashiorkor]

25.30 What are some conditions that require a positive nitrogen balance? [App: Nitrogen Balance and Kwashiorkor]

25.31 What is negative nitrogen balance? [App: Nitrogen Balance and Kwashiorkor]

25.32 What are some conditions that cause a negative nitrogen balance to occur? [App: Nitrogen Balance and Kwashiorkor]

25.33 How do marasmus and kwashiorkor differ in cause and in symptoms? [App: Nitrogen Balance and Kwashiorkor]

25.34 What does the term "xenobiotic" mean? [Int: Xenobiotics]

25.35 In what two ways does a body respond to a xenobiotic agent? [Int: Xenobiotics]

Additional Questions and Problems

25.36 Why is the formation of urea an energy-expensive process?

25.37 How can amino acids "pay" for the disposing of extra nitrogen?

25.38 Write the equation for the transamination reaction that occurs between valine and pyruvate. What do you think is the name of the associated enzyme?

25.39 Detail the five points at which amino acids can enter the citric acid cycle.

25.40 Can an amino acid be both glucogenic and ketogenic? Why or why not?

25.41 Briefly explain how the carbons from amino acids can end up in adipose tissue.

25.42 Consider all the metabolic processes that we have studied. Why do we say that tissue is dynamic? Detail some examples of dynamic relationships.

CHAPTER

26
Nucleic Acids and Protein Synthesis

We've had Superman and Superwoman in comic strips. Maybe, thanks to genetic engineering, someday we'll have real Supercattle that produce low-fat milk and low-cholesterol beef.

How does a seed "know" what kind of plant to become? How does a fertilized ovum know how to grow into a human being? And how does a cell know what part of the body it's in, whether brain or big toe, so that it can produce the right chemicals necessary for sustaining life? The answers to these and myriad other questions about all living organisms reside in the biological molecules called *nucleic* (nu-**clay**-ic) *acids*.

The nucleic acids, *deoxyribonucleic acid (DNA)* and *ribonucleic acid (RNA),* are the chemical carriers of an organism's genetic information. Coded in an organism's DNA is all the information that determines the nature of the organism, whether dandelion, goldfish, or human being. Coded also in DNA are all the instructions needed for cellular functioning, and all the directions needed for producing the tens or hundreds of thousands of different proteins required by the organism. In this chapter, we'll answer the following questions about nucleic acids:

1. *What is the composition of the nucleic acids, DNA and RNA?* The goal: Be able to describe and identify the components of nucleosides, nucleotides, DNA, and RNA.
2. *What is the structure of DNA?* The goal: Be able to describe the double helix and base pairing in DNA.
3. *How is DNA reproduced?* The goal: Be able to explain the process of DNA replication.
4. *What are the functions of RNA?* The goal: Be able to list the types of RNA, their locations in the cell, and their functions.
5. *How do organisms synthesize RNA?* The goal: Be able to explain the process of transcription.
6. *How does RNA control protein synthesis?* The goal: Be able to explain the genetic code and describe each step in translation.
7. *What are mutations?* Be able to describe mutations, how they might occur, and their role in hereditary disease.

26.1 DNA, CHROMOSOMES, AND GENES

The terms "chromosome" and "gene" were coined long before the chemical nature of these cell components was understood. A "chromosome" was a structure in the cell nucleus thought to be the carrier of *genetic information*—all the information needed by an organism to duplicate itself. A "gene" was the portion of a chromosome that controlled a specific inheritable trait such as brown eyes or red hair. We've come a long way since these terms were introduced, with especially dramatic progress since the 1970s thanks to a number of new techniques we'll discuss shortly.

When a cell is not actively dividing, its nucleus is occupied by *chromatin,* a tangle of fibers composed of protein and **DNA (deoxyribonucleic acid),** the

DNA (deoxyribonucleic acid) The nucleic acid that stores the genetic information.

Chromosome A complex between proteins and a DNA molecule that is visible during cell division.

Gene All segments of DNA needed to direct the synthesis of a protein with a specific function.

Nucleic acid A biological polymer made by the linking together of nucleotide units.

Nucleotide A building block for nucleic acid synthesis, consisting of a five-carbon sugar bonded to a cyclic amine base and to phosphoric acid.

Heterocycle A ring that contains nitrogen or some other atom in addition to carbon.

RNA (ribonucleic acid) Nucleic acid responsible for putting the genetic information to use in protein synthesis.

nucleic acid molecule that stores genetic information. Just prior to cell division, the DNA is duplicated so that each new cell can receive a complete copy. During cell division chromatin organizes itself into **chromosomes**, each of which contains a DNA molecule (Figure 26.1). Each DNA molecule, in turn, is made up of many **genes**—individual segments of the DNA that contain the instructions needed to direct the synthesis of a protein with a specific function.

Different organisms differ in their complexity and therefore have different numbers of chromosomes and genes. A frog, for example, has 26 chromosomes (13 pairs), whereas a human has 46 chromosomes (23 pairs). Each chromosome contains a single immense molecule of DNA that, in humans, has a molecular weight of up to 150 *billion* amu and a length of up to *12 centimeters* when stretched out. The 23 pairs of human chromosomes are estimated to include about 100,000 genes.

26.2 COMPOSITION OF NUCLEIC ACIDS

Like proteins and carbohydrates, **nucleic acids** are polymers made up of individual molecules linked together in long chains. Proteins are polypeptides, carbohydrates are polysaccharides, and nucleic acids are poly*nucleotides*. A **nucleotide** can itself be further broken down by hydrolysis to yield three components: a simple aldopentose sugar, a **heterocyclic** amine base, and phosphoric acid.

$$\text{Nucleic acid} \xrightarrow{\text{hydrolysis}} \text{many nucleotides} \xrightarrow{\text{hydrolysis}} \begin{array}{c} \text{aldopentose sugars} \\ + \\ \text{amine bases} \\ + \\ H_3PO_4 \end{array}$$

Figure 26.1
Human chromosomes. Chromosomes are roughly X-shaped during cell division and have four banded arms connected at a centromere.

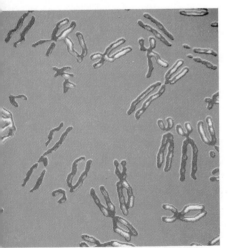

There are two types of nucleic acids: DNA and RNA. DNA stores genetic information, and **RNA (ribonucleic acid)** makes possible the use of that information. Before we discuss the overall structure of the nucleic acids, it's important to understand how their component parts are joined together and how DNA and RNA differ.

Nucleosides In RNA, the sugar component is D-ribose, as indicated by the name "*ribo*nucleic acid." In DNA, or *deoxy*ribonucleic acid, the sugar component is 2-D-*deoxy*ribose. (The prefix "2-deoxy-" means that an oxygen atom is missing from the C2 position of ribose.)

Ribose

2-Deoxyribose — Oxygen missing

Nucleoside A compound consisting of a five-carbon sugar bonded to a cyclic amine base; like a nucleotide but missing the phosphate group.

The compounds formed by bonding of either of these sugars to a heterocyclic amine are known as **nucleosides.** As illustrated in the following general formula, the sugar and the amine are connected by a β-glycosidic bond between a heterocyclic ring nitrogen and the sugar's anomeric carbon atom (C1).

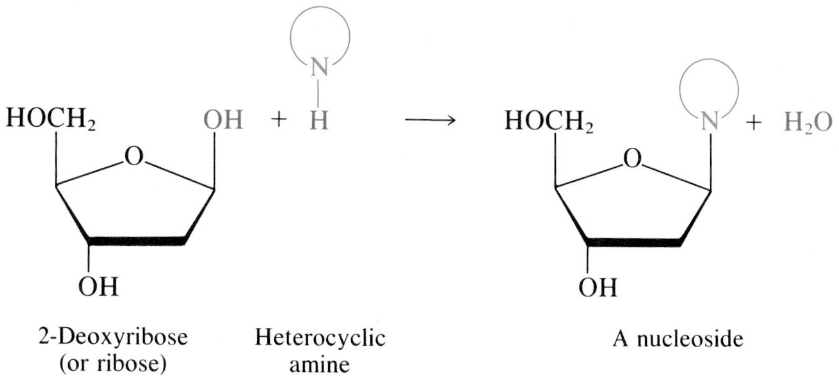

2-Deoxyribose (or ribose)   Heterocyclic amine   A nucleoside

Five heterocyclic amine bases are found in DNA and RNA. Two are derivatives of purine, and three are derivatives of pyrimidine (Section 15.3). The hydrogen atoms lost in the formation of nucleosides are shown in color in the following structures. Adenine, guanine, and cytosine are included in both DNA and RNA, thymine is part of only DNA molecules, and uracil is part of only RNA molecules.

Bases in DNA and RNA

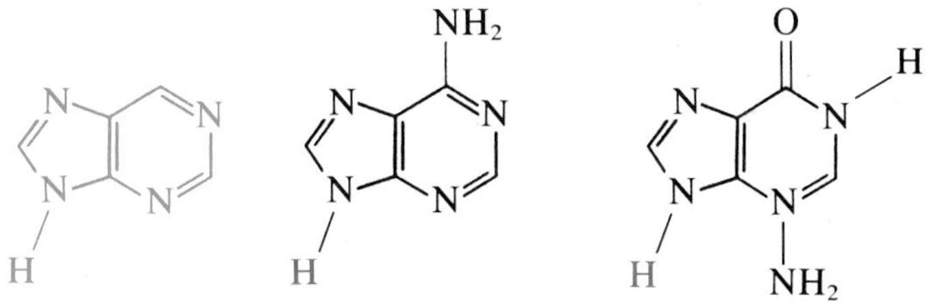

Purine   Adenine (DNA, RNA)   Guanine (DNA, RNA)

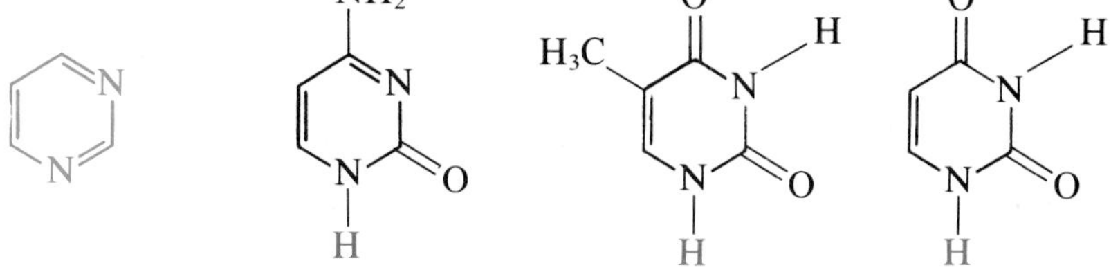

Pyrimidine   Cytosine (DNA, RNA)   Thymine (DNA)   Uracil (RNA)

Nucleosides are named with the base name modified by the ending -*osine* for the purine bases and -*idine* for the pyrimidine bases. No prefix is used for names of nucleosides containing ribose, but the prefix "deoxy-" is added for those that contain deoxyribose. To distinguish between them, numbers without primes are used for the nitrogen base, and numbers with primes are used for the sugar ring.

Two nucleosides

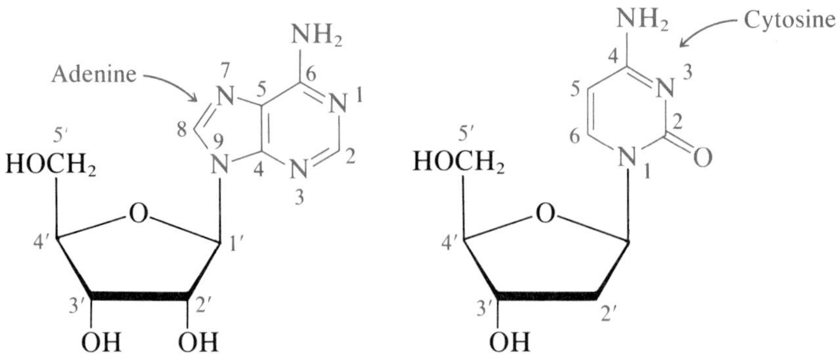

Adenosine

Deoxycytidine

Nucleotides   The nucleotides that are the building blocks of nucleic acids are monophosphate esters of nucleosides, formed by reaction of phosphoric acid with the alcohol —OH group on C5′ of the sugar. The phosphate group is ionized in body fluids and is often represented as $-OPO_3^{2-}$.

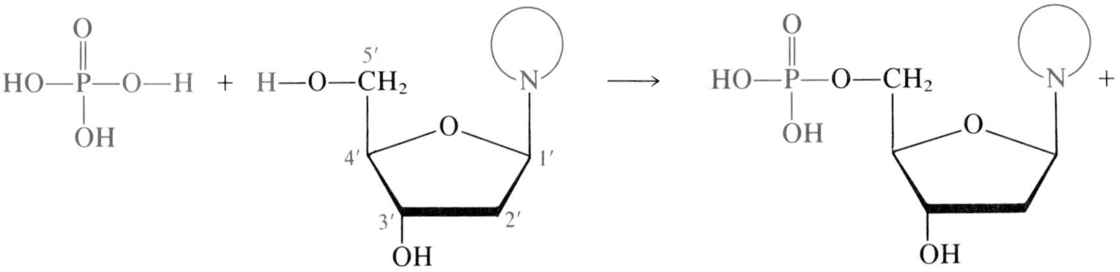

Phosphoric acid

A deoxyribonucleoside

A deoxyribonucleotide

Nucleotides are named by adding *5′-monophosphate* to the name of the nucleoside. The nucleotides corresponding to adenosine and deoxycytidine are thus adenosine 5′-monophosphate (AMP) and deoxycytidine 5′-monophosphate (dCMP). Nucleotides containing D-ribose are known as **ribonucleotides,** and those containing D-deoxyribose are known as **deoxyribonucleotides.**

Ribonucleotide A nucleotide containing D-2-ribose.

Deoxyribonucleotide A nucleotide containing D-deoxyribose.

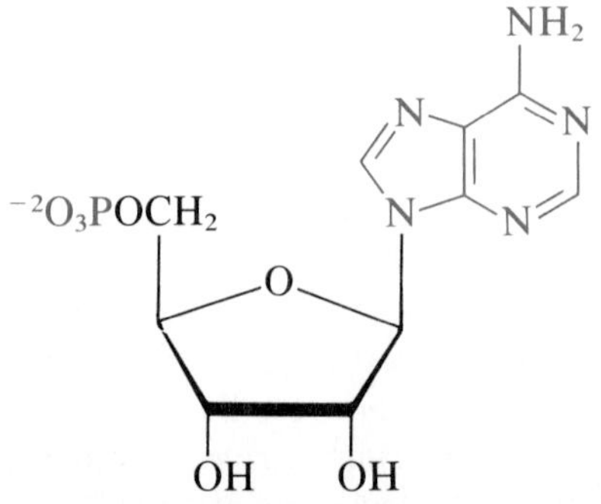

Adenosine 5′-monophosphate (AMP)
(a ribonucleotide)

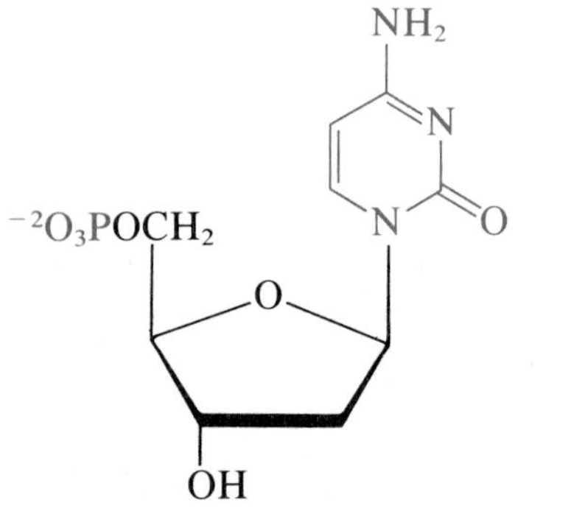

Deoxycytidine 5′-monophosphate (dCMP)
(a deoxyribonucleotide)

The names of the bases, nucleosides, and nucleotides are summarized in Table 26.1 together with their abbreviations. We rely heavily on abbreviations in biochemistry, so it's important to be familiar with them. Any of the nucleosides or nucleotides can add phosphate groups to form diphosphate or triphosphate esters like the guanosine phosphates. As you've seen on numerous occasions, a number of these phosphates—particularly adenosine triphosphate, ATP—play essential roles in biochemistry.

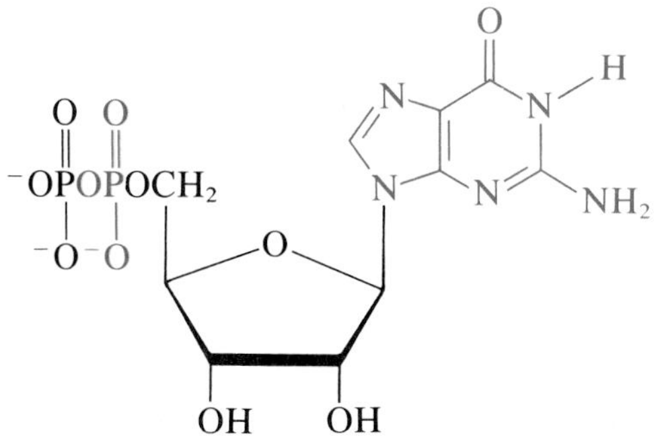

Guanosine diphosphate (GDP) 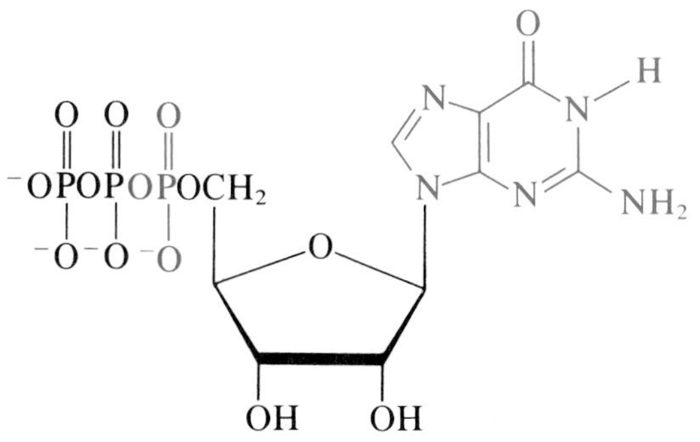   Guanosine triphosphate (GTP)

Summary of nucleic acid composition

DNA (deoxyribonucleic acid)

- Sugar is D-2-deoxyribose.
- DNA is a polymer of deoxyribonucleotides.
- Bases are adenine, guanine, cytosine, and *thymine*.

RNA (ribonucleic acid)

- Sugar is D-ribose.
- RNA is a polymer of ribonucleotides.
- Bases are adenine, guanine, cytosine, and *uracil* (instead of thymine).

Table 26.1   Names of Bases, Nucleosides, and Nucleotides in DNA and RNA

Bases	Nucleosides	Nucleotides	
DNA			
Adenine (A)	Deoxyadenosine	Deoxyadenosine 5-monophosphate	dAMP
Guanine (G)	Deoxyguanosine	Deoxyguanosine 5'-monophosphate	dGMP
Cytosine (C)	Deoxycytidine	Deoxycytidine 5'-monophosphate	dCMP
Thymine (T)	Deoxythymidine	Deoxythymidine 5'-monophosphate	dTMP
RNA			
Adenine (A)	Adenosine	Adenosine 5-monophosphate	AMP
Guanine (G)	Guanosine	Guanosine 5'-monophosphate	GMP
Cytosine (C)	Cytidine	Cytidine 5'-monophosphate	CMP
Uracil (U)	Uridine	Uridine 5'-monophosphate	UMP

Solved Problem 26.1 Classify the following structure as that of a nucleoside or a nucleotide, identify its sugar and base components, name it, and give its abbreviation.

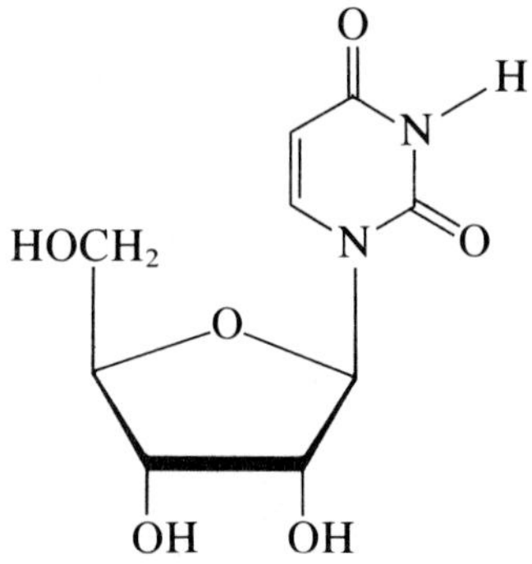

Solution The compound contains a sugar and a nitrogen base, but no phosphate group, so it is a *nucleoside*. The sugar has an —OH in the 2' position and is therefore *ribose*. Checking the base structures given earlier shows that this one is *uracil*, a pyrimidine base. The nucleoside is therefore named *uridine*.

Practice Problems

26.1 Draw the structure of 2'-deoxythymidine 5'-phosphate.

26.2 Draw the structure of the triphosphate of deoxycytidine.

26.3 Write the full names of dTMP, AMP, ADP, and ATP.

26.3 THE STRUCTURE OF NUCLEIC ACID CHAINS

Nucleotides are joined together in DNA and RNA by phosphate ester bonds between the phosphate component of one nucleotide and the carbon 3' —OH group on the sugar component of the next nucleotide:

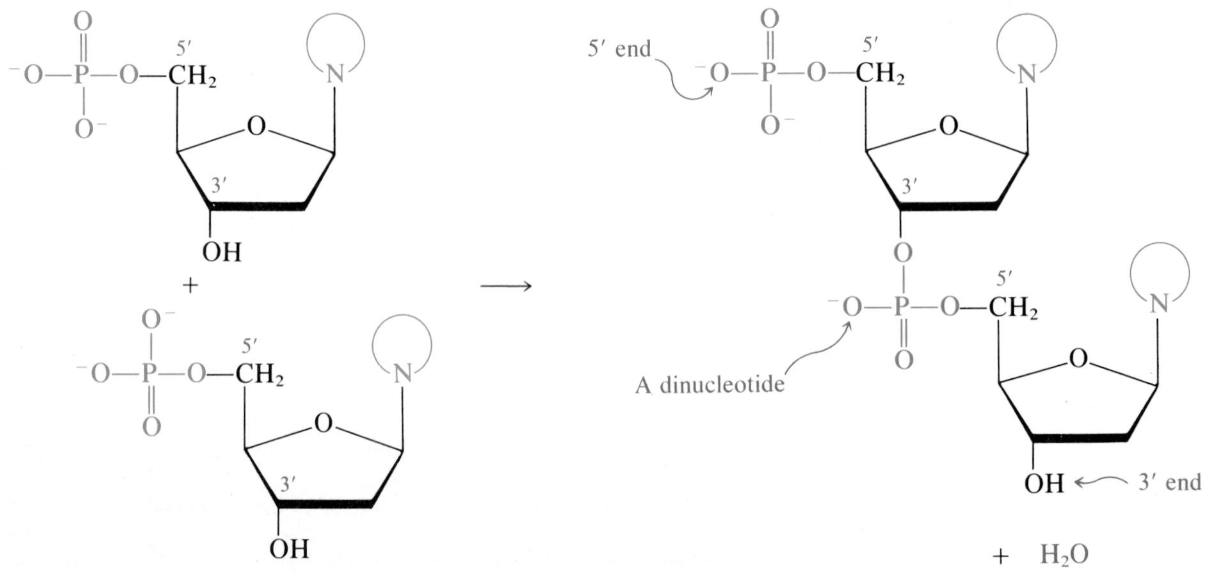

A dinucleotide

+ H₂O

Additional nucleotides join by formation of still more phosphate ester bonds until ultimately an immense chain is formed (Figure 26.2). Regardless of how long a polynucleotide chain is, one end of the nucleic acid molecule always has a free —OH group at C3′ (called the *3′ end*), and the other end of the molecule always has a phosphoric acid group at C5′ (the *5′ end*).

Just as the exact structure of a protein depends on the sequence in which the individual amino acids are connected (Section 18.6), the exact structure of a nucleic acid molecule depends on the sequence in which individual nucleotides are connected. To carry the analogy even further, just as a protein has a polyamide backbone with different side chains attached to it at regular intervals, a nucleic acid has an alternating sugar-phosphate backbone with different heterocyclic amine bases attached to it at regular intervals.

The sequence of nucleotides in a chain is described by starting at the 5′ end and identifying the bases in order of occurrence. Rather than write the full name of each nucleotide or each base, however, it's more convenient to use the simple one-letter abbreviations of the bases: A for adenine, G for guanine, C for cytosine, T for thymine (and U for uracil in RNA). Thus, a typical sequence in DNA might be written as -T-A-G-G-C-T-.

Figure 26.2
Generalized structure of a nucleic acid chain. Phosphate ester bonds link the individual nucleotides together.

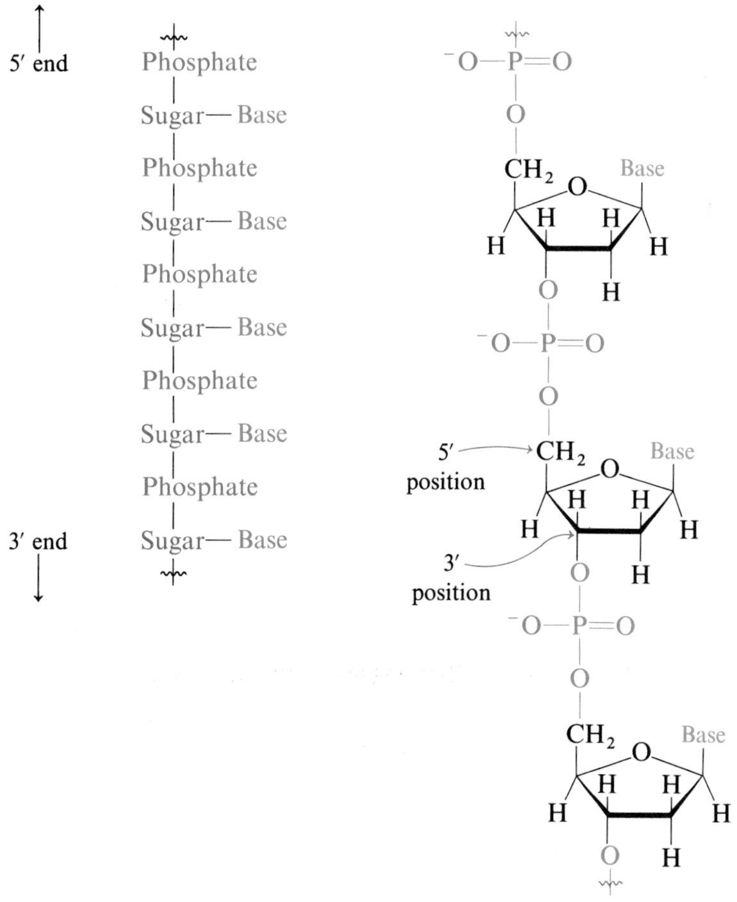

Comparison of protein and nucleic acid backbones and side chains

A protein: N terminus — Different side-chains — C terminus

R^1 O R^2 O R^3 O R^4 O R^5 O

NH—CH—C—NH—CH—C—NH—CH—C—NH—CH—C—NH—CH—C

Amide bonds

A nucleic acid: 5′ End — Different bases — 3′ End

Base^1 Base^2 Base^3

Phosphate—Sugar—Phosphate—Sugar—Phosphate—Sugar

Phosphate ester bonds

Practice Problem   **26.4**   Write the full structure of the DNA dinucleotide A-G. Identify the 5′ and 3′ ends of the dinucleotide.

26.4   BASE PAIRING IN DNA: THE WATSON-CRICK MODEL

DNA samples from different cells of the same species have the same proportions of the four heterocyclic bases, but samples from different species can have quite different proportions of bases. For example, human DNA contains about 30% each of adenine and thymine, and 20% each of guanine and cytosine. The bacterium *Escherichia coli,* however, contains only about 24% each of adenine and thymine, and 25% each of guanine and cytosine. Note that in both cases, the bases occur in *pairs.* A and T are usually present in equal amounts, as are G and C. Why should this be?

In 1953, James Watson and Francis Crick proposed a structure for DNA that not only accounts for the pairing of bases but also provides a simple method for the storage and transfer of genetic information. According to the *Watson-Crick model* (Figure 26.3), a DNA molecule consists of *two* polynucleotide strands coiled around each other in a helical, screw-like fashion. The sugar-

Figure 26.3
The double-helical structure of DNA. The sugar-phosphate backbone (blue) runs along the outside of the helix, while amine bases (purines in red and pyrimidines in green) hydrogen-bond to one another on the inside. A section of the wider major groove is at the bottom in the picture, with a section of the minor groove immediately above it.

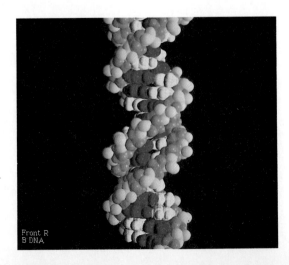

Front R
B DNA

Double helix Two strands coiled around each other in a screw-like fashion, as in the two polynucleotide strands in DNA.

phosphate backbone is on the outside of the so-called **double helix,** and the heterocyclic bases are on the inside, so that a base on one strand points directly toward a base on the second strand (Figure 26.4).

The two strands of the DNA double helix run in opposite directions, one in the $5' \rightarrow 3'$ direction, the other in the $3' \rightarrow 5'$ direction, as shown in Figure 26.4. The strands are held together by hydrogen bonds between bases. This hydrogen bonding is not random, but rather is the basis for the structure of DNA, just as hydrogen bonding interactions determine the secondary structure of proteins. In the double helix, adenine and thymine form two hydrogen bonds to each other but not to cytosine or guanine. Similarly, cytosine and guanine form three hydrogen bonds to each other in the double helix, but not to adenine or thymine.

Hydrogen-bonding base pairs in DNA

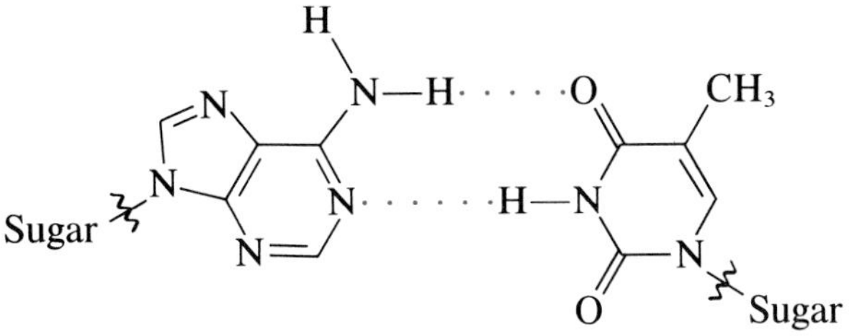

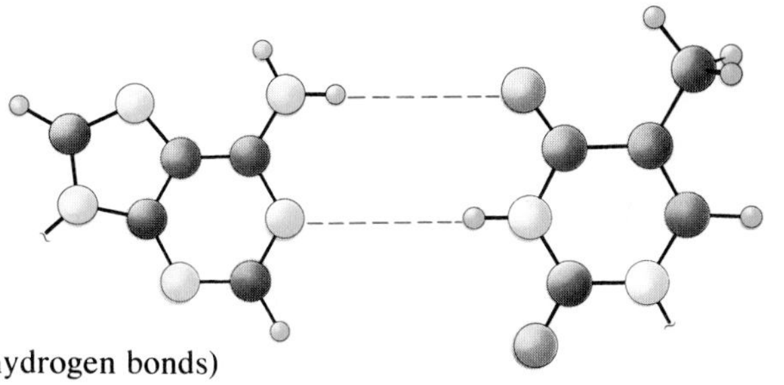

Adenine-thymine (two hydrogen bonds)

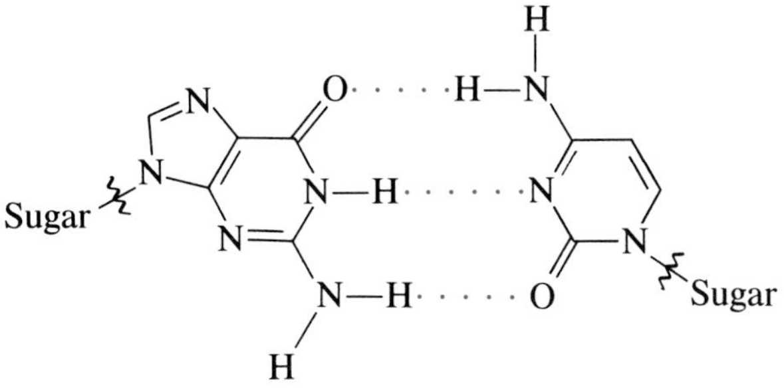

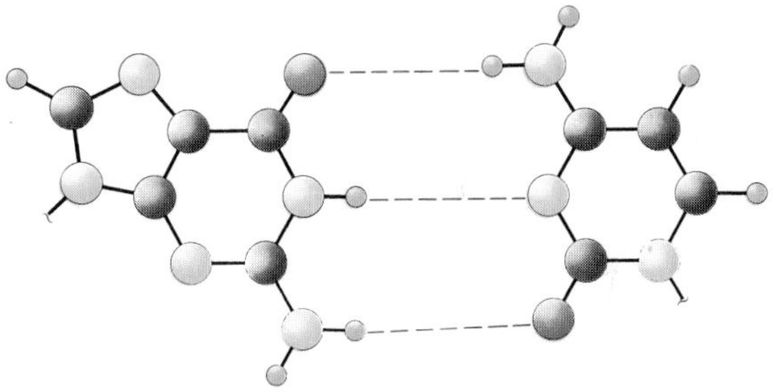

Guanine-cytosine (three hydrogen bonds)

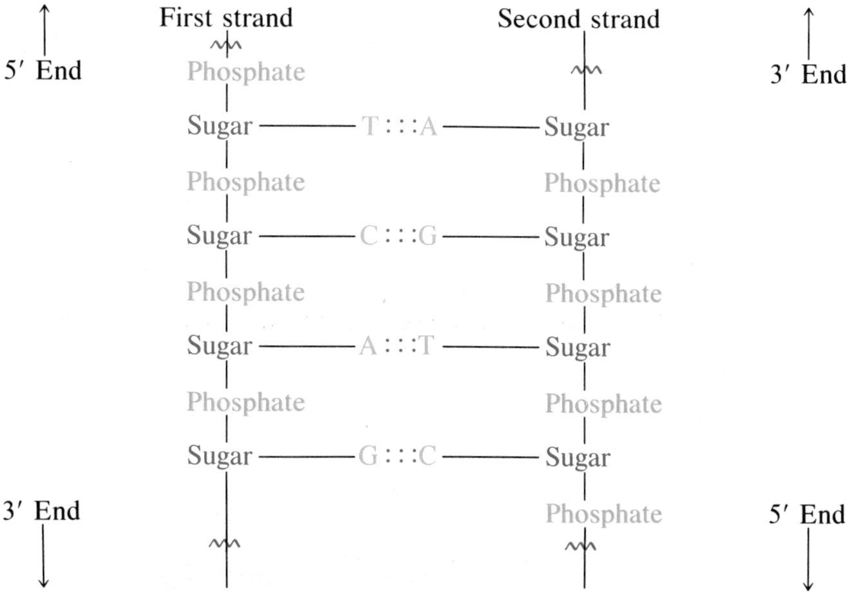

Figure 26.4
Complementary base pairing in the DNA double helix.

The two strands of the DNA double helix are *complementary* rather than identical. Whenever an A base occurs in one strand, a T base occurs opposite it in the other strand; when a C base occurs in one strand, a G base occurs in the other. This complementary *base pairing* in the two strands explains why A/T and C/G always occur in equal amounts.

As indicated in Figure 26.3, the DNA double helix is like a twisted ladder, with the sugar-phosphate backbone making up the sides and the hydrogen-bonded base pairs the rungs. X-ray measurements show that the helix is 2.0 nm wide, that there are exactly 10 nucleotides in each full turn, and that each turn is 3.4 nm long. A helpful way to remember the base pairing in DNA is to memorize the phrase "Pure silver taxi."

Pure	Silver	Taxi
Pur	**AG**	**TC**
The purines	A and G	pair with T and C

Notice in Figure 26.3 that the two strands of the double helix coil in such a way that two kinds of "grooves" result, a *major groove* and a *minor groove*. The major groove is slightly deeper than the minor groove, and both are lined by potential hydrogen-bond donors and acceptors. Thus, a variety of molecules are able to *intercalate*, or fit, into one of the grooves between the strands. A large number of cancer-causing and cancer-preventing agents are thought to function by interacting with DNA in this way.

Solved Problem 26.2 What sequence of bases of one strand of DNA is complementary to the sequence T-A-T-G-C-A-G on the other strand?

Solution Remembering that A and G (silver) bond to T and C (taxi), respectively, go through the original sequence replacing each A by T, each G by C, each T by A, and each C by G.

Original: T-A-T-G-C-A-G
              : : : : : : :
Complement: A-T-A-C-G-T-C

Practice Problems **26.5** What sequences of bases on one DNA strand are complementary to these sequences on another strand?
(a) G-C-C-T-A-G-T   (b) A-A-T-G-G-C-T-C-A

26.6 Why doesn't DNA include any adenine-guanine base pairs?

26.7 Draw the structures of adenine and uracil (which replaces thymine in RNA) and show the hydrogen bonding between them.

26.5 NUCLEIC ACIDS AND HEREDITY

How do organisms use nucleic acids to store and express genetic information? If the information is to be preserved and passed on from generation to generation, a mechanism must exist for copying DNA. If the information is to be used, mechanisms must exist for decoding the information and for carrying out the instructions therein.

According to what has been called the *central dogma of molecular genetics,* the function of DNA is to store information and pass it on to RNA, while the function of RNA is to read, decode, and use the information received from DNA to make proteins. Each of the hundreds or thousands of individual genes on a DNA molecule contains the instructions necessary to make a specific protein that is in turn needed for a specific biological purpose. The primary structure of each enzyme, for example, is coded for by a different gene. By decoding the right genes at the right time in the right place, an organism can use genetic information to synthesize the many thousands of proteins necessary to carry out the biochemical reactions required for smooth functioning.

Three fundamental processes take place in the transfer and use of genetic information:

$$\text{DNA} \xrightarrow{\text{replication}} \text{DNA} \xrightarrow{\text{transcription}} \text{mRNA} \xrightarrow{\text{translation}} \text{Proteins}$$

Replication The process by which copies of DNA are made in the cell.

1. Replication is the process by which a replica, or identical copy, of DNA is made. Replication occurs every time a cell divides so that information can be preserved and handed down to offspring.

Transcription The process by which the information in DNA is read and used to synthesize mRNA.

2. Transcription is the process by which the genetic messages contained in DNA are "read," or transcribed. The product of transcription, known as *messenger RNA* (mRNA), leaves the cell nucleus and carries the message to the sites of protein synthesis.

Translation The process by which mRNA directs protein synthesis.

3. Translation is the process by which the genetic messages carried by mRNA are decoded and used to build proteins.

In the following sections, we'll look at these important processes, focusing on what happens to DNA, the RNAs, and the proteins being synthesized. The complex processes of replication, transcription, and translation proceed with great accuracy and require participation by many other types of molecules. Energy-supplying nucleoside triphosphates and many different enzymes play essential roles, some understood, some not yet fully understood, and most likely some not yet discovered. At this point, however, our goal is to present an overview of how the genetic information is duplicated and put to work.

26.6 REPLICATION OF DNA

The Watson-Crick double-helix model of DNA does more than just explain base pairing, it also provides for an explanation of the ingenious way that DNA molecules reproduce exact copies of themselves. DNA replication begins with a partial unwinding of the double helix (Figure 26.5). As the two DNA strands separate and the bases are exposed, the enzyme *DNA polymerase* moves into

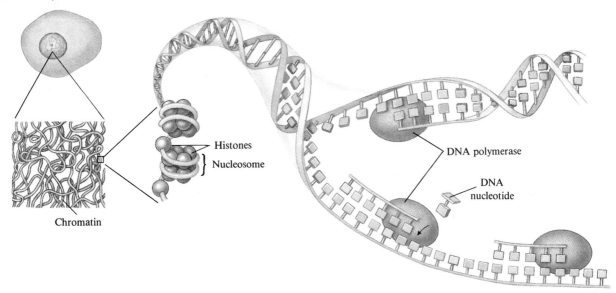

Histones

Nucleosome

Chromatin

DNA polymerase

DNA nucleotide

Figure 26.5
DNA replication. In a nondividing cell, chromatin consists of the DNA double helix coiled around *histones* (proteins) in a series of beads known as *nucleosomes*. When replication is about to occur, the DNA unwinds from the histones and the hydrogen bonds of the double helix open up, allowing DNA polymerase to move along the strands.

position at the point where synthesis will begin. One by one, nucleotide triphosphates approach and hydrogen-bond to the bases in each strand in an exactly complementary manner, A to T, and G to C. DNA polymerase then catalyzes bond formation between each arriving nucleotide and its neighbors, accompanied by removal of the extra phosphate groups, and the two new strands begin to grow.

Since each new strand is complementary to its old template strand, two identical new copies of the DNA double helix are produced during replication. In each new helix, one strand is the old template and the other is newly synthesized, a result described by saying that the replication is *semi-conservative*. The process is shown schematically in Figure 26.6.

Crick probably described the DNA replication process best when he described the fitting together of two DNA strands as being like a hand in a glove. The hand and glove separate, a new hand forms inside the old glove, and a new glove forms around the old hand. Two identical copies now exist where only one existed before.

DNA polymerase catalyzes the reaction between the 5′ phosphate on an incoming nucleotide and the free 3′ OH on the growing polynucleotide. As a result, the new DNA strands can grow only in the 5′ → 3′ direction, and strand growth must begin at the 3′ end of the template. Because the original DNA strands are complementary, though, only one new strand can begin at the 3′ end and grow continuously as the point of replication (the *replication fork*) moves along the original DNA. The other strand must grow in the opposite direction. The result is production of a series of short sections of DNA, as shown schematically in Figure 26.7. To complete formation of this strand, the sections are joined by the action of a *DNA ligase* enzyme.

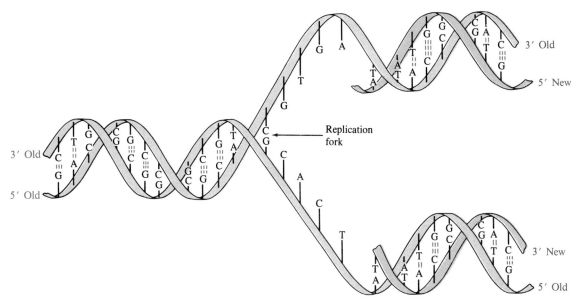

Figure 26.6
A schematic representation of DNA replication. The original DNA double helix partially unzips at a point called the *replication fork*, and complementary new nucleotides line up on each strand. When the new nucleotides are joined by DNA polymerase, two identical new DNA molecules result.

Figure 26.7
Replication at a fork. Because the polynucleotide must grow in the 5′ → 3′ direction, one strand (top) grows continuously toward the replication fork and the other grows in segments as the fork moves. The segments are later joined by a ligase enzyme.

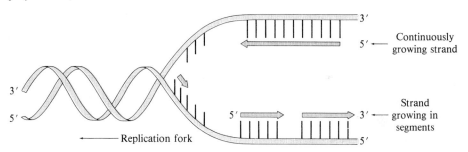

Genome The sum of all genes in an organism.

It's difficult to conceive of the magnitude of the replication process. The sum of all genes in a human cell—the human **genome**—is estimated to be approximately 3 billion base pairs, and a single DNA chain might contain up to 250 million pairs of bases. Regardless of the size of these enormous molecules, their base sequence is faithfully copied during replication, and an error occurs only about once in each 10–100 billion bases. The complete copying process in human cells takes several hours. To replicate such huge molecules as human DNA at this speed requires not one, but many replication forks, producing many segments of DNA strands that ultimately are united to form the new double helix.

Growth of new
DNA strands
from numerous
replication
forks

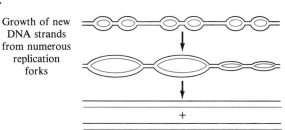

AN APPLICATION: THE HUMAN GENOME PROJECT

To "map the human genome" means to identify the location on the chromosomes of our 100,000 genes (a few thousand have been located so far) and to find the complete sequence of the 3 billion base pairs that make up human DNA. Ever since techniques to make this possible began to appear, debate has raged about whether the mapping should be done. Is it worth the huge expense of time and money? Who should do it? How should it be done? The technology is changing so rapidly, it's hard to make plans.

Perhaps the toughest question is, What will be done with the results? The benefits in understanding the molecular basis of inherited diseases are clear, for improved diagnosis and treatment will certainly result. Identification of individuals especially at risk to environmental factors that cause cancer or heart disease will also become possible. Many scientists look forward to intellectual benefits in their varied fields. But tough emotional and ethical problems are bound to arise when an individual can know many years in advance that he or she is destined to develop a degenerative and costly disease. Should this information be available to potential employers, insurers, or spouses?

Whatever the outcomes, the course toward mapping the human genome seems to have been set in the late 1980s. The Human Genome Project was given initial funding by the U.S. government, and James Watson, codiscoverer of the DNA double helix 36 years earlier, was appointed to coordinate efforts at research centers throughout the United States and, perhaps, the world. The project is predicted to take 15 years and cost at least $3 billion, or $1 per nucleotide. Genome mapping is a topic that's bound to show up with regularity in newspapers and magazines for many years to come.

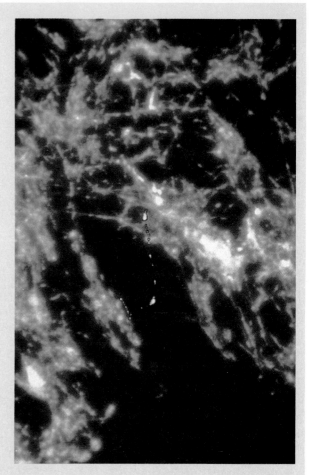

The picture shows in blue the DNA released onto a microscope slide from a single human sperm cell. Fluorescently labeled DNA sequences complementary to those in genes under study have been added to the slide, a technique that will be used to map the human genome. The bright dots identify locations where the complementary fluorescent DNA probes have attached to the genes in question. The highlighted genes are on chromosome 19, are expressed during embryo development, and have been implicated in formation of certain types of tumors.

26.7 STRUCTURE AND FUNCTION OF RNA

RNA is structurally similar to DNA—both are sugar-phosphate polymers and both have heterocyclic bases attached—but there are important differences. For one thing, they differ in composition: The sugar in RNA is ribose rather than the 2-deoxyribose in DNA, and the base uracil is present in RNA instead of thymine. They also differ in size and structure—RNA molecules are smaller than DNA molecules, and RNA is single-stranded, not double-stranded like DNA.

Another difference between RNA and DNA is in function. DNA has only one function—storing genetic information—but there are three main kinds of ribonucleic acid, each of which has a specific function.

Ribosome A structure in the cell where protein synthesis occurs.

Ribosomal RNA (rRNA) The type of RNA complexed with proteins in ribosomes.

Messenger RNA (mRNA) The RNA whose function is to carry genetic messages transcribed from DNA and to direct protein synthesis.

Transfer RNA (tRNA) The RNA whose function is to transport specific amino acids into position for protein synthesis.

● **Ribosomal RNAs** Outside the nucleus but within the cytoplasm of a cell are the **ribosomes,** small, granular structures where protein synthesis takes place. Each ribosome is a complex consisting of about 60% **ribosomal RNA (rRNA)** and 40% protein and has a total molecular weight of approximately 5,000,000 amu.

● **Messenger RNAs** The nucleic acids that record information from DNA in the cell nucleus and carry it to the ribosomes are known as **messenger RNAs (mRNA).** Protein synthesis can be controlled by variation of the rates at which mRNAs are synthesized and then destroyed once the protein they produce is no longer needed.

● **Transfer RNAs** The function of **transfer RNAs (tRNA)** is to deliver amino acids one by one to protein chains growing at ribosomes.

26.8 RNA SYNTHESIS: TRANSCRIPTION

The process of converting the information contained in a DNA segment into proteins begins with the synthesis of mRNA molecules containing anywhere from several hundred to several thousand ribonucleotides, depending on the size of the protein to be made. Each of the 100,000 or so proteins in the human body is synthesized from a different mRNA that has been transcribed from a specific gene on DNA.

Messenger RNA is synthesized in the cell nucleus by *transcription* of DNA, a process similar to DNA replication. As in replication, a small section of the DNA double helix unwinds, and the bases on the two strands are exposed. RNA nucleotides (*ribonucleotides*) line up in the proper order by hydrogen-bonding to their complementary bases on DNA, the nucleotides are joined together by an *RNA polymerase* enzyme, and mRNA results.

Unlike what happens in DNA replication, where both strands are copied, only one of the two DNA strands is transcribed into mRNA. The DNA strand that is transcribed is called the *template strand,* while its complement is called the *informational strand.* Since the template strand and the informational strand are complementary, and since the template strand and the mRNA molecule are also complementary, it follows that *the messenger RNA molecule produced during transcription is a copy of the DNA informational strand.* The only difference is that the mRNA molecule has a U base everywhere that the DNA informational strand has a T base. The transcription process is shown schematically in Figure 26.8, and an electron micrograph of transcription taking place is shown in Figure 26.9.

Transcription of DNA by the process just discussed raises as many questions as it answers. How does the DNA know where to unwind? Where along the chain does one gene stop and the next one start? How do the ribonucleotides know exactly the right place along the template strand to begin lining up and the right place to stop? These and many other questions about both replication and

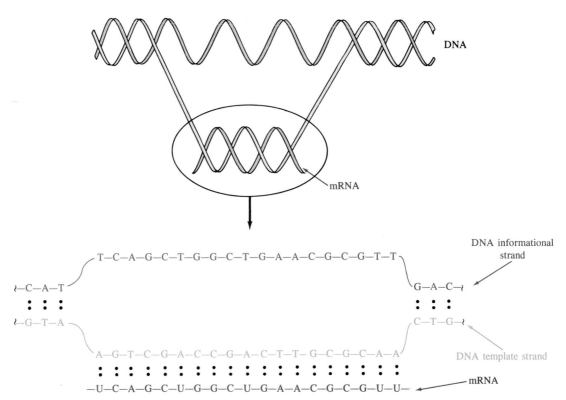

—C—A—T
: : :
—G—T—A

T—C—A—G—C—T—G—G—C—T—G—A—A—C—G—C—G—T—T

G—A—C—
: : :
C—T—G—

DNA informational strand

A—G—T—C—G—A—C—C—G—A—C—T—T—G—C—G—C—A—A
: : : : : : : : : : : : : : : : : : : :
—U—C—A—G—C—U—G—G—C—U—G—A—A—C—G—C—G—U—U—

DNA template strand

mRNA

Figure 26.8
The transcription of DNA to synthesize mRNA. The mRNA molecule produced is complementary to the template strand from which it is transcribed and is identical to the informational strand except for the replacement of T by U.

Figure 26.9
Electron micrograph of DNA undergoing transcription. The backbone of the feather-like structure running down the image is the DNA strand, and the clusters along the backbone are the mRNA molecules, which increase in length as they move along the DNA.

transcription are extremely difficult to answer. The current picture is that a DNA chain contains certain base sequences called *promoter sites,* which bind to the RNA polymerase enzyme that actually carries out RNA synthesis, thus signaling the beginning of a gene. Similarly, there are other base sequences at the end of a gene that signal a stop to mRNA synthesis.

Exon A DNA segment in a gene that codes for part of a protein molecule.

Intron A seemingly nonsensical segment of DNA that does not code for part of a protein.

Another part of the picture is the recent discovery that genes are not necessarily continuous segments of the DNA chain. Often a gene will begin in one small section of DNA called an **exon,** then be interrupted by a seemingly nonsensical section called an **intron** that does not code for any part of the protein to be synthesized, and then take up again farther down the chain in another exon. The final mRNA molecule to be released from the nucleus results only after the nonsense sections are cut out by appropriate enzymes and the remaining pieces are spliced together. Current evidence is that up to 90% of human DNA is made up of introns and only about 10% of DNA actually contains genetic instructions. The possible functions of introns and how they came to be there in the first place are, as you might guess, lively areas of study and speculation.

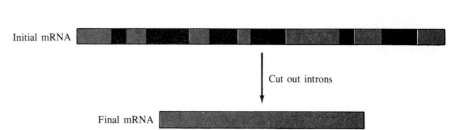

rRNA and tRNA are synthesized in the nucleus in essentially the same manner as mRNA, and before leaving the nucleus, all three types of RNA molecules are modified in various ways needed for their different functions.

Practice Problem   ***26.8***   What mRNA base sequences are complementary to each of the following DNA template sequences?
(a) -G-A-T-T-A-C-C-G-T-A-   (b) -T-A-T-G-G-C-T-A-G-G-C-A-

26.9 THE GENETIC CODE

Since all biological processes are ultimately regulated by proteins, the primary function of mRNA is to direct the biosynthesis of the thousands of diverse peptides and proteins required by an organism. The ribonucleotide sequence in an mRNA chain acts like a long, coded sentence to specify the order in which different amino acid residues should be joined to form a protein. Each "word"

Codon A sequence of three ribonucleotides in the RNA chain that codes for a specific amino acid.

or **codon** in the mRNA sentence consists of a series of three ribonucleotides that is specific for a particular amino acid. For example, the series cytosine-uracil-guanine (C-U-G) on an mRNA chain is a codon directing incorporation of the amino acid leucine into a growing protein. Similarly, the sequence guanine-adenine-uracil (G-A-U) codes for aspartic acid.

AN APPLICATION: VIRUSES AND AIDS

Viruses are small structures containing a piece of nucleic acid wrapped in a protective coat of protein. The nucleic acid may be either DNA or RNA, it may be either single-stranded or double-stranded, and it may consist either of a single piece or of several pieces. Many hundreds of different viruses are known, each of which can infect a particular plant or animal cell.

Viruses occupy the gray area between living and nonliving. By itself, a virus has none of the cellular machinery necessary for replication. Once it enters a living cell, though, a virus can take over the host cell and force it to produce virus copies. Some of the infected cells eventually die, but others continue to produce viral copies, which then leave the host and spread the infection to other cells.

Viral infection begins when a virus particle enters a host cell, loses its protein coat, and releases its nucleic acid. What happens next de-

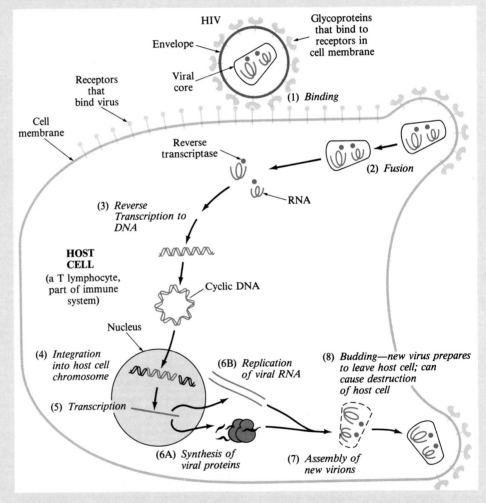

Interaction of HIV virus with host cell, a T lymphocyte. The parenthetical numbers show the events beginning with binding and cell entry by the virus, proceeding to reproduction of the viral RNA and proteins by reverse transcription, and ending with release of a new virus particle from the host cell.

pends on whether the virus is based on DNA or RNA. If the infectious agent is a DNA virus, the host cell first replicates the viral DNA and then decodes it in the normal way. The viral DNA is transcribed to produce RNA, the RNA is translated to synthesize viral protein coating, and the newly formed virus copies are released from the cell. If the infectious agent is an RNA virus, however, a problem exists. Either the cell must transcribe and produce proteins directly from the viral *RNA* template, or else it must first produce DNA from the viral RNA by a process called *reverse transcription*. Viruses that follow the reverse transcription route are called *retroviruses*. The human immunodeficiency virus (HIV-1) responsible for most cases of AIDS is an example; its mode of action is shown in the accompanying diagram.

Viral infections, whether HIV infections or the common cold, are difficult to treat with chemical agents. The challenge is to design a drug that can act on viruses within cells without damaging the cells and their genetic machinery. AZT, the first drug to be licensed for treatment of AIDS, is similar enough in structure to the nucleoside thymidine to halt the process of reverse transcription. Though not a cure for full-blown AIDS, AZT is effective at slowing the progress of the

disease in individuals infected with HIV but as yet without symptoms. Another approach to AIDS therapy is the search for a chemical agent that could inhibit binding of the virus to the cell membrane (step 1 in the diagram).

AZT (3′-azido-3′-deoxythymidine)

One possible defense against viral infections is the use of vaccines, which cause the human immune system to produce antibodies that will inactivate a virus. Advances have been made toward developing an HIV vaccine, but the task is especially challenging because HIV undergoes frequent mutations—variations in its nucleotide structure. The result is that an HIV vaccine active against one strain of HIV might soon encounter a mutant it can't react with.

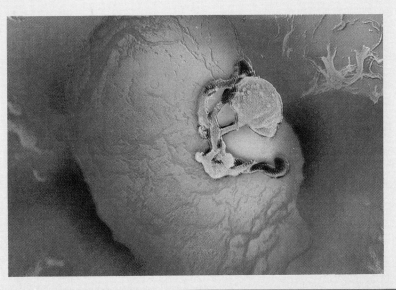

An electron micrograph of a cell that has been invaded by HIV-1 and is about to release a newly formed virus particle (green).

Genetic code The complete assignment of the 64 mRNA codons to specific amino acids (or stop signals).

Of the 64 possible three-base combinations in RNA, 61 code for specific amino acids, and 3 code for chain termination (the "stop codes"). The meaning of each codon—the **genetic code** universal to all but a few living organisms—is shown in Table 26.2. Note that most amino acids are specified by more than one codon and that codons are always written in the $5' \rightarrow 3'$ direction.

Practice Problems **26.9** List possible codon sequences for the following amino acids:
(a) Ala   (b) Phe   (c) Leu   (d) Val   (e) Tyr

26.10 What amino acids do the following sequences code for?
(a) A-U-U   (b) G-C-G   (c) C-G-A   (d) A-A-C

26.10   RNA AND PROTEIN BIOSYNTHESIS: TRANSLATION

The message contained in mRNA is decoded by tRNA molecules. Each different tRNA acts as a carrier to transport a specific amino acid into place on the ribosome so that it can be incorporated into the growing protein chain. A typical tRNA molecule is a single chain held together by regions of base-pairing in a structure something like a cloverleaf (Figure 26.10a) and containing about 70–100 ribonucleotides.

In three dimensions, a tRNA is L-shaped, as shown in Figures 26.10b and c. Every tRNA is bonded to its specific amino acid by an ester linkage between the —COOH of the amino acid and the free —OH group at the C3' position of ribose on the 3' end of the tRNA chain. At the other end of the L, near the

Table 26.2   Codon Assignments of Base Triplets

First Base (5' end)	Second Base	Third Base (3' end) U	C	A	G
U	U	Phe	Phe	Leu	Leu
	C	Ser	Ser	Ser	Ser
	A	Tyr	Tyr	Stop	Stop
	G	Cys	Cys	Stop	Trp
C	U	Leu	Leu	Leu	Leu
	C	Pro	Pro	Pro	Pro
	A	His	His	Gln	Gln
	G	Arg	Arg	Arg	Arg
A	U	Ile	Ile	Ile	Met
	C	Thr	Thr	Thr	Thr
	A	Asn	Asn	Lys	Lys
	G	Ser	Ser	Arg	Arg
G	U	Val	Val	Val	Val
	C	Ala	Ala	Ala	Ala
	A	Asp	Asp	Glu	Glu
	G	Gly	Gly	Gly	Gly

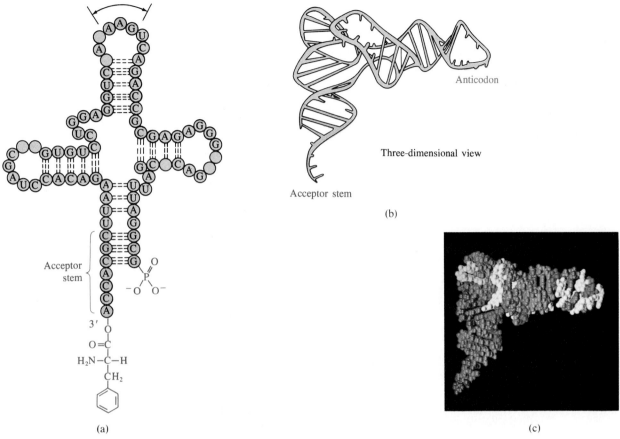

Figure 26.10

Structure of a tRNA molecule. (a) Schematic, flattened tRNA molecule. The cloverleaf-shaped tRNA contains an anticodon triplet on one "leaf" and a covalently bonded amino acid at its 3′ end. The example shown is a yeast tRNA that codes for phenylalanine. (The nucleotides not specifically identified are slightly altered analogs of the four normal ribonucleotides.) (b) The three-dimensional shape (the tertiary structure) of a tRNA molecule. Note how the anticodon is at one end of the molecule and the "acceptor stem" where the amino acid bonds is at the other end. (c) A computer-generated space-filling model of a tRNA.

Anticodon A sequence of three ribonucleotides on tRNA that recognizes the complementary sequence on mRNA.

molecule's halfway point, each tRNA contains a section called an **anticodon,** a sequence of three ribonucleotides complementary to certain codon sequences on mRNA. For example, the codon sequence C-U-G on mRNA can match up with the complementary anticodon sequence G-A-C on a leucine-bearing tRNA.

The three main steps in protein synthesis are initiation, elongation, and termination.

Initiation   Protein synthesis is initiated when an mRNA, a ribosome, and the first tRNA come together (1, Figure 26.11). The first A-U-G codon on the 5′ end of mRNA acts as a "start" signal for the translation machinery and codes for the introduction of a methionine unit. Initiation is completed when the methionine tRNA occupies one of the two binding sites on the ribosome. Since this is the site where the growing peptide will reside, it's known as the *active site,* or

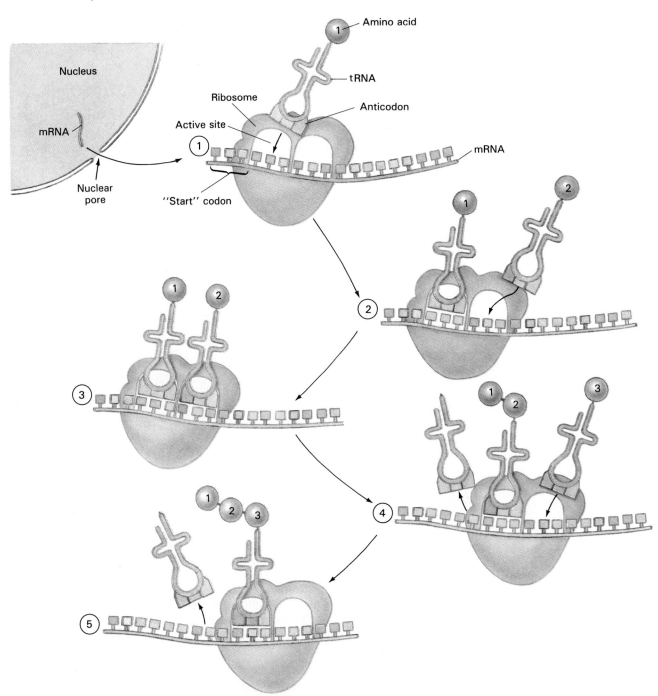

Figure 26.11
Initiation and elongation steps in protein synthesis. *Initiation:* An mRNA molecule
joins with the two ribosomal subunits and a methionine tRNA at the active, or P
binding site (①). *Elongation:* The second tRNA approaches the ribosome and binds
at the A site (②, ③). A peptide bond then forms between the first two amino acids.
Then translocation shifts the entire ribosome one codon further down the mRNA
chain as the first tRNA is freed from the P site, the peptide-carrying tRNA
previously at the A site shifts to the P site, and the A site is available for the next
tRNA (④). With the second peptide bond formed, the ribosome translocates again
and is ready for the cycle to repeat as the peptide chain grows (⑤).

the *P binding site*. (Clearly, not all proteins have methionine at one end—if it's not needed, the methionine from chain initiation is removed before the new protein goes to work.)

Elongation   Next to the active site on the ribosome is the *A binding site*, the point where the next codon on mRNA is exposed and where the tRNA carrying the next amino acid will be attached (hence, the "A" binding site). All available tRNA molecules can approach and try to fit, but only one with an appropriate anticodon sequence can bind (2, Figure 26.11). Let's assume that the second codon in our mRNA is C-U-G, which codes for leucine, and a leucine-bearing tRNA binds to the A site (3, Figure 26.11). Next, a peptide bond forms between methionine and leucine, and the bond to the first tRNA breaks.

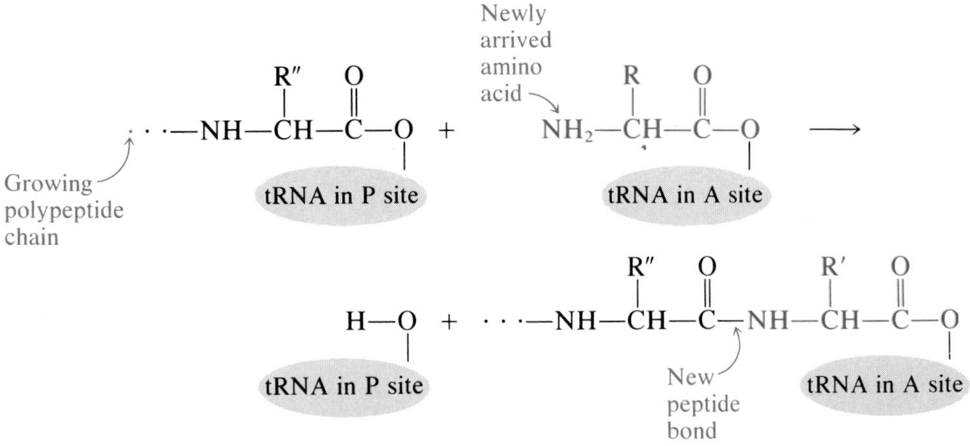

Formation of peptide bond between amino acids during translation

With the first peptide bond formed, the entire ribosome shifts three positions (one codon) along the mRNA chain, a process called *translocation*. At the same time, the empty methionine tRNA previously at the P site is freed from the ribosome, the leucine tRNA bearing the new polypeptide chain is moved from the A site to the active site, and the A site is opened up to accept the tRNA carrying the next amino acid (4, Figure 26.11).

The three elongation steps now repeat:

1. The next appropriate tRNA binds to the A site.

2. Peptide bond formation occurs to attach the newly arrived amino acid to the growing chain.

3. Translocation takes place to free the A site for the next tRNA (5, Figure 26.11).

A single RNA molecule can be "read" simultaneously by many ribosomes, as shown schematically in the following diagram. The growing polypeptides increase in length as the ribosomes move down the mRNA strand.

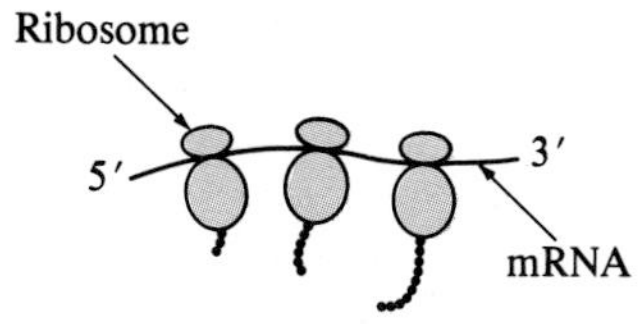

Termination   When synthesis of the proper protein is completed, a "stop" codon (U-A-A, U-G-A, or U-A-G) signals the end of the process. An enzyme called a *releasing factor* then frees the polypeptide chain from the last tRNA, and the mRNA molecule is released from the ribosome.

In our discussion in this and the preceding sections, we've left many questions about replication, transcription, and translation unanswered. What tells a cell when to start replication? Since each cell contains the entire genome and since cells differ widely in their function, what keeps genes for unneeded proteins turned off? What turns on genes for proteins needed by the cell? Are there mechanisms to repair damaged DNA or correct errors in its structure? Exactly how do mRNA and a ribosome unite in the proper orientation? These are just a few examples of further questions. Some can be answered with details beyond what we can cover here. Others can't yet be fully answered, but they may be soon, because molecular genetics is one of today's most active areas of scientific research.

Solved Problem 26.3   What amino acid sequence is coded for by the mRNA base sequence AUC-GGU?

Solution   Table 26.1 indicates that AUC codes for isleucine and that GGU codes for glycine. Thus, AUC-GGU codes for Ile-Gly.

Practice Problems   26.11   What amino acid sequence is coded for by the following mRNA base sequence? CUU-AUG-GCU-UGG-CCC-UAA

26.12   What anticodon sequences of tRNAs can match the codons in the mRNA in Problem 26.11?

26.13   What was the base sequence in the original DNA template strand on which the mRNA sequence in Problem 26.11 was made?

26.11   GENE MUTATION AND HEREDITARY DISEASE

The base-pairing mechanism of DNA replication and RNA transcription is an extremely efficient and accurate method for preserving and using genetic information, but it's not perfect. Occasionally an error occurs, resulting in the incorporation of an incorrect base at some point.

If an occasional error occurs during the transcription of an mRNA molecule, the problem is not too serious. After all, large numbers of mRNA molecules are continually being produced, and an error that occurs only one out of a million or so times would hardly be noticed. If an occasional error occurs during the replication of a *DNA* molecule, however, the consequences can be far more damaging. Each chromosome in a cell contains only *one* DNA molecule, and if that molecule is miscopied during replication, then the error is passed on when the cell divides.

Mutation An error in base sequence occurring during DNA replication.

Mutagen A substance that causes mutations.

An error in base sequence that occurs during DNA replication is called a **mutation.** Some mutations occur spontaneously, others are caused by exposure to ionizing radiation such as cosmic rays and gamma rays, and still others are caused by exposure to certain chemicals called **mutagens.** Some, perhaps most, mutations are harmless, but others can be devastating. Imagine, for example, that the sequence A-T-G on the informational strand of DNA is miscopied as A-C-G during replication. The mRNA transcribed from the corresponding template strand will then have the incorrect codon sequence A-C-G rather than the correct sequence A-U-G. But since A-C-G codes for threonine, whereas A-U-G codes for methionine, an incorrect amino acid will be inserted into the corresponding protein during translation. Furthermore, *every* copy made of the protein will have the same error.

The biological effects of incorporating an incorrect amino acid into a protein can range from negligible to catastrophic. If the incorrect amino acid occurs at some unimportant site, there may be little or no change in the biological properties of the protein. If, however, the error occurs at an important point, the biological activity of the protein can be completely changed.

Somatic cell Any cell other than a reproductive one.

Germ cell A reproductive (sperm or egg) cell.

When mutation occurs in a **somatic cell,** meaning any cell other than a sperm or egg cell, that cell might undergo uncontrolled growth leading to cancer. When mutation occurs in a **germ cell** (sperm or egg), then the genetic alteration is passed on to the offspring where it might show up as a hereditary disease. There are some 2000 known hereditary diseases that affect humans, with consequences that are sometimes fatal. Some of the more common ones are listed in Table 26.3.

The biochemical nature of the genetic defects in some hereditary diseases is well understood, but the nature of many others has eluded investigation. The key to understanding hereditary diseases is to identify the defective gene and its protein product. Cystic fibrosis, a hereditary disease that affects one of every 1600 Caucasian children, resisted the search for its cause until 1989. After narrowing in on a particular chromosome, researchers scanned the DNA in that chromosome and finally located the faulty gene. Its mutation results in production of a protein that normally aids in transporting sodium and chloride ions across cell membranes but is missing one phenylalanine. The next step toward a treatment will be to discover exactly how this defective protein causes the disease symptoms. A more immediate benefit is development of a screening test to identify carriers of this gene.

Table 26.3   Some Common Hereditary Diseases and Their Causes

Name	Nature and Cause of Defect
Phenylketonuria	Brain damage in infants caused by the defective enzyme phenylalanine hydroxylase
Albinism	Absence of skin pigment caused by the defective enzyme tyrosinase
Tay-Sachs disease	Mental retardation caused by a defect in production of the enzyme hexosaminidase A
Cystic fibrosis	Bronchopulmonary, liver, and pancreatic obstructions by thickened mucus; defective gene and protein identified
Sickle-cell anemia	Anemia and obstruction of blood flow caused by a defect in hemoglobin

Practice Problems **26.14** If the sequence T-G-G in a gene on a DNA informational strand mutated and became T-G-A, the protein encoded by that gene would stop short of completion. Explain.

26.15 What changes would result in the proteins made from genes in which the following mutations had occurred?
(a) A-T-C becomes A-T-G   (b) C-C-T becomes C-G-T

INTERLUDE: RECOMBINANT DNA

The revolution in molecular biology that began in the late 1970s owes its existence to the discovery of powerful techniques for manipulating the transfer and expression of genetic information. Using what has come to be called **recombinant DNA** technology, it is actually possible to cut a specific gene out of one organism and recombine it into the genetic machinery of a second organism. Normally, a specific gene from a higher organism is spliced into the DNA of a bacterium, and the bacterium then manufactures the specified protein, perhaps insulin or some other valuable material.

Bacterial cells, unlike the cells of higher organisms, contain part of their DNA in circular pieces called *plasmids,* each of which carries just a few genes. Plasmids are extremely easy to isolate, several copies of each plasmid are present in a cell, and each plasmid replicates through the normal base-pairing pathway.

To prepare a plasmid for insertion of a foreign gene, the plasmid is first cut open by use of a restriction endonuclease—one of over 200 enzymes known to cleave a DNA molecule between two nucleotides where a specific base sequence occurs. For example, the much-used restriction endonuclease *Eco*RI is capable of cutting a plasmid between G and A in the sequence G-A-A-T-T-C. The cut is not straight, however, but is slightly offset so that both DNA strands are left with a few unpaired bases on each end—so-called *sticky ends.*

(a)   (b)   (c)

A schematic representation for preparing recombinant DNA. (a) A bacterial plasmid and a chromosome of a higher organism are cut with the same restriction enzyme, leaving "sticky ends." (b) The opened plasmid and the desired gene fragment are then joined by DNA ligase enzyme to yield (c) an altered plasmid that is reinserted into a bacterial cell.

Cut here

$$\text{-G-A-A-T-T-C-} \xrightarrow[\text{(a)}]{Eco\text{RI}}$$
$$\text{-C-T-T-A-A-G-}$$

Cut here

Sticky ends

$$\text{-G} \qquad\qquad \text{A-A-T-T-C-}$$
$$\text{-C-T-T-A-A} \qquad + \qquad \text{G-}$$

SUMMARY

Deoxyribonucleic acid (DNA) and **ribonucleic acid (RNA)** are the carriers of an organism's genetic information. DNA and protein are combined in chromatin in cell nuclei and during cell division in higher organisms are visible as **chromosomes.** Each chromosome contains one DNA molecule, which contains several thousand **genes**—sections of the DNA chain that encode instructions for constructing a single protein.

Nucleic acids are polymers of **nucleotides.** Each nucleotide consists of a **heterocyclic amine base** linked to C1 of an aldopentose sugar, with the sugar in turn linked through its C5 —OH group to phosphoric acid. The sugar component in RNA is ribose, while that in DNA is 2-deoxyribose. DNA contains two purine amine bases (adenine and guanine) and two pyrimidine bases (cytosine and thymine); RNA contains adenine, guanine, cytosine, and

Once the plasmid has been cut, the specific gene to be inserted is then cut from its chromosome using the same restriction endonuclease so that the sticky ends on the gene fragment are complementary to the sticky ends on the opened plasmid. The gene fragment and opened plasmid are then mixed in the presence of a DNA ligase enzyme that joins them together and reconstitutes the now altered plasmid. The entire sequence of events is shown schematically in the accompanying figure.

Once the altered plasmid has been made, it can be inserted back into a bacterial cell where the normal processes of transcription and translation take place to synthesize the protein encoded by the inserted gene. Since bacteria multiply rapidly, there are soon a large number of them, all containing the altered plasmid and all manufacturing the desired protein. In a sense, the bacterium has been harnessed and put to work as a protein factory.

Among the present commercial uses of recombinant DNA technology are the syntheses of human insulin, human growth hormone, and interferon, an antiviral protein that is being tested for cancer treatment and has already proven effective in protecting individuals with an immune system disorder (chronic granulomatous disease) from microbial infections.

A landmark in the rapid growth of biotechnology was reached in the early 1990s when gene therapy, the injection into humans of genetically altered copies of their own cells, had its first clinical trials. A child with a rare hereditary immune disorder was treated with her own white blood cells to which the missing gene had been added and her condition apparently improved. In a larger trial, white blood cells altered to produce tumor necrosis factor, which kills cancerous cells, were administered to a group of skin cancer patients. These trials are just the beginning of what promises to be a long series of studies leading to new therapies based on manipulation of DNA.

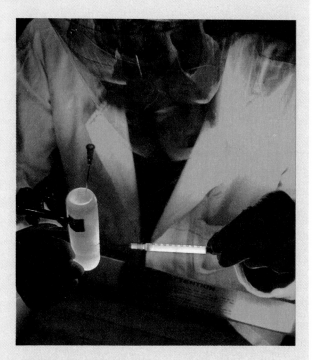

The scientist is withdrawing a sample of human DNA from a tube containing centrifuged lymphocytes marked with a dye that fluoresces yellow under ultraviolet light.

a second pyrimidine base called uracil in place of thymine. Nucleotides join together to yield nucleic acid chains by formation of an ester link between the phosphate component of one nucleotide and the —OH group at C3′ on the sugar of the neighboring nucleotide.

Molecules of DNA consist of two polynucleotide strands held together by hydrogen bonds between bases on the two strands and coiled into a **double helix.** Adenine (A) and thymine (T) form hydrogen bonds only to each other, as do cytosine (C) and guanine (G). Thus, the two strands are complementary rather than identical. Whenever an A occurs in one strand, a T occurs opposite it in the other strand; when a G occurs in one strand, a C occurs opposite it.

Three main processes take place in the transfer of genetic information:

1. Replication of DNA is the process by which identical copies of DNA are made and genetic information is preserved. The DNA double helix partially unwinds, complementary deoxyribonucleotides line up in order, DNA polymerase enzyme forms bonds between them, and two new DNA molecules are produced.

2. Transcription is the process by which RNA is produced. A segment of the DNA double helix unwinds, complementary ribonucleotides line up on the **template strand** of DNA, and an RNA polymerase enzyme forms bonds between them. The RNA produced is an exact copy of the **informational strand** of the DNA.

There are three types of RNA: **Ribosomal RNA (rRNA)** combines with proteins to form **ribosomes,** structures in the cytoplasm that are the sites of protein synthesis. **Messenger RNA (mRNA)** duplicates the information from DNA needed for synthesis of a given protein, carries it out of the nucleus to ribosomes, and directs protein synthesis. **Transfer RNA (tRNA)** delivers amino acids in the correct order for protein synthesis.

3. Translation is the process by which mRNA directs protein synthesis. Each mRNA has three-base segments called **codons** along its chain. Each codon is recognized by the **anticodon** of the tRNA that delivers the appropriate amino acid to the ribosome during protein synthesis.

Mutations occur when an error is made during DNA replication, an incorrect base is placed into the DNA chain, and a protein containing an incorrect amino acid sequence is produced. If the genetic alteration occurs in a sperm or egg cell, the mutation is passed on to offspring.

REVIEW PROBLEMS

Structure and Function of Nucleic Acids

26.16 What is a nucleotide, and what three kinds of components does it contain?

26.17 What are the names of the sugars in DNA and RNA, and how do they differ?

26.18 What does the prefix *deoxy-* mean when used in DNA?

26.19 What are the names of the four heterocyclic bases in DNA?

26.20 What are the names of the four heterocyclic bases in RNA?

26.21 Where in the cell is most DNA found?

26.22 What are the three main kinds of RNA, and what is the function of each?

26.23 What is meant by these terms?
(a) base pairing
(b) replication (of DNA)
(c) translation
(d) transcription

26.24 What is the difference between a gene and a chromosome?

26.25 What genetic information does a single gene contain?

26.26 Approximately how many genes are in human DNA?

26.27 What kind of bonding holds the DNA double helix together?

26.28 What does it mean to speak of bases as being "complementary"?

26.29 Rank the following kinds of nucleic acids in order of size: DNA, mRNA, tRNA.

26.30 What is the difference between a nucleotide's 3′ end and its 5′ end?

26.31 Show by drawing structures how the phosphate and sugar components of a nucleic acid are joined.

26.32 Show by drawing structures how the sugar and heterocyclic base components of a nucleic acid are joined.

26.33 Draw the complete structure of deoxycytidine 5′-phosphate, one of the four deoxyribonucleotides.

26.34 Draw the full structure of the RNA dinucleotide U-C. Identify the 5′ and 3′ ends of the dinucleotide.

Nucleic Acids and Heredity

26.35 What are the names of the two DNA strands, and how do they differ?

26.36 What is a codon, and on what kind of nucleic acid is it found?

26.37 What is an anticodon, and on what kind of nucleic acid is it found?

26.38 What are the general shape and structure of a tRNA molecule?

26.39 What are exons and introns? Which of the two carries genetic information? Which of the two is more abundant in DNA?

26.40 How can the phrase "Pure silver taxi" help you remember the nature of base pairing in DNA?

26.41 The DNA from sea urchins contains about 32% A and about 18% G. What percentages of T and C would you expect in sea urchin DNA? Explain.

26.42 Look at Table 26.2 and find codons for these amino acids:
(a) Pro   (b) Lys   (c) Met

26.43 What amino acids are specified by these codons?
(a) A-C-U   (b) G-G-A   (c) C-U-U

26.44 What anticodon sequences on tRNA are complementary to the codons in Problem 26.43?

26.45 If the sequence T-A-C-C-G-A appeared on the information strand of DNA, what sequence would appear opposite it on the template strand?

26.46 What sequence would appear on the mRNA molecule transcribed from the DNA in Problem 26.45?

26.47 What dipeptide would be synthesized from the gene sequence in Problem 26.45?

26.48 What is a mutation, and how can it be caused?

26.49 Why does a mutation in RNA have a much smaller effect on an organism than a mutation in DNA?

26.50 If the gene sequence A-T-T-G-G-C-C-T-A on the informational strand of DNA mutated and became A-C-T-G-G-C-C-T-A, what effect would the mutation have on the sequence of the protein produced?

26.51 What kind of cell must undergo mutation in order for the genetic error to be passed down to future generations?

26.52 What problem would you foresee if codons were made of only two nucleotides rather than three? How many codons are possible using combinations of two bases?

26.53 In general terms, what is the cause of hereditary diseases?

26.54 A substantial percentage of chemical agents that cause mutations also cause cancer. Explain in general terms why this might be.

26.55 Metenkephalin is a small peptide having morphine-like properties and found in animal brains. Give an mRNA sequence that will code for the synthesis of metenkephalin:

Tyr-Gly-Gly-Phe-Met

26.56 Give a DNA gene sequence that will code for metenkephalin (Problem 26.55).

Applications

26.57 Do you think that the presence of introns will make the process of genome mapping more difficult? Why or why not? [App: Human Genome Project]

26.58 Discuss the implications of knowing your own genetic map. [App: Human Genome Project]

26.59 How do viruses differ from normal living organisms? [App: Viruses and AIDS]

26.60 Explain the process of reverse transcription. What is the name given to viruses that use this process? [App: Viruses and AIDS]

26.61 How do vaccines work? Why will it be difficult to design a vaccine to counteract the virus responsible for AIDS? [App: Viruses and AIDS]

26.62 Give two reasons why bacterial cells are used for recombinant DNA techniques. [Int: Recombinant DNA]

26.63 Discuss the general procedure for recombinant DNA techniques, including the purposes of plasmids, restriction endonucleases, and DNA ligases. [Int: Recombinant DNA]

26.64 What types of compounds are produced by genetically altered bacteria? [Int: Recombinant DNA]

Additional Questions and Problems

26.65 Discuss how human genome mapping might be combined with recombinant DNA techniques to alleviate or eliminate many hereditary diseases.

26.66 Normal hemoglobin has a glutamic acid at the same site in which sickle-cell hemoglobin has a valine. List a possible mRNA sequence that could be present in each type of hemoglobin.

26.67 List a DNA sequence that could account for the amino acids in normal and sickle-cell hemoglobin (Problem 26.66).

26.68 Would you expect a mutation in the DNA from TAA to TAG to produce a hazardous effect? Why or why not? What about a change from TAA to GAA?

26.69 Human and horse insulin are both composed of two polypeptide chains with one chain containing 21 amino acids and the other containing 30 amino acids. How many nitrogen bases are present in the DNA coding for each chain?

26.70 Human and horse insulin (Problem 26.69) differ in primary structure at two amino acids: at the ninth position in one chain (human has serine and horse has glycine) and in the thirtieth position on the other chain (human has threonine and horse has alanine). How must the DNA differ to account for this?

26.71 Explain how a tRNA delivers the correct amino acid to the growing protein chain.

26.72 The initiation code for all proteins is AUG. What happens if a protein does not include methionine as its first amino acid?

C H A P T E R

27

Body Fluids

Blood is a water solution (the plasma) that also contains (left to right) red cells (5 billion/mL), white cells (7 million/mL), and platelets (250 million/mL). In this chapter you'll discover the functions of these four components of blood.

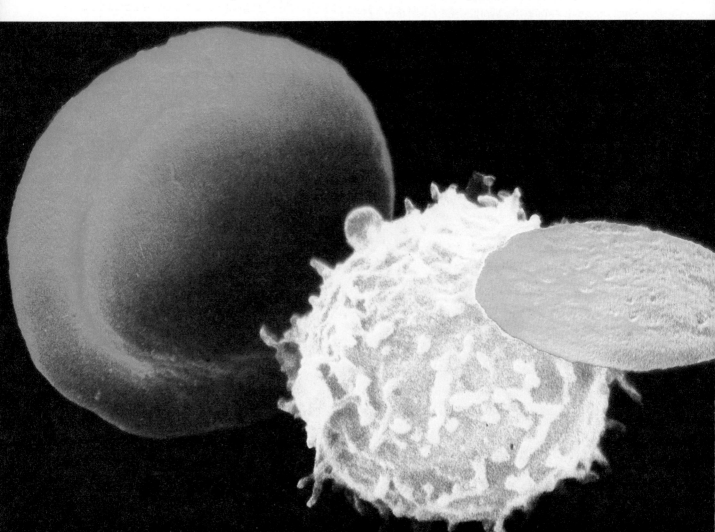

Just about every major aspect of chemistry you've studied so far applies to the subject of this chapter—body fluids. Electrolytes, nutrients and waste products, metabolism intermediates, and chemical messengers flow through the body in blood, and wastes exit in the urine. The chemical composition of blood and urine therefore mirrors chemical reactions throughout the body. Fortunately, samples of these fluids are easily collected. Many advances in understanding biological chemistry have been based on information from blood and urine analysis. In addition, studies of blood and urine chemistry provide information essential for the diagnosis and treatment of disease.

The questions answered in this chapter demonstrate the central role of chemistry in understanding physiology:

1. **How are body fluids classified?** The goal: Be able to describe the major categories of body fluids, their general composition, and the exchange of solutes between them.
2. **What are the roles of blood in maintaining homeostasis?** The goal: Be able to explain the composition and functions of blood.
3. **How do red blood cells participate in the transport of blood gases?** The goal: Be able to explain the relationships among O_2 and CO_2 transport, acid–base balance, and temperature.
4. **How do blood components participate in the body's defense mechanisms?** The goal: Be able to identify and describe the roles of blood components that participate in inflammation, the immune response, and blood clotting.
5. **How does the kidney function?** The goal: Be able to describe the process of urine formation.
6. **How does the composition of urine vary?** The goal: Be able to explain the influences on urine volume and acid elimination.

27.1   BODY WATER AND ITS SOLUTES

Intracellular fluid Fluid inside cells.

Extracellular fluid Fluid outside cells.

Blood plasma Liquid portion of blood; an extracellular fluid.

Interstitial fluid Fluid surrounding cells; an extracellular fluid.

Water is the solvent in all body fluids, and the water content of the human body averages about 60% (Figure 27.1). Physiologists describe body water as occupying two different ''compartments''—the intracellular and the extracellular compartments. Thus far we have looked mainly at chemical reactions occurring in the **intracellular fluid,** the fluid inside cells, which includes about two-thirds of all body water. The remaining one-third of body water is **extracellular fluid,** which includes mainly **blood plasma** (the fluid portion of blood) and **interstitial fluid** (the fluid that fills the spaces between cells).

To be soluble in water a substance must be an ion, a gas, a small polar molecule, or a large molecule with many polar or ionic groups on its surface. All four types of solutes are present in body fluids. The major components of body fluids are inorganic ions and ionized biomolecules, mainly proteins, as shown in

743

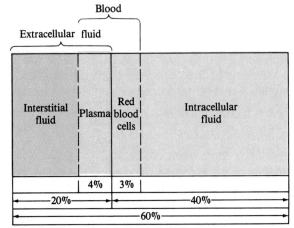

Figure 27.1
Body water as a percent of total body weight. The body averages 60% water by weight, which is divided between extracellular and intracellular fluid.

Electrolyte A substance that conducts electricity when dissolved in water; specifically, ions in body fluids.

Osmolarity The number of moles of dissolved solute particles (ions or molecules) per liter of solution.

the comparison of blood plasma, interstitial fluid, and intracellular fluid in Figure 27.2.

Inorganic ions, known collectively as **electrolytes**, have many specific roles in the reactions of metabolism, are major contributors to the **osmolarity** of body fluids, and move about as necessary to maintain charge balance inside and outside cells. Water-soluble proteins, which are negatively charged at physiological pH, make up a large proportion of the mass of solutes in blood plasma and intracellular fluid. One hundred milliliters of blood contains about 7 g of protein. Blood proteins transport lipids and other molecules, and play essential roles in the protection afforded by blood clotting (Section 27.6) and the immune response (Section 27.5). The blood gases oxygen and carbon dioxide, along with

Figure 27.2
The distribution of cations and anions in body fluids. Outside cells, Na^+ is the major cation and Cl^- is the major anion. Inside cells, K^+ is the major cation and HPO_4^{2-} is the major anion. Note that at physiological pH, proteins are negatively charged.

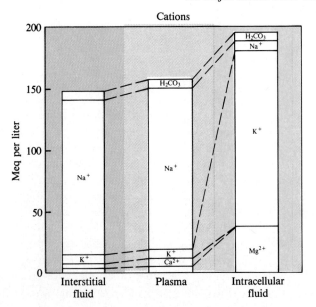

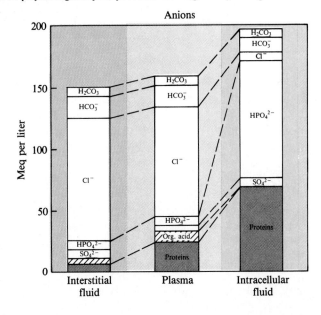

glucose, amino acids, and the nitrogen-containing by-products of protein catabolism, are the major small molecules in body fluids.

Blood travels through peripheral tissue in a network of tiny, hair-like capillaries that connect the arterial and venous parts of the circulatory system (Figure 27.3a). It is here that the exchange of nutrients and end products of metabolism occurs. Capillary walls consist of a single layer of loosely spaced cells (Figure 27.3b). Water and many small solutes move freely across the walls in response to differences in fluid pressure and concentration.

On the arterial ends of capillaries (connected to *arteries,* which carry blood from the heart), blood pressure is higher than interstitial fluid pressure and pushes solutes and water into interstitial fluid. On the venous ends (connected to *veins,* which carry blood to the heart), blood pressure is lower, and water and solutes reenter the plasma. Solutes that can cross membranes passively by diffusion move from regions of high concentration to regions of low concentration. The combined result of water and solute exchange at capillaries is that, except for proteins, blood plasma and interstitial fluid are very similar in composition (see Figure 27.2).

In addition to blood capillaries, peripheral tissue is networked with lymph capillaries that terminate in blind pockets. The lymphatic system collects excess interstitial fluid, debris from cellular breakdown, and proteins and lipid droplets too large to pass through capillary walls. Interstitial fluid and the substances that accompany it into the lymphatic system are referred to as *lymph,* and the walls of lymph capillaries are constructed so that lymph cannot return to the surrounding tissue. Ultimately lymph enters the bloodstream at the thoracic duct.

Active transport   Energy-requiring transport of molecules or ions across cell membranes against a concentration gradient.

Solutes in the interstitial fluid and the intracellular fluid are exchanged by crossing cell membranes (Figure 27.4). Here, major differences in concentration are maintained by **active transport** against concentration gradients (from regions of low concentration to regions of high concentration) and by the impermeability of cell membranes to certain solutes, notably Na^+.

Figure 27.3
(a) The capillary network. Solute exchange between blood and interstitial fluid occurs across capillary walls. (b) Cross section of a blood capillary. The walls are made of individual cells lying parallel to the capillary blood channel. The nucleus of a single cell appears as a green semicircle on the right. This falsely colored electron micrograph enlarges the original tissue more than 2500 times.

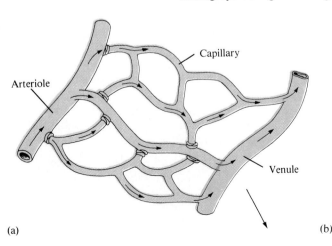

Arteriole

Capillary

Venule

(a)

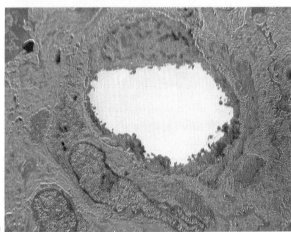

(b)

Figure 27.4
Exchange among body fluids. Water exchanges freely in most tissues, with the result that the osmolarity of blood plasma, interstitial fluid, and intracellular fluid is the same. Large proteins cross neither capillary walls nor cell membranes, leaving the interstitial fluid protein concentration low. Concentration differences between interstitial fluid and intracellular fluid are maintained by active transport.

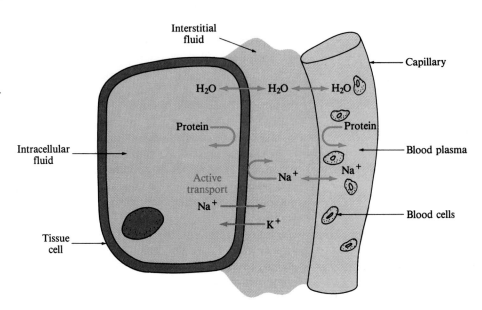

AN APPLICATION: ORGAN CRYOPRESERVATION

Organ cryopreservation is the freezing of live tissue for storage and subsequent use. To make this type of preservation successful, the molecular mechanisms essential to life in the frozen state must be understood. Freezing unprotected body fluids causes formation of ice crystals that rip through cell membranes, damage subcellular organelles, and allow the cell contents to spill out. Much of what has been attempted in medical cryopreservation is based on strategies observed in the survival of insects, mammals, and fish in subfreezing temperatures.

Successful cryopreservation of human sperm frozen in a glycerol solution first occurred in 1949. Since then, techniques have been developed for freezing many single-cell suspensions (sperm, red and white blood cells, platelets) and simple tissues (embryos, skin, cornea, pancreatic islets). Cryopreservation of larger organs for transplantation and the return of the frozen organs to a fully viable state is more difficult and is not yet possible.

To survive freezing, tissues must satisfy three biochemical conditions. First, ice formation in extracellular fluids must be controlled so that freezing occurs slowly and ice crystals are small.

To promote small crystal formation, freeze-tolerant animals use blood proteins as nuclei around which large numbers of very small crystals form. Since small ice crystals are thermodynamically unstable, however, and like to form larger and larger crystals (much as sizable ice crystals appear in ice cream kept too long after opening), antifreeze compounds that slow crystal growth are also needed.

Secondly, cell structure and the cell membrane must be protected because water tends to exit cells by osmosis when the extracellular fluids freeze. Freeze-tolerant animals use low-molecular-weight compounds (trehalose, proline, glycerol, sorbitol, glucose) to prevent the injuries that would result during freezing.

Finally, cell viability must be maintained, which means that metabolism cannot cease. Although the lowered temperature automatically lowers the metabolism rate, there must be some production of ATP. (In freeze-tolerant animals, it's as low as 1–10% of the normal resting rate.) Since there is no blood supply, the fuel must be present in the cells. Also, since there isn't any oxygen supply, the metabolism must be anaerobic, and the metabolites must not be toxic be-

27.2 FLUID BALANCE

The body is kept in overall fluid balance by roughly equal daily intake and output of water as summarized in the following table. In a hot environment or when doing strenuous work, the intake of drinking water and loss in sweat and exhaled gases both increase greatly.

Water intake (mL/day)		Water output (mL/day)	
Drinking water	1200	Urine	1400
Water from food	1000	Skin	400
Water from metabolic		Lungs	400
oxidation of food	300	Sweat	100
		Feces	200
Total	2500		2500

cause there is no mechanism for their elimination.

Large-organ cryopreservation must meet the same conditions as the survival of freeze-tolerant animals. Ice-nucleating compounds and/or antifreeze compounds must be provided (and then removed while the tissue is thawed if the compounds are potentially toxic). Also, membranes must be stabilized and metabolism inhibited, but not so much so that ATP production ceases. Further studies of freeze-tolerant animals are expected to contribute to the eventual successful cryopreservation of organs for transplantation.

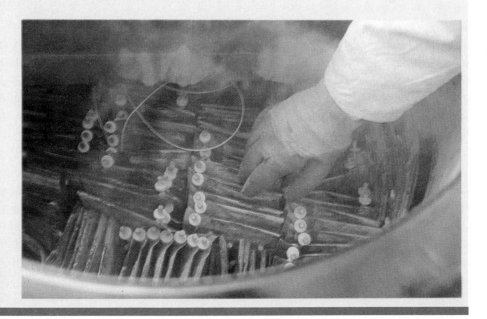

Bags of cryopreserved blood for transfusions or for controlled thawing that allows separation of useful components such as a clotting agent for treatment of hemophilia.

One important role of the kidneys is to keep water and electrolytes in balance by increasing or decreasing the amounts eliminated. The intake and output of water are, in turn, controlled by hormones. Receptors in the hypothalamus monitor the concentration of solutes in blood plasma, and as little as a 2% change in osmolarity can cause an adjustment in hormone secretion. For example, when a rise in osmolarity indicates an increased concentration of solutes in blood and therefore a shortage of water, secretion of *antidiuretic hormone* (also known as vasopressin) increases. In the kidney, antidiuretic hormone causes a decrease in the water content of the urine. At the same time, the thirst mechanism is activated to cause increased water intake.

A long list of abnormal conditions cause what physicians refer to as the *syndrome of inappropriate antidiuretic hormone secretion (SIADH),* the result of excess secretion of the hormone. When this occurs, the kidney excretes too little water, the water content of body compartments increases, and serum concentrations of electrolytes drop to dangerously low levels. Among the causes of SIADH are regional low blood volume due to decreased return of blood to the heart (caused by, for example, asthma, pneumonia, pulmonary obstruction, or heart failure) and misinterpretation by the hypothalamus of osmolarity (due, for example, to central nervous system disorders, barbiturates, or morphine).

The reverse problem, inadequate secretion of antidiuretic hormone, often a result of injury to the hypothalamus, causes *diabetes insipidus* (unrelated to diabetes mellitus). In this condition, up to 15 L of dilute urine is excreted each day. Administration of synthetic hormone controls the problem.

27.3 BLOOD

Blood flows through the body in the circulatory system, which in the absence of trauma or disease is an essentially closed system. About 55% of blood is plasma, which contains the solutes shown in Figure 27.5, and the remaining 45% is a mixture of red blood cells (erythrocytes), platelets, and white blood cells suspended in the plasma.

Whole blood Blood plasma plus blood cells.

The plasma and cells together make up **whole blood** (Figure 27.6), which is usually collected for clinical laboratory analysis directly into evacuated tubes. An anticoagulant must be added to whole blood to prevent clotting, which otherwise will occur in 20–60 min at room temperature. Often the anticoagulant is supplied right in the collection tube. *Heparin,* a common anticoagulant, is a natural polysaccharide that interferes with the action of enzymes needed for clotting. Other anticoagulants, such as citrate ion and oxalate ion, form precipitates with calcium ion, which is also needed for blood clotting. Plasma is separated from blood cells by spinning the sample in a centrifuge.

Blood serum Fluid portion of blood remaining after clotting has occurred.

Many laboratory analyses are performed on **blood serum,** the fluid remaining after blood has completely clotted. When a serum sample is desired, whole blood is collected in the presence of an agent that hastens clotting, often thrombin, a natural component of the clotting system (Section 27.6). Centrifugation separates the clot and cells to leave behind the serum.

The functions of blood fall into three major categories:

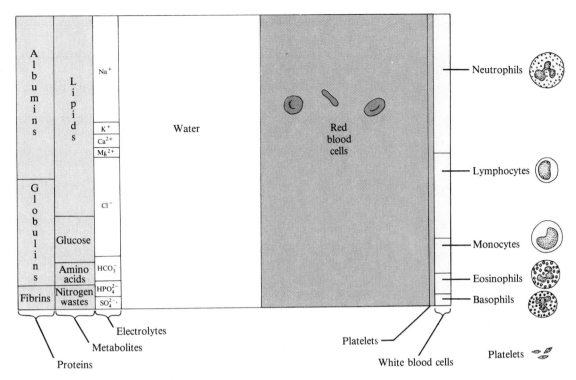

Figure 27.5
The composition of whole blood.

Figure 27.6
Whole blood for a blood count analysis. The number of red cells, white cells, and platelets per liter will be determined by an electronic cell counter.

Erythrocytes Red blood cells.

● **Transport** The circulatory system is the body's equivalent of the interstate highway network, transporting materials from where they are produced to where they are used or disposed of.

● **Regulation** Blood redistributes body heat as it flows along, thereby participating in the regulation of body temperature. It also picks up or delivers water and electrolytes as they are needed to maintain fluid and electrolyte balance. In addition, blood buffers are essential to the maintenance of acid–base balance.

● **Defense** Blood carries the molecules and cells needed for two major defense mechanisms: (1) blood clotting, which prevents loss of blood and begins healing of wounds; (2) the immune response, which destroys foreign invaders.

27.4 RED BLOOD CELLS AND BLOOD GASES

Red blood cells, or **erythrocytes,** have one major purpose: to transport blood gases. Erythrocytes have no nuclei or ribosomes and cannot replicate themselves. In addition, they have no mitochondria or glycogen and must obtain glucose from the surrounding plasma. Their enormous number—about 250 million in a single drop of blood—and their large surface area provide for rapid exchange of gases throughout the body. Because they are small and flexible, erythrocytes can squeeze through the tiniest capillaries one at a time.

Ninety-five percent of the protein in an erythrocyte is hemoglobin, the transporter of oxygen and carbon dioxide. Hemoglobin (Hb) is composed of four protein chains with the quaternary structure shown earlier in Figure 18.16. Each protein chain has a heme molecule embedded in its nonpolar interior, and each of the four hemes can combine with one O_2 molecule.

Oxygen Transport   The iron(II) ion, Fe^{2+}, in the center of each heme molecule is the site of the action in transporting oxygen, which is held to the iron by bonding through an unshared electron pair. In contrast to the cytochromes of the respiratory chain, where iron cycles between Fe^{2+} and Fe^{3+}, heme iron must remain in the reduced Fe^{2+} state or it loses its oxygen-carrying ability. Hemoglobin carrying four oxygens (*oxyhemoglobin*) is bright red, hemoglobin that has lost one or more oxygens (*deoxyhemoglobin*) is dark red-purple and accounts for the darker color of venous blood, and dried blood is brown because exposure to oxygen has oxidized the heme iron.

At normal physiological conditions, the percentage of heme molecules that carry oxygen, known as the *percent saturation,* is dependent on the **partial pressure** of oxygen in surrounding tissues as shown in Figure 27.7. The relation is not a simple equilibrium, as shown by the S shape of the curve, but is determined by allosteric interaction (Section 19.9). Binding each O_2 causes changes in hemoglobin structure that enhance binding of the next O_2, and releasing each oxygen enhances release of the next. As a result, oxygen is more readily released to tissue where the partial pressure of oxygen is low. The average oxygen partial pressure in peripheral tissue is 40 mm Hg, a pressure at which Hb remains 75% saturated by oxygen, leaving a large amount of O_2 in reserve for emergencies. Note, however, the rapid drop in the curve between 40 mm Hg and 20 mm Hg, the oxygen pressure in tissue where metabolism is occurring rapidly.

Carbon Dioxide Transport, H^+ Transport, and Temperature Effects   Carbon dioxide from metabolism in peripheral cells diffuses into interstitial fluid and then into capillaries, where it is transported in three ways: (1) dissolved, (2) bonded to Hb, or (3) as HCO_3^-. About 7% of the CO_2 dissolves in blood plasma. The rest enters erythrocytes, where some of it bonds to hemoglobin, not in the heme portion of the molecule but by reaction with nonionized amino acid —NH_2 groups,

Partial pressure   The contribution to total gas pressure contributed by each individual component of a mixture of gases.

Figure 27.7
Oxygen saturation of hemoglobin at normal physiologic conditions. Oxygen pressure is about 100 mm Hg in arteries and 20 mm Hg in active muscles. Note the big release of oxygen between 40 mm Hg and 20 mm Hg.

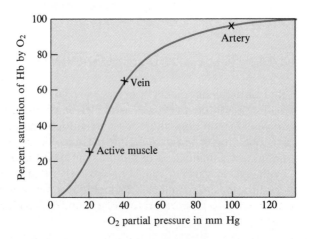

$$Hb\text{---}NH_2 + CO_2 \longrightarrow Hb\text{---}NHCOO^- + H^+$$

Most of the CO_2 is rapidly converted to bicarbonate ion within erythrocytes, which contain a large concentration of carbonic anhydrase:

$$H_2O + CO_2 \xrightarrow{\text{carbonic anhydrase}} H^+ + HCO_3^-$$

Without some compensating change, the result of the two reactions of carbon dioxide shown above would be an unacceptably large increase in blood acidity. To cope with the increased acidity, hemoglobin responds by reversibly binding hydrogen ions:

$$Hb \cdot 4\,O_2 + 2\,H^+ \rightleftharpoons Hb \cdot 2\,H^+ + 4\,O_2$$

The release of oxygen is enhanced by allosteric effects when the hydrogen ion binding increases, and oxygen is held more firmly when the hydrogen ion binding decreases. The result of changes in H^+ concentration on the oxygen affinity of hemoglobin is shown in the following diagram. Where H^+ concentration is high in peripheral tissues, Hb releases O_2 and picks up H^+. The Hb then carries H^+ to the lungs, where it is released as O_2 is picked up.

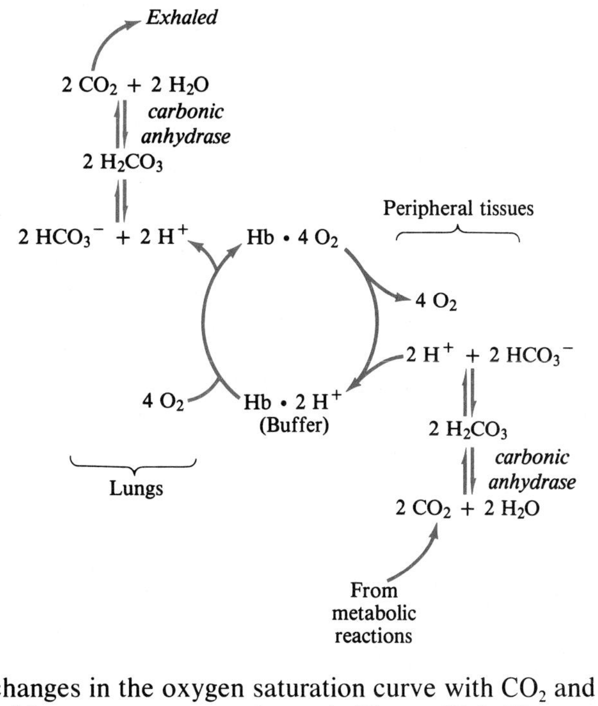

The changes in the oxygen saturation curve with CO_2 and H^+ concentrations and with temperature are shown in Figure 27.8. The curve shifts to the right, indicating decreased affinity of Hb for O_2, when the H^+ and CO_2 concentrations increase and when the temperature increases. These are exactly the conditions in muscles that are working hard and need more oxygen. The curve shifts to the left, indicating increased affinity of Hb for oxygen, under the opposite conditions of decreased H^+ and CO_2 concentrations and lower temperature.

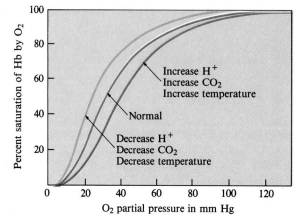

Figure 27.8
Changes in oxygen affinity of hemoglobin with changing conditions. The normal curve of Figure 27.7 is shown in red.

The intimate relationships among H^+ and HCO_3^- concentrations and O_2 and CO_2 partial pressures are essential to maintaining electrolyte and acid–base balance (Section 11.15). Clinical laboratory measurements of these parameters are often used in diagnosis. Table 27.1 lists a few common disorders that cause acidosis and their effects on blood concentrations.

27.5   PLASMA PROTEINS, WHITE BLOOD CELLS, AND IMMUNITY

Having examined the function of red blood cells, we now want to take a look at the roles of plasma proteins and other types of blood cells (Table 27.2) in the *defense* functions of blood: the immune response and blood clotting. Each of these processes is complex, and we can only take an overview of them. In this section we will focus on the immune function, and in the next section we will do the same for blood clotting.

Antigen A substance that stimulates the immune response.

A foreign invader, known as an **antigen,** is any substance recognized by the body as not part of itself. The invader might be a foreign molecule or a microorganism. Antigens can also be small molecules known as *haptens* that are recognized as antigens only after they have bonded to carrier proteins. Haptens include some antibiotics, environmental pollutants, and allergens from plants and animals.

Table 27.1   Effects on Blood Gas Concentrations of Some Common Disorders That Cause Acidosis

Disorder	H^+ Concentration	CO_2 Partial Pressure	O_2 Partial Pressure	HCO_3^- Concentration
Kidney failure	Increase	Initially normal, then decrease	Normal	Decrease
Diabetic ketoacidosis	Increase	Initially normal, then decrease	Normal	Decrease
Emphysema	. Increase	Increase	Decrease	Normal
Respiratory distress syndrome	Increase	Increase	Decrease	Normal

Table 27.2   Protein and Cellular Components of Blood

Blood Component	Function
Proteins	
Albumins	Transport lipids, hormones, drugs; major contributor to plasma osmolarity
Globulins	
Immunoglobulins (γ-globulins, antibodies)	Identify antigens (microorganisms and other foreign invaders) and initiate their destruction
Transport globulins	Transport lipids and metal ions
Fibrinogen	Forms fibrin, the basis of blood clots
Blood cells	
Red blood cells	Transport O_2, CO_2, H^+
White blood cells	
Lymphocytes (T cells and B cells)	Defend against specific pathogens and foreign substances
Neutrophils, eosinophils, and monocytes	Carry out phagocytosis—engulf foreign invaders
Basophils	Release histamine during inflammatory response of injured tissue
Platelets	Help to initiate blood clotting

Inflammation   A nonspecific defense mechanism triggered by antigens or tissue damage; includes swelling, redness, warmth, and pain.

Figure 27.9
Red blood cells squeezing through capillary wall (top left to bottom right). When inflammation reduces the permeability of a capillary wall, red and white cells, plus plasma, cross into the surrounding tissue.

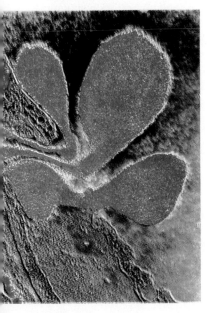

The recognition of an antigen can initiate three different responses, the first, inflammation, being nonspecific and the other two being specific to a given antigen.

Inflammatory Response   Cell damage due to infection or injury initiates **inflammation.** For example, the swollen, painful, red bump that develops around a splinter in your finger is an inflammation. Chemical messengers released at the site of the injury direct the inflammatory response. One important such messenger is *histamine,* which is synthesized from the amino acid histidine.

$$\underset{\text{Histidine}}{\boxed{\text{N}\,\text{N}\,\text{H}}-\text{CH}_2\text{CH}{-}{-}\text{CO}^- \atop {}^+\text{NH}_3} \quad \xrightarrow{\text{histidine} \atop \text{decarboxylase}} \quad \underset{\text{Histamine}}{\boxed{\text{N}\,\text{N}\,\text{H}}-\text{CH}_2\text{CH}_2{-}\text{NH}_3{}^+} \;+\; \text{CO}_2$$

Histamine sets off dilation of capillaries and increases the permeability of capillary walls (Figure 27.9). The resulting increased blood flow into the damaged area reddens and warms the skin, and swelling occurs as plasma carrying blood-clotting factors and defensive proteins enters the intercellular space. At the same time, white blood cells cross capillary walls to attack invaders.

Bacteria or other antigens at the site of inflammation are attacked by **phagocytosis:** White blood cells (*phagocytes*) surround the bacteria and destroy them by enzyme-catalyzed hydrolysis reactions (Figure 27.10). Phagocytes that

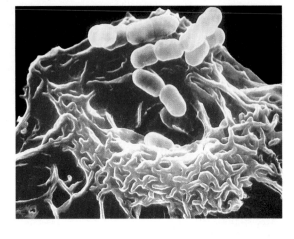

Figure 27.10
Phagocytosis. The green
Escherichia coli bacteria,
normal residents of the
human intestine that can
cause infection under some
conditions, are about to be
engulfed by a white blood
cell.

Phagocytosis Engulfing
and digestion of a particle by
a cell.

Immune response The
recognition of and attack on
antigens, including viruses,
bacteria, toxic substances,
and infected cells.

Antibody (immunoglobulin)
Protein molecule that
identifies antigens.

have encountered antigens also emit a variety of chemical messengers that help
to direct the inflammatory response and initiate specific immune responses.

Yet another defense system consisting of a large family of plasma proteins
known as the *complement* system is activated by inflammation. The complement proteins assemble into a *membrane attack complex* able to destroy bacteria directly by creating holes in their cell walls.

An inflammation caused by a wound will heal completely only if all
infectious agents have been removed, with dead cells and other debris absorbed
into the lymph system. The inflammatory response is dependent upon normal
numbers of white blood cells, and if the white blood cell count falls below 1000
per mL of blood, any infection can be life-threatening.

Specific Immune Responses   As opposed to the inflammatory response, an
immune response depends on recognition of specific invaders. At the molecular
level, an antigen is detected by an interaction very much like that between an
enzyme and its substrate. Noncovalent attraction allows a spatial fit between
the antigen, or more likely a small region in an antigen, and an **antibody** specific
to it. The antigen and the antibody may be independent molecules or they may
be portions of proteins or glycoproteins protruding from the surfaces of cell
membranes.

Before an immune response is triggered, an antigen often undergoes
phagocytosis in such a manner that a segment of the antigen molecule is
transported to the surface of the phagocyte. At the surface, it will be recognized
by a lymphocyte, a white blood cell responsible for one of the two antigen-
specific responses: the antibody-directed or the cell-directed immune response.
This recognition can take place anywhere in the body but often occurs in lymph
nodes, the tonsils, or the spleen, which have large concentrations of lymphocytes.

Antibody-directed Immune Response   In the antibody-directed immune response, antigens are first detected by binding to antibodies on the surfaces of B
lymphocytes, known as B cells. The B cells divide into *plasma cells* that
produce identical non-cell-bound antibodies specific for the antigen they have
encountered.

AN APPLICATION: THE BLOOD–BRAIN BARRIER

Nowhere in human beings is the maintenance of a constant internal environment more important than in the brain. If the brain were exposed to the fluctuations in concentrations of hormones, amino acids, neurotransmitters, and potassium that occur elsewhere in the body, inappropriate nervous activity would result. Therefore, the brain must be rigorously isolated from variations in blood composition.

How can the brain receive nutrients from the blood in capillaries and yet be protected? The answer lies in the unique structure of the *endothelial* cells that form the walls of brain capillaries. Unlike the cells in other capillaries, those in brain capillaries form a series of continuous tight junctions so that nothing can pass between them. To reach the brain, therefore, a substance must cross this *blood–brain barrier* by crossing the endothelial cell membranes.

The brain, of course, can't be completely isolated or it would die from lack of nourishment. Glucose, the main source of energy for brain cells, and certain amino acids the brain can't manufacture are recognized and brought across the cell membranes by transporters specific to each nutrient. Similar specific transporters move surplus substances out of the brain.

An asymmetric (one-way) transport system exists for glycine, a small amino acid that is a potent neurotransmitter. Glycine inhibits rather than activates transmission of nerve signals, and its concentration must be held at a lower level in the brain than in the blood. To accomplish this, there is a glycine transport system in the cell membrane closest to the brain, but no matching transport system on the other side. Thus, glycine can be transported out of the brain but not into it.

The brain is also protected by the "metabolic" blood–brain barrier. In this case, a compound that gets into an endothelial cell is converted there to a metabolite that is unable to enter the brain. A striking demonstration of the metabolic brain barrier is provided by dopamine, a neurotransmitter that is deficient in Parkinson's

disease, and L-dopa, a metabolic precursor of dopamine.

L-Dopa can both enter and leave the brain because it is recognized by an amino acid transporter. The brain is protected from an entering excess of L-dopa, however, by its conversion to dopamine *within* the endothelial cell. Dopamine, which is also produced from L-dopa within the brain (thereby allowing treatment of Parkinson's disease with L-dopa), can, like glycine, leave the brain but cannot enter it.

Since crossing the endothelial cell membrane is the necessary route into the brain, substances soluble in the membrane lipids readily breach the blood–brain barrier. Among such substances are nicotine, caffeine, codeine, diazepam (Valium, a tranquilizer), and heroin. Heroin differs from morphine in having two nonpolar acetyl groups where the morphine molecule has polar hydroxyl groups. The resulting difference in lipid solubility allows heroin to enter the brain much more efficiently than morphine. Once heroin is inside the brain, enzymes remove the acetyl groups to give morphine, in essence trapping it in the brain. When developing therapeutic drugs, it is possible to take advantage of such trapping so that a carefully controlled dosage can have a potent effect.

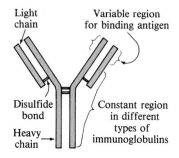

Figure 27.11
The basic structure of an immunoglobulin.

Antibodies, known also as **immunoglobulins,** make up 20% of the plasma proteins. The body may contain up to 10,000 different immunoglobulins at any one time, and we have the capacity to make more than one hundred million others. The immunoglobulins are glycoproteins that have in common two *heavy* polypeptide chains and two *light* polypeptide chains joined by disulfide bonds, as shown schematically in Figure 27.11. The antigen-binding sites are composed of antigen-specific sequences of amino acids. Once synthesized, antibodies spread out to mark their antigens for destruction by phagocytosis or by the complement system.

B-cell division also yields *memory cells* that remain on guard and quickly produce more plasma cells if the same antigen reappears in the future.

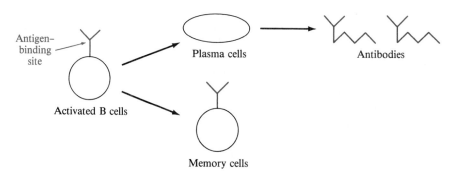

The long-lived memory cells are responsible for long-term immunity to diseases after the first illness or after a vaccination.

Immunoglobulin G antibodies (known as gamma globulins) protect against viruses and bacteria. Allergies and asthma are caused by an oversupply of *immunoglobulin E.* Numerous disorders result from the mistaken identification of normal body constituents as foreign and the overproduction of antibodies to combat them. These *autoimmune diseases* include attack on connective tissue at joints in rheumatoid arthritis, attack on pancreatic islet cells in some forms of diabetes mellitus, and a generalized attack on nucleic acids and blood components in systemic lupus erythematosus.

Cell-directed Immune Response   In the cell-directed immune response, antigenic regions on the surface of infected body cells are detected by compatible regions on the surface of T lymphocytes, or T cells. The T cells themselves destroy the infected cells by releasing a toxic protein that kills by perforating the cell membranes. Cell-directed immunity is responsible for protection against cancer cells and virus-infected cells and also causes the rejection of transplanted organs.

27.6 BLOOD CLOTTING

Fibrin   Insoluble protein that forms the fiber framework of a blood clot.

A blood clot consists of blood cells trapped in a mesh of the insoluble fibrous protein known as **fibrin** (Figure 27.12). Formation of a clot is a many-step process requiring participation of 12 clotting factors. The calcium ion is one such clotting factor, and vitamin K is necessary for synthesis in the liver of other clotting factors, most of which are glycoproteins. Therefore, a deficiency of vitamin K, the presence of a competitive inhibitor of vitamin K, or a defi-

Figure 27.12
Red blood cells trapped in a fibrin clot.

Hemostasis The stopping of bleeding.

Blood clotting Formation of a structure of fibrin and blood cells to close the rupture of a blood vessel.

Zymogen A compound that becomes an active enzyme after undergoing a chemical change.

ciency of a clotting factor can cause excessive bleeding, sometimes even from minor tissue damage. *Hemophilia* is a disorder caused by an inherited genetic defect that results in the absence of one or another of the clotting factors.

The body's mechanism for halting blood loss from even the tiniest capillary is referred to as **hemostasis.** The first events in hemostasis are constriction of surrounding blood vessels and formation at the site of tissue damage of a plug composed of the blood cells known as *platelets.*

Next, **blood clotting,** which may be triggered by two different pathways, swings into action. One pathway (the *intrinsic pathway*) begins when blood makes contact with the negatively charged surface of the fibrous protein collagen exposed at the site of tissue damage. Glass is also negatively charged, and clotting is activated in exactly the same manner when blood is placed in a glass tube. The other pathway (the *extrinsic pathway*) begins with release by tissue damage of an integral membrane glycoprotein known as *tissue factor.*

The result in either case is a cascade of reactions in which several inactive clotting factors that are zymogens are activated one after another by cleavage of polypeptides to give active clotting factors, many of which are enzymes that then catalyze activation of the next factor in the cascade. The two pathways merge and, in the final step of the *common pathway,* the factor *thrombin* catalyzes cleavage of small polypeptides from the soluble plasma protein *fibrinogen* (Figure 27.13). Negatively charged groups in these polypeptides make

Figure 27.13
Schematic drawing of fibrinogen (mol wt 340,000) and a fibrin clot, which is essentially a polymer of fibrin. In fibrinogen there are three globular protein regions connected by fibrous protein. Removal of ionic polypeptides from the central region gives fibrin, in which central sites are free to associate with compatible sites on the ends of the molecule, forming a half-staggered arrangement which is then stabilized by covalent cross-links.

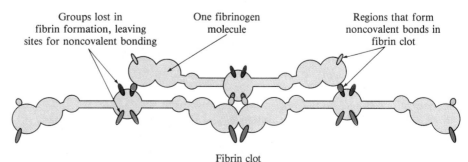

Groups lost in fibrin formation, leaving sites for noncovalent bonding

One fibrinogen molecule

Regions that form noncovalent bonds in fibrin clot

Fibrin clot

fibrinogen soluble and keep the molecules apart. Once these polypeptides are removed, the resulting insoluble fibrin molecules immediately associate with each other by noncovalent interactions. Then, they are bound into fibers by formation of peptide bond cross-links between lysine and glutamine side chains in a reaction catalyzed by another of the clotting factors:

$$Gln-CH_2CH_2-\overset{\overset{\displaystyle O}{\|}}{C}-NH_2 \quad + \quad {}^+H_3NCH_2CH_2CH_2CH_2-Lys \quad \text{(Protein chain)}$$

$$\downarrow$$

$$Gln-CH_2CH_2-\overset{\overset{\displaystyle O}{\|}}{C}-NHCH_2CH_2CH_2CH_2-Lys \quad + \quad NH_4{}^+$$

Cross-link between protein chains

The final stage of hemostasis is breakdown of the clot by hydrolysis of peptide bonds after it has done its job of preventing blood loss and binding together damaged surfaces as they heal.

27.7   THE KIDNEY AND URINE FORMATION

The kidneys bear the major responsibility for maintaining a constant internal environment in the body. By managing the elimination of appropriate amounts of water, electrolytes, hydrogen ions, and nitrogen-containing wastes, the kidneys respond to changes in health, diet, and physical activity.

About 25% of the blood pumped from the heart goes directly to the kidneys, where the functional units are the *nephrons*. Each kidney contains over a million of them. Blood enters a nephron at a *glomerulus* (Figure 27.14), a tangle of capillaries surrounded by a fluid-filled space. **Filtration,** the first of three essential kidney functions, occurs here. The pressure of blood pumped into the glomerulus directly from the heart is high enough to push plasma and all its solutes except large proteins across the capillary membrane into the surrounding fluid, the **glomerular filtrate.** The filtrate flows from the capsule into the tubule that makes up the rest of the nephron, and the blood enters the network of capillaries intertwined with the tubule.

Since about 125 mL of filtrate per minute enters the kidneys, they produce a total of 180 L of filtrate per day. This filtrate contains not only waste products but also many solutes the body can't afford to lose, such as glucose and electrolytes. Since we excrete only about 1.4 L of urine each day, you can see that another important function of the kidneys is **reabsorption:** the recapture of water and essential solutes by moving them *out of* the tubule. Reabsorption alone, however, is not sufficient to provide the kind of control over urine composition that is needed. Suppose, for example, more of a particular solute

Filtration, kidney   Filtration of blood plasma through a glomerulus and into a kidney nephron.

Glomerular filtrate   Fluid that enters the nephron from the glomerulus; filtered blood plasma.

Reabsorption, kidney Movement of solutes out of filtrate in the kidney tubule.

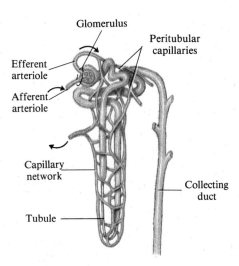

Figure 27.14
Structure of a nephron.

Secretion   Movement of solutes into filtrate in a kidney tubule.

must be excreted than is present in the filtrate. This situation is dealt with by **secretion:** the transfer of solutes *into* the kidney tubule.

Reabsorption and secretion require the transfer of solutes and water among the filtrate, the interstitial fluid surrounding the tubule, and blood in the capillaries. Solutes may cross the tubule and capillary membranes by passive diffusion in response to concentration or ionic charge differences, or by active transport. Water moves in response to differences in the osmolarity of the fluids on the two sides of the membranes. Solute and water movement is also controlled by hormone-directed variations in the permeability of the tubule membrane. The changes in filtrate composition during urine formation are diagrammed in Figure 27.15.

27.8   URINE COMPOSITION AND FUNCTION

Urine contains the products of glomerular filtration, minus the substances reabsorbed in the tubules, plus the substances secreted in the tubules. The actual concentration of these substances in urine at any time is determined by the amount of water being excreted, which can vary significantly with water intake, exercise, temperature, and state of health.

About 50 g of solids in solution are excreted every day. The major components of these solids are about 20 g of electrolytes and 30 g of nitrogen-containing wastes, which include urea and ammonia from amino acid catabolism, creatinine from breakdown of creatine phosphate in muscles, and uric acid from purine catabolism. Normal urine composition is usually reported as amount excreted per day, and laboratory urinalysis often requires collection of a 24-hr urine sample.

The following paragraphs briefly describe a few of the mechanisms that control the composition of urine.

Fluid and Na$^+$ Balance   The higher the concentration of solutes in urine, the less water is reabsorbed and the greater the volume of urine. The elevated concentration of glucose in the urine of individuals with diabetes mellitus, for example, is responsible for their increased urine volume.

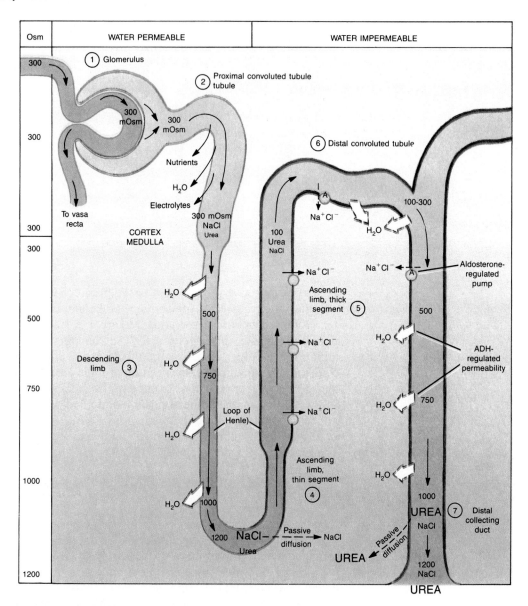

Figure 27.15
Formation of urine in the kidney. The numerical scale at the left shows the osmolarity of the outer (cortex) and inner (medulla) regions of the kidney. The high osmolarity of the medulla is maintained by removal of water via the capillary network (not shown). After filtration in the *glomerulus* (①), about 80% of the Na^+, 100% of the glucose, and significant amounts of other electrolytes and nutrients are reabsorbed by active transport in the *proximal convoluted tubule* (②). Next, in the *descending limb* of the loop of Henle (③), which is not permeable to Na^+ and carries out no active transport, water passively exits the tubule because the surrounding interstitial fluid has a higher osmolarity. Then, in the *ascending limb*, the situation reverses. The tubule here is not very permeable to water, but Na^+ departs by passive diffusion at the lower end (④) and active transport at the upper end (⑤). To maintain charge balance, Cl^- moves with Na^+. Finally, in the *distal convoluted tubule* (⑥) and *collecting duct*, the filtrate concentration receives its final adjustment.

The amount of water reabsorbed is in turn dependent on both the antidiuretic-hormone-controlled permeability of the collecting duct membrane and the amount of sodium actively reabsorbed. Increased sodium reabsorption means higher interstitial osmolarity, greater water reabsorption, and decreased urine volume. In the opposite condition of decreased sodium reabsorption, less water is reabsorbed and urine volume increases. Diuretic drugs such as furosemide (Endural or Lasix), which is used in treating hypertension and congestive heart failure, act by inhibiting the active transport of Na^+ out of the loop of Henle. Caffeine has a similar effect.

The reabsorption of Na^+ is normally under the control of the steroid hormone aldosterone. The arrival of chemical messengers signaling a decrease in total blood plasma volume accelerates the secretion of aldosterone. The result is increased Na^+ reabsorption in the kidney tubules accompanied by increased water reabsorption.

Acid–base Balance   Respiration, buffering of body fluids, and excretion of hydrogen ions in the urine combine to maintain acid–base balance. Metabolism normally produces an excess of hydrogen ions, 50–100 mEq of which must be excreted each day to prevent acidosis. In the process of eliminating H^+, the kidneys also regulate the HCO_3^- concentration of body fluids. Very little free hydrogen ion exists in blood plasma, and therefore very little enters the glomerular filtrate. Instead, the H^+ to be eliminated is produced from CO_2 in the cells lining the proximal and distal convoluted tubules.

$$CO_2 + H_2O \xrightarrow{\text{carbonic anhydrase}} H^+ + HCO_3^-$$

The HCO_3^- ions return to the bloodstream, and the H^+ ions enter the filtrate.

The urine must carry away the necessary quantity of H^+ without becoming excessively acid. To accomplish this, the H^+ is tied up by reaction with HPO_4^{2-} that is absorbed at the glomerulus or by reaction with NH_3 produced in the distal tubule cells by deamination of glutamate.

$$H^+ + HPO_4^{2-} \longrightarrow H_2PO_4^-$$

$$H^+ + NH_3 \longrightarrow NH_4^+$$

When acidosis occurs, the kidney responds by synthesizing more ammonia, thereby increasing the quantity of H^+ eliminated.

A further outcome of H^+ production in tubule cells is the net reabsorption of HCO_3^- that entered the filtrate at the glomerulus. The body cannot afford to lose HCO_3^-: The result would be production of additional acid from carbon dioxide by reaction with water. Instead, H^+ secreted into the tubules combines with HCO_3^- in the filtrate to produce CO_2 and water. Upon returning to the bloodstream, the CO_2 is reconverted to HCO_3^-.

In summary, acid–base reactions in the kidneys have the following results:

- Secreted H^+ is eliminated in the urine as NH_4^+ or $H_2PO_4^-$.
- Secreted H^+ combines with filtered HCO_3^-, producing CO_2 that returns to the bloodstream and again is converted to HCO_3^-.

INTERLUDE: AUTOMATED CLINICAL LABORATORY ANALYSIS

What happens when a physician orders chemical tests of blood, urine, or spinal fluid? The sample goes to a clinical chemistry laboratory, often in a hospital, where most tests are done by automated clinical chemistry analyzers. There are basically two types of analysis, one for the quantity of a chemical component and the other for the quantity of an enzyme with a specific metabolic activity.

Many chemical components are measured by mixing a reagent with the sample—the analyte—and determining the quantity of a colored product with a *photometer,* a unit that measures the absorption of light of a wavelength specific to the product. For each test specified, a portion of the sample is mixed with the appropriate reagent and the photometer is adjusted to the exact wavelength necessary.

When it's not possible to directly form a product that can be seen by a photometer, other types of reactions are used. Since most analytes are important in body fluids, they are usually substrates for enzyme-catalyzed reactions that can often be used in analysis. Glucose is determined in this manner by a pair of enzyme-catalyzed reactions: The glucose is converted to glucose 6-phosphate by the hexokinase-catalyzed reaction with ATP, the phosphate is then oxidized by $NADP^+$, and the quantity of NADPH produced is measured photometrically.

The second type of analysis, determination of the quantity of a specific enzyme or the ratio of two or more enzymes, is valuable in detecting organ damage that allows enzymes to leak into body fluids. For example, elevation of both ALT (alanine aminotransferase) and AST (aspartate aminotransferase) with an AST/ALT ratio greater than 1.0 is characteristic of liver disease. If, however, the AST is greatly elevated and the AST/ALT ratio is higher than 1.5, a myocardial infarction (heart attack) is likely to have occurred. When the analyte is an enzyme, advantage is often taken of its action on a substrate. ALT, for example, is determined by photometrically monitoring the disappearance of NADH in the following pair of coupled reactions (where LD = lactate dehydrogenase):

$$\text{L-Alanine} + \alpha\text{-ketoglutarate} \xrightarrow{\text{ALT}} \text{pyruvate} + \text{L-glutamate}$$

SUMMARY

Two-thirds of body fluid is **intracellular fluid** and one-third is **extracellular fluid,** mainly **blood plasma** and **interstitial fluid.** Daily water intake and output are kept in balance by hormonal control of water in urine and thirst. At capillaries throughout the body, nutrients and metabolism end products carried in blood are exchanged by passage across capillary walls, through interstitial fluid, and across cell membranes of peripheral tissue.

Blood functions fall into three major categories: transport, regulation, and defense. Oxygen and carbon dioxide are transported in **erythrocytes** by hemoglobin, with the amounts of bound gases under allosteric control such that more oxygen is released to tissues where metabolism is active.

Plasma proteins and white blood cells are responsible for defense mechanisms. Infection or cell damage initiates **inflammation,** a nonspecific response that includes release of histamine, increased blood flow, and phagocytosis of **antigens** by white blood cells. In the antibody-directed and cell-directed **immune responses,** specific antigens are recognized by noncovalent interactions between protein amino acid sequences like those in enzyme-substrate interaction. The antibody-directed immune response occurs when B lymphocytes detect an antigen and divide into plasma cells that produce **immunoglobulins,** glycoprotein **antibodies** that mark that antigen for destruction. B cell division also yields memory cells that produce antibody if their antigen reappears. The cell-directed immune response occurs when T lymphocytes detect antigenic regions on the surface of infected body cells and release a toxic protein that perforates the cell membrane and destroys the infected cells.

Hemostasis, the halt of blood loss, begins with formation of a platelet plug at the site of tissue damage. **Blood clotting,** a process requiring stepwise activation of a series of 12 clotting factors, follows. The end result is removal

$$\text{Pyruvate} + \text{NADH/H}^+ \xrightarrow{\text{LD}} \text{lactate} + \text{NAD}^+$$

Automated analyzers rely on premixed reagents and automatic division of a fluid sample into small portions for each test. A low-volume analyzer that can provide rapid results for a few tests accepts a bar-coded serum or plasma sample cartridge (about the size of a small cassette tape cartridge) followed by bar-coded reagent cartridges. The instrument software reads the bar codes and directs an automatic pipettor to transfer the appropriate volume of sample to each test cartridge. The instrument then moves the test cartridge along as the sample and reagents are mixed, the reaction takes place for a measured amount of time, and the photometer reading is taken and converted to the test result.

A high-volume analyzer with more-complex software can randomly access 40 or more tests, can do STAT tests in less than 5 minutes, and can run over 400 tests per hour at a cost of less than 10 cents per test. The end result is a printed report on each sample listing the types of tests, the sample values, and a normal range for each test.

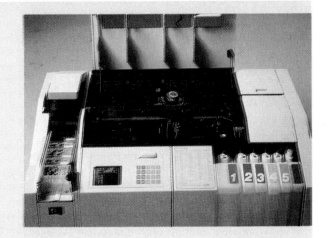

The small automated analyzer shown is used mainly for specialized tests such as those for cardiac enzymes, drugs of abuse, and therapeutic drugs. Eighty-two different analyses are available and up to 10 can be chosen for a single blood or urine sample and completed in about 30 minutes. The sample cup and a reagent pack for each test are loaded at the left. All results are read by a photometer and printed out on a slip that emerges just above the push-button control panel. The numbered bottles at the right contain water automatically dispensed to dissolve and dilute reagents.

by thrombin of solubilizing polypeptides from fibrinogen, followed by clot formation by cross-linking of the resulting **fibrin** fibers.

The kidney maintains homeostasis by controlled elimination of water, electrolytes, hydrogen ions, and nitrogen-containing wastes. The three essential kidney functions are **filtration** of all but large proteins from blood, **reabsorption** of water and essential solutes, and **secretion** of solutes to be eliminated. These functions are controlled by passive diffusion between regions of different **osmolarity, active transport,** and hormone-directed variation in the permeability of the kidney tubule membrane. The water and Na^+ content of urine are interrelated and controlled by antidiuretic hormone and aldosterone. Acid–base balance is maintained by elimination of H^+ as NH_4^+ or $H_2PO_4^-$ and net reabsorption of HCO_3^-.

REVIEW PROBLEMS

Body Fluids

27.1 What are the three principal body fluids?

27.2 Which body fluid contains the majority of the body's water?

27.3 What characteristics are needed for a substance to be soluble in body fluids?

27.4 How does blood pressure compare to interstitial fluid pressure in arterial capillaries? In venous capillaries? What effects do these pressure differences have on solute crossing cell membranes?

27.5 What is the purpose of the lymphatic system?

27.6 What is another name for antidiuretic hormone, and what is the purpose of this hormone?

27.7   What is plasma?

27.8   What are the three main types of blood cells?

27.9   What is the major function of each type of blood cell?

27.10   Heparin used for commercial purposes is commonly isolated from beef lung or pork intestines. Why do these tissues contain this substance?

27.11   What is blood serum?

27.12   How many O_2 molecules can be bonded by a hemoglobin molecule?

27.13   What must be the oxidation state of the iron in hemoglobin for it to perform its function?

27.14   How do deoxyhemoglobin and oxyhemoglobin differ in color?

27.15   How does the degree of saturation of hemoglobin vary with the partial pressure of O_2 in the tissues?

27.16   Oxygen has an allosteric interaction with hemoglobin. What are the results of this interaction as oxygen is bonded and as it is released?

27.17   What are the three ways of transporting CO_2 in the body?

27.18   Use Figure 27.7 to estimate the partial pressure at which hemoglobin is 50% saturated with oxygen.

27.19   Explain two ways in which the CO_2 released by oxidation of glucose tends to raise the H^+ concentration of the blood?

27.20   Does each of the following effects cause hemoglobin to release more O_2 to the tissues or to absorb more O_2?

(a) raising the temperature

(b) production of CO_2

(c) increase of the H^+ concentration

27.21   What are the three types of responses the body makes upon exposure to an antigen?

27.22   Antihistamines are often prescribed to counteract the effects of allergies. Propose one way in which these drugs might work to stop production of histamine from histidine.

27.23   How are specific immune responses similar to the enzyme–substrate interaction?

27.24   What types of cells are associated with antibody-directed immune response and how do they work?

27.25   What are memory cells?

27.26   T cells are often discussed in conjunction with the disease AIDS, in which a virus destroys these cells. How do T cells work to combat disease?

27.27   What is a blood clot?

27.28   What vitamin and what mineral are specifically associated with the clotting process?

27.29   What two pathways trigger blood clotting?

27.30   Why do you suppose that many of the enzymes involved in blood clotting are secreted by the body as zymogens?

27.31   Kidneys are often referred to as filters that purify the blood. What other two essential functions do the kidneys perform to help maintain homeostasis?

27.32   Write the reactions by which HPO_4^{2-} and HCO_3^- absorb excess H^+ from the urine before elimination.

Applications

27.33   What is cryopreservation? [App: Organ Cryopreservation]

27.34   Why do cells rupture as the water in them freezes? [App: Organ Cryopreservation]

27.35   What two steps are taken in cryopreservation to ensure that small, rather than large, ice crystals form and that the crystals do not merge to make larger crystals upon standing? [App: Organ Cryopreservation]

27.36   What problems are associated with the need to maintain metabolism while tissues are frozen? [App: Organ Cryopreservation]

27.37   How do endothelial cells in brain capillaries differ from those in other capillary systems? [App: Blood–Brain Barrier]

27.38   What is meant by an asymmetric transport system? Give one specific example of such a system. [App: Blood–Brain Barrier]

27.39   What type of substance is likely to breach the blood–brain barrier? Would ethanol be likely to cross this barrier? Why or why not? [App: Blood–Brain Barrier]

27.40   How are photometers used in automated analysis? [Int: Automated Clinical Chemistry Analysis]

27.41   Why is it useful to test for enzyme levels in body fluids? [Int: Automated Clinical Chemistry Analysis]

27.42   What are some advantages of using automated analyzers compared to using technicians? [Int: Automated Clinical Chemistry Analysis]

Additional Questions

27.43   Why is ethanol soluble in blood?

27.44   Nursing mothers are said to impart immunity to their infants. Why do you think this is so?

27.45   Many people find they retain water, with fingers and ankles swelling after eating salty food. Explain this in terms of the method of operation of the kidneys.

27.46   How does active transport differ from osmosis?

27.47   When is active transport necessary to move substances through cell walls?

27.48   Discuss the importance of the CO_2/HCO_3^- equilibrium in blood and in urine.

27.49   Name the three principal functions of blood and give an example of each type of function.

Exponential Notation

Numbers that are either very large or very small are usually represented in *exponential notation* as a number between 1 and 10 multiplied by a power of 10. In this kind of expression, the small raised number to the right of the 10 is the *exponent*.

Number	Exponential Form	Exponent
1,000,000	1×10^6	6
100,000	1×10^5	5
10,000	1×10^4	4
1,000	1×10^3	3
100	1×10^2	2
10	1×10^1	1
1		
0.1	1×10^{-1}	-1
0.01	1×10^{-2}	-2
0.001	1×10^{-3}	-3
0.000 1	1×10^{-4}	-4
0.000 01	1×10^{-5}	-5
0.000 001	1×10^{-6}	-6
0.000 000 1	1×10^{-7}	-7

Numbers greater than 1 have *positive* exponents, which tell how many times a number must be *multiplied* by 10 to obtain the correct value. For example, the expression 5.2×10^3 means that 5.2 must be multiplied by 10 three times:

$$5.2 \times 10^3 = 5.2 \times 10 \times 10 \times 10 = 5.2 \times 1000 = 5200$$

Note that doing this means moving the decimal point three places to the right:

$$5\ 2\ 0\ 0.$$

1 2 3

The value of a positive exponentindicates *how many places to the right the decimal point must be moved* to give the correct number in ordinary decimal notation.

Numbers less than 1 have *negative exponents,* which tell how many times a number must be *divided* by 10 (or multiplied by one-tenth) to obtain the correct value. Thus, the expression 3.7×10^{-2} means that 3.7 must be divided by 10 two times:

$$3.7 \times 10^{-2} = \frac{3.7}{10 \times 10} = \frac{3.7}{100} = 0.037$$

Note that doing this means moving the decimal point two places to the left:

$$0.\ 0\ 3\ 7$$

$$2\ \ 1$$

The value of a negative exponent indicates *how many places to the left the decimal point must be moved* to give the correct number in ordinary decimal notation.

CONVERTING DECIMAL NUMBERS TO EXPONENTIAL NOTATION

To convert a number greater than 1 from decimal notation to exponential notation, first move the decimal point to the *left* until there is only a single digit to the left of the decimal point. The *positive* exponent needed for exponential notation is the same as *the number of places the decimal point was moved*.

$$6\ 3\ 5\ 7\ 8\ 1. = 6.35781 \times 10^5$$

$$5\ 4\ 3\ 2\ 1$$

To convert a number smaller than 1 from decimal notation to exponential notation, first move the decimal point to the *right* until there is *a single nonzero digit* to the left of the decimal point. The *negative* exponent needed for exponential notation is the same as *the number of places the decimal point was moved*.

$$0.\ 0\ 0\ 0\ 4\ 2\ 6 = 4.26 \times 10^{-4}$$

$$1\ 2\ 3\ 4$$

MULTIPLYING EXPONENTIAL NUMBERS

To multiply two numbers in exponential form, the exponents are *added*. For example:

$$
\begin{aligned}
(3.5 \times 10^3) \times (4.2 \times 10^4) &= 3.5 \times 4.2 \times 10^{(3\,+\,4)} \\
&= 14.7 \times 10^7 \\
&= 1.47 \times 10^8 = 1.5 \times 10^8 \text{ (rounded off)} \\
(5.2 \times 10^4) \times (4.6 \times 10^{-3}) &= 5.2 \times 4.6 \times 10^{[4\,+\,(-3)]} \\
&= 23.92 \times 10^1 \\
&= 2.392 \times 10^2 = 2.4 \times 10^2 \text{ (rounded off)}
\end{aligned}
$$

DIVIDING EXPONENTIAL NUMBERS

To divide two numbers in exponential form, the exponents are *subtracted*. For example:

$$\frac{4.1 \times 10^4}{6.2 \times 10^6} = \frac{4.1}{6.2} \times 10^{(4-6)} = 0.6613 \times 10^{-2} = 6.6 \times 10^{-3}$$

$$\frac{6.6 \times 10^3}{8.4 \times 10^{-2}} = \frac{6.6}{8.4} \times 10^{[3-(-2)]} = 0.7857 \times 10^5 = 7.9 \times 10^4$$

Conversion Factors

Length SI Unit; Meter (m)

1 meter = 0.001 kilometer (km)
= 100 centimeters (cm)
= 1.0936 yards (yd)

1 centimeter = 10 millimeters (mm)
= 0.3937 inch (in.)

1 nanometer = 1×10^{-9} meter

1 Angstrom (Å) = 1×10^{-10} meter

1 inch = 2.54 centimeters

1 mile = 1.6094 kilometers

Volume SI Unit: Cubic meter (m^3)

1 cubic meter = 1000 liters (L)

1 liter = 1000 cubic centimeters (cm^3)
= 1000 milliliters (mL)
= 1.056710 quarts (qt)

1 cubic inch = 16.4 cubic centimeters

Temperature SI Unit: Kelvin (K)

0 K = $-273.15°C$
= $-459.67°F$

$°F = (9/5)°C + 32°; (1.8 \times °C) + 32°$

$°C = (5/9)(°F - 32°); \dfrac{(°F - 32°)}{1.8}$

K = $°C + 273.15°$

Mass SI Unit: Kilogram (kg)

1 kilogram = 100 grams (g)
= 2.205 pounds (lb)

1 gram = 1000 milligrams (mg)
= 0.03527 ounce (oz)

1 pound = 453.6 grams

1 atomic mass unit = 1.66054×10^{-24} gram

Pressure SI Unit: Pascal (Pa)

1 pascal = 9.869×10^{-7} atmospheres

1 atmosphere = 101,325 pascals
= 760 mm Hg (Torr)
= 14.70 lb/in^2

Energy SI Unit: Joule (J)

1 joule = 0.23901 calorie (cal)

1 calorie = 4.184 joules

G L O S S A R Y

Note: This glossary is alphabetized according to the letter-by-letter method, as are most dictionaries. All the letters of compound terms up to the first comma are taken into account in the alphabeticization. For example, "Acid, Brønsted-Lowry" comes before "Acid anhydride," which comes before "Acid–base indicator."

Acetal A compound that has two —OR groups bonded to the same carbon atom.

Acetyl coenzyme A (acetyl SCoA) Acetyl-substituted coenzyme A that is the common intermediate carrying acetyl groups into the citric acid cycle.

Acetyl group The $CH_3C=O$ group.

Achiral The opposite of chiral; not having (right or left) handedness.

Acid, Brønsted-Lowry A substance that is able to donate a hydrogen ion, H^+.

Acid anhydride A compound that has a carbonyl group bonded to a carbon substituent and an —OCOR group, RCO_2COR.

Acid–base indicator A dye that changes color to indicate the acidity or alkalinity of a solution (the pH).

Acid–base reaction, Brønsted-Lowry The transfer of a proton from an acid to a base.

Acid dissociation constant (K_a) The equilibrium constant for the dissociation of an acid (HA), equal to $[H^+][A^-]/[HA]$.

Acidic solution A solution in which pH is less than 7 and hydronium ion concentration is greater than hydroxide ion concentration.

Acidosis The medical condition that results when blood pH drops below 7.35.

Activation energy (E_{act}) The amount of energy reactants must have in order to surmount the energy barrier to reaction.

Activation of an enzyme Any process that initiates or increases the action of an enzyme.

Active site A small, three-dimensional portion of an enzyme with the specific shape and structure necessary to bind a substrate.

Active transport Energy-requiring transport of molecules or ions across cell membranes against a concentration gradient.

Acyclic alkane An alkane that contains no rings.

Acyl carrier protein Protein that carries acyl groups during fatty acid synthesis.

Addition reaction, aldehydes and ketones Addition of an alcohol or other reagent to the carbon–oxygen double bond to give a carbon–oxygen single bond.

Addition reaction, alkenes and alkynes Addition of a reactant of the general form X-Y to the multiple bond of an unsaturated compound to yield a saturated product.

Adenosine triphosphate (ATP) The triphosphate of adenosine; the "energy currency" molecule of metabolism.

Aerobic In the presence of oxygen.

Alcohol A compound that contains an —OH functional group covalently bonded to an alkane-like carbon atom, R—OH.

Aldehyde A compound that has a carbonyl group bonded to one carbon atom and one hydrogen atom, RCHO.

Aldol reaction The reaction of a ketone or aldehyde to form a hydroxy ketone product on treatment with base catalyst.

Aldose A monosaccharide that contains an aldehyde carbonyl group.

Alkali metal An element in Group 1A of the periodic table (Li, Na, K, Rb, Cs, Fr).

Alkaline earth metal An element in Group 2A of the periodic table (Be, Mg, Ca, Sr, Ba, Ra).

Alkaline solution A solution in which pH is greater than 7 and hydroxide ion concentration is greater than hydronium ion concentration.

Alkaloid A naturally occurring nitrogen-containing compound isolated from a plant; usually basic, bitter, and poisonous.

Alkalosis The medical condition that results when blood pH rises above 7.45.

Alkane A compound that contains only carbon and hydrogen and has only single bonds.

Alkene A compound that contains only carbon and hydrogen and has a carbon–carbon double bond.

Alkoxy group An RO— group.

Alkyl group The part of an alkane that remains when one hydrogen atom is removed.

Alkyl halide A compound that contains an alkyl group bonded to a halogen atom, RX.

Alkyne A compound that contains only carbon and hydrogen and has a carbon–carbon triple bond.

Allosteric control Cooperative interaction by which binding a substrate or a regulator at one site in an enzyme influences shape and therefore binding at other sites.

Allosteric enzyme Enzyme with two or more protein chains (quaternary structure) and two or more interactive binding sites for substrate and regulators.

Alloy A solution or mixture of metals.

Alpha (α) amino acid An amino acid in which the amino group is bonded to the carbon atom next to the —COOH group.

Alpha (α) helix A common secondary protein structure in which a protein chain wraps into a coil stabilized by hydrogen bonds between peptide links in the backbone.

Alpha (α) radiation Emission of helium nuclei, $^4_2$He.

Amide A compound that has a carbonyl group bonded to one carbon atom and one nitrogen atom group, $RCONR_2'$, where R' may be an H atom.

Amide bond The bond between a carbonyl group and the nitrogen atom in an amide.

Amine A compound that has one or more organic groups bonded to nitrogen, RNH_2, R_2NH, or R_3N.

Amino acid A molecule that contains both an amino group and a carboxylic acid functional group; proteins are polymers of amino acids.

Amino acid pool The entire collection of free amino acids in the body.

Amino group The —NH_2 functional group.

Ammonium salt An ionic substance formed by reaction of ammonia or an amine with an acid; an ammonium ion, NH_4^+, and an anion or an amine salt.

Amorphous solid A solid in which the atoms, molecules, or ions do not have a regular, repeating arrangement as in a crystalline solid.

Amphoteric Able to react as both an acid and a base.

Anabolism Metabolic reactions that build larger biological molecules from smaller pieces.

Anaerobic In the absence of oxygen.

Anion A negatively charged ion.

Anomeric carbon atom The hemiacetal C atom in a cyclic sugar; the C atom bonded to an —OH group and an —OR group (R is the ring).

Anomers Cyclic sugars that differ only in positions of substituents at the hemiacetal carbon (the anomeric carbon); the α form has the —OH on the opposite side from the —CH_2OH; the β form has the —OH on the same side as the —CH_2OH.

Antibody (immunoglobulin) Protein molecule that identifies antigens.

Anticodon A sequence of three ribonucleotides on tRNA that recognizes the complementary sequence (the codon) on mRNA.

Antigen A substance that stimulates the immune response.

Antiseptic An agent that can be used to destroy or prevent the growth of harmful microorganisms on or in the body.

Apoenzyme The protein portion of an enzyme.

Aqueous solution A solution in which water is the solvent.

Aromatic compound A compound that contains a six-membered ring of carbon atoms with three double bonds or a similarly stable ring.

Artificial radioisotope A manmade radioactive isotope.

Artificial transmutation The change of one element into another by bombardment with a high-energy particle.

Aryl group General name for any substituent derived from an aromatic compound.

Atom The smallest and simplest particle of an element.

Atomic mass unit (amu) Unit of the relative atomic mass scale, equal to 1/12 the mass of a carbon atom, which is 1.6606×10^{-24} g.

Atomic number The primary characteristic that distinguishes atoms of different elements; equal to the number of protons in an atom's nucleus.

Atomic radius The radius of an atom; one-half the distance between the nuclei of adjacent identical atoms in a metal or molecule.

Atomic weight The average mass (in amu) of atoms in the naturally occurring isotopic mixture of an element.

Avogadro's law Equal volumes of gases at the same temperature and pressure contain equal numbers of molecules (V/n = constant, or $V_1/n_1 = V_2/n_2$).

Avogadro's number The number of particles (atoms, formula units, ions, or molecules) in one mole (6.02×10^{23}).

Background radiation The sum of low-level radiation to which all individuals are exposed.

Balanced chemical equation A chemical equation in which the number of atoms of each kind is the same in the reactants and products and, for a net ionic equation, charge is also balanced.

Basal metabolic rate The minimum amount of energy required per unit time to stay alive.

Base, Brønsted-Lowry A substance that can accept a hydrogen ion (H^+) from an acid.

Benedict's reagent A reagent (Cu^{2+} in basic solution) that converts an aldehyde into a carboxylic acid and yields a brick-red precipitate of Cu_2O.

Beta (β) radiation Emission of electrons.

β-Pleated sheet A common secondary protein structure in which segments of a protein chain fold back on themselves to form parallel strands held together by hydrogen bonds between peptide links in the backbone.

Bile Fluid secreted by the liver and released into the small intestine during digestion; contains bile salts, bicarbonate ion, and other electrolytes.

Bile acids Steroid acids derived from cholesterol that are secreted in bile.

Binary compound A compound composed of atoms of two elements.

Biochemistry The study of the chemistry of substances that occur in living matter.

Biomolecule A molecule of a compound that occurs naturally and plays a role in the chemistry of living things.

Blood clotting Formation of a structure of fibrin and blood cells to close the rupture of a blood vessel.

Blood plasma Liquid portion of blood; an extracellular fluid.

Blood serum Fluid portion of blood remaining after clotting has occurred.

Boiling point (bp) The temperature at which the vapor pressure of a liquid is equal to atmospheric pressure and the liquid boils; the temperature at which a liquid begins to boil.

Bond angle The angle formed between any two adjacent covalent bonds.

Bond energy The energy needed to separate two covalently bonded atoms.

Boyle's law The pressure of a gas at constant temperature is inversely proportional to its volume (PV = constant, or $P_1V_1 = P_2V_2$).

Branched-chain alkane An alkane that has a branching connection of carbon atoms along its chain.

Buffer A combination of substances, usually a weak acid and its anion, that act together to prevent a large change in the pH of a solution.

C-Terminal amino acid The amino acid with the free —COOH group at the end of a polypeptide.

Calorie (cal) A common unit of energy equal to exactly 4.184 J.

Carbocation A polyatomic ion with a positively charged carbon atom.

Carbohydrate A member of a large class of naturally occurring polyhydroxy ketones and aldehydes.

Carbonyl group A functional group that has a carbon atom joined to an oxygen atom by a double bond, C=O.

Carbonyl-group substitution reaction A reaction in which a new group X replaces (substitutes for) a group Y attached to a carbonyl-group carbon.

Carboxylate anion The anion that results from dissociation of a carboxylic acid, $RCOO^-$.

Carboxyl group The —COOH functional group.

Carboxylic acid A compound that has a carbonyl group bonded to one carbon substituent and one —OH group, RCOOH.

Carboxylic acid salt An ionic compound containing a carboxylate ion.

Carcinogenic Cancer-causing.

Catabolism Metabolic reactions that break down molecules into smaller pieces.

Catalyst A substance that speeds up a chemical reaction without itself undergoing any permanent chemical change.

Catalytic cracking A procedure for converting the alkane mixture in kerosene into the smaller, branched-chain alkanes needed for gasoline.

Cation A positively charged ion.

Cell membrane (plasma membrane) The membrane surrounding a cell (a lipid bilayer).

Celsius degree (C°) The metric unit of temperature; equal to 1.8 Fahrenheit degrees.

Centimeter (cm) A unit of length equal to 1/100 m, or about 0.3937 in.

Chain-growth polymer Polymer formed by the addition of monomer molecules one by one to the end of a growing chain.

Chain reaction A self-sustaining reaction that continues unchecked once it has been started.

Change of state The conversion of a substance from one state to another, for example, from a liquid to a gas.

Charles' law The volume of a gas at constant pressure is directly proportional to its kelvin temperature (V/T = constant, or V_1/T_1 = V_2/T_2).

Chemical bonds The forces that hold atoms together in chemical compounds.

Chemical compound A pure substance that can be broken down into simpler substances only by chemical reactions; composed of atoms of different elements.

Chemical element A fundamental substance that cannot be chemically broken down into any simpler substance.

Chemical energy The potential energy stored in chemical compounds.

Chemical equation A written expression in which symbols and formulas are used to represent reactants and products, and which is balanced.

Chemical equilibrium The point in a reversible reaction at which forward and reverse reactions take place at the same rate, so that the concentrations of products and reactants no longer change.

Chemical formula Representation with symbols and subscript numbers of the atoms combined in a chemical compound.

Chemical property A property that involves a change in identity of a substance.

Chemical reaction (chemical change) A process in which the identity and composition of one or more substances are changed.

Chemical reactivity The tendency of an element or chemical compound to undergo chemical reactions.

Chemiosmotic hypothesis An explanation of how the establishment of a pH difference across the mitochondrial membrane by the respiratory chain is coupled to the synthesis of ATP.

Chemistry The study of the nature, properties, and transformations of matter.

Chiral Having (right or left) handedness; able to have two different mirror-image forms.

Chiral carbon atom A carbon atom bonded to four different groups.

Chromosome A complex between proteins and a DNA molecule that is visible during cell division.

Chyme The semisolid mixture produced by digestion in the stomach.

Chylomicron Spherical, low-density lipoprotein particles that transport triacylglycerols.

Cis isomer Isomer with a specific pair of atoms or groups on same side of a double bond.

Citric acid cycle The series of biochemical reactions that break down acetyl groups to carbon dioxide and energy stored in reduced coenzymes.

Claisen condensation reaction A reaction that joins two ester molecules together to yield a keto ester product.

Clinical chemistry Chemical analysis of body tissues and fluids.

Codon A sequence of three ribonucleotides in the RNA chain that codes for a specific amino acid.

Coefficient A number placed before a formula in a chemical equation to show how many units of that substance are required to balance the equation.

Coenzyme A small, organic molecule that acts as an enzyme cofactor.

Coenzyme A Coenzyme that functions as a carrier of acetyl groups and other acyl (R—C=O) groups.

Cofactor A small, nonprotein part of an enzyme that is essential to the enzyme's catalytic activity.

Colloid (colloidal dispersion) A mixture of a dispersing medium and dispersed particles larger than most molecules but too small to be seen by the naked eye.

Combination reaction A reaction in which two reactants combine to give a single product: $A + B \rightarrow C$.

Combined gas law The product of the pressure and volume of a gas is inversely proportional to its temperature (PV/T = constant, or P_1V_1/T_1 = P_2V_2/T_2).

Combustion A chemical reaction in which heat and often light are produced; usually refers to burning in the presence of oxygen.

Competitive enzyme inhibition Enzyme regulation in which an inhibitor competes with a similarly shaped substrate for binding to the enzyme active site.

Concentration The quantity of a solute per unit quantity of a solution or solvent.

Condensed structure A structure in which central atoms and the atoms bonded to them are written as groups, e.g., $CH_3CH_2CH_3$.

Conformation The exact three-dimensional shape of a molecule at any given instant.

Conjugate acid–base pair An acid and base related to each other by the loss and gain of a proton.

Conjugated protein A protein that yields one or more other substances in addition to amino acids when hydrolyzed.

Conversion factor A fraction that states a relationship between two different units of measure.

Coordinate covalent bond A covalent bond in which both shared electrons are contributed by the same atom.

Cosmic rays Energetic particles, primarily protons, from space.

Coupled reactions Combination of an energy-requiring and energy-releasing reaction that occur together to give an overall energy-releasing reaction.

Covalent bond A bond that results when two atoms share one or more pairs of electrons.

Crenation The shriveling of red blood cells due to a loss of water.

Crystal A solid with planar faces meeting at fixed angles, and in which atoms, ions, or molecules are arranged in a regular, repeating pattern.

Crystalline solid A solid composed of crystals, in which atoms, molecules, or ions have a regular, repeating arrangement.

Cubic centimeter (cc or cm^3) An alternative name for a milliliter; volume of a 1 cm cube.

Cubic meter (m^3) The SI unit of volume; equal to 264.2 gal.

Curie A unit for measuring the number of radioactive disintegrations per second.

Cycloalkane An alkane that contains a ring of carbon atoms.

Cytochromes Heme-containing coenzymes that function as electron carriers.

Cytoplasm The region between the cell membrane and the nuclear membrane in a eukaryotic cell.

Cytosol The contents of the cytoplasm surrounding the organelles.

D sugar Monosaccharide with —OH group on chiral carbon atom farthest from the carbonyl group pointing to the right in the Fischer projection.

Dalton An alternate name for atomic mass unit.

Dalton's law of partial pressure The total pressure exerted by a mixture of gases is equal to the sum of the partial pressures exerted by each individual gas.

Decomposition reaction A reaction in which a single compound breaks down into other substances, often after energy has been supplied as heat: $C \rightarrow A + B$.

Dehydration Loss of water, as from an alcohol to yield an alkene.

Denaturation The loss of secondary, tertiary, or quaternary protein structure due to disruption of noncovalent interactions that leaves peptide bonds intact.

Density The mass of an object per unit of volume.

Deoxyribonucleotide A nucleotide containing D-deoxyribose.

Diabetes mellitus A chronic condition due to either insufficient insulin or failure of insulin to cross cell membranes.

Dialysis The passage through a semipermeable membrane of solvent and small solute molecules, but not of large solute molecules.

Diastereomers Stereoisomers that are not mirror images of each other.

Diatomic molecule A molecule that consists of two atoms bonded together.

Diffusion The mixing of atoms, ions, or molecules by random motion.

Digestion A general term for the breakdown of food into small molecules.

Dilution A decrease in the concentration of a solution caused by addition of solvent.

Dilution factor The ratio of original to final volumes of a solution being diluted.

Dimer A unit formed by the joining together of two identical molecules.

Dipole–dipole attraction Attraction between polar molecules.

Diprotic acid A substance that has two acidic hydrogen atoms.

Disaccharide A carbohydrate that yields two monosaccharides on hydrolysis.

Disinfectant An agent that can be used to destroy or prevent the growth of harmful microorganisms on inanimate objects only.

Displacement reaction A reaction in which one element takes the place of a less reactive element in a compound: $A + BC \rightarrow B + AC$.

Dissociation The splitting apart of a substance to yield two or more ions.

Distillation The process by which a mixture is separated according to boiling points.

Disulfide A compound that contains a sulfur–sulfur single bond, $R—S—S—R$.

Disulfide bridge An S—S bond formed between two cysteine residues that can join two peptide chains together or cause a loop in a peptide chain.

DNA (deoxyribonucleic acid) The nucleic acid that stores the genetic information.

Double covalent bond A covalent bond that results from sharing of two electron pairs between atoms.

Double helix Two strands coiled around each other in a screw-like fashion, as in the two polynucleotide strands in DNA.

Dynamic equilibrium An equilibrium in which the rates of forward and reverse changes are equal.

Electrode Conductor through which electrical current enters or leaves a conducting medium.

Electrolyte A substance that conducts electricity when dissolved in water; specifically, ions in body fluids.

Electron A negatively charged (-1) subatomic particle of very small mass that moves in the large volume of space surrounding nuclei.

Electron configuration The specific arrangement of an atom's electrons in shells and subshells; represented by notation such as $1s^2 2s^2$.

Electron-dot symbol An atomic symbol surrounded by dots that show the number of valence electrons.

Electronegativity The ability of an atom in a molecule to attract electrons; a numerical value on the electronegativity scale.

Endergonic Describes a process that absorbs free energy from the surroundings (positive ΔG).

Endocrine system A system of specialized cells, tissues, and ductless glands that excrete hormones and share with the nervous system the responsibility for maintaining constant internal body conditions and responding to changes in the environment.

End point The point at which neutralization is made visible in a titration.

Endothermic Describes a process that absorbs heat from the surroundings.

Energy The capacity to do work or to cause change.

Enthalpy change The amount of heat released (negative ΔH) or absorbed (positive ΔH) in a specific chemical or physical change.

Entropy (S) A measure of the amount of molecular disorder of a substance.

Enzyme A protein or other molecule that acts as a catalyst for biological reactions.

Enzyme–substrate complex A complex of enzyme and substrate in which the two are bound together by noncovalent interactions with the substrate in position to react.

Equilibrium constant (K) Value of the equilibrium constant expression for a given reaction.

Equilibrium constant expression Expression showing the concentrations of reactants and products in a reaction at equilibrium; equal to a constant

$$K = \frac{[M]^m [N]^n \cdots}{[A]^a [B]^b \cdots}$$

Equivalent, acid (base) The amount in grams of an acid (or base) that can donate one mole of H^+ (or OH^-) ions.

Equivalent, ion (Eq) The amount of an ion in grams that contains Avogadro's number of charges.

Erythrocytes Red blood cells.

Essential amino acid An amino acid that cannot be synthesized by the body and so must be obtained in the diet.

Ester A compound that has a carbonyl group bonded to one carbon atom and one —OR group, RCOOR'.

Esterification reaction The reaction between an alcohol and a carboxylic acid to yield an ester plus water.

Ether A compound that has an oxygen atom bonded to two carbon atoms, R—O—R.

Ethyl group —CH_2CH_3, the alkyl group derived from ethane.

Eukaryotic cell Cell with a membrane-enclosed nucleus; found in all higher organisms.

Evaporation The gradual escape of molecules from a liquid into the gaseous state.

Exchange reaction A reaction in which two compounds exchange partners: $AB + XY \rightarrow AY + XB$.

Exergonic Describes a process that releases free energy to the surroundings (negative ΔG).

Exon A DNA segment in a gene that codes for part of a protein molecule.

Exothermic Describes a process that gives off heat to the surroundings.

Extracellular fluid Fluid outside cells.

Factor-label method A method of problem solving in which equations are set up so that unwanted units cancel and only the correct units remain.

Fat A mixture of triacylglycerols that is solid because it contains a high proportion of saturated fatty acids.

Fatty acid A long-chain carboxylic acid; those in animal fats and vegetable oils often have 12 to 22 carbon atoms.

Fatty-acid spiral (β-oxidation) A repetitive series of biochemical reactions that degrade fatty acids to acetyl SCoA.

Feedback control Activation or inhibition of the first reaction in a sequence by a product of the sequence.

Fermentation The breakdown of glucose to ethanol plus carbon dioxide by the action of yeast enzymes.

Fibrin Insoluble protein that forms the fiber framework of a blood clot.

Fibrous protein A tough, insoluble protein whose peptide chains are arranged in long filaments.

Filtration, kidney Filtration of blood plasma through glomerulus and into kidney nephron.

Fischer projection Structure that represents a chiral carbon atom as the intersection of two lines. The horizontal line represents bonds pointing out of the page, and the vertical line represents bonds pointing behind the page. For sugars, the aldehyde or ketone is at the top.

Flavin adenine dinucleotide (FAD) Coenzyme that functions as an oxidizing agent and forms $FADH_2$.

Formula unit The group of atoms or ions represented by the formula of a chemical compound.

Formula weight The sum of the individual atomic weights for all atoms represented in the formula of a chemical compound or polyatomic ion.

Free energy change (ΔG) A quantity that is a function of energy change and entropy change and is the criterion for spontaneous change (negative ΔG; $\Delta G = \Delta H - T\Delta S$).

Free radical An atom or group that has an unpaired electron.

Functional group A part of a larger molecule composed of an atom or group of atoms that has characteristic chemical behavior.

Gamma (γ) radiation Emission of high-energy light waves.

Gas A substance that has neither a definite volume nor a definite shape.

Gas laws A series of laws that describe the behavior of gases under conditions of differing pressure, volume, and temperature.

Gene All segments of DNA needed to direct the synthesis of a protein with a specific function.

Genetic code The complete assignment of the 64 mRNA codons to specific amino acids (or stop signals).

Genetic enzyme control Regulation of enzyme activity by control of the synthesis of enzymes.

Genetic mutation An error in the code that determines inherited traits.

Genome The sum of all genes in an organism.

Germ cell A reproductive (sperm or egg) cell.

Globular protein A water-soluble protein that adopts a compact, coiled-up shape.

Glomerular filtrate The fluid that enters the nephron from the glomerulus; filtered blood plasma.

Glucogenic amino acid An amino acid catabolized to a compound that can enter the citric acid cycle for conversion to glucose.

Gluconeogenesis The biochemical pathway for the synthesis of glucose from noncarbohydrates such as lactate, amino acids, and glycerol.

Glycerophospholipid (phosphoglyceride) A lipid in which glycerol is linked by ester bonds to two fatty acids and one phosphate, which is in turn linked by another ester bond to an amino alcohol (or other alcohol); a derivative of phosphatidic acid.

Glycogenesis The biochemical pathway for synthesis of glycogen.

Glycogenolysis The biochemical pathway for breakdown of glycogen to free glucose.

Glycol A dialcohol; a compound that contains two —OH groups.

Glycolipid A sphingolipid with a fatty acid bonded to the C2 —NH$_2$ and a sugar bonded to the C1 —OH group of sphingosine.

Glycolysis The biochemical pathway that breaks down a molecule of glucose into two molecules of pyruvate plus energy.

Glycoside A cyclic acetal formed by reaction of a monosaccharide with an alcohol, accompanied by loss of H$_2$O.

Glycosidic bond Bond between the anomeric carbon atom of a monosaccharide and an —OR group.

Grain alcohol A common name for ethyl alcohol, CH$_3$CH$_2$OH.

Gram (g) The metric unit of mass; equal to 1/1000 kg.

Group A vertical column of elements in the periodic table; elements in a group have similar properties.

Half-life ($t_{1/2}$) The amount of time required for 50% of a sample to undergo radioactive decay.

Halogen An element in Group 7A of the periodic table (F, Cl, Br, I, At).

Halogenation, alkane The substitution of one or more hydrogen atoms in an alkane by halogen atoms.

Halogenation, alkene The reaction of an alkene with a halogen (Cl$_2$ or Br$_2$) to yield a 1,2-dihaloalkane product.

Heat A form of energy transferred from a hotter object to a colder one.

Heat of fusion The amount of heat necessary to convert one gram (or one mole) of solid into a liquid.

Heat of reaction The amount of heat released or absorbed during a chemical reaction.

Heat of vaporization The amount of heat necessary to convert one gram (or one mole) of liquid into a gas.

Hemiacetal A compound that has both an alcohol-like —OH group and an ether-like —OR group bonded to the same carbon atom.

Hemodialysis The use of dialysis to purify blood.

Hemolysis Bursting of red blood cells due to a buildup of pressure in the cell.

Hemostasis The stopping of bleeding.

Henderson-Hasselbalch equation Equation that gives pH of buffer solution containing weak acid (HA) and its anion (A$^-$); pH = pK_a + log([A$^-$]/[HA]).

Henry's law The solubility of a gas in a liquid is directly proportional to its partial pressure over the liquid at constant temperature.

Heterocycle A ring that contains nitrogen or some other atom in addition to carbon.

Heterogeneous mixture A mixture that is visually nonuniform.

Holoenzyme The combination of apoenzyme and cofactor that is active as a biological catalyst.

Homeostasis Maintenance of unchanging internal conditions by living things.

Homogeneous mixture A mixture that is uniform to the naked eye.

Hormone A chemical messenger, usually secreted by an endocrine gland and transported through the bloodstream to elicit response from a specific target tissue.

Hydrate A solid substance that has water molecules included in its crystals.

Hydration The surrounding of a solute ion or molecule by water molecules.

Hydration, alkene The reaction of an alkene with water to yield an alcohol.

Hydrocarbon A compound that contains only carbon and hydrogen.

Hydrogenation The reaction of an alkene (or alkyne) with H$_2$ to yield an alkane product.

Hydrogen bonding Intermolecular force due to attraction between an electronegative atom (O, N, or F) and a hydrogen atom bonded to an electronegative atom (usually O or N).

Hydrohalogenation The reaction of an alkene with HCl or HBr to yield an alkyl halide product.

Hydrolysis The breakdown of a compound, such as an acetal, by reaction with water; the H's and O from water add to the atoms in the broken bond.

Hydrolysis reaction, ion The reaction of an ion with water.

Hydrometer A weighted bulb used to measure the specific gravity of a liquid.

Hydronium ion H_3O^+; the species formed when an acid is dissolved in water.

Hydrophilic Water-loving; a hydrophilic substance dissolves in water.

Hydrophobic Water-fearing; a hydrophobic substance does not dissolve in water.

Hydroxyl group A name for the —OH group in an organic compound.

Hygroscopic Having the ability to attract water vapor from the air.

Hyperglycemia Higher than normal blood glucose concentration.

Hypertonic Referring to a solution that has higher osmolarity than another, usually blood.

Hyperventilation Heavy breathing that depletes carbon dioxide from the lungs and blood.

Hypoglycemia Lower than normal blood glucose concentration.

Hypotonic Referring to a solution that has lower osmolarity than another, usually blood.

Ideal gas A gas that obeys all the assumptions of the kinetic theory of gases.

Immiscible Mutually insoluble.

Immune response The recognition and attack on antigens, including viruses, bacteria, toxic substance, and infected cells.

Induced-fit model A model that pictures an enzyme with a conformationally flexible active site that can change shape to accommodate a range of different substrate molecules.

Inflammation A nonspecific defense mechanism triggered by antigens or tissue damage; includes swelling, redness, warmth, and pain.

Inhibition of an enzyme Any process that slows down or stops the action of an enzyme.

Inorganic compound A chemical compound composed of elements other than carbon; not an organic compound based on a carbon structure.

Intermolecular forces Forces of attraction and repulsion between partial charges in different molecules.

Interstitial fluid Fluid surrounding cells; an extracellular fluid.

Intracellular fluid Fluid inside cells.

Intron A seemingly nonsensical segment of DNA that does not code for part of a protein.

Ion An electrically charged atom or group of atoms.

Ionic bond The force of electrical attraction between an anion and a cation in a crystal.

Ionic compound A chemical compound composed of positively and negatively charged ions held together by ionic bonds.

Ionization energy The energy required to remove one electron from a single atom in the gaseous state.

Ionizing radiation Radiation capable of dislodging an electron from (ionizing) an atom or a chemical compound it strikes.

Ion product constant of water (K_w) The product of H_3O^+ and OH^- molar concentrations in water or any aqueous solution ($K_w = [H_3O^+][OH^-] = 1 \times 10^{-14}$).

Irreversible enzyme inhibition A mechanism of enzyme deactivation in which an inhibitor forms covalent bonds to the active site.

Isoelectric point The pH at which a large sample of amino acid molecules has equal numbers of + and – charges.

Isomers, cis–trans Alkenes that have the same formula and connections between atoms but differ in having pairs of groups on opposite sides of the double bond.

Isomers, constitutional Compounds with the same molecular formula but with different connections between atoms.

Isomers, optical The two mirror-image forms of a chiral molecule.

Isopropyl group —$CH(CH_3)_2$, the alkyl group derived by removing a hydrogen atom from the central carbon of propane.

Isotonic Referring to a solution that has the same osmolarity as another, usually blood.

Isotopes Atoms of the same element that have different numbers of neutrons in their nuclei and therefore different mass numbers.

Joule (J) The SI unit of energy.

Kelvin (K) The SI unit of temperature; $1 \text{ K} = 1 \text{ C}°$.

Ketoacidosis Lowered blood pH due to accumulation of ketone bodies.

Ketogenesis The synthesis of ketone bodies from acetyl SCoA.

Ketogenic amino acid An amino acid catabolized to a compound that can be converted to ketone bodies (also fatty acids).

Ketone A compound that has a carbonyl group bonded to two carbon atoms, $R_2C{=}O$.

Ketone bodies Compounds produced in the liver that can be used as fuel by muscle and brain tissue; β-hydroxybutyrate, acetoacetate, and acetone.

Ketose A monosaccharide that contains a ketone carbonyl group.

Kilocalorie (kcal) A unit of energy equal to 1000 cal.

Kilogram (kg) The SI unit of mass; equal to 2.205 lb.

Kinetic energy The energy of an object in motion.

Kinetic theory of gases A set of assumptions for explaining the general behavior of gases.

L sugar Monosaccharide with —OH group on chiral carbon atom farthest from the carbonyl group pointing to the left in the Fischer projection.

Law of conservation of energy Energy can be neither created nor destroyed in any physical or chemical change (first law of thermodynamics).

Law of conservation of mass Matter can be neither created nor destroyed in any physical or chemical change.

Le Chatelier's principle When a stress is applied to a system in equilibrium, the equilibrium shifts so that the stress is relieved.

Length The distance an object extends in a given direction.

Leukotriene A simple lipid derived from a C_{20} carboxylic acid like a prostaglandin but with no cyclopentane ring.

Lewis structure A structural formula that uses dots to represent valence electrons.

Limiting reactant The reactant present in an amount that limits the extent to which a reaction can occur.

Line structure A shorthand way of representing ring structures as polygons without showing individual carbon atoms.

1,4 Link An acetal link between the hydroxyl group at C1 of one sugar and the hydroxyl group at C4 of another sugar.

Lipid A naturally occurring molecule found in plant and animal sources that is soluble in nonpolar organic solvents.

Lipid bilayer The basic structural unit of cell membranes; composed of two parallel sheets of membrane lipid molecules arranged tail to tail.

Lipogenesis The biochemical pathway for synthesis of fatty acids from acetyl SCoA.

Lipoprotein A lipid–protein complex that transports lipids.

Liquid A substance that has a definite volume but that changes shape to fill its container.

Liter (L) The metric unit of volume; equal to 1.057 qt.

Lock-and-key model A model for enzyme specificity that pictures an enzyme as a large molecule with a cleft into which only specific substrate molecules can fit.

London forces Intermolecular forces of attraction between short-lived, temporary dipoles.

Lone pair A pair of outer-shell electrons not used by an atom in forming bonds.

Main group An element group on the far right (Groups 3A–8A) or far left (Groups 1A–2A) of the periodic table.

Manometer A mercury-filled U-tube used to measure pressure in a closed system.

Markovnikov's rule In the addition of HX to an alkene, the H becomes attached to the carbon that already has the most H's, and the X becomes attached to the carbon that has fewer H's.

Mass The amount of matter in an object.

Mass number The sum of an atom's protons and neutrons.

Matter The physical material that makes up the universe; anything that has mass and volume.

Melting point (mp) The temperature at which a solid turns into a liquid.

Mercaptan An alternate name for a thiol, R—SH.

Mercury barometer A mercury-filled glass tube used to measure atmospheric pressure.

Messenger RNA (mRNA) The RNA whose function is to carry genetic messages transcribed from DNA and direct protein synthesis.

Meta Indicates 1,3 substituents on a benzene ring.

Metabolic disease A disease due to disruption of metabolism caused by a genetically determined defect.

Metabolism The overall sum of the many reactions taking place in an organism.

Metal A malleable element with a lustrous appearance that is a good conductor of heat and electricity.

Metalloids Elements with properties intermediate between those of metals and nonmetals.

Meter (m) The SI unit of length; equal to 1.0936 yd.

Methyl group —CH_3, the alkyl group derived from methane.

Metric units Units of a common system of measure used throughout the world.

Micelle A spherical cluster formed by the aggregation of soap molecules (or other molecules with hydrophilic and hydrophobic ends) in water.

Microgram (μg) A unit of mass equal to 1/1000 mg.

Millicurie A unit 1/1000 the size of a curie.

Milliequivalent (mEq) The amount of an ion equal to 1/1000 of an equivalent.

Milligram (mg) A unit of mass equal to 1/1000 g.

Milliliter (mL) A unit of volume equal to 1/1000 L.

Millimeter (mm) A unit of length equal to 1/1000 m.

Millimeter of mercury (mm Hg) A unit of pressure equal to the force exerted by a 1 mm column of mercury.

Mirror image The reverse image produced when a object is reflected in a mirror.

Miscible Two substances mutually soluble in all proportions without limit.

Mitochondrial matrix The space surrounded by the inner membrane.

Mitochondrion An egg-shaped organelle where small molecules are broken down to provide the energy to power an organism.

Mixture A physical blend of two or more substances, each of which retains its chemical identity; can be separated by physical changes.

Mobilization (of triacylglycerols) Hydrolysis of triacylglycerols in adipose tissue and release of fatty acids into bloodstream.

Molarity (M) Concentration expressed as the number of moles of solute per liter of solution.

Molar mass The mass, usually in grams, of one mole of a substance.

Mole Amount of substance containing an Avogadro's number of particles; in practice, usually the quantity in grams equal to the formula weight of the substance.

Molecular compound A chemical compound composed of molecules.

Molecular formula A formula that shows by subscripts the numbers and kinds of atoms in one molecule.

Molecule A group of atoms held together by covalent bonds in a discrete unit capable of independent existence.

Mole ratio A ratio of the molar amounts of substances shown by the coefficients in a balanced chemical equation.

Monatomic ion An ion formed from a single atom.

Monomer Small molecule combined to form a polymer.

Monoprotic acid A substance that has one acidic hydrogen atom.

Monosaccharide (simple sugar) A carbohydrate that can't be chemically broken down into a smaller sugar by hydrolysis with aqueous acid.

Mutagen A substance that causes mutations.

Mutarotation Change in rotation of plane-polarized light resulting from the equilibrium between cyclic anomers and the open-chain form of a sugar.

Mutation An error in base sequence occurring during DNA replication.

*n***-Propyl group** —$CH_2CH_2CH_3$, the alkyl group derived by removing a hydrogen atom from an end carbon of propane.

N-Terminal amino acid The amino acid with the free —NH_2 group at the end of a protein.

Natural gas A gaseous mixture of small alkanes, primarily methane.

Natural radioisotope A radioactive isotope found in nature.

Net ionic equation A chemical equation including only the species that undergo change in a reaction involving ions in solution; must be balanced for atoms and charge.

Neurotransmitter A chemical messenger that transmits a nerve impulse between neighboring nerve cells.

Neutralization The exchange reaction between an acid and a base to yield water and a salt.

Neutral solution A solution in which pH is equal to 7 and hydronium ion and hydroxide ion concentrations are equal.

Neutron An electrically neutral subatomic particle found in the nuclei of atoms; mass = 1 amu.

Nicotinamide adenine dinucleotide (NAD⁺) Coenzyme that functions as an oxidizing agent and forms NADH/H⁺.

Nitrate ester A compound formed by reaction of an alcohol with nitric acid.

Noble gas An element in Group 8A of the periodic table (He, Ne, Ar, Kr, Xe, Rn).

Nomenclature A system for naming chemical compounds.

Noncompetitive enzyme inhibition Enzyme regulation in which an inhibitor binds to an enzyme elsewhere than at the active site, thereby changing the shape of the enzyme's active site.

Nonelectrolyte A substance that does not conduct electricity when dissolved in water.

Nonessential amino acid One of 10 amino acids that are synthesized in the body and are therefore not necessary in the diet.

Nonmetal An element that is a poor conductor of heat and electricity.

Nonpolar covalent bond A covalent bond in which the electrons are shared equally.

Nonredox reaction A reaction in which no oxidation numbers change.

Normality A measure of acid (or base) concentration expressed as the number of acid (or base) equivalents per liter of solution.

Nuclear chemistry The study of atomic nuclei and their reactions.

Nuclear decay The emission of a particle from the nucleus of an element.

Nuclear fission The splitting apart of an atom by neutron bombardment to give smaller atoms.

Nuclear fusion The combination of two light nuclei to give a heavier nucleus.

Nucleic acid A biological polymer made by the linking together of nucleotide units.

Nucleoside A compound consisting of a five-carbon sugar bonded to a cyclic amine base; like a nucleotide but missing the phosphate group.

Nucleotide A building block for nucleic acid synthesis, consisting of a five-carbon sugar bonded to a cyclic amine base and to phosphoric acid.

Nucleus The dense, positively charged mass at the center of an atom where protons and neutrons are located.

Octane number A measure of the antiknock properties of a fuel.

Octet rule Atoms of main-group elements tend to combine in chemical compounds by gaining, losing, or sharing electrons so that they attain eight outer-shell electrons.

Oil A mixture of triacylglycerols that is liquid because it contains a high proportion of unsaturated fatty acids.

Optical isomers (enantiomers) The two mirror-image forms of a chiral molecule.

Orbitals Regions of specific shapes occupied by electrons in s, p, d, and f subshells.

Organelle A small, organized unit in the cell that performs a specific function.

Organic compounds Compounds containing carbon and hydrogen, and derivatives of these compounds.

Ortho Indicates 1,2 substituents on a benzene ring.

Osmolarity (Osmol) The number of moles of dissolved solute particles (ions or molecules) per liter of solution.

Osmosis The passage of solvent molecules across a semipermeable membrane from a more dilute solution to a more concentrated solution.

Osmotic membrane A membrane that allows only the passage of solvent molecules.

Osmotic pressure The amount of external pressure that must be applied to halt the passage of solvent molecules across an osmotic membrane.

Oxidation The loss of electrons or increase in oxidation number of a reactant in a chemical reaction; in organic chemistry, the removal of hydrogen from a molecule or the addition of oxygen to a molecule.

Oxidation number (oxidation state) The charge on an ion formed from a single atom or a number assigned according to a fixed set of rules and indicating relative electron ownership in bonds.

Oxidation–reduction (redox) reaction A reaction in which oxidation and reduction occur, shown by electron gain and loss, oxidation number changes, or transfer of hydrogen or oxygen.

Oxidative deamination Conversion of an amino acid —NH_2 group to an α-keto group, with removal of NH_4^+.

Oxidative phosphorylation The synthesis of ATP from ADP using energy released in the respiratory chain.

Oxidized Indicates a substance that has lost hydrogen or gained oxygen, or a substance containing atoms that have lost electrons or increased in oxidation number.

Oxidizing agent The reactant that causes an oxidation by taking electrons or decreasing in oxidation number.

Para Indicates 1,4 substituents on a benzene ring.

Paraffin A mixture of waxy alkanes having 20 to 36 carbon atoms.

Partial pressure The contribution to total gas pressure of each individual component of a mixture of gases.

Parts per million (ppm) Number of parts per one million (10^6) parts.

Pascal (Pa) The SI unit of pressure; equal to 0.007500 mm Hg.

Pentose phosphate pathway The biochemical pathway that produces ribose (a pentose), NADPH, and other sugar phosphates from glucose; an alternative to glycolysis.

Peptide bond An amide bond that links two amino acids together.

Percent yield The percent of the theoretical yield actually obtained from a chemical reaction.

Period A horizontal row of elements in the periodic table; elements are listed in order of increasing atomic number across a period.

Periodic table A table displaying the elements in order of increasing atomic number so that elements with similar properties fall into the same column.

Petroleum A complex mixture of hydrocarbons, primarily of marine origin.

pH A number usually between 0 and 14 that describes the acidity of an aqueous solution; mathematically, the negative common logarithm of a solution's H_3O^+ concentration.

Phagocytosis Engulfing and digestion of a particle by a cell.

Phenol A compound that has an —OH functional group bonded directly to an aromatic, benzene-like ring.

Phenyl group The name of the C_6H_5— unit when a benzene ring is considered a substituent group.

Phosphate ester A compound formed by reaction of an alcohol with phosphoric acid.

Phosphate group A —PO_3^{2-} group in an organic molecule.

Phosphatidylcholine (lecithin) A glycerophospholipid containing choline.

Phospholipid A lipid that has an ester link between phosphoric acid and an alcohol.

Phosphoric anhydride A —P—O—P— group, like those in ADP and ATP.

Phosphorylation A reaction that results in addition of a phosphate group (—PO_3^{2-}).

Physical change A change in which no change in identity occurs.

Physical property A property that can be determined without a change in identity.

Physical quantity A physical property that can be measured.

Polar covalent bond A covalent bond in which one atom attracts bonding electrons more strongly than the other atom.

Polyatomic ion An ion that contains two or more atoms.

Polycyclic aromatic compound A substance that has two or more benzene-like rings fused together along their edges.

Polymer Very large molecule composed of identical repeating units and formed by combination of small molecules.

Polymerization A reaction in which monomers combine to form a polymer.

Polypeptide A molecule composed of roughly 10 to 100 amino acids linked by peptide bonds.

Polysaccharide (complex carbohydrate) A carbohydrate composed of many monosaccharides bonded together.

Polyunsaturated fatty acid (PUFA) A long-chain fatty acid that has two or more carbon–carbon double bonds.

Potential energy Energy that is stored because of position, composition, or shape.

Precipitate A solid that forms during a reaction in solution.

Precipitation The formation of a solid during a reaction in solution.

Precursor A compound necessary for the synthesis of another compound.

Pressure The force per unit area exerted on a surface.

Primary alcohol An alcohol in which the OH-bearing carbon atom is bonded to one other carbon (and two hydrogens), RCH_2OH.

Primary amine An amine that has one organic group bonded to nitrogen, RNH_2.

Primary (1°) carbon A carbon atom that is bonded to one other carbon atom.

Primary protein structure The sequence in which amino acids are linked together in a protein.

Product A substance formed as the result of a chemical reaction.

Prokaryotic cell Cell that has no nucleus; found in bacteria and algae.

Property A characteristic useful for identifying a substance or object.

Prostaglandin A lipid derived from a C_{20} carboxylic acid and containing a cyclopentane ring with two long side chains.

Protease An enzyme for hydrolysis of the peptide bonds in proteins.

Protein A large biological molecule made of many amino acids linked together through amide bonds.

Proton A positively charged (+1) subatomic particle found in the nuclei of atoms; mass = 1 amu.

Pure substance Any type of matter that has unvarying properties and constant composition, no matter what its source.

Quaternary ammonium salt An ammonium salt with four organic groups bonded to the nitrogen atom.

Quaternary (4°) carbon A carbon atom that is bonded to four other carbons.

Quaternary protein structure The way in which two or more protein chains aggregate to form large, ordered structures.

R— The general symbol for an alkyl group.

Rad A unit for measuring the amount of radiation energy absorbed per gram of tissue.

Radioactive tracer A radioisotope used to track the location of a substance to which it is attached.

Radioactivity Spontaneous emission of radiation (alpha, beta, or gamma rays) from an unstable atomic nucleus.

Radioisotope A radioactive isotope.

Reabsorption, kidney Movement of solutes out of filtrate in kidney tubule.

Reactant A starting substance that undergoes change in a chemical reaction.

Reaction energy diagram A plot representing the energy changes that occur during a chemical reaction.

Reaction mechanism A complete description of how a reaction occurs, including the details of each individual step in the overall process.

Reaction rate The rate at which a chemical reaction occurs, as measured by the change of a reactant or product concentration per unit time.

Reagent A reactant used to bring about a specific chemical reaction.

Reduced Indicates a substance that has gained hydrogen or lost oxygen, or a substance containing atoms that have gained electrons or decreased in oxidation number.

Reducing agent The reactant that causes a reduction by giving up electrons or increasing in oxidation number.

Reducing sugar A carbohydrate that reacts with an oxidizing agent such as Benedict's reagent.

Reduction The gain of electrons or decrease in oxidation number of a reactant in a chemical reaction; in organic chemistry, the addition of hydrogen or the removal of oxygen.

Reductive amination Conversion of an α-keto acid to an amino acid by reaction with NH_4^+.

Refining The process by which petroleum is converted into gasoline and other useful products.

Rem A unit for measuring the amount of tissue damage caused by radiation.

Replication The process by which copies of DNA are made in the cell.

Residue, amino acid An alternative name for an amino acid unit in a polypeptide or protein.

Resonance The existence of a molecule in a single structure intermediate among two or more correct possible double-bond-containing structures that can be drawn.

Respiratory chain (electron transport chain) The series of biochemical reactions that passes electrons from reduced coenzymes to oxygen and is coupled to ATP formation.

Reversible reaction A reaction that can proceed in either the forward or the reverse direction.

Ribonucleotide A nucleotide containing D-ribose.

Ribosomal RNA (rRNA) The type of RNA complexed with proteins in ribosomes.

Ribosome A structure in the cell where protein synthesis occurs.

RNA (ribonucleic acid) Nucleic acid responsible for putting the genetic information to use in protein synthesis.

Roentgen A unit for measuring the ionizing intensity of radiation.

Rounding off The procedure for expressing values with the correct number of significant figures.

Salt An ionic substance composed of the cation of a base and the anion of an acid, such as NaCl, $BaSO_4$.

Saponification Reaction of a fat or oil with aqueous hydroxide ion to yield glycerol and carboxylate salts of fatty acids.

Saponification reaction The reaction of an ester with aqueous hydroxide ion to yield an alcohol and the metal salt of a carboxylic acid.

Saturated Containing only single bonds between carbon atoms and thus unable to accommodate additional hydrogen atoms.

Saturated solution A solution in which the solute has reached its solubility limit.

Scientific notation Representation of a number as the product of a number (usually between 1 and 10) and 10 with an exponent.

Secondary alcohol An alcohol in which the OH-bearing carbon atom is bonded to two other carbons (and one hydrogen), R_2CHOH.

Secondary amine An amine that has two organic groups bonded to nitrogen, R_2NH.

Secondary (2°) carbon A carbon atom that is bonded to two other carbons.

Secondary protein structure The way in which nearby segments of a protein chain are oriented into a regular pattern, for example, an α-helix or a β-pleated sheet.

Secretion, kidney Movement of solutes into filtrate in kidney tubule.

Semipermeable membrane A thin membrane that allows water and other small solvent molecules to pass through but blocks the flow of larger solute molecules and ions.

Shells Specific regions surrounding a nucleus that are occupied by electrons of increasing energy; main energy levels.

Significant figures In describing a quantity, the total of the number of digits whose values are known with certainty, plus one estimated digit.

Simple protein A protein that yields only amino acids when hydrolyzed.

Simplest formula A chemical formula giving the lowest possible ratio of atoms combined in a chemical compound.

Single covalent bond A covalent bond that results from sharing two electrons between atoms.

SI Unit An internationally agreed on unit of measure (in the *Système International d'Unités*) derived from the metric system.

Soap The mixture of carboxylate salts of fatty acids formed on saponification of animal fat.

Solid A substance that has a definite shape and volume.

Solubility The amount of a substance that can be dissolved in a given volume of solvent.

Solute The dissolved substance in a homogeneous mixture; often a solid or gas in a liquid solution.

Solution A homogeneous mixture in which atoms, molecules, or ions are uniformly mixed together.

Solvation The surrounding of a solute ion or molecule by solvent molecules.

Solvent The major substance in a homogeneous mixture; often a liquid in which a gas or solid is dissolved.

Somatic cell Any cell other than a reproductive one.

Specific gravity The density of a substance divided by the density of water at the same temperature.

Specific heat The amount of heat that will raise the temperature of one gram of a substance by one Celsius degree.

Specificity The extent to which an enzyme reacts with different substrates.

Spectator ion An ion that undergoes no change in a reaction in aqueous solution.

Sphingolipid A lipid derived from the amino alcohol sphingosine (or related amino alcohols), rather than glycerol.

Sphingomyelin A sphingolipid with a fatty acid bonded to the C2 —NH_2 and a phosphate bonded to the C1 —OH group of sphingosine.

Spontaneous process A process that, once begun, proceeds without any external influence.

Standard molar volume The volume of one mole of a gas at standard temperature and pressure (22.4 L).

Standard temperature and pressure (STP) Standard conditions for a gas, defined as 0°C (273 K) and 1 atm (760 mm Hg) pressure.

State of matter The physical state of a substance as a solid, a liquid, or a gas.

Stereoisomers Isomers that have the same molecular and structural formulas but different arrangements of their atoms in space.

Steroid A lipid whose structure is based on a tetracyclic (four-ring) carbon skeleton.

Straight-chain alkane An alkane that has all its carbon atoms connected in a row.

Strong acid An acid that is a strong electrolyte.

Strong base A hydroxide ion containing base that is a strong electrolyte.

Strong electrolyte A substance that is completely converted to ions when dissolved in water.

Structural formula A formula that shows how atoms are connected to each other.

Subatomic particle An elementary particle smaller than an atom.

Subshells Subdivisions within shells of atoms occupied by electrons of different energies; designated by s, p, d, and f.

Substituent A group attached to a root compound.

Substitution reaction An organic reaction in which two reactants exchange atoms or groups, AB + XY → AY + XB.

Substitution reaction, aromatic Substitution of an atom or group for one of the hydrogens on an aromatic ring.

Substrate The reactant in an enzyme-catalyzed reaction.

Substrate-level phosphorylation Formation of ATP by transfer of a phosphate to ADP from another phosphate-containing compound.

Supersaturated solution A solution containing solute in greater concentration than the normal solubility limit of the solute.

Surface tension Result of intermolecular forces that pull molecules at a liquid surface inward, causing the surface to act like a stretched membrane.

Suspension A mixture containing particles just large enough to be visible to the naked eye and which settle out on standing.

Symmetry plane An imaginary plane cutting through the middle of an object so that one half of the object is a mirror image of the other half.

Synapse Narrow gap between nerve cells across which a signal is carried by a neurotransmitter.

Temperature Relative measure of how hot or cold something is; heat flows between objects or materials that are at different temperatures and are in contact.

Tertiary alcohol An alcohol in which the OH-bearing carbon atom is bonded to three other carbons (and no hydrogens), R_3COH.

Tertiary amine An amine that has three organic groups bonded to nitrogen, R_3N.

Tertiary (3°) carbon A carbon atom that is bonded to three other carbons.

Tertiary protein structure The way in which an entire protein chain is coiled and folded into its specific three-dimensional shape.

Tetrahedron A geometrical figure with four identical triangular faces.

Theoretical yield The calculated maximum amount of product that can be formed from a given amount of reactant.

Thiol A compound that contains the —SH functional group, R—SH.

Titration An experimental method for determining acid (or base) concentration by neutralizing a sample with a base (or acid) of known concentration.

Tollens' reagent A reagent ($AgNO_3$ in aqueous NH_3) that converts an aldehyde into a carboxylic acid and deposits a silver mirror on the inside surface of the reaction flask.

Torr Alternate name for the pressure unit mm Hg.

Transamination The interchange of the amino group of an amino acid and the keto group of an α-keto acid.

Transcription The process by which the information in DNA is read and used to synthesize mRNA.

Transfer RNA (tRNA) The RNA whose function is to transport specific amino acids into position for protein synthesis.

Trans isomer Isomer with a specific pair of atoms or groups on opposite sides of double bond.

Transition metal group A short element group in the middle of the periodic table, where d subshells are filling.

Transition metals Metals that fall in the 10 short groups in the center of the periodic table.

Translation The process by which mRNA directs protein synthesis.

Transmutation The change of one element into another brought about by a nuclear reaction.

Transuranium elements Elements of higher atomic number than uranium (at no 92) and known only as artificial radioisotopes.

Triacylglycerol (triglyceride) A triester of glycerol with three fatty acids.

Triple covalent bond A covalent bond that results from sharing of three electron pairs between atoms.

Triple helix The secondary protein structure of tropocollagen in which three protein chains coil around each other to form a rod.

Triprotic acid A substance that has three acidic hydrogen atoms.

Turnover number The number of substrate molecules acted on by one molecule of enzyme per unit time.

Unit A defined quantity adopted as a standard of measurement.

Universal gas law (ideal gas law) A law that relates the temperature, pressure, volume, and molar amount of a gas sample ($PV = nRT$).

Unsaturated Containing one or more double or triple bonds between carbon atoms and thus able to add more hydrogen atoms.

Unsaturated solution A solution that contains less than the maximum amount of solute.

Urea cycle The cyclic biochemical pathway that produces urea for excretion.

Valence electrons Electrons in the highest occupied main energy level of an atom.

Valence-shell electron-pair repulsion (VSEPR) Repulsion by valence-shell electron pairs that maintains them as far apart as possible and determines molecular shape.

Vapor The gaseous state of a substance that is normally a liquid.

Vapor pressure The pressure of a vapor at equilibrium with its liquid.

Variable A measurable property that is not constant.

Vinyl monomer A compound with the structure $CH_2{=}CHZ$ that undergoes polymerization to give a polymer with the repeating unit —CH_2—CHZ—.

Vitamin A small, organic molecule that must be obtained in the diet and that is essential in trace amounts for proper biological functioning.

Volatile Evaporates readily.

Volume The amount of space occupied by an object or a sample of a substance.

Volumetric flask A flask whose volume has been precisely calibrated.

Volume/volume percent concentration [(v/v)%] Concentration expressed as the number of milliliters of solute per 100 mL of solution.

Wax A mixture of esters of long-chain carboxylic acids with long-chain alcohols.

Weak acid An acid that, because it holds its proton strongly, is not fully dissociated in aqueous solution and establishes an equilibrium with water.

Weak base A base that, because it has low affinity for a proton, establishes an equilibrium with water.

Weak electrolyte A substance that is partly ionized when dissolved in water and establishes equilibrium between its ionized and nonionized forms.

Weight The measure of the gravitational force exerted on an object by the earth, the moon, or another massive body.

Weight/volume percent concentration [(w/v)%] Concentration expressed as the number of grams of solute dissolved in 100 mL of solution.

Weight/weight percent concentration [(w/w)%] Concentration expressed as the number of grams of solute per 100 g of solution.

Whole blood Blood plasma plus blood cells.

Wood alcohol A common name for methyl alcohol, CH_3OH.

X rays High-energy electromagnetic radiation.

Zwitterion A neutral dipolar compound that contains both + and − charges in its structure.

Zymogen (proenzyme) A compound that becomes an active enzyme after undergoing a chemical change.

A N S W E R S

Selected Answers to Problems

Answers are given for most in-chapter problems and even-numbered end-of-chapter problems.

Chapter 1

1.1 physical: **(a)**, **(c)**; chemical: **(b)**, **(d)** **1.2** all **1.3** gas **1.4** solid **1.5** mixture: **(a)**, **(d)**; pure: **(b)**, **(c)** **1.6** physical: **(a)**, **(c)**; chemical: **(b)** **1.7 (a)** U **(b)** Ti **(c)** W **1.8 (a)** sodium **(b)** calcium **(c)** palladium **(d)** potassium **(e)** strontium **(f)** tin **1.9** kinetic: **(b)**; potential: **(a)**, **(c)**, **(d)** **1.10** all **1.12 (a)**, **(c)**, **(e)** **1.13 (a)**, **(b)**, **(e)** **1.14** physical: **(a)**, **(b)**, **(c)**; chemical **(d)** **1.18** gas **1.20 (a)** solid **(b)** liquid **1.22** mixture: **(a)**, **(c)**, **(d)**; pure: **(b)**, **(e)**, **(f)** **1.24** element: **(a)**; compound: **(b)**, **(c)**; mixture: **(d)**, **(e)**, **(f)** **1.26 (a)** reactant: hydrogen peroxide; products: water, oxygen **(b)** hydrogen peroxide and water are compounds **1.30 (a)** nitrogen **(b)** potassium **(c)** chlorine **(d)** calcium **(e)** phosphorus **(f)** magnesium **(g)** manganese **1.32 (a)** Br **(b)** Mn **(c)** C **(d)** K **1.34 (a)** Fe **(b)** Cu **(c)** Co **(d)** Mo **(e)** Cr **(f)** F **(g)** S **1.36 (a)** magnesium, sulfur, oxygen **(b)** iron, bromine **(c)** cobalt, phosphorus **(d)** arsenic, hydrogen **(e)** calcium, chromium, oxygen **1.38** carbon, hydrogen, nitrogen, oxygen; 10 **1.40** $C_{16}H_{18}N_2O_5S$ **1.42** potential: **(a)**, **(b)**, **(c)** **1.44** food **1.52** false: **(a)**, **(b)**, **(c)** **1.54** compounds: white solid; elements: metal, brown gas **1.56** Use a magnet to attract iron, add water to dissolve salt, filter off sand, and then evaporate.

Chapter 2

2.1 centiliter **2.2 (a)** milliliter **(b)** kilogram **(c)** centimeter **(d)** kilometer **(e)** microgram **2.3 (a)** L **(b)** μL **(c)** nm **(d)** Mm **2.4 (a)** 0.000 000 001 m **(b)** 0.1 g **(c)** 1 000 m **(d)** 0.001 L **(e)** 0.000 000 001 g **2.5 (a)** 5.8×10 g **(b)** 4.6792×10^4 m **(c)** 6.720×10^{-4} cm **(d)** 3.453×10^2 kg **2.6 (a)** 48,850 mg **(b)** 0.000 008 3 m **(c)** 0.0400 m **2.7** 2.78×10^{-10} m, 2.78×10^2 pm **2.8 (a)** 3 **(b)** 4 **(c)** 5 **(d)** exact **2.9 (a)** 6.0×10^5 **(b)** 1.300×10^3 **(c)** 7.942×10^{11} **2.10 (a)** 2.30 g **(b)** 188.38 mL **(c)** 0.009 L **(d)** 1.000 kg **2.11 (a)** 50.9 mL **(b)** 0.078 g **(c)** 11.9 m **(d)** 51 mg **(e)** 103 **2.12 (a)** 1 L = 1000 mL **(b)** 1 g = 0.0353 oz **(c)** 1 L = 1.057 qt **2.13 (a)** 454 g **(b)** 2.5 L **(c)** 105 qt **2.14** 795 mL **2.15 (a)** 3.4 kg **(b)** 120 mL **2.16 (a)** 10.6 mg/kg **(b)** 36 mg/kg **2.17** 39.4°C **2.18** −38.0°F **2.19** 7700 cal **2.20** 0.21 cal/g · °C **2.21** float: ice, human fat, cork, balsa wood; sink: gold, table sugar, earth **2.22** 8.392 mL **2.23** 2.2 g/cm³ **2.28 (a)** pg **(b)** cm **(c)** dL **(d)** μL **(e)** mL **2.30** 1 L = 10^6 μL, 20 mL = 2×10^4 μL **2.32 (a)** 9.457×10^3 **(b)** 7×10^{-5} **(c)** 2×10^{10} **(d)** 1.2345×10^{-2} **(e)** 6.5238×10^2 **2.34 (a)** 6 **(b)** 3 **(c)** 3 **(d)** 4 **(e)** 1–5 **(f)** 3 **2.36 (a)** 5 **(b)** 3 **(c)** 4 **(d)** 1 **(e)** 3 **2.38 (a)** 12.1 g **(b)** 96.19 cm **(c)** 263 mL **(d)** 20.9 mg **2.40 (a)** 0.3614 cg **(b)** 0.0120 ML **(c)** 0.0144 mm **(d)** 60.3 ng **2.42 (a)** 97.8 kg **(b)** 0.133 mL **(c)** 0.46 ng **2.44 (a)** 62.1 mi/hr **(b)** 91.1 ft/s **2.46** $1.20/L, $0.80/L **2.48** 4,370,000 ft² **2.50** 10 g **2.52** 4.1 cups **2.54** 102.7°F **2.56** 537 cal, 0.537 kcal **2.58** 39°C **2.60** 0.178 cm³ **2.62** 901.9 mL, 817.1 mL **2.66** shock absorber, insulator, energy storehouse **2.68** 2.4 L **2.70** 177°C **2.72** 8.5×10^{-3} g/kg; 1.2 tablets **2.74** 7.8×10^6 mL/day **2.76 (a)** 13.6 **(b)** float **(c)** 60.0 lb **2.78** 600 000 cal/day **2.80** 2.8 mL

Chapter 3

3.1 14 amu **3.2** 3.27×10^{-17} g **3.3** 6.02×10^{23} atoms in all cases **3.5** rhenium, lithium, tellurium **3.6** 92 protons, 143 neutrons **3.7** ^{35}Cl: 17 protons, 18 neutrons; ^{37}Cl: 17 protons, 20 neutrons **3.8** ${}^{35}_{17}$Cl and ${}^{37}_{17}$Cl **3.9** ${}^{11}_{5}$B, ${}^{56}_{26}$Fe **3.10** 6, 2, 6 **3.11** 10, neon **3.12** 12, magnesium **3.13 (a)** $1s^2\, 2s^2\, 2p^2$ **(b)** $1s^2\, 2s^2\, 2p^6\, 3s^1$ **(c)** $1s^2\, 2s^2\, 3s^2\, 3p^5$ **(d)** $1s^2\, 2s^2\, 2p^6\, 3s^2\, 3p^6\, 4s^2$ **3.14** $1s^2\, 2s^2\, 2p^6\, 3s^1$; $1s^2\, 2s^2\, 2p^6\, 3s^2\, 3p^6\, 4s^2\, 3d^{10}\, 4p^6$

3.15 $4p^3$; all are unpaired **3.16 (a)** Group 3A **(b)** period 3 **3.17 (a)** nonmetal, main-group, noble gas **(b)** metal, main-group **(c)** nonmetal, main-group **(d)** metal, transition element **3.18** $1s^2\,2s^2\,2p^6\,3s^2\,3p^6\,4s^2\,3d^{10}\,4p^3$ **3.19** Group 2A **3.20** Group 7A, $1s^2\,2s^2\,2p^6\,3s^2\,3p^5$ **3.22 (a)** 3.4703×10^{-22} g **(b)** 2.1802×10^{-22} g **(c)** 6.6467×10^{-24} g **3.24** 14.01 g **3.28** 18, 20, 22 **3.30 (a)** and **(c)** **3.32 (a)** $^{14}_{6}\text{C}$ **(b)** $^{39}_{19}\text{K}$ **(c)** $^{20}_{10}\text{Ne}$ **3.34 (a)** 122 **(b)** 136 **(c)** 118 **(d)** 48 **3.36** $^{131}_{53}\text{I}$ **3.38** 63.55 **3.40** 2, 8, 18 **3.42** 10, neon **3.44 (a)** two unpaired **(b)** one unpaired **(c)** two unpaired **3.46** 2 **3.48** s, p, and d subshells are full **3.50 (a–b)** transition metals **(c)** $3d$ **3.52 (a)** metal, main-group, alkaline earth **(b)** metal, transition **(c)** nonmetal, main-group **(d)** nonmetal, main-group, noble gas **3.54** Group 6A **3.56 (a)** 8 **(b)** 4 **(c)** 2 **(d)** 1 **(e)** 3 **(f)** 7 **3.58** F, Cl, Br, I, At **3.60** Xenon has a filled-shell configuration. **3.62** Cesium is mistaken for potassium. **3.64 (a)** cosmic rays **(b)** gamma rays **(c)** cosmic rays **3.66** ultraviolet **3.70 (a)** $3p$ **(b)** $3s$ **(c)** $4d$ **(d)** $3d$ **3.72** 24.31 **3.74** carbon **3.76** 23 g **3.78** $4s$ **3.80** 1.29×10^{20} **3.82 (a)** Cu **(b)** Mo

Chapter 4

4.1 $^{218}_{84}\text{Po}$ **4.2** $^{226}_{88}\text{Ra}$ **4.3** $^{14}_{6}\text{C} \rightarrow\ ^{0}_{-1}\text{e} +\ ^{14}_{7}\text{N}$ **4.4 (a)** $^{3}_{1}\text{H} \rightarrow\ ^{0}_{-1}\text{e} +\ ^{3}_{2}\text{He}$ **(b)** $^{210}_{82}\text{Pb} \rightarrow\ ^{0}_{-1}\text{e} +\ ^{210}_{83}\text{Bi}$ **4.5** 12% **4.6** 13 m **4.7** 4% **4.8** 4.0 mL **4.9** $^{237}_{93}\text{Np}$ **4.10** $^{241}_{95}\text{Am} +\ ^{4}_{2}\text{He} \rightarrow 2\ ^{1}_{0}\text{n} +\ ^{243}_{97}\text{Bk}$ **4.11** $^{40}_{18}\text{Ar} +\ ^{1}_{1}\text{H} \rightarrow\ ^{1}_{0}\text{n} +\ ^{40}_{19}\text{K}$ **4.12** $^{235}_{92}\text{U} +\ ^{1}_{0}\text{n} \rightarrow\ ^{137}_{52}\text{Te} + 2\ ^{1}_{0}\text{n} +\ ^{97}_{40}\text{Zr}$ **4.14** transmutation **4.16** $^{4}_{2}\text{He}$ **4.18** Gamma is highest, alpha lowest. **4.22** A neutron decays. **4.24** Half of a sample decays in that time. **4.36** $^{75}_{34}\text{Se} \rightarrow\ ^{0}_{-1}\text{e} +\ ^{75}_{35}\text{Br}$ **4.38** 2.3 yr **4.40 (a)** $^{140}_{55}\text{Cs}$ **(b)** $^{246}_{96}\text{Cm}$ **4.42 (a)** $^{113}_{49}\text{In}$ **(b)** $^{13}_{7}\text{N}$ **4.44** $^{198}_{80}\text{Hg} +\ ^{1}_{0}\text{n} \rightarrow\ ^{198}_{79}\text{Au} +\ ^{1}_{1}\text{H}$, a proton **4.46** 22%, 0.1% **4.48** 0.26 rem **4.50** 2.4 mL **4.54** 17,200 yr **4.56** Not enough time has passed. **4.60 (a)** beta decay **(b)** Mo-98 **4.62 (a)** U-234 **(b)** radiation shielding

Chapter 5

5.1 $\cdot\ddot{\text{X}}\cdot$ **5.2** $:\ddot{\text{Rn}}:$ $\cdot\ddot{\text{Pb}}\cdot$ $:\ddot{\text{Br}}\cdot$ $\cdot\text{Ra}\cdot$ **5.3** $\text{K} \rightarrow \text{K}^+ + \text{e}^-$; $\text{I} + \text{e}^- \rightarrow \text{I}^-$ **5.4 (a)** $\text{Se} + 2\,\text{e}^- \rightarrow \text{Se}^{2-}$ **(b)** $\text{Ba} \rightarrow \text{Ba}^{2+} + 2\,\text{e}^-$ **(c)** $\text{Br} + \text{e}^- \rightarrow \text{Br}^-$ **5.5** $1s^2\,2s^2\,2p^6\,3s^2\,3p^6\,4s^1$; lose one electron **5.6** two **5.7** gain two electrons; $1s^2\,2s^2\,2p^6$; neon **5.8** cation **5.9** chromium **5.10 (a), (b), (d)**

5.11 (a) copper(II) ion **(b)** fluoride ion **(c)** magnesium ion **(d)** sulfide ion **5.12** Ag^+, Fe^{2+}, Cu^+, Te^{2-} **5.13** Na^+, sodium ion; K^+, potassium ion; Ca^{2+}, calcium ion; Cl^-, chloride ion **5.14** nitrate ion, cyanide ion, hydroxide ion, hydrogen phosphate ion **5.15** HCO_3^-, OH^-, NO_2^-, MnO_4^- **5.16 (a)** AgI **(b)** Ag_2O **(c)** Ag_3PO_4 **5.17 (a)** Na_2SO_4 **(b)** FeSO_4 **(c)** $\text{Cr}_2(\text{SO}_4)_3$ **5.18** $(\text{NH}_4)_2\text{CO}_3$ **5.19** $\text{Al}_2(\text{SO}_4)_3$, $\text{Al}(\text{CH}_3\text{CO}_2)_3$ **5.20** BaSO_4 **5.21** silver(I) sulfide **5.22 (a)** copper(II) oxide **(b)** calcium cyanide **(c)** sodium nitrate **(d)** copper(I) sulfate **(e)** lithium phosphate **(f)** ammonium chloride **5.23 (a)** Ba(OH)_2 **(b)** CuCO_3 **(c)** $\text{Mg(HCO}_3)_2$ **(d)** CuF **(e)** $\text{Fe}_2(\text{SO}_4)_3$ **(f)** $\text{Fe(NO}_3)_2$ **5.24 (a)** 148.3 **(b)** 136.1 **(c)** 74.1 **(d)** 278.0 **5.25 (a)** 2 **(b)** 3 **(c)** 2 **(d)** 3 **(e)** 4 **(f)** 2 **5.26 (a)** 40.3 **(b)** 79.5 **(c)** 151.0 **5.28 (a)** $\text{H}\cdot$ **(b)** $\text{He}:$ **(c)** $\text{Li}\cdot$ **5.30 (a)** $\text{Ca} \rightarrow \text{Ca}^{2+} + 2\,\text{e}^-$ **(b)** $\text{Au} \rightarrow \text{Au}^+ + \text{e}^-$ **(c)** $\text{F} + \text{e}^- \rightarrow \text{F}^-$ **(d)** $\text{Cr} \rightarrow \text{Cr}^{3+} + 3\,\text{e}^-$ **5.32** true: **(c)**; false: **(a), (b), (d)** **5.34** $2-$ **5.36** 2 **5.38 (a)**, **(b)** $1s^2\,2s^2\,2p^6\,3s^2\,3p^6\,4s^2\,3d^{10}\,4p^6$ **(c)** $1s^2\,2s^2\,2p^6\,3s^2\,3p^6$ **(d)** $1s^2\,2s^2\,2p^6\,3s^2\,3p^6\,4s^2\,3d^{10}\,4p^6\,5s^2\,4d^{10}\,5p^6$ **(e)** $1s^2\,2s^2\,2p^6$ **5.42** low ionization energies **5.44 (a)** 1A, 2A, 3A **(b)** 6A, 7A **5.46 (c)** **5.48** greater **5.50 (a)** sulfide **(b)** tin(II) **(c)** strontium **(d)** magnesium **(e)** gold(I) **5.52 (a)** Se^{2-} **(b)** O^{2-} **(c)** Ag^+ **(d)** Co^{2+} **5.54 (a)** ammonium **(b)** cyanide **(c)** carbonate **(d)** sulfite **5.56** Mg(OH)_2, Al(OH)_3 **5.58 (a)** Ca(OCl)_2 **(b)** CuSO_4 **(c)** Na_3PO_4 **5.60 (a)** magnesium carbonate **(b)** calcium acetate **(c)** silver cyanide **(d)** sodium dichromate **5.62** Ag^+ and Cl^-, Ca^{2+} and $2\,\text{F}^-$, $2\,\text{Ag}^+$ and SO_4^{2-} **5.64 (a)** $(\text{NH}_4)_2\text{Cr}_2\text{O}_7$, 252.1 **(b)** $\text{Cr(NO}_3)_3$, 238.0 **(c)** LiHSO_4, 104.0 **5.66** 6×10^{-4} g/L **5.70** hyper: too high; hypo: too low **5.72** calcium ion (Ca^{2+}), phosphate ion (PO_4^{3-}), hydroxide ion (OH^-) **5.74** H^- has a helium configuration **5.76** $\text{Pb}_3(\text{AsO}_4)_2$, 899.4 **5.78 (a)** copper(I) phosphate **(b)** sodium sulfite **(c)** manganese(IV) oxide **(d)** gold(III) chloride **(e)** lead(IV) carbonate **(f)** nickel(III) sulfide **5.80 (a)** 8 p, 8 n, 10 e **(b)** 39 p, 50 n, 36 e **(c)** 55 p, 78 n, 54 e **(d)** 35 p, 46 n, 36 e

Chapter 6

6.1 helium **6.2** $:\ddot{\text{I}}:\ddot{\text{I}}:$, xenon **6.3 (a)** P 3, H 1 **(b)** Se 2, H 1 **(c)** H 1, Cl 1 **(d)** Si 4, F 1 **6.4** PbCl_4 **6.5** ionic and covalent: N, O, S, F, Cl, Br, I; covalent: B, C, Si, P **6.6** H 2; C 8; O 8 **6.7 (a)** H_2O_2

(b) 34.0   **6.8** 165.4   **6.10 (a)**   :C≡O:
:O:
(b) :Cl̈—S̈—Cl̈:   **6.14** H—C—H   **6.16** chloroform: tetrahedral; dichloroethylene: planar
6.17 tetrahedral   **6.18** Both are bent.   **6.19** H
< P < S < N < O   **6.20 (a)** polar covalent   **(b)**
ionic   **(c)** covalent   **(d)** polar covalent   **6.21 (a)**
$^{\delta+}S—F^{\delta-}$ **(b)** $^{\delta+}P—O^{\delta-}$ **(c)** $^{\delta+}As—Cl^{\delta-}$ **6.22** All
are polar.   **6.23 (a)** disulfur dichloride   **(b)** iodine
chloride   **(c)** iodine trichloride   **6.24 (a)** SeF_4 **(b)**
P_2O_5 **(c)** BrF_3 **6.26 (a)** diatomic, both ionic and
covalent   **(b)** ionic   **(c)** covalent   **(d)** diatomic, covalent and ionic   **(e)** diatomic, covalent and ionic   **(f)**
ionic   **6.28** two bonds   **6.30** H, B, P, S   **6.32 (b)**,
(c), **(e)**, **(f)**   **6.38 (a)**   **6.40 (a)** $CH_3CH_2CH_3$ **(b)**
$H_2C=CHCH_3$ **(c)** CH_3CH_2Cl **6.44** CH_3CH_2OH
6.46 H_2NNH_2 **6.48** S=C=S   **6.52** pyramidal,
tetrahedral, angular   **6.54 (a)** 4 bonding pairs, tetrahedral   **(b)** 3 bonding pairs, planar   **(c)** 4 bonding
pairs, tetrahedral   **(d)** 3 bonding pairs, planar triangular   **(e)** 3 bonding pairs, planar triangular   **(f)** 3
bonding pairs, 1 lone pair, pyramidal   **6.58** ionic:
(a); polar covalent: **(b)**, **(c)**, **(d)**   **6.60** no   **6.62 (a)**
phosphorus triiodide   **(b)** arsenic trichloride   **(c)**
tetraphosphorus trisulfide   **(d)** dialuminum hexafluoride   **(e)** nitrogen triiodide   **(f)** iodine heptafluor-
ide   **6.66** Mg, Fe, Co, Cu, Zn   **6.68** both are carbon; have different structures   **6.72** 13 p, 14 e
6.74 (b) tetrahedral   **(c)** coordinate covalent bond
(d) 19 p, 18 e   **6.76 (a)** 111.0   **(b)** 198.5   **(c)** 67.8   **(d)**
120.4   **(e)** 94.2   **(f)** 112.8   **(g)** 88.0   **6.78** polar

Chapter 7

7.1 balanced: **(a)**, **(c)**   **7.2 (a)** Solid cobalt chloride
plus gaseous hydrogen fluoride gives solid cobalt
fluoride plus gaseous hydrogen chloride.   **(b)** Aqueous lead(II) nitrate plus aqueous potassium iodide
gives solid lead(II) iodide plus aqueous potassium
nitrate.   **7.3** $2 Na + Cl_2 \rightarrow 2 NaCl$ **7.4** $2 O_3 \rightarrow 3$
O_2 **7.5 (a)** $Ca(OH)_2 + 2 HCl \rightarrow CaCl_2 + 2 H_2O$
(b) $4 Al + 3 O_2 \rightarrow 2 Al_2O_3$ **(c)** $2 CH_3CH_3 + 7 O_2 \rightarrow 4$
$CO_2 + 6 H_2O$ **(d)** $2 AgNO_3 + MgCl_2 \rightarrow 2 AgCl +$
$Mg(NO_3)_2$ **7.6** 0.217 mol; 4.6 g   **7.7** C: 1.4×10^{23};
H: 1.2×10^{23}; O: 6.0×10^{22} **7.8** 60 g   **7.9 (a)** Ni
$+ 2 HCl \rightarrow NiCl_2 + H_2$; 4.90 mol   **(b)** 3.0 mol
7.10 $6 CO_2 + 6 H_2O \rightarrow C_6H_{12}O_6 + 6 O_2$; 90 mol CO_2
7.11 (a) 39.6 mol   **(b)** 14 g   **7.12** 88.2%   **7.13 (a)**
combination   **(b)** displacement   **(c)** combination

7.14 $Mg(s) + ZnCl_2(aq) \rightarrow Zn(s) + MgCl_2(aq)$
7.15 $CuCO_3(s) \rightarrow CuO(s) + CO_2(g)$ **7.16** $2 HgO(s)$
$\rightarrow 2 Hg(l) + O_2(g)$ **7.17 (a)** $2 Al(NO_3)_3(aq) + 3$
$Na_2S(aq) \rightarrow Al_2S_3(s) + 6 NaNO_3(aq)$ **(b)** $CaCO_3(s)$
$+ 2 HCl(aq) \rightarrow CaCl_2(aq) + CO_2(g) + H_2O(l)$
7.18 (a) displacement   **(b)** exchange   **(c)** decomposition   **7.19** $NH_4NO_3(s) \rightarrow N_2O(g) + 2 H_2O(l)$; decomposition   **7.20 (a)** $2 K(s) + Pb^{2+} \rightarrow 2 K^+ +$
$Pb(s)$ **(b)** $OH^- + H^+ \rightarrow H_2O$ **(c)** $CuS(s) + 2 H^+ \rightarrow$
$Cu^{2+} + H_2S(g)$ **7.21 (a)** oxidizing: Cu^{2+}; reducing:
Fe   **(b)** oxidizing: Cl_2; reducing: Mg   **(c)** oxidizing:
Cr_2O_3; reducing: Al   **7.22** $2 K(s) + Br_2(l) \rightarrow 2$
$KBr(s)$; oxidizing: Br_2; reducing: K   **7.23 (a)** V(III)
(b) Mg(II)   **(c)** Sn(IV)   **(d)** Cr(VI)   **(e)** Cu(II)   **(f)**
Ni(II)   **7.24 (a)** exchange, not redox   **(b)** displacement, redox   **7.25** oxidized: S^{2-}; reduced: O_2
7.26 redox: **(a)**, **(c)**, **(d)**   **7.28 (a)** $2 C_2H_6 + 7 O_2 \rightarrow 4$
$CO_2 + 6 H_2O$ **(b)** balanced   **(c)** $2 Mg + O_2 \rightarrow 2 MgO$
(d) $(NH_4)_3P + 3 NaOH \rightarrow Na_3P + 3 NH_3 + 3 H_2O$
(e) $2 K + 2 H_2O \rightarrow 2 KOH + H_2$ **7.30 (a)**
$Hg(NO_3)_2 + 2 LiI \rightarrow 2 LiNO_3 + HgI_2$ **(b)** $I_2 + 5 Cl_2$
$\rightarrow 2 ICl_5$ **(c)** $4 Al + 3 O_2 \rightarrow 2 Al_2O_3$ **(d)** $CuSO_4 + 2$
$AgNO_3 \rightarrow Ag_2SO_4 + Cu(NO_3)_2$ **(e)** $2 Mn(NO_3)_3 + 3$
$Na_2S \rightarrow Mn_2S_3 + 6 NaNO_3$ **(f)** $4 NO_2 + O_2 \rightarrow 2$
N_2O_5 **(g)** $P_4O_{10} + 6 H_2O \rightarrow 4 H_3PO_4$ **7.32** C_2H_6O
$+ O_2 \rightarrow C_2H_4O_2 + H_2O$ **7.34 (a)** $2 C_4H_{10} + 13 O_2$
$\rightarrow 8 CO_2 + 10 H_2O$ **(b)** $C_2H_6O + 3 O_2 \rightarrow 2 CO_2 + 3$
H_2O **(c)** $2 C_8H_{18} + 25 O_2 \rightarrow 16 CO_2 + 18 H_2O$
7.36 2.43×10^{23} **7.38** 6.44×10^{-4} mol
7.40 284.75   **7.42 (a)** 0.0641 mol   **(b)** 0.0595 mol
(c) 0.0418 mol   **(d)** 0.0143 mol   **7.44** 151.8; 1.98
mmol   **7.46 (a)** $N_2 + O_2 \rightarrow 2 NO$ **(b)** 7.50 mol   **(c)**
7.62 mol   **(d)** 0.125 mol   **(e)** O_2 is limiting; 24 mol NO
can form.   **7.48 (a)** $N_2 + 3 H_2 \rightarrow 2 NH_3$ **(b)** 0.471
mol   **(c)** 16.1 g   **(d)** H_2 is limiting; 15.1 g NH_3
7.50 (a) 52.4 kg Fe   **(b)** 97.9%   **7.52 (a)** $FeCl_2(aq)$
$+ 2 Ag(s) \rightarrow 2 AgCl(s) + Fe(s)$ **(b)** $Ba(s) + 2$
$H_2O(l) \rightarrow Ba(OH)_2(aq) + H_2(g)$ **(c)** $CuSO_4(aq) +$
$Zn(s) \rightarrow ZnSO_4(aq) + Cu(s)$ **(d)** $Mg(NO_3)_2(aq) +$
$Pb(s) \rightarrow Pb(NO_3)_2(aq) + Mg(s)$ **7.54 (a)** $Mg(s) +$
$Cu^{2+} \rightarrow Mg^{2+} + Cu(s)$ **(b)** $2 Cl^- + Pb^{2+} \rightarrow PbCl_2(s)$
(c) $2 Cr^{3+} + 3 S^{2-} \rightarrow Cr_2S_3(s)$ **(d)** $2 Au^{3+} + 3 Sn(s)$
$\rightarrow 3 Sn^{2+} + 2 Au(s)$ **(e)** $2 I^- + Br_2(l) \rightarrow 2 Br^- +$
$I_2(s)$ **7.56 (a)** Co(III)   **(b)** Fe(II)   **(c)** U(VI)   **(d)**
Cu(II)   **(e)** Ti(IV)   **(f)** Sn(II)   **7.58** oxidation: addition of oxygen or removal of hydrogen; reduction:
addition of hydrogen or removal of oxygen
7.60 production of electricity from redox reactions
7.62 (a) 0.459 g   **(b)** displacement   **(c)** H^+ is oxidizing agent; Zn is reducing agent.   **7.64** 5.6×10^{14}

7.66 (a) $C_{12}H_{22}O_{11} \rightarrow 12\,C + 11\,H_2O$ **(b)** 25.2 g **(c)** 8.94 g **7.68 (a)** oxidized **(b)** 6.40 g **(c)** 104 g **7.70 (a)** exchange **(b)** combination **(c)** combination **(d)** exchange **(e)** exchange **(f)** combination **(g)** combination

Chapter 8

8.1 (a) endothermic **(b)** $\Delta H = +678$ kcal **(c)** $C_6H_{12}O_6(aq) + 6\,O_2(g) \rightarrow 6\,CO_2(g) + 6\,H_2O(l) + 678$ kcal **8.2 (a)** endothermic **(b)** 200 kcal **(c)** 74 kcal **8.3** 100 kcal **8.4 (a)** decrease **(b)** increase **(c)** decrease **8.5** yes; entropy increases

8.7 (a) $K = \dfrac{[NO_2]^2}{[N_2O_4]}$ **(b)** $K = \dfrac{[CH_3Cl]\,[HCl]}{[CH_4]\,[Cl_2]}$

(c) $K = \dfrac{[Br_2]\,[F_2]^5}{[BrF_5]^2}$ **8.8 (a)** products **(b)** reactants **(c)** products **8.9** reaction favored by high pressure **8.10 (a)** decrease product amount **(b)** increase product amount **(c)** increase product amount **8.14 (a)** positive **(b)** 48 kcal **(c)** 3.5 kcal **8.16 (a)** $C_2H_5OH + 3\,O_2 \rightarrow 2\,CO_2 + 3\,H_2O$ **(b)** negative **(c)** 327 kcal **(d)** 1.38 mol; 63.5 g **(e)** 71.0 kcal **(f)** 5.63 g **8.18** increased disorder: **(a)**; decreased disorder: **(b)**, **(c)**, **(d)**, **(e)** **8.20** A reaction with negative ΔG is spontaneous. **8.22 (a)** endothermic **(b)** increase **(c)** ΔG is positive, so ΔS is more important than ΔH **(d)** more soluble in hot water **8.24 (a)** $H_2(g) + Br(l) \rightarrow 2\,HBr(g)$ **(b)** increase **(c)** ΔH is negative and ΔS is positive. **8.26** $E_{act} = 5$ kcal **8.30** Collisions increase, and a larger fraction of particles move faster. **8.32** Catalysts increase reaction rate by lowering activation energy. **8.34 (a)** yes **(b)** reaction rate is slow **8.36** Amounts of reactants diminish. **8.38 (a)** $K = \dfrac{[CO_2]^2}{[CO]^2[O_2]}$

(b) $K = \dfrac{[HCl]^2[C_2H_4Cl_2]}{[Cl_2]^2[C_2H_6]}$ **(c)** $K = \dfrac{[CaO]\,[CO_2]}{[CaCO_3]}$

(d) $K = \dfrac{[H_3O^+]\,[F^-]}{[HF]\,[H_2O]}$ **(e)** $K = \dfrac{[O_3]^2}{[O_2]^3}$

8.40 (a) $K = 140$ **(b)** 0.054 mol/L **8.42** more product **8.44 (a)** endothermic **(b)** reactants are favored **(c)** (1) favors ozone; (2) favors ozone; (3) favors O_2; (4) does not affect equilibrium; (5) favors O_3 **8.46** conversion into useful compounds **8.48** thyroid and hypothalamus **8.50** epinephrine **8.52** stimulates hemoglobin production **8.54 (a)** CO reacts with Hb, leaving less Hb for O_2 to react with. **(b)** O_2 shifts equilibrium, displacing CO from Hb. **8.56 (a)** 5.40 kcal **(b)** 5.40 kcal **8.58 (a)** $2\,CH_3OH + 3\,O_2 \rightarrow 2\,CO_2 + 4\,H_2O$ **(b)** 272 kcal **8.60** no

Chapter 9

9.1 0.289 atm; 4.25 psi; 29,300 Pa **9.2** 9.3 atm He; 0.19 atm O_2 **9.3** approximately identical **9.4** 75.4% N_2, 13.2% O_2, 5.3% CO_2, 6.2% H_2O **9.5** 90.4 mm Hg **9.6** 450 L **9.7** 1.2 atm, 62 atm **9.8** 0.38 L; 0.85 L **9.9** 410 K = 137°C **9.10** 209 K **9.11** 467 mL **9.12** 4460 mol; 7.14×10^4 g CH_4; 1.96×10^5 g CO_2 **9.13** 5.0 atm **9.14** 1100 mol; 4400 g **9.15 (a)** decrease **(b)** increase **9.16 (a)**, **(c)** **9.17 (a)** London forces **(b)** hydrogen bonds, dipole–dipole forces **(c)** dipole–dipole forces **9.18** the amount of pressure needed to hold a column of mercury 760 mm high **9.20** The total pressure exerted by a gas mixture is the sum of the individual pressures of the components in the mixture. **9.24 (a)** 0.579 atm **(b)** 17.3 in. Hg **9.26** $P_1V_1 = P_2V_2$ at constant temperature and number of moles **9.28** 1.75 atm **9.30** 73.5 psi **9.32** $V_1/T_1 = V_2/T_2$ at constant pressure and number of moles **9.34** 130 mL **9.36** amount of gas **9.38** 493 K **9.40** 49.5 mL **9.42** 29 atm **9.46** equal numbers **9.48** 63 mL **9.50** $PV = nRT$ **9.52** CO_2 has fewer molecules but weighs more. **9.54** 41 K **9.56** 0.048 mol **9.58** 1 atm **9.60** the amount of heat needed to vaporize one gram of a liquid at its boiling point **9.62 (a)** all molecules **(b)** dipole moment **(c)** —OH or —NH bonds **9.64 (b)**, **(e)** **9.66 (a)** 29.2 kcal **(b)** 138 kcal **9.68 (a)** ionic **(b)** intermolecular **(c)** metallic **(d)** intermolecular **(e)** ionic **(f)** intermolecular **9.72** 180/110 is high; the first number is systolic. **9.74** Bioglass chemically bonds to bone. **9.76** Ice has more space between molecules than water has. **9.80** 0.13 mol; 4.0 L **9.82** 600 g/day **9.84 (a)** 0.714 g/L **(b)** 1.96 g/L **(c)** 1.42 g/L **(d)** 15.7 g/L **9.86 (a)** CH_3OH: 9.04 atm; H_2O: 9.64 atm **(b)** 18.68 atm

Chapter 10

10.1 (b), **(d)** **10.2 (c)**, **(d)** **10.3** $Na_2SO_4 \cdot 10H_2O$ **10.4** 322 g **10.5** 5.56 g/dL **10.6** 0.0086% (w/v) **10.7** 6.6% (w/v) **10.8** Yes. Concentration depends on solubility. **10.9 (a)** 12 g **(b)** 1.5 g **10.10** Place 38 mL acetic acid in flask and dilute to 500 mL. **10.11 (a)** 22 mL **(b)** 18 mL **10.12** 9000 ppm **10.13** 1.6 ppm **10.14** 0.93 M **10.15 (a)** 25 g **(b)** 68 g **10.16 (a)** 0.025 mol **(b)** 1.6 mol **10.17** 695 mL   **10.18** 1.5 g   **10.19** 0.38 g   **10.20** 2.4 M **10.21** 39.1 mL   **10.22** (a) 39.1 g   (b) 79.9 g   (c) 12.2 g   (d) 48.0 g   **10.23** (a) 39.1 mg   (b) 79.9 mg (c) 12.2 mg   (d) 48.0 mg   **10.24** 9 mg   **10.25** $CaCl_2$ gives three solute particles; NaCl gives only two.   **10.26** (a) $-1.86°C$ (b) 101.2°C   **10.27** (a) 0.70 Osmol   (b) 0.15 Osmol   **10.28** A homogeneous mixture is uniform; a heterogeneous mixture is nonuniform.   **10.30** polarity   **10.32** Add water, filter off the insoluble aspirin, and recover the salt by evaporation of the water.   **10.34** all **10.36** 2.1 g/dL   **10.38** Molarity is the number of moles of solute per liter of solution.   **10.40** 400 mL **10.42** 2.5 g   **10.44** Dissolve 1.5 g NaCl in water to a final volume of 250 mL.   **10.46** (a) 5% (w/v) (b) 2.5% (w/v)   **10.48** (a) 4.0 g   (b) 15 g **10.50** 0.090% (w/v)   **10.52** 10. ppm   **10.54** (a) 0.425 M   (b) 1.53 M   (c) 1.02 M   **10.56** (a) 3.6 g   (b) 15 g   (c) 118 g   **10.58** 37 g   **10.60** (a) 0.00520 M (b) 62 mL   **10.62** 375 mL   **10.64** Ca^{2+} concentration is 0.0015 M   **10.66** 0.355 g   **10.68** Vapor pressure of the syrup is lower.   **10.70** 102.74°C **10.72** The inside of the cell has higher osmolarity than water.   **10.74** Water passes through the membrane to the NaCl solution.   **10.76** Water passes through the membrane from the glucose solution to the NaCl solution.   **10.78** Water is drawn out by osmosis.   **10.80** Electrolyte imbalances can be controlled.   **10.82** 9.4 mL   **10.84** 0.57 M **10.86** 0.31 Osmol, nearly isotonic with blood plasma **10.90** (b) 3.05 M

Chapter 11

11.1 strong: (b), (e); weak: (a), (c), (d)   **11.2** (a), (b) **11.3** (a), (c)   **11.4** (a) base   (b) acid   (c) base **11.5** (a) $HNO_3 + KOH \rightarrow KNO_3 + H_2O$ (b) $H_2SO_4 + 2\ LiOH \rightarrow 2\ H_2O + Li_2SO_4$ (c) $2\ HCl + Mg(OH)_2 \rightarrow 2\ H_2O + MgCl_2$ **11.6** $Al(OH)_3 + 3\ HCl \rightarrow AlCl_3 + 3\ H_2O$ **11.7** (a) $KHCO_3 + HNO_3 \rightarrow KNO_3 + CO_2 + H_2O$ (b) $MgCO_3 + H_2SO_4 \rightarrow MgSO_4 + CO_2 + H_2O$ (c) $HI + LiHCO_3 \rightarrow LiI + CO_2 + H_2O$ **11.8** $H_2SO_4 + 2\ NH_3 \rightarrow (NH_4)_2SO_4$ **11.9** $CH_3CH_2NH_2 + HCl \rightarrow CH_3CH_2NH_3^+Cl^-$ **11.10** (a) H_2S (b) HPO_4^{2-} (c) HCO_3^- (d) NH_3 **11.11** (a) NH_4^+ (b) H_2SO_4 (c) H_2CO_3 **11.12** (a) F^- (b) OH^- **11.13** $HPO_4^{2-} + OH^- \rightleftarrows PO_4^{3-} + H_2O$ **11.14** (a) acidic, $[OH^-] = 3.1 \times 10^{-10}$ M   (b) basic, $[OH^-] = 3.2 \times 10^{-3}$ M   **11.15** (a) 5   (b) 5 **11.16** (a) 1×10^{-13} M   (b) 1×10^{-3} M   (c) 1×10^{-8} M; (b) is most acidic; (a) is least acidic   **11.17** (a) acidic   (b) basic   (c) acidic   (d) acidic; (b) is least

acidic and (d) wine is most acidic   **11.18** (a) $[H_3O^+]$ $= 3 \times 10^{-7}$ M   (b) $[H_3O^+] = 1 \times 10^{-8}$ M   (c) $[H_3O^+] = 2.0 \times 10^{-4}$ M   (d) $[H_3O^+] = 3.2 \times 10^{-4}$ M   **11.19** (a) 8.28   (b) 5.05   **11.20** Chloride ion is too weak a base to neutralize any added acid. **11.21** pH = 7.2   **11.22** (a) 0.079 Eq   (b) 0.338 Eq (c) 0.14 Eq   **11.23** (a) 0.26 N   (b) 1.13 N   (c) 0.47 N **11.24** 0.019 Eq   (b) 0.019 Eq   **11.25** 0.730 M **11.26** 133 mL   **11.27** 0.22 M   **11.28** none **11.29** (b), (d)   **11.30** (a) H_2SO_4 (b) HNO_3 (c) $Mg(OH)_2$ (d) $Al(OH)_3$ (e) HF   (f) KOH **11.32** CH_3COOH dissociates only to the extent of about 1%.   **11.34** A monoprotic acid can give up only one proton; a diprotic acid can give up two. **11.36** (a), (c), (e)   **11.38** (a) acid   (b) base   (c) neither   (d) acid   (e) neither   (f) acid   **11.40** (a) $CH_2ClCOOH$ (b) $C_5H_5NH^+$ (c) $HSeO_4^-$ (d) $(CH_3)_3NH^+$ **11.42** (a) $HCO_3^- + HCl \rightarrow H_2O + CO_2 + Cl^-$; $HCO_3^- + NaOH \rightarrow H_2O + Na^+ + CO_3^{2-}$ (b) $H_2PO_4^- + HCl \rightarrow H_3PO_4 + Cl^-$; $H_2PO_4^- + 2\ NaOH \rightarrow 2\ H_2O + 2\ Na^+ + PO_4^{3-}$ **11.44** (a) $HCl + PO_4^{3-} \rightleftarrows HPO_4^{2-} + Cl^-$ (b) $HCN + SO_4^{2-} \leftrightarrows HSO_4^- + CN^-$ (c) $HClO_4 + NO_2^- \rightleftarrows HNO_2 + ClO_4^-$ (d) $CH_3O^- + HF \rightleftarrows CH_3OH + F^-$ **11.46** Citric acid reacts with sodium bicarbonate to release CO_2 bubbles: $C_6H_5O_7H_3 + 3\ NaHCO_3 \rightarrow C_6H_5O_7Na_3 + 3\ H_2O + 3\ CO_2$ **11.48** (a) HF   (b) HSO_4^- (c) $H_2PO_4^-$ (d) CH_3COOH **11.50** pH is the negative logarithm of H_3O^+ concentration. **11.52** weakly basic   **11.54** pH 1.0   **11.56** (a) 1×10^{-4} M   (b) 1×10^{-11} M   (c) 1 M   (d) 4.2×10^{-2} M   (e) 1.1×10^{-8} M   **11.58** 4×10^{-8} M **11.60** (a) 4.0×10^{-7} (b) 2.0×10^{-11} (c) 4.3×10^{-3}; (c) < (a) < (b)   **11.62** $CH_3COOH + CH_3COO^-$ Na^+ **11.66** The normality of an acid or base solution is the number of equivalents per liter of solution.   **11.68** 0.035 Eq   **11.70** Dilute 2.1 mL of 12.0 M HCl to a final volume of 250 mL.   **11.72** 0.56 L **11.74** (a) 1.5 N   (b) 0.26 N   (c) 1.4 N   **11.76** 0.17 M, 0.51 N   **11.78** 0.075 M   **11.80** (a) $NaAl(OH)_2CO_3 + 4\ HCl \rightarrow CO_2 + 3\ H_2O + NaCl + AlCl_3$ (b) 51.5 mg   **11.82** (a) 0.03 M   (b) 50 g **11.84** 16 mL   **11.86** 0.35 M   **11.88** Both contain the same number of moles of acid; the HCl solution has an H_3O^+ concentration of 0.20 N and has a lower pH.   **11.90** (a), (c)   **11.92** pH = 11.70

Chapter 12

12.1 (a) alcohol, carboxylic acid   (b) double bond, ester   **12.2** (a) CH_3CHO (b) CH_3CH_2COOH **12.3** $CH_3CH_2CH_2CH_2CH_2CH_2CH_3$ **12.6** Structures **(a)** and **(c)** are identical and are isomers of structure **(b)**. **12.9 (a)** 2,6-dimethyloctane **(b)** 3,3-diethylheptane **12.11** CH_3's are primary, CH_2's are secondary, CH's are tertiary, and C's are quaternary. **12.12 (a)** 2-methylbutane **(b)** 2,2,3-trimethylbutane **12.13** $2\ C_2H_6 + 7O_2 \rightarrow 4\ CO_2 + 6\ H_2O$ **12.15 (a)** 1-ethyl-4-methylcyclohexane **(b)** 1-ethyl-3-isopropylcyclopentane **12.20** Add more water and see which is miscible. **12.22 (a)** CH_3OH **(b)** CH_3NH_2 **(c)** CH_3COOH **(d)** CH_3OCH_3 **12.26** A straight-chain alkane has all its carbons in a row. **12.28** No. Isomers must have the same formula. **12.30** Carbon forms only four bonds. **12.32** $CH_3CH_2CH_2OH$, $CH_3CH(OH)CH_3$, and $CH_3CH_2OCH_3$ **12.34** There are too many hydrogens. **12.36** identical: **(a)**; isomers: **(b)**, **(d)**, **(e)**; unrelated: **(c)** **12.38 (a)** First and second are identical. **(b)** First and second are identical. **12.42** hexane, 2-methylpentane, 3-methylpentane, 2,2-dimethylbutane, 2,3-dimethylbutane **12.44 (a)** 1-isopropyl-1-methylcyclopentane **(b)** 1,1,3,3-tetramethylcyclopentane **(c)** propylcyclohexane **(d)** 4-butyl-1,1-dimethylcyclohexane **(e)** ethylcyclooctane **(f)** 1,2-diethyl-3-methylcyclopropane **(g)** 2-ethyl-1-methyl-3-propylcyclopentane **12.46** heptane, 2-methylhexane, 3-methylhexane, 2,2-dimethylpentane, 2,3-dimethylpentane, 2,4-dimethylpentane, 3,3-dimethylpentane, 3-ethylpentane, 2,2,3-trimethylbutane **12.48** $2\ C_8H_{18} + 25\ O_2 \rightarrow 16\ CO_2 + 18\ H_2O$ **12.50** hydrogen **12.54** A semisynthetic compound is one derived by laboratory manipulation of a naturally occurring substance. **12.56** Shape affects reactivity and biological properties. **12.58** branched-chain alkanes **12.60** On a percentage basis, the molecular weights of hexadecane and heptadecane are closer than those of methane and ethane. **12.62** CH_3's are primary, CH_2's are secondary, CH's are tertiary, and C's are quaternary. **12.64** Compounds with similar structures are often miscible. **12.68** ethylcyclopropane, 1,1-dimethylcyclopropane, 1,2-dimethylcyclopropane, methylcyclobutane, cyclopentane.

Chapter 13

13.1 (a) 2-methyl-3-heptene **(b)** 2-methyl-1,5-hexadiene **13.3** Compounds **(a)** and **(c)** can exist as cis–trans isomers. **13.5 (a)**, **(b)**, **(c)** butane; **(d)** methylcyclohexane **13.8 (a)** 3-ethyl-2-pentene **(b)** 2,3-dimethyl-2-butene or 2,3-dimethyl-1-butene **13.10** 2-ethyl-1-butene or 3-methyl-2-pentene

13.11 $(CH_3)_3C^+$ **13.13 (a)** *ortho*-bromochlorobenzene **(b)** butylbenzene **(c)** *ortho*-bromomethylbenzene or *ortho*-bromotoluene **13.18** The word "aromatic" refers only to chemical structure. **13.20** alkene: *-ene*; alkyne: *-yne*; aromatic: *-benzene* **13.22 (a)** 1-pentene **(b)** 5-methyl-2-pentyne **(c)** 2,3-dimethyl-2-butene **(d)** 2-ethyl-3-methyl-1,3-pentadiene **(e)** 4-ethyl-3,5-dimethylcyclohexene **(f)** 3,3-diethylcyclobutene **13.24** 1-pentyne, 2-pentyne, 3-methyl-1-butyne **13.30** 1,2-pentadiene, 1,3-pentadiene, 1,4-pentadiene, 3-methyl-1,2-butadiene, 2-methyl-1,3-butadiene **13.32** A triple bond is linear. **13.34** 2-pentene **13.36** identical: **(a)**, **(b)** **13.44** $2\ C_2H_2 + 5\ O_2 \rightarrow 4\ CO_2 + 2\ H_2O$ **13.50 (b)** bromobenzene **13.52** cyclohexane **13.56** The cis double bond in rhodopsin isomerizes to trans. **13.60** ultraviolet **13.62** They have no double bonds. **13.64** Cyclohexene reacts with Br_2. **13.70** 2-bromopentane and 3-bromopentane **13.74** hydration of 3,4,4-trimethyl-2-pentene or of 2-ethyl-3,3-dimethyl-1-butene

Chapter 14

14.1 (a) alcohol **(b)** alcohol **(c)** phenol **(d)** alcohol **(e)** ether **(f)** ether **14.2** A hydroxyl group is a part of a larger molecule. **14.4 (a)** 3-pentanol **(b)** 2-ethyl-1-pentanol **(c)** 5-bromo-2-ethyl-1-pentanol **(d)** 4,4-dimethylcyclohexanol **14.5** 14.3: tertiary **(a)**; secondary **(b)**, **(c)**, **(d)**; primary **(e)**; 14.4: secondary **(a)**, **(d)**; primary **(b)**, **(c)** **14.6 (a)** **14.7 (b)** **14.8 (a)** propene **(b)** cyclohexene **(c)** 4-methyl-1-pentene or 4-methyl-2-pentene **14.9 (a)** 2,3-dimethyl-2-butanol **(b)** 1-butanol or 2-butanol **14.11 (a)** 2-propanol **(b)** cycloheptanol **(c)** 3-methyl-1-butanol **14.13 (a)** *p*-chlorophenol **(b)** 4-bromo-3-methylphenol **14.16** A primary alcohol has one, a secondary alcohol has two, and a tertiary alcohol has three organic R groups attached to the OH-bearing carbon. **14.18** phenol **14.20** phenol, ether **14.24 (a)** *o*-ethylphenol **(b)** isopropyl methyl ether **(c)** methyl *p*-nitrophenyl ether **(d)** cyclopentyl methyl ether **(e)** *o*-butylphenol **(f)** dipropyl ether **14.26 (a)** < **(c)** < **(b)** **14.28** a ketone **14.30** aldehyde or carboxylic acid **14.32** Phenols dissolve in aqueous NaOH; alcohols don't. **14.38** odor **14.42** Alcohols can form hydrogen bonds; thiols and alkyl chlorides can't. **14.44** depressant **14.48** alcohol dehydrogenase **14.50** sterilant, anesthetic **14.52** vitamin E **14.54** Chlorofluorocarbons destroy ozone.

14.58 Neither is similar to water. **14.60** Antiseptics can be used on living tissue. **14.62** flammability **14.66** CH_3COOH **14.68** Sugar hydroxyl groups form hydrogen bonds with water.

Chapter 15

15.1 primary (a), (c); secondary (b), (d); tertiary (e) **15.2** (a) propylamine (b) dimethylamine (c) N-ethylaniline **15.4** $(CH_3)_2NH + H_2O \rightleftarrows (CH_3)_2NH_2^+ + OH^-$ **15.5** (a) $CH_3CH_2CH(CH_3)NH_2 + HBr(aq) \rightleftarrows CH_3CH_2CH(CH_3)NH_3^+Br^-(aq)$ (b) $C_6H_5NH_2 + HCl(aq) \rightleftarrows C_6H_5NH_3^+Cl^-(aq)$ (c) $CH_3CH_2NH_2 + CH_3COOH(aq) \rightleftarrows CH_3CH_2NH_3^+CH_3COO^-(aq)$ (d) $CH_3NH_3^+Cl^- + NaOH(aq) \rightleftarrows CH_3NH_2 + H_2O + NaCl(aq)$ **15.6** (a) sec-butylammonium bromide (b) anilinium chloride (c) ethylammonium acetate **15.7** (a) ethylamine (b) triethylamine **15.9** (a) tertiary (b) primary **15.10** N-ethyl-N,N-dimethylcyclohexylammonium chloride **15.12** -amine **15.16** Problem 15.14: primary (a), (f); secondary (b), (c), (e); tertiary (d). Problem 15.15: primary (a), (b); secondary (c); tertiary (d). **15.18** weaker **15.20** (a) $CH_3CH_2NH_2 + H_2O \rightleftarrows CH_3CH_2NH_3 + OH^-$; ethylamine is predominant (b) ethylammonium ion is predominant **15.24** diethylamine **15.26** tertiary **15.28** Quinine dissolves in aqueous acid. **15.30** does not react **15.34** The salts are more soluble in body fluids. **15.36** $C_{17}H_{20}N_2S$ **15.40** Decylamine has a larger organic part. **15.44** (a) purine **15.48** Amines are (a) stronger smelling, (b) more basic, (c) lower-boiling.

Chapter 16

16.1 (a) ester, carboxylic acid (b) ketone (c) aldehyde (d) ketone (e) aldehyde (f) ester **16.3** (a) pentanal (b) 3-pentanone (c) 4-methylhexanal **16.5** (a) 2-methyl-1-propanol (b) m-chlorobenzyl alcohol (c) cyclopentanol **16.6** (a) 4,4-dimethylcyclohexanone (b) 4-methylpentanal (c) 2-methylpentanal **16.9** hemiacetal: (b); acetal (a), (d) **16.10** (a) $C_6H_5CH_2COCH_2CH_3$, methanol (b) formaldehyde, propanol **16.12** (a), (b) **16.14** -al, -one **16.16** (a) cyclopentanone (b) octanal (c) $CH_3COCH_2CH_2CH_2CHO$ (d) 2-hydroxycyclopentanone **16.18** (a), (b), (d), (e) **16.22** (a) 2-methylbutanal (b) 2-methylpentanal (c) 2,2-dimethylpropanal (d) 3-methyl-2-nitrobenzaldehyde (e) 3-ethyl-3-methylpentanal (f) 3,3-dibromobutanal **16.24** (a) butanal (b) butanal (c) 2-bu-

tanone (d) impossible structure (e) 2-butanone **16.26** acetal **16.28** (a) $CH_3COCH_3 \rightarrow CH_3CH(OH)CH_3$ (b) $CH_3CHO \rightarrow CH_3COOH$ (c) $CH_3CHO + CH_3OH \rightarrow CH_3CH(OCH_3)_2$ **16.30** (a) cyclopentanol (b) 1-hexanol (c) 2-ethyl-1-pentanol (d) benzyl alcohol (e) 2-butanol (f) 2,2-dichloroethanol **16.32** A Tollen's test is positive for pentanal. **16.34** (a) p-methylbenzyl alcohol (b) 2-ethyl-4-methyl-1-pentanol (c) 2-buten-1-ol (d) m-hydroxybenzyl alcohol **16.38** (a) butanal, methanol, and ethanol (b) acetone and ethylene glycol (c) cyclohexanone and methanol (d) benzaldehyde and ethanol **16.40** $HOCH_2CH_2CH_2CH_2CHO$ and CH_3OH **16.42** The hemiacetal and aldehyde forms of glucose are in equilibrium. **16.46** (a) cyclohexanone (b) 3-cyclohexenol (c) 3,4-dibromocyclohexanone **16.50** 2-methylpropanal **16.56** HCN is a gas that reacts with NaOH. **16.58** HCN is generated outside the insect. **16.60** no **16.64** (a) 2-methyl-3-pentanone (b) 1,5-hexadiene (c) m-bromotoluene (d) 4,5,5-trimethyl-3-hexanone (e) o-methoxyisopropylbenzene (f) 5,5-diethyl-3-heptyne

Chapter 17

17.1 (b) > (c) > (a) **17.2** highest (a); lowest (c) **17.3** (a) $CH_3CH_2CONH_2$ (b) $CH_3CH_2CH_2COOH$ (c) $(CH_3)_2CHCOOH$ **17.4** (a) 4-methylpentanoic acid (b) isopropyl butanoate (c) N-methyl-p-chlorobenzamide **17.6** (a) potassium butanoate (b) barium 2-methylpentanoate **17.7** (a) $2 \ HCOO^-Ca^{2+}$ (b) $H_2C=CHCOO^-Na^+$ **17.8** $HCOOCH_2CH(CH_3)_2$ **17.9** (a) cyclohexanol and 4-methylpentanoic acid (b) 2-propanol and pentanoic acid **17.10** (a) 2-propanol and 2-methylpropanoic acid (b) ethanol and 2-butenoic acid (c) 1-propanol and p-bromobenzoic acid **17.13** acetic acid and p-ethoxyaniline **17.14** (a) 2-butenoic acid and methylamine (b) p-chlorobenzoic acid and diethylamine **17.15** (a) amide; acetic acid and ammonia (b) phosphate ester; ethanol and phosphoric acid (c) ester; propanoic acid and methanol **17.18** The carbonyl carbon is bonded to an electronegative atom. **17.22** (a–b) CH_3CH_2COOH (c) $CH_3CH_2COO^-$ **17.24** (a) potassium 3-ethylpentanoate (b) ammonium benzoate (c) calcium propanoate **17.26** heptanoic acid, 2-methylhexanoic acid, 3-methylhexanoic acid **17.30** (a) 9.34 g (b) 28.1 g **17.32** 190 mL **17.34** chloroacetic acid **17.36** (a) 3-methylbutyl acetate (b) methyl 4-methylpentanoate (c) ethyl 2,2-dimethylpropanoate (d) ethyl benzoate (e) cyclopentyl propanoate 17.38 (a) methanol and pentanoic acid (b) 2-propanol and 2-methylbutanoic acid (c) cyclohexanol and acetic acid (d) phenol and o-hydroxybenzoic acid 17.40 (a) 2-ethylbutanoic acid and ammonia (b) benzoic acid and aniline (c) formic acid and dimethylamine (d) propanoic acid and isopropylamine 17.42 (a) 3-methylpentanoic acid and ammonia (b) acetic acid and aniline (c) methylethylamine and benzoic acid (d) 2,3-dibromohexanoic acid and ammonia 17.44 o-aminobenzoic acid and methanol 17.46 $HOCH_2CH_2CH_2COOH$ 17.50 $H_2NCH_2CH_2CH_2CH_2CH_2COOH$ 17.52 acetic anhydride and p-ethoxyaniline 17.54 a hydrogen on carbon next to the ester 17.58 They react with water to give phosphoric acid. 17.60 (a) addition (b) addition (c) elimination 17.62 molds and fungi 17.64 CH_3COSH 17.70 Propanamide forms hydrogen bonds. 17.72 ethyl butanoate 17.74 (a) 2-chloro-3,4-dimethyl-3-hexene (b) N-methyl-N-phenylpropanamide (c) phenyl 2,2-diethylbutanoate (d) N-ethyl-o-nitrobenzamide

Chapter 18

18.1 aromatic ring: phenylalanine, tryptophan, tyrosine; sulfur: cysteine, methionine; alcohols: serine, threonine; alkyl groups: alanine, isoleucine, leucine, valine 18.3 low pH, protonated; isoelectric point, neutral; high pH, deprotonated 18.4 hydrophilic 18.5 (a), (c) 18.6 handed: shoe, shirt, coin; not handed: pencil, paper clip, drinking glass 18.7 2-aminobutane has four different groups attached to C2 18.8 (b), (c) 18.9 isoleucine, threonine 18.10 Val-Cys, Cys-Val 18.11 Val-Tyr-Gly, Val-Gly-Tyr, Tyr-Gly-Val, Tyr-Val-Gly, Gly-Val-Tyr, Gly-Tyr-Val 18.12 (a) Leu-Asp (b) Tyr-Ser-Lys 18.14 (a) hydrogen bonds (b) hydrophobic (c) salt bridge (d) hydrophobic (e) hydrogen bonds 18.15 eight amino acids 18.16 a molecule with both an amino group and a carboxylic acid group 18.18 (a) serine (b) threonine (c) proline (d) phenylalanine (e) cysteine 18.20 (a) valine (b) threonine (c) cysteine (d) tyrosine 18.22 (a) basic (b) neutral (c) acidic (d) neutral 18.26 Chiral means handed; achiral means not handed. 18.28 (a), (c), (e), (f) 18.30 (a) is chiral 18.32 a large amino acid polymer 18.34 A simple protein

contains only amino acids; a conjugated protein contains other kinds of compounds as well. 18.36 A fibrous protein is thread-like and water-insoluble; a globular protein is coiled up and water-soluble. 18.38 (a) amino acid sequence (b) helical or pleated sheet segments (c) overall three-dimensional coiling (d) aggregate structure of several chains 18.40 It forms disulfide links. 18.42 (a) Hydrophobic amino acids stay on inside of globular proteins. (b) Salt bridges hold charged groups near each other. (c) Hydrogen bonding holds groups together. 18.44 Tertiary structure is disrupted. 18.46 N-terminus: Asp; C-terminus: Phe 18.50 Proteins are hydrolyzed in the stomach. 18.52 N-terminus: Tyr; C-terminus: Met 18.54 Glycine is an internal salt. 18.56 Val-Gly-Ser-Met-Ala-Asp 18.58 They have no net charge at this pH. 18.60 Asp, Phe 18.62 94% 18.64 one that does not contain all essential amino acids 18.66 identity and amounts of amino acids in a protein 18.68 Enzyme is denatured during processing. 18.70 Tropocollagen has a complex quaternary structure. 18.72 basic pH, perhaps 9.0

Chapter 19

19.2 (a) hydration of fumaric acid (b) oxidation of squalene (c) phosphorylation of glucose (d) hydrolysis of cellulose 19.3 ligase 19.4 Vitamin A is a hydrocarbon; vitamin C has many hydroxyl groups. 19.6 An enzyme is a biological catalyst, usually a protein. 19.8 by lowering the activation energy 19.10 (a) breaking a bond by addition of water (b) isomerization of a substrate (c) addition or elimination of a small molecule 19.12 A cofactor is a small, nonprotein part of an enzyme; a coenzyme is an organic cofactor. 19.14 Enzymes bind substrates in their active sites and hold them in the correct position for reaction. 19.16 (a) hydrolysis of a protein (b) bond formation in DNA (c) transfer of a methyl group 19.18 competitive inhibition 19.20 hydrolases 19.24 by acting as a noncompetitive inhibitor for bacterial enzymes 19.26 Reaction rates increase with substrate concentration and then level off. 19.28 (a) decrease (b) probable decrease (c) probable stop (d) increase 19.30 competitive, noncompetitive, irreversible 19.32 competitive and noncompetitive: noncovalent bonds; irreversible: covalent bonds 19.34 Papain hydrolyzes peptide bonds in meat.

19.36 The product of an enzyme-catalyzed reaction series is an inhibitor for an earlier step in the series. **19.38** A positive regulator changes the shape of an active site so that it accepts substrate more readily. A negative regulator changes the shape of an active site so that it accepts substrate less readily. **19.40** fewer calories **19.42** A hormone is a chemical messenger produced in the endocrine glands. A neurotransmitter transmits chemical messages in the nervous system. **19.44** vitamins **19.46** Vitamin C is not stored in the body because it is water-soluble. **19.48** hypothalamus **19.52** Enzymes are usually proteins; hormones can have any structure. **19.54** cAMP is a second messenger released in a cell's interior. **19.56** central and peripheral **19.58** cholinergic and adrenergic **19.60** autonomic: unconsciously controlled functions; somatic: consciously controlled functions **19.62** CK(MB), AST, and LDH$_1$ **19.64** Water-soluble vitamins are not stored in the body. **19.66** by preventing cell wall synthesis **19.70** Pyroglutamic acid is a cyclic amide of glutamic acid. **19.72** Sulfa drugs mimic PABA.

Chapter 20

20.1 exergonic (a), (c); endergonic (b); (a) proceeds furthest **20.3** What is favorable in one direction is unfavorable in the other direction. **20.4** Acetyl phosphate + ADP → acetate + ATP; favorable by −3.0 kcal **20.5** (a) NAD$^+$ (b) FAD **20.7** Citric acid, isocitric acid **20.8** Acetyl SCoA + 2 O$_2$ → 2 CO$_2$ + HSCoA + H$_2$O **20.9** A stable carboxylate ion is produced. **20.10** NADH/H$^+$, FMNH$_2$, FeSP(red), CoQ **20.11** It has a large hydrocarbon portion. **20.12** exergonic: releases energy; endergonic: absorbs energy **20.14** Both ΔH and ΔS are important. **20.16** exergonic: (a), (c); endergonic: (b); furthest toward products: (a) **20.18** Prokaryotic: bacteria and algae; eukaryotic: higher organisms **20.20** cytoplasm: everything between the cell membrane and the nuclear membrane in a eukaryotic cell; cytosol: material that fills the interior of the cell **20.22** ATP is produced there. **20.24** digestion: breakdown of bulk food; metabolism: total of all cellular reactions **20.26** acetyl SCoA **20.28** adenosine triphosphate **20.30** Energy is released when a phosphate group is transferred. **20.34** exergonic by 4.5 kcal **20.36** No. Reaction is unfavorable by 4.0 kcal/mol. **20.38** in mitochondria **20.40** oxaloacetic acid

20.42 formation of ATP by transfer of a phosphate to ADP from another phosphate-containing compound such as GTP **20.44** 3 NADH, 1 FADH$_2$ **20.46** the electron transport system **20.48** NADH or FADH$_2$ is oxidized; ADP is phosphorylated. **20.50** H$_2$O and energy **20.52** iron **20.54** FADH$_2$, CoQ, 2 Fe^{3+} **20.56** It would stop. **20.58** the minimum energy expenditure to keep functioning **20.60** Energy is needed above the minimum. **20.62** They prevent electron transport. **20.64** 2,4-dinitrophenol **20.68** oxidoreductases **20.70** cytochrome oxidase

Chapter 21

21.1 (a) aldopentose (b) ketotriose (c) aldotetrose **21.4** eight **21.5** (a) D-ribose (b) L-mannose **21.8** β anomer **21.12** reducing because the right-hand sugar has a hemiacetal linkage **21.13** a β 1,4 glycoside link **21.14** two β-D-glucoses **21.15** nonreducing **21.16** a polyhydroxy aldehyde or ketone **21.18** aldose: aldehyde; ketose: ketone **21.22** glucose: in most food, fruits, and vegetables; galactose: in brain tissue and lactose; fructose: in honey and fruits **21.24** Enantiomers are mirror-image stereoisomers; diastereomers are non-mirror-image stereoisomers. **21.26** no **21.28** The product from erythrose has a plane of symmetry. **21.30** It reduces oxidizing agents like Benedict's reagent. **21.32** α form **21.40** A new chiral center is formed when fructose is reduced. **21.42** A hemiacetal has a carbon atom bonded to one —OH group and one —OR group. An acetal has a carbon bonded to two —OR groups. **21.46** Maltose occurs in fermenting grains; lactose occurs in milk; sucrose occurs in many plants. **21.48** Both are α-glucose polymers; amylopectin has branches. **21.50** Both monosaccharide units are acetals. **21.52** Gentiobiose is reducing and has a hemiacetal linkage on the right-hand sugar. **21.54** Trehalose is nonreducing and has an acetal linkage. **21.56** one that rotates plane-polarized light **21.58** D-Fructose rotates light more strongly than D-glucose. **21.60** It is dietary fiber. **21.62** Insulin regulates blood sugar. **21.66** It causes the production of antibodies. **21.68** Their caloric value is low relative to their sweetness. **21.70** They are diastereomers; their properties have no relation. **21.76** Saliva hydrolyzes the glycoside links, yielding glucose.

Chapter 22

22.1 (a) glucose + galactose **(b)** glucose + fructose   **22.2 (a)** glycogenolysis **(b)** gluconeogenesis **(c)** glycogenesis   **22.3** glycolysis, glycogenesis, pentose phosphate pathway   **22.4** steps 6, 7, and 9, 10   **22.5** steps 2, 5, and 8   **22.6** at step 3   **22.7** They differ in the stereochemistry of the hydroxyl group at C4.   **22.8** Glucose → 2 pyruvate → 2 CO_2 + 2 acetyl SCoA; 2 acetyl SCoA → 4 CO_2 **22.9** Because NADH can't cross the inner mitochondrial membrane but must be converted into $FADH_2$ which produces only 2 ATPs.   **22.10** PP_i; $\Delta G = -6.9$ kcal; favorable   **22.11** energy cost of 2 ATP + 2 GTP   **22.12** NAD^+; oxidation   **22.13** 2 fructose 6-phosphate + 3 CO_2 + glyceraldehyde 3-phosphate   **22.14** mouth, stomach, and small intestines; hydrolysis of carbohydrates, fats, and proteins   **22.16** anaerobic: lactate; aerobic: acetyl SCoA; fermentation: ethanol   **22.18** glycolysis: breakdown of glucose to pyruvate; gluconeogenesis: synthesis of glucose from pyruvate; glycogenesis: synthesis of glycogen from glucose; glycogenolysis: breakdown of glycogen to glucose   **22.20** pyruvate   **22.22 (a)** steps 1, 3, 6 **(b)** step 6 **(c)** step 9   **22.24 (a)** 6 mol **(b)** 2 × 3 mol **(c)** 12 mol   **22.26** 72 mol   **22.28** 12 mol   **22.30** allows synthesis of glucose from fat or protein   **22.32** triggers breakdown of glycogen   **22.34** ketone bodies   **22.36** liver   **22.38** Some steps are too exergonic to be reversed.   **22.40** lactate, glycerol   **22.42** Some steps are too exergonic to be reversed.   **22.44** fructose 6-phosphate, glyceraldehyde 3-phosphate, or ribose 5-phosphate   **22.46** deficiency of enzyme for breakdown of muscle glycogen   **22.50** Energy for sprinting comes from creatine phosphate.   **22.52** Enzyme for galactose catabolism is missing.   **22.54** hosphorylation   **22.56** Glucose from glycogen is phosphorylated directly with inorganic phosphate rather than with ATP.

Chapter 23

23.1 $CH_3(CH_2)_{18}COOCH_2(CH_2)_{30}CH_3$
23.5 $[CH_3(CH_2)_7CH=CH(CH_2)_7CO_2^-]_2Ca^{2+}$
23.9 double bond, ketone   **23.10** An aromatic A ring makes them estrogens.   **23.12** a naturally occurring molecule that dissolves in nonpolar solvents   **23.14** A long, straight-chain carboxylic acid   **23.18** insulation, long-term fuel storage   **23.20** $CH_3(CH_2)_{14}COOCH_2(CH_2)_{14}CH_3$
23.22 The two stearic acids can be bonded to adjacent or nonadjacent oxygens.

23.24 (a) $CH_3(CH_2)_{14}COO^-Na^+$
(b) $CH_3(CH_2)_7CH=CH(CH_2)_7COOCH_2(CH_2)_8CH_3$
23.26 glycerol, stearic acid, oleic acid, linolenic acid   **23.28** higher   **23.30** A fat has three fatty acids bonded to glycerol; a phospholipid has a phosphate group.   **23.32** The charged head group is necessary.   **23.34** A soap micelle is a monolayer.   **23.36** sphingomyelins and spingoglycolipids   **23.40** four fatty acids, two phosphoric acids, three glycerols   **23.42** The A ring of estradiol is a phenol; the A ring of testosterone has a double bond and a ketone.   **23.44** regulates membrane fluidity   **23.48** arachidonic acid   **23.52** They are not biodegradable.   **23.54** active transport   **23.56** It inhibits an enzyme.   **23.58** They are species-specific.   **23.60 (b)** choline, glycerol, phosphoric acid, two fatty acids **(d)** sphingosine, fatty acid, phosphoric acid, choline **(e)** sphingosine, fatty acid, monosaccharide **(f)** three oleic acids, glycerol   **23.62** lower melting   **23.64** $CH_3(CH_2)_{16}COOCH_2(CH_2)_{20}CH_3$ **23.66** to send chemical messages in the body   **23.68** Like soaps, lecithins are attracted to both water and fats.   **23.70** 11–12 g

Chapter 24

24.1 step 5 of glycolysis   **24.2 (a)** 6 acetyl CoA, 5 turns **(b)** 10 acetyl CoA, 9 turns   **24.4** steps 6–8 of the citric acid cycle   **24.5** 129 ATP   **24.6** 260 g   **24.7** (i), (iii)   **24.8** small intestine   **24.10** to emulsify lipids   **24.12** lipoproteins that aid in lipid transport   **24.14** in association with albumins   **24.16** 5 ATP   **24.18** 20 ATP   **24.20** in mitochondrial matrix   **24.22** A thiol ester is formed with HSCoA.   **24.24** A different product results from each turn.   **24.26 (d)** > **(b)** > **(c)** > **(a)**   **24.28** $CH_3COCH_2COSCoA$ + HSCoA → 2 $CH_3COSCoA$ **24.30 (a)** three turns **(b)** six turns   **24.32** Acetyl SCoA   **24.34** They are made from a two-carbon unit.   **24.36** seven rounds   **24.38** deposition of cholesterol in the arteries   **24.40** high HDL/LDL   **24.42** ^1H **24.44** Carbohydrates are metabolized to yield acetyl SCoA, which is converted into lipids.   **24.46** acetone, 3-hydroxybutyrate, acetoacetate   **24.48** 22   **24.50** Fatty acids yield more than twice as much energy per gram as carbohydrates do.

Chapter 25

25.1 $(CH_3)_2CHCH_2COCOO^-$
25.2 $C_6H_5CH_2COCOO^-$ **25.3** fumarate → malate → oxaloacetate → aspartate   **25.5** transamina-

tions: (1), (2), (3); hydrolysis: (3); oxidation: (3)
25.6 in the stomach   **25.8** food, tissue breakdown
25.10 An amino group of an amino acid exchanges
with the keto group of an α-keto acid.   **25.12 (a)**
CH₃CH₂CH(CH₃)COCOO⁻   **(b)** HSCH₂COCOO⁻
(c) C₆H₅CH₂COCOO⁻   **25.14** An amino group is
replaced by a keto group.   **25.16** a dehydrogenase
25.18 ammonia   **25.20** an amino acid that can be
converted into a ketone body   **25.22** Ornithine is
the reactant in step 1 and the product of step 4.
25.24 Nonessential amino acids require 1–3 steps;
essential amino acids require 7–10 steps.
25.26 reductive amination; oxidative deamination
25.28 a defect in tyrosine biosynthesis
25.30 growing children, pregnant women, those with
healing wounds   **25.32** inadequate protein intake,
kidney disease   **25.34** a compound that is foreign
to the body   **25.36** Two molecules of ATP are consumed.   **25.38** (CH₃)₂CHCH(NH₂)COOH +
CH₃COCOO⁻ → (CH₃)₂CHCOCOO⁻ +
CH₃CH(NH₂)COOH; valine aminotransferase
25.40 yes   **25.42** Tissue is constantly undergoing
degradation and resynthesis.

Chapter 26

26.3 deoxythymidine 5′-monophosphate, adenosine
5′-monophosphate, adenosine 5′-diphosphate,
adenosine 5′-triphosphate
26.5 (a) C-G-G-A-T-C-A   **(b)** T-T-A-C-C-G-A-G-T
26.6 A and G don't hydrogen-bond to each other.
26.8 (a) C-U-A-A-U-G-G-C-A-U
(b) A-U-A-C-C-G-A-U-C-C-G-U   **26.9 (a)** GCU,
GCC, GCA, GCG **(b)** UUU, UUC **(c)** UUA,
UUG, CUU, CUC, CUA, CUG **(d)** GUU, GUC,
GUA, GUG **(e)** UAU, UAC   **26.10 (a)** Ile **(b)** Ala
(c) Arg **(d)** Asn   **26.11** Leu-Met-Ala-Trp-Pro
26.12 GAA, UAC, CGA, ACC, GGG, AUU
26.13 GAA-TAC-CGA-ACC-GGG-ATT
26.14 TGA codes for UGA, which is a stop codon.
26.15 (a) Stop becomes Tyr. **(b)** Gly becomes Ala.
26.16 A nucleotide consists of a heterocyclic amine
base, an aldopentose, and a phosphoric acid.
26.18 oxygen atom missing   **26.20** adenine,
cytosine, guanine, and uracil   **26.22** Messenger
RNA carries genetic message from DNA to ribosomes; ribosomal RNA makes up ribosomes; transfer RNA transports amino acids to ribosomes.
26.24 A chromosome is a large molecule of DNA; a
gene is a part of the chromosome that codes for
a single protein.   **26.26** approximately 100,000
26.28 They hydrogen-bond to each other.

26.30 The 5′ end has a free phosphoric acid group;
the 3′ end has a free —OH group.   **26.36** a sequence of three nucleotides on mRNA that codes
for a specific amino acid   **26.38** cloverleaf; contains 70–100 nucleotides   **26.40** The purines A and
G pair with T and C, respectively.   **26.42 (a)** CCU,
CCC, CCA, CCG **(b)** AAA, AAG **(c)** AUG
26.44 (a) UGA **(b)** CCU **(c)** GAA
26.46 U-A-C-C-G-A   **26.48** an error in base sequence that occurs during DNA replication
26.50 Thr replaces Ile.   **26.52** If codons were made
up of 2 nucleotides, only 4² = 16 combinations
would be possible, and it would not be possible to
code for all 20 amino acids.   **26.54** A mutation
changes the biological activity of a protein and
can lead to uncontrollable cell growth.
26.56 TAT-GGT-GGT-TTT-ATG-TAA
26.60 Reverse transcriptases give DNA from RNA.
26.62 They are small and contain plasmids.
26.64 proteins   **26.66** Glu: GAA, GAG; Val:
GUU, GUC, GUA, GUG   **26.68** TAA → TAG
would have no effect; TAA → GAA would change
Ile to Leu.   **26.70** At least two bases in DNA must
differ.   **26.72** It is removed.

Chapter 27

27.2 intracellular fluid   **27.4** higher in arterial;
lower in venous   **27.6** vasopressin; to control water–electrolyte balance   **27.8** red blood cells,
platelets, white blood cells   **27.10** Clotting would
be dangerous in these organs.   **27.12** four
27.14 Oxyhemoglobin is red; deoxyhemoglobin is
purple.   **27.16** The binding and release of each successive oxygen is enhanced.   **27.18** approximately
30 mm Hg   **27.20** releases O₂: **(a)**, **(b)**, **(c)**
27.22 by inhibiting the decarboxylase enzyme
27.24 lymphocytes cause phagocytosis   **27.26** T
cells release a toxin that destroys infected cells.
27.28 Ca²⁺, vitamin K   **27.30** so that they do not
trigger unwanted clots   **27.32** H⁺ + HPO₄²⁻ →
H₂PO₄⁻; H⁺ + HCO₃⁻ → H₂O + CO₂   **27.34** Ice
crystals disrupt cell membranes.   **27.36** ATP
production must continue, and metabolism must
be anaerobic.   **27.38** Transport occurs in only
one direction.   **27.40** to measure amounts of
colored products   **27.44** Antibodies are passed
from mother to child.   **27.46** Active transport
involves an expenditure of energy to carry a substance across a membrane. Osmosis is spontaneous.
27.48 The equilibrium is necessary for pH control.

CREDITS

Photo/Illustration Credits

All experiments and chemistry demonstrations not otherwise credited were photographed by Tom Bochsler, Photography Limited; page v, Dr. Tony Brain/David Parker/Science Photo Library/Photo Researchers; Page vi, CNRI/Science Photo Library/Photo Researchers, John Gillmoure/The Stock Market; vii, U.S. Navy/Science Photo Library/Photo Researchers, John Kaprielian/Photo Researchers; vii, Richard Megna, Fundamental Photos; ix, G. Bordessoule/Agence Vandystadt/Photo Researchers, David Muench/The Image Bank; x, Will McIntyre/Photo Researchers, Paul Lowe/Matrix; xi, Hans Reinhard/Okapia/Photo Researchers, Richard Megna/Fundamental Photographs; xii, J.P. Ferrero Jacana/Photo Researchers, CNRI/Science Photo Library/Photo Researchers; xiv, Peter G. Aitken/Photo Researchers, Lennart Nillsson: *BEHOLD MAN;* xv, P. Durand/Sygma; xvi, M.I. Walker/Photo Researchers, CNRI/Science Photo Library/Photo Researchers.

1, Courtesy Lawrence Livermore National Laboratory; 3, Richard Megna/Fundamental Photos; 6, Richard Megna, Fundamental Photographs; 9, Tom Bochsler (N, Ph, S) and Department of Mineral Sciences, Museum of Natural History, Smithsonian Institution (Au, Zn); 11, Norm Thomas/Photo Researchers; 13, Omni Photo and Tom Bochsler; 14, Paul Chauncey/The Stock Market, John Dommers/Photo Researchers; 18, Photo courtesy of IBM®; 21, CNRI/Science Photo Library/Photo Researchers; 24, Sartorius Balances and Scales and Brinkman Instruments, Inc.; 27, The Granger Collection; 34, C.C. Duncan/Medical Images, Inc.; 41, Nathan Benn/Woodfin Camp & Associates; 43, Dr. Tony Brain/David Parker/Science Photo Library/Photo Researchers and CNRI, Science Photo Library/Photo Researchers; 47, IBM Research Division, Yorktown Heights; 54, Gamma-Liaison; 55, Diane Schiumo/Fundamental Photographs; 61, John Gillmoure/The Stock Market; 65, Tom Bochsler; 67, courtesy IBM®; 71, UCLA and Mallinckrodt; 79, Grace Moore/Medichrome/The Stock Shop; 84, CNRI/Science Photo Library/Photo Researchers; 87, Ander Tsiara/Science Source/Photo Researchers; 91, U.S. Navy/Science Photo Library/Photo Researchers; 92, The Granger Collection; 96, Jeremy Burgess/Science Photo Library/Photo Researchers; 98, Tom Bochsler.

101, Sidney Moulds/Science Photo Library/Photo Researchers; 118, Paul Silverman/Fundamental Photographs (malachite) and Ralph Wetmore/Photo Researchers (topaz); 122, Ted Horowitz/The Stock Market; 131, John Kaprielian/Photo Researchers; 132, Joe Munroe/Photo Researchers; 146, Richard Megna/Fundamental Photographs; 149, Sinclair Stammers/Science Photo Library/Photo Researchers; 153, Chris Jones/The Stock Market; 159, Richard Megna/Fundamental Photographs; 160, Tom Bochsler; 168, Fundamentals Photographs; 170, Hoffman-Spooner/Gamma-Liaison; 171, Clyde Metz; 172, Tom Bochsler (left), Grant Heilman Photography (right); 177, Richard Megna/Fundamental Photographs; 178, Richard Megna/Fundamental Photographs; 183, Richard Megna/Fundamental Photographs; 187, Chris Collins/The Stock Market; 196, Tom Bochsler; 198, Richard Megna/Fundamental Photographs (left), Eastman Chemical Company.

211, John Roshelley/Photo Researchers; 216, Kip Peticolas/Fundamental Photographs; 225, G. Bordessoule/Agence Vandystadt/Photo Researchers; 228, David Guyon/The BOC Group PLC/Science Photo Library/Photo Researchers; 235, Peter Steiner/The Stock Market; 237, Arthur Sirdoisky/Medical Images, Inc.; 245, David Muench/The Image Bank; 247, Richard Megna/Fundamental Photographs; 253, The Granger Collection; 254, Richard Megna/Fundamental Photographs; 258, Donald Clegg and Roxy Wilson; 270, Tom Bochsler; 271, Walter Dawn/Photo Researchers (a) and Biophoto Associates/Photo Researchers (b, c); 278, Paul Silverman/Fundamental Photographs; 281, Richard Megna/Fundamental Photographs (Fig. 11.1) and Tom Bochsler (Fig. 11.2); 286 and 299, Tom Bochsler; 297, Will McIntyre/Photo Researchers; 298, Tom Bochsler.

308, Richard Megna/Fundamental Photographs; 310, Hank Morgan/Photo Researchers; 314, Paul Lowe/Matrix; 317, Sygma; 325, Gary Retherford/Photo Researchers (vitamins) and Any Levin/Photo Researchers (vegetables); 328, image generated on the Evans & Sutherland PS 390 computer graphics system using Tripos® SYBYL computational chemistry software; 345, Hans Reinhard/Okapia/Photo Researchers;

357, © Joel Gordon 1988. Courtesy West Publishing Co./ CHEMISTRY by Radel/Navidi; **362,** Luiz Claudio Marigo/Peter Arnold; **363,** © Christo, 1983; Photo: Wolfgang Volz; **364,** Richard Megna/Fundamental Photographs; **365,** Michael English/Medical Images Inc.; **370,** Dr. E.R. Degginger; **373,** Tom Bochsler; **378,** Philippe Plailly/Science Photo Library/Photo Researchers; **382,** Donald Clegg and Roxy Wilson; **383,** Richard Megna/Fundamental Photographs; **393,** Donald Clegg and Roxy Wilson; **394,** John Bova/Photo Researchers; **398,** Anaquest Pharmaceutical Products.

401, NASA/Science Source/Photo Researchers; **412,** reproduced from *The Merck Index*, Eleventh Edition (1989), S. Budavari, M.J. O'Neil, A. Smith, P.E. Heckelman, Eds., by permission of the copyright owner, Merck & Co., Inc., Rahway, NJ, U.S.A. © Merck & Co., Inc., 1989; **413,** reproduced with permission of Medical Economics Co., Oradell, N.J. (No Doz) and Dean: Lange's Handbook of Chemistry, 13/E, © 1987, McGraw-Hill. Reprinted with permission of McGraw-Hill; **407,** Susan Lapides/Woodfin Camp & Associates; **420,** Donald Clegg and Roxy Wilson; **422,** Jerome Wexler/ Photo Researchers; **423,** Dr. E.R. Degginger (opium poppy) and Hans Pfletschinger/Peter Arnold (mosquito); **427,** Will and Deni McIntyre/Photo Researchers; **431,** T. Eisner and D. Aneshanslet, Cornell University; **437,** J.P. Ferrero Jacana/Photo Researchers; **441,** S. Varnedoe; **442,** © Joel Gordon, 1988. Courtesy West Publishing Co./ CHEMISTRY by Radel and Navidi (Tollens Test) and Joel Gordon (Benedict's test); **453,** adapted with permission of Academic Press Inc., San Diego, Calif.; **458,** Bridgeman/ Art Resource; **464,** Tom Bochsler/Photography Unlimited; **471,** Nobel Proctor/Photo Researchers; **475,** The Granger Collection; **481,** CNRI/Science Photo Library/ Photo Researchers; **488,** Tom Bochsler/Photography Unlimited; **494,** Jean Marc Barey/Agence Vandystadt/Photo Researchers.

509, Bill Longcore/Photo Researchers; **510, 511,** adapted from Annino and Grese, *Clinical Chemistry,* 4th Ed., copyright 1976, Little, Brown & Company, used with permission, **515,** William Ober and Claire Garrison; **518,** after Jane Richardson (Fig. 18.14); **519,** Science Photo Library/Photo Researchers; **528,** CNRI/Science Photo Library/Photo Researchers; **531,** © Richard Megna/ Fundamental Photographs; **537,** from Mathews and van Holde, *Biochemistry*, copyright 1990, with permission of The Benjamin/Cummings Publishing Company; **551,** Dr. E.R. Degginger; **552,** William Ober and Claire Garrison (nerve cell); **555,** Dr. Edward J. Bottone, Dept. of Microbiology, Mount Sinai Hospital, New York; **559,** Globus Brothers/The Stock Market; **561,** Michael English/Medical Images; **567,** Richard Megna/Fundamental Photographs; **570,** Peter G. Aitken/Photo Researchers; **581,** image generated on the Evans & Sutherland PS 390 computer graphics system using Tripos® SYBYL com-

putational chemistry software; **584,** Norman R. Lightfoot/Photo Researchers; **588,** CNRI/Science Photo Library/Photo Researchers; **595,** James H. Carmichael, Jr./Photo Researchers.

600, Bonnie Rauch/Photo Researchers; **604,** Don W. Fawcett/Ito/Photo Researchers; **605,** Rainer Berg/ Okapia/Science Source/Photo Researchers; **608,** George Haling/Photo Researchers; **609,** M.I. Walker/Photo Researchers; **612,** United Nations Photo; **616,** Lennart Nilsson: *BEHOLD MAN;* **619,** CNRI/Science Photo Library/Photo Researchers; **620,** from Mathews and van Holde, *Biochemistry,* copyright 1990, with permission of The Benjamin/Cummings Publishing Company; **627,** Jack Fields/Photo Researchers; **629,** Frederick Martini; **632,** Eli Lilly and Company; **642,** Peter Aprahamian/ Science Library/Photo Researchers; **646,** F. Durand/ Sygma; **651,** Patrick Donehue/Photo Researchers; **656,** CNRI/Science Photo Library/Photo Researchers; **659,** William Ober and Claire Garrison (cell membrane) and Don W. Fawcett/Photo Researchers; **660,** William Ober and Claire Garrison; **665,** SIU/Photo Researchers; **667,** John Serrao/Photo Researchers; **671,** Tripos Associates, St. Louis, MO, U.S.A.; **674,** Craig Luce (liver); **675,** William Ober and Claire Garrison; **679,** Biophoto Associates, Science Source/Photo Researchers; **681,** Lennart Nilsson from *The Body Victorious,* © Delacorte Press; **691,** Toshiba America Medical Systems, Inc.; **694,** M.I. Walker/Photo Researchers; **696,** John Colwell from Grant Heilman.

701, Andrew J. Martinez/Photo Researchers; **709,** Lennart Nilsson: *BEHOLD MAN;* **712,** Land O'Lakes, Inc.; **714,** Howard Sochurek/Medical Images; **720,** Richard Feldman/National Institute of Health; **724,** William Ober and Claire Garrison; **725,** Kuchel and Ralston: *Schaum's Outline on Theory and Problems of Biochemistry*, copyright 1988, McGraw-Hill. Reprinted with permission of McGraw-Hill (growth of new DNA strands); **726,** Lawrence Livermore National Laboratory; **728,** Professor Oscar Miller/Science Photo Library/Photo Researchers; **731,** CDC/Science Source/Photo Researchers; **733,** (c) image generated on the Evans & Sutherland PS 390 computer graphics system using Tripos® SYBYL computational chemistry software; William Ober and Claire Garrison; **739,** Philippe Plailly/Science Photo Library/Photo Researchers; **742,** NIBSC/Science Photo Library/Photo Researchers; **745,** CNRI/Science Photo Library/Photo Researchers; **747,** Philippe Plailly/Science Photo Library/Photo Researchers; **749,** Sheila Terr/Science Photo Library/Photo Researchers; **754,** CNRI/ Science Photo Library (red blood cells) and TEM; Boerhinger-Ingelheim/Lennart Nilsson (phagocytosis); **757,** Tony Brain/Science Photo Library/Photo Researchers; **759,** William Ober and Claire Garrison; **760,** William Ober and Claire Garrison.

I N D E X

SOME IMPORTANT FAMILIES OF ORGANIC MOLECULES

Family Name	Functional Group Structure[a]	Simple Example	Name Ending
Alkane	(contains only C—H and C—C single bonds)	CH_3CH_3 ethane	-ane
Alkene	$\diagdown C=C \diagup$	$H_2C=CH_2$ ethylene	-ene
Alkyne	—C≡C—	H—C≡C—H acetylene (ethyne)	-yne
Arene	(benzene ring structure)	benzene	none
Alkyl halide[b]	—C—X	CH_3—Cl methyl chloride	none
Alcohol	—C—O—H	CH_3—OH methyl alcohol (methanol)	-ol
Ether	—C—O—C—	CH_3—O—CH_3 dimethyl ether	none
Amine	—N—H, —N—H, —N— with H	CH_3—NH_2 methylamine	-amine
Aldehyde	—C(=O)—H	CH_3—C(=O)—H acetaldehyde (ethanal)	-al
Ketone	—C—C(=O)—C—	CH_3—C(=O)—CH_3 acetone	-one
Carboxylic acid	—C(=O)—OH	CH_3—C(=O)—OH acetic acid	-ic acid
Anhydride	—C(=O)—O—C(=O)—	CH_3—C(=O)—O—C(=O)—CH_3 acetic anhydride	none
Ester	—C(=O)—O—	CH_3—C(=O)—O—CH_3 methyl acetate	-ate
Amide	—C(=O)—NH_2, —C(=O)—N—H, —C(=O)—N—	CH_3—C(=O)—NH_2 acetamide	-amide

[a] The bonds whose connections aren't specified are assumed to be attached to carbon or hydrogen atoms in the rest of the molecule.
[b] X = F, Cl, Br, or I.